Elements of Physical Chemistry

Peter Atkins
University of Oxford

Julio de Paula
Lewis & Clark College

With contributions from

David Smith
University of Bristol

SIXTH EDITION

OXFORD
UNIVERSITY PRESS

OXFORD
UNIVERSITY PRESS

Great Clarendon Street, Oxford, OX2 6DP,
United Kingdom

Oxford University Press is a department of the University of Oxford.
It furthers the University's objective of excellence in research, scholarship,
and education by publishing worldwide. Oxford is a registered trade mark of
Oxford University Press in the UK and in certain other countries

Impression: 2

British Library Cataloguing in Publication Data
Data available

Library of Congress Cataloging in Publication Data
Library of Congress Control Number: 2012940889

ISBN 978–0–19–960811–9

Printed and bound by Bell and Bain Ltd, Glasgow

Fundamental constants

Quantity	Symbol	Value	Power of ten	Units
Speed of light	c	2.997 924 58	10^{8}	$\mathrm{m\ s^{-1}}$
Elementary charge	e	1.602 176 565	10^{-19}	C
Planck's constant	h	6.626 069 57	10^{-34}	J s
	$\hbar = h/2\pi$	1.054 571 726	10^{-34}	J s
Boltzmann's constant	k	1.380 6488	10^{-23}	$\mathrm{J\ K^{-1}}$
Avogadro's constant	N_A	6.022 141 29	10^{23}	$\mathrm{mol^{-1}}$
Gas constant	$R = N_A k$	8.314 4621		$\mathrm{J\ K^{-1}\ mol^{-1}}$
Faraday's constant	$F = N_A e$	9.648 533 65	10^{4}	$\mathrm{C\ mol^{-1}}$
Mass				
electron	m_e	9.109 382 91	10^{-31}	kg
proton	m_p	1.672 621 777	10^{-27}	kg
neutron	m_n	1.674 927 351	10^{-27}	kg
atomic mass constant	m_u	1.660 538 921	10^{-27}	kg
Vacuum permeability	μ_0	4π	10^{-7}	$\mathrm{J\ s^2\ C^{-2}\ m^{-1}}$
Vacuum permittivity	$\varepsilon_0 = 1/\mu_0 c^2$	8.854 187 817	10^{-12}	$\mathrm{J^{-1}\ C^2\ m^{-1}}$
	$4\pi\varepsilon_0$	1.112 650 056	10^{-10}	$\mathrm{J^{-1}\ C^2\ m^{-1}}$
Bohr magneton	$\mu_B = e\hbar/2m_e$	9.274 009 68	10^{-24}	$\mathrm{J\ T^{-1}}$
Nuclear magneton	$\mu_N = e\hbar/2m_p$	5.050 783 53	10^{-27}	$\mathrm{J\ T^{-1}}$
Proton magnetic moment	μ_p	1.410 606 743	10^{-26}	$\mathrm{J\ T^{-1}}$
g-value of electron	g_e	2.002 319 304		
Magnetogyric ratio				
electron	$\gamma_e = -g_e e/2m_e$	−1.001 159 652	10^{10}	$\mathrm{C\ kg^{-1}}$
proton	$\gamma_p = 2\mu_p/\hbar$	2.675 222 004	10^{8}	$\mathrm{C\ kg^{-1}}$
Bohr radius	$a_0 = 4\pi\varepsilon_0\hbar^2/e^2 m_e$	5.291 772 109	10^{-11}	m
Rydberg constant	$R_\infty = m_e e^4/8h^3 c\varepsilon_0^2$	1.097 373 157	10^{5}	$\mathrm{cm^{-1}}$
	hcR_∞/e	13.605 692 53		eV

* The values quoted here were extracted in December 2011 from the National Institute of Standards and Technology (NIST) website (search term: physical constants).

Using this book to master physical chemistry

This icon indicates where material can be found on the accompanying Online Resource Centre.

Grasping the main concepts

The 'Foundations' chapter

You should review this chapter first, as it introduces basic concepts that are used throughout the text.

Annotated equations and equation labels

We have annotated many equations so that you can follow how they are developed. A green annotation takes you across the equals sign: it is a reminder of the substitution used, an approximation made, the terms that have been assumed constant, and so on. A red annotation is a reminder of the significance of an individual term in an expression. Many of the equations are labelled to highlight their significance. We sometimes colour a collection of numbers or symbols to show how they carry from one line to the next.

$$\Delta V = \overbrace{V_f}^{n_fRT/p_{ex}} - \overbrace{V_i}^{n_iRT/p_{ex}} \overset{\Delta n_g = n_f - n_i}{=} \frac{RT\Delta n_g}{p_{ex}}$$

Derivations

Mathematical development is an intrinsic part of physical chemistry, and to achieve full understanding you need to see how a particular expression is obtained and if any assumptions have been made. The 'Derivations' are set off from the text to let you adjust the level of detail that you require for your current needs and make it easier to review material. All the calculus in the book is confined within these 'Derivations'.

Derivation 2.3

Heat transfers at constant pressure

Consider a system open to the atmosphere, so that its pressure p is constant and equal to the external pressure p_{ex}. From eqn 2.13b we can write

$\Delta H = \Delta U + p\Delta V = \Delta U + p_{ex}\Delta V$

The chemist's toolkit

To make progress with a derivation you might need to be reminded of some mathematical, physical, or chemical concepts and techniques. The toolkits occur just where they are needed.

The chemist's toolkit 2.1 Integration

The area under a graph of any function f is found by the techniques of integration. For instance, the area under

A note on good practice

Our 'notes on good practice' will help you to avoid making common mistakes. They encourage conformity to the international language of science by setting out the language and procedures adopted by the International Union of Pure and Applied Chemistry (IUPAC).

A note on good practice Keep track of signs by considering whether the stored energy has decreased when the system does work (w is then negative) or has increased when work has been done on the system (w is then positive).

Artwork

Many concepts in physical chemistry are best understood by visualizing what is taking place. Our diagrams and graphs are meant to help you to answer the important question 'What is this equation telling me?'

Checklist of key concepts

The principal concepts introduced in each chapter are summarized in a checklist at the end of the chapter. Check off the box that precedes each entry when you feel confident about the topic.

Checklist of key concepts

☐ 1 A system is classified as open, closed, or isolated.

☐ 2 The surroundings remain at constant temperature

Roadmap of key equations

You don't have to memorize every equation in the text. A roadmap at the end of the chapter summarizes the relations between key equations and, where appropriate, indicates the approximations you have to make in order to transform an equation into a simpler form.

Together, the 'Checklists' and 'Roadmaps' will help you to see how the concepts and equations of physical chemistry come together, forming a network of ideas rather than an overwhelming, scattered collection.

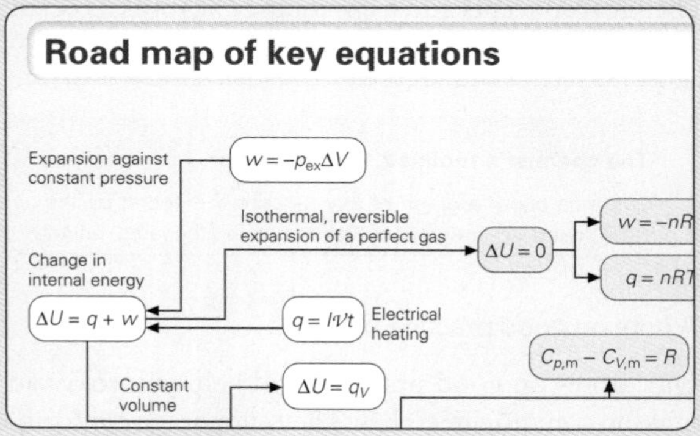

Resource section

Thermodynamic, kinetic, and spectroscopic data help you to understand trends in chemical and physical properties and become familiar with the magnitudes of properties. But long tables can break up the flow of the text, so where appropriate they have been collected at the end of the book.

Becoming a problem solver

Brief illustrations

Brief illustrations are short examples of how to use equations that have been introduced in the text. They will teach you how to use data, manipulate units correctly, and become familiar with the implications of an equation.

> **Brief illustration 2.3** The energy of heating
>
> If the heat capacity of a beaker of water is 0.50 kJ K^{-1}, and we observe a temperature rise of 4.0 K, then we can infer that the heat transferred to the water is
>
> $$q = (0.50\ \text{kJ K}^{-1}) \times (4.0\ \text{K}) = +2.0\ \text{kJ}$$

Examples

Working through problems in physical chemistry is a highly personal business, but in each 'Example' we suggest a 'Strategy' for organizing the information in a problem and then finding its solution. Following this section is the worked-out 'Answer', where again we note the importance of using units correctly.

> **Example 1.2**
>
> Calculating mole fractions
>
> A mass of 100.0 g of dry air consists of 75.5 g of N_2, 23.2 g of O_2, and 1.3 g of Ar. Express the composition of dry air in terms of mole fractions.

Webcasts

These online videos walk you through the solutions of a selection of 'Examples', just as your tutor might do in class.

Testing yourself

Self-tests

It is important to monitor your own progress, so you should use the 'Self-tests' to check that you have mastered the concepts and procedures in the 'Examples' and 'Brief illustrations'.

> **Self-test 0.4**
>
> Determine the pressure exerted by someone of mass 64 kg whose shoes have a combined surface area of 480 cm^2.
>
> *Answer:* 13 kPa

Discussion questions

The end-of-chapter material starts with a short set of questions that are intended to encourage reflection on the material and to view it conceptually before tackling it mathematically.

Exercises

The 'Exercises' allow you to assess your grasp of the material covered in the chapter.

Multiple choice questions

A bank of self-assessing multiple choice questions, with worked-out solutions, is provided for each chapter. Make the most of these questions to cement your knowledge and use them to draw your attention to areas where you need further study.

> Question 1
>
> A cylinder of volume 1810 dm^3 contains gas at a pressure of 197 atm of 25°C. Assuming that the gas behaves as a perfect gas, calculate th contained in the cylinder.
>
> a) 298 mol
>
> b) 1720 mol

Taking the material further

Impact sections

The 'Impact' sections show how the principles developed in a chapter are being applied to a selection of modern problems in a variety of disciplines, especially biology and materials science.

> **Impact on biology 4.2**
>
> Life and the Second Law
>
> Every chemical reaction that is spontaneous under conditions of constant temperature and pressure, including those driving the processes of growth, learning, and reproduction, is a reaction that proceeds in the direction of lower Gibbs energy, or—another way of expressing the same thing—

Further information

Sometimes a derivation is too long, too detailed, or too different in level for it to be included in the text. Where this is the case, they can be found less obtrusively at the end of the chapter.

Projects

At the end of a chapter you will find a small collection of 'Projects', which challenge you, sometimes through the use of calculus, to weave together many threads of foregoing material.

Using related resources

Solutions manual

The accompanying Solutions Manual (ISBN: 978-019-967449-7) provides full, detailed solutions to the 'Discussion questions', 'Exercises', and 'Projects'. As well as being a place to check your answers, it is also a very helpful learning aid.

Explorations in Physical Chemistry

Explorations in Physical Chemistry consists of interactive Mathcad® worksheets and interactive Excel® workbooks, complete with thought-stimulating exercises. You can simulate chemical phenomena by changing parameters in over 75 graphs.

Figures with associated interactive graphs have this icon next to the figure legend.

This resource can be purchased as an online resource or as a CD-ROM (ISBN: 978-019-928894-6).

Using this package in the classroom

The following password protected resources are available for registered adopters:

Figures and Tables of data

If, as an instructor using this book, you wish to use the figures or tables in a lecture you may do so without charge (but not for commercial purposes without specific permission). Almost all are available in PowerPoint® format.

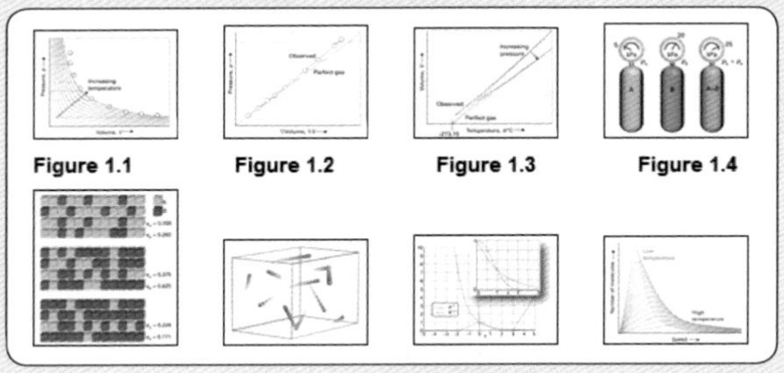

Test bank

A ready-made, electronic testing resource. It is fully customizable and contains feedback for students.

Preface

There is always room for improvement: the emphasis on one topic wanes and another becomes important; ways to clarify an explanation or improve the presentation become apparent either to the authors or at the suggestion of users; ideas emerge about how to give assistance with that ever-present challenge, the mathematics. Once again, therefore, we are pleased to have the opportunity to prepare a new edition with a variety of enhancements.

The structure of this new edition remains much the same as in the fifth edition. Although the structure is the same, there are many new features in this edition and we have paid special attention to the presentation of the mathematics:

- The new *Foundations* chapter replaces the Introduction of the fifth edition. In it we have tried to identify the topics, especially but not only from physics, that are needed to understand the principles later in the text, such as classical mechanics and electromagnetism.
- Chemists, and scientists in general, need a variety of mathematical techniques and concepts from physics and introductory chemistry to set up and explore their models, their data, and their theories. We have recognized this need in this edition by providing about 20 *Chemist's toolkits* at relevant points throughout the text. These short accounts will remind you of or provide some background to you about a variety of techniques as they become necessary as the text is developed.
- The *annotated equations* were popular in the fifth edition, and we have developed their use further.
- We have attached *labels* to all the main equations: that should help you to recall their meaning and relevance, and remind you of any conditions for them to be valid.
- In place of the lists of key equations in the fifth edition, we have provided a much more structured summary of the equations in the form of *Road maps*. These maps summarize the relations between expressions and, where appropriate, indicate the approximations you have to make in order to transform an equation into something simpler. They should also help you grasp the intellectual unity of the subject, showing that it is a network of equations rather than an overwhelming collection.
- The *Impact* sections, about one in each chapter, replaces the Boxes of earlier editions and show how the principles developed in the chapter are currently being applied in a variety of modern contexts, especially biology and materials science.
- The *Brief illustrations* are now numbered and there are more of them. Each one shows how to use an equation that has just been introduced in the text, paying particular attention to the correct use of units.
- We have increased the number of the all-important worked *Examples*, all of which, as always, have strategy sections to help you collect your thoughts.
- Almost all the *Brief illustrations* are now followed (like many of the worked *Examples*) by a *Self-test*, to enable you to assess your grasp of the calculation or topic. We consider it very important that you become familiar with the magnitudes of properties and appreciate that equations make useful, practical predictions.
- We have added more steps to the *Derivations* where we were advised it was appropriate in order to make the flow of mathematics easier to follow.
- The Appendices have been eliminated with most of their material present either in the *Foundations* chapter or at the point of use in *The chemist's toolkits*.
- There is a new *Resource section*, which now consolidates material on symbols and units and the *Data section*.

As always in the preparation of a new edition we have relied heavily on advice from users throughout the world, our numerous translators into other languages, and colleagues who have given their time in the reviewing process. Although we have always had sporadic feedback from student readers from around the world, for this edition we have collected comments from student panels, and we are very grateful to them for putting us so closely in touch with their attitudes, difficulties, and needs. We would particularly like to thank David Smith, of Bristol University, for the care with which he reviewed the drafts and his extensive work on the end-of-chapter Exercises, which he reviewed in detail and augmented where he felt it appropriate.

Our publishers have, as always, been a pleasure to work with and supportive throughout.

PWA
JdeP

About the authors

Peter Atkins is a fellow of Lincoln College in the University of Oxford and the author of about seventy books for students and a general audience. His texts are market leaders around the globe. A frequent lecturer in the United States and throughout the world, he has held visiting professorships in France, Israel, Japan, China, and New Zealand. He was the founding chairman of the Committee on Chemistry Education of the International Union of Pure and Applied Chemistry and was a member of IUPAC's Physical and Biophysical Chemistry Division.

Julio de Paula is Professor of Chemistry at Lewis & Clark College. A native of Brazil, Professor de Paula received a B.A. degree in chemistry from Rutgers, The State University of New Jersey, and a Ph.D. in biophysical chemistry from Yale University. His research activities encompass the areas of molecular spectroscopy, biophysical chemistry, and nanoscience. He has taught courses in general chemistry, physical chemistry, biophysical chemistry, instrumental analysis, and writing.

Acknowledgements

The authors have received a great deal of help during the preparation and production of this text and wish to thank all their colleagues who have made such thought-provoking and useful suggestions. In particular, we wish to record publicly our thanks to:

Hashim M. Ali, Arkansas State University
Chris Amodio, University of Surrey
Teemu Arppe, University of Helsinki
Jochen Autschbach, State University Of New York–Buffalo
Anil C. Banerjee, Columbus State University
Simon Biggs, University of Leeds
Timothy Brewer, Eastern Michigan University
Jorge Chacón, University of the West of Scotland
Anders Ericsson, Uppsala University
Stefan Franzen, North Carolina State University
Qingfeng Ge, Southern Illinois University
Fiona Gray, University of St Andrews
Ron Haines, University of New South Wales
Grant Hill, Glasgow University
Meez Islam, Teesside University
Emily A. A. Jarvis, Loyola Marymount University
Peter B. Karadakov, University of York
Peter Kroll, University of Texas at Arlington
Yu Kay Law, Fort Hays State University
Kristi Lazar, Westmont College
Mike Lyons, Trinity College Dublin
Alexandra J. MacDermott, University of Houston-Clear Lake
Michael D. McCorcle, Evangel University
Katie Mitchell-Koch, Emporia State University
Damien M. Murphy, Cardiff University
Nixon O. Mwebi, Jacksonville State University
Martin J. Paterson, Heriot-Watt University
Greg Van Patten, Ohio University
Julia Percival, University of Surrey
Patricia Redden, St. Peter's College
Juliana Serafin, University of Charleston
Susan Sinnott, University of Florida
Alyssa C. Thomas, Utica College
Harald Walderhaug, University of Oslo
T. Ffrancon Williams, The University of Tennessee
Christopher A. Wilson, Lewis University

Student reviewers

Frances Anastassacos, University of St Andrews
Jonathan Booth, University of York
Sinead Brady, University of St Andrews
Gareth Davis, University of Newcastle
Kate Horner, University of York
Sinead Keaveney, University of New South Wales
Emily McHale, University of St Andrews
James McManus, University of Newcastle
Maria O'Brien, Trinity College Dublin
Riccardo Serreli, Heriot-Watt University
Kristina Sladekova, University of the West of Scotland
Patrycja Stachelek, University of Newcastle
Eden Tanner, University of New South Wales
Jay Pritchard, University of St Andrews
Christopher Redford, University of New South Wales
Lisa Russell, Trinity College Dublin
Matthew Ryder, Heriot-Watt University

Brief contents

Foundations 1

1 The properties of gases 18

2 Thermodynamics: the First Law 47

3 Thermodynamics: applications of the First Law 69

4 Thermodynamics: the Second Law 90

5 Physical equilibria: pure substances 113

6 Physical equilibria: the properties of mixtures 133

7 Chemical equilibria: the principles 165

8 Chemical equilibria: solutions 186

9 Chemical equilibria: electrochemistry 209

10 Chemical kinetics: the rates of reactions 235

11 Chemical kinetics: accounting for the rate laws 261

12 Quantum theory 287

13 Quantum chemistry: atomic structure 315

14 Quantum chemistry: the chemical bond 344

15 Molecular interactions 376

16 Macromolecules and aggregates 394

17 Metallic, ionic, and covalent solids 420

18 Solid surfaces 449

19 Spectroscopy: molecular rotations and vibrations 477

20 Spectroscopy: electronic transitions 503

21 Spectroscopy: magnetic resonance 529

22 Statistical thermodynamics 551

Resource section 569

1 Quantities and units 570

2 Data section 572

Full contents

List of The chemist's toolkits xx
List of tables xxi

Foundations **1**

Matter 1
0.1 Mass and amount of substance 2
0.2 Volume 4
0.3 Density 4
0.4 Extensive and intensive properties 4

Energy 5
0.5 Velocity and momentum 5
0.6 Acceleration 6
0.7 Force 6
0.8 Pressure 6
0.9 Work and energy 8
0.10 Temperature 10
0.11 Quantization and the Boltzmann distribution 10
0.12 Equipartition 12

Electromagnetic radiation 12
0.13 Electromagnetic waves 12
0.14 Photons 13

Building on the foundations 14
QUESTIONS AND EXERCISES 15

Chapter 1
The properties of gases **18**

Equations of state 18
1.1 The perfect gas equation of state 19
1.2 Using the perfect gas law 22
1.3 Mixtures of gases: partial pressures 23
Impact on environmental science 1.1: The gas laws and the weather 25

The molecular model of gases 26
1.4 The pressure of a gas according to the kinetic model 27
1.5 The average speed of gas molecules 27
1.6 The Maxwell distribution of speeds 28
1.7 Diffusion and effusion 30
1.8 Molecular collisions 32

Real gases 33
1.9 Molecular interactions 33
1.10 The critical temperature 34
1.11 The compression factor 36
1.12 The virial equation of state 36
1.13 The van der Waals equation of state 37
1.14 The liquefaction of gases 40
FURTHER INFORMATION 1.1: KINETIC MOLECULAR THEORY 41
CHECKLIST OF KEY CONCEPTS 42
ROAD MAP OF KEY EQUATIONS 43
QUESTIONS AND EXERCISES 43

Chapter 2
Thermodynamics: the First Law **47**

The conservation of energy 48
2.1 Systems and surroundings 48
2.2 Work and heat 49
2.3 The molecular interpretation of work, heat, and temperature 50
2.4 The measurement of work 51
2.5 The measurement of heat 55
2.6 Heat influx during expansion 58

Internal energy and enthalpy 58
2.7 The internal energy 59
2.8 Internal energy as a state function 60
2.9 The enthalpy 62
2.10 The temperature variation of the enthalpy 63
CHECKLIST OF KEY CONCEPTS 65
ROAD MAP OF KEY EQUATIONS 66
QUESTIONS AND EXERCISES 66

Chapter 3
Thermodynamics: applications of the First Law **69**

Physical change 69
3.1 The enthalpy of phase transition 70

	Impact on biochemistry 3.1: Differential scanning calorimetry	73
3.2	Atomic and molecular change	74
Chemical change		**78**
3.3	Enthalpies of combustion	78
	Impact on technology 3.2: Fuels	79
	Impact on biochemistry 3.3: Food and energy reserves	80
3.4	The combination of reaction enthalpies	80
3.5	Standard enthalpies of formation	82
3.6	The variation of reaction enthalpy with temperature	83
	CHECKLIST OF KEY CONCEPTS	85
	ROAD MAP OF KEY EQUATIONS	86
	QUESTIONS AND EXERCISES	86

Chapter 4
Thermodynamics: the Second Law **90**

Entropy		**91**
4.1	The direction of spontaneous change	91
4.2	Entropy and the Second Law	92
	Impact on technology 4.1: Heat engines, refrigerators, and heat pumps	93
4.3	The entropy change accompanying expansion	94
4.4	The entropy change accompanying heating	95
4.5	The entropy change accompanying a phase transition	97
4.6	Entropy changes in the surroundings	99
4.7	The molecular interpretation of entropy	100
4.8	Absolute entropies and the Third Law of thermodynamics	101
4.9	The molecular interpretation of Third-Law entropies	103
4.10	The standard reaction entropy	105
4.11	The spontaneity of chemical reactions	105
The Gibbs energy		**106**
4.12	Focussing on the system	106
4.13	Properties of the Gibbs energy	107
	Impact on biology 4.2: Life and the Second Law	108
	CHECKLIST OF KEY CONCEPTS	109
	ROAD MAP OF KEY EQUATIONS	109
	QUESTIONS AND EXERCISES	110

Chapter 5
Physical equilibria: pure substances **113**

The thermodynamics of transition		**113**
5.1	The condition of stability	114
5.2	The variation of Gibbs energy with pressure	114
5.3	The variation of Gibbs energy with temperature	116
Phase diagrams		**118**
5.4	Phase boundaries	119
5.5	The location of phase boundaries	120
5.6	Characteristic points	124
	Impact on technology 5.1: Supercritical fluids	125
5.7	The phase rule	126
5.8	Phase diagrams of typical materials	127
5.9	The molecular structure of liquids	129
	CHECKLIST OF KEY CONCEPTS	130
	ROAD MAP OF KEY EQUATIONS	131
	QUESTIONS AND EXERCISES	131

Chapter 6
Physical equilibria: the properties of mixtures **133**

The thermodynamic description of mixtures		**133**
6.1	Measures of concentration	134
6.2	Partial molar properties	135
6.3	Spontaneous mixing	138
6.4	Ideal solutions	139
6.5	Ideal–dilute solutions	142
	Impact on biology 6.1: Gas solubility and breathing	145
6.6	Real solutions: activities	146
Colligative properties		**146**
6.7	The modification of boiling and freezing points	147
6.8	Osmosis	149
Phase diagrams of mixtures		**152**
6.9	Mixtures of volatile liquids	153
6.10	Liquid–liquid phase diagrams	155
6.11	Liquid–solid phase diagrams	156
	Impact on technology 6.2: Ultrapurity and controlled impurity	159

6.12 The Nernst distribution law 159
CHECKLIST OF KEY CONCEPTS 160
ROAD MAP OF KEY EQUATIONS 160
QUESTIONS AND EXERCISES 161

Chapter 7
Chemical equilibria: the principles 165

Thermodynamic background 165
7.1 The reaction Gibbs energy 166
7.2 The variation of $\Delta_r G$ with composition 167
7.3 Reactions at equilibrium 169
7.4 The standard reaction Gibbs energy 170
Impact on biochemistry 7.1: Coupled reactions in biochemical processes 172
7.5 The equilibrium composition 173
7.6 The equilibrium constant in terms of concentration 175
7.7 The molecular interpretation of equilibrium constants 176

The response of equilibria to the conditions 177
7.8 The effect of temperature 177
7.9 The effect of compression 179
7.10 The presence of a catalyst 181
Impact on biochemistry 7.2: Binding of oxygen to myoglobin and haemoglobin 181
CHECKLIST OF KEY CONCEPTS 182
ROAD MAP OF KEY EQUATIONS 183
QUESTIONS AND EXERCISES 183

Chapter 8
Chemical equilibria: solutions 186

Proton transfer equilibria 186
8.1 Brønsted–Lowry theory 186
8.2 Protonation and deprotonation 187
8.3 Polyprotic acids 192
8.4 Amphiprotic systems 195

Salts in water 196
8.5 Acid–base titrations 196
8.6 Buffer action 198
Impact on medicine 8.1: Buffer action in blood 200
8.7 Indicators 200

Solubility equilibria 202
8.8 The solubility constant 202
8.9 The common-ion effect 203
8.10 The effect of added salts on solubility 204
CHECKLIST OF KEY CONCEPTS 205
ROAD MAP OF KEY EQUATIONS 206
QUESTIONS AND EXERCISES 206

Chapter 9
Chemical equilibria: electrochemistry 209

Ions in solution 209
9.1 The Debye–Hückel theory 210
9.2 The migration of ions 213
Impact on biochemistry 9.1: Ion channels and pumps 216

Electrochemical cells 217
9.3 Half-reactions and electrodes 217
9.4 Reactions at electrodes 219
9.5 Varieties of cell 221
9.6 The cell reaction 222
9.7 The cell potential 222
9.8 Cells at equilibrium 224
9.9 Standard potentials 225
9.10 The variation of potential with pH 225
9.11 The determination of pH 227

Applications of standard potentials 228
9.12 The electrochemical series 228
9.13 The determination of thermodynamic functions 228
Impact on technology 9.2: Fuel cells 230
CHECKLIST OF KEY CONCEPTS 231
ROAD MAP OF KEY EQUATIONS 231
QUESTIONS AND EXERCISES 232

Chapter 10
Chemical kinetics: the rates of reactions 235

Empirical chemical kinetics 236
10.1 Spectrophotometry 236
10.2 Experimental techniques 237

Reaction rates 238
10.3 The definition of rate 238
10.4 Rate laws and rate constants 239

10.5 Reaction order 240
10.6 The determination of the rate law 241
10.7 Integrated rate laws 243
10.8 Half-lives and time constants 247

The temperature dependence of reaction rates 248

10.9 The Arrhenius parameters 249
10.10 Collision theory 251
10.11 Transition-state theory 253
CHECKLIST OF KEY CONCEPTS 256
ROAD MAP OF KEY EQUATIONS 257
QUESTIONS AND EXERCISES 257

Chapter 11
Chemical kinetics: accounting for the rate laws **261**

Reaction schemes 261

11.1 The approach to equilibrium 261
11.2 Relaxation methods 263
11.3 Consecutive reactions 265

Reaction mechanisms 266

11.4 Elementary reactions 266
11.5 The formulation of rate laws 267
11.6 The steady-state approximation 268
11.7 The rate-determining step 269
11.8 Kinetic control 270
11.9 Unimolecular reactions 270

Reactions in solution 272

11.10 Activation control and diffusion control 272
11.11 Diffusion 273

Homogeneous catalysis 276

11.12 Acid and base catalysis 276
11.13 Enzymes 277

Chain reactions 280

11.14 The structure of chain reactions 280
11.15 The rate laws of chain reactions 281
FURTHER INFORMATION 11.1: FICK'S LAWS OF DIFFUSION 282
CHECKLIST OF KEY CONCEPTS 283
ROAD MAP OF KEY EQUATIONS 283
QUESTIONS AND EXERCISES 284

Chapter 12
Quantum theory **287**

The emergence of quantum theory 287

12.1 Atomic and molecular spectra: discrete energies 288
12.2 The photoelectric effect: light as particles 289
12.3 Electron diffraction: electrons as waves 290

The dynamics of microscopic systems 292

12.4 The Schrödinger equation 292
12.5 The Born interpretation 293
12.6 The uncertainty principle 295

Applications of quantum mechanics 297

12.7 Translation 298
12.8 Rotational motion 303
12.9 Vibration: the harmonic oscillator 307
FURTHER INFORMATION 12.1: THE SEPARATION OF VARIABLES PROCEDURE 310
CHECKLIST OF KEY CONCEPTS 311
ROAD MAP OF KEY EQUATIONS 311
QUESTIONS AND EXERCISES 312

Chapter 13
Quantum chemistry: atomic structure **315**

Hydrogenic atoms 315

13.1 The spectra of hydrogenic atoms 316
13.2 The permitted energies of hydrogenic atoms 316
13.3 Quantum numbers 318
13.4 The wavefunctions: s orbitals 321
13.5 The wavefunctions: p and d orbitals 324
13.6 Electron spin 325
13.7 Spectral transitions and selection rules 326

The structures of many-electron atoms 327

13.8 The orbital approximation 327
13.9 The Pauli principle 328
13.10 Penetration and shielding 328
13.11 The building-up principle 329
13.12 The occupation of d orbitals 330
13.13 The configurations of cations and anions 331
13.14 Self-consistent field orbitals 331

Periodic trends in atomic properties 332
13.15 Atomic radius 332
13.16 Ionization energy and electron affinity 333
The spectra of complex atoms 335
13.17 Term symbols 335
13.18 Spin–orbit coupling 338
13.19 Selection rules 338
Impact on astronomy 13.1: The spectroscopy of stars 338
FURTHER INFORMATION 13.1: THE PAULI PRINCIPLE 339
CHECKLIST OF KEY CONCEPTS 340
ROAD MAP OF KEY EQUATIONS 341
QUESTIONS AND EXERCISES 341

Chapter 14
Quantum chemistry: the chemical bond 344

Introductory concepts 345
14.1 The classification of bonds 345
14.2 Potential energy curves 345
Valence bond theory 346
14.3 Diatomic molecules 346
14.4 Polyatomic molecules 348
14.5 Promotion and hybridization 349
14.6 Resonance 352
14.7 The language of valence bonding 353
Molecular orbitals 353
14.8 Linear combinations of atomic orbitals 353
14.9 Bonding and antibonding orbitals 355
14.10 The structures of homonuclear diatomic molecules 356
14.11 The structures of heteronuclear diatomic molecules 363
14.12 The structures of polyatomic molecules 365
14.13 The Hückel method 366
Computational chemistry 369
14.14 Techniques 369
14.15 Graphical output 370
14.16 Applications 371
CHECKLIST OF KEY CONCEPTS 372
ROAD MAP OF KEY EQUATIONS 373
QUESTIONS AND EXERCISES 373

Chapter 15
Molecular interactions 376

van der Waals interactions 376
15.1 Interactions between partial charges 377
15.2 Electric dipole moments 377
15.3 Interactions between dipoles 380
15.4 Induced dipole moments 382
15.5 Dispersion interactions 384
The total interaction 385
15.6 Hydrogen bonding 385
15.7 The hydrophobic effect 386
15.8 Modelling the total interaction 387
Impact on medicine 15.1: Molecular recognition and drug design 389
Molecules in motion 390
CHECKLIST OF KEY CONCEPTS 390
ROAD MAP OF KEY EQUATIONS 391
QUESTIONS AND EXERCISES 391

Chapter 16
Macromolecules and aggregates 394

Biological and synthetic macromolecules 395
16.1 Models of structure 395
Impact on biochemistry 16.1: The prediction of protein structure 399
16.2 Mechanical properties of polymers 401
16.3 Electrical properties of polymers 403
Mesophases and disperse systems 403
16.4 Liquid crystals 403
16.5 Classification of disperse systems 404
16.6 Surface, structure, and stability 405
Impact on biochemistry 16.2: Biological membranes 407
16.7 The electric double layer 408
16.8 Liquid surfaces and surfactants 409
Determination of size and shape 411
16.9 Average molar masses 412
16.10 Mass spectrometry 413
16.11 Ultracentrifugation 414
16.12 Electrophoresis 415
16.13 Laser light scattering 415

CHECKLIST OF KEY CONCEPTS 416
ROAD MAP OF KEY EQUATIONS 417
QUESTIONS AND EXERCISES 417

Chapter 17
Metallic, ionic, and covalent solids **420**

Bonding in solids 420
17.1 The band theory of solids 421
17.2 The occupation of bands 422
17.3 The optical properties of junctions 424
17.4 Superconductivity 424
17.5 The ionic model of bonding 425
17.6 Lattice enthalpy 425
17.7 The origin of lattice enthalpy 427
17.8 Covalent networks 429
Impact on technology 17.1: Nanowires 429
17.9 Magnetic properties of solids 430

Crystal structure 432
17.10 Unit cells 432
17.11 The identification of crystal planes. 433
17.12 The determination of structure 435
17.13 Bragg's law 437
17.14 Experimental techniques 438
17.15 Metal crystals 440
17.16 Ionic crystals 442
17.17 Molecular crystals 443
Impact on biochemistry 17.2: X-ray crystallography of biological macromolecules 444
CHECKLIST OF KEY CONCEPTS 445
ROAD MAP OF KEY EQUATIONS 446
QUESTIONS AND EXERCISES 446

Chapter 18
Solid surfaces **449**

The growth and structure of surfaces 449
18.1 Surface growth 450
18.2 Surface composition and structure 450

The extent of adsorption 455
18.3 Physisorption and chemisorption 455
18.4 Adsorption isotherms 456
18.5 The rates of surface processes 461

Catalytic activity at surfaces 463
18.6 Unimolecular reactions 463
18.7 The Langmuir–Hinshelwood mechanism 464
18.8 The Eley–Rideal mechanism 464
Impact on technology 18.1: Examples of heterogeneous catalysis 465

Processes at electrodes 467
18.9 The electrode–solution interface 467
18.10 The rate of electron transfer 468
18.11 Voltammetry 470
18.12 Electrolysis 472
CHECKLIST OF KEY CONCEPTS 473
ROAD MAP OF KEY EQUATIONS 474
QUESTIONS AND EXERCISES 474

Chapter 19
Spectroscopy: molecular rotations and vibrations **477**

Rotational spectroscopy 478
19.1 The rotational energy levels of molecules 478
19.2 Forbidden and allowed rotational states 482
19.3 Populations at thermal equilibrium 483
19.4 Rotational transitions: microwave spectroscopy 484
19.5 Linewidths 486
19.6 Rotational Raman spectra 488

Vibrational spectroscopy 489
19.7 The vibrations of molecules 489
19.8 Vibrational transitions 490
19.9 Anharmonicity 491
19.10 Vibrational Raman spectra of diatomic molecules 492
19.11 The vibrations of polyatomic molecules 492
19.12 Vibration–rotation spectra 495
19.13 Vibrational Raman spectra of polyatomic molecules 496
Impact on the environment 19.1: Climate change 497
CHECKLIST OF KEY CONCEPTS 498
ROAD MAP OF KEY EQUATIONS 499
QUESTIONS AND EXERCISES 499

Chapter 20
Spectroscopy: electronic transitions **503**

Ultraviolet and visible spectra 503
20.1 Practical considerations 505
20.2 Absorption intensities 505
20.3 The Franck–Condon principle 507
20.4 Specific types of transitions 508
Impact on biochemistry 20.1: Vision 509

Radiative and non-radiative decay 510
20.5 Fluorescence 511
20.6 Phosphorescence 511
20.7 Quenching 512
Impact on biochemistry 20.2: Photosynthesis 517
20.8 Lasers 518

Photoelectron spectroscopy 522
FURTHER INFORMATION 20.1: THE BEER–LAMBERT LAW 524
FURTHER INFORMATION 20.2: THE EINSTEIN TRANSITION PROBABILITIES 524
CHECKLIST OF KEY CONCEPTS 525
ROAD MAP OF KEY EQUATIONS 526
QUESTIONS AND EXERCISES 526

Chapter 21
Spectroscopy: magnetic resonance **529**

Nuclear magnetic resonance 529
21.1 Nuclei in magnetic fields 530
21.2 The technique 532

The information in NMR spectra 533
21.3 The chemical shift 533
21.4 The fine structure 535
21.5 Spin relaxation 539
21.6 Conformational conversion and chemical exchange 541
21.7 Two-dimensional NMR 542
Impact on medicine 21.1: Magnetic resonance imaging 542

Electron paramagnetic resonance 543
21.8 The *g*-value 544
21.9 Hyperfine structure 545
CHECKLIST OF KEY CONCEPTS 547
ROAD MAP OF KEY EQUATIONS 548
QUESTIONS AND EXERCISES 549

Chapter 22
Statistical thermodynamics **551**

The Boltzmann distribution 551
22.1 The general form of the Boltzmann distribution 552
22.2 The origins of the Boltzmann distribution 553

The partition function 553
22.3 The interpretation of the partition function 554
22.4 Examples of partition functions 555
22.5 The molecular partition function 557

Thermodynamic properties 557
22.6 The internal energy 557
22.7 The heat capacity 559
22.8 The entropy 560
22.9 The Gibbs energy 560
22.10 The equilibrium constant 561
FURTHER INFORMATION 22.1: THE CALCULATION OF PARTITION FUNCTIONS 563
FURTHER INFORMATION 22.2: THE EQUILIBRIUM CONSTANT FROM THE PARTITION FUNCTION 564
CHECKLIST OF KEY CONCEPTS 564
ROAD MAP OF KEY EQUATIONS 565
QUESTIONS AND EXERCISES 565

Resource section **569**

1 Quantities and units 570
2 Data section 572
1 Thermodynamic data 572
2 Standard potentials 579

Index 581

List of The chemist's toolkits

0.1	Quantities and units	2
1.1	Graphs	20
1.2	Exponential and Gaussian functions	29
1.3	Differentiation	40
2.1	Integration	53
2.2	Logarithms	55
2.3	Electrical charge, current, power, and energy	57
6.1	Power series and expansions	149
7.1	Quadratic equations	175
9.1	The Coulomb interaction	210
9.2	Electric current	213
9.3	Oxidation numbers	218
10.1	Ordinary differential equations	244
11.1	Differential equations for kinetics	265
12.1	Vectors	306
12.2	Partial differential equations	310
13.1	Addition and subtraction of vectors	336
14.1	The Lewis theory of covalent bonding	345
14.2	The VSEPR model	345
14.3	Simultaneous equations	367

List of tables

0.1	Pressure units and conversion factors	7
1.1	The gas constant in various units	19
1.2	The molar volumes of gases at SATP	22
1.3	The composition of the Earth's atmosphere	25
1.4	Collision cross-sections of atoms and molecules	32
1.5	The critical temperatures of gases	35
1.6	van der Waals parameters of gases	38
2.1	Heat capacities of common materials	56
2.2	Temperature dependence of heat capacities	64
3.1	Standard enthalpies of transition at the transition temperature	71
3.2	First and second (and some higher) standard enthalpies of ionization	75
3.3	Standard electron gain enthalpies of the main-group elements	76
3.4	Selected bond enthalpies	76
3.5	Mean bond enthalpies	77
3.6	Standard enthalpies of combustion	79
3.7	Thermochemical properties of some fuels	80
3.8	Reference states of some elements	82
3.9	Standard enthalpies of formation	83
4.1	Entropies of vaporization at 1 atm and the normal boiling point	98
4.2	Standard molar entropies of some substances	103
5.1	Vapour pressure	124
5.2	Critical constants	124
6.1	Henry's law constants for gases dissolved in water	144
6.2	Activities and standard states	146
6.3	Cryoscopic and ebullioscopic constants	147
7.1	Thermodynamic criteria of spontaneity	170
7.2	Standard Gibbs energies of formation	171
8.1	Acidity and basicity constants	189
8.2	Successive acidity constants of polyprotic acids	192
8.3	Indicator colour changes	201
8.4	Solubility constants	203
9.1	Ionic conductivities	214
9.2	Ionic mobilities in water	215
9.3	Standard potentials	226

10.1	Kinetic techniques for fast reactions	236
10.2	Kinetic data for first-order reactions	244
10.3	Kinetic data for second-order reactions	246
10.4	Integrated rate laws	247
10.5	Arrhenius parameters	249
11.1	Diffusion coefficients	273
13.1	Hydrogenic wavefunctions	319
13.2	Atomic radii of main-group elements	333
13.3	First ionization energies of main-group elements	334
13.4	Electron affinities of main-group elements	335
14.1	Hybrid orbitals	351
14.2	Electronegativities of main-group elements	363
14.3	Summary of *ab initio* calculations and spectroscopic data for linear polyenes	372
15.1	Partial charges in polypeptides	377
15.2	Dipole moments, mean polarizabilities, and polarizability volumes	378
15.3	Potential energy of molecular interactions	386
15.4	Lennard-Jones parameters for the (12,6) potential	389
16.1	Surface tensions of liquids	409
17.1	Lattice enthalpies	426
17.2	Madelung constants	428
17.3	Magnetic susceptibilities	431
17.4	The essential symmetries of the seven crystal systems	433
17.5	Radius ratio and crystal type	443
17.6	Ionic radii	443
18.1	Maximum observed enthalpies of physisorption	455
18.2	Enthalpies of chemisorption	456
18.3	Properties of catalysts	465
18.4	Chemisorption abilities	466
18.5	Exchange current densities and transfer coefficients	470
19.1	Moments of inertia	480
19.2	Properties of diatomic molecules	492
19.3	Typical vibrational wavenumbers	494
20.1	Colour, frequency, and energy of light	504
20.2	Values of R_0 for some donor–acceptor pairs	516
21.1	Nuclear constitution and the nuclear spin quantum number	530
21.2	Nuclear spin properties	530

Resource section

A1.1	The SI base units	570
A1.2	A selection of derived units	570
A1.3	Common SI prefixes	570
A1.4	Some common units	571
A2.1	Thermodynamic data for organic compounds	572
A2.2	Thermodynamic data for elements and inorganic compounds	573
A2.3a	Standard potentials in electrochemical order	579
A2.3b	Standard potentials in alphabetical order	580

Foundations

Physical chemistry provides a link between the properties of bulk matter and the behaviour of the particles—atoms, molecules, or ions—of which matter is composed. Physical chemists are concerned about the structure of matter and how and why it changes. They formulate theories to understand and explain chemical phenomena. These theories typically result in mathematical models that may be tested by comparison with experimental data.

Physical chemistry draws on two of the great pillars of modern physical science: **thermodynamics** and **quantum theory**. In thermodynamics, the focus is on the macroscopic, bulk properties of a system; quantum theory is commonly applied to the study of individual atoms and molecules. The two pillars are intimately connected, for the bulk properties of systems are determined and explained by quantum-mechanical effects.

This chapter is the foundation for everything else that follows. In it, we describe the fundamental principles that underpin physical chemistry. We also cover many of these principles in *The chemist's toolkits* within the individual chapters. However, these fundamentals are integral to an understanding of physical chemistry and it is useful to introduce them before we proceed. We start by describing the physical properties that characterize the states of matter. Although the term 'energy' is widely used in everyday language, in science it has a precise meaning that we describe in Section 0.9. Finally, we shall discuss electromagnetic radiation, which is central to many of the phenomena that we observe and interpret.

Matter 1

0.1 Mass and amount of substance 2

0.2 Volume 4

0.3 Density 4

0.4 Extensive and intensive properties 4

Energy 5

0.5 Velocity and momentum 5

0.6 Acceleration 6

0.7 Force 6

0.8 Pressure 6

0.9 Work and energy 8

0.10 Temperature 10

0.11 Quantization and the Boltzmann distribution 10

0.12 Equipartition 12

Electromagnetic radiation 12

0.13 Electromagnetic waves 12

0.14 Photons 13

Building on the foundations 14

QUESTIONS AND EXERCISES 15

Matter

Matter may exist as a solid, a liquid, or a gas. These different states are distinguished by their behaviour when enclosed within a container:

A **solid** retains its shape regardless of the shape of the container it occupies.

A **liquid** is a fluid form of matter that possesses a well-defined surface and fills the lower part of the container it occupies.

A **gas** is a fluid form of matter that fills any container it occupies.

The different states of matter result from the strength of the interactions between their atoms, ions, or molecules and hence their freedom to move past one another. In a solid, the atoms, ions, or molecules interact so strongly that they are locked together rigidly into either ordered crystalline arrays or disordered amorphous structures. They oscillate around their mean position and only rarely are they able to move past one another. The interactions are weaker in a liquid and the atoms, ions, or molecules possess sufficient energy to overcome them. As a result, they are able to move past one another in a restricted manner. They are in a continuous state of motion but travel only a fraction of a diameter before colliding with a neighbour. A gas is composed of much more widely separated atoms or molecules that are in continuous, rapid, and random motion. They are usually so far apart that they interact with one another only very weakly and may travel several, often many, diameters before colliding with another particle.

The transition from solid to liquid to gas results from the increased freedom of their constituent particles. As a sample is heated, the increased energy allows the atoms, ions, or molecules to overcome the attractive interactions that otherwise would tend to hold them rigidly, and they are able to move more freely as a liquid. The supply of more energy ultimately results in the molecules escaping from one another completely, and the liquid vaporizes and becomes a gas.

The chemist's toolkit 0.1 Quantities and units

The result of a measurement is a **physical quantity** that is reported as a numerical multiple of a unit:

physical quantity = numerical value × unit

It follows that units may be treated like algebraic quantities and may be multiplied, divided, and cancelled. Thus, the expression (physical quantity)/unit is the numerical value (a dimensionless quantity) of the measurement in the specified units. For instance, the mass m of an object could be reported as $m = 2.5$ kg or m/kg = 2.5. See the *Resource section* for a list of units. Although it is good practice to use only SI units, there will be occasions where accepted practice is so deeply rooted that physical quantities are expressed using other, non-SI units. By international convention, all physical quantities are represented by sloping symbols; all units are roman (upright).

Units may be modified by a prefix that denotes a factor of a power of 10. Among the most common SI prefixes are those listed in Table A1.3 in the *Resource section*. Examples of the use of these prefixes are:

$1\ \text{nm} = 10^{-9}\ \text{m} \quad 1\ \text{ps} = 10^{-12}\ \text{s} \quad 1\ \mu\text{mol} = 10^{-6}\ \text{mol}$

Powers of units apply to the prefix as well as the unit they modify. For example, $1\ \text{cm}^3 = 1\ (\text{cm})^3$, and $(10^{-2}\ \text{m})^3 = 10^{-6}\ \text{m}^3$. Note that $1\ \text{cm}^3$ does not mean $1\ \text{c(m}^3)$. When carrying out numerical calculations, it is usually safest to write out the numerical value of an observable in scientific notation (as $n.nnn \times 10^n$).

There are seven SI base units, which are listed in Table A1.1 in the *Resource section*. All other physical quantities may be expressed as combinations of these base units (see Table A1.2). **Molar concentration** (more formally, but very rarely, *amount of substance concentration*), for example, which is an amount of substance divided by the volume it occupies, can be expressed using the derived units of mol dm^{-3} as a combination of the base units for amount of substance and length. A number of these derived combinations of units have special names and symbols and we shall highlight them as they arise.

0.1 Mass and amount of substance

Mass, m, is a measure of the quantity of matter of a sample regardless of its chemical identity. Thus, 2 kg of lead contains twice as much matter as 1 kg of lead and indeed twice as much matter as 1 kg of anything. The *Système International* (SI) unit of mass is the **kilogram** (kg). Since December 2011, 1 kg has been defined in an abstract manner in terms of fundamental constants, although prior to that, the unit was defined as the mass of a certain block of platinum–iridium alloy preserved at Sèvres, outside Paris. For typical laboratory-sized samples it is usually more convenient to use a smaller unit and to express mass in grams (g), where 1 kg = 10^3 g. The chemist's toolkit 0.1 is a brief review of the SI.

A note on good practice Be sure to distinguish mass and weight. Mass is a measure of the quantity of matter, and is independent of location. Weight is the gravitational force exerted by an object, and depends upon the effect of gravity. An astronaut has a different weight on the Earth and the Moon, but the same mass.

In chemistry it is often more useful to know the numbers of each specific type of atom, ion, or molecule in a sample than the mass of each substance. However, a sample of water of mass 10 g consists of about 10^{23} H_2O molecules, and it is clearly not convenient to report the number of molecules. Instead, we report the **amount of substance**, which is also known as the **chemical amount**, n, in a sample. The amount of substance is expressed in terms of the SI unit **mole**, which is abbreviated as mol. The name mole is derived, ironically, from the Latin word meaning 'massive heap'. The mole is currently defined in terms of the number of ^{12}C atoms in exactly 12 g of carbon-12, which is close to 6.022×10^{23}. More specifically, **Avogadro's constant**, N_A, is the number of entities per mole, and is

$$N_A = 6.022\ 141\ 29 \times 10^{23}\ \text{mol}^{-1}$$ Avogadro's constant

Unless we require greater precision, we shall commonly approximate this value to 6.022×10^{23} mol^{-1}. We use the term 'entities' because the concept of chemical amount may apply to atoms, molecules, or formula units. Thus, a sample of hydrogen gas that consists of 1 mol H_2 contains $6.022\ldots \times 10^{23}$ hydrogen molecules, and a sample of water that consists of 2 mol H_2O contains $2 \times 6.022\ldots \times 10^{23} = 1.2\ldots \times 10^{24}$ water molecules.

A note on good practice Always specify the identity of the entities when using the unit mole to avoid any ambiguity. If, improperly, we report that a sample consisted of 1 mol of hydrogen, it would not be clear whether it consisted of 6×10^{23} hydrogen atoms (1 mol H) or 6×10^{23} hydrogen molecules (1 mol H_2).

Avogadro's constant is used to calculate the number of particles N in a sample from the chemical amount n:

Number of particles = chemical amount
× number of particles per mole

That is,

$$N = n \times N_A \qquad (0.1)$$ Relation between number and chemical amount

Avogadro's constant has units of mol^{-1} and the chemical amount is measured in units of mol. The units of these two quantities cancel when multiplied together and the number of entities is correctly expressed as a dimensionless quantity without units.

Brief illustration 0.1 Number and amount of atoms[1]

From eqn 0.1 rearranged into $n = N/N_A$, and being careful to label the chemical amount as n_{Cu}, so that it refers unambiguously to the amount of Cu atoms, a sample of copper containing 8.8×10^{22} Cu atoms corresponds to

$$n_{Cu} = \frac{N}{N_A} = \frac{8.8 \times 10^{22}}{6.022 \times 10^{23}\ \text{mol}^{-1}} = 0.15\ \text{mol}$$

Notice how much easier it is to report the amount of Cu atoms present rather than their actual number.

Molar mass, M, is the mass of the sample, m, divided by the amount of substance, n:

$$M = \frac{m}{n} \qquad (0.2)$$ Definition Molar mass

Molar mass is the mass per mole of atoms, molecules, or formula units and therefore has SI units of kilograms per mole (kg mol^{-1}), or, more commonly, grams per mole (g mol^{-1}). When we refer to the molar mass of an element we always mean the mass per mole of its atoms. When we refer to the molar mass of a compound, we always mean the molar mass of its molecules or, in the case of solid compounds in general, the mass per mole of its formula units, such as NaCl for sodium chloride and Cu_2Au for a specific alloy of copper and gold.

Care has to be taken to allow for the isotopic composition of an element, so we must use a suitably weighted mean of the masses of the isotopes present. The molar mass of a typical sample of carbon, the mass per mole of carbon atoms, with carbon-12 and carbon-13 atoms in their typical abundances, is 12.01 g mol^{-1}. The molar mass of water is the mass per mole of H_2O molecules, with the isotopic abundances of hydrogen and oxygen those of typical samples of the elements, and is 18.02 g mol^{-1}. The values obtained in this way are shown on the periodic table inside the back cover.

The molar mass of a compound of known composition is calculated by taking a sum of the molar masses of its constituent atoms. The molar mass of a compound of unknown composition is determined experimentally by using mass spectrometry in a similar way to the determination of atomic masses.

Brief illustration 0.2 Mass and amount of atoms

To find the amount of C atoms present in 21.5 g of carbon, given that the molar mass of carbon is 12.01 g mol^{-1}, from eqn 0.2 in the form $n = m/M$ we write (taking care to specify the species)

$$n_C = \frac{m}{M_C} = \frac{21.5\ \text{g}}{12.01\ \text{g mol}^{-1}} = 1.79\ \text{mol}$$

That is, the sample contains 1.79 mol C atoms.

[1] Throughout this text, we use Brief illustrations to show how an equation or a concept is used. Worked examples are used when an equation needs to be developed substantially or a problem analysed before it is solved.

Self-test 0.1

What amount of H_2O molecules is present in 10.0 g of water?

Answer: 0.555 mol H_2O

The terms **atomic weight** (AW) or **relative atomic mass** (RAM or A_r) and **molecular weight** (MW) or **relative molar mass** (RMM or M_r) are still commonly used to signify the numerical value of the molar mass of an element or compound, respectively. In practice the atomic or molecular weight is the molar mass with the units g mol^{-1} struck out. Thus, the atomic weight (or RAM) of a natural sample of carbon is 12.01 and the molecular weight (or RMM) of water is 18.02.

A note on good practice The terms 'atomic weight' and 'molecular weight' are ingrained in chemists' practice, and still acknowledged by the *International Union of Pure and Applied Chemistry* (IUPAC) as fixtures in the language despite the terms not referring to what is correctly regarded as 'weight' (the gravitational force on a body). In an attempt to encourage better usage, we largely avoid them.

The actual mass of an atom or molecule is denoted m and must be distinguished from its molar mass, M. The two quantities are related by $m = M/N_A$. Atomic and molecular masses, like any masses, are reported in kilograms and are the actual masses of the entities. To avoid the awkwardly small numbers that are typical of atomic and molecular masses, they are often reported as a multiple of the atomic mass constant, $m_u = 1.660\ 54 \times 10^{-27}$ kg.

Brief illustration 0.3 Molecular mass

The molar mass of ethane is 30.07 g mol^{-1}. The actual mass of a C_2H_6 molecule is

$$m = \frac{30.07\ \text{g mol}^{-1}}{6.022 \times 10^{23}\ \text{mol}^{-1}} = 4.993 \times 10^{-23}\ \text{g}$$

or 4.993×10^{-26} kg. In terms of the atomic mass constant,

$$\frac{m}{m_u} = \frac{4.993 \times 10^{-26}\ \text{kg}}{1.660\,54 \times 10^{-27}\ \text{kg}} = 30.07$$

That is, $m = 30.07m_u$.

0.2 Volume

The **volume**, V, of a sample is the extent of three-dimensional space that it occupies. Volume is expressed in cubic metres, m^3, and its sub-multiples, such as cubic decimetres, dm^3 (1 dm^3 = 10^{-3} m^3), and cubic centimetres (1 cm^3 = 10^{-6} m^3). It is also common to encounter the non-SI unit litre (1 L = 1 dm^3) and its submultiple, the millilitre (1 mL = 1 cm^3).

Brief illustration 0.4 Volume units

To carry out simple unit conversions, simply replace the fraction of the unit (such as cm) by its definition (in this case, 10^{-2} m). Thus, to convert 100 cm^3 to cubic decimetres (litres), use 1 cm = 10^{-1} dm, in which case 100 cm^3 = 100 (10^{-1} dm)^3, which is the same as 0.100 dm^3.

Self-test 0.2

Express a volume of 100 mm^3 in units of cm^3.

Answer: 0.100 cm^3

0.3 Density

Mass density, or, more commonly, just density, ρ (rho) is the mass of a sample, m, divided by its volume, V:

$$\rho = \frac{m}{V} \qquad \text{Definition} \quad \text{Mass density} \qquad (0.3)$$

Dense materials have a lot of matter packed into a small volume. With the mass measured in kilograms and the volume in cubic metres, density is reported in kilograms per cubic metre (kg m^{-3}); however, it is equally acceptable and often more convenient to report mass density in grams per cubic centimetre (g cm^{-3}). The relation between these units is

$$1\ \text{g cm}^{-3} = 10^3\ \text{kg m}^{-3}$$

The density of mercury, for example, may be reported as either 13.6 g cm^{-3} or as 1.36×10^4 kg m^{-3}.

0.4 Extensive and intensive properties

Properties such as mass and volume that depend upon the amount of substance in the sample are known as **extensive properties:** they depend on the 'extent' of the sample. In contrast, **intensive properties,** such as pressure and temperature do not depend upon the amount of substance present in a sample. For example, density is an intensive property, because its value does not depend upon the amount of substance present; doubling the amount of substance results in a doubling of both the mass and the volume, so their ratio remains the same. Density is thus an example of an intensive property that is a ratio of two extensive properties. Temperature is an intensive property because the temperature of a sample is independent of the size of the sample.

A **molar quantity** is the value of a property of a sample divided by the amount of substance in the sample:

$$X_m = \frac{X}{n} \qquad \text{Definition} \quad \text{Molar quantity} \qquad (0.4)$$

An example is molar volume, V_m, the volume occupied per mole of entities. By convention, the subscript m denotes a molar quantity. We have, however, already seen that the molar mass is denoted simply *M*.

A note on good practice Distinguish a molar quantity, such as the molar volume, with units of cubic metres per mole ($m^3\ mol^{-1}$), from the quantity *for* 1 mole, such as the volume occupied by 1 mole of the substance, with units cubic metres (m^3).

Energy

To appreciate much of what is to follow, it is essential to understand the concept of energy. To do so, we must first consider how objects such as atoms or molecules move under the influence of forces. We start by considering **classical mechanics**, the system of mechanics devised by Isaac Newton in the seventeenth century, which is appropriate for macroscopic particles (particles visible to the naked eye), and use it to describe the relation between the concepts of velocity, momentum, acceleration, force, work, and energy.

0.5 Velocity and momentum

Translation is the motion of a particle through space. The **speed**, v, of a body is defined as the rate of change of position. Speed is usually measured in SI units of metres per second ($m\ s^{-1}$). The concept of **velocity** is related to that of speed. The terms are not, however, synonymous: velocity defines the direction as well as the rate of motion, and particles travelling at the same speed but in different directions have different velocities.

The concepts of classical mechanics, and quantum mechanics too, are commonly expressed in terms of the **linear momentum**, p, which is defined as

$$p = mv \qquad \text{Definition} \quad \text{Linear momentum} \qquad (0.5)$$

Its units are kilogram metres per second ($kg\ m\ s^{-1}$). Momentum also mirrors velocity in having a sense of direction, and bodies of the same mass and moving at the same speed but in different directions have different linear momenta.

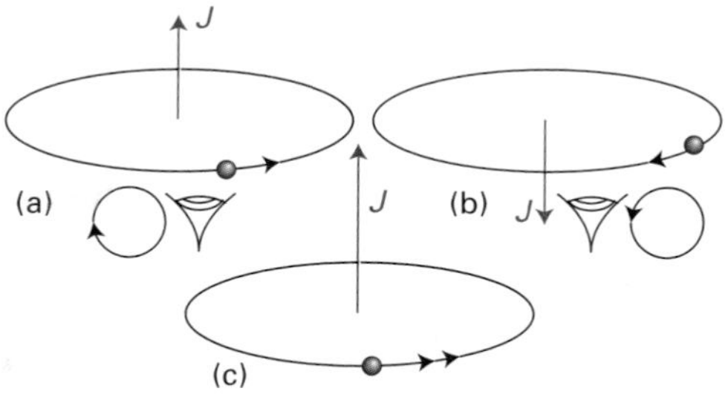

Fig. 0.1 Angular momentum has a sense of direction as well as a magnitude and can be represented by an arrow. The direction of the arrow indicates sense of rotation and its length represents the rate of rotation. (a) Low momentum, clockwise rotation (as seen from below); (b) low momentum, anticlockwise rotation; (c) high momentum, clockwise rotation.

We shall also need to consider the *rotation* of bodies, such as the motion of electrons around nuclei in atoms and the rotation of entire molecules. **Angular velocity**, ω (omega), is the rate of change of *angular* position; it is reported in radians per second ($rad\ s^{-1}$). There are 2π radians in a circle, so 1 cycle per second is the same as 2π radians per second. In that case, we would write $\omega = 2\pi\ s^{-1}$. Expressions for other angular properties then follow by analogy with the corresponding equations for linear motion. The **angular momentum**, J (Fig. 0.1), is, for example, defined, by analogy with eqn 0.5 as

$$J = I\omega \qquad \text{Definition} \quad \text{Angular momentum} \qquad (0.6)$$

The quantity I is the **moment of inertia** of the body, the sum of the product of the mass of each atom multiplied by the square of the distance from the axis of rotation. The moment of inertia represents the resistance of the body to a change in the state of rotation in the same way that mass represents the resistance of the body to a change in the state of translation.

● Brief illustration 0.5 The moment of inertia

There are two possible axes of rotation in a $C^{16}O_2$ molecule, each passing through the C atom and perpendicular to the axis of the molecule and to each other. Each O atom is at a distance R from the axis of rotation, where R is the length of a CO bond, 116 pm. The mass of each ^{16}O atom is $16.00m_u$. The C atom is stationary (it lies on the axis of rotation) and does not contribute to the moment of inertia. Therefore, the moment of inertia of the molecule around the rotation axis is

$$I = 2m(^{16}\text{O})R^2$$

$$= 2 \times \left(\overbrace{16.00 \times \overbrace{1.660\,54 \times 10^{-27}\,\text{kg}}^{m_u}}^{m(^{16}\text{O})}\right) \times \left(\overbrace{1.16 \times 10^{-10}\,\text{m}}^{R}\right)^2$$

$$= 7.15 \times 10^{-46}\ \text{kg m}^2$$

Note that the units of moments of inertia are kilograms metre squared (kg m^2).

0.6 Acceleration

Acceleration, *a*, is the rate of change of velocity. A body accelerates if its speed changes. A body also accelerates if its speed remains unchanged but its direction of motion changes. For example, a charged molecular fragment in a mass spectrometer accelerates when its speed increases as it flies in a straight line towards a detector. An electron moving with a constant speed in a circular trajectory in an electrostatic analyser is also accelerating, because the direction of its motion, and therefore its velocity, is continuously changing even though its speed is constant. Acceleration, is measured in SI units of metres per second squared (m s^{-2}).

0.7 Force

According to Newton's **second law of motion**, the acceleration *a* of a body of mass *m* is proportional to the force, *F*, acting on it:

$$F = ma \qquad \text{Force} \quad (0.7)$$

Eqn 0.7 implies that the SI units of force are those of the product of mass and acceleration, namely kilogram metres per second squared (kg m s^{-2}). Force is, however, such an important quantity that it is often expressed in terms of the newton, N, where 1 N = 1 kg m s^{-2}.

Brief illustration 0.6 The force on a freely falling body

The acceleration, *g*, of a freely falling body may be taken to be constant at the Earth's surface and has a value g = 9.81 m s^{-2}. The magnitude of the gravitational force acting on a body of mass *m* is therefore $F_{\text{gravitational}} = mg$, and for a mass of 1.0 kg at the surface of the Earth,

$$F_{\text{gravitational}} = (1.0\ \text{kg}) \times (9.81\ \text{m s}^{-2}) = 9.8\ \text{kg m s}^{-2}$$
$$= 9.8\ \text{N}$$

This force is directed towards the centre of the Earth. We call this force the *weight* of the body. It might be helpful to note that a force of 1 N is approximately the gravitational force exerted on a small apple (of mass 100 g).

Self-test 0.3

Calculate the gravitational force acting on a mass of 1.00 kg at the surface of the Moon, where the acceleration due to gravity is 1.63 m s^{-2}.

Answer: 1.63 N

Force is a quantity that has a sense of direction, and Newton's law indicates that the acceleration occurs in the same direction in which the force acts. If, for an isolated system, no external force acts, then there is no acceleration. This statement is the **law of conservation of momentum,** that the momentum of a body is constant in the absence of a force acting on the body.

0.8 Pressure

Pressure, *p*, is the ratio of the force, *F*, to the area, *A*, over which the force is exerted:

$$p = \frac{F}{A} \qquad \text{Definition Pressure} \quad (0.8)$$

Although both pressure and linear momentum are denoted by *p*, the context should always make it clear which is meant. The force may arise in many ways, including the result of the gravitational pull of the Earth on a body resting on a piston and the action of atoms or molecules colliding with the walls of a container.

The SI unit of pressure is called the **pascal** (Pa), where 1 Pa = 1 N m^{-2} = 1 kg m^{-1} s^{-2}. The pascal is approximately the same as the pressure exerted by a mass of 10 mg spread over 1 cm^2, so it is actually rather a small unit. However, it often proves convenient to use other units. One of the most commonly used alternatives is the **bar** where 1 bar = 10^5 Pa; the bar is not an SI unit, but it is an accepted and widely used abbreviation for 10^5 Pa. The atmospheric pressure that we normally experience is close to 1 bar. In due course, when we encounter various standard conditions, we shall refer to a **standard pressure,** $p^{\ominus}$ = 1 bar exactly. Some of the other non-SI units that are commonly used in reporting pressures are shown in Table 0.1.

The pressure resulting from the weight of a solid object at the surface of the Earth is

$$p = \frac{F_{\text{gravitational}}}{A} = \frac{mg}{A} \qquad (0.9)$$

Table 0.1
*Pressure units and conversion factors**

pascal, Pa	1 Pa = 1 N m^{-2}
bar	1 bar = 10^5 Pa
atmosphere, atm	1 atm = **101.325** kPa = **1.013 25** bar
torr, Torr†	**760** Torr = 1 atm 1 Torr = 133.32 Pa

* Values in bold are exact.
† The name of the unit is torr, its symbol is Torr.

An ice skater exerts a high pressure on the ice because the blades of the skates have a very small contact area with the surface of the ice. The pressure exerted by the skater wearing normal shoes would be much lower, because even though the mass of the skater remains the same, the contact area with the ice is greater.

Brief illustration 0.7 The pressure resulting from the weight of an object

The pressure exerted by a mass of 10 g resting on a surface area of 1.0 cm^2 at the surface of the Earth is, from eqn 0.9,

$$p = \frac{F_{\text{gravitational}}}{A} = \frac{mg}{A} = \frac{\overbrace{(10\text{ g})}^{m}\times\overbrace{(9.81\text{ m s}^{-2})}^{g}}{\underbrace{(1.0\text{ cm}^2)}_{A}}$$

$$= \frac{\overbrace{(1.0\times10^{-2}\text{ kg})}^{10\text{ g}}\times(9.81\text{ m s}^{-2})}{\underbrace{(1.0\times10^{-4}\text{ m}^2)}_{1.0\text{ cm}^2}}$$

$$= 9.8\times10^2\ \overbrace{\text{kg m}^{-1}\text{ s}^{-2}}^{\text{Pa}} = 0.98\text{ kPa}$$

This result is about 1/100 of the pressure exerted by the atmosphere at the Earth's surface (100 kPa).

Self-test 0.4

Determine the pressure exerted by someone of mass 64 kg whose shoes have a combined surface area of 480 cm^2.

Answer: 13 kPa

An incompressible fluid exerts a pressure at its base because of the gravitational force exerted by the mass of fluid above. A **hydrostatic pressure** of this kind differs from that exerted by a solid weight in the sense that a body immersed in the liquid at the base of the column experiences the same pressure on all its faces, even the underside, because the pressure is transmitted through the fluid to each face.

A body immersed in a gas, even one in a closed container and therefore not directly subject to gravitational effects on a column of fluid above it or in empty space, also experiences a pressure on each of its faces. This pressure is due to the impact of gas molecules, with each collision giving rise to a force and hence a pressure. The individual impacts are so numerous that the force is effectively constant. Atmospheric pressure results in this way from the collisions of the molecules in air and is greatest at the Earth's surface where the density of air is highest.

When a gas is confined to a cylinder fitted with a movable piston, the position of the piston adjusts until the pressure of the gas inside the cylinder is equal to that exerted by the atmosphere. When the pressures on either side of the piston are the same, we say that the two regions on either side are in **mechanical equilibrium** (Fig. 0.2).

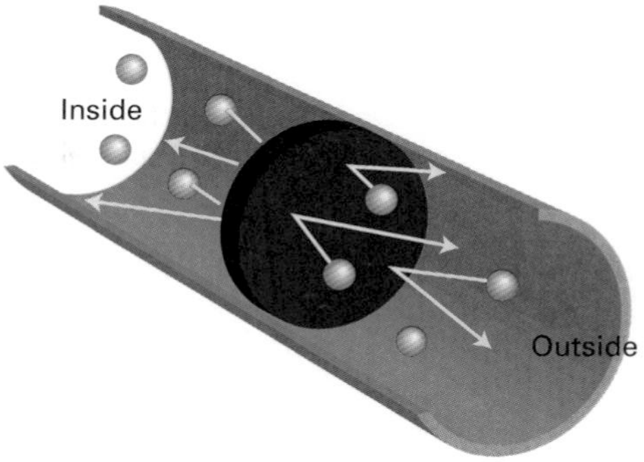

Fig. 0.2 A system is in mechanical equilibrium with its surroundings if it is separated from them by a movable wall and the external pressure is equal to the pressure of the gas in the system.

Example 0.1

Converting between units

A scientist was exploring the effect of atmospheric pressure on the rate of growth of a lichen, and measured a pressure of 1.115 bar. What is the pressure in atmospheres?

Strategy Write the relation between the 'old units' (the units to be replaced) and the 'new units' (the units required) in the form

1 old unit = x new units

then replace the 'old unit' everywhere it occurs by '*x* new units', and multiply out the numerical expression.

Solution From Table 0.1 we have 1.013 25 bar = 1 atm, with atm the 'new unit' and bar the 'old unit'. As a first step we write

$$1\ \text{bar} = \frac{1}{1.01325}\ \text{atm}$$

Then we replace bar wherever it appears by (1/1.013 25) atm:

$$p = 1.115\ \text{bar} = 1.115 \times \frac{1}{1.01325}\ \text{atm} = 1.100\ \text{atm}$$

A note on good practice The number of significant figures in the answer (four in this instance) is the same as the number of significant figures in the data; the relation between old and new numbers in this case is exact.

Self-test 0.5

The pressure in the eye of a hurricane was recorded as 723 Torr. What is the pressure in kilopascals?

Answer: 96.4 kPa

0.9 Work and energy

Work, w, is done when a body is moved against an opposing force. For example, work is done when a gas at high pressure expands, moving a piston against the force exerted by the external pressure.

If the opposing force is constant, then the work done is given by the product of the magnitude of the opposing force and the distance, d, through which the body is moved. The magnitude of this work (we worry about signs in Chapter 2) is

$$w = Fd \qquad \text{Work done against a force} \qquad (0.10)$$

Clearly, more work is done against a strong opposing force than when the force is weak. Mechanical work is done on a body when it is raised through a vertical distance against the force of gravity. The magnitude of the gravitational force acting on the body is $F_{\text{gravitational}} = mg$ (see Brief illustration 0.6). The magnitude of the work done in raising the object through a height h is therefore

$$w = \overbrace{mg}^{F} \times \overbrace{h}^{d} = mgh \qquad \text{At the surface of the Earth} \quad \text{Mechanical work} \qquad (0.11)$$

Brief illustration 0.8 The work of raising a weight

To raise a body of mass 1.0 kg on the surface of the Earth through a vertical distance (against the direction of the force) of 1.0 m requires the following work to be done:

$$w = (1.0\ \text{kg}) \times (9.81\ \text{m s}^{-2}) \times (1.0\ \text{m})$$

$$= 9.8\ \overbrace{\text{kg m}^2\ \text{s}^{-2}}^{\text{N m}} = 9.8\ \text{N m}$$

As we see more formally in the next section, the unit 1 N m (or, in terms of base units, 1 kg m^2 s^{-2}) is called 1 joule (1 J). So, 9.8 J is needed to raise a mass of 1.0 kg through 1.0 m on the surface of the Earth.

Self-test 0.6

An engine does work of 0.12 kJ in raising a mass of 250 g vertically from the surface of the Earth. Through what distance is the mass raised?

Answer: 49 m

Energy is defined as the capacity to do work. For example, a system such as a steam engine, which is connected through pulleys to a weight, has energy because it may do work in raising the weight against the force of gravity. An electrochemical cell also has energy because it may be used to drive an electrical motor and hence move a body against an opposing force.

A body may have energy by virtue of either its motion or its position. **Kinetic energy, E_k,** is the energy of a body due to its motion. For a body of mass m moving at a speed v,

$$E_k = \tfrac{1}{2}mv^2 \qquad \text{Linear motion} \quad \text{Kinetic energy} \qquad (0.12a)$$

That is, a heavy object moving at the same speed as a light object has a higher kinetic energy. We may also use the definition of linear momentum, eqn 0.5, to rewrite our expression for kinetic energy as

$$E_k = \frac{p^2}{2m} \qquad \text{Kinetic energy in terms of the linear momentum} \qquad (0.12b)$$

The kinetic energy of a rotating body is

$$E_k = \tfrac{1}{2}I\omega^2 = \frac{J^2}{2I} \qquad \text{Angular motion} \quad \text{Kinetic energy} \qquad (0.13)$$

by analogy with eqn 0.12a and 0.12b, with J the angular momentum and I the moment of inertia.

Brief illustration 0.9 Kinetic energy

The kinetic energy of a body of mass 1.0 kg travelling at 1.0 m s^{-1} is

$$E_k = \tfrac{1}{2} \times (1.0\ \text{kg}) \times (1.0\ \text{m s}^{-1})^2 = 0.5\ \overbrace{\text{kg m}^2\ \text{s}^{-2}}^{\text{J}} = 0.5\ \text{J}$$

The moment of inertia of a typical bicycle wheel is about 0.050 kg m^2. When travelling at 15 mph (24 km h^{-1}, 6.7 m s^{-1}) it makes 3.6 revolutions per second, corresponding to 2π × 3.6 radians per second and therefore to $\omega = 2\pi \times 3.6\ \text{s}^{-1}$. Its rotational kinetic energy is therefore

$$E_k = \tfrac{1}{2} \times (0.050\ \text{kg m}^2) \times (2\pi \times 3.6\ \text{s}^{-1})^2$$

$$= 1.3 \times 10^2\ \overbrace{\text{kg m}^2\ \text{s}^{-2}}^{\text{J}}$$

or 0.13 kJ.

The **potential energy,** E_p, of a body is the energy it possesses due to its position. The precise dependence on position depends on the type of force acting on the body. For a body of mass m on the surface of the Earth, the **gravitational potential energy** relative to its value at the surface itself depends on its height, h, above the surface as

$$E_p = mgh \qquad \text{Gravitational potential energy} \qquad (0.14)$$

A force acts between charges and results in an **electrostatic potential energy.** The SI unit of charge is the coulomb, C. The fundamental charge e is

$$e = 1.602\ 176\ 565 \times 10^{-19}\ \text{C}$$

An electron has a charge of $-e$ and a proton a charge of $+e$. When we do not require great precision, we shall use $e = 1.602 \times 10^{-19}$ C. The coulomb is thus a relatively large unit with a charge of 1 C corresponding to 6×10^{18} electrons. For two electric charges Q_1 and Q_2 separated by a distance r the electrostatic potential energy, which is known as the **Coulomb potential energy,** is

$$E_p = \frac{Q_1 Q_2}{4\pi\varepsilon r} \qquad \text{Coulomb potential energy} \qquad (0.15)$$

The quantity ε (epsilon) is the **permittivity;** its value depends upon the nature of the medium between the charges. If the charges are separated by a vacuum, then the constant is known as the **vacuum permittivity,** ε_0 (epsilon zero), which has the value $8.854 \times 10^{-12}\ \text{J}^{-1}\ \text{C}^2\ \text{m}^{-1}$. The permittivity is greater for other media, such as air, water, or oil.

The Coulomb potential energy is inversely proportional to the separation of the charges and is zero when the charges are infinitely far apart. The potential energy of two charges with the same sign is positive and rises as they are brought together. In other words, work must be done to bring like charges together from infinite separation. In contrast, the potential energy is negative for two charges with opposing signs. Their potential energy falls as they are brought together. As we shall see as the text develops, most contributions to the potential energy that we need to consider in chemistry are due to the Coulombic interaction.

Brief illustration 0.10 The Coulomb potential energy

The Coulomb potential energy resulting from the electrostatic interaction between a positively charged sodium cation, Na^+, and a negatively charged chloride anion, Cl^-, at a distance of 0.28 nm, which is the separation between ions in the lattice of a sodium chloride crystal, is

$$E_p = \frac{\overbrace{(-1.602 \times 10^{-19}\ \text{C})}^{Q(\text{Cl}^-)} \times \overbrace{(1.602 \times 10^{-19}\ \text{C})}^{Q(\text{Na}^+)}}{4\pi \times \underbrace{(8.854 \times 10^{-12}\ \text{C}^2\ \text{J}^{-1}\text{m}^{-1})}_{\varepsilon_0} \times \underbrace{(0.28 \times 10^{-9}\ \text{m})}_{r}}$$

$$= -8.2 \times 10^{-19}\ \text{J}$$

This value is equivalent to a molar energy of

$$E_p \times N_A = (-8.2 \times 10^{-19}\ \text{J}) \times (6.022 \times 10^{23}\ \text{mol}^{-1})$$
$$= -490\ \text{kJ mol}^{-1}$$

A note on good practice Write units at *every* stage of a calculation and do not simply attach them to a final numerical value. Also, it is often sensible to express all numerical quantities in scientific notation using exponential format rather than SI prefixes to denote powers of ten.

Self-test 0.7

The centres of neighbouring cations and anions in magnesium oxide crystals are separated by 0.21 nm. Determine the molar Coulomb potential energy resulting from the electrostatic interaction between a Mg^{2+} and a O^{2-} ion in such a crystal.

Answer: -2600 kJ mol^{-1}

The **total energy,** E, of a body is the sum of its kinetic and potential energies:

$$E = E_k + E_p \qquad \text{Total energy} \qquad (0.16)$$

Provided no external forces are acting on the body, its total energy is constant. This central statement of physics is known as the **law of the conservation of energy.** Potential and kinetic energy may be freely interchanged, but their sum remains constant. Thus, a falling ball loses potential energy but gains an equivalent amount of kinetic energy as it accelerates. A mass on a pendulum continually exchanges kinetic and potential energy as it swings. In all cases, however, the total energy remains constant provided the body is isolated from external influences. We use the law of the conservation of energy many times in the following chapters.

The SI unit for energy may be deduced from the equations for kinetic and potential energy. In all cases the combination of the SI base units of the quantities on the right-hand side of the equation is kg m^2 s^{-2}. The concept of energy occurs so often in science, and especially in physical chemistry, that energy is given its own derived SI unit called the **joule** (J), where 1 kg m^2 s^{-2} = 1 J. A joule is quite a small unit, and in chemistry we often deal with energies of the order of kilojoules (1 kJ = 10^3 J). We may also demonstrate that, because 1 N = 1 kg m s^{-2}, then 1 J = 1 kg m^2 s^{-2} = 1 N m, and hence show (as must be the case) that energy and work may be expressed in the same units.

0.10 Temperature

Temperature is the property of an object that determines in which direction energy will flow as heat when it is in contact with another object: energy flows from higher temperature to lower temperature. When two bodies have the same temperature, there is no net flow of energy between them. In that case we say that the bodies are in **thermal equilibrium** (Fig. 0.3).

A note on good practice Never confuse temperature with heat. Everyday language comes close to confusing them, by equating 'high temperature' with 'hot', but they are entirely different concepts. Heat—as we shall see in detail in Chapter 2—is a mode of transfer of energy; temperature is an intensive property that determines the direction of flow of energy as heat.

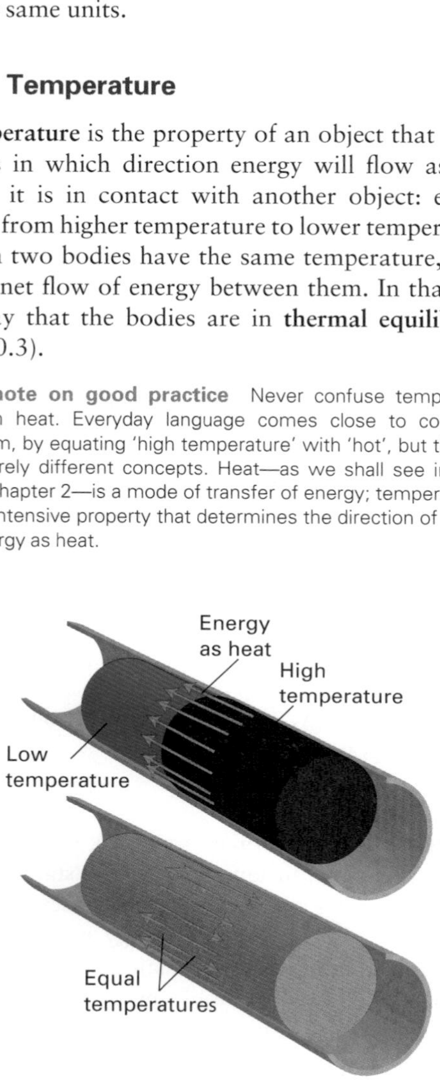

Fig. 0.3 The temperatures of two objects act as a signpost showing the direction in which energy will flow as heat through a thermally conducting wall. (a) Energy flows as heat from high temperature to low temperature. (b) When the two objects are at the same temperature, although there is still energy transfer in both directions, there is no net flow of energy as heat.

In science, we use the **thermodynamic temperature**, T. This 'absolute' scale sets $T = 0$ as the lowest possible temperature. Temperatures are then reported on the **Kelvin scale**. The scale is defined in terms of the 'triple point' of water, the temperature at which water, ice, and water vapour are in mutual equilibrium. That temperature is defined as 273.16 K, where K (not °K) denotes kelvin. On this scale, water freezes at very close to 273.15 K. In practice, scientists also make use of the **Celsius scale** and denote temperatures θ (theta). The Celsius scale is defined in terms of the Kelvin scale by

$$\theta/°\text{C} = T/\text{K} - 273.15 \quad \text{Definition Celsius scale} \quad (0.17)$$

The magnitude of 1 kelvin is the same as 1 degree Celsius. On the Celsius scale, water at 1 atm freezes at close to 0 °C and boils at close to 100 °C.

0.11 Quantization and the Boltzmann distribution

As we shall see in Chapter 12, experiments reveal that, contrary to the laws of classical mechanics and everyday experience, microscopic objects, such as atoms and molecules, exist in distinct, well-defined **states** with characteristic energies. Energy is therefore said to be **quantized**, meaning that it is allowed only certain discrete values. The separation in energy between these allowed states depends upon the characteristics of the motion and the mass of the particle. The separation in energy is greatest for objects of low mass (like electrons) that are confined to small regions of space (like atoms and molecules). Although the states resulting from electronic motion are, in general, widely separated, those arising from vibrational, rotational, and translational motion are progressively closer together. In practice, the separation in energy between the allowed translational states of atoms and molecules is so small that it may be ignored and the effects of quantization are insignificant. We therefore often treat the translational energy of atoms and molecules as if it is continuous. The quantization of the other modes of motion, however, cannot be ignored. Figure 0.4 shows the typical relative energy separations for each of these types of motion.

The atoms or molecules in a sample do not all have the same energy. Some have a lot of energy in the sense that they occupy high-energy states. Others have little energy in the sense that they occupy low-energy states. That is, the atoms or molecules of a sample are distributed over many different states. Moreover, the ceaseless exchange of energy as the

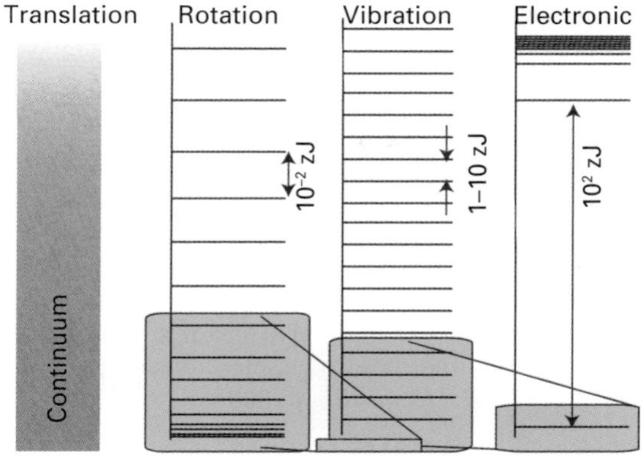

Fig. 0.4 The energy level separations typical of translational, rotational, vibrational, and electronic motion. Note that 1 zJ = 10^{-21} J (corresponding to about 0.6 kJ mol^{-1}).

particles in the sample collide with one another means that the atoms or molecules jump between the different states. Some are excited to higher levels and others relax to lower energy. It turns out that even though there are many possible arrangements of atoms or molecules over the states, one of them, the **Boltzmann distribution**, is overwhelmingly more likely to occur than any of the others. We therefore proceed as if the Boltzmann distribution is the only arrangement of the system.

According to the Boltzmann distribution, which is discussed in detail in Chapter 22, the relative populations, N_1 and N_2, of two states depends upon the temperature T and the difference in their energies, ε_1 and ε_2:

$$\frac{N_2}{N_1} = e^{-(\varepsilon_2-\varepsilon_1)/kT} \qquad \text{Boltzmann distribution} \qquad (0.18a)$$

The constant k is **Boltzmann's constant:**

$$k = 1.380\,6488 \times 10^{-23}\ \mathrm{J\,K^{-1}}$$

We shall normally use the approximation $k = 1.381 \times 10^{-23}\ \mathrm{J\,K^{-1}}$. In chemical applications it is common to use not the individual energies but energies per mole of molecules, E_i, with $E_i = N_A\varepsilon_i$, where N_A is Avogadro's constant. When both the numerator and denominator in the exponential are multiplied by N_A, eqn 0.18a becomes

$$\frac{N_2}{N_1} = e^{-(E_2-E_1)/RT} \qquad \text{Boltzmann distribution in terms of molar energies} \qquad (0.18b)$$

where $R = N_Ak$:

$$R = 8.314\,4621\ \mathrm{J\,K^{-1}\,mol^{-1}}$$

When such precision is not required, we shall normally approximate this value to 8.3145 J K^{-1} mol^{-1}.

The constant R is called the (molar) **gas constant** because it originally arose in relation to the properties of gases (Section 1.1), but the fact that it is related to Boltzmann's constant and the Boltzmann distribution means that it has a much broader range of application than to gases alone, as we shall see throughout the text.

The population, N, in a state with an energy $\Delta\varepsilon$ relative to the lowest state, which has a population N_0, is given by

$$\frac{N}{N_0} = e^{-\Delta\varepsilon/kT} \qquad \text{Boltzmann distribution} \qquad (0.18c)$$

This difference in the populations of the lower and higher energy states becomes more marked as the separation, $\Delta\varepsilon$, between the states increases. For example, because the separation in energy between different electronic states is usually large, most atoms or molecules are found in the lowest electronic energy state and the population of excited electronic states is usually very low. In contrast, because the rotational energy states of molecules lie much closer in energy, the populations of excited rotational states can be significant.

We can now see, also through eqn 0.18, the true significance of temperature: for a given set of quantum states, *the temperature is the single parameter that controls the relative population of the states.* The effect is shown in Fig. 0.5. When the temperature is low, there is only a small population of molecules in high-energy states. When the temperature is greater, many high-energy states are populated. We draw on the Boltzmann distribution many times in the following chapters because it gives great insight into the properties of matter and how they vary with temperature. The single point to keep in mind at this

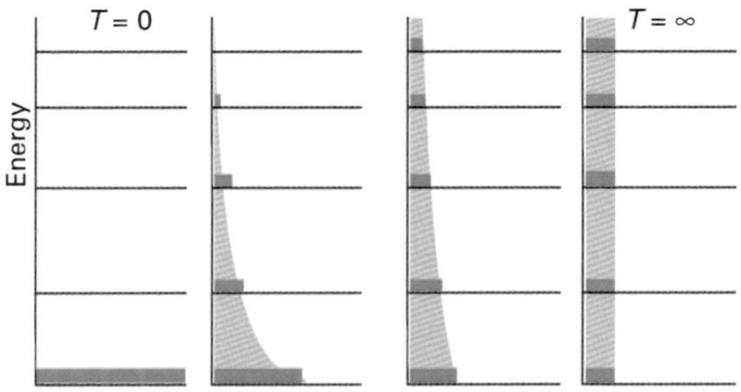

Fig. 0.5 The Boltzmann distribution of populations for a system of five states as the temperature is increased. At low temperature, the population of molecules in high-energy states is small. At higher temperature, more high-energy states are populated.

stage is that as the temperature increases, more states of higher energy become populated.

● Brief illustration 0.11 Relative populations

Methylcyclohexane molecules may exist in one of two conformations, with the methyl group in either an equatorial or axial position. The equatorial form is lower in energy with the axial form being 6.0 kJ mol^{-1} higher in energy. At a temperature of 300 K, this difference in energy implies that the relative populations of molecules in the axial and equatorial states is

$$\frac{N_a}{N_e} = e^{-(E_a - E_e)/RT}$$

$$= e^{-(6.0\times10^3 \text{ J mol}^{-1})/(8.3145 \text{ J K}^{-1}\text{ mol}^{-1}\times 300 \text{ K})} = 0.090$$

The number of molecules in an axial conformation is therefore just 9 per cent of those in the equatorial conformation.

Self-test 0.8

Determine the temperature at which the relative proportion of molecules in axial and equatorial conformations in a sample of methylcyclohexane is 0.30 or 30 per cent.

Answer: 600 K

0.12 Equipartition

Although the Boltzmann distribution can be used to calculate the average energy associated with each mode of motion of an atom or molecule in a sample at a given temperature, there is a much simpler shortcut. When the temperature is so high that many energy levels are occupied, we can use the **equipartition theorem:**

> For a sample at thermal equilibrium the average value of each quadratic contribution to the energy is $\frac{1}{2}kT$.

By a 'quadratic contribution' we mean a term that is proportional to the square of the momentum (as in the expression for the kinetic energy, $E_k = p^2/2m$) or the displacement from an equilibrium position (as for the potential energy of a harmonic oscillator, $E_p = \frac{1}{2}k_f x^2$). The theorem is strictly valid only at high temperatures or if the separation between energy levels is small because under these conditions many states are populated. The equipartition theorem is most reliable for translational and rotational modes of motion. The separation between vibrational and electronic states is typically greater than for rotation or translation, and so the equipartition theorem is unreliable for these types of motion.

● Brief illustration 0.12 Average molecular energies

An atom or molecule may move in three dimensions and its translational kinetic energy is therefore the sum of three quadratic contributions:

$$E_{\text{trans}} = \tfrac{1}{2}mv_x^2 + \tfrac{1}{2}mv_y^2 + \tfrac{1}{2}mv_z^2$$

The equipartition theorem predicts that the average energy for each of these quadratic contributions is $\frac{1}{2}kT$. Thus, the average kinetic energy is $E_{\text{trans}} = 3 \times \frac{1}{2}kT = \frac{3}{2}kT$. The molar translational energy is thus $E_{\text{trans,m}} = \frac{3}{2}kT \times N_A = \frac{3}{2}RT$. At 300 K:

$$E_{\text{trans,m}} = \tfrac{3}{2} \times (8.3145 \text{ J K}^{-1}\text{ mol}^{-1}) \times (300 \text{ K})$$

$$= 3700 \text{ J mol}^{-1} = 3.7 \text{ kJ mol}^{-1}$$

Self-test 0.9

A linear molecule may rotate about two axes in space, each of which counts as a quadratic contribution. Calculate the rotational contribution to the molar energy of a collection of linear molecules at 500 K.

Answer: 4.2 kJ mol^{-1}

Electromagnetic radiation

Many of the techniques, such as spectroscopy and X-ray diffraction, which are used by physical chemists to investigate and understand the structure and properties of matter, involve **electromagnetic radiation.** For many purposes, electromagnetic radiation may be treated as consisting of oscillating electric and magnetic disturbances that propagate as waves. However, there are other occasions where this wave picture cannot explain all the properties of electromagnetic radiation. In these cases, electromagnetic radiation must be treated as if it is a stream of particles called **photons.**

0.13 Electromagnetic waves

A **wave** is a periodic disturbance that propagates through matter or space. An electromagnetic wave consists of oscillating electric and magnetic fields. The two components of an electromagnetic wave are mutually perpendicular and perpendicular to the direction of propagation (Fig. 0.6). Both components act on charged particles: the electric field acts on both stationary and moving charged particles but the magnetic field acts only on charged particles that are moving. No medium is required for propagation of electromagnetic waves and they travel through a

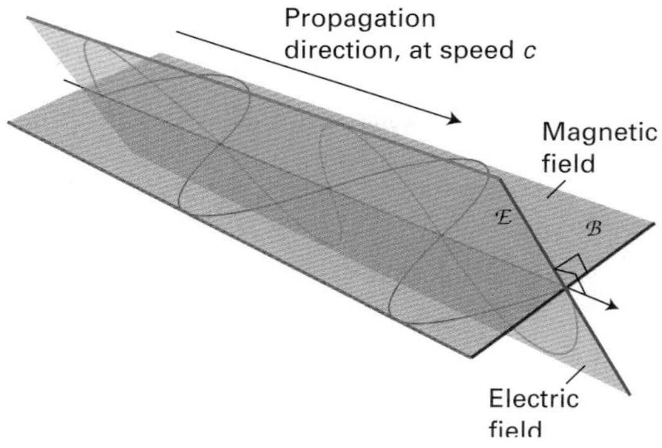

Fig. 0.6 The electric and magnetic fields oscillate in mutually perpendicular planes that are also perpendicular to the direction of propagation.

vacuum at a constant speed called the **speed of light**, c, which has the defined value of exactly

$$c = 2.997\ 924\ 58 \times 10^8 \text{ m s}^{-1}$$

and which we shall usually quote as 2.998×10^8 m s^{-1}. Electromagnetic waves propagate more slowly in media such as air, water, and glass.

A wave is characterized by its **wavelength,** λ (lambda), which is the distance between consecutive peaks of the wave (Fig. 0.7). Light, which is electromagnetic radiation that is visible to the human eye, has a wavelength in the range 420 to 700 nm. The properties of a wave may also be expressed in terms of its **frequency,** ν (nu), which is the rate at which the fields change direction. Frequency is thus the number of oscillations per second and is measured in units of reciprocal seconds (s^{-1}) and is often quoted as the derived SI unit the hertz, symbol Hz, where 1 s^{-1} = 1 Hz.

The **period** for each oscillation is the reciprocal of the frequency, $1/\nu$. During this period the wave propagates by one complete wavelength, λ. It follows that the speed of propagation is

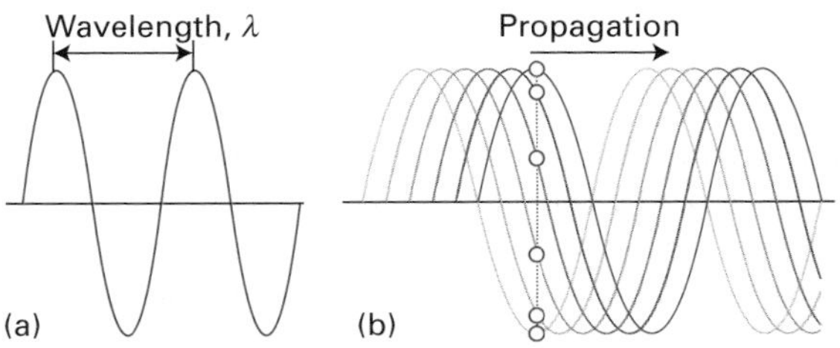

Fig. 0.7 (a) The wavelength, λ, of a wave is the peak-to-peak distance. (b) The frequency, ν, is the number of cycles per second that occur as the wave travels past a particular point.

$$c = \frac{\text{distance of propagation during one period of oscillation}}{\text{period of one oscillation}} = \frac{\lambda}{1/\nu}$$

or

$$c = \lambda\nu \qquad \text{The relation between wavelength and frequency} \qquad (0.19)$$

The frequency of an electromagnetic wave is therefore inversely proportional to its wavelength: a wave with a high frequency has a short wavelength, and a wave with a low frequency has a long wavelength. The properties of an electromagnetic wave are also expressed in terms of its **wavenumber,** $\tilde{\nu}$ (nu tilde), which is defined as

$$\tilde{\nu} = \frac{\nu}{c} = \frac{1}{\lambda} \qquad \text{Wavenumber} \qquad (0.20)$$

Thus, wavenumber is the reciprocal of the wavelength and can be interpreted as the number of wavelengths in a given distance. In spectroscopy, for historic reasons, wavenumber is usually reported in units of reciprocal centimetres (cm^{-1}). Visible light therefore corresponds to electromagnetic radiation with a wavenumber of 14 000 to 24 000 cm^{-1}.

Brief illustration 0.13 Wavenumbers

The wavenumber of electromagnetic radiation of wavelength 660 nm is

$$\tilde{\nu} = \frac{1}{\lambda} = \frac{1}{660 \times 10^{-9}\text{ m}} = 1.5 \times 10^6 \text{ m}^{-1} = 15\,000 \text{ cm}^{-1}$$

You can avoid errors in converting between units of m^{-1} and cm^{-1} by remembering that wavenumber represents the number of wavelengths in a given distance. Thus, a wavenumber expressed as the number of waves per centimetre and hence in units of cm^{-1} must be 100 times less than the equivalent quantity expressed per metre in units of m^{-1}.

The classification of electromagnetic radiation according to its wavelength is shown in Fig. 0.8. Electromagnetic radiation that consists of a ray of a single frequency (and therefore single wavelength) is known as **monochromatic**, because it represents a single colour. In contrast, **white light** consists of electromagnetic waves with a continuous, but not necessarily uniform, spread of frequencies throughout the visible region of the spectrum.

0.14 Photons

A wave interpretation of electromagnetic radiation cannot explain certain experimental observations.

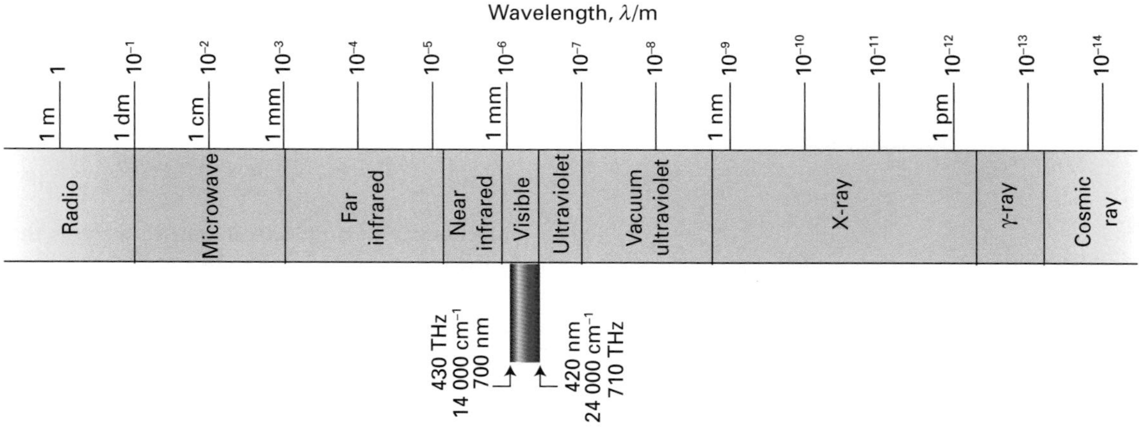

Fig. 0.8 The electromagnetic spectrum and the classification of the spectral regions.

Instead, we need to consider an alternative interpretation in which electromagnetic radiation is viewed as a stream of particle-like photons. These photons represent discrete packets of energy, and for a given frequency of radiation, each one has exactly the same energy given by

$$E = h\nu \qquad \text{Energy of a photon in terms of frequency} \qquad (0.21a)$$

The constant of proportionality, h, is known as **Planck's constant:**

$$h = 6.626\ 069\ 57 \times 10^{-34}\ \text{J s}$$

which we shall normally quote as 6.626×10^{-34} J s. The energy varies with wavelength and wavenumber as

$$E = \frac{hc}{\lambda} = hc\tilde{\nu} \qquad \text{Energy of a photon in terms of the wavelength and wavenumber} \qquad (0.21b)$$

The energy of a photon of short-wavelength X-ray radiation is thus higher than a photon of long-wavelength microwave radiation.

Example 0.2

Determining the number of photons emitted by a source

A pulsed laser emits near-infrared radiation of wavelength 1064 nm. Each pulse has a total energy of 2.5 kJ. How many photons are emitted in each pulse?

Strategy Determine the energy of each photon using eqn 0.21b. The number of photons emitted is then the total energy of each pulse divided by the energy of one photon.

Solution A wavelength of 1064 nm corresponds to 1064×10^{-9} m. Thus, from eqn 0.21b, the energy of each photon is

$$E = \frac{hc}{\lambda} = \frac{\overbrace{(6.626\times10^{-34}\ \text{J s})}^{h}\times\overbrace{(2.998\times10^{8}\ \text{m s}^{-1})}^{c}}{\underbrace{1064\times10^{-9}\ \text{m}}_{\lambda}}$$

$$= 1.867\times10^{-19}\ \text{J}$$

The number of photons emitted in each pulse is then the ratio of the energy of each pulse to the energy of an individual photon

$$N = \frac{\overbrace{2500\ \text{J}}^{\text{Total energy of pulse}}}{\underbrace{1.867\times10^{-19}\ \text{J}}_{\text{Energy of one photon}}} = 1.3\times10^{22}$$

Self-test 0.10

Determine the number of photons emitted per second by monochromatic mercury vapour lamp of wavelength 404.7 nm that emits energy at a rate of 80 J s^{-1}.

Answer: 1.6×10^{20} s^{-1}

Building on the foundations

You now have enough basic information to move on to physical chemistry itself, but you should be prepared to return to this chapter whenever necessary when you need some background or a quick reminder of a concept. You will come to see that the concepts presented here underlie the whole of the subject. Thermodynamics, which opens the main body of the text, draws heavily on concepts relating to work and energy. Quantum theory, which deals among other things with the structures of atoms and molecules, draws heavily on concepts relating to energy and momentum. Spectroscopy draws heavily on concepts relating to the electromagnetic field, with photons acting as messengers from inside atoms and molecules, bringing information about their identities and structures. Throughout the text you will see how the Boltzmann distribution acts as a bridge between atomic and molecular structure on one hand and thermodynamic and kinetic properties on the other. It will always be useful to keep the Boltzmann distribution in mind when thinking about an atomic interpretation of a bulk property.

This chapter is by no means exhaustive: if it were, you would be overwhelmed. It is here to act as an introduction to the most fundamental concepts that you need. Throughout the text you will find panels labelled 'The chemist's toolkit'. These items—you have already seen one example in this chapter—provide a deeper discussion of a concept that needs to be elaborated and which should form a part of the intellectual equipment of every chemist. They are presented at precisely the point where the concept is first needed, and as you work through the text you will gradually acquire all the techniques that you need to master the subject.

Questions and exercises

Discussion questions

0.1 Explain the differences in observable properties between gases, liquids, and solids and account for these differences in molecular terms.

0.2 Distinguish between the terms: force, work, energy, kinetic energy, and potential energy.

0.3 Distinguish between mechanical and thermal equilibrium. In what sense are these equilibria dynamic?

0.4 Identify and define the various uses of the term 'state' in chemistry.

0.5 Explain the role of temperature in determining the populations of the states in a quantum system.

Exercises

0.1 Calculate the amount of $C_6H_{12}O_6$ molecules, in moles, in 10.0 g of glucose.

0.2 Calculate the molecular mass of buckminsterfullerene, C_{60}, from its molar mass.

0.3 The molar mass of the oxygen-storage protein myoglobin is 16.1 kg mol^{-1}. How many myoglobin molecules are present in 1.0 g of the compound?

0.4 The mass of a red blood cell is about 33 pg (where 1 pg = 10^{-12} g), and it contains typically 3×10^8 haemoglobin molecules. Each haemoglobin molecule is a tetramer of myoglobin (see preceding exercise). What fraction of the mass of the cell is due to haemoglobin?

0.5 What mass of carbon monoxide is needed to reduce 1.0 t (1 t = 10^3 kg) of iron(III) oxide to the metal?

0.6 The internal volume of a particular cylinder of oxygen gas is given as 50 dm^3. Express this volume in (a) cubic metres, (b) cubic centimetres.

0.7 A spherical drop of oil has a mass of 20.4 μg and a radius of 1.74 mm. Given that the volume of a sphere is $\frac{4}{3}\pi r^3$, what is the mass density of the oil in (a) grams per cubic centimetre, (b) kilograms per cubic metre?

0.8 The density of octane (which we take to model gasoline) is 0.703 g cm^{-3}; what amount (in moles) of octane molecules do you get when you buy 1.00 dm^3 (1.00 litre) of gasoline?

0.9 Calculate the mass of carbon dioxide produced by the combustion of 1.00 dm^3 of gasoline treated as octane of mass density 0.703 g cm^{-3}.

0.10 Identify whether the following properties are extensive or intensive: (a) volume, (b) mass density, (c) temperature, (d) molar volume, (e) amount of substance.

0.11 A sample of methane gas, CH_4, with a mass of 45.2 g occupies 2.19 cm^3 at 19.7 MPa and 298 K. Determine the molar volume of methane gas under these conditions, expressing the answer in cubic decimetres per mole.

0.12 The (molar) ionization energy of atomic sodium is 495.8 kJ mol^{-1}. Calculate the energy required to ionize a single sodium atom.

0.13 Oxygen molecules in air move with an average speed of 482 m s^{-1} at 298 K. Calculate the magnitude of the average linear momentum of an oxygen molecule at this temperature.

0.14 The moment of inertia for rotation of a hydrogen molecule, $^1H_2$, about an axis perpendicular to its bond is 4.61×10^{-48} kg m^2. What is the bond length of H_2? The mass of a ^{1}H atom is $1.01 m_u$.

0.15 The electronic structure of atoms and molecules may be investigated using photoelectron spectroscopy. An electron in a photoelectron spectrometer is accelerated from rest by a uniform electric field to a speed of 420 km s^{-1} in 10 μs. Determine (a) the magnitude of the acceleration of the electron, (b) the force acting on the electron (c) the kinetic energy of the electron, and (d) the magnitude of the work done on the electron.

0.16 What is the gravitational force that you are currently experiencing?

0.17 Calculate the percentage change in your weight as you move from the North Pole, where $g = 9.832$ m s^{-2}, to the equator, where $g = 9.789$ m s^{-2}.

0.18 Express (a) 108 kPa in torr, (b) 0.975 bar in atmospheres, (c) 22.5 kPa in atmospheres, (d) 770 Torr in pascals.

0.19 A centrifuge used in the separation of solids from suspensions consists of a rotor arm that rotates at high angular speed. The centrifuge rotates at an angular speed of 400 rad s^{-1}. The mass of the rotor arm is concentrated at the outer edge, so that effectively it may be considered as a mass of 2.0 kg rotating at a distance of 20 cm from the axis of rotation. What is the kinetic energy of the rotating centrifuge arm?

0.20 Calculate the work that a person of mass 65 kg must do to climb between two floors of a building separated by 4.0 m.

0.21 What is the kinetic energy of a tennis ball of mass 58 g served at 35 m s^{-1}?

0.22 A car of mass 1.5 t (1 t = 10^3 kg) travelling at 50 km h^{-1} must be brought to a stop. How much kinetic energy must be dissipated?

0.23 Consider a region of the atmosphere of volume 25 dm^3, which at 20 °C contains about 1.0 mol of molecules. Take the average molar mass of the molecules as 29 g mol^{-1} and their average speed as about 400 m s^{-1}. Estimate the energy stored as molecular translational kinetic energy in this volume of air.

0.24 Calculate the minimum energy that a bird of mass 25 g must expend in order to reach a height of 50 m.

0.25 The most probable distance of an electron from the nucleus in the lowest energy state of a hydrogen atom is called the Bohr radius: $a_0 = 52.9$ pm. What is the electrostatic potential energy of interaction between the nucleus and the electron at this separation?

0.26 Chemists often quote quantities in terms of the electronvolt, symbol eV, which is the energy acquired by an electron when it moves through a potential difference of 1 V: 1 eV = 1.602×10^{-19} J. The work function of potassium, which is the energy required to eject an electron from the surface of a metal, is 2.30 eV. Express this energy as an equivalent molar quantity in units of kilojoules per mole.

0.27 Given that the Celsius and Fahrenheit temperature scales are related by $\theta_{\text{Celsius}}/°\text{C} = \frac{5}{9}(\theta_{\text{Fahrenheit}}/°\text{F} - 32)$, what is the temperature of absolute zero ($T = 0$) on the Fahrenheit scale?

0.28 In his original formulation, Anders Celsius identified 0 with the boiling point of water and 100 with its freezing point. Find a relation (expressed like eqn 0.17) between this original scale (denote it $\theta'/°\text{C}'$) and (a) the current Celsius scale ($\theta/°\text{C}$), (b) the Fahrenheit scale.

0.29 Imagine that Pluto is inhabited and that its scientists use a temperature scale in which the freezing point of liquid nitrogen is 0 °P (degrees Plutonium) and its boiling point is 100 °P. The inhabitants of Earth report these temperatures as −209.9 °C and −195.8 °C, respectively. What is the relation between temperatures on (a) the Plutonium and Kelvin scales, (b) the Plutonium and Fahrenheit scales?

0.30 The *Rankine scale* is used in some engineering applications. On it, the absolute zero of temperature is set at zero but the size of the Rankine degree (°R) is the same as that of the Fahrenheit degree (°F). What is the boiling point of water on the Rankine scale?

0.31 A hypothetical system consists of quantum states with energies of 0, ε, 2ε, . . . , with the populations of the states determined by the Boltzmann distribution. Show that, at a temperature of $T = \varepsilon/k$, the populations of the first four excited states relative to that of the lowest state are 37 per cent, 14 per cent, 5 per cent, and 1 per cent, with the populations of all higher states being negligible.

0.32 What is the total energy of a quantum system such as that described in the previous exercise that consists of 100 molecules?

0.33 Use the equipartition theorem to calculate the average translational kinetic energy of a nitrogen, N_2, molecule at a temperature of 298 K. Hence determine the average translational speed of a N_2 molecule at this temperature.

0.34 Nonlinear molecules, such as ammonia, NH_3, may rotate about three axes. Determine the rotational contribution to the molar energy of a sample of ammonia gas at 298 K.

0.35 A helium–neon laser emits red light of wavelength 632.8 nm. What is (a) the wavenumber, (b) the frequency of the radiation, and (c) the energy of a photon of this radiation?

0.36 Calculate the energy per photon and the energy per mole of photons for radiation of wavelength (a) 600 nm (red), (b) 550 nm (yellow), (c) 400 nm (violet), (d) 200 nm (ultraviolet), (e) 150 pm (X-ray), (f) 1.0 cm (microwave).

0.37 A photodetector produces energy at a rate of 0.68 μW, where 1 W = 1 J s^{-1} when exposed to radiation of wavelength 245 nm. How many photons does it detect per second?

0.38 A certain lamp emits blue light of wavelength 380 nm. How many photons does it emit each second if its power is (a) 1.00 W, (b) 100 W?

0.39 For how long must a sodium lamp rated at 100 W operate to generate 1.00 mol of photons of wavelength 590 nm? Assume all the power is used to generate those photons.

0.40 A solid-state InGaAsP diode laser produces monochromatic radiation of wavelength 1.13 μm. The energy is emitted at a rate of 1.05 mJ s^{-1}. How many photons are emitted per second?

1

The properties of gases

Equations of state 18

1.1 The perfect gas equation of state 19

1.2 Using the perfect gas law 22

1.3 Mixtures of gases: partial pressures 23

The molecular model of gases 26

1.4 The pressure of a gas according to the kinetic model 27

1.5 The average speed of gas molecules 27

1.6 The Maxwell distribution of speeds 28

1.7 Diffusion and effusion 30

1.8 Molecular collisions 32

Real gases 33

1.9 Molecular interactions 33

1.10 The critical temperature 34

1.11 The compression factor 36

1.12 The virial equation of state 36

1.13 The van der Waals equation of state 37

1.14 The liquefaction of gases 40

FURTHER INFORMATION 1.1 41

CHECKLIST OF KEY CONCEPTS 42

ROAD MAP OF KEY EQUATIONS 43

QUESTIONS AND EXERCISES 43

Although gases are simple, both to describe and in terms of their internal structure, they are of immense importance. We spend our whole lives surrounded by gas in the form of air and the local variation in its properties is what we call the 'weather'. To understand the atmospheres of this and other planets we need to understand gases. As we breathe, we pump gas in and out of our lungs, where it changes composition and temperature. Many industrial processes involve gases, and both the outcome of the reaction and the design of the reaction vessels depend on knowledge of their properties.

Equations of state

We can specify the state of any sample of substance by giving the values of the following properties (all of which are defined in the Foundations section):

V, the volume of the sample
p, the pressure of the sample
T, the temperature of the sample
n, the amount of substance in the sample

However, an astonishing experimental fact is that *these four quantities are not independent of one another*. For instance, we cannot arbitrarily choose to have a sample of 0.555 mol H_2O in a volume of 10 dm^3 at 100 kPa and 500 K: it is found *experimentally* that that state simply does not exist. If we select the amount, the volume, and the temperature, then we find that we have to accept a particular pressure (in this case, close to 230 kPa). This experimental generalization is summarized by saying the substance obeys an **equation of state**, an equation of the form

$$p = f(n,V,T) \qquad \text{Equation of state} \quad (1.1)$$

This expression tells us that the pressure depends on ('is a function of') the amount, volume, and temperature and that if we know those three variables, then the pressure can have only one value.

The equations of state of most substances are not known, so in general we cannot write down an explicit expression for the pressure in terms of the other variables. However, certain equations of state are known. In particular, the equation of state of a low-pressure gas is known, and proves to be very simple and very useful. This equation is used to describe the behaviour of gases taking part in reactions, the behaviour of the atmosphere, as a starting point for problems in chemical engineering, and even in the description of the structures of stars.

1.1 The perfect gas equation of state

The equation of state of a low-pressure gas was among the first results to be established in physical chemistry. The original experiments were carried out by Robert Boyle in the seventeenth century and there was a revival of interest later in the century when people began to fly in balloons. This technological progress demanded more knowledge about the response of gases to changes of pressure and temperature and, like technological advances in other fields today, that interest stimulated a lot of experiments.

The experiments of Boyle and his successors led to the formulation of the following **perfect gas equation of state**:

$$pV = nRT \qquad \text{Perfect gas equation of state} \qquad (1.2a)$$

This equation has the form of eqn 1.1 when rearranged into

$$p = \frac{nRT}{V} \qquad (1.2b)$$

The **gas constant** R is currently an experimentally determined quantity obtained by evaluating $R = pV/nRT$ as the pressure is allowed to approach zero. For calculations, we shall normally use the approximate value 8.3145 J K^{-1} mol^{-1}. Values of R in a variety of units are given in Table 1.1.

The perfect gas equation of state—more briefly, the 'perfect gas law'—is so called because it is an idealization of the equations of state that gases actually obey. Specifically, it is found that all gases obey the equation ever more closely as the pressure is reduced towards zero. That is, eqn 1.2 is an example of a **limiting law**, a law that becomes increasingly valid in a certain limit, in this case as the pressure is reduced to zero. The law is obeyed exactly in the limit of zero pressure.

Table 1.1
The gas constant in various units

R = 8.314 47	J K^{-1} mol^{-1}
8.314 47	dm^3 kPa K^{-1} mol^{-1}
8.205 74 × 10^{-2}	dm^3 atm K^{-1} mol^{-1}
62.364	dm^3 Torr K^{-1} mol^{-1}
1.987 21	cal K^{-1} mol^{-1}

1 dm^3 = 10^{-3} m^3

A hypothetical substance that obeys eqn 1.2 at *all* pressures, not just at very low pressures, is called a **perfect gas**. From what has just been said, an actual gas, which is termed a **real gas**, behaves more and more like a perfect gas as its pressure is reduced towards zero. In practice, normal atmospheric pressure at sea level ($p \approx 100$ kPa) is already low enough for most real gases to behave almost perfectly, and unless stated otherwise we shall always assume in this text that the gases we encounter behave like a perfect gas. The reason why a real gas behaves differently from a perfect gas can be traced to the attractions and repulsions that exist between actual molecules and which are absent in a perfect gas (Chapter 15).

A note on good practice A perfect gas is widely called an 'ideal gas' and the perfect gas equation of state is commonly called 'the ideal gas equation'. We use 'perfect gas' to imply the absence of molecular interactions; we use 'ideal'; in Chapter 6 to denote mixtures in which all the molecular interactions are the same but not necessarily zero.

The perfect gas law is based on and summarizes three sets of experimental observations. One is **Boyle's law**:

> At constant temperature, the pressure of a fixed amount of gas is inversely proportional to its volume.

Mathematically:

$$p \propto \frac{1}{V} \qquad \text{At constant temperature} \quad \text{Boyle's law} \qquad (1.3)$$

We can easily verify that eqn 1.2 is consistent with Boyle's law: by treating n and T as constants, the perfect gas law becomes pV = constant, and hence $p \propto 1/V$. Boyle's law implies that if we compress (reduce the volume of) a fixed amount of gas at constant temperature into half its original volume, then its pressure will double. Figure 1.1 shows the graph obtained by plotting experimental values of p against

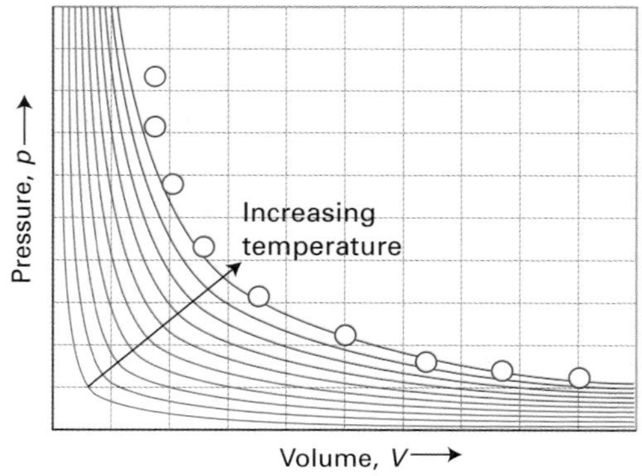

Fig. 1.1 The volume of a gas decreases as the pressure on it is increased. For a sample that obeys Boyle's law and that is kept at constant temperature, the graph showing the dependence is a hyperbola, as shown here. Each curve corresponds to a single temperature, and hence is an isotherm. The isotherms are hyperbolas (The chemist's toolkit 1.1).

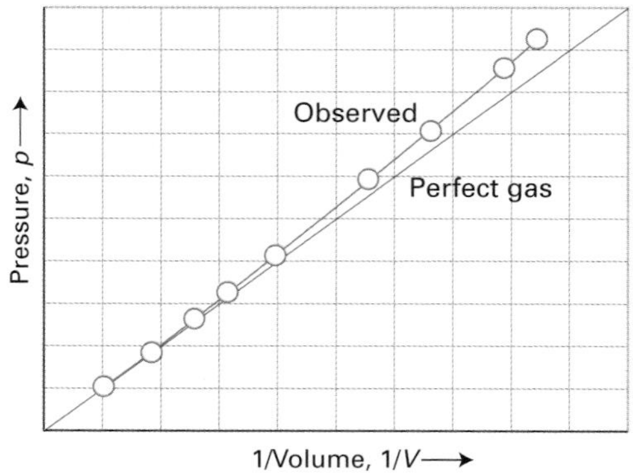

Fig. 1.2 A good test of Boyle's law is to plot the pressure against $1/V$ (at constant temperature), when a straight line should be obtained. This diagram shows that the observed pressures approach a straight line (the blue line) as the volume is increased and the pressure reduced. A perfect gas would follow the straight line at all pressures; real gases obey Boyle's law in the limit of low pressures.

V for a fixed amount of gas at different temperatures and the curves predicted by Boyle's law. Each curve is a hyperbola (see The chemist's toolkit 1.1 for a discussion of graphs) and called an **isotherm** because it depicts the variation of a property (in this case, the pressure) at a single constant temperature. It is hard, from this graph, to judge how well Boyle's law is obeyed. However, when we plot p against $1/V$, we get straight lines, just as we would expect from Boyle's law (Fig. 1.2).

The chemist's toolkit 1.1 Graphs

The graphs of $xy = a$, with a a constant, or $y = a/x$ are *hyperbolas*. Examples are the isotherms in Fig. 1.1. The graph of $y = ax^2$ is a *parabola*; it plays a role in the discussion of molecular vibrations. These two 'conic sections' are illustrated in Sketch 1.1.

The graph of $y = mx + b$ is a straight line of slope m and passing through $y = b$ at $x = 0$, the 'intercept with the y-axis' (Sketch 1.2). In the special case $b = 0$, $y = mx$ and the straight line passes through the origin $y = 0$ when $x = 0$. In that case we say that 'y is proportional to x' ($y \propto x$) the constant of proportionality being m. The intercept with the x-axis occurs when $x = -b/m$ (because then $y = m(-b/m) + b = 0$).

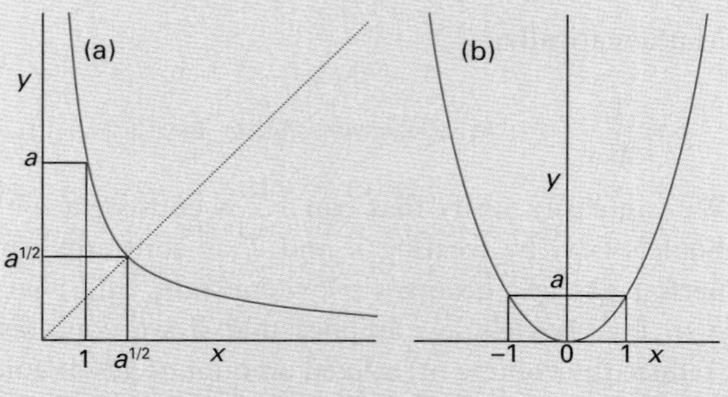

Sketch 1.1 The characteristics of (a) hyperbolas and (b) parabolas.

The horizontal and vertical axes of graphs are labelled with pure numbers. Thus, if the units of x and y are denoted 'units of x' and 'units of y', the plot should be of y/(units of y) against x/(units of x). It follows that the slope of a graph is dimensionless. For a graph of the form y/(units of y) = $b + mx$/(units of x), both m and b are pure numbers. To interpret them in terms of physical quantities we multiply both sides of this expression by 'units of y', and obtain

$$y = \overbrace{b \times (\text{units of } y)}^{\text{Interpretation of the intercept}} + \overbrace{m \times \frac{\text{units of } y}{\text{units of } x}}^{\text{Interpretation of the slope}} \times x$$

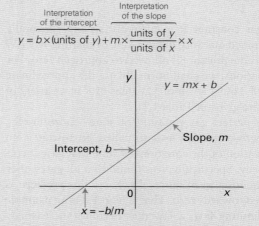

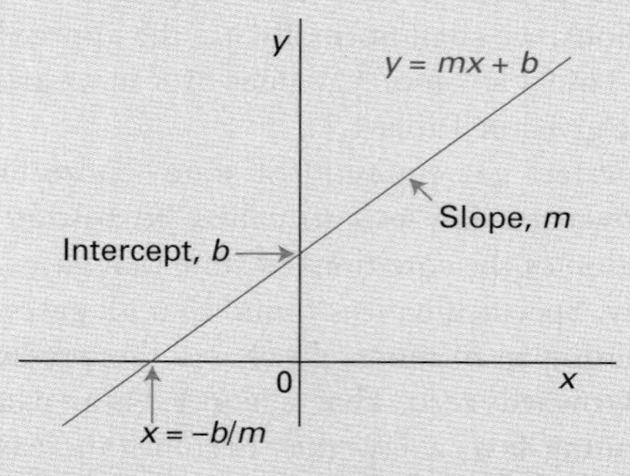

Sketch 1.2 The characteristics of a straight-line graph of y against x with $y = mx + b$.

A note on good practice In general, it is best to plot data as straight lines to test the validity of a theory because it is difficult to distinguish between subtly different shapes of a curve by eye. Instead, the mathematical function that represents the theory is rearranged so that the resulting graph has the form of a straight line, $y = mx + b$.

The second experimental observation summarized by eqn 1.2 is **Charles's law**:

> At constant pressure, the volume of a fixed amount of gas varies linearly with the temperature.

The equation for this linear dependence is

$$V = A + B\theta \qquad \text{Constant pressure} \quad \text{Charles's law} \quad (1.4a)$$

where θ (theta) is the temperature on the Celsius scale and A and B are constants that depend on the amount of gas and the pressure. Figure 1.3 shows typical plots of volume against temperature for a series of samples of gases at different pressures and confirms that (at low pressures, and for temperatures that are not too low) the volume varies linearly with the Celsius temperature. We also see that all the volumes extrapolate to zero as θ approaches the same very low temperature (−273.15 °C, in fact), regardless of the identity of the gas. Because a volume cannot be negative, this common temperature must represent the **absolute zero** of temperature, a temperature below which it is impossible to cool an object. Indeed, as explained in Foundations 0.10, the 'thermodynamic' scale ascribes the value $T = 0$ to this absolute zero of temperature. In terms of the thermodynamic temperature, therefore, Charles's law takes the simpler form

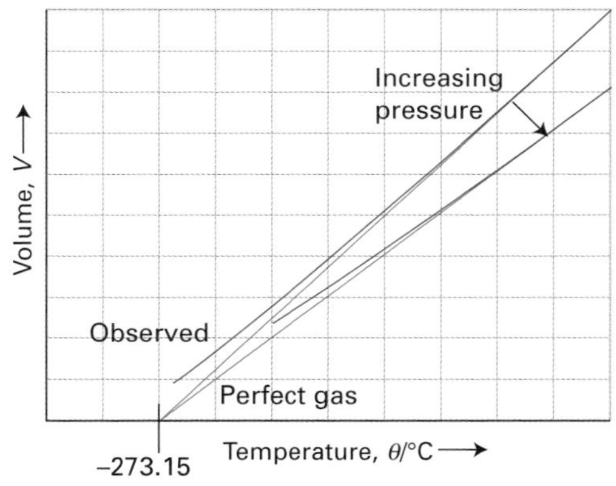

Fig. 1.3 This diagram illustrates the content and implications of Charles's law, which asserts that the volume occupied by a gas (at constant pressure) varies linearly with the temperature. When plotted against Celsius temperatures (as here), all gases give straight lines that extrapolate to $V = 0$ at −273.15 °C. This extrapolation suggests that −273.15 °C is the lowest attainable temperature.

$$V \propto T \qquad \text{Constant pressure} \quad \text{Charles's law} \quad (1.4b)$$

It follows that doubling the temperature (such as from 300 K to 600 K, corresponding to an increase from 27 °C to 327 °C) doubles the volume, provided the pressure remains the same. Now we can see that eqn 1.2 is consistent with Charles's law. First, we rearrange it into $V = nRT/p$, and then note that when the amount n and the pressure p are both constant, we can write $V \propto T$, as required.

The third feature of gases summarized by eqn 1.2 is **Avogadro's principle:**

> At a given temperature and pressure, equal volumes of gas contain the same numbers of molecules.

That is, 1.00 dm^3 of oxygen at 100 kPa and 300 K contains the same number of molecules as 1.00 dm^3 of carbon dioxide, or any other gas, at the same temperature and pressure. The principle implies that if we double the number of molecules, but keep the temperature and pressure constant, then the volume of the sample will double. We can therefore write

$$V \propto n \qquad \text{Constant temperature and pressure} \quad \text{Avogadro's principle} \quad (1.5)$$

This result follows easily from eqn 1.2 if we treat p and T as constants. Avogadro's suggestion is a principle rather than a law (a direct summary of experience), because it is based on a model of a gas, in this case as a collection of molecules. Even though there is no longer any doubt that molecules exist, this relation remains a principle rather than a law.

The molar volume, V_m, the volume a substance occupies per mole of molecules, was introduced in Foundations 0.4:

$$V_m = \frac{V}{n} \qquad \text{Definition} \quad \text{Molar volume} \quad (1.6a)$$

For a perfect gas, $n = pV/RT$, so in this case

$$V_m \overset{n=pV/RT}{=} \frac{V}{pV/RT} \overset{\text{Cancel }V}{=} \frac{1}{p/RT} \overset{1/(a/x)=x/a}{=} \frac{RT}{p} \qquad \text{Perfect gas} \quad \text{Molar volume} \quad (1.6b)$$

To interpret this expression, note that

- Because eqn 1.6b makes no reference to the identity of the gas, provided the gas is behaving perfectly, its molar volume is the same whatever its chemical identity at the same temperature and pressure.

The data in Table 1.2 show that this conclusion is approximately true for most gases under normal conditions (normal atmospheric pressure of about 100 kPa and room temperature).

Table 1.2
The molar volumes of gases at standard ambient temperature and pressure (SATP: 298.15 K and 1 bar)

Gas	$V_m/(dm^3\ mol^{-1})$
Perfect gas	24.7896*
Ammonia	24.8
Argon	24.4
Carbon dioxide	24.6
Nitrogen	24.8
Oxygen	24.8
Hydrogen	24.8
Helium	24.8

*At STP (0 °C, 1 atm), $V_m = 24.4140\ dm^3\ mol^{-1}$.

Chemists have found it convenient to report much of their data on gases at a particular set of 'standard' conditions.[1] By **standard ambient temperature and pressure** (SATP) they mean a temperature of 25 °C (more precisely, 298.15 K) and a pressure of exactly 1 bar (100 kPa). The **standard pressure** is denoted $p^\ominus$, so $p^\ominus$ = 1 bar exactly. The molar volume of a perfect gas at SATP is 24.79 $dm^3\ mol^{-1}$, as can be verified by substituting the values of the temperature and pressure into eqn 1.6b. This value implies that at SATP, 1 mol of perfect gas molecules occupies about 25 dm^3 (a cube of about 30 cm on a side). An earlier set of standard conditions, which is still encountered, is **standard temperature and pressure** (STP), namely 0 °C and 1 atm. The molar volume of a perfect gas at STP is 22.41 $dm^3\ mol^{-1}$.

1.2 Using the perfect gas law

Here we review two elementary applications of the perfect gas equation of state:

- The prediction of the pressure of a gas given its temperature, its chemical amount, and the volume it occupies.
- The prediction of the change in pressure arising from changes in the conditions.

Calculations like these underlie more advanced considerations, including the way that meteorologists understand the changes in the atmosphere that we call the weather (see Impact 1.1 following Section 1.3).

[1] Be very careful not to confuse these 'standard conditions' for gases with the 'standard states' of substances, which are introduced in Chapter 3.

Example 1.1

Predicting the pressure of a sample of gas

A chemist is investigating the conversion of atmospheric nitrogen to usable form by the bacteria that inhabit the root systems of certain legumes, and needs to know the pressure in kilopascals exerted by 1.25 g of nitrogen gas in a flask of volume 250 cm^3 at 20 °C.

Strategy For this calculation, arrange eqn 1.2a ($pV = nRT$) into a form that gives the unknown (the pressure, p) in terms of the information supplied, which in this case is eqn 1.2b ($p = nRT/V$). To use this expression, we need to know the amount of molecules (in moles) in the sample, which we can obtain from the mass and the molar mass (by using $n = m/M$) and to convert the temperature to the Kelvin scale (by adding 273.15 to the Celsius temperature.

Solution The amount of N_2 molecules (of molar mass 28.02 g mol^{-1}) present is

$$n(N_2) = \frac{m}{M(N_2)} = \frac{1.25\ g}{28.02\ g\ mol^{-1}} = \frac{1.25}{28.02}\ mol$$

The temperature of the sample is

$$T/K = 20 + 273.15,\ so\ T = (20 + 273.15)\ K$$

Therefore, from $p = nRT/V$,

$$p = \frac{\overbrace{(1.25/28.02)\ mol}^{n} \times \overbrace{(8.3145\ J\ K^{-1}\ mol^{-1})}^{R} \times \overbrace{(20+273.15)\ K}^{T}}{\underbrace{(2.50\times 10^{-4})\ m^3}_{V}}$$

$$= \frac{(1.25/28.02)\times(8.3145)\times(20+273.15)}{2.50\times 10^{-4}}\ \frac{J}{m^3}$$

$$\overset{1\ J\ m^{-3}=1\ Pa}{=}\ 4.35\times 10^5\ Pa\ \overset{1\ kPa=10^3\ Pa}{=}\ 435\ kPa$$

Note how the units cancel like ordinary numbers. Had we required the pressure in different units we could, instead, have used Table 1.1 to select the appropriate value of R to match the information required.

A note on good practice It is best to postpone a numerical calculation to the last possible stage, and carry it out in a single step. This procedure avoids rounding errors.

Self-test 1.1

Calculate the pressure exerted by 1.22 g of carbon dioxide confined to a flask of volume 500 dm^3 (ie, 5.00 × 10^2 dm^3) at 37 °C.

Answer: 143 Pa

In some cases, we are given the pressure under one set of conditions and are asked to predict the pressure of the same sample under a different set of conditions. In that case, we use the perfect gas law as follows. Suppose the initial pressure is p_1, the initial temperature is T_1, and the initial volume is V_1. Then by dividing both sides of eqn 1.2a by the temperature, which gives $pV/T = nR$ in general, we can write in this instance

$$\frac{p_1V_1}{T_1} = nR$$

Suppose now that the conditions are changed to T_2 and V_2, and the pressure changes to p_2 as a result. Then under the new conditions eqn 1.2a tells us that

$$\frac{p_2V_2}{T_2} = nR$$

The nR on the right of these two equations is the same in each case, because R is a constant and the amount of gas molecules has not changed. It follows that we can combine the two equations into a single equation:

$$\frac{p_1V_1}{T_1} = \frac{p_2V_2}{T_2} \qquad \text{Constant amount of gas} \quad \text{Combined gas equation} \qquad (1.7)$$

This equation can be rearranged to calculate any one unknown (such as p_2 for instance) in terms of the other variables.

Brief Illustration 1.1 The combined gas equation

Consider a sample of gas with an initial volume of 15 cm^3 that has been heated from 25 °C to 1000 °C and had its pressure increased from 10.0 kPa to 150.0 kPa. We may rearrange eqn 1.7 and use it to calculate the final volume

$$V_2 = \frac{p_1V_1}{T_1} \times \frac{T_2}{p_2} = \frac{p_1}{p_2} \times \frac{T_2}{T_1} \times V_1$$

$$= \overbrace{\frac{10.0\ \text{kPa}}{150.0\ \text{kPa}}}^{p_1/p_2} \times \overbrace{\frac{(1000+273.15)\ \text{K}}{(25+273.15)\ \text{K}}}^{T_2/T_1} \times \overbrace{15\ \text{cm}^3}^{V_1} = 4.3\ \text{cm}^3$$

Self-test 1.2

Calculate the final pressure of a gas that is compressed from a volume of 20.0 dm^3 to 10.0 dm^3 and cooled from 100 °C to 25 °C if the initial pressure is 1.00 bar.

Answer: 1.60 bar

1.3 Mixtures of gases: partial pressures

Scientists are often concerned with mixtures of gases, such as when they are considering the properties of the atmosphere in meteorology, the composition of exhaled air in medicine, or the mixtures of hydrogen and nitrogen used in the industrial synthesis of ammonia. They need to be able to assess the contribution that each component of a gaseous mixture makes to the total pressure.

In the early nineteenth century, John Dalton carried out a series of experiments that led him to formulate what has become known as **Dalton's law:**

> The pressure exerted by a mixture of perfect gases is the sum of the pressures that each gas would exert if it were alone in the container at the same temperature:

$$p = p_A + p_B + \cdots \qquad \text{Perfect gases} \quad \text{Dalton's law} \qquad (1.8)$$

In this expression, p_J is the pressure that the gas J would exert if it were alone in the container at the same temperature. Dalton's law is strictly valid only for mixtures of perfect gases (or for real gases at such low pressures that they are behaving perfectly), but it can be treated as valid under most conditions we encounter.

Brief illustration 1.2 Dalton's law

Suppose we were interested in the composition of inhaled and exhaled air, and we knew that a certain mass of carbon dioxide exerts a pressure of 5 kPa when present alone in a container, and that a certain mass of oxygen exerts 20 kPa when present alone in the same container at the same temperature. Then, when both gases are present in the container, the carbon dioxide in the mixture contributes 5 kPa to the total pressure and oxygen contributes 20 kPa; according to Dalton's law, the total pressure of the mixture is the sum of these two pressures, or 25 kPa (Fig. 1.4).

Self-test 1.3

Suppose some nitrogen, enough to generate a pressure of 10 kPa when there alone, is introduced into the mixture. What is the new total pressure?

Answer: 35 kPa

For any type of gas (real or perfect) in a mixture, the **partial pressure**, p_J, of the gas J is defined as

$$p_J = x_J p \qquad \text{Definition} \quad \text{Partial pressure} \qquad (1.9)$$

where x_J is the **mole fraction** of the gas J in the mixture. The mole fraction of J is the amount of J

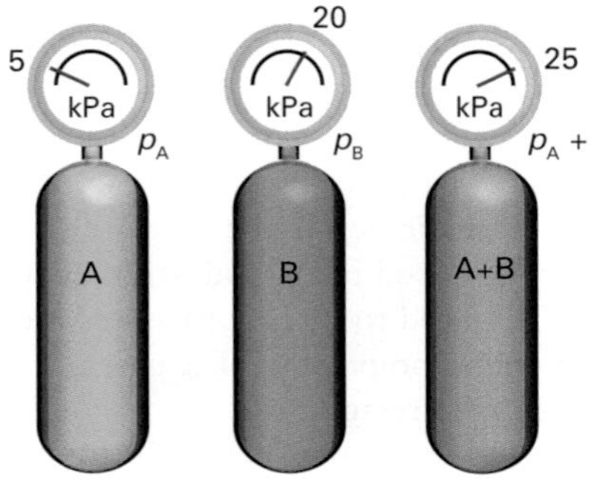

Fig. 1.4 The partial pressure p_A of a perfect gas A is the pressure that the gas would exert if it occupied a container alone; similarly, the partial pressure p_B of a perfect gas B is the pressure that the gas would exert if it occupied the same container alone. The total pressure p when both perfect gases simultaneously occupy the container is the sum of their partial pressures.

molecules expressed as a fraction of the total amount of molecules in the mixture. In a mixture that consists of n_A A molecules, n_B B molecules, and so on (where the n_J are amounts in moles), the mole fraction of J (where J = A, B, . . .) is

$$x_J = \frac{n_J}{n} \qquad n = n_A + n_B + \cdots \qquad \text{Definition} \quad \text{Mole fraction} \tag{1.10a}$$

Mole fractions are unitless because the unit mole in numerator and denominator cancel. For a **binary mixture**, one that consists of two species, this general expression becomes

$$x_A = \frac{n_A}{n_A + n_B} \qquad x_B = \frac{n_B}{n_A + n_B} \qquad x_A + x_B = 1 \tag{1.10b}$$

When only A is present, $x_A = 1$ and $x_B = 0$. When only B is present, $x_B = 1$ and $x_A = 0$. When both are present in the same amounts, $x_A = 1/2$ and $x_B = 1/2$ (Fig. 1.5).

When it is typographically more convenient, we write amounts as $n(\text{J})$, mole fractions as $x(\text{J})$, and partial pressures as $p(\text{J})$.

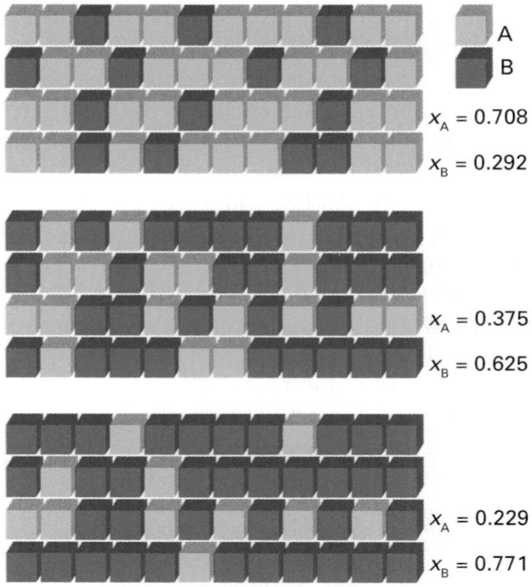

Fig. 1.5 A representation of the meaning of mole fraction. In each case, a small square represents one molecule of A (yellow squares) or B (green squares). There are 48 squares in each sample.

Example 1.2

Calculating mole fractions

A mass of 100.0 g of dry air consists of 75.5 g of N_2, 23.2 g of O_2, and 1.3 g of Ar. Express the composition of dry air in terms of mole fractions.

Strategy Begin by converting each mass to an amount in moles. Determine the mole fractions using eqn 1.10b as a ratio of the amount of each constituent to the total amount of substance.

Solution From eqn 0.2, the amounts of N_2, O_2, and Ar are

$$n_{N_2} = \frac{m_{N_2}}{M_{N_2}} = \frac{75.5\text{ g}}{28.0\text{ g mol}^{-1}} = 2.70\text{ mol}$$

$$n_{O_2} = \frac{m_{O_2}}{M_{O_2}} = \frac{23.2\text{ g}}{32.0\text{ g mol}^{-1}} = 0.725\text{ mol}$$

$$n_{Ar} = \frac{m_{Ar}}{M_{Ar}} = \frac{1.3\text{ g}}{39.9\text{ g mol}^{-1}} = 0.033\text{ mol}$$

The mole fraction of N_2 molecules is, from eqn 1.10b,

$$x_{N_2} = \frac{n_{N_2}}{\underbrace{n_{N_2} + n_{O_2} + n_{Ar}}_{n}} = \frac{2.70\text{ mol}}{2.70\text{ mol} + 0.725\text{ mol} + 0.033\text{ mol}}$$

$$= 0.780$$

By repeating the calculation for the other constituents, we find that the mole factions of O_2 and Ar in dry air are 0.210 and 0.009, respectively.

Self-test 1.4

The mole fraction of NH_3 in a sample of 10.0 mol of gas is 0.285. What mass of NH_3 is present in the sample?

Answer: 79.8 g

For a mixture of perfect gases, we can identify the partial pressure of J with the contribution that J makes to the total pressure. Thus, if we introduce $p = nRT/V$ into eqn 1.9, we get

$$p_J = \frac{n_J}{n} \times \overbrace{\frac{nRT}{V}}^{p} = \frac{n_J RT}{V}$$

The quantity $n_J RT/V$ is the pressure that an amount n_J of perfect gas J would exert in the otherwise empty container. That is, *for a perfect gas*, the partial pressure of a gas in a mixture is the pressure it would exert if it were alone in the container (at the same temperature). If the gases are real, then their partial pressures are still given by eqn 1.9, for that definition applies to all gases, and the sum of these partial pressures is the total pressure (because the sum of all the mole fractions is 1); however, we can no longer say that each partial pressure is the pressure that the gas would exert when it is alone in the container.

Brief illustration 1.3 Partial pressures

From Example 1.2, we have $x(N_2) = 0.780$, $x(O_2) = 0.210$, and $x(Ar) = 0.009$ for dry air at sea level. It then follows from eqn 1.9 that when the total atmospheric pressure is 100 kPa, the partial pressure of nitrogen is

$$p(N_2) = x(N_2)p = 0.780 \times (100\text{ kPa}) = 78.0\text{ kPa}$$

Similarly, for the other two components we find $p(O_2) = 21.0$ kPa and $p(Ar) = 0.9$ kPa. Provided the gases are perfect, these partial pressures are the pressures that each gas would exert if it were separated from the mixture and put in the same container on its own.

Self-test 1.5

The partial pressure of oxygen in air plays an important role in the aeration of water, to enable aquatic life to thrive, and in the absorption of oxygen by blood in our lungs (see Section 6.4). Calculate the partial pressures of a sample of gas consisting of 2.50 g of oxygen and 6.43 g of carbon dioxide with a total pressure of 88 kPa.

Answer: 31 kPa, 57 kPa

Impact on environmental science 1.1

The gas laws and the weather

The biggest sample of gas readily accessible to us is the atmosphere, a mixture of gases with the composition summarized in Table 1.3. The composition is maintained moderately constant by diffusion and convection (winds, particularly the local turbulence called **eddies**) but the pressure and temperature vary with altitude and with the local conditions, particularly in the troposphere (the 'sphere of change'), the layer extending up to about 11 km.

Table 1.3
The composition of the Earth's atmosphere

Substance		Percentage	
		By volume	By mass
Nitrogen, N_2		78.08	75.53
Oxygen, O_2		20.95	23.14
Argon, Ar		0.93	1.28
Carbon dioxide, CO_2		0.031	0.047
Hydrogen, H_2		5.0×10^{-3}	2.0×10^{-4}
Neon, Ne		1.8×10^{-3}	1.3×10^{-3}
Helium, He		5.2×10^{-4}	7.2×10^{-5}
Methane, CH_4		2.0×10^{-4}	1.1×10^{-4}
Krypton, Kr		1.1×10^{-4}	3.2×10^{-4}
Nitric oxide, NO		5.0×10^{-5}	1.7×10^{-6}
Xenon, Xe		8.7×10^{-6}	3.9×10^{-5}
Ozone, O_3	Summer:	7.0×10^{-6}	1.2×10^{-5}
	Winter:	2.0×10^{-6}	3.3×10^{-6}

One of the most variable constituents of air is water vapour, and the humidity it causes. The presence of water vapour results in a *lower* density of air at a given temperature and pressure, as we may conclude from Avogadro's principle. The numbers of molecules in 1 m^3 of moist air and dry air are the same (at the same temperature and pressure), but the mass of an H_2O molecule is less than that of all the other major constituents of air (the molar mass of H_2O is 18 g mol^{-1}, the average molar mass of air molecules is 29 g mol^{-1}), so the density of the moist sample is less than that of the dry sample.

The pressure and temperature vary with altitude. In the troposphere the average temperature is 15 °C at sea level, falling to −57 °C at the bottom of the tropopause at 11 km. This variation is much less pronounced when expressed on the Kelvin scale, ranging from 288 K to 216 K, an average of 268 K. If we suppose that the temperature has its average value all the way up to the edge of the troposphere, then the pressure varies with altitude, h, according to the **barometric formula**:

$$p = p_0 e^{-h/H}$$

where p_0 is the pressure at sea level and H is a constant approximately equal to 8 km. More specifically, $H = RT/Mg$, where M is the average molar mass of air, T is the temperature, and g is the acceleration of free fall. The barometric formula fits the observed pressure distribution quite well even for regions well above the troposphere (Fig. 1.6). It implies that the pressure of the air and its density fall to half their sea-level value at $h = H \ln 2$, or 6 km.

Local variations of pressure, temperature, and composition in the troposphere are manifest as 'weather'. A small region of air is termed a **parcel**. First, we note that a parcel of warm

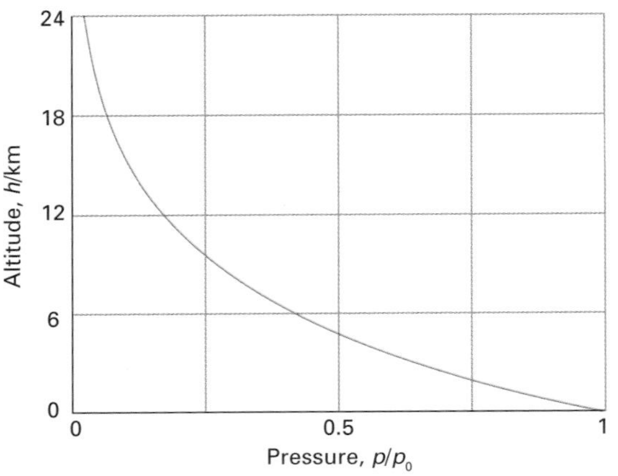

Fig. 1.6 The variation of atmospheric pressure with altitude as predicted by the barometric formula.

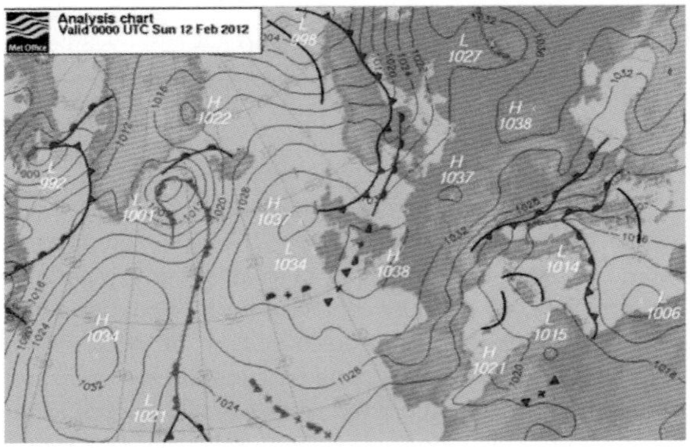

Fig. 1.7 A typical weather map; this one for the North Atlantic in Feburary 2012. Regions of high pressure are denoted *H* and those of low pressure *L*. Pressures are in millibars (1 mbar = 100 Pa).

air is less dense than the same parcel of cool air. As a parcel rises, it expands without transfer of heat from its surroundings: it uses energy to push back the surrounding atmosphere and as a result it cools. Cool air can absorb lower concentrations of water vapour than warm air, so the moisture forms clouds. Cloudy skies can therefore be associated with rising air and clear skies are often associated with descending air.

The motion of air in the upper altitudes may lead to an accumulation in some regions and a loss of molecules from other regions. The former result in the formation of regions of high pressure ('highs' or anticyclones) and the latter result regions of low pressure ('lows', depressions, or cyclones). These regions are shown as H and L on the weather map in Fig. 1.7. The lines of constant pressure—differing by 4 mbar (400 Pa, about 3 Torr)—marked on it are called **isobars**. The elongated regions of high and low pressure are known, respectively, as **ridges** and **troughs**.

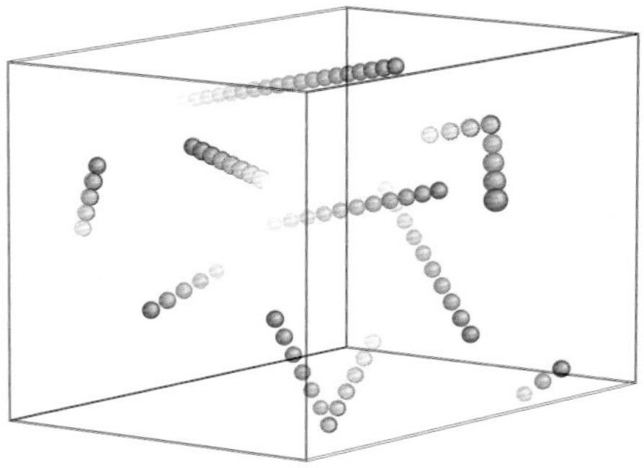

Fig. 1.8 The model used for discussing the molecular basis of the physical properties of a perfect gas. The pointlike molecules move randomly with a wide range of speeds and in random directions, both of which change when they collide with the walls or with other molecules.

The molecular model of gases

We remarked in Foundations (in the section on matter) that a gas may be pictured as a collection of particles in ceaseless, random motion (Fig. 1.8). Now we develop this model of the gaseous state of matter to see how it accounts for the perfect gas law. One of the most important functions of physical chemistry is to convert qualitative notions into quantitative statements that can be tested experimentally by making measurements and comparing the results with predictions. Indeed, an important component of science as a whole is its technique of proposing a qualitative model and then expressing that model mathematically. The 'kinetic model' (or the 'kinetic molecular theory', KMT) of gases is an excellent example of this procedure: the model is very simple, and the quantitative prediction (the perfect gas law) is experimentally verifiable.

The **kinetic model of gases** is based on three assumptions:

1. A gas consists of molecules in ceaseless random motion.
2. The size of the molecules is negligible in the sense that their diameters are much smaller than the average distance travelled between collisions.
3. The molecules do not interact, except during collisions.

The assumption that the molecules do not interact unless they are in contact implies that the potential energy of the molecules (their energy due to their position) is independent of their separation and may be set equal to zero. The total energy of a sample of gas is therefore the sum of the kinetic energies (the energy due to motion) of all the molecules present in it. It follows that the faster the molecules travel (and hence the greater their kinetic energy), the greater the total energy of the gas.

1.4 The pressure of a gas according to the kinetic model

The kinetic model accounts for the steady pressure exerted by a gas in terms of the collisions the molecules make with the walls of the container. Each impact gives rise to a brief force on the wall, but as billions of collisions take place every second, the walls experience a virtually constant force, and hence the gas exerts a steady pressure. On the basis of this model, the pressure exerted by a gas of molar mass M in a volume V is

$$p = \frac{nMv_{\text{rms}}^2}{3V} \qquad \text{Pressure, according to KMT} \qquad (1.11)$$

See Further information 1.1 for a derivation of this equation. Here v_{rms} is the **root-mean-square speed** (r.m.s. speed) of the molecules. This quantity is defined as the square root of the mean value of the squares of the speeds, v, of the molecules. That is, for a sample consisting of N molecules with speeds v_1, v_2, . . . , v_N, we square each speed, add the squares together, divide by the total number of molecules (to get the mean, denoted by ⟨. . .⟩), and finally take the square root of the result:

$$v_{\text{rms}} = \langle v^2 \rangle^{1/2} \qquad \text{Definition} \quad \text{Root-mean-square speed} \qquad (1.12)$$

$$= \left(\frac{v_1^2 + v_2^2 + \cdots + v_N^2}{N} \right)^{1/2}$$

The r.m.s. speed might at first sight seem to be a rather peculiar measure of the mean speeds of the molecules, but its significance becomes clear when we make use of the fact that the kinetic energy of a molecule of mass m travelling at a speed v is $E_k = \frac{1}{2}mv^2$, which implies that the mean kinetic energy of a collection of molecules, $\langle E_k \rangle$, is the average of this quantity, or $\frac{1}{2}mv_{\text{rms}}^2$. It follows from the relation $\frac{1}{2}mv_{\text{rms}}^2 = \langle E_k \rangle$, that

$$v_{\text{rms}} = \left(\frac{2\langle E_k \rangle}{m} \right)^{1/2} \qquad (1.13)$$

Therefore, wherever v_{rms} appears, we can think of it as a measure of the mean kinetic energy of the molecules of the gas. The r.m.s. speed is quite close in value to another and more readily visualized measure of molecular speed, the **mean speed**, $\bar{v}$, of the molecules:

$$\bar{v} = \frac{v_1 + v_2 + \cdots + v_N}{N} \qquad \text{Definition} \quad \text{Mean speed} \qquad (1.14)$$

For samples consisting of large numbers of molecules, the mean speed is slightly smaller than the r.m.s. speed. The precise relation is

$$\bar{v} = \left(\frac{8}{3\pi} \right)^{1/2} v_{\text{rms}} = 0.921 v_{\text{rms}} \qquad (1.15)$$

For elementary purposes, and for qualitative arguments, we don't need to distinguish between the two measures of average speed, but for precise work the distinction is important.

Brief illustration 1.4 Root-mean-square values

Cars pass a point travelling at 45.0 (5), 47.0 (7), 50.0 (9), 53.0 (4), 57.0 (5) km h^{-1}, where the number of cars is given in parentheses. The r.m.s. speed of the cars is given by eqn 1.12 in the form

$$v_{\text{rms}} = \left\{ \frac{5 \times (45.0\ \text{km h}^{-1})^2 + 7 \times (47.0\ \text{km h}^{-1})^2 + \cdots 5 \times (57.0\ \text{km h}^{-1})^2}{5 + 7 + \cdots 5} \right\}^{1/2}$$

$$= 50.2\ \text{km h}^{-1}$$

Self-test 1.6

Calculate the mean speed of the cars (from eqn 1.14).

Answer: 50.0 km h^{-1}

Eqn 1.11 can be rearranged into

$$pV = \tfrac{1}{3}nMv_{\text{rms}}^2 \qquad (1.16)$$

which is starting to resemble the equation $pV = nRT$. This conclusion is a major success of the kinetic model, for the model implies an experimentally verified result.

1.5 The average speed of gas molecules

We now suppose that the expression for pV derived from the kinetic model, eqn 1.16, is indeed the equation of state of a perfect gas, $pV = nRT$. That being so, we can equate the expression on the right to nRT, which gives

$$\tfrac{1}{3}nMv_{\text{rms}}^2 = nRT$$

The n on each side now cancel to give

$$\tfrac{1}{3}Mv_{\text{rms}}^2 = RT$$

The great usefulness of this expression is that we can rearrange it into a formula for the r.m.s. speed of the gas molecules at any temperature, first by writing

$v_{rms}^2 = 3RT/M$ and then by taking the square root of both sides:

$$v_{rms} = \left(\frac{3RT}{M}\right)^{1/2} \quad \text{r.m.s. speed of molecules} \quad (1.17)$$

Brief illustration 1.5 The r.m.s. speed of molecules

Substitution of the molar mass of O_2 (32.0 g mol^{-1}, corresponding to 3.20×10^{-2} kg mol^{-1}) and a temperature corresponding to 25 °C (that is, 298 K) into eqn 1.17 gives an r.m.s. speed for these molecules of

$$v_{rms} = \left\{\frac{3 \times \overbrace{(8.3145\ \text{J K}^{-1}\ \text{mol}^{-1})}^{R} \times \overbrace{(298\ \text{K})}^{T}}{\underbrace{3.20 \times 10^{-2}\ \text{kg mol}^{-1}}_{M}}\right\}^{1/2} = 482\ \text{m s}^{-1}$$

(We used 1 J = 1 kg m^2 s^{-2} to cancel units.) The same calculation for nitrogen molecules gives 515 m s^{-1}. Both these values are not far off the speed of sound in air (346 m s^{-1} at 25 °C). That similarity is reasonable, because sound is a wave of pressure variation transmitted by the movement of molecules, so the speed of propagation of a wave should be approximately the same as the speed at which molecules can adjust their locations.

Self-test 1.7

Calculate the rms speed of H_2 molecules at 25 °C.

Answer: 1920 m s^{-1}

The important conclusion to draw from eqn 1.17 is that

> The r.m.s. speed of molecules in a gas is proportional to the square root of the temperature: $v_{rms} \propto T^{1/2}$.

Because the mean speed is proportional to the r.m.s. speed, the same is true of the mean speed too. Therefore, doubling the thermodynamic temperature (that is, doubling the temperature on the Kelvin scale) increases the mean and the r.m.s. speed of molecules by a factor of $2^{1/2} = 1.414\ldots$.

Brief illustration 1.6 Molecular speeds

Cooling a sample of air from 25 °C (298 K) to 0 °C (273 K) reduces the original r.m.s. speed of the molecules by a factor of

$$\frac{v_{rms}(273\ \text{K})}{v_{rms}(298\ \text{K})} \overset{v_{rms} \propto T^{1/2}}{=} \left(\frac{273\ \text{K}}{298\ \text{K}}\right)^{1/2} = 0.957$$

So, on a cold day, the average speed of air molecules (which is changed by the same factor) is about 4 per cent less than on a warm day.

1.6 The Maxwell distribution of speeds

So far, we have dealt only with the *average* speed of molecules in a gas. Not all molecules, however, travel at the same speed: some move more slowly than the average (until they collide, and get accelerated to a high speed, like the impact of a bat on a ball), and others may briefly move at much higher speeds than the average, but be brought to a sudden stop when they collide. There is a ceaseless redistribution of speeds among molecules as they undergo collisions. Each molecule collides about once every nanosecond (1 ns = 10^{-9} s) or so in a gas under normal conditions.

The mathematical expression that tells us the probability, P, that the molecules in a sample of gas have a speed that lies in a particular range at any instant is called the **distribution of molecular speeds.** Thus, the distribution might tell us that at 20 °C 19 out of 1000 O_2 molecules, corresponding to P = 0.019, have a speed in the range between 300 and 310 m s^{-1}, that 21 out of 1000 have a speed in the range 400 to 410 m s^{-1}, corresponding to $P = 0.021$, and so on. The precise form of the distribution was worked out by James Clerk Maxwell towards the end of the nineteenth century, and his expression is known as the **Maxwell distribution of speeds.** According to Maxwell, the probability $P(v,v + \Delta v)$ that the molecules have a speed in a narrow range between v and $v + \Delta v$ (for example, between 300 m s^{-1} and 310 m s^{-1}, corresponding to v = 300 m s^{-1} and Δv = 10 m s^{-1}) is

$$P(v,v + \Delta v) = \rho(v)\Delta v \text{ with}$$

$$\rho(v) = 4\pi\left(\frac{M}{2\pi RT}\right)^{1/2} v^2\, e^{-Mv^2/2RT} \quad \text{Maxwell distribution of speeds} \quad (1.18)$$

(ρ is the Greek letter rho; for a review of exponential functions, see The chemist's toolkit 1.2.) This formula was used to calculate the numbers quoted above. Its origin can be traced to the Boltzmann distribution (Foundations 0.11), because the fraction of molecules that have a particular speed v have a kinetic energy $E_k = \tfrac{1}{2}mv^2$ (there is no contribution from potential energy) and according to Boltzmann, that fraction is proportional to $e^{-E_k/kT}$, which becomes first $e^{-mv^2/2kT}$ and then $e^{-Mv^2/2RT}$ when m and k are both multiplied by Avogadro's constant, N_A:

$$\rho(v) \propto e^{-\overbrace{mv^2/2kT}^{E_k/kT}} \overset{mN_A=M,\ kN_A=R}{=} e^{-Mv^2/2RT}$$

The chemist's toolkit 1.2 Exponential and Gaussian functions

In preparation for their occurrence throughout the text (and throughout physical chemistry), it will be useful to know the shapes of the functions e^{-ax} and e^{-ax^2} (Sketch 1.3).

An **exponential function**, a function of the form e^{-ax}, starts off at 1 when $x = 0$ and decays toward zero, which it reaches as x approaches infinity. This function approaches zero more rapidly when a is large than when it is small.

A **Gaussian function**, a function of the form e^{-ax^2}, also starts off at 1 when $x = 0$ and decays to zero as x increases, however, its decay is initially slower but then plunges down to zero more rapidly than an exponential function.

The illustration also shows the behaviour of the two functions for negative values of x. The exponential function e^{-ax} rises rapidly to infinity as $x \to -\infty$, but the Gaussian function is symmetrical about $x = 0$ and traces out a bell-shaped curve that falls to zero as $x \to \pm\infty$.

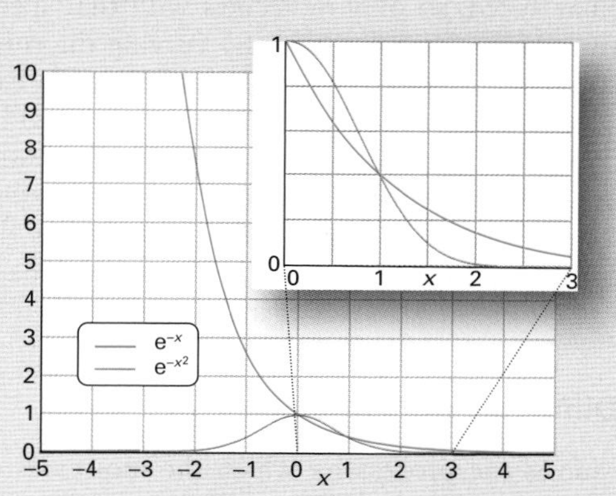

Sketch 1.3 The exponential function, e^{-x}, and the bell-shaped Gaussian function, e^{-x^2}. Note that both are equal to 1 at $x = 0$ but the exponential function rises to infinity as $x \to -\infty$. The enlargement in the right shows the behaviour for $x > 0$ in more detail.

Note that the function $\rho(v)$ is a 'probability density', not the probability itself, in the sense that the actual probability of molecules having a speed in the range v to $v + \Delta v$ is given by the product of $\rho(v)$ and the width of the range. (This relation is like the mass of a volume of material being given by the product of its mass density and the size of the region, the volume.) We shall encounter several other examples of probability densities elsewhere in the text.

Although eqn 1.18 looks complicated, its features can be picked out quite readily. One of the skills to develop in physical chemistry is the ability to interpret the message carried by equations. Equations convey information, and it is far more important to be able to read that information than simply to remember the equation. Let's read the information in eqn 1.18 piece by piece.

- Because $P(v,v + \Delta v)$ is proportional to the range of speeds Δv, we see that the probability of the speed lying in the range Δv increases in proportion to the width of the range. If at a given speed we double the range of interest (but still keep it narrow), then the probability of molecules having speeds lying in that range doubles too.
- Equation 1.18 includes a decaying exponential function, the term $4\pi\left(\frac{M}{2\pi RT}\right)^{1/2} v^2 e^{-Mv^2/2RT}$. Its presence implies that the probability that molecules will be found with very high speeds is very small because e^{-ax^2} becomes very small when ax^2 is large.
- The factor $M/2RT$ multiplying v^2 in the exponent, $4\pi\left(\frac{M}{2\pi RT}\right)^{1/2} v^2 e^{-Mv^2/2RT}$, is large when the molar mass, M, is large, so the exponential factor goes most rapidly towards zero when M is large. That tells us that heavy molecules are unlikely to be found with very high speeds.
- The opposite is true when the temperature, T, is high: then the factor $M/2RT$ in the exponent is small, so the exponential factor falls towards zero relatively slowly as v increases. This tells us that at high temperatures, there is a greater probability of the molecules having high speeds than at low temperatures.
- A factor v^2 in $4\pi\left(\frac{M}{2\pi RT}\right)^{1/2} v^2 e^{-Mv^2/2RT}$ multiplies the exponential. This factor, which comes from taking into account that high velocities can be achieved in more ways than low velocities, goes to zero as v goes to zero, so the probability of finding molecules with very low speeds will also be very small.
- The remaining factors (the term $4\pi(M/2\pi RT)^{1/2}v^2 e^{-Mv^2/2RT}$) simply ensure that when we add together the probabilities over the entire range of speeds from zero to infinity, then we get 1.

Figure 1.9 is a graph of the Maxwell distribution, and shows these features pictorially for the same gas (the same value of M) but different temperatures. As we deduced from the equation, we see that there is only a small probability that molecules in the sample have very low or very high speeds. However, the probability of molecules having very high speeds increases sharply as the temperature is raised, as the tail of the distribution reaches up to higher speeds.

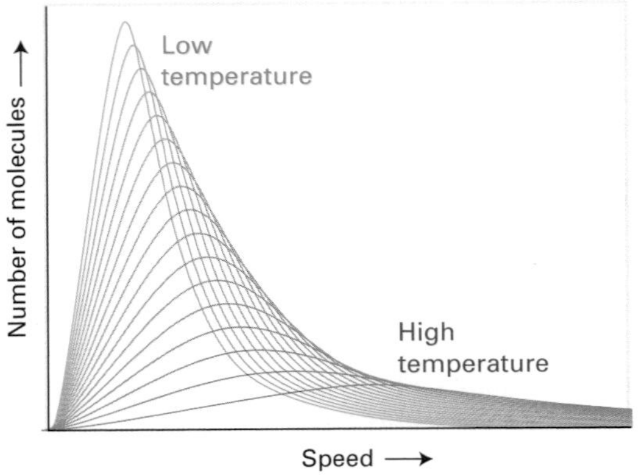

Fig. 1.9 The Maxwell distribution of speeds and its variation with the temperature. Note the broadening of the distribution and the shift of the r.m.s. speed to higher values as the temperature is increased.

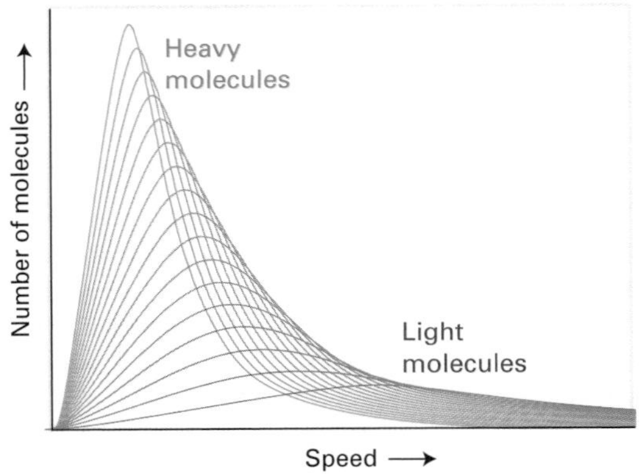

Fig. 1.10 The Maxwell distribution of speeds also depends on the molar mass of the molecules. Molecules of low molar mass have a broad spread of speeds, and a significant fraction may be found travelling much faster than the r.m.s. speed. The distribution is much narrower for heavy molecules, and most of them travel with speeds close to the r.m.s. value.

This feature plays an important role in the rates of gas-phase chemical reactions, for (as we shall see in Section 10.10), the rate of a reaction in the gas phase depends on the energy with which two molecules crash together, which in turn depends on their speeds.

Figure 1.10 is a plot of the Maxwell distribution for molecules with different molar masses at the same temperature. As can be seen, not only do heavy molecules have lower average speeds than light molecules at a given temperature, but they also have a significantly narrower spread of speeds. That narrow spread means that most molecules will be found with speeds close to the average. In contrast, light molecules (such as H_2) have high average speeds and a wide spread of speeds: many molecules will be found travelling either much more slowly or much more quickly than the average. This feature plays an important role in determining the composition of planetary atmospheres, because it means that a significant fraction of light molecules travel at sufficiently high speeds to escape from the planet's gravitational attraction. The ability of light molecules to escape is one reason why hydrogen (molar mass 2.02 g mol^{-1}) and helium (4.00 g mol^{-1}) are very rare in the Earth's atmosphere.

The Maxwell distribution has been verified experimentally by passing a beam of molecules from an oven at a given temperature through a series of coaxial slotted disks. The speed of rotation of the disks brings the slots into line for molecules travelling at a particular speed, so only molecules with that speed pass through and are detected. By varying the rotation speed, the shape of the speed distribution can be explored and is found to match that predicted by eqn 1.18. Although the selector measures the distribution of speeds in one dimension, collisions within the beam ensure that that distribution matches the distribution in three dimensions and that the experiment monitors the Maxwell distribution.

1.7 Diffusion and effusion

Diffusion is the process by which the molecules of different substances mingle with each other. The atoms of two solids diffuse into each other when the two solids are in contact, but the process is very slow. The diffusion of a solute through a liquid solvent is much faster but mixing needs to be encouraged by stirring or shaking (the process is then no longer pure diffusion). The diffusion of one gas into another is much faster. It accounts for the largely uniform composition of the atmosphere, for if a gas is produced by a localized source (such as carbon dioxide from the respiration of animals, oxygen from photosynthesis by green plants, and pollutants from vehicles and industrial sources), then the molecules of gas will diffuse from the source and in due course be distributed throughout the atmosphere. In practice, the process of mixing is accelerated by the bulk motion we experience as winds. The process of **effusion** is the escape of a gas through a small hole, as in a puncture in an inflated balloon or tyre (Fig. 1.11).

The rates of diffusion and effusion of gases increase with increasing temperature, for both processes

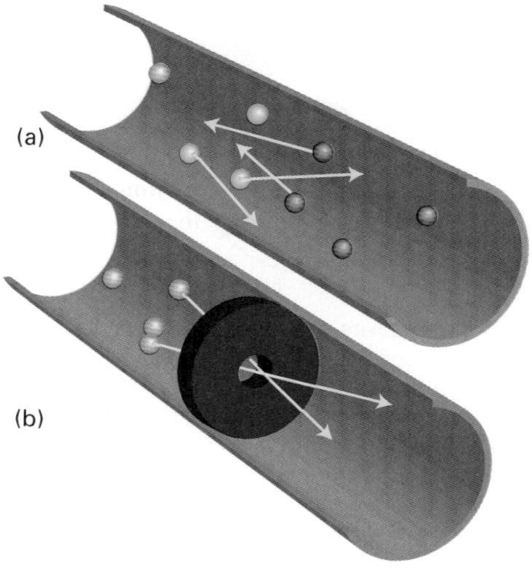

Fig. 1.11 (a) Diffusion is the spreading of the molecules of one substance into the region initially occupied by another species. Note that molecules of both substances move, and each substance diffuses into the other. (b) Effusion is the escape of molecules through a small hole in a confining wall.

depend on the motion of molecules, and molecular speeds increase with temperature. The rates also decrease with increasing molar mass, for molecular speeds decrease with increasing molar mass. The dependence on molar mass, however, is simple only in the case of effusion. In effusion, only a single substance is in motion, not the two or more intermingling gases involved in diffusion.

The experimental observations on the dependence of the rate of effusion of a gas on its molar mass are summarized by **Graham's law of effusion**, proposed by Thomas Graham in 1833:

> At a given pressure and temperature, the rate of effusion of a gas is inversely proportional to the square root of its molar mass:

$$\text{Rate of effusion} \propto \frac{1}{M^{1/2}} \qquad \text{Graham's law} \quad (1.19)$$

Rate in this context means the number (or amount of molecules, in moles) of molecules that escape per second.

Brief illustration 1.7 Graham's law

The rates (in terms of amounts of molecules) at which hydrogen (molar mass 2.02 g mol^{-1}) and carbon dioxide (44.01 g mol^{-1}) effuse under the same conditions of pressure and temperature are in the ratio

$$\frac{\text{Rate of effusion of } H_2}{\text{Rate of effusion of } CO_2} \overset{\text{Rate}\propto 1/M^{1/2}}{=} \left(\frac{M(CO_2)}{M(H_2)}\right)^{1/2} = \left(\frac{44.01\text{ g mol}^{-1}}{2.016\text{ g mol}^{-1}}\right)^{1/2} = \left(\frac{44.01}{2.016}\right)^{1/2} = 4.672$$

The *mass* of carbon dioxide that escapes in a given interval is greater than the mass of hydrogen, because although nearly 5 times as many hydrogen molecules escape, each carbon dioxide molecule has over 20 times the mass of a molecule of hydrogen.

Self-test 1.8

Suppose 5.0 g of argon escapes by effusion; what mass of nitrogen would escape under the same conditions?

Answer: 4.2 g

Note on good practice Always make it clear what terms mean: in this instance 'rate' alone is ambiguous; you need to specify that it is the rate in terms of amount of molecules.

The high rate of effusion of hydrogen and helium is one reason why these two gases leak from containers and through rubber diaphragms so readily. The different rates of effusion through a porous barrier are employed in the separation of uranium-235 from the more abundant and less useful uranium-238 in the processing of nuclear fuel. The process depends on the formation of uranium hexafluoride, a volatile solid. However, because the ratio of the molar masses of $^{238}UF_6$ and $^{235}UF_6$ is only 1.008, the ratio of the rates of effusion is only $(1.008)^{1/2} = 1.004$. Thousands of successive effusion stages are therefore required to achieve a significant separation. The rate of effusion of gases was once used to determine molar mass by comparison of the rate of effusion of a gas or vapour with that of a gas of known molar mass. However, there are now much more precise methods available, such as mass spectrometry.

Graham's law is explained by noting that the r.m.s. speed of molecules in a gas is inversely proportional to the square root of their molar mass (eqn 1.17). Because the rate of effusion through a hole in a container is proportional to the rate at which molecules pass through the hole, it follows that the rate should be inversely proportional to $M^{1/2}$, which is in accord with Graham's law.

1.8 Molecular collisions

The average distance that a molecule travels between collisions is called its **mean free path, λ** (lambda). The mean free path in a liquid is less than the diameter of the molecules, because a molecule in a liquid meets a neighbour even if it moves only a fraction of a diameter. However, in gases, the mean free paths of molecules can be several hundred molecular diameters. If we think of a molecule as the size of a tennis ball, then the mean free path in a typical gas would be about the length of a tennis court.

The **collision frequency, z**, is the average rate of collisions made by one molecule. Specifically, z is the average number of collisions one molecule makes in a given time interval divided by the length of the interval. It follows that the inverse of the collision frequency, $1/z$, is the **time of flight**, the average time that a molecule spends in flight between two collisions (for instance, if there are 10 collisions per second, so the collision frequency is 10 s^{-1}, then the average time between collisions is $^1/_{10}$ of a second and the time of flight is $^1/_{10}$ s). As we shall see, the collision frequency in a typical gas is about 10^9 s^{-1} at 1 atm and room temperature, so the time of flight in a gas is typically 1 ns.

Because speed is distance travelled divided by the time taken for the journey, the r.m.s. speed v_{rms}, which we can loosely think of as the average speed, is the average length of the flight of a molecule between collisions (that is, the mean free path, λ) divided by the time of flight ($1/z$). It follows that the mean free path and the collision frequency are related by

$$v_{rms} = \frac{\text{distance between collisions}}{\text{time between collisions}} = \frac{\overbrace{\text{mean free path}}^{\lambda}}{\underbrace{\text{time of flight}}_{1/z}}$$

$$= \frac{\lambda}{1/z} = \lambda z \qquad (1.20)$$

Therefore, if we can calculate either λ or z, then we can find the other from this equation and the value of v_{rms} given in eqn 1.17.

To find expressions for λ and z we need a slightly more elaborate version of the kinetic model. The basic kinetic model supposes that the molecules are effectively point-like; however, to obtain collisions, we need to assume that two 'points' score a hit whenever they come within a certain range d of each other, where d can be thought of as the diameter of the molecules (Fig. 1.12). The **collision cross-section, σ** (sigma), the target area presented by one molecule to another, is therefore the area of a circle of radius d, so $\sigma = \pi d^2$. When this quantity is built into the kinetic model, it is possible to show[2] that

$$\lambda = \frac{kT}{\sigma p} \qquad \text{Mean free path} \quad (1.21)$$

where k is Boltzmann's constant. Table 1.4 lists the collision cross-sections of some common atoms and molecules. Similarly, by combining this expression with that in eqn 1.20,

$$z \overset{z=v_{rms}/\lambda}{=} \frac{\sigma v_{rms} p}{kT} \qquad \text{Collision frequency} \quad (1.22)$$

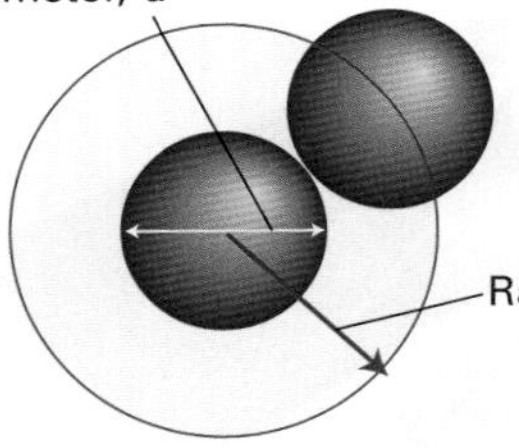

Fig. 1.12 To calculate features of a perfect gas that are related to collisions, a point is regarded as being surrounded by a sphere of diameter d. A molecule will hit another molecule if the centre of the former lies within a circle of radius d. The collision cross-section is the target area, πd^2.

Table 1.4
Collision cross-sections of atoms and molecules

Species	σ/nm^2
Argon, Ar	0.36
Benzene, C_6H_6	0.88
Carbon dioxide, CO_2	0.52
Chlorine, Cl_2	0.93
Ethene, C_2H_4	0.64
Helium, He	0.21
Hydrogen, H_2	0.27
Methane, CH_4	0.46
Nitrogen, N_2	0.43
Oxygen, O_2	0.40
Sulfur dioxide, SO_2	0.58

1 nm^2 = 10^{-18} m^2.

[2] See, for instance, our *Physical Chemistry* (2010).

Brief illustration 1.8 Mean free path

From the information in Table 1.4 we can calculate that the mean free path of O_2 molecules in a sample of oxygen at SATP (25 °C, 1 bar) is

$$\lambda = \frac{\overbrace{(1.381\times10^{-23}\ \text{J K}^{-1})}^{k}\times\overbrace{(298\ \text{K})}^{T}}{\underbrace{(0.40\times10^{-18}\ \text{m}^2)}_{\sigma}\times\underbrace{(1\times10^5\ \text{Pa})}_{p}}$$

$$= \frac{(1.381\times10^{-23})\times(298)}{(0.40\times10^{-18})\times(1\times10^5)}\ \frac{\text{J}}{\text{Pa m}^2}$$

$$\overset{1\ \text{J}=1\ \text{Pa m}^{-3}}{=}\ 1.0\times10^{-7}\ \text{m}\ \overset{1\ \text{nm}=10^{-9}\ \text{m}}{=}\ 100\ \text{nm}$$

Under the same conditions, the collision frequency is $6.2\times10^9\ \text{s}^{-1}$, so each molecule makes 6.2 billion collisions each second.

Self-test 1.9

Use eqns 1.17 and 1.22, together with the information in Table 1.4, to determine the collision frequency for Cl_2 molecules in a sample of chlorine gas under the same conditions.

Answer: $7.3\times10^9\ \text{s}^{-1}$

Once again, we should *interpret the essence* of eqns 1.21 and 1.22 rather than trying to remember them.

- Because $\lambda \propto 1/p$, the mean free path decreases as the pressure increases.

This decrease is a result of the increase in the number of molecules present in a given volume as the pressure is increased, so each molecule travels a shorter distance before it collides with a neighbour. For example, the mean free path of an O_2 molecule decreases from 73 nm to 36 nm when the pressure is increased from 1.0 bar to 2.0 bar at 25 °C.

- Because $\lambda \propto 1/\sigma$, the mean free path is shorter for molecules with large collision cross-sections.

For instance, the collision cross-section of a benzene molecule (0.88 nm^2) is about four times greater than that of a helium atom (0.21 nm^2), and at the same pressure and temperature its mean free path is four times shorter.

- Because $z \propto p$, the collision frequency increases with the pressure of the gas.

This dependence follows from the fact that, provided the temperature is the same, each molecule takes less time to travel to its neighbour in a denser, higher-pressure gas. For example, although the collision frequency for an O_2 molecule in oxygen gas at SATP is $6.2\times10^9\ \text{s}^{-1}$, at 2.0 bar and the same temperature the collision frequency is doubled, to $1.2\times10^{10}\ \text{s}^{-1}$.

- Because eqn 1.22 shows that $z \propto v_{\text{rms}}$, and we know that $v_{\text{rms}} \propto 1/M^{1/2}$, heavy molecules have lower collision frequencies than light molecules, providing their collision cross-sections are the same.

Heavy molecules travel more slowly on average than light molecules do (at the same temperature), so they collide with other molecules less frequently.

Real gases

So far, everything we have said applies to perfect gases, in which the average separation of the molecules is so great that they move independently of one another. In terms of the quantities introduced in the previous section, a perfect gas is a gas for which the mean free path, λ, of the molecules in the sample is much greater than d, the separation at which they are regarded as being in contact:

$$\text{Condition for perfect-gas behaviour: } \lambda \gg d$$

As a result of this large average separation, a perfect gas is a gas in which the only contribution to the energy comes from the kinetic energy of the motion of the molecules and there is no contribution to the total energy from the potential energy arising from the interaction of the molecules with one another. However, in fact all molecules do interact with one another provided they are close enough together, so the 'kinetic energy only' model is only an approximation. Nevertheless, under most conditions the criterion $\lambda \gg d$ (the separation of the molecules is much greater than their diameters; think tennis court compared with tennis ball) is satisfied and the gas can be treated as though it is perfect.

1.9 Molecular interactions

There are two types of contribution to the potential energy of interaction between molecules. At relatively large separations (a few molecular diameters), molecules attract each other. This attraction is responsible for the condensation of gases into liquids at low temperatures. At low enough temperatures the molecules of a gas have insufficient kinetic energy to escape from each other's attraction and they stick

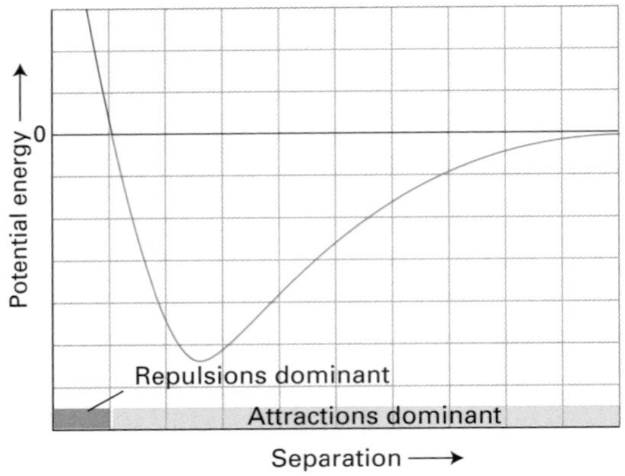

Fig. 1.13 The variation of the potential energy of two molecules with their separation. High positive potential energy (at very small separations) indicates that the interactions between them are strongly repulsive at these distances. At intermediate separations, where the potential energy is negative, the attractive interactions dominate. At large separations (on the right) the potential energy is zero and there is no interaction between the molecules.

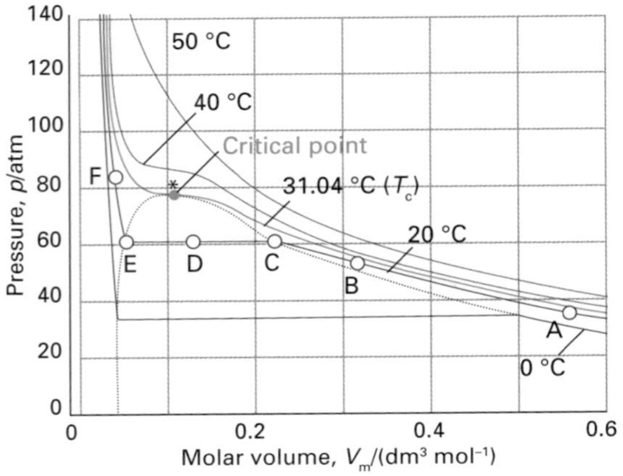

Fig. 1.14 The experimental isotherms of carbon dioxide at several temperatures. The critical isotherm is at 31.04 °C.

together. Second, although molecules attract each other when they are a few diameters apart, as soon as they come into contact they repel each other. This repulsion is responsible for the fact that liquids and solids have a definite bulk and do not collapse to an infinitesimal point.

Molecular interactions—the attractions and repulsions between molecules—give rise to a potential energy that contributes to the total energy of a gas. Because attractions correspond to a lowering of total energy as molecules get closer together, they make a *negative* contribution to the potential energy. On the other hand, repulsions make a positive contribution to the total energy as the molecules squash together. Figure 1.13 illustrates the general form of the variation of the intermolecular potential energy. At large separations, the energy-lowering interactions are dominant, but at short distances the energy-raising repulsions dominate.

Molecular interactions affect the bulk properties of a gas and, in particular, their equations of state. For example, the isotherms of real gases have shapes that differ from those implied by Boyle's law, particularly at high pressures and low temperatures when the interactions are most important. Figure 1.14 shows a set of experimental isotherms for carbon dioxide. They should be compared with the perfect-gas isotherms shown in Fig. 1.1. Although the experimental isotherms resemble the perfect-gas isotherms at high temperatures (and at low pressures, off the scale on the right of the graph), there are very striking differences between the two at temperatures below about 50 °C and at pressures above about 1 bar.

1.10 The critical temperature

To understand the significance of the isotherms in Fig. 1.14, let's begin with the isotherm at 20 °C and follow the changes from A to F:

- At point A the sample of carbon dioxide is a gas.
- As the sample is compressed to B by pressing in a piston, the pressure increases broadly in agreement with Boyle's law.
- The increase continues until the sample reaches point C.
- Beyond this point, we find that the piston can be pushed in without any further increase in pressure, through D to E.
- The reduction in volume from E to F requires a very large increase in pressure.

This variation of pressure with volume is exactly what we expect if the gas at C condenses to a compact liquid at E. Indeed, if we could observe the sample we would see that:

- At C the liquid begins to condense to a liquid.
- The condensation is complete when the piston is pushed in to E.
- At E, the piston is resting on the surface of the liquid.
- The subsequent reduction in volume, from E to F, corresponds to the very high pressure needed to compress a liquid into a smaller volume.

In terms of intermolecular interactions:

- The step from C to E corresponds to the molecules being so close on average that they attract each other and cohere into a liquid.
- The step from E to F represents the effect of trying to force the molecules even closer together when they are already in contact, and hence trying to overcome the strong repulsive interactions between them.

If we could look inside the container at point D, we would see a liquid separated from the remaining gas by a sharp surface (Fig. 1.15). At a slightly higher temperature (at 30 °C, for instance), a liquid forms, but a higher pressure is needed to produce it. It might be difficult to make out the surface because the remaining gas is at such a high pressure that its density is similar to that of the liquid. At the special temperature of 31.04 °C (304.19 K) the gaseous state of carbon dioxide appears to transform continuously into the condensed state and at no stage is there a visible surface between the two states of matter. At this temperature (which is 304.19 K for carbon dioxide but varies from substance to substance), called the **critical temperature**, T_c, and at all higher temperatures, a single form of matter fills the container at all stages of the compression and there is no separation of a liquid from the gas. We have to conclude that *a gas cannot be condensed to a liquid by the application of pressure unless the temperature is below the critical temperature.*

Figure 1.14 also shows that in the **critical isotherm**, the isotherm at the critical temperature, the volumes at each end of the horizontal part of the isotherm have merged to a single point, the **critical point** of the gas. The pressure and molar volume at the critical point are called the **critical pressure,** p_c, and **critical molar volume,** V_c, of the substance. Collectively, p_c, V_c, and T_c are the **critical constants** of a substance. Table 1.5 lists the critical temperatures of some common gases. The data there imply, for example, that liquid nitrogen cannot be formed by the application of pressure unless the temperature is below 126 K (−147 °C). The critical temperature is sometimes used to distinguish the terms 'vapour' and 'gas':

a **vapour** is the gaseous phase of a substance below its critical temperature (and which can therefore be liquefied by compression alone)

a **gas** is the gaseous phase of a substance above its critical temperature (and which cannot therefore be liquefied by compression alone)

Oxygen at room temperature is therefore a true gas; the gaseous phase of water at room temperature is a vapour.

Table 1.5
The critical temperatures of gases

	Critical temperature/°C
Noble gases	
Helium, He	−268 (5.2 K)
Neon, Ne	−229
Argon, Ar	−123
Krypton, Kr	−64
Xenon, Xe	17
Halogens	
Chlorine, Cl_2	144
Bromine, Br_2	311
Small inorganic molecules	
Ammonia, NH_3	132
Carbon dioxide, CO_2	31
Hydrogen, H_2	−240
Nitrogen, N_2	−147
Oxygen, O_2	−118
Water, H_2O	374
Organic compounds	
Benzene, C_6H_6	289
Methane, CH_4	−83
Tetrachloromethane, CCl_4	283

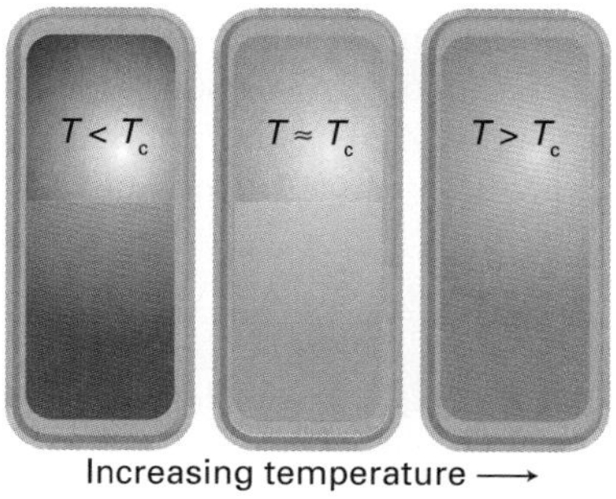

Fig. 1.15 When a liquid is heated in a sealed container, the density of the vapour phase increases and that of the liquid phase decreases, as depicted here by the changing density of shading. There comes a stage at which the two densities are equal and the interface between the two fluids disappears. This disappearance occurs at the critical temperature. The container needs to be strong: the critical temperature of water is at 373 °C and the vapour pressure is then 218 atm.

The dense fluid obtained by compressing a gas when its temperature is higher than its critical temperature is not a true liquid, but it behaves like a liquid in many respects: it has a density similar to that of a liquid, for instance, and can act as a solvent. However, despite its density, the fluid is not strictly a liquid because it never possesses a surface that separates it from a vapour phase. Nor is it much like a gas, because it is so dense. It is an example of a **supercritical fluid.** Supercritical fluids are currently being used as solvents. For example, supercritical carbon dioxide is used to extract caffeine in the manufacture of decaffeinated coffee where, unlike organic solvents, it does not result in the formation of an unpleasant and possibly toxic residue. Supercritical fluids are also currently of great interest in industrial processes, for they can be used instead of chlorofluorocarbons (CFCs) and hence avoid the environmental damage that CFCs are known to cause. Because supercritical carbon dioxide is obtained either from the atmosphere or from renewable organic sources (by fermentation) its use does not increase the net load of atmospheric carbon dioxide.

1.11 The compression factor

A useful quantity for discussing the properties of real gases is the **compression factor**, Z, which is the ratio of the actual experimentally determined molar volume of a gas, V_m, to the molar volume of a perfect gas, $V_m^{perfect}$, under the same conditions:

$$Z = \frac{V_m}{V_m^{perfect}} \quad \text{Definition} \quad \text{Compression factor} \qquad (1.23a)$$

For a perfect gas, $V_m = V_m^{perfect}$, so $Z = 1$ and deviations of Z from 1 are a measure of how far a real gas departs from behaving perfectly. The molar volume of a perfect gas is RT/p (recall eqn 1.6b), so we can rewrite the definition of Z as

$$Z \overbrace{=}^{V_m^{perfect}=RT/p} \frac{V_m}{RT/p} = \frac{pV_m}{RT} \qquad (1.23b)$$

When Z is measured for real gases, it is found to vary with pressure, as shown in Fig. 1.16. At low pressures, some gases (methane, ethane, and ammonia, for instance) have $Z < 1$. That is, their molar volumes are smaller than that of a perfect gas, suggesting that the molecules are pulled together slightly. We can conclude that for these molecules and these conditions, the attractive interactions are dominant. The compression factor rises above 1 at high pressures, whatever the identity of the gas, and for some

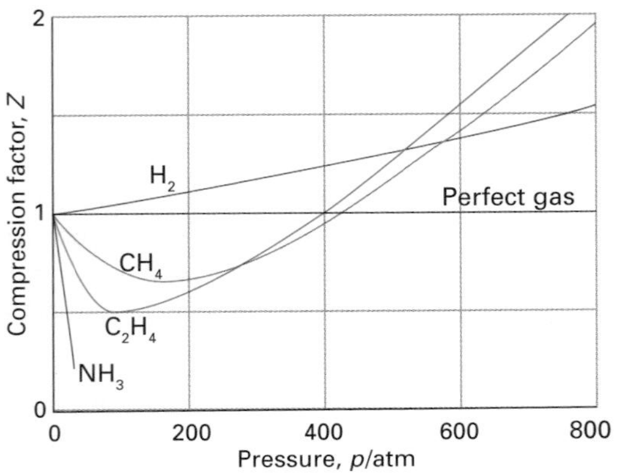

Fig. 1.16 The variation of the compression factor, Z, with pressure for several gases at 0 °C. A perfect gas has $Z = 1$ at all pressures. Of the gases shown, hydrogen shows positive deviations at all pressures (at this temperature); all the other gases show negative deviations initially but positive deviations at high pressures. The negative deviations are a result of the attractive interactions between molecules and the positive deviations are a result of the repulsive interactions.

gases (hydrogen in the illustration) $Z > 1$ at all pressures. The type of behaviour exhibited depends on the temperature. The observation that $Z > 1$ tells us that the molar volume of the gas is now greater than that expected for a perfect gas of the same temperature and pressure, so the molecules are pushed apart slightly. This behaviour indicates that the repulsive forces are dominant. For hydrogen, the attractive interactions are so weak that the repulsive interactions dominate even at low pressures.

1.12 The virial equation of state

We can use the deviation of Z from its 'perfect' value of 1 to construct an *empirical* (observation based) equation of state. To do so we suppose that, for real gases, the relation $Z = 1$ is only the first term of a lengthier expression and write instead

$$Z = 1 + \frac{B}{V_m} + \frac{C}{V_m^2} + \cdots \qquad (1.24a)$$

The coefficients $B, C, \ldots$ are called **virial coefficients;** B is the second virial coefficient, C, the third, and so on; the unwritten $A = 1$ is the first. The word 'virial' comes from the Latin word for force, and it reflects the fact that intermolecular forces are now significant. The virial coefficients, which are also denoted B_2, B_3, etc. in place of B, C, etc., vary from gas to gas and depend on the temperature. This technique, of taking a limiting expression (in this case, $Z = 1$, which applies to gases at very large molar volumes) and supposing that it is the first term of a more complicated expression, is quite common in physical chemistry. The limiting expression is the first approximation to the true expression, whatever that may be, and the additional terms progressively take into account the secondary effects that the limiting expression ignores.

The most important additional term on the right in eqn 1.24a is the one proportional to B (because under most conditions $C/V_m^2 \ll B/V_m$ and C/V_m^2 can be neglected). In that case

$$Z \approx 1 + \frac{B}{V_m} \tag{1.24b}$$

From the graphs in Fig. 1.16, it follows that, for the temperature to which the data apply, B must be positive for hydrogen (so that $Z > 1$) but negative for methane, ethane, and ammonia (so that for them $Z < 1$). However, in all cases Z rises again as the gas is compressed further (corresponding to small values of V_m and high pressures in Fig. 1.16), indicating that C/V_m^2 is positive. The values of the virial coefficients for many gases are known from measurements of Z over a range of molar volumes and using mathematical software to fit the data to eqn 1.24a by varying the coefficients until a good match is obtained.

To convert eqn 1.24a into an equation of state, we combine it with eqn 1.23b ($Z = pV_m/RT$), which gives

$$\frac{pV_m}{RT} = 1 + \frac{B}{V_m} + \frac{C}{V_m^2} + \cdots$$

Now multiply both sides by RT/V_m and obtain

$$p = \frac{RT}{V_m}\left(1 + \frac{B}{V_m} + \frac{C}{V_m^2} + \cdots\right)$$

Next, replace V_m by V/n throughout to get p as a function of n, V, and T:

$$p = \frac{nRT}{V}\left(1 + \frac{nB}{V} + \frac{n^2C}{V^2} + \cdots\right) \quad \text{Virial equation of state} \tag{1.25}$$

Equation 1.25 is the **virial equation of state.** When the molar volume is very large, the terms B/V_m and C/V_m^2 are both very small, and only the 1 inside the parentheses survives. In this limit, the virial equation of state approaches that of a perfect gas.

Brief illustration 1.9 The virial equation of state

The molar volume of NH_3 is 1.00 dm^3 mol^{-1} at 36.2 bar and 473 K. Assuming that under these conditions the virial equation of state may be written as $p = (RT/V_m) \times (1 - B/V_m)$, so

$$B = \left(\frac{pV_m}{RT} - 1\right)V_m$$

The value of the second virial coefficient at this temperature is therefore

$$B = \left(\frac{\overbrace{(36.2 \times 10^5\ \text{Pa})}^{36.2\ \text{bar}} \times \overbrace{(1.00 \times 10^{-3}\ \text{m}^3\ \text{mol}^{-1})}^{1.00\ \text{dm}^3\ \text{mol}^{-1}}}{(8.3145\ \text{J K}^{-1}\ \text{mol}^{-1}) \times (473\ \text{K})} - 1\right)$$
$$\times 1.00 \times 10^{-3}\ \text{m}^3\ \text{mol}^{-1}$$
$$= -79.5 \times 10^{-6}\ \text{m}^3\ \text{mol}^{-1} = -79.5\ \text{cm}^3\ \text{mol}^{-1}$$

Self-test 1.10

The second virial coefficient for NH_3 is -45.6 cm^3 mol^{-1} at a temperature of 573 K. Determine the pressure at which the molar volume is 1.00 dm^3 mol^{-1} at this temperature.

Answer: 45.6 bar

1.13 The van der Waals equation of state

Although it is the most reliable equation of state, the virial equation does not give us much immediate insight into the behaviour of gases and their condensation to liquids. The **van der Waals equation,** which was proposed in 1873 by the Dutch physicist Johannes van der Waals, is only an approximate equation of state but it has the advantage of showing how the intermolecular interactions contribute to the deviations of a gas from the perfect gas law. We can view the van der Waals equation as another example of taking a soundly based qualitative idea and building up a mathematical expression that can be tested quantitatively.

The repulsive interaction between two molecules implies that they cannot come closer than a certain distance. Therefore, instead of being free to travel anywhere in a volume V, the actual volume in which the molecules can travel is reduced to an extent proportional to the number of molecules present and the volume they each exclude (Fig. 1.17). We can

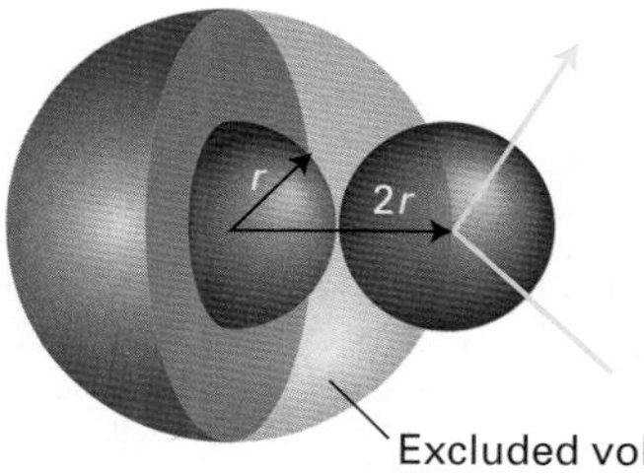

Fig. 1.17 When two molecules, each of radius r and volume $V_{mol} = \frac{4}{3}\pi r^3$ approach each other, the centre of one of them cannot penetrate into a sphere of radius $2r$ and therefore volume $8V_{mol}$ surrounding the other molecule.

therefore model the effect of the repulsive, volume-excluding forces by changing V in the perfect gas equation to $V - nb$, where b is the proportionality constant between the reduction in volume and the amount of molecules present in the container (see the following Derivation).

Derivation 1.1

The molar volume of a gas described by the van der Waals equation

The volume of a sphere of radius R is $\frac{4}{3}\pi R^3$. Figure 1.17 shows that the closest distance of two hard-sphere molecules of radius r, and volume $V_{\text{molecule}} = \frac{4}{3}\pi r^3$, is $2r$. Therefore, the excluded volume is $\frac{4}{3}\pi(2r)^3 = 8(\frac{4}{3}\pi r^3)$, or $8V_{\text{molecule}}$. The volume excluded per molecule is one-half this volume, or $4V_{\text{molecule}}$, so $b \approx 4V_{\text{molecule}}N_A$.

So far, the perfect gas equation of state changes from $p = nRT/V$ to

$$p = \frac{nRT}{V - nb}$$

This equation of state—it is not yet the full van der Waals equation—should describe a gas in which repulsions are important. Note that when the pressure is low, the volume is large compared with the volume excluded by the molecules (which we write $V \gg nb$). The nb can then be ignored in the denominator and the equation reduces to the perfect gas equation of state. It is always a good plan to verify that an equation reduces to a known form when a plausible physical approximation is made.

The effect of the attractive interactions between molecules is to reduce the pressure that the gas exerts. We can model the effect by supposing that the attraction experienced by a given molecule is proportional to the concentration, n/V, of molecules in the container. Because the attractions slow the molecules down, the molecules strike the walls less frequently *and* strike it with a weaker impact. (This slowing does not mean that the gas is cooler close to the walls: the simple relation between T and r.m.s. speed in eqn 1.17 is valid only in the absence of intermolecular forces.) We can therefore expect the reduction in pressure to be proportional to the *square* of the molar concentration, one factor of n/V reflecting the reduction in frequency of collisions and the other factor the reduction in the strength of their impulse. If the constant of proportionality is written a, we can write

$$\text{Reduction in pressure} = a \times \left(\frac{n}{V}\right)^2$$

It follows that the equation of state allowing for both repulsions and attractions is

$$p = \frac{nRT}{V - nb} - a\left(\frac{n}{V}\right)^2 \quad \text{van der Waals equation of state} \quad (1.26a)$$

This expression is the **van der Waals equation of state.** To show its resemblance to the perfect gas equation $pV = nRT$, eqn 1.26a is sometimes rearranged by bringing the term in a to the left, to give $p + an^2/V^2$, and then multiplying both sides by $V - nb$:

$$\left(p + \frac{an^2}{V^2}\right)(V - nb) = nRT \quad (1.26b)$$

We have built the van der Waals equation by using physical arguments about the volumes of molecules and the effects of forces between them. It can be derived in other ways, but the present method has the advantage of showing how to derive the form of an equation out of general ideas. The derivation also has the advantage of keeping imprecise the significance of the **van der Waals parameters,** the constants a and b: they are much better regarded as empirical parameters than as precisely defined molecular properties. The van der Waals parameters depend on the gas, but are taken as independent of temperature (Table 1.6). It follows from the way we have constructed the equation that a (the parameter representing the role of attractions) can be expected to be large when the molecules attract each other strongly, whereas b (the parameter representing the role of

Table 1.6
van der Waals parameters of gases

Substance	$a/(10^2$ kPa dm^6 mol$^{-2})$	$b/(10^{-2}$ dm^3 mol$^{-1})$
Air	1.4	0.039
Ammonia, NH_3	4.114	3.71
Argon, Ar	1.320	3.20
Carbon dioxide, CO_2	3.119	4.29
Ethane, C_2H_6	5.435	6.51
Ethene, C_2H_4	4.545	5.82
Helium, He	0.0337	2.38
Hydrogen, H_2	0.2388	2.65
Nitrogen, N_2	1.347	3.87
Oxygen, O_2	1.359	3.19
Xenon, Xe	4.135	5.16

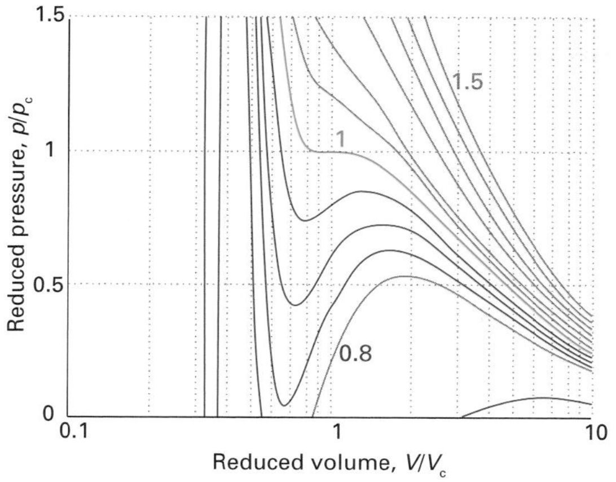

Fig. 1.18 Isotherms calculated by using the van der Waals equation of state. The axes are labelled with the 'reduced pressure', p/p_c, and 'reduced volume', V/V_c, where $p_c = a/27b^2$ and $V_c = 3b$. The individual isotherms are labelled with the 'reduced temperature', T/T_c, where $T_c = 8a/27Rb$. The isotherm labelled 1 is the critical isotherm (the isotherm at the critical temperature).

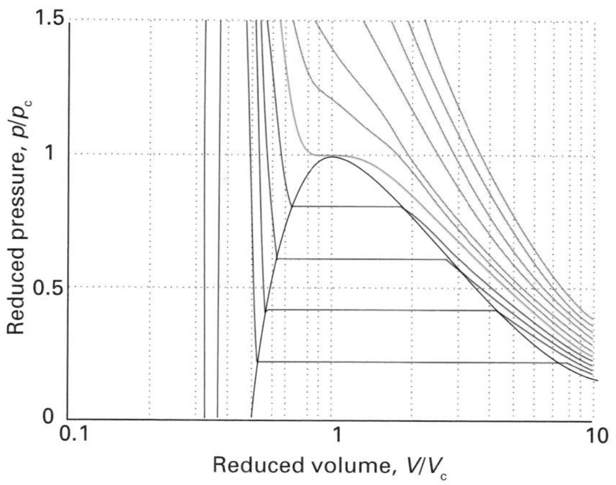

Fig. 1.19 The unphysical van der Waals loops are eliminated by drawing straight lines that divide the loops into areas of equal size. With this procedure, the isotherms strongly resemble the observed isotherms.

repulsions) can be expected to be large when the molecules are large.

The reliability of the van der Waals equation can be judged by comparing the isotherms it predicts, which are shown in Fig. 1.18, with the experimental isotherms already shown in Fig. 1.14. Apart from the waves below the critical temperature they do resemble experimental isotherms quite well. The waves, which are called **van der Waals loops**, are unrealistic because they suggest that under some conditions compression results in a decrease of pressure. The loops are therefore trimmed away and replaced by horizontal lines (Fig. 1.19). The van der Waals parameters in Table 1.6 were found by fitting the calculated curves to experimental isotherms.

Two important features of the van der Waals equation should be noted. First, perfect-gas isotherms are obtained from the van der Waals equation at high temperatures and low pressures. To confirm this remark, we need to note that when the temperature is high, RT may be so large that the first term on the right in eqn 1.26a greatly exceeds the second, so the latter may be ignored. Furthermore, at low pressures, the molar volume is so large that $V - nb$ can be replaced by V. Hence, under these conditions (of high temperature and low pressure), eqn 1.26a reduces to $p = nRT/V$, the perfect gas equation. Second, and as shown in the following Derivation, the critical constants are related to the van der Waals coefficients as follows:

$$V_c = 3b, \; T_c = \frac{8a}{27Rb}, \; p_c = \frac{a}{27b^2} \qquad (1.27)$$

The first of these relations shows that the critical volume is about three times the volume occupied by the molecules themselves.

Example 1.3

Estimating the critical constants of a gas

Estimate the critical constants of carbon dioxide.

Strategy Treat carbon dioxide as a van der Waals gas and use eqn 1.27 with values of parameters taken from Table 1.6. Convert the parameters to base units before using them.

Answer The van der Waals parameters for CO_2 are $a = 3.460$ dm^6 bar mol^{-2} and $b = 0.004\,267$ dm^3 mol^{-1}. To convert them to base units we write

$$a = 3.658\ \overbrace{\mathrm{dm}^6}^{(10^{-1}\,\mathrm{m})^6}\ \overbrace{\mathrm{bar}}^{10^5\,\mathrm{Pa}}\ \mathrm{mol}^{-2} = 0.3658\ \mathrm{m}^6\ \mathrm{Pa\ mol}^{-2}$$

$$b = 0.0429\ \overbrace{\mathrm{dm}^3}^{(10^{-1}\,\mathrm{m})^3}\ \mathrm{mol}^{-1} = 4.29\times10^{-5}\ \mathrm{m}^3\ \mathrm{mol}^{-1}$$

Then, by using eqn 1.27, we predict the values of the critical constants for CO_2 to be

$$V_c = 3b = 3\times(4.29\times10^{-5}\ \mathrm{m}^3\ \mathrm{mol}^{-1}) = 1.29\times10^{-4}\ \mathrm{m}^3\ \mathrm{mol}^{-1}, \text{ or } 0.129\ \mathrm{dm}^3\ \mathrm{mol}^{-1}$$

$$T_c = \frac{8a}{27Rb} = \frac{8\times(0.3658\ \mathrm{m}^6\ \mathrm{Pa\ mol}^{-2})}{27\times(8.3145\ \mathrm{J\ K}^{-1}\ \mathrm{mol}^{-1})\times(4.29\times10^{-5}\ \mathrm{m}^3\ \mathrm{mol}^{-1})}$$

$$= 304\ \mathrm{K, or}\ 31\ ^\circ\mathrm{C}$$

$$p_c = \frac{a}{27b^2} = \frac{0.3658\ \mathrm{m}^6\ \mathrm{Pa\ mol}^{-2}}{27\times(4.29\times10^{-5}\ \mathrm{m}^3\ \mathrm{mol}^{-1})^2} = 7.36\ \mathrm{MPa}$$

The experimental values are 0.094 $dm^3\ mol^{-1}$, 304 K, and 7.375 MPa, respectively.

Self-test 1.11

The critical pressure and temperature of CH_4 are 46.1 bar and 191 K, respectively. Determine the value of the van der Waals parameter *b*.

Answer: 0.0431 $dm^3\ mol^{-1}$

Derivation 1.2

Relating the critical constants to the van der Waals parameters

For this derivation we need know some of the rules of calculus related to differentiation as summarized in The chemist's toolkit 1.3. We see from Fig. 1.18 that, for $T < T_c$, the calculated isotherms oscillate, and each one passes through a minimum followed by a maximum. These extrema converge as $T \rightarrow T_c$ and coincide at $T = T_c$; at the critical point the curve has a flat inflexion (**1**). From the properties of curves, we know that an inflexion of this type occurs when both the first and second derivatives are zero. Hence, we can find the critical temperature by calculating these derivatives and setting them equal to zero. First, we use $V_m = V/n$ to write eqn 1.26a as

$$p = \frac{RT}{V_m - b} - \frac{a}{V_m^2}$$

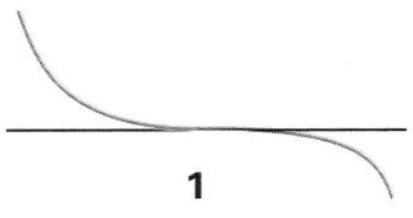

1

The first and second derivatives of p with respect to V_m (with p playing the role of y in The chemist's toolkit 1.3, and V_m playing the role of x) are, respectively:

$$\frac{dp}{dV_m} = -\frac{RT}{(V_m - b)^2} + \frac{2a}{V_m^3} \qquad \frac{d^2p}{dV_m^2} = \frac{2RT}{(V_m - b)^3} - \frac{6a}{V_m^4}$$

At the critical point $T = T_c$, $V_m = V_c$, and both derivatives are equal to zero:

$$-\frac{RT_c}{(V_c - b)^2} + \frac{2a}{V_c^3} = 0 \qquad \frac{2RT_c}{(V_c - b)^3} - \frac{6a}{V_c^4} = 0$$

Solving this pair of equations gives (as you should verify) the expressions for V_c and T_c in eqn 1.27. When they are inserted in the van der Waals equation itself, we find the expression for p_c given there too.

1.14 The liquefaction of gases

A gas may be liquefied by cooling it below its boiling point at the pressure of the experiment. For example, chlorine at 1 atm can be liquefied by cooling it to below −34 °C in a bath cooled with dry ice (solid carbon dioxide). For gases with very low boiling points (such as oxygen and nitrogen, at −183 °C and −186 °C, respectively), such a simple technique is not practicable unless an even colder bath is available.

One alternative and widely used commercial technique makes use of the forces that act between molecules. We saw earlier that the r.m.s. speed of molecules in a gas is proportional to the square root of the

The chemist's toolkit 1.3 Differentiation

An important result from calculus is that

$$\frac{dx^n}{dx} = nx^{n-1}$$

Thus, if $y = mx^2 + b$, then, because m and b are both constants, $dy/dx = 2mx$. In this instance, we see that the slope (which is equal to dy/dx) increases with x (Sketch 1.4). The expression for dx^n/dx also applies when n is negative. So, for instance,

$$\frac{d}{dx}\frac{1}{x} = \frac{d(1/x)}{dx} = \frac{d(x^{-1})}{dx} \overset{n=-1}{=} -x^{-2} = -\frac{1}{x^2}$$

Two very important relations are

$$\frac{d}{dx}e^{ax} = ae^{ax} \qquad \frac{d}{dx}e^{f(x)} = \left(\frac{df(x)}{dx}\right)e^{f(x)}$$

Two other fundamental results are

$$\frac{d}{dx}\frac{1}{a+bx} = -\frac{b}{(a+bx)^2} \qquad \frac{d}{dx}\frac{1}{(a+bx)^2} = -\frac{2b}{(a+bx)^3}$$

The 'second derivative' of a function $d(dy/dx)/dx$ is denoted d^2y/dx^2, or equivalently $(d^2/dx^2)y$, and is the result of taking the derivative a second time by using the same rule given above. For instance, because the first derivative of $y = mx^2 + b$ is $2mx$, the second derivative of the same function is $2m$. For Derivation 1.2 you also need to know that

$$\frac{d^2}{dx^2}\frac{1}{a+bx} = \frac{d}{dx}\left\{-\frac{b}{(a+bx)^2}\right\} = \frac{2b^2}{(a+bx)^3}$$

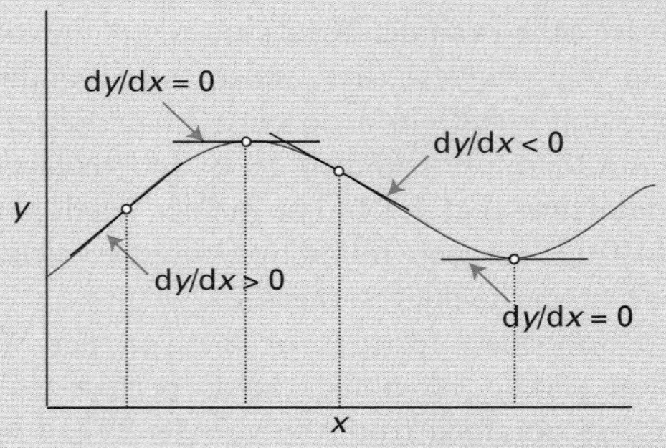

Sketch 1.4 A function and its slope; the slope at any point x is given by the derivative of the function at that point.

temperature (eqn 1.17). It follows that reducing the r.m.s. speed of the molecules is equivalent to cooling the gas. If the speed of the molecules can be reduced to the point that neighbours can capture each other by their intermolecular attractions, then the cooled gas will condense to a liquid.

To slow the gas molecules, we make use of an effect similar to that seen when a ball is thrown into the air: as it rises it slows in response to the gravitational attraction of the Earth and its kinetic energy is converted into potential energy. Molecules attract each other, as we have seen (the attraction is not gravitational, but the effect is the same), and if we can cause them to move apart from each other, like a ball rising from a planet, then they should slow. It is very easy to move molecules apart from each other: we simply allow the gas to expand, which increases the average separation of the molecules. To cool a gas, therefore, we allow it to expand without allowing any heat to enter from outside. As it does so, the molecules move apart to fill the available volume, struggling as they do so against the attraction of their neighbours. Because some kinetic energy must be converted into potential energy to reach greater separations, the molecules travel more slowly as their separation increases. Therefore, because the average speed of the molecules has been reduced, the gas is now cooler than before the expansion. This process of cooling a real gas by expansion through a narrow opening called a 'throttle' is called the **Joule–Thomson effect**. The effect was first observed and analysed by James Joule (whose name is commemorated in the unit of energy) and William Thomson (who later became Lord Kelvin). The procedure works only for real gases in which the attractive interactions are dominant, because the molecules have to climb apart against the attractive force in order for them to travel more slowly. For molecules under conditions when repulsions are dominant (corresponding to $Z > 1$, where Z is the compression factor), the Joule–Thomson effect results in the gas becoming warmer.

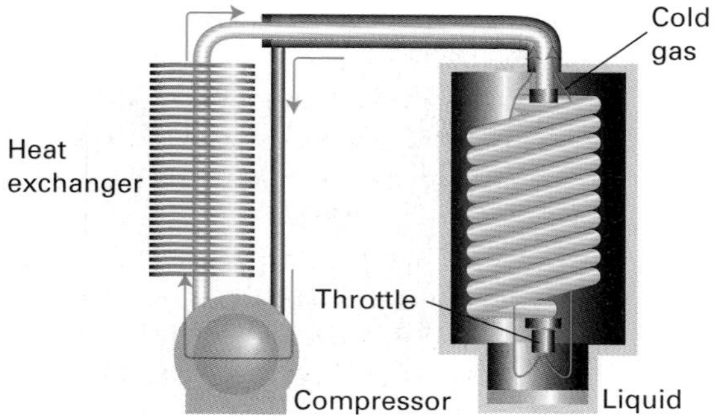

Fig. 1.20 The principle of the Linde refrigerator. The gas is recirculated and cools the gas that is about to undergo expansion through the throttle. The expanding gas cools still further. Eventually, liquefied gas drips from the throttle.

In practice, the gas is allowed to expand several times by recirculating it through a device called a **Linde refrigerator** (Fig. 1.20). On each successive expansion the gas becomes cooler, and as it flows past the incoming gas, the latter is cooled further. After several successive expansions, the gas becomes so cold that it condenses to a liquid.

Further information 1.1

Kinetic molecular theory

One of the essential skills of a physical chemist is the ability to turn simple, qualitative ideas into rigid, testable, quantitative theories. The kinetic model of gases is an excellent example of this technique, as it takes the concepts set out in the text and turns them into precise expressions. As usual in model building, there are a number of steps, but each one is motivated by a clear appreciation of the underlying physical picture, in this case a swarm of mass points in ceaseless random motion. The key quantitative ingredients we need are the equations of classical mechanics, which are covered in Foundations.

We begin the derivation of eqn 1.11, $p = nMv_{\rm rms}^2/3V$, by considering the arrangement in Fig. 1.21. When a particle of mass m that is travelling with a component of velocity v_x parallel to the x-axis ($v_x > 0$ corresponding to motion to the right and $v_x < 0$ to motion to the left) collides with the wall on the right and is reflected, its linear momentum changes from $+m|v_x|$ before the collision to $-m|v_x|$ after the collision (when it is travelling in the opposite direction at the same speed; $|x|$ means the value of x with the sign discarded, so $|-3| = 3$, for instance). The x-component of the momentum therefore changes by $2m|v_x|$ on each collision (the y- and z-components are unchanged). Many molecules collide with the wall in an interval Δt, and the total change of

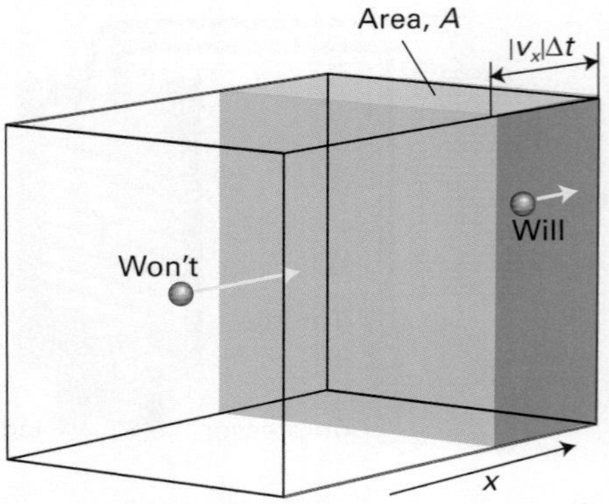

Fig. 1.21 The model used for calculating the pressure of a perfect gas according to the kinetic molecular theory. Here, for clarity, we show only the x-component of the velocity (the other two components are not changed when the molecule collides with the wall). All molecules within the shaded area will reach the wall in an interval Δt provided they are moving towards it.

momentum is the product of the change in momentum of each molecule multiplied by the number of molecules that reach the wall during the interval.

We need to calculate that number. Because a molecule with velocity component v_x can travel a distance $|v_x|\Delta t$ along the x-axis in an interval Δt, all the molecules within a distance $|v_x|\Delta t$ of the wall will strike it if they are travelling towards it. It follows that if the wall has area A, then all the particles in a volume $A \times |v_x|\Delta t$ will reach the wall (if they are travelling towards it). The number density, the number of particles divided by the total volume, is nN_A/V (where n is the total amount in moles of molecules in the container of volume V and N_A is Avogadro's constant), so the number of molecules in the volume $A|v_x|\Delta t$ is $(nN_A/V) \times A|v_x|\Delta t$. At any instant, half the particles are moving to the right and half are moving to the left. Therefore, the average number of collisions with the wall during the interval Δt is $nN_AA|v_x|\Delta t/2V$.

The total momentum change in the interval Δt is the product of the number of collisions we have just calculated and the change of momentum on each collision, $2m|v_x|$:

$$\text{Momentum change} = \overbrace{\frac{nN_AA|v_x|\Delta t}{2V}}^{\text{Number of collisions}} \times \overbrace{2m|v_x|}^{\text{Momentum change on one collision}}$$

$$= \frac{nmN_AAv_x^2\Delta t}{V} \overset{M=mN_A}{=} \frac{nMAv_x^2\Delta t}{V}$$

Next, to find the force, we calculate the rate of change of momentum:

$$\text{Force} = \frac{\overbrace{nMAv_x^2\Delta t/V}^{\text{Momentum change}}}{\underbrace{\Delta t}_{\text{Interval}}} = \frac{nMAv_x^2}{V}$$

It follows that the pressure, the force divided by the area, is

$$\text{Pressure} = \frac{\overbrace{nMAv_x^2/V}^{\text{Force}}}{\underbrace{A}_{\text{Area}}} = \frac{nMv_x^2}{V}$$

Not all the molecules travel with the same velocity, so the detected pressure, p, is the average (denoted $\langle\ldots\rangle$) of the quantity just calculated:

$$p = \frac{nM\langle v_x^2\rangle}{V}$$

To write an expression of the pressure in terms of the root-mean-square speed, v_{rms}, we begin by writing the speed of a single molecule, v, as $v^2 = v_x^2 + v_y^2 + v_z^2$. Because the root-mean-square speed, v_{rms}, is defined as $v_{rms} = \langle v^2\rangle^{1/2}$ (eqn 1.12), it follows that

$$v_{rms}^2 = \langle v^2\rangle = \langle v_x^2\rangle + \langle v_y^2\rangle + \langle v_z^2\rangle$$

However, because the molecules are moving randomly, all three averages are the same. It follows that $v_{rms}^2 = 3\langle v_x^2\rangle$. Equation 1.11 now follows by substituting $\langle v_x^2\rangle = \frac{1}{3}v_{rms}^2$ into $p = nM\langle v_x^2\rangle/V$.

Checklist of key concepts

☐ 1 An equation of state is an equation relating pressure, volume, temperature, and amount of a substance.

☐ 2 The perfect-gas equation of state is based on Boyle's law ($p \propto 1/V$), Charles's law ($V \propto T$), and Avogadro's principle ($V \propto n$).

☐ 3 Dalton's law states that the total pressure of a mixture of perfect gases is the sum of the pressures that each gas would exert if it were alone in the container at the same temperature.

☐ 4 The partial pressure of any gas is defined as $p_J = x_Jp$, where x_J is its mole fraction in a mixture and p is the total pressure.

☐ 5 The kinetic model of gases expresses the properties of a perfect gas in terms of a collection of mass points in ceaseless random motion.

☐ 6 The mean speed and root-mean-square speed of molecules are proportional to the square root of the (absolute) temperature and inversely proportional to the square root of the molar mass.

☐ 7 The properties of the Maxwell distribution of speeds are summarized in Figs 1.9 and 1.10.

☐ 8 Diffusion is the spreading of one substance through another; effusion is the escape of a gas through a small hole.

☐ 9 Graham's law states that the rate of effusion is inversely proportional to the square root of the molar mass.

☐ 10 The Joule–Thomson effect is the cooling of gas that occurs when it expands through a throttle without the influx of heat.

Road map of key equations

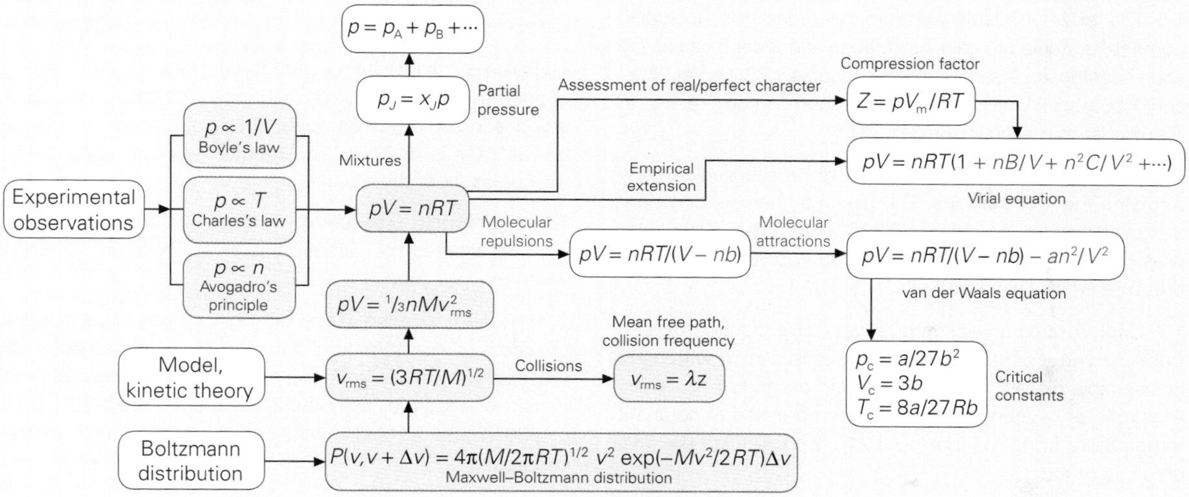

Blue boxes denote limitation to perfect gases.

Questions and exercises

Discussion questions

1.1 Explain how the experiments of Boyle, Charles, and Avogadro led to the formulation of the perfect gas equation of state.

1.2 Explain the term 'partial pressure' and why Dalton's law is a limiting law.

1.3 Use the kinetic model of gases to explain why light gases, such as H_2 and He, are rare in the Earth's atmosphere but heavier gases, such as O_2, CO_2, and N_2, are abundant.

1.4 Provide a molecular interpretation for the variation of the rates of diffusion and effusion of gases with temperature.

1.5 Explain how the compression factor varies with pressure and temperature and describe how it reveals information about intermolecular interactions in real gases.

1.6 What is the significance of the critical constants of a gas?

1.7 Justify the formulation of the van der Waals equation of state.

Exercises

Treat all gases as perfect unless instructed otherwise.

1.1 What pressure is exerted by a sample of nitrogen gas of mass 3.055 g in a container of volume 3.00 dm^3 at 32 °C?

1.2 A sample of neon of mass 425 mg occupies 6.00 dm^3 at 77 K. What pressure does it exert?

1.3 Much to everyone's surprise, nitrogen monoxide (NO) has been found to act as a neurotransmitter. To prepare to study its effect, a sample was collected in a container of volume 300.0 cm^3. At 14.5 °C its pressure is found to be 34.5 kPa. What amount (in moles) of NO has been collected?

1.4 A domestic water-carbonating kit uses steel cylinders of carbon dioxide of volume 250 cm^3. They weigh 1.04 kg when full and 0.74 kg when empty. What is the pressure of gas in the cylinder at 20 °C?

1.5 The effect of high pressure on organisms, including humans, is studied to gain information about deep-sea diving and anaesthesia. A sample of air occupies 1.00 dm^3 at 25 °C and 1.00 atm. What pressure is needed to compress it to 100 cm^3 at this temperature?

1.6 You are warned not to dispose of pressurized cans by throwing them on to a fire. The gas in an aerosol container exerts a pressure of 125 kPa at 18 °C. The container is thrown on a fire, and its temperature rises to 700 °C. What is the pressure at this temperature?

1.7 Until we find an economical way of extracting oxygen from sea water or lunar rocks, we have to carry it with us to inhospitable places, and do so in compressed form in tanks. A sample of oxygen at 101 kPa is compressed at constant temperature from 7.20 dm^3 to 4.21 dm^3. Calculate the final pressure of the gas.

1.8 To what temperature must a sample of helium gas be cooled from 22.2 °C to reduce its volume from 1.00 dm^3 to 100 cm^3?

1.9 Hot-air balloons gain their lift from the lowering of density of air that occurs when the air in the envelope is heated. To what temperature should you heat a sample of air, initially at 315 K, to increase its volume by 25 per cent?

1.10 At sea level, where the pressure was 104 kPa and the temperature 21.1 °C, a certain mass of air occupied 2.0 m^3. To what volume will the region expand when it has risen to an altitude where the pressure and temperature are (a) 52 kPa, –5.0 °C, (b) 880 Pa, –52.0 °C?

1.11 A diving bell has an air space of 3.0 m^3 when on the deck of a boat. What is the volume of the air space when the bell has been lowered to a depth of 50 m? Take the mean density of sea water to be 1.025 g cm^{-3} and assume that the temperature is the same as on the surface.

1.12 Balloons were used to obtain much of the early information about the atmosphere and continue to be used today to obtain weather information. In 1783, Jacques Charles used a hydrogen-filled balloon to fly from Paris 36 km into the French countryside. What is the mass density of hydrogen relative to air at the same temperature and pressure? What mass of payload can be lifted by 10 kg of hydrogen, neglecting the mass of the balloon?

1.13 Atmospheric pollution is a problem that has received much attention. Not all pollution, however, is from industrial sources. Volcanic eruptions can be a significant source of air pollution. The Kilauea volcano in Hawaii emits 200–300 t of SO_2 per day. If this gas is emitted at 800 °C and 1.0 atm, what volume of gas is emitted?

1.14 A meteorological balloon had a radius of 1.5 m when released at sea level at 20 °C and expanded to a radius of 3.5 m when it had risen to its maximum altitude where the temperature was –25 °C. What is the pressure inside the balloon at that altitude?

1.15 A gas mixture being used to simulate the atmosphere of another planet consists of 320 mg of methane, 175 mg of argon, and 225 mg of nitrogen. The partial pressure of nitrogen at 300 K is 15.2 kPa. Calculate (a) the volume and (b) the total pressure of the mixture.

1.16 The vapour pressure of water at blood temperature is 47 Torr. What is the partial pressure of dry air in our lungs when the total pressure is 760 Torr?

1.17 A determination of the density of a gas or vapour can provide a quick estimate of its molar mass even though for practical work mass spectrometry is far more precise. The density of a gaseous compound was found to be 1.23 g dm^{-3} at 330 K and 25.5 kPa. What is the molar mass of the compound?

1.18 In an experiment to measure the molar mass of a gas, 250 cm^3 of the gas was confined in a glass vessel. The pressure was 152 Torr at 298 K and the mass of the gas was 33.5 mg. What is the molar mass of the gas?

1.19 A vessel of volume 22.4 dm^3 contains 2.0 mol H_2 and 1.0 mol N_2 at 273.15 K. Calculate (a) their partial pressures and (b) the total pressure.

1.20 Use the kinetic model to calculate the root-mean-square speed of (a) N_2, (b) H_2O molecules in the Earth's atmosphere at 273 K.

1.21 Use the equipartition theorem, which is outlined in Foundations 0.12, to derive eqn 1.17.

1.22 Calculate the mean speed of (a) He atoms, (b) CH_4 molecules at (i) 79 K, (ii) 315 K, (iii) 1500 K.

1.23 A 1.0 dm^3 glass bulb contains 1.0×10^{23} H_2 molecules. If the pressure exerted by the gas is 100 kPa, what is (a) the temperature of the gas, (b) the root-mean-square speeds of the molecules? (c) Would the temperature be different if they were O_2 molecules?

1.24 Synthesis gas (also known as 'syngas') consists of a mixture of hydrogen, H_2, and carbon monoxide, CO. Calculate the relative rate, as molecules per second, at which the hydrogen and carbon monoxide escape from a leaking cylinder of synthesis gas.

1.25 A cylinder containing gas for a carbon dioxide laser contains equal amounts of carbon dioxide, nitrogen, and helium. If 1.0 g of carbon dioxide leaks from the cylinder by effusion, what mass of nitrogen and helium escapes?

1.26 At what pressure does the mean free path of argon at 25 °C become comparable to the diameter of a spherical vessel of volume 1.0 dm^3 that contains it? Take $\sigma = 0.36\ nm^2$.

1.27 At what pressure does the mean free path of argon at 25 °C become comparable to 10 times the diameters of the atoms themselves? Take $\sigma = 0.36\ nm^2$.

1.28 When studying the photochemical processes that can occur in the upper atmosphere, it is important to know how often atoms and molecules collide. At an altitude of 20 km the temperature is 217 K and the pressure 0.050 atm. What is the mean free path of N_2 molecules? Take $\sigma = 0.43\ nm^2$.

1.29 How many collisions does a single Ar atom make in 1.0 s when the temperature is 25 °C and the pressure is (a) 10 bar, (b) 100 kPa, (c) 1.0 Pa?

1.30 Calculate the total number of collisions per second in 1.0 dm^3 of argon under the same conditions as in Exercise 1.29.

1.31 How many collisions per second does an N_2 molecule make at an altitude of 20 km? (See Exercise 1.28 for data.)

1.32 How does the mean free path in a sample of a gas vary with temperature in a constant-volume container?

1.33 The spread of pollutants through the atmosphere is governed partly by the effects of winds but also by the natural tendency of molecules to diffuse. The latter depends on how far a molecule can travel before colliding with another molecule. Calculate the mean free path of diatomic molecules in air using $\sigma = 0.43\ nm^2$ at 25 °C and (a) 10 bar, (b) 103 kPa, (c) 1.0 Pa.

1.34 The critical point of ammonia, NH_3, gas occurs at 111.3 atm, 72.5 $cm^3\ mol^{-1}$, and 405.5 K. Calculate the compression factor at the critical point. What can you infer?

1.35 The virial equation of state may also be written as an expansion in terms of pressure: $Z = 1 + B'p + \cdots$. The critical constants for water, H_2O, are 218.3 atm, 55.3 $cm^3\ mol^{-1}$, and 647.4 K. Assuming that the expansion may be truncated after the second term, calculate the value of the second virial coefficient B' at the critical temperature.

1.36 Calculate the pressure exerted by 1.0 mol C_2H_6 behaving as (a) a perfect gas, (b) a van der Waals gas when it is confined under the following conditions: (i) at 273.15 K in 22.414 dm^3, (ii) at 1000 K in 100 cm^3. Use the data in Table 1.6.

1.37 How reliable is the perfect gas law in comparison with the van der Waals equation? Calculate the difference in pressure of 10.00 g of carbon dioxide confined to a container of volume 100 cm^3 at 25.0 °C between treating it as a perfect gas and a van der Waals gas.

1.38 A certain gas obeys the van der Waals equation with $a = 0.50\ m^6\ Pa\ mol^{-2}$. Its molar volume is found to be $5.00 \times 10^{-4}\ m^3\ mol^{-1}$ at 273 K and 3.0 MPa. From this information calculate the van der Waals constant b. What is the compression factor for this gas at the prevailing temperature and pressure?

1.39 Express the van der Waals equation of state as a virial expansion in powers of $1/V_m$ and obtain expressions for B and C in terms of the parameters a and b. *Hint*: The expansion you will need is $(1 - x)^{-1} = 1 + x + x^2 + \cdots$.

1.40 Measurements on argon gave $B = -21.7\ cm^3\ mol^{-1}$ and $C = 1200\ cm^6\ mol^{-2}$ for the virial coefficients at 273 K. What are the values of a and b in the corresponding van der Waals equation of state? *Hint*: Use the expression for B derived in Exercise 1.39.

1.41 Show that there is a temperature at which the second virial coefficient, B, is zero for a van der Waals gas, and calculate its value for carbon dioxide. *Hint*: Use the expression for B derived in Exercise 1.39.

1.42 The critical constants of ethane are $p_c = 48.20$ atm, $V_c = 148\ cm^3\ mol^{-1}$, and $T_c = 305.4$ K. Calculate the van der Waals parameters of the gas and estimate the radius of the molecules.

Projects

The symbol ‡ indicates that calculus is required.

1.43‡ In the following you are invited to explore the Maxwell distribution of speeds in more detail. (a) Confirm that the mean speed of molecules of molar mass M at a temperature T is equal to $(8RT/\pi M)^{1/2}$. *Hint*: You will need an integral of the form $\int_0^\infty x^3 e^{-ax^2} dx = n!/2a^2$. (b) Confirm that the root-mean-square speed of molecules of molar mass M at a temperature T is equal to $(3RT/M)^{1/2}$ and hence confirm eqn 1.17. *Hint*: You will need an integral of the form $\int_0^\infty x^4 e^{-ax^2} dx = (3/8a^2)(\pi/a)^{1/2}$. (c) Find an expression for the most probable speed of molecules of molar mass M at a temperature T. *Hint:* Look for a maximum in the Maxwell distribution (the maximum occurs as $dF/ds = 0$). (d) Estimate the fraction of N_2 molecules at 500 K that has speeds in the range 290 to 300 $m\ s^{-1}$.

1.44‡ Here we explore the van der Waals equation of state. Using the language of calculus, the critical point of a van der

Waals gas occurs where the isotherm has a flat inflexion, which is where $dp/dV_m = 0$ (zero slope) and $d^2p/dV_m^2 = 0$ (zero curvature). (a) Evaluate these two expressions using eqn 1.26b, and find expressions for the critical constants in terms of the van der Waals parameters. (b) Show that the value of the compression factor at the critical point is $^3/_8$.

1.45 The kinetic model of gases is valid when the size of the particles is negligible compared with their mean free path. It may seem absurd, therefore, to expect the kinetic theory and, as a consequence, the perfect gas law, to be applicable to the dense matter of stellar interiors. In the Sun, for instance, the density is 150 times that of liquid water at its centre and comparable to that of water about halfway to its surface. However, we have to realize that the state of matter is that of a *plasma*, in which the electrons have been stripped from the atoms of hydrogen and helium that make up the bulk of the matter of stars. As a result, the particles making up the plasma have diameters comparable to those of nuclei, or about 10 fm. Therefore, a mean free path of only 0.1 pm satisfies the criterion for the validity of the kinetic model and the perfect gas law. We can therefore use $pV = nRT$ as the equation of state for the stellar interior. (a) Calculate the pressure halfway to the centre of the Sun, assuming that the interior consists of ionized hydrogen atoms, the temperature is 3.6 MK, and the mass density is 1.20 g cm^{-3} (slightly higher than the density of water). (b) Combine the result from part (a) with the expression for the pressure from kinetic model to show that the pressure of the plasma is related to its **kinetic** *energy density* $\rho_k = E_k/V$, the kinetic energy of the molecules in a region divided by the volume of the region, by $p = (2/3)\rho_k$. (c) What is the kinetic energy density halfway to the centre of the Sun? Compare your result with the (translational) kinetic energy density of the Earth's atmosphere on a warm day (25 °C): 1.5×10^5 J m^{-3} (corresponding to 0.15 J cm^{-3}). (d) A star eventually depletes some of the hydrogen in its core, which contracts and results in higher temperatures. The increased temperature results in an increase in the rates of nuclear reaction, some of which result in the formation of heavier nuclei, such as carbon. The outer part of the star expands and cools to produce a red giant. Assume that halfway to the centre a red giant has a temperature of 3500 K, is composed primarily of fully ionized carbon atoms and electrons, and has a mass density of 1200 kg m^{-3}. What is the pressure at this point? (e) If the red giant in part (d) consisted of neutral carbon atoms, what would be the pressure at the same point under the same conditions?

2

Thermodynamics: the First Law

The branch of physical chemistry known as **thermodynamics** is concerned with the study of the transformations of energy and, in particular, the transformation of heat into work and vice versa. This concern might seem remote from chemistry; indeed, thermodynamics was originally formulated by physicists and engineers interested in the efficiency of steam engines. However, thermodynamics has proved to be of immense importance in chemistry. Not only does it deal with the energy output of chemical reactions but it also helps to answer questions that lie right at the subject's heart, such as why reactions reach equilibrium, their composition at equilibrium, and how reactions in electrochemical (and biological) cells can be used to generate electricity.

Chemical thermodynamics is a tree with many branches. **Thermochemistry** is the branch that deals with the heat output of chemical reactions. As we elaborate the content of thermodynamics, we shall see that we can also discuss the output of energy in the form of work. This connection leads us into the fields of **electrochemistry**, the interaction between electricity and chemistry, and **bioenergetics**, the deployment of energy in living organisms. The whole of equilibrium chemistry—the formulation of equilibrium constants, and the very special case of the equilibrium composition of solutions of acids and bases—is an aspect of thermodynamics.

Classical thermodynamics, the thermodynamics developed during the nineteenth century, stands aloof from any models of the internal constitution of matter: we could develop and use thermodynamics without ever mentioning atoms and molecules. However, the subject is greatly enriched by acknowledging that atoms and molecules do exist and interpreting thermodynamic properties and relations in terms of them. As we already saw in Foundations, we shall often cross back and forth between thermodynamics, which provides useful relations between observable properties of bulk matter, and the properties of atoms and molecules, which are ultimately

The conservation of energy 48

2.1 Systems and surroundings 48

2.2 Work and heat 49

2.3 The molecular interpretation of work, heat, and temperature 50

2.4 The measurement of work 51

2.5 The measurement of heat 55

2.6 Heat influx during expansion 58

Internal energy and enthalpy 58

2.7 The internal energy 59

2.8 Internal energy as a state function 60

2.9 The enthalpy 62

2.10 The temperature variation of the enthalpy 63

CHECKLIST OF KEY CONCEPTS 65

ROAD MAP OF KEY EQUATIONS 66

QUESTIONS AND EXERCISES 66

responsible for these bulk properties. The theory of the connection between atomic and bulk thermodynamic properties is called **statistical thermodynamics** and is treated in Chapter 22.

The conservation of energy

Almost every argument and explanation in chemistry boils down to a consideration of some aspect of a single property: the energy. Energy determines what molecules may form, what reactions may occur, how fast they may occur, and—with a refinement in our conception of energy that we explore in Chapter 4—in which direction a reaction has a tendency to occur.

As we saw in Foundations 0.9:

Energy is the capacity to do work.

Work is done to achieve motion against an opposing force.

These definitions imply that a raised weight of a given mass has more energy than one of the same mass resting on the ground because the former has a greater capacity to do work: it can do work as it falls to the level of the lower weight. The definition also implies that a gas at high temperature has more energy than the same gas at a low temperature: the hot gas has a higher pressure and can do more work in driving out a piston.

People struggled for centuries to create energy from nothing, for they believed that if they could create energy, then they could produce work (and wealth) endlessly. However, without exception, despite strenuous efforts, many of which degenerated into deceit, they failed. As a result of their failed efforts, we have come to recognize that energy can be neither created nor destroyed but merely converted from one form into another or moved from place to place. This 'law of the conservation of energy' is of great importance in chemistry. Most chemical reactions release energy or absorb it as they occur; so according to the law of the conservation of energy, we can be confident that all such changes must result only in the conversion of energy from one form into another or its transfer from place to place, not its creation or annihilation. The detailed study of that conversion and transfer is the domain of thermodynamics.

2.1 Systems and surroundings

In thermodynamics, a **system** is the part of the world in which we have a special interest. The **surroundings** are where we make our observations (Fig. 2.1). The surroundings, which can be modelled as a large water bath, remain at constant temperature regardless of how much energy flows into or out of them. They are so huge that they also have either constant volume or constant pressure regardless of any changes that take place to the system. Thus, even though the system might expand, the surroundings remain effectively the same size.

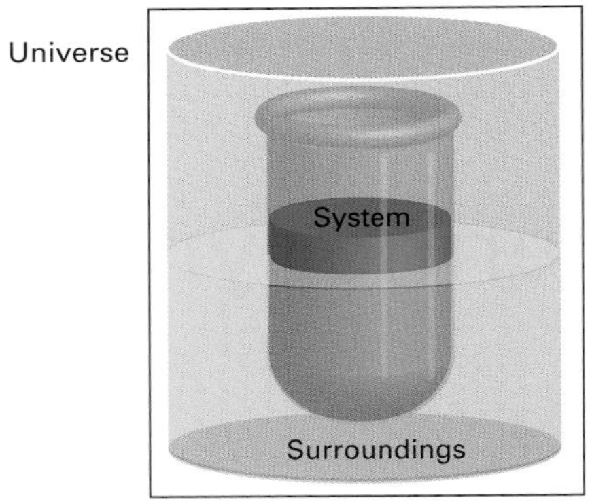

Fig. 2.1 The sample is the system of interest; the rest of the world is its surroundings. The surroundings are where observations are made on the system. They can often be modelled, as here, by a large water bath. The universe consists of the system and surroundings.

We need to distinguish three types of system (Fig. 2.2):

An **open system** can exchange both energy and matter with its surroundings.

A **closed system** can exchange energy but not matter with its surroundings.

An **isolated system** can exchange neither matter nor energy with its surroundings.

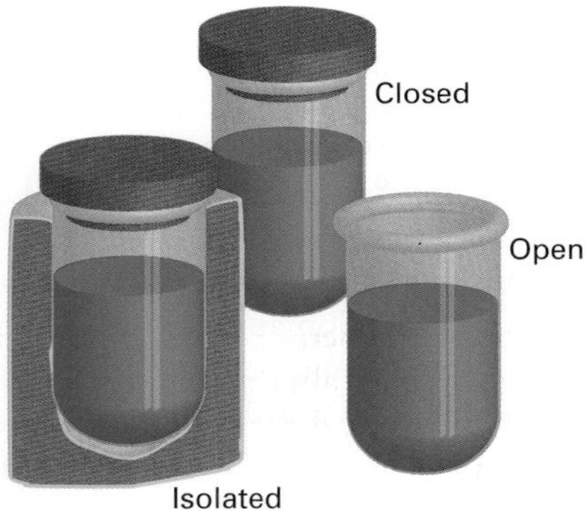

Fig. 2.2 A system is *open* if it can exchange energy and matter with its surroundings, *closed* if it can exchange energy but not matter, and *isolated* if it can exchange neither energy nor matter.

An example of an open system is a flask that is not stoppered and to which various substances can be added. A biological cell is an open system because nutrients and waste can pass through the cell wall. You and I are open systems: we ingest, respire, perspire, and excrete. An example of a closed system is a stoppered flask: energy can be exchanged with the contents of the flask because the walls may be able to conduct heat. An example of an isolated system is a sealed flask that is thermally, mechanically, and electrically insulated from its surroundings.

2.2 Work and heat

Energy can be exchanged between a closed system and its surroundings by doing work or by the process called 'heating'. A system does work when it causes motion against an opposing force. As we saw in Foundations 0.9, the magnitude of the work done in moving against a constant opposing force is the product of the distance moved and the strength of the force; we develop that relation shortly.

Heating is the process of transferring energy as a result of a temperature difference between the system and its surroundings. To avoid a lot of awkward circumlocution, it is common to say that 'energy is transferred as work' when the system does work, and 'energy is transferred as heat' when the system is heated by its surroundings. However, we should always remember that 'work' and 'heat' are *modes of transfer* of energy, not *forms* of energy.

Although in everyday language the terms 'temperature' and 'heat' are sometimes not distinguished, they are entirely different concepts:

> **Heat**, q, is energy in transit as a result of a temperature difference.
>
> **Temperature**, T, is an intensive property (a property that does not depend on the amount of substance in the sample, Foundations 0.4) that is used to define the state of a system and determines the direction in which energy flows as heat.

Walls that permit heating as a mode of transfer of energy are called **diathermic** (Fig. 2.3). A metal container is diathermic. Walls that do not permit heating even though there is a difference in temperature are called **adiabatic** (from the Greek words for 'not passing through'). The double walls of a vacuum flask are adiabatic to a good approximation.

As an example of the different ways of transferring energy, consider a chemical reaction that produces gases, such as the reaction of an acid with zinc, $Zn(s) + 2\,HCl(aq) \rightarrow ZnCl_2(aq) + H_2(g)$. Suppose first that the reaction takes place inside a cylinder fitted with a piston, then the gas produced drives out the piston and raises a weight in the surroundings (Fig. 2.4). In this case, energy has migrated to the surroundings as a result of the system doing work because a weight has been raised in the surroundings: that weight can now do more work, so it possesses more energy. Some energy also migrates into the surroundings as heat. We can detect that transfer of energy by immersing the reaction vessel in an ice bath (a bath containing a mixture of ice and water at 0 °C) and noting how much ice melts. Alternatively, we could let the same reaction take place in a vessel with a

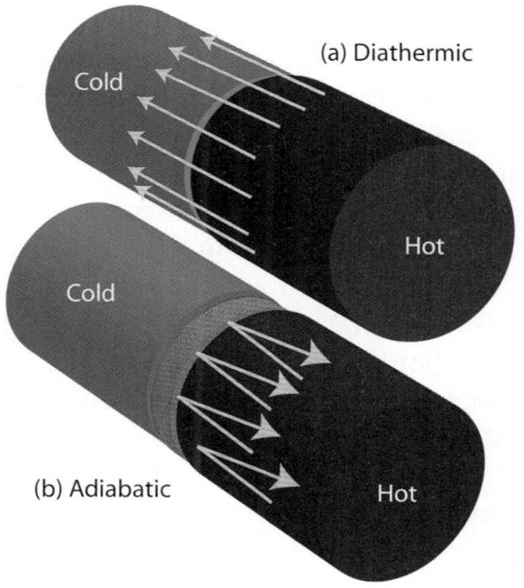

Fig. 2.3 (a) A diathermic wall permits the passage of energy as heat; (b) an adiabatic wall does not, even if there is a temperature difference across the wall.

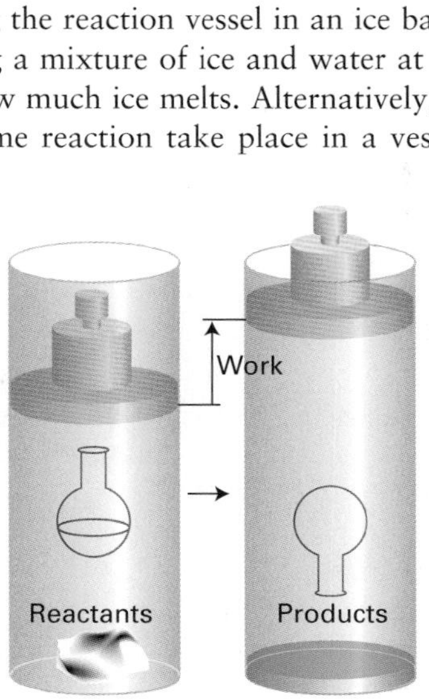

Fig. 2.4 When hydrochloric acid reacts with zinc, the hydrogen gas produced must push back the surrounding atmosphere (represented by the weight resting on the piston), and hence must do work on its surroundings. This is an example of energy leaving a system as work.

piston locked in position. No work is done, because no weight is raised. However, because it is found that more ice melts than in the first experiment, we can conclude that more energy has migrated to the surroundings as heat.

A process in a system that releases energy as heat is called **exothermic.** A process in a system that absorbs energy as heat is called **endothermic.** An example of an exothermic reaction is any combustion of an organic compound. Endothermic reactions are much less common. The endothermic dissolution of ammonium nitrate in water is the basis of the instant cold-packs that are included in some first-aid kits. They consist of a plastic envelope containing water dyed blue (for psychological reasons) and a small tube of ammonium nitrate, which is broken when the pack is to be used.

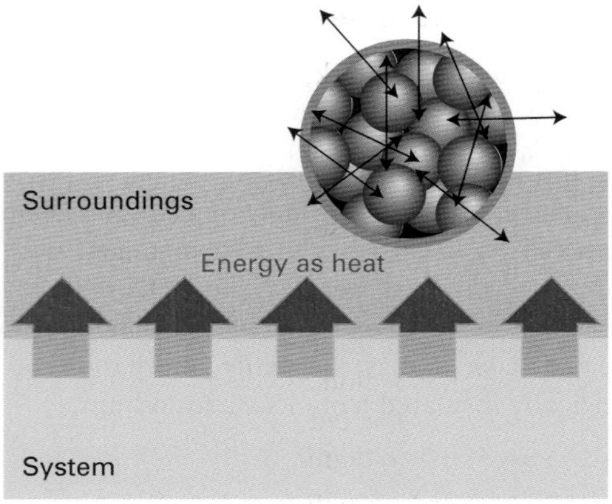

Fig. 2.6 Heat is the transfer of energy that causes or utilizes random motion in the surroundings. When energy leaves the system (the pale green region), it generates random motion in the surroundings (shown magnified).

2.3 The molecular interpretation of work, heat, and temperature

The clue to the molecular nature of work comes from thinking about the motion of a weight in terms of its component atoms. When a weight is raised, all its atoms move in the same direction. This observation suggests that:

> Work is the mode of transfer of energy that achieves or utilizes uniform motion in the surroundings (Fig. 2.5).

Whenever we think of work, we can always think of it in terms of uniform motion of some kind. Electrical work, for instance, corresponds to electrons being pushed in the same direction through a circuit. Mechanical work corresponds to atoms being pushed in the same direction against an opposing force.

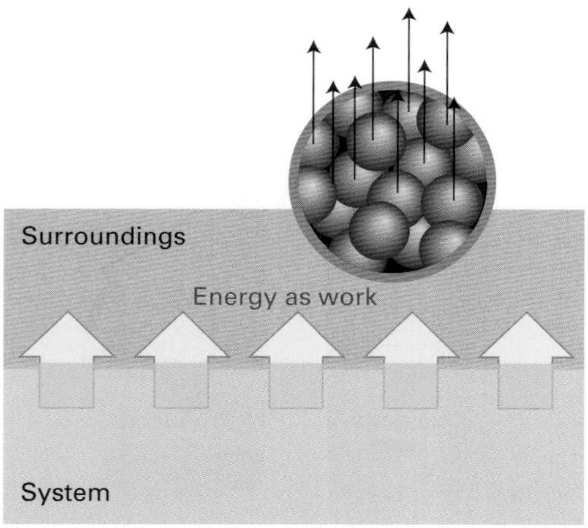

Fig. 2.5 Work is transfer of energy that causes or utilizes uniform motion of atoms in the surroundings. For example, when a weight is raised, all the atoms of the weight (shown magnified) move in unison in the same direction.

Now consider the molecular nature of heat. When energy is transferred as heat to the surroundings, the atoms and molecules oscillate more vigorously around their positions or move more rapidly from place to place. The key point is that motion stimulated by the arrival of energy from the system as heat is random, not uniform as in the case of doing work. This observation suggests that:

> Heat is the mode of transfer of energy that achieves or utilizes random motion in the surroundings (Fig. 2.6).

A fuel burning, for example, generates disorderly molecular motion in its vicinity.

An interesting historical point is that the molecular difference between work and heat correlates with the chronological order of their application. The release of energy when a fire burns is a relatively unsophisticated procedure because the energy emerges in a disordered fashion from the burning fuel. It was developed—stumbled upon—early in the history of civilization. The generation of work by a burning fuel, in contrast, relies on a carefully controlled transfer of energy so that vast numbers of molecules move in unison. Apart from Nature's achievement of work through the evolution of muscles, the large-scale transfer of energy by doing work was achieved thousands of years later than the transfer of energy by heating, for it had to await the development of the steam engine.

2.4 The measurement of work

When a system does work, such as by raising a weight in the surroundings or forcing an electric current through a circuit, the energy transferred, which is denoted w, is reported as a negative quantity. For instance, if a system raises a weight in the surroundings and in the process does 100 J of work (that is, 100 J of energy leaves the system by doing work), then we write $w = -100$ J. When work is done on the system, as when we wind a spring inside a clockwork mechanism, w is reported as a positive quantity. We write $w = +100$ J to signify that 100 J of work has been done on the system (that is, 100 J of energy had been transferred to the system by doing work). The sign convention is easy to follow if we think of changes to the energy of the system: its energy decreases (w is negative) if energy leaves it and its energy increases (w is positive) if energy enters it (Fig. 2.7).

We use the same convention for energy transferred as heat, q. We write $q = -100$ J if 100 J of energy leaves the system as heat, so reducing the energy of the system, and $q = +100$ J if 100 J of energy enters the system as heat.

Because many chemical reactions produce gas, one very important type of work in chemistry is **expansion work**, the work done when a system expands against an opposing pressure. The action of acid on zinc illustrated in Fig. 2.4 is an example of a reaction in which expansion work is done in the process of making room for the gaseous product, hydrogen in this case. We show in the following Derivation that when a system expands through a volume ΔV against a constant external pressure p_{ex} the work done is

$$w = -p_{ex}\Delta V \qquad \text{Constant pressure} \qquad \text{Work of expansion} \qquad (2.1)$$

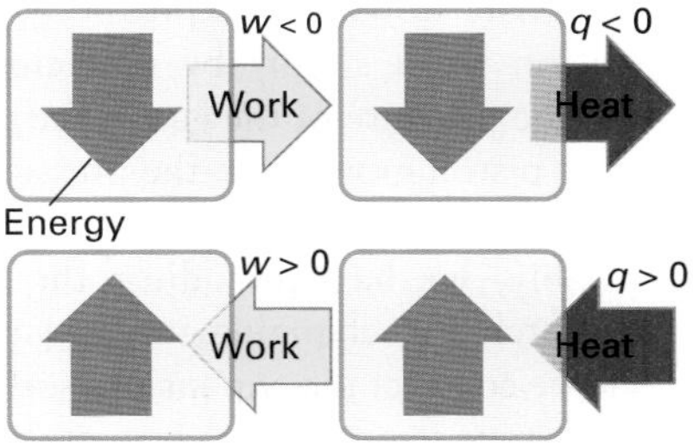

Fig. 2.7 The sign convention in thermodynamics: w and q are positive if energy enters the system (as work and heat, respectively), but negative if energy leaves the system.

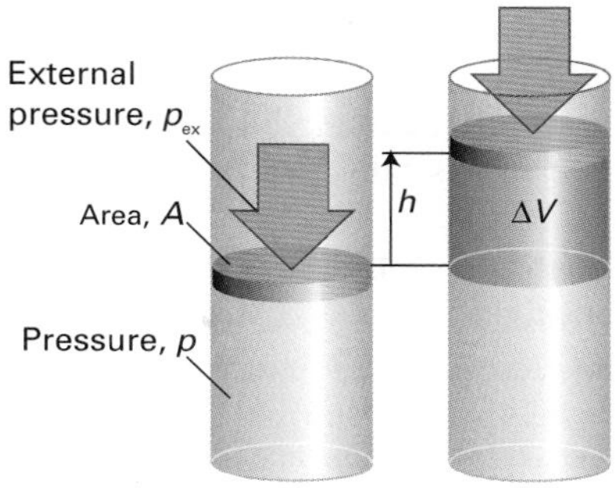

Fig. 2.8 When a piston of area A moves out through a distance h, it sweeps out a volume $\Delta V = Ah$. The external pressure p_{ex} opposes the expansion with a force $p_{ex}A$.

Derivation 2.1

Expansion work

To calculate the work done when a system expands from an initial volume V_i to a final volume V_f, a change $\Delta V = V_f - V_i$, we consider a piston of area A moving out through a distance h (Fig. 2.8). There need not be an actual piston: we can think of the piston as representing the boundary between the expanding gas and the surrounding atmosphere. However, there may be an actual piston, such as when the expansion takes place inside an internal combustion engine.

The force opposing the expansion is the constant external pressure p_{ex} multiplied by the area of the piston (because force is pressure times area, Foundations 0.8), so $F = p_{ex}A$. From eqn 0.10, the work done is therefore

$$\text{Work done} \overset{\text{eqn 0.10}}{=} (\overbrace{p_{ex}A}^{F}) \times \overbrace{h}^{d} = p_{ex} \times \overbrace{hA}^{\Delta V} = p_{ex} \times \Delta V$$

The last equality follows from the fact that hA is the volume of the cylinder swept out by the piston as the gas expands, so we can write $hA = \Delta V$. That is, for expansion work,

$$\text{Work done} = p_{ex}\Delta V$$

Now consider the sign. A system does work and thereby loses energy (that is, w is negative) when it expands (when ΔV is positive). Therefore, we need a negative sign in the equation to ensure that w is negative when ΔV is positive), so we obtain eqn 2.1.

A note on good practice Keep track of signs by considering whether the stored energy has decreased when the system does work (w is then negative) or has increased when work has been done on the system (w is then positive).

According to eqn 2.1, the *external* pressure determines how much work a system does when it expands through a given volume: the greater the external pressure, the greater the opposing force and the greater the work that a system does. When the external pressure is zero, $w = 0$. In this case, the system does no work as it expands because it has nothing to push against. Expansion against zero external pressure is called **free expansion**.

Brief illustration 2.1 The work done by a chemical reaction

Consider a reaction that results in the formation of perfect gases. To calculate the work done by this reaction at a specified temperature and external pressure, we first need to calculate ΔV from the perfect gas law

$$\Delta V = \overbrace{V_f}^{n_f RT/p_{ex}} - \overbrace{V_i}^{n_i RT/p_{ex}} \overset{\Delta n_g = n_f - n_i}{=} \frac{RT\Delta n_g}{p_{ex}}$$

where Δn_g is the change in the amount of gas molecules. It follows from eqn 2.1 that the work is

$$w = -p_{ex}\Delta V = -p_{ex}\frac{RT\Delta n_g}{p_{ex}} \overset{\text{cancel } p_{ex}}{=} -RT\Delta n_g$$

and does not depend on the external pressure, if it remains constant during the reaction. If the reaction results in the net production of gases, then $\Delta n_g > 0$ and $w < 0$. Conversely, if there is net consumption of gases, $\Delta n_g < 0$ and $w > 0$. For example, the work of a chemical reaction that forms 1.0 mol CO_2(g), treated as a perfect gas, at 25 °C is

$$\begin{aligned} w &= -\overbrace{(8.3145\ \text{J K}^{-1}\ \text{mol}^{-1})}^{R}\times\overbrace{(298\ \text{K})}^{T}\times\overbrace{(1.0\ \text{mol})}^{\Delta n_g} \\ &= -2.5\times10^{3}\ \text{J} = -2.5\ \text{kJ} \end{aligned}$$

Self-test 2.1

Determine the work done when 1.0 mol of propane gas molecules react with oxygen to produce carbon dioxide and liquid water at 25 °C: $C_3H_8(g) + 5\ O_2(g) \rightarrow 3\ CO_2(g) + 4\ H_2O(l)$

Answer: +7.4 kJ

Equation 2.1 shows us how to get the *least* expansion work from a system: we just reduce the external pressure to zero. But how can we achieve the *greatest* work for a given change in volume? According to eqn 2.1, the system does maximum work when the external pressure has its maximum value. The force opposing the expansion is then the greatest and the system must exert most effort to push the piston out. However, that external pressure cannot be greater than the pressure, p, of the gas inside the system, for otherwise the external pressure would compress the gas instead of allowing it to expand. Therefore, *maximum work is obtained when the external pressure is only infinitesimally less than the pressure of the gas in the system.* In effect, the two pressures must be adjusted to be the same at all stages of the expansion. In Foundations 0.8 we called this balance of pressures a state of mechanical equilibrium. Therefore, we can conclude that

> A system that remains in mechanical equilibrium with its surroundings at all stages of the expansion does maximum expansion work.

There is another way of expressing this condition. Because the external pressure is infinitesimally less than the pressure of the gas at some stage of the expansion, the piston moves out. However, suppose we increase the external pressure so that it became infinitesimally greater than the pressure of the gas; now the piston moves in. That is, *when a system is in a state of mechanical equilibrium, an infinitesimal change in the pressure results in opposite directions of motion.*

A process that can be reversed by an *infinitesimal* change in a variable—in this case, the pressure—is said to be **reversible**. In everyday life 'reversible' means a process that can be reversed; in thermodynamics it has a stronger meaning: it means that a process can be reversed by an *infinitesimal* modification in some variable (such as the pressure).

We can summarize this discussion by the following remarks:

- A system does maximum expansion work when the external pressure is equal to that of the system at every stage of the expansion ($p_{ex} = p$).
- A system does maximum expansion work when it is in mechanical equilibrium with its surroundings at every stage of the expansion.
- Maximum expansion work is achieved in a reversible change.

All three statements are equivalent, but they reflect different degrees of sophistication in the way the point is expressed.

We cannot write down the expression for maximum expansion work simply by replacing p_{ex} in eqn 2.1 by p (the pressure of the gas in the cylinder) because, as the piston moves out, the pressure inside the system falls. To make sure the entire process occurs reversibly, we have to adjust the external pressure to match the changing internal pressure. Suppose that we conduct the expansion isothermally (that is, at constant temperature) by immersing the system in a water bath held at a specified temperature. As we show in the following Derivation, the work of isothermal, reversible expansion of a perfect gas from an initial volume V_i to a final volume V_f at a temperature T is

$$w = -nRT\ln\frac{V_f}{V_i}$$ Perfect gas, isothermal conditions; Work of reversible expansion (2.2)

where n is the amount of gas molecules in the system. As explained in Derivation 2.2 and The chemist's toolkit 2.1, this result is equal to the area beneath a graph of $p = nRT/V$ between the limits V_i and V_f (Fig. 2.9).

The chemist's toolkit 2.1 Integration

The area under a graph of any function f is found by the techniques of integration. For instance, the area under the graph of the function $f(x)$ between $x = a$ and $x = b$ (Sketch 2.1) is expressed by the relation

$$\text{Area between } a \text{ and } b = \int_a^b f(x)\mathrm{d}x$$

The expression on the right is called the **integral** of the function f. When written as $\int$ alone, it is the **indefinite integral** of the function. When written with limits (as in the expression above), it becomes the **definite integral** of the function. The definite integral is the indefinite integral evaluated at the upper limit (b) minus the indefinite integral evaluated at the lower limit (a).

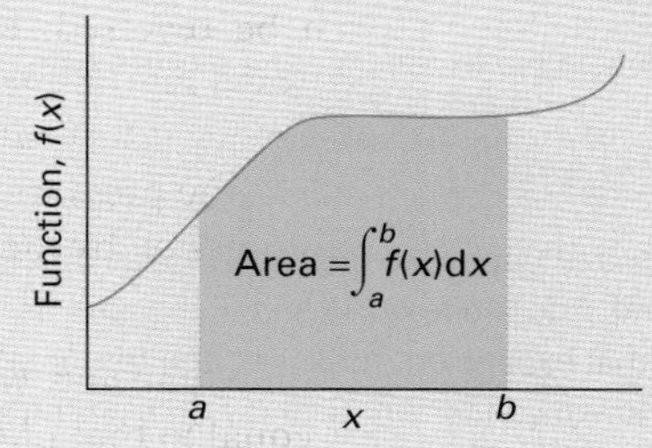

Sketch 2.1 Graphical representation of the integral of $f(x)$ between $x = a$ and $x = b$.

The indefinite integral of x^n is

$$\int x^n \mathrm{d}x = \frac{x^{n+1}}{n+1} + \text{constant}$$

It follows that the definite integral of x^n evaluated between a and b is

$$\int_a^b x^n\,\mathrm{d}x = \left(\frac{x^{n+1}}{n+1} + \text{constant}\right)\Bigg|_a^b = \left(\frac{b^{n+1}}{n+1} + \text{constant}\right) - \left(\frac{a^{n+1}}{n+1} + \text{constant}\right) = \frac{1}{n+1}(b^{n+1} - a^{n+1})$$

We see that the constant cancels. As an example, it follows that the area under the graph of x^2 lying between $a = 2$ and $b = 3$ (Sketch 2.2) is

$$\int_2^3 x^2\,\mathrm{d}x \overset{n=2}{=} \frac{1}{\underset{n+1}{3}}\left(3^{\overset{n+1}{3}} - 2^3\right) = \frac{1}{3}(27-8) = \frac{19}{3}$$

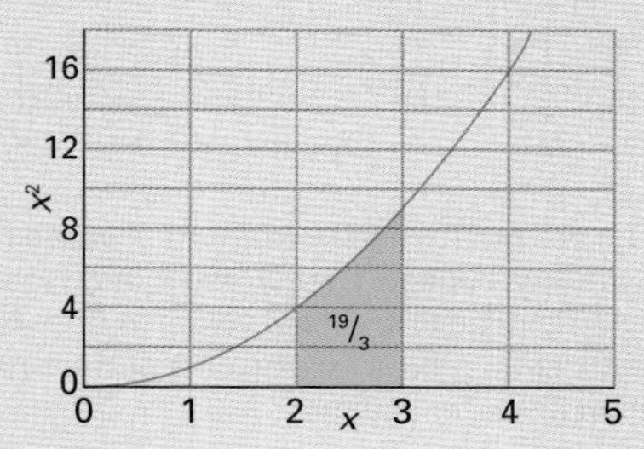

Sketch 2.2 Graphical representation of the integral of x^2 between $x = 2$ and $x = 3$.

A common integral in physical chemistry is

$$\int \frac{\mathrm{d}x}{x} = \ln x + \text{constant}$$

where $\ln x$ is the natural logarithm of x. To evaluate the integral between the limits $x = a$ and $x = b$, we write

$$\int_a^b \frac{\mathrm{d}x}{x} = (\ln x + \text{constant})\Big|_a^b = (\ln b + \text{constant}) - (\ln a + \text{constant}) = \ln b - \ln a = \ln\frac{b}{a}$$

where we have used a standard mathematical relation between logarithms (see The chemist's toolkit 2.2) to simplify the final expression. For instance, the area under the graph of $1/x$ lying between $a = 2$ and $b = 3$ (Sketch 2.3) is $\ln(3/2) = 0.41$.

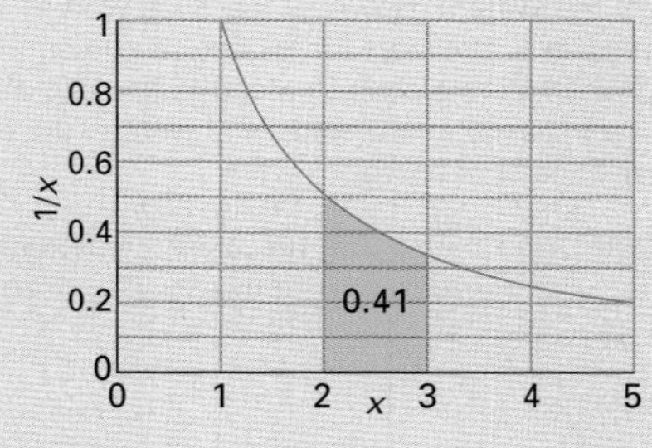

Sketch 2.3 Graphical representation of the integral of $1/x$ between $x = 2$ and $x = 3$.

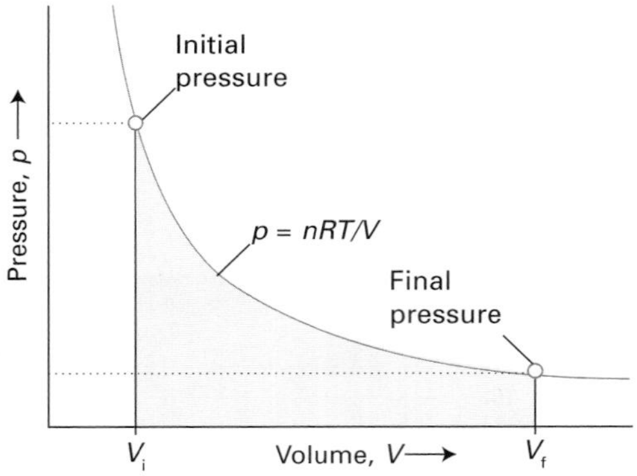

Fig. 2.9 The work of reversible isothermal expansion of a gas is equal to the area beneath the corresponding isotherm evaluated between the initial and final volumes (the yellow area). The isotherm shown here is that of a perfect gas, but the same relation holds for any gas.

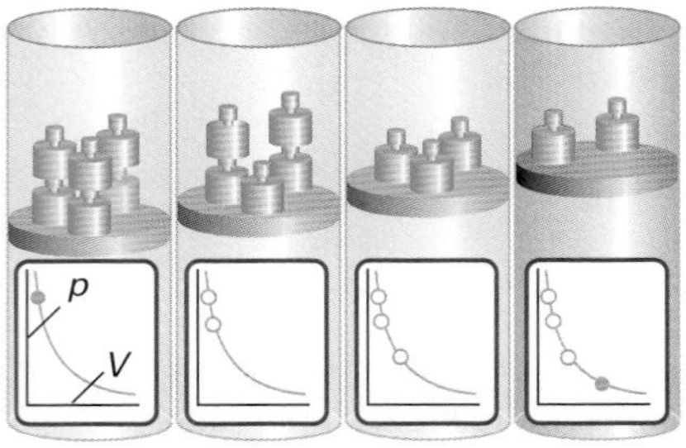

Fig. 2.10 For a gas to expand reversibly, the external pressure must be adjusted to match the internal pressure at each stage of the expansion. This matching is represented in this illustration by gradually unloading weights from the piston as the piston is raised and the internal pressure falls. The procedure results in the extraction of the maximum possible work of expansion.

Derivation 2.2

Reversible, isothermal expansion work

Because (to ensure reversibility) the external pressure must be adjusted in the course of the expansion, we have to think of the process as taking place in a series of small steps during each one of which the external pressure is constant. We calculate the work done in each step for the prevailing external pressure, and then add all these values together. To ensure that the overall result is accurate, we have to make the steps as small as possible—infinitesimal, in fact—so that the pressure is truly constant during each one. In other words, we have to use the calculus, in which case the sum over an infinite number of infinitesimal steps becomes an integral (The chemist's toolkit 2.1).

When the system expands through an infinitesimal volume dV, the infinitesimal work, dw, done is the infinitesimal version of eqn 2.1 ($w = -p_{ex}\Delta V$):

$$dw = -p_{ex}dV$$

A note on good practice The replacement of Δ by d always indicates an infinitesimal change. In the equation above, dw denotes an infinitesimal quantity of energy transferred as work, and dV is the resulting infinitesimal change in volume of the system. However, because work is a process and volume is a property, the symbol d carries a different meaning when it is attached to w and V. Namely, dV denotes a very small *change in the state of a system*: dw denotes a very small transfer of energy as work.

At each stage, we ensure that the external pressure is the same as the current pressure, p, of the gas (Fig. 2.10). So, we set $p_{ex} = p$ and obtain

$$dw = -pdV$$

The total work when the system expands from V_i to V_f is the sum (integral) of all the infinitesimal changes between the limits V_i and V_f, which we write (see The chemist's toolkit 2.1)

For a reversible expansion: $w = -\int_{V_i}^{V_f} p\,dV$

To evaluate the integral, we need to know how p, the pressure of the gas in the system, changes as it expands. For this step, we suppose that the gas is perfect, in which case we can use the perfect gas $pV = nRT$ in the form $p = nRT/V$. At this stage we have

For the reversible expansion of a perfect gas:

$$w = -\int_{V_i}^{V_f} \frac{nRT}{V}dV$$

In general, the temperature might change as the gas expands, so in general T depends on V, and T changes as V changes. For isothermal expansion, however, the temperature is held constant and we can take n, R, and T outside the integral and write

For the isothermal, reversible expansion of a perfect gas:

$$w = -nRT\int_{V_i}^{V_f} \frac{dV}{V}$$

It follows from the techniques introduced in The chemist's toolkits 2.1 and 2.2 that

$$\int_{V_i}^{V_f} \frac{dV}{V} = |\ln V|_{V_i}^{V_f} = \ln V_f - \ln V_i = \ln\frac{V_f}{V_i}$$

When we insert this result into the preceding expression ($w = -nRT\int_{V_i}^{V_f} dV/V$), we obtain eqn 2.2. The interpretation of eqn 2.2 as an area follows from the fact, as explained in The chemist's toolkit 2.1, that a definite integral is equal to the area beneath a graph of the function lying between the two limits of the integral.

A note on good practice Introduce (and keep note of) the restrictions (in this case, in succession: reversible process, perfect gas, isothermal) only as they prove necessary, as you might be able to use an intermediate formula without needing to restrict it in the way that was necessary to achieve the final expression.

The chemist's toolkit 2.2 Logarithms

Some equations are most readily solved by using logarithms and related functions. The **natural logarithm** of a number x is denoted $\ln x$, and is defined as the power to which a number designated e = 2.718 ... must be raised for the result to be equal to x. It follows from the definitions of logarithms that

$$\ln x + \ln y = \ln xy$$
$$\ln x - \ln y = \ln x/y$$
$$a \ln x = \ln x^a$$

Some useful points about logarithms are summarized in Sketch 2.4:

- Logarithms increase very slowly as x increases.
- The logarithm of 1 is 0: $\ln 1 = 0$.
- The logarithms of numbers less than 1 are negative.
- In elementary mathematics the logarithms of negative numbers are not defined.

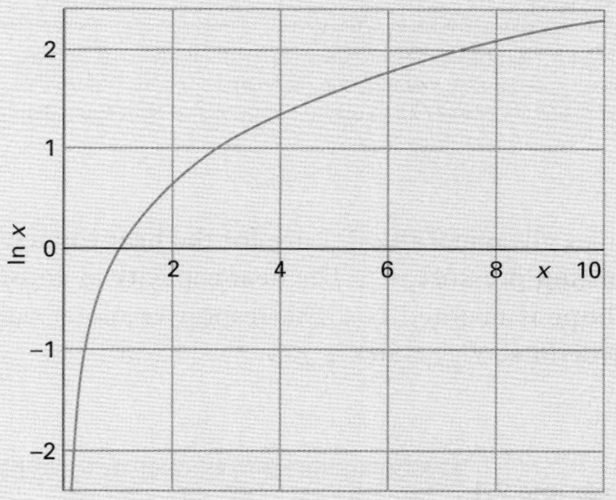

Sketch 2.4 The graph of $\ln x$ for $0 < x \le 10$.

We also encounter the **common logarithm** of a number, the logarithm compiled with 10 in place of e; they are denoted $\log x$. Common logarithms follow the same rules of addition and subtraction as natural logarithms. Common and natural logarithms (log and ln, respectively) are related by

$$\ln x = \ln 10 \times \log x = (2.303\ldots) \times \log x$$

Equation 2.2 will turn up in various disguises throughout this text. Once again, it is important to be able to interpret it rather than just remember it:

- In an expansion $V_f > V_i$, so $V_f/V_i > 1$ and the logarithm is positive ($\ln x > 0$ if $x > 1$). Therefore, in an expansion, w is negative. That is what we should expect: energy *leaves* the system as the system does expansion work.
- For a given change in volume, we get more work the higher the temperature of the confined gas (Fig. 2.11). That is also what we should expect: at high temperatures, the pressure of the gas is high, so we have to use a high external pressure, and therefore a stronger opposing force, to match the internal pressure at each stage.

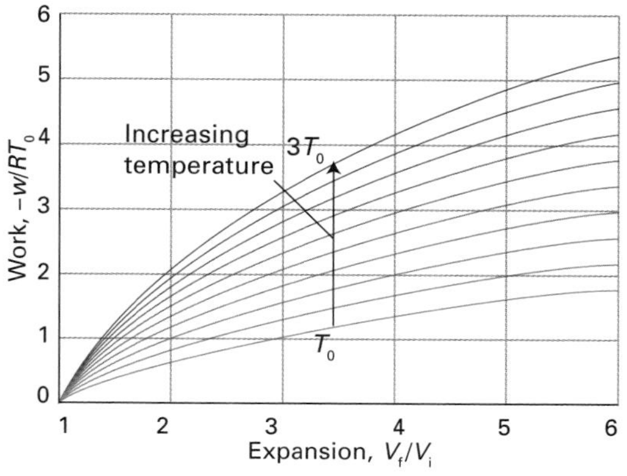

Fig. 2.11 The work of reversible, isothermal expansion of a perfect gas. Note that for a given change of volume and fixed amount of gas, the work is greater the higher the temperature.

Brief illustration 2.2 The work of reversible isothermal expansion

The work done when 1.0 mol Ar(g) confined in a cylinder of volume 1.0 dm³ at 25 °C expands isothermally and reversibly to 2.0 dm³ is

$$w = -\overbrace{(1.0\ \text{mol})}^{n} \times \overbrace{(8.3145\ \text{J K}^{-1}\ \text{mol}^{-1})}^{R} \times \overbrace{(298\ \text{K})}^{T} \ln \overbrace{\frac{2.0\ \text{dm}^3}{1.0\ \text{dm}^3}}^{V_f/V_i}$$
$$= -1.7 \times 10^3\ \text{J} = -1.7\ \text{kJ}$$

We have seen that the work done depends upon the manner in which an expansion occurs. Work is an example of a **path function** because its value depends upon the path taken rather than the state of the system. Thus, the magnitude of the work done when a gas expands reversibly is greater than if it expands by a different path between the same initial and final states.

2.5 The measurement of heat

Upon heating, the temperature of a substance typically rises. We say 'typically' because the temperature does not always rise. The temperature of boiling water, for instance, remains unchanged as energy is supplied as heat (see Chapter 5).

For a specified energy, q, transferred by heating, the size of the resulting temperature change, ΔT, depends on the 'heat capacity' of the substance. The **heat capacity**, C, is defined as

$$C = \frac{q}{\Delta T} \qquad \text{Definition} \quad \text{Heat capacity} \qquad (2.3a)$$

It follows that we have a simple way of measuring the heat absorbed or released by a system: we measure a temperature change and then use the appropriate value of the heat capacity of the system and eqn 2.3a rearranged into

$$q = C\Delta T \qquad (2.3b)$$

Brief illustration 2.3 The energy of heating

If the heat capacity of a beaker of water is 0.50 kJ K^{-1}, and we observe a temperature rise of 4.0 K, then we can infer that the heat transferred to the water is

$$q = (0.50 \text{ kJ K}^{-1}) \times (4.0 \text{ K}) = +2.0 \text{ kJ}$$

Heat capacities occur frequently in the following sections and chapters, and we need to be aware of their properties and how their values are reported. First, we note that the heat capacity is an extensive property (a property that depends on the amount of substance in the sample, Foundations 0.4): 2 kg of iron has twice the heat capacity of 1 kg of iron, so twice as much heat is required to raise its temperature by a given amount. It is more convenient to report the heat capacity of a substance as an intensive property (a property that is independent of the amount of substance in the sample). We therefore use either the **specific heat capacity**, C_s, the heat capacity divided by the mass of the sample ($C_s = C/m$, in joules per kelvin per gram, J K^{-1} g^{-1}) or the **molar heat capacity**, C_m, the heat capacity divided by the amount of substance ($C_m = C/n$, in joules per kelvin per mole, J K^{-1} mol^{-1}). In common usage, the specific heat capacity is often called the *specific heat*. To obtain the heat capacity of a sample of known mass or that contains a known amount of substance, we use these definitions in the form $C = mC_s$ or $C = nC_m$, respectively.

For reasons that will be explained shortly, the heat capacity of a substance depends on whether the sample is maintained at constant volume (like a gas in a sealed, rigid vessel) as it is heated, or whether the sample is maintained at constant pressure (like water in an open container), and free to change its volume. The latter is a more common arrangement, and the values given in Table 2.1 are for the **heat capacity at constant pressure**, C_p. The **heat capacity at constant volume** is denoted C_V. The respective molar values are denoted $C_{p,m}$ and $C_{V,m}$.

Table 2.1
Heat capacities of common materials

Substance	Heat capacity	
	Specific, $C_{p,s}$/(J K^{-1} g^{-1})	Molar, $C_{p,m}$/(J K^{-1} mol^{-1})*
Air	1.01	29
Benzene, C_6H_6(l)	1.05	136.1
Brass (Cu/Zn)	0.37	
Copper, Cu(s)	0.38	24.44
Ethanol, C_2H_5OH(l)	2.42	111.46
Glass (Pyrex)	0.78	
Granite	0.80	
Marble	0.84	
Polyethylene	2.3	
Stainless steel	0.51	
Water, H_2O(s)	2.03	37
H_2O(l)	4.18	75.29
H_2O(g)	2.01	33.58

*Molar heat capacities are given only for air and well-defined pure substances; see also the *Resource section*. The text's website contains links to online databases of heat capacities.

Example 2.1

Calculating a temperature change from the heat capacity

Suppose a 1.0 kW kettle contains 1.0 kg of water and is turned on for 100 s. By how much does the temperature of the water change?

Strategy We calculate the energy supplied as heat to the water in the kettle from the power P (in watts; 1 W = 1 J s^{-1}) and the time interval t (in seconds) by using $q = Pt$. Then we calculate the temperature change ΔT by using eqn 2.3, rewritten as

$$\Delta T = \frac{q}{C_p} = \frac{q}{nC_{p,m}}$$

where $n = m/M$ is the amount of material in the kettle (the amount, in moles, of H_2O molecules, in our case) and $C_{p,m}$ is the constant-pressure molar heat capacity of the material.

Solution The energy supplied as heat to 1.0 kg of water is

$$q = (1.0 \text{ kW}) \times (100 \text{ s}) = (1.0 \times 10^3 \text{ J s}^{-1}) \times (100 \text{ s})$$
$$= +1.0 \times 10^5 \text{ J}$$

From $m = 1.0$ kg and $M = 18.0$ g mol^{-1} we obtain

$$n = \frac{1.0\times10^3 \text{ g}}{18.0 \text{ g mol}^{-1}} = \frac{1.0\times10^3}{18.0} \text{ mol}$$

Now we use $C_{p,m} = 75$ J K^{-1} mol^{-1} for the constant-pressure molar heat capacity of liquid water (Table 2.1) to calculate the temperature change:

$$\Delta T = \frac{q}{nC_{p,m}} = \frac{\overbrace{1.0\times10^5 \text{ J}}^{q}}{\underbrace{(1.0\times10^3/18.0 \text{ mol})}_{n}\times\underbrace{(75 \text{ J K}^{-1}\text{ mol}^{-1})}_{C_{p,m}}} = +24 \text{ K}$$

Self-test 2.2

What energy is needed to increase the temperature of a sample of 250 g of water by 40 °C?

Answer: 42 kJ

One way to measure the energy transferred as heat in a process is to use a **calorimeter** (Fig. 2.12), which consists of a container in which the reaction or physical process occurs, a thermometer, and a surrounding water bath. The entire assembly is insulated from the rest of the world. The principle of a calorimeter is to use the rise in temperature to determine the energy released as heat by the process occurring inside it. To interpret the rise in temperature, we must first calibrate the calorimeter by comparing the observed change in temperature with a change in temperature brought about by the transfer of a known quantity of energy as heat. One procedure is to heat the calorimeter electrically by passing a known current for a measured time through a heater, and record the increase in temperature. The energy provided electrically is (see The chemist's toolkit 2.3)

$$q = I\mathcal{V}t \qquad (2.4)$$

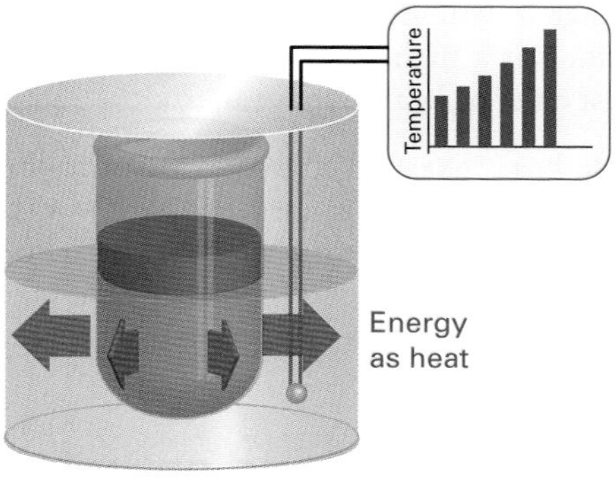

Fig. 2.12 The loss of energy into the surroundings can be detected by noting the change in temperature as the process takes place.

The chemist's toolkit 2.3 Electrical charge, current, power, and energy

Electrical charge is measured in **coulombs**, C. The elementary charge, the magnitude of charge carried by a single electron or proton is approximately 1.6×10^{-19} C. The motion of charge gives rise to an electric current, I, measured in coulombs per second, or **amperes**, A, where 1 A = 1 C s^{-1}. If the electric charge is that of electrons (as it is through metals and semiconductors), then a current of 1 A represents the flow of 6×10^{18} electrons per second.

The rate of supply of energy is the power. If a constant current I flows through a potential difference $\mathcal{V}$ (measured in volts, V), the power, P, is given by

$$P = I\mathcal{V}$$

It follows that the energy supplied in an interval t is

$$E = Pt = I\mathcal{V}t$$

Because 1 A V s = 1 (C s^{-1}) V s = 1 C V = 1 J, the energy is obtained in joules with the current in amperes, the potential difference in volts, and the time in seconds. If the energy is supplied as heat, then we obtain eqn 2.4.

where I is the current (in amperes, A), $\mathcal{V}$ is the potential difference of the supply (in volts, V), and t is the time (in seconds, s) for which the current flows.

Brief illustration 2.4 Electrical heating

If we pass a current of 10.0 A from a 12 V supply for 300 s, then the energy supplied as heat is

$$q = (10.0 \text{ A})\times(12 \text{ V})\times(300 \text{ s})$$
$$= 3.6\times10^4 \overbrace{\text{A V s}}^{1 \text{ A V s}=1 \text{ J}} = 36 \text{ kJ}$$

The observed rise in temperature lets us calculate the heat capacity of the calorimeter (which in this context is also called the **calorimeter constant**) from eqn 2.3. Then we use this heat capacity to interpret a temperature rise due to a reaction in terms of the heat released or absorbed. An alternative procedure is to calibrate the calorimeter by using a reaction of known heat output, such as the combustion of benzoic acid (C_6H_5COOH), for which the heat output is 3227 kJ per mole of C_6H_5COOH consumed.

Example 2.2

Calibrating a calorimeter and measuring a heat transfer

In an experiment to measure the heat released by the combustion of a sample of nutrient, the compound was burned in a calorimeter and the temperature rose by 3.22 °C. When a current of 1.23 A from a 12.0 V source flows through a heater in the same calorimeter for 156 s, the temperature rose by 4.47 °C. What is the heat released by the combustion reaction?

Strategy We calculate the heat supplied electrically by using eqn 2.4 and 1 A V s = 1 J. Then we use the observed rise in temperature to find the heat capacity of the calorimeter. Finally, we use this heat capacity to convert the temperature rise observed for the combustion into a heat output by writing $q = C\Delta T$ (or $q = C\Delta\theta$ if the temperature is given on the Celsius scale).

Solution The heat supplied during the calibration step is

$$\begin{aligned} q = I\mathcal{V}t &= (1.23\ \text{A}) \times (12.0\ \text{V}) \times (156\ \text{s}) \\ &= 1.23 \times 12.0 \times 156\ \text{A V s} \\ &= 1.23 \times 12.0 \times 156\ \text{J} \end{aligned}$$

This product works out as 2.30 kJ, but to avoid rounding errors we save the numerical work to the final stage. The heat capacity of the calorimeter is

$$C = \frac{q}{\Delta\theta} = \frac{1.23\times12.0\times156\ \text{J}}{4.47\ \text{°C}} = \frac{1.23\times12.0\times156}{4.47}\ \text{J °C}^{-1}$$

The numerical value of C is 515 J °C^{-1}, but we don't evaluate it yet in the actual calculation. The heat output of the combustion is therefore

$$q = C\Delta\theta = \left(\frac{1.23\times12.0\times156}{4.47}\ \text{J °C}^{-1}\right)\times(3.22\ \text{°C}) = 1.66\ \text{kJ}$$

A note on good practice As well as keeping the numerical evaluation to the final stage, show the units at each stage of the calculation.

Self-test 2.3

In an experiment to measure the heat released by the combustion of a sample of fuel, the compound was burned in an oxygen atmosphere inside a calorimeter and the temperature rose by 2.78 °C. When a current of 1.12 A from an 11.5 V source flows through a heater in the same calorimeter for 162 s, the temperature rose by 5.11 °C. What is the heat released by the combustion reaction?

Answer: 1.1 kJ

2.6 Heat influx during expansion

In certain cases, we can relate the value of q to the change in volume of a system, and so can calculate, for instance, the flow of energy as heat into the system when a gas expands. The simplest case is that of a perfect gas undergoing isothermal expansion. We can use a molecular interpretation to guide our thoughts. Because the expansion is isothermal, the temperature of the gas is the same at the end of the expansion as it was initially. Therefore, the mean speed of the molecules of the gas is the same before and after the expansion. That implies in turn that the total kinetic energy of the molecules is the same. But for a perfect gas, the *only* contribution to the energy is the kinetic energy of the molecules (recall Section 1.4), so we have to conclude that the *total* energy of the gas is the same before and after the expansion. Energy has left the system as work; therefore, a compensating amount of energy must have entered the system as heat. We can therefore write

$$q = -w \qquad \text{Perfect gas, isothermal conditions} \qquad \text{Relation between heat and work} \qquad (2.5)$$

Brief illustration 2.5 Heating during expansion

If we find that $w = -100$ J for a particular expansion (meaning that 100 J has left the system as a result of the system doing work), then we can conclude that $q = +100$ J (that is, 100 J must enter as heat). For free expansion, $w = 0$, so we conclude that $q = 0$ too: there is no influx of energy as heat when a perfect gas expands against zero pressure.

If the isothermal expansion is also reversible, we can use eqn 2.2 for the work in eqn 2.5, and write

$$q = nRT\ln\frac{V_f}{V_i} \qquad \text{Isothermal, reversible, perfect gas} \qquad \text{Heat transfer during expansion} \qquad (2.6)$$

We interpret this expression as follows:

- When $V_f > V_i$, as in an expansion, the logarithm is positive and we conclude that $q > 0$, as expected: flows as heat into the system to make up for the energy lost as work.
- The greater the ratio of the final and initial volumes, the greater the influx of energy as heat.
- The higher the temperature, the greater the quantity of energy as heat that must enter for a given change in volume: we have seen that more work is done at a higher temperature, so more energy must enter as heat to make up for the energy lost.

Because heat, like work, is a path function, when considering the transfer of energy as heat it is necessary to define how the expansion occurs. The energy transferred as heat depends upon the path taken and is not simply a function of the state of the system.

Internal energy and enthalpy

Heat and work are *equivalent* ways of transferring energy into or out of a system in the sense that once the energy is inside, it is stored simply as 'energy': regardless of how the energy was supplied, as work or as heat, it can be released in either form. In the next few sections we explore this **equivalence of heat and work** in more detail.

2.7 The internal energy

We need some way of keeping track of the energy changes in a system. This tracking is the job of the property called the **internal energy**, U, of the system, the sum of all the kinetic and potential contributions to the energy of all the atoms, ions, and molecules in the system. The internal energy is the grand total energy of the system. It has a value that depends on the temperature and, in general, the pressure. The internal energy is an extensive property because 2 kg of iron at a given temperature and pressure, for instance, has twice the internal energy of 1 kg of iron under the same conditions. The **molar internal energy**, $U_m = U/n$, the internal energy per mole of material, is an intensive property.

In practice, we do not know and cannot measure the total energy of a sample, because it includes the kinetic and potential energies of all the electrons and all the components of the atomic nuclei. Nevertheless, there is no problem with dealing with the *changes* in internal energy, ΔU, because we can determine those changes by monitoring the energy supplied or lost as heat or as work. All practical applications of thermodynamics deal with ΔU, not with U itself. A change in internal energy is written

$$\Delta U = w + q \quad \text{Change in internal energy in terms of work and heat} \quad (2.7)$$

where w is the energy transferred to the system by doing work and q the energy transferred to it by heating.

A note on good practice We write ΔU for the change in internal energy because it is the difference between the final and initial values. We do not write Δq or Δw because it is meaningless to refer to a 'difference of heat' or a 'difference of work': q and w are the quantities of energy transferred as heat and work, respectively, which result in the change ΔU. However, as we saw in Section 2.4, we do use the symbol d to denote an infinitesimally small transfer of energy as work (as in dw) and heat (dq).

Brief illustration 2.6 Changes in internal energy

When a system releases 10 kJ of energy into the surroundings by doing work (that is, when $w = -10$ kJ), the internal energy of the system decreases by 10 kJ, and we write $\Delta U = -10$ kJ. The minus sign signifies the reduction in internal energy that has occurred. If the system loses 20 kJ of energy by heating its surroundings (so $q = -20$ kJ), we write $\Delta U = -20$ kJ. If the system loses 10 kJ as work *and* 20 kJ as heat, as in an inefficient internal combustion engine, the internal energy falls by a total of 30 kJ, and we write $\Delta U = -30$ kJ. On the other hand, if we do 10 kJ of work on the system ($w = +10$ kJ), for instance, by winding a spring it contains or pushing in a piston to compress a gas (Fig. 2.13), then the internal energy of the system increases by 10 kJ, and we write $\Delta U = +10$ kJ. Likewise, if we supply 20 kJ of energy by heating the system ($q = +20$ kJ), then the internal energy increases by 20 kJ, and we write $\Delta U = +20$ kJ.

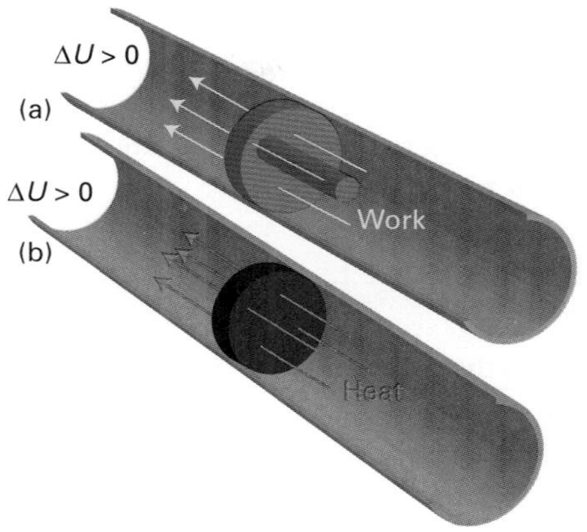

Fig. 2.13 (a) Provided there are no other transfers of energy, when work is done on a system, its internal energy rises ($\Delta U > 0$). (b) Likewise, provided there are no other transfers of energy, the internal energy also rises when energy is transferred into the system as heat.

Another note on good practice Notice that ΔU always carries a sign explicitly, even if it is positive: we never write $\Delta U = 20$ kJ, for instance, but always +20 kJ.

We have seen that a feature of a perfect gas is that for any *isothermal* expansion, the total energy of the sample remains the same, and therefore, because $\Delta U = 0$, that $q = -w$. That is, any energy lost as work is restored by an influx of energy as heat. We can express this property in terms of the internal energy, for it implies that the internal energy remains constant when a perfect gas expands isothermally: from eqn 2.7 we can write

$$\Delta U = 0 \quad \text{Isothermal, perfect gas} \quad \text{Change of internal energy on expansion} \quad (2.8)$$

In other words, *the internal energy of a sample of perfect gas at a given temperature is independent of the volume it occupies*. We can understand this independence by realizing that when a perfect gas expands isothermally the only feature that changes is the average distance between the molecules; their average speed and therefore total kinetic energy remains the same. However, as there are no intermolecular interactions, the total energy is independent of the average separation, so the internal energy is unchanged by expansion.

Example 2.3

Calculating the change in internal energy

Nutritionists are interested in the use of energy by the human body and we can consider our own body as a thermodynamic 'system'. Calorimeters have been constructed that can accommodate a person to measure (non-destructively!) their net energy output. Suppose in the course of an experiment someone does 622 kJ of work on an exercise bicycle and loses 82 kJ of energy as heat. What is the change in internal energy of the person? Disregard any matter loss by perspiration.

Strategy This example is an exercise in keeping track of signs correctly. When energy is lost from the system, w or q is negative. When energy is gained by the system, w or q is positive.

Solution To take note of the signs we write $w = -622$ kJ (622 kJ is lost by doing work) and $q = -82$ kJ (82 kJ is lost by heating the surroundings). Then eqn 2.7 gives us

$$\Delta U = w + q = (-622 \text{ kJ}) + (-82 \text{ kJ}) = -704 \text{ kJ}$$

We see that the person's internal energy falls by 704 kJ. Later, that energy will be restored by eating.

A note on good practice Always attach the correct signs: use a positive sign when there is a flow of energy into the system and a negative sign when there is a flow of energy out of the system.

Self-test 2.4

An electric battery is charged by supplying 250 kJ of energy to it as electrical work (by driving an electric current through it), but in the process it loses 25 kJ of energy as heat to the surroundings. What is the change in internal energy of the battery?

Answer: +225 kJ

2.8 Internal energy as a state function

An important characteristic of the internal energy is that it is a **state function**, a physical property that depends only on the present state of the system and is independent of the path by which that state was reached. Unlike work and heat, a change in internal energy does not depend upon the path taken. If we were to change the temperature of the system, then change the pressure, then adjust the temperature and pressure back to their original values, the internal energy would return to its original value too.

A state function is very much like altitude: each point on the surface of the Earth can be specified by quoting its latitude and longitude, and (on land areas, at least) there is a unique property, the altitude, that has a fixed value at that point. Regardless of the path taken to arrive at the point with a specific latitude and longitude, the value of the altitude is always the same at that point. In thermodynamics, the role of latitude and longitude is played by the pressure and temperature (and any other variables needed to specify the state of the system), and the internal energy plays the role of the altitude, with a single, fixed value for each state of the system, regardless of the changes in pressure and temperature (and any other variables) that brought the system to that state.

The fact that U is a state function implies that *a change, ΔU, in the internal energy depends only upon the initial and final states of a system* (Fig. 2.14). Once again, the altitude is a helpful analogy. If we climb a mountain between two fixed points, we make the same change in altitude regardless of the path we take between the two points. Likewise, if we compress a sample of gas until it reaches a certain pressure and then cool it to a certain temperature, the change in internal energy has a particular value. If, on the other hand, we change the temperature and then the pressure, but ensure that the two final values are the same as in the first experiment, then the overall change in internal energy is exactly the same as

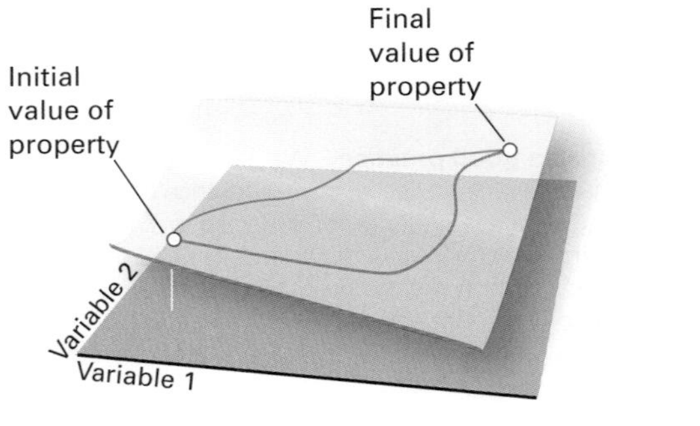

Fig. 2.14 The curved sheet shows how a property (for example, the altitude) changes as two variables (for example, latitude and longitude) are changed. The altitude is a state property, because it depends only on the current state of the system. The change in the value of a state property is independent of the path between the two states. For example, the difference in altitude between the initial and final states shown in the diagram is the same whatever path (as depicted by the red and green lines) is used to travel between them.

before. This path independence of the value of ΔU is of the greatest importance in chemistry, as we shall soon see.

Suppose we now consider an isolated system. Because an isolated system can neither do work nor heat its surroundings (or acquire energy by either process), it follows that its internal energy cannot change. That is,

> The internal energy of an isolated system is constant.

This statement is the **First Law of thermodynamics.** It is closely related to the law of conservation of energy, but allows for transfers of energy as heat as well as by doing work. Unlike thermodynamics, mechanics does not deal with the concept of heat.

The experimental evidence for the First Law is the impossibility of making a 'perpetual motion machine', a device for producing work without consuming fuel. As we have already remarked, try as people might, they have never succeeded. No device has ever been made that creates internal energy to replace the energy drawn off as work. We cannot extract energy as work, leave the system isolated for some time, and hope that when we return the internal energy will have become restored to its original value.

The definition of ΔU in terms of w and q points to a very simple method for measuring the change in internal energy of a system when a reaction takes place. We have seen already that the work done by a system when it pushes against a fixed external pressure is proportional to the change in volume. Therefore, if we carry out a reaction in a container of constant volume, the system can do no expansion work and provided it can do no other kind of work (so-called 'non-expansion work', such as electrical work) we can set $w = 0$. Then eqn 2.7 simplifies to

$$\Delta U = q \quad \text{Constant volume, no non-expansion work} \quad \text{Internal energy change} \qquad (2.9a)$$

This relation is commonly written

$$\Delta U = q_V \qquad (2.9b)$$

The subscript V signifies that the volume of the system is constant. An example of a chemical system that can be approximated as a constant-volume container is an individual biological cell.

To measure a change in internal energy, we should use a calorimeter that has a fixed volume and monitor the energy released as heat ($q < 0$) or supplied as heat ($q > 0$). A **bomb calorimeter** is an example of a constant-volume calorimeter: it consists of a sturdy, sealed, constant-volume vessel in which the reaction takes place, and a surrounding water bath (Fig. 2.15). To ensure that no heat escapes unnoticed from the calorimeter, it is immersed in a water bath with a temperature adjusted to match the rising temperature of the calorimeter. The fact that the temperature of the bath is the same as that of the calorimeter ensures that no heat flows from one to the other. That is, the arrangement is adiabatic.

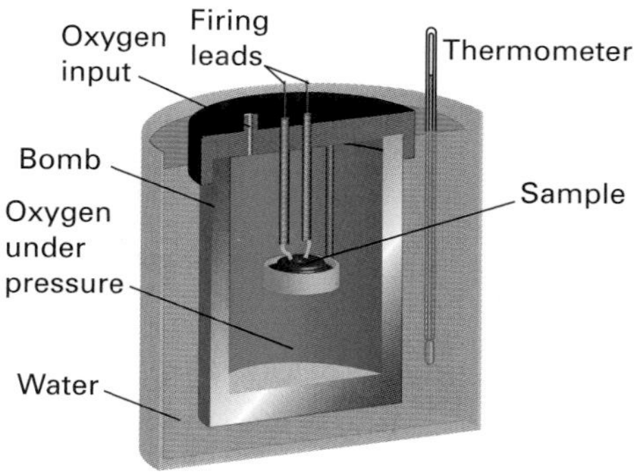

Fig. 2.15 A constant-volume bomb calorimeter. The 'bomb' is the central, sturdy vessel, which is strong enough to withstand moderately high pressures. The calorimeter is the entire assembly shown here. To ensure that no heat escapes into the surroundings, the calorimeter may be immersed in a water bath with a temperature that is continuously adjusted to that of the calorimeter at each stage of the combustion.

We can use eqn 2.9 to obtain more insight into the heat capacity of a substance. The definition of heat capacity is given in eqn 2.3 ($C = q/\Delta T$). At constant volume, q may be replaced by the change in internal energy of the substance, so

$$C_V = \frac{\Delta U}{\Delta T} \quad \text{Definition} \quad \text{Constant-volume heat capacity} \qquad (2.10)$$

The expression on the right is the slope of the graph of internal energy plotted against temperature, with the volume of the system held constant, so C_V tells us how the internal energy of a constant-volume system varies with temperature. If, as is generally the case, the graph of internal energy against temperature is not a straight line, we interpret C_V as the slope of the tangent to the curve at the temperature of interest (Fig. 2.16).

A note on good practice In the language of the calculus, the constant-volume heat capacity is the derivative of the function U with respect to the variable T at a specified volume (see The chemist's toolkit 1.3). As shown in that toolkit, the slope of a function is the first derivative, which in this context is dU/dT. When it is important to specify that one variable is kept fixed, as the volume is in this case, we attach that fixed quantity as a suffix to the differential and write $C_V = (\partial U/\partial T)_V$. Note the 'curly d' in this expression, which is the convention when one or more possible variables are kept constant.

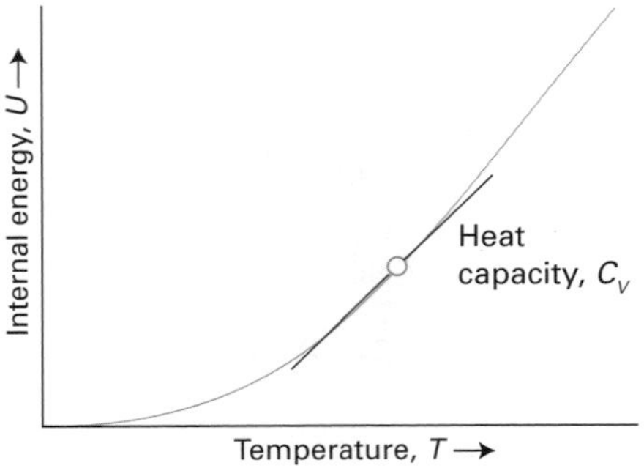

Fig. 2.16 The constant-volume heat capacity is the slope of a curve showing how the internal energy varies with temperature. The slope, and therefore the heat capacity, may be different at different temperatures.

2.9 The enthalpy

Much of chemistry, and most of biology, takes place in vessels that are open to the atmosphere and subjected to constant pressure, not constrained to constant volume in a rigid, sealed container. In general, when a change takes place in a system open to the atmosphere, the volume of the system changes. For example, the thermal decomposition of 1.0 mol $CaCO_3$(s) at 1 bar results in an increase in volume of nearly 90 dm^3 at 800 °C on account of the carbon dioxide gas produced. To create this large volume for the carbon dioxide to occupy, the surrounding atmosphere must be pushed back. That is, the system must perform expansion work of the kind treated in Section 2.4. Therefore, although a certain quantity of heat may be supplied to bring about the endothermic decomposition, the increase in internal energy of the system is not equal to the energy supplied as heat because some energy has been used to do work of expansion (Fig. 2.17). In other words, because the volume has increased, some of the heat supplied to the system has leaked back into the surroundings as work.

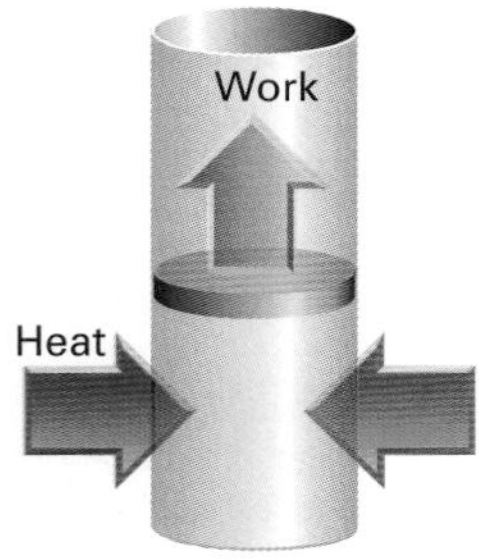

Fig. 2.17 The change in internal energy of a system that is free to expand or contract is not equal to the energy supplied as heat because some energy may escape back into the surroundings as work. However, the change in enthalpy of the system under these conditions is equal to the energy supplied as heat.

Another example is the oxidation of a fat, such as tristearin, to carbon dioxide in the body. The overall reaction is $2\,C_{57}H_{110}O_6(s) + 163\,O_2(g) \rightarrow 114\,CO_2(g) + 110\,H_2O(l)$. In this exothermic reaction there is a net *decrease* in volume equivalent to the elimination of (163 − 114) mol = 49 mol of gas molecules for every 2 mol of tristearin molecules that react. The decrease in volume at 25 °C is about 600 cm^3 for the consumption of 1 g of fat. Because the volume of the system decreases, the atmosphere does work *on* the system as the reaction proceeds. That is, energy is transferred *to* the system as it contracts. In effect, a weight has been lowered in the surroundings, so the surroundings can do less work after the reaction has occurred. Some of their energy has been transferred into the system. For this reaction, the decrease in the internal energy of the system is less than the energy released as heat because some energy has been restored by doing work.

We can avoid the complication of having to take into account the work of expansion by introducing a new property that will be at the centre of our attention throughout the rest of the chapter and will recur throughout the book. The **enthalpy**, H, of a system is defined as

$$H = U + pV \qquad \text{Definition} \quad \text{Enthalpy} \quad (2.11)$$

That is, the enthalpy differs from the internal energy by the addition of the product of the pressure, p, and the volume, V, of the system. This expression applies to *any* system or individual substance: don't be misled by the 'pV' term into thinking that eqn 2.11 applies only to a perfect gas.

Enthalpy is an extensive property. The **molar enthalpy**, $H_m = H/n$, of a substance, an intensive property, differs from the molar internal energy by an amount proportional to the molar volume, V_m, of the substance:

$$H_m = U_m + pV_m \qquad \text{Definition} \quad \text{Molar enthalpy} \quad (2.12a)$$

This relation is valid for all substances. For a perfect gas we can go on to write $pV_m = RT$, and obtain

$$H_m = U_m + RT \qquad \text{Perfect gas} \quad \text{Molar enthalpy} \quad (2.12b)$$

At 25 °C, $RT = 2.5$ kJ mol^{-1}, so the molar enthalpy of a perfect gas is greater than its molar internal energy by 2.5 kJ mol^{-1}. Because the molar volume of a solid or liquid is typically about a thousand times less than that of a gas, we can also conclude that the molar enthalpy of a solid or liquid is only about 2.5 J mol^{-1}

(note: joules, not kilojoules) more than its molar internal energy, so the numerical difference is negligible.

A change in enthalpy (the only quantity we can measure in practice) arises from a change in the internal energy and a change in the product pV:

$$\Delta H = \Delta U + \Delta(pV) \qquad (2.13a)$$

where $\Delta(pV) = p_f V_f - p_i V_i$. If the change takes place at constant pressure p, the second term on the right simplifies to

$$\Delta(pV) = pV_f - pV_i = p(V_f - V_i) = p\Delta V$$

and we can write

$$\Delta H = \Delta U + p\Delta V \quad \text{Constant pressure} \quad \text{Enthalpy change} \qquad (2.13b)$$

We shall often make use of this important relation for processes occurring at constant pressure, such as chemical reactions taking place in containers open to the atmosphere.

Although the enthalpy and internal energy of a sample may have similar numerical values, the introduction of the enthalpy has very important consequences in thermodynamics. First, notice that because H is defined in terms of state functions (U, p, and V), *the enthalpy is a state function*. The implication is that the change in enthalpy, ΔH, when a system changes from one state to another is independent of the path between the two states. Secondly, we show in the following Derivation that the change in enthalpy of a system can be identified with the heat transferred to it at constant pressure:

$$\Delta H = q \qquad (2.14a)$$

This relation is commonly written

$$\Delta H = q_p \quad \text{Constant pressure, no non-expansion work} \quad \text{Enthalpy change} \qquad (2.14b)$$

The subscript p signifies that the pressure is held constant.

Derivation 2.3

Heat transfers at constant pressure

Consider a system open to the atmosphere, so that its pressure p is constant and equal to the external pressure p_{ex}. From eqn 2.13b we can write

$$\Delta H = \Delta U + p\Delta V = \Delta U + p_{ex}\Delta V$$

However, we know that the change in internal energy is given by eqn 2.7 ($\Delta U = w + q$) with $w = -p_{ex}\Delta V$ (provided the system does no other kind of work). When we substitute that expression into this one we obtain

$$\Delta H = (-p_{ex}\Delta V + q) + p_{ex}\Delta V = q$$

which is eqn 2.14.

The result expressed by eqn 2.14, that *at constant pressure, with no non-expansion work, we can identify the energy transferred by heating with a change in enthalpy of the system*, is enormously powerful. It relates a quantity we can measure (the energy transferred as heat at constant pressure) to the change in a state function (the enthalpy). Dealing with state functions greatly extends the power of thermodynamic arguments, because we don't have to worry about how we get from one state to another: all that matters is the initial and final states.

Brief illustration 2.7 Enthalpy changes

Equation 2.14 implies that if 10 kJ of energy is supplied as heat to the system that is free to change its volume at constant pressure, then the enthalpy of the system increases by 10 kJ regardless of how much energy enters or leaves by doing work, and we write $\Delta H = +10$ kJ. On the other hand, if the reaction is exothermic and releases 10 kJ of energy as heat when it occurs, then $\Delta H = -10$ kJ regardless of how much work is done. For the particular case of the combustion of tristearin mentioned at the beginning of the section, in which 90 kJ of energy is released as heat, we would write $\Delta H = -90$ kJ.

An endothermic reaction ($q > 0$) taking place at constant pressure results in an increase in enthalpy ($\Delta H > 0$) because energy enters the system as heat. On the other hand, an exothermic process ($q < 0$) taking place at constant pressure corresponds to a decrease in enthalpy ($\Delta H < 0$) because energy leaves the system as heat. All combustion reactions, including the controlled combustions that contribute to respiration, are exothermic and are accompanied by a decrease in enthalpy. These relations are consistent with the name 'enthalpy', which is derived from the Greek words meaning 'heat inside': the 'heat inside' the system is increased if the process is endothermic and absorbs energy as heat from the surroundings; it is decreased if the process is exothermic and releases energy as heat into the surroundings. However, never forget that heat does not actually 'exist' inside: only energy exists in a system; heat is a means of recovering or increasing that energy.

2.10 The temperature variation of the enthalpy

We have seen that the internal energy of a system rises as the temperature is increased. The same is true of the enthalpy, which also rises when the temperature is increased (Fig. 2.18). For example, the enthalpy of 100 g of water is greater at 80 °C than at

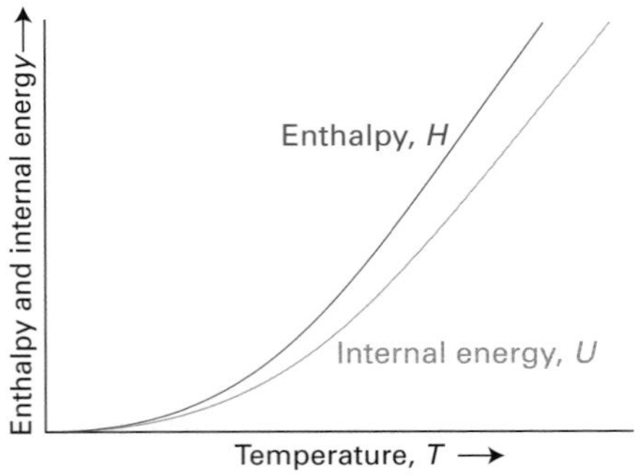

Fig. 2.18 The enthalpy of a system increases as its temperature is raised. Note that the enthalpy is always greater than the internal energy of the system, and that the difference increases with temperature.

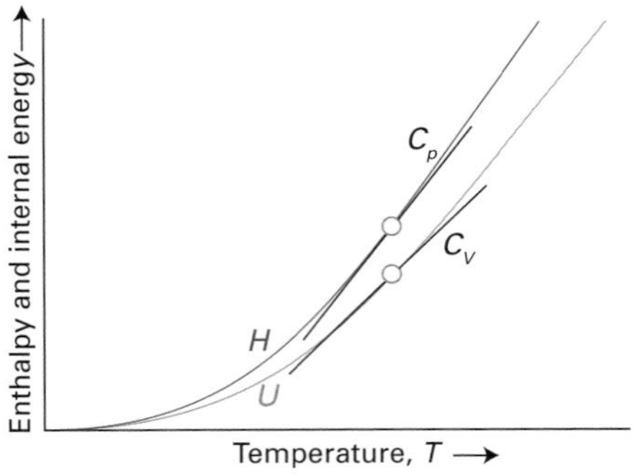

Fig. 2.19 The heat capacity at constant pressure is the slope of the curve showing how the enthalpy varies with temperature; the heat capacity at constant volume is the corresponding slope of the internal energy curve. Note that the heat capacity varies with temperature (in general) and that C_p is greater than C_V.

20 °C. We can measure the change by monitoring the energy that we must supply as heat to raise the temperature through 60 °C when the sample is open to the atmosphere (or subjected to some other constant pressure); it is found that $\Delta H \approx +25$ kJ in this instance.

Just as we saw that the constant-volume heat capacity tells us about the temperature dependence of the internal energy at constant volume, so the constant-pressure heat capacity tells us how the enthalpy of a system changes as its temperature is raised at constant pressure. To derive the relation, we combine the definition of heat capacity in eqn 2.3 ($C = q/\Delta T$) with eqn 2.14 and obtain

$$C_p = \frac{\Delta H}{\Delta T} \quad \text{Definition} \quad \text{Constant-pressure heat capacity} \quad (2.15)$$

A note on good practice As for C_V, the formal definition of this heat capacity is written in terms of derivatives, but in this case with p held constant: $C_p = (\partial H/\partial T)_p$.

The constant-pressure heat capacity is the slope of a plot of enthalpy against temperature of a system kept at constant pressure (Fig. 2.19). In general, the constant-pressure heat capacity depends on the temperature, and some values at 298 K are given in Table 2.1 (many more will be found in the *Resource Section*). Values at other temperatures not too different from room temperature are commonly estimated from the expression

$$C_{p,\mathrm{m}} = a + bT + \frac{c}{T^2} \quad (2.16a)$$

with values of the constants a, b, and c obtained by fitting this expression to experimental data. Some values of the constants are given in Table 2.2 and a typical temperature variation is shown in Fig. 2.20. At very low temperatures, non-metallic solids are

Table 2.2
*Temperature dependence of heat capacities**

Substance	a/ (J K^{-1} mol^{-1})	b/ (J K^{-2} mol^{-1})	c/ (J K mol^{-1})
C(s, graphite)	16.86	4.77×10^{-3}	-8.54×10^5
CO_2(g)	44.22	8.79×10^{-3}	-8.62×10^5
H_2O(l)	75.29	0	0
N_2(g)	28.58	3.77×10^{-3}	-5.0×10^4
Cu(s)	22.64	6.28×10^{-3}	0
NaCl(s)	45.94	16.32×10^{-3}	0

*The constants are for use in the expression $C_{p,\mathrm{m}} = a + bT + c/T^2$.

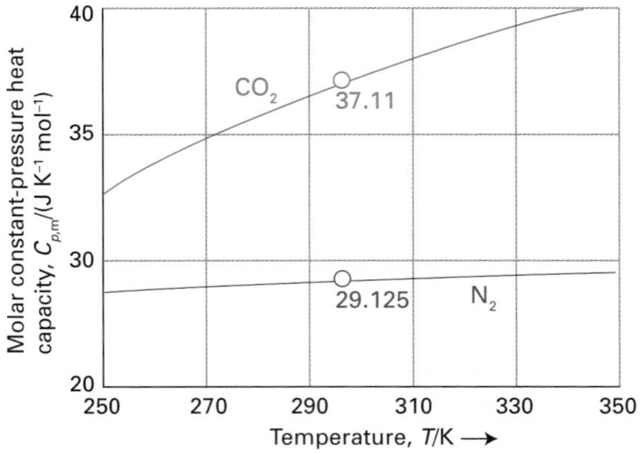

Fig. 2.20 The variation of heat capacity with temperature as expressed by the empirical formula in eqn 2.16a. For this plot we have used the values for carbon dioxide and nitrogen: the circles show the measured values at 298 K.

found to have heat capacities that are proportional to T^3:

$$C_{p,\mathrm{m}} = aT^3 \qquad (2.16b)$$

where a is another constant (not the same as in eqn 2.16a). The reason for this behaviour was unclear until quantum mechanics provided an explanation.

Brief illustration 2.8 Temperature dependence of the enthalpy

Provided the heat capacity is constant over the range of temperatures of interest, we can write eqn 2.15 as $\Delta H = C_p\Delta T$. This relation means that when the temperature of 100 g of water (5.55 mol H_2O) is raised from 20 °C to 80 °C (so $\Delta T = +60$ K) at constant pressure, the enthalpy of the sample changes by

$$\begin{aligned}\Delta H = C_p\Delta T &= nC_{p,\mathrm{m}}\Delta T\\ &= (5.55\ \mathrm{mol}) \times (75.29\ \mathrm{J\ K^{-1}\ mol^{-1}}) \times (60\ \mathrm{K}) = +25\ \mathrm{kJ}\end{aligned}$$

The calculation in the illustration is only approximate, because the heat capacity depends on the temperature, and we have used an average value for the temperature range of interest. In Project 2.29 you are invited to explore a case in which this approximation is removed.

Because we know that the difference between the enthalpy and internal energy of a perfect gas depends in a very simple way on the temperature (eqn 2.12b), we can suspect that there is a simple relation between the heat capacities at constant volume and constant pressure. We show in the following Derivation that, in fact,

$$C_{p,\mathrm{m}} - C_{V,\mathrm{m}} = R \qquad \text{Perfect gas} \quad \text{Difference between the molar heat capacities} \qquad (2.17)$$

Derivation 2.4

The relation between heat capacities

The molar internal energy and enthalpy of a perfect gas are related by eqn 2.12b ($H_\mathrm{m} = U_\mathrm{m} + RT$) which we can write as $H_\mathrm{m} - U_\mathrm{m} = RT$. When the temperature increases by ΔT, the molar enthalpy increases by ΔH_m and the molar internal energy increases by ΔU_m, so

$$\Delta H_\mathrm{m} - \Delta U_\mathrm{m} = R\Delta T$$

Now divide both sides by ΔT, which gives

$$\frac{\Delta H_\mathrm{m}}{\Delta T} - \frac{\Delta U_\mathrm{m}}{\Delta T} = R$$

We recognize the first term on the left as the molar constant-pressure heat capacity, $C_{p,\mathrm{m}}$, and the second term as the molar constant-volume heat capacity, $C_{V,\mathrm{m}}$. Therefore, this relation can be written as in eqn 2.17.

Equation 2.17 shows that the molar heat capacity of a perfect gas is greater at constant pressure than at constant volume. This difference is what we should expect. At constant volume, all the energy supplied as heat to the system remains inside and the temperature rises accordingly. At constant pressure, though, some of the energy supplied as heat escapes back into the surroundings when the system expands and does work. As less energy remains in the system, the temperature does not rise as much, which corresponds to a greater heat capacity. The difference is significant for gases (for oxygen, $C_{V,\mathrm{m}} = 20.8\ \mathrm{J\ K^{-1}\ mol^{-1}}$ and $C_{p,\mathrm{m}} = 29.1\ \mathrm{J\ K^{-1}\ mol^{-1}}$), which undergo large changes of volume when heated, but is negligible for most solids and liquids under normal conditions, for they expand much less and therefore lose less energy to the surroundings as work.

Checklist of key concepts

- ☐ 1 A system is classified as open, closed, or isolated.
- ☐ 2 The surroundings remain at constant temperature and either constant volume or constant pressure when processes occur in the system.
- ☐ 3 An exothermic process releases energy as heat; an endothermic process absorbs energy as heat.
- ☐ 4 Maximum expansion work is achieved in a reversible change.
- ☐ 5 The First Law of thermodynamics states that the internal energy of an isolated system is constant.
- ☐ 6 Energy transferred as heat at constant volume is equal to the change in internal energy of the system.
- ☐ 7 Energy transferred as heat at constant pressure is equal to the change in enthalpy.
- ☐ 8 The heat capacity at constant volume is the slope of the internal energy with respect to temperature.
- ☐ 9 The heat capacity at constant pressure is the slope of the enthalpy with respect to temperature.

Road map of key equations

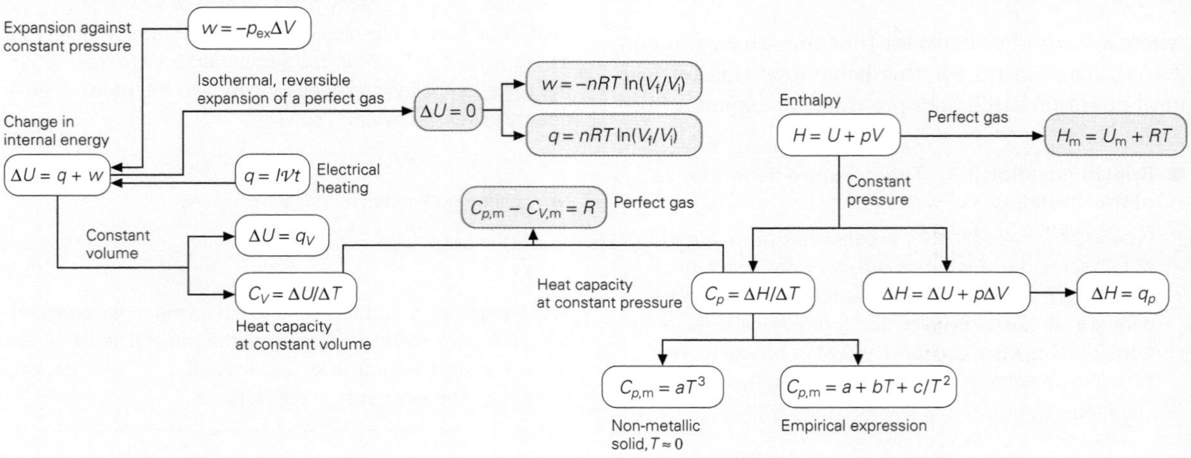

Blue boxes are relations for perfect gases.

Questions and exercises

Discussion questions

2.1 Discuss the statement that a system and its surroundings are distinguished by specifying the properties of the boundary that separates them.

2.2 Give examples of chemical systems that are (a) open, (b) closed, (c) isolated.

2.3 What is (a) temperature, (b) heat, (c) work, (d) energy?

2.4 Provide molecular interpretations of work and heat.

2.5 Are the law of conservation of energy in dynamics and the First Law of thermodynamics identical? If not, describe the difference.

2.6 Explain the difference between expansion work against constant pressure and work of reversible expansion and their consequences.

2.7 Distinguish between reversible and irreversible expansions.

2.8 Explain the difference between the change in internal energy and the change in enthalpy of a chemical or physical process.

2.9 Specify and explain the limitations of the following expressions: (a) $q = nRT\ln(V_f/V_i)$; (b) $\Delta H = \Delta U + p\Delta V$; (c) $C_{p,m} - C_{V,m} = R$.

Exercises

Assume all gases are perfect unless stated otherwise.

2.1 Calculate the work done by a gas when it expands through (a) 1.0 cm^3, (b) 1.0 dm^3 against an atmospheric pressure of 100 kPa. What work must be done to compress the gas back to original state in each case?

2.2 Calculate the work done by a gas consisting of 2.0 mol of molecules when it expands reversibly and isothermally from 1.0 dm^3 to 3.0 dm^3 at 300 K.

2.3 A sample of methane of mass 4.50 g occupies 12.7 dm^3 at 310 K. (a) Calculate the work done when the gas expands isothermally against a constant external pressure of 30.0 kPa until its volume has increased by 3.3 dm^3. (b) Calculate the work that would be done if the same expansion occurred isothermally and reversibly.

2.4 In the isothermal reversible compression of 52.0 mmol of perfect-gas molecules at 260 K, the volume of the gas is reduced from 300 cm^3 to 100 cm^3. Calculate the magnitude of the work done for this process.

2.5 A sample of blood plasma occupies 0.550 dm^3 at 0 °C and 1.03 bar, and is compressed isothermally by 57 per cent by being subjected to a constant external pressure of 95.2 bar. Calculate the magnitude of the work done.

2.6 A strip of magnesium metal of mass 12.5 g is dropped into a beaker of dilute hydrochloric acid. Given that the magnesium is the limiting reactant, calculate the work done by the system as a result of the reaction. The atmospheric pressure is 1.00 atm and the temperature 20.2 °C.

2.7 Calculate the work of expansion accompanying the complete combustion of 10.0 g of sucrose ($C_{12}H_{22}O_{11}$) to carbon dioxide and (a) liquid water, (b) water vapour at 20 °C when the external pressure is 1.20 atm.

2.8 We are all familiar with the general principles of operation of an internal combustion reaction: the combustion of fuel drives out the piston. It is possible to imagine engines that use reactions other than combustions, and we need to assess the work they can do. A chemical reaction takes place in a container of cross-sectional area 100 cm^2; the container has a piston at one end. As a result of the reaction, the piston is pushed out through 10.0 cm against a constant external pressure of 100 kPa. Calculate the work done by the system.

2.9 What is the heat capacity of a sample of liquid that rose in temperature by 5.23 °C when supplied with 124 J of energy as heat?

2.10 A cube of iron was heated to 70 °C and transferred to a beaker containing 100 g of water at 20 °C. The final temperature of the water and the iron was 23 °C. What is (a) the heat capacity, (b) the specific heat capacity, and (c) the molar heat capacity of the iron cube? Ignore heat losses from the assembly.

2.11 The high heat capacity of water is ecologically benign because it stabilizes the temperatures of lakes and oceans: a large quantity of energy must be lost or gained before there is a significant change in temperature. Conversely, it means that a lot of heat must be supplied to achieve a large rise in temperature. The molar heat capacity of water is 75.3 J K^{-1} mol^{-1}. What energy is needed to heat 250 g of water (a cup of coffee, for instance) through 40 °C?

2.12 A current of 1.55 A from a 110 V source was passed through a heater for 8.5 min. The heater was immersed in a water bath. What quantity of energy was transferred to the water as heat?

2.13 When 229 J of energy is supplied as heat to 3.00 mol Ar(g), the temperature of the sample increases by 2.55 K. Calculate the molar heat capacities at constant volume and constant pressure of the gas.

2.14 The heat capacity of air is much smaller than that of water, and relatively modest amounts of heat are needed to change its temperature. This is one of the reasons why desert regions, though very hot during the day, are bitterly cold at night. The heat capacity of air at room temperature and pressure is approximately 21 J K^{-1} mol^{-1}. How much energy is required to raise the temperature of a room of dimensions 5.5 m × 6.5 m × 3.0 m by 10 °C? If losses are neglected, how long will it take a heater rated at 1.5 kW to achieve that increase given that 1 W = 1 J s^{-1}?

2.15 The transfer of energy from one region of the atmosphere to another is of great importance in meteorology for it affects the weather. Calculate the heat needed to be supplied to a parcel of air containing 1.00 mol air molecules to maintain its temperature at 300 K when it expands reversibly and isothermally from 22 dm^3 to 30.0 dm^3 as it ascends.

2.16 The temperature of a block of iron ($C_{V,m}$ = 25.1 J K^{-1} mol^{-1}) of mass 1.4 kg fell by 65 °C as it cooled to room temperature. What is its change in internal energy?

2.17 In an experiment to determine the calorific value of a food, a sample of the food was burned in an oxygen atmosphere and the temperature rose by 2.89 °C. When a current of 1.27 A from a 12.5 V source flowed through the same calorimeter for 157 s, the temperature rose by 3.88 °C. What energy is released as heat by the combustion?

2.18 In preparation for a study of the metabolism of an organism, a small, sealed calorimeter was prepared. In the initial phase of the experiment, a current of 22.22 mA from a 11.8 V source was passed for 162 s through a heater inside the calorimeter. What is the change in internal energy of the calorimeter?

2.19 In a computer model of the atmosphere, 20 kJ of energy was transferred as heat to a parcel of air of initial volume 1.0 m^3 at 1.0 atm. What is the change in enthalpy of the parcel of air?

2.20 The internal energy of a perfect gas does not change when the gas undergoes isothermal expansion from V_i to V_f. What is the change in enthalpy?

2.21 Carbon dioxide, although only a minor component of the atmosphere, plays an important role in determining the weather and the composition and temperature of the atmosphere. Calculate the difference between the molar enthalpy and the molar internal energy of carbon dioxide regarded as a real gas at 298.15 K. For this calculation treat carbon dioxide as a van der Waals gas and use the data in Table 1.6.

2.22 A sample of a serum of mass 25 g is cooled from 290 K to 275 K at constant pressure by the extraction of 1.2 kJ of energy as heat. Calculate q and ΔH and estimate the heat capacity of the sample.

2.23 When 3.0 mol O_2(g) is heated at a constant pressure of 3.25 atm, its temperature increases from 260 K to 285 K. Given that the molar heat capacity of O_2 at constant pressure is 29.4 J K^{-1} mol^{-1}, calculate q, ΔH, and ΔU.

2.24 The molar heat capacity at constant pressure of carbon dioxide is 29.14 J K^{-1} mol^{-1}. What is the value of its molar heat capacity at constant volume?

2.25 Use the information in the preceding exercise to calculate the change in (a) molar enthalpy, (b) molar internal energy when carbon dioxide is heated from 15 °C (the temperature when air is inhaled) to 37 °C (blood temperature, the temperature in our lungs).

Projects

The symbol ‡ indicates that calculus is required.

2.26‡ Here we explore the van der Waals equation of state in more detail by deriving expressions for the work done in a reversible isothermal expansion for two different equations of state. (a) Following the method in Derivation 2.2, derive the work done for a gas that obeys the equation of state $p = nRT/(V - nb)$, which is appropriate when molecular repulsions are important. Does the gas do more or less work than a perfect gas for the same change of volume? (b) Now derive an equivalent expression for a gas that obeys the equation of state $p = nRT/V - n^2a/V^2$, which is appropriate when molecular attractions are important. Does the gas do more or less work than a perfect gas for the same change of volume?

2.27‡ Derivation 2.2 showed how to calculate the work of reversible, isothermal expansion of a perfect gas. Suppose that the expansion is reversible but not isothermal and that the temperature decreases as the expansion proceeds. (a) Find an expression for the work when $T = T_i - c(V - V_i)$, with c a positive constant. (b) Is the work greater or smaller than for isothermal expansion?

2.28‡ We now explore the effect of the temperature dependence of the heat capacity on the internal energy. (a) The heat capacity of a non-metallic solid at very low temperatures (close to $T = 0$) typically varies as aT^3, where a is a constant. How does its internal energy vary? (b) Suppose that the molar internal energy of a substance over a limited temperature range can be expressed as a polynomial in T as $U_m(T) = a + bT + cT^2$. Find an expression for the constant-volume molar heat capacity at a temperature T.

2.29‡ Now we explore the implications of the temperature dependence of the heat capacity for the enthalpy. (a) The heat capacity of a substance is often reported in the form $C_{p,m} = a + bT + c/T^2$. Use this expression to make a more accurate estimate of the change in molar enthalpy of carbon dioxide when it is heated from 15 °C to 37 °C (as in the preceding exercise), given $a = 44.22$ J K^{-1} mol^{-1}, $b = 8.79 \times 10^{-3}$ J K^{-2} mol^{-1}, and $c = -8.62 \times 10^5$ J K mol^{-1}. You will need to integrate $dH = C_p dT$. (b) Use the expression from part (a) to determine how the molar enthalpy of the substance change over that limited temperature range. Plot the molar enthalpy as a function of temperature.

2.30‡ The exact expression for the relation between the heat capacities at constant volume and constant pressure is $C_p - C_V = \alpha^2 TV/\kappa$, where α is the expansion coefficient, $\alpha = (dV/dT)/V$ at constant pressure and κ (kappa) is the isothermal compressibility, $\kappa = -(dV/dp)/V$. Confirm that this general expression reduces to that in eqn 2.17 for a perfect gas.

3

Thermodynamics: applications of the First Law

Physical change 69

3.1 The enthalpy of phase transition 70

3.2 Atomic and molecular change 74

Chemical change 78

3.3 Enthalpies of combustion 78

3.4 The combination of reaction enthalpies 80

3.5 Standard enthalpies of formation 82

3.6 The variation of reaction enthalpy with temperature 83

CHECKLIST OF KEY CONCEPTS 85

ROAD MAP OF KEY EQUATIONS 86

QUESTIONS AND EXERCISES 86

This chapter is an extended illustration of the role of enthalpy in chemistry. There are three properties of enthalpy to keep in mind:

1. A change in enthalpy can be identified with the heat supplied at constant pressure ($\Delta H = q_p$).
2. Enthalpy is a state function, so the change in its value ($\Delta H = H_f - H_i$) between two specified initial and final states can be calculated by selecting the most convenient path between them.
3. The slope of a plot of enthalpy against temperature is the constant-pressure heat capacity of the system ($C_p = \Delta H/\Delta T$).

All the material in this chapter is based on these three properties.

Physical change

First, we consider physical change, such as when one form of a substance changes into another form of the same substance, as when ice melts to water. We shall also include changes of a particularly simple kind, such as the ionization of an atom or the breaking of a bond in a molecule.

The numerical value of a thermodynamic property depends on the conditions, such as the states of the substances involved, the pressure, and the temperature. Chemists have therefore found it convenient to report their data for a set of standard conditions at the temperature of their choice:

> The **standard state** of a substance is the pure substance at exactly 1 bar.

(Recall that 1 bar = 10^5 Pa.) We denote the standard state value by the superscript $^\ominus$ on the symbol for the property, as in $H_m^\ominus$ for the standard molar enthalpy of a substance and $p^\ominus$ for the standard pressure of 1 bar. For example, the standard state of hydrogen gas is the pure gas at 1 bar and the standard state of solid calcium carbonate is the pure solid at 1 bar, with either the calcite or aragonite form specified. The physical state needs to be specified because we can speak of the standard states of the solid, liquid, and vapour forms of water, for instance, which are the pure solid, the pure liquid, and the pure vapour, respectively, at 1 bar in each case.

In older texts you might come across a standard state defined for 1 atm (101.325 kPa) in place of 1 bar. That is the old convention. In most cases, data for 1 atm differ only a little from data for 1 bar. You might also come across standard states defined as referring to 298.15 K. That is incorrect: temperature is not a part of the definition of standard state, and standard states may refer to any temperature (but it should be specified). Thus, it is possible to speak of the standard state of water vapour at 100 K, 273.15 K, or any other temperature. It is conventional, though, for data to be reported at the so-called 'conventional temperature' of 298.15 K (25.00 °C), and from now on, unless specified otherwise, all data will be for that temperature. For simplicity, we shall often refer to 298.15 K as '25 °C'. Finally, a standard state need not be a stable state and need not be realizable in practice. Thus, the standard state of water vapour at 25 °C is the vapour at 1 bar, but water vapour at that temperature and pressure would immediately condense to liquid water.

3.1 The enthalpy of phase transition

A **phase** is a specific state of matter that is uniform throughout in composition and physical state. The liquid and vapour states of water are two of its phases. The term 'phase' is more specific than 'state of matter' because a substance may exist in more than one solid form, each of which is a solid phase. Thus, the element sulfur may exist as a solid. However, as a solid it may be found as rhombic sulfur or as monoclinic sulfur; these two solid phases differ in the manner in which the crown-like S_8 molecules stack together. No substance has more than one gaseous phase, so 'gas phase' and 'gaseous state' are effectively synonyms. The only substance that exists in more than one liquid phase is helium. Many substances exist in a variety of solid phases. Carbon, for instance, exists as graphite, diamond, and a variety of forms based on fullerene structures; calcium carbonate exists as calcite and aragonite; there are at least twelve forms of ice; one more was discovered in 2006.

The conversion of one phase of a substance to another phase is called a **phase transition**. Thus, vaporization (liquid → gas) is a phase transition, as is a transition between solid phases (such as rhombic sulfur → monoclinic sulfur). Most phase transitions are accompanied by a change of enthalpy because the rearrangement of atoms or molecules usually requires or releases energy. 'Evaporation' is virtually synonymous with vaporization, but commonly denotes vaporization that continues until all the liquid has disappeared.

The vaporization of a liquid, such as the conversion of liquid water to water vapour when a pool of water evaporates at 20 °C or a kettle boils at 100 °C is an endothermic process ($\Delta H > 0$), because heat must be supplied to bring about the change. At a molecular level, molecules are being driven apart from the grip they exert on one another, and this process requires energy. One of the body's strategies for maintaining its temperature at about 37 °C is to use the endothermic character of the vaporization of water, because the evaporation of perspiration requires heat and withdraws it from the skin.

The energy that must be supplied as heat at constant pressure per mole of molecules that are vaporized under standard conditions (that is, pure liquid at 1 bar changing to pure vapour at 1 bar) is called the **standard enthalpy of vaporization** of the liquid, and is denoted $\Delta_{vap}H^{\ominus}$ (Table 3.1). For example, 44 kJ of heat is required to vaporize 1 mol $H_2O(l)$ at 1 bar and 25 °C, so $\Delta_{vap}H^{\ominus} = 44\ \text{kJ mol}^{-1}$. All enthalpies of vaporization are positive, so the sign is not normally given. Alternatively, the same information can be reported by writing the **thermochemical equation**

$$H_2O(l) \rightarrow H_2O(g) \qquad \Delta H^{\ominus} = +44\ \text{kJ}$$

A thermochemical equation shows the standard enthalpy change (including the sign) that accompanies the conversion of an amount of reactant equal to its stoichiometric coefficient in the accompanying chemical equation (in this case, 1 mol H_2O). If the stoichiometric coefficients in the chemical equation are multiplied through by 2, then the thermochemical equation would be written

$$2\ H_2O(l) \rightarrow 2\ H_2O(g) \qquad \Delta H^{\ominus} = +88\ \text{kJ}$$

This equation signifies that 88 kJ of heat is required to vaporize 2 mol $H_2O(l)$ at 1 bar and (recalling our convention) at 298.15 K. Unless otherwise stated, all data in this text are for 298.15 K.

A note on good practice The attachment of the subscript vap to the Δ is the modern convention; however, the older convention in which the subscript is attached to the *H*, as in ΔH_{vap}, is still widely used.

Table 3.1
*Standard enthalpies of transition at the transition temperature**

Substance	Freezing point, T_f/K	$\Delta_{fus}H^⦵$/ (kJ mol^{-1})	Boiling point, T_b/K	$\Delta_{vap}H^⦵$/ (kJ mol^{-1})
Ammonia, NH_3	195.3	5.65	239.7	23.4
Argon, Ar	83.8	1.2	87.3	6.5
Benzene, C_6H_6	278.7	9.87	353.3	30.8
Ethanol, C_2H_5OH	158.7	4.60	351.5	43.5
Helium, He	3.5	0.02	4.22	0.08
Mercury, Hg	234.3	2.292	629.7	59.30
Methane, CH_4	90.7	0.94	111.7	8.2
Methanol, CH_3OH	175.5	3.16	337.2	35.3
Propanone, CH_3COCH_3	177.8	5.72	329.4	29.1
Water, H_2O	273.15	6.01	373.2	40.7

*For values at 298.15 K, use the information in the *Data section*. The text's website also contains links to online databases of thermochemical data.

Example 3.1

Determining the enthalpy of vaporization of a liquid

Ethanol, C_2H_5OH, is brought to the boil at 1 atm. When an electric current of 0.682 A from a 12.0 V supply is passed for 500 s through a heating coil immersed in the boiling liquid, it is found that the temperature remains constant but 4.33 g of ethanol is vaporized. What is the enthalpy of vaporization of ethanol at its boiling point at 1 atm?

Strategy Because the heat is supplied at constant pressure, the heat supplied, q_p, can be identified with the change in enthalpy of the ethanol when it vaporizes. We need to calculate the heat supplied and the amount of ethanol molecules vaporized. Then the enthalpy of vaporization is the heat supplied divided by the amount: $\Delta_{vap}H = \Delta H/n$. The heat supplied is given by eqn 2.4 ($q = I\mathcal{V}t$; recall that 1 A V s = 1 J). The amount of ethanol molecules is determined by dividing the mass of ethanol vaporized by its molar mass: $n = m/M$.

Solution The energy supplied by heating is

$$q_p = I\mathcal{V}t = (0.682\text{ A}) \times (12.0\text{ V}) \times (500\text{ s})$$
$$= 0.682 \times 12.0 \times 500\text{ J}$$

This value is the change in enthalpy of the sample. The amount of ethanol molecules (of molar mass 46.07 g mol^{-1}) vaporized is

$$n = \frac{m}{M} = \frac{4.33\text{ g}}{46.07\text{ g mol}^{-1}} = \frac{4.33}{46.07}\text{ mol}$$

The molar enthalpy change is therefore

$$\Delta_{vap}H = \frac{\overbrace{0.682 \times 12.0 \times 500\text{ J}}^{q_p}}{\underbrace{(4.33/46.07)\text{ mol}}_{n}} = 4.35 \times 10^4\text{ J mol}^{-1}$$

corresponding to 43.5 kJ mol^{-1}.

Because the pressure is 1 atm, not 1 bar, the enthalpy of vaporization calculated here is not the standard value. However, 1 atm differs only slightly from 1 bar, so we can expect the standard enthalpy of vaporization at the boiling point of ethanol, which is 78 °C (351 K), can be expected to be similar.

A note on good practice Molar quantities are expressed as a quantity per mole (as in kilojoules per mole, kJ mol^{-1}). Distinguish them from the magnitude of a property *for* 1 mol of substance, which is expressed as the quantity itself (as in kilojoules, kJ). All enthalpies of transition, denoted $\Delta_{trs}H$, are molar quantities.

Self-test 3.1

In a similar experiment, it was found that 1.36 g of boiling benzene, C_6H_6, is vaporized when a current of 0.835 A from a 12.0 V source is passed for 53.5 s. What is the enthalpy of vaporization of benzene at its normal boiling point?

Answer: 30.8 kJ mol^{-1}

There are some striking differences in standard enthalpies of vaporization: although the value for water is 44 kJ mol^{-1}, that for methane, CH_4, at its boiling point is only 8 kJ mol^{-1}. Even allowing for the fact that vaporization is taking place at different temperatures, the difference between the enthalpies of vaporization signifies that water molecules are held together in the bulk liquid much more tightly than methane molecules are in liquid methane. We shall see in Chapter 15 that the interaction responsible for the low volatility of water is the hydrogen bond. The high enthalpy of vaporization of water has profound ecological consequences, for it is partly responsible for the survival of the oceans and the generally low humidity of the atmosphere. If only a small amount of heat had to be supplied to vaporize the oceans, the atmosphere would be much more heavily saturated with water vapour than is in fact the case.

Another common phase transition is **fusion**, or melting, as when ice melts to water or iron becomes molten. The change in molar enthalpy that accompanies fusion under standard conditions (pure solid at 1 bar changing to pure liquid at 1 bar) is called the **standard enthalpy of fusion, $\Delta_{fus}H^{\ominus}$**. Its value for water at 0 °C is 6.01 kJ mol^{-1} (all enthalpies of fusion are positive, and the sign need not be given), which signifies that 6.01 kJ of energy is needed to melt 1 mol $H_2O(s)$ at 0 °C and 1 bar. Notice that the enthalpy of fusion of water is much less than its enthalpy of vaporization. In vaporization the molecules become completely separated from each other, whereas in melting the molecules are merely loosened without separating completely (Fig. 3.1).

The reverse of vaporization is **condensation** and the reverse of fusion (melting) is **freezing**. The molar enthalpy changes are, respectively, the negative of the enthalpies of vaporization and fusion, because the heat that is supplied to vaporize or melt the substance is released when it condenses or freezes. The fact that enthalpy is a state function implies that it is always the case that, under the same conditions of temperature and pressure, *the enthalpy change of a reverse transition is the negative of the enthalpy change of the forward transition*:

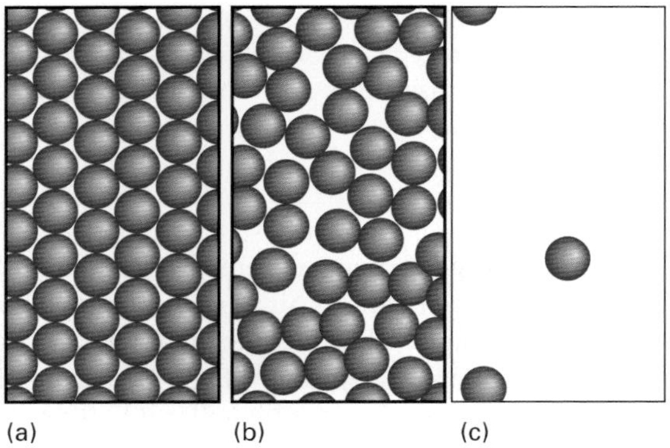

Fig. 3.1 When a solid (a) melts to a liquid (b), the molecules separate from one another only slightly, the intermolecular interactions are reduced only slightly, and there is only a small change in enthalpy. When a liquid vaporizes (c), the molecules are separated by a considerable distance, the intermolecular forces are reduced almost to zero, and the change in enthalpy is much greater. The text's website contains links to animations illustrating this point.

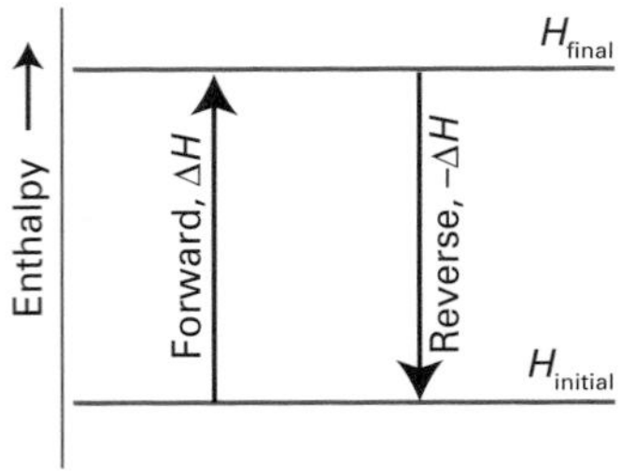

Fig. 3.2 An implication of the First Law is that the enthalpy change accompanying a reverse process is the negative of the enthalpy change for the forward process.

$$H_2O(s) \rightarrow H_2O(l) \qquad \Delta H^{\ominus} = +6.01 \text{ kJ}$$
$$H_2O(l) \rightarrow H_2O(s) \qquad \Delta H^{\ominus} = -6.01 \text{ kJ}$$

and in general, under the same conditions,

$$\Delta_{forward}H = -\Delta_{reverse}H \qquad (3.1)$$

This relation follows directly from the fact that H is a state property, for H must return to the same value if a forward change is followed by the reverse of that change (Fig. 3.2). The high standard enthalpy of vaporization of water (44 kJ mol^{-1}), signifying a strongly endothermic process, implies that the condensation of water (−44 kJ mol^{-1}) is a strongly exothermic process. That exothermicity is in part the origin of the ability of steam to scald severely, because the energy is passed on to the skin. We write 'in part' because steam is typically hot, and gives up the high kinetic energy of its molecular motion when it comes in contact with a cooler body as well as releasing the energy of condensation.

The direct conversion of a solid to a vapour is called **sublimation**. Sublimation occurs when a molecule is so weakly bound to its neighbours that it escapes into the vapour as soon as it has enough energy to move past them. (The thermodynamic explanation is given in Section 5.3.) The reverse process is called **vapour deposition**. Sublimation can be observed on a cold, frosty morning, when frost vanishes as vapour without first melting. The frost itself forms by vapour deposition from cold, damp air. The vaporization of solid carbon dioxide ('dry ice') is another example of sublimation. The standard molar enthalpy change accompanying sublimation is called the **standard enthalpy of sublimation, $\Delta_{sub}H^{\ominus}$**. Because enthalpy is a state function, the same change in enthalpy must be obtained both in the *direct* conversion of solid to vapour and in the *indirect* conversion, in which the solid first melts to the liquid and then that liquid vaporizes (Fig. 3.3):

$$\Delta_{sub}H^{\ominus} = \Delta_{fus}H^{\ominus} + \Delta_{vap}H^{\ominus} \qquad (3.2)$$

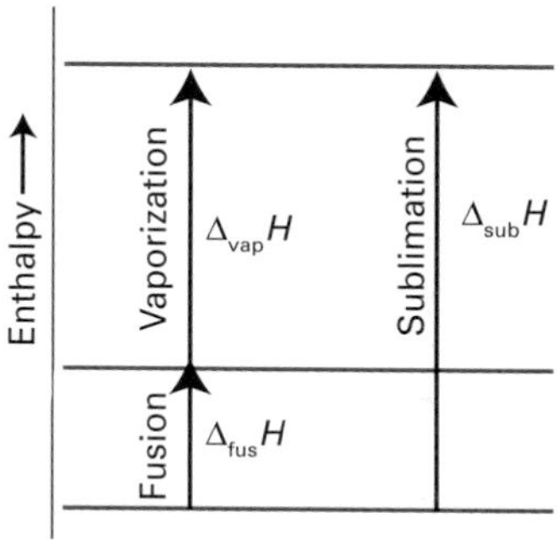

Fig. 3.3 The enthalpy of sublimation at a given temperature is the sum of the enthalpies of fusion and vaporization at that temperature. Another implication of the First Law is that the enthalpy change of an overall process is the sum of the enthalpy changes for the possibly hypothetical steps into which it may be divided.

The two enthalpies that are added together must be for the same temperature, so to get the enthalpy of sublimation of water at 0 °C we must add together the enthalpies of fusion and vaporization for this temperature. Adding together enthalpies of transition for different temperatures gives a meaningless result. The result expressed by eqn 3.2 is an example of a more general statement that will prove useful time and again during our study of thermochemistry:

> The enthalpy change of an overall process is the sum of the enthalpy changes for the steps (actual or hypothetical) into which it may be divided.

Brief illustration 3.1 Enthalpy of sublimation

The standard enthalpy of fusion of ice at 0 °C is 6.01 kJ mol^{-1} and the standard enthalpy of vaporization of water at 0 °C is 45.07 kJ mol^{-1}. The standard enthalpy of sublimation of ice at 0 °C is therefore given by eqn 3.2 as

$$\begin{aligned}\Delta_{sub}H^{\ominus} &= \Delta_{fus}H^{\ominus} + \Delta_{vap}H^{\ominus}\\ &= 6.01\ \text{kJ mol}^{-1} + 45.07\ \text{kJ mol}^{-1}\\ &= 51.08\ \text{kJ mol}^{-1}\end{aligned}$$

Self-test 3.2

Deduce the standard enthalpy of vaporization of liquid carbon dioxide at 298 K from the standard enthalpy of sublimation, 25.23 kJ mol^{-1}, and the standard enthalpy of fusion, 8.33 kJ mol^{-1}, at that temperature.

Answer: 16.90 kJ mol^{-1}

Very large molecules, such as synthetic and biological polymers, and molecular assemblies, such as biological membranes, also undergo a different type of phase transition: the disruption of intra- and intermolecular interactions that help the polymers or assemblies attain complex three-dimensional structures. This transition is an endothermic process that can be studied by **differential scanning calorimetry** (see Impact 3.1).

Impact on biochemistry 3.1

Differential scanning calorimetry

A differential scanning calorimeter, DSC, consists of two small compartments that are heated electrically at a constant rate—the 'scanning' part of the technique (Fig. 3.4). The temperature, T, at time t during a linear scan is $T = T_0 + \alpha t$, where T_0 is the initial temperature and α is the temperature scan rate (in kelvin per second, K s^{-1}). Note that the rate of increase of temperature is $dT/dt = \alpha$. A computer controls the electrical power output to each compartment in order to maintain the same temperature in the two compartments throughout the analysis. The term 'differential' refers to the fact that the behaviour of the sample is compared to that of a reference material that does not undergo a physical or chemical change during the analysis. The temperature of the sample would change relative to that of the reference material if a chemical or physical process involving heat transfer occurs in the sample during the scan. To maintain the same temperature in both compartments, excess heat is transferred to the sample during the process.

If an endothermic process occurs in the sample at a particular temperature, we have to supply additional 'excess' heat, dq_{ex}, to the sample to achieve the same temperature as the reference. We can express this excess heat in terms of an 'excess' heat capacity at each stage of the scan, C_{ex}, by writing $dq_{ex} = C_{ex}dT$. It follows that, because $dT = \alpha dt$, then

$$C_{ex} = \frac{dq_{ex}}{dT} = \frac{dq_{ex}}{\alpha dt} = \frac{P_{ex}}{\alpha}$$

where $P_{ex} = dq_{ex}/dt$ is the excess electrical power (the rate of supply of energy, in watts, where 1 W = 1 J s^{-1}) necessary to equalize the temperature of the sample and reference

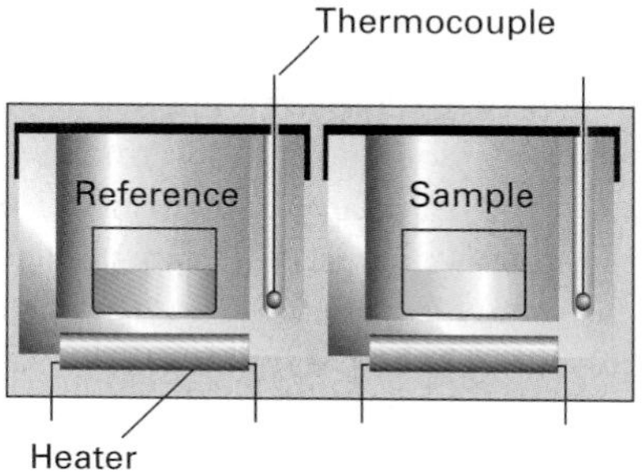

Fig. 3.4 A differential scanning calorimeter. The sample and a reference material are heated in separate but identical compartments. The output is the difference in power needed to maintain the compartments at equal temperatures as the temperature rises.

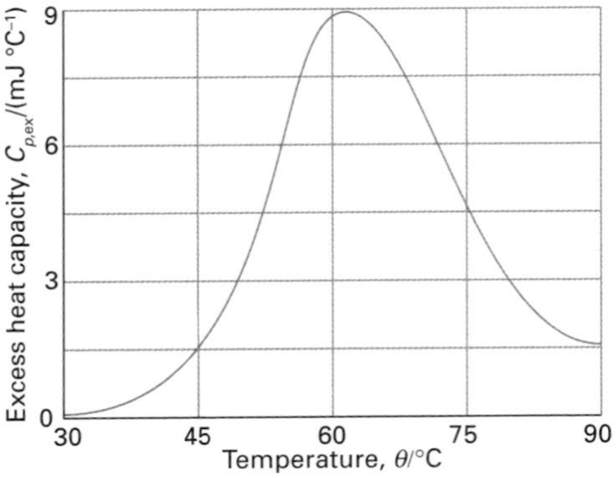

Fig. 3.5 A thermogram for the protein ubiquitin. The protein retains its native structure up to about 45 °C and then undergoes an endothermic conformational change. (Adapted from B. Chowdhry and S. LeHarne, *J. Chem. Educ.* 74, 236 (1997).)

compartments at each stage of the scan. A **thermogram** is a plot of C_{ex} (which is obtained from the value of P_{ex}/α and the power monitored throughout the scan) against T. Figure 3.5 is a thermogram that indicates that the protein ubiquitin retains its native structure up to about 45 °C. At higher temperatures, the protein undergoes an endothermic conformational change that results in the loss of its three-dimensional structure.

To find the total heat transferred, we need to integrate both sides of $dq_{ex} = C_{ex}dT$ from an initial temperature T_1 to a final temperature T_2 (see The chemist's toolkit 2.1):

$$q_{ex} = \int_{T_1}^{T_2} C_{ex}dT$$

The integral is the area under the thermogram between T_1 and T_2. Because the apparatus is at constant pressure, we can identify q_{ex} with the change in enthalpy accompanying the process.

3.2 Atomic and molecular change

A group of enthalpy changes we often employ in the following pages are those accompanying changes to individual atoms and molecules. Among the most important is the **standard enthalpy of ionization,** $\Delta_{ion}H^{\ominus}$, the standard molar enthalpy change accompanying the removal of an electron from a gas-phase atom (or ion). For example, because

$$H(g) \rightarrow H^+(g) + e^-(g) \qquad \Delta H^{\ominus} = +1312\text{ kJ}$$

the standard enthalpy of ionization of hydrogen atoms is reported as 1312 kJ mol^{-1}. This value signifies that 1312 kJ of energy must be supplied as heat to ionize 1 mol H(g) at 1 bar (and 298.15 K). Table 3.2 gives values of the ionization enthalpies for a number of elements; note that all enthalpies of ionization of neutral atoms are positive. The enthalpy of ionization is closely related to the 'ionization energy', the energy of ionization at $T = 0$ (see Chapter 13).

We often need to consider a succession of ionizations, such as the conversion of magnesium atoms to Mg^+ ions followed by the ionization of these Mg^+ ions to Mg^{2+} ions. The successive molar enthalpy changes are called, respectively, the **first ionization enthalpy,** the **second ionization enthalpy,** and so on. For magnesium, these enthalpies refer to the processes

$$Mg(g) \rightarrow Mg^+(g) + e^-(g) \qquad \Delta H^{\ominus} = +738\text{ kJ}$$
$$Mg^+(g) \rightarrow Mg^{2+}(g) + e^-(g) \qquad \Delta H^{\ominus} = +1451\text{ kJ}$$

Note that the second ionization enthalpy is larger than the first: more energy is needed to separate an electron from a positively charged ion than from the neutral atom. Note also that enthalpies of ionization refer to the ionization of the gas phase atom or ion, not to the ionization of an atom or ion in a solid. To determine the latter, we need to combine two or more enthalpy changes.

Example 3.2

Combining enthalpy changes

The standard enthalpy of sublimation of magnesium at 25 °C is 148 kJ mol^{-1}. How much energy as heat (at that temperature and 1 bar) must be supplied to 1.00 g of solid magnesium metal to produce a gas composed of Mg^{2+} ions and electrons?

Strategy The enthalpy change for the overall process is a sum of the steps, sublimation followed by the two stages of ionization, into which it can be divided. Then the heat required for the specified process is the product of the overall molar enthalpy change and the amount of atoms; the latter is calculated from the given mass and the molar mass of the substance.

Solution The overall process is

$$Mg(s) \rightarrow Mg^{2+}(g) + 2\,e^-(g)$$

The thermochemical equation for this process is the sum of the following thermochemical equations:

		$\Delta H^{\ominus}$/kJ
Sublimation:	$Mg(s) \rightarrow Mg(g)$	+148
First ionization:	$Mg(g) \rightarrow Mg^+(g) + e^-(g)$	+738
Second ionization:	$Mg^+(g) \rightarrow Mg^{2+}(g) + e^-(g)$	+1451
Overall (sum):	$Mg(s) \rightarrow Mg^{2+}(g) + 2\,e^-(g)$	+2337

These processes are illustrated diagrammatically in Fig. 3.6. It follows that the overall enthalpy change per mole of Mg is +2337 kJ mol^{-1}. Because the molar mass of magnesium atoms is 24.31 g mol^{-1}, 1.0 g of magnesium corresponds to

Table 3.2

*First and second (and some higher) standard enthalpies of ionization, $\Delta_{ion}H^{\ominus}/(kJ\ mol^{-1})$**

1	2	13	15	15	16	17	18
H 1312							He 2370 5250
Li 519 7300	Be 900 1760	B 799 2420 14 800	C 1090 2350 3660 25 000	N 1400 2860	O 1310 3390	F 1680 3370	Ne 2080 3950
Na 494 4560	Mg 738 1451 7740	Al 577 1820 2740 11 600	Si 786	P 1060	S 1000	Cl 1260	Ar 1520
K 418 3070	Ca 590 1150 4940	Ga 577	Ge 762	As 966	Se 941	Br 1140	Kr 1350
Rb 402 2650	Sr 548 1060 4120	In 556	Sn 707	Sb 833	Te 870	I 1010	Xe 1170
Cs 376 2420 3300	Ba 502 966 3390	Tl 812	Pb 920	Bi 1040	Po 812	At 920	Rn 1040

*Strictly, these are the values of $\Delta_{ion}U(0)$. For more precise work, use $\Delta_{ion}H(T) = \Delta_{ion}U(0) + \frac{5}{2}RT$, with $\frac{5}{2}RT = 6.20\ J\ mol^{-1}$ at 298 K.

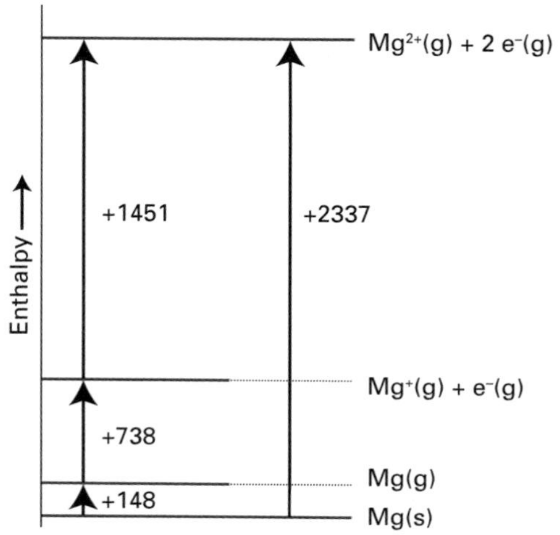

Fig. 3.6 The contributions to the enthalpy change treated in Example 3.2.

$$n(\mathrm{Mg}) = \frac{m(\mathrm{Mg})}{M(\mathrm{Mg})} = \frac{1.00\ \mathrm{g}}{24.31\ \mathrm{g\ mol^{-1}}} = \frac{1.00}{24.31}\ \mathrm{mol}$$

Therefore, the energy that must be supplied as heat (at constant pressure) to ionize 1.00 g of magnesium metal is

$$q_p = \left(\frac{1.00}{24.31}\ \mathrm{mol}\right) \times (2337\ \mathrm{kJ\ mol^{-1}}) = +96.1\ \mathrm{kJ}$$

To put this value into perspective, it is approximately the same as the energy needed to vaporize about 43 g of boiling water.

Self-test 3.3

The enthalpy of sublimation of aluminium is 326 kJ mol^{-1}. Use this information and the ionization enthalpies in Table 3.2 to calculate the energy that must be supplied as heat (at constant pressure) to convert 1.00 g of solid aluminium metal to a gas of Al^{3+} ions and electrons at 25 °C.

Answer: +203 kJ

Table 3.3
*Standard electron gain enthalpies of the main-group elements, $\Delta_{eg}H^{\ominus}/(\text{kJ mol}^{-1})$**

1	2	13	14	15	16	17	18
H −73							He +21
Li −60	Be +18	B −27	C −122	N +7	O −141 +844	F −328	Ne +29
Na −53	Mg +232	Al −43	Si −134	P −44	S −200 +532	Cl −349	Ar +35
K −48	Ca +186	Ga −29	Ge −116	As −78	Se −195	Br −325	Kr +39
Rb −47	Sr +146	In −29	Sn −116	Sb −103	Te −190	I −295	Xe +41
Cs −46	Ba +46	Tl −19	Pb −35	Bi −91	Po −183	At −270	Ra

*Where two values are given, the first refers to the formation of the ion X^- from the neutral atom X; the second, to the formation of X^{2-} from X^-.

The reverse of ionization is **electron gain**, and the corresponding molar enthalpy change under standard conditions is called the **standard electron-gain enthalpy, $\Delta_{eg}H^{\ominus}$**. This quantity is closely related to the electron affinity, as we shall see in Section 13.16. For example, because experiments show that

$$\text{Cl(g)} + \text{e}^-\text{(g)} \rightarrow \text{Cl}^-\text{(g)} \qquad \Delta H^{\ominus} = -349 \text{ kJ}$$

it follows that the electron-gain enthalpy of Cl atoms is −349 kJ mol^{-1}. Notice that electron gain by Cl is an exothermic process, so heat is released when a Cl atom captures an electron and forms an ion. It can be seen from Table 3.3, which lists a number of electron gain enthalpies, that some electron gains are exothermic and others are endothermic, so we need to include their sign. For example, electron gain by an O^- ion is strongly endothermic because it takes energy to push an electron on to an already negatively charged species:

$$\text{O}^-\text{(g)} + \text{e}^-\text{(g)} \rightarrow \text{O}^{2-}\text{(g)} \qquad \Delta H^{\ominus} = +844 \text{ kJ}$$

The final atomic and molecular process to consider at this stage is the **dissociation**, or breaking, of a chemical bond, as in the process

$$\text{HCl(g)} \rightarrow \text{H(g)} + \text{Cl(g)} \qquad \Delta H^{\ominus} = +431 \text{ kJ}$$

The corresponding standard molar enthalpy change is called the **bond enthalpy**, so we would report the H—Cl bond enthalpy as 431 kJ mol^{-1}. All bond enthalpies are positive.

Table 3.4
Selected bond enthalpies, $\Delta H(\text{A—B})/(\text{kJ mol}^{-1})$

Diatomic molecules

H—H	43	O=O	497	F—F	155	H—F	565
		N≡N	945	Cl—Cl	242	H—Cl	431
		O—H	428	Br—Br	193	H—Br	366
		C≡O	1074	I—I	151	H—I	299

Polyatomic molecules

H—CH_3	435	H—NH_2	431	H—OH	492
H—C_6H_5	469	O_2N—NO_2	57	HO—OH	213
H_3C—CH_3	368	O=CO	531	HO—CH_3	377
H_2C=CH_2	699	Cl—CH_3	452	HC≡CH	962
		Br—CH_3	293	I—CH_3	234

Some bond enthalpies are given in Table 3.4. Note that the nitrogen–nitrogen bond in molecular nitrogen, N_2, is very strong, at 945 kJ mol^{-1}, which helps to account for the chemical inertness of nitrogen and its ability to dilute the oxygen in the atmosphere without reacting with it. In contrast, the fluorine–fluorine bond in molecular fluorine, F_2, is relatively weak, at 155 kJ mol^{-1}; the weakness of this bond contributes to the high reactivity of elemental fluorine. However, bond enthalpies alone do not account for reactivity because, although the bond in molecular iodine is even weaker, I_2 is less reactive than F_2,

Table 3.5
*Mean bond enthalpies, $\Delta H_B/(\text{kJ mol}^{-1})$**

	H	C	N	O	F	Cl	Br	I	S	P	Si
H	436										
C	412	348 (1) 612 (2) 838 (3) 518 (a)									
N	388	305 (1) 613 (2) 890 (3)	163 (1) 409 (2) 945 (3)								
O	463	360 (1) 743 (2)	157	146 (1) 497 (2)							
F	565	484	270	185	155						
Cl	431	338	200	203	254	242					
Br	366	276				219	193				
I	299	238				210	178	151			
S	338	259			496	250	212		264		
P	322									200	
Si	318		374	466							226

*Values are for single bonds except where otherwise stated (in parentheses). (a) Denotes aromatic.

and the bond in CO is stronger than the bond in N_2, but CO forms many carbonyl compounds, such as $Ni(CO)_4$. The types and strengths of the bonds that the elements can make to other elements are additional factors.

A complication when dealing with bond enthalpies is that their values depend on the molecule in which the two linked atoms occur. For instance, the total standard enthalpy change for the atomization (the complete dissociation into atoms) of water

$$H_2O(g) \rightarrow 2\,H(g) + O(g) \qquad \Delta H^{\ominus} = +927\text{ kJ}$$

is not twice the O—H bond enthalpy in H_2O even though two O—H bonds are dissociated. There are in fact two different dissociation steps. In the first step, an O—H bond is broken in an H_2O molecule:

$$H_2O(g) \rightarrow HO(g) + H(g) \qquad \Delta H^{\ominus} = +499\text{ kJ}$$

In the second step, the O—H bond is broken in an OH radical:

$$HO(g) \rightarrow H(g) + O(g) \qquad \Delta H^{\ominus} = +428\text{ kJ}$$

The sum of the two steps is the atomization of the molecule. As can be seen from this example, the O—H bonds in H_2O and HO have similar but not identical bond enthalpies.

Although accurate calculations must use bond enthalpies for the molecule in question and its successive fragments, when such data are not available there is no choice but to make estimates by using **mean bond enthalpies,** ΔH_B, which are the averages of bond enthalpies over a related series of compounds (Table 3.5). For example, the mean HO bond enthalpy, $\Delta H_B(\text{H—O}) = 463\text{ kJ mol}^{-1}$, is the mean of the O—H bond enthalpies in H_2O and several other similar compounds, including methanol, CH_3OH.

Example 3.3

Using mean bond enthalpies

Estimate the standard enthalpy change for the reaction

$$C(s, \text{graphite}) + 2\,H_2(g) + \tfrac{1}{2}\,O_2(g) \rightarrow CH_3OH(l)$$

in which liquid methanol is formed from its elements at 25 °C. Use information from the *Data section* and bond enthalpy data from Tables 3.4 and 3.5.

Strategy In calculations of this kind, the procedure is to break the overall process down into a sequence of steps such that their sum is the chemical equation required. Always ensure, when using bond enthalpies, that all the species are in the gas phase. That may mean including the appropriate enthalpies of vaporization or sublimation. One approach is to atomize all the reactants and then to build the products from the atoms so produced. When explicit bond enthalpies are available (that is, data are given in the tables available), use them; otherwise, use mean bond enthalpies to obtain estimates. It is often helpful to display the enthalpy changes diagrammatically.

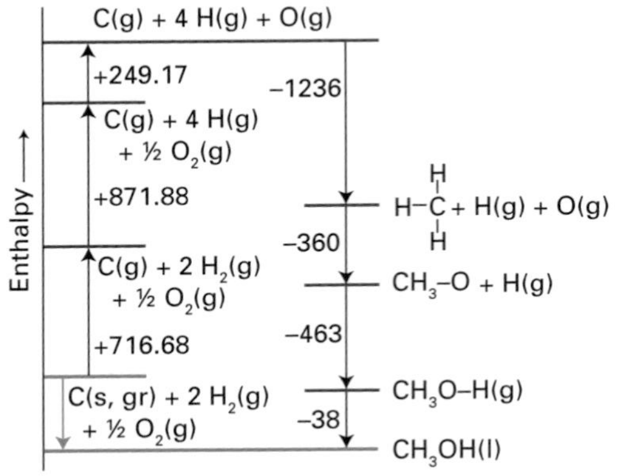

Fig. 3.7 The enthalpy changes used to estimate the enthalpy change accompanying the formation of liquid methanol from its elements. The bond enthalpies are mean values, so the final value is only approximate.

Solution The following steps are required (Fig. 3.7):

		$\Delta H^{\ominus}$/ kJ
Atomization of graphite:	C(s, graphite) → C(g)	+716.68
Dissociation of 2 mol $H_2(g)$:	$2\,H_2(g) \rightarrow 4\,H(g)$	+871.88
Dissociation of $^1/_2$ $O_2(g)$:	$^1/_2\,O_2(g) \rightarrow O(g)$	+249.17
Overall, so far:	$C(s) + 2\,H_2(g) + {}^1/_2\,O_2(g) \rightarrow C(g) + 4\,H(g) + O(g)$	+1837.73

These values are accurate. In the second step, three CH bonds, one CO bond, and one OH bond are formed, and we estimate their enthalpies from mean values. The standard enthalpy change for bond formation (the reverse of dissociation) is the negative of the mean bond enthalpy (obtained from Table 3.5):

	$\Delta H^{\ominus}$/kJ
Formation of 3 C—H bonds:	−1236
Formation of 1 C—O bond:	−360
Formation of 1 O—H bond:	−463
Overall, in this step: $C(g) + 4\,H(g) + O(g) \rightarrow CH_3OH(g)$	−2059

These values are estimates. The final stage of the reaction is the condensation of methanol vapour:

$$CH_3OH(g) \rightarrow CH_3OH(l) \qquad \Delta H^{\ominus} = -38.00\ \text{kJ}$$

The sum of the enthalpy changes is

$$\Delta H^{\ominus} = (+1837.73\ \text{kJ}) + (-2059\ \text{kJ}) + (-38.00\ \text{kJ}) = -259\ \text{kJ}$$

The experimental value is −239 kJ.

Self-test 3.4

Estimate the enthalpy change for the combustion of liquid ethanol to carbon dioxide and liquid water under standard conditions by using the bond enthalpies, mean bond enthalpies, and the appropriate standard enthalpies of vaporization.

Answer: −1348 kJ; the experimental value is −1368 kJ

Chemical change

In the remainder of this chapter we concentrate on enthalpy changes accompanying chemical reactions, such as the hydrogenation of ethene:

$$CH_2{=}CH_2(g) + H_2(g) \rightarrow CH_3CH_3(g) \qquad \Delta H^{\ominus} = -137\ \text{kJ}$$

The value of $\Delta H^{\ominus}$ given here signifies that the enthalpy of the system decreases by 137 kJ (and, if the reaction takes place at constant pressure, that 137 kJ of energy is released by heating the surroundings) when 1 mol $CH_2{=}CH_2(g)$ at 1 bar combines with 1 mol $H_2(g)$ at 1 bar to give 1 mol $CH_3CH_3(g)$ at 1 bar, all at 25 °C.

3.3 Enthalpies of combustion

One commonly encountered reaction is **combustion**, the complete reaction of a compound, most commonly an organic compound, with oxygen, as in the combustion of methane in a natural gas flame:

$$CH_4(g) + 2\,O_2(g) \rightarrow CO_2(g) + 2\,H_2O(l) \qquad \Delta H^{\ominus} = -890\ \text{kJ}$$

By convention, combustion of an organic compound results in the formation of carbon dioxide gas, liquid water, and—if the compound contains nitrogen—nitrogen gas. The **standard enthalpy of combustion**, $\Delta_c H^{\ominus}$, is the change in standard enthalpy per mole of combustible substance. In this example, we would write $\Delta_c H^{\ominus}(CH_4,g) = -890\ \text{kJ mol}^{-1}$. Some typical values are given in Table 3.6. Note that $\Delta_c H^{\ominus}$ is a molar quantity, and is obtained from the value of $\Delta H^{\ominus}$ by dividing by the amount of organic reactant consumed (in this case, by 1 mol CH_4). We see in Impact 3.2 that the enthalpy of combustion is a useful measure of the efficiency of fuels. It also determines the utility of foods, such as carbohydrates, fats, and proteins, as sources of energy for organisms (see Impact 3.3).

Enthalpies of combustion are commonly measured by using a **bomb calorimeter**, a device in which energy is transferred as heat at constant volume. According to the discussion in Section 2.8 and the relation $\Delta U = q_V$, the energy transferred as heat at constant volume is equal to the change in internal energy, ΔU, not ΔH. To convert from ΔU, to ΔH we need to note that the molar enthalpy of a substance is related to its molar internal energy by $H_m = U_m + pV_m$ (eqn 2.12a). As explained in Section 2.9, for condensed phases, but not for gases, pV_m is so small it may be ignored. For gases treated as perfect, pV_m

Table 3.6
Standard enthalpies of combustion

Substance	$\Delta_c H^\ominus/(\text{kJ mol}^{-1})$
Benzene, $C_6H_6(l)$	−3268
Carbon monoxide, $CO(g)$	−283
Carbon, C(s,graphite)	−394
Ethanol, $C_2H_5OH(l)$	−1368
Ethyne, $C_2H_2(g)$	−1300
Glucose, $C_6H_{12}O_6(s)$	−2808
Hydrogen, $H_2(g)$	−286
Iso-octane*, $C_8H_{18}(l)$	−5461
Methane, $CH_4(g)$	−890
Methanol, $CH_3OH(l)$	−726
Methylbenzene, $C_6H_5CH_3(l)$	−3910
Octane, $C_8H_{18}(l)$	−5471
Propane, $C_3H_8(g)$	−2220
Sucrose, $C_{12}H_{22}O_{11}(s)$	−5645
Urea, $CO(NH_2)_2(s)$	−632

*2,2,4-Trimethylpentane.

may be replaced by RT. Therefore, if in the chemical equation the difference (products − reactants) in the stoichiometric coefficients of *gas phase* species is $\Delta\nu_{\text{gas}}$, we can write

$$\Delta_c H = \Delta_c U + \Delta\nu_{\text{gas}} RT \qquad (3.3)$$

Relation between ΔH and ΔU

Note that $\Delta\nu_{\text{gas}}$ (where ν is nu) is a dimensionless quantity. This expression applies to any reaction that involves gases, not just to combustions (see Section 3.4).

Brief illustration 3.2 The difference between ΔH and ΔU for a reaction involving gases

The energy released as heat when glycine is burned in a bomb calorimeter is 969.6 kJ mol^{-1} at 298.15 K, so $\Delta_c U = -969.6$ kJ mol^{-1}. From the chemical equation

$$NH_2CH_2COOH(s) + \tfrac{9}{4}\,O_2(g) \rightarrow 2\,CO_2(g) + \tfrac{5}{2}\,H_2O(l) + \tfrac{1}{2}\,N_2(g)$$

we find that $\Delta\nu_{\text{gas}} = (2 + \tfrac{1}{2}) - \tfrac{9}{4} = \tfrac{1}{4}$. Therefore,

$$\begin{aligned}\Delta_c H - \Delta_c U &= \tfrac{1}{4}RT \\ &= \tfrac{1}{4} \times (8.3145 \times 10^{-3}\ \text{J K}^{-1}\ \text{mol}^{-1}) \times (298.15\ \text{K}) \\ &= 0.62\ \text{kJ mol}^{-1}\end{aligned}$$

This difference, though small, is significant.

Self-test 3.5

Determine the difference $\Delta_c H - \Delta_c U$ for the reaction $N_2(g) + 3\,H_2(g) \rightarrow 2\,NH_3(g)$ at 500 °C.

Answer: −13 kJ mol^{-1}

Impact on technology 3.2

Fuels

We shall see in Chapter 4 that the best assessment of the ability of a compound to act as a fuel to drive many of the processes occurring in the body makes use of the 'Gibbs energy'. However, a useful guide to the resources provided by a fuel, and the only one that matters when its heat output is being considered, is the enthalpy, particularly the enthalpy of combustion. The thermochemical properties of fuels and foods are commonly discussed in terms of their **specific enthalpy**, the enthalpy of combustion divided by the mass of material (typically in kilojoules per gram), or the **enthalpy density**, the magnitude of the enthalpy of combustion divided by the volume of material (typically in kilojoules per cubic decimetre or kilojoules per litre). Thus, if the standard enthalpy of combustion is $\Delta_c H^\ominus$ and the molar mass of the compound is M, then the specific enthalpy is $\Delta_c H^\ominus/M$. Similarly, the enthalpy density is $\Delta_c H^\ominus/V_m$, where V_m is the molar volume of the material under the same conditions of pressure and temperature.

Table 3.7 lists the specific enthalpies and enthalpy densities of several fuels. The most suitable fuels may be those with high specific enthalpies, as the advantage of a high molar enthalpy of combustion may be eliminated if a large mass of fuel is to be transported. We see that H_2 gas compares very well with more traditional fuels such as methane (natural gas), iso-octane (gasoline), and methanol. Furthermore, the combustion of H_2 gas does not generate CO_2. As a result, H_2 gas has been proposed as an efficient, clean alternative to fossil fuels, such as natural gas and petroleum. However, we also see that H_2 gas has a very low enthalpy density, which arises from the fact that hydrogen is a very light gas. So, the advantage of a high specific enthalpy is undermined by the large volume of fuel to be transported and stored. Strategies are being developed to solve the storage problem. For example, the small H_2 molecules can travel through holes in the crystalline lattice of a sample of metal, such as titanium, where they bind as metal hydrides. In this way it is possible to increase the effective density of hydrogen atoms to a value that is higher than that of liquid H_2. The fuel can be released on demand by heating the metal.

We now assess the factors that optimize the heat output of carbon-based fuels. Let's consider the combustion of 1 mol $CH_4(g)$, the main constituent of natural gas. From the thermochemical equation, we see that the combustion of 1 mol $CH_4(g)$ releases 890 kJ of energy as heat. Now consider the combustion of 1 mol $CH_3OH(g)$:

$$CH_3OH(g) + \tfrac{3}{2}\,O_2(g) \rightarrow CO_2(g) + 2\,H_2O(l) \qquad \Delta H^\ominus = -764\ \text{kJ}$$

This reaction is also exothermic, but now only 764 kJ of energy is released as heat per mole of molecules. The replacement of a C—H bond by a C—O bond renders the carbon in methanol more highly oxidized than the carbon in methane, so it is reasonable to expect that less energy is released to complete the oxidation of carbon to CO_2 in methanol.

Table 3.7
Thermochemical properties of some fuels

Fuel	Combustion equation	$\Delta_c H^{\ominus}$/(kJ mol^{-1})	Specific enthalpy/(kJ g^{-1})	Enthalpy density*/(kJ dm^{-3})
Hydrogen	$2\,H_2(g) + O_2(g) \rightarrow 2\,H_2O(l)$	−286	142	13
Methane	$CH_4(g) + 2\,O_2(g) \rightarrow CO_2(g) + 2\,H_2O(l)$	−890	55	40
Iso-octane†	$2\,C_8H_{18}(l) + 25\,O_2(g) \rightarrow 16\,CO_2(g) + 18\,H_2O(l)$	−5461	48	3.3×10^4
Methanol	$2\,CH_3OH(l) + 3\,O_2(g) \rightarrow 2\,CO_2(g) + 4\,H_2O(l)$	−726	23	1.8×10^4

*At atmospheric pressures and room temperature.
†2,2,4-Trimethylpentane.

Another factor that determines the heat output of combustion reactions is the number of carbon atoms in hydrocarbon compounds. For example, from the value of the standard enthalpy of combustion for methane we know that for each mole of CH_4 supplied to a furnace, 890 kJ of heat can be released, whereas for each mole of iso-octane molecules (C_8H_{18}, 2,2,4-trimethylpentane, a typical component of gasoline) supplied to an internal combustion engine, 5461 kJ of heat is released. The much larger value for iso-octane is a consequence of each molecule having eight C atoms to contribute to the formation of carbon dioxide whereas methane has only one.

Impact on biochemistry 3.3

Food and energy reserves

A typical 18–20 year old man requires a daily input of about 12 MJ (1 MJ = 10^6 J); a woman of the same age needs about 9 MJ. If the entire consumption were in the form of glucose, which has a specific enthalpy of 16 kJ g^{-1}, meeting energy needs would require the consumption of 750 g of glucose by a man and 560 g by a woman. In fact, the complex carbohydrates (polymers of carbohydrate units, such as starch) more commonly found in our diets have slightly higher specific enthalpies (17 kJ g^{-1}) than glucose itself, so a carbohydrate diet is slightly less daunting than a pure glucose diet, as well as being more appropriate in the form of fibre, the indigestible cellulose that helps move digestion products through the intestine.

The specific enthalpy of fats, which are long-chain esters like tristearin (beef fat), is much greater than that of carbohydrates, at around 38 kJ g^{-1}, slightly less than the value for the hydrocarbon oils used as fuel (48 kJ g^{-1}). The reason for this difference lies in the fact that many of the carbon atoms in carbohydrates are bonded to oxygen atoms and are already partially oxidized whereas most of the carbon atoms in fats are bonded to hydrogen and other carbon atoms and hence have lower oxidation numbers. As we saw above, the presence of partially oxidized carbons lowers the heat output of a fuel.

Fats are commonly used as an energy store, to be used only when the more readily accessible carbohydrates have fallen into short supply. In Arctic species, the stored fat also acts as a layer of insulation; in desert species (such as the camel), the fat is also a source of water, one of its oxidation products.

Proteins are also used as a source of energy, but their components, the amino acids, are often too valuable to squander in this way, and are used to construct other proteins instead. When proteins are oxidized (to urea, $CO(NH_2)_2$), the equivalent enthalpy density is comparable to that of carbohydrates.

We have already remarked that not all the energy released by the oxidation of foods is converted to work. The heat that is also released needs to be discarded in order to maintain body temperature within its typical range of 35.6–37.8 °C. A variety of mechanisms contribute to this aspect of homeostasis, the ability of an organism to counteract environmental changes with physiological responses. The general uniformity of temperature throughout the body is maintained largely by the flow of blood. When heat needs to be dissipated rapidly, warm blood is allowed to flow through the capillaries of the skin, so producing flushing. Radiation is one means of discarding heat; another is evaporation and the energy demands of the enthalpy of vaporization of water. Evaporation removes about 2.4 kJ per gram of water perspired. When vigorous exercise promotes sweating (through the influence of heat selectors on the hypothalamus), 1–2 dm^3 of perspired water can be produced per hour, corresponding to a heat loss of 2.4–5.0 MJ h^{-1}.

3.4 The combination of reaction enthalpies

The **reaction enthalpy** (or the 'enthalpy of reaction'), $\Delta_r H$, is the change in enthalpy that accompanies a chemical reaction: the enthalpy of combustion is just a special case. The reaction enthalpy is the difference between the molar enthalpies of the reactants and the products, with each term weighted by the stoichiometric coefficient, ν (nu), in the chemical equation

$$\Delta_r H = \Sigma \nu H_m(\text{products}) - \Sigma \nu H_m(\text{reactants}) \quad \text{Reaction enthalpy} \quad (3.4a)$$

where the Σ (uppercase sigma) are instructions to form the sum of the following terms. The **standard reaction enthalpy** (the 'standard enthalpy of reaction') $\Delta_r H^\ominus$, is the value of the reaction enthalpy when all the reactants and products are in their standard states:

$$\Delta_r H^\ominus = \Sigma \nu H^\ominus_m(\text{products}) - \Sigma \nu H^\ominus_m(\text{reactants}) \quad \text{Standard reaction enthalpy} \quad (3.4b)$$

Because the $H^\ominus_m$ are molar quantities and the stoichiometric coefficients are pure numbers, the units of $\Delta_r H^\ominus$ are kilojoules per mole. The standard reaction enthalpy is the change in enthalpy of the system when the reactants in their standard states (pure, 1 bar) are completely converted into products in their standard states (pure, 1 bar), with the change expressed in kilojoules per mole of reaction as written. Thus, if for the reaction $2\ H_2(g) + O_2(g) \rightarrow 2\ H_2O(l)$ we report that $\Delta_r H^\ominus = -572\ \text{kJ mol}^{-1}$, then the 'per mole' means that the reaction releases 572 kJ of heat per mole of O_2 consumed or per 2 mol H_2O formed (and therefore 286 kJ per mole of H_2O formed).

It is often the case that a reaction enthalpy is needed but is not available in tables of data. Now the fact that enthalpy is a state function comes in handy, because it implies that we can construct the required reaction enthalpy from the reaction enthalpies of known reactions. We have already seen a primitive example when we calculated the enthalpy of sublimation from the sum of the enthalpies of fusion and vaporization. The only difference is that we now apply the technique to a sequence of chemical reactions. The procedure is summarized by **Hess's law**:

> The standard enthalpy of a reaction is the sum of the standard enthalpies of the reactions into which the overall reaction may be divided.

Although the procedure is given the status of a law, it hardly deserves the title because it is nothing more than a consequence of enthalpy being a state function, which implies that an overall enthalpy change can be expressed as a sum of enthalpy changes for each step in an indirect path. The individual steps need not be actual reactions that can be carried out in the laboratory—they may be entirely hypothetical reactions, the only requirement being that their equations should balance. Each step must correspond to the same temperature.

Example 3.4

Using Hess's law

Given the thermochemical equations

$C_3H_6(g) + H_2(g) \rightarrow C_3H_8(g) \qquad \Delta H^\ominus = -124\ \text{kJ}$

$C_3H_8(g) + 5\ O_2(g) \rightarrow 3\ CO_2(g) + 4\ H_2O(l) \qquad \Delta H^\ominus = -2220\ \text{kJ}$

where C_3H_6 is propene and C_3H_8 is propane, calculate the standard enthalpy of combustion of propene.

Strategy We need to add or subtract the thermochemical equations, together with any others that are needed (from the *Data section*), so as to reproduce the thermochemical equation for the reaction required. In calculations of this type, it is often necessary to use the synthesis of water to balance the hydrogen or oxygen atoms in the overall equation. Once again, it may be helpful to express the changes diagrammatically.

Solution The overall reaction is

$C_3H_6(g) + \tfrac{9}{2}\ O_2(g) \rightarrow 3\ CO_2(g) + 3\ H_2O(l) \qquad \Delta H^\ominus$

We can recreate this thermochemical equation from the following sum (Fig. 3.8):

	$\Delta H^\ominus$/kJ
$C_3H_6(g) + H_2(g) \rightarrow C_3H_8(g)$	−124
$C_3H_8(g) + 5\ O_2(g) \rightarrow 3\ CO_2(g) + 4\ H_2O(l)$	−2220
$H_2O(l) \rightarrow H_2(g) + \tfrac{1}{2}\ O_2(g)$	+286
Overall: $C_3H_6(g) + \tfrac{9}{2}\ O_2(g) \rightarrow 3\ CO_2(g) + 3\ H_2O(l)$	−2058

It follows that the standard enthalpy of combustion of propene is $-2058\ \text{kJ mol}^{-1}$.

Self-test 3.6

Calculate the standard enthalpy of the reaction $C_6H_6(l) + 3\ H_2(g) \rightarrow C_6H_{12}(l)$ from the standard enthalpies of combustion of benzene (Table 3.6) and cyclohexane ($-3930\ \text{kJ mol}^{-1}$).

Answer: −196 kJ

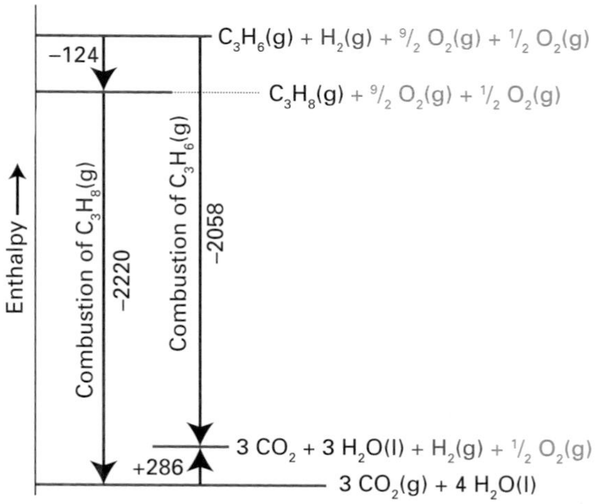

Fig. 3.8 The enthalpy changes used in Example 3.4 to illustrate Hess's law.

3.5 Standard enthalpies of formation

The problem with eqn 3.4 is that we have no way of knowing the absolute enthalpies of the substances. To avoid this problem, we can imagine the reaction as taking place by an indirect route, in which the reactants are first broken down into the elements and then the products are formed from the elements (Fig. 3.9). Specifically, the **standard enthalpy of formation**, $\Delta_f H^{\ominus}$, of a substance is the standard enthalpy (per mole of the substance) for its formation from its elements in their reference states. The **reference state** of an element is its most stable form under the prevailing conditions (Table 3.8). Don't confuse 'reference state' with 'standard state': the reference state of carbon at 25 °C is graphite; the standard state of carbon is any specified phase of the element at 1 bar. For example, the standard enthalpy of formation of liquid water (at 25 °C, as always in this text) is obtained from the thermochemical equation

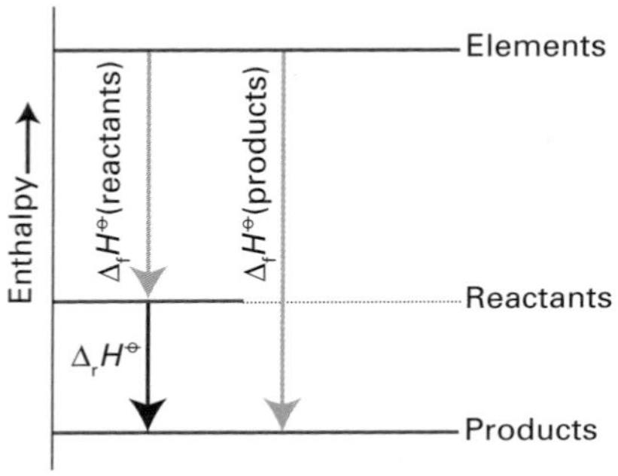

Fig. 3.9 An enthalpy of reaction may be expressed as the difference between the enthalpies of formation of the products and the reactants.

Table 3.8
Reference states of some elements

Element	Reference state
Arsenic	grey arsenic
Bromine	liquid
Carbon	graphite
Hydrogen	gas
Iodine	solid
Mercury	liquid
Nitrogen	gas
Oxygen	gas
Phosphorus	white phosphorus
Sulfur	rhombic sulfur
Tin	white tin, α-tin

$$H_2(g) + \tfrac{1}{2} O_2(g) \rightarrow H_2O(l) \qquad \Delta H^{\ominus} = -286 \text{ kJ}$$

and is $\Delta_f H^{\ominus}(H_2O,l) = -286 \text{ kJ mol}^{-1}$. Note that enthalpies of formation are molar quantities, so to go from $\Delta H^{\ominus}$ in a thermochemical equation to $\Delta_f H^{\ominus}$ for that substance, divide by the amount of substance formed (in this instance, by 1 mol H_2O).

With the introduction of standard enthalpies of formation, we can write

$$\Delta_r H^{\ominus} = \sum \nu \Delta_f H^{\ominus}(\text{products}) - \sum \nu \Delta_f H^{\ominus}(\text{reactants}) \quad (3.5)$$

Practical implementation Standard reaction enthalpy

The first term on the right is the enthalpy of formation of all the products from their elements; the second term on the right is the enthalpy of formation of all the reactants from their elements. The fact that the enthalpy is a state function means that a reaction enthalpy calculated in this way is identical to the value that would be calculated from eqn 3.4 if absolute enthalpies were available.

The values of some standard enthalpies of formation at 25 °C are given in Table 3.9, and a longer list is given in the *Data section*. Most are obtained by calorimetric measurements, but the computational methods described in Chapter 14 are now reasonably reliable and can be used to estimate data when experimental information is unavailable. The standard enthalpies of formation of elements in their reference states are zero by definition (because their formation is the null reaction: element → element). Note, however, that the standard enthalpy of formation of an element in a state other than its reference state is not zero:

$$C(s, \text{graphite}) \rightarrow C(s, \text{diamond}) \qquad \Delta H^{\ominus} = +1.895 \text{ kJ}$$

Therefore, although $\Delta_f H^{\ominus}(C, \text{graphite}) = 0$, $\Delta_f H^{\ominus}(C, \text{diamond}) = +1.895 \text{ kJ mol}^{-1}$.

Example 3.5

Using standard enthalpies of formation

Calculate the standard enthalpy of combustion of liquid benzene from the standard enthalpies of formation of the reactants and products.

Strategy Write the chemical equation, identify the stoichiometric numbers of the reactants and products, and then use eqn 3.5. Note that the expression has the form 'products − reactants'. Numerical values of standard enthalpies of formation are given in the *Data section*. The standard enthalpy of combustion is the enthalpy change per mole of substance, so we need to interpret the enthalpy change accordingly.

Table 3.9

*Standard enthalpies of formation at 298.15 K**

Substance	$\Delta_f H^\ominus$/(kJ mol^{-1})
Inorganic compounds	
Ammonia, NH_3(g)	−46.11
Ammonium nitrate, NH_4NO_3(s)	−365.56
Carbon monoxide, CO(g)	−110.53
Carbon disulfide, CS_2(l)	+89.70
Carbon dioxide, CO_2(g)	−393.51
Dinitrogen tetroxide, N_2O_4(g)	+9.16
Dinitrogen monoxide, N_2O(g)	+82.05
Hydrogen chloride, HCl(g)	−92.31
Hydrogen fluoride, HF(g)	−271.1
Hydrogen sulfide, H_2S(g)	−20.63
Nitric acid, HNO_3(l)	−174.10
Nitrogen dioxide, NO_2(g)	+33.18
Nitrogen monoxide, NO(g)	+90.25
Sodium chloride, NaCl(s)	−411.15
Sulfur dioxide, SO_2(g)	−296.83
Sulfur trioxide, SO_3(g)	−395.72
Sulfuric acid, H_2SO_4(l)	−813.99
Water, H_2O(l)	−285.83
H_2O(g)	−241.82
Organic compounds	
Benzene, C_6H_6(l)	+49.0
Ethane, C_2H_6(g)	−84.68
Ethanol, C_2H_5OH(l)	−277.69
Ethene, C_2H_4(g)	+52.26
Ethyne, C_2H_2(g)	+226.73
Glucose, $C_6H_{12}O_6$(s)	−1268
Methane, CH_4(g)	−74.81
Methanol, CH_3OH(l)	−238.86
Sucrose, $C_{12}H_{22}O_{11}$(s)	−2222

*A longer list is given in the *Data section* at the end of the book. The text's website also contains links to additional data.

Solution The chemical equation is

$$C_6H_6(l) + {}^{15}\!/_2\, O_2(g) \rightarrow 6\, CO_2(g) + 3\, H_2O(l)$$

It follows that

$$\begin{aligned}\Delta_r H^\ominus &= \{6\Delta_f H^\ominus(CO_2,g) + 3\Delta_f H^\ominus(H_2O,l)\} - \{\Delta_f H^\ominus(C_6H_6,l) + {}^{15}\!/_2\Delta_f H^\ominus(O_2,g)\}\\ &= \{6 \times (-393.51\ \text{kJ mol}^{-1}) + 3 \times (-285.83\ \text{kJ mol}^{-1})\} - \{(49.0\ \text{kJ mol}^{-1}) + 0\}\\ &= -3268\ \text{kJ mol}^{-1}\end{aligned}$$

Inspection of the chemical equation shows that, in this instance, the 'per mole' is per mole of C_6H_6, which is exactly what we need for an enthalpy of combustion. It follows that the standard enthalpy of combustion of liquid benzene is −3268 kJ mol^{-1}.

A note on good practice The standard enthalpy of formation of an element in its reference state (oxygen gas in this example) is written 0 not 0 kJ mol^{-1}, because it is zero whatever units we happen to be using.

Self-test 3.7

Use standard enthalpies of formation to calculate the enthalpy of combustion of propane gas to carbon dioxide and water vapour.

Answer: −2220 kJ mol^{-1}

The reference states of the elements define a thermochemical 'sea level', and enthalpies of formation can be regarded as thermochemical 'altitudes' above or below sea level (Fig. 3.10). It is sometimes useful to classify compounds according to their standard enthalpies of formation:

- **Exothermic compounds** ($\Delta_f H^\ominus < 0$) have a lower enthalpy than their component elements.
- **Endothermic compounds** ($\Delta_f H^\ominus > 0$) have a higher enthalpy than their component elements.

Water is an exothermic compound; carbon disulfide is an endothermic compound.

3.6 The variation of reaction enthalpy with temperature

It often happens that you have access to data at one temperature but need it at another temperature. For example, you might want to know the enthalpy of a particular reaction at body temperature, 37 °C, but

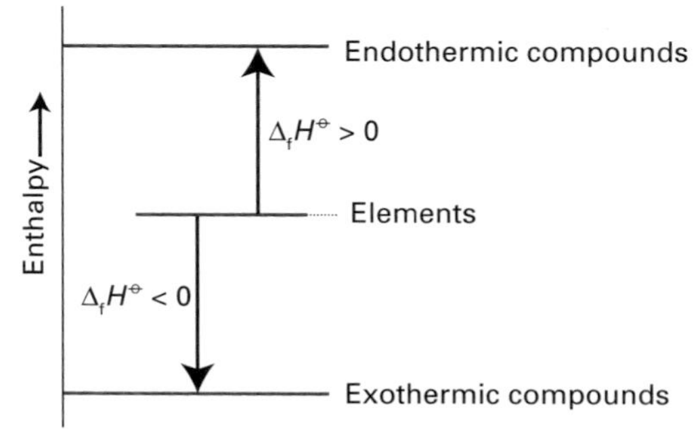

Fig. 3.10 The enthalpy of formation acts as a kind of thermochemical 'altitude' of a compound with respect to the 'sea level' defined by the elements from which it is made. Endothermic compounds have positive enthalpies of formation; exothermic compounds have negative enthalpies of formation.

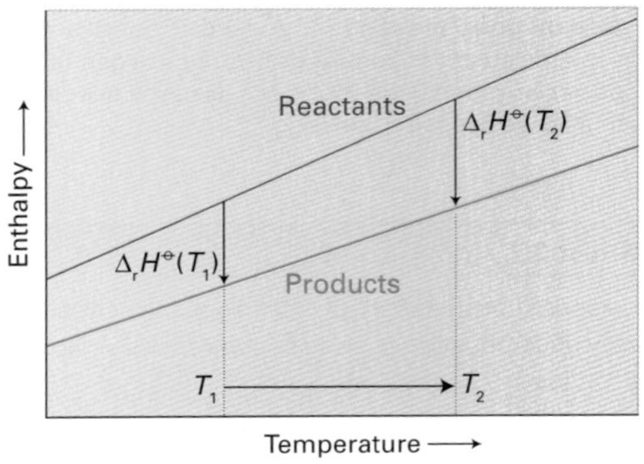

Fig. 3.11 The enthalpy of a substance increases with temperature. Therefore, if the total enthalpy of the reactants increases by a different amount from that of the products, the reaction enthalpy will change with temperature. The change in reaction enthalpy depends on the relative slopes of the two lines and hence on the heat capacities of the substances.

may have data available for 25 °C. Another type of question that might arise might be whether the oxidation of glucose is more exothermic when it takes place inside an Arctic fish that inhabits water at 0 °C than when it takes place at mammalian body temperatures. Similarly, you may be asked to predict whether the synthesis of ammonia is more exothermic at a typical industrial temperature of 450 °C than at 25 °C. In precise work, every attempt would be made to measure the reaction enthalpy at the temperature of interest, but it is useful to have a quick way of estimating the direction of change and even a moderately reliable numerical value and, perhaps more importantly, understand the origin of the change.

Figure 3.11 illustrates the technique we use. As we have seen, the enthalpy of a substance increases with temperature; therefore, the total enthalpy of the reactants and the total enthalpy of the products increase as shown in the illustration. Provided the two total enthalpy increases are different, the standard reaction enthalpy (their difference at a given temperature) will change as the temperature is changed. The change in the enthalpy of a substance depends on the slope of the graph and therefore on the constant-pressure heat capacities of the substances. We can therefore expect the temperature dependence of the reaction enthalpy to be related to the difference in heat capacities of the products and the reactants.

As a simple example, consider the reaction

$$2\,H_2(g) + O_2(g) \rightarrow 2\,H_2O(l)$$

where the standard enthalpy of reaction is known at one temperature (for example, at 25 °C from the tables in this book). According to eqn 3.4, we can write

$$\Delta_r H^{\ominus}(T) = 2H^{\ominus}_m(H_2O,l) - \{2H^{\ominus}_m(H_2,g) + H^{\ominus}_m(O_2,g)\}$$

for the reaction at a temperature T. If the reaction takes place at a higher temperature T', the molar enthalpy of each substance is increased because it stores more energy and the standard reaction enthalpy becomes

$$\Delta_r H^{\ominus}(T') = 2H^{\ominus\prime}_m(H_2O,l) - \{2H^{\ominus\prime}_m(H_2,g) + H^{\ominus\prime}_m(O_2,g)\}$$

where the primes signify the values at the new temperature. Equation 2.15 ($C_p = \Delta H/\Delta T$) implies that the increase in molar enthalpy of a substance when the temperature is changed from T to T' is $C^{\ominus}_{p,m} \times (T' - T)$, where $C^{\ominus}_{p,m}$ is the standard molar constant-pressure heat capacity of the substance, its molar heat capacity measured at 1 bar. For example, the molar enthalpy of water changes to

$$H^{\ominus\prime}_m(H_2O,l) = H^{\ominus}_m(H_2O,l) + C^{\ominus}_{p,m}(H_2O,l) \times (T' - T)$$

if $C^{\ominus}_{p,m}(H_2O,l)$ is constant over the temperature range. When we substitute terms like this into the expression above, we arrive at **Kirchhoff's law:**

$$\Delta_r H^{\ominus}(T') = \Delta_r H^{\ominus}(T) + \Delta_r C^{\ominus}_p (T' - T) \qquad \text{Kirchhoff's law} \qquad (3.6)$$

where

$$\Delta_r C^{\ominus}_p = 2C^{\ominus}_{p,m}(H_2O,l) - \{2C^{\ominus}_{p,m}(H_2,g) + C^{\ominus}_{p,m}(O_2,g)\}$$

Note that this combination has the same pattern as the reaction enthalpy and the stoichiometric numbers occur in the same way. In general, $\Delta_r C^{\ominus}_p$ is the difference between the weighted sums of the standard molar heat capacities of the products and the reactants:

$$\Delta_r C^{\ominus}_p = \sum \nu C^{\ominus}_{p,m}(\text{products}) - \sum \nu C^{\ominus}_{p,m}(\text{reactants}) \qquad (3.7)$$

We see that, just as we anticipated, the standard reaction enthalpy at one temperature can be calculated from the standard reaction enthalpy at another temperature provided we know the standard molar constant-pressure heat capacities of all the substances. These values are given in the *Data section*. The derivation of Kirchhoff's law supposes that the heat capacities are constant over the range of temperature of interest, so the law is best restricted to small temperature differences (of no more than 100 K or so).

Example 3.6

Using Kirchhoff's law

The standard enthalpy of formation of gaseous water at 25 °C is −241.82 kJ mol^{-1}. Estimate its value at 100 °C.

Strategy First, write the chemical equation and identify the stoichiometric coefficients. Then calculate the value of $\Delta_r C_p^\ominus$ from the data in the *Data section* by using eqn 3.7 and use the result in eqn 3.6.

Solution The chemical equation is

$$H_2(g) + \tfrac{1}{2} O_2(g) \rightarrow H_2O(g)$$

and the molar constant-pressure heat capacities of $H_2O(g)$, $H_2(g)$, and $O_2(g)$ are 33.58 J K^{-1} mol^{-1}, 28.84 J K^{-1} mol^{-1}, and 29.37 J K^{-1} mol^{-1}, respectively. It follows that

$$\begin{aligned}\Delta_r C_p^\ominus &= C_{p,m}^\ominus(H_2O,g) - \{C_{p,m}^\ominus(H_2,g) + \tfrac{1}{2}C_{p,m}^\ominus(O_2,g)\}\\ &= (33.58\ \text{J K}^{-1}\ \text{mol}^{-1}) - \{(28.84\ \text{J K}^{-1}\ \text{mol}^{-1})\\ &\quad + \tfrac{1}{2} \times (29.37\ \text{J K}^{-1}\ \text{mol}^{-1})\}\\ &= -9.95\ \text{J K}^{-1}\ \text{mol}^{-1} = -9.95 \times 10^{-3}\ \text{kJ K}^{-1}\ \text{mol}^{-1}\end{aligned}$$

Then, because $T' - T = +75$ K, from eqn 3.6 we find

$$\begin{aligned}\Delta_r H^\ominus(T') &= (-241.82\ \text{kJ mol}^{-1}) + (-9.95 \times 10^{-3}\ \text{kJ K}^{-1}\ \text{mol}^{-1})\\ &\quad \times (75\ \text{K})\\ &= (-241.82\ \text{kJ mol}^{-1}) - (0.75\ \text{kJ mol}^{-1})\\ &= -242.57\ \text{kJ mol}^{-1}\end{aligned}$$

The experimental value is −242.58 kJ mol^{-1}.

Self-test 3.8

Estimate the standard enthalpy of formation of $NH_3(g)$ at 400 K from the data in the *Data section*.

Answer: −48.4 kJ mol^{-1}

The calculation in Example 3.6 shows that the standard reaction enthalpy at 100 °C is only slightly different from that at 25 °C. The reason is that the change in reaction enthalpy is proportional to the *difference* between the molar heat capacities of the products and the reactants, which is usually not very large. It is generally the case that, provided the temperature range is not too wide, enthalpies of reactions vary only slightly with temperature. A reasonable first approximation is that standard reaction enthalpies are independent of temperature. When the temperature range is too wide for it to be safe to assume that heat capacities are constant, the empirical temperature dependence of each heat capacity given in eqn 3.6 may be used: the resulting expression is best developed by using mathematical software.

Checklist of key concepts

- ☐ **1** The standard state of a substance is the pure substance at 1 bar.
- ☐ **2** The standard enthalpy of transition, $\Delta_{trs}H^\ominus$, is the change in molar enthalpy when a substance in one phase changes into another phase, both phases being in their standard states.
- ☐ **3** The enthalpy of the reverse of a process is the negative of the enthalpy of the forward process under the same conditions, $\Delta_{reverse}H = -\Delta_{forward}H$.
- ☐ **4** The standard enthalpy of a process is the sum of the standard enthalpies of the individual processes into which it may be regarded as divided, as in $\Delta_{sub}H^\ominus = \Delta_{fus}H^\ominus + \Delta_{vap}H^\ominus$.
- ☐ **5** Hess's law states that the standard enthalpy of a reaction is the sum of the standard enthalpies of the reactions into which the overall reaction may be divided.
- ☐ **6** The standard enthalpy of formation of a compound, $\Delta_f H^\ominus$, is the standard reaction enthalpy for the formation of the compound from its elements in their reference states.
- ☐ **7** At constant pressure, exothermic compounds are those for which $\Delta_f H^\ominus < 0$; endothermic compounds are those for which $\Delta_f H^\ominus > 0$.

Road map of key equations

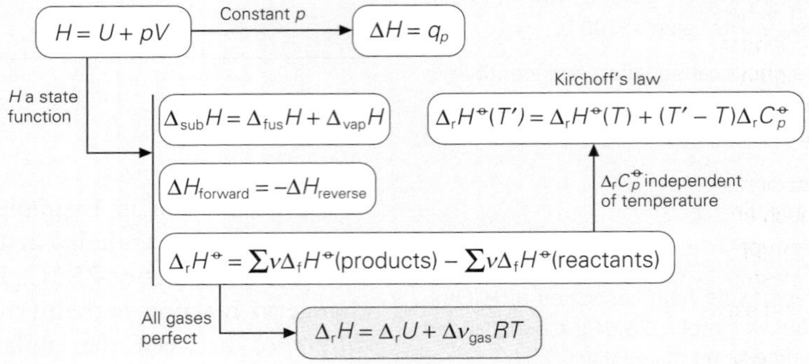

A blue box indicates that the expression applies to perfect gases.

Questions and exercises

Discussion questions

3.1 Define the terms (a) standard enthalpy of vaporization, (b) standard enthalpy of fusion, (c) standard enthalpy of sublimation, (d) standard enthalpy of ionization, (e) standard electron-gain enthalpy, (f) mean bond enthalpy, and identify an application for each one.

3.2 Define the terms (a) standard reaction enthalpy, (b) standard enthalpy of combustion, (c) standard enthalpy of formation, and identify an application for each one.

3.3 A primitive air-conditioning unit for use in places where electrical power is not available can be made by hanging up strips of linen soaked in water. Explain why this strategy is effective.

3.4 Why is it important to distinguish between the standard state and the reference state of an element?

3.5 Discuss the limitations of the expressions: (a) $\Delta_r H = \Delta_r U + \Delta \nu_{gas} RT$, (b) $\Delta_r H^\ominus(T') = \Delta_r H^\ominus(T) + \Delta_r C_p^\ominus \times (T' - T)$.

3.6 In the earlier literature, and still not uncommonly, you will find the expressions 'heat of combustion' and 'heat of vaporization'. Why are the thermodynamic expressions 'enthalpy of combustion' and 'enthalpy of vaporization' superior?

Exercises

Assume all gases are perfect unless stated otherwise. All thermochemical data are for 298.15 K.

3.1 Estimate the difference between the standard enthalpy of formation of CO_2(g) as currently defined (at 1 bar) and its value using the former definition (at 1 atm).

3.2 Liquid mixtures of sodium and potassium are used in some nuclear reactors as coolants that can survive the intense radiation inside reactor cores. Calculate the energy required as heat to melt 250 kg of sodium metal at 371 K.

3.3 Calculate the energy that must be transferred as heat to evaporate 1.00 kg of water at (a) 25 °C, (b) 100 °C.

3.4 2-Propanol (isopropanol) is commonly used as 'rubbing alcohol' to relieve sprain injuries in sport: its action is due to the cooling effect that accompanies its rapid evaporation when applied to the skin. In an experiment to determine the enthalpy of vaporization of 2-propanol, a sample was brought to the boil. It was then found that when an electric current of 0.812 A from a 11.5 V supply was passed for 303 s, then 4.27 g of the alcohol was vaporized. What is the (molar) enthalpy of vaporization of isopropanol at its boiling point?

3.5 Refrigerators make use of the heat absorption required to vaporize a volatile liquid. A fluorocarbon liquid being investigated to replace a chlorofluorocarbon has $\Delta_{vap}H^\ominus =$

+32.0 kJ mol^{-1}. Calculate q, w, ΔH, and ΔU when 2.50 mol is vaporized at 250 K and 750 Torr.

3.6 Use the information in Table 3.1 to calculate the energy that must be transferred as heat to melt 100 g of ice at 0 °C, increase the sample temperature to 100 °C, and then vaporize it at that temperature. Sketch a graph of temperature against time on the assumption that the sample is heated at a constant rate.

3.7 The enthalpy of sublimation of calcium at 25 °C is 178.2 kJ mol^{-1}. How much energy (at constant temperature and pressure) must be supplied as heat to 5.0 g of solid calcium to produce a plasma (a gas of charged particles) composed of Ca^{2+} ions and electrons?

3.8 Estimate the difference between the standard enthalpy of ionization of Ca(g) to Ca^{2+}(g) and the accompanying change in internal energy at 25 °C.

3.9 Estimate the difference between the standard electron-gain enthalpy of Br(g) and the corresponding change in internal energy at 25 °C.

3.10 How much energy (at constant temperature and pressure) must be supplied as heat to 10.0 g of chlorine gas (as Cl_2) to produce a plasma (a gas of charged particles, in this case ions) composed of Cl^- and Cl^+ ions? The enthalpy of ionization of Cl(g) is +1257.5 kJ mol^{-1} and its electron-gain enthalpy is −354.8 kJ mol^{-1}.

3.11 Use the data in the preceding exercise to identify (a) the standard enthalpy of ionization of Cl^-(g) and (b) the accompanying change in molar internal energy.

3.12 The enthalpy changes accompanying the dissociation of successive bonds in NH_3(g) are 460, 390, and 314 kJ mol^{-1}, respectively. (a) What is the mean enthalpy of an N—H bond? (b) Do you expect the mean bond internal energy to be larger or smaller than the mean bond enthalpy?

3.13 Use bond enthalpies and mean bond enthalpies to estimate (a) the enthalpy of the glycolysis reaction adopted by anaerobic bacteria as a source of energy, $C_6H_{12}O_6$(aq) → 2 $CH_3CH(OH)COOH$(aq), lactic acid, which is produced via the formation of pyruvic acid, $CH_3COCOOH$, and the action of lactate dehydrogenase and (b) the enthalpy of combustion of glucose. Ignore the contributions of enthalpies of fusion and vaporization.

3.14 Ethane is flamed off in abundance from oil wells, because it is unreactive and difficult to use commercially. But would it make a good fuel? The standard enthalpy of reaction for 2 C_2H_6(g) + 7 O_2(g) → 4 CO_2(g) + 6 H_2O(l) is −3120 kJ mol^{-1}. (a) What is the standard enthalpy of combustion of ethane? (b) What is the specific enthalpy of combustion of ethane? (c) Is ethane a more or less efficient fuel than methane?

3.15 Standard enthalpies of formation are widely available, but we might need a standard enthalpy of combustion instead. The standard enthalpy of formation of ethylbenzene is −12.5 kJ mol^{-1}. Calculate its standard enthalpy of combustion.

3.16 Combustion reactions are relatively easy to carry out and study, and their data can be combined to give enthalpies of other types of reaction. As an illustration, calculate the standard enthalpy of hydrogenation of cyclohexene to cyclohexane given that the standard enthalpies of combustion of the two compounds are −3752 kJ mol^{-1} (cyclohexene) and −3953 kJ mol^{-1} (cyclohexane).

3.17 The efficient design of chemical plants depends on the designer's ability to assess and use the heat output in one process to supply another process. The standard enthalpy of reaction for N_2(g) + 3 H_2(g) → 2 NH_3(g) is −92.22 kJ mol^{-1}. What is the change in enthalpy when (a) 1.00 t of N_2(g) is consumed, (b) 1.00 t of NH_3(g) is formed?

3.18 Estimate the standard internal energy of formation of liquid methyl acetate (methyl ethanoate, CH_3COOCH_3) at 298 K from its standard enthalpy of formation, which is −442 kJ mol^{-1}.

3.19 The standard enthalpy of combustion of anthracene is −7163 kJ mol^{-1}. Calculate its standard enthalpy of formation.

3.20 When 320 mg of naphthalene, $C_{10}H_8$(s), was burned in a bomb calorimeter, the temperature rose by 3.05 K. Calculate the heat capacity of the calorimeter. By how much will the temperature rise when 100 mg of phenol, C_6H_5OH(s), is burned in the calorimeter under the same conditions?

3.21 The energy resources of glucose are of major concern for the assessment of metabolic processes. When 0.3212 g of glucose was burned in a bomb calorimeter of heat capacity 641 J K^{-1} the temperature rose by 7.793 K. Calculate (a) the standard molar enthalpy of combustion, (b) the standard internal energy of combustion, and (c) the standard enthalpy of formation of glucose.

3.22 The complete combustion of fumaric acid in a bomb calorimeter released 1333 kJ mol^{-1} of HOOCCH=CHCOOH(s) at 298 K. Calculate (a) the internal energy of combustion, (b) the enthalpy of combustion, (c) the enthalpy of formation of lactic acid.

3.23 The mean bond enthalpies of the C–C, C—H, C–O, C=O, and O—H bonds are 348, 412, 360, 743, and 463 kJ mol^{-1}, respectively. The combustion of a fuel such as octane is exothermic because relatively weak bonds break to form relatively strong bonds. Use this information to justify why glucose has a lower specific enthalpy than the lipid decanoic acid ($C_{10}H_{20}O_2$) even though these compounds have similar molar masses.

3.24 Calculate the standard enthalpy of solution of AgI(s) in water from the standard enthalpies of formation of the solid and the aqueous ions.

3.25 The standard enthalpy of decomposition of the yellow complex NH_3SO_2 into NH_3 and SO_2 is +40 kJ mol^{-1}. Calculate the standard enthalpy of formation of NH_3SO_2.

3.26 Given that the enthalpy of combustion of graphite is −393.5 kJ mol^{-1} and that of diamond is −395.41 kJ mol^{-1}, calculate the standard enthalpy of the C(s, graphite) → C(s, diamond) transition.

3.27 The pressures deep within the Earth are much greater than those on the surface, and to make use of thermochemical data in geochemical assessments we need to take the differences into account. Use the information in the preceding exercise, together with the densities of graphite (2.250 g cm^{-3}) and diamond (3.510 g cm^{-3}), to calculate the internal energy of the transition when the sample is under a pressure of 150 kbar.

3.28 A typical human produces about 10 MJ of energy transferred as heat each day through metabolic activity. The main mechanism of heat loss is through the evaporation of water. (a) If a human body were an isolated system of mass 65 kg with the heat capacity of water, what temperature rise would the body experience? (b) Human bodies are actually open systems. What mass of water should be evaporated each day to maintain constant temperature?

3.29 Camping gas is typically propane. The standard enthalpy of combustion of propane gas is −2220 kJ mol^{-1} and the standard enthalpy of vaporization of the liquid is +15 kJ mol^{-1}. Calculate (a) the standard enthalpy and (b) the standard internal energy of combustion of the liquid.

3.30 Classify as endothermic or exothermic (a) a combustion reaction for which $\Delta_r H^\ominus = -2020$ kJ mol^{-1}, (b) a dissolution for which $\Delta H^\ominus = +4.0$ kJ mol^{-1}, (c) vaporization, (d) fusion, (e) sublimation.

3.31 Standard enthalpies of formation are of great usefulness, for they can be used to calculate the standard enthalpies of a very wide range of reactions of interest in chemistry, biology, geology, and industry. Use information in the *Data section* to calculate the standard enthalpies of the following reactions:

(a) $2\,NO_2(g) \rightarrow N_2O_4(g)$

(b) $NO_2(g) \rightarrow \frac{1}{2}\,N_2O_4(g)$

(c) $3\,NO_2(g) + H_2O(l) \rightarrow 2\,HNO_3(aq) + NO(g)$

(d) Cyclopropane(g) → propene(g)

(e) $HCl(aq) + NaOH(aq) \rightarrow NaCl(aq) + H_2O(l)$

3.32 Calculate the standard enthalpy of formation of N_2O_5 from the following data:

$2\,NO(g) + O_2(g) \rightarrow 2\,NO_2(g) \qquad \Delta_r H^\ominus = -114.1$ kJ mol^{-1}

$4\,NO_2(g) + O_2(g) \rightarrow 2\,N_2O_5(g) \qquad \Delta_r H^\ominus = -110.2$ kJ mol^{-1}

$N_2(g) + O_2(g) \rightarrow 2\,NO(g) \qquad \Delta_r H^\ominus = +180.5$ kJ mol^{-1}

3.33 Heat capacity data can be used to estimate the reaction enthalpy at one temperature from its value at another. Use the information in the *Data section* to predict the standard reaction enthalpy of $2\,NO_2(g) \rightarrow N_2O_4(g)$ at 100 °C from its value at 25 °C.

3.34 Estimate the enthalpy of vaporization of water at 100 °C from its value at 25 °C (44.01 kJ mol^{-1}) given the constant-pressure heat capacities of 75.29 J K^{-1} mol^{-1} and 33.58 J K^{-1} mol^{-1} for liquid and gas, respectively.

3.35 It is often useful to be able to anticipate, without doing a detailed calculation, whether an increase in temperature will result in a raising or a lowering of a reaction enthalpy. The constant-pressure molar heat capacity of a gas of linear molecules is approximately $\frac{7}{2}R$ whereas that of a gas of nonlinear molecules is approximately $4R$. Decide whether the standard enthalpies of the following reactions will increase or decrease with increasing temperature:

(a) $2\,H_2(g) + O_2(g) \rightarrow 2\,H_2O(g)$

(b) $N_2(g) + 3\,H_2(g) \rightarrow 2\,NH_3(g)$

(c) $CH_4(g) + 2\,O_2(g) \rightarrow CO_2(g) + 2\,H_2O(g)$

3.36 The molar heat capacity of liquid water is approximately $9R$. Decide whether the standard enthalpy of the reactions (a) and (c) in Exercise 3.35 will increase or decrease with a rise in temperature if the water is produced as a liquid.

Projects

The symbol ‡ indicates that calculus is required.

3.37‡ This exercise explores differential scanning calorimetry in more detail. (a) In many experimental thermograms, such as that shown in Fig. 3.5, the baseline on the left of the transition is at a different level from that on the right. Explain this observation. (b) You have at your disposal a sample of pure polymer P and a sample of P that has just been synthesized in a large chemical reactor and which might contain impurities. Describe how you would use differential scanning calorimetry to determine the mole percentage composition of P in the allegedly impure sample.

3.38‡ Here we explore Kirchhoff's law (eqn 3.6) in greater detail. (a) Derive a version of Kirchhoff's law for the temperature dependence of the internal energy of reaction. (b) The formulation of Kirchhoff's law given in eqn 3.6 is valid when the difference in heat capacities is independent of temperature over the temperature range of interest. Suppose instead that $\Delta_r C_p^\ominus = a + bT + c/T^2$. Derive a more accurate form of Kirchhoff's law in terms of the parameters a, b, and c. Start by expressing the change in the reaction enthalpy for an infinitesimal change in temperature as $dH = \Delta_r C_p dT$ and substitute the expression for the change in the heat capacities on reaction. Then integrate this expression between the two temperatures of interest.

3.39 In this project we explore the thermodynamics of carbohydrates as biological fuels. It is useful to know that glucose and fructose are simple sugars with the molecular formula $C_6H_{12}O_6$. Sucrose (cane sugar) is a complex sugar with molecular formula $C_{12}H_{22}O_{11}$ that consists of a glucose unit covalently bound to a fructose unit (a water molecule is eliminated as a result of the reaction between glucose and fructose to form sucrose). There are no dietary recommendations for consumption of carbohydrates. Some nutritionists recommend diets that are largely devoid of carbohydrates, with most of the energy needs being met by fats. However, the most common diets are those in which at least 65 per cent of our food calories come from carbohydrates. (a) A ¾-cup serving of pasta contains 40 g of carbohydrates. What percentage of the daily calorie requirement for a person on a 2200 Calorie diet (1 Cal = 1 kcal) does this serving represent? (b) The mass of a typical glucose tablet is 2.5 g. Calculate the energy released as heat when a glucose tablet is burned in air. (c) To what height could you climb on the energy a glucose tablet provides assuming 25 per cent of the energy is available for work? (d) Is the standard enthalpy of combustion of glucose likely to be higher or lower at blood temperature than at 25 °C? (e) Calculate the energy released as heat when a typical sugar cube of mass 1.5 g is burned in air. (f) To what height could you climb on the energy a table sugar cube provides assuming 25 per cent of the energy is available for work?

4

Thermodynamics: the Second Law

Entropy 91

4.1 The direction of spontaneous change 91

4.2 Entropy and the Second Law 92

4.3 The entropy change accompanying expansion 94

4.4 The entropy change accompanying heating 95

4.5 The entropy change accompanying a phase transition 97

4.6 Entropy changes in the surroundings 99

4.7 The molecular interpretation of entropy 100

4.8 Absolute entropies and the Third Law of thermodynamics 101

4.9 The molecular interpretation of Third-Law entropies 103

4.10 The standard reaction entropy 105

4.11 The spontaneity of chemical reactions 105

The Gibbs energy 106

4.12 Focussing on the system 106

4.13 Properties of the Gibbs energy 107

CHECKLIST OF KEY CONCEPTS 109

ROAD MAP OF KEY EQUATIONS 109

QUESTIONS AND EXERCISES 110

Some things happen; some things don't. A gas expands to fill the vessel it occupies; a gas that already fills a vessel does not suddenly contract into a smaller volume. A hot object cools to the temperature of its surroundings; a cool object does not suddenly become hotter than its surroundings. Hydrogen and oxygen combine explosively (once their ability to do so has been liberated by a spark) and form water; water left standing in oceans and lakes does not gradually decompose into hydrogen and oxygen. These everyday observations suggest that changes can be divided into two classes. A **spontaneous change** is a change that has a tendency to occur without work having to be done to bring it about. A spontaneous change has a natural tendency to occur. A **non-spontaneous change** is a change that can be brought about only by doing work. A non-spontaneous change has no natural tendency to occur. Non-spontaneous changes can be *made* to occur by doing work: gas can be compressed into a smaller volume by pushing in a piston, the temperature of a cool object can be raised by forcing an electric current through a heater attached to it, and water can be decomposed by the passage of an electric current. However, in each case we need to act in some way on the system to bring about the non-spontaneous change. There must be some feature of the world that accounts for the distinction between the two types of change.

Throughout the chapter we shall use the terms 'spontaneous' and 'non-spontaneous' in their thermodynamic sense. That is, we use them to signify that a change does or does not have a natural *tendency* to occur. In thermodynamics the term spontaneous has nothing to do with speed. Some spontaneous changes are very fast, such as the precipitation reaction that occurs when solutions of sodium chloride and silver nitrate are mixed. However, some spontaneous changes are so slow that there may be no observable change even after millions of years. For example, although the decomposition of benzene into carbon and hydrogen is spontaneous, it does not occur at a measurable rate under normal conditions, and benzene is a common laboratory commodity with a shelf life of (in principle) millions of years. Thermodynamics deals with the tendency to change; it is silent on the rate at which that tendency is realized.

Entropy

A few moments' thought is all that is needed to identify the reason why some changes are spontaneous and others are not. That reason is *not* the tendency of the system to move towards lower energy. This point is easily established by identifying an example of a spontaneous change in which there is no change in energy. The isothermal expansion of a perfect gas into a vacuum is spontaneous, but the total energy of the gas does not change because the molecules continue to travel at the same average speed and so keep their same total kinetic energy. Even in a process in which the energy of a system does decrease (as in the spontaneous cooling of a block of hot metal), the First Law requires the total energy of the system and the surroundings to be constant. Therefore, in this case the energy of another part of the world must increase if the energy decreases in the part that interests us. For instance, a hot block of metal in contact with a cool block cools and loses energy; however, the second block becomes warmer, and increases in energy. It is equally valid to say that the second block moves spontaneously to higher energy as it is to say that the first block has a tendency to go to lower energy!

4.1 The direction of spontaneous change

We shall now show that *the apparent driving force of spontaneous change is the tendency of energy to disperse and matter to become disordered.* For example, the molecules of a gas may all be in one region of a container initially, but their ceaseless random motion ensures that they spread rapidly throughout the entire volume of the container (Fig. 4.1). Because their motion is so disorderly, there is a negligibly small probability that all the molecules will find their way back simultaneously into the region of the container they occupied initially. In this instance, the natural direction of change corresponds to the disorderly dispersal of matter.

A similar explanation accounts for spontaneous cooling, but now we need to consider the dispersal of energy rather than that of matter. In a block of hot metal, the atoms are oscillating vigorously and the hotter the block the more vigorous their motion. The cooler surroundings also consist of oscillating atoms, but their motion is less vigorous. The vigorously oscillating atoms of the hot block jostle their neighbours in the surroundings, and the energy of the atoms in the block is handed on to the atoms in the surroundings (Fig. 4.2). The process continues until the vigour with which the atoms in the system are oscillating has fallen to that of the surroundings. The opposite flow of energy is very unlikely. It is highly improbable that there will be a net flow of energy into the system as a result of jostling from less vigorously oscillating molecules in the surroundings. In this case, the natural direction of change corresponds to the dispersal of energy.

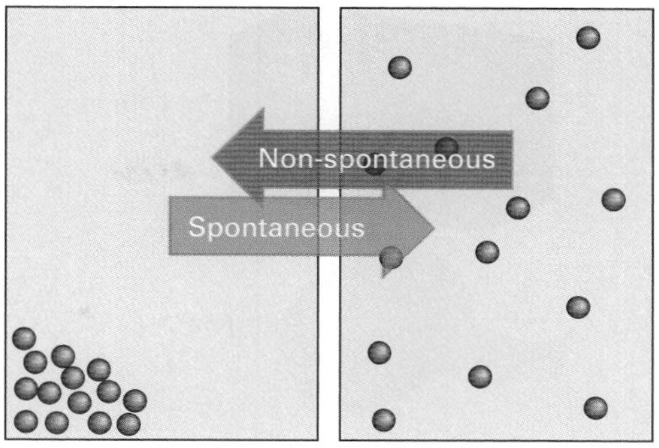

Fig. 4.1 One fundamental type of spontaneous process is the dispersal of matter. This tendency accounts for the spontaneous tendency of a gas to spread into and fill the container it occupies. It is extremely unlikely that all the particles will collect into one small region of the container. (In practice, the number of particles is of the order of 10^{23}.)

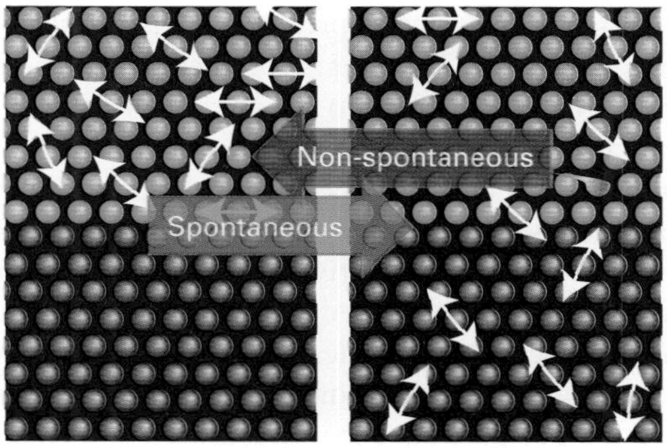

Fig. 4.2 Another fundamental type of spontaneous process is the dispersal of energy (represented by the small arrows). In these diagrams, the yellow spheres represent the system and the purple spheres represent the surroundings. The double headed arrows represent the thermal motion of the atoms.

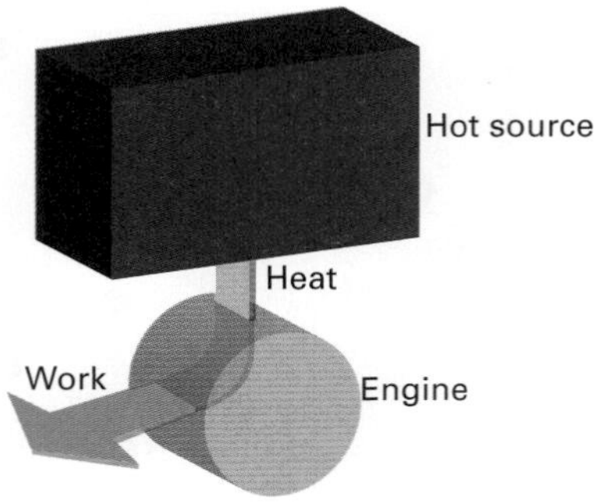

Fig. 4.3 The Second Law denies the possibility of the process illustrated here, in which heat is changed completely into work, there being no other change. The process is not in conflict with the First Law, because the energy is conserved.

The tendency of energy to disperse also explains the fact that, despite numerous attempts, it has proved impossible to construct an engine like that shown in Fig. 4.3, in which heat, perhaps from the combustion of a fuel, is drawn from a hot reservoir and completely converted into work, such as the work of moving an automobile. All actual heat engines have both a hot region, the 'source', and a cold region, the 'sink', and it has been found that some energy must be discarded into the cold sink as heat and not used to do work. In molecular terms, only some of the energy stored in the atoms and molecules of the hot source can be used to do work and transferred to the surroundings in an orderly way. For the engine to do work, some energy must be transferred to the cold sink as heat, to stimulate disorderly motion of its atoms and molecules.

In summary, we have identified a single basic type of spontaneous physical process:

Matter and energy tend to disperse.

By 'disperse' we mean spread in a disorderly way through space or, if matter is confined to a particular region, for its structure to become disorganized, as when a solid melts. We must now see how this natural process results in some chemical reactions being spontaneous and others not. It may seem very puzzling that such a simple principle can account for the formation of such ordered systems as proteins and biological cells. Nevertheless, in due course we shall see that organized structures can emerge as energy and matter disperse. We shall see, in fact, that collapse into disorder accounts for change in all its forms.

4.2 Entropy and the Second Law

The measure of disorderly dispersal used in thermodynamics is called the **entropy**, S. Initially, we can take entropy to be a synonym for the extent of disorderly dispersal, but shortly we shall see that it can be defined precisely and quantitatively, measured, and then applied to chemical reactions. At this point, all we need to know is that when matter and energy become dispersed, the entropy increases. That being so, we can express the basic principle underlying change as the **Second Law of thermodynamics:**

The entropy of an isolated system tends to increase. — The Second Law

The 'isolated system' may consist of a system in which we have a special interest (a beaker containing reagents) and that system's surroundings: the two components jointly form a little 'universe' in the thermodynamic sense.

To make progress and turn the Second Law into a quantitatively useful statement, we need to define entropy precisely. We shall use the following definition of a *change* in entropy for a system maintained at constant temperature:

$$\Delta S = \frac{q_{\text{rev}}}{T} \qquad \text{Definition} \quad \text{Entropy change} \qquad (4.1)$$

That is, the change in entropy of a substance is equal to the energy transferred as heat to it *reversibly* divided by the temperature at which the transfer takes place. We need to understand three points about the definition in eqn 4.1: the significance of the term 'reversible', why heat (not work) appears in the numerator, and why temperature appears in the denominator.

- *Why reversible?* We met the concept of reversibility in Section 2.4, where we saw that it refers to the ability of an infinitesimal change in a variable to change the direction of a process. Mechanical reversibility refers to the equality of pressure acting on either side of a movable wall. Thermal reversibility, the type involved in eqn 4.1, refers to the equality of temperature on either side of a thermally conducting wall. Reversible transfer of heat is smooth, careful, restrained transfer between two bodies at the same temperature. By making the transfer reversible we ensure that there are no hot spots generated in the object that later disperse spontaneously and hence add to the entropy.
- *Why heat and not work in the numerator?* Now consider why heat and not work appears in eqn 4.1. Recall from Section 2.3 that to transfer energy as heat we make use of the disorderly motion of molecules whereas to transfer energy as work we

make use of orderly motion. It should be plausible that the change in entropy—the change in the degree of disorder—is proportional to the energy transfer that takes place by making use of disorderly motion rather than orderly motion.

- *Why temperature in the denominator?* The presence of the temperature in the denominator in eqn 4.1 takes into account the disorder that is already present. If a given quantity of energy is transferred as heat to a hot object (one in which the atoms have a lot of disorderly thermal motion), then the additional disorder generated is less significant than if the same quantity of energy is transferred as heat to a cold object in which the atoms have less thermal motion. The difference is like sneezing in a busy street (an environment analogous to a high temperature) and sneezing in a quiet library (an environment analogous to a low temperature).

Brief illustration 4.1 Entropy changes

The transfer of 100 kJ of energy as heat to a large mass of water at 0 °C (273 K) results in a change in entropy of

$$\Delta S = \frac{q_{rev}}{T} = \frac{100 \times 10^3 \text{ J}}{273 \text{ K}} = +366 \text{ J K}^{-1}$$

We use a large mass of water to ensure that the temperature of the sample does not change as heat is transferred. The same transfer at 100 °C (373 K) results in

$$\Delta S = \frac{100 \times 10^3 \text{ J}}{373 \text{ K}} = +268 \text{ J K}^{-1}$$

The increase in entropy is greater at the lower temperature.

A note on good practice The units of entropy are joules per kelvin (J K^{-1}). Entropy is an extensive property. When we deal with molar entropy, an intensive property, the units will be joules per kelvin per mole (J K^{-1} mol^{-1}).

The entropy is a state function, a property with a value that depends only on the present state of the system.[1] The entropy is a measure of the current state of disorder of the system, and how that disorder was achieved is not relevant to its current value. A sample of liquid water of mass 100 g at 60 °C and 98 kPa has exactly the same degree of molecular disorder—the same entropy—regardless of what has happened to it in the past. The implication of entropy being a state function is that a change in its value when a system undergoes a change of state is independent of how the change of state is brought about.

[1] For a proof of this statement, see our *Physical Chemistry* (2010).

Impact on technology 4.1

Heat engines, refrigerators, and heat pumps

One practical application of entropy is to the discussion of the efficiencies of heat engines, refrigerators, and heat pumps. As remarked in the text, to achieve spontaneity—an engine is less than useless if it has to be driven—some energy must be discarded as heat into the cold sink. It is quite easy to calculate the minimum energy that must be discarded in this way by thinking about the flow of energy and the changes in entropy of the hot source and cold sink. To simplify the discussion, we shall express it in terms of the magnitudes of the heat and work transactions, which we write as $|q|$ and $|w|$, respectively (so, if $q = -100$ J, $|q| = 100$ J). Maximum work—and therefore maximum efficiency—is achieved if all energy transactions take place reversibly, so we assume that to be the case in the following.

Suppose that the hot source is at a temperature T_{hot}. Then when energy $|q|$ is released from it reversibly as heat, its entropy changes by $-|q|/T_{hot}$. Suppose that we allow an energy $|q'|$ to flow reversibly as heat into the cold sink at a temperature T_{cold}. Then the entropy of that sink changes by $+|q'|/T_{cold}$ (Fig. 4.4). The total change in entropy is therefore

$$\Delta S_{total} = \underbrace{-\frac{|q|}{T_{hot}}}_{\substack{\text{Reduction of}\\ \text{entropy of the}\\ \text{hot source}}} \underbrace{+\frac{|q'|}{T_{cold}}}_{\substack{\text{Increase of}\\ \text{entropy of the}\\ \text{cold sink}}}$$

The engine will not operate spontaneously if this change in entropy is negative, and just becomes spontaneous as ΔS_{total} becomes positive. This change of sign occurs when $\Delta S_{total} = 0$, which is achieved when

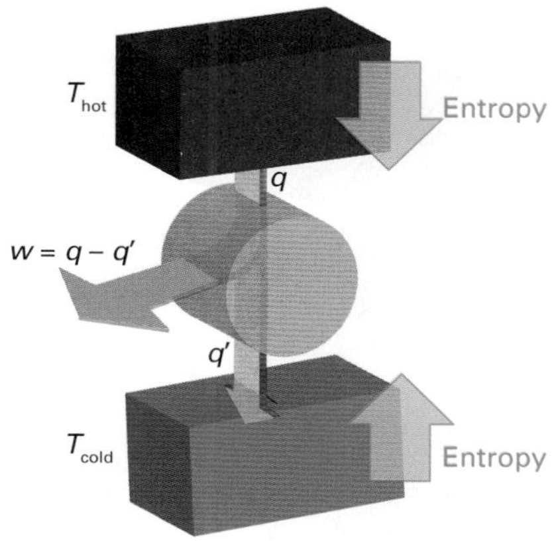

Fig. 4.4 The flow of energy in a heat engine. For the process to be spontaneous, the decrease in entropy of the hot source must be offset by the increase in entropy of the cold sink. However, because the latter is at a lower temperature, not all the energy removed from the hot source need be deposited in it, leaving the difference available as work.

$$|q'| = \frac{T_{\text{cold}}}{T_{\text{hot}}} \times |q|$$

If we have to discard an energy $|q'|$ into the cold sink, the maximum energy that can be extracted as work is $|q| - |q'|$. It follows that the **efficiency**, η (eta), of the engine, the ratio of the work produced to the heat absorbed by the engine, is

$$\eta = \frac{\text{work produced}}{\text{heat absorbed}} = \frac{|q|-|q'|}{|q|} = 1-\frac{|q'|}{|q|} = 1-\frac{T_{\text{cold}}}{T_{\text{hot}}}$$

This remarkable result tells us that the efficiency of a perfect heat engine (one working reversibly and without mechanical defects such as friction) depends only on the temperatures of the hot source and cold sink. It shows that maximum efficiency (closest to $\eta = 1$) is achieved by using a sink that is as cold as possible and a source that is as hot as possible.

Brief illustration 4.2 The maximum efficiency

The maximum efficiency of an electrical power station using steam at 200 °C (473 K) and discharging at 20 °C (293 K) is

$$\eta = 1-\frac{\overbrace{293\ \text{K}}^{T_{\text{cold}}}}{\underbrace{473\ \text{K}}_{T_{\text{hot}}}} = 1-\frac{293}{473} = 0.381$$

or 38.1 per cent.

A refrigerator can be analysed similarly (Fig. 4.5). The entropy change when an energy $|q|$ is withdrawn reversibly as heat from the cold interior at a temperature T_{cold} is $-|q|/T_{\text{cold}}$. The entropy change when an energy $|q'|$ is deposited reversibly as heat in the outside world at a temperature T_{hot} is $+|q'|/T_{\text{hot}}$. The total change in entropy would be negative if $|q'| = |q|$, and the refrigerator would not work. However, if we increase the flow of energy into the warm exterior by doing work on the refrigerator, then the entropy change of the warm exterior can be increased to the point at which it overcomes the decrease in entropy of the cold interior, and the refrigerator operates. The calculation of the maximum efficiency of this process is left as an exercise (see Project 4.34).

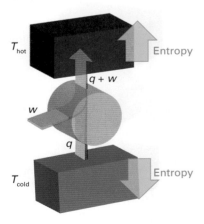

Fig. 4.5 The flow of energy as heat from a cold source to a hot sink becomes feasible if work is provided to add to the energy stream. Then the increase in entropy of the hot sink can be made to cancel the entropy decrease of the cold source.

A heat pump is simply a refrigerator, but in which we are more interested in the supply of heat to the exterior than the cooling achieved in the interior. You are invited to show (see Project 4.34) that the efficiency of a perfect heat pump, as measured by the heat produced divided by the work done, also depends on the ratio of the two temperatures.

4.3 The entropy change accompanying expansion

We can often rely on intuition to judge whether the entropy increases or decreases when a substance undergoes a physical change. For instance, the entropy of a sample of gas increases as it expands because the molecules get to move in a greater volume and so have a greater degree of disorderly dispersal. However, the advantage of eqn 4.1 is that it lets us express the increase *quantitatively* and make numerical calculations. For instance, as shown in the following Derivation, we can use the definition to calculate the change in entropy when a perfect gas expands isothermally from a volume V_{i} to a volume V_{f}, and obtain

$$\Delta S = nR\ln\frac{V_{\text{f}}}{V_{\text{i}}} \qquad \text{Perfect gas} \qquad \text{Change in entropy on isothermal expansion} \qquad (4.2)$$

We have already stressed the importance of reading equations for their physical content. In this case:

- If $V_{\text{f}} > V_{\text{i}}$, as in an expansion, then $V_{\text{f}}/V_{\text{i}} > 1$ and the logarithm is positive. Consequently, eqn 4.2 predicts a positive value for ΔS, corresponding to an increase in entropy, just as we anticipated (Fig. 4.6).
- The change in entropy is independent of the temperature at which the isothermal expansion occurs. More work is done if the temperature is high (because the external pressure must be matched to a higher value of the pressure of the gas), so more energy must be supplied as heat to maintain the temperature. The temperature in the denominator of eqn 4.1 is higher, but the 'sneeze' (in terms of the analogy introduced earlier) is greater too, and the two effects cancel.

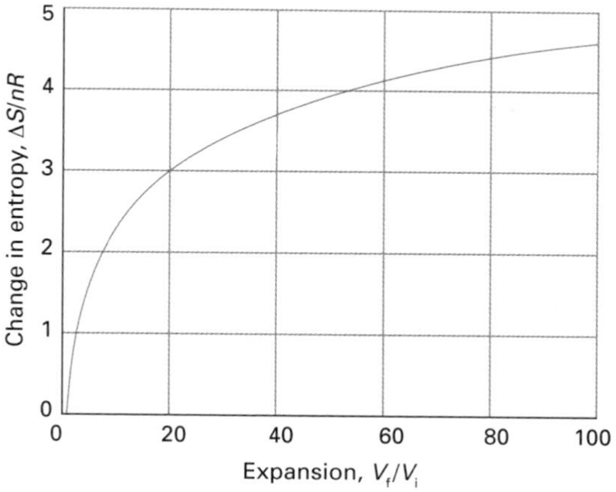

Fig. 4.6 The entropy of a perfect gas increases logarithmically (as ln V) as the volume is increased.

Derivation 4.1

The variation of the entropy of a perfect gas with volume

We need to know q_{rev}, the energy transferred as heat in the course of a *reversible* change at the temperature T. From eqn 2.6 we know that the energy transferred as heat to a perfect gas when it undergoes reversible, isothermal expansion from a volume V_i to a volume V_f at a temperature T is

$$q_{rev} = nRT \ln \frac{V_f}{V_i}$$

It follows that

$$\Delta S = \frac{q_{rev}}{T} = \frac{nRT \ln(V_f/V_i)}{T} \overset{\text{cancel } T}{=} nR \ln \frac{V_f}{V_i}$$

which is eqn 4.2.

Brief illustration 4.3 The entropy change accompanying expansion

When the volume occupied by 1.00 mol of any perfect gas molecules is doubled at any constant temperature, $V_f/V_i = 2$ and

$$\Delta S = (1.00 \text{ mol}) \times (8.3145 \text{ J K}^{-1} \text{ mol}^{-1}) \times \ln 2$$
$$= +5.76 \text{ J K}^{-1}$$

Self-test 4.1

What is the change in entropy when the pressure of a perfect gas is changed isothermally from p_i to p_f?

Answer: $\Delta S = nR \ln(p_i/p_f)$

Here is a subtle but important point. The definition in eqn 4.1 makes use of a *reversible* transfer of heat, and that is what we used in the derivation of eqn 4.2. However, entropy is a state function, so its value is independent of the path between the initial and final states. This independence of path means that although we have used a reversible path to calculate ΔS, the same value applies to an irreversible change (for instance, free expansion) between the same two states. We cannot use an irreversible path to calculate ΔS, but the value calculated for a reversible path applies however the path is traversed in practice between the specified initial and final states. You may have noticed that in Brief illustration 4.3 we did not specify how the expansion took place other than that it is isothermal.

4.4 The entropy change accompanying heating

We should expect the entropy of a sample to increase as the temperature is raised from T_i to T_f, because the thermal disorder of the system is greater at the higher temperature, when the molecules move more vigorously. To calculate the change in entropy, we go back to the definition in eqn 4.1 and, as shown in the following Derivation, find that, provided the heat capacity is constant over the range of temperatures of interest,

$$\Delta S = C \ln \frac{T_f}{T_i} \quad \text{Constant heat capacity} \quad \text{Change in entropy on heating} \qquad (4.3)$$

where C is the heat capacity of the system; if the pressure is held constant during the heating, we use the constant-pressure heat capacity, C_p, and if the volume is held constant, we use the constant-volume heat capacity, C_V.

Once more, we interpret the equation:

- When $T_f > T_i$, $T_f/T_i > 1$, which implies that the logarithm is positive, that $\Delta S > 0$, and therefore that the entropy increases as the temperature is raised (Fig. 4.7).
- The higher the heat capacity of the substance, the greater the change in entropy for a given rise in temperature. A high heat capacity implies that a lot of heat is required to produce a given change in temperature, so the 'sneeze' must be more powerful than for when the heat capacity is low, and the entropy increase is correspondingly high.

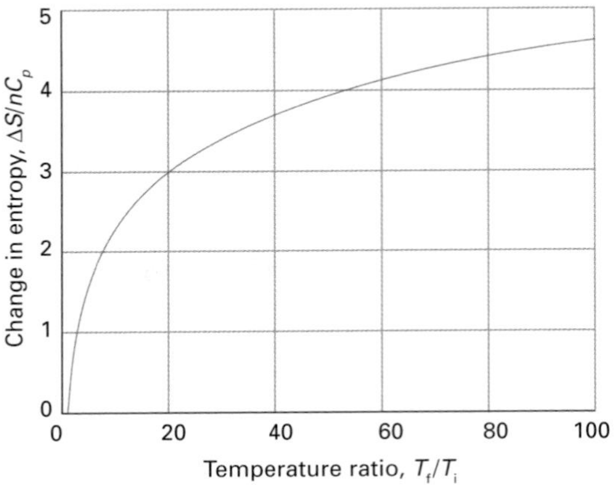

Fig. 4.7 The entropy of a sample with a heat capacity that is independent of temperature, such as a monatomic perfect gas, increases logarithmically (as ln T) as the temperature is increased. The increase is proportional to the heat capacity of the sample.

Derivation 4.2

The variation of entropy with temperature

Equation 4.1 refers to the transfer of heat to a system at a temperature T. In general, the temperature changes as we heat a system, so we cannot use eqn 4.1 directly. Suppose, however, that we transfer only an infinitesimal energy, dq, to the system, then there is only an infinitesimal change in temperature and we introduce negligible error if we keep the temperature in the denominator of eqn 4.1 equal to T during that transfer. As a result, the entropy increases by an infinitesimal amount dS given by

$$dS = \frac{dq_{rev}}{T}$$

To calculate dq, we recall from Section 2.5 that the heat capacity $C = q/\Delta T$, is where ΔT is macroscopic change in temperature. For the infinitesimal change dT brought about by an infinitesimal transfer of heat we write $C = dq/dT$. This relation also applies when the transfer of energy is carried out reversibly. It follows that $dq_{rev} = CdT$ and therefore that

$$dS = \frac{CdT}{T}$$

The total change in entropy, ΔS, when the temperature changes from T_i to T_f is the sum (integral; see The chemist's toolkit 2.1) of all such infinitesimal terms with T in general different for each of the infinitesimal steps:

$$\Delta S = \int_{T_i}^{T_f} \frac{CdT}{T} \qquad (4.4)$$

For many substances and for small temperature ranges we may take C to be constant. This assumption is strictly true for a monatomic perfect gas. Then C may be taken outside the integral, and the latter evaluated as follows:

$$\Delta S = \int_{T_i}^{T_f} \frac{CdT}{T} = C\int_{T_i}^{T_f} \frac{dT}{T} = C \ln \frac{T_f}{T_i}$$

We have used the same standard integral as in Derivation 2.2, and evaluated the limits similarly.

Brief illustration 4.4 The entropy change upon heating

When calculating entropy changes at constant volume, the heat capacity to be used in eqn 4.3 is $C_{V,m}$. For example, the change in molar entropy when hydrogen gas is heated from 20 °C to 30 °C at constant volume is (assuming that $C_{V,m} = 22.44$ J K^{-1} mol^{-1} is constant over this temperature range):

$$\Delta S_m = C_{V,m} \ln \frac{T_f}{T_i} = (22.44 \text{ J K}^{-1} \text{ mol}^{-1}) \times \ln \frac{\overbrace{303 \text{ K}}^{30\,°\text{C}}}{\underbrace{293 \text{ K}}_{20\,°\text{C}}}$$

$$= +0.75 \text{ J K}^{-1} \text{ mol}^{-1}$$

When we cannot assume that the heat capacity is constant over the temperature range of interest, which is the case for all solids at low temperatures, we have to allow for the variation of C with temperature. As we show in the following very brief Derivation, the result is

$$\Delta S = \text{area under the graph of } C/T \text{ plotted against } T, \text{ between } T_i \text{ and } T_f \qquad (4.5)$$

Experimental basis of determining an entropy change

Derivation 4.3

The entropy change when the heat capacity varies with temperature

In Derivation 4.2 we found, before making the assumption that the heat capacity is constant, that

$$\Delta S = \int_{T_i}^{T_f} \frac{CdT}{T}$$

This, which is eqn 4.4, is our starting point. All we need recognize is the standard result from calculus, illustrated in Derivation 2.2 and The chemist's toolkit 2.1, that the integral of a function between two limits is the area under the graph of the function between the two limits. In this case, the function is C/T, the heat capacity at each temperature divided by that temperature.

The process of calculating an entropy change is illustrated in Fig. 4.8:

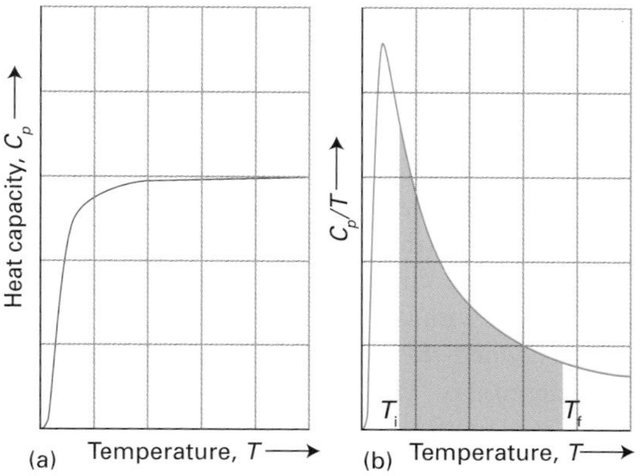

Fig. 4.8 The experimental determination of the change in entropy of a sample that has a heat capacity that varies with temperature involves: (a) measuring the heat capacity over the range of temperatures of interest, then plotting C_p/T against T and (b) determining the area under the curve (the tinted area shown here). The heat capacity of all solids decreases toward zero as the temperature is reduced.

- First we measure and tabulate the heat capacity C throughout the range of temperatures of interest (Fig. 4.8a).
- Then we divide each one of the values of C by the corresponding temperature, to get C/T at each temperature, and plot these C/T against T (Fig. 4.8b).
- Lastly, we evaluate the area under the graph between the temperatures T_i and T_f (Fig. 4.8b). The only way to proceed reliably is to fit the data to a polynomial in T and then to use a computer to evaluate the integral.

4.5 The entropy change accompanying a phase transition

We can suspect that the entropy of a substance increases when it melts and when it boils because its molecules become more disordered as it changes from solid to liquid and from liquid to vapour. The transfer of energy as heat occurs reversibly when a solid is at its melting temperature. If the temperature of the surroundings is infinitesimally lower than that of the system, then energy flows out of the system as heat and the substance freezes. If the temperature is infinitesimally higher, then energy flows into the system as heat and the substance melts. Moreover, because the transition occurs at constant pressure, we can identify the heat transferred per mole of substance with the enthalpy of fusion (melting). Therefore, the **entropy of fusion**, $\Delta_{fus}S$, the change

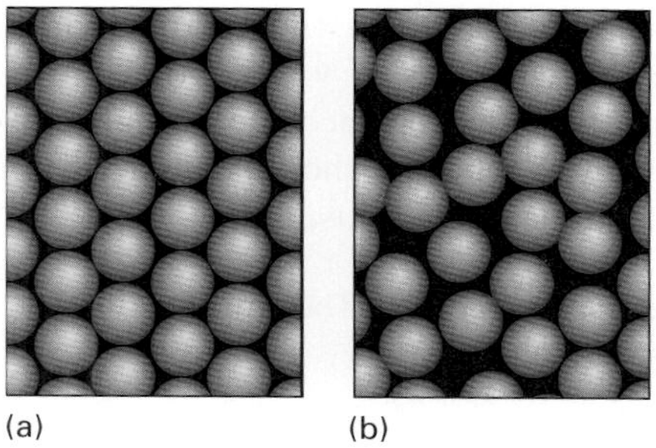

Fig. 4.9 When a solid, depicted by the orderly array of spheres (a), melts, the molecules form a liquid, the disorderly array of spheres (b). As a result, the entropy of the sample increases.

of entropy per mole of substance, at the melting temperature, T_f (with f now denoting fusion), is

$$\Delta_{fus}S = \frac{\Delta_{fus}H(T_f)}{T_f} \qquad \text{At the melting temperature} \qquad \text{Entropy of fusion} \qquad (4.6)$$

Notice how we must use the enthalpy of fusion *at the melting temperature*. To get the standard entropy of fusion, $\Delta_{fus}S^{\ominus}$, at the melting temperature we use the melting temperature at 1 bar and the corresponding standard enthalpy of fusion at that temperature. All enthalpies of fusion are positive (melting is endothermic: it requires heat), so all entropies of fusion are positive too: disorder increases on melting. The entropy of water, for example, increases when it melts because the orderly structure of ice collapses as the liquid forms (Fig. 4.9).

Brief illustration 4.5 The entropy of fusion

From eqn 4.6 and the information in Table 3.1, the entropy of fusion of ice at 0 °C is

$$\Delta_{fus}S = \frac{\overbrace{6.01\ \text{kJ mol}^{-1}}^{\Delta_{fus}H(T_f)\text{ of water}}}{\underbrace{273.15\ \text{K}}_{T_f\text{ of water}}} = +2.20\times10^{-2}\ \overbrace{\text{kJ}}^{10^3\,\text{J}}\ \text{K}^{-1}\ \text{mol}^{-1}$$

$$= +22.0\ \text{J K}^{-1}\ \text{mol}^{-1}$$

The entropy of other types of transition may be discussed similarly. Thus, the entropy of vaporization, $\Delta_{vap}S$, at the boiling temperature, T_b, of a liquid is related to its enthalpy of vaporization at that temperature by

$$\Delta_{vap}S = \frac{\Delta_{vap}H(T_b)}{T_b} \qquad \text{At the boiling temperature} \qquad \text{Entropy of vaporization} \qquad (4.7)$$

To use this formula, we use the enthalpy of vaporization at the boiling temperature. For the standard value, $\Delta_{vap}S^{\ominus}$, we use data corresponding to 1 bar.

Because vaporization is endothermic for all substances, all entropies of vaporization are positive. The increase in entropy accompanying vaporization is in line with what we should expect when a compact liquid turns into a gas.

Brief illustration 4.6 The entropy of vaporization

From eqn 4.7 and the information in Table 3.1, the entropy of vaporization of water at 100 °C is

$$\Delta_{\text{vap}}S = \frac{\overbrace{40.7\ \text{kJ mol}^{-1}}^{\Delta_{\text{vap}}H(T_b)\text{ of water}}}{\underbrace{373.2\ \text{K}}_{T_b\text{ of water}}} = +1.09\times10^{-1}\ \text{kJ K}^{-1}\ \text{mol}^{-1}$$

$$= +109\ \text{J K}^{-1}\ \text{mol}^{-1}$$

Entropies of vaporization shed light on an empirical relation known as **Trouton's rule:**

The quantity $\Delta_{\text{vap}}H^{\ominus}(T_b)/T_b$ is approximately the same (and equal to about 85 J K^{-1} mol^{-1}) for all liquids except when hydrogen bonding or some other kind of specific molecular interaction is present. Trouton's rule

The data in Table 4.1 provide support for this rule. We know that the quantity $\Delta_{\text{vap}}H^{\ominus}(T_b)/T_b$, however, is the entropy of vaporization of the liquid at its boiling point, so Trouton's rule is explained if all liquids have approximately the same entropy of vaporization at their boiling points. This near equality is to be expected because when a liquid vaporizes, the compact condensed phase changes into a widely dispersed gas that occupies approximately the same volume whatever its identity. To a good approximation, therefore, we expect the increase in disorder, and therefore the entropy of vaporization, to be almost the same for all liquids at their boiling temperatures.

The exceptions to Trouton's rule include liquids in which the interactions between molecules result in the liquid being less disordered than a random jumble of molecules. For example, the high value for water implies that the H_2O molecules are linked together in some kind of ordered structure by hydrogen bonding, with the result that the entropy change is greater when this relatively ordered liquid forms a disordered gas. The high value for mercury has a similar explanation but stems from the presence of metallic bonding in the liquid, which organizes the atoms into more definite patterns than would be the case if such bonding were absent.

Table 4.1

Entropies of vaporization at 1 atm and the normal boiling point

	$\Delta_{\text{vap}}S/(\text{J K}^{-1}\ \text{mol}^{-1})$
Ammonia, NH_3	97.4
Benzene, C_6H_6	87.2
Bromine, Br_2	88.6
Carbon tetrachloride, CCl_4	85.9
Cyclohexane, C_6H_{12}	85.1
Hydrogen sulfide, H_2S	87.9
Mercury, Hg(l)	94.2
Water, H_2O	109.1

Brief illustration 4.7 Trouton's rule

We can estimate the enthalpy of vaporization of liquid bromine from its boiling temperature, 59.2 °C. No hydrogen bonding or other kind of special interaction is present, so we use the rule after converting the boiling point to 332.4 K:

$$\Delta_{\text{vap}}H^{\ominus} \approx (332.4\ \text{K}) \times (85\ \text{J K}^{-1}\ \text{mol}^{-1}) = 28\ \text{kJ mol}^{-1}$$

The experimental value is 29 kJ mol^{-1}.

Self-test 4.2

Estimate the enthalpy of vaporization of ethane from its boiling point, which is –88.6 °C.

Answer: 16 kJ mol^{-1}

To calculate the entropy of phase transition at a temperature other than the transition temperature, we have to do additional calculations, as shown in the following Example.

Example 4.1

Calculating the entropy of vaporization

Calculate the entropy of vaporization of water at 25 °C from thermodynamic data and its enthalpy of vaporization at its normal boiling point.

Strategy The most convenient way to proceed is to perform three calculations. First, calculate the entropy change for heating liquid water from 25 °C to 100 °C (using eqn 4.3 with data for the liquid from Table 2.1). Then use eqn 4.7 and data from Table 3.1 to calculate the entropy of transition at 100 °C. Next, calculate the change in entropy for cooling the vapour from 100 °C to 25 °C (using eqn 4.3 again, but now with data for the vapour from Table 2.1). Finally, add the three contributions together. The steps may be hypothetical.

Answer From eqn 4.3 with data for the liquid from Table 2.1:

$$\Delta S_1 = C_{p,\mathrm{m}}(\mathrm{H_2O,l})\ \ln\frac{T_\mathrm{f}}{T_\mathrm{i}}$$

$$= (75.29\ \mathrm{J\ K^{-1}\ mol^{-1}}) \times \ln\frac{373\ \mathrm{K}}{298\ \mathrm{K}}$$

$$= +16.9\ \mathrm{J\ K^{-1}\ mol^{-1}}$$

From eqn 4.7 and data from Table 3.1:

$$\Delta S_2 = \frac{\Delta_\mathrm{vap}H(T_\mathrm{b})}{T_\mathrm{b}} = \frac{4.07\times10^4\ \mathrm{J\ K^{-1}\ mol^{-1}}}{373\ \mathrm{K}}$$

$$= +109\ \mathrm{J\ K^{-1}\ mol^{-1}}$$

From eqn 4.3 with data for the vapour from Table 2.1:

$$\Delta S_1 = C_{p,\mathrm{m}}(\mathrm{H_2O,g})\ \ln\frac{T_\mathrm{f}}{T_\mathrm{i}}$$

$$= (33.58\ \mathrm{J\ K^{-1}\ mol^{-1}}) \times \ln\frac{298\ \mathrm{K}}{373\ \mathrm{K}}$$

$$= -7.54\ \mathrm{J\ K^{-1}\ mol^{-1}}$$

The sum of the three entropy changes is the entropy of transition at 25 °C:

$$\Delta_\mathrm{vap}S\ (298\ \mathrm{K}) = \Delta S_1 + \Delta S_2 + \Delta S_3 = +118\ \mathrm{J\ K^{-1}\ mol^{-1}}$$

Self-test 4.3

Calculate the entropy of vaporization of benzene at 25 °C from the following data: $T_\mathrm{b} = 353.2$ K, $\Delta_\mathrm{vap}H^\ominus(T_\mathrm{b}) = 30.8$ kJ mol^{-1}, $C_{p,\mathrm{m}}(\mathrm{l}) = 136.1$ J K^{-1} mol^{-1}, $C_{p,\mathrm{m}}(\mathrm{g}) = 81.6$ J K^{-1} mol^{-1}.

Answer: 96.4 J K^{-1} mol^{-1}

4.6 Entropy changes in the surroundings

We can use the definition of entropy in eqn 4.1 to calculate the entropy change of the surroundings in contact with the system at the temperature T:

$$\Delta S_\mathrm{sur} = \frac{q_\mathrm{sur}}{T}$$

The surroundings are so extensive that they remain at constant pressure regardless of any events taking place in the system, so $q_\mathrm{sur,rev} = \Delta H_\mathrm{sur}$. The enthalpy is a state function, so a change in its value is independent of the path and we get the same value of ΔH_sur regardless of how the heat is transferred. Therefore, we can drop the label 'rev' from q and write

$$\Delta S_\mathrm{sur} = \frac{q_\mathrm{sur}}{T} \qquad (4.8)$$

Entropy change of the surroundings in terms of heating of the surroundings

This formula can be used to calculate the entropy change of the surroundings regardless of whether the change in the system is reversible or not.

Example 4.2

Estimating the entropy change of the surroundings

A typical resting person heats the surroundings at a rate of about 100 W. Estimate the entropy you generate in the surroundings in the course of a day at 20 °C.

Strategy We can estimate the approximate change in entropy from eqn 4.8 once we have calculated the energy transferred as heat. To find this quantity, we use 1 W = 1 J s^{-1} and the fact that there are 86 400 s in a day. Convert the temperature to kelvins.

Solution The heat transferred to the surroundings in the course of a day is

$$q_\mathrm{sur} = (86\,400\ \mathrm{s}) \times (100\ \mathrm{J\ s^{-1}}) = 86\,400 \times 100\ \mathrm{J}$$

The increase in entropy of the surroundings is therefore

$$\Delta S_\mathrm{sur} = \frac{q_\mathrm{sur}}{T} = \frac{86\,400\times100\ \mathrm{J}}{293\ \mathrm{K}} = +2.95\times10^4\ \mathrm{J\ K^{-1}}$$

That is, the entropy production is about 30 kJ K^{-1}. Just to stay alive, each person on the planet contributes about 30 kJ K^{-1} each day to the entropy of their surroundings. The use of transport, machinery, and communications generates far more in addition.

Self-test 4.4

Suppose a small reptile operates at 0.50 W. What entropy does it generate in the course of a day in the water in the lake that it inhabits, where the temperature is 15 °C?

Answer: +150 J K^{-1}

Equation 4.8 is expressed in terms of the energy supplied to the *surroundings* as heat, q_sur. Normally, we have information about the energy supplied to or escaping from the *system* as heat, q. The two quantities are related by $q_\mathrm{sur} = -q$. For instance, if $q = +100$ J, an influx of 100 J, then $q_\mathrm{sur} = -100$ J, indicating that the surroundings have lost that 100 J. Therefore, at this stage we can replace q_sur in eqn 4.8 by $-q$ and write

$$\Delta S_\mathrm{sur} = -\frac{q}{T} \qquad (4.9)$$

Entropy change of the surroundings in terms of heating of the system

This expression is in terms of the properties of the system. Moreover, it applies whether or not the process taking place in the system is reversible.

To gain insight into eqn 4.9, let's consider two scenarios:

- Suppose a perfect gas expands isothermally and reversibly from V_i to V_f. The entropy change of the gas itself (the system) is given by eqn 4.2. To calculate the entropy change in the surroundings, we note that q, the heat required to keep the temperature constant, is given in Derivation 4.1. Therefore,

$$\Delta S_{\text{sur}} = -\frac{q}{T} = -\frac{nRT\ln(V_f/V_i)}{T} \overset{\text{cancel } T}{=} -nR\ln\frac{V_f}{V_i}$$

The change of entropy in the surroundings is therefore the negative of the change in entropy of the system, and the total entropy change for the reversible process is zero.

- Now suppose that the gas expands isothermally but freely ($p_{ex} = 0$) between the same two volumes. The change in entropy of the system is the same, because entropy is a state function. However, because $\Delta U = 0$ for the isothermal expansion of a perfect gas and no work is done, no heat is taken in from the surroundings. Because $q = 0$, it follows from eqn 4.9 (which, remember, can be used for either reversible or irreversible heat transfers), that $\Delta S_{\text{sur}} = 0$. The total change in entropy is therefore equal to the change in entropy of the system, which is positive. We see that for this irreversible process, the entropy of the universe has increased, in accord with the Second Law.

If a chemical reaction or a phase transition takes place at constant pressure, we can identify q in eqn 4.9 with the change in enthalpy of the system, and obtain

$$\Delta S_{\text{sur}} = -\frac{\Delta H}{T} \qquad \text{Constant pressure} \quad \text{Entropy change of the surroundings} \qquad (4.10)$$

This enormously important expression will lie at the heart of our discussion of chemical equilibrium. We see that it is consistent with common sense: if the process is exothermic, ΔH is negative and therefore ΔS_{sur} is positive. The entropy of the surroundings increases if heat is released into them. If the process is endothermic ($\Delta H > 0$), then the entropy of the surroundings decreases.

4.7 The molecular interpretation of entropy

We have referred frequently to 'molecular disorder' and have interpreted the thermodynamic quantity of entropy in terms of this so far ill-defined concept. The concept of disorder, however, can be expressed precisely and used to calculate entropies. The procedures required will be described in Chapter 22, for they draw on information that we have not yet encountered. However, it is possible to understand the basis of the approach that we use there, and see how it illuminates what we have achieved so far.

The fundamental equation that we need is the **Boltzmann formula**, which was originally proposed by Ludwig Boltzmann towards the end of the nineteenth century (and is carved as his epitaph on his tombstone):

$$S = k \ln W \qquad \text{Boltzmann formula for the entropy} \qquad (4.11)$$

where $k = 1.381 \times 10^{-23}$ J K^{-1} is Boltzmann's constant (Foundations 0.11). The quantity W is the number of ways that the molecules of the sample can be arranged yet correspond to the same total energy and formally is called the 'weight' of a 'configuration' of the sample. Each way to arrange the molecules in a system (under the constraint of keeping the total energy constant) is also called a 'microstate' of the system.

Brief illustration 4.8 The weight of a configuration

Suppose we had a tiny system of four molecules A, B, C, and D that could occupy three equally spaced levels of energies 0, ε, and 2ε, and we know that the total energy is 4ε. The 19 arrangements shown in Fig. 4.10 are possible, so W = 19 and the system has 19 microstates.

Self-test 4.5

How many arrangements are possible for three molecules with a total energy 3ε that could occupy equally spaced levels of energies 0, ε, and 2ε?

Answer: 7

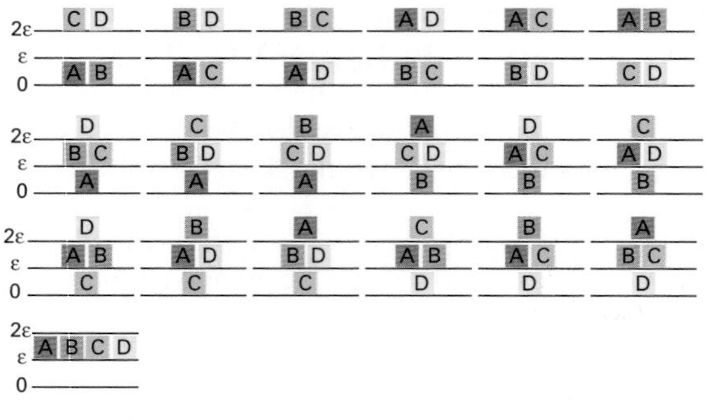

Fig. 4.10 The 19 arrangements of four molecules (represented by the blocks) in a system with three energy levels and a total energy of 4ε.

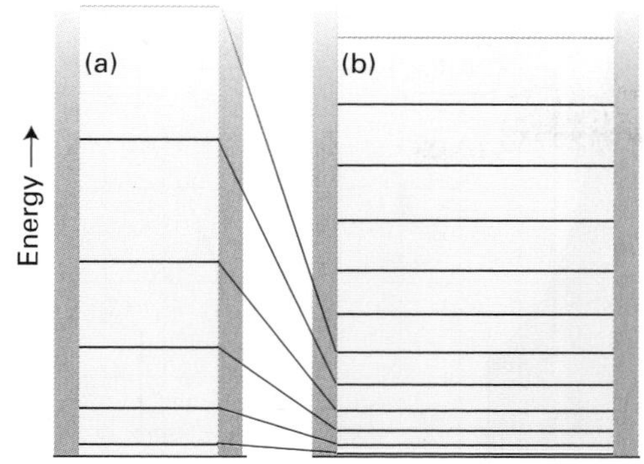

Fig. 4.11 As a box expands, the energy levels of the particles inside it come closer together. At a given temperature, the number of arrangements corresponding to the same total energy is greater when the energy levels are closely spaced than when they are far apart.

The entropy calculated from the Boltzmann formula is sometimes called the **statistical entropy**. We see that, if $W = 1$, which corresponds to one microstate (only one way of achieving a given energy; all molecules in exactly the same state), then $S = 0$ because $\ln 1 = 0$. However, if the system can exist in more than one microstate, then $W > 1$ and $S > 0$. These are important results, to which we shall return in Section 4.9. For now, we note that if the molecules in the system have access to a greater number of energy levels, then there may be more ways of achieving a given total energy. That is, there are more microstates for a given total energy, W is greater, and the entropy is greater than when fewer levels accessible. Therefore, the statistical view of entropy summarized by the Boltzmann formula is consistent with our previous statement that the entropy is related to the dispersal of energy. For instance, when a perfect gas expands, the available translational energy levels get closer together (Fig. 4.11; this is a conclusion from quantum theory that we verify in Chapter 12), so it is possible to distribute the molecules over them in more ways than when the volume of the container is small and the energy levels are further apart. Therefore, as the container expands, W and therefore S increase. It is no coincidence that the thermodynamic expression for ΔS (eqn 4.2) is proportional to a logarithm: the logarithm in Boltzmann's formula turns out to lead to the same logarithmic expression (see Chapter 22).

Brief illustration 4.9 The Boltzmann formula

Suppose that a large, flexible polymer can adopt 1.0×10^{31} different conformations of the same energy. It follows from eqn 4.11 that the entropy of the polymer is

$$S = \overbrace{(1.381\times10^{-23}\ \text{J K}^{-1})}^{k}\times\ln\overbrace{(1.0\times10^{31})}^{W} = 9.9\times10^{-22}\ \text{J K}^{-1}$$

Multiplication by Avogadro's constant gives the molar entropy: $S_m = 6.0 \times 10^2\ \text{J K}^{-1}\ \text{mol}^{-1}$.

The Boltzmann formula also illuminates the thermodynamic definition of entropy (eqn 4.1) and in particular the role of the temperature. Molecules in a system at high temperature can occupy a large number of the available energy levels, so a small additional transfer of energy as heat will lead to a relatively small change in the number of accessible energy levels. Consequently, the number of microstates does not increase appreciably and nor does the entropy of the system. In contrast, the molecules in a system at low temperature have access to far fewer energy levels (at $T = 0$, only the lowest level is accessible), and the transfer of the same quantity of energy by heating will increase the number of accessible energy levels and the number of microstates significantly. Hence, the change in entropy upon heating will be greater when the energy is transferred to a cold body than when it is transferred to a hot body. This argument suggests that the change in entropy should be inversely proportional to the temperature at which the transfer takes place, as in eqn 4.1.

4.8 Absolute entropies and the Third Law of thermodynamics

The graphical procedure summarized by Fig. 4.8 for the determination of the difference in entropy of a substance at two temperatures has a very important application. If $T_i = 0$, then the area under the graph between $T = 0$ and some temperature T gives us the value of $\Delta S = S(T) - S(0)$. We are supposing that there are no phase transitions below the temperature T. If there are any phase transitions (for example, melting) in the temperature range of interest, then the entropy of each transition at the transition temperature is calculated using an equation like eqn 4.6. In any case, at $T = 0$, all the motion of the atoms has been eliminated, and there is no thermal disorder. Moreover, if the substance is perfectly crystalline, with every atom in a well-defined location, then there is no spatial disorder either. We can therefore suspect that at $T = 0$, the entropy is zero. When we have done some quantum mechanics, we shall see that molecules cannot lose all their vibrational energy, so they retain some motion even at $T = 0$. However, they are then all in the same state (their lowest energy state), and so in this sense lack any thermal disorder.

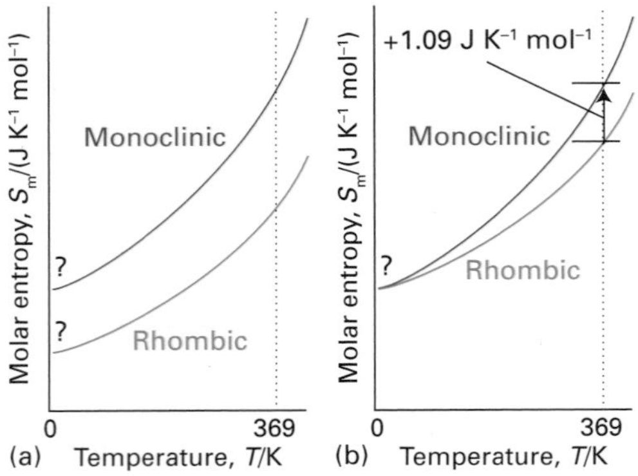

Fig. 4.12 (a) The molar entropies of monoclinic and rhombic sulfur vary with temperature as shown here. At this stage we do not know their values at $T = 0$. (b) When we slide the two curves together by matching their separation to the measured entropy of transition at the transition temperature, we find that the entropies of the two forms are the same at $T = 0$.

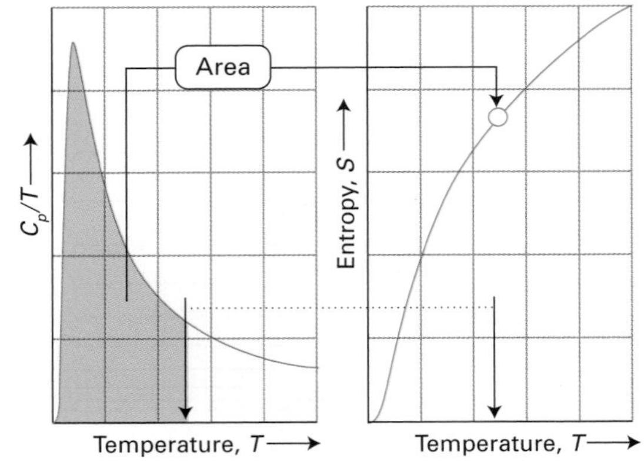

Fig. 4.13 The absolute entropy (or Third-Law entropy) of a substance is calculated by extending the measurement of heat capacities down to $T = 0$ (or as close to that value as possible), and then determining the area of the graph of C_p/T against T up to the temperature of interest. The area is equal to the absolute entropy at the temperature T.

One example of the thermodynamic evidence for the conclusion that $S(0) = 0$ is as follows. Sulfur undergoes a phase transition from rhombic to monoclinic at 96 °C (369 K) and the enthalpy of transition is +402 J mol^{-1}. The entropy of transition is therefore $\Delta S = (+402\ \mathrm{J\ K^{-1}\ mol^{-1}})/(369\ \mathrm{K}) = +1.09\ \mathrm{J\ K^{-1}\ mol^{-1}}$ at this temperature. We can also measure the molar entropy of each phase relative to its value at $T = 0$ by determining the heat capacity from $T = 0$ up to the transition temperature (Fig. 4.12). At this stage, we do not know the values of the entropies at $T = 0$. However, as we see from the illustration, to match the observed entropy of transition at 369 K, *the molar entropies of the two crystalline forms must be the same at $T = 0$*. We cannot say that the entropies are zero at $T = 0$, but from the experimental data we do know that they are the same. This observation is generalized into the **Third Law of thermodynamics:**

The entropies of all perfectly crystalline substances are the same at $T = 0$. The Third Law

For convenience (and in accord with our understanding of entropy as a measure of disorder), we take this common value to be zero. Then, with this convention, according to the Third Law,

$S(0) = 0$ for all perfectly ordered crystalline materials.

The **Third-Law entropy** at any temperature, $S(T)$, is equal to the area under the graph of C/T between $T = 0$ and the temperature T (Fig. 4.13). If there are any phase transitions (for example, melting) in the temperature range of interest, then the entropy of each transition at the transition temperature is calculated like that in eqn 4.6 and its contribution added to the contributions from each of the phases, as shown in Fig. 4.14. The Third-Law entropy, which is commonly called simply 'the entropy', of a substance depends on the pressure; we therefore select a standard pressure (1 bar) and report the **standard molar entropy**, $S_m^{\ominus}$, the molar entropy of a substance in its standard state at the temperature of interest. Some values at 298.15 K (the conventional temperature for reporting data) are given in Table 4.2.

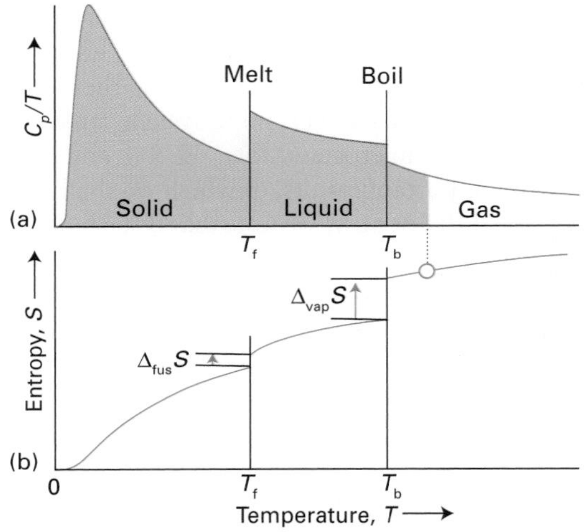

Fig. 4.14 The determination of entropy from heat capacity data. (a) Variation of C_p/T with the temperature of the sample. (b) The entropy, which is equal to the area beneath the upper curve up to the temperature of interest plus the entropy of each phase transition between $T = 0$ and the temperature of interest.

Table 4.2

*Standard molar entropies of some substances at 298.15 K**

Substance	$S_m^⊖/(J\ K^{-1}\ mol^{-1})$
Gases	
Ammonia, NH_3	192.5
Carbon dioxide, CO_2	213.7
Helium, He	126.2
Hydrogen, H_2	130.7
Neon, Ne	146.3
Nitrogen, N_2	191.6
Oxygen, O_2	205.1
Water vapour, H_2O	188.8
Liquids	
Benzene, C_6H_6	173.3
Ethanol, CH_3CH_2OH	160.7
Water, H_2O	69.9
Solids	
Calcium oxide, CaO	39.8
Calcium carbonate, $CaCO_3$	92.9
Copper, Cu	33.2
Diamond, C	2.4
Graphite, C	5.7
Lead, Pb	64.8
Magnesium carbonate, $MgCO_3$	65.7
Magnesium oxide, MgO	26.9
Sodium chloride, NaCl	72.1
Sucrose, $C_{12}H_{22}O_{11}$	360.2
Tin, Sn (white)	51.6
Sn (grey)	44.1

*See the *Data section* for more values.

Heat capacities can be measured only with great difficulty at very low temperatures, particularly close to $T = 0$. However, as remarked in Section 2.10, it has been found that many non-metallic substances have a heat capacity that obeys the **Debye T^3 law**:

At temperatures close to $T = 0$, $C_{p,\mathrm{m}} = aT^3$ Debye T^3 law   (4.12a)

where a is an empirical constant that depends on the substance and is found by fitting eqn 4.12a to a series of measurements of the heat capacity close to $T = 0$. With a determined, it is easy to deduce, as we show in the Derivation below, the molar entropy at low temperatures:

At temperatures close to $T = 0$, $S_\mathrm{m}(T) = \frac{1}{3}C_{p,\mathrm{m}}(T)$ Entropy at low temperatures   (4.12b)

That is, the molar entropy at the low temperature T is equal to one-third of the constant-pressure heat capacity at that temperature. The Debye T^3 law strictly applies to C_p, but C_V and C_p converge as $T \to 0$, so we can use it for estimating C_V too without significant error at low temperatures.

Derivation 4.4

Entropies close to $T = 0$

Once again, we use the general expression, eqn 4.4, for the entropy change accompanying a change of temperature deduced in Derivation 4.2, with ΔS interpreted as $S(T_\mathrm{f}) - S(T_\mathrm{i})$, taking molar values, and supposing that the heating takes place at constant pressure:

$$S_\mathrm{m}(T_\mathrm{f}) - S_\mathrm{m}(T_\mathrm{i}) = \int_{T_\mathrm{i}}^{T_\mathrm{f}} \frac{C_{p,\mathrm{m}}}{T}\,\mathrm{d}T$$

If we set $T_\mathrm{i} = 0$ and T_f some general temperature T, we transform this expression into

$$S_\mathrm{m}(T) - \overbrace{S_\mathrm{m}(0)}^{0} = \int_0^T \frac{C_{p,\mathrm{m}}}{T}\,\mathrm{d}T$$

According to the Third Law, $S(0) = 0$, and according to the Debye T^3 law, $C_{p,\mathrm{m}} = aT^3$, so

$$S_\mathrm{m}(T) = \int_0^T \frac{aT^3}{T}\,\mathrm{d}T = a\int_0^T T^2\,\mathrm{d}T$$

At this point we can use the standard integral

$$\int x^2\mathrm{d}x = \tfrac{1}{3}x^3 + \text{constant}$$

to write

$$\int_0^T T^2\mathrm{d}T = (\tfrac{1}{3}T^3 + \text{constant})\Big|_0^T$$
$$= (\tfrac{1}{3}T^3 + \text{constant}) - \text{constant} = \tfrac{1}{3}T^3$$

We can conclude that

$$S_\mathrm{m}(T) = \tfrac{1}{3}aT^3 = \tfrac{1}{3}C_{p,\mathrm{m}}(T)$$

as in eqn 4.12b.

4.9 The molecular interpretation of Third-Law entropies

We can readily verify that the Boltzmann formula (eqn 4.11) agrees with the Third-Law value $S(0) = 0$. When $T = 0$, all the molecules must be in the lowest possible energy level. Because there is just a single arrangement of the molecules, $W = 1$, and as $\ln 1 = 0$, eqn 4.11 gives $S = 0$ too. We can also see that the Boltzmann formula is consistent with the entropy of

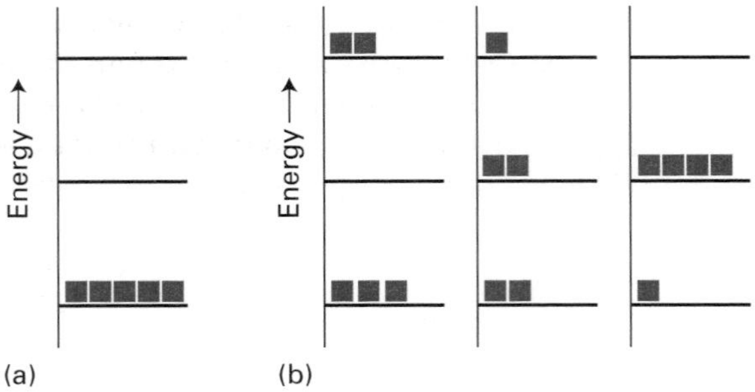

Fig. 4.15 The arrangements of molecules over the available energy levels determines the value of the statistical entropy. (a) At $T = 0$, there is only one arrangement possible: all the molecules must be in the lowest energy state. (b) When $T > 0$ several arrangements may correspond to the same total energy. In this simple case, $W = 3$.

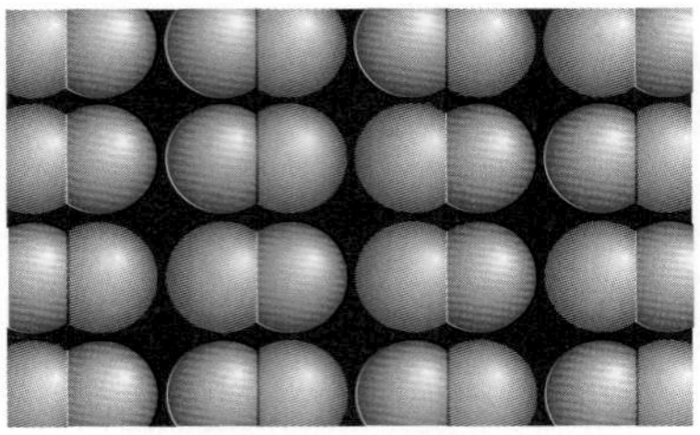

Fig. 4.16 The positional disorder of a substance that accounts for the residual entropy of molecules that can adopt either of two orientations at $T = 0$ (in this case, CO). If there are N molecules in the sample, there are 2^N possible arrangements with the same energy.

a substance always being positive (because $W \geq 1$, the natural logarithm term in eqn 4.11 is never negative) and increasing with temperature. When $T > 0$, the molecules of a sample can occupy energy levels above the lowest one, and now many different arrangements of the molecules will correspond to the same total energy (Fig. 4.15). That is, when $T > 0$, $W > 1$ and according to eqn 4.11 the entropy rises above zero (because $\ln W > 0$ when $W > 1$).

It is worth spending a moment to look at the values in Table 4.2 to see that they are consistent with our molecular interpretation of entropy. For example, the standard molar entropy of diamond (2.4 J K^{-1} mol^{-1}) is lower than that of graphite (5.7 J K^{-1} mol^{-1}). This difference is consistent with the atoms being linked less rigidly in graphite than in diamond and their thermal disorder being correspondingly greater. The standard molar entropies of ice, water, and water vapour at 25 °C are, respectively, 45, 70, and 189 J K^{-1} mol^{-1}, and the increase in values corresponds to the increasing disorder on going from a solid to a liquid and then to a gas.

The Boltzmann formula also provides an explanation of a rather startling observation: the entropy of some substances is greater than zero at $T = 0$, apparently contrary to the Third Law. When the entropy of carbon monoxide gas is measured thermodynamically (from heat capacity and boiling point data, and assuming from the Third Law that the entropy is zero at $T = 0$), it is found that $S_m^{\ominus}(298\text{ K}) = 192$ J K^{-1} mol^{-1}. However, when Boltzmann's formula is used and the relevant molecular data included, the standard molar entropy is calculated as 198 J K^{-1} mol^{-1}. One explanation might be that the thermodynamic calculation failed to take into account a phase transition in solid carbon monoxide, which could have contributed the missing 6 J K^{-1} mol^{-1}. An alternative explanation is that the CO molecules are disordered in the solid, even at $T = 0$, and that there is a contribution to the entropy at $T = 0$ from positional disorder that is frozen in. This contribution is called the **residual entropy** of a solid.

We can estimate the value of the residual entropy by using the Boltzmann formula and supposing that at $T = 0$ each CO molecule can lie in either of two orientations (Fig. 4.16). Then the total number of ways of arranging N molecules is $(2 \times 2 \times 2 \ldots)_{N\text{ times}} = 2^N$. Then

$$S = k \ln 2^N = Nk \ln 2$$

(We used $\ln x^a = a \ln x$; see The chemist's toolkit 2.2.) The molar residual entropy is obtained by replacing N by Avogadro's constant:

$$S_m = N_A k \ln 2 = R \ln 2$$

This expression evaluates to 5.8 J K^{-1} mol^{-1}, in good agreement with the value needed to bring the thermodynamic value into line with the statistical value, for instead of taking $S_m(0) = 0$ in the thermodynamic calculation, we should take $S_m(0) = 5.8$ J K^{-1} mol^{-1}.

Brief illustration 4.10 Residual entropy

Ice has a residual entropy of 3.4 J K^{-1} mol^{-1}. This value can be traced to the positional disorder of the location of the H atoms that lie between neighbouring molecules. Thus, although each H_2O molecule has two short O—H covalent bonds and two long O···H—O bonds, there is a randomness in which bonds are long and which are short (Fig. 4.17). When the statistics of the disorder are analysed for a sample that contains N molecules, it turns out that $W = (3/2)^N$. It follows that the residual entropy is expected to be $S(0) = k \ln (3/2)^N = Nk \ln 3/2$, and therefore the molar residual entropy is $S_m(0) = R \ln 3/2$, which evaluates to 3.4 J K^{-1} mol^{-1}, in agreement with the experimental value.

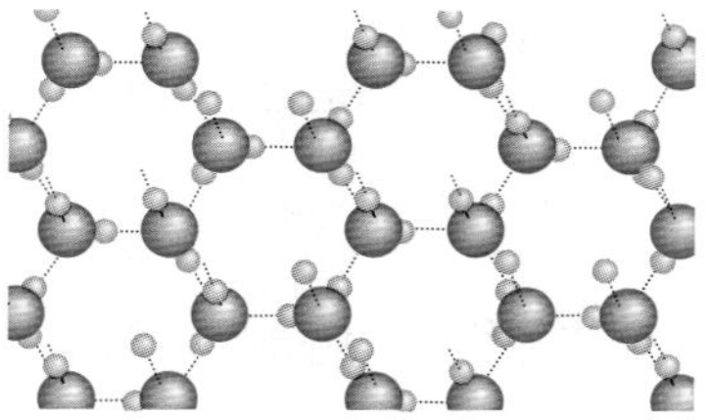

Fig. 4.17 The origin of the residual entropy of ice is the randomness in the location of the hydrogen atom in the O—H···O bonds between neighbouring molecules. Note that each molecule has two short O—H bonds and two long O···H hydrogen bonds. This schematic diagram shows one possible arrangement.

Self-test 4.6

Calculate the residual molar entropy of a solid in which the molecules can adopt six orientations of equal energy at $T = 0$.

Answer: $S_m(0) = 14.9\ \text{J K}^{-1}\ \text{mol}^{-1}$

4.10 The standard reaction entropy

Now we move into the arena of chemistry, where reactants are transformed into products. When there is a net formation of a gas in a reaction, as in combustion, we can usually anticipate that the entropy increases. When there is a net consumption of gas, as in photosynthesis, it is usually safe to predict that the entropy decreases. However, for a quantitative value of the change in entropy, and to predict the sign of the change when no gases are involved, we need to make an explicit calculation.

The difference in molar entropy between the products and the reactants in their standard states is called the **standard reaction entropy**, $\Delta_r S^\ominus$. It can be expressed in terms of the molar entropies of the substances in much the same way as we have already used for the standard reaction enthalpy:

$$\Delta_r S^\ominus = \sum \nu S_m^\ominus(\text{products}) - \sum \nu S_m^\ominus(\text{reactants}) \qquad \text{The standard reaction entropy} \qquad (4.13)$$

where the ν are the stoichiometric coefficients in the chemical equation.

Brief illustration 4.11 Standard reaction entropy

For the reaction $2\ H_2(g) + O_2(g) \rightarrow 2\ H_2O(l)$ we expect a negative entropy of reaction as gases are consumed. To find the explicit value we use the values in the *Data section* to write

$$\begin{aligned}\Delta_r S^\ominus &= 2S_m^\ominus(H_2O,l) - \{2S_m^\ominus(H_2,g) + S_m^\ominus(O_2,g)\}\\ &= 2(70\ \text{J K}^{-1}\ \text{mol}^{-1}) - \{2(131\ \text{J K}^{-1}\ \text{mol}^{-1})\\ &\quad + (205\ \text{J K}^{-1}\ \text{mol}^{-1})\}\\ &= -327\ \text{J K}^{-1}\ \text{mol}^{-1}\end{aligned}$$

A note on good practice Do not make the mistake of setting the standard molar entropies of elements equal to zero: they have nonzero values (provided $T > 0$), as we have already discussed.

Self-test 4.7

(a) Calculate the standard reaction entropy for $N_2(g) + 3\ H_2(g) \rightarrow 2\ NH_3(g)$ at 25 °C. (b) What is the change in entropy when 2 mol H_2 reacts?

Answer: (a) (Using values from Table 4.2) $-198.7\ \text{J K}^{-1}\ \text{mol}^{-1}$; (b) $-132.5\ \text{J K}^{-1}$

4.11 The spontaneity of chemical reactions

The result of the calculation in Brief illustration 4.11 should be rather surprising at first sight. We know that the reaction between hydrogen and oxygen is spontaneous and, once initiated, that it proceeds with explosive violence. Nevertheless, the entropy change that accompanies it is negative: the reaction results in less disorder, yet it is spontaneous!

The resolution of this apparent paradox underscores a feature of entropy that recurs throughout chemistry: *it is essential to consider the entropy of both the system and its surroundings when deciding whether a process is spontaneous or not.* The reduction in entropy by 327 J K^{-1} mol^{-1} in the reaction $2\ H_2(g) + O_2(g) \rightarrow 2\ H_2O(l)$ relates only to the system, the reaction mixture. To apply the Second Law correctly, we need to calculate the *total* entropy, the sum of the changes in the system and the surroundings that jointly compose the 'isolated system' referred to in the Second Law. It may well be the case that the entropy of the system decreases when a change takes place, but there may be a more than compensating increase in entropy of the surroundings so that overall the entropy change is positive. The opposite may also be true: a large decrease in entropy of the surroundings may occur when the entropy of the system increases. In that case we would be wrong to conclude from the increase of the system alone that the change is spontaneous. *Whenever considering the implications of entropy, we must always consider the total change of the system and its surroundings.*

To calculate the entropy change in the surroundings when a reaction takes place at constant pressure, we use eqn 4.10, interpreting the ΔH in that expression as the standard reaction enthalpy, $\Delta_r H^\ominus$.

Brief illustration 4.12 Entropy change in the surroundings resulting from a chemical reaction

Consider the water formation reaction, $2\ H_2(g) + O_2(g) \rightarrow 2\ H_2O(l)$, for which $\Delta_r H^{\ominus} = -572$ kJ mol^{-1}. The change in entropy of the surroundings (which are maintained at 25 °C, the same temperature as the reaction mixture) is

$$\Delta_r S_{sur} = -\frac{\Delta_r H^{\ominus}}{T} = -\frac{(-572\times10^3\ \text{J mol}^{-1})}{298\ \text{K}}$$

$$= +1.92\times10^3\ \text{J K}^{-1}\ \text{mol}^{-1}$$

Now we can see that the total entropy change is positive:

$$\Delta_r S_{total} = (-327\ \text{J K}^{-1}\ \text{mol}^{-1}) + (1.92\times10^3\ \text{J K}^{-1}\ \text{mol}^{-1})$$

$$= +1.59\times10^3\ \text{J K}^{-1}\ \text{mol}^{-1}$$

This calculation confirms that the reaction is spontaneous under standard conditions. In this case, the spontaneity is a result of the considerable disorder that the reaction generates in the surroundings: water is dragged into existence, even though $H_2O(l)$ has a lower entropy than the gaseous reactants, by the tendency of energy to disperse into the surroundings.

Self-test 4.8

Calculate the entropy change in the surroundings for the reaction $N_2(g) + 3\ H_2(g) \rightarrow 2\ NH_3(g)$ at 25 °C, given the standard enthalpy of reaction $\Delta_r H^{\ominus} = -92.2$ kJ mol^{-1} and standard entropy of reaction $\Delta_r S^{\ominus} = -199$ J K^{-1} mol^{-1} at this temperature.

Answer: +111 J K^{-1} mol^{-1}

The Gibbs energy

One of the problems with entropy calculations is already apparent: we have to work out two entropy changes, the change in the system and the change in the surroundings, and then consider the sign of their sum. The great American theoretician J.W. Gibbs (1839–1903), who laid the foundations of chemical thermodynamics towards the end of the nineteenth century, discovered how to combine the two calculations into one. The combination of the two procedures in fact turns out to be of much greater relevance than just saving a little labour, and throughout this text we shall see consequences of the procedure he developed.

4.12 Focussing on the system

The total entropy change that accompanies a process is $\Delta S_{total} = \Delta S + \Delta S_{sur}$, where ΔS is the entropy change for the system; for a spontaneous change, $\Delta S_{total} > 0$. If the process occurs at constant pressure and temperature, we can use eqn 4.10 to express the change in entropy of the surroundings in terms of the enthalpy change of the system, ΔH. When the resulting expression is inserted into this one, we obtain

$$\Delta S_{total} = \Delta S - \frac{\Delta H}{T} \quad \text{Constant temperature and pressure} \quad \text{Total entropy change} \qquad (4.14)$$

The great advantage of this formula is that it expresses the total entropy change of the system and its surroundings in terms of properties of the system alone. The only restriction is to changes at constant pressure and temperature.

Now we take a very important step. First, we introduce the **Gibbs energy**, G, which is defined as

$$G = H - TS \quad \text{Definition} \quad \text{Gibbs energy} \qquad (4.15)$$

The Gibbs energy is commonly referred to as the 'free energy' and the 'Gibbs free energy'. Because H, T, and S are state functions, G is a state function too. A change in Gibbs energy, ΔG, at constant temperature arises from changes in enthalpy and entropy, and is

$$\Delta G = \Delta H - T\Delta S \quad \text{Constant temperature} \quad \text{Gibbs energy change} \qquad (4.16)$$

By comparing eqns 4.14 and 4.16 we obtain

$$\Delta G = -T\Delta S_{total} \quad \text{Constant temperature and pressure} \quad \text{Gibbs energy change} \qquad (4.17)$$

We see that at constant temperature and pressure, the change in Gibbs energy of a system is proportional to the overall change in entropy of the system plus its surroundings.

The difference in sign between ΔG and ΔS_{total} implies that the condition for a process being spontaneous changes from $\Delta S_{total} > 0$ in terms of the total entropy (which is universally true) to $\Delta G < 0$ in terms of the Gibbs energy (for processes occurring at constant temperature and pressure). That is, *in a spontaneous change at constant temperature and pressure, the Gibbs energy decreases* (Fig. 4.18).

It may seem more natural to think of a system as falling to a lower value of some property. However, it must never be forgotten that to say that a system tends to fall to lower Gibbs energy is only a modified way of saying that a system and its surroundings jointly tend towards greater total entropy. The *only* criterion of spontaneous change is the total entropy of the system and its surroundings; the Gibbs energy merely contrives a way of expressing that total change in terms of the properties of the system alone, and is valid only for processes that occur at constant

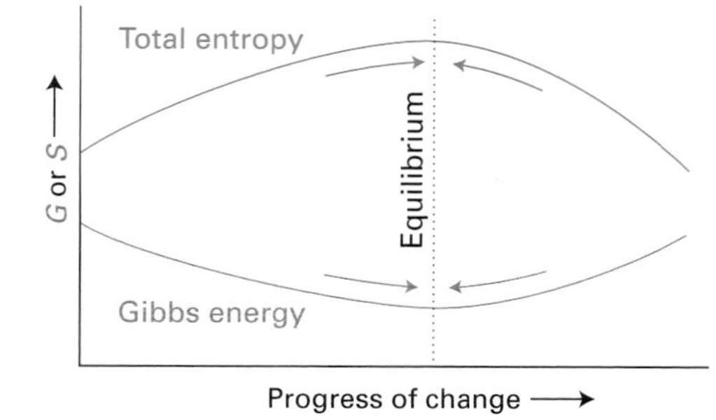

Fig. 4.18 (a) The criterion of spontaneous change is the increase in total entropy of the system and its surroundings. (b) Provided we accept the limitation of working at constant pressure and temperature, we can focus entirely on properties of the system, and express the criterion as a tendency to move to lower Gibbs energy.

temperature and pressure. Every chemical reaction that is spontaneous under conditions of constant temperature and pressure, including those that drive the processes of growth, learning, and reproduction, are reactions that change in the direction of lower Gibbs energy, or—another way of expressing the same thing—result in the overall entropy of the system and its surroundings becoming greater.

4.13 Properties of the Gibbs energy

As well as providing a criterion of spontaneous change, another feature of the Gibbs energy is that *the value of ΔG for a process gives the maximum non-expansion work that can be extracted from the process at constant temperature and pressure.* By **non-expansion work**, w', we mean any work other than that arising from the expansion of the system. It may include electrical work, if the process takes place inside an electrochemical or biological cell, or other kinds of mechanical work, such as the winding of a spring or the contraction of a muscle (we saw examples in Chapter 2). To demonstrate this property, we need to combine the First and Second Laws, and as shown in the following Derivation we find

$$\Delta G = w'_{\text{max}} \qquad \text{Constant temperature and pressure} \qquad \text{Gibbs energy and non-expansion work} \qquad (4.18)$$

Brief illustration 4.13 Non-expansion work

Experiments show that for the formation of 1 mol $H_2O(l)$ at 25 °C and 1 bar, $\Delta H = -286$ kJ and $\Delta G = -237$ kJ. It follows that up to 237 kJ of non-expansion work can be extracted from the reaction between hydrogen and oxygen to produce 1 mol $H_2O(l)$ at 25 °C. If the reaction takes place in a fuel cell—a device for using a chemical reaction to produce an electric current, like those used on the space shuttle—then up to 237 kJ of electrical energy can be generated for each mole of H_2O produced. This energy is enough to keep a 60 W light bulb shining for about 1.1 h. If no attempt is made to extract any energy as work, then 286 kJ (in general, ΔH) of energy will be produced as heat. If some of the energy released is used to do work, then up to 237 kJ (in general, ΔG) of non-expansion work can be obtained.

Derivation 4.5

Maximum non-expansion work

We need to consider infinitesimal changes, because dealing with reversible processes is then much easier. Our aim is to derive the relation between the infinitesimal change in Gibbs energy, dG, accompanying a process and the maximum amount of non-expansion work that the process can do, dw'. We start with the infinitesimal form of eqn 4.16:

At constant temperature: $dG = dH - TdS$

where, as usual, d denotes an infinitesimal difference. A good rule in the manipulation of thermodynamic expressions is to feed in definitions of the terms that appear. We do this twice. First, we use the expression for the change in enthalpy at constant pressure (eqn 2.13; $dH = dU + pdV$), and obtain

At constant temperature and pressure:
$dG = dH - TdS = dU + pdV - TdS$

Then we replace dU in terms of infinitesimal contributions from work and heat ($dU = dw + dq$):

$$dG = dU + pdV - TdS = dw + dq + pdV - TdS$$

A note on good practice Recall from Chapter 2 that, because heat and work are processes and the internal energy is a property, the symbol d carries a different meaning when it is attached to q, w, and U. Namely, dU denotes a very small *change* in the internal energy of a system whereas dq and dw denote very small *transfers* of energy as heat and work.

The work done on the system consists of expansion work, $-p_{ex}dV$, and non-expansion work, dw'. Therefore,

$$\begin{aligned} dG &= dw + dq + pdV - TdS \\ &= -p_{ex}dV + dw' + dq + pdV - TdS \end{aligned}$$

This derivation is valid for any process taking place at constant temperature and pressure.

Now we specialize to a reversible change. For expansion work to be reversible, we need to match p and p_{ex}, in which case the first and fourth terms on the right cancel. Moreover, because the heat transfer is also reversible, we can replace dq by TdS, in which case the third and fifth terms also cancel:

$$dG = -pdV + dw' + TdS + pdV - TdS$$

We are left with

At constant temperature and pressure, for a reversible process: $dG = dw'_{rev}$

Maximum work is done during a reversible change (Section 2.4), so another way of writing this expression is

At constant temperature and pressure: $dG = dw'_{max}$

Because this relation holds for each infinitesimal step between the specified initial and final states, it applies to the overall change too. Therefore, we obtain eqn 4.18.

Example 4.3

Estimating a change in Gibbs energy

Suppose a certain small bird has a mass of 30 g. What is the minimum mass of glucose that it must consume to fly to a branch 10 m above the ground? The change in Gibbs energy that accompanies the oxidation of 1.0 mol $C_6H_{12}O_6$(s) to carbon dioxide and water vapour at 25 °C is −2828 kJ.

Strategy First, we need to calculate the work needed to raise a mass *m* through a height *h* on the surface of the Earth. As we saw in eqn 0.11, this work is equal to *mgh*, where *g* is the acceleration of free fall. This work, which is non-expansion work, can be identified with ΔG. We need to determine the amount of substance that corresponds to the required change in Gibbs energy, and then convert that amount to a mass by using the molar mass of glucose.

Solution The non-expansion work to be done is

$$w' = mgh = (30 \times 10^{-3}\ \text{kg}) \times (9.81\ \text{m s}^{-2}) \times (10\ \text{m})$$
$$= 3.0 \times 9.81 \times 1.0 \times 10^{-1}\ \text{J}$$

(because 1 kg m^2 s^{-2} = 1 J). The amount, *n*, of glucose molecules required for oxidation to give a change in Gibbs energy of this value given that 1 mol provides 2828 kJ is

$$n = \frac{3.0\times 9.81\times 1.0\times 10^{-1}\ \text{J}}{2.828\times 10^{6}\ \text{J mol}^{-1}} \overset{\text{cancel J}}{=} \frac{3.0\times 9.81\times 1.0\times 10^{-7}}{2.828}\text{mol}$$

Therefore, because the molar mass, *M*, of glucose is 180 g mol^{-1}, the mass, *m*, of glucose that must be oxidized is

$$m = nM = \left(\frac{3.0\times 9.81\times 1.0\times 10^{-7}}{2.828}\text{mol}\right)\times(180\ \text{g mol}^{-1}) \overset{\text{cancel mol}}{=} 1.9\times 10^{-4}\ \text{g}$$

That is, the bird must consume at least 0.19 mg of glucose for the mechanical effort (and more if it thinks about it).

Self-test 4.9

A hard-working human brain, perhaps one that is grappling with physical chemistry, operates at about 25 W (1 W = 1 J s^{-1}). What mass of glucose must be consumed to sustain that power output for an hour?

Answer: 5.7 g

The great importance of the Gibbs energy in chemistry is becoming apparent. At this stage, we see that it is a measure of the non-expansion work resources of chemical reactions: if we know ΔG, then we know the maximum non-expansion work that we can obtain by harnessing the reaction in some way. In some cases, the non-expansion work is extracted as electrical energy. This is the case when the reaction takes place in an electrochemical cell, of which a fuel cell is a special case, as we see in Chapter 9. In other cases, the reaction may be used to build other molecules. This is the case in biological cells, where the Gibbs energy available from the hydrolysis of ATP (adenosine triphosphate) to ADP is used to build proteins from amino acids, to power muscular contraction, and to drive the neuronal circuits in our brains. Indeed, simple arguments involving entropy and the Gibbs energy lead easily to an explanation of why life is consistent with the Second Law (see Impact 4.2).

This interpretation can be expressed in terms of $\Delta G = \Delta H - T\Delta S$. Thus, we saw in Section 2.9 that *in the absence of non-expansion work* ΔH can be identified with the energy transferred as heat (at constant pressure). But in a reversible process $T\Delta S$ is also equal to the energy transferred as heat even if other processes, including non-expansion work, are taking place. Therefore, the difference between these two quantities must represent the energy transfer due to these other processes, and specifically to the non-expansion work that the system does.

Impact on biology 4.2

Life and the Second Law

Every chemical reaction that is spontaneous under conditions of constant temperature and pressure, including those driving the processes of growth, learning, and reproduction, is a reaction that proceeds in the direction of lower Gibbs energy, or—another way of expressing the same thing—results in the overall entropy of the system and its surroundings becoming greater. With these ideas in mind, it is possible to explain why life, which can be regarded as a collection of biological processes taking place in a highly organized body, proceeds in accord with the Second Law of thermodynamics.

Conditions in a cell ensure that many of the reactions that lead to the breakdown of food in organisms are spontaneous. For instance, the dissociation of large molecules, such as sugars and lipids, into smaller molecules leads to the dispersal of matter in the cell. Energy is also dispersed, as it is released upon reorganization of bonds in foods when they are oxidized. More difficult to rationalize is life's requirement of organization of a very large number of molecules into biological cells, which in turn assemble into organisms. To be sure, the entropy of the system—the organism—is very low because matter becomes less dispersed when molecules assemble to form cells, tissues, organs, and so on. However, the lowering of the system's entropy comes at the expense of an increase in the entropy of the surroundings.

To understand this point, we need to know that cells grow and act by converting energy from the Sun or oxidation of foods partially into work. The remaining energy is released as heat into the surroundings, so $q_{sur} > 0$ and $\Delta S_{sur} > 0$. As with any process, life is spontaneous and organisms thrive as long as the increase in the entropy of the organism's environment compensates for decreases in the entropy arising from the assembly of the organism. Alternatively, we may say that $\Delta G < 0$ for the overall sum of physical and chemical changes that we call life.

Checklist of key concepts

- ☐ **1** A spontaneous change is a change that has a tendency to occur without work having to be done to bring it about.
- ☐ **2** Matter and energy tend to disperse.
- ☐ **3** The Second Law states that the entropy of an isolated system tends to increase.
- ☐ **4** In general, the entropy change accompanying the heating of a system is equal to the area under the graph of C/T against T between the two temperatures of interest.
- ☐ **5** The Third Law of thermodynamics states that the entropies of all perfectly crystalline substances are the same at $T = 0$ (and may be taken to be zero).
- ☐ **6** The residual entropy of a substance is its entropy at $T = 0$ due to any positional disorder that remains.
- ☐ **7** At constant temperature and pressure, a system tends to change in the direction of decreasing Gibbs energy.
- ☐ **8** At constant temperature and pressure, the change in Gibbs energy accompanying a process is equal to the maximum non-expansion work the process can do.

Road map of key equations

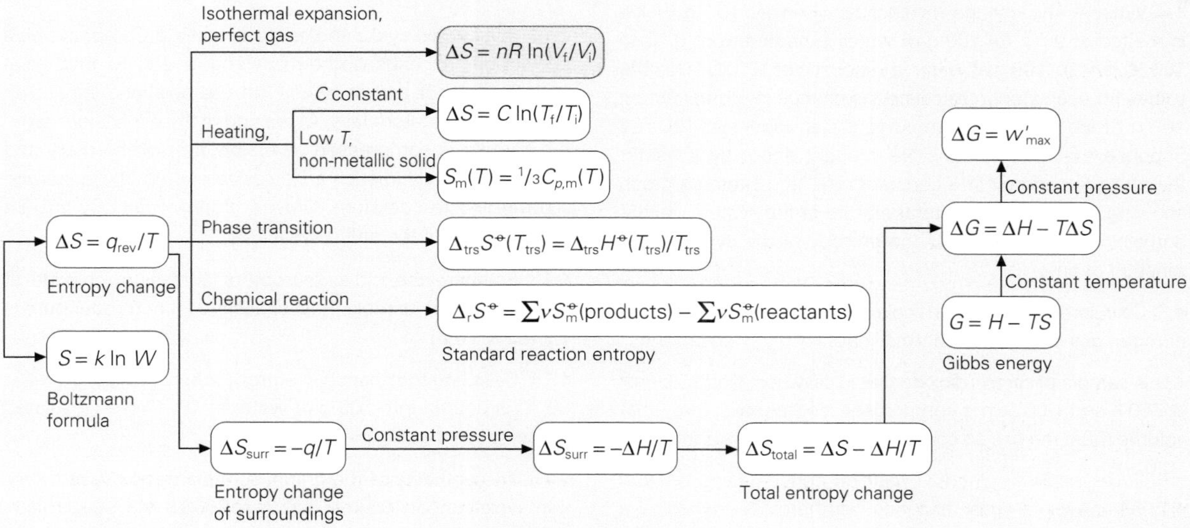

A blue box indicates a relation valid only for perfect gases.

Questions and exercises

Discussion questions

4.1 Explain the single criterion for a physical process to be classified as spontaneous.

4.2 Explain in molecular terms why the entropy of a gas increases (a) with volume, (b) with temperature.

4.3 Justify Trouton's rule. What are the sources of discrepancies?

4.4 Justify the identification of the statistical entropy with the thermodynamic entropy.

4.5 Under what circumstances may the properties of the system alone be used to identify the direction of spontaneous change?

4.6 The evolution of life requires the organization of a very large number of molecules into biological cells. Does the formation of living organisms violate the Second Law of thermodynamics? State your conclusion clearly and present detailed arguments to support it.

Exercises

Assume all gases are perfect unless stated otherwise. All thermochemical data are for 298.15 K.

4.1 A goldfish swims in a bowl of water at 20 °C. Over a period of time, the fish transfers 120 J to the water as a result of its metabolism. What is the change in entropy of the water, assuming no losses to the environment?

4.2 Suppose you put a cube of ice of mass 100 g into a glass of water at just above 0 °C. When the ice melts, about 33 kJ of energy is absorbed from the surroundings as heat. What is the change in entropy of (a) the sample (the ice), (b) the surroundings (the glass of water)?

4.3 A sample of aluminium of mass 1.00 kg is cooled at constant pressure from 300 K to 250 K. Calculate the energy that must be removed as heat and the change in entropy of the sample. The molar heat capacity of aluminium is 24.35 J K^{-1} mol^{-1}.

4.4 What is the change in entropy when (a) 100 g of ice is melted at 0 °C, (b) 100 g of water is heated from 0 °C to 100 °C, and (c) 100 g of water is vaporized at 100 °C? Use the values for each step to calculate the change in entropy when 100 g of ice at is transformed to water vapour at 100 °C? Suppose that the changes are brought about by a heater that supplies energy at a constant rate, and sketch a graph showing (a) the change in temperature of the system, (b) the enthalpy of the system, (c) the entropy of the system as a function of time.

4.5 Calculate the change in molar entropy when a sample of nitrogen gas expands isothermally from 1.0 dm^3 to 5.5 dm^3.

4.6 A sample of carbon dioxide that initially occupies 15.0 dm^3 at 250 K and 1.00 atm is compressed isothermally. Into what volume must the gas be compressed to reduce its entropy by 10.0 J K^{-1}?

4.7 Whenever a gas expands isothermally—when we exhale, when a flask is opened, and so on—the gas undergoes an increase in entropy. A sample of methane gas of mass 15 g at 260 K and 105 kPa expands isothermally and (a) reversibly, (b) irreversibly until its pressure is 1.50 kPa. Calculate the change in entropy of the gas.

4.8 What is the change in entropy of 100 g of water when it is heated from room temperature (20 °C) to body temperature (37 °C)? Use $C_{p,m}$ = 75.5 J K^{-1} mol^{-1}.

4.9 Calculate the change in entropy of 1.0 kg of lead when it cools from 500 °C to 100 °C. Take $C_{p,m}$ = 26.44 J K^{-1} mol^{-1}.

4.10 Calculate the change in molar entropy when a sample of argon is compressed from 2.0 dm^3 to 500 cm^3 and simultaneously heated from 300 K to 400 K. Take $C_{V,m} = \frac{3}{2}R$.

4.11 A monatomic perfect gas at a temperature T_i is expanded isothermally to twice its initial volume. To what temperature should it be cooled to restore its entropy to its initial value? Take $C_{V,m} = \frac{3}{2}R$.

4.12 In a certain cyclic engine (technically, a *Carnot cycle*), a perfect gas expands isothermally and reversibly, then adiabatically ($q = 0$) and reversibly. In the adiabatic expansion step the temperature falls. At the end of the expansion stage, the sample is compressed reversibly first isothermally and then adiabatically in such a way as to end up at the starting volume and temperature. Draw a graph of entropy against temperature for the entire cycle.

4.13 Estimate the molar entropy of potassium chloride at 5.0 K given that its molar heat capacity at that temperature is 1.2 mJ K^{-1} mol^{-1}.

4.14 Calculate the change in entropy when 100 g of water at 80 °C is poured into 100 g of water at 10 °C in an insulated vessel given that $C_{p,m}$ = 75.5 J K^{-1} mol^{-1}.

4.15 The enthalpy of the graphite → diamond phase transition, which under 100 kbar occurs at 2000 K, is +1.9 kJ mol^{-1}. Calculate the entropy of transition at this temperature.

4.16 The enthalpy of vaporization of chloroform (trichloromethane), $CHCl_3$, is 29.4 kJ mol^{-1} at its normal boiling point of 334.88 K. (a) Calculate the entropy of vaporization of chloroform at this temperature. (b) What is the entropy change in the surroundings?

4.17 Calculate the entropy of fusion of a compound at 25 °C given that its enthalpy of fusion is 36 kJ mol^{-1} at its melting point of 151 °C and the molar heat capacities (at constant pressure) of the liquid and solid forms are 33 J K^{-1} mol^{-1} and 17 J K^{-1} mol^{-1}, respectively.

4.18 Octane is typical of the components of gasoline. Estimate (a) the entropy of vaporization, (b) the enthalpy of vaporization of octane, which boils at 126 °C at 1 atm.

4.19 Suppose that the weight of a configuration of N molecules in a gas of volume V is proportional to V^N. Use Boltzmann's formula to deduce the change in entropy when the gas expands isothermally from V_i to V_f.

4.20 An $FClO_3$ molecule can adopt four orientations in the solid with negligible difference in energy. What is its residual molar entropy?

4.21 Without performing a calculation, estimate whether the standard entropies of the following reactions are positive or negative:

(a) Ala–Ser–Thr–Lys–Gly–Arg–Ser $\xrightarrow{\text{trypsin}}$ Ala–Ser–Thr–Lys + Gly–Arg

(b) $N_2(g) + 3\ H_2(g) \rightarrow 2\ NH_3(g)$

(c) $ATP^{4-}(aq) + 2\ H_2O(l) \rightarrow ADP^{3-}(aq) + HPO_4^{2-}(aq) + H_3O^+(aq)$

4.22 Use the information on standard molar entropies in the *Data section* to calculate the standard reaction entropy at 298 K of

(a) $2\ CH_3CHO(g) + O_2(g) \rightarrow 2\ CH_3COOH(l)$

(b) $2\ AgCl(s) + Br_2(l) \rightarrow 2\ AgBr(s) + Cl_2(g)$

(c) $Hg(l) + Cl_2(g) \rightarrow HgCl_2(s)$

(d) $Zn(s) + Cu^{2+}(aq) \rightarrow Zn^{2+}(aq) + Cu(s)$

(e) $C_{12}H_{22}O_{11}(s) + 12\ O_2(g) \rightarrow 12\ CO_2(g) + 11\ H_2O(l)$

4.23 Suppose that when you exercise, you consume 100 g of glucose and that all the energy released as heat remains in your body at 37 °C. What is the change in entropy of your body?

4.24 Calculate the standard reaction entropy and the change in entropy of the surroundings (at 298 K) of the reaction $N_2(g) + 3\ H_2(g) \rightarrow 2\ NH_3(g)$.

4.25 The constant-pressure molar heat capacities of linear gaseous molecules are approximately $\frac{7}{2}R$ and those of nonlinear gaseous molecules are approximately $4R$. Estimate the change in standard reaction entropy of the following two reactions when the temperature is increased by 10 K at constant pressure:

(a) $2\ H_2(g) + O_2(g) \rightarrow 2\ H_2O(g)$

(b) $CH_4(g) + 2\ O_2(g) \rightarrow CO_2(g) + 2\ H_2O(g)$

4.26 Use the information you deduced in Exercise 4.24 to calculate the standard Gibbs energy of reaction of $N_2(g) + 3\ H_2(g) \rightarrow 2\ NH_3(g)$.

4.27 In a particular biological reaction taking place in the body at 37 °C, the change in enthalpy was −135 kJ mol^{-1} and the change in entropy was −136 J K^{-1} mol^{-1}. (a) Calculate the change in Gibbs energy. (b) Is the reaction spontaneous? (c) Calculate the total change in entropy of the system and the surroundings.

4.28 The change in Gibbs energy that accompanies the oxidation of $C_6H_{12}O_6(s)$ to carbon dioxide and water vapour at 25 °C is −2828 kJ mol^{-1}. How much glucose does a person of mass 65 kg need to consume to climb through 10 m?

4.29 Fuel cells are being developed that make use of organic fuels; in due course they might be used to power tiny intravenous machines for carrying out repairs on diseased tissue. What is the maximum non-expansion work that can be obtained from the metabolism of 1.0 mg of sucrose to carbon dioxide and water?

4.30 The formation of glutamine from glutamate and ammonium ions requires 14.2 kJ mol^{-1} of energy input. It is driven by the hydrolysis of ATP to ADP mediated by the enzyme glutamine synthetase. (a) Given that the change in Gibbs energy for the hydrolysis of ATP corresponds to $\Delta G = -31$ kJ mol^{-1} under the conditions prevailing in a typical cell, can the hydrolysis drive the formation of glutamine? (b) How many moles of ATP must be hydrolysed to form 1 mol glutamine?

4.31 The hydrolysis of acetyl phosphate has $\Delta G = -42$ kJ mol^{-1} under typical biological conditions. If acetyl phosphate were to be synthesized by coupling to the hydrolysis of ATP, what is the minimum number of ATP molecules that would need to be involved?

4.32 Suppose that the radius of a typical cell is 10 μm and that inside it 106 ATP molecules are hydrolysed each second. What is the power density of the cell in watts per cubic metre (1 W = 1 J s^{-1})? A computer battery delivers about 15 W and has a volume of 100 cm^3. Which has the greater power density, the cell or the battery? (For data, see Exercise 4.31.)

Projects

The symbol ‡ indicates that calculus is required.

4.33‡ Equation 4.3 is based on the assumption that the heat capacity is independent of temperature. Suppose, instead, that the heat capacity depends on temperature as $C = a + bT + c/T^2$. Find an expression for the change of entropy accompanying heating from T_i to T_f. *Hint:* See Derivation 4.2.

4.34 Here we explore the thermodynamics of refrigerators and heat pumps. (a) Follow the arguments used in the definition of the efficiency of a heat engine in Impact on technology *4.1*, to show that the best *coefficient of cooling performance*, c_{cool}, the ratio of the energy extracted as heat at T_{cold} to the energy supplied as work in a perfect refrigerator, is $c_{cool} = T_{cold}/(T_{hot} - T_{cold})$. What is the maximum rate of extraction of energy as heat in a domestic refrigerator rated at 200 W operating at 5.0 °C in a room at 22 °C? (b) Show that the best *coefficient of heating performance*, c_{warm}, the ratio of the energy produced as heat at T_{hot} to the energy supplied as work in a perfect heat pump, is $c_{warm} = T_{hot}/(T_{hot} - T_{cold})$. What is the maximum power rating of a heat pump that consumes power at 2.5 kW operating at 18.0 °C and warming a room at 22 °C?

4.35‡ The temperature dependence of the heat capacity of non-metallic solids is found to follow the Debye T^3 law at very low temperatures, with $C_{p,m} = aT^3$. (a) Derive an expression for the change in molar entropy on heating for such a solid. (b) For solid nitrogen $a = 6.15 \times 10^{-3}$ J K^{-4} mol^{-1}. What is the molar entropy of solid nitrogen at 5 K?

5

Physical equilibria: pure substances

Boiling, freezing, and the conversion of graphite to diamond are all examples of **phase transitions**, or changes of phase without change of chemical composition. Many phase changes are common everyday phenomena and their description is an important part of physical chemistry. They occur whenever a solid changes into a liquid, as in the melting of ice, or a liquid changes into a vapour, as in the vaporization of water in our lungs. They also occur when one solid phase changes into another, as in the conversion of graphite into diamond under high pressure, or the conversion of one phase of iron into another as it is heated in the process of steelmaking. The tendency of a substance to form a liquid crystal, a distinct phase with properties intermediate between those of solid and liquid, guides the design of displays for electronic devices. Phase changes are important geologically too; for example, calcium carbonate is typically deposited as aragonite, but then gradually changes into another crystal form, calcite.

This chapter deals with two aspects of phase transitions. In the first part, we explore what thermodynamics has to say about the response of phase transitions to changes in pressure and temperature. In the second part, we see how regions of phase stability can be reported diagrammatically.

The thermodynamics of transition 113

5.1 The condition of stability 114

5.2 The variation of Gibbs energy with pressure 114

5.3 The variation of Gibbs energy with temperature 116

Phase diagrams 118

5.4 Phase boundaries 119

5.5 The location of phase boundaries 120

5.6 Characteristic points 124

5.7 The phase rule 126

5.8 Phase diagrams of typical materials 127

5.9 The molecular structure of liquids 129

CHECKLIST OF KEY CONCEPTS 130

ROAD MAP OF KEY EQUATIONS 131

QUESTIONS AND EXERCISES 131

The thermodynamics of transition

The Gibbs energy, $G = H - TS$, of a substance, with H its enthalpy, T its temperature, and S its entropy, will be at centre stage in all that follows. We need to know how its value depends on the pressure and temperature. As we work out these dependencies, we shall acquire deep insight into the thermodynamic properties of matter and the transitions it can undergo.

5.1 The condition of stability

First, we need to establish the importance of the *molar* Gibbs energy, $G_m = G/n$, in the discussion of phase transitions of a pure substance. The molar Gibbs energy, an intensive property, depends on the phase of the substance. For instance, the molar Gibbs energy of liquid water is in general different from that of water vapour at the same temperature and pressure. When an amount n of the substance changes from phase 1 (for instance, liquid), with molar Gibbs energy $G_m(1)$ to phase 2 (for instance, vapour) with molar Gibbs energy $G_m(2)$, the change in Gibbs energy is

$$\Delta G = nG_m(2) - nG_m(1) = n\{G_m(2) - G_m(1)\}$$

We know that a spontaneous change at constant temperature and pressure is accompanied by a negative value of ΔG. This expression shows, therefore, that a change from phase 1 to phase 2 is spontaneous if the molar Gibbs energy of phase 2 is lower than that of phase 1 because then $G_m(2) - G_m(1) < 0$. In other words:

> A substance has a spontaneous tendency to change into the phase with the lowest molar Gibbs energy.

If at a certain temperature and pressure the solid phase of a substance has a lower molar Gibbs energy than its liquid phase, then the solid phase is thermodynamically more stable and the liquid will (or at least has a tendency to) freeze. If the opposite is true, the liquid phase is thermodynamically more stable and the solid will melt. For example, at 1 atm, ice has a lower molar Gibbs energy than liquid water when the temperature is below 0 °C, and under these conditions the conversion of water into ice is spontaneous.

Brief illustration 5.1 Thermodynamic stability

The Gibbs energy of transition from metallic white tin (α-Sn) to non-metallic grey tin (β-Sn) is +0.13 kJ mol^{-1} at 298 K. We saw in Section 3.5 that the reference state of an element is defined as its most stable form under the prevailing conditions. The molar Gibbs energy of metallic white tin (α-Sn) is lower than that of non-metallic grey tin (β-Sn) by 0.13 kJ mol^{-1}. The thermodynamically most stable form, and therefore the reference state at 298 K, is therefore metallic white tin (α-Sn).

5.2 The variation of Gibbs energy with pressure

To discuss how phase transitions depend on the pressure, we need to know how the molar Gibbs energy varies with pressure. We show in the following Derivation that when the temperature is held constant and the pressure is changed by a small amount Δp, the molar Gibbs energy of a substance changes by

$$\Delta G_m = V_m \Delta p \qquad (5.1)$$

where V_m is the molar volume of the substance. This expression is valid when the molar volume is constant in the pressure range of interest.

Derivation 5.1

The variation of *G* with pressure

We start with the definition of Gibbs energy, $G = H - TS$, and change the temperature, volume, and pressure by an infinitesimal amount. As a result, H changes to $H + dH$. T changes to $T + dT$, S changes to $S + dS$, and G changes to $G + dG$. After the change

$$G + dG = H + dH - (T + dT)(S + dS)$$
$$= H + dH - TS - TdS - SdT - dTdS$$

The G on the left cancels the $H - TS$ (in blue) on the right, the doubly infinitesimal $dTdS$ is so small that it can be neglected, and we are left with

$$dG = dH - TdS - SdT$$

To make progress, we need to know how the enthalpy changes. From its definition $H = U + pV$, in a similar way (letting U change to $U + dU$, and so on:

$$H + dH = U + dU + (p + dp)(V + dV)$$
$$= U + dU + pV + pdV + Vdp + dpdV$$

Then, neglecting the doubly infinitesimal term $dpdV$ and letting the H on the left cancel the $U + pV$ on the right (in blue) we can write

$$dH = dU + pdV + Vdp$$

On substituting this expression into the previous one, we obtain

$$dG = dU + pdV + Vdp - TdS - SdT$$

At this point we need to know how the internal energy changes, dU, when infinitesimal quantities of energy as heat and work, dq and dw, are supplied, and write

$$dU = dq + dw$$

If initially we consider only reversible changes, we can replace dq by TdS (because $dS = dq_{rev}/T$) and dw by $-pdV$ (because $dw = -p_{ex}dV$ and $p_{ex} = p$ for a reversible change), and obtain

$$dU = TdS - pdV$$

Now we substitute this expression into the expression for dG and obtain

$$dG = TdS - pdV + pdV + Vdp - TdS - SdT$$

We are left with the important result that

$$dG = Vdp - SdT \tag{5.2}$$

Now here is a subtle but important point. To derive this result we have supposed that the changes in conditions have been made reversibly. However, G is a state function, and so the change in its value is independent of path. Therefore, eqn 5.2 is valid for any change, not just a reversible change.

At this point we decide to keep the temperature constant, and set $dT = 0$ in eqn 5.2; this leaves

$$dG = Vdp$$

and, for molar quantities, $dG_m = V_m dp$. This expression is exact, but applies only to an infinitesimal change in the pressure. For an observable change, we replace dG_m and dp by ΔG_m and Δp, respectively, and obtain eqn 5.1, provided the molar volume is constant over the range of interest.

A note on good practice When confronted with a proof in thermodynamics, go back to fundamental definitions (as we did three times in succession in this derivation: first of G, then of H, and finally of U).

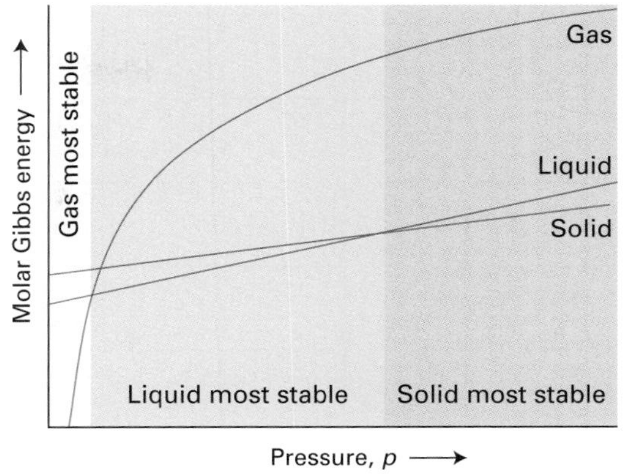

Fig. 5.1 The variation of molar Gibbs energy with pressure. The region where the molar Gibbs energy of a particular phase is least is shown by a green line and the corresponding region of stability of each phase is indicated in the coloured bands across the illustration.

What does eqn 5.1 tell us? First of all, note it tells us that, because all molar volumes are positive:

- An increase in pressure ($\Delta p > 0$) results in an increase in molar Gibbs energy ($\Delta G_m > 0$).

Next, we also see that:

- For a given change in pressure, the resulting change in molar Gibbs energy is greatest for substances with large molar volumes.

Therefore, because the molar volume of a gas is much larger than that of a condensed phase (a liquid or a solid), the dependence of G_m on p is much greater for a gas than for a condensed phase. For most substances (water is an important exception), the molar volume of the liquid phase is greater than that of the solid phase. Therefore, for most substances, the slope of a graph of G_m against p is greater for a liquid than for a solid. These characteristics are illustrated in Fig. 5.1.

As we see from Fig. 5.1, when we increase the pressure on a substance, the molar Gibbs energy of the gas phase rises above that of the liquid, then the molar Gibbs energy of the liquid rises above that of the solid. Because the system has a tendency to convert into the state of lowest molar Gibbs energy, the graphs show that at low pressures the gas phase is the most stable, then at higher pressures the liquid phase becomes the most stable, followed by solid phase. In other words, under pressure the substance condenses to a liquid, and then further pressure can result in the formation of a solid.

We can use eqn 5.1 to predict the actual shape of graphs like those in Fig. 5.1. For a solid or liquid, the molar volume is almost independent of pressure, so eqn 5.1 is an excellent approximation to the change in molar Gibbs energy, and after writing $\Delta G_m = G_m(p_f) - G_m(p_i)$ and $\Delta p = p_f - p_i$ we obtain

$$G_m(p_f) = G_m(p_i) + V_m(p_f - p_i) \quad \text{Liquid or solid} \quad \text{Pressure dependence of } G_m \tag{5.3a}$$

This equation shows that the molar Gibbs energy of a solid or liquid increases linearly with pressure. However, because the molar volume of a condensed phase is so small, the dependence is very weak, and for the typical ranges of pressure normally of interest to us we can ignore the pressure dependence of G. In contrast, the molar Gibbs energy of a gas does depend on the pressure, and because the molar volume of a gas is large, the dependence is significant. We show in the following Derivation that

$$G_m(p_f) = G_m(p_i) + RT \ln\frac{p_f}{p_i} \quad \text{Perfect gas} \quad \text{Pressure dependence of } G_m \tag{5.3b}$$

This equation shows that the molar Gibbs energy increases logarithmically (as $\ln p$) with the pressure (Fig. 5.2). The flattening of the curve at high pressures reflects the fact that as V_m gets smaller, G_m becomes less responsive to pressure.

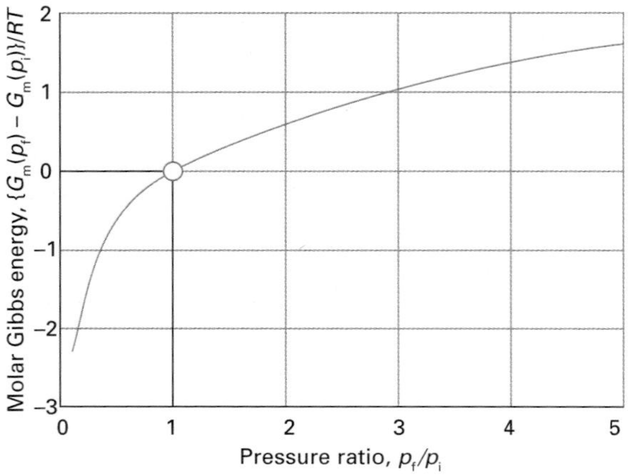

Fig. 5.2 The variation of the molar Gibbs energy of a perfect gas with pressure.

Derivation 5.2

The pressure variation of the Gibbs energy of a perfect gas

We start with the exact expression for the effect of an infinitesimal change in pressure obtained in Derivation 5.1, that $dG_m = V_m dp$. For a change in pressure from p_i to p_f, we need to add together (integrate) all these infinitesimal changes, and write

$$\Delta G_m = \int_{p_i}^{p_f} V_m dp$$

To evaluate the integral, we must know how the molar volume depends on the pressure. For a perfect gas $V_m = RT/p$. Then

$$\Delta G_m = \int_{p_i}^{p_f} V_m dp \overset{V_m=RT/p}{\overbrace{=}} \int_{p_i}^{p_f} \frac{RT}{p} dp \overset{RT \text{ is constant}}{\overbrace{=}} RT \int_{p_i}^{p_f} \frac{1}{p} dp$$
$$= RT \ln \frac{p_f}{p_i}$$

In the last line we have used the standard integral, which we discussed in The chemist's toolkit 2.1:

$$\int \frac{1}{x} dx = \ln x + \text{constant}$$

Finally, with $\Delta G_m = G_m(p_f) - G_m(p_i)$, we get eqn 5.3b.

Example 5.1

Assessing the variation of Gibbs energy with pressure

At 0 °C liquid water and ice are in equilibrium, so they have the same molar Gibbs energy. What is the effect on the difference in molar Gibbs energy when the pressure is increased from 1.0 bar to 100 bar? The mass densities of the two phases at 0 °C are $\rho(\text{ice}) = 0.9150\ \text{g cm}^{-3}$ and $\rho(\text{liquid}) = 0.9999\ \text{g cm}^{-3}$.

Strategy Begin by setting up an expression for the difference in molar Gibbs energies at a given pressure, and then use eqn 5.3a to express how that difference changes when the pressure is changed. The molar volume is related to the mass density by $V_m = M/\rho$, where M is the molar mass of water (18.02 g mol^{-1}).

Solution The difference in molar Gibbs energies at a pressure p_f is

$$\begin{aligned}\Delta G_m(p_f) &= G_m(\text{liquid},p_f) - G_m(\text{ice},p_f)\\ &= \{G_m(\text{liquid},p_i) + V_m(\text{liquid})(p_f - p_i)\} - \{G_m(\text{ice},p_i) \\ &\quad + V_m(\text{ice})(p_f - p_i)\}\\ &= \Delta G_m(p_i) + \{V_m(\text{liquid}) - V_m(\text{ice})\}(p_f - p_i)\end{aligned}$$

Because the difference is zero at the initial pressure (because the phases are in equilibrium),

$$\Delta G_m(p_f) = \{V_m(\text{liquid}) - V_m(\text{ice})\}(p_f - p_i)$$
$$\overset{V_m=M/\rho}{\overbrace{=}} M\left\{\frac{1}{\rho(\text{liquid})} - \frac{1}{\rho(\text{ice})}\right\}(p_f - p_i)$$

Now insert the data:

$$\Delta G_m = \overbrace{(18.02\ \text{g mol}^{-1})}^{M}$$
$$\times \left\{\underbrace{\frac{1}{0.9999\ \text{g cm}^{-3}}}_{\rho(\text{liquid})} - \underbrace{\frac{1}{0.9150\ \text{g cm}^{-3}}}_{\rho(\text{ice})}\right\} \times \overbrace{(99\ \text{kPa})}^{(p_f - p_i)}$$
$$= -165\ \text{kPa cm}^3\ \text{mol}^{-1} = -165 \times \left(\overbrace{10^3\ \overbrace{\text{N m}^{-2}}^{\text{Pa}}}^{\text{kPa}}\right)$$
$$\times \left(\overbrace{10^{-6}\ \text{m}^3}^{1\ \text{cm}^3}\right)\text{mol}^{-1}$$
$$= -165 \times \overbrace{10^{-3}\ \overbrace{\text{N m}}^{\text{J}}}^{\text{mJ}}\ \text{mol}^{-1} = -165\ \text{mJ mol}^{-1}$$

The difference is negative, so the transition from ice to liquid is spontaneous under this increased pressure.

Self-test 5.1

Calculate the difference in molar Gibbs energies of liquid water and its vapour at 100 °C when the pressure is increased from 1.00 bar to 2.00 bar. What change becomes spontaneous? You can ignore the tiny effect of pressure on the liquid phase.

Answer: +2.15 kJ mol^{-1}; condensation becomes spontaneous

5.3 The variation of Gibbs energy with temperature

Now we consider how the molar Gibbs energy varies with temperature. For small changes in temperature, we show in the following Derivation that the change in molar Gibbs energy at constant pressure is

$$\Delta G_m = -S_m \Delta T \qquad \text{Dependence of } G_m \text{ on temperature} \qquad (5.4)$$

where $\Delta G_m = G_m(T_f) - G_m(T_i)$ and $\Delta T = T_f - T_i$. This expression is valid provided the entropy of the substance is unchanged over the range of temperatures of interest.

Derivation 5.3

The variation of the Gibbs energy with temperature

The starting point for this short derivation is eqn 5.2 ($dG = Vdp - SdT$), the expression obtained in Derivation 5.1 for the change in molar Gibbs energy when both the pressure and the temperature are changed by infinitesimal amounts. If we hold the pressure constant, $dp = 0$, and eqn 5.2 becomes (for molar quantities)

$$dG_m = -S_m dT$$

This expression is exact. If we suppose that the molar entropy is unchanged in the range of temperatures of interest, we can replace the infinitesimal changes by observable changes, and so obtain eqn 5.4.

Equation 5.4 tells us that, because molar entropy is positive,

An increase in temperature ($\Delta T > 0$) results in a decrease in G_m ($\Delta G_m < 0$).

We see that, for a given change of temperature, the change in molar Gibbs energy is proportional to the molar entropy. For a given substance, there is more spatial disorder in the gas phase than in a condensed phase, so the molar entropy of the gas phase is greater than that for a condensed phase. It follows that the molar Gibbs energy falls more steeply with temperature for a gas than for a condensed phase. The molar entropy of the liquid phase of a substance is greater than that of its solid phase, so the slope is least steep for a solid. Figure 5.3 summarizes these characteristics.

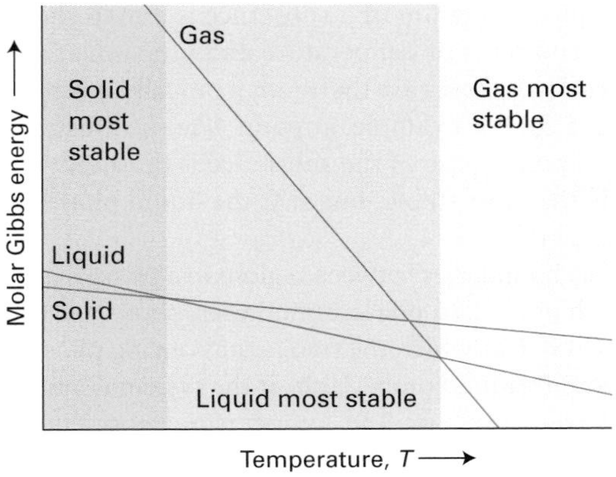

Fig. 5.3 The variation of molar Gibbs energy with temperature. All molar Gibbs energies decrease with increasing temperature. The regions of temperature over which the solid, liquid, and gaseous forms of a substance have the lowest molar Gibbs energy are indicated by the coloured bands and the green line.

Example 5.2

Calculating the effect of temperature on the Gibbs energy

Liquid water and ice are in equilibrium at 0 °C under a pressure of 1 bar. What is the effect on the difference of molar Gibbs energies of the two phases when the temperature is increased to 1 °C?

Strategy Set up an expression for the difference in molar Gibbs energies at the second temperature, and then use eqn 5.4 to relate that difference to the difference at the equilibrium temperature, which we know to be 0. The relevant molar entropies are $S_m(\text{ice}) = 37.99\ \text{J K}^{-1}\ \text{mol}^{-1}$ and $S_m(\text{liquid}) = 69.91\ \text{J K}^{-1}\ \text{mol}^{-1}$ at 25 °C, but they differ little from the values at 0 °C.

Solution The difference in molar Gibbs energies at the final temperature T_f is

$$\begin{aligned}\Delta G_m(T_f) &= G_m(\text{liquid}, T_f) - G_m(\text{ice}, T_f)\\ &= \{G_m(\text{liquid}, T_i) - S_m(\text{liquid})(T_f - T_i)\} - \{G_m(\text{ice}, T_i) - S_m(\text{ice})(T_f - T_i)\}\\ &= \Delta G_m(T_i) - \{S_m(\text{liquid}) - S_m(\text{ice})\}(T_f - T_i)\end{aligned}$$

The difference in Gibbs energies is zero at the initial temperature ($\Delta G_m(T_i) = 0$), so

$$\Delta G_m(T_f) = -\{S_m(\text{liquid}) - S_m(\text{ice})\}(T_f - T_i)$$

Now insert the data:

$$\begin{aligned}\Delta G_m(T_f) &= -\{\overbrace{69.91\ \text{J K}^{-1}\ \text{mol}^{-1}}^{S_m(\text{liquid})} - \overbrace{37.99\ \text{J K}^{-1}\ \text{mol}^{-1}}^{S_m(\text{ice})}\} \times \overbrace{(1\ \text{K})}^{T_f - T_i}\\ &= -31.92\ \text{J mol}^{-1}\end{aligned}$$

The difference is negative, signifying that the transition from solid to liquid becomes spontaneous at this higher temperature.

Self-test 5.2

Water vapour and liquid are in equilibrium at 100 °C; what is the effect on the difference (liquid → vapour) of molar Gibbs energies of lowering the temperature to 98 °C? Take the relevant molar entropies to be those at 25 °C; see the *Data section*.

Answer: $+238\ \text{J mol}^{-1}$; condensation becomes spontaneous

Figure 5.3 also reveals the thermodynamic reason why substances melt and vaporize as the temperature is raised. At low temperatures, the solid phase has the lowest molar Gibbs energy and is therefore the most stable. However, as the temperature is raised, the molar Gibbs energy of the liquid phase falls below that of the solid phase, and the substance melts. At even higher temperatures, the molar Gibbs energy of the gas phase plunges down below that of the liquid phase, and the gas becomes the most stable phase. In other words, above a certain temperature, the liquid vaporizes to a gas.

We can also start to understand why some solid substances, such as solid carbon dioxide, sublime, that is, convert directly to a vapour without first forming a liquid. There is no fundamental requirement for the three lines to lie exactly in the positions we have drawn them in Fig. 5.3: the liquid line, for instance, could lie where we have drawn it in Fig. 5.4. Now we see that at no temperature (at the given pressure) does the liquid phase have the lowest molar Gibbs energy. Such a substance converts spontaneously directly from the solid to the vapour. That is, the substance sublimes.

The **transition temperature**, T_{trs}, between two phases, such as between liquid and solid or between ordered and disordered states of a protein, is the temperature, at a given pressure, at which the molar Gibbs energies of the two phases are equal. Above the solid–liquid transition temperature the liquid phase is thermodynamically more stable; below it, the solid phase is more stable.

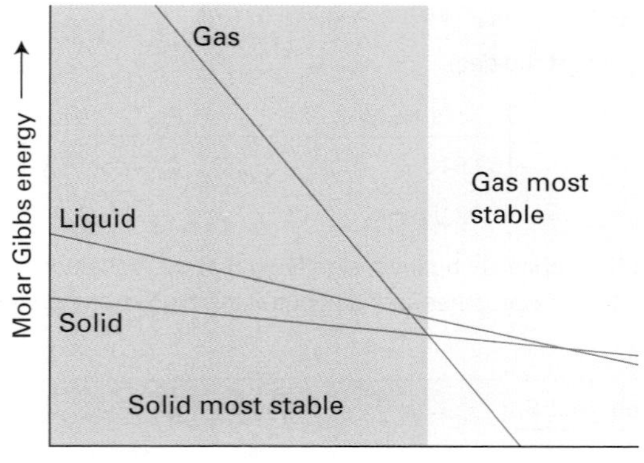

Fig. 5.4 If the line for the Gibbs energy of the liquid phase does not cut through the line for the solid phase (at a given pressure) before the line for the gas phase cuts through the line for the solid, the liquid is not stable at any temperature at that pressure. Such a substance sublimes.

Brief illustration 5.2 Phase transitions

At 1 atm, the transition temperature for ice and liquid water is 0 °C and that for grey and white tin is 13 °C. At the transition temperature itself, the molar Gibbs energies of the two phases are identical and there is no tendency for either phase to change into the other. At this temperature, therefore, the two phases are in equilibrium. At 1 atm, ice and liquid water are in equilibrium at 0 °C and the two allotropes of tin are in equilibrium at 13 °C.

As always when using thermodynamic arguments, it is important to keep in mind the distinction between the spontaneity of a phase transition and its rate. *Spontaneity is a tendency, not necessarily an actuality.* A phase transition predicted to be spontaneous may occur so slowly as to be unimportant in practice. For instance, at normal temperatures and pressures the molar Gibbs energy of graphite is 3 kJ mol^{-1} lower than that of diamond, so there is a thermodynamic tendency for diamond to convert into graphite. However, for this transition to take place, the carbon atoms of diamond must change their locations, and because the bonds between the atoms are so strong and large numbers of bonds must change simultaneously, this process is unmeasurably slow except at high temperatures. In gases and liquids the mobilities of the molecules normally allow phase transitions to occur rapidly, but in solids thermodynamic instability may be frozen in and a thermodynamically unstable phase may persist for thousands of years.

Phase diagrams

The **phase diagram** of a substance is a map showing the conditions of temperature and pressure at which its various phases are thermodynamically most stable (Fig. 5.5). For example, at point A in the illustration, the vapour phase of the substance is thermodynamically the most stable, but at C the liquid phase is the most stable.

The boundaries between regions in a phase diagram, which are called **phase boundaries**, show the values of p and T at which the two neighbouring phases are in equilibrium. For example, if the system is arranged to have a pressure and temperature represented by point B, then the liquid and its vapour are in equilibrium (like liquid water and water vapour at 1 atm and 100 °C). If the temperature is reduced at constant pressure, the system moves to point C where the liquid is the only stable phase (like water at 1 atm and at temperatures between 0 °C and 100 °C). If the

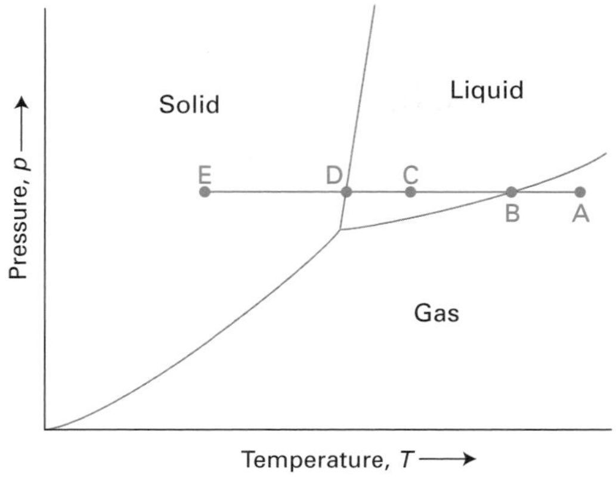

Fig. 5.5 A typical phase diagram, showing the regions of pressure and temperature at which each phase is the most stable. The phase boundaries (three are shown here) show the values of pressure and temperature at which the two phases separated by the line are in equilibrium. The significance of the letters A, B, C, D, and E (also referred to in Fig. 5.8) is explained in the text.

temperature is reduced still further to D, then the solid and the liquid phases are in equilibrium (like ice and water at 1 atm and 0 °C). A further reduction in temperature takes the system into the region at E where the solid is the stable phase.

5.4 Phase boundaries

The pressure of the vapour in equilibrium with its condensed phase is called the **vapour pressure** of the substance. The liquid–vapour boundary in a phase diagram, the line showing the conditions under which vapour and liquid are in equilibrium, is therefore a plot of the vapour pressure of the liquid against temperature. Vapour pressure increases with temperature because according to the Boltzmann distribution, as the temperature is raised more molecules have sufficient energy to leave their neighbours in the liquid.

To determine the liquid–vapour boundary, we simply measure the vapour pressure as a function of temperature. One procedure is to introduce a liquid into the near vacuum at the top of a mercury barometer and measure by how much the column is depressed (Fig. 5.6). To ensure that the pressure exerted by the vapour is truly the vapour pressure, we have to add enough liquid for some to remain after the vapour forms, for only then are the liquid and vapour phases in equilibrium. We change the temperature and determine another point on the curve, and so on (Fig. 5.7).

Now suppose we have a liquid in a cylinder sealed with a piston. If we apply a pressure greater than the vapour pressure of the liquid, the vapour is eliminated,

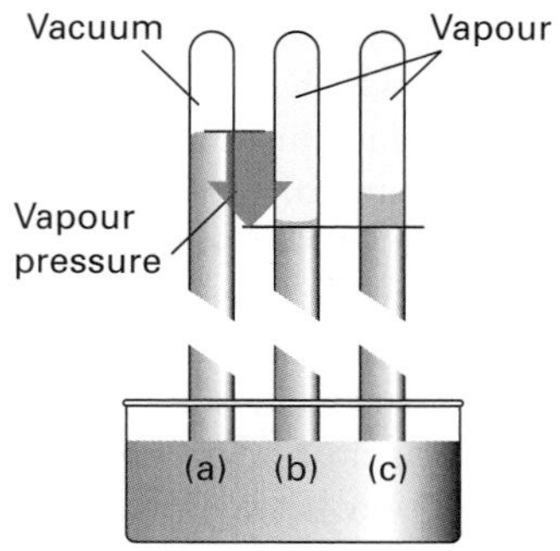

Fig. 5.6 When a small volume of water is introduced into the vacuum above the mercury in a barometer (a), the mercury is depressed (b) by an amount that is proportional to the vapour pressure of the liquid. (c) The same pressure is observed however much liquid is present (provided some is present).

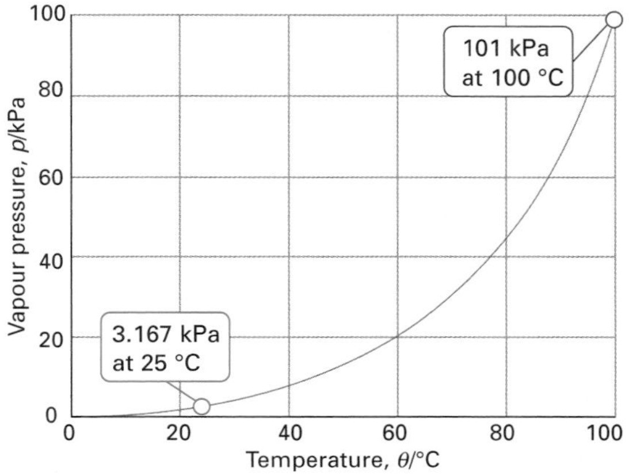

Fig. 5.7 The experimental variation of the vapour pressure of water with temperature.

the piston rests on the surface of the liquid, and the system moves to one of the points in the 'liquid' region of the phase diagram. Only a single phase is present. If instead we reduce the pressure on the system to a value below the vapour pressure, the system moves to one of the points in the 'vapour' region of the diagram. Reducing the pressure will involve pulling out the piston a long way, so that all the liquid evaporates; while any liquid is present, the pressure in the system remains constant at the vapour pressure of the liquid.

Brief Illustration 5.3 The effect of pressure on a phase transition

At 25 °C liquid water coexists in equilibrium with water vapour with a vapour pressure of 3.2 kPa. If the pressure is increased to, for example, 7.0 kPa, the Gibbs energy of the vapour increases by more than that of the liquid, and the vapour ceases to be thermodynamically stable. The vapour therefore condenses entirely to liquid water.

The same approach can be used to plot the solid–vapour boundary, which is a graph of the vapour pressure of the solid against temperature. The **sublimation vapour pressure** of a solid, the pressure of the vapour in equilibrium with a solid at a particular temperature, is usually much lower than that of a liquid.

A more sophisticated procedure is needed to determine the locations of solid–solid phase boundaries like that between calcite and aragonite, for instance, because the transition between two solid phases is more difficult to detect. One approach is to use **thermal analysis**, which takes advantage of the heat released during a transition. In a typical thermal analysis experiment, a sample is allowed to cool and its temperature is monitored. When the transition occurs, energy is released as heat and the cooling stops until the transition is complete (Fig. 5.8). The transition temperature is obvious from the shape of the graph and is used to mark a point on the phase diagram. The pressure can then be changed, and the corresponding transition temperature determined.

Any point lying on a phase boundary represents a pressure and temperature at which there is a 'dynamic equilibrium' between the two adjacent phases. A state of **dynamic equilibrium** is one in which a reverse process is taking place at the same rate as the forward process. Although there may be a great deal of activity at a molecular level, there is no net change in the bulk properties or appearance of the sample. For example, any point on the liquid–vapour boundary represents a state of dynamic equilibrium in which vaporization and condensation continue at matching rates. Molecules are leaving the surface of the liquid at a certain rate, and molecules already in the gas phase are returning to the liquid at the same rate; as a result, there is no net change in the number of molecules in the vapour and hence no net change in its pressure. Similarly, a point on the solid–liquid curve represents conditions of pressure and temperature at which molecules are ceaselessly breaking away from the surface of the solid and contributing to the liquid. However, they are doing so at a rate that exactly matches that at which molecules already in the liquid are settling on to the surface of the solid and contributing to the solid phase.

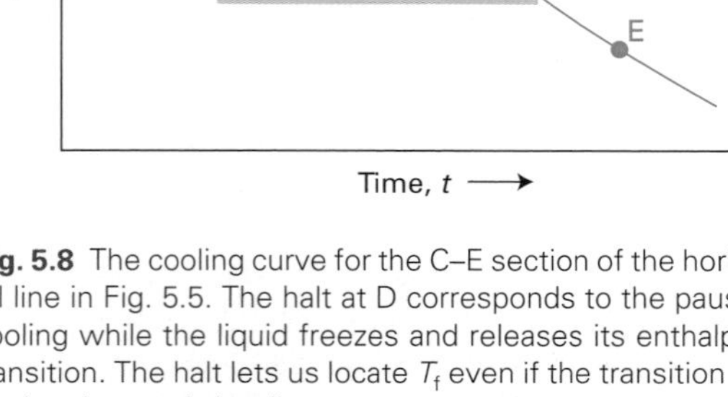

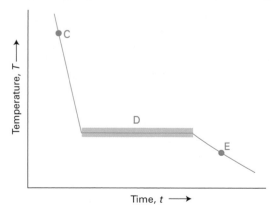

Fig. 5.8 The cooling curve for the C–E section of the horizontal line in Fig. 5.5. The halt at D corresponds to the pause in cooling while the liquid freezes and releases its enthalpy of transition. The halt lets us locate T_f even if the transition cannot be observed visually.

5.5 The location of phase boundaries

Thermodynamics provides us with a way of predicting the location of the phase boundaries. Suppose two phases are in equilibrium at a given pressure and temperature. Then, if we change the pressure, we must adjust the temperature to a different value to ensure that the two phases remain in equilibrium. In other words, there must be a relation between the change in pressure, Δp, that we exert and the change in temperature, ΔT, we must make to ensure that the two phases remain in equilibrium. We show in the following Derivation that the relation between the change in temperature and the change in pressure needed to maintain equilibrium is given by the **Clapeyron equation:**

$$\Delta p = \frac{\Delta_{trs}H}{T\Delta_{trs}V} \times \Delta T \qquad \text{Clapeyron equation} \qquad (5.5a)$$

where $\Delta_{trs}H$ is the enthalpy of transition and $\Delta_{trs}V$ is the volume of transition (the change in molar volume when the transition occurs); these quantities are specified in the following Derivation. This form of the Clapeyron equation is valid for small changes in pressure and temperature, because only then can $\Delta_{trs}H$, $\Delta_{trs}V$, and T be taken as constant across the range.

Derivation 5.4

The Clapeyron equation

This derivation is based on eqn 5.2 ($dG = Vdp - SdT$), the relation obtained in Derivation 5.1.

Consider two phases 1 (for instance, a liquid) and 2 (a vapour). At a certain pressure and temperature the two phases are in equilibrium and $G_m(1) = G_m(2)$, where $G_m(1)$ is the molar Gibbs energy of phase 1 and $G_m(2)$ that of phase 2 (Fig. 5.9). Now change the pressure by an infinitesimal amount dp and the temperature by dT. The molar Gibbs energies of each phase change as follows:

Phase 1: $dG_m(1) = V_m(1)dp - S_m(1)dT$

Phase 2: $dG_m(2) = V_m(2)dp - S_m(2)dT$

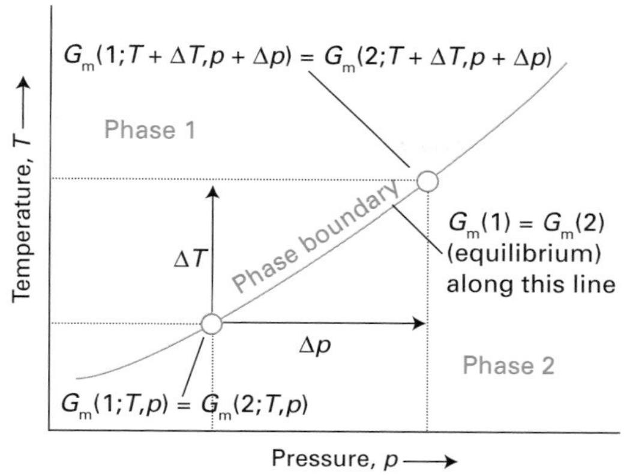

Fig. 5.9 At equilibrium, two phases have the same molar Gibbs energy. When the temperature is changed, for the two phases to remain in equilibrium, the pressure must be changed so that the Gibbs energies of the two phases remain equal.

where $V_m(1)$ and $S_m(1)$ are the molar volume and molar entropy of phase 1 and $V_m(2)$ and $S_m(2)$ are those of phase 2. The two phases were in equilibrium before the change, so the two molar Gibbs energies were equal. The two phases are still in equilibrium after the pressure and temperature are changed, so their two molar Gibbs energies remain equal. Therefore, the two *changes* in molar Gibbs energy must be equal, $dG_m(2) = dG_m(1)$, and we can write

$$V_m(2)dp - S_m(2)dT = V_m(1)dp - S_m(1)dT$$

This equation can be rearranged to

$$\{V_m(2) - V_m(1)\}dp = \{S_m(2) - S_m(1)\}dT$$

The entropy of transition, $\Delta_{trs}S$, is the difference between the two molar entropies, and the volume of transition, $\Delta_{trs}V$, is the difference between the molar volumes of the two phases:

$$\Delta_{trs}V = V_m(2) - V_m(1); \ \Delta_{trs}S = S_m(2) - S_m(1)$$

We can therefore write

$$\Delta_{trs}V dp = \Delta_{trs}S dT$$

or

$$dp = \frac{\Delta_{trs}S}{\Delta_{trs}V} dT$$

We saw in Chapter 4 that the transition entropy is related to the enthalpy of transition by $\Delta_{trs}S = \Delta_{trs}H/T_{trs}$, so we may also write

$$dp = \frac{\Delta_{trs}H}{T\Delta_{trs}V} dT \qquad (5.5b)$$

We have dropped the 'trs' subscript from the temperature because all the points on the phase boundary—the only points we are considering—are transition temperatures. This expression is exact. For variations in pressure and temperature small enough for $\Delta_{trs}H$, $\Delta_{trs}V$, and T to be treated as constant, the infinitesimal changes dp and dT can be replaced by observable changes, Δp and ΔT, and we obtain eqn 5.5a.

The Clapeyron equation tells us the slope (the value of $\Delta p/\Delta T$) of any phase boundary in terms of the enthalpy and volume of transition. For the solid–liquid phase boundary, the enthalpy of transition is the enthalpy of fusion, which is positive because melting is always endothermic. For most substances, the molar volume increases slightly on melting, so $\Delta_{trs}V$ is positive but small. It follows that the slope of the phase boundary is large and positive (up from left to right), and therefore that a large increase in pressure brings about only a small increase in melting temperature. Water, though, is quite different, for although its melting is endothermic, its molar volume *decreases* on melting (liquid water is denser than ice at 0 °C, which is why ice floats on water), so $\Delta_{trs}V$ is small but negative. Consequently, the slope of the ice–water phase boundary is steep but negative (down from left to right). Now a large increase in pressure brings about a small lowering of the melting temperature of ice.

Example 5.3

Estimating the effect of pressure on the boiling temperature

Estimate the typical size of the effect of increasing pressure on the boiling point of a liquid.

Strategy We begin by calculating a typical value of dp/dT for the case of vaporization. First, we rewrite eqn 5.5 as

$$\frac{dp}{dT} = \frac{\Delta_{trs}H}{T\Delta_{trs}V} \overset{trs\rightarrow vap}{=} \frac{\Delta_{vap}H}{T\Delta_{vap}V}$$

Second, we need to estimate the right-hand side. At the boiling point, the term $\Delta_{vap}H/T$ is given by Trouton's rule: 85 J K^{-1} mol^{-1} (Section 4.5). Because the molar volume of a gas is so much greater than the molar volume of a liquid, we can write $\Delta_{vap}V = V_m(g) - V_m(l) \approx V_m(g)$ and take for $V_m(g)$ the molar volume of a perfect gas (at low pressures, at least). Finally, we invert the estimated value of dp/dT to obtain dT/dp and, consequently, an estimate of the extent of change in the boiling point with pressure.

Solution From the perfect gas equation, $V_m(g) = RT/p$ is about 25 dm^3 mol^{-1} = 2.5 × 10^{-2} m^3 mol^{-1} at 1 atm and near but above room temperature, T = 298 K. Therefore,

$$\frac{dp}{dT} \approx \frac{\overbrace{85 \text{ J K}^{-1}\text{ mol}^{-1}}^{\Delta_{vap}H/T}}{\underbrace{2.5\times10^{-2}\text{ m}^3\text{ mol}^{-1}}_{V_m}} \overset{1\text{ J}=1\text{ Pa m}^3}{=} \overbrace{3.4\times10^3\text{ Pa K}^{-1}}^{0.034\text{ atm}}$$

where we have used 1 J = 1 Pa m^3. This value corresponds to 0.034 atm K^{-1} and hence to dT/dp = 29 K atm^{-1}. Therefore, a change of pressure of +0.1 atm can be expected to change a boiling temperature by about +3 K.

Self-test 5.3

Estimate dT/dp for water at its normal boiling point using the information in Table 3.1.

Answer: 28 K atm^{-1}

We cannot use eqn 5.5 to discuss the liquid–vapour phase boundary, except over very small ranges of temperature and pressure, because we cannot assume that the volume of the vapour, and therefore the volume of transition, is independent of pressure. However, if we suppose that the vapour behaves as a perfect gas, then it turns out (see the following Derivation) that the relation between a change in pressure and a change in temperature is given by the **Clausius–Clapeyron equation:**

$$\Delta(\ln p) = \frac{\Delta_{vap}H}{RT^2} \times \Delta T \quad \text{Vapour a perfect gas} \quad \text{Clausius–Clapeyron equation} \quad (5.6)$$

Equation 5.6 shows that as the temperature of a liquid is raised ($\Delta T > 0$) its vapour pressure increases (an increase in the logarithm of p, $\Delta(\ln p) > 0$, implies that p increases).

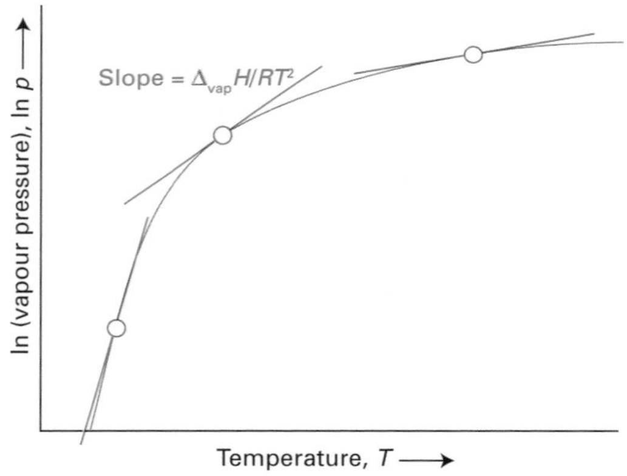

Fig. 5.10 The Clausius–Clapeyron equation gives the slope of a plot of the logarithm of the vapour pressure of a substance against the temperature. That slope at a given temperature is proportional to the enthalpy of vaporization of the substance.

Derivation 5.5

The Clausius–Clapeyron equation

For the liquid–vapour boundary the 'trs' label in the exact form of the Clapeyron equation, eqn 5.5b in Derivation 5.4, becomes 'vap' and that equation can be written

$$\frac{dp}{dT} \overset{\text{trs}\to\text{vap}}{=} \frac{\Delta_{vap}H}{T\Delta_{vap}V}$$

Because the molar volume of a gas is much larger than the molar volume of a liquid, the volume of vaporization, $\Delta_{vap}V = V_m(g) - V_m(l)$, is approximately equal to the molar volume of the gas itself. Therefore, to a good approximation,

$$\frac{dp}{dT} = \frac{\Delta_{vap}H}{T\Delta_{vap}V} = \frac{\Delta_{vap}H}{T\{V_m(g) - V_m(l)\}} \overset{V_m(g) \gg V_m(l)}{\approx} \frac{\Delta_{vap}H}{TV_m(g)}$$

To make further progress, we can treat the vapour as a perfect gas and write its molar volume as $V_m = RT/p$. Then

$$\frac{dp}{dT} = \frac{\Delta_{vap}H}{TV_m(g)} \overset{V_m(g)=RT/p}{=} \frac{\Delta_{vap}H}{T(RT/p)} = \frac{p\Delta_{vap}H}{RT^2}$$

and therefore

$$\frac{dp}{p} = \frac{\Delta_{vap}H}{RT^2}dT$$

A standard result of calculus, which we covered in The chemist's toolkit 2.1, is $d(\ln x)/dx = 1/x$, and therefore (by multiplying both sides by dx), $dx/x = d(\ln x)$. It follows that we may write the last equation as the Clausius–Clapeyron equation:

$$d(\ln p) = \frac{\Delta_{vap}H}{RT^2}dT$$

Provided the range of temperature and pressure is small, the infinitesimal changes $d(\ln p)$ and dT can be replaced by measurable changes, and we obtain eqn 5.6.

One important application of the Clausius–Clapeyron equation is to derive an expression that shows how the vapour pressure varies with temperature and thus show how the variation depends on molecular properties. It follows from eqn 5.6, as we also show in the following Derivation, that the vapour pressure p' at a temperature T' is related to the vapour pressure p at a temperature T by

$$\ln p' = \ln p + \frac{\Delta_{vap}H}{R}\left(\frac{1}{T} - \frac{1}{T'}\right) \quad \text{Temperature dependence of vapour pressure} \quad (5.7)$$

Equation 5.7 lets us calculate the vapour pressure at one temperature provided we know it at another temperature. The equation tells us that:

> For a given change in temperature, the greater the enthalpy of vaporization, the greater the change in vapour pressure.

Water, for instance, has a high enthalpy of vaporization on account of the strong hydrogen bonds that hold the molecules together in the liquid, so we can expect its vapour pressure to increase rapidly as the temperature is increased. Benzene lacks hydrogen bonds, has a much smaller enthalpy of vaporization, and so has a vapour pressure that varies much more weakly with temperature.

Derivation 5.6

Temperature dependence of the vapour pressure

To obtain the explicit expression for the vapour pressure at any temperature (eqn 5.7) we take the equation at the end of the preceding Derivation,

$$d(\ln p) = \frac{\Delta_{vap}H}{RT^2}dT$$

and integrate both sides. If the vapour pressure is p at a temperature T and p' at a temperature T', this integration takes the form

$$\int_{\ln p}^{\ln p'} d(\ln p) = \int_{T}^{T'} \frac{\Delta_{vap}H}{RT^2}dT$$

A note on good practice When setting up an integration, make sure the limits match on each side of the expression. Here, the lower limits are ln p on the left and T on the right, and the upper limits are ln p' and T', respectively.

The integral on the left evaluates as follows:

$$\int_{\ln p}^{\ln p'} d(\ln p) \overbrace{=}^{\int_a^b dx=b-a} \ln p' - \ln p \overbrace{=}^{\ln x - \ln y = \ln \frac{x}{y}} \ln\frac{p'}{p}$$

To evaluate the integral on the right, we suppose that the enthalpy of vaporization is constant over the temperature range of interest, so together with R it can be taken outside the integral:

$$\int_{T}^{T'} \frac{\Delta_{vap}H}{RT^2}dT \overbrace{=}^{\Delta_{vap}H/R \text{ a constant}} \frac{\Delta_{vap}H}{R}\int_{T}^{T'}\frac{1}{T^2}dT$$

$$= \frac{\Delta_{vap}H}{R}\left(\frac{1}{T} - \frac{1}{T'}\right)$$

To obtain this result we have used the standard integral (The chemist's toolkit 2.1)

$$\int \frac{1}{x^2} = -\frac{1}{x} + \text{constant}$$

By equating the preceding two lines we obtain eqn 5.7.

A note on good practice Keep a note of any approximations made in a derivation. In this pair of derivations we have made three: (1) the molar volume of a gas is much greater than that of a liquid, (2) the vapour behaves as a perfect gas, and (3) the enthalpy of vaporization is independent of temperature in the range of interest. Approximations limit the ways in which an expression may be used to solve problems.

Note too that we can write eqn 5.7 as

$$\ln p = \overbrace{\ln p' + \frac{\Delta_{vap}H}{RT'}}^{A} - \overbrace{\frac{\Delta_{vap}H}{RT}}^{B/T}$$

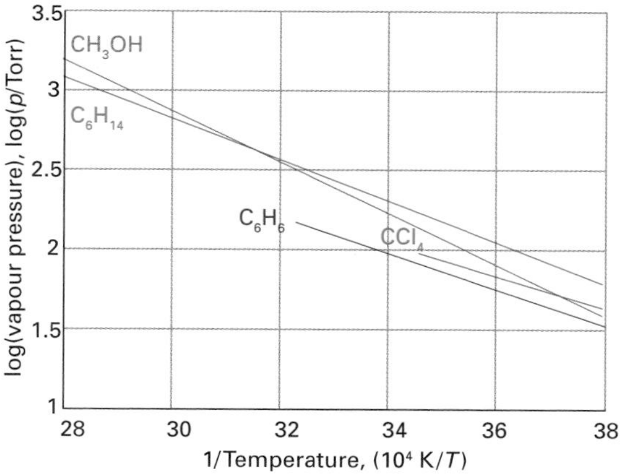

Fig. 5.11 The vapour pressures of some substances based on the data in Table 5.1. The plot is of the common logarithm, not the natural logarithm (ln x = ln 10 × log x).

This expression has the form

$$\ln p = A - \frac{B}{T} \tag{5.8}$$

where A and B are constants and the value of A depends on the units adopted for p. This is the form in which vapour pressures are commonly reported (Table 5.1 and Fig. 5.11).

Brief illustration 5.4 Temperature variation of vapour pressure

The vapour pressure of benzene in the range 0–42 °C can be expressed in the form of eqn 5.8:

$$\ln(p/\text{kPa}) = 16.319 - \frac{4110\text{ K}}{T}$$

because B = 4110 K. It follows from the preceding discussion that, because $B = \Delta_{vap}H/R$, then

$$\Delta_{vap}H = BR = (4110\text{ K}) \times (8.3145\text{ J K}^{-1}\text{ mol}^{-1})$$
$$= 34.2\text{ kJ mol}^{-1}$$

A note on good practice You will sometimes see eqn 5.8 written without units, or with the units in parentheses. It is much better practice to include the units in such a way as to make all the quantities unitless, as in this Brief illustration: p/kPa is a dimensionless number.

Self-test 5.4

For benzene in the range 42–100 °C, ln(p/kPa) = 15.61 – (3884 K)/T. Estimate the normal boiling point of benzene. (The normal boiling point is the temperature at which the vapour pressure is 1 atm; see below.)

Answer: 80.2 °C; the actual value is 80.1 °C

Table 5.1
*Vapour pressure**

Substance	A	B/K	Temperature range (°C)
Benzene, C_6H_6(l)	16.32	4110	0 to +42
	15.61	3884	42 to 100
Hexane, C_6H_{14}(l)	15.77	3811	−10 to +90
Methanol, CH_3OH(l)	18.25	4610	−10 to +80
Methylbenzene, $C_6H_5CH_3$(l)	17.17	4713	−92 to +15
Phosphorus, P_4(s, white)	20.21	7592	20 to 44
Sulfur trioxide, SO_3(l)	21.06	5225	24 to 48
Tetrachloromethane, CCl_4(l)	16.42	4078	−19 to +20

*A and B are the constants in the expression $\ln(p/\text{kPa}) = A - B/T$.

5.6 **Characteristic points**

As we have just seen, as the temperature of a liquid is raised, its vapour pressure increases. First, consider what we would observe when we heat a liquid in an open vessel. At a certain temperature, the vapour pressure becomes equal to the external pressure. At this temperature, the vapour can drive back the surrounding atmosphere and expand indefinitely. Moreover, because there is no constraint on expansion, bubbles of vapour can form throughout the body of the liquid, a condition known as **boiling**. The temperature at which the vapour pressure of a liquid is equal to the external pressure is called the **boiling temperature**. When the external pressure is 1 atm, the boiling temperature is called the **normal boiling point**, T_b. It follows that we can predict the normal boiling point of a liquid by noting the temperature on the phase diagram at which its vapour pressure is 1 atm. The use of 1 atm in the definition of normal boiling point rather than 1 bar is historical: the boiling temperature at 1 bar is called the **standard boiling point**.

Now consider what happens when we heat the liquid in a closed vessel. Because the vapour cannot escape, its density increases as the vapour pressure rises and in due course the density of the vapour becomes equal to that of the remaining liquid. At this stage the surface between the two phases disappears, as was depicted in Fig. 1.15. The temperature at which the surface disappears is the critical temperature, T_c, which we first encountered in Section 1.10. The vapour pressure at the critical temperature is called the **critical pressure**, p_c, and the critical temperature and critical pressure together identify the **critical point** of the substance (see Table 5.2). If we exert pressure on a sample that is above its critical temperature, we produce a denser fluid. However, no surface appears to separate the two parts of the sample and a single uniform phase, a **supercritical fluid**, continues to fill the container. That is, we have to conclude that *a liquid cannot be produced by the application of pressure to a substance if it is at or above its critical temperature*. That is why the liquid–vapour boundary in a phase diagram terminates at the critical point (Fig. 5.12).

The temperature at which the liquid and solid phases of a substance coexist in equilibrium at a specified pressure is called the **melting temperature**

Table 5.2
*Critical constants**

	p_c/atm	V_c/(cm^3 mol^{-1})	T_c/K
Ammonia, NH_3	111	73	406
Argon, Ar	48	75	151
Benzene, C_6H_6	49	260	563
Bromine, Br_2	102	135	584
Carbon dioxide, CO_2	73	94	304
Chlorine, Cl_2	76	124	417
Ethane, C_2H_6	48	148	305
Ethene, C_2H_4	51	124	283
Hydrogen, H_2	13	65	33
Methane, CH_4	46	99	191
Oxygen, O_2	50	78	155
Water, H_2O	218	55	647

*The critical volume V_c is the molar volume at the critical pressure and critical temperature.

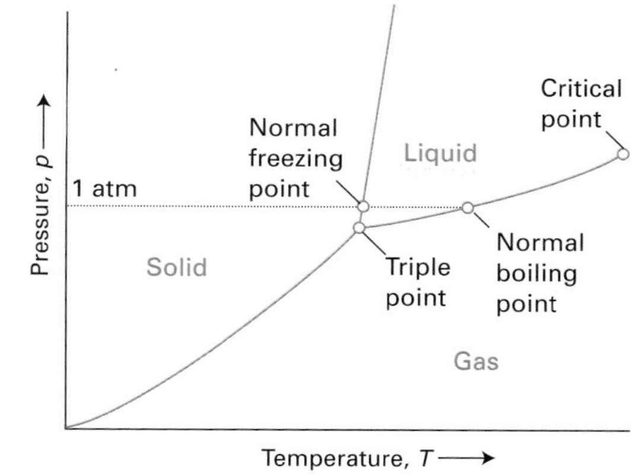

Fig. 5.12 The significant points of a phase diagram. The liquid–vapour phase boundary terminates at the *critical point*. At the *triple point*, solid, liquid, and vapour are in dynamic equilibrium. The *normal freezing point* is the temperature at which the liquid freezes when the pressure is 1 atm; the *normal boiling point* is the temperature at which the vapour pressure of the liquid is 1 atm.

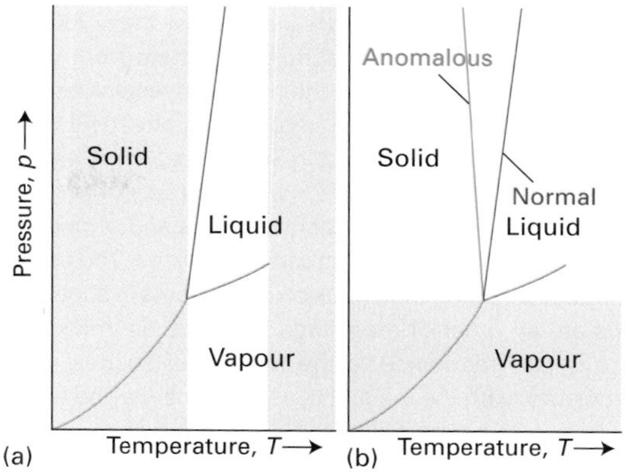

Fig. 5.13 (a) For substances that have phase diagrams resembling the one shown here (which is common for most substances, with the important exception of water), the triple point and the critical point mark the range of temperatures over which the substance may exist as a liquid. The shaded areas show the regions of temperature in which a liquid cannot exist as a stable phase. (b) A liquid cannot exist as a stable phase if the pressure is below that of the triple point for normal or anomalous liquids.

of the substance. Because a substance melts at the same temperature as it freezes, 'melting temperature' is synonymous with **freezing temperature**. The solid–liquid boundary therefore shows how the melting temperature of a solid varies with pressure. The melting temperature when the pressure on the sample is 1 atm is called the **normal melting point** or the **normal freezing point,** T_f. A liquid freezes when the energy of the molecules in the liquid is so low that they cannot escape from the attractive forces of their neighbours and lose their mobility.

There is a set of conditions under which three different phases (typically solid, liquid, and vapour) all coexist in equilibrium simultaneously. It is represented by the **triple point**, where the three phase boundaries meet. The triple point of a pure substance is a characteristic, unchangeable physical property of the substance. For water the triple point lies at 273.16 K and 611 Pa, and ice, liquid water, and water vapour coexist in equilibrium at no other combination of pressure and temperature. At the triple point, the rates of each forward and reverse process are equal (but the three individual rates are not necessarily the same).

The triple point and the critical point are important features of a substance because they act as frontier posts for the existence of the liquid phase. As we see from Fig. 5.13a, if the slope of the solid–liquid phase boundary is as shown in the diagram:

> The triple point marks the lowest temperature at which the liquid can exist.

> The critical point marks the highest temperature at which the liquid can exist.

We shall see in Section 5.8 that for a few materials (most notably water) the solid–liquid phase boundary slopes in the opposite direction, and then only the second of these conclusions is relevant (see Fig. 5.13b).

Impact on technology 5.1

Supercritical fluids

Supercritical carbon dioxide, $scCO_2$, is the centre of attention for an increasing number of solvent-based processes. The critical temperature, 31.0 °C (304.2 K), and pressure, 72.9 atm (7.39 MPa), are readily accessible and carbon dioxide is cheap. The mass density of $scCO_2$ at its critical point is 0.45 g cm^{-3}. However, the transport properties of any supercritical fluid depend strongly on its density, which in turn is sensitive to the pressure and temperature. For instance, densities may be adjusted from a gas-like 0.1 g cm^{-3} to a liquid-like 1.2 g cm^{-3}. A useful rule of thumb is that the solubility of a solute is an exponential function of the density of the supercritical fluid, so small increases in pressure, particularly close to the critical point, can have very large effects on solubility.

A great advantage of $scCO_2$ is that there are no noxious residues once the solvent has been allowed to evaporate; so, coupled with its low critical temperature, $scCO_2$ is ideally suited to food processing and the production of pharmaceuticals. It is used, for instance, to remove caffeine from coffee. The supercritical fluid is also increasingly being used for dry cleaning, which avoids the use of carcinogenic and environmentally deleterious chlorinated hydrocarbons.

Supercritical CO_2 has been used since the 1960s as a mobile phase in **supercritical fluid chromatography** (SFC), but it fell out of favour when the more convenient technique of high-performance liquid chromatography (HPLC) was introduced. However, interest in SFC has returned, and there are separations possible in SFC that cannot easily be achieved by HPLC, such as the separation of lipids and of phospholipids. Samples as small as 1 pg can be analysed. The essential advantage of SFC is that diffusion coefficients in supercritical fluids are an order of magnitude greater than in liquids, so there is less resistance to the transfer of solutes through the column, with the result that separations may be effected rapidly or with high resolution.

The principal problem with $scCO_2$ is that the fluid is not a very good solvent and surfactants are needed to induce many potentially interesting solutes to dissolve. Indeed, $scCO_2$-based dry cleaning depends on the availability of cheap surfactants, so too does the use of $scCO_2$ as a solvent for homogeneous catalysts, such as metal complexes. There appear to be two principal approaches to solving the solubilizing problem. One solution is to use fluorinated and siloxane-based polymeric stabilizers, which allow polymerization reactions to proceed in $scCO_2$. The disadvantage of these stabilizers for commercial use is their great expense. An alternative and much cheaper approach is poly(ether-carbonate) copolymers. The copolymers can be made more soluble in $scCO_2$ by adjusting the ratio of ether and carbonate groups.

The critical temperature of water is 374 °C (647 K) and its pressure is 218 atm (221 MPa). The conditions for using scH_2O are therefore much more demanding than for $scCO_2$ and the properties of the fluid are highly sensitive to pressure. Thus, as the density of scH_2O decreases, the characteristics of a solution change from those of an aqueous solution through those of a non-aqueous solution and eventually to those of a gaseous solution. One consequence is that reaction mechanisms may change from ionic to radical.

5.7 The phase rule

At this point you might wonder whether *four* phases of a single substance could ever be in equilibrium (such as the two solid forms of tin, liquid tin, and tin vapour). To explore this question we think about the thermodynamic criterion for four phases to be in equilibrium. For equilibrium, the four molar Gibbs energies would all have to be equal and we could write

$$G_m(1) = G_m(2) \quad G_m(2) = G_m(3) \quad G_m(3) = G_m(4)$$

(The other equalities $G_m(1) = G_m(4)$, and so on, are implied by these three equations.) Each Gibbs energy is a function of the pressure and temperature, so we should think of these three relations as three equations for the two unknowns p and T. In general, three equations for two unknowns have no solution. For instance, the three equations $5x + 3y = 4$, $2x + 6y = 5$, and $x + y = 1$ have no solutions (try it). Therefore, we have to conclude that the four molar Gibbs energies cannot all be equal. In other words, *four phases of a single substance cannot coexist in mutual equilibrium.*

The conclusion we have reached is a special case of one of the most elegant results of chemical thermodynamics. The **phase rule**, which is derived in the following Derivation, was deduced by Gibbs and states that, for a system at equilibrium,

$$F = C - P + 2 \qquad \text{Phase rule} \quad (5.9)$$

Here F is the number of degrees of freedom, C is the number of components, and P is the number of phases:

- The **number of components**, C, in a system is the minimum number of independent species necessary to define the composition of all the phases present in the system.

The definition is easy to apply when the species present in a system do not react, for then we simply count their number. For instance, pure water is a one-component system ($C = 1$) and a mixture of ethanol and water is a two-component system ($C = 2$).

- The **number of degrees of freedom**, F, of a system is the number of intensive variables (such as the pressure, temperature, or mole fractions: they are independent of the amount of material in the sample) that can be changed independently without disturbing the number of phases in equilibrium.

Derivation 5.7

The phase rule

We begin by counting the total number of intensive variables. The pressure, p, and temperature, T, count as 2. We can specify the composition of a phase by giving the mole fractions of $C - 1$ components. We need specify only $C - 1$ and not all C mole fractions because $x_1 + x_2 + \cdots + x_C = 1$, and all mole fractions are known if all except one are specified. Because there are P phases, the total number of composition variables is $P(C - 1)$. At this stage, the total number of intensive variables is $P(C - 1) + 2$.

At equilibrium, the chemical potential of a component J must be the same in every phase:

$$\mu_J(1) = \mu_J(2) = \ldots = \mu_J(P) \text{ for } P \text{ phases}$$

That is, there are $P - 1$ equations of this kind to be satisfied for each component J. As there are C components, the total number of equations is $C(P - 1)$. Each equation reduces our freedom to vary one of the $P(C - 1) + 2$ intensive variables. It follows that the total number of degrees of freedom is

$$F = P(C - 1) + 2 - C(P - 1) = C - P + 2$$

which is eqn 5.9.

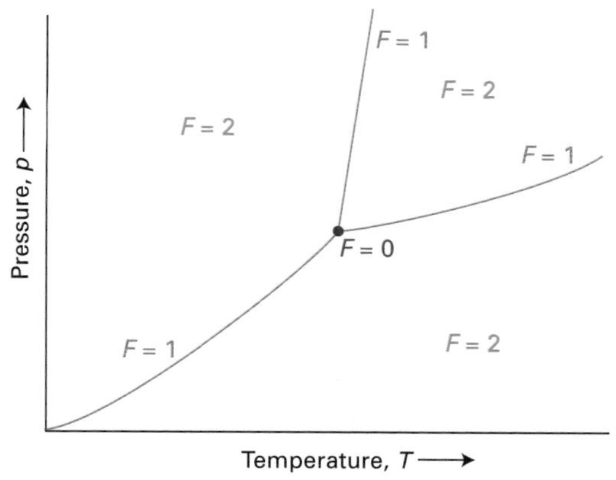

Fig. 5.14 The features of a phase diagram represent different degrees of freedom. When only one phase is present, $F = 2$ and the pressure and temperature can be varied at will. When two phases are present in equilibrium, $F = 1$: now, if the temperature is changed, the pressure must be changed by a specific amount. When three phases are present in equilibrium, $F = 0$ and there is no freedom to change either variable.

For a one-component system, such as pure water, we set $C = 1$ and the phase rule simplifies to $F = 3 - P$. When only one phase is present, $F = 2$, which implies that p and T can be varied independently. In other words, a single phase is represented by an *area* on a phase diagram. When two phases are in equilibrium $F = 1$; that is, the equilibrium of two phases is represented by a *line* in a phase diagram: a line in a graph shows how one variable must change if another variable is varied (Fig. 5.14). This constraint implies that pressure is not freely variable if the temperature has been set. Instead of selecting the temperature, we can select the pressure, but having done so the two phases come into equilibrium at a single definite temperature. Therefore, freezing (or any other phase transition of a single substance) occurs at a definite temperature at a given pressure. When three phases are in equilibrium $F = 0$. This special 'invariant condition' can therefore be established only at a definite temperature and pressure. The equilibrium of three phases is therefore represented by a *point*, the triple point, on the phase diagram. If we set $P = 4$, we get the absurd result that F is negative; that result is in accord with the conclusion at the start of this section that four phases cannot be in equilibrium in a one-component system.

Example 5.4

Using the phase rule

A saturated solution of copper(II) sulfate, with excess of the solid, is present in equilibrium with its vapour in a closed vessel. (a) How many phases and components are present? (b) How many degrees of freedom are available, and what are they?

Strategy Use the definitions of C and P given in the text, and then the phase rule to determine F.

Solution (a) The system consists of two components, water and copper(II) sulfate, so $C = 2$. We could consider it to be composed of water and Cu^{2+} and SO_4^{2-} ions, but the ion concentrations are not independent. There are three phases present (the liquid solution, the excess solid, and the vapour), so $P = 3$. (b) It then follows from the phase rule that

$$F = 2 - 3 + 2 = 1$$

The single degree of freedom can be taken to be the temperature; if it is varied the vapour pressure changes. The two cannot be varied independently.

Self-test 5.5

Repeat the question for the case in which there is no excess solid solute present.

Answer: (a) $C = 2$, $P = 2$; (b) $F = 2$, temperature and composition are variable

5.8 Phase diagrams of typical materials

We shall now see how these general features appear in the phase diagrams of a selection of pure substances.

Figure 5.15 is the phase diagram for water. The liquid–vapour phase boundary shows how the vapour pressure of liquid water varies with temperature. We can use this curve, which is shown in more detail in Fig. 5.7, to decide how the boiling temperature varies with changing external pressure.

Brief illustration 5.5 The phases of water

When the external pressure is 19.9 kPa (at an altitude of 12 km), water boils at 60 °C because that is the temperature at which the vapour pressure is 19.9 kPa. The solid–liquid boundary line in Fig. 5.15, which is shown in more detail in Fig. 5.16, shows how the melting temperature of water depends on pressure. For example, although ice melts at 0 °C at 1 atm, it melts at –1 °C when the pressure is 130 bar. The very steep slope of the boundary indicates that enormous pressures are needed to bring about significant changes. Notice that the line slopes down from left to right, which—as we anticipated—means that the melting temperature of ice falls as the pressure is raised.

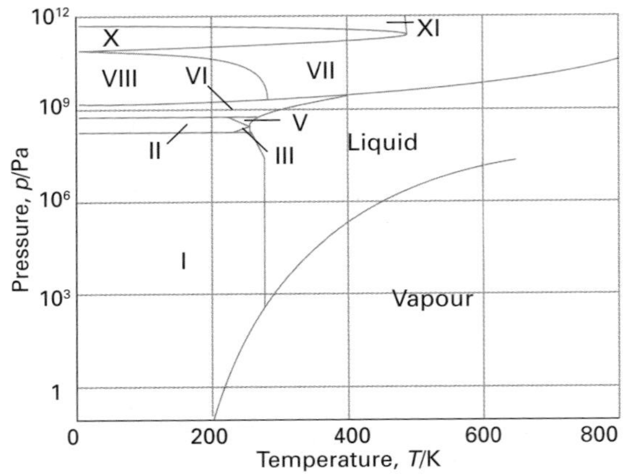

Fig. 5.15 The phase diagram for water showing the different solid phases.

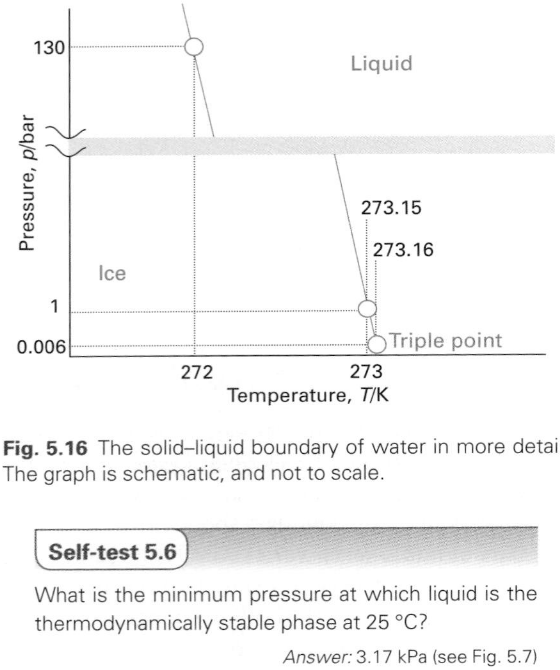

Fig. 5.16 The solid–liquid boundary of water in more detail. The graph is schematic, and not to scale.

Self-test 5.6

What is the minimum pressure at which liquid is the thermodynamically stable phase at 25 °C?

Answer: 3.17 kPa (see Fig. 5.7)

The reason for the unusual behaviour of water pointed out in the *Brief illustration*, that the melting temperature of ice falls as the pressure is raised, can be ascribed to the decrease in volume that occurs when ice melts: it is favourable for the solid to transform into the denser liquid as the pressure is raised. The decrease in volume is a result of the very open structure of the crystal structure of ice: as shown in Fig. 5.17, the water molecules are held apart, as well as together, by the hydrogen bonds between them but the structure partially collapses on melting and the liquid is denser than the solid.

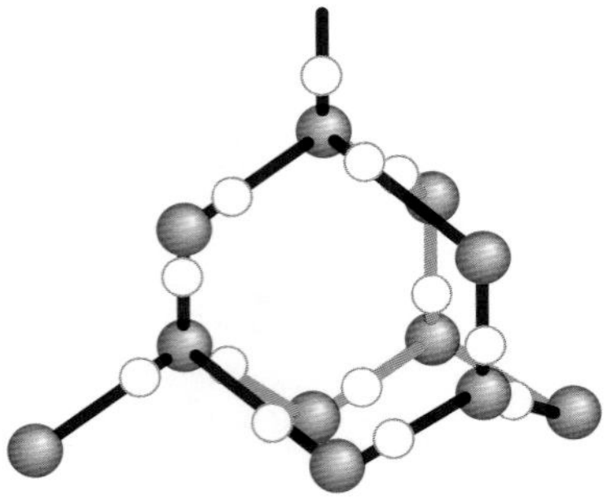

Fig. 5.17 The structure of ice-I. Each O atom is at the centre of a tetrahedron of four O atoms at a distance of 276 pm. The central O atom is attached by two short O—H bonds to two H atoms and by two long hydrogen bonds to the H atoms of two of the neighbouring molecules. Overall, the structure consists of planes of puckered hexagonal rings of H_2O molecules (like the chair form of cyclohexane). This structure collapses partially on melting, leading to a liquid that is denser than the solid.

Figure 5.15 shows that water has many different solid phases other than ordinary ice ('ice-I', shown in Fig. 5.17). These solid phases differ in the arrangement of the water molecules: under the influence of very high pressures, hydrogen bonds buckle and the H_2O molecules adopt different arrangements. These **polymorphs**, or different solid phases, of ice may be responsible for the advance of glaciers, for ice at the bottom of glaciers experiences very high pressures where it rests on jagged rocks. The sudden apparent explosion of Halley's comet in 1991 may have been due to the conversion of one form of ice into another in its interior.

Figure 5.18 shows the phase diagram for carbon dioxide. The features to notice include the slope of the solid–liquid boundary: this positive slope is typical of almost all substances. The slope indicates that the melting temperature of solid carbon dioxide rises as the pressure is increased. As the triple point (217 K, 5.11 bar) lies well above ordinary atmospheric pressure, liquid carbon dioxide does not exist at normal atmospheric pressures whatever the temperature, and the solid sublimes when left in the open (hence the name 'dry ice'). To obtain liquid carbon dioxide, it is necessary to exert a pressure of at least 5.11 bar.

Cylinders of carbon dioxide generally contain the liquid or compressed gas; if both gas and liquid are present inside the cylinder, then at 20 °C the pressure must be about 65 atm. When the gas squirts through the throttle it cools by the Joule–Thomson effect, so when it emerges into a region where the pressure is only 1 atm, it condenses into a finely divided snow-like solid.

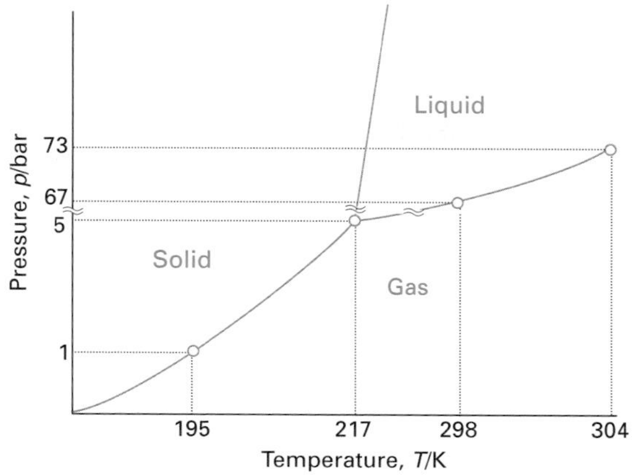

Fig. 5.18 The phase diagram of carbon dioxide. Note that, as the triple point lies well above atmospheric pressure, liquid carbon dioxide does not exist under normal conditions (a pressure of at least 5.11 bar must be applied). The text's website contains links to online databases of data on phase transitions, including phase diagrams.

Figure 5.19 shows the phase diagram of helium. Helium behaves unusually at low temperatures. For instance, the solid and gas phases of helium are never in equilibrium however low the temperature: the atoms are so light that they vibrate with a large-amplitude motion even at very low temperatures and the solid simply shakes itself apart. Solid helium can be obtained, but only by holding the atoms together by applying pressure. A second unique feature of helium is that pure helium-4 has two liquid phases.

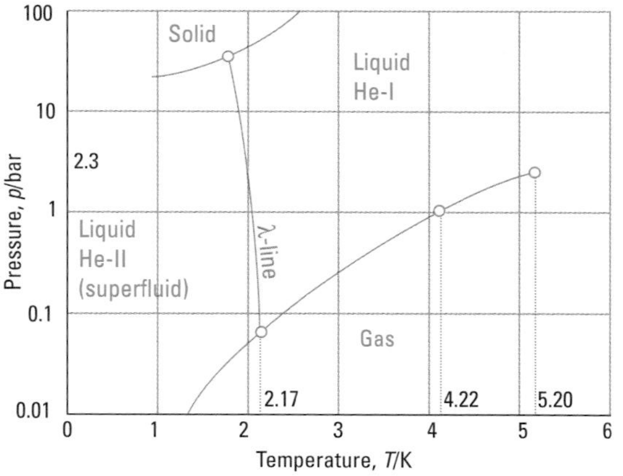

Fig. 5.19 The phase diagram for helium-4. The 'λ-line' marks the conditions under which the two liquid phases are in equilibrium. Helium-I is a conventional liquid and helium-II is a superfluid. Note that a pressure of at least 20 bar must be exerted before solid helium can be obtained.

The phase marked He-I in the diagram behaves like a normal liquid; the other phase, He-II, is a **superfluid**; it is so called because it flows without viscosity. Helium is the only known substance with a liquid–liquid boundary in its phase diagram, although recent work has suggested that water may also have a superfluid liquid phase.

5.9 The molecular structure of liquids

One question that should arise in your mind is the molecular basis of the material we have been discussing and, in particular, the molecular nature of a pure liquid phase. That is the question we address here.

The starting point for the discussion of gases is the totally random distribution of the molecules of a perfect gas. The starting point for the discussion of solids is the well-ordered structure of perfect crystals (Chapter 17). The liquid state is between these extremes: there is some structure and some disorder. The particles of a liquid are held together by intermolecular forces of the kind we discuss in Chapter 15, but their kinetic energies are comparable to their potential energies. As a result, although the molecules are not free to escape completely from the bulk, the whole structure is very mobile. The flow of molecules is like a crowd of spectators leaving a stadium.

In a crystal, particles lie at definite locations (in the absence of defects and thermal motion). This regularity continues out to large distances (to the edge of the crystal, billions of molecules away), so we say that crystals have **long-range order**. When the crystal melts, the long-range order is lost and wherever we look at long distances from a given particle there is equal probability of finding a second particle. Close to the first particle, though, there may be a remnant of order. Its nearest neighbours might still adopt approximately their original positions, and even if they are displaced by newcomers the new particles might adopt their vacated positions. The existence of this **short-range order** is due largely to intermolecular forces that exert their influence over short distances. For example, in liquid water any given H_2O molecule is surrounded by other molecules at the corners of a tetrahedron, similar to the arrangement in ice. The intermolecular forces (in this case, largely hydrogen bonds) are strong enough to affect the local structure right up to the boiling point.

The 'structure', such as it is, of a liquid is reported in terms of the **pair distribution function**, $g(r)$. This function is defined so that $g(r)\Delta r$ is the probability that a molecule will be found anywhere in a thin shell of radius r and thickness Δr. In a crystal, the pair distribution function is a series of spikes corresponding to molecules being found at fixed positions relative to any selected molecule, nearest neighbours, next-nearest neighbours, and so on. In a liquid, these spikes are blurred because the molecules are not in precisely fixed locations (Fig. 5.20). At small distances from a given molecule there is some short-range order, and the pair distribution function shows a series of oscillations. At greater distances, where there is no long-range order, these oscillations disappear and there is a uniform probability of finding molecules.

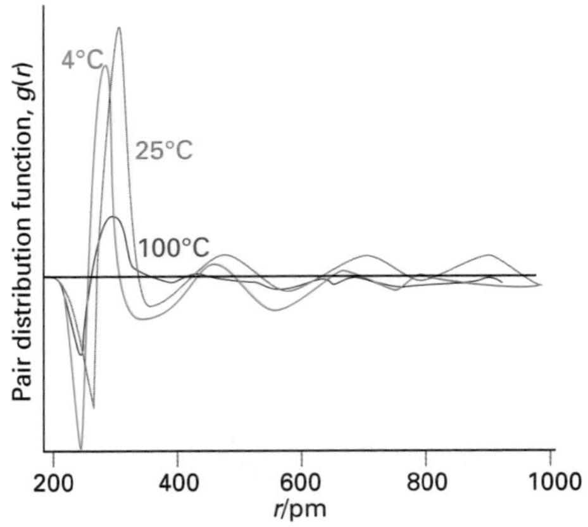

Fig. 5.20 The pair distribution function for the oxygen atoms in liquid water at three temperatures. Note the expansion as the temperature is raised.

Checklist of key concepts

☐ 1 The molar Gibbs energy of a liquid or a solid is almost independent of pressure.

☐ 2 A phase diagram of a substance shows the conditions of pressure and temperature at which its various phases are most stable.

☐ 3 A phase boundary depicts the pressures and temperatures at which two phases are in equilibrium.

☐ 4 The slope of a phase boundary is given by the Clapeyron equation; see the following *Road map*.

☐ 5 The slope of the liquid–vapour phase boundary is given by the Clausius–Clapeyron equation; see the following *Road map*.

☐ 6 The vapour pressure of a liquid is the pressure of the vapour in equilibrium with the liquid; it depends on temperature as $\ln p = A - B/T$.

☐ 7 The boiling temperature is the temperature at which the vapour pressure is equal to the external pressure; the normal boiling point is the temperature at which the vapour pressure is 1 atm.

☐ 8 The critical temperature is the temperature above which a substance does not form a liquid.

☐ 9 The triple point is the condition of pressure and temperature at which three phases are in mutual equilibrium.

☐ 10 The structure of a liquid is characterized by short-range order that is due largely to intermolecular forces that exert their influence over short distances.

☐ 11 Short-range order is displayed as broad oscillations in the plot of the pair distribution function against distance from a given molecule.

Road map of key equations

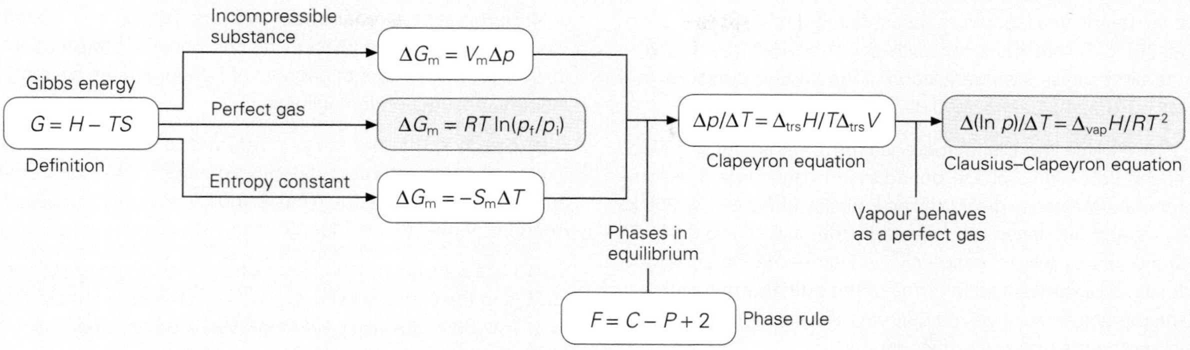

A blue box indicates a relation valid for a perfect gas.

Questions and exercises

Discussion questions

5.1 Why does the molar Gibbs energy vary with (a) temperature, (b) pressure?

5.2 Discuss the implication for phase stability of the variation of molar Gibbs energy with temperature and pressure.

5.3 Without doing a calculation, decide whether the presence of (a) attractive, (b) repulsive interactions between gas molecules will raise of lower the molar Gibbs energy of a gas relative to its 'perfect' value.

5.4 Explain the significance of the Clapeyron equation and of the Clausius–Clapeyron equation.

5.5 Use the phase rule to discuss the form of the phase diagram of sulfur, which has two solid phases, one liquid phase. Identify the number of degrees of freedom for each possible combination of phase equilibrium.

5.6 Explain what is meant by the 'structure of a liquid'.

5.7 The use of supercritical fluids for the extraction of a component from a complicated mixture is not restricted to the decaffeination of coffee. Consult library and internet resources and prepare a discussion of the principles, advantages, disadvantages, and current uses of supercritical fluid extraction technology.

Exercises

Assume all gases are perfect unless stated otherwise. All thermochemical data are for 298.15 K.

5.1 The standard Gibbs energy of formation of rhombic sulfur is zero and that of monoclinic sulfur is +0.33 kJ mol^{-1} at 25 °C. Which polymorph is the more stable at that temperature?

5.2 Carbon may exist as diamond, graphite, and other allotropes. The standard Gibbs energy of formation of diamond is 2.900 kJ mol^{-1} greater than that of graphite at 298 K. Predict which of these two allotropes is the more stable at this temperature.

5.3 The density of rhombic sulfur is 2.070 g cm^{-3} and that of monoclinic sulfur is 1.957 g cm^{-3}. Can the application of pressure be expected to make monoclinic sulfur more stable than rhombic sulfur?

5.4 Calculate the change in molar Gibbs energy of carbon dioxide (treated as a perfect gas) at 20 °C when its pressure is changed isothermally from 1.0 bar to (a) 3.0 bar, (b) 2.7 × 10^{-4} atm, its partial pressure in dry air at sea level.

5.5 A sample of water vapour at 200 °C is compressed isothermally from 350 cm^3 to 120 cm^3. What is the change in its molar Gibbs energy?

5.6 The standard molar entropy of rhombic sulfur is 31.80 J K^{-1} mol^{-1} and that of monoclinic sulfur is 32.6 J K^{-1} mol^{-1}. (a) Can an increase in temperature be expected to make monoclinic sulfur more stable than rhombic sulfur? (b) If so, at what temperature will the transition occur at 1 bar? (See Exercise 5.1 for data.)

5.7 The standard molar entropy of benzene is 173.3 J K^{-1} mol^{-1}. Calculate the change in its standard molar Gibbs energy when benzene is heated from 25 °C to 45 °C.

5.8 The standard molar entropies of water ice, liquid, and vapour are 37.99, 69.91, and 188.83 J K^{-1} mol^{-1}, respectively.

On a single graph, show how the Gibbs energy of each of these phases varies with temperature.

5.9 An open vessel containing (a) water, (b) benzene, (c) mercury stands in a laboratory measuring 5.0 m × 4.3 m × 2.2 m at 25 °C. What mass of each substance will be found in the air if there is no ventilation? (The vapour pressures are (a) 2.3 kPa, (b) 10 kPa, (c) 0.30 Pa.)

5.10 (a) Use the Clapeyron equation to estimate the slope of the solid–liquid phase boundary of water on a pressure–temperature phase diagram. The enthalpy of fusion is 6.008 kJ mol^{-1} and the densities of ice and water at 0 °C are 0.916 71 and 0.999 84 g cm^{-3}, respectively. Determine the entropy of fusion by expressing it in terms of the enthalpy of fusion and the melting point of ice. (b) Estimate the pressure required to lower the melting point of ice by 1 °C.

5.11 The normal boiling temperature of diethyl ether, $(C_2H_5)_2O$, is 307.7 K and the standard enthalpy of vaporization is 27.4 kJ mol^{-1}. Use the Clausius–Clapeyron equation to predict the vapour pressure of liquid diethyl ether at 298.15 K.

5.12 The average of the enthalpy of vaporization of propanone, C_3H_6O, over the temperature range 280 to 340 K is 30.2 kJ mol^{-1}. The vapour pressure is 30.600 kPa at 298.15 K. Estimate the normal boiling temperature of propanone.

5.13 Given the parameterization of the vapour pressure in eqn 5.8 and Table 5.1, what is (a) the enthalpy of vaporization, (b) the normal boiling point of hexane?

5.14 Suppose we wished to express the vapour pressure in eqn 5.8 in torr. What would be the values of A and B for methylbenzene? See Table 5.1 for data.

5.15 The vapour pressure of mercury is at 20 °C is 160 mPa; what is its vapour pressure at 40 °C given that its enthalpy of vaporization is 59.30 kJ mol^{-1}?

5.16 The vapour pressure of pyridine is 50.0 kPa at 365.7 K and the normal boiling point is 388.4 K. What is the enthalpy of vaporization of pyridine?

5.17 Estimate the normal boiling point of benzene given that its vapour pressure is 20 kPa at 35 °C and 50.0 kPa at 58.8 °C.

5.18 A saturated solution of Na_2SO_4, with excess of the solid, is present at equilibrium with its vapour in a closed vessel. (a) How many phases and components are present? (b) What is the number of degrees of freedom of the system? Identify the independent variables.

5.19 Suppose that the solution referred to in the preceding exercise is not saturated. (a) How many phases and components are present? (b) What is the number of degrees of freedom of the system? Identify the independent variables.

5.20 On a cold, dry morning after a frost, the temperature was –5 °C and the partial pressure of water in the atmosphere fell to 2 Torr. Will the frost sublime? What partial pressure of water would ensure that the frost remained?

5.21 (a) Refer to Fig. 5.15 and describe the changes that would be observed when water vapour at 1.0 bar and 400 K is cooled at constant pressure to 260 K. (b) Suggest the appearance of a plot of temperature against time if energy is removed at a constant rate. To judge the relative slopes of the cooling curves, you need to know that the constant-pressure molar heat capacities of water vapour, liquid, and solid are approximately $4R$, $9R$, and $4.5R$; the enthalpies of transition are given in Table 3.1.

5.22 Refer to Fig. 5.16 and describe the changes that would be observed when cooling takes place at the pressure of the triple point.

5.23 Use the phase diagram in Fig. 5.18 to state what would be observed when a sample of carbon dioxide, initially at 1.0 atm and 298 K, is subjected to the following cycle: (a) constant-pressure heating to 320 K, (b) isothermal compression to 100 atm, (c) constant-pressure cooling to 210 K, (d) isothermal decompression to 1.0 atm, (e) constant-pressure heating to 298 K.

5.24 Infer from the phase diagram for helium in Fig. 5.19 whether helium-I is more dense or less dense than helium-II.

Projects

The symbol ‡ indicates that calculus is required.

5.24‡ Here we explore the thermodynamic properties of a gas and its ability to condense to a liquid. Suppose that a gas obeys the van der Waals equation of state with the repulsive effects much greater than the attractive effects (that is, neglect the parameter a). (a) Find an expression for the change in molar Gibbs energy when the pressure is changed from p_i to p_f by following the method outlined in Derivation 5.2. (b) Is the change greater or smaller than for a perfect gas? (c) Estimate the percentage difference between the van der Waals and perfect gas calculations for carbon dioxide undergoing a change from 1.0 atm to 10.0 atm. (d) At the critical point of the gas, $dp/dV = 0$ and $d^2p/dV^2 = 0$. These relations identify the point of inflexion of the isotherms in Fig. 1.18. Show that a gas described by the equation of state $p = nRT/V - an^2/V^2 + bn^3/V^3$ shows critical behaviour, and express the critical constants in terms of the parameters a and b.

5.25‡ The Clausius–Clapeyron equation is derived in Section 5.5 on the assumption that the enthalpy of vaporization is independent of temperature in the range of interest. (a) The enthalpy of vaporization of water is 40.656 kJ mol^{-1} at its normal boiling temperature, 373 K. Assuming that the enthalpy of vaporization does not vary with temperature, use eqn 5.7 to calculate the vapour pressure of water at 308 K. (b) Derive an improved version of the equation on the basis that the enthalpy of vaporization has the form $\Delta_{vap}H = a + bT$. (c) For water in the range 298–373 K, a = 57.373 kJ mol^{-1} and b = –44.801 J K^{-1} mol^{-1}. Use your improved version of the Clausius–Clapeyron equation to calculate a more reliable value for the vapour pressure of water at 308 K.

6

Physical equilibria: the properties of mixtures

We now leave pure materials and the limited but important changes they can undergo and examine mixtures. We shall consider only **homogeneous mixtures**, or solutions, in which the composition is uniform however small the sample. The component in smaller abundance is called the **solute** and that in larger abundance is the **solvent**. These terms, however, are normally but not invariably reserved for solids dissolved in liquids; one liquid mixed with another is normally called simply a 'mixture' of the two liquids. In this chapter we consider mainly **non-electrolyte solutions**, where the solute is not present as ions. Examples are sucrose dissolved in water, sulfur dissolved in carbon disulfide, and a mixture of ethanol and water. We delay until Chapter 9 the special problems of **electrolyte solutions**, in which the solute consists of ions that interact strongly with one another.

The thermodynamic description of mixtures 133

6.1 Measures of concentration 134

6.2 Partial molar properties 135

6.3 Spontaneous mixing 138

6.4 Ideal solutions 139

6.5 Ideal–dilute solutions 142

6.6 Real solutions: activities 146

Colligative properties 146

6.7 The modification of boiling and freezing points 147

6.8 Osmosis 149

Phase diagrams of mixtures 152

6.9 Mixtures of volatile liquids 153

6.10 Liquid–liquid phase diagrams 155

6.11 Liquid–solid phase diagrams 156

6.12 The Nernst distribution law 159

CHECKLIST OF KEY CONCEPTS 160

ROAD MAP OF KEY EQUATIONS 160

QUESTIONS AND EXERCISES 161

The thermodynamic description of mixtures

We need a set of concepts that enable us to apply thermodynamics to mixtures of variable composition. First, we need to be able to define the composition of the mixture. We have already encountered mole fraction in Section 1.3 in our discussion of mixtures of gases but will now need to introduce the concepts of molar concentration and molality. Second, we also need to describe the properties of the mixture. We have already seen how to use the partial pressure, the contribution of one component in a gaseous mixture to the total pressure, to discuss the properties of mixtures of gases. For a more general description of the thermodynamics of mixtures we have to introduce other 'partial' properties, each one being the contribution that a particular component makes to the mixture.

6.1 Measures of concentration

The **molar concentration,** c_J or [J], of a solute J dissolved in a solvent is defined as the ratio of the chemical amount of J, n_J, divided by the volume of the solution, V:

$$c_J = \frac{n_J}{V} \quad \text{Definition} \quad \text{Molar concentration} \quad (6.1)$$

A note on good practice The informal term *molarity* is used widely, though not always properly. We speak of the molar concentration of a solute or the molarity of a solution, but not the molarity of a solute.

Molar concentration is typically reported in units of moles per cubic decimetre (mol dm^{-3}, which chemists commonly denote as M and read as 'molar'). We define the **standard molar concentration** as $c^{\ominus}$ = 1 mol dm^{-3} exactly. To prepare a solution of known molar concentration, a known amount of solute is dissolved in some solvent, and then more solvent is added to reach the desired total volume V. The volume V in the definition of molar concentration is the volume of the solution and not the volume of the solvent used to make up the solution.

Example 6.1

Making up a solution

A chemist wishes to make up an aqueous solution of copper sulfate of volume 100 cm^3 and molar concentration 0.250 mol dm^{-3}. What mass of hydrated copper sulfate, $CuSO_4{\cdot}5H_2O$, is required?

Strategy Use eqn 6.1 in the form $n_J = c_J V$ to determine the amount of $CuSO_4{\cdot}5H_2O$ required. Then, use the relation between molar mass and amount that was introduced in Foundations ($n = m/M$, in the form $m = nM$) to deduce the mass that corresponds to this amount.

Solution Because 1 cm^3 = 10^{-3} dm^3, 100 cm^3 corresponds to 0.100 dm^3. From eqn 6.1, the amount of $CuSO_4{\cdot}5H_2O$ required is therefore

$$n = cV = 0.250 \text{ mol dm}^{-3} \times 0.100 \text{ dm}^3 = 2.50 \times 10^{-4} \text{ mol}$$

The molar mass of $CuSO_4{\cdot}5H_2O$ is M = 249.68 g mol^{-1}, so the mass required is

$$m = n \times M = 2.50 \times 10^{-4} \text{ mol} \times 249.68 \text{ g mol}^{-1} = 0.0624 \text{ g}$$

which is equivalent to 62.4 mg.

Self-test 6.1

Determine the concentration of solute in an aqueous solution of volume 250 cm^3 that contains a mass of 13.2 g of phenol, C_6H_5OH.

Answer: 0.561 mol dm^{-3}

The **mass concentration** (which we also denote c, but always draw attention to it, in some cases by writing c_{mass}), is the mass of solute divided by the volume of the solution. Mass concentration is reported in grams (or kilograms) per cubic decimetre (g dm^{-3}) and is related to the molar concentration by

$$c_{mass} = \frac{[J]}{M} \quad \text{Mass concentration} \quad (6.2)$$

where M is the molar mass of the solute J.

The **molality,** b_J, of a solute is defined as the ratio of the chemical amount of the solute J divided by the mass of the solvent used to make up the solution

$$b_J = \frac{n_J}{m_{solvent}} \quad \text{Definition} \quad \text{Molality} \quad (6.3)$$

The molality of a solute is typically reported in units of moles per kilogram (mol kg^{-1}, which chemists commonly denote as m and read as 'molal'). The **standard molality** of the solution is defined as $b^{\ominus}$ = 1 mol kg^{-1} exactly. An important distinction between molar concentration and molality is that whereas the former is defined in terms of the volume of the *solution*, the molality is defined in terms of the mass of *solvent* used to prepare the solution. The molar concentration of a solute varies with temperature as the solution expands and contracts but the molality remains constant.

For dilute solutions in water, the numerical values of the molality and molar concentration differ very little because 1 dm^3 of solution is mostly water and has a mass close to 1 kg; for concentrated aqueous solutions and for all non-aqueous solutions with densities different from 1 g cm^{-3}, the two values are very different.

Example 6.2

Relating mole fraction and molality

What is the mole fraction of glucose molecules, $C_6H_{12}O_6$, in 0.140 m $C_6H_{12}O_6$(aq)?

Strategy We consider a sample that contains (exactly) 1 kg of solvent, and hence an amount $n_G = b_G \times$ (1 kg) of solute molecules. The amount of water molecules in exactly 1 kg of water is n_W = (1 kg)/M_W, where M_W is the molar mass of water. We refer to *exactly* 1 kg of water to avoid problems with significant figures. We can then calculate the mole fraction of glucose molecules as the ratio of the amount of solute, n_{solute}, to the total amount of water and solute $n = n_{solute} + n_W$.

Solution The amount of glucose molecules in exactly 1 kg of water is

$$n_{solute} = (0.140 \text{ mol kg}^{-1}) \times (1 \text{ kg}) = 0.140 \text{ mol}$$

The amount of water molecules in exactly 1 kg (10^3 g) of water is

$$n_W = \frac{10^3\ \text{g}}{18.02\ \text{g mol}^{-1}} = \frac{10^3}{18.02}\ \text{mol}$$

The total amount of molecules present is

$$n = n_{\text{solute}} + n_W = 0.140\ \text{mol} + \frac{10^3}{18.02}\ \text{mol}$$

The mole fraction of glucose molecules is therefore

$$x_{\text{solute}} = \frac{n_{\text{solute}}}{n} = \frac{0.140\ \text{mol}}{0.140 + (10^3/18.02)\ \text{mol}} = 2.52 \times 10^{-3}$$

Self-test 6.2

Calculate the mole fraction of sucrose molecules, $C_{12}H_{22}O_{11}$, in 1.22 m $C_{12}H_{22}O_{11}$(aq).

Answer: 2.15×10^{-2}

6.2 Partial molar properties

A **partial molar property** is the contribution (per mole) that a substance makes to an overall property of a mixture. The easiest partial molar property to visualize is the **partial molar volume**, V_J, of a substance J, the contribution J makes to the total volume of a mixture. Partial molar quantities are also commonly denoted by a bar over the symbol, as in $\bar{V}_J$. We have to be alert to the fact that although 1 mol of a substance has a characteristic volume when it is pure, 1 mol of that substance can make different contributions to the total volume of a mixture because molecules pack together in different ways in the pure substances and in mixtures.

Brief illustration 6.1 Partial molar volume

Imagine a huge volume of pure water. When a further 1 mol H_2O is added, the volume increases by 18 cm^3. However, when we add 1 mol H_2O to a huge volume of pure ethanol, the volume increases by only 14 cm^3. The quantity 18 cm^3 mol^{-1} is the volume occupied per mole of water molecules in pure water; 14 cm^3 mol^{-1} is the volume occupied per mole of water molecules in virtually pure ethanol. In other words, the partial molar volume of water in pure water is 18 cm^3 mol^{-1} and the partial molar volume of water in pure ethanol is 14 cm^3 mol^{-1}. In the latter case there is so much ethanol present that each H_2O molecule is surrounded by ethanol molecules and the packing of the molecules results in the water molecules occupying only 14 cm^3.

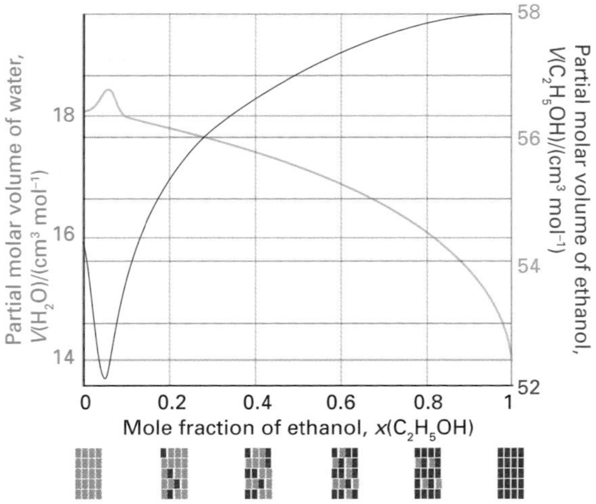

Fig. 6.1 The partial molar volumes of water and ethanol at 25 °C. Note the different scales (water on the left, ethanol on the right).

The partial molar volume at an intermediate composition of the water/ethanol mixture is an indication of the volume the H_2O molecules occupy when they are surrounded by a mixture of molecules representative of the overall composition (half water, half ethanol, for instance, when the mole fractions are both 0.5). The partial molar volume of ethanol varies as the composition of the mixture is changed, because the environment of an ethanol molecule changes from pure ethanol to pure water as the proportion of water increases and the volume occupied by the ethanol molecules varies accordingly. Figure 6.1 shows the variation of the two partial molar volumes across the full composition range at 25 °C.

Once we know the partial molar volumes V_A and V_B of the two components A and B of a mixture at the composition (and temperature) of interest, then we show in the following Derivation that we can state the total volume V of the mixture by using

$$V = n_A V_A + n_B V_B \qquad (6.4)$$

Derivation 6.1

Total volume and partial molar volume

Consider a very large sample of the mixture of the specified composition. Then, when an amount n_A of A is added, the composition remains virtually unchanged but the volume of the sample increases by $n_A V_A$. Similarly, when an amount n_B of B is added, the volume increases by $n_B V_B$. The total increase in volume is $n_A V_A + n_B V_B$. The mixture now occupies a larger volume but the proportions of the components are still the same. Next, scoop out of this enlarged volume a sample containing n_A of A and n_B of B. Its volume is $n_A V_A + n_B V_B$. Because volume is a state function, the same sample could have been prepared simply by mixing the appropriate amounts of A and B.

If you prefer to use mathematical arguments, then consider the following. When the composition of the mixture at constant temperature and pressure is changed by the addition of dn_A of A and dn_B of B, then the total volume the mixture changes by

$$dV = V_A dn_A + V_B dn_B$$

Provided the relative proportions of A and B are held constant as they are added, the mixture has the same composition and the two partial volumes are therefore constant. Therefore, to calculate the total volume, we integrate dV as n_A and n_B are raised simultaneously from 0 to their final values with the V_J treated as constants:

$$V = \overbrace{\int_0^{n_A} V_A dn_A + \int_0^{n_B} V_B dn_B}^{n_A:n_B \text{ constant}} = V_A \int_0^{n_A} dn_A + V_B \int_0^{n_B} dn_B$$

$$= V_A n_A + V_B n_B$$

as in eqn 6.4. Although we have envisaged the two integrations as being linked (in order to preserve constant composition), because V is a state function the final result in eqn 6.4 is valid however the solution is in fact prepared.

Example 6.3

Using partial molar volumes

What is the total volume of a mixture of 50.0 g of ethanol and 50.0 g of water at 25 °C?

Strategy To use eqn 6.4, we need the mole fractions of each substance and the corresponding partial molar volumes. We calculate the mole fractions in the same way as in Example 1.2 by using the molar masses of the components to calculate the amounts by using $n_J = m_J/M_J$. Then find the partial molar volumes corresponding to these mole fractions by referring to Fig. 6.1.

Solution The molar masses of CH_3CH_2OH and H_2O are 46.07 g mol^{-1} and 18.02 g mol^{-1}, respectively. Therefore the amounts present in the mixture are

$$n_{\text{ethanol}} = \frac{50.0\text{ g}}{46.07\text{ g mol}^{-1}} = 1.09\text{ mol}$$

$$n_{\text{water}} = \frac{50.0\text{ g}}{18.02\text{ g mol}^{-1}} = 2.77\text{ mol}$$

for a total of 3.86 mol. Hence $x_{\text{ethanol}} = 0.282$ and $x_{\text{water}} = 0.718$. According to Fig. 6.1, the partial molar volumes of the two substances in a mixture of this composition are 56 cm^3 mol^{-1} and 18 cm^3 mol^{-1}, respectively, so from eqn 6.4 the total volume of the mixture is

$$V = \overbrace{(1.09\text{ mol}) \times (56\text{ cm}^3\text{ mol}^{-1})}^{\text{Contribution from ethanol}} + \overbrace{(2.77\text{ mol}) \times (18\text{ cm}^3\text{ mol}^{-1})}^{\text{Contribution from water}}$$

$$\underbrace{=}_{\text{cancel mol}} 1.09 \times 56 + 2.77 \times 18\text{ cm}^3 = 110\text{ cm}^3$$

Self-test 6.3

Use Fig. 6.1 to calculate the mass density of a mixture of 20 g of water and 100 g of ethanol.

Answer: 0.84 g cm^{-3}

Now we extend the concept of a partial molar quantity to other state functions. The most important for our purposes is the **partial molar Gibbs energy**, G_J, of a substance J, which is the contribution of J (per mole of J) to the total Gibbs energy of a mixture. It follows, in the same way as for volume, that if we know the partial molar Gibbs energies of two substances A and B in a mixture of a given composition, then we can calculate the total Gibbs energy of the mixture by using an expression like eqn 6.4:

$$G = n_A G_A + n_B G_B \qquad (6.5a)$$

The partial molar Gibbs energy has exactly the same significance as the partial molar volume. For instance, ethanol has a certain partial molar Gibbs energy when it is pure (and every molecule is surrounded by other ethanol molecules), and it has a different partial molar Gibbs energy when it is in an aqueous solution of a certain composition (because then each ethanol molecule is surrounded by a mixture of ethanol and water molecules).

The partial molar Gibbs energy is so important in chemistry that it is given a special name and symbol. From now on, we shall call it the **chemical potential** and denote it μ (mu). Then, eqn 6.5a becomes

$$G = n_A \mu_A + n_B \mu_B \qquad \text{The total Gibbs energy of a mixture} \qquad (6.5b)$$

where μ_A is the chemical potential of A in the mixture and μ_B is the chemical potential of B. Formally, the chemical potential is the slope of a graph of the total Gibbs energy plotted against the amount of substance J (which may be either A or B) present in the mixture, with the temperature, pressure, and amounts of other components held constant.[1] In the course of this chapter and the next we shall see that the name 'chemical potential' is very appropriate, for

[1] Using proper mathematical notation of the type introduced in Section 2.8, the chemical potential is then written $\mu_J = (\partial G/\partial n_J)_{T,p,n_B}$.

it will become clear that μ_J is a measure of the ability of J to bring about physical and chemical change. A substance with a high chemical potential has a high ability, in a sense we shall explore, to drive a reaction or some other physical process forward.

To make progress, we need an explicit expression for the variation of the chemical potential of a substance with the composition of the mixture. Our starting point is eqn 5.3b:

$$G_m(p_f) = G_m(p_i) + RT\ln\frac{p_f}{p_i}$$

which shows how the molar Gibbs energy of a perfect gas depends on pressure. First, we set $p_f = p$, the pressure of interest, and $p_i = p^{\ominus}$, the standard pressure (1 bar). At the latter pressure, the molar Gibbs energy has its standard value, $G_m^{\ominus}$, so we can write

$$\overbrace{G_m(p)}^{p_f} \overset{\text{let } p_i=p^{\ominus}}{=} \overbrace{G_m(p_i)}^{G_m^{\ominus}} + RT\ln\frac{\overbrace{p}^{p_f}}{\underbrace{p_i}_{p^{\ominus}}}$$

$$= G_m^{\ominus} + RT\ln\frac{p}{p^{\ominus}} \qquad (6.6)$$

Next, for a *mixture* of perfect gases, we interpret p as the *partial* pressure of the gas (Section 1.3), and the G_m is the *partial* molar Gibbs energy, the chemical potential. Therefore, for a mixture of perfect gases, for each component J present at a partial pressure p_J,

$$\mu_J = \mu_J^{\ominus} + RT\ln\frac{p_J}{p^{\ominus}} \quad \text{Perfect gas} \quad \text{Chemical potential} \qquad (6.7a)$$

In this expression, $\mu_J^{\ominus}$ is **standard chemical potential** of the gas J, which is identical to its standard molar Gibbs energy, the value of G_m for the pure gas at 1 bar. If we adopt the convention that, whenever p_J appears in a formula it is to be interpreted as $p_J/p^{\ominus}$ (so, if the pressure is 2.0 bar, $p_J = 2.0$), we can write eqn 6.7a more simply as

$$\mu_J = \mu_J^{\ominus} + RT\ln p_J \qquad (6.7b)$$

Figure 6.2 illustrates the pressure dependence of the chemical potential of a perfect gas predicted by this equation. The chemical potential becomes negatively infinite as the pressure tends to zero. As the pressure is increased from zero, the chemical potential rises to its standard value at 1 bar (because ln 1 = 0), and then increases slowly (logarithmically, as ln p) as the pressure is increased further.

As always, we can become familiar with an equation by listening to what it tells us. In this case:

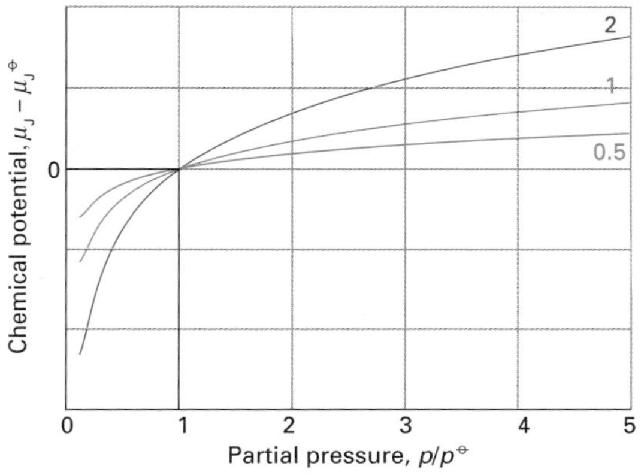

Fig. 6.2 The variation with partial pressure of the chemical potential of a perfect gas at three different temperatures (in the ratios 0.5:1:2). Note that the chemical potential increases with pressure and, at a given pressure, with temperature.

- As p_J increases, so does ln p_J. Therefore, eqn 6.7 tells us that the higher the partial pressure of a gas, the higher its chemical potential.

This conclusion is consistent with the interpretation of the chemical potential as an indication of the potential of a substance to be active chemically: the higher the partial pressure, the more active chemically the species. In this instance the chemical potential represents the tendency of the substance to react when it is in its standard state (the significance of the term $\mu^{\ominus}$) plus an additional tendency that reflects whether it is at a different pressure. For a given amount of substance, a higher partial pressure gives that substance more chemical 'punch', just like winding a spring gives a spring more physical punch (that is, enables it to do more work).

Brief illustration 6.2 The chemical potential of a perfect gas

Suppose that the partial pressure of a perfect gas in a mixture is reduced. According to eqn 6.7a, the change in chemical potential of the gas is

$$\Delta\mu = \left(\mu^{\ominus} + RT\ln\frac{p_f}{p^{\ominus}}\right) - \left(\mu^{\ominus} + RT\ln\frac{p_i}{p^{\ominus}}\right)$$

$$= RT\left(\ln\frac{p_f}{p^{\ominus}} - \ln\frac{p_i}{p^{\ominus}}\right)$$

$$\underbrace{=}_{\ln a - \ln b = \ln(a/b)} RT\ln\frac{p_f/p^{\ominus}}{p_i/p^{\ominus}} = RT\ln\frac{p_f}{p_i}$$

For example, if the pressure of a gas fall from 100 kPa to 50 kPa as it is consumed in a reaction at 298 K, the change in chemical potential of the gas is

$$\Delta\mu = (8.3145\ \text{J K}^{-1}\ \text{mol}^{-1}) \times (298\ \text{K}) \times \ln\frac{50\ \text{kPa}}{100\ \text{kPa}}$$

$$= -1.7 \times 10^{3}\ \text{J mol}^{-1} = -1.7\ \text{kJ mol}^{-1}$$

We saw in Chapter 5 that the molar Gibbs energy of a pure substance is the same in all the phases at equilibrium. We use the same argument in the following Derivation to show that:

> A system is at equilibrium when the chemical potential of each substance has the same value in every phase in which it occurs.

We can think of the chemical potential as the pushing power of each substance, and equilibrium is reached only when each substance pushes with the same strength in any phase it occupies.

Derivation 6.2

The uniformity of chemical potential

Suppose a substance J occurs in different phases in different regions of a system. For instance, we might have a liquid mixture of ethanol and water and a mixture of their vapours. Let the substance J have chemical potential $\mu_J(\text{l})$ in the liquid mixture and $\mu_J(\text{g})$ in the vapour. We could imagine an infinitesimal amount, dn_J, of J migrating from the liquid to the vapour. As a result, the Gibbs energy of the liquid phase falls by $\mu_J(\text{l})dn_J$ and that of the vapour rises by $\mu_J(\text{g})dn_J$. The net change in Gibbs energy is

$$dG = \mu_J(\text{g})dn_J - \mu_J(\text{l})dn_J = \{\mu_J(\text{g}) - \mu_J(\text{l})\}dn_J$$

At equilibrium, there is no tendency for this migration (or the reverse process, migration from the vapour to the liquid), so $dG = 0$, which is true if $\mu_J(\text{g}) = \mu_J(\text{l})$. The argument applies to each component of the system. Therefore, *for a substance to be at equilibrium throughout the system, its chemical potential must be the same everywhere.*

6.3 Spontaneous mixing

All gases mix spontaneously with one another because the molecules of one gas can mingle with the molecules of the other gas. But how can we show *thermodynamically* that mixing is spontaneous? At constant temperature and pressure, we need to show that $\Delta G < 0$. The first step is therefore to find an expression for ΔG when two gases mix, and then to decide whether it is negative. As we see in the following Derivation, when an amount n_A of A and n_B of B of two gases mingle at a temperature T,

$$\Delta G = nRT\{x_A \ln x_A + x_B \ln x_B\} \qquad \text{Perfect gases} \quad \text{Gibbs energy of mixing} \qquad (6.8)$$

with $n = n_A + n_B$ and the x_J the mole fractions of the components J in the mixture.

Derivation 6.3

The Gibbs energy of mixing

Suppose we have an amount n_A of a perfect gas A at a certain temperature T and pressure p, and an amount n_B of a perfect gas B at the same temperature and pressure. The two gases are in separate compartments initially (Fig. 6.3). The Gibbs energy of the system (the two unmixed gases) is the sum of their individual Gibbs energies:

$$G_i \overset{\text{eqn 6.5b}}{=} n_A\mu_A + n_B\mu_B \overset{\text{eqn 6.7b}}{=} n_A\{\mu_A^{\ominus} + RT\ln p\} + n_B\{\mu_B^{\ominus} + RT\ln p\}$$

The chemical potentials are those for the two gases, each at a pressure p. When the partition is removed, the total pressure remains the same, but, according to Dalton's law (Section 1.3), the partial pressures fall to $p_A = x_A p$ and $p_B = x_B p$, where the x_J are the mole fractions of the two gases in the mixture ($x_J = n_J/n$, with $n = n_A + n_B$). The final Gibbs energy of the system is therefore

$$G_f = n_A\{\mu_A^{\ominus} + RT\ln p_A\} + n_B\{\mu_B^{\ominus} + RT\ln p_B\}$$
$$= n_A\{\mu_A^{\ominus} + RT\ln x_A p\} + n_B\{\mu_B^{\ominus} + RT\ln x_B p\}$$

The difference $G_f - G_i$ is the change in Gibbs energy that accompanies mixing. The standard chemical potentials cancel, and by making use of the relation (see The chemist's toolkit 2.2):

$$\ln x_J p - \ln p \overset{\ln a - \ln b = \ln(a/b)}{=} \ln\frac{x_J p}{p} \overset{\text{Cancel } p}{=} \ln x_J$$

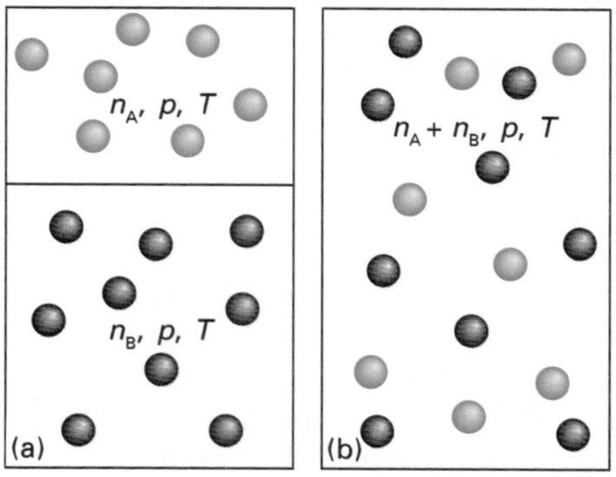

Fig. 6.3 The (a) initial and (b) final states of a system in which two perfect gases mix. The molecules do not interact, so the enthalpy of mixing is zero. However, because the final state is more disordered than the initial state, there is an increase in entropy.

for each gas, we obtain

$$\Delta G = RT\{n_A \ln x_A + n_B \ln x_B\} \overset{n_A=nx_A,\ n_B=nx_B}{=} nRT\{x_A \ln x_A + x_B \ln x_B\}$$

which is eqn 6.8.

Equation 6.8 tells us the change in Gibbs energy when two gases mix at constant temperature and pressure (Fig. 6.4). The crucial feature is that because x_A and x_B are both less than 1, the two logarithms are negative ($\ln x < 0$ if $x < 1$), so $\Delta G < 0$ at all compositions. Therefore, *perfect gases mix spontaneously in all proportions*. Furthermore, if we compare eqn 6.8 (written in a slightly different way) with $\Delta G = \Delta H - T\Delta S$,

$$\Delta G = \overset{\Delta H}{0} + \overbrace{T[nR\{x_A \ln x_A + x_B \ln x_B\}]}^{-T\Delta S}$$

we can conclude that

$$\Delta H = 0 \quad \text{Perfect gases} \quad \text{Enthalpy of mixing} \quad (6.9a)$$

$$\Delta S = -nR\{x_A \ln x_A + x_B \ln x_B\} \quad \text{Perfect gases} \quad \text{Entropy of mixing} \quad (6.9b)$$

That is, there is no change in enthalpy when two perfect gases mix, which reflects the fact that there are no interactions between the molecules. There is an increase in entropy, because the mixed gas is more disordered than the unmixed gases (Fig. 6.5). The entropy of the surroundings is unchanged because the enthalpy of the system is constant, so no energy escapes as heat into the surroundings. It follows that the increase in entropy of the system is the 'driving force' of the mixing.

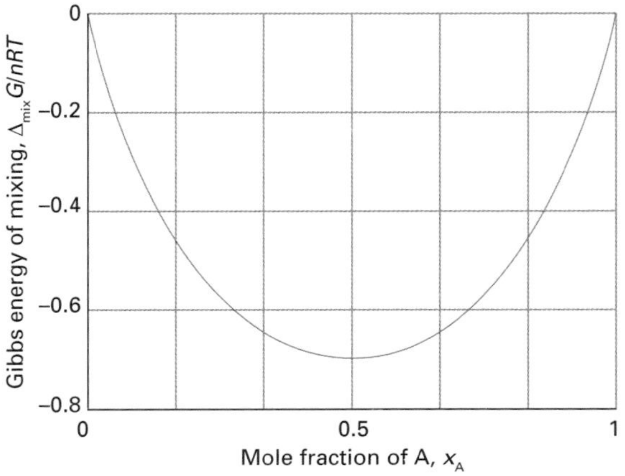

Fig. 6.4 The variation of the Gibbs energy of mixing with composition for two perfect gases at constant temperature and pressure. Note that $\Delta G < 0$ for all compositions, which indicates that two gases mix spontaneously in all proportions.

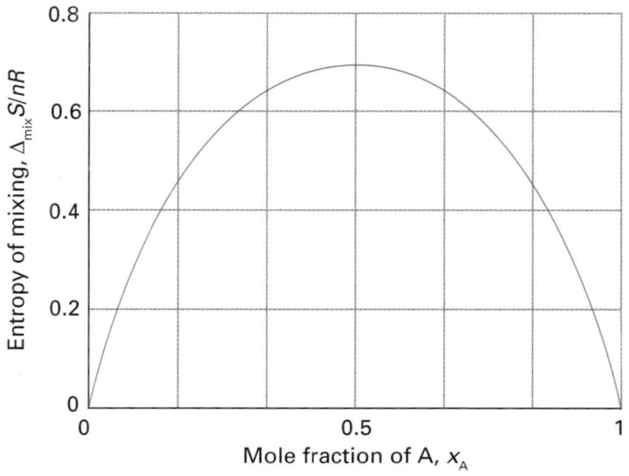

Fig. 6.5 The variation of the entropy of mixing with composition for two perfect gases at constant temperature and pressure.

6.4 Ideal solutions

In chemistry we are concerned with liquids as well as gases, so we need an expression for the chemical potential of a substance in a liquid solution. In the following, we use the following notation:

J denotes a substance in general
A denotes a solvent
B denotes a solute

We can anticipate that the chemical potential of a species ought to increase with concentration, because the higher its concentration the greater its chemical 'punch'.

(a) Raoult's law

The key to setting up an expression for the chemical potential of a solute is the work done by the French chemist François Raoult (1830–1901), who spent most of his life measuring the vapour pressures of solutions. In general, the vapour above a mixture is also a mixture, so the total vapour pressure of the mixture is the sum of the **partial vapour pressure,** p_J, the contribution to the total vapour pressure of each component in the mixture. Raoult measured the partial pressure of the vapour of each component in dynamic equilibrium with the liquid mixture, and established what is now called **Raoult's law:**

> The partial vapour pressure of a substance in a liquid mixture is proportional to its mole fraction in the mixture and its vapour pressure when pure: $p_J = x_J p_J^*$ Raoult's law (6.10)

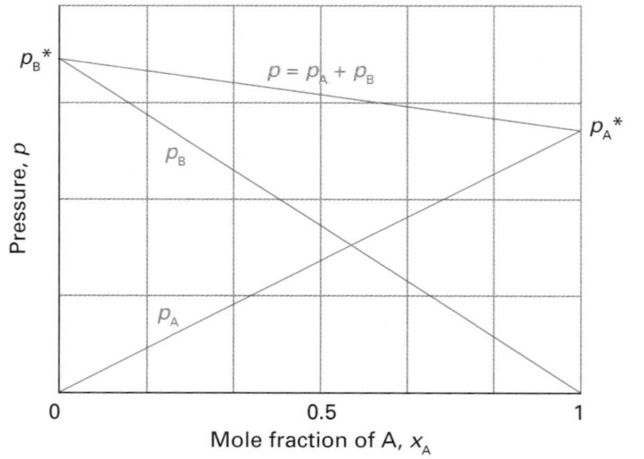

Fig. 6.6 The partial vapour pressures of the two components of an ideal binary mixture are proportional to the mole fractions of the components in the liquid. The total pressure of the vapour is the sum of the two partial vapour pressures.

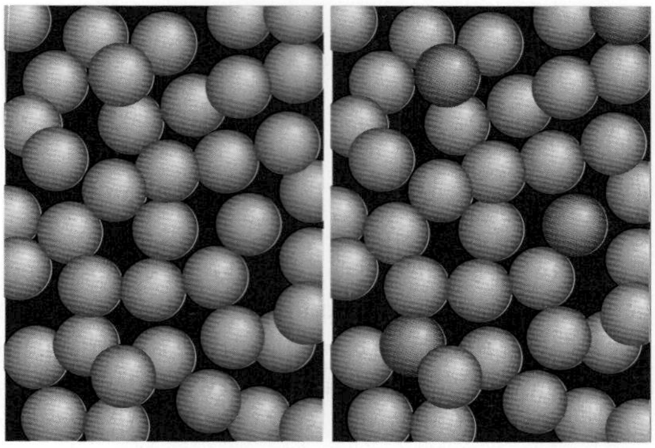

Fig. 6.7 (a) In a pure liquid, we can be confident that any molecule selected from the sample is a solvent molecule. (b) When a solute is present, we cannot be sure that blind selection will give a solvent molecule, so the entropy of the system is greater than in the absence of the solute.

In this expression, p_J^* is the vapour pressure of the pure substance. For example, when the mole fraction of water in an aqueous solution is 0.90, then, provided Raoult's law is obeyed, the partial vapour pressure of the water in the solution is 90 per cent that of pure water. This conclusion is approximately true whatever the identity of the solute and the solvent (Fig. 6.6).

Brief illustration 6.3 Raoult's law

A solution is prepared by dissolving 1.5 mol $C_{10}H_8$ (naphthalene) in 1.00 kg of benzene (C_6H_6). To calculate the partial vapour pressure of benzene in the solution, we express the mole fraction, $x_{benzene}$, of benzene as

$$x_{benzene} = \frac{n_{benzene}}{n_{benzene} + n_{naphthalene}} \overset{n=m/M}{=} \frac{m_{benzene}/M_{benzene}}{(m_{benzene}/M_{benzene}) + n_{naphthalene}}$$

where $m_{benzene}$ and $M_{benzene}$ are the mass and molar mass of benzene, respectively. It follows from eqn 6.10 and the vapour pressure of pure benzene, $p^*_{benzene} = 12.6$ kPa at 25 °C that

$$p_{benzene} \overset{\text{Raoult's law}}{=} x_{benzene}p^*_{benzene}$$

$$= \frac{m_{benzene}/M_{benzene}}{(m_{benzene}/M_{benzene}) + n_{naphthalene}} p^*_{benzene}$$

$$= \frac{(1.00\times10^3\ \text{g})/(78.54\ \text{g mol}^{-1})}{\{(1.00\times10^3\ \text{g})/(78.54\ \text{g mol}^{-1})\}+1.5\ \text{mol}} \times (12.6\ \text{kPa})$$

$$\overset{\text{cancel g and mol}}{=} 11.3\ \text{kPa}$$

The molecular origin of Raoult's law is the effect of the solute on the entropy of the solution. In the pure solvent, the molecules are dispersed and have a corresponding entropy; the vapour pressure then represents the tendency of the system and its surroundings to reach a higher entropy. When a solute is present, the solution has a greater disorder than the pure solvent because we cannot be sure that a molecule chosen at random will be a solvent molecule (Fig. 6.7). Because the entropy of the solution is higher than that of the pure solvent, the solution has a lower tendency to acquire an even higher entropy by the solvent vaporizing. In other words, the vapour pressure of the solvent in the solution is lower than that of the pure solvent.

An **ideal solution** is a hypothetical solution of a solute B in a solvent A that obeys Raoult's law throughout the composition range from pure A to pure B. The law is most reliable when the components of a mixture have similar molecular shapes and are held together in the liquid by similar types and strengths of intermolecular forces. An example is a mixture of two structurally similar hydrocarbons. A mixture of benzene and methylbenzene (toluene) is a good approximation to an ideal solution, for the partial vapour pressure of each component satisfies Raoult's law reasonably well throughout the composition range from pure benzene to pure methylbenzene (Fig. 6.8).

No mixture is truly ideal and all real mixtures show deviations from Raoult's law. However, the deviations are small for the component of the mixture that is in large excess (the solvent) and become smaller as the concentration of solute decreases

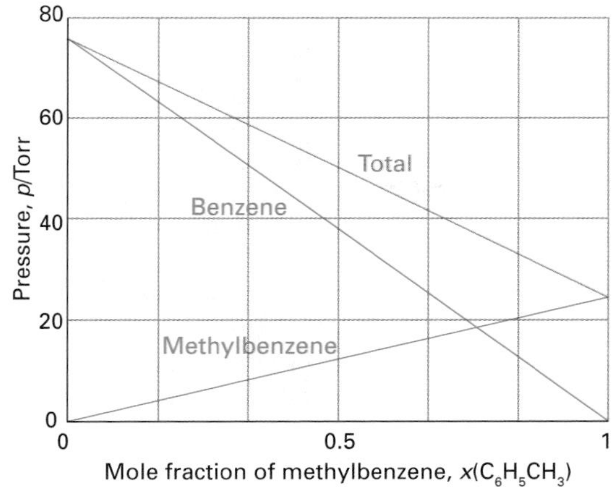

Fig. 6.8 Two similar substances, in this case benzene and methylbenzene (toluene), behave almost ideally and have vapour pressures that closely resemble those for the ideal case depicted in Fig. 6.6.

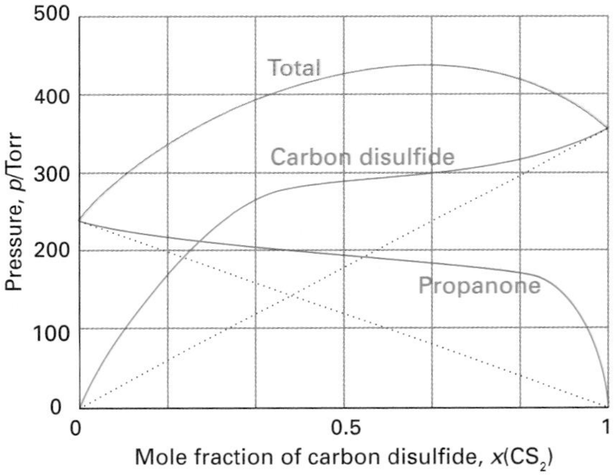

Fig. 6.9 Strong deviations from ideality are shown by dissimilar substances, in this case carbon disulfide and acetone (propanone). Note, however, that Raoult's law is obeyed by propanone when only a small amount of carbon disulfide is present (on the left) and by carbon disulfide when only a small amount of propanone is present (on the right).

(Fig. 6.9). We can usually be confident that Raoult's law is reliable for the solvent when the solution is very dilute. More formally, Raoult's law is a *limiting law* (like the perfect gas law), and is strictly valid only in the limit of zero concentration of solute.

(b) The chemical potential of the solvent

The theoretical importance of Raoult's law is that, because it relates vapour pressure to composition, and we know how to relate pressure to chemical potential, we can use the law to relate chemical potential to the composition of a solution. As we show in the following Derivation, the chemical potential of a solvent A present in solution at a mole fraction x_A is

$$\mu_A = \mu_A^* + RT \ln x_A \quad \text{Ideal solution} \quad \text{Chemical potential of the solvent} \quad (6.11)$$

where μ_A^* is the chemical potential of pure A. This expression is valid throughout the concentration range for either component of a binary ideal solution. It is valid for the solvent of a real solution the closer the composition approaches pure solvent (pure A).

A note on good practice An asterisk (*) is used to denote a pure substance, but not one that is necessarily in its standard state. Only if the pressure is 1 bar would μ_A^* be the standard chemical potential of A, and it would then be written $\mu_A^\ominus$.

Figure 6.10 shows the variation of chemical potential of the solvent predicted by eqn 6.11. The essential feature is as follows:

- Because $x_A < 1$ implies that $\ln x_A < 0$, the chemical potential of a solvent is lower in a solution than when it is pure (when $x_A = 1$).

Provided the solution is almost ideal, a solvent in which a solute is present has less chemical 'punch' (including a lower ability to generate a vapour pressure) than when it is pure.

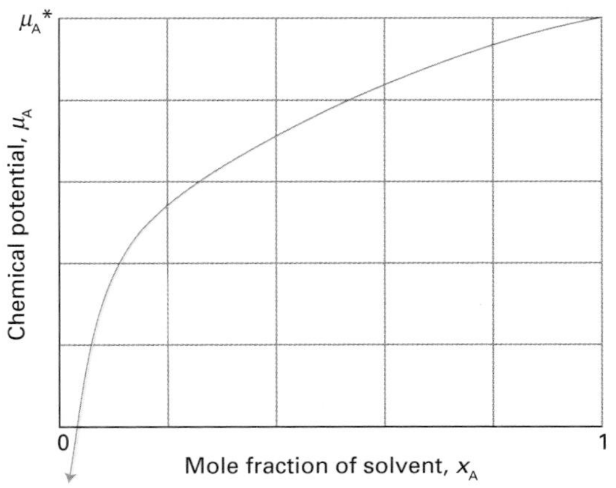

Fig. 6.10 The variation of the chemical potential of the solvent with the composition of the solution. Note that the chemical potential of the solvent is lower in the mixture than for the pure liquid (for an ideal system). This behaviour is likely to be shown by a dilute solution in which the solvent is almost pure (and obeys Raoult's law).

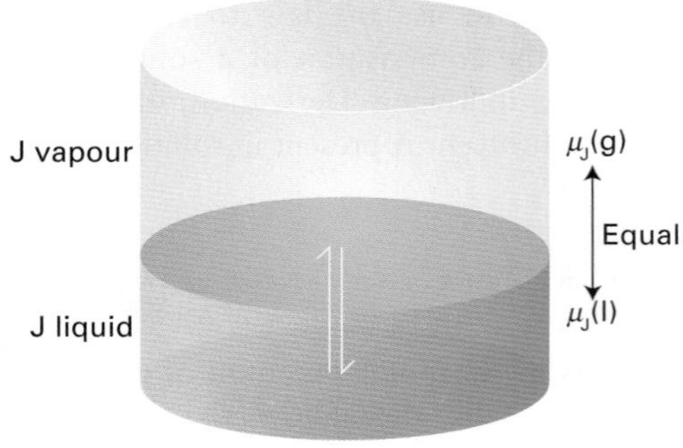

Fig. 6.11 At equilibrium, the chemical potential of a substance in its liquid phase is equal to the chemical potential of the substance in its vapour phase.

Derivation 6.4

The chemical potential of a solvent

We have seen that when a liquid A in a mixture is in equilibrium with its vapour at a partial pressure p_A, the chemical potentials of the two phases are equal (Fig. 6.11), and we can write $\mu_A(l) = \mu_A(g)$. However, we already have an expression for the chemical potential of a vapour, eqn 6.7b; so at equilibrium

$$\mu_A(l) = \mu_A^{\ominus}(g) + RT \ln p_A$$

According to Raoult's law, $p_A = x_A p_A^*$, so we can write

$$\mu_J(l) = \mu_A^{\ominus}(g) + RT \ln x_A p_A^* \overset{\ln(ab) = \ln a + \ln b}{=} \mu_A^{\ominus}(g) + RT \ln p_A^* + RT \ln x_A$$

The first two terms on the right, $\mu_A^{\ominus}(g)$ and $RT \ln p_A^*$, are independent of the composition of the mixture. We can write them as the constant μ_A^*, the chemical potential of pure liquid A. Then eqn 6.11 follows.

Brief illustration 6.4 The chemical potential of the solvent

The change in chemical potential of benzene caused by a solute is given by

$$\Delta\mu_{\text{benzene}} = \overbrace{\mu_{\text{benzene}}}^{\mu^*_{\text{benzene}} + RT \ln x_{\text{benzene}}} - \mu^*_{\text{benzene}} = RT \ln x_{\text{benzene}}$$

If a solute is present at a mole fraction of 0.10, then $x_{\text{benzene}} = 0.90$, and at 298 K the change in chemical potential is

$$\Delta\mu_{\text{benzene}} = (8.3145 \text{ J K}^{-1} \text{ mol}^{-1}) \times (298 \text{ K}) \times \ln 0.90$$
$$\overset{\text{cancel K}}{=} -2.6 \times 10^2 \text{ J mol}^{-1}$$
$$= -0.26 \text{ kJ mol}^{-1}$$

Is dissolving a solute to form an ideal solution spontaneous? To answer this question, we need to discover whether ΔG is negative for dissolving. The calculation is essentially the same as for the mixing of two perfect gases, and we conclude that

$$\Delta G = nRT\{x_A \ln x_A + x_B \ln x_B\}$$

Ideal solution Gibbs energy of dissolving (6.12)

exactly as for two perfect gases. As for perfect gases, the enthalpy and entropy of dissolving are

$$\Delta H = 0$$ Ideal solution Enthalpy of dissolving (6.13a)

$$\Delta S = -nR\{x_A \ln x_A + x_B \ln x_B\}$$

Ideal solution Entropy of dissolving (6.13b)

The value of ΔH indicates that although (unlike for perfect gases) there are interactions between the molecules, the average solute–solute, solvent–solvent, and solute–solvent interactions are all the same, so the solute slips into solution without a change in enthalpy. The driving force for dissolving is the increase in entropy of the system as one component mingles with the other (as in Fig. 6.5).

A note on good practice 'Ideality' implies that the average interactions are all the same. A 'perfect' gas is a special case of an ideal system in which the average intermolecular interactions are not merely the same but are in fact zero. Most scientists do not make this helpful distinction, and refer to an 'ideal gas' rather than a 'perfect gas'.

6.5 Ideal–dilute solutions

Raoult's law provides a good description of the vapour pressure of the *solvent* in a very dilute solution, because the solvent A is almost pure. Here we develop a thermodynamic description of the solute.

We cannot in general expect Raoult's law to be a good description of the vapour pressure of the solute B because a solute in dilute solution is very far from being pure. In a dilute solution, each solute molecule is surrounded by nearly pure solvent, so its environment is quite unlike that in the pure solute and except when solute and solvent are very similar (such as benzene and methylbenzene) it is very unlikely that its vapour pressure will be related in a simple manner to that of the pure solute.

(a) Henry's law

It is found experimentally that in dilute solutions the vapour pressure of the solute is in fact proportional to its mole fraction, just as for the solvent. Unlike the solvent, though, the constant of proportionality is not in general the vapour pressure of the pure solute. This linear but different dependence was discovered by the English chemist William Henry (1775–1836), and is summarized as **Henry's law**:

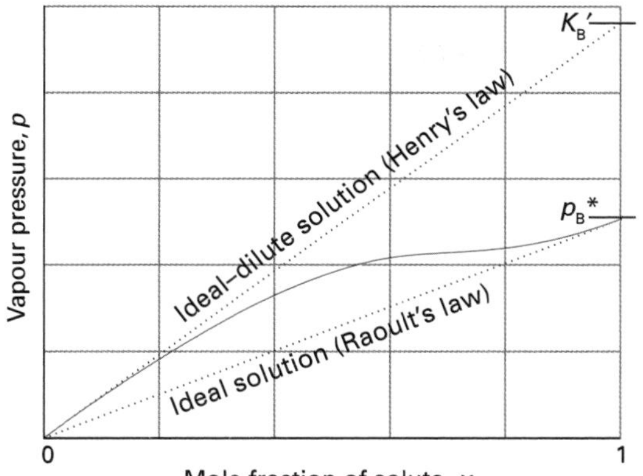

Fig. 6.12 When a component (the solvent) is almost pure, it behaves in accord with Raoult's law and has a vapour pressure that is proportional to the mole fraction in the liquid mixture, and a slope p^*, the vapour pressure of the pure substance. When the same substance is the minor component (the solute), its vapour pressure is still proportional to its mole fraction, but the constant of proportionality is now K_H' (that is, for the solute B, specifically K_B').

The vapour pressure of a volatile solute B is proportional to its mole fraction in a solution:

$$p_B = K_H'x_B \quad \text{Henry's law} \quad (6.14)$$

Here K_H', which is called the **Henry's law constant,** is characteristic of the solute and chosen so that the straight line predicted by eqn 6.14 is the tangent of the experimental curve at $x_B = 0$ (Fig. 6.12). Expressed in this way, the Henry's law constant has units of pressure. Henry's law is usually obeyed only at low concentrations of the solute (close to $x_B = 0$). Solutions that are dilute enough for the solute to obey Henry's law are called **ideal–dilute solutions.**

Example 6.4

Verifying Raoult's and Henry's laws

The partial vapour pressures of each component in a mixture of propanone (acetone, A) and trichloromethane (chloroform, C) were measured at 35 °C with the following results:

x_C	0	0.20	0.40	0.60	0.80	1
p_C/kPa	0	4.7	11	18.9	26.7	36.4
p_A/kPa	43.3	33.3	23.3	12.3	4.9	0

Confirm that the mixture conforms to Raoult's law for the component in large excess and to Henry's law for the minor component. Find the Henry's law constants.

Strategy The original procedure was to plot the partial vapour pressures against mole fraction. Raoult's law is verified by comparing the data to the straight line $p_J = x_Jp_J^*$ for each component in the region in which it is in excess and therefore acting as the solvent. Henry's law is verified by finding a straight line $p_J = K_H'x_J$ that is tangent to each partial vapour pressure at low x_J where the component can be treated as the solute.

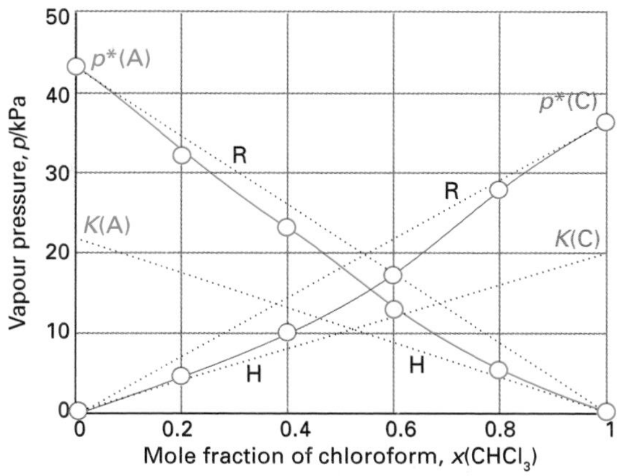

Fig. 6.13 The experimental partial vapour pressures of a mixture of trichloromethane, $CHCl_3$ (C), and propanone, CH_3COCH_3 (acetone, A), based on the data in Example 6.4. Henry's and Raoult's law behaviour are denoted H and R, respectively.

Solution The data are plotted in Fig. 6.13, together with the Raoult's law lines. Henry's law requires $K_H' = 20.0$ kPa for propanone and $K_H' = 22.0$ kPa for trichloromethane. Notice how the data deviate from both Raoult's and Henry's laws for even quite small departures from $x = 1$ and $x = 0$, respectively.

Self-test 6.4

The vapour pressure of chloromethane at various mole fractions in a mixture at 25 °C was found to be as follows:

x	0.005	0.009	0.019	0.024
p/kPa	27.3	48.4	101	126

Estimate the Henry's law constant.

Answer: 5 MPa

Henry's law is commonly written to show how the molar concentration of the dissolved gas depends on its partial pressure in the vapour phase above the solvent:

$$[B] = K_Hp_B \quad \text{Another version of Henry's law} \quad (6.15)$$

Written in this way, and with the pressure in kilopascals, the Henry's law constant, K_H, is reported in moles per cubic decimetre per kilopascal (mol dm^{-3} kPa^{-1}). The Henry's law constants of some gases are listed in Table 6.1. This form of the law and these

Table 6.1

Henry's law constants for gases dissolved in water at 25 °C

	K_H/(mol m^{-3} kPa^{-1})
Carbon dioxide, CO_2	3.39×10^{-1}
Hydrogen, H_2	7.78×10^{-3}
Methane, CH_4	1.48×10^{-2}
Nitrogen, N_2	6.48×10^{-3}
Oxygen, O_2	1.30×10^{-2}

units make it very easy to calculate the molar concentration of the dissolved gas, simply by multiplying the partial pressure of the gas (in kilopascals) by the appropriate constant. Equation 6.15 is used, for instance, to estimate the concentration of O_2 in natural waters (see Example 6.5). Knowledge of Henry's law constants for gases in fats and lipids is important for the discussion of respiration, especially when the partial pressure of oxygen is abnormal, as in diving and mountaineering (see Impact 6.1).

Example 6.5

Determining whether natural water can support aquatic life

The concentration of O_2 in water required to support aerobic aquatic life is about 4.0 mg dm^{-3}. What is the minimum partial pressure of oxygen in the atmosphere that can achieve this concentration?

Strategy The strategy of the calculation is to determine the partial pressure of oxygen that, according to Henry's law (written as eqn 6.15), corresponds to the concentration specified.

Solution Equation 6.15 becomes

$$p_{O_2} = \frac{[O_2]}{K_H}$$

We note that the molar concentration of O_2 is

$$[O_2] \overset{[J]=c_{mass}/M}{=} \frac{4.0\times10^{-3}\ \text{g dm}^{-3}}{32\ \text{g mol}^{-1}} = \frac{4.0\times10^{-3}}{32}\frac{\text{mol}}{\text{dm}^3}$$

$$= \frac{4.0\times10^{-3}}{32}\frac{\text{mol}}{10^{-3}\ \text{m}^3} = \frac{4.0}{32}\ \text{mol m}^{-3}$$

From Table 6.1, K_H for oxygen in water is 1.30×10^{-2} mol m^{-3} kPa^{-1}; therefore the partial pressure needed to achieve the stated concentration is

$$p_{O_2} = \frac{(4.0/32)\ \text{mol m}^{-3}}{1.30\times10^{-2}\ \text{mol m}^{-3}\ \text{kPa}^{-1}} = 9.6\ \text{kPa}$$

The partial pressure of oxygen in air at sea level is 21 kPa (158 Torr), which is greater than 9.6 kPa (72 Torr), so the required concentration can be maintained under normal conditions.

A note on good practice The number of significant figures in the result of a calculation should not exceed the number in the data.

Self-test 6.5

What partial pressure of methane is needed to dissolve 21 mg of methane in 100 g of benzene at 25 °C ($K_H = 5.69 \times 10^4$ kPa, for Henry's law in the form given in eqn 6.15)?

Answer: 57 kPa (4.3×10^2 Torr)

(b) The chemical potential of the solute

Henry's law lets us write an expression for the chemical potential of a solute in a dilute solution. We show in the following Derivation that the chemical potential of the solute when it is present at a mole fraction x_B is

$$\mu_B = \mu_B^* + RT \ln x_B \qquad (6.16)$$

Chemical potential of the solute in terms of the mole fraction

This expression, which is illustrated in Fig. 6.14, applies when Henry's law is valid, in an ideal–dilute solution. The chemical potential of the solute has its pure value when it is present alone ($x_B = 1$, $\ln 1 = 0$) and a smaller value when dissolved (when $x_B < 1$, $\ln x_B < 0$).

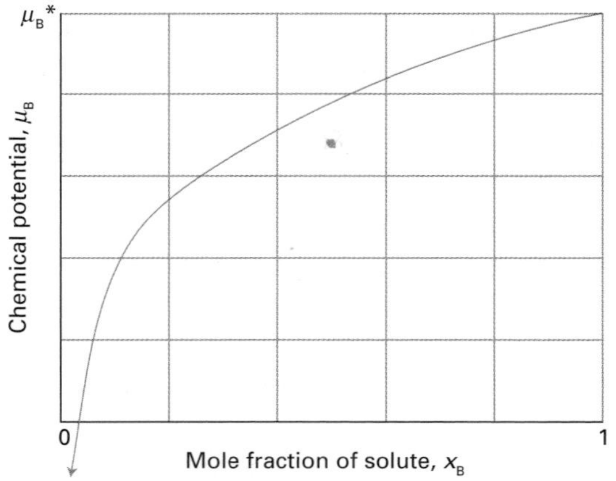

Fig. 6.14 The variation of the chemical potential of the solute with the composition of the solution expressed in terms of the mole fraction of solute. Note that the chemical potential of the solute is lower in the mixture than for the pure solute (for an ideal system). This behaviour is likely to be shown by a dilute solution in which the solvent is almost pure and the solute obeys Henry's law.

Derivation 6.5

The chemical potential of the solute

We apply the same reasoning as in Derivation 6.4. When a solute B in a solution is in equilibrium with its vapour at a partial pressure p_B, we can write $\mu_B(l) = \mu_B(g)$ and (from eqn 6.7)

$$\mu_B(l) = \mu_B^{\ominus}(g) + RT \ln p_B$$

According to Henry's law, $p_B = K'_H x_B$, so it follows that

$$\mu_B(l) = \mu_B^{\ominus}(g) + RT \ln K'_H x_B = \overbrace{\mu_B^{\ominus}(g) + RT \ln K'_H}^{\mu_B^*} + RT \ln x_B$$

The terms $\mu_B^{\ominus}(g)$ and $RT \ln K'_H$ are independent of the composition of the mixture and can be combined into the constant μ_B^*, the chemical potential of pure liquid B. Equation 6.16 then follows.

It is often more natural to express the composition of a solution in terms of the molar concentration of the solute, [B], rather than as a mole fraction. The mole fraction of the solute and the molar concentration are proportional to each other in dilute solutions, so we write x_B = constant × $[B]/c^{\ominus}$, where the standard molar concentration is introduced to ensure that the constant is dimensionless. Then eqn 6.16 becomes

$$\mu_B = \mu_B^* + RT \ln[\text{constant} \times ([B]/c^{\ominus})]$$

$$\underbrace{=}_{\ln ab = \ln a + \ln b} \overbrace{\mu_B^* + RT \ln(\text{constant})}^{\mu_B^{\ominus}} + RT \ln([B]/c^{\ominus})$$

A note on good practice It is meaningless to take logarithms of quantities with units, so always ensure that the x of $\ln x$ is a pure number.

Provided that the pressure is 1 bar we can combine the first two terms into a single constant, $\mu_B^{\ominus}$, and write this relation as

$$\mu_B = \mu_B^{\ominus} + RT \ln([B]/c^{\ominus}) \quad \text{Chemical potential of the solute in terms of the molar concentration} \quad (6.17a)$$

This equation is the best way to write the relation; however, it is cumbersome, and for the rest of the chapter we shall write $[B]/c^{\ominus}$ simply as [B] and—to conform to the requirement stated in the note on good practice—interpret [B] as the molar concentration in moles per cubic decimetre with the units deleted (we treated pressure similarly earlier in the chapter). Thus, if in fact [B] = 0.1 mol dm^{-3}, so $[B]/c^{\ominus} = 0.1$, from now on we shall write [B] = 0.1 and use eqn 6.17a in the form

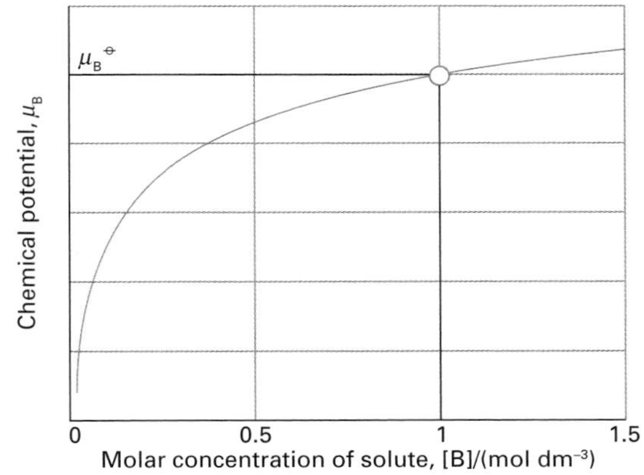

Fig. 6.15 The variation of the chemical potential of the solute with the composition of the solution that obeys Henry's law expressed in terms of the molar concentration of solute. The chemical potential has its standard value at [B] = 1 mol dm^{-3}.

$$\mu_B = \mu_B^{\ominus} + RT \ln[B] \quad \text{Simplified version of eqn 6.17a} \quad (6.17b)$$

Figure 6.15 illustrates the variation of chemical potential with concentration predicted by this equation. The chemical potential of the solute has its standard value when the molar concentration of the solute is 1 mol dm^{-3} (that is, $c^{\ominus}$).

Impact on biology 6.1

Gas solubility and breathing

We inhale about 500 cm^3 of air with each breath we take. The influx of air is a result of changes in volume of the lungs as the diaphragm is depressed and the chest expands, which results in a decrease in pressure of about 100 Pa relative to atmospheric pressure. Expiration occurs as the diaphragm rises and the chest contracts and gives rise to a differential pressure of about 100 Pa above atmospheric pressure. The total volume of air in the lungs is about 6 dm^3, and the additional volume of air that can be exhaled forcefully after normal expiration is about 1.5 dm^3. Some air remains in the lungs at all times to prevent the collapse of the alveoli.

The effect of gas exchange between blood and air inside the alveoli of the lungs means that the composition of the air in the lungs is different from that in the atmosphere, and changes throughout the breathing cycle. Alveolar gas is in fact a mixture of newly inhaled air and air about to be exhaled. The concentration of oxygen present in arterial blood is equivalent to a partial pressure of about 40 Torr (5.3 kPa), whereas the partial pressure of freshly inhaled air in the alveoli of the lungs is about 100 Torr (13.3 kPa). Arterial blood remains in the capillary passing through the wall of an alveolus for about 0.75 s, but such is the steepness of the pressure gradient that it becomes fully saturated with oxygen in about 0.25 s. If the lungs collect fluids (as in pneumonia), then the respiratory membrane thickens, diffusion is greatly slowed, and body tissues begin to suffer from oxygen starvation. Carbon dioxide moves in the opposite direction across the respiratory tissue, but the partial pressure gradient is much less, corresponding to about 45 Torr (6.0 kPa) in blood and 40 Torr (5.3 kPa) in air at equilibrium in the alveoli of the lungs. However, because carbon dioxide is much more soluble in the alveolar fluid than oxygen is, equal amounts of oxygen and carbon dioxide are exchanged in each breath.

A hyperbaric oxygen chamber, in which oxygen is at an elevated partial pressure, is used to treat certain types of disease. Carbon monoxide poisoning can be treated in this way, as can the consequences of shock. Diseases that are caused by anaerobic bacteria, such as gas gangrene and tetanus, can also be treated because the bacteria cannot thrive in high oxygen concentrations.

6.6 **Real solutions: activities**

No actual solutions are ideal, and many solutions deviate from ideal–dilute behaviour as soon as the concentration of solute rises above a small value. In thermodynamics we try to preserve the form of equations developed for ideal systems so that it becomes easy to step between the two types of system. This is the thought behind the introduction of the **activity**, a_J, of a substance, which is a kind of effective concentration. The activity is defined so that the expression

$$\mu_J = \mu_J^{\ominus} + RT \ln a_J \qquad (6.18)$$

The chemical potential in terms of the activity

is true at *all* concentrations and for both the solvent and the solute.

For ideal solutions, $a_J = x_J$, and the activity of each component is equal to its mole fraction. For ideal–dilute solutions comparison of eqn 6.18 with eqn 6.17a gives $a_B = [B]/c^{\ominus}$, and the activity of the solute is equal to the numerical value of its molar concentration. For real solutions we write

$$\text{For the solvent:}\quad a_A = \gamma_A x_A$$
$$\text{For the solute:}\quad a_B = \gamma_B [B]/c^{\ominus} \qquad (6.19)$$

The activity in terms of the activity coefficient

where the γ (gamma) in each case is the **activity coefficient**. Both activities and activity coefficients are dimensionless. Activity coefficients depend on the composition of the solution and we should note the following:

- Because the solvent behaves more in accord with Raoult's law as it becomes pure, $\gamma_A \to 1$ as $x_A \to 1$.
- Because the solute behaves more in accord with Henry's law as the solution becomes very dilute, $\gamma_B \to 1$ as $[B] \to 0$.

Table 6.2
*Activities and standard states**

Substance	Standard state	Activity, *a*
Solid	Pure solid, 1 bar	1
Liquid	Pure liquid, 1 bar	1
Gas	Pure gas, 1 bar	$p/p^{\ominus}$
Solute	Molar concentration of 1 mol dm^{-3}	$[J]/c^{\ominus}$

$p^{\ominus}$ = 1 bar (= 10^5 Pa), $c^{\ominus}$ = 1 mol dm^{-3}

*Activities are for perfect gases and ideal–dilute solutions; all activities are dimensionless.

These conventions and relations are summarized in Table 6.2.

Activities and activity coefficients are often branded as 'fudge factors'. To some extent that is true. However, their introduction does allow us to derive thermodynamically exact expressions for the properties of non-ideal solutions. Moreover, in a number of cases it is possible to calculate or measure the activity coefficient of a species in solution. In this text we shall normally derive thermodynamic relations in terms of activities, but when we want to make contact with actual measurements, we shall set the activities equal to the 'ideal' values in Table 6.2.

Colligative properties

An ideal solute has no effect on the enthalpy of a solution in the sense that the enthalpy of mixing is zero. However, it does affect the entropy by introducing a degree of disorder that is not present in the pure solvent, and we found in eqn 6.13b that $\Delta S > 0$ when two components mix to give an ideal solution. We can therefore expect a solute to modify the physical properties of the solution. Apart from lowering the vapour pressure of the solvent, which we have already considered, a non-volatile solute has three main effects:

- It raises the boiling point of a solution.
- It lowers the freezing point.
- It gives rise to an osmotic pressure.

(The meaning of the last will be explained shortly.) Because these properties all stem from changes in the disorder of the solvent, and the increase in disorder is independent of the identity of the species we use to bring it about, for a given solvent all of them depend only on the number of solute particles present, not their chemical identity. For this reason they are called **colligative properties,** with 'colligative' denoting 'depending on the collection'. Thus, a 0.01 mol kg^{-1} aqueous solution of any non-electrolyte should have the same boiling point, freezing point, and osmotic pressure.

6.7 The modification of boiling and freezing points

As we have indicated, the effect of a solute is to raise the boiling point and to lower the freezing point of a solvent. It is found empirically, and is justified by the calculation in the following Derivation, that the **elevation of boiling point,** ΔT_b and the **depression of freezing point,** ΔT_f, are both proportional to the molality, b_B, of the solute:

$$\Delta T_b = K_b b_B \quad \text{Elevation of boiling point} \quad (6.20a)$$

$$\Delta T_f = K_f b_B \quad \text{Depression of freezing point} \quad (6.20b)$$

K_b is the **ebullioscopic constant** and K_f is the **cryoscopic constant** of the solvent. They are also called the 'boiling-point constant' and the 'freezing-point constant', respectively. The two constants can be estimated from other properties of the solvent, but both are best treated as empirical constants (Table 6.3).

Brief illustration 6.5 Depression of the freezing point

To estimate the lowering of the freezing point of a solution made by dissolving 3.0 g (about one cube) of sucrose ($C_{12}H_{22}O_{11}$) in 100 g of water, we must first express the molality of sucrose as

$$b_{\text{sucrose}} = \frac{n_{\text{sucrose}}}{m_{\text{water}}} \overset{n=m/M}{=} \frac{m_{\text{sucrose}}/M_{\text{sucrose}}}{m_{\text{water}}} = \frac{m_{\text{sucrose}}}{M_{\text{sucrose}} m_{\text{water}}}$$

where m_{sucrose} and M_{sucrose} are the mass and molar mass of sucrose, respectively. It follows from eqn 6.20b and Table 6.3 that

$$\Delta T_f \overset{\text{eqn 6.20b}}{=} K_{f,\text{water}} \frac{m_{\text{sucrose}}}{M_{\text{sucrose}} m_{\text{water}}}$$

$$= (1.86\ \text{K kg mol}^{-1}) \times \frac{3.0\ \text{g}}{(343.88\ \text{g mol}^{-1}) \times (0.100\ \text{kg})}$$

$$\overset{\text{cancel kg, g, and mol}}{=} 0.16\ \text{K}$$

Table 6.3
Cryoscopic and ebullioscopic constants

Solvent	K_f/(K kg mol^{-1})	K_b/(K kg mol^{-1})
Acetic acid	3.90	3.07
Benzene	5.12	2.53
Camphor	40	
Carbon disulfide	3.8	2.37
Naphthalene	6.94	5.8
Phenol	7.27	3.04
Tetrachloromethane	30	4.95
Water	1.86	0.51

To understand the origin of these effects we shall make two simplifying assumptions:

- The solute is not volatile, and therefore does not appear in the vapour phase.
- The solute is insoluble in the solid solvent, and therefore does not appear in the solid phase.

For example, a solution of sucrose in water consists of a solute (sucrose) that is not volatile and therefore never appears in the vapour, which is therefore pure water vapour. The sucrose is also left behind in the liquid solvent when ice begins to form, so the ice remains pure.

The origin of colligative properties is the lowering of chemical potential of the solvent by the presence of a solute, as expressed by eqn 6.11. We saw in Section 5.3 that the freezing and boiling points correspond to the temperatures at which the graph of the molar Gibbs energy of the liquid intersects the graphs of the molar Gibbs energy of the solid and vapour phases, respectively. Because we are now dealing with mixtures, we have to think about the *partial* molar Gibbs energy (the chemical potential) of the solvent. The presence of a solute lowers the chemical potential of the liquid but, because the vapour and solid remain pure, their chemical potentials remain unchanged. As a result, we see from Fig. 6.16 that the freezing point moves to lower values; likewise, from Fig. 6.17 we see that the boiling point moves to higher values. In other words, the freezing point is depressed, the boiling point is elevated, and the liquid phase exists over a wider range of temperatures. The following Derivation shows how to express these changes quantitatively.

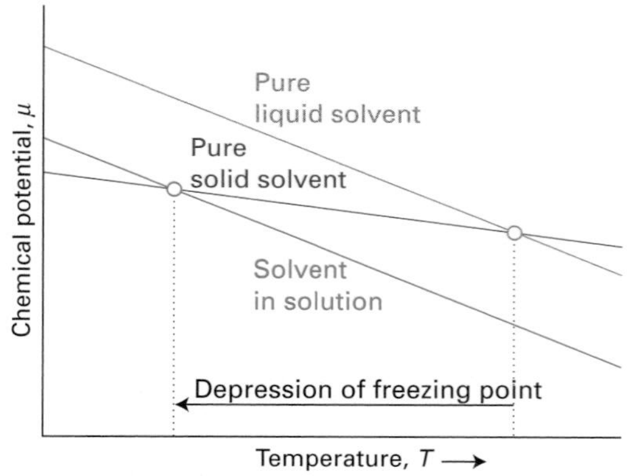

Fig. 6.16 The chemical potentials of pure solid solvent and pure liquid solvent also decrease with temperature, and the point of intersection, where the chemical potential of the liquid rises above that of the solid, marks the freezing point of the pure solvent. A solute lowers the chemical potential of the solvent but leaves that of the solid unchanged. As a result, the intersection point lies further to the left and the freezing point is therefore lowered.

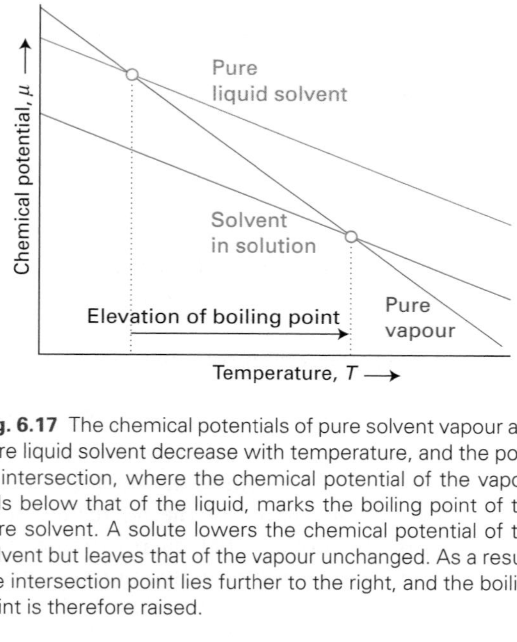

Fig. 6.17 The chemical potentials of pure solvent vapour and pure liquid solvent decrease with temperature, and the point of intersection, where the chemical potential of the vapour falls below that of the liquid, marks the boiling point of the pure solvent. A solute lowers the chemical potential of the solvent but leaves that of the vapour unchanged. As a result, the intersection point lies further to the right, and the boiling point is therefore raised.

Derivation 6.6

The modification of transition temperatures

To derive an expression for the elevation of boiling point, we note that at the normal (that is, at 1 atm) boiling point, T_b^*, of the pure solvent A, the solvent vapour and liquid are in equilibrium at 1 atm, so their chemical potentials are equal:

$$\mu_A^*(\text{g}, 1\text{ atm}, T_b^*) = \mu_A^*(\text{l}, 1\text{ atm}, T_b^*)$$

For simplicity, we no longer specify the pressure as 1 atm, but remember that that is its value. In the presence of a solute B, the mole fraction of the solvent A is reduced from 1 to $x_A = 1 - x_B$ and the boiling point is T_b; then, because the solvent vapour and liquid remain in equilibrium under these new conditions,

$$\mu_A^*(\text{g}, T_b) = \mu_A(\text{l}, x_A, T_b)$$

This equality is depicted in Fig. 6.18a. According to eqn 6.11, the chemical potential of the solvent in the solution is related to its mole fraction by

$$\mu_A(\text{l}, x_A, T_b) = \mu_A^*(\text{l}, T_b) + RT_b \ln x_A$$

Therefore, the last equation becomes

$$\mu_A^*(\text{g}, T_b) = \mu_A^*(\text{l}, T_b) + RT_b \ln x_A$$

which rearranges into

$$\ln x_A = \frac{\mu_A^*(\text{g}, T_b) - \mu_A^*(\text{l}, T_b)}{RT_b}$$

The chemical potential of a pure substance is the same as the molar Gibbs energy of the substance, so this expression is the same as

$$\ln x_A = \frac{\overbrace{G_m(\text{g}, T_b) - G_m(\text{l}, T_b)}^{\Delta_{vap}G(T_b)}}{RT_b} = \frac{\Delta_{vap}G(T_b)}{RT_b}$$

When $x_A = 1$ (so $\ln x_A = 0$), the pure liquid solvent, the boiling point is T_b^*, and we can write

$$0 = \frac{\Delta_{vap}G(T_b^*)}{RT_b^*}$$

The difference of these two equations is

$$\ln x_A = \frac{\Delta_{vap}G(T_b)}{RT_b} - \frac{\Delta_{vap}G(T_b^*)}{RT_b^*}$$

Now we use $\Delta G = \Delta H - T\Delta S$ and the approximate temperature-independence of ΔH and ΔS to turn this equation into

$$\ln x_A \overset{\Delta_{vap}G(T) = \Delta_{vap}H(T) - T\Delta_{vap}S(T)}{=} \frac{\Delta_{vap}H(T_b) - T_b\Delta_{vap}S(T_b)}{RT_b} - \frac{\Delta_{vap}H(T_b^*) - T_b^*\Delta_{vap}S(T_b^*)}{RT_b^*}$$

Now assume that in the range of temperatures of interest, the enthalpy and entropy of vaporization are independent of temperature, so each can be written as a constant term:

$$\ln x_A = \frac{\Delta_{vap}H}{RT_b} - \frac{\Delta_{vap}S}{R} - \frac{\Delta_{vap}H}{RT_b^*} + \frac{\Delta_{vap}S}{R} \overset{\text{Cancel blue terms}}{=} \frac{\Delta_{vap}H}{R}\left(\frac{1}{T_b} - \frac{1}{T_b^*}\right)$$

Now we use $\ln x_A = \ln(1 - x_B) \approx -x_B$ (see The chemist's toolkit 6.1) to express this equation as

$$-x_B \approx \frac{\Delta_{vap}H}{R}\left(\frac{1}{T_b} - \frac{1}{T_b^*}\right)$$

or

$$x_B \approx -\frac{\Delta_{vap}H}{R}\left(\frac{1}{T_b} - \frac{1}{T_b^*}\right) = \frac{\Delta_{vap}H}{R}\left(\frac{1}{T_b^*} - \frac{1}{T_b}\right) = \frac{\Delta_{vap}H}{R}\left(\frac{T_b - T_b^*}{T_b^* T_b}\right)$$

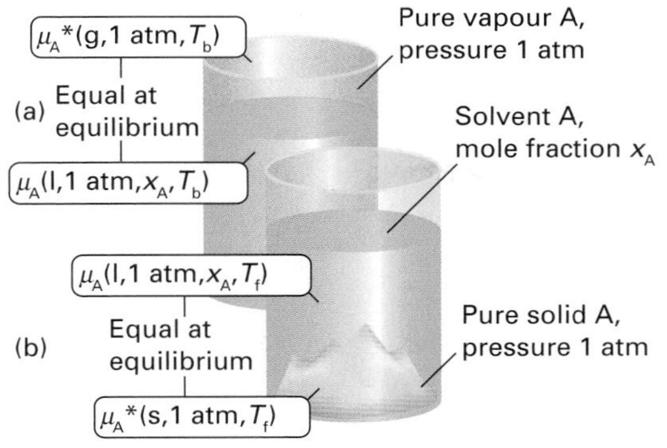

Fig. 6.18 The equilibria between phases and the corresponding relation between chemical potentials of the solvent in solution at (a) the normal boiling point and (b) the normal freezing point.

We are almost there. First, we note that the elevation of boiling point is $\Delta T_b = T_b - T_b^*$. Then we note that because the value of T_b is very close to T_b^*, little error is introduced by replacing $T_b^* T_b$ by T_b^{*2}. In this way we arrive at

$$x_B \approx \frac{\Delta_{vap}H}{R} \times \frac{\Delta T_b}{T_b^{*2}}$$

which we can rearrange into

$$\Delta T_b \approx \frac{RT_b^{*2}}{\Delta_{vap}H} \times x_B$$

At this point, we see that the elevation of boiling point is proportional to the mole fraction of solute B and independent of its identity (the $\Delta_{vap}H$ and T_b are properties of the solvent). The mole fraction of the solute is proportional to its molality, b_B, so the equation we have derived has the form $\Delta T_b = K_b b_B$, as in eqn 6.20.

The calculation of the depression of freezing point starts with the equilibrium condition depicted in Fig. 6.18b, which implies

$$\mu_A^*(s, T_f) = \mu_A(l, x_A, T_f) \quad \text{at 1 atm}$$

The calculation then proceeds in exactly the same way, and we arrive at

$$\Delta T_f \approx \frac{RT_f^{*2}}{\Delta_{fus}H} \times x_B$$

with $\Delta T_f = T_f^* - T_f$, as in eqn 6.20. The explicit calculation is left as an exercise.

The elevation of boiling point is too small to have any practical significance. A practical consequence of the lowering of freezing point, and hence the lowering of the melting point of the pure solid, is its employment in organic chemistry to judge the purity of a sample, for any impurity lowers the melting point of a substance from its accepted value. The salt water of the oceans freezes at temperatures lower than that of fresh water and salt is spread on highways to delay the onset of freezing. The addition of 'antifreeze' to car engines and, by natural processes, to arctic fish, is commonly held up as an example of the lowering of freezing point, but the concentrations are far too high for the arguments we have used here to be applicable. The 1,2-ethanediol ('glycol') used as antifreeze simply interferes with bonding between water molecules. Likewise, the antifreeze proteins of arctic fish act by binding to small ice crystals and preventing larger crystals from forming.

The chemist's toolkit 6.1 Power series and expansions

It is possible, and often useful, to express a function $f(x)$ as a special sum of terms called a **power series** of the form

$$f(x) = c_0 + c_1 x + c_2 x^2 + \cdots$$

where c_n are constants. The following series are used frequently in physical chemistry. The approximations are valid if $x \ll 1$.

$$(1 + x)^{-1} = 1 - x + x^2 - \cdots \approx 1 - x$$

$$e^x = 1 + x + \tfrac{1}{2}x^2 + \tfrac{1}{6}x^3 + \cdots \approx 1 + x$$

$$\ln x = (x - 1) - \tfrac{1}{2}(x - 1)^2 + \tfrac{1}{3}(x - 1)^3 - \cdots$$

$$\ln(1 + x) = x - \tfrac{1}{2}x^2 + \tfrac{1}{3}x^3 - \cdots \approx x$$

$$(1 + x)^{1/2} = 1 + \tfrac{1}{2}x - \tfrac{1}{8}x^2 + \cdots$$

6.8 Osmosis

The phenomenon of **osmosis** (from the Greek word for 'push') is the passage of a pure solvent into a solution separated from it by a semipermeable membrane. A **semipermeable membrane** is a membrane that is permeable to the solvent but not to the solute. The membrane might have microscopic holes that are large enough to allow water molecules to pass through, but not ions or carbohydrate molecules with their bulky coating of hydrating water molecules. The **osmotic pressure,** Π (uppercase pi), is the pressure that must be applied to the solution to stop the inward flow of solvent.

In the simple arrangement shown in Fig. 6.19, the pressure opposing the passage of solvent into the solution arises from the hydrostatic pressure of the column of solution that the osmosis itself produces. This column is formed when the pure solvent flows through the membrane into the solution and pushes the column of solution higher up the tube. Equilibrium is reached when the downward pressure exerted by the column of solution is equal to the upward osmotic pressure. A complication of this arrangement is that the entry of solvent into the solution results in dilution of the latter, so it is more

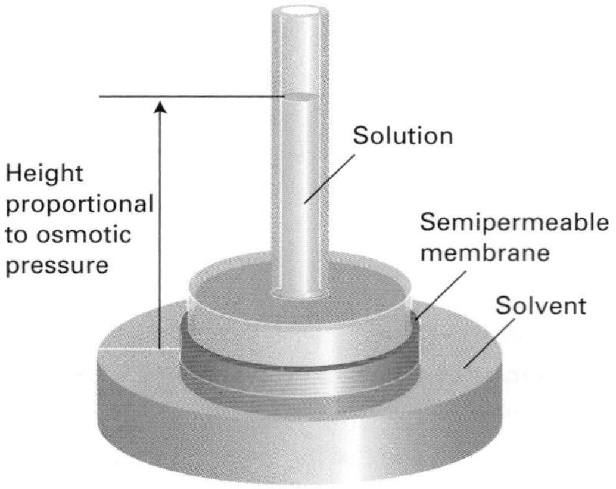

Fig. 6.19 In a simple osmosis experiment, a solution is separated from the pure solvent by a semipermeable membrane. Pure solvent passes through the membrane and the solution rises in the inner tube. The net flow ceases when the pressure exerted by the column of liquid is equal to the osmotic pressure of the solution.

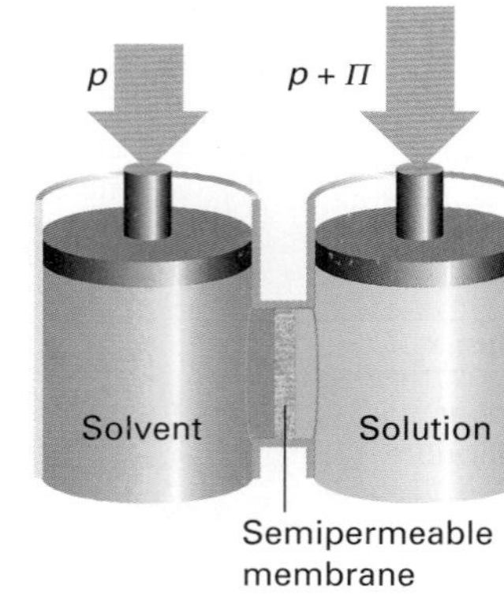

Fig. 6.20 The basis of the calculation of osmotic pressure. The presence of a solute lowers the chemical potential of the solvent in the right-hand compartment, but the application of pressure raises it. The osmotic pressure is the pressure needed to equalize the chemical potential of the solvent in the two compartments.

difficult to treat mathematically than an arrangement in which an externally applied pressure opposes any flow of solvent into the solution.

The osmotic pressure of a solution is proportional to the concentration of solute. In fact, we show in the following Derivation that the expression for the osmotic pressure of an ideal solution, which is called the **van 't Hoff equation**, bears an uncanny resemblance to the expression for the pressure of a perfect gas:

$$\Pi V \approx n_B RT \quad \text{Ideal solution} \quad \text{van 't Hoff equation} \qquad (6.21a)$$

Because $n_B/V = [B]$, the molar concentration of the solute, a simpler form of this equation is

$$\Pi \approx [B]RT \quad \text{Ideal solution} \quad \text{Another version of the van 't Hoff equation} \qquad (6.21b)$$

This equation applies only to solutions that are sufficiently dilute to behave as ideal–dilute solutions.

Derivation 6.7

The van 't Hoff equation

The thermodynamic treatment of osmosis makes use of the fact that, at equilibrium, the chemical potential of the solvent A is the same on each side of the membrane (Fig. 6.20). The starting relation is therefore

$$\mu_A(\text{pure solvent at pressure } p) = \mu_A(\text{solvent in the solution at pressure } p + \Pi)$$

The pure solvent is at atmospheric pressure, p, and the solution is at a pressure $p + \Pi$ on account of the additional pressure, Π, that has to be exerted on the solution to establish equilibrium. We shall write the chemical potential of the pure solvent at the pressure p as $\mu_A^*(p)$. The chemical potential of the solvent in the solution is lowered by the solute but it is raised on account of the greater pressure, $p + \Pi$, acting on the solution. We denote this chemical potential by $\mu_A(x_A, p + \Pi)$. Our task is to find the extra pressure Π needed to balance the lowering of chemical potential caused by the solute.

The condition for equilibrium written above is

$$\mu_A^*(p) = \mu_A(x_A, p + \Pi)$$

We take the effect of the solute into account by using eqn 6.11:

$$\mu_A(x_A, p + \Pi) \overset{\text{eqn 6.11}}{=} \mu_A^*(p + \Pi) + RT \ln x_A$$

The effect of pressure on an (assumed incompressible) liquid is given by eqn 5.1 ($\Delta G_m = V_m \Delta p$) but now expressed in terms of the chemical potential and the partial molar volume of the solvent:

$$\mu_A^*(p + \Pi) = \mu_A^*(p) + V_A \Delta p$$

At this point we identify the difference in pressure Δp as Π, so

$$\mu_A^*(p + \Pi) = \mu_A^*(p) + V_A \Pi$$

When we combine this relation with $\mu_A^*(p) = \mu_A^*(p + \Pi) + RT \ln x_A$ we get

$$\mu_A^*(p) = \mu_A^*(p) + V_A \Pi + RT \ln x_A$$

and therefore, after cancelling the $\mu_A^*(p)$,

$$-RT \ln x_A = \Pi V_A$$

The mole fraction of the solvent x_A is equal to $1 - x_B$, where x_B is the mole fraction of solute molecules. In dilute solution,

$\ln(1 - x_B)$ is approximately equal to $-x_B$ (The chemist's toolkit 6.1), so this equation becomes

$$RTx_B \approx \Pi V_A$$

When the solution is dilute, $x_B = n_B/n \approx n_B/n_A$, so

$$RTn_B \approx n_A\Pi V_A \overset{n_AV_A=V}{=} \Pi V$$

where V is the volume of the solvent, which is eqn 6.21a.

Osmosis helps biological cells maintain their structure. Cell membranes are semipermeable and allow water, small molecules, and hydrated ions to pass, while blocking the passage of biopolymers synthesized inside the cell. The difference in concentrations of solutes inside and outside the cell gives rise to an osmotic pressure, and water passes into the more concentrated solution in the interior of the cell, carrying small nutrient molecules. The influx of water also keeps the cell swollen, whereas dehydration causes the cell to shrink.

One of the most common applications of osmosis is **osmometry**, the measurement of molar masses of proteins and synthetic polymers from the osmotic pressure of their solutions. As these huge molecules dissolve to produce solutions that are far from ideal, we assume that the van 't Hoff equation is only the first term of an expansion:

$$\Pi = [B]RT\{1 + B[B] + \cdots\} \quad \text{Expanded van 't Hoff equation} \quad (6.22a)$$

Exactly the same strategy was used in Section 1.12 to extend the perfect gas equation to real gases and there it led to the virial equation of state. The empirical parameter B in this expression is called the **osmotic virial coefficient**. To use eqn 6.22a, we rearrange it into a form that gives a straight line by dividing both sides by [B]:

$$\overbrace{\frac{\Pi}{[B]}}^{y} = \overbrace{RT}^{=\text{intercept}} + \overbrace{BRT}^{+\text{slope}}\overbrace{[B]}^{\times x} \quad (6.22b)$$

As we illustrate in the following example, we can find the molar mass of the solute B by measuring the osmotic pressure at a series of mass concentrations and making a plot of $\Pi/[B]$ against [B] (Fig. 6.21).

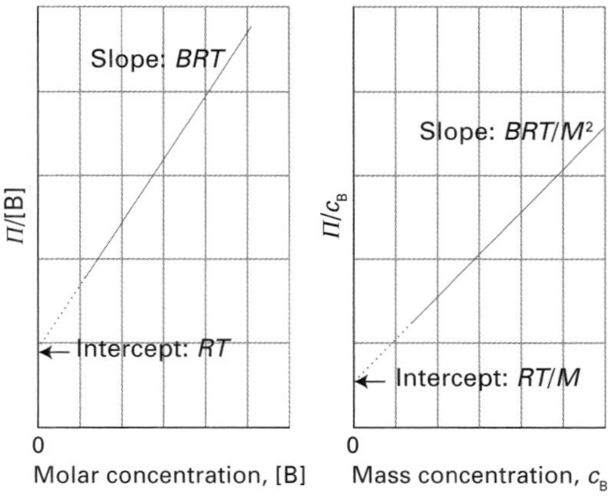

Fig. 6.21 The plot and extrapolation made to analyse the results of an osmometry experiment.

Example 6.6

Using osmometry to determine molar mass

The osmotic pressures of solutions of an enzyme, denoted B, in water at 298 K are given below. Determine the molar mass of the enzyme.

$c_B/(\text{g dm}^{-3})$	1.00	2.00	4.00	7.00	9.00
Π/Pa	27	70	197	500	785

Strategy First, we need to express eqn 6.22b in terms of the mass concentration, c_B. The molar concentration [B] of the solute is related to the mass concentration $c_B = m_B/V$ by

$$c_B = \frac{m_B}{V} = \overbrace{\frac{m_B}{n_B}}^{M} \times \overbrace{\frac{n_B}{V}}^{[B]} = M \times [B]$$

where M is the molar mass of the solute, so $[B] = c_B/M$. With this substitution, eqn 6.22b becomes

$$\overbrace{\frac{\Pi M}{c_B}}^{\Pi/[B]} = RT + \overbrace{\frac{BRTc_B}{M}}^{BRT[B]} + \cdots$$

Division through by M gives

$$\overbrace{\frac{\Pi}{c_B}}^{y} = \overbrace{\frac{RT}{M}}^{=\text{intercept}} + \overbrace{\left(\frac{BRT}{M^2}\right)}^{+\text{slope}}\overbrace{c_B}^{\times x} + \cdots$$

It follows that, by plotting Π/c_B against c_B, the results should fall on a straight line with intercept RT/M on the vertical axis at $c_B = 0$. Therefore, by locating the intercept by extrapolation of the data to $c_B = 0$, we can find the molar mass of the solute.

Solution The following values of Π/c_B can be calculated from the data:

$c_B/(\text{g dm}^{-3})$	1.00	2.00	4.00	7.00	9.00
$(\Pi/\text{Pa})/(c_B/\text{g dm}^{-3})$	27	35	49.2	71.4	87.2

The points are plotted in Fig. 6.22. The intercept with the vertical axis at $c_B = 0$ (which is best found by using linear regression and mathematical software, as we have done) is at

$$\frac{\Pi/\text{Pa}}{c_B/(\text{g dm}^{-3})} = 19.8$$

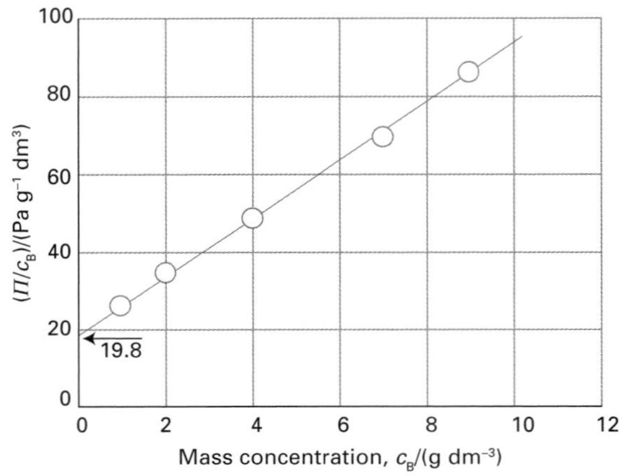

Fig. 6.22 The plot of the data in Example 6.6. The molar mass is determined from the intercept at $c_B = 0$.

which we can rearrange into

$$\Pi/c_B = 19.8 \text{ Pa g}^{-1} \text{ dm}^3$$

Therefore, because this intercept is equal to RT/M, we can write

$$M = \frac{RT}{19.8 \text{ Pa g}^{-1} \text{ dm}^3}$$

It follows that

$$M = \frac{\overbrace{(8.3145\times10^3 \text{ J K}^{-1} \text{ mol}^{-1})}^{R}\times\overbrace{(298 \text{ K})}^{T}}{19.8 \text{ Pa g}^{-1} \text{ dm}^3}$$

$$\underset{\text{Cancel K and J (1 J = 1 Pa m}^3\text{)}}{=} 1.25\times10^5 \text{ g mol}^{-1}$$

The molar mass of the enzyme is therefore close to 125 kg mol^{-1}.

A note on good practice Graphs should be plotted on axes labelled with pure numbers. Note how the plotted quantities are divided by their units, so that c_B/(g dm^{-3}), for instance, is a dimensionless number. By carrying the units through every stage of the calculation, we end up with the correct units for M. It is far better to proceed systematically in this way than to try to guess the units at the end of the calculation.

Self-test 6.6

Measurements of the osmotic pressure of a solution of poly(vinyl chloride), PVC, in dioxane at 25 °C were as follows:

c/(g dm^{-3})	0.50	1.00	1.50	2.00	2.50
Π/Pa	33.6	35.2	36.8	38.4	40.0

Determine the molar mass of the polymer.

Answer: 77 kg mol^{-1}

When pressure greater than the osmotic pressure is applied to the solution, there is a thermodynamic tendency for the solvent to flow out of the solution and into the pure solvent. This process is called **reverse osmosis.** Reverse osmosis is of great importance for the purification of sea water so that it is potable (drinkable) and can be used for irrigation, and many reverse osmosis plants are in operation around the world to supply fresh water to arid or water-deficient regions. The principal technical problem is to manufacture semipermeable membranes that are strong enough to withstand the high pressures required but still allow an economic flow.

Phase diagrams of mixtures

As in the discussion of pure substances (Chapter 5), the phase diagram of a mixture shows which phase is most stable for the given conditions. However, for mixtures composition is a variable in addition to the pressure and temperature.

It will be useful to keep in mind the implications of the phase rule ($F = C - P + 2$, eqn 5.9). We shall consider only **binary mixtures**, which are mixtures of two components (such as ethanol and water) and may therefore set $C = 2$. Then $F = 4 - P$. For simplicity we keep the pressure constant (at 1 atm, for instance), which uses up one of the degrees of freedom, and write $F' = 3 - P$ for the number of degrees of freedom remaining. One of these degrees of freedom is the temperature, the other is the composition. Hence we should be able to depict the phase equilibria of the system on a **temperature–composition diagram** in which one axis is the temperature and the other axis is the mole fraction. In a region where there is only one phase, $F' = 2$ and both the temperature and the composition can be varied (Fig. 6.23). If two phases are present at equilibrium, $F' = 1$, and only one of the two variables may be changed at will. For example, if we change the composition, then to maintain equilibrium between the two phases we have to adjust the temperature too. Such two-phase equilibria therefore define a line in the phase diagram. If three phases are present, $F' = 0$ and there is no degree of freedom for the system. To establish equilibrium between three phases we must adopt a specific temperature and composition. Such a condition is therefore represented by a point on the phase diagram.

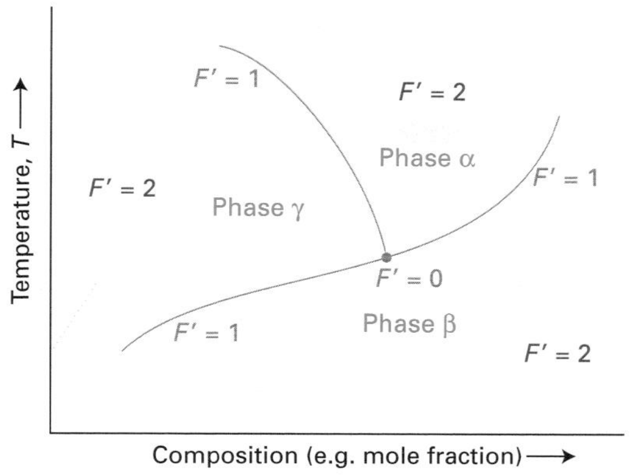

Fig. 6.23 The interpretation of a temperature–composition phase diagram at constant pressure. In a region where only one phase is present, $F' = 2$ and both composition and temperature can be varied. On a phase boundary, where two phases are in equilibrium, $F' = 1$ and only one variable can be changed independently. At a point where three phases are present in equilibrium, $F' = 0$, and the temperature and composition is fixed.

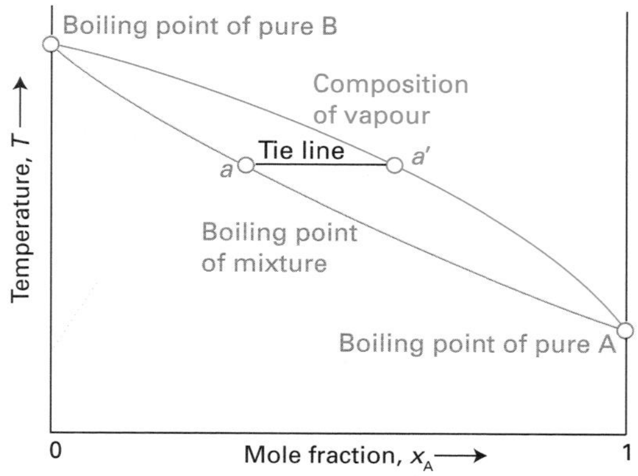

Fig. 6.24 A temperature–composition diagram for a binary mixture of volatile liquids. The tie line connects the points that represent the compositions of liquid and vapour that are in equilibrium at each temperature. The lower curve is a plot of the boiling point of the mixture against composition.

6.9 Mixtures of volatile liquids

First, we consider the phase diagram of a binary mixture of two volatile components. This kind of system is important for understanding fractional distillation, which is a widely used technique in industry and the laboratory. Intuitively, we might expect the boiling point of a mixture of two volatile liquids to vary smoothly from the boiling point of one pure component when only that liquid is present to the boiling point of the other pure component when only that liquid is present. This expectation is often borne out in practice, and Fig. 6.24 shows a typical plot of boiling point against composition (the lower curve).

The vapour in equilibrium with the boiling mixture is also a mixture of the two components. We should expect the vapour to be richer than the liquid mixture in the more volatile of the two substances. This difference is also often found in practice, and the upper curve in the illustration shows the composition of the vapour in equilibrium with the boiling liquid. To identify the composition of the vapour, we note the boiling point of the liquid mixture (point *a*, for instance) and draw a horizontal **tie line**, a line joining two phases that are in equilibrium with each other, across to the upper curve. Its point of intersection (a') gives the composition of the vapour. In this example, we see that the mole fraction of A in the vapour is about 0.6. As expected, the vapour is richer than the liquid in the more volatile component. Graphs like these are determined empirically, by measuring the boiling points of a series of mixtures

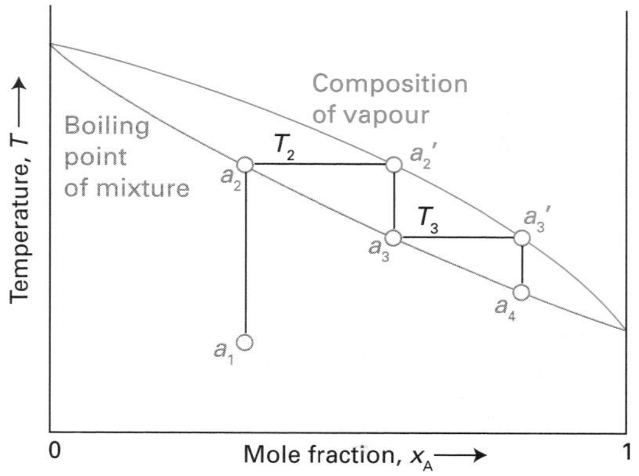

Fig. 6.25 The process of fractional distillation can be represented by a series of steps on a temperature–composition diagram like that in Fig. 6.24. The initial liquid mixture may be at a temperature and have a composition like that represented by point a_1. It boils at the temperature T_2, and the vapour in equilibrium with the boiling liquid has composition a_2'. If that vapour is condensed (to a_3 or below), the resulting condensate boils at T_3 and gives rise to a vapour of composition represented by a_3'. As the succession of vaporizations and condensations is continued, the composition of the distillate moves towards pure A (the more volatile component).

(to plot the lower curve of boiling point against composition), and measuring the composition of the vapour in equilibrium with each boiling mixture (to plot the corresponding points of the vapour-composition curve).

We can follow the changes that occur during the fractional distillation of a mixture of volatile liquids by noting what happens when a mixture of composition a_1 is heated (Fig. 6.25). The mixture boils at a_2 and its vapour has composition a_2'. This vapour condenses to a liquid of the same composition when it has risen to a cooler part of the 'fractionating column', a vertical column packed with glass rings or beads to give a large surface area. This condensate boils at the temperature corresponding to the point a_3 and yields a vapour of composition a'_3. This vapour is even richer in the more volatile component. That vapour condenses to a liquid which boils at the temperature corresponding to the point a_4. The cycle is repeated until almost pure A emerges from the top of the column.

Whereas many binary liquid mixtures do have temperature–composition diagrams resembling that shown in Fig. 6.25, in a number of important cases there are marked differences. For example, a maximum in the boiling point curve is sometimes found (Fig. 6.26). This behaviour is a sign that favourable interactions between the molecules of the two components reduce the vapour pressure of the mixture below the ideal value. Examples of this behaviour include trichloromethane/propanone and nitric acid/water mixtures. Temperature–composition curves are also found that pass through a minimum (Fig. 6.27). This behaviour indicates that the (A,B) interactions are unfavourable and hence that the mixture is more volatile than expected on the basis of simple mingling of the two species. Examples include dioxane/water and ethanol/water.

There are important consequences for distillation when the temperature–composition diagram has a maximum or a minimum. Consider a liquid of composition a_1 on the right of the maximum in Fig. 6.26. It boils at a_2 and its vapour (of composition a'_2) is richer in the more volatile component A. If that vapour is removed, the composition of the remaining liquid moves towards a_3. The vapour in equilibrium with this boiling liquid has composition a'_3: note that the two compositions are more similar than the original pair (a_3 and a'_3 are closer together than a_2 and a'_2). If that vapour is removed, the composition of the boiling liquid shifts towards a_4 and the vapour of that boiling mixture has a composition identical to that of the liquid. At this stage, evaporation occurs without change of composition. The mixture is said to form an **azeotrope** (from the Greek words for 'boiling without changing'). When the azeotropic composition has been reached, distillation cannot separate the two liquids because the condensate retains the composition of the liquid. One example of azeotrope formation is hydrochloric acid/water, which is azeotropic at 80 per cent water (by mass) and boils unchanged at 108.6 °C.

The system shown in Fig. 6.27 is also azeotropic, but shows this character in a different way. Suppose we start with a mixture of composition a_1 and follow the changes in the composition of the vapour that rises through a fractionating column. The mixture boils at a_2 to give a vapour of composition a'_2. This vapour condenses in the column to a liquid of the same composition (now marked a_3). That liquid reaches equilibrium with its vapour at a'_3, which condenses higher up the tube to give a liquid of the same composition. The fractionation therefore shifts the vapour towards the azeotropic composition at a_4,

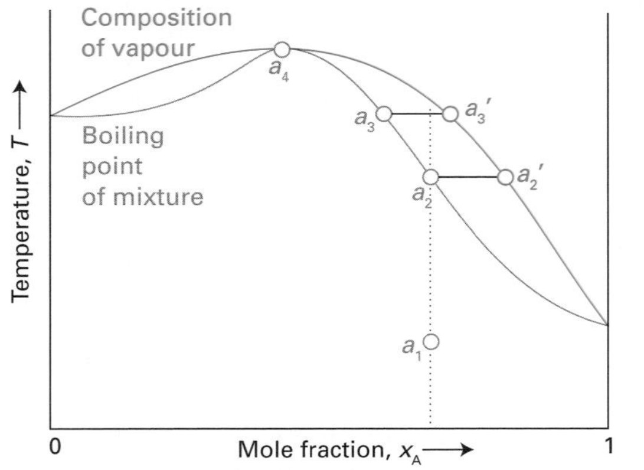

Fig. 6.26 The temperature–composition diagram for a high-boiling azeotrope. As fractional distillation proceeds, the composition of the remaining liquid moves towards a_4; however, once there, the vapour in equilibrium with that liquid has the same composition, so the mixture evaporates with an unchanged composition and no further separation can be achieved.

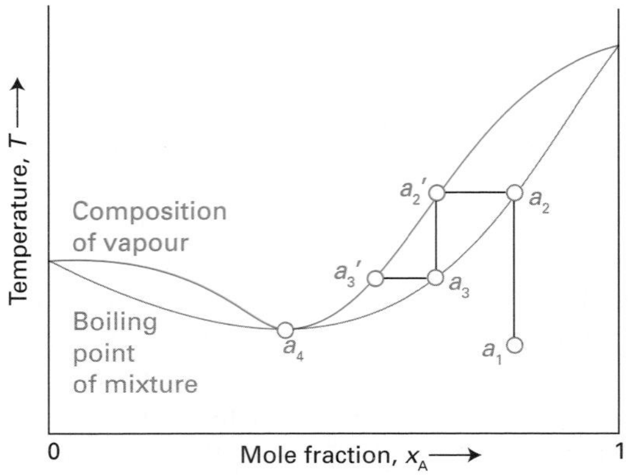

Fig. 6.27 The temperature–composition diagram for a low-boiling azeotrope. As fractional distillation proceeds, the composition of the vapour moves towards a_4; however, once there, the vapour in equilibrium with that liquid has the same composition, so no further separation of the distillate can be achieved.

but the composition cannot move beyond a_4 because now the vapour and the liquid have the same composition. Consequently, the azeotropic vapour emerges from the top of the column. An example is ethanol/water, which boils unchanged when the water content is 4 per cent and the temperature is 78 °C.

6.10 Liquid–liquid phase diagrams

Partially miscible liquids are liquids that do not mix together in all proportions. An example is a mixture of hexane and nitrobenzene: when the two liquids are shaken together, the liquid consists of two liquid phases, one is a saturated solution of hexane in nitrobenzene and the other is a saturated solution of nitrobenzene in hexane. Because the two solubilities vary with temperature, the compositions and proportions of the two phases change as the temperature is changed. We can use a temperature–composition diagram to display the composition of the system at each temperature.

Suppose we add a small amount of nitrobenzene to hexane at a temperature T'. The nitrobenzene dissolves completely; however, as more nitrobenzene is added, a stage comes when no more dissolves. The sample now consists of two phases in equilibrium with each other, the more abundant one consisting of hexane saturated with nitrobenzene, the less abundant one a trace of nitrobenzene saturated with hexane. In the temperature–composition diagram drawn in Fig. 6.28, the composition of the former is represented by the point a' and that of the latter by the point a''. The relative abundances of the two phases are given by the **lever rule** (Fig. 6.29):

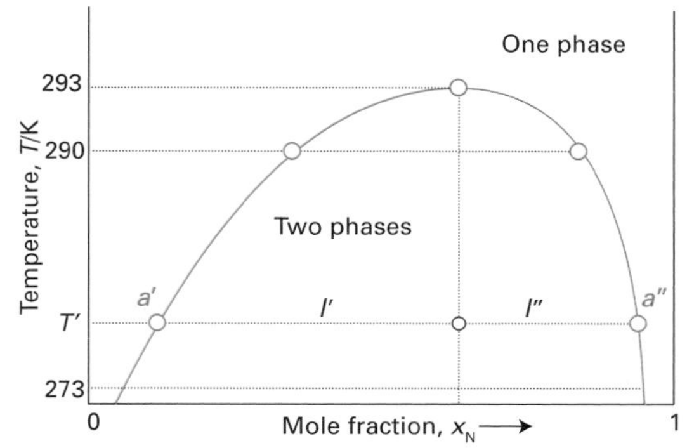

Fig. 6.28 The temperature–composition diagram for hexane and nitrobenzene at 1 atm. The upper critical solution temperature, T_{uc}, is the temperature above which no phase separation occurs. For this system it lies at 293 K (when the pressure is 1 atm).

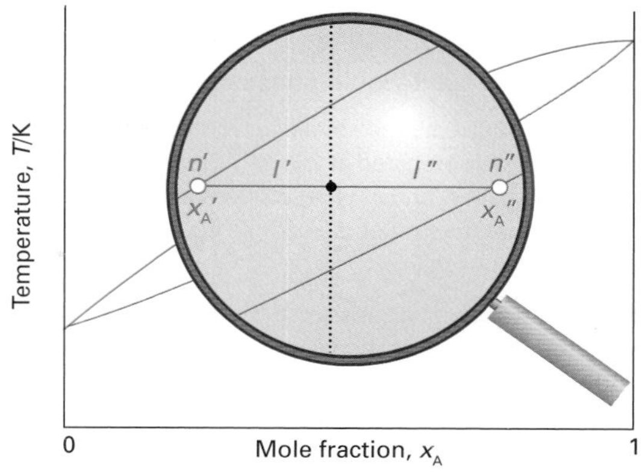

Fig. 6.29 The coordinates and compositions referred to by the lever rule.

$$\frac{\text{Amount of phase of composition } a''}{\text{Amount of phase of composition } a'} = \frac{l'}{l''}$$

The lever rule (6.23)

Derivation 6.8

The lever rule

We write $n = n' + n''$, where n' is the total amount of molecules in the one phase, n'' is the total amount in the other phase, and n is the total amount of molecules in the sample. The total amount of A in the sample is nx_A, where x_A is the overall mole fraction of A in the sample (this is the quantity plotted along the horizontal axis). The overall amount of A is also the sum of its amounts in the two phases, where it has the mole fractions x'_A and x''_A, respectively:

$$nx_A = n'x'_A + n''x''_A$$

We can also multiply each side of the relation $n = n' + n''$ by x_A and obtain

$$nx_A = n'x_A + n''x_A$$

Then, by equating these two expressions we get first

$$n'x'_A + n''x''_A = n'x_A + n''x_A$$

and then after a slight rearrangement

$$n'(x'_A - x_A) = n''(x_A - x''_A)$$

or (as can be seen by referring to Fig. 6.29)

$$n'l' = n''l''$$

which is eqn 6.23.

Example 6.7

Interpreting a liquid–liquid phase diagram

A mixture of 50 g (0.59 mol) of hexane and 50 g (0.41 mol) of nitrobenzene was prepared at 290 K. What are the compositions of the phases, and in what proportions do they occur? To what temperature must the sample be heated in order to obtain a single phase?

Strategy The answer is based on Fig. 6.28. First, we need to identify the tie-line corresponding to the temperature specified: the points at its two ends give the compositions of the two phases in equilibrium. Next, we identify the location on the horizontal axis corresponding to the overall composition of the system and draw a vertical line. Where that line cuts the tie line it divides it into the two lengths needed to use the lever rule, eqn 6.23. For the final part, we note the temperature at which the same vertical line cuts through the phase boundary: at that temperature and above, the system consists of a single phase.

Solution We denote hexane by H and nitrobenzene by N. The horizontal tie-line at 290 K cuts the phase boundary at $x_N = 0.37$ and at $x_N = 0.83$, so those mole fractions are the compositions of the two phases. The overall composition of the system corresponds to $x_N = 0.41$, so we draw a vertical line at that mole fraction. The lever rule then gives the ratio of amounts of each phase as

$$\frac{l'}{l''} = \frac{0.41-0.37}{0.83-0.41} = \frac{0.04}{0.42} = 0.1$$

We conclude that the hexane-rich phase is ten times more abundant than the nitrobenzene-rich phase at this temperature. Heating the sample to 292 K takes it into the single phase region.

Self-test 6.7

Repeat the problem for 50 g hexane and 100 g nitrobenzene at 273 K.

Answer: $x_N = 0.09$ and 0.95 in the ratio 1:1.3; 290 K

When more nitrobenzene is added to the two-phase mixture at the temperature T', hexane dissolves in it slightly. The overall composition moves to the right in the phase diagram, but the compositions of the two phases in equilibrium remain a' and a''. The difference is that the amount of the second phase increases at the expense of the first. A stage is reached when so much nitrobenzene is present that it can dissolve all the hexane, and the system reverts to a single phase. Now the point representing the overall composition and temperature lies to the right of the phase boundary in the illustration and the system is a single phase.

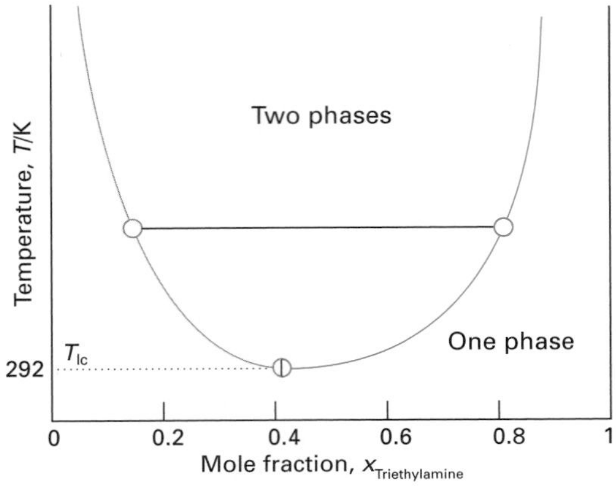

Fig. 6.30 The temperature–composition diagram for water and triethylamine. The lower critical solution temperature, T_{lc}, is the temperature below which no phase separation occurs. For this system it lies at 292 K (when the pressure is 1 atm).

The **upper critical solution temperature,** T_{uc} (which is also called the *upper consolute temperature*), is the upper limit of temperatures at which phase separation occurs. Above the upper critical solution temperature the two components are fully miscible. In molecular terms, this temperature arises because the greater thermal motion of the molecules leads to greater miscibility of the two components. In thermodynamic terms, the Gibbs energy of mixing becomes negative above a certain temperature, regardless of the composition.

Some systems show a **lower critical solution temperature,** T_{lc} (which is also called the *lower consolute temperature*), below which they mix in all proportions and above which they form two phases. An example is water and triethylamine (Fig. 6.30). In this case, at low temperatures the two components are more miscible because they form a weak complex; at higher temperatures the complexes break up and the two components are less miscible.

A few systems have both upper and lower critical temperatures. The reason can be traced to the fact that, as the temperature is increased, the weak complexes are disrupted, leading to partial miscibility, whereas the thermal motion at higher temperatures homogenizes the mixture again, just as in the case of ordinary partially miscible liquids. One example is nicotine and water, which are partially miscible between 61 °C and 210 °C (Fig. 6.31).

6.11 Liquid–solid phase diagrams

Phase diagrams are also used to show the regions of temperature and composition at which solids and

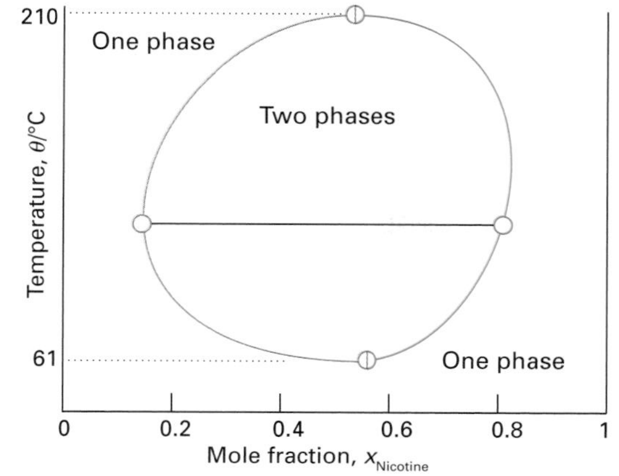

Fig. 6.31 The temperature–composition diagram for water and nicotine, which has both upper and lower critical solution temperatures. Note the high temperatures on the graph: the diagram corresponds to a sample under pressure.

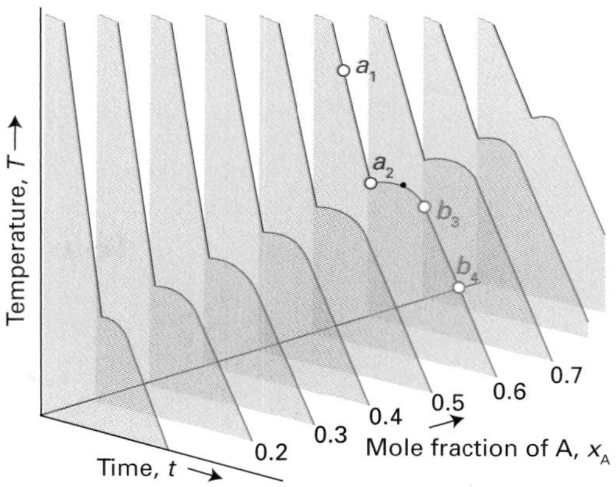

Fig. 6.33 The cooling curves for the alloy shown in Fig. 6.32; the labelled points correspond to those in that phase diagram.

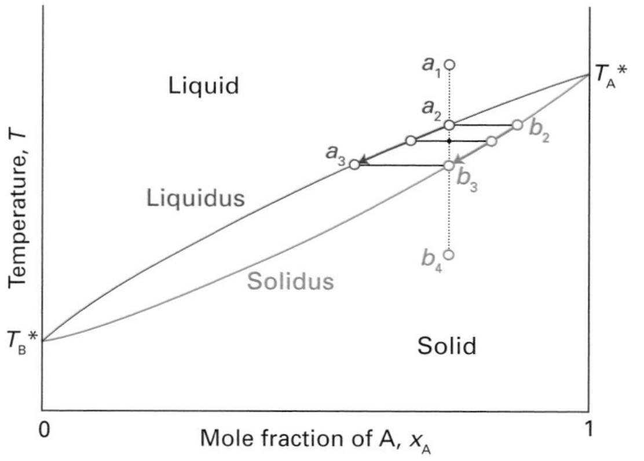

Fig. 6.32 The phase diagram for an alloy formed from two metals (with normal melting points T_A^* and T_B^* in their pure form) that are miscible in all proportions in the liquid and solid phases.

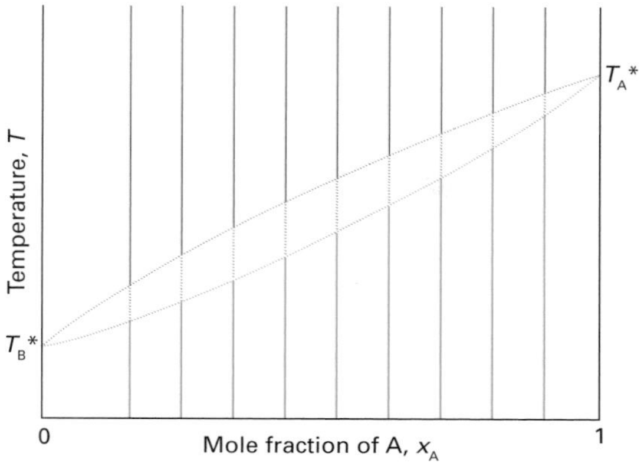

Fig. 6.34 When the time dependence of the cooling curves in Fig. 6.33 is removed and each curve is plotted vertically against the mole fraction of A, the liquidus and solidus can be identified and the phase diagram constructed.

liquids exist in binary systems. Such diagrams are useful for discussing the techniques that are used to prepare the high-purity materials used in the electronics industry and are also of great importance in metallurgy.

Figure 6.32 shows a simple phase diagram for an alloy of two metals that are miscible in all proportions. The **liquidus** is the line above which the entire sample is liquid; the **solidus** is the line below which the sample is entirely solid. When a sample of composition and temperature a_1 is cooled, at a_2 it initially deposits a solid of composition b_2. As the temperature is lowered, the equilibrium composition of the deposited solid moves towards b_3 and that of the remaining liquid moves towards a_3. Below the solidus, only solid of the original composition is present.

Phase diagrams such as these are constructed by monitoring the **cooling curve** at a series of compositions (Fig. 6.33). The different slopes of the liquid phase and solid phase cooling curves is due to their different heat capacities: the rate at which energy is lost as heat from a sample is proportional to the temperature difference between it and its surroundings, and the corresponding change in temperature of the sample will be large if the heat capacity is small (from $\Delta T = q/C$) and small if the heat capacity is high. Cooling to the temperature of the surroundings also slows if the surroundings are at a constant temperature. The changing slope between the temperatures corresponding to the liquidus and the solidus is due to the exothermic character of the phase transition: the progressive release of heat as the solid forms retards the cooling. Figure 6.34 shows a portrayal of

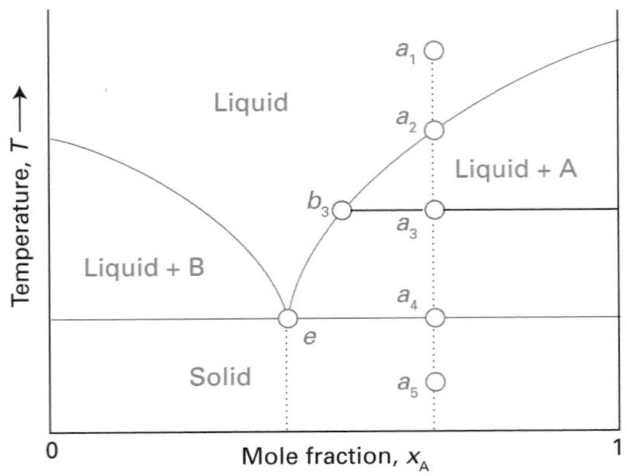

Fig. 6.35 The temperature–composition diagram for two almost immiscible solids and their completely immiscible liquids. The vertical line through e corresponds to the eutectic composition, the mixture with lowest melting point.

the sequence of cooling curves plotted against the initial composition of the liquid but with the time dependence removed. By joining the points that terminate each cooling region the liquidus and solidus can be constructed.

Figure 6.35 shows the phase diagram for a system composed of two metals that are almost completely immiscible right up to their melting points (such as antimony and bismuth). Consider the molten liquid of composition a_1. When the liquid is cooled to a_2, the system enters the two-phase region labelled 'Liquid + A'. Almost pure solid A begins to come out of solution and the remaining liquid becomes richer in B. On cooling to a_3, more of the solid forms, and the relative amounts of the solid and liquid (which are in equilibrium) are given by the lever rule: at this stage there are nearly equal amounts of each. The liquid phase is richer in B than before (its composition is given by b_3) because A has been deposited. At a_4 there is less liquid than at a_3 and its composition is given by e. This liquid now freezes to give a two-phase system of almost pure A and almost pure B and cooling down to a_5 leads to no further change in composition.

The vertical line through e in Fig. 6.35 corresponds to the **eutectic composition** (from the Greek words for 'easily melted'). A solid with the eutectic composition melts, without change of composition, at the lowest temperature of any mixture. Solutions of composition to the right of e deposit A as they cool, and solutions to the left deposit B: only the eutectic mixture (apart from pure A or pure B) solidifies at a single definite temperature without gradually unloading one or other of the components from the liquid.

One technologically important eutectic is solder, which typically consists of about 67 per cent tin and 33 per cent lead by mass and melts at 183 °C. Eutectic formation occurs in the great majority of binary alloy systems. It is of great importance for the microstructure of solid materials, for although a eutectic solid is a two-phase system, it crystallizes out in a nearly homogeneous mixture of microcrystals. The two microcrystalline phases can be distinguished by microscopy and structural techniques such as X-ray diffraction (Chapter 17).

Cooling curves are used to detect eutectics. We can see how it is used by considering the rate of cooling down the vertical line at a_1 in Fig. 6.35. The liquid cools steadily until it reaches a_2, when A begins to be deposited. Cooling is now slower because the solidification of A is exothermic and retards the cooling (Fig. 6.36). When the remaining liquid reaches the eutectic composition, the temperature remains constant until the whole sample has solidified: this pause in the decrease in temperature is known as the **eutectic halt**. If the liquid has the eutectic composition e initially, then the liquid cools steadily down to the freezing temperature of the eutectic, when there is a long eutectic halt as the entire sample solidifies just like the freezing of a pure liquid.

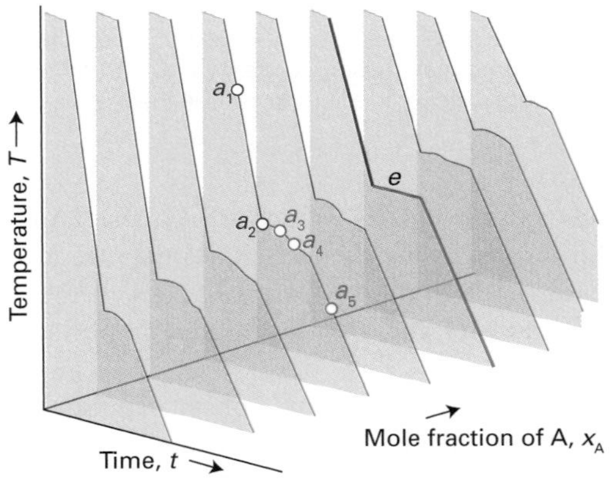

Fig. 6.36 The cooling curves for the system shown in Fig. 6.35. For a sample of composition represented by the vertical line through a_1 to a_5, the rate of cooling decreases at a_2 because solid A comes out of solution until the eutectic composition is reached, when there is a plateau. The cooling curve at the eutectic composition e has a complete halt at e when the eutectic solidifies without change of composition.

Impact on technology 6.2

Ultrapurity and controlled impurity

Advances in technology have called for materials of extreme purity. For example, semiconductor devices consist of almost perfectly pure silicon or germanium doped to a precisely controlled extent. For these materials to operate successfully, the impurity level must be kept down to less than 1 in 10^9. The technique of **zone refining** makes use of the non-equilibrium properties of mixtures. It relies on the impurities being more soluble in the molten sample than in the solid, and sweeps them up by passing a molten zone repeatedly from one end to the other along a sample (Fig. 6.37). In practice, a train of hot and cold zones are swept repeatedly from one end to the other. The zone at the end of the sample is the impurity dump: when the heater has gone by, it cools to a dirty solid that can be discarded.

We can use a phase diagram to discuss zone refining, but we have to allow for the fact that the molten zone moves along the sample and the sample is uniform in neither temperature nor composition. Consider a liquid (which represents the molten zone) on the vertical line at a_1 in Fig. 6.38, and let it cool without the entire sample coming to overall equilibrium. If the temperature falls to a_2, a solid of composition b_2 is deposited and the remaining liquid (the zone where the heater has moved on) is at a_2'. Cooling that liquid down a vertical line passing through a_2' deposits solid of composition b_3 and leaves liquid at a_3'. The process continues until the last drop of liquid to solidify is heavily contaminated with A. There is plenty of everyday evidence that impure liquids freeze in this way. For example, an ice cube is clear near the surface but misty in the core. The water used to make ice normally contains dissolved air; freezing proceeds from the outside, and air is accumulated in the retreating liquid phase. The air cannot escape from the interior of the cube, so when that freezes the air is trapped in a mist of tiny bubbles.

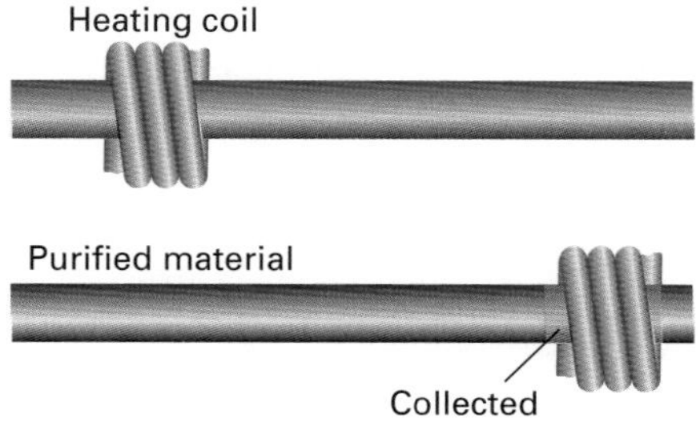

Fig. 6.37 In the zone refining procedure, a heater is used to melt a small region of a long cylindrical sample of the impure solid, and that zone is swept to the other end of the rod. As it moves, it collects impurities. If a series of passes are made, the impurities accumulate at one end of the rod and can be discarded.

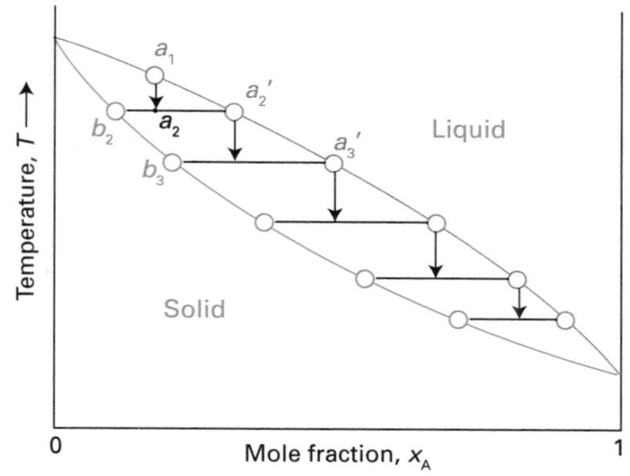

Fig. 6.38 A binary temperature–composition diagram can be used to discuss zone-refining, as explained in the text.

A modification of zone refining is **zone levelling**. This technique is used to introduce controlled amounts of impurity (for example, of indium into germanium). A sample rich in the required dopant is put at the head of the main sample, and made molten. The zone is then dragged repeatedly in alternate directions through the sample, where it deposits a uniform distribution of the impurity.

6.12 The Nernst distribution law

Suppose we shake a compound up with a mixture of two immiscible liquids, and allow the two phases to separate into layers. What can be said about the relative concentrations of the compound, which we denote C, in the two layers?

This question can be answered by making use of the chemical potential, for at equilibrium the chemical potential of the compound, μ_C, must be the same in each phase: $\mu_C(1) = \mu_C(2)$. If we suppose that each solution is an ideal–dilute solution, it follows that

$$\mu_C^{\ominus}(1) + RT\ln x_C(1) = \mu_C^{\ominus}(2) + RT\ln x_C(2)$$

The two standard states are different because we are using Henry's law to define the chemical potentials and the constants that occur in that law vary between solvents. It follows that

$$\mu_C^{\ominus}(1) - \mu_C^{\ominus}(2) = RT\ln x_C(2) - RT\ln x_C(1)$$

and therefore that

$$\ln\left(\frac{x_C(2)}{x_C(1)}\right) = \frac{\mu_C^{\ominus}(1) - \mu_C^{\ominus}(2)}{RT}$$

The right-hand side is a constant for a given pair of liquids, so we arrive at the **Nernst distribution law:**

$$\frac{x_C(2)}{x_C(1)} = \text{constant} \qquad \text{The Nernst distribution law} \qquad (6.24)$$

That is, regardless of the overall concentration (but providing the solutions are both behaving ideally), the ratio of mole fractions in the two phases is the same.

Brief illustration 6.6 The Nernst distribution law

Suppose a certain amount of benzoic acid, C_6H_5COOH, is shaken up with a mixture of benzene and water; then the acid distributes itself between the two phases such that the ratio of mole fractions is equal to a constant. If twice that amount of acid is shaken up in the mixture, the ratio of mole fractions will be the same.

Checklist of key concepts

☐ 1 A partial molar quantity is the contribution of a component (per mole) to the overall property of a mixture.

☐ 2 The chemical potential of a component is the partial molar Gibbs energy of that component in a mixture.

☐ 3 An ideal solution is one in which both components obey Raoult's law over the entire composition range.

☐ 4 An ideal–dilute solution is one in which the solute obeys Henry's law.

☐ 5 The activity of a substance is an effective concentration; see Table 6.2.

☐ 6 A colligative property is a property that depends on the number of solute particles, not their chemical identity; they arise from the effect of a solute on the entropy of the solution.

☐ 7 Colligative properties include lowering of vapour pressure, depression of freezing point, elevation of boiling point, and osmotic pressure.

☐ 8 The equilibria between phases (at constant pressure) are represented by lines on a temperature–composition phase diagram.

☐ 9 The relative abundance of phases is obtained by using the lever rule (see the *Road map*).

☐ 10 An azeotrope is a mixture that vaporizes and condenses without a change of composition.

☐ 11 A eutectic is a mixture that freezes and melts without change of composition.

Road map of key equations

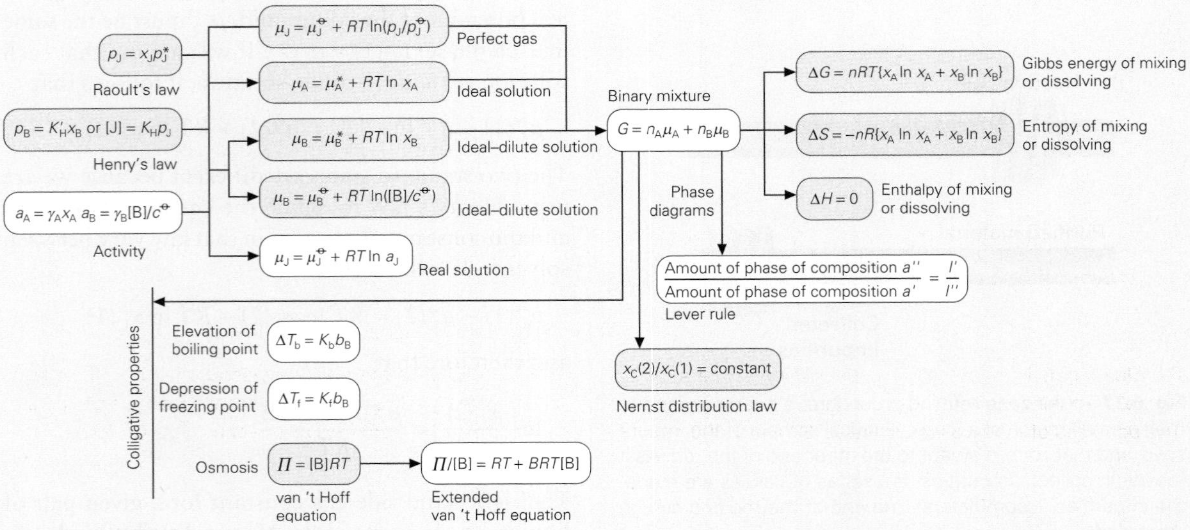

Blue boxes indicate relations valid for perfect gases, ideal solutions, and ideal–dilute solutions.

Questions and exercises

Discussion questions

6.1 Explain the significance of a partial molar quantity and how it depends on composition.

6.2 Define and describe the applications of the chemical potential of a substance.

6.3 State and justify the thermodynamic criterion for solution–vapour equilibrium.

6.4 Justify Raoult's and Henry's laws in terms of the molecular interactions in a mixture.

6.5 Explain the origin of the colligative properties.

6.6 What is meant by the activity of a solute?

6.7 Explain the origin of osmosis in terms of the thermodynamic and molecular properties of a mixture.

6.8 Explain how osmotic pressure measurements can be used to determine the molar mass of a polymer.

Exercises

6.1 What mass of sodium chloride, NaCl, should be dissolved in enough water to make up 300 cm^3 of 1.00 M NaCl(aq)?

6.2 A mass of 5.96 g of hydrated sodium dihydrogenphosphate, $Na_2HPO_4{\cdot}7H_2O$, was dissolved in 250 cm^3 water at 298 K. Calculate the molality of the solute in the resulting solution, given that the mass density of water at this temperature is 0.997 g cm^{-3}.

6.3 An aqueous solution was made up by dissolving 34.5 g of sucrose, $C_{12}H_{22}O_{11}$, in enough water to make 250 cm^3 of solution. The mass density of the resulting solution was 1040 kg m^{-3}. Calculate the molar concentration and the molality of sucrose in the solution.

6.4 Calculate the mole fractions of the molecules of a mixture that contains 56 g of benzene (C_6H_6) and 120 g of toluene ($C_6H_5CH_3$).

6.5 Dry air may be considered as consisting of 75.53 per cent by mass of nitrogen, N_2, and 23.14 per cent by mass of oxygen, O_2. The remainder is made up principally of argon. What are the mole fractions of these three principal substances?

6.6 Determine the mole fraction of glycine molecules, NH_2CH_2COOH, in an aqueous solution in which the molality of glycine is 0.100 mol kg^{-1}.

6.7 The partial molar volumes of propanone and trichloromethane in a mixture in which the mole fraction of $CHCl_3$ is 0.4693 are 74.166 cm^3 mol^{-1} and 80.235 cm^3 mol^{-1}, respectively. What is the volume of a solution of total mass 1.000 kg?

6.8 Use Fig. 6.1 to estimate the total volume of a solution formed by mixing 50.0 cm^3 of ethanol with 50.0 cm^3 of water. The densities of the two liquids are 0.789 and 1.000 g cm^{-3}, respectively.

6.9 By how much does the chemical potential of carbon dioxide at 310 K and 2.0 bar differ from its standard value at that temperature?

6.10 The standard state of a substance used to be defined as 1 atm at the specified temperature. By how much does the current definition of standard chemical potential (at 1 bar) differ from its former value at 298.15 K?

6.11 Calculate (a) the (molar) Gibbs energy of mixing, (b) the (molar) entropy of mixing when the two major components of air (nitrogen and oxygen) are mixed to form air at 298 K. The mole fractions of N_2 and O_2 are 0.78 and 0.22, respectively. Is the mixing spontaneous?

6.12 Suppose now that argon is added to the mixture in Exercise 6.11 to bring the composition closer to real air, with mole fractions 0.780, 0.210, and 0.0096, respectively. What is the additional change in molar Gibbs energy and entropy at 298 K? Is the mixing spontaneous?

6.13 Derive eqn 6.12, which shows how the Gibbs energy of mixing varies with the mole fractions of two components in an ideal solution. Follow the method used in Derivation 6.3. The initial Gibbs energy of the unmixed components is $G_i = n_A\mu_A^* + n_B\mu_B^*$; after mixing, use the chemical potentials in eqn 6.11.

6.14 A solution is prepared by dissolving 2.33 g of C_{60} (buckminsterfullerene) in 100.0 g of toluene (methylbenzene). Given that the vapour pressure of pure toluene is 5.00 kPa at 30 °C, what is the vapour pressure of toluene in the solution?

6.15 Estimate the vapour pressure of sea water at 20 °C given that the vapour pressure of pure water is 2.338 kPa at that temperature and the solute is largely Na^+ and Cl^- ions, each present at about 0.50 mol dm^{-3}.

6.16 A solution is prepared by dissolving iodine, I_2, in tetrachloromethane, CCl_4, at 25 °C. What are the mole fractions of the solute and solvent for a solute of molality 0.100 mol kg^{-1}? Hence calculate the change in the chemical potential of the solvent caused by the solute.

6.17 At 300 K, the vapour pressure of dilute solutions of HCl in liquid $GeCl_4$ are as follows:

x(HCl)	0.005	0.012	0.019
p/kPa	32.0	76.9	121.8

Show that the solution obeys Henry's law in this range of mole fractions and calculate the Henry's law constant at 300 K.

6.18 Calculate the concentration of carbon dioxide in fat given that the Henry's law constant is 8.6×10^4 Torr and the partial pressure of carbon dioxide is 55 kPa.

6.19 What partial pressure of hydrogen results in a molar concentration of 1.0 mmol dm^{-3} in water at 25 °C?

6.20 The rise in atmospheric carbon dioxide results in higher concentrations of dissolved carbon dioxide in natural waters. Use Henry's law and the data in Table 6.1 to calculate the solubility of CO_2 in water at 25 °C when its partial pressure is (a) 3.8 kPa, (b) 50.0 kPa.

6.21 The mole fractions of N_2 and O_2 in air at sea level are approximately 0.78 and 0.21. Calculate the molalities of the solution formed in an open flask of water at 25 °C.

6.22 A water-carbonating plant is available for use in the home and operates by providing carbon dioxide at 1.0 atm. Estimate the molar concentration of the CO_2 in the soda water it produces.

6.23 At 90 °C the vapour pressure of toluene (methylbenzene) is 53 kPa and that of *o*-xylene (1,2-dimethylbenzene) is 20 kPa. What is the composition of the liquid mixture that boils at 25 °C when the pressure is 0.50 atm? What is the composition of the vapour produced?

6.24 The vapour pressures of the two components A and B of a binary mixture varied as follows:

$$p_A/\text{Torr} = 861x - 790x^2 + 245x^3 + 100x^4$$

$$p_B/\text{Torr} = 749 - 749x + 404x^2 - 404x^3$$

Confirm that the mixture conforms to Raoult's law for the component in large excess and to Henry's law for the minor component. Find the Henry's law constants.

6.25 The vapour pressures of the two components A and B of a binary mixture varied as follows:

x_A	0	0.20	0.40	0.60	0.80	1
p_A/Torr	0	70	173	295	422	539
p_B/Torr	701	551	391	237	101	0

Confirm that the mixture conforms to Raoult's law for the component in large excess and to Henry's law for the minor component. Find the Henry's law constants.

6.26 What is the change in chemical potential of glucose when its concentration in water at 20.0 °C is changed from 0.10 mol dm^{-3} to 1.00 mol dm^{-3}?

6.27 The vapour pressure of a sample of benzene is 53.0 kPa at 60.6 °C, but it fell to 51.2 kPa when 0.133 g of an organic compound was dissolved in 5.00 g of the solvent. Calculate the molar mass of the compound.

6.28 Estimate the freezing point of 200 cm^3 of water sweetened by the addition of 2.5 g of sucrose. Treat the solution as ideal.

6.29 Estimate the freezing point of 200 cm^3 of water to which 2.5 g of sodium chloride has been added. Treat the solution as ideal.

6.30 The addition of 28.0 g of a compound to 750 g of tetrachloromethane, CCl_4, lowered the freezing point of the solvent by 5.40 K. Calculate the molar mass of the compound.

6.31 A compound A existed in equilibrium with its dimer, A_2, in propanone solution. Derive an expression for the equilibrium constant $K = [A_2]/[A]^2$ in terms of the depression in vapour pressure caused by a given concentration of compound. *Hint*: Suppose that a fraction *f* of the A molecules are present as the dimer. The depression of vapour pressure is proportional to the total concentration of A and A_2 molecules regardless of their chemical identities.

6.32 The osmotic pressure of an aqueous solution of urea at 300 K is 150 kPa. Calculate the freezing point of the same solution.

6.33 The osmotic pressure of a solution of polystyrene in toluene (methylbenzene) was measured at 25 °C with the following results:

c/(g dm^{-3})	2.042	6.613	9.521	12.602
Π/Pa	58.3	188.2	270.8	354.6

Determine the molar mass of the polymer.

6.34 The molar mass of an enzyme was determined by dissolving it in water, and measuring the height, h, of the resulting solution drawn up a capillary tube at 20 °C. The following data were obtained.

c/(mg cm^{-3})	3.221	4.618	5.112	6.722
h/cm	5.746	8.238	9.119	11.990

The osmotic pressure may be calculated from the height of the column, $\Pi = h\rho g$, taking the mass density of the solution as ρ = 1.000 g cm^{-3} and the acceleration of free fall as g = 9.81 m s^{-2}. Determine the molar mass of the enzyme.

6.35 The following temperature–composition data were obtained for a mixture of octane (O) and toluene (T) at 760 Torr, where x is the mole fraction in the liquid and y the mole fraction in the vapour at equilibrium.

θ/°C	110.9	112.0	114.0	115.8	117.3	119.0	120.0	123.0
x_T	0.908	0.795	0.615	0.527	0.408	0.300	0.203	0.097
y_T	0.923	0.836	0.698	0.624	0.527	0.410	0.297	0.164

The boiling points are 110.6 °C for toluene and 125.6 °C for octane. Plot the temperature–composition diagram of the mixture. What is the composition of the vapour in equilibrium with the liquid of composition (a) x_T = 0.250 and (b) x_O = 0.250?

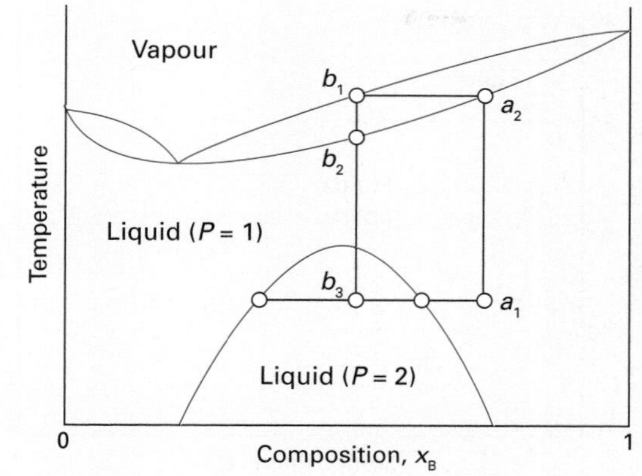

Fig. 6.39 A phase diagram for two partially miscible liquids.

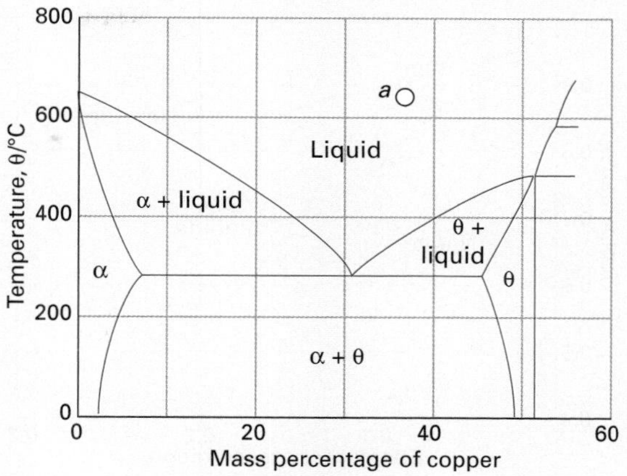

Fig. 6.41 A part of the phase diagram of an alloy of copper and aluminium.

6.36 Figure 6.39 shows the phase diagram for two partially miscible liquids, which can be taken to be that for water (A) and 2-methyl-1-propanol (B). Describe what will be observed when a mixture of composition b_3 is heated, at each stage giving the number, composition, and relative amounts of the phases present.

6.37 Sketch the phase diagram of the system NH_3/N_2H_4 given that the two substances do not form a compound with each other, that NH_3 freezes at −78 °C and N_2H_4 freezes at +2 °C, and that a eutectic is formed when the mole fraction of N_2H_4 is 0.07 and that the eutectic melts at −80 °C.

6.38 Figure 6.40 is the phase diagram for silver/tin. Label the regions, and describe what will be observed when liquids of compositions a and b are cooled to 200 °C.

6.39 Sketch the cooling curves for the compositions a and b in Fig. 6.40.

6.40 Use the phase diagram in Fig. 6.40 to determine (a) the solubility of silver in tin at 800 °C, (b) the solubility of Ag_3Sn in silver at 460 °C, and (c) the solubility of Ag_3Sn in silver at 300 °C.

6.41 Figure 6.41 shows a part of the phase diagram of an alloy of copper and aluminium. Describe what you will observe as a melt of composition a is cooled. What is the solubility of copper in aluminium at 500 °C?

6.42 Figure 6.42 shows a part of the phase diagram typical of a simple steel. Describe what you will observe as the melt at a is allowed to cool to room temperature.

6.43 Hexane and perfluorohexane (C_6F_{14}) show partial miscibility below 22.70 °C. The critical concentration at the upper critical temperature is x = 0.355, where x is the mole fraction of C_6F_{14}. At 22.0 °C the two solutions in equilibrium have x = 0.24 and x = 0.48 respectively, and at 21.5 °C the mole fractions are 0.22 and 0.51. Sketch the phase diagram. Describe the phase changes that occur when perfluorohexane is added to a fixed amount of hexane at (a) 23 °C, (b) 25 °C.

6.44 In a theoretical study of protein-like polymers, the phase diagram shown in Fig. 6.43 was obtained. It shows

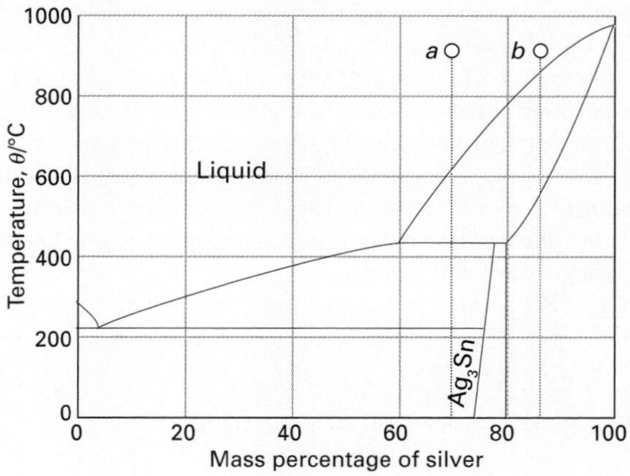

Fig. 6.40 The phase diagram for silver/tin.

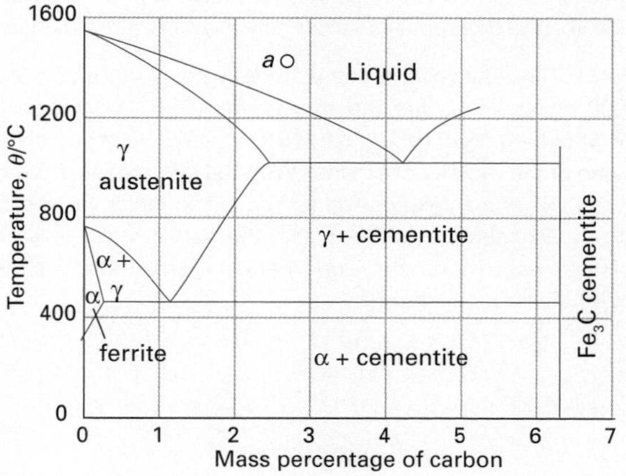

Fig. 6.42 A part of the phase diagram typical of a simple steel.

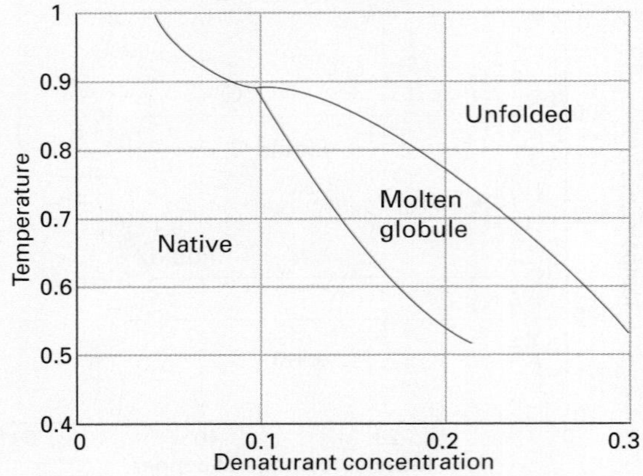

Fig. 6.43 A theoretical phase diagram for a model protein.

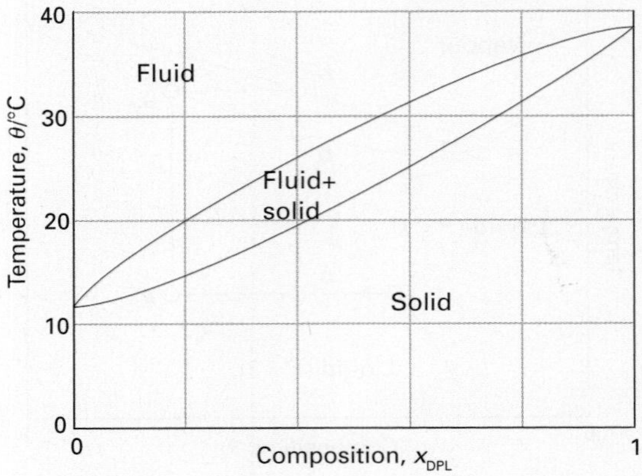

Fig. 6.44 The phase diagram for a membrane-forming two-component system.

three structural regions: the native form, the unfolded form, and a 'molten globule' form. (a) Is the molten-globule form ever stable when the denaturant concentration is below 0.1? (b) Describe what happens to the polymer as the native form is heated in the presence of denaturant at concentration 0.15.

6.45 In an experimental study of membrane-like assemblies of synthetic materials, a phase diagram like that shown in Fig. 6.44 was obtained. The two components are dielaidoylphosphatidylcholine (DEL) and dipalmitoylphosphatidylcholine (DPL). Explain what happens as a liquid mixture of composition $x_{DEL} = 0.5$ is cooled from 45 °C.

6.46 When 2.0 g of aspirin was shaken up in a flask containing two immiscible liquids it was found that the mole factions in the two liquids were 0.11 and 0.18. When a further 1.0 g of aspirin was added, it was found that the mole fraction in the first liquid increased to 0.15. What would you expect the mole fraction the second liquid to become?

Projects

The symbol ‡ indicates that the exercise requires calculus.

6.47‡ (a) The partial molar volume of ethanol in a mixture with water at 25 °C is $V_{ethanol}/(\text{cm}^3\ \text{mol}^{-1}) = 54.6664 - 0.727\,88b + 0.084\,768b^2$, where b is the numerical value of the molality of ethanol. Plot this quantity as a function of b and identify the composition at which the partial molar volume is a minimum. Express that composition as a mole fraction. (b) Use differentiation to identify the minimum in part (a).

6.48‡ The total volume of a water/ethanol mixture of mass 1.00 kg at 25 °C fits the expression $V/\text{cm}^3 = 1002.93 + 54.6664b - 0.363\,94b^2 + 0.028\,256b^3$, where b is the numerical value of the molality of ethanol. With the information in Exercise 6.47, find an expression for the partial molar volume of water. Plot the curve. Show that the partial molar volume of water has a maximum value where the partial molar volume of ethanol is a minimum.

6.49 Haemoglobin, the red blood protein responsible for oxygen transport, binds about 1.34 cm^3 of oxygen per gram. Normal blood has a haemoglobin concentration of 150 g dm^{-3}. Haemoglobin in the lungs is about 97 per cent saturated with oxygen, but in the capillary is only about 75 per cent saturated. (a) What volume of oxygen is given up by 100 cm^3 of blood flowing from the lungs in the capillary? Breathing air at high pressures, such as in scuba diving, results in an increased concentration of dissolved nitrogen. The Henry's law constant in the form $c = Kp$ for the solubility of nitrogen is 0.18 μg/(g H_2O atm). (b) What mass of nitrogen is dissolved in 100 g of water saturated with air at 4.0 atm and 20 °C? Compare your answer to that for 100 g of water saturated with air at 1.0 atm. (Air is 78.08 mole per cent N_2.) (c) If nitrogen is four times as soluble in fatty tissues as in water, what is the increase in nitrogen concentration in fatty tissue in going from 1 atm to 4 atm.

7

Chemical equilibria: the principles

Now we arrive at the point where real chemistry begins. Chemical thermodynamics is used to predict whether a mixture of reactants has a spontaneous tendency to change into products, to predict the composition of the reaction mixture at equilibrium, and to predict how that composition will be modified by changing the conditions. Although reactions in industry are rarely allowed to reach equilibrium, knowing whether equilibrium lies in favour of reactants or products under certain conditions is a good indication of the feasibility of a process. Much the same is true of biochemical reactions, where the avoidance of equilibrium is life and the attainment of equilibrium is death.

There is one word of warning that is essential to remember: *thermodynamics is silent about the rates of reaction*. All it can do is to identify whether a particular reaction mixture has a tendency to form products, it cannot say whether that tendency will ever be realized. Chapters 10 and 11 explore what determines the rates of chemical reactions. However, although thermodynamics has nothing to say about rates, it is important to keep in mind the physical picture that chemical equilibria are *dynamic*, and that at equilibrium the rates of the forward reaction and its reverse are equal, with products decaying at the same rate as they are formed. This dynamic character means that chemical reactions are responsive to changes in the conditions, and although thermodynamics can say nothing about how quickly they respond, it can be used to predict how they respond, such as by tending to form more products as the temperature is raised.

Thermodynamic background 165

7.1 The reaction Gibbs energy 166

7.2 The variation of $\Delta_r G$ with composition 167

7.3 Reactions at equilibrium 169

7.4 The standard reaction Gibbs energy 170

7.5 The equilibrium composition 173

7.6 The equilibrium constant in terms of concentration 175

7.7 The molecular interpretation of equilibrium constants 176

The response of equilibria to the conditions 177

7.8 The effect of temperature 177

7.9 The effect of compression 179

7.10 The presence of a catalyst 181

CHECKLIST OF KEY CONCEPTS 182

ROAD MAP OF KEY EQUATIONS 183

QUESTIONS AND EXERCISES 183

Thermodynamic background

The thermodynamic criterion for spontaneous change at constant temperature and pressure is $\Delta G < 0$. The principal idea behind this chapter, therefore, is that

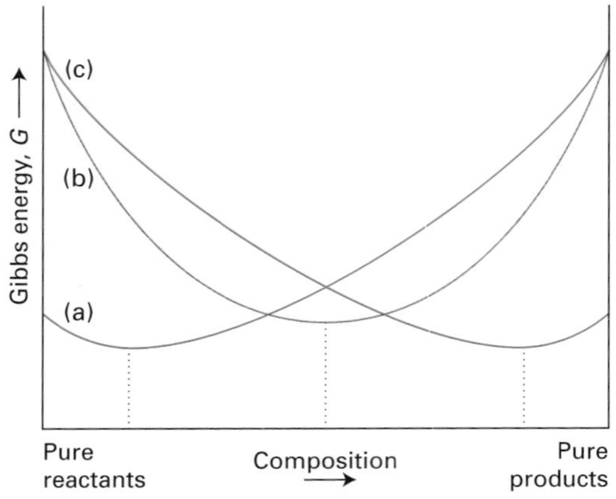

Fig. 7.1 The variation of Gibbs energy of a reaction mixture with progress of the reaction, pure reactants on the left and pure products on the right. (a) This reaction 'does not go': the minimum in the Gibbs energy occurs very close to reactants. (b) This reaction reaches equilibrium with approximately equal amounts of reactants and products present in the mixture. (c) This reaction goes almost to completion, as the minimum in Gibbs energy lies very close to pure products.

At constant temperature and pressure, a reaction mixture tends to adjust its composition until its Gibbs energy is a minimum.

If the Gibbs energy of a mixture varies as shown in Fig. 7.1a, very little of the reactants convert into products before G has reached its minimum value and the reaction 'does not go'. If G varies as shown in Fig. 7.1c, then a high proportion of products must form before G reaches its minimum and the reaction 'goes'. In many cases, the equilibrium mixture contains almost no reactants or almost no products. Many reactions have a Gibbs energy that varies as shown in Fig. 7.1b, and at equilibrium the reaction mixture contains substantial amounts of both reactants and products. One of our tasks is to see how to use thermodynamic data to predict the equilibrium composition and to see how that composition depends on the conditions.

7.1 The reaction Gibbs energy

To keep our ideas in focus, we consider two important reactions. One is the isomerism of glucose-6-phosphate (**1**, G6P) to fructose-6-phosphate (**2**, F6P), which is an early step in the anaerobic breakdown of glucose:

$$\text{G6P(aq)} \rightleftharpoons \text{F6P(aq)} \qquad \text{(R1)}$$

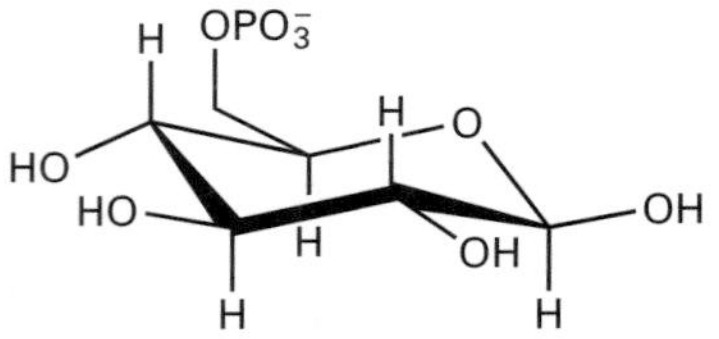

1 Glucose-6-phosphate

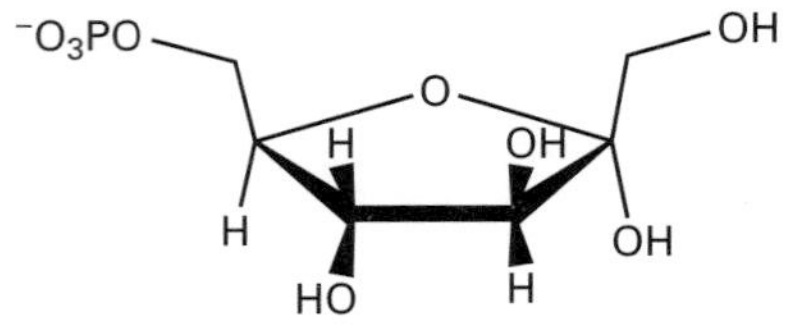

2 Fructose-6-phosphate

This reaction takes place in the aqueous environment of the cell. The second is the synthesis of ammonia, which is of crucial importance for industry and agriculture:

$$N_2(g) + 3\,H_2(g) \rightleftharpoons 2\,NH_3(g) \qquad \text{(R2)}$$

These two reactions are specific examples of a general reaction of the form

$$a\,\text{A} + b\,\text{B} \rightleftharpoons c\,\text{C} + d\,\text{D} \qquad \text{(R3)}$$

with arbitrary physical states.

First, consider reaction R1. Suppose that in a short interval while the reaction is in progress, the amount of G6P changes by $-\Delta n$. As a result of this change, the contribution of G6P to the total Gibbs energy of the system changes by $-\mu_{\text{G6P}}\Delta n$, where μ_{G6P} is the chemical potential (the partial molar Gibbs energy) of G6P in the reaction mixture. In the same interval, the amount of F6P changes by $+\Delta n$, so its contribution to the total Gibbs energy changes by $+\mu_{\text{F6P}}\Delta n$, where μ_{F6P} is the chemical potential of F6P. Provided Δn is small enough to leave the composition virtually unchanged, the net change in Gibbs energy of the system is

$$\Delta G = \mu_{\text{F6P}}\Delta n - \mu_{\text{G6P}}\Delta n$$

If we divide through by Δn, we obtain a special case of the **reaction Gibbs energy**, $\Delta_r G$:

$$\Delta_r G = \frac{\Delta G}{\Delta n} = \mu_{\text{F6P}} - \mu_{\text{G6P}} \qquad (7.1a)$$

There are two ways to interpret $\Delta_r G$. First, it is the difference of the chemical potentials of the products and reactants *at the composition of the reaction mixture*. Second, because $\Delta_r G$ is the change in G divided by the change in composition, we can think of $\Delta_r G$ as

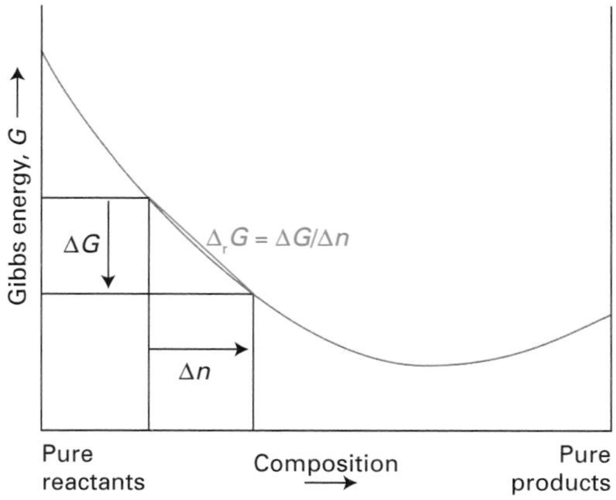

Fig. 7.2 The variation of Gibbs energy with progress of reaction showing how the reaction Gibbs energy, $\Delta_r G$, is related to the slope of the curve at a given composition.

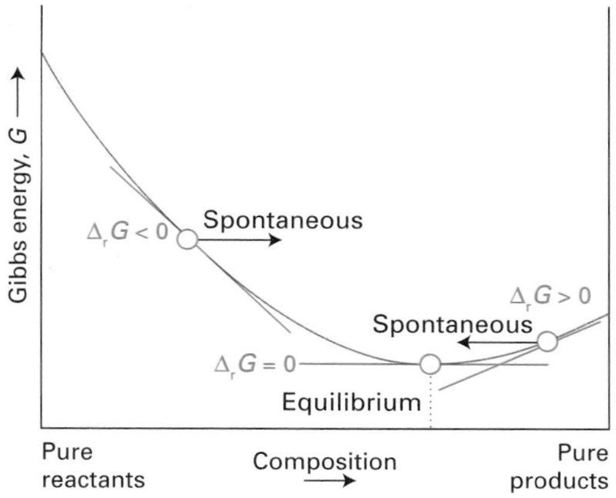

Fig. 7.3 At the minimum of the curve, corresponding to equilibrium, $\Delta_r G = 0$. To the left of the minimum, $\Delta_r G < 0$ and the forward reaction is spontaneous. To the right of the minimum, $\Delta_r G > 0$ and the reverse reaction is spontaneous.

being the slope of the graph of G plotted against the changing composition of the system (Fig. 7.2).

The synthesis of ammonia, reaction R2, provides a slightly more complicated example. If the amount of N_2 changes by $-\Delta n$, then from the reaction stoichiometry we know that the change in the amount of H_2 will be $-3\Delta n$ and the change in the amount of NH_3 will be $+2\Delta n$. Each change contributes to the change in the total Gibbs energy of the mixture, and the overall change is

$$\begin{aligned}\Delta G &= \mu_{NH_3} \times 2\Delta n - \mu_{N_2} \times \Delta n - \mu_{H_2} \times 3\Delta n \\ &= (2\mu_{NH_3} - \mu_{N_2} - 3\mu_{H_2})\Delta n\end{aligned}$$

where the μ_J are the chemical potentials of the species in the reaction mixture. In this case, therefore, the reaction Gibbs energy is

$$\Delta_r G = \frac{\Delta G}{\Delta n} = 2\mu_{NH_3} - (\mu_{N_2} + 3\mu_{H_2}) \qquad (7.1b)$$

Note that each chemical potential is multiplied by the corresponding stoichiometric coefficient and that reactants are subtracted from products. For the general case, reaction R3,

$$\Delta_r G = (c\mu_C + d\mu_D) - (a\mu_A + b\mu_B) \qquad \text{Reaction Gibbs energy} \qquad (7.1c)$$

The chemical potential of a substance depends on the composition of the mixture in which it is present, and is high when its concentration or partial pressure is high. Therefore, $\Delta_r G$ changes as the composition changes (Fig. 7.3). Remember that $\Delta_r G$ is the *slope* of G plotted against composition. We see that $\Delta_r G < 0$ and the slope of G is negative (down from left to right, from reactants towards products) when the mixture is rich in the reactants A and B because μ_A and μ_B are then high. Conversely, $\Delta_r G > 0$ and the slope of G is positive (up from left to right) when the mixture is rich in the products C and D because μ_C and μ_D are then high.

At compositions corresponding to $\Delta_r G < 0$ the *forward* reaction is spontaneous and the reactants tend to form more products. At composition corresponding to $\Delta_r G > 0$, the *reverse* reaction is spontaneous, and the products tend to decompose into reactants. Where $\Delta_r G = 0$ (at the minimum of the graph where the slope is zero), the reaction has no tendency to form either products or reactants. In other words, the reaction is at equilibrium. That is, *the criterion for chemical equilibrium at constant temperature and pressure* is

$$\Delta_r G = 0 \qquad \text{Constant temperature and pressure} \qquad \text{Condition of equilibrium} \qquad (7.2)$$

7.2 The variation of $\Delta_r G$ with composition

Our next step is to find how $\Delta_r G$ varies with the composition of the system. Once we know that, we shall be able to identify the composition corresponding to $\Delta_r G = 0$. Our starting point is the general expression for the composition dependence of the chemical potential derived in Section 6.6:

$$\mu_J = \mu_J^{\ominus} + RT \ln a_J \qquad (7.3)$$

where a_J is the activity of the species J. When we are dealing with ideal systems, which will be the case in this chapter, we use the identifications given in Table 6.2:

For solutes in an ideal solution, $a_J = [J]/c^{\ominus}$, the molar concentration of J relative to the standard value $c^{\ominus} = 1\ \text{mol dm}^{-3}$.

For perfect gases, $a_J = p_J/p^{\ominus}$, the partial pressure of J relative to the standard pressure $p^{\ominus} = 1$ bar.

For pure solids and liquids, $a_J = 1$.

As in Chapter 6, to simplify the appearance of expressions in what follows, we shall not write $c^{\ominus}$ and $p^{\ominus}$ explicitly except when we need to make a particular point.

Substitution of eqn 7.3 into eqn 7.1c gives

$$\Delta_r G = \left\{c\Big(\overbrace{\mu_C^{\ominus} + RT\ln a_C}^{\mu_C}\Big) + d\Big(\overbrace{\mu_D^{\ominus} + RT\ln a_D}^{\mu_D}\Big)\right\}$$

$$- \left\{a\Big(\overbrace{\mu_A^{\ominus} + RT\ln a_A}^{\mu_A}\Big) + b\Big(\overbrace{\mu_B^{\ominus} + RT\ln a_B}^{\mu_B}\Big)\right\}$$

$$= \{(c\mu_C^{\ominus} + d\mu_D^{\ominus}) - (a\mu_A^{\ominus} + b\mu_B^{\ominus})\}$$
$$+ RT\{c \ln a_C + d \ln a_D - a \ln a_A - b \ln a_B\}$$

The first term on the right in the second equality (in blue) is the **standard reaction Gibbs energy**, $\Delta_r G^{\ominus}$:

$$\Delta_r G^{\ominus} = (c\mu_C^{\ominus} + d\mu_D^{\ominus}) - (a\mu_A^{\ominus} + b\mu_B^{\ominus}) \quad (7.4a)$$

Because the standard states refer to the pure materials, the standard chemical potentials in this expression are the standard molar Gibbs energies of the (pure) species. Therefore, eqn 7.4a is the same as

$$\Delta_r G^{\ominus} = \{cG_m^{\ominus}(C) + dG_m^{\ominus}(D)\} - \{aG_m^{\ominus}(A) + bG_m^{\ominus}(B)\}$$

Standard reaction Gibbs energy (7.4b)

We consider this important quantity in more detail shortly. At this stage, therefore, we know that

$$\Delta_r G = \Delta_r G^{\ominus} + RT\{c \ln a_C + d \ln a_D - a \ln a_A - b \ln a_B\}$$

and the expression for $\Delta_r G$ is beginning to look much simpler.

To make further progress, we rearrange the remaining terms on the right (in blue) as follows by using the properties of logarithms set out in The chemist's toolkit 2.2:

$$c \ln a_C + d \ln a_D - a \ln a_A - b \ln a_B$$
$$\overset{a\ln x = \ln x^a}{=} \ln a_C^c + \ln a_D^d - \ln a_A^a - \ln a_B^b$$
$$\overset{\ln x + \ln y = \ln xy}{=} \ln a_C^c a_D^d - \ln a_A^a a_B^b$$
$$\overset{\ln x - \ln y = \ln \frac{x}{y}}{=} \ln \frac{a_C^c a_D^d}{a_A^a a_B^b}$$

At this point, we have deduced that

$$\Delta_r G = \Delta_r G^{\ominus} + RT\ln \frac{a_C^c a_D^d}{a_A^a a_B^b}$$

To simplify the appearance of this expression still further we introduce the (dimensionless) **reaction quotient**, Q, for reaction C:

$$Q = \frac{a_C^c a_D^d}{a_A^a a_B^b}$$

Definition | Reaction quotient (7.5)

Note that Q has the form of products divided by reactants, with the activity of each species raised to a power equal to its stoichiometric coefficient in the reaction. We can now write the overall expression for the reaction Gibbs energy at any composition of the reaction mixture as

$$\Delta_r G = \Delta_r G^{\ominus} + RT \ln Q$$

Dependence of $\Delta_r G$ on composition (7.6)

This simple but hugely important equation will occur several times in different disguises.

Example 7.1

Writing a reaction quotient

Write the reaction quotients for reactions R1 and R2.

Strategy Identify the appropriate expressions for the activities of the reactants and products, and insert them into the definition of Q in eqn 7.5. Simplify the appearance of the expressions by using the convention that neither $c^{\ominus}$ nor $p^{\ominus}$ is written explicitly.

Solution The reaction quotient for reaction A is

$$Q = \frac{a_{F6P}}{a_{G6P}}$$

For dilute solutions, we can express the activity of the solute as $a_B = [B]/c^{\ominus}$, and the reaction quotient therefore is

$$Q = \frac{[F6P]/c^{\ominus}}{[G6P]/c^{\ominus}} = \frac{[F6P]}{[G6P]}$$

In this example, the $c^{\ominus}$ cancel. For reaction R2, the synthesis of ammonia, a gas-phase reaction, the reaction quotient is

$$Q = \frac{a_{NH_3}^2}{a_{N_2} a_{H_2}^3} = \frac{(p_{NH_3}/p^{\ominus})^2}{(p_{N_2}/p^{\ominus})(p_{H_2}/p^{\ominus})^3} = \frac{p_{NH_3}^2 p^{\ominus 2}}{p_{N_2} p_{H_2}^3}$$

When the standard pressure is not written explicitly, this expression simplifies to

$$Q = \frac{p_{NH_3}^2}{p_{N_2} p_{H_2}^3}$$

with p_J interpreted as the numerical value of the partial pressure of J in bar (so, if $p_{NH_3} = 2$ bar, we just write $p_{NH_3} = 2$ when using this expression).

Self-test 7.1

Write the reaction quotient for the esterification reaction $CH_3COOH + C_2H_5OH \rightarrow CH_3COOC_2H_5 + H_2O$. (All four components are present in the reaction mixture as liquids: the mixture is not an aqueous solution.)

Answer: $Q = [CH_3COOC_2H_5][H_2O]/[CH_3COOH][C_2H_5OH]$

7.3 Reactions at equilibrium

When the reaction has reached equilibrium, the composition has no further tendency to change because $\Delta_r G = 0$ and the reaction is spontaneous in neither direction. At equilibrium, the reaction quotient has a certain value called the **equilibrium constant**, K, of the reaction:

$$K = Q_{\text{equilibrium}} = \left(\frac{a_C^c a_D^d}{a_A^a a_B^b}\right)_{\text{equilibrium}} \quad \text{Definition} \quad \text{Equilibrium constant} \quad (7.7)$$

We shall not normally write 'equilibrium'; the context will always make it clear that Q refers to an *arbitrary* stage of the reaction whereas K, the value of Q at equilibrium, is calculated from the *equilibrium* composition. It now follows from eqn 7.6 that at equilibrium

$$0 = \Delta_r G^{\ominus} + RT \ln K$$

and therefore that

$$\Delta_r G^{\ominus} = -RT \ln K \quad \text{Relation between } \Delta_r G^{\ominus} \text{ and } K \quad (7.8)$$

This is one of the most important equations in the whole of chemical thermodynamics. Its principal use is to predict the value of the equilibrium constant of any reaction from tables of thermodynamic data, like those in the *Data section*. Alternatively, we can use it to determine $\Delta_r G^{\ominus}$ by measuring the equilibrium constant of a reaction.

Brief illustration 7.1 The equilibrium constant

For the reaction $H_2(g) + I_2(s) \rightarrow 2\,HI(g)$, $\Delta_r G^{\ominus} = +3.40$ kJ mol^{-1} at 25 °C. To calculate the equilibrium constant we write

$$\ln K = -\frac{\Delta_r G^{\ominus}}{RT} = -\frac{3.40 \times 10^3 \text{ J mol}^{-1}}{(8.3145 \text{ J K}^{-1}\text{mol}^{-1}) \times (298.15 \text{ K})} = -\frac{3.40 \times 10^3}{8.3145 \times 298.15}$$

This expression evaluates to $\ln K = -1.37$, but to avoid rounding errors, we leave the evaluation until the next step, which is to use the relation $e^{\ln x} = x$ with $x = K$ to write

$$K = e^{-\frac{3.40 \times 10^3}{8.3145 \times 298.15}} = 0.25$$

A note on good practice All equilibrium constants and reaction quotients are dimensionless numbers.

Self-test 7.2

The equilibrium constant of the reaction $N_2(g) + 3\,H_2(g) \rightarrow 2\,NH_3(g)$ at 25 °C is 5.8×10^5. What is (a) the reaction Gibbs energy at equilibrium, (b) the standard reaction Gibbs energy?

Answer: (a) 0, (b) −32.90 kJ mol^{-1}

An important feature of eqn 7.8 is that it tells us that $K > 1$ if $\Delta_r G^{\ominus} < 0$. Broadly speaking, $K > 1$ implies that products are dominant at equilibrium (Fig. 7.4), so we can conclude that

A reaction is thermodynamically feasible (in the sense $K > 1$) if $\Delta_r G^{\ominus} < 0$.

Reactions for which $\Delta_r G^{\ominus} < 0$ are called **exergonic.** Conversely, because eqn 7.8 tells us that $K < 1$ if $\Delta_r G^{\ominus} > 0$, then we know that the reactants will be

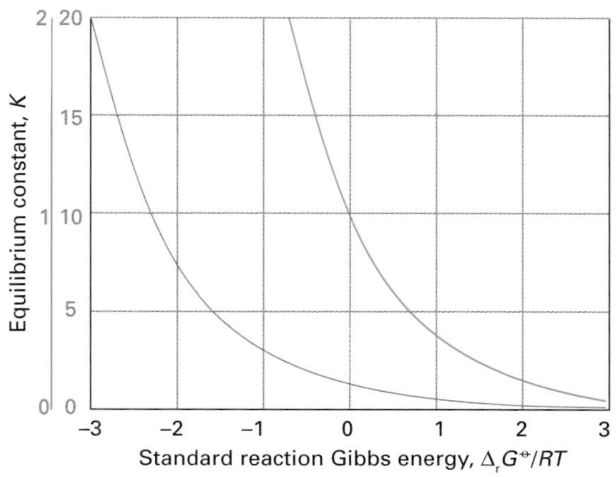

Fig. 7.4 The relation between standard reaction Gibbs energy and the equilibrium constant of the reaction. Note the different scales for the red and green lines.

Table 7.1
Thermodynamic criteria of spontaneity

(1) If the reaction is exothermic ($\Delta_r H^\ominus < 0$) and $\Delta_r S^\ominus > 0$
$\Delta_r G^\ominus < 0$ and $K > 1$ at all temperatures
(2) If the reaction is exothermic ($\Delta_r H^\ominus < 0$) and $\Delta_r S^\ominus < 0$
$\Delta_r G^\ominus < 0$ and $K > 1$ provided that $T < \Delta_r H^\ominus/\Delta_r S^\ominus$
(3) If the reaction is endothermic ($\Delta_r H^\ominus > 0$) and $\Delta_r S^\ominus > 0$
$\Delta_r G^\ominus < 0$ and $K > 1$ provided that $T > \Delta_r H^\ominus/\Delta_r S^\ominus$
(4) If the reaction is endothermic ($\Delta_r H^\ominus > 0$) and $\Delta_r S^\ominus < 0$
$\Delta_r G^\ominus < 0$ and $K > 1$ at no temperature

dominant in a reaction mixture at equilibrium if $\Delta_r G^\ominus > 0$. In other words,

> A reaction is not thermodynamically feasible (in the sense $K < 1$) if $\Delta_r G^\ominus > 0$.

Reactions for which $\Delta_r G^\ominus > 0$ are called **endergonic.** Some care must be exercised with these rules, however, because the products will be significantly more abundant than reactants only if $K \gg 1$ (more than about 10^3) and even a reaction with $K < 1$ may have a reasonable abundance of products at equilibrium.

Table 7.1 summarizes the conditions under which $\Delta_r G^\ominus < 0$ and $K > 1$. Because $\Delta_r G^\ominus = \Delta_r H^\ominus - T\Delta_r S^\ominus$, the standard reaction Gibbs energy is certainly negative if both $\Delta_r H^\ominus < 0$ (an exothermic reaction) and $\Delta_r S^\ominus > 0$ (a reaction system that becomes more disorderly, such as by forming a gas). The standard reaction Gibbs energy is also negative if the reaction is endothermic ($\Delta_r H^\ominus > 0$) and $T\Delta_r S^\ominus$ is sufficiently large and positive. Note that for an endothermic reaction to have $\Delta_r G^\ominus < 0$, its standard reaction

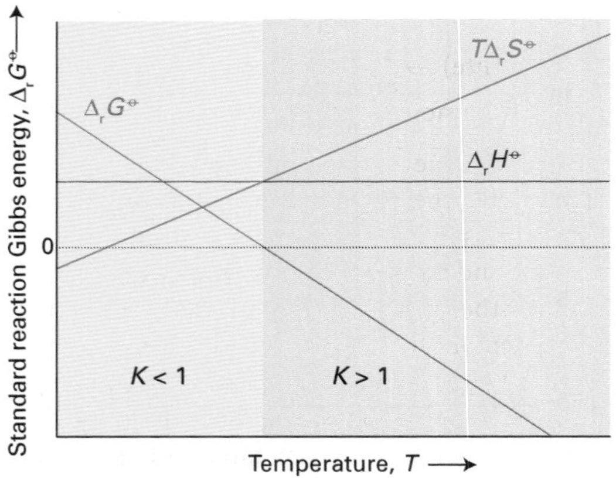

Fig. 7.5 An endothermic reaction may have $K > 1$ provided the temperature is high enough for $T\Delta_r S^\ominus$ to be large enough that, when subtracted from $\Delta_r H^\ominus$, the result is negative.

entropy *must* be positive. Moreover, the temperature must be high enough for $T\Delta_r S^\ominus$ to be greater than $\Delta_r H^\ominus$ (Fig. 7.5). The switch of $\Delta_r G^\ominus$ from positive to negative, corresponding to the switch from $K < 1$ (the reaction 'does not go') to $K > 1$ (the reaction 'goes'), occurs at a temperature given by equating $\Delta_r H^\ominus - T\Delta_r S^\ominus$ to 0, which gives:

$$T = \frac{\Delta_r H^\ominus}{\Delta_r S^\ominus} \quad \text{Minimum temperature for spontaneity of endothermic reaction} \qquad (7.9)$$

Brief illustration 7.2 Spontaneity of an endothermic reaction

Consider the (endothermic) thermal decomposition of calcium carbonate:

$$CaCO_3(s) \rightarrow CaO(s) + CO_2(g)$$

For this reaction $\Delta_r H^\ominus = +178$ kJ mol^{-1} and $\Delta_r S^\ominus = +161$ J K^{-1} mol^{-1}. The decomposition temperature, the temperature at which the decomposition becomes spontaneous, is

$$T = \frac{\overbrace{1.78 \times 10^5\,\text{J mol}^{-1}}^{\Delta_r H^\ominus}}{\underbrace{161\,\text{J K}^{-1}\,\text{mol}^{-1}}_{\Delta_r S^\ominus}} = 1.11 \times 10^3\,\text{K}$$

or about 832 °C. Because the entropy of decomposition is similar for all such reactions (they all involve the decomposition of a solid into a gas), we can conclude that the decomposition temperatures of solids increase as their enthalpy of decomposition increases.

7.4 The standard reaction Gibbs energy

The standard reaction Gibbs energy, $\Delta_r G^\ominus$, is central to the discussion of chemical equilibria and the calculation of equilibrium constants. We have seen that it is defined as the difference in standard molar Gibbs energies of the products and the reactants weighted by the stoichiometric coefficients, ν, in the chemical equation

$$\Delta_r G^\ominus = \sum \nu G_m^\ominus(\text{products}) - \sum \nu G_m^\ominus(\text{reactants}) \quad \text{Definition} \quad \text{Standard reaction Gibbs energy} \qquad (7.10)$$

where, as usual, Σ instructs us to take the sum of the items that follow it. For example, the standard reaction Gibbs energy for reaction R1 is the difference between the molar Gibbs energies of fructose-6-phosphate and glucose-6-phosphate in solution at 1 mol dm^{-3} and 1 bar.

We cannot calculate $\Delta_r G^\ominus$ from the standard molar Gibbs energies themselves, because these quantities are not known. One practical approach is to calculate the standard reaction enthalpy from standard enthalpies of formation (Section 3.5), the standard reaction entropy from Third-Law entropies (Section 4.10), and then to combine the two quantities by using

$$\Delta_r G^\ominus = \Delta_r H^\ominus - T\Delta_r S^\ominus \tag{7.11}$$

Brief illustration 7.3 The standard reaction Gibbs energy from thermodynamic data

To evaluate the standard reaction Gibbs energy at 25 °C for the reaction $H_2(g) + \frac{1}{2} O_2(g) \rightarrow H_2O(l)$, we note that

$$\Delta_r H^\ominus = \Delta_f H^\ominus(H_2O,l) - \overbrace{0}^{\Delta_f H^\ominus(H_2)} - \overbrace{0}^{\frac{1}{2}\Delta_f H^\ominus(O_2)}$$
$$= -285.83 \text{ kJ mol}^{-1}$$

The standard reaction entropy, calculated in Chapter 4, is $\Delta_r S^\ominus = -163.34 \text{ J K}^{-1} \text{ mol}^{-1}$, which, because 163.34 J is the same as 0.163 34 kJ, corresponds to $-0.163\,34 \text{ kJ K}^{-1} \text{ mol}^{-1}$. Therefore, from eqn 7.11,

$$\Delta_r G^\ominus = (-285.83 \text{ kJ mol}^{-1}) - (298.15 \text{ K}) \times (-0.163\,34 \text{ kJ K}^{-1} \text{ mol}^{-1})$$
$$= -237.13 \text{ kJ mol}^{-1}$$

Self-test 7.3

Use the information in the *Data section* to determine the standard reaction Gibbs energy for $3\,O_2(g) \rightarrow 2\,O_3(g)$ from standard enthalpies of formation and standard entropies.

Answer: $+326.4 \text{ kJ mol}^{-1}$

We saw in Section 3.5 how to use standard enthalpies of formation of substances to calculate standard reaction enthalpies. We can use the same technique for standard reaction Gibbs energies. To do so, we list the **standard Gibbs energy of formation**, $\Delta_f G^\ominus$, of a substance:

> The standard Gibbs energy of formation is the standard reaction Gibbs energy (per mole of the species) for its formation from the elements in their reference states.

The concept of reference state was introduced in Section 3.5; the temperature is arbitrary, but we shall almost always take it to be 25 °C (298 K, more precisely 298.15 K). For example, the standard Gibbs energy of formation of liquid water, $\Delta_f G^\ominus(H_2O, l)$, is the standard reaction Gibbs energy for

$$H_2(g) + \tfrac{1}{2} O_2(g) \rightarrow H_2O(l)$$

Table 7.2
*Standard Gibbs energies of formation at 298.15 K**

Substance	$\Delta_f G^\ominus/(\text{kJ mol}^{-1})$
Gases	
Ammonia, NH_3	−16.45
Carbon dioxide, CO_2	−394.36
Dinitrogen tetroxide, N_2O_4	+97.89
Hydrogen iodide, HI	+1.70
Nitrogen dioxide, NO_2	+51.31
Sulfur dioxide, SO_2	−300.19
Water, H_2O	−228.57
Liquids	
Benzene, C_6H_6	+124.3
Ethanol, CH_3CH_2OH	−174.78
Water, H_2O	−237.13
Solids	
Calcium carbonate, $CaCO_3$	−1128.8
Iron(III) oxide, Fe_2O_3	−742.2
Silver bromide, AgBr	−96.90
Silver chloride, AgCl	−109.79

*Additional values are given in the *Data section* and the text's website.

and is -237 kJ mol^{-1} at 298 K. Some standard Gibbs energies of formation are listed in Table 7.2 and more can be found in the *Data section*. It follows from the definition that the standard Gibbs energy of formation of an element in its reference state is zero because reactions such as C(s, graphite) → C(s, graphite) are null (that is, nothing happens). The standard Gibbs energy of formation of an element in a phase different from its reference state is nonzero:

$$\text{C(s, graphite)} \rightarrow \text{C(s, diamond)}$$
$$\Delta_f G^\ominus(\text{C, diamond}) = +2.90 \text{ kJ mol}^{-1}$$

Many of the values in the tables have been compiled by combining the standard enthalpy of formation of the species with the standard entropies of the compound and the elements, as illustrated above, but there are other sources of data and we encounter some of them later.

Standard Gibbs energies of formation can be combined to obtain the standard Gibbs energy of almost any reaction. We use the now familiar expression

$$\Delta_r G^\ominus = \sum \nu \Delta_f G^\ominus(\text{products}) - \sum \nu \Delta_f G^\ominus(\text{reactants}) \tag{7.12}$$

Practical implementation | Standard reaction Gibbs energy

Brief illustration 7.4 The standard reaction Gibbs energy from Gibbs energies of formation

To determine the standard reaction Gibbs energy for

$$2\,CO(g) + O_2(g) \rightarrow 2\,CO_2(g)$$

we carry out the following calculation:

$$\begin{aligned}\Delta_r G^\ominus &= 2\Delta_f G^\ominus(CO_2, g) - \{2\Delta_f G^\ominus(CO, g) + \Delta_f G^\ominus(O_2, g)\} \\ &= 2 \times (-394\ \text{kJ mol}^{-1}) - \{2 \times (-137\ \text{kJ mol}^{-1}) + 0\} \\ &= -514\ \text{kJ mol}^{-1}\end{aligned}$$

Self-test 7.4

Calculate the standard reaction Gibbs energy of the oxidation of ammonia to nitric oxide according to the equation $4\,NH_3(g) + 5\,O_2(g) \rightarrow 4\,NO(g) + 6\,H_2O(g)$.

Answer: -959.42 kJ mol^{-1}

Standard Gibbs energies of formation of compounds have their own significance as well as being useful in calculations of K. They are a measure of the 'thermodynamic altitude' of a compound above or below a 'sea level' of stability represented by the elements in their reference states (Fig. 7.6). If the standard Gibbs energy of formation is positive and the compound lies above 'sea level', then the compound has a spontaneous tendency to sink towards thermodynamic sea level and decompose into the elements. That is, $K < 1$ for their formation reaction. We say that a compound with $\Delta_f G^\ominus > 0$ is **thermodynamically unstable** with respect to its elements or that it is an **endergonic compound**. Thus, ozone, for which $\Delta_f G^\ominus = +163$ kJ mol^{-1}, has a spontaneous tendency to decompose into oxygen under standard conditions at 25 °C. More precisely, the equilibrium constant for the reaction $\tfrac{3}{2}\,O_2(g) \rightleftharpoons O_3(g)$ is less than 1 (much less in fact, for $K = 2.7 \times 10^{-29}$). However, although ozone is thermodynamically unstable, it can survive if the reactions that convert it into oxygen are slow. That is the case in the upper atmosphere, and the O_3 molecules in the ozone layer survive for long periods. Benzene ($\Delta_f G^\ominus = +124$ kJ mol^{-1}) is also thermodynamically unstable with respect to its elements ($K = 1.8 \times 10^{-22}$). However, the fact that bottles of benzene are everyday laboratory commodities also reminds us of the point made at the start of the chapter, that *spontaneity is a thermodynamic tendency that might not be realized at a significant rate in practice.*

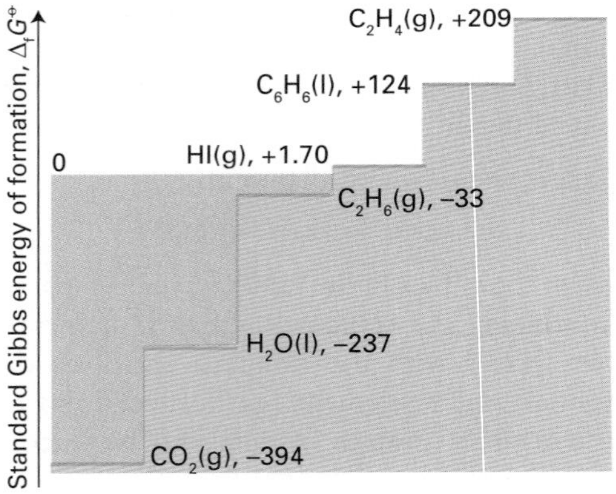

Fig. 7.6 The standard Gibbs energy of formation of a compound is like a measure of the compound's altitude above (or below) sea level: compounds that lie above sea level have a spontaneous tendency to decompose into the elements (and to revert to sea level). Compounds that lie below sea level are stable with respect to decomposition into the elements. The numerical values are in kilojoules per mole.

Another useful point that can be made about standard Gibbs energies of formation is that there is no point in searching for *direct* syntheses of a thermodynamically unstable compound from its elements (under standard conditions, at the temperature to which the data apply), because the reaction does not occur in the required direction: the *reverse* reaction, decomposition, is spontaneous. Endergonic compounds must be synthesized by alternative routes or under conditions for which their Gibbs energy of formation is negative and they lie beneath thermodynamic sea level.

Compounds with $\Delta_f G^\ominus < 0$ (corresponding to $K > 1$ for their formation reactions) are said to be **thermodynamically stable** with respect to their elements or are called **exergonic compounds**. Exergonic compounds lie below the thermodynamic sea level of the elements (under standard conditions). An example is the exergonic compound ethane, with $\Delta_f G^\ominus = -33$ kJ mol^{-1}: the negative sign shows that the formation of ethane gas is spontaneous in the sense that $K > 1$ (in fact, $K = 7.1 \times 10^5$ at 25 °C).

Impact on biochemistry 7.1

Coupled reactions in biochemical processes

A reaction that is not spontaneous may be driven forward by coupling it to a reaction that is spontaneous. A simple mechanical analogy is a pair of weights joined by a string (Fig. 7.7): the lighter of the pair of weights will be pulled up as the heavier weight falls. Although the lighter weight has a natural tendency to move downwards, its coupling to the heavier weight results in it being raised. The thermodynamic analogue is an endergonic reaction (the analogue of the lighter weight) being forced to occur by coupling it to an exergonic reaction (the analogue of the heavier weight falling to the

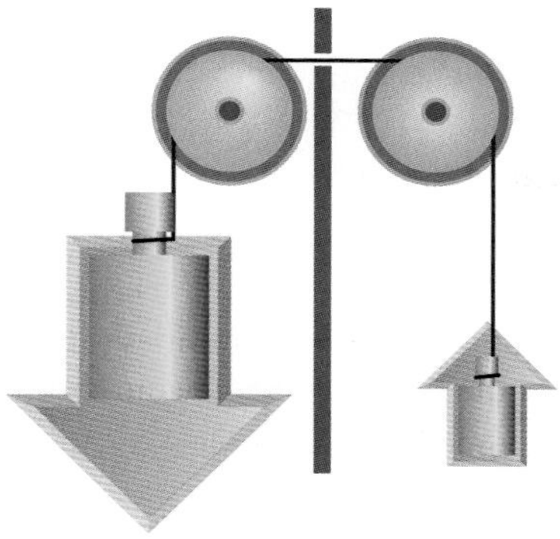

Fig. 7.7 If two weights are coupled as shown here, then the heavier weight will move the lighter weight in its non-spontaneous direction: overall, the process is still spontaneous. The weights are the analogues of two chemical reactions: a reaction with a large negative ΔG can force another reaction with a smaller ΔG to run in its non-spontaneous direction.

ground). The overall reaction is spontaneous because the sum $\Delta_r G + \Delta_r G'$ is negative. The whole of life's activities depend on coupling of this kind, for the oxidation reactions of food act as the heavy weights that drive other reactions forward and result in the formation of proteins from amino acids, the actions of muscles for propulsion, and even the activities of the brain for reflection, learning, and imagination.

The function of adenosine triphosphate, ATP (**3**), for instance, is to store the energy made available when food is oxidized and then to supply it on demand to a wide variety of processes, including muscular contraction, reproduction, and vision. The essence of ATP's action is its ability to lose its terminal phosphate group by hydrolysis and to form adenosine diphosphate, ADP (**4**):

$$\text{ATP(aq)} + \text{H}_2\text{O(l)} \rightarrow \text{ADP(aq)} + \text{P}_i^-\text{(aq)} + \text{H}^+\text{(aq)}$$

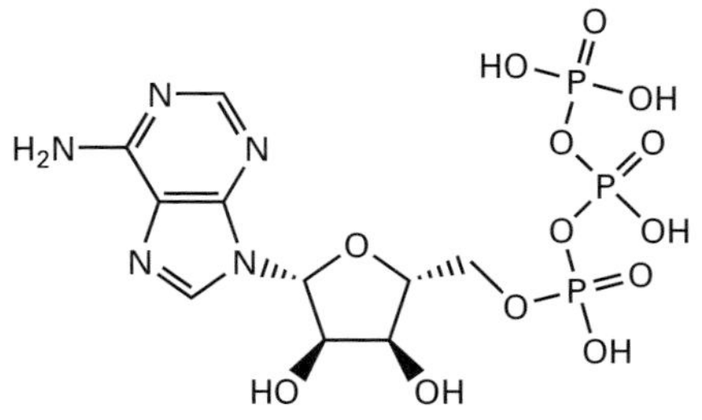

3 ATP

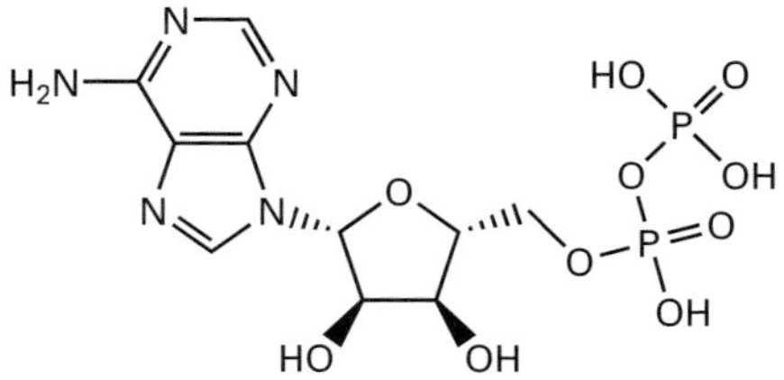

4 ADP

P_i^- denotes an inorganic phosphate group, such as $H_2PO_4^-$. This reaction is exergonic under the conditions prevailing in cells and can drive an endergonic reaction forward if suitable enzymes are available to couple the reactions. For example, the endergonic phosphorylation of glucose is coupled to the hydrolysis of ATP in the cell, so the net reaction

$$\text{glucose(aq)} + \text{ATP(aq)} \rightarrow \text{G6P(aq)} + \text{ADP(aq)}$$

is exergonic and initiates glycolysis.

Before discussing the hydrolysis of ATP quantitatively, we need to note that the conventional standard state of hydrogen ions ($a = 1$, corresponding to pH = 0, a strongly acidic solution; recall from introductory chemistry that pH = $-\log a_{H_3O^+} \approx -\log [H_3O^+]$) is not appropriate to normal biological conditions inside cells, where the pH is close to 7. Therefore, in biochemistry it is common to adopt the **biological standard state**, in which pH = 7, a neutral solution. We shall adopt this convention in this section, and label the corresponding standard quantities as $G^{\oplus}$, $H^{\oplus}$, and $S^{\oplus}$. Another convention to denote the biological standard state is to write $X^{\circ\prime}$ or $X^{\ominus\prime}$.

The biological standard values for the hydrolysis of ATP at 37 °C (310 K, body temperature) are

$$\Delta_r G^{\oplus} = -31 \text{ kJ mol}^{-1} \qquad \Delta_r H^{\oplus} = -20 \text{ kJ mol}^{-1}$$
$$\Delta_r S^{\oplus} = +34 \text{ kJ mol}^{-1}$$

The hydrolysis is therefore exergonic ($\Delta_r G < 0$) under these conditions, and 31 kJ mol^{-1} is available for driving other reactions. On account of its exergonic character, the ADP–phosphate bond has been called a 'high-energy phosphate bond'. The name is intended to signify a high tendency to undergo reaction and should not be confused with 'strong' bond in its normal chemical sense (that of a high bond enthalpy). In fact, even in the biological sense it is not of very 'high energy'. The action of ATP depends on the bond being intermediate in strength. Thus ATP acts as a phosphate donor to a number of acceptors (such as glucose), but is recharged with a new phosphate group by more powerful phosphate donors in the phosphorylation steps in the respiration cycle.

7.5 The equilibrium composition

The magnitude of an equilibrium constant is a good *qualitative* indication of the feasibility of a reaction regardless of whether the system is ideal or not. Broadly speaking, if $K \gg 1$ (typically $K > 10^3$, corresponding to $\Delta_r G^{\ominus} < -17$ kJ mol^{-1} at 25 °C), then the reaction has a strong tendency to form products. If $K \ll 1$ (that is, for $K < 10^{-3}$, corresponding to $\Delta_r G^{\ominus} > +17$ kJ mol^{-1} at 25 °C), then the equilibrium composition will consist of largely unchanged reactants. If K is comparable to 1 (typically lying between 10^{-3} and 10^3), then significant amounts of both reactants and products will be present at equilibrium.

An equilibrium constant expresses the composition of an equilibrium mixture as a ratio of products of activities. Even if we confine our attention to ideal systems it is still necessary to do some work to extract the actual equilibrium concentrations or partial pressures of the reactants and products given their initial values.

Example 7.2

Calculating an equilibrium composition 1

Estimate the fraction *f* of F6P in a solution, where *f* is defined as

$$f = \frac{[\text{F6P}]}{[\text{F6P}]+[\text{G6P}]}$$

in which G6P and F6P are in equilibrium at 25 °C (reaction R1) given that $\Delta_r G^\ominus = +1.7\ \text{kJ mol}^{-1}$ at that temperature.

Strategy Express *f* in terms of *K*. To do so, recognize that if the numerator and denominator in the expression for *f* are both divided by [G6P], then the ratios [F6P]/[G6P] can be replaced by *K*. Calculate the value of *K* by using eqn 7.8.

Solution Division of the numerator and denominator by [G6P] gives

$$f = \frac{\overbrace{[\text{F6P}]/[\text{G6P}]}^{K}}{\underbrace{[\text{F6P}]/[\text{G6P}]}_{K}+1} = \frac{K}{K+1}$$

We find the equilibrium constant by using $K = e^{\ln K}$ and rearranging eqn 7.8 into

$$K = e^{-\Delta_r G^\ominus/RT}$$

First, note that, because $+1.7\ \text{kJ mol}^{-1}$ is the same as $+1.7 \times 10^3\ \text{J mol}^{-1}$,

$$\frac{\Delta_r G^\ominus}{RT} = \frac{1.7\times10^3\ \text{J mol}^{-1}}{(8.3145\ \text{J K}^{-1}\ \text{mol}^{-1})\times(298\ \text{K})} = \frac{1.7\times10^3}{8.3145\times298}$$

Therefore,

$$K = e^{-\frac{1.7\times10^3}{8.3145\times298}} = 0.50$$

and

$$f = \frac{0.50}{0.50+1} = 0.33$$

That is, at equilibrium, 33 per cent of the solute is F6P and 67 per cent is G6P.

Self-test 7.5

Estimate the composition of a solution in which two isomers A and B are in equilibrium ($A \rightleftharpoons B$) at 37 °C and $\Delta_r G^\ominus = -2.2\ \text{kJ mol}^{-1}$.

Answer: The fraction of B at equilibrium is $f = 0.70$

In more complicated cases it is best to organize the necessary work into a systematic procedure resembling a spreadsheet by constructing a table with columns headed by the species and, in successive rows:

1. The initial molar concentrations of solutes or partial pressures of gases.
2. The changes in these quantities that must take place for the system to reach equilibrium.
3. The resulting equilibrium values.

In most cases, we do not know the change that must occur for the system to reach equilibrium, so the change in the concentration or partial pressure of one species is written as *x* and the reaction stoichiometry is used to write the corresponding changes in the other species. When the values at equilibrium (the last row of the table) are substituted into the expression for the equilibrium constant, we obtain an equation for *x* in terms of *K*. This equation can be solved for *x*, and hence the concentrations of all the species at equilibrium may be found.

Example 7.3

Calculating an equilibrium composition 2

Suppose that in an industrial process, N_2 at 1.00 bar is mixed with H_2 at 3.00 bar and the two gases are allowed to come to equilibrium with the product ammonia in a reactor of constant volume (in the presence of a catalyst, so the reaction proceeds quickly). At the temperature of the reaction, it has been determined experimentally that $K = 977$ for reaction R2. What are the equilibrium partial pressures of the three gases?

Strategy Proceed as set out above. Write down the chemical equation of the reaction and the expression for *K*. Set up the equilibrium table, express *K* in terms of *x*, and solve the equation for *x*. Because the volume of the reaction vessel is constant, each partial pressure is proportional to the amount of its molecules present ($p_J = n_J RT/V$), so the stoichiometric relations apply to the partial pressures directly. In general, solution of the equation for *x* results in several mathematically possible values of *x*. Select the chemically acceptable solution by considering the signs of the predicted concentrations or partial pressures: they must be positive. Confirm the accuracy of the calculation by substituting the calculated equilibrium partial pressures into the expression for the equilibrium constant to verify that the value so calculated is equal to the experimental value used in the calculation.

Solution The chemical equation is reaction R2, $N_2(g) + 3\,H_2(g) \rightarrow 2\,NH_3(g)$, and the equilibrium constant is

$$K = \frac{p_{NH_3}^2}{p_{N_2}\,p_{H_2}^3}$$

with the partial pressures those at equilibrium (and, as usual, relative to $p^{\ominus}$). The equilibrium table is

Species:	N_2	H_2	NH_3
Initial partial pressure/bar	1.00	3.00	0
Change/bar	$-x$	$-3x$	$+2x$
Equilibrium partial pressure/bar	$1.00-x$	$3.00-3x$	$2x$

The equilibrium constant for the reaction is therefore

$$K=\frac{\overbrace{(2x)^2}^{p_{NH_3}}}{\underbrace{(1.00-x)}_{p_{N_2}}\underbrace{(3.00-3x)^3}_{p_{H_2}}}=\frac{4x^2}{27(1.00-x)^4}$$

Our task is to solve this equation for x. Because $K=977$, this equation rearranges first to

$$977=\frac{4}{27}\left(\frac{x}{(1.00-x)^2}\right)^2$$

and then, after multiplying both sides by $^{27}/_4$ and taking the square root of both sides, to

$$\overbrace{\sqrt{\frac{27\times977}{4}}}^{g}=\frac{x}{(1.00-x)^2}$$

To keep the appearance of this equation simple, we write $g=(27\times977/4)^{1/2}$, so it becomes

$$g=\frac{x}{(1.00-x)^2}=\frac{x}{1.00-2.00x+x^2}$$

This expression can now be rearranged into

$$gx^2-2.00gx+1.00g=x$$

and then into

$$\overbrace{gx^2}^{ax^2}\overbrace{-(2.00g+1)x}^{+bx}\overbrace{+1.00g}^{+c}=0$$

This equation has the form of a quadratic equation (see The chemist's toolkit 7.1) with $a=g$, $b=-(2.00g+1)$, and $c=1.00g$. Its solutions are $x=1.12$ and $x=0.895$. Because p_{N_2} cannot be negative, and $p_{N_2}=1.00-x$ (from the equilibrium table), we know that x cannot be greater than 1.00; therefore, we select $x=0.895$ as the acceptable solution. It then follows from the last line of the equilibrium table that (with the units bar restored):

$$p_{N_2}=0.10\text{ bar}\qquad p_{H_2}=0.32\text{ bar}\qquad p_{NH_3}=1.8\text{ bar}$$

This is the composition of the reaction mixture at equilibrium. Note that, because K is large (of the order of 10^3), the products dominate. To verify the result, we calculate

$$\frac{p_{NH_3}^2}{p_{N_2}p_{H_2}^3}=\frac{1.8^2}{0.10\times0.32^3}=9.9\times10^2$$

The result is close to the experimental value (the discrepancy stems from rounding errors).

The chemist's toolkit 7.1 Quadratic equations

A quadratic equation is an equation of the form

$$ax^2+bx+c=0$$

It has the solutions ('roots')

$$x=\frac{-b\pm\sqrt{b^2-4ac}}{2a}$$

There are often physical reasons for choosing one root rather than the other. For instance, if x represents a concentration or a pressure, then it must be positive.

It is sometimes helpful to display the roots graphically. A graph of $y=ax^2+bx+c$ is a parabola (Sketch 7.1), which intersects the horizontal x-axis (where $y=0$) at the two roots of the equation.

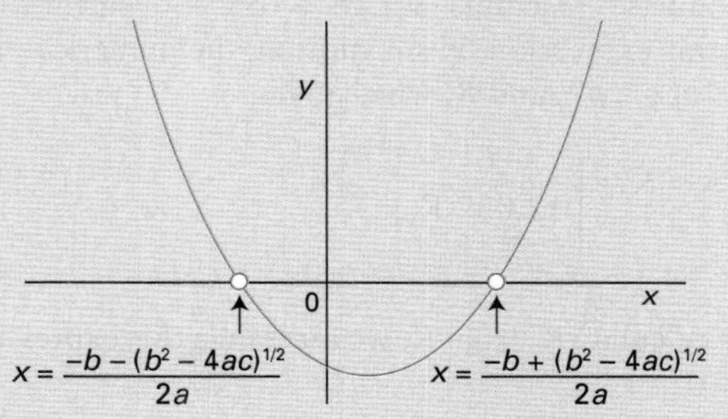

Sketch 7.1 The graphical representation of the roots of a quadratic equation.

Self-test 7.6

In an experiment to study the formation of nitrogen oxides in jet exhausts, N_2 at 0.100 bar is mixed with O_2 at 0.200 bar and the two gases are allowed to come to equilibrium with the product NO in a reactor of constant volume. Take $K=3.4\times10^{-21}$ at 800 K. What is the equilibrium partial pressure of NO?

Answer: 8.2 pbar

7.6 The equilibrium constant in terms of concentration

An important point to appreciate is that the equilibrium constant K calculated from thermodynamic data refers to activities. For gas-phase reactions, that means partial pressures (and explicitly, $p_J/p^{\ominus}$). This requirement is sometimes emphasized by writing K as K_p, but the practice is unnecessary if the thermodynamic origin of K is remembered. In practical applications, however, we might wish to discuss gas-phase reactions in terms of molar concentrations.

The equilibrium constant is then denoted K_c, and for reaction R2 is

$$K_c = \frac{[NH_3]^2}{[N_2][H_2]^3}$$

with, as usual, the molar concentration [J] interpreted as $[J]/c^{\ominus}$ with $c^{\ominus} = 1$ mol dm^{-3}. To obtain the value of K_c from thermodynamic data, we must first calculate K and then convert K to K_c by using, as shown in the following Derivation, the relation

$$K = K_c \times \left(\frac{c^{\ominus}RT}{p^{\ominus}}\right)^{\Delta\nu_{gas}} \quad \text{Relation between } K \text{ and } K_c \quad (7.13a)$$

In this expression, $\Delta\nu_{gas}$ is the difference in the stoichiometric coefficients of the gas-phase species, products – reactants. We get a very convenient form of this expression by substituting the numerical values of $c^{\ominus}$, $p^{\ominus}$, and R, which gives

$$K = K_c \times \left(\frac{T}{\underbrace{12.027 \text{ K}}_{p^{\ominus}/c^{\ominus}R}}\right)^{\Delta\nu_{gas}} \quad (7.13b)$$

To calculate K_c from K, we rearrange this expression into

$$K_c = \frac{K}{(T/12.027 \text{ K})^{\Delta\nu_{gas}}} \quad (7.13c)$$

Derivation 7.1

The relation between K and K_c

In this derivation, we need to be fussy about units, and will write the equilibrium constants of reaction R3 in all their glory as

$$K = \frac{a_C^c a_D^d}{a_A^a a_B^b} = \frac{(p_C/p^{\ominus})^c(p_D/p^{\ominus})^d}{(p_A/p^{\ominus})^a(p_B/p^{\ominus})^b} \qquad K_c = \frac{([C]/c^{\ominus})^c([D]/c^{\ominus})^d}{([A]/c^{\ominus})^a([B]/c^{\ominus})^b}$$

The inclusion of $p^{\ominus}$ and $c^{\ominus}$ ensures that the two equilibrium constants are dimensionless. Now we use the perfect gas law to replace each partial pressure by

$$p_J = n_J RT/V = [J]RT$$

(because $[J] = n_J/V$). This substitution turns the expression for K into

$$K = \frac{\left(\overbrace{[C]RT}^{p_C}/p^{\ominus}\right)^c\left(\overbrace{[D]RT}^{p_D}/p^{\ominus}\right)^d}{\left(\underbrace{[A]RT}_{p_A}/p^{\ominus}\right)^a\left(\underbrace{[B]RT}_{p_B}/p^{\ominus}\right)^b} = \frac{[C]^c[D]^d}{[A]^a[B]^b} \times \left(\frac{RT}{p^{\ominus}}\right)^{(c+d)-(a+b)}$$

Next, we recognize that

$$K_c = \frac{[C]^c[D]^d}{[A]^a[B]^b} \times \left(\frac{1}{c^{\ominus}}\right)^{(c+d)-(a+b)},$$

$$\text{so} \quad \frac{[C]^c[D]^d}{[A]^a[B]^b} = K_c \times (c^{\ominus})^{(c+d)-(a+b)}$$

and therefore (by replacing the blue expression in the equation for K) conclude that

$$K = K_c \times \left(\frac{c^{\ominus}RT}{p^{\ominus}}\right)^{(c+d)-(a+b)}$$

We obtain eqn 7.13 by writing $(c + d) - (a + b) = \Delta\nu_{gas}$.

Brief illustration 7.5 Calculating K_c from K

For reaction R2 we have $\Delta\nu_{gas} = 2 - (1 + 3) = -2$; therefore, from eqn 7.13c,

$$K_c = \frac{K}{(T/12.027 \text{ K})^{-2}} = K \times \left(\frac{T}{12.027 \text{ K}}\right)^2$$

At 298 K, $K = 5.8 \times 10^5$, so at this temperature

$$K_c = 5.8 \times 10^5 \times \left(\frac{298 \text{ K}}{12.027 \text{ K}}\right)^2 = 3.6 \times 10^8$$

7.7 The molecular interpretation of equilibrium constants

We can obtain a deeper insight into the origin and significance of the equilibrium constant K by considering the Boltzmann distribution (Foundations 0.10) of molecules over the available states of a system composed of reactants and products. When atoms can exchange partners, as in a reaction, the available states of the system include arrangements in which the atoms are present both in the form of reactants and in the form of products: these arrangements have their characteristic sets of energy levels, but the Boltzmann distribution does not distinguish between their identities, only their energies. The atoms distribute themselves over both sets of energy levels in accord with the Boltzmann distribution (Fig. 7.8).

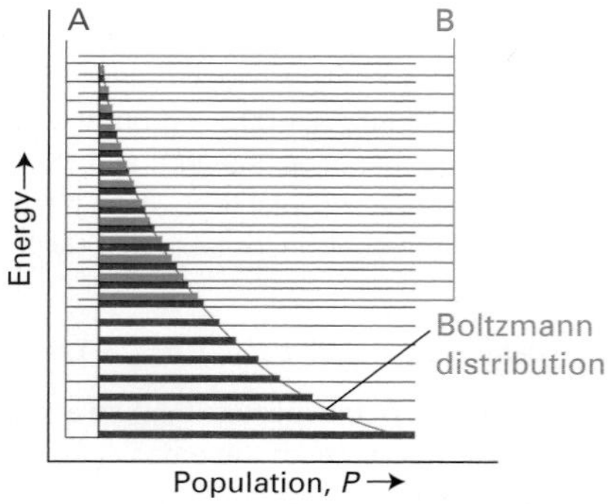

Fig. 7.8 The Boltzmann distribution of populations over the energy levels of two species A and B with similar densities of energy levels; the reaction A → B is endothermic in this example. The bulk of the population is associated with the species A, so that species is dominant at equilibrium.

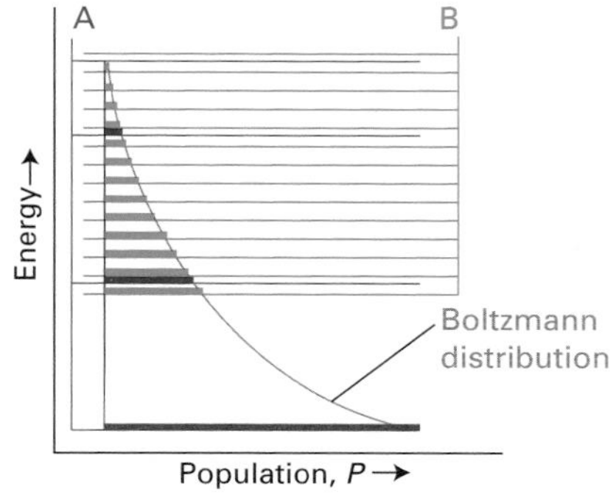

Fig. 7.9 Even though the reaction A → B is endothermic, the density of energy levels in B is so much greater than that in A, that the population associated with B is greater than that associated with A, so B is dominant at equilibrium.

The amount of reactants is the sum of all their populations, and the amount of products likewise the sum of all their populations. At a given temperature, there is a specific distribution of populations, and hence a specific composition of the reaction mixture.

It can be appreciated from Fig. 7.8 that if the reactants and products both have similar arrays of molecular energy levels, then the dominant species in a reaction mixture at equilibrium will be the species with the lower set of energy levels. However, the fact that the equilibrium constant is related to the Gibbs energy is a signal that entropy plays a role as well as energy. Its role can be appreciated by referring to Fig. 7.9. We see that although the B energy levels lie higher than the A energy levels, in this instance they are much more closely spaced. As a result, their total population may be considerable and B could even dominate in the reaction mixture at equilibrium. Closely spaced energy levels correlate with a high entropy, so in this case we see that entropy effects dominate adverse energy effects. That is, a positive reaction enthalpy results in a lowering of the equilibrium constant (that is, an endothermic reaction can be expected to have an equilibrium composition that favours the reactants). However, if there is positive reaction entropy, then the equilibrium composition may favour products, despite the endothermic character of the reaction.

The response of equilibria to the conditions

In introductory chemistry, we meet the empirical rule of thumb known as **Le Chatelier's principle**:

> When a system at equilibrium is subjected to a disturbance, the composition of the system adjusts so as to tend to minimize the effect of the disturbance.

For instance, if a system is compressed, then the equilibrium position can be expected to shift in the direction that leads to a reduction in the number of molecules in the gas phase, for that tends to minimize the effect of compression. Le Chatelier's principle, though, is only a rule of thumb, and to understand why reactions respond as they do, and to calculate the new equilibrium composition, we need to use thermodynamics. We need to keep in mind that some changes in conditions affect the value of $\Delta_r G^\ominus$ and therefore of K (temperature is the only instance) whereas others change the consequences of K having a particular fixed value without changing the value of K (the pressure, for instance).

7.8 The effect of temperature

According to Le Chatelier's principle, we can expect a reaction to respond to a lowering of temperature by releasing heat and to respond to an increase of temperature by absorbing heat. That is:

> When the temperature is raised, the equilibrium composition of an exothermic reaction will tend to shift towards reactants; the equilibrium composition of an endothermic reaction will tend to shift towards products.

In each case, the response tends to minimize the effect of raising the temperature. But *why* do reactions at equilibrium respond in this way? Le Chatelier's principle is only a rule of thumb, and gives no clue to the reason for this behaviour. As we shall now see, the origin of the effect is the dependence of $\Delta_r G^\ominus$, and therefore of K, on the temperature.

First, we consider the effect of temperature on $\Delta_r G^\ominus$. We use the relation $\Delta_r G^\ominus = \Delta_r H^\ominus - T\Delta_r S^\ominus$ and make the assumption that neither the reaction enthalpy nor the reaction entropy varies much with temperature (over small ranges, at least). It follows that

$$\text{Change in } \Delta_r G^\ominus = -(\text{change in } T) \times \Delta_r S^\ominus \quad (7.14)$$

This expression is easy to apply when there is a consumption or formation of gas because, as we have seen (Section 4.5), gas formation dominates the sign of the reaction entropy.

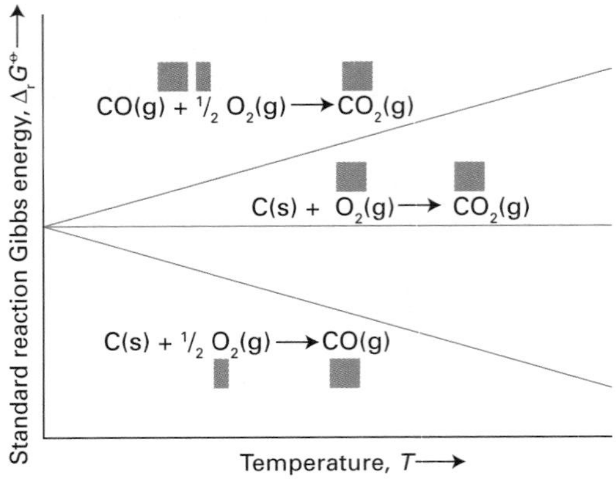

Fig. 7.10 The variation of reaction Gibbs energy with temperature depends on the reaction entropy and therefore on the net production or consumption of gas in a reaction (as indicated by the blue boxes, which show the relative amounts of gas on each side of the equations). The Gibbs energy of a reaction that produces gas decreases with increasing temperature. The Gibbs energy of a reaction that results in a net consumption of gas increases with temperature.

Brief illustration 7.6 The effect of temperature

Consider the three reactions

(i) $\frac{1}{2}$ C(s) + $\frac{1}{2}$ O_2(g) → $\frac{1}{2}$ CO_2(g)

(ii) C(s) + $\frac{1}{2}$ O_2(g) → CO(g)

(iii) CO(g) + $\frac{1}{2}$ O_2(g) → CO_2(g)

all of which are important in the discussion of the extraction of metals from their ores. In reaction (i), the amount of gas is constant, so the reaction entropy is small and $\Delta_r G^{\ominus}$ for this reaction changes only slightly with temperature. Because in reaction (ii) there is a net increase in the amount of gas molecules, from $\frac{1}{2}$ mol to 1 mol, the reaction entropy is large and positive; therefore, $\Delta_r G^{\ominus}$ for this reaction decreases sharply with increasing temperature. In reaction (iii), there is a similar net decrease in the amount of gas molecules, from $\frac{3}{2}$ mol to 1 mol, so $\Delta_r G^{\ominus}$ for this reaction increases sharply with increasing temperature. These remarks are summarized in Fig. 7.10.

Now consider the effect of temperature on K itself. The value of K changes, because $-\Delta_r G^{\ominus}/RT$ changes as T changes. At first, this problem looks troublesome, because both T and $\Delta_r G^{\ominus}$ appear in the expression for K. However, as we show in the following Derivation, the effect of temperature can be expressed very simply as the **van 't Hoff equation:**[1]

$$\ln K' = \ln K + \frac{\Delta_r H^{\ominus}}{R}\left(\frac{1}{T} - \frac{1}{T'}\right) \quad \text{van 't Hoff equation} \quad (7.15)$$

where K is the equilibrium constant at the temperature T and K' is its value when the temperature is T'. All we need to know to calculate the temperature dependence of an equilibrium constant, therefore, is the standard reaction enthalpy.

Derivation 7.2

The van 't Hoff equation

As before, we use the approximation that the standard reaction enthalpy and entropy are independent of temperature over the range of interest, so the entire temperature dependence of $\Delta_r G^{\ominus}$ stems from the T in $\Delta_r G^{\ominus} = \Delta_r H^{\ominus} - T\Delta_r S^{\ominus}$. At a temperature T,

$$\ln K = -\frac{\Delta_r G^{\ominus}}{RT} \overset{\Delta_r G^{\ominus}=\Delta_r H^{\ominus}-T\Delta_r S^{\ominus}}{=} -\frac{\Delta_r H^{\ominus}}{RT} + \frac{\Delta_r S^{\ominus}}{R}$$

At another temperature T', when $\Delta_r G^{\ominus\prime} = \Delta_r H^{\ominus} - T'\Delta_r S^{\ominus}$ and the equilibrium constant is K', a similar expression holds:

$$\ln K' = -\frac{\Delta_r G^{\ominus}}{RT'} \overset{\Delta_r G^{\ominus}=\Delta_r H^{\ominus}-T'\Delta_r S^{\ominus}}{=} -\frac{\Delta_r H^{\ominus}}{RT'} + \frac{\Delta_r S^{\ominus}}{R}$$

The difference between the two (noting that the two terms in blue cancel) is

$$\ln K' - \ln K = \frac{\Delta_r H^{\ominus}}{R}\left(\frac{1}{T} - \frac{1}{T'}\right)$$

which is the van 't Hoff equation, eqn 7.15.

Let's explore the information in the van 't Hoff equation. Consider the case when $T' > T$. Then the term in parentheses in eqn 7.15 is positive. If $\Delta_r H^{\ominus} > 0$, corresponding to an endothermic reaction, the entire term on the right is positive. In this case, therefore, $\ln K' > \ln K$. That being so, we conclude that $K' > K$ for an endothermic reaction. In general:

> The equilibrium constant of an endothermic reaction increases with temperature.

The opposite is true when $\Delta_r H^{\ominus} < 0$, so we can conclude in this case that:

> The equilibrium constant of an exothermic reaction decreases with an increase in temperature.

The Boltzmann distribution gives us insight into these conclusions. The typical arrangement of energy levels for an endothermic reaction is shown in

[1] To distinguish this van 't Hoff equation from the van 't Hoff equation for osmotic pressure, Section 6.8, it is sometimes called the *van 't Hoff isochore*.

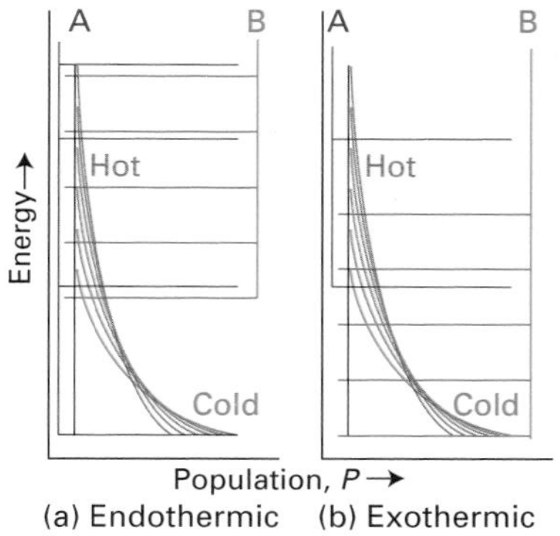

Fig. 7.11 The effect of temperature on a chemical equilibrium can be interpreted in terms of the change in the Boltzmann distribution with temperature and the effect of that change in the population of the species. (a) In an endothermic reaction, the population of B increases at the expense of A as the temperature is raised. (b) In an exothermic reaction, the opposite happens.

Fig. 7.11a. When the temperature is increased, the Boltzmann distribution adjusts and the populations change as shown. The change corresponds to an increased population of the higher energy states at the expense of the population of the lower energy states. We see that the states that arise from the B molecules become more populated at the expense of the A molecules. Therefore, the total population of B states increases, and B becomes more abundant in the equilibrium mixture. Conversely, if the reaction is exothermic (Fig. 7.11b), then an increase in temperature increases the population of the A states (which start at higher energy) at the expense of the B states, so the reactants become more abundant.

The effect of temperature on K is of considerable commercial and environmental significance. For example, the synthesis of ammonia is exothermic, so its equilibrium constant decreases as the temperature is increased; in fact, K falls below 1 when the temperature is raised to above 200 °C. Unfortunately, the reaction is slow at low temperatures and is commercially feasible only if the temperature exceeds about 750 °C even in the presence of a catalyst; but then K is very small. We shall see shortly how Fritz Haber, the inventor of the Haber process for the commercial synthesis of ammonia, was able to overcome this difficulty. Another example is the oxidation of nitrogen:

$$N_2(g) + O_2(g) \rightarrow 2\,NO(g)$$

This reaction is endothermic ($\Delta_r H^{\ominus} = +180\ \mathrm{kJ\ mol^{-1}}$) largely as a consequence of the very high bond enthalpy of N_2, so its equilibrium constant increases with temperature. It is for this reason that nitrogen monoxide (nitric oxide) is formed in significant quantities in the hot exhausts of jet engines and in the hot exhaust manifolds of internal combustion engines, and then goes on to contribute to the problems caused by acid rain.

A further point in this connection is that to use the van 't Hoff equation for the temperature dependence of K_c, we first convert K_c to K by using eqn 7.13 at the temperature to which it applies, use eqn 7.15 to convert K to the new temperature, and then use eqn 7.13 again, but with the new temperature, to convert the new K to K_c. As you might appreciate, on the whole it is better to stick to using K.

7.9 The effect of compression

We have seen that Le Chatelier's principle suggests that the effect of compression (decrease in volume) on a gas-phase reaction at equilibrium is as follows:

> When a system at equilibrium is compressed, the composition of a gas-phase equilibrium adjusts so as to reduce the number of molecules in the gas phase.

For example, in the synthesis of ammonia, reaction B, four reactant molecules give two product molecules, so compression favours the formation of ammonia. Indeed, this is the key to resolving Haber's dilemma, for by working with highly compressed gases he was able to increase the yield of ammonia. Pressure plays an important role in governing the uptake and release of oxygen from oxygen transport and storage proteins.

Brief Illustration 7.7 The effect of compression

For the reaction $2\,NO_2(g) \rightleftharpoons N_2O_4(g)$ the formation of products is accompanied by a decrease in the number of molecules in the gas phase: $\Delta\nu_{gas} = 1 - 2 = -1$. Le Chatelier's principle suggests that if the reaction vessel is compressed, then the equilibrium composition of the mixture adjusts so as to reduce the number of molecules in the gas phase. Therefore, for this reaction, compression favours the formation of products.

Self-test 7.7

Is the formation of products in the reaction $4\,NH_3(g) + 5\,O_2(g) \rightleftharpoons 4\,NO(g) + 6\,H_2O(g)$ favoured by compression or expansion of the reaction vessel?

Answer: expansion

Let's explore the thermodynamic basis of this dependence. First, we note that $\Delta_r G^\ominus$ is defined as the difference between the Gibbs energies of substances in their standard states and therefore at 1 bar. It follows that $\Delta_r G^\ominus$ has the same value whatever the actual pressure used for the reaction. Therefore, because ln K is proportional to $\Delta_r G^\ominus$:

> K is independent of the pressure at which the reaction is carried out.

Thus, if the reaction mixture in which ammonia is being synthesized is compressed isothermally, the equilibrium constant remains unchanged.

This rather startling conclusion should not be misinterpreted. The value of K is independent of the pressure to which the system is subjected, but because partial pressures occur in the expression for K in a rather complicated way, the pressure-independence of K does not mean that the *individual* partial pressures or concentrations are unchanged. Suppose, for example, the volume of the reaction vessel in which the reaction $H_2(g) + I_2(s) \rightarrow 2\ HI(g)$ has reached equilibrium is reduced by a factor of 2 and the system is allowed to reach equilibrium again. If the partial pressures were simply to double (that is, there is no adjustment of composition by further reaction), the equilibrium constant would change from

$$K = \frac{p_{HI}^2}{p_{H_2}} \overset{p_J \to 2p_J}{\frown} \quad \text{to} \quad K' = \frac{(2p_{HI})^2}{2p_{H_2}} = 2K$$

However, we have seen that compression leaves K unchanged. Therefore, the two partial pressures must adjust by different amounts. In this instance, K' will remain equal to K if the partial pressure of HI changes by a factor of less than 2 and the partial pressure of H_2 increases by more than a factor of 2. In other words, the equilibrium composition must shift in the direction of the reactants in order to preserve the equilibrium constant.

We can express this effect quantitatively by expressing the partial pressures in terms of the mole fractions and the total pressure. For the reaction above, we find

$$K = \frac{p_{HI}^2}{p_{H_2}} \overset{p_J = x_J p}{=} \frac{x_{HI}^2 p^2}{x_{H_2} p} = \frac{x_{HI}^2 p}{x_{H_2}}$$

For K to remain constant as the pressure is increases, the ratio of mole fractions must decrease, implying that the proportion of HI in the mixture must increase. Because $x_{HI} + x_{H_2} = 1$, the explicit dependence of the mole fractions can be found by substituting $x_{H_2} = 1 - x_{HI}$, which gives

$$K = \frac{x_{HI}^2 p}{1 - x_{HI}}$$

and solving the resulting quadratic equation

$$\overbrace{p x_{HI}^2}^{ax^2} \overbrace{+ K x_{HI}}^{+bx} \overbrace{-K}^{+c} = 0$$

for x_{HI} by using the formula in The chemist's toolkit 7.1:

$$x_{HI} = \frac{\overbrace{-K}^{-b} \pm \left(\overbrace{K^2}^{b^2} + \overbrace{4Kp}^{-4ac}\right)^{1/2}}{\underbrace{2p}_{2a}} = \frac{K}{2p}\left\{-1 \pm \left(1 + \frac{4p}{K}\right)^{1/2}\right\}$$

Because a mole fraction must be positive, we select the following solution:

$$x_{HI} = \left(\frac{K}{2p}\right)\left\{-1 + \left(1 + \frac{4p}{K}\right)^{1/2}\right\} \qquad (7.16)$$

The dependence of the mole fractions implied by this expression is shown in Fig. 7.12. When the pressure is very low in the sense that $4p/K \ll 1$, we can replace the square root by using the information in The chemist's toolkit 6.1 to write

$$\left(1 + \frac{4p}{K}\right)^{1/2} \overset{(1+x)^{1/2} = 1 + \frac{1}{2}x - \frac{1}{8}x^2 + \cdots}{=} 1 + \frac{2p}{K} - \frac{2p^2}{K^2} + \cdots$$

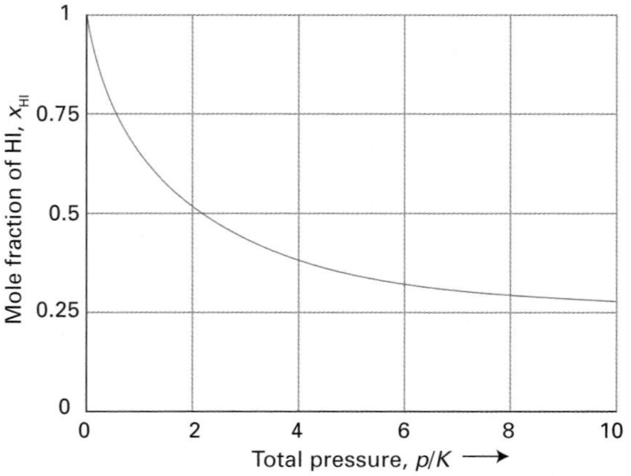

Fig.7.12 The mole fraction of HI molecules in a gas-phase reaction mixture of H_2 and HI as a function of pressure (expressed as $4p/K$); the I_2 is present as a solid throughout.

Then

$$x_{HI} = \frac{K}{2p}\left\{-1+1+\frac{2p}{K}-\frac{2p^2}{K^2}+\cdots\right\} = 1-\frac{p}{K}+\cdots$$

and as p approaches zero, x_{HI} approaches 1 and the equilibrium lies entirely in favour of HI.

A note on good practice When one factor increases and another decreases, always evaluate the limit of an expression in this way: never rely on simply setting a term equal to zero.

Compression has no effect on the composition when the number of gas phase molecules is the same in the reactants as in the products. An example is the synthesis of hydrogen iodide in which all three substances are present in the gas phase and the chemical equation is $H_2(g) + I_2(g) \rightarrow 2\,HI(g)$ in place of $H_2(g) + I_2(s) \rightarrow 2\,HI(g)$.

A more subtle example is the effect of the addition of an inert gas to a reaction mixture contained inside a vessel of constant volume. The overall pressure increases as the gas (such as argon) is added, but the addition of a foreign gas does not affect the *partial* pressures of the other gases present: the partial pressure of a perfect gas (Section 1.3), the pressure a gas would exert if it alone occupied the vessel, is independent of the presence or absence of any other gases. Therefore, under these circumstances, not only does the equilibrium constant remain unchanged, but the partial pressures of the reactants and products remain the same whatever the stoichiometry of the reaction.

7.10 The presence of a catalyst

A catalyst is a substance that accelerates a reaction without itself appearing in the overall chemical equation. Enzymes are biological versions of catalysts. We study the action of catalysts in Section 11.12, and at this stage do not need to know in detail how they work other than that they provide an alternative, faster route from reactants to products.

Although the new route from reactants to products is faster, the initial reactants and the final products are the same. The quantity $\Delta_r G^{\ominus}$ is defined as the difference of the standard molar Gibbs energies of the reactants and products, so it is independent of the path linking the two. It follows that an alternative pathway between reactants and products leaves $\Delta_r G^{\ominus}$ and therefore K unchanged. That is:

> The presence of a catalyst does not change the equilibrium constant of a reaction and has no effect on the equilibrium composition of the reaction mixture.

Impact on biochemistry 7.2

Binding of oxygen to myoglobin and haemoglobin

The protein myoglobin (Mb) stores O_2 in muscle and the protein haemoglobin (Hb) transports O_2 in blood; haemoglobin is composed of four myoglobin-like molecules. In each protein, the O_2 molecule attaches to an iron ion in a haem group, and each myoglobin-like component of haemoglobin responds to the change in shape of the others when O_2 binds to them.

First, consider the equilibrium between Mb and O_2:

$$Mb(aq) + O_2(g) \rightleftharpoons MbO_2(aq) \qquad K = \frac{[MbO_2]}{p[Mb]}$$

where p is the numerical value of the partial pressure (in bar) of O_2 gas. It follows that the **fractional saturation**, s, the fraction of Mb molecules that are oxygenated, is

$$s = \frac{\overbrace{[MbO_2]}^{\text{Concentration of oxygenated Mb}}}{\underbrace{[Mb]+[MbO_2]}_{\text{Total Mb concentration}}} = \frac{\overbrace{[MbO_2]/[Mb]}^{Kp}}{1+\underbrace{[MbO_2]/[Mb]}_{Kp}} = \frac{Kp}{1+Kp}$$

The dependence of s on p is shown in Fig. 7.13.

Now consider the equilibria between Hb and O_2:

$$Hb(aq) + O_2(g) \rightleftharpoons HbO_2(aq) \qquad K_1 = \frac{[HbO_2]}{p[Hb]}$$

$$HbO_2(aq) + O_2(g) \rightleftharpoons Hb(O_2)_2(aq) \qquad K_2 = \frac{[Hb(O_2)_2]}{p[HbO_2]}$$

$$Hb(O_2)_2(aq) + O_2(g) \rightleftharpoons Hb(O_2)_3(aq) \qquad K_3 = \frac{[Hb(O_2)_3]}{p[Hb(O_2)_2]}$$

$$Hb(O_2)_3(aq) + O_2(g) \rightleftharpoons Hb(O_2)_4(aq) \qquad K_4 = \frac{[Hb(O_2)_4]}{p[Hb(O_2)_3]}$$

To develop an expression for s, we express $[Hb(O_2)_2]$ in terms of $[HbO_2]$ by using K_2, then express $[HbO_2]$ in terms of

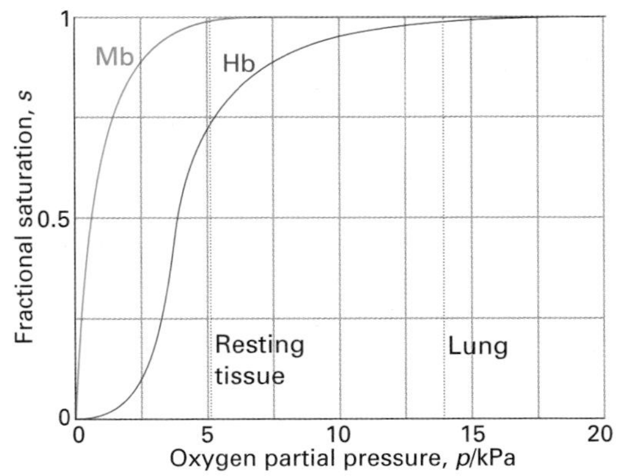

Fig.7.13 The variation of the fractional saturation of myoglobin and haemoglobin molecules with the partial pressure of oxygen. The different shapes of the curves account for the different biological functions of the two proteins.

[Hb] by using K_1, and likewise for all the other concentrations of $Hb(O_2)_3$ and $Hb(O_2)_4$. It follows that

$$[HbO_2] = K_1p[Hb] \qquad [Hb(O_2)_2] = K_1K_2p^2[Hb]$$
$$[Hb(O_2)_3] = K_1K_2K_3p^3[Hb] \qquad [Hb(O_2)_4] = K_1K_2K_3K_4p^4[Hb]$$

Because HbO_2 accounts for one O_2 molecule, $Hb(O_2)_2$ accounts for two, and so on, the total concentration of bound O_2 is

$$[O_2]_{bound} = [HbO_2] + 2[Hb(O_2)_2] + 3[Hb(O_2)_3] + 4[Hb(O_2)_4]$$
$$= (1 + 2K_2p + 3K_2K_3p^2 + 4K_2K_3K_4p^3)K_1p[Hb]$$

and the total concentration of haemoglobin is

$$[Hb]_{total} = (1 + K_1p + K_1K_2p^2 + K_1K_2K_3p^3 + K_1K_2K_3K_4p^4)[Hb]$$

Because each Hb molecule has four sites at which O_2 can attach, the fractional saturation is

$$s = \frac{[O_2]_{bound}}{4[Hb]_{total}}$$
$$= \frac{(1+2K_2p+3K_2K_3p^2+4K_2K_3K_4p^3)K_1p}{4(1+K_1p+K_1K_2p^2+K_1K_2K_3p^3+K_1K_2K_3K_4p^4)}$$

A reasonable fit of the experimental data can be obtained with $K_1 = 0.01$, $K_2 = 0.02$, $K_3 = 0.04$, and $K_4 = 0.08$ when p is expressed in kilopascals (kPa).

The binding of O_2 to haemoglobin is an example of **cooperative binding**, in which the binding of a ligand (in this case O_2) to a biopolymer (in this case Hb) becomes more favourable thermodynamically (that is, the equilibrium constant increases) as the number of bound ligands increases up to the maximum number of binding sites. We see the effect of cooperativity in Fig. 7.13. Unlike the myoglobin saturation curve, the haemoglobin saturation curve is *sigmoidal* (S-shaped): the fractional saturation is small at low ligand concentrations, increases sharply at intermediate ligand concentrations, and then levels off at high ligand concentrations. Cooperative binding of O_2 by haemoglobin is explained by an **allosteric effect**, in which an adjustment of the conformation of a molecule when one substrate binds affects the ease with which a subsequent substrate molecule binds.

The differing shapes of the saturation curves for myoglobin and haemoglobin have important consequences for the way O_2 is made available in the body: in particular, the greater sharpness of the Hb saturation curve means that Hb can load O_2 more fully in the lungs and unload it more fully in different regions of the organism. In the lungs, where $p \approx 14$ kPa, $s \approx 0.98$, representing almost complete saturation. In resting muscular tissue, p is equivalent to about 5 kPa, corresponding to $s \approx 0.75$, implying that sufficient O_2 is still available should a sudden surge of activity take place. If the local partial pressure falls to 3 kPa, s falls to about 0.1. Note that the steepest part of the curve falls in the range of typical tissue oxygen partial pressure. Myoglobin, on the other hand, begins to release O_2 only when p has fallen below about 3 kPa, so it acts as a reserve to be drawn on only when the Hb oxygen has been used up.

Checklist of key concepts

- ☐ 1 The reaction Gibbs energy, Δ_rG, is the slope of a plot of Gibbs energy against composition.
- ☐ 2 The condition of chemical equilibrium at constant temperature and pressure is $\Delta_rG = 0$.
- ☐ 3 The equilibrium constant is the value of the reaction quotient at equilibrium.
- ☐ 4 Reactions for which $\Delta_rG^{\ominus} < 0$ are exergonic; reactions for which $\Delta_rG^{\ominus} > 0$ are endergonic.
- ☐ 5 A compound is thermodynamically stable with respect to its elements if $\Delta_fG^{\ominus} < 0$.
- ☐ 6 The equilibrium constant of a reaction is independent of the presence of a catalysts and independent of the pressure.
- ☐ 7 The variation of an equilibrium constant with temperature is expressed by the van 't Hoff equation.
- ☐ 8 The equilibrium constant K increases with temperature if $\Delta_rH^{\ominus} > 0$ (an endothermic reaction), and decreases if $\Delta_rH^{\ominus} < 0$ (an exothermic reaction).
- ☐ 9 When a system at equilibrium is compressed, the composition of a gas-phase equilibrium adjusts so as to reduce the number of molecules in the gas phase.

Road map of key equations

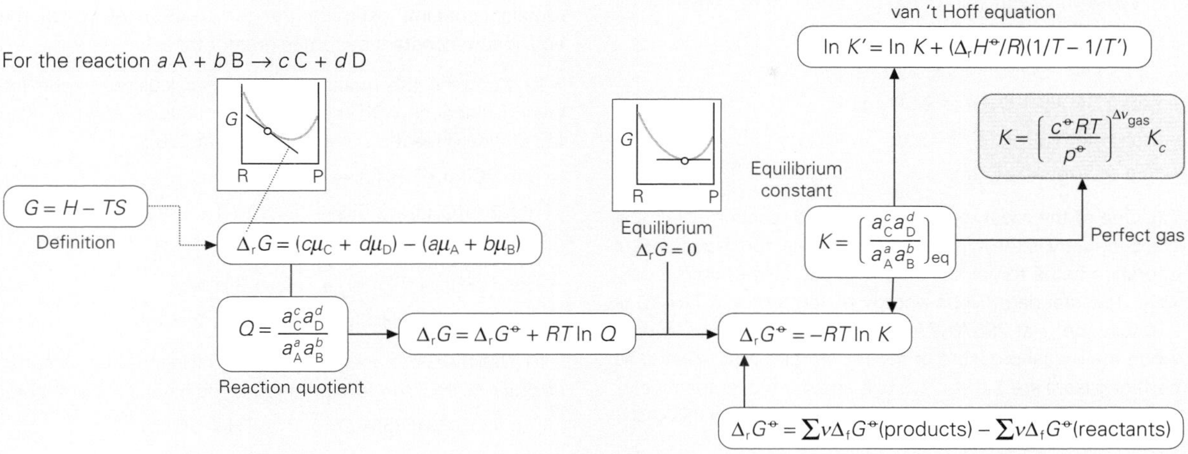

A blue box indicates a relation valid only for perfect gases.

Questions and exercises

Discussion questions

7.1 Explain how the mixing of reactants and products affects the position of chemical equilibrium.

7.2 Explain how a reaction that is not spontaneous may be driven forward by coupling it to a spontaneous reaction.

7.3 State and explain Le Chatelier's principle in terms of thermodynamic quantities. Could there be exceptions to Le Chatelier's principle?

7.4 Suggest how the thermodynamic equilibrium constant and the equilibrium constant expressed in terms of partial pressures may respond differently to changes in pressure and temperature.

7.5 Identify and justify the approximations made in the derivation of the van 't Hoff equation, eqn 7.15.

Exercises

7.1 Write the reaction quotients for the following reactions making the approximation of replacing activities by molar concentrations or partial pressures:

(a) $2\,CH_3COCOOH(aq) + 5\,O_2(g) \rightleftharpoons 6\,CO_2(g) + 4\,H_2O(l)$

(b) $Fe(s) + PbSO_4(aq) \rightleftharpoons FeSO_4(aq) + Pb(s)$

(c) $Hg_2Cl_2(s) + H_2(g) \rightleftharpoons 2\,HCl(aq) + 2\,Hg(l)$

(d) $2\,CuCl(aq) \rightleftharpoons Cu(s) + CuCl_2(aq)$

7.2 Borneol is a pungent compound obtained from the camphorwood tree of Borneo and Sumatra. The standard reaction Gibbs energy of the isomerization of borneol (**5**) to isoborneol (**6**) in the gas phase at 503 K is +9.4 kJ mol^{-1}. Calculate the reaction Gibbs energy in a mixture consisting of 0.15 mol of borneol and 0.30 mol of isoborneol when the total pressure is 600 Torr.

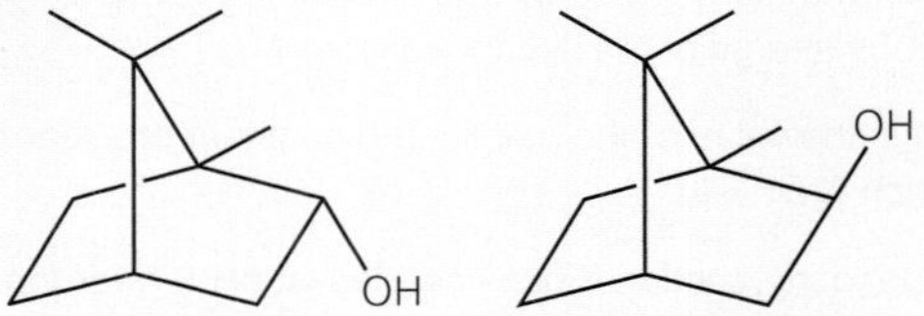

5 d-Borneol **6** d-Isoborneol

7.3 The standard Gibbs energy for the hydrolysis of ATP to ADP given in Impact on biochemistry 7.1 is −30.5 kJ mol^{-1}; what is the Gibbs energy of reaction in an environment at 37 °C in which the ATP, ADP, and P_i concentrations are all (a) 1.0 mmol dm^{-3}, (b) 1.0 μmol dm^{-3}?

7.4 The distribution of Na^+ ions across a typical biological membrane is 10 mmol dm^{-3} inside the cell and 140 mmol dm^{-3} outside the cell. At equilibrium the concentrations are equal. What is the Gibbs energy difference across the membrane at 37 °C?

7.5 Write the expressions for the equilibrium constants of the following reactions:

(a) $CO(g) + Cl_2(g) \rightleftharpoons COCl(g) + Cl(g)$

(b) $2\ SO_2(g) + O_2(g) \rightleftharpoons 2\ SO_3(g)$

(c) $H_2(g) + Br_2(g) \rightleftharpoons 2\ HBr(g)$

(d) $2\ O_3(g) \rightleftharpoons 3\ O_2(g)$

7.6 One of the most extensively studied reactions of industrial chemistry is the synthesis of ammonia, for its successful operation helps to govern the efficiency of the entire economy. The standard Gibbs energy of formation of NH_3(g) is −16.5 kJ mol^{-1} at 298 K. What is the reaction Gibbs energy when the partial pressure of the N_2, H_2, and NH_3 (treated as perfect gases) are 3.0 bar, 1.0 bar, and 4.0 bar, respectively? What is the spontaneous direction of the reaction in this case?

7.7 If the equilibrium constant for the reaction $A + B \rightleftharpoons C$ is reported as 0.432, what would be the equilibrium constant for the reaction written as $C \rightleftharpoons A + B$?

7.8 The equilibrium constant for the reaction $A + B \rightleftharpoons 2\ C$ is reported as 7.2×10^5. What would it be for the reaction written as (a) $2\ A + 2\ B \rightleftharpoons 4\ C$, (b) $\frac{1}{2} A + \frac{1}{2} B \rightleftharpoons C$?

7.9 The equilibrium constant for the isomerization of *cis*-2-butene to *trans*-2-butene is $K = 2.07$ at 400 K. Calculate the standard reaction Gibbs energy for the isomerization.

7.10 The standard reaction Gibbs energy of the isomerization of *cis*-2-pentene to *trans*-2-pentene at 400 K is −3.67 kJ mol^{-1}. Calculate the equilibrium constant of the isomerization.

7.11 One reaction has a standard Gibbs energy of −320 kJ mol^{-1} and a second reaction has a standard Gibbs energy of −55 kJ mol^{-1}. What is the ratio of their equilibrium constants at 300 K?

7.12 One enzyme-catalysed reaction in a biochemical cycle has an equilibrium constant that is 8.4 times the equilibrium constant of a second reaction. If the standard Gibbs energy of the former reaction is −250 kJ mol^{-1}, what is the standard reaction Gibbs energy of the second reaction?

7.13 What is the value of the equilibrium constant of a reaction for which $\Delta_r G^\ominus = 0$?

7.14 The standard reaction Gibbs energies (at pH = 7) for the hydrolysis of glucose-1-phosphate, glucose-6-phosphate, and glucose-3-phosphate are −21, −14, and −9.2 kJ mol^{-1}. Calculate the equilibrium constants for the hydrolyses at 37 °C.

7.15 Use the information in the *Data section* to classify the following compounds as endergonic or exergonic: (a) glucose, (b) methylamine, (c) octane, (d) ethanol.

7.16 Use the information in the *Data section* to estimate the temperature at which (a) $CaCO_3$ decomposes spontaneously and (b) $CuSO_4.5H_2O$ undergoes dehydration.

7.17 The standard reaction enthalpy of $Zn(s) + H_2O(g) \rightarrow ZnO(s) + H_2(g)$ is approximately constant at +224 kJ mol^{-1} from 920 K up to 1280 K. The standard reaction Gibbs energy is +33 kJ mol^{-1} at 1280 K. Assuming that both quantities remain constant, estimate the temperature at which the equilibrium constant becomes greater than 1.

7.18 Combine the reaction entropies calculated in the following reactions with the reaction enthalpies and calculate the standard reaction Gibbs energies at 298 K.

(a) $HCl(g) + NH_3(g) \rightarrow NH_4Cl(s)$

(b) $2\ Al_2O_3(s) + 3\ Si(s) \rightarrow 3\ SiO_2(s) + 4\ Al(s)$

(c) $Fe(s) + H_2S(g) \rightarrow FeS(s) + H_2(g)$

(d) $FeS_2(s) + 2\ H_2(g) \rightarrow Fe(s) + 2\ H_2S(g)$

(e) $2\ H_2O_2(l) + H_2S(g) \rightarrow H_2SO_4(l) + 2\ H_2(g)$

7.19 Use the Gibbs energies of formation in the *Data section* to decide which of the following reactions have $K > 1$ at 298 K.

(a) $2\ CH_3CHO(g) + O_2(g) \rightleftharpoons 2\ CH_3COOH(l)$

(b) $2\ AgCl(s) + Br_2(l) \rightleftharpoons 2\ AgBr(s) + Cl_2(g)$

(c) $Hg(l) + Cl_2(g) \rightleftharpoons HgCl_2(s)$

(d) $Zn(s) + Cu^{2+}(aq) \rightleftharpoons Zn^{2+}(aq) + Cu(s)$

(e) $C_{12}H_{22}O_{11}(s) + 12\ O_2(g) \rightleftharpoons 12\ CO_2(g) + 11\ H_2O(l)$

7.20 The standard enthalpy of combustion of solid phenol, C_6H_5OH, is −3054 kJ mol^{-1} at 298 K and its standard molar entropy is 144.0 J K^{-1} mol^{-1}. Calculate the standard Gibbs energy of formation of phenol at 298 K.

7.21 The standard reaction Gibbs energy for the hydrolysis of ATP to ADP is +10 kJ mol^{-1} at 298 K. What is the biological standard state value?

7.22 The reaction

$$\text{Pyruvate}^-(aq) + NADH(aq) + H^+(aq) \rightarrow \text{lactate}^-(aq) + NAD^+(aq)$$

where NAD^+ is the oxidized form of nicotinamide dinucleotide, occurs in muscle cells deprived of oxygen during strenuous exercise and can lead to cramp. Calculate the standard biological Gibbs energy for the reaction at 310 K give that $\Delta_r G^\ominus = -66.6$ kJ mol^{-1}.

7.23 The standard biological reaction Gibbs energy for the removal of the phosphate group from adenosine monophosphate is −14 kJ mol^{-1} at 298 K. Following Impact on biochemistry 7.1, what is the value of the thermodynamic standard reaction Gibbs energy?

7.24 The overall reaction for the glycolysis reaction is $C_6H_{12}O_6(aq) + 2\ NAD^+(aq) + 2\ ADP(aq) + 2\ P_i^-(aq) + 2\ H_2O(l) \rightarrow 2\ CH_3COCO_2^-(aq) + 2\ NADH(aq) + 2\ ATP(aq) + 2\ H_3O^+(aq)$. For this reaction, $\Delta_r G^\oplus = -80.6$ kJ mol^{-1} at 298 K. What is the value of $\Delta_r G^\ominus$? Use the relation $\Delta_r G = \Delta_r G^\ominus + RT \ln Q$ with the appropriate value of Q for the presence of H_3O^+ at pH = 7.

7.25 The second step in glycolysis is the isomerization of glucose-6-phosphate (G6P to fructose-6-phosphate (F6P). Example 7.2 considered the equilibrium between F6P and G6P. Draw a graph to show how the reaction Gibbs energy varies with the fraction f of F6P in solution. Label the regions of the graph that correspond to the formation of F6P and G6P being spontaneous, respectively.

7.26 The equilibrium constant for the gas-phase isomerization of borneol, $C_{10}H_{17}OH$, to isoborneol (see Exercise 7.2) at 503 K is 0.106. A mixture consisting of 6.70 g of borneol and 12.5 g of isoborneol in a container of volume 5.0 dm^3 is heated to 503 K and allowed to come to equilibrium. Calculate the mole fractions of the two substances at equilibrium.

7.27 Calculate the composition of a system in which nitrogen and hydrogen are mixed at partial pressures of 1.00 bar and 4.00 bar and allowed to reach equilibrium with their product, ammonia, under conditions when $K = 89.8$.

7.28 In a gas-phase equilibrium mixture of $SbCl_5$, $SbCl_3$, and Cl_2 at 500 K, $p_{SbCl_5} = 0.17$ bar and $p_{SbCl_3} = 0.22$ bar. Calculate the equilibrium partial pressure of Cl_2 given that $K = 3.5 \times 10^{-4}$ for the reaction $SbCl_5(g) \rightleftharpoons SbCl_3(g) + Cl_2(g)$.

7.29 The equilibrium constant $K = 0.36$ for the reaction $PCl_5(g) \rightleftharpoons PCl_3(g) + Cl_2(g)$ at 400 K. (a) Given that 1.5 g of PCl_5 was initially placed in a reaction vessel of volume 250 cm^3, determine the molar concentrations in the mixture at equilibrium. (b) What is the percentage of PCl_5 decomposed at 400 K?

7.30 In the Haber process for ammonia, $K = 0.036$ for the reaction $N_2(g) + 3\,H_2(g) \rightleftharpoons 2\,NH_3(g)$ at 500 K. If a reactor is charged with partial pressures of 0.020 bar of N_2 and 0.020 bar of H_2, what will be the equilibrium partial pressure of the components?

7.31 The equilibrium pressure of H_2 over a mixture of solid uranium and solid uranium hydride at 500 K is 1.04 Torr. Calculate the standard Gibbs energy of formation of $UH_3(s)$ at 500 K.

7.32 The equilibrium constant for the reaction $I_2(g) \rightleftharpoons 2\,I(g)$ is 0.26 at 1000 K. What is the corresponding value of K_c?

7.33 The standard reaction Gibbs energy for the reaction $H_2(g) + \tfrac{1}{2}\,O_2(g) \rightarrow H_2O\,(l)$ is -237.13 kJ mol^{-1} at 25 °C. Determine the equilibrium constant in terms of concentration, K_c, at this temperature.

7.34 What is the standard enthalpy of a reaction for which the equilibrium constant is (a) doubled, (b) halved when the temperature is increased by 10 K at 298 K?

7.35 The dissociation vapour pressure (the pressure of gaseous products in equilibrium with the solid reactant) of NH_4Cl at 427 °C is 608 kPa but at 459 °C it has risen to 1115 kPa. Calculate (a) the equilibrium constant, (b) the standard reaction Gibbs energy, (c) the standard enthalpy, (d) the standard entropy of dissociation, all at 427 °C. Assume that the vapour behaves as a perfect gas and that $\Delta H^{\ominus}$ and $\Delta S^{\ominus}$ are independent of temperature in the range given.

7.36 Use the following data on the reaction $H_2(g) + Cl_2(g) \rightleftharpoons 2\,HCl(g)$ to determine the standard reaction enthalpy:

T/K	300	500	1000
K	4.0×10^{31}	4.0×10^{18}	5.1×10^{8}

7.37 The equilibrium constant of the reaction $2\,C_3H_6(g) \rightleftharpoons C_2H_4(g) + C_4H_8(g)$ is found to fit the expression

$$\ln K = -1.04 - \frac{1088\text{ K}}{T} + \frac{1.51\times10^{-2}\text{ K}^2}{T^2}$$

between 300 K and 600 K. Calculate the standard reaction enthalpy and standard reaction entropy at 400 K. *Hint*: Begin by calculating ln K at 390 K and 410 K; then use eqn 7.15 to determine the standard reaction enthalpy. To determine the standard reaction entropy, calculate the values for the standard reaction Gibbs energy at each temperature using eqn 7.8 and then apply eqn 7.11.

7.38 Express the equilibrium constant for $N_2O_4(g) \rightleftharpoons 2\,NO_2(g)$ in terms of the fraction α of N_2O_4 that has dissociated and the total pressure p of the reaction mixture, and show that when the extent of dissociation is small ($\alpha \ll 1$), α is inversely proportional to the square root of the total pressure ($\alpha \propto p^{-1/2}$).

Projects

The symbol ‡ indicates that calculus is required.

7.39‡ Here we explore the van 't Hoff equation in more detail. (a) The bond in molecular iodine is weak, and hot iodine vapour contains a proportion of iodine atoms. When 1.00 g of I_2 is heated to 1000 K in a sealed container of volume 1.00 dm^3, the resulting equilibrium mixture contains 0.830 g of I_2. Calculate K for the dissociation equilibrium $I_2(g) \rightleftharpoons 2\,I(g)$. (b) The thermodynamically exact form of the van 't Hoff equation (eqn 7.15) is $\mathrm{d}(\ln K)/\mathrm{d}T = -\Delta_r H^{\ominus}/RT^2$. Use the data in part (a) to deduce an expression for the temperature dependence of the standard reaction enthalpy for the reaction treated there, and draw a graph to show the variation. (c) The van 't Hoff equation (eqn 7.15) applies to K, not to K_c. Find the corresponding expression for K_c.

7.40 The saturation curves shown in Impact on biochemistry 7.2 may also be modelled mathematically by the equation

$$\log \frac{s}{1-s} = v \log p - v \log K$$

where s is the saturation, p is the partial pressure of O_2 (specifically, $p/p^{\ominus}$), K is a constant (not the binding constant for one ligand), and v is the *Hill coefficient*, which varies from 1, for no cooperativity, to N for all-or-none binding of N ligands ($N = 4$ in Hb). The Hill coefficient for myoglobin is 1, and for haemoglobin it is 2.8. (a) Determine the constant K for both Mb and Hb from the graph of fractional saturation (at $s = 0.5$) and then calculate the fractional saturation of Mb and Hb for the following values of p/kPa: 1.0, 1.5, 2.5, 4.0, 8.0. (b) Use the information from part (a) to calculate the value of s at the same p values assuming that v has the theoretical maximum value of 4.

8

Chemical equilibria: solutions

Proton transfer equilibria 186

8.1 Brønsted–Lowry theory 186

8.2 Protonation and deprotonation 187

8.3 Polyprotic acids 192

8.4 Amphiprotic systems 195

Salts in water 196

8.5 Acid–base titrations 196

8.6 Buffer action 198

8.7 Indicators 200

Solubility equilibria 202

8.8 The solubility constant 202

8.9 The common-ion effect 203

8.10 The effect of added salts on solubility 204

CHECKLIST OF KEY CONCEPTS 205

ROAD MAP OF KEY EQUATIONS 206

QUESTIONS AND EXERCISES 206

In this chapter we examine some consequences of dynamic chemical equilibria. We concentrate on the equilibria that exist in solutions of acids, bases, and their salts in water, where rapid proton transfer between species ensures that equilibrium is maintained at all times. The link between Chapter 7 and the discussion here is that, provided the temperature is held constant, *an equilibrium constant retains its value even though the individual activities may change*. So, if one substance is added to a mixture at equilibrium, the other substances adjust their abundances to restore the value of K.

Proton transfer equilibria

The reaction of acids and bases are central to chemistry and its applications, such as chemical analysis and synthesis. One particularly important application of proton transfer equilibrium is in living cells, for even small drifts in the equilibrium concentration of hydrogen ions can result in disease, cell damage, and death. Throughout this chapter, keep in mind that a free hydrogen ion (H^+, a proton) does not exist in water: it is always attached to a water molecule and exists as H_3O^+, a hydronium ion.

8.1 Brønsted–Lowry theory

According to the **Brønsted–Lowry theory** of acids and bases, an **acid** is a proton donor and a **base** is a proton acceptor.[1] The proton, which in this context means a hydrogen ion, H^+, is highly mobile and acids and bases in water are always in equilibrium with their deprotonated and protonated counterparts and

[1] There are many different definitions of acids and bases, but in this text we discuss only Brønsted–Lowry acids and bases.

hydronium ions (H_3O^+). Thus, an acid HA, such as HCN, immediately establishes the equilibrium

$$HA(aq) + H_2O(l) \rightleftharpoons H_3O^+(aq) + A^-(aq)$$

$$K = \frac{a_{H_3O^+}a_{A^-}}{a_{HA}a_{H_2O}} \qquad (8.1a)$$

A base B (such as NH_3) immediately establishes the equilibrium

$$B(aq) + H_2O(l) \rightleftharpoons BH^+(aq) + OH^-(aq)$$

$$K = \frac{a_{BH^+}a_{OH^-}}{a_B a_{H_2O}} \qquad (8.1b)$$

In these equilibria, A^- is the **conjugate base** of the acid HA and BH^+ is the **conjugate acid** of the base B. Even in the absence of added acids and bases, proton transfer occurs between water molecules and the **autoprotolysis equilibrium**

$$2\,H_2O(l) \rightleftharpoons H_3O^+(aq) + OH^-(aq)$$

$$K = \frac{a_{H_3O^+}a_{OH^-}}{a_{H_2O}^2} \qquad \text{Autoprotolysis equilibrium} \qquad (8.2)$$

is always present. Autoprotolysis is also called *autoionization*.

As will be familiar from introductory chemistry, the hydronium ion concentration is commonly expressed in terms of the pH, which is defined formally as

$$pH = -\log a_{H_3O^+} \qquad \text{Definition} \quad \text{The pH scale} \qquad (8.3)$$

where the logarithm is to base 10 (see The chemist's toolkit 2.2). In elementary work, the hydronium ion activity is replaced by the numerical value of its molar concentration, $[H_3O^+]$, which is equivalent to setting the activity coefficient γ equal to 1 and writing $a_{H_3O^+} = [H_3O^+]/c^{\ominus}$ with $c^{\ominus} = 1\ mol\ dm^{-3}$. However, it should never be forgotten that the replacement of activities by molar concentration is invariably hazardous. Because ions interact over long distances, the replacement is unreliable for all but the most dilute solutions (less than about $10^{-3}\ mol\ dm^{-3}$).

Brief illustration 8.1 The pH scale

If the molar concentration of H_3O^+ is 2.0 mmol dm^{-3} (where 1 mmol = 10^{-3} mol), then

$$pH \approx -\log(2.0 \times 10^{-3}) = 2.70$$

If the molar concentration were ten times less, at 0.20 mmol dm^{-3}, then the pH would be 3.70. Notice that *the higher the pH, the lower the concentration of hydronium ions in the solution* and that a change in pH by 1 unit corresponds to a tenfold change in their molar concentration.

Self-test 8.1

Death is likely if the pH of human blood plasma changes by more than ±0.4 from its normal value of 7.4. What is the approximate range of molar concentrations of hydrogen ions for which life can be sustained?

Answer: 16 nmol dm^{-3} to 100 nmol dm^{-3} (1 nmol = 10^{-9} mol)

8.2 Protonation and deprotonation

All the solutions we consider in this chapter are so dilute that we can regard the water present as being a nearly pure liquid and therefore as having unit activity (see Table 6.2). When we set $a_{H_2O} = 1$ for all the solutions we consider, the resulting equilibrium constant is called the **acidity constant,** K_a, of the acid HA:

$$K_a = \frac{a_{H_3O^+}a_{A^-}}{a_{HA}} \approx \frac{([H_3O^+]/c^{\ominus})([A^-]/c^{\ominus})}{[HA]/c^{\ominus}} \qquad \text{Definition} \quad \text{Acidity constant} \qquad (8.4a)$$

This rather cumbersome expression is normally written

$$K_a = \frac{[H_3O^+][A^-]}{[HA]} \qquad (8.4b)$$

with '[J]' interpreted as $[J]/c^{\ominus}$ (that is, as the numerical value of the molar concentration of J with the units mol dm^{-3} struck out). Acidity constants are also called *acid ionization constants* and, less appropriately (because deprotonation is not a simple fragmentation into atoms), *dissociation constants*. Data are widely reported in terms of the negative common logarithm of this quantity:

$$pK_a = -\log K_a \qquad (8.5)$$

It follows from eqn 7.8, in the form $\ln K = -\Delta_r G^{\ominus}/RT$, that pK_a is proportional to $\Delta_r G^{\ominus}$ for the proton transfer reaction (see Exercise 8.8). Therefore, manipulations of pK_a and related quantities are actually manipulations of standard reaction Gibbs energies in disguise. Note that if $\Delta_r G^{\ominus} > 0$, corresponding to a deprotonation equilibrium lying strongly in favour of reactants (the original acid molecules), then $pK_a > 0$ too.

The value of the acidity constant indicates the extent to which proton transfer occurs at equilibrium in aqueous solution. The smaller the value of K_a, and therefore the larger the value of pK_a, the lower is the concentration of deprotonated molecules. In short, the higher the value of pK_a, the weaker the acid. Most acids have $K_a < 1$ (and usually much less than 1), with $pK_a > 0$, indicating only a small extent of deprotonation in water. These acids are classified as **weak acids**. A few acids, most notably, in aqueous solution, HCl, HBr, HI, HNO_3, H_2SO_4, and $HClO_4$, are classified as **strong acids**, and are commonly regarded as being completely deprotonated in aqueous solution. Sulfuric acid, H_2SO_4, is strong with respect only to its first deprotonation; the second deprotonation (the deprotonation of HSO_4^-), is weak.

The corresponding expression for a base is called the **basicity constant**, K_b:

$$K_b = \frac{a_{BH^+}a_{OH^-}}{a_B} \approx \frac{[BH^+][OH^-]}{[B]}$$
$$pK_b = -\log K_b \qquad \text{Definition} \quad \text{Basicity constant} \qquad (8.6)$$

where we have used the same convention for interpreting '[J]' as in eqn 8.5. A **strong base** is fully protonated in solution in the sense that $K_b > 1$. One example is the oxide ion, O^{2-}, which cannot survive in water but is immediately and fully converted into its conjugate acid OH^-. A **weak base** is not fully protonated in water in the sense that $K_b < 1$ (and usually much less than 1). Ammonia, NH_3, and its organic derivatives the amines are all weak bases in water, and only a small proportion of their molecules exist as the conjugate acid (NH_4^+ or RNH_3^+).

The **autoprotolysis constant** for water, K_w, is the equilibrium constant in eqn 8.2 with the activity of water set equal to 1:

$$K_w = a_{H_3O^+}a_{OH^-}$$
$$pK_w = -\log K_w \qquad \text{Definition} \quad \text{Autoprotolysis constant} \qquad (8.7)$$

At 25 °C, the only temperature we consider in this chapter, $K_w = 1.0 \times 10^{-14}$ and $pK_w = 14.00$. As may be confirmed by multiplying the two constants together, the acidity constant of the conjugate acid, BH^+, of a base B (the equilibrium constant for the reaction $BH^+ + H_2O \rightleftharpoons H_3O^+ + B$) is related to the basicity constant of B (the equilibrium constant for the reaction $B + H_2O \rightleftharpoons BH^+ + OH^-$) by

$$K_aK_b = \frac{a_{H_3O^+}a_B}{a_{BH^+}} \times \frac{a_{BH^+}a_{OH^-}}{a_B} \overset{\text{cancel highlighted terms}}{=} a_{H_3O^+}a_{OH^-} = K_w \qquad (8.8a)$$

The implication of this relation is that K_a increases as K_b decreases to maintain a product equal to the constant K_w. That is, *as the strength of a base decreases, the strength of its conjugate acid increases*, and vice versa. On taking the common logarithm of both sides of eqn 8.8a, we obtain (by using the expressions given in The chemist's toolkit 2.2)

$$\log K_aK_b \overset{\log xy = \log x + \log y}{=} \underbrace{\log K_a}_{-pK_a} + \underbrace{\log K_b}_{-pK_b} = \underbrace{\log K_w}_{-pK_w}$$

and therefore

$$pK_a + pK_b = pK_w \qquad \text{Relation between } pK_a \text{ and } pK_b \qquad (8.8b)$$

The great advantage of this relation is that the pK_b values of bases may be expressed as the pK_a of their conjugate acids, so the strengths of all weak acids and bases may be listed in a single table (Table 8.1).

Brief illustration 8.2 Acidity and basicity constants

If the acidity constant of the conjugate acid ($CH_3NH_3^+$) of the base methylamine (CH_3NH_2) is reported as $pK_a = 10.56$,

$$CH_3NH_3^+(aq) + H_2O(l) \rightleftharpoons H_3O^+(aq) + CH_3NH_2(aq)$$
$$pK_a = 10.56$$

we can infer that the basicity constant of methylamine itself, the equilibrium constant for

$$CH_3NH_2(aq) + H_2O(l) \rightleftharpoons CH_3NH_3^+(aq) + OH^-(aq)$$

is

$$pK_b = pK_w - pK_a = 14.00 - 10.56 = 3.44$$

Another useful relation is obtained by taking the common logarithm of both sides of the definition of K_w in eqn 8.7, which gives

$$\log a_{H_3O^+}a_{OH^-} \overset{\log xy = \log x + \log y}{=} \underbrace{\log a_{H_3O^+}}_{-pH} + \underbrace{\log a_{OH^-}}_{-pOH} = \underbrace{\log K_w}_{-pK_w}$$

and therefore

$$pH + pOH = pK_w \qquad \text{Relation between pH and pOH} \qquad (8.9)$$

where $pOH = \log a_{OH^-}$. This enormously important relation means that the activities (in elementary work, the molar concentrations) of hydronium and hydroxide ions are related by a seesaw relation: as one goes up, the other goes down to preserve the value of pK_w.

The extent of deprotonation of a weak acid in solution depends on the acidity constant and the initial concentration of the acid, its concentration as prepared. The **fraction deprotonated**, $f_{\text{deprotonated}}$, the fraction of acid molecules HA that have donated a proton, is

Table 8.1
Acidity and basicity constants at 298.15 K*

Acid/Base	K_b	pK_b	K_a	pK_a
Strongest weak acids				
Trichloroacetic acid, CCl_3COOH	3.3×10^{-14}	13.48	3.0×10^{-1}	0.52
Benzenesulfonic acid, $C_6H_5SO_3H$	5.0×10^{-14}	13.30	2×10^{-1}	0.70
Iodic acid, HIO_3	5.9×10^{-14}	13.23	1.7×10^{-1}	0.77
Sulfurous acid, H_2SO_3	6.3×10^{-13}	12.19	1.6×10^{-2}	1.81
Chlorous acid, $HClO_2$	1.0×10^{-12}	12.00	1.0×10^{-2}	2.00
Phosphoric acid, H_3PO_4	1.3×10^{-12}	11.88	7.6×10^{-3}	2.12
Chloroacetic acid, $CH_2ClCOOH$	7.1×10^{-12}	11.15	1.4×10^{-3}	2.85
Lactic acid, $CH_3CH(OH)COOH$	1.2×10^{-11}	10.92	8.4×10^{-4}	3.08
Nitrous acid, HNO_2	2.3×10^{-11}	10.63	4.3×10^{-4}	3.37
Hydrofluoric acid, HF	2.9×10^{-11}	10.55	3.5×10^{-4}	3.45
Formic acid, HCOOH	5.6×10^{-11}	10.25	1.8×10^{-4}	3.75
Benzoic acid, C_6H_5COOH	1.5×10^{-10}	9.81	6.5×10^{-5}	4.19
Acetic acid, CH_3COOH	5.6×10^{-10}	9.25	1.8×10^{-5}	4.75
Carbonic acid, H_2CO_3	2.3×10^{-8}	7.63	4.3×10^{-7}	6.37
Hypochlorous acid, HClO	3.3×10^{-7}	6.47	3.0×10^{-8}	7.53
Hypobromous acid, HBrO	5.0×10^{-6}	5.31	2.0×10^{-9}	8.69
Boric acid, $B(OH)_3H$†	1.4×10^{-5}	4.86	7.2×10^{-10}	9.14
Hydrocyanic acid, HCN	2.0×10^{-5}	4.69	4.9×10^{-10}	9.31
Phenol, C_6H_5OH	7.7×10^{-5}	4.11	1.3×10^{-10}	9.89
Hypoiodous acid, HIO	4.3×10^{-4}	3.36	2.3×10^{-11}	10.64
Weakest weak acids				
Weakest weak bases				
Urea, $CO(NH_2)_2$	1.3×10^{-14}	13.90	7.7×10^{-1}	0.10
Aniline, $C_6H_5NH_2$	4.3×10^{-10}	9.37	2.3×10^{-5}	4.63
Pyridine, C_5H_5N	1.8×10^{-9}	8.75	5.6×10^{-6}	5.25
Hydroxylamine, NH_2OH	1.1×10^{-8}	7.97	9.1×10^{-7}	6.03
Nicotine, $C_{10}H_{11}N_2$	1.0×10^{-6}	5.98	1.0×10^{-8}	8.02
Morphine, $C_{17}H_{19}O_3N$	1.6×10^{-6}	5.79	6.3×10^{-9}	8.21
Hydrazine, NH_2NH_2	1.7×10^{-6}	5.77	5.9×10^{-9}	8.23
Ammonia, NH_3	1.8×10^{-5}	4.75	5.6×10^{-10}	9.25
Trimethylamine, $(CH_3)_3N$	6.5×10^{-5}	4.19	1.5×10^{-10}	9.81
Methylamine, CH_3NH_2	3.6×10^{-4}	3.44	2.8×10^{-11}	10.56
Dimethylamine, $(CH_3)_2NH$	5.4×10^{-4}	3.27	1.9×10^{-11}	10.73
Ethylamine, $C_2H_5NH_2$	6.5×10^{-4}	3.19	1.5×10^{-11}	10.81
Triethylamine, $(C_2H_5)_3N$	1.0×10^{-3}	2.99	1.0×10^{-11}	11.01
Strongest weak bases				

*Values for polyprotic acids—those capable of donating more than one proton—refer to the first deprotonation.

†The proton transfer equilibrium is $B(OH)_3(aq) + 2\,H_2O(l) \rightleftharpoons H_3O^+(aq) + B(OH)_4^-(aq)$

Fraction deprotonated

$$= \frac{\text{equilibrium molar concentration of conjugate base}}{\text{initial concentration of acid}}$$

$$f_{\text{deprotonated}} = \frac{[A^-]_{\text{equilibrium}}}{[HA]_{\text{as prepared}}} \quad \text{Fraction of deprotonated acid} \quad (8.10a)$$

The extent to which a weak base B is protonated is reported in terms of $f_{\text{protonated}}$, the **fraction protonated:**

Fraction protonated

$$= \frac{\text{equilibrium molar concentration of conjugate acid}}{\text{initial concentration of base}}$$

$$f_{\text{protonated}} = \frac{[BH^+]_{\text{equilibrium}}}{[B]_{\text{as prepared}}} \quad \text{Fraction of protonated acid} \quad (8.10b)$$

We can estimate the pH of a solution of a weak acid or a weak base and calculate either of these fractions by using the equilibrium-table technique described in Section 7.5.

Example 8.1

Assessing the extent of deprotonation of a weak acid

Estimate the pH and the fraction of CH_3COOH molecules deprotonated in 0.15 M $CH_3COOH(aq)$.

Strategy The aim is to calculate the equilibrium composition of the solution. To do so, we use the technique illustrated in Example 7.3, with x the change in molar concentration of H_3O^+ ions required to reach equilibrium. Ignore the tiny concentration of hydronium ions present in pure water. Once x has been found, calculate pH = −log x. Because we can anticipate that the extent of deprotonation is small (the acid is weak), use the approximation that x is very small to simplify the equations.

Solution We draw up the following equilibrium table:

Species:	CH_3COOH	H_3O^+	$CH_3CO_2^-$
Initial concentration/(mol dm^{-3})	0.15	0	0
Change to reach equilibrium/(mol dm^{-3})	$-x$	$+x$	$+x$
Equilibrium concentration/(mol dm^{-3})	$0.15-x$	x	x

The value of x is found by inserting the equilibrium concentrations into the expression for the acidity constant:

$$K_a = \frac{\overbrace{[H_3O^+]}^{x}\overbrace{[CH_3CO_2^-]}^{x}}{\underbrace{[CH_3COOH]}_{0.15-x}} = \frac{x \times x}{0.15-x}$$

> **A note on good practice** Acetic acid (ethanoic acid) is written CH_3COOH because the two O atoms are inequivalent; its conjugate base, the acetate ion (ethanoate ion) is written $CH_3CO_2^-$ because the two O atoms are now equivalent (by resonance).

We could arrange the expression into a quadratic equation and use the solution in Example 7.3, as explained in The chemist's toolkit 7.1. However, it is more instructive to make use of the smallness of x to replace $0.15 - x$ by 0.15 (this approximation is valid if $x \ll 0.15$, which is likely because the acid is weak, but should be verified at the end of the calculation once x has been calculated). Then the simplified equation $K_a = x^2/0.15$ rearranges first to $0.15 \times K_a = x^2$ and then to

$$x = (0.15 \times K_a)^{1/2} = (0.15 \times 1.8 \times 10^{-5})^{1/2} = 1.6 \times 10^{-3}$$

where we have taken the value for the acidity constant, K_a, from Table 8.1. Therefore,

$$\text{pH} = -\log(1.6 \times 10^{-3}) = 2.80$$

Calculations of this kind are rarely accurate to more than one decimal place in the pH (and even that may be too optimistic) because the effects of ion–ion interactions have been ignored, so this answer would be reported as pH = 2.8. The fraction deprotonated, $f_{\text{deprotonated}}$, is

$$f_{\text{deprotonated}} = \frac{[CH_3CO_2^-]_{\text{equilibrium}}}{[CH_3COOH]_{\text{as prepared}}} = \frac{x}{0.15} = \frac{1.6\times10^{-3}}{0.15} = 0.011$$

That is, only 1.1 per cent of the acetic acid molecules have donated a proton.

> **A note on good practice** When an approximation has been assumed, verify at the end of the calculation that the approximation is consistent with the result obtained. In this case, we assumed that $x \ll 0.15$ and have found that $x = 0.011$, which is consistent.

Self-test 8.2

Estimate the pH of 0.010 M $CH_3CH(OH)COOH(aq)$ (lactic acid) from the data in Table 8.1. Before carrying out the numerical calculation, decide whether you expect the pH to be higher or lower than that calculated for the same concentration of acetic acid.

Answer: 2.5

Example 8.2

Estimating the pH of a dilute solution of a weak acid

Estimate the pH of 1.5×10^{-4} M $CH_3COOH(aq)$, being careful to treat this solution as dilute, and not amenable to the approximations used in Example 8.1.

Strategy Proceed as outlined in Example 8.1, but because the solution is much more dilute, and consequently the extent of deprotonation may not be small (even though the acid is weak), be prepared to solve the quadratic equation that results from manipulation of the expression for K_a exactly (see The chemist's toolkit 7.1).

Solution As in Example 8.1, we draw up the following equilibrium table:

Species:	CH_3COOH	H_3O^+	$CH_3CO_2^-$
Initial concentration/(mol dm^{-3})	1.5×10^{-4}	0	0
Change to reach equilibrium/(mol dm^{-3})	$-x$	$+x$	$+x$
Equilibrium concentration/(mol dm^{-3})	$1.5\times10^{-4}-x$	x	x

Had we proceeded as in Example 8.1, we would calculate $x = 5.2\times10^{-5}$, which although less than the initial concentration is not much less. Therefore, we must solve the quadratic equation

$$\overset{a}{1}x^2+\overbrace{K_a}^{b}x\overbrace{-(1.5\times10^{-4})K_a}^{c}=0$$

with, from Table 8.1, $K_a = 1.8\times10^{-5}$. Therefore, from the expression in The chemist's toolkit 7.1,

$$x=\frac{\overbrace{-1.8\times10^{-5}}^{b=K_a}\pm\left\{\overbrace{(-1.8\times10^{-5})^2}^{b^2}-\overbrace{4\overbrace{(1)}^{a}\times\overbrace{(-1.5\times10^{-4}\times\underbrace{1.8\times10^{-5}}_{K_a})}^{c}}^{4ac}\right\}^{1/2}}{2\times\underset{a}{1}}$$

$= 4.4\times10^{-5}$ or -6.2×10^{-5}

Because x is equal to the concentration of H_3O^+, it cannot be negative, so we select $x = 4.4\times10^{-5}$. It follows that pH = 4.4. (The illegal calculation, which would have assumed from the outset that $x \ll 1.5\times10^{-4}$, would have given 4.3.)

Self-test 8.3

Estimate the pH of 1.5×10^{-4} M $CH_3CH(OH)COOH(aq)$ (lactic acid) from the data in Table 8.1.

Answer: 3.9

Example 8.3

Assessing the extent of protonation of a weak base

The conjugate acid of the base quinoline (**1**) has $pK_a = 4.88$. Estimate the pH and the fraction of molecules protonated in a 0.010 M aqueous solution of quinoline.

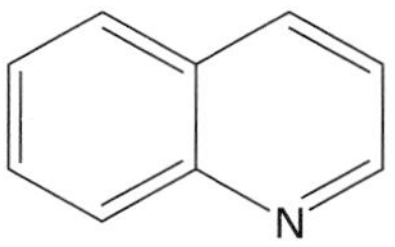

1 Quinoline

Strategy The calculation of the pH of a solution of a base involves one more step than that for the pH of a solution of an acid. The first step is to calculate the concentration of OH^- ions in the solution by using the equilibrium-table technique, and to express it as the pOH of the solution. The additional step is to convert that pOH into a pH by using the water autoprotolysis equilibrium, eqn 8.2, in the form $pH = pK_w - pOH$, with $pK_w = 14.00$ at 25 °C. We also need to compute $pK_b = pK_w - pK_a$.

Solution First, write

$pK_b = 14.00 - 4.88 = 9.12$, corresponding to

$K_b = 10^{-9.12} = 7.6\times10^{-10}$

Now draw up the following equilibrium table, denoting quinoline by Q and its conjugate acid by QH^+:

Species:	Q	OH^-	QH^+
Initial concentration/(mol dm^{-3})	0.010	0	0
Change to reach equilibrium/(mol dm^{-3})	$-x$	$+x$	$+x$
Equilibrium concentration/(mol dm^{-3})	$0.010-x$	x	x

The value of x is found by inserting the equilibrium concentrations into the expression for the basicity constant:

$$K_b=\frac{\overbrace{[OH^-]}^{x}\overbrace{[QH^+]}^{x}}{\underbrace{[Q]}_{0.010-x}}=\frac{x\times x}{\underbrace{0.010-x}_{\approx0.010}}\approx\frac{x^2}{0.010}$$

We have supposed that $x \ll 0.010$. Then the simplified equation $K_b = x^2/0.010$ rearranges to

$x = (0.010\times K_b)^{1/2} = (0.010\times7.6\times10^{-10})^{1/2} = 2.8\times10^{-6}$

This value is consistent with the assumption that $x \ll 0.010$. Therefore,

$$pOH=-\log\overbrace{[OH^-]}^{x}=-\log(2.8\times10^{-6})=5.55$$

and consequently pH = 14.00 − 5.55 = 8.45, or about 8.4. The fraction protonated, $f_{protonated}$, is

$$f_{protonated}=\frac{[QH^+]_{equilibrium}}{[Q]_{as\ prepared}}=\frac{x}{0.010}=\frac{2.8\times10^{-6}}{0.010}=2.8\times10^{-4}$$

or 1 molecule in about 3500.

Self-test 8.4

The pK_a for the first protonation of nicotine (**2**) is 8.02. What is the pH and the fraction of molecules protonated in a 0.015 M aqueous solution of nicotine?

Answer: 10.1; 1/120

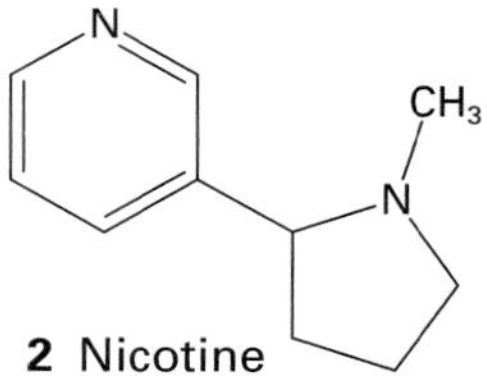

2 Nicotine

8.3 Polyprotic acids

A **polyprotic acid** is a molecular compound that can donate more than one proton. Two examples are sulfuric acid, H_2SO_4, which can donate up to two protons, and phosphoric acid, H_3PO_4, which can donate up to three. A polyprotic acid is best considered to be a molecular species that can give rise to a series of Brønsted acids as it donates a succession of protons. Thus, sulfuric acid is the parent of two Brønsted acids, H_2SO_4 itself and HSO_4^-, and phosphoric acid is the parent of three Brønsted acids, namely H_3PO_4, $H_2PO_4^-$, and HPO_4^{2-}.

For a species H_2A with two acidic protons (such as H_2SO_4), the successive equilibria we need to consider are

$$H_2A(aq) + H_2O(l) \rightleftharpoons H_3O^+(aq) + HA^-(aq)$$

$$K_{a1} = \frac{a_{H_3O^+}a_{HA^-}}{a_{H_2A}}$$

$$HA^-(aq) + H_2O(l) \rightleftharpoons H_3O^+(aq) + A^{2-}(aq)$$

$$K_{a2} = \frac{a_{H_3O^+}a_{A^{2-}}}{a_{HA^-}}$$

In the first of these equilibria, HA^- is the conjugate base of H_2A. In the second, HA^- acts as the acid and A^{2-} is its conjugate base. Values of successive acidity constants of polyprotic acids are given in Table 8.2. In all cases, K_{a2} is smaller than K_{a1}, typically by three orders of magnitude for small molecular species, because the second proton is more difficult to remove, partly on account of the negative charge on HA^-, which attracts the positive charge of H^+ and hinders its escape. Enzymes are polyprotic acids, for they possess many protons that can be donated to a substrate molecule or to the surrounding aqueous medium of the cell. For them, successive acidity constants vary much less because the molecules are so large that the loss of a proton from one part of the molecule has little effect on the ease with which another some distance away may be lost.

Example 8.4

Calculating the concentration of carbonate ion in carbonic acid

Groundwater contains dissolved carbon dioxide, carbonic acid, hydrogencarbonate ions, and a very low concentration of carbonate ions. Estimate the molar concentration of CO_3^{2-} ions in a solution in which water and CO_2(g) are in equilibrium.

Strategy We must be very cautious in the interpretation of calculations involving carbonic acid because equilibrium between dissolved CO_2 and H_2CO_3 is achieved only very slowly. In organisms, attainment of equilibrium is facilitated by the enzyme carbonic anhydrase. We start with the equilibrium that produces the ion of interest (such as A^{2-}) and write its activity in terms of the acidity constant for its formation (K_{a2}). That expression will contain the activity of the conjugate acid (HA^-), which we can express in terms of the activity of *its* conjugate acid (H_2A) by using the appropriate acidity constant (K_{a1}). This equilibrium dominates all the rest provided the molecule is small and there are marked differences between its acidity constants, so it may be possible to make an approximation at this stage.

Solution The CO_3^{2-} ion, the conjugate base of the acid HCO_3^-, is produced in the equilibrium

$$HCO_3^-(aq) + H_2O(l) \rightleftharpoons H_3O^+(aq) + CO_3^{2-}(aq)$$

$$K_{a2} = \frac{a_{H_3O^+}a_{CO_3^{2-}}}{a_{HCO_3^-}}$$

Hence,

$$a_{CO_3^{2-}} = \frac{a_{HCO_3^-}K_{a2}}{a_{H_3O^+}}$$

Table 8.2

Successive acidity constants of polyprotic acids at 298.15 K

Acid	K_{a1}	pK_{a1}	K_{a2}	pK_{a2}	K_{a3}	pK_{a3}
Carbonic acid, H_2CO_3	4.3×10^{-7}	6.37	5.6×10^{-11}	10.25		
Hydrosulfuric acid, H_2S	1.3×10^{-7}	6.88	7.1×10^{-15}	14.15		
Oxalic acid, $(COOH)_2$	5.9×10^{-2}	1.23	6.5×10^{-5}	4.19		
Phosphoric acid, H_3PO_4	7.6×10^{-3}	2.12	6.2×10^{-8}	7.21	2.1×10^{-13}	12.67
Phosphorous acid, H_2PO_3	1.0×10^{-2}	2.00	2.6×10^{-7}	6.59		
Sulfuric acid, H_2SO_4	Strong		1.2×10^{-2}	1.92		
Sulfurous acid, H_2SO_3	1.5×10^{-2}	1.81	1.2×10^{-7}	6.91		
Tartaric acid, $C_2H_4O_2(COOH)_2$	6.0×10^{-4}	3.22	1.5×10^{-5}	4.82		

The HCO_3^- ions are produced in the equilibrium

$$H_2CO_3(aq) + H_2O(l) \rightleftharpoons H_3O^+(aq) + HCO_3^-(aq)$$

One H_3O^+ ion is produced for each HCO_3^- ion produced. These two concentrations are not exactly the same, because a little HCO_3^- is lost in the second deprotonation and the amount of H_3O^+ has been increased by it. Also, HCO_3^- is a weak base and abstracts a proton from water to generate H_2CO_3. However, those secondary changes can safely be ignored in an approximate calculation. Because the molar concentrations of HCO_3^- and H_3O^+ are approximately the same, we can suppose that their activities are also approximately the same, and set $a_{HCO_3^-} \approx a_{H_3O^+}$, which implies that $a_{CO_3^{2-}} \approx K_{a2}$. Then, when we make the approximation that $a_{CO_3^{2-}} = [CO_3^{2-}]/c^{\ominus}$, we obtain

$$[CO_3^{2-}] \approx K_{a2}c^{\ominus}$$

Because we know from Table 8.2 that $pK_{a2} = 10.25$, it follows that $[CO_3^{2-}] = 5.6 \times 10^{-11}c^{\ominus}$, and therefore, if equilibrium has in fact been achieved, that the molar concentration of CO_3^{2-} ions is 5.6×10^{-11} mol dm^{-3} and (within the approximations we have made) independent of the concentration of H_2CO_3 present initially.

Self-test 8.5

Calculate the molar concentration of S^{2-} ions in $H_2S(aq)$.

Answer: 7.1×10^{-15} mol dm^{-3}

Example 8.5

Calculating the fractional composition of a solution of a diprotic acid

Oxalic acid, $H_2C_2O_4$ (ethanedioic acid, HOOC–COOH), exists in solution in equilibrium with $HC_2O_4^-$ and $C_2O_4^{2-}$. Show how the composition of an aqueous solution that contains 0.010 mol dm^{-3} of oxalic acid varies with pH.

Strategy We expect the fully protonated species ($H_2C_2O_4$) at low pH, the partially deprotonated species ($HC_2O_4^-$) at intermediate pH, and the fully deprotonated species ($C_2O_4^{2-}$) at high pH. Set up the expressions for the two acidity constants, treating $H_2C_2O_4$ as the parent acid, and an expression for the total concentration of oxalic acid. Solve the resulting expressions for the fraction of each species in terms of the hydronium ion concentration.

Solution The two acidity constants are

$$H_2C_2O_4(aq) + H_2O(l) \rightleftharpoons H_3O^+(aq) + HC_2O_4^-(aq)$$

$$K_{a1} = \frac{[H_3O^+][HC_2O_4^-]}{[H_2C_2O_4]}$$

$$HC_2O_4^-(aq) + H_2O(l) \rightleftharpoons H_3O^+(aq) + C_2O_4^{2-}(aq)$$

$$K_{a2} = \frac{[H_3O^+][C_2O_4^{2-}]}{[HC_2O_4^-]}$$

We also know that the total concentration of oxalic acid in all its forms, its initial concentration $O = [H_2C_2O_4]_{\text{as prepared}}$, is

$$[H_2C_2O_4] + [HC_2O_4^-] + [C_2O_4^{2-}] = O$$

We now have three equations for three unknown concentrations: $[H_2C_2O_4]$, $[HC_2O_4^-]$, and $[C_2O_4^{2-}]$. To solve the equations, we proceed systematically first by using K_{a1} to express $[HC_2O_4^-]$ in terms of $[H_2C_2O_4]$:

$$[HC_2O_4^-] = \frac{K_{a1}[H_2C_2O_4]}{[H_3O^+]}$$

Then, use K_{a2} to express $[C_2O_4^{2-}]$ in terms of $[HC_2O_4^-]$:

$$[C_2O_4^{2-}] \overbrace{=}^{\text{expression for } K_{a2}} \frac{K_{a2}[HC_2O_4^-]}{[H_3O^+]} \overbrace{=}^{\text{expression for } [HC_2O_4^-]} \frac{K_{a1}K_{a2}[H_2C_2O_4]}{[H_3O^+]^2}$$

The expression for the total concentration O can now be written in terms of $[H_2C_2O_4]$ and $[H_3O^+]$:

$$O = [H_2C_2O_4] + \overbrace{\frac{K_{a1}[H_2C_2O_4]}{[H_3O^+]}}^{[HC_2O_4^-]} + \overbrace{\frac{K_{a1}K_{a2}[H_2C_2O_4]}{[H_3O^+]^2}}^{[C_2O_4^{2-}]}$$

$$= \left\{1 + \frac{K_{a1}}{[H_3O^+]} + \frac{K_{a1}K_{a2}}{[H_3O^+]^2}\right\}[H_2C_2O_4]$$

$$= \frac{1}{[H_3O^+]^2}\{[H_3O^+]^2 + [H_3O^+]K_{a1} + K_{a1}K_{a2}\}[H_2C_2O_4]$$

where we have removed a factor $1/[H_3O^+]^2$ to simplify the next steps of the calculation. It follows that the fractions f of each species present in the solution are

$$f(H_2C_2O_4) = \frac{[H_2C_2O_4]}{O}$$

$$= \frac{[H_2C_2O_4]}{(1/[H_3O^+])^2\{[H_3O^+]^2 + [H_3O^+]K_{a1} + K_{a1}K_{a2}\}[H_2C_2O_4]}$$

$$\overbrace{=}^{\text{cancel } [H_2C_2O_4]} \frac{[H_3O^+]^2}{[H_3O^+]^2 + [H_3O^+]K_{a1} + K_{a1}K_{a2}} \quad (8.11a)$$

and similarly

$$f(HC_2O_4^-) = \frac{[HC_2O_4^-]}{O}$$

$$= \frac{[H_3O^+]K_{a1}}{[H_3O^+]^2 + [H_3O^+]K_{a1} + K_{a1}K_{a2}} \quad (8.11b)$$

$$f(C_2O_4^{2-}) = \frac{[C_2O_4^{2-}]}{O}$$

$$= \frac{K_{a1}K_{a2}}{[H_3O^+]^2 + [H_3O^+]K_{a1} + K_{a1}K_{a2}} \quad (8.11c)$$

These fractions are plotted against pH $= -\log[H_3O^+]$ in Fig. 8.1 (so that $[H_3O^+]$ in each expression is given by 10^{-pH}). Note from the graphs that:

- $H_2C_2O_4$ is dominant for pH < pK_{a1},
- $H_2C_2O_4$ and $HC_2O_4^-$ have the same concentration at pH = pK_{a1}, and
- $HC_2O_4^-$ is dominant for pH > pK_{a1}, until $C_2O_4^{2-}$ becomes dominant.

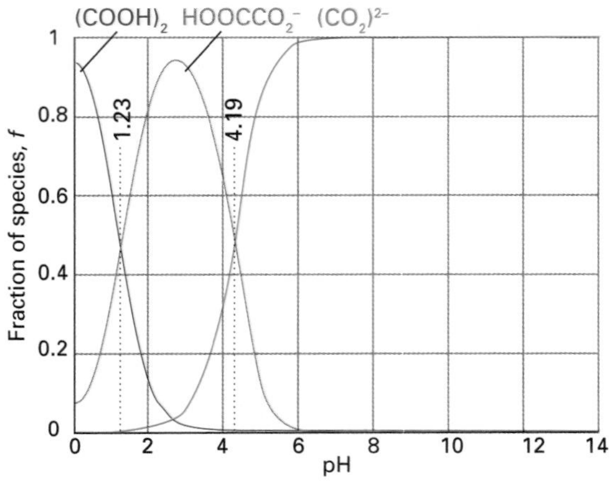

Fig. 8.1 The fractional composition of the protonated and deprotonated forms of oxalic acid in aqueous solution as a function of pH. Note that conjugate pairs are present at equal concentrations when the pH is equal to the pK_a of the acid member of the pair.

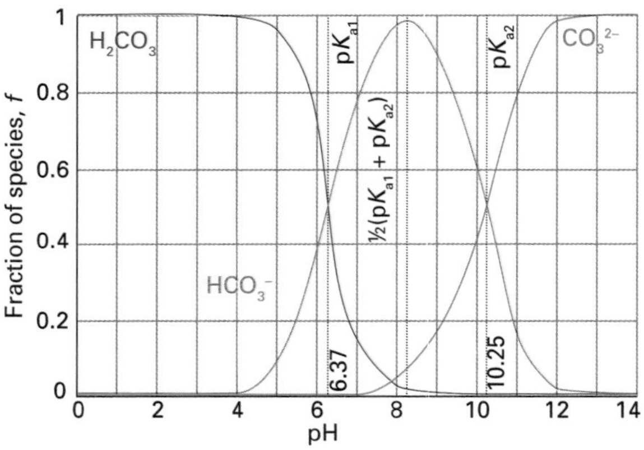

Fig. 8.2 The fractional composition of the protonated and deprotonated forms of carbonic acid in aqueous solution as a function of pH.

Self-test 8.6

Construct the diagram for the fraction of protonated species in an aqueous solution of carbonic acid.

Answer: Fig. 8.2

Be ready to take advantage of symmetries in the expressions: inspection of the three expressions for the fractions of the species present in Example 8.5 shows a symmetry in the appearance of $[H_3O^+]$ and the Ks. By noting this symmetry, it is possible to write down the expression for the species present in a solution of a triprotic acid without further calculation.

Example 8.6

Calculating the fractional composition of a solution of a triprotic acid

Phosphoric acid, H_3PO_4, exists in solution in equilibrium with $H_2PO_4^-$, HPO_4^{2-}, and PO_4^{3-}. Write expressions for the fractional compositions of these four species.

Strategy In an aqueous solution of phosphoric acid, H_3PO_4, the following equilibria must be considered:

$$H_3PO_4(aq) + H_2O(l) \rightleftharpoons H_3O^+(aq) + H_2PO_4^-(aq)$$

$$K_{a1} = \frac{[H_3O^+][H_2PO_4^-]}{[H_3PO_4]}$$

$$H_2PO_4^-(aq) + H_2O(l) \rightleftharpoons H_3O^+(aq) + HPO_4^{2-}(aq)$$

$$K_{a2} = \frac{[H_3O^+][HPO_4^{2-}]}{[H_2PO_4^-]}$$

$$HPO_4^{2-}(aq) + H_2O(l) \rightleftharpoons H_3O^+(aq) + PO_4^{3-}(aq)$$

$$K_{a3} = \frac{[H_3O^+][PO_4^{3-}]}{[HPO_4^{2-}]}$$

Proceed by using the approach of Example 8.5:

- Begin by writing an expression for $P = [H_3PO_4]_{\text{as prepared}}$ and H.
- Then write expressions for $f(H_3PO_4)$, $f(H_2PO_4^-)$, $f(HPO_4^{2-})$, and $f(PO_4^{3-})$. The task is made simple by extending the expressions in Example 8.5 to the case of a triprotic acid.

Solution Write

$$P = [H_3PO_4]_{\text{as prepared}}$$
$$= [H_3PO_4] + [H_2PO_4^-] + [HPO_4^{2-}] + [PO_4^{3-}]$$
$$H = [H_3O^+]^3 + K_{a1}[H_3O^+]^2 + K_{a1}K_{a2}[H_3O^+] + K_{a1}K_{a2}K_{a3}$$

By direct extension of the expressions written in Example 8.5, we write

$$f(H_3PO_4) = \frac{[H_3O^+]^3}{H} \qquad f(H_2PO_4^-) = \frac{K_{a1}[H_3O^+]^2}{H}$$

$$f(HPO_4^{2-}) = \frac{K_{a1}K_{a2}[H_3O^+]}{H} \qquad f(PO_4^{3-}) = \frac{K_{a1}K_{a2}K_{a3}}{H}$$

Self-test 8.7

Plot the diagram for the fraction of protonated species in an aqueous solution of phosphoric acid.

Answer: Fig. 8.3

We can summarize the behaviour found in Examples 8.5 and 8.6 (and illustrated in Figs. 8.1–8.3) as follows. Consider each conjugate acid–base pair, with acidity constant K_a; then:

- The acid form is dominant for pH < pK_a
- The conjugate pair have equal concentrations at pH = pK_a
- The base form is dominant for pH > pK_a

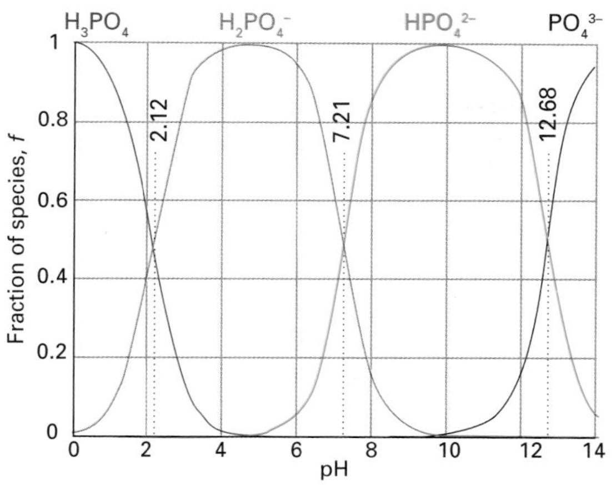

Fig. 8.3 The fractional composition of the protonated and deprotonated forms of phosphoric acid in aqueous solution as a function of pH.

In each case, the other possible forms of a polyprotic system can be ignored, provided the pK_a values are not too close together.

8.4 Amphiprotic systems

An **amphiprotic** species is a molecule or ion that can both accept and donate protons. For instance, HCO_3^- can act as an acid (to form CO_3^{2-}) and as a base (to form H_2CO_3). We need to calculate the pH of a solution of a salt with an amphiprotic anion, such as a solution of $NaHCO_3$. Is the solution acidic on account of the acid character of HCO_3^-, or is it basic on account of the anion's basic character? As we show in the following Derivation, the pH of such a solution is given by

$$\mathrm{pH} = \tfrac{1}{2}(\mathrm{p}K_{a1} + \mathrm{p}K_{a2}) \qquad \text{pH of amphiprotic salt solution} \qquad (8.12)$$

This expression is valid provided the molar concentration of the salt is high in the sense (using the notation in the following Derivation) that $S/c^{\ominus} \gg K_w/K_{a2}$ and $S/c^{\ominus} \gg K_{a1}$, where S is the initial concentration of the salt. If these conditions are not satisfied, a much more complicated expression must be used.[2]

Brief illustration 8.3 The pH of a solution of an amphiprotic species

If the dissolved salt is sodium hydrogencarbonate, we can immediately conclude from the pK_a values for the successive equilibria that the pH of the solution *of any concentration* (provided the approximations remain valid) is

$$\mathrm{pH} = \tfrac{1}{2}(\underbrace{6.37}_{\mathrm{p}K_{a1}} + \underbrace{10.25}_{\mathrm{p}K_{a2}}) = 8.31$$

The solution is basic. Because $K_w/K_{a2} = 2 \times 10^{-4}$ and $K_{a1} = 4.3 \times 10^{-7}$, this result is reliable provided $[NaHCO_3]_{\text{as prepared}}/c^{\ominus} \gg 2 \times 10^{-4}$ (that is, $[NaHCO_3]_{\text{as prepared}} \gg 0.2$ mmol dm^{-3}).

We can treat a solution of potassium dihydrogenphosphate in the same way, taking into account only the second and third acidity constants of H_3PO_4 because protonation as far as H_3PO_4 is negligible:

$$\mathrm{pH} = \tfrac{1}{2}(\underbrace{7.21}_{\mathrm{p}K_{a2}} + \underbrace{12.67}_{\mathrm{p}K_{a3}}) = 9.94$$

Because $K_w/K_{a3} = 0.05$ and $K_{a2} = 6.2 \times 10^{-8}$, this expression is reliable provided $[KH_2PO_4]_{\text{as prepared}}/c^{\ominus} \gg 0.05$ (that is, $[KH_2PO_4]_{\text{as prepared}} \gg 0.05$ mol dm^{-3}).

[2] See Chapter 4 of our *Physical Chemistry for the Life Sciences* (2011).

Derivation 8.1

The pH of an amphiprotic salt solution

Let's suppose that we make up a solution of the salt MHA with initial concentration S, where HA^- is the amphiprotic anion (such as HCO_3^-) and M^+ is a cation (such as Na^+). The equilibrium table is as follows:

Species:	H_2A	HA^-	A^{2-}	H_3O^+
Initial molar concentration/(mol dm^{-3})	0	S	0	0
Change to reach equilibrium/(mol dm^{-3})	$+x$	$-(x+y)$	$+y$	$+(y-x)$
Equilibrium concentration/(mol dm^{-3})	x	$S-x-y$	y	$y-x$

The two acidity constants are

$$K_{a1} = \frac{\overbrace{[H_3O^+]}^{y-x}\overbrace{[HA^-]}^{S-x-y}}{\underbrace{[H_2A]}_{x}} = \frac{(y-x)(S-x-y)}{x}$$

$$K_{a2} = \frac{\overbrace{[H_3O^+]}^{y-x}\overbrace{[A^{2-}]}^{y}}{\underbrace{[HA^-]}_{S-x-y}} = \frac{(y-x)y}{S-x-y}$$

Multiplication of these two expressions, noting from the equilibrium table that at equilibrium $y - x = [H_3O^+]$, gives

$$K_{a1}K_{a2} = \frac{(y-x)(S-x-y)}{x}\,\frac{(y-x)y}{S-x-y} \overset{\text{cancel terms}}{=} \frac{\overbrace{(y-x)^2}^{[H_3O^+]^2}y}{x}$$

$$= [H_3O^+]^2 \times \frac{y}{x}$$

Next, we show that, to a good approximation, $y/x \approx 1$ and therefore that $[H_3O^+] = (K_{a1}K_{a2})^{1/2}$. For this step we rearrange the expression for K_{a1} as follows:

$$xK_{a1} = (y - x)(S - x - y) = Sy - y^2 - Sx + x^2$$

Because xK_{a1}, x^2, and y^2 (in blue) are all very small compared with terms that have S in them, this expression reduces to

$$0 \approx Sy - Sx$$

We conclude that $x \approx y$, and therefore that $y/x \approx 1$, as required. Equation 8.12 now follows by taking the common logarithm of both sides of $[H_3O^+] = (K_{a1}K_{a2})^{1/2}$.

Salts in water

The ions present when a salt is added to water may themselves be either acids or bases and consequently affect the pH of the solution. For example, when ammonium chloride is added to water, it provides both an acid (NH_4^+) and a base (Cl^-). The solution consists of a weak acid (NH_4^+) and a very weak base (Cl^-). The net effect is that the solution is acidic. Similarly, a solution of sodium acetate consists of a neutral ion (the Na^+ ion) and a base ($CH_3CO_2^-$). The net effect is that the solution is basic, and its pH is greater than 7.

To estimate the pH of the solution, we proceed in exactly the same way as for the addition of a 'conventional' acid or base, for in the Brønsted–Lowry theory, there is no distinction between 'conventional' acids like acetic acid and the conjugate acids of bases (like NH_4^+).

Brief illustration 8.4 The pH of a salt solution

To calculate the pH of 0.010 M NH_4Cl(aq) at 25 °C, we proceed exactly as in Example 8.1, taking the initial concentration of the acid (NH_4^+) to be 0.010 mol dm^{-3}. The K_a to use is the acidity constant of the acid NH_4^+, which is listed in Table 8.1. Alternatively, we use K_b for the conjugate base (NH_3) of the acid and convert that quantity to K_a by using eqn 8.8a ($K_aK_b = K_w$). We find pH = 5.63, which is on the acid side of neutral. Exactly the same procedure is used to find the pH of a solution of a salt of a weak acid, such as sodium acetate. The equilibrium table is set up by treating the anion $CH_3CO_2^-$ as a base (which it is), and using for K_b the value obtained from the value of K_a for its conjugate acid (CH_3COOH).

Self-test 8.8

Estimate the pH of 0.0025 M $NH(CH_3)_3Cl$(aq) at 25 °C.

Answer: 6.2

8.5 Acid–base titrations

Acidity constants play an important role in acid–base titrations, for we can use them to decide the value of the pH that signals the **stoichiometric point,** the stage at which a stoichiometrically equivalent amount of acid has been added to a given amount of base. For historical reasons, the stoichiometric point is widely called the 'equivalence point' of a titration. (The meaning of the related term 'end point' is explained in Section 8.7.) The plot of the pH of the **analyte,** the solution being analysed, against the volume of **titrant,** the solution in the burette, added is called the **pH curve.** It shows a number of features that are still of interest even nowadays when many titrations are carried out in automatic titrators with the pH monitored electronically: automatic titration equipment is built to make use of the concepts we describe here.

First, consider the titration of a strong acid with a strong base, such as the titration of hydrochloric acid with aqueous sodium hydroxide. The reaction is

$$HCl(aq) + NaOH(aq) \rightarrow NaCl(aq) + H_2O(l)$$

Initially, the analyte (hydrochloric acid) has a low pH. The ions present at the stoichiometric point (the Na^+ ions from the strong base and the Cl^- ions from the strong acid) barely affect the pH, so the pH is that of almost pure water, namely pH = 7. After the stoichiometric point, when base is added to a neutral solution, the pH rises sharply to a high value. The pH curve for such a titration is shown in Fig. 8.4.

Figure 8.5 shows the pH curve for the titration of a weak acid (such as CH_3COOH) with a strong base (NaOH). At the stoichiometric point the solution

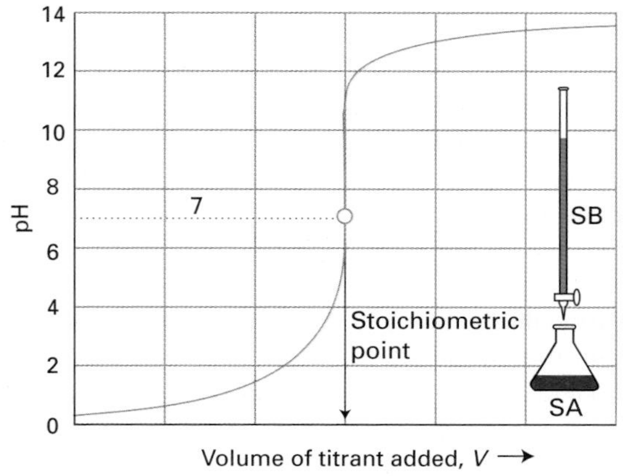

Fig. 8.4 The pH curve for the titration of a strong acid (the analyte) with a strong base (the titrant). There is an abrupt change in pH near the stoichiometric point at pH = 7. The final pH of the medium approaches that of the titrant.

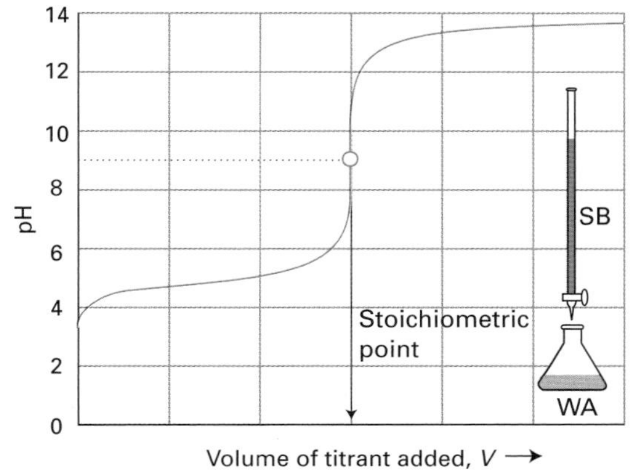

Fig. 8.5 The pH curve for the titration of a weak acid (the analyte) with a strong base (the titrant). Note that the stoichiometric point occurs at pH > 7 and that the change in pH near the stoichiometric point is less abrupt than in Fig. 8.4. The pK_a of the acid is equal to the pH halfway to the stoichiometric point.

contains $CH_3CO_2^-$ ions and Na^+ ions together with any ions stemming from autoprotolysis. The presence of the Brønsted base $CH_3CO_2^-$ in the solution means that we can expect pH > 7. In a titration of a weak base (such as NH_3) and a strong acid (HCl), the solution contains NH_4^+ ions and Cl^- ions at the stoichiometric point. Because Cl^- is only a very weak Brønsted base and NH_4^+ is a weak Brønsted acid the solution is acidic and its pH will be less than 7.

Now we consider the shape of the pH curve in Fig. 8.5 in terms of the acidity constants of the species involved. The approximations we make are based on the fact that the acid is weak, and therefore that HA is more abundant than any A^- ions in the solution. Furthermore, when HA is present, it provides so many H_3O^+ ions, even though it is a weak acid, that they greatly outnumber any H_3O^+ ions that come from the very feeble autoprotolysis of water. Finally, when excess base is present after the stoichiometric point has been passed, the OH^- ions it provides dominate any that come from the water autoprotolysis.

To be specific, let's suppose that we are titrating 25.00 cm^3 of 0.10 M $CH_3COOH(aq)$ with 0.20 M NaOH(aq) at 25 °C. We can calculate the pH at the start of a titration of a weak acid with a strong base as explained in Example 8.1, and find pH = 2.9. The addition of titrant converts some of the acid to its conjugate base in the reaction

$$CH_3COOH(aq) + OH^-(aq) \rightarrow H_2O(l) + CH_3CO_2^-(aq)$$

Suppose we add enough titrant to produce a concentration [base] of the conjugate base and simultaneously reduce the concentration of acid to [acid]. Then, because the acid and its conjugate base remain at equilibrium:

$$CH_3COOH(aq) + H_2O(l) \rightleftharpoons H_3O^+(aq) + CH_3CO_2^-(aq)$$

we can write

$$K_a = \frac{a_{H_3O^+}a_{CH_3CO_2^-}}{a_{CH_3COOH}} \approx \frac{a_{H_3O^+}[\text{base}]}{[\text{acid}]}$$

This expression rearranges first to

$$a_{H_3O^+} \approx \frac{K_a[\text{acid}]}{[\text{base}]}$$

and then, by taking common logarithms, we obtain

$$\overbrace{\log a_{H_3O^+}}^{-\text{pH}} \approx \log\frac{K_a[\text{acid}]}{[\text{base}]} \overbrace{=}^{\log xy=\log x+\log y} \overbrace{\log K_a}^{-\text{p}K_a} + \log\frac{[\text{acid}]}{[\text{base}]}$$

which can be written as the **Henderson–Hasselbalch equation**

$$\text{pH} \approx \text{p}K_a - \log\frac{[\text{acid}]}{[\text{base}]} \quad \text{Henderson–Hasselbalch equation} \quad (8.13)$$

Example 8.7

Estimating the pH at an intermediate stage in a titration

Calculate the pH of the solution after the addition of 5.00 cm^3 of the titrant to the analyte in the titration described above.

Strategy The first step involves deciding the amount of OH^- ions added in the titrant, and then to use that amount to calculate the amount of CH_3COOH remaining. Notice that because the ratio of acid to base molar concentrations occurs in eqn 8.13, the volume of solution cancels, and we can equate the ratio of concentrations to the ratio of amounts:

$$\frac{[\text{acid}]}{[\text{base}]} = \frac{n_{\text{acid}}/V}{n_{\text{base}}/V} = \frac{n_{\text{acid}}}{n_{\text{base}}}$$

Solution The addition of 5.00 cm^3, or 5.00×10^{-3} dm^3 (because 1 cm^3 = 10^{-3} dm^3), of titrant corresponds to the addition of

$$n_{OH^-} = (5.00 \times 10^{-3}\ \text{dm}^3) \times (0.200\ \text{mol dm}^{-3}) = 1.00 \times 10^{-3}\ \text{mol}$$

This amount of OH^- (1.00 mmol) converts 1.00 mmol CH_3COOH to the base $CH_3CO_2^-$. The initial amount of CH_3COOH in the analyte is

$$n_{CH_3COOH} = (25.00 \times 10^{-3}\ \text{dm}^3) \times (0.200\ \text{mol dm}^{-3}) = 2.50 \times 10^{-3}\ \text{mol}$$

so the amount remaining after the addition of titrant is 1.50 mmol. It then follows from the Henderson–Hasselbalch equation that

$$\mathrm{pH} \approx 4.75 - \log\frac{1.50\times10^{-3}}{1.00\times10^{-3}} = 4.6$$

As expected, the addition of base has resulted in an increase in pH from 2.9. You should not take the value 4.6 too seriously because we have already pointed out that such calculations are approximate. However, it is important to note that the pH has increased from its initial acidic value.

Self-test 8.9

Calculate the pH after the addition of a further 5.00 cm^3 of titrant.

Answer: 5.4

Halfway to the stoichiometric point, when enough base has been added to neutralize half the acid, the concentrations of acid and base are equal and because log 1 = 0 the Henderson–Hasselbalch equation gives

$$\mathrm{pH} \approx \mathrm{p}K_a \qquad \text{pH halfway to the stoichiometric point} \qquad (8.14)$$

In the present titration, we see that at this stage of the titration, pH ≈ 4.75. Note from the pH curve in Fig. 8.5 how much more slowly the pH is changing compared with initially: this point will prove important shortly. Equation 8.14 implies that we can determine the pK_a of the acid directly from the pH of the mixture. Indeed, an approximate value of the pK_a may be calculated by recording the pH during a titration and then examining the record for the pH halfway to the stoichiometric point.

At the stoichiometric point, enough base has been added to convert all the acid to its base, so the solution consists primarily of $CH_3CO_2^-$ ions. These ions are Brønsted bases, so we can expect the solution to be basic with a pH of well above 7. We have already seen how to estimate the pH of a solution of a weak base in terms of its concentration (Example 8.3), so all that remains to be done is to calculate the concentration of $CH_3CO_2^-$ at the stoichiometric point.

Brief illustration 8.5 The pH at the stoichiometric point

Because the analyte initially contained 2.50 mmol CH_3COOH, the volume of titrant needed to neutralize it is the volume that contains the same amount of base:

$$V_{\text{base}} = \frac{2.50\times10^{-3}\,\text{mol}}{0.200\,\text{mol dm}^{-3}} = 1.25\times10^{-2}\,\text{dm}^3$$

or 12.5 cm^3. The total volume of the solution at this stage is therefore 37.5 cm^3, so the concentration of base is

$$[CH_3CO_2^-] = \frac{2.50\times10^{-3}\,\text{mol}}{37.5\times10^{-3}\,\text{dm}^3} = 6.67\times10^{-2}\,\text{mol dm}^{-3}$$

It then follows from a calculation similar to that in Example 8.3 (with pK_b = 9.25 for $CH_3CO_2^-$) that the pH of the solution at the stoichiometric point is 8.8.

It is very important to note that the pH at *the stoichiometric point of a weak acid–strong base titration is on the basic side of neutrality* (pH > 7). At the stoichiometric point, the solution consists of a weak base (the conjugate base of the weak acid, here the $CH_3CO_2^-$ ions) and neutral cations (the Na^+ ions from the titrant).

The general form of the pH curve suggested by these estimates throughout a weak acid–strong base titration is illustrated in Fig. 8.5. The pH rises slowly from its initial value, passing through the values given by the Henderson–Hasselbalch equation when the acid and its conjugate base are both present, until the stoichiometric point is approached. It then changes rapidly to and through the value characteristic of a solution of a salt, which takes into account the effect on the pH of a solution of a weak base, the conjugate base of the original acid. The pH then climbs less rapidly towards the value corresponding to a solution consisting of excess base, and finally approaches the pH of the original base solution when so much titrant has been added that the solution is virtually the same as the titrant itself. The stoichiometric point is detected by observing where the pH changes rapidly through the value calculated in Brief illustration 8.5.

A similar sequence of changes occurs when the analyte is a weak base (such as ammonia) and the titrant is a strong acid (such as hydrochloric acid). In this case the pH curve is like that shown in Fig. 8.6: the pH falls as acid is added, plunges through the pH corresponding to a solution of the conjugate acid of the original base, in this case NH_4^+, and then slowly approaches the pH of the original strong acid which in this example is HCl. The pH of the stoichiometric point is that of a solution of a weak acid, and is calculated as illustrated in Example 8.1.

8.6 Buffer action

The slow variation of the pH when the concentrations of the conjugate acid and base are nearly equal, when pH ≈ pK_a, is the basis of **buffer action**, the ability

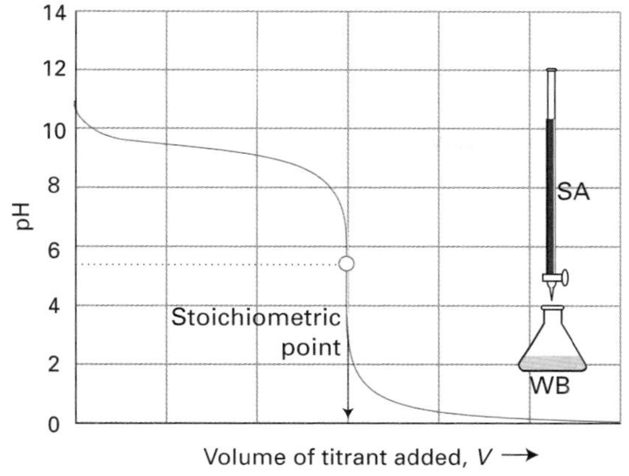

Fig. 8.6 The pH curve for the titration of a weak base (the analyte) with a strong acid (the titrant). The stoichiometric point occurs at pH < 7. The final pH of the solution approaches that of the titrant.

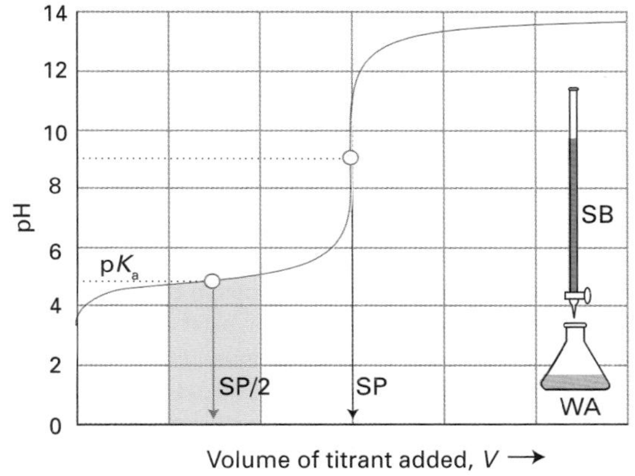

Fig. 8.7 The pH of a solution changes only slowly in the region of halfway to the stoichiometric point. In this region the solution is buffered to a pH close to pK_a.

of a solution to oppose changes in pH when small amounts of strong acids and bases are added (Fig. 8.7). An **acid buffer** solution, one that stabilizes the solution at a pH below 7, is typically prepared by making a solution of a weak acid (such as acetic acid) and a salt that supplies its conjugate base (such as sodium acetate). A **base buffer**, one that stabilizes a solution at a pH above 7, is prepared by making a solution of a weak base (such as ammonia) and a salt that supplies its conjugate acid (such as ammonium chloride).

Brief illustration 8.6 The pH of a buffer solution

Suppose we need to estimate the pH of a buffer formed from equal amounts of KH_2PO_4(aq) and K_2HPO_4(aq). We note that the two anions present are $H_2PO_4^-$ and HPO_4^{2-}. The former is the conjugate acid of the latter:

$$H_2PO_4^-(aq) + H_2O(l) \rightleftharpoons H_3O^+(aq) + HPO_4^{2-}(aq)$$

so we need the pK_a of the acid form, $H_2PO_4^-$. In this case we can take it from Table 8.1, or recognize it as the pK_{a2} of phosphoric acid, and take it from Table 8.2 instead. In either case, $pK_a = 7.21$. Hence, the solution should buffer close to pH = 7.

Self-test 8.10

Calculate the pH of an aqueous buffer solution that contains equal amounts of NH_3 and NH_4Cl.

Answer: 9.25; more realistically: 9

An acid buffer stabilizes the pH of a solution because the abundant supply of A^- ions (from the salt) can remove any H_3O^+ ions brought by additional acid; furthermore, the abundant supply of HA molecules can provide H_3O^+ ions to react with any base that is added. Similarly, in a base buffer the weak base B can accept protons when an acid is added and its conjugate acid BH^+ can supply protons if a base is added.

These abilities can be expressed quantitatively by considering the changes in equilibrium composition of hydronium ions in the presence of a buffer. This behaviour is best illustrated with a specific example.

Example 8.8

Illustrating the effect of a buffer

When a drop (0.20 cm^3, say) of 1.0 mol dm^{-3} HCl(aq) is added to 25 cm^3 of pure water, the resulting hydronium ion concentration rises to 0.0080 mol dm^{-3} and so the pH changes from 7.0 to 2.1, a big change. Now suppose the drop is added to 25 cm^3 of an acetate buffer solution that is 0.040 mol dm^{-3} $NaCH_3CO_2$(aq) and 0.080 mol dm^{-3} CH_3COOH(aq). What will be the change in pH?

Strategy The presence of the acid tells us that this mixture will be an acid buffer. Use the Henderson–Hasselbalch equation (or, better, first principles), to estimate the initial pH. Then calculate the amount of H_3O^+ added in the drop and the consequent changes to the amounts of acetic acid and acetate ions in the solution. Use the Henderson–Hasselbalch equation to estimate the pH of the resulting solution.

Solution The initial pH of the buffer solution is

$$\mathrm{pH} = \overbrace{4.75}^{pK_a} - \log \frac{\overbrace{0.080}^{[CH_3COOH]}}{\underbrace{0.040}_{[CH_3CO_2^-]}} = 4.45$$

The drop of HCl(aq) contains

$$n(H_3O^+) = (0.20 \times 10^{-3}\ dm^3) \times (1.0\ mol\ dm^{-3}) = 0.20\ mmol$$

The buffer solution contains

$$n(CH_3CO_2^-) = (25 \times 10^{-3}\ dm^3) \times (0.040\ mol\ dm^{-3}) = 1.0\ mmol$$

$$n(CH_3COOH) = (25 \times 10^{-3}\ dm^3) \times (0.080\ mol\ dm^{-3}) = 2.0\ mmol$$

The acetate ion is protonated by the incoming acid, and so its amount is reduced to 0.8 mmol. As a result, the amount of CH_3COOH rises from 2.0 mmol to 2.2 mmol. The volume of the solution barely changes, so the two concentrations become 0.032 mol dm^{-3} $NaCH_3CO_2$(aq) and 0.088 mol dm^{-3} CH_3COOH(aq). It then follows from the Henderson–Hasselbalch equation that

$$pH = \overbrace{4.75}^{pK_a} - \log \overbrace{\frac{0.088}{0.032}}^{[CH_3COOH]/[CH_3CO_2^-]} = 4.31$$

The change in pH is from 4.45 to 4.31, far smaller than in the absence of the buffer.

Self-test 8.11

Estimate the change in pH when 0.20 cm^3 of 1.5 mol dm^{-3} NaOH(aq) is added to 30 cm^3 of (a) pure water and (b) a phosphate buffer that is 0.20 mol dm^{-3} KH_2PO_4(aq) and 0.30 mol dm^{-3} K_2HPO_4(aq).

Answer: (a) 7.00 to 9.00; (b) 7.39 to 4.42

Impact on medicine 8.1

Buffer action in blood

Physiological buffers are responsible for maintaining the pH of blood within a narrow range of 7.37 to 7.43, thereby stabilizing the active conformations of biological macromolecules and optimizing the rates of biochemical reactions. There are two buffer systems that help maintain the pH of blood relatively constant: one arising from a carbonic acid/bicarbonate (hydrogencarbonate) ion equilibrium and another involving protonated and deprotonated forms of haemoglobin, the protein responsible for the transport of O_2 in blood (Impact 7.1).

Carbonic acid forms in blood from the reaction between water and CO_2 gas, which comes from inhaled air and is also a by-product of metabolism:

$$CO_2(g) + H_2O(l) \rightleftharpoons H_2CO_3(aq)$$

In red blood cells, this reaction is catalysed by the enzyme carbonic anhydrase. Aqueous carbonic acid then deprotonates to form bicarbonate (hydrogencarbonate) ion:

$$H_2CO_3(aq) + H_2O(l) \rightleftharpoons H_3O^+(aq) + HCO_3^-(aq)$$

The fact that the pH of normal blood is approximately 7.4 implies that $[HCO_3^-]/[H_2CO_3] \approx 20$. The body's control of the pH of blood is an example of *homeostasis*, the ability of an organism to counteract environmental changes with physiological responses. For instance, the concentration of carbonic acid can be controlled by respiration: exhaling air depletes the system of CO_2(g) and H_2CO_3(aq) so the pH of blood rises when air is exhaled. Conversely, inhalation increases the concentration of carbonic acid in blood and lowers its pH. The kidneys also play a role in the control of the concentration of hydronium ions. There, ammonia formed by the release of nitrogen from some amino acids (such as glutamine) combines with excess hydronium ions and the ammonium ion is excreted through urine.

The condition known as *alkalosis* occurs when the pH of blood rises above about 7.45. *Respiratory alkalosis* is caused by hyperventilation, or excessive respiration. The simplest remedy consists of breathing into a paper bag in order to increase the levels of inhaled CO_2. *Metabolic alkalosis* may result from illness, poisoning, repeated vomiting, and overuse of diuretics. The body may compensate for the increase in the pH of blood by decreasing the rate of respiration.

Acidosis occurs when the pH of blood falls below about 7.35. In *respiratory acidosis*, impaired respiration increases the concentration of dissolved CO_2 and lowers the blood's pH. The condition is common in victims of smoke inhalation and patients with asthma, pneumonia, and emphysema. The most efficient treatment consists of placing the patient in a ventilator. *Metabolic acidosis* is caused by the release of large amounts of lactic acid or other acidic by-products of metabolism, which react with hydrogencarbonate ion to form carbonic acid, thus lowering the blood's pH. The condition is common in patients with diabetes and severe burns.

The concentration of hydronium ions in blood is also controlled by haemoglobin, which can exist in deprotonated (basic) or protonated (acidic) forms, depending on the state of protonation of several amino acid residues on the protein's surface. The carbonic acid/bicarbonate ion equilibrium and proton equilibria in haemoglobin also regulate the oxygenation of blood. The key to this regulatory mechanism is the *Bohr effect*, the observation that haemoglobin binds O_2 strongly when it is deprotonated and releases O_2 when it is protonated. It follows that when dissolved CO_2 levels are high and the pH of blood falls slightly, haemoglobin becomes protonated and releases bound O_2 to tissue. Conversely, when CO_2 is exhaled and the pH rises slightly, haemoglobin becomes deprotonated and binds O_2.

8.7 Indicators

The rapid change of pH near the stoichiometric point of an acid–base titration is the basis of indicator detection. An **acid–base indicator** is a water-soluble organic molecule with acid (HIn) and conjugate base (In^-) forms that differ in colour. The two forms are in equilibrium in solution:

$$HIn(aq) + H_2O(l) \rightleftharpoons H_3O^+(aq) + In^-(aq)$$

$$K_{In} = \frac{a_{H_3O^+}a_{In^-}}{a_{HIn}} \approx \frac{a_{H_3O^+}[In^-]}{[HIn]}$$

Table 8.3
Indicator colour changes

Indicator	Acid colour	pH range of colour change	pK_{In}	Base colour
Thymol blue	Red	1.2 to 2.8	1.7	Yellow
Methyl orange	Red	3.2 to 4.4	3.4	Yellow
Bromophenol blue	Yellow	3.0 to 4.6	3.9	Blue
Bromocresol green	Yellow	4.0 to 5.6	4.7	Blue
Methyl red	Red	4.8 to 6.0	5.0	Yellow
Bromothymol blue	Yellow	6.0 to 7.6	7.1	Blue
Litmus	Red	5.0 to 8.0	6.5	Blue
Phenol red	Yellow	6.6 to 8.0	7.9	Red
Thymol blue	Yellow	9.0 to 9.6	8.9	Blue
Phenolphthalein	Colourless	8.2 to 10.0	9.4	Pink
Alizarin yellow	Yellow	10.1 to 12.0	11.2	Red
Alizarin	Red	11.0 to 12.4	11.7	Purple

The pK_{In} of some indicators are listed in Table 8.3. The ratio of the concentrations of the conjugate acid and base forms of the indicator is

$$\frac{[\mathrm{In}^-]}{[\mathrm{HIn}]} \approx \frac{K_{\mathrm{In}}}{a_{\mathrm{H_3O^+}}}$$

This expression can be rearranged (after taking common logarithms and using the methods given in The chemist's toolkit 2.2) to

$$\log\frac{[\mathrm{In}^-]}{[\mathrm{HIn}]} \approx \log\frac{K_{\mathrm{In}}}{a_{\mathrm{H_3O^+}}} \overset{\log(x/y)=\log x-\log y}{=} \overbrace{\log K_{\mathrm{In}}}^{-\mathrm{p}K_{\mathrm{In}}} - \overbrace{\log a_{\mathrm{H_3O^+}}}^{-\mathrm{pH}}$$

and written as

$$\log\frac{[\mathrm{In}^-]}{[\mathrm{HIn}]} \approx \mathrm{pH} - \mathrm{p}K_{\mathrm{In}} \qquad (8.15)$$

Brief illustration 8.7 Indicators

The pH swings from higher than pK_{In} to lower than pK_{In} as acid is added to the solution; the ratio of In^- to HIn swings from well above 1 to well below 1 (Fig. 8.8). For instance, if $pH = pK_{In} - 1$, then $[In^-]/[HIn] = 10$, but if $pH = pK_{In} + 1$, then $[In^-]/[HIn] = 10^{-1}$, a decrease of two orders of magnitude.

Self-test 8.12

What is the ratio of the yellow and blue forms of bromocresol green in solution of pH (a) 3.7, (b) 4.7, and (c) 5.7?

Answer: (a) 10:1, (b) 1:1, (c) 1:10

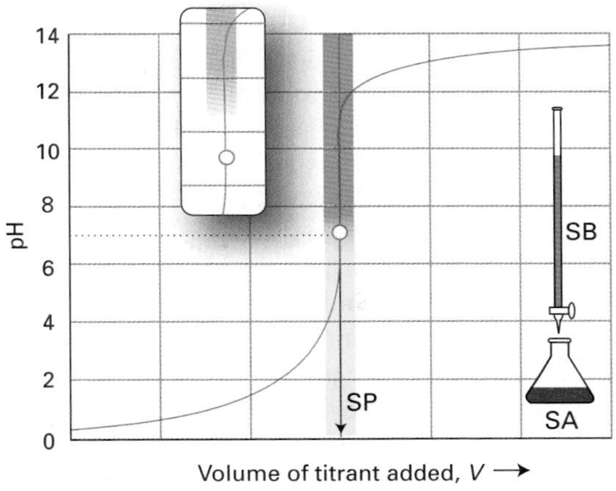

Fig. 8.8 The range of pH over which an indicator changes colour is depicted by the tinted band. For a strong acid–strong base titration, the stoichiometric point is indicated accurately by an indicator that changes colour at pH = 7 (such as bromothymol blue). However, the change in pH is so sharp that accurate results are also obtained even if the indicator changes colour in neighbouring values. Thus, phenolphthalein (which has $pK_{In} = 9.4$, see Table 8.3) is also often used.

At the stoichiometric point, the pH changes sharply through several pH units, so the molar concentration of H_3O^+ changes through several orders of magnitude. The indicator equilibrium changes so as to accommodate the change of pH, with HIn the dominant species on the acid side of the stoichiometric point, when H_3O^+ ions are abundant, and In^- dominant on the basic side, when the base can remove protons from HIn. The accompanying colour change signals the stoichiometric point of the

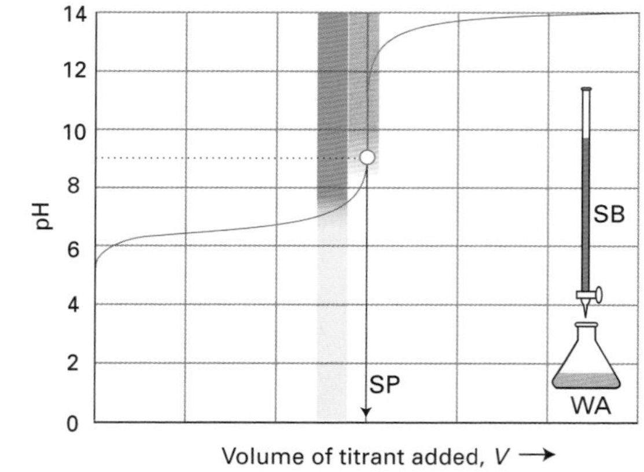

Fig. 8.9 In a weak acid–strong base titration, an indicator with $pK_{In} \approx 7$ (the lower band, like bromothymol blue) would give a false indication of the stoichiometric point; it is necessary to use an indicator that changes colour close to the pH of the stoichiometric point. If that lies at about pH = 9, then phenolphthalein would be appropriate.

titration. The colour in fact changes over a range of pH, typically from $pH \approx pK_{In} - 1$, when HIn is ten times as abundant as In^-, to $pH \approx pK_{In} + 1$, when In^- is ten times as abundant as HIn. The pH halfway through a colour change, when $pH \approx pK_{In}$ and the two forms, HIn and In^-, are in equal abundance, is the **end point** of the indicator. With a well-chosen indicator, the end point of the indicator coincides with the stoichiometric point of the titration.

Care must be taken to use an indicator that changes colour at the pH appropriate to the type of titration. Specifically, we need to match the end point to the stoichiometric point, and therefore select an indicator for which pK_{In} is close to the pH of the stoichiometric point. Thus, in a weak acid–strong base titration, the stoichiometric point lies at pH > 7, and we should select an indicator that changes at that pH (Fig. 8.9). Similarly, in a strong acid–weak base titration, we need to select an indicator with an end point at pH < 7. Qualitatively, we should choose an indicator with $pK_{In} \approx 7$ for strong acid–strong base titrations, one with $pK_{In} < 7$ for strong acid–weak base titrations, and one with $pK_{In} > 7$ for weak acid–strong base titrations.

Solubility equilibria

A solid dissolves in a solvent until the solution and the solid solute are in equilibrium. At this stage, the solution is said to be **saturated**, and its molar concentration is the **molar solubility** of the solid. That the two phases—the solid solute and the solution—are in dynamic equilibrium implies that we can use equilibrium concepts to discuss the composition of the saturated solution. You should note that a solubility equilibrium is an example of a *heterogeneous* equilibrium in which the species are in different phases (the solid solute and the solution) The properties of aqueous solutions of electrolytes are commonly treated in terms of equilibrium constants, and in this section we shall confine our attention to them. We shall also limit our attention to **sparingly soluble** compounds, which are compounds that dissolve only slightly in water. This restriction is applied because the effects of ion–ion interactions are a complicating feature of more concentrated solutions and more advanced techniques are then needed before the calculations are reliable. Once again, we shall concentrate on general trends and properties rather than expecting to obtain numerically precise results and confine numerical calculations to systems at 298 K.

8.8 The solubility constant

The heterogeneous equilibrium between a sparingly soluble ionic compound, such as calcium hydroxide, $Ca(OH)_2$, and its ions in aqueous solution is

$$Ca(OH)_2(s) \rightleftharpoons Ca^{2+}(aq) + 2\,OH^-(aq)$$

$$K_s = \frac{a_{Ca^{2+}}a^2_{OH^-}}{\underbrace{a_{Ca(OH)_2}}_{1}} = a_{Ca^{2+}}a^2_{OH^-}$$

The equilibrium constant for an ionic equilibrium such as this, bearing in mind that the solid does not appear in the equilibrium expression because its activity is 1, is called the **solubility constant** (which is also called the *solubility product constant* or simply the *solubility product*). For very dilute solutions, $a_J \approx [J]/c^{\ominus}$ and as usual we can replace the activity a_J of a species J by the numerical value of its molar concentration. Experimental values for solubility constants are given in Table 8.4.

We can interpret the solubility constant in terms of the numerical value of the **molar solubility**, s, of a sparingly soluble substance. For an ionic compound of the form A_xB_y made up of A^{a+} and B^{b-} ions, the molar concentration of cations in solution is $[A^{a+}] = xs$ and that of anions is $[B^{b-}] = ys$. Provided it is permissible to replace activities by molar concentrations, the solubility constant is then

$$K_s = [A^{a+}]^x[B^{b-}]^y = (xs)^x(ys)^y = x^xy^ys^{(x+y)}$$

Table 8.4
Solubility constants at 298.15 K

Compound	Formula	K_s
Aluminium hydroxide	$Al(OH)_3$	1.0×10^{-33}
Antimony sulfide	Sb_2S_3	1.7×10^{-93}
Barium carbonate	$BaCO_3$	8.1×10^{-9}
fluoride	BaF_2	1.7×10^{-6}
sulfate	$BaSO_4$	1.1×10^{-10}
Bismuth sulfide	Bi_2S_3	1.0×10^{-97}
Calcium carbonate	$CaCO_3$	8.7×10^{-9}
fluoride	CaF_2	4.0×10^{-11}
hydroxide	$Ca(OH)_2$	5.5×10^{-6}
sulfate	$CaSO_4$	2.4×10^{-5}
Copper(I) bromide	CuBr	4.2×10^{-8}
chloride	CuCl	1.0×10^{-6}
iodide	CuI	5.1×10^{-12}
sulfide	Cu_2S	2.0×10^{-47}
Copper(II) iodate	$Cu(IO_3)_2$	1.4×10^{-7}
oxalate	CuC_2O_4	2.9×10^{-8}
sulfide	CuS	8.5×10^{-45}
Iron(II) hydroxide	$Fe(OH)_2$	1.6×10^{-14}
sulfide	FeS	6.3×10^{-18}
Iron(III) hydroxide	$Fe(OH)_3$	2.0×10^{-39}
Lead(II) bromide	$PbBr_2$	7.9×10^{-5}
chloride	$PbCl_2$	1.6×10^{-5}
fluoride	PbF_2	3.7×10^{-8}
iodate	$Pb(IO_3)_2$	2.6×10^{-13}
iodide	PbI_2	1.4×10^{-8}
sulfate	$PbSO_4$	1.6×10^{-8}
sulfide	PbS	3.4×10^{-28}
Magnesium ammonium phosphate	$MgNH_4PO_4$	2.5×10^{-13}
carbonate	$MgCO_3$	1.0×10^{-5}
fluoride	MgF_2	6.4×10^{-9}
hydroxide	$Mg(OH)_2$	1.1×10^{-11}
Mercury(I) chloride	Hg_2Cl_2	1.3×10^{-18}
iodide	Hg_2I_2	1.2×10^{-28}
Mercury(II) sulfide	HgS black:	1.6×10^{-52}
	red:	1.4×10^{-53}
Nickel(II) hydroxide	$Ni(OH)_2$	6.5×10^{-18}
Silver bromide	AgBr	7.7×10^{-13}
carbonate	Ag_2CO_3	6.2×10^{-12}
chloride	AgCl	1.6×10^{-10}
hydroxide	AgOH	1.5×10^{-8}
iodide	AgI	1.5×10^{-16}
sulfide	Ag_2S	6.3×10^{-51}
Zinc hydroxide	$Zn(OH)_2$	2.0×10^{-17}
sulfide	ZnS	1.6×10^{-24}

Brief illustration 8.8 The solubility of a salt

It follows from the stoichiometry of calcium hydroxide that the molar concentration of Ca^{2+} ions in solution is equal to that of the $Ca(OH)_2$ dissolved in solution, so $[Ca^{2+}] = s$. Likewise, because the concentration of OH^- ions is twice that of $Ca(OH)_2$ formula units, it follows that $[OH^-] = 2s$. Therefore,

$$K_s \approx s \times (2s)^2 = 4s^3 \quad \text{and} \quad s \approx (\tfrac{1}{4}K_s)^{1/3}$$

Thus, from Table 8.4, $K_s = 5.5 \times 10^{-6}$, so $s \approx 1 \times 10^{-2}$ and the molar solubility is 1×10^{-2} mol dm^{-3}.

Self-test 8.13

Copper occurs in many minerals, one of which is chalcocite, Cu_2S. What is the approximate solubility of this compound in water at 25 °C? Use the data for Cu_2S in Table 8.4.

Answer: 1.7×10^{-16} mol dm^{-3}

The result obtained in Brief illustration 8.8 is only approximate because ion–ion interactions have been ignored. However, because the solid is sparingly soluble, the concentrations of the ions are low and the inaccuracy is moderately low. Solubility constants (which are determined by electrochemical measurements of the kind described in Chapter 9) provide a more accurate way of measuring solubilities of very sparingly soluble compounds than the direct measurement of the mass that dissolves.

Because solubility products are equilibrium constants, they may be calculated from thermodynamic data, particularly the standard Gibbs energies of formation of ions in solution and the relation ($\Delta_r G^{\ominus} = -RT \ln K$). The direct determination of solubilities is very difficult for almost insoluble salts. Another application is to the discussion of qualitative analysis, where judicious choice of concentrations guided by the values of K_s can result in the successive precipitation of compounds (sulfides, for instance) and the recognition of the heavy elements (barium, for instance), present in a mixture.

8.9 The common-ion effect

The principle that an equilibrium constant remains unchanged whereas the individual concentrations of species may change is applicable to solubility constants, and may be used to assess the effect of the addition of species to solutions. An example of particular importance is the effect on the solubility of a compound of the presence of another freely soluble solute that provides an ion in common with the sparingly soluble compound already present. For example, we may consider the effect on the solubility of adding sodium chloride to a saturated solution of silver chloride, the common ion in this case being Cl^-.

We know what to expect from Le Chatelier's principle (Chapter 7): when the concentration of the common ion is increased, we can expect the equilibrium to respond by tending to minimize that increase. As a result, the solubility of the original salt can be expected to decrease. To treat the effect quantitatively, we note that the molar solubility of silver chloride in pure water is related to its solubility constant by $s \approx K_s^{1/2}$. To assess the effect of the common ion, we suppose that Cl^- ions are added to a concentration C mol dm^{-3}, which greatly exceeds the concentration of the same ion that stems from the presence of the silver chloride. Therefore, we can write

$$K_s = a_{Ag^+}a_{Cl^-} \approx ([Ag^+]/c^{\ominus})(C/c^{\ominus})$$

It is very dangerous to neglect deviations from ideal behaviour in ionic solutions, so from now on the calculation will only be indicative of the kinds of changes that occur when a common ion is added to a solution of a sparingly soluble salt: the qualitative trends are reproduced, but the quantitative calculations are unreliable. With these remarks in mind, it follows that the solubility s' of silver chloride in the presence of added chloride ions is

$$\frac{s'}{c^{\ominus}} = \frac{K_s}{C/c^{\ominus}} \quad \text{or} \quad s' = \frac{K_s \times c^{\ominus 2}}{C} \qquad (8.16)$$

The solubility is greatly reduced by the presence of the common ion. The reduction of the solubility of a sparingly soluble salt by the presence of a common ion is called the **common-ion effect**.

Brief illustration 8.9 The common-ion effect

The solubility of silver chloride in water is 1.3×10^{-5} mol dm^{-3}. The solubility constant is therefore

$$K_s \approx ([Ag^+]/c^{\ominus})([Cl^-]/c^{\ominus}) = (s/c^{\ominus}) \times (s/c^{\ominus}) = (s/c^{\ominus})^2$$

In the presence of 0.10 M NaCl(aq)

$$s' = \frac{K_s \times c^{\ominus 2}}{C} \approx \frac{(s/c^{\ominus})^2 \times c^{\ominus 2}}{C} = \frac{s^2}{C}$$

$$= \frac{(1.3 \times 10^{-5}\ \text{mol dm}^{-3})^2}{0.10\ \text{mol dm}^{-3}} = 1.7 \times 10^{-9}\ \text{mol dm}^{-3}$$

which is nearly ten thousand times less.

Self-test 8.14

Estimate the molar solubility of calcium fluoride, CaF_2, in (a) water, (b) 0.010 M NaF(aq).

Answer: (a) 2.2×10^{-4} mol dm^{-3}; (b) 4.0×10^{-7} mol dm^{-3}

8.10 The effect of added salts on solubility

Even a salt that has no ion in common with the sparingly soluble salt can affect the latter's solubility. At low concentrations of the added salt, the solubility of the sparingly soluble salt is increased. The explanation lies in a phenomenon that moves to centre stage in Chapter 9, where we see that in aqueous solution cations tend to be found near anions and anions tend to be found near cations. That is, each ion is in an environment, called an 'ionic atmosphere', of opposite charge. The charge imbalance is not great, because the ions are ceaselessly churned around by thermal motion, but it is enough to lower the energy of the central ion slightly.

When a soluble salt is added to the solution of a sparingly soluble salt, the abundant ions of the latter form ionic atmospheres around the ions of the sparingly soluble salt. As a result of the lowering of energy that results, the sparingly soluble salt has a greater tendency to go into solution. That is, the presence of the added salt raises the solubility of the sparingly soluble salt.

To estimate the effect of an added salt MX on a sparingly soluble salt AB, we write the solubility constant for AB in terms of activities:

$$K_s = a_A a_B = \gamma_A\gamma_B[A][B] = \gamma_A\gamma_B s^2$$

It follows that $s = (K_s/\gamma_A\gamma_B)^{1/2}$ and therefore, from The chemist's toolkit 2.2, that

$$\log s = \tfrac{1}{2}\log(K_s/\gamma_A\gamma_B) \overset{\log(x/y)=\log x-\log y}{=} \tfrac{1}{2}\log K_s - \tfrac{1}{2}\log \gamma_A\gamma_B$$

We shall see in Chapter 9 that the logarithm of the product of activity coefficients is proportional to the square root of the concentration, C, of the added salt; with $\log \gamma_A\gamma_B = -2AC^{1/2}$ where A is a constant that depends on the identity of the solvent and the temperature; for water at 25 °C, $A = 0.51$. It follows that

$$\underbrace{\log s}_{\text{AB in the presence of MX}} = \tfrac{1}{2}\log K_s + \underbrace{AC^{1/2}}_{\text{Effect of added salt}} \qquad (8.17)$$

Note that $AC^{1/2}$ increases with increasing concentration of added salt, so $\log S$, and therefore S itself, also increases, as we anticipated. The linear dependence of $\log S$ on $C^{1/2}$ is observed, but only for low concentrations of added salt.

Brief illustration 8.10 The effect of added salts on solubility

The solubility of AgCl in water at 25 °C is $s = K_s^{1/2}$; because $K_s = 1.6 \times 10^{-10}$, it follows that $s = 1.3 \times 10^{-5}$

mol dm^{-3}. In the presence of 0.10 mol dm^{-3} $KNO_3(aq)$ its solubility increases to

$$\log s = \tfrac{1}{2}\times\log\underbrace{(1.6\times10^{-10})}_{K_s} + \underbrace{0.5100}_{A}\times\underbrace{(0.10)^{1/2}}_{C^{1/2}} = -4.74$$

corresponding to $s = 1.8\times10^{-5}$ mol dm^{-3}.

Self-test 8.15

Determine the concentration of KNO_3 that results in an increase in the solubility of AgCl in water to 2.1×10^{-5} mol dm^{-3}.

Answer: 0.19 mol dm^{-3}

Checklist of key concepts

☐ **1** In the Brønsted–Lowry theory of acids and bases, an acid is a proton donor and a base is a proton acceptor.

☐ **2** The strength of an acid HA is reported in terms of its acidity constant, K_a, and that of a base B in terms of its basicity constant, K_b.

☐ **3** Weak acids have $K_a < 1$, with $pK_a > 0$. Strong acids are commonly regarded as being completely deprotonated in aqueous solution.

☐ **4** A strong base is fully protonated in solution ($K_b > 1$). A weak base is not fully protonated in water ($K_b < 1$).

☐ **5** The acid form of a species is dominant if $pH < pK_a$ and the base form is dominant if $pH > pK_a$.

☐ **6** A polyprotic acid is a molecular compound that can donate more than one proton.

☐ **7** An amphiprotic species is a molecule or ion that can both accept and donate protons.

☐ **8** The pH of a mixed solution of a weak acid and its conjugate base is given by the Henderson–Hasselbalch equation.

☐ **9** The pH of a buffer solution containing equal concentrations of a weak acid and its conjugate base is $pH = pK_a$.

☐ **10** An acid buffer solution stabilizes the solution at a pH below 7, and is prepared by making a solution of a weak acid and a salt that supplies its conjugate base.

☐ **11** A base buffer stabilizes a solution at a pH above 7, and is prepared by making a solution of a weak base and a salt that supplies its conjugate acid.

☐ **12** The end point of the colour change of an acid–base indicator occurs at $pH = pK_{In}$; in a titration, choose an indicator with an end point that coincides with the stoichiometric point.

☐ **13** A solid dissolves in a solvent until the solution and the solid solute are in equilibrium. The molar solubility of the solid is its molar concentration in this saturated solution.

☐ **14** The common-ion effect is the reduction in solubility of a sparingly soluble salt by the presence of a common ion.

☐ **15** At low concentrations of and added salt that does not provide common ions, the solubility of a sparingly soluble salt increases.

Road map of key equations

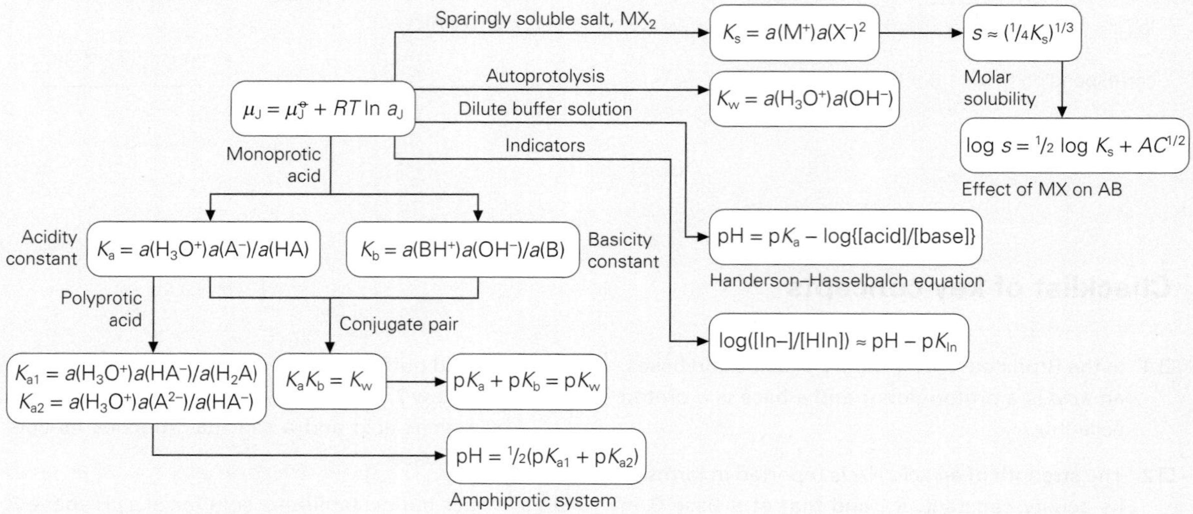

Questions and exercises

Discussion questions

8.1 Describe the changes in pH that take place during the titration of: (a) a weak acid with a strong base, (b) a weak base with a strong acid.

8.2 Describe the basis of buffer action and indicator detection.

8.3 Explain the difference between 'stoichiometric (equivalence) point' and 'end point' in the context of a titration.

8.4 Outline the change in composition of a solution of the salt of a triprotic acid as the pH is changed from 1 to 14.

8.5 State the limits to the generality of the expression for estimating the pH of an amphiprotic salt solution. Suggest reasons for why these limitations exist.

8.6 Describe and justify the approximations used in the derivation of the Henderson–Hasselbalch equation?

8.7 Explain the common-ion effect.

Exercises

8.1 Write the proton transfer equilibria for the following acids in aqueous solution and identify the conjugate acid–base pairs in each one: (a) H_2SO_4, (b) HF (hydrofluoric acid), (c) $C_6H_5NH_3^+$ (anilinium ion), (d) $H_2PO_4^-$ (dihydrogenphosphate ion), (e) HCOOH (formic acid), (f) $NH_2NH_3^+$ (hydrazinium ion).

8.2 Numerous acidic species are found in living systems. Write the proton transfer equilibria for the following biochemically important acids in aqueous solution: (a) lactic acid ($CH_3CHOHCOOH$), (b) glutamic acid ($HOOCCH_2CH_2CH(NH_2)COOH$), (c) glycine (NH_2CH_2COOH), (d) oxalic acid (HOOCCOOH).

8.3 The molar concentration of H_3O^+ ions in the following solutions was measured at 25 °C. Calculate the pH and pOH of the solution: (a) 1.5×10^{-5} mol dm^{-3} (a sample of rain water), (b) 1.5 mmol dm^{-3}, (c) 5.1×10^{-14} mol dm^{-3}, (d) 5.01×10^{-5} mol dm^{-3}.

8.4 Calculate the molar concentration of H_3O^+ ions and the pH of the following solutions: (a) 25.0 cm^3 of 0.144 M HCl(aq) was added to 25.0 cm^3 of 0.125 M NaOH(aq), (b) 25.0 cm^3 of 0.15 M HCl(aq) was added to 35.0 cm^3 of 0.15 M KOH(aq), (c) 21.2 cm^3 of 0.22 M HNO_3(aq) was added to 10.0 cm^3 of 0.30 M NaOH(aq).

8.5 For biological and medical applications we often need to consider proton transfer equilibria at body temperature (37 °C). The value of K_w for water at body temperature is 2.5×10^{-14}. (a) What is the value of $[H_3O^+]$ and the pH of neutral water at 37 °C? (b) What is the molar concentration of OH^- ions and the pOH of neutral water at 37 °C?

8.6 Suppose that something had gone wrong in the Big Bang, and instead of ordinary hydrogen there was an abundance of deuterium in the universe. There would be many subtle changes in equilibria, particularly the deuteron transfer equilibria of heavy atoms and bases. The K_w for D_2O, heavy water, at 25 °C is 1.35×10^{-15}. (a) Write the chemical equation for the autoprotolysis (more precisely, autodeuterolysis) of D_2O. (b) Evaluate pK_w for D_2O at 25 °C. (c) Calculate the molar concentrations of D_3O^+ and OD^- in neutral heavy water at 25 °C. (d) Evaluate the pD and pOD of neutral heavy water at 25 °C. (e) Formulate the relation between pD, pOD, and $pK_w(D_2O)$.

8.7 Estimate the pH of a solution of 0.50 M HCl(aq), assuming ideal behaviour. The mean activity coefficient at this concentration is 0.769. Calculate a more reliable value of the pH.

8.8 Use the van 't Hoff equation (eqn 7.15) to derive an expression for the slope of a plot of pK_a against temperature.

8.9 The pK_w of water varies with temperature as follows:

θ/°C	10	15	20	25	30	35
pK_w	14.5346	14.3463	14.1669	13.9965	13.8330	13.6801

Determine the standard enthalpy of deprotonation of water.

8.10 The pK_b of ammonia in water varies with temperature as follows:

θ/°C	10	15	20	25	30	35
pK_b	4.804	4.782	4.767	4.751	4.740	4.733

Deduce as much information as you can from these values.

8.11 The pK_b of the organic base nicotine (denoted Nic) is 5.98. Write the corresponding protonation reaction, the deprotonation reaction of the conjugate acid, and find the value of pK_a for nicotine.

8.12 Determine the fraction of solute deprotonated or protonated in (a) 0.25 M C_6H_5COOH(aq), (b) 0.150 M NH_2NH_2(aq) (hydrazine), (c) 0.112 M $(CH_3)_3N$(aq) (trimethylamine).

8.13 Calculate the pH, pOH, and fraction of solute protonated or deprotonated in the following aqueous solutions: (a) 0.150 M $CH_3CH(OH)COOH$(aq) (lactic acid), (b) 2.4×10^{-4} M $CH_3CH(OH)COOH$(aq), (c) 0.25 M $C_6H_5SO_3H$(aq) (benzenesulfonic acid).

8.14 The amino acid tyrosine has $pK_a = 2.20$ for deprotonation of its carboxylic acid group. What are the relative concentrations of tyrosine and its conjugate base at a pH of (a) 7, (b) 2.2, (c) 1.5?

8.15 Calculate the pH of the following acid solutions at 25 °C: (a) 1.0×10^{-4} M H_3BO_3(aq) (boric acid acts as a monoprotic acid), (b) 0.015 M H_3PO_4(aq), (c) 0.10 M H_2SO_3(aq).

8.16 Calculate the molar concentrations of $(COOH)_2$, $HOOCCO_2^-$, $(CO_2)_2^{2-}$, H_3O^+, and OH^- in 0.15 M $(COOH)_2$(aq).

8.17 Calculate the molar concentrations of H_2S, HS^-, S^{2-}, H_3O^+, and OH^- in 0.065 M H_2S(aq).

8.18 Show how the composition of an aqueous solution that contains 20 mmol dm^{-3} glycine varies with pH.

8.19 The hydrogensulfite ion, HSO_3^- is amphiprotic. Use the values for the successive acidity constants for sulfurous acid, H_2SO_3, in Table 8.2 to calculate the pH of an aqueous solution of sodium hydrogensulfite, $NaHSO_3$.

8.20 Determine whether aqueous solutions of the following salts have a pH equal to, greater than, or less than 7; if pH > 7 or pH < 7, write a chemical equation to justify your answer. (a) NH_4Br, (b) Na_2CO_3, (c) KF, (d) KBr, (e) $AlCl_3$, (f) $Co(NO_3)_2$.

8.21 Sodium acetate, $NaCH_3CO_2$, of mass 7.4 g is used to prepare 250 cm^3 of aqueous solution. What is the pH of the solution?

8.22 What is the pH of a solution when 2.75 g of ammonium chloride, NH_4Cl, is used to make 100 cm^3 of aqueous solution?

8.23 An aqueous solution of volume 1.0 dm^3 contains 10.0 g of potassium bromide. What is the percentage of Br^- ions that are protonated?

8.24 Sketch reasonably accurately the pH curve for the titration of 25.0 cm^3 of 0.15 M $Ba(OH)_2$(aq) with 0.22 M HCl(aq). Mark on the curve (a) the initial pH, (b) the pH at the stoichiometric point.

8.25 Show how the composition of an aqueous solution that contains 30 mmol dm^{-3} tyrosine varies with pH.

8.26 Estimate the pH of an aqueous solution of sodium hydrogenoxalate. Under what conditions is this estimate reasonably reliable?

8.27 Calculate the pH of (a) 0.10 M NH_4Cl(aq), (b) 0.25 M $NaCH_3CO_2$(aq), (c) 0.200 M CH_3COOH(aq).

8.28 A sample of 0.10 M CH_3COOH(aq) of volume 25.0 cm^3 is titrated with 0.10 M NaOH(aq). The K_a for CH_3COOH is 1.8×10^{-5}. (a) What is the pH of 0.10 M CH_3COOH(aq)? (b) What is the pH after the addition of 10.0 cm^3 of 0.10 M NaOH(aq)? (c) What volume of 0.10 M NaOH(aq) is required to reach halfway to the stoichiometric point? (d) Calculate the pH at that halfway point. (e) What volume of 0.10 M NaOH(aq) is required to reach the stoichiometric point? (f) Calculate the pH at the stoichiometric point.

8.29 At the halfway point in the titration of a weak acid with a strong base the pH was measured as 5.16. What is the acidity constant and the pK_a of the acid? What is the pH of the solution that is 0.025 M in the acid?

8.30 Calculate the pH at the stoichiometric point of the titration of 25.00 cm^3 of 0.150 M lactic acid with 0.188 M NaOH(aq).

8.31 Sketch the pH curve of a solution containing 0.10 M $NaCH_3CO_2(aq)$ and a variable amount of acetic acid.

8.32 There are many organic acids and bases in our cells, and their presence modifies the pH of the fluids inside them. It is useful to be able to assess the pH of solutions of acids and bases and to make inferences from measured values of the pH. A solution of equal concentrations of lactic acid and sodium lactate was found to have pH = 3.08. (a) What are the values of pK_a and K_a of lactic acid? (b) What would the pH be if the acid had twice the concentration of the salt?

8.33 The weak base colloquially known as Tris, and more precisely as tris(hydroxymethyl)aminomethane (**3**), has $pK_a = 8.3$ at 20 °C and is commonly used to produce a buffer for biochemical applications. At what pH would you expect Tris to act as a buffer in a solution that has equal molar concentrations of Tris and its conjugate acid?

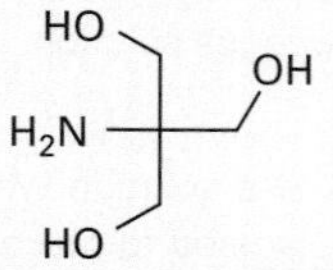

3 Tris(hydroxymethyl)aminomethane

8.34 A buffer solution of volume 100 cm³ consists of 0.10 M $CH_3COOH(aq)$ and 0.10 M $Na(CH_3CO_2)(aq)$. (a) What is its pH? (b) What is the pH after the addition of 3.3 mmol NaOH to the buffer solution? (c) What is the pH after the addition of 6.0 mmol HNO_3 to the initial buffer solution?

8.35 Predict the pH region in which each of the following buffers will be effective, assuming equal molar concentrations of the acid and its conjugate base: (a) sodium lactate and lactic acid, (b) sodium benzoate and benzoic acid, (c) potassium hydrogenphosphate and potassium phosphate, (d) potassium hydrogenphosphate and potassium dihydrogenphosphate, (e) hydroxylamine and hydroxylammonium chloride.

8.36 From the information in Tables 8.1 and 8.2, select suitable buffers for (a) pH = 2.2 and (b) pH = 7.0.

8.37 Write the expression for the solubility constants of the following compounds: (a) AgI, (b) Hg_2S, (c) $Fe(OH)_3$, (d) Ag_2CrO_4.

8.38 Use the data in Table 8.4 to estimate the molar solubilities of (a) $BaSO_4$, (b) Ag_2CO_3, (c) $Fe(OH)_3$, (d) Hg_2Cl_2 in water.

8.39 Use the data in Table 8.4 to estimate the solubility in water of each sparingly soluble substance in its respective solution: (a) silver bromide in 1.4×10^{-3} M NaBr(aq), (b) magnesium carbonate in 1.1×10^{-5} M $Na_2CO_3(aq)$, (c) lead(II) sulfate in a 0.10 M $CaSO_4(aq)$, (d) nickel(II) hydroxide in 2.7×10^{-5} M $NiSO_4(aq)$.

8.40 The solubility of mercury(I) iodide is 5.5 fmol dm^{-3} (1 fmol = 10^{-15} mol) in water at 25 °C. What is the standard Gibbs energy of dissolution of the salt?

8.41 Thermodynamic data can be used to predict the solubilities of compounds that would be very difficult to measure directly. Calculate the solubility of mercury(II) chloride in water at 25 °C from standard Gibbs energies of formation.

8.42 (a) Derive an expression for the ratio of solubilities of AgCl at two different temperatures; assume that the standard enthalpy of solution of AgCl is independent of temperature in the range of interest. (b) Do you expect the solubility of AgCl to increase or decrease as the temperature is raised?

Projects

8.43 Deduce expressions for the fractions of each type of species present in an aqueous solution of lysine (**4**) as a function of pH and plot the appropriate speciation diagram. Use the following values of the acidity constants: $pK_a(H_3Lys^{2+}) = 2.18$, $pK_a(H_2Lys^+) = 8.95$, $pK_a(HLys) = 10.53$.

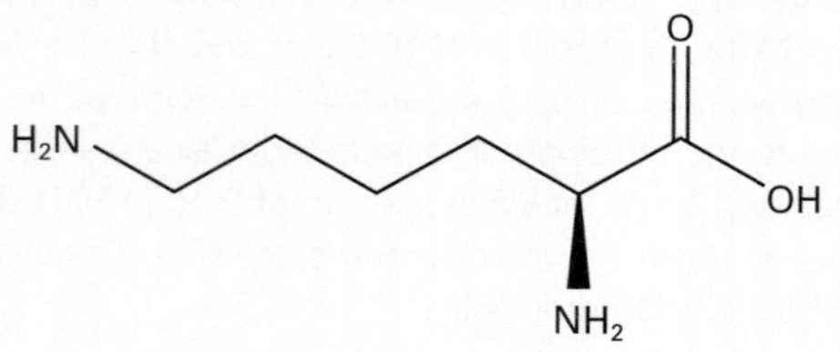

4 Lysine (Lys)

8.44 Using the insights gained through your work on Exercise 8.43, and *without doing a calculation*, sketch the speciation diagram for histidine (**5**) in water and label the axes with the significant values of pH. Use $pK_a(H_3His^{2+}) = 1.77$, $pK_a(H_2His^+) = 6.10$, $pK_a(HHis) = 9.18$.

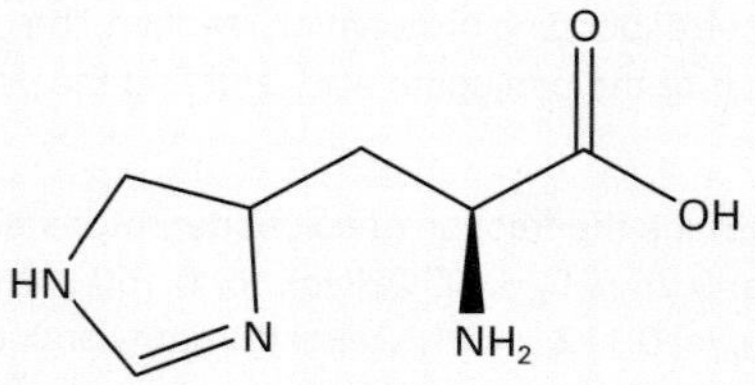

5 Histidine (His)

8.45 Here we explore buffer action in blood more quantitatively. (a) What are the values of the ratio $[HCO_3^-]/[H_2CO_3]$ at the onset of acidosis and alkalosis? (b) The *Bohr effect* may be understood in terms of a dependence on pH of the degree of cooperativity in the binding of O_2 by haemoglobin. Based on the description of the Bohr effect given here and the information provided in Exercise 7.40, does the Hill coefficient of haemoglobin increase or decrease with pH?

9

Chemical equilibria: electrochemistry

Such apparently unrelated processes as combustion, respiration, photosynthesis, and corrosion are actually all closely related, for in each of them an electron, sometimes accompanied by a group of atoms, is transferred from one species to another. Indeed, together with the proton transfer typical of acid–base reactions, processes in which electrons are transferred, the so-called **redox reactions**, account for many of the reactions encountered in chemistry. Redox reactions—the principal topic of this chapter—are of immense practical significance, not only because they underlie many biochemical and industrial processes, but also because they are the basis of the generation of electricity by chemical reactions and the investigation of reactions by making electrical measurements.

Measurements like the ones we describe in this chapter lead to a collection of data that are very useful for discussing the characteristics of electrolyte solutions and of a wide range of different types of equilibria in solution. They are also used throughout inorganic chemistry to assess the thermodynamic feasibility of reactions and the stabilities of compounds. They are used in physiology to discuss the details of the propagation of signals in neurons.

This chapter considers both electrolytes and non-electrolytes. It should be familiar from introductory chemistry that an **electrolyte** (or *electrolyte solution*) is an electrically conducting solution. Most electrolytes are ionic solutions, with the ions acting as the carriers of charge. A **non-electrolyte** (or *non-electrolyte solution*) is a solution that does not conduct electricity because no ions are present.

Ions in solution 209

9.1 The Debye–Hückel theory 210

9.2 The migration of ions 213

Electrochemical cells 217

9.3 Half-reactions and electrodes 217

9.4 Reactions at electrodes 219

9.5 Varieties of cell 221

9.6 The cell reaction 222

9.7 The cell potential 222

9.8 Cells at equilibrium 224

9.9 Standard potentials 225

9.10 The variation of potential with pH 225

9.11 The determination of pH 227

Applications of standard potentials 228

9.12 The electrochemical series 228

9.13 The determination of thermodynamic functions 228

CHECKLIST OF KEY CONCEPTS 231

ROAD MAP OF KEY EQUATIONS 231

QUESTIONS AND EXERCISES 232

Ions in solution

The most significant difference between electrolyte and non-electrolyte solutions is that there are long-range Coulombic interactions between the ions in the former. As a result, electrolyte solutions exhibit non-ideal behaviour even at very low concentrations because the solute particles, the ions, do not move independently of one another. Some idea of the importance of ion–ion interactions is obtained by noting their average separations in solutions of different molar concentration c and, to appreciate the scale, the typical number of H_2O molecules that can fit between them:

$c/(\text{mol dm}^{-3})$	0.001	0.01	0.1	1	10
Separation/nm	12	5	3	1	0.5
Number of H_2O molecules	40	18	8	4	2

We see how to take the interactions between ions into account—which become very important for concentrations of 0.01 mol dm^{-3} and more—in the first part of this chapter. A second difference is that an ion in solution responds to the presence of an electric field, migrates through the solution, and carries charge from one location to another. Our bodies are electric conductors and some of the thoughts you are currently having as you read this sentence can be traced to the migration of ions through membranes in the enormously complex electrical circuits of your brain.

Much of this chapter depends on the energy of the interaction between ions through the Coulomb interaction between charges. The properties of that interaction are reviewed in The chemist's toolkit 9.1.

The chemist's toolkit 9.1 The Coulomb interaction

As explained in Foundations, the Coulomb interaction between two charges Q_1 and Q_2 separated by a distance r in a vacuum is described by the *Coulombic potential energy*:

$$E_p = \frac{Q_1 Q_2}{4\pi\varepsilon_0 r}$$

where $\varepsilon_0 = 8.854 \times 10^{-12}\ \text{J}^{-1}\ \text{C}^2\ \text{m}^{-1}$ is the vacuum permittivity. Note that the interaction is attractive ($E_p < 0$) when Q_1 and Q_2 have opposite signs and repulsive ($E_p > 0$) when their signs are the same. The potential energy of a charge is zero when it is at an infinite distance from the other charge. When the charges are separated by a medium (such as water), the vacuum permittivity is replaced by the permittivity of the medium, ε:

$$E_p = \frac{Q_1 Q_2}{4\pi\varepsilon r}$$

The permittivity is related to the vacuum permittivity by writing $\varepsilon = \varepsilon_r\varepsilon_0$, where ε_r is the empirically determined *relative permittivity* (formerly, the 'dielectric constant').

9.1 The Debye–Hückel theory

We have seen (in Section 6.6) that the thermodynamic properties of solutes are expressed in terms of their activities, a_J, which is a kind of dimensionless effective concentration, and that activities are related to concentrations by multiplication by an activity coefficient, γ_J. There are various ways of expressing concentration; in the first part of this chapter we use the molality, b_J, and write

$$a_J = \gamma_J b_J / b^{\ominus} \qquad (9.1a)$$

where $b^{\ominus} = 1\ \text{mol kg}^{-1}$. Because the solution becomes more ideal as the molality approaches zero, we know that $\gamma_J \to 1$ as $b_J \to 0$. For notational simplicity, we shall replace $b_J/b^{\ominus}$ by b_J itself, treat b as the numerical value of the molality, and write

$$a_J = \gamma_J b_J \qquad (9.1b)$$

All this convention means is that to use expressions in which b_J appears, express it in moles per kilogram (such as 0.02 mol kg^{-1}), discard those units (mol kg^{-1}), and use only its numerical value (0.02) in calculations.

Once we know the activity of the species J, we can write its chemical potential by using

$$\mu_J = \mu_J^{\ominus} + RT \ln a_J \qquad (9.2)$$

The thermodynamic properties of the solution—such as the equilibrium constants of reactions involving ions—can then be derived in the same way as for ideal solutions but with activities in place of concentrations. However, when we want to relate the results we derive to observations, we need to know how to relate those activities to concentrations. We ignored that problem when discussing acids and bases, and simply assumed that all activity coefficients were 1. In this chapter, we see how to improve that approximation.

One problem that confronts us from the outset is that cations and anions always occur together in solution. Therefore, there is no experimental procedure for distinguishing the deviations from ideal behaviour due to the cations from those of the anions: we cannot measure the activity coefficients of cations and anions separately. The best we can do experimentally is to ascribe deviations from ideal behaviour equally to each kind of ion and to talk in terms of a **mean activity coefficient**, $\gamma_\pm$. For a salt MX, such as NaCl, we show in the following Derivation that the mean activity coefficient is related to the activity coefficients of the individual ions as follows:

$$\gamma_\pm = (\gamma_+\gamma_-)^{1/2} \qquad \text{MX salt} \quad \text{Mean activity coefficient} \qquad (9.3a)$$

For a salt M_pX_q, the mean activity coefficient is related to the activity coefficients of the individual ions as follows:

$$\gamma_\pm = (\gamma_+^p \gamma_-^q)^{1/s} \quad s = p + q \qquad \text{M}_p\text{X}_q \text{ salt} \quad \text{Mean activity coefficient} \qquad (9.3b)$$

Thus, for $Mg_3(PO_4)_2$, where $p = 3$, $q = 2$, and $s = 5$, the mean activity coefficient for each type of ion is $\gamma_\pm = (\gamma_+^3\gamma_-^2)^{1/5}$.

Brief illustration 9.1 Mean activity coefficients 1

Suppose we found a way to calculate the actual activity coefficients of Na^+ and SO_4^{2-} ions in 0.010 M Na_2SO_4(aq) and found them to be 0.65 and 0.84, respectively (these values are invented). The mean activity coefficient would be

$$\gamma_\pm = \{(0.65)^2 \times (0.84)\}^{1/3} = 0.71$$

because $p = 2$ and $q = 1$ and $s = 3$ (0.71 is the real experimental value). We would then write the activities of the two ions as

$$a_+ = \gamma_\pm b_+/b^\ominus = 0.71 \times (2 \times 0.010) = 0.014$$
$$a_- = \gamma_\pm b_-/b^\ominus = 0.71 \times (0.010) = 0.0071$$

Self-test 9.1

The experimental value of the mean activity coefficient for 0.020 m $CaCl_2$(aq) is 0.664. Suppose we knew that the activity coefficient of the Cl^- ion is 0.811. What would be the activity coefficient of the Ca^{2+} ion in this solution?

Answer: 0.445

Derivation 9.1

Mean activity coefficients

For a salt MX that dissociates completely in solution, the molar Gibbs energy of the ions is

$$G_m = \mu_+ + \mu_-$$

where μ_+ and μ_- are the chemical potentials of the cations and anions, respectively. Each chemical potential can be expressed in terms of a molality b and an activity coefficient γ by using eqn 9.2 ($\mu = \mu^\ominus + RT\ln a$) and then eqn 9.1 ($a = \gamma b$), which gives

$$G_m = \{\mu_+^\ominus + RT\ln a_+\} + \{\mu_-^\ominus + RT\ln a_-\}$$
$$= \{\mu_+^\ominus + RT\ln \gamma_+ b_+\} + \{\mu_-^\ominus + RT\ln \gamma_- b_-\}$$
$$\overset{\ln xy = \ln x + \ln y}{=} \{\mu_+^\ominus + RT\ln \gamma_+ + RT\ln b_+\} + \{\mu_-^\ominus + RT\ln \gamma_- + RT\ln b_-\}$$

where we have used the rules for the manipulation of logarithms in The chemist's toolkit 2.2. We do so again to combine the two terms involving the activity coefficients (blue) as

$$G_m = \{\mu_+^\ominus + RT\ln b_+\} + \{\mu_-^\ominus + RT\ln b_-\} + RT\ln \gamma_+\gamma_-$$

Next, write the term inside the final (blue) logarithm as $\gamma_\pm^2$, to obtain

$$G_m = \{\mu_+^\ominus + RT\ln b_+\} + \{\mu_-^\ominus + RT\ln b_-\} + RT\ln \gamma_\pm^2$$
$$\overset{\ln x^2 = 2\ln x}{=} \{\mu_+^\ominus + RT\ln b_+\} + \{\mu_-^\ominus + RT\ln b_-\} + 2RT\ln \gamma_\pm$$
$$= \{\mu_+^\ominus + RT\ln b_+ + RT\ln \gamma_\pm\} + \{\mu_-^\ominus + RT\ln b_- + RT\ln \gamma_\pm\}$$
$$\overset{\ln x + \ln y = \ln xy}{=} \{\mu_+^\ominus + RT\ln \gamma_\pm b_+\} + \{\mu_-^\ominus + RT\ln \gamma_\pm b_-\}$$

We see that, with the mean activity coefficient defined as in eqn 9.3a, the deviation from ideal behaviour (as expressed by the activity coefficient) is now shared equally between the two types of ion. In exactly the same way, the Gibbs energy of a salt M_pX_q can be written

$$G_m = p(\mu_+^\ominus + RT\ln \gamma_\pm b_+) + q(\mu_-^\ominus + RT\ln \gamma_\pm b_-)$$

with the mean activity coefficient defined as in eqn 9.3b.[1]

The question still remains, however, about how the mean activity coefficients may be estimated. A theory that accounts for their values in very dilute solutions was developed by Peter Debye and Erich Hückel in 1923. They supposed that each ion in solution is surrounded by an **ionic atmosphere** of counter charge. This 'atmosphere' is actually the slight imbalance of charge arising from the competition between thermal motion, which tends to keep all the ions distributed uniformly throughout the solution, and the Coulombic interaction between ions, which tends to attract counter-ions (ions of opposite charge) into each other's vicinity and repel ions of like charge (Fig. 9.1). As a result of this competition, there is a slight excess of cations near any anion, giving a positively charged ionic atmosphere around the anion, and a slight excess of anions near any cation, giving a negatively charged ionic atmosphere around the cation. Because each ion is in an atmosphere of opposite charge, its energy is lower than in a uniform, ideal solution, and therefore its chemical potential is lower than in an ideal solution. A lowering of the chemical potential of an ion below its ideal solution value is equivalent to the activity coefficient of the ion being less than 1 (because $\ln \gamma$ is negative when $\gamma < 1$). Debye and Hückel were able to derive an expression that is a limiting law in the sense that it becomes increasingly valid as the concentration of ions approaches zero. The **Debye–Hückel limiting law**[2] is

$$\log \gamma_\pm = -A|z_+z_-|I^{1/2} \qquad \text{Debye–Hückel limiting law} \qquad (9.4)$$

[1] For the details of this general case, see our *Physical Chemistry* (2010); see also Exercise 9.3.

[2] For a derivation of the Debye–Hückel limiting law, see our *Physical Chemistry* (2010).

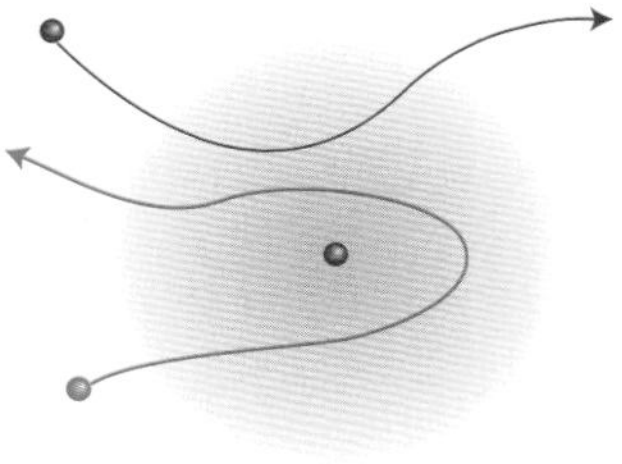

Fig. 9.1 The ionic atmosphere surrounding an ion consists of a slight excess of opposite charge as ions move through the vicinity of the central ion, with counter-ions lingering longer than ions of the same charge. The ionic atmosphere lowers the energy of the central ion.

(Note the common logarithm.) In this expression, A is a constant which for water at 25 °C works out as 0.509. The z_J are the charge numbers of the ions (so $z_+=+1$ for Na^+ and $z_-=-2$ for SO_4^{2-}); the vertical bars means that we ignore the sign of the product. The quantity I is the **ionic strength** of the solution, which is defined in terms of the molalities of the ions as

$$I = \tfrac{1}{2}(z_+^2 b_+ + z_-^2 b_-)/b^{\ominus} \qquad \text{Definition} \quad \text{Ionic strength} \qquad (9.5a)$$

Brief illustration 9.2 Mean activity coefficients 2

To estimate the mean activity coefficient for the ions in 0.0010 M Na_2SO_4(aq) at 25 °C, we first evaluate the ionic strength of the solution from eqn 9.5a by using $b_+/b^{\ominus}=2\times 0.0010$ and $z_+=+1$ for Na^+ and $b_-/b^{\ominus}=0.0010$ and $z_-=-2$ for SO_4^{2-}:

$$I = \tfrac{1}{2}\{(+1)^2 \times (2\times 0.0010) + (-2)^2 \times (0.0010)\} = 0.0030$$

Then we use the Debye–Hückel limiting law, eqn 9.4, to write

$$\begin{aligned}\log \gamma_\pm &= -0.509 \times |(+1)||(-2)| \times (0.0030)^{1/2}\\ &= -2\times 0.509 \times (0.0030)^{1/2} = -0.056\end{aligned}$$

Thus, taking antilogarithms,

$$\gamma_\pm \overset{x=10^{\log x}}{=} 0.88$$

The experimental value is 0.886.

Self-test 9.2

Estimate the mean activity coefficient for the ions in 0.0005 M $Al_2(SO_4)_3$ at 25 °C.

Answer: 0.77

When calculating the ionic strength it is important to include all the ions present in the solution, not just those of interest. For instance, when calculating the ionic strength of a solution of silver chloride and potassium nitrate, there are contributions to the ionic strength from all four types of ion. When more than two ions contribute to the ionic strength, we write

$$I = \frac{1}{2}\overbrace{\sum_i z_i^2 b_i/b^{\ominus}}^{\text{Sum over all ions present}} \qquad \text{General case} \quad \text{Ionic strength} \qquad (9.5b)$$

where z_i is the charge number of an ion i (positive for cations and negative for anions) and b_i is its molality.

As we have stressed, eqn 9.4 is a *limiting* law and is reliable only in very dilute solutions. For solutions more concentrated than about 10^{-3} mol dm^{-3} ion–ion interactions become even more important and it is better to use an empirical modification known as the **extended Debye–Hückel law:**

$$\log \gamma_\pm = -\frac{A|z_+z_-|I^{1/2}}{1+BI^{1/2}} + CI \qquad \text{Extended Debye–Hückel law} \qquad (9.6)$$

where B and C are dimensionless constants (Fig. 9.2). Although B can be interpreted as a measure of the closest approach of the ions, it (like C) is best regarded as an adjustable empirical parameter. As usual, it is better to interpret this equation than to try to remember it:

- When $BI^{1/2} \ll 1$ and CI is small, the extended law reduces to the limiting law.
- The term proportional to I becomes more important than the terms proportional to $I^{1/2}$ at higher ionic strengths; because CI is positive, it increases the value of $\log \gamma_\pm$ above its limiting law value, as shown in Fig. 9.2.

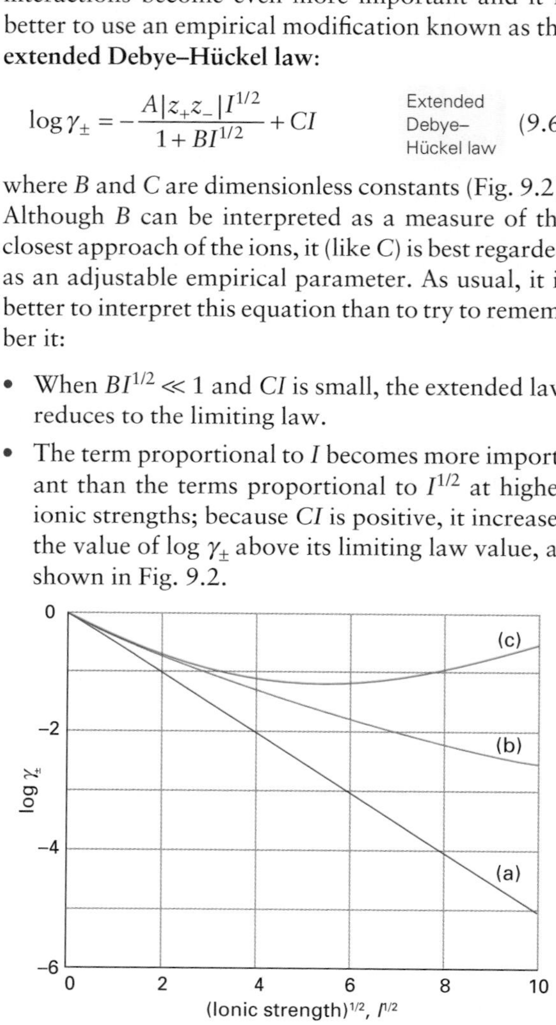

Fig. 9.2 The variation of the activity coefficient with ionic strength according to the extended Debye–Hückel theory. (a) The limiting law for a 1,1-electrolyte. (b) The extended law with $B = 0.5$. (c) The extended law, extended further by the addition of a term CI, in this case with $C=0.2$. The last form of the law reproduces the observed behaviour reasonably well.

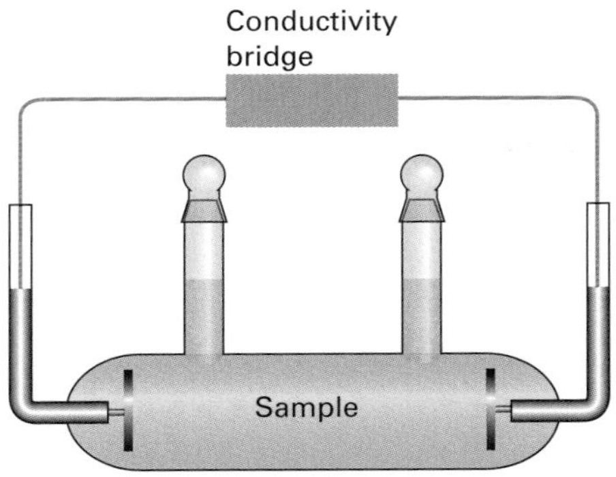

Fig. 9.3 A typical conductivity cell. The cell is made part of a 'bridge' and its resistance is measured. The conductivity is normally determined by comparison of its resistance to that of a solution of known conductivity. An alternating current is used to avoid the formation of decomposition products at the electrodes.

9.2 The migration of ions

Ions are mobile in solution, and the study of their motion down a potential gradient gives an indication of their size, the effect of solvation, and details of the type of motion they undergo. The migration of ions in solution is studied by measuring the electrical resistance of a solution of known concentration in a cell like that in Fig. 9.3. Certain technicalities must be dealt with in practice, such as using an alternating current to minimize the effects of electrolysis, but the essential point is the determination of R and from R and the dimensions of the sample, the conductivity κ (kappa); see The chemist's toolkit 9.2.

The chemist's toolkit 9.2 Electric current

The resistance, R (in ohms, Ω), of a sample controls the current, I (in amperes, A), that flows when a potential difference, $\mathcal{V}$ (in volts, V), is applied between the two electrodes. According to Ohm's law: $I = \mathcal{V}/R$.

The resistance depends on the identity and temperature of the substance; it is proportional to the length, L, and inversely proportional to the cross-sectional area, A, of the sample: $R \propto L/A$. The constant of proportionality is called the **resistivity**, ρ (rho), and we write $R = \rho L/A$. The units of resistivity are ohm metre (Ω m). The reciprocal of the resistivity, $1/\rho$, is called the **conductivity**, κ (kappa), which is expressed in $\Omega^{-1}\,m^{-1}$. Reciprocal ohms are given their own name, siemens (S, with $1\,S = 1\,\Omega^{-1}$); then conductivities are expressed in siemens per metre ($S\,m^{-1}$).

The ampere, A, is one of the SI base units. Charge is reported in coulombs, with $1\,C = 1\,A\,s$. For potential and potential difference the volt, V, is defined as $1\,V = 1\,J\,C^{-1}$ (or, in base units, $1\,V = 1\,kg\,m^2\,A^{-1}\,s^{-3}$) and for resistance the ohm, Ω, is defined as $1\,\Omega = 1\,V\,A^{-1}$ (in base units, $1\,\Omega = 1\,kg\,m^2\,A^{-2}\,s^{-3}$), so $1\,S = 1\,A\,V^{-1}$. For most applications, it is possible to work in amperes, volts, and coulombs.

Once we have determined κ (in practice, by calibrating the cell with a solution of known conductivity), we find the **molar conductivity**, Λ_m (lambda), when the solute molar concentration is c by forming

$$\Lambda_m = \frac{\kappa}{c} \qquad \text{Definition} \quad \text{Molar conductivity} \qquad (9.7)$$

With molar concentration in moles per cubic decimetre, molar conductivity is expressed in siemens per metre per (moles per cubic decimetre), or $S\,m^{-1}\,(mol\,dm^{-3})^{-1}$. These awkward units are useful in practical applications but can be simplified to siemens metre squared per mole ($S\,m^2\,mol^{-1}$). Specifically, the relation between units is

$$1\ S\,m^{-1}\,(mol\,dm^{-3})^{-1} = 1\ mS\,m^2\,mol^{-1}$$

where $1\,mS = 10^{-3}\,S$.

The molar conductivity of a *strong* electrolyte (a solution in which the solute is fully dissociated into ions in solution, such as the solution of a salt) varies with molar concentration in accord with the empirical law discovered by Friedrich Kohlrausch in 1876:

$$\Lambda_m = \Lambda_m^\circ - \mathcal{K}c^{1/2} \qquad \text{Strong electrolyte} \quad \text{Kohlrausch law} \qquad (9.8)$$

The constant Λ_m°, the **limiting molar conductivity**, is the molar conductivity in the limit of such low concentration that the ions no longer interact with one another. The constant $\mathcal{K}$ takes into account the effect of these interactions when the concentration is nonzero. The fact that the interactions give rise to a square-root dependence on the concentration suggests that they arise from effects like those responsible for activity coefficients in the Debye–Hückel theory, and in particular the effect of an ionic atmosphere on the mobilities of ions. The fact that the molar conductivity *decreases* with increasing concentration can be traced to the retarding effect of the ions on the motion of one another. We shall concentrate on the limiting conductivity.

When the ions are so far apart that their interactions can be ignored, we can suspect that the molar conductivity is due to the independent migration of cations in one direction and of anions in the opposite direction, and write

$$\Lambda_m^\circ = \lambda_+ + \lambda_- \qquad \text{Definition} \quad \text{Ionic conductivities} \qquad (9.9)$$

where λ_+ and λ_- are the ionic conductivities of the individual cations and anions (Table 9.1).

The molar conductivity of a *weak* electrolyte varies in a more complex way with the concentration of the

Table 9.1
*Ionic conductivities, λ/(mS m^2 mol^{-1})**

Cations		Anions	
H^+ (H_3O^+)	34.96	OH^-	19.91
Li^+	3.87	F^-	5.54
Na^+	5.01	Cl^-	7.64
K^+	7.35	Br^-	7.81
Rb^+	7.78	I^-	7.68
Cs^+	7.72	CO_3^{2-}	13.86
Mg^{2+}	10.60	NO_3^-	7.15
Ca^{2+}	11.90	SO_4^{2-}	16.00
Sr^{2+}	11.89	$CH_3CO_2^-$	4.09
NH_4^+	7.35	HCO_2^-	5.46
$[N(CH_3)_4]^+$	4.49		
$[N(CH_2CH_3)_4]^+$	3.26		

*The same numerical values apply when the units are S m^{-1} (mol dm^{-3})$^{-1}$.

solute. This variation reflects the fact that the degree of ionization (or, in the case of weak acids and bases, the degree of deprotonation or protonation) varies with the concentration, with relatively more ions present at low concentrations than at high. Because we can use simple equilibrium-table techniques to relate the ion concentrations to the nominal (initial) concentration, we can use measurements of molar conductivity to determine acidity constants. The same kind of measurements can also be used to monitor the progress of reactions in solution, provided that they involve ions.

Example 9.1

Determining the acidity constant from the conductivity of a weak acid

The molar conductivity of 0.010 m CH_3COOH(aq) is 1.65 mS m^2 mol^{-1}. What is the acidity constant, K_a, of the acid?

Strategy Acidity constants were introduced in Section 8.2. Because acetic acid is weak, it is only partly deprotonated in aqueous solution. Only the fraction of acid molecules present as ions contributes to the conduction, so we need to express Λ_m in terms of the fraction deprotonated. To do so, we set up an equilibrium table, find the molar concentration of H_3O^+ and $CH_3CO_2^-$ ions, and relate those concentrations to the observed molar conductivity.

Answer The equilibrium table for $CH_3COOH(aq) + H_2O(l) \rightleftharpoons H_3O^+(aq) + CH_3CO_2^-(aq)$ is

Species:	CH_3COOH	H_3O^+	$CH_3CO_2^-$
Initial molar concentration/(mol dm^{-3})	0.010	0	0
Change/(mol dm^{-3})	$-x$	$+x$	$+x$
Equilibrium molar concentration/(mol dm^{-3})	$0.010 - x$	x	x

The value of x is found by substituting the entries in the last line into the expression for K_a:

$$K_a = \frac{\overbrace{[H_3O^+]}^{x}\overbrace{[CH_3CO_2^-]}^{x}}{\underbrace{[CH_3COOH]}_{0.010-x}} = \frac{x^2}{0.010 - x}$$

On the assumption that x is small, we replace $0.010 - x$ by 0.010 and find that $x = (0.010K_a)^{1/2}$. (If you prefer not to make an assumption, you could proceed as in Example 8.2.) The fraction, α, of CH_3COOH molecules present as ions is therefore $x/0.010$, or $\alpha = (K_a/0.010)^{1/2}$. The molar conductivity of the solution is therefore this fraction multiplied by the molar conductivity of acetic acid calculated on the assumption that deprotonation is complete:

$$\Lambda_m = \alpha\Lambda_m^\circ = \alpha(\lambda_{H_3O^+} + \lambda_{CH_3CO_2^-})$$

where $\Lambda_m^\circ = \lambda_{H_3O^+} + \lambda_{CH_3CO_2^-}$. Because

$$\lambda_{H_3O^+} + \lambda_{CH_3CO_2^-} = 34.96 + 4.09 \text{ mS m}^2 \text{ mol}^{-1} = 39.05 \text{ mS m}^2 \text{ mol}^{-1}$$

it follows that α = (1.65 mS m^2 mol^{-1})/(39.05 mS m^2 mol^{-1}) = 0.0423. Therefore,

$$K_a = 0.010\alpha^2 = 0.010 \times (0.0423)^2 = 1.8 \times 10^{-5}$$

This value corresponds to pK_a = 4.75.

Self-test 9.3

The molar conductivity of 0.0250 M HCOOH(aq) is 4.61 mS m^2 mol^{-1}. What is the pK_a of formic acid?

Answer: 3.49

The ability of an ion to conduct electricity depends on its ability to move through the solution. When an ion is subjected to an electric field $\mathcal{E}$, it accelerates. However, the faster it travels through the solution, the greater the retarding force it experiences from the viscosity of the medium. As a result, it settles down into a limiting speed called its **drift speed**, s, which is proportional to the strength of the applied field:

$$s = u\mathcal{E} \qquad \text{Definition} \quad \text{Mobility} \quad (9.10)$$

The **mobility**, u, depends on the radius, a, of the ion and the viscosity, and, as we show in the following Derivation, on the viscosity, η (eta), of the solution:

$$u = \frac{e|z|}{6\pi\eta a} \qquad (9.11a)$$

where z is the charge number of the ion. The mobility of an ion determines the rate at which it can transport charge through a solution and therefore its molar conductivity. The relation between the two is[3]

$$\lambda_{\pm} = |z|u_{\pm}F \qquad \text{Conductivity and mobility} \quad (9.11b)$$

Derivation 9.2

The ionic mobility

An *electric field* is an influence that accelerates a charged particle. An ion of charge ze in an electric field $\mathcal{E}$ (typically, in volts per metre, V m^{-1}) experiences a force of magnitude $|z|e\mathcal{E}$, which accelerates it. However, the ion experiences a frictional force due to its motion through the medium, which increases the faster the ion travels. The retarding force due to the viscosity on a spherical particle of radius a travelling at a speed s is given by 'Stokes's law':

$$F_{\text{viscous}} = 6\pi\eta a s \qquad \text{Stokes's law} \quad (9.12)$$

When the particle has reached its drift speed, the accelerating and viscous retarding forces are equal, so we can write

$$\overbrace{e|z|\mathcal{E}}^{\text{Accelerating force}} = \overbrace{6\pi\eta a s}^{\text{Retarding force}}$$

and solve this expression for s:

$$s = \frac{e|z|\mathcal{E}}{6\pi\eta a}$$

At this point we can compare this expression for the drift speed with eqn 9.10, and hence find the expression for mobility given in eqn 9.11.

Table 9.2

Ionic mobilities in water at 298 K, $u/(10^{-8}$ m^2 s^{-1} V$^{-1})$

Cations		Anions	
H^+ (H_3O^+)	36.23	OH^-	20.64
Li^+	4.01	F^-	5.74
Na^+	5.19	Cl^-	7.92
K^+	7.62	Br^-	8.09
Rb^+	8.06	I^-	7.96
Cs^+	8.00	CO_3^{2-}	7.18
Mg^{2+}	5.50	NO_3^-	7.41
Ca^{2+}	6.17	SO_4^{2-}	8.29
Sr^{2+}	6.16		
NH_4^+	7.62		
$[N(CH_3)_4]^+$	4.65		
$[N(CH_2CH_3)_4]^+$	3.38		

Equation 9.11 tells us that the mobility of an ion is high if it is highly charged, is small, and if it is in a solution with low viscosity. These features appear to contradict the trends in Table 9.2, which lists the mobilities of a number of ions. For instance, the mobilities of the Group 1 cations *increase* down the group despite their increasing radii. The explanation is that the radius to use in eqn 9.11 is the **hydrodynamic radius**, the *effective* radius for the migration of the ions taking into account the entire object that moves. When an ion migrates, it carries its hydrating water molecules with it, and as small ions are more extensively hydrated than large ions (because they give rise to a stronger electric field in their vicinity), ions of small radius actually have a large hydrodynamic radius. Thus, hydrodynamic radius *decreases* down Group 1 because the extent of hydration decreases with increasing ionic radius.

One significant deviation from this trend is the very high mobility of the proton in water. It is believed that this high mobility reflects an entirely different mechanism for conduction, the **Grotthus mechanism**, in which the proton on one H_2O molecule migrates to its neighbour, the proton on that H_2O molecule migrates to its neighbour, and so on along a chain (Fig. 9.4). The motion is therefore an *effective* motion of a proton, not the actual motion of a single proton.

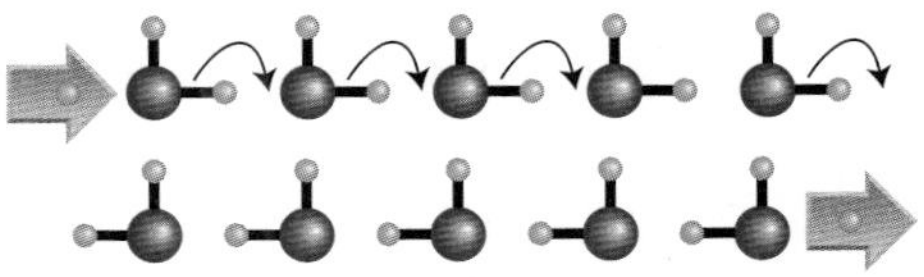

Fig. 9.4 A simplified version of the 'Grotthus mechanism' of proton conduction through water. The proton leaving the chain on the right is not the same as the proton entering the chain on the left.

[3] For a derivation, see our *Physical Chemistry* (2010).

Impact on biochemistry 9.1

Ion channels and pumps

The controlled transport of molecules and ions across biological membranes is at the heart of a number of key cellular processes, such as the transmission of nerve impulses, the transfer of glucose into red blood cells, and the synthesis of ATP. Here we examine in some detail the various ways in which ions cross the alien environment of the lipid bilayer.

Suppose that a membrane provides a barrier that slows down the transfer of molecules or ions into or out of the cell. The thermodynamic tendency to transport a species A through the membrane is partially determined by a concentration gradient (more precisely, an activity gradient) across the membrane, which results in a difference in molar Gibbs energy between the inside and the outside of the cell

$$\Delta G_m = \overbrace{G_m(\text{in})}^{\mu^\ominus + RT\ln a_{in}} - \overbrace{G_m(\text{out})}^{\mu^\ominus + RT\ln a_{out}} = RT\ln\frac{a_{in}}{a_{out}}$$

The equation implies that transport into the cell of either neutral or charged species is thermodynamically favourable if $a_{in} < a_{out}$ or, if we set the activity coefficients to 1, if $[A]_{in} < [A]_{out}$. If A is an ion, there is a second contribution to ΔG_m that is due to the different potential energy of the ions on each side of the bilayer, where the difference in electrostatic potential is $\Delta\phi = \phi_{in} - \phi_{out}$ and the difference in molar energy is $zF\Delta\phi$ where z is the ion charge number and F is Faraday's constant, the magnitude of electric charge per mole of electrons:

$$F = eN_A = 96\,485\ \text{C mol}^{-1} = 96.485\ \text{kC mol}^{-1}$$

The final expression for ΔG is then

$$\Delta G_m = RT\ \ln\frac{[A]_{in}}{[A]_{out}} + zF\Delta\phi$$

This equation implies that there is a tendency, called **passive transport**, for a species to move down concentration and membrane potential gradients. It is also possible to move a species against these gradients, but now the flow must be driven by an exergonic process, such as the hydrolysis of ATP. This process is called **active transport**.

The transport of ions into or out of a cell needs to be mediated (that is, facilitated by other species) because the hydrophobic environment of the membrane is inhospitable to ions. There are two mechanisms for ion transport: mediation by a carrier molecule and transport through a *channel former*, a protein that creates a hydrophilic pore through which the ion can pass. An example of a channel former is the polypeptide gramicidin A, which increases the membrane permeability to cations such as H^+, K^+, and Na^+.

Ion channels are proteins that effect the movement of specific ions down a membrane potential gradient. They are highly selective, so there is a channel protein for Ca^{2+}, another for Cl^-, and so on. The opening of the gate may be triggered by potential differences between the two sides of the membrane or by the binding of an *effector molecule* to a specific receptor site on the channel.

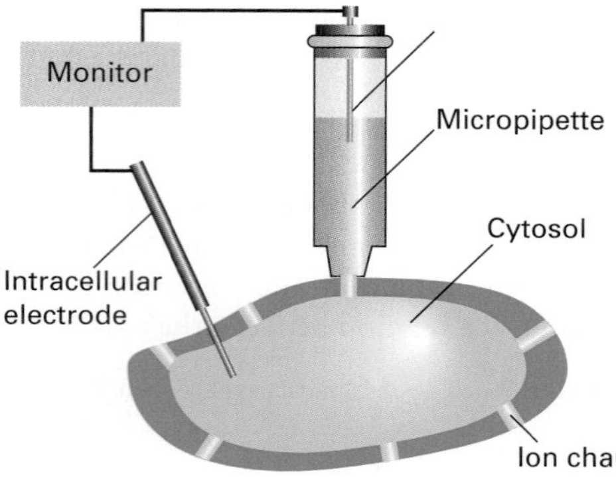

Fig. 9.5 A representation of the patch clamp technique for the measurement of ionic currents through membranes in intact cells. A section of membrane containing an ion channel is in tight contact with the tip of a micropipette containing an electrolyte solution and the patch electrode. An intracellular electrode is inserted into the cytosol of the cell and the two electrodes are connected to a power supply and current measuring device.

The **patch clamp technique** can be used to measure the transport of ions across cell membranes. One of many possible experimental arrangements is shown in Fig. 9.5. With mild suction, a 'patch' of membrane from a whole cell or a small section of a broken cell can be attached tightly to the tip of a micropipette filled with an electrolyte solution and containing an electrode, the **patch electrode**. A potential difference (the 'clamp') is applied between the patch electrode and an intra-cellular electrode in contact with the cytosol of the cell. If the membrane is permeable to ions at the applied potential difference, a current flows through the completed circuit. Using sufficiently narrow micropipette tips with diameters of less than 1 μm, ion currents of a few picoamperes (1 pA = 10^{-12} A) have been measured across sections of membranes containing only one ion channel protein.

A striking example of the importance of ion channels is their role in the propagation of impulses by neurons, the fundamental units of the nervous system. The cell membrane of a neuron is more permeable to K^+ ions than to either Na^+ or Cl^- ions. The key to the mechanism of action of a nerve cell is its use of Na^+ and K^+ channels to move ions across the membrane, modulating its potential. For example, the concentration of K^+ inside an inactive nerve cell is about 20 times that on the outside, whereas the concentration of Na^+ outside the cell is about 10 times that on the inside. The difference in concentrations of ions results in a transmembrane potential difference of about −62 mV, with the negative sign denoting that the inside has a lower potential. This potential difference is also called the *resting potential* of the cell membrane.

The transmembrane potential difference plays a particularly interesting role in the transmission of nerve impulses. Upon receiving an impulse, which is called an **action potential**, a site in the nerve cell membrane becomes transiently permeable to Na^+ and the transmembrane potential changes. To propagate along a nerve cell, the action potential must change the transmembrane potential by at least 20 mV, to values that are less negative than −40 mV. Propagation occurs when an action potential in one site of the membrane triggers an action potential in an adjacent site, with sites behind the moving action potential returning to the resting potential.

Ions such as H^+, Na^+, K^+, and Ca^{2+} are often transported actively across membranes by integral proteins called *ion pumps*. Ion pumps are molecular machines that work by adopting conformations that are permeable to one ion but not others depending on the state of phosphorylation of the protein. Because protein phosphorylation requires dephosphorylation of ATP, the conformational change that opens or closes the pump is endergonic and requires the use of energy stored during metabolism.

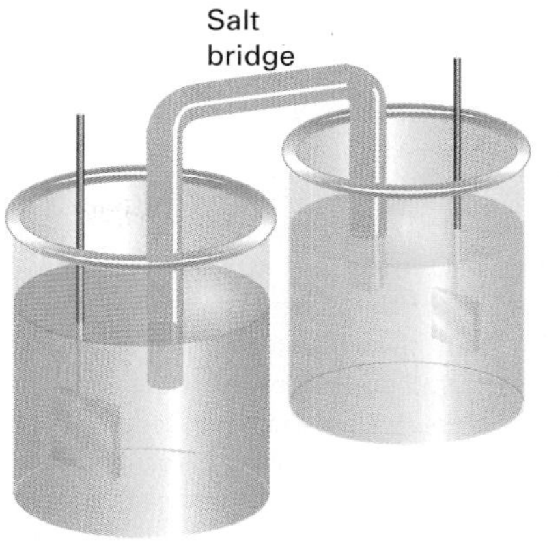

Fig. 9.7 When the electrolytes in the electrode compartments of a cell are different, they need to be joined so that ions can travel from one compartment to another. One device for joining the two compartments is a salt bridge.

Electrochemical cells

An **electrochemical cell** consists of two electronic conductors (metal or graphite, for instance) dipping into an electrolyte (an ionic conductor), which may be a solution, a liquid, or a solid. The electronic conductor and its surrounding electrolyte is an **electrode.** The physical structure containing them is called an **electrode compartment.** The two electrodes may share the same compartment (Fig. 9.6). If the electrolytes are different, then the two compartments may be joined by a **salt bridge,** which is an electrolyte solution that completes the electrical circuit by permitting ions to move between the compartments (Fig. 9.7). Alternatively, the two solutions may be in direct physical contact (for example, through a porous membrane) and form a **liquid junction.** However, a liquid junction introduces complications into the interpretation of measurements, and we do not consider it further.

A **galvanic cell** (also called a *voltaic cell*) is an electrochemical cell that produces electricity as a result of the spontaneous reaction occurring inside it. An **electrolytic cell** is an electrochemical cell in which a non-spontaneous reaction is driven by an external source of direct current. The commercially available dry cells, mercury cells, nickel–cadmium ('nicad'), and lithium ion cells used to power electrical equipment are all galvanic cells and produce electricity as a result of the spontaneous chemical reaction between the substances built into them at manufacture. A **fuel cell** is a galvanic cell in which the reagents, such as hydrogen and oxygen or methane and oxygen, are supplied continuously from outside. Fuel cells are used on manned spacecraft, are beginning to be considered for use in terrestrial vehicles, and gas supply companies hope that one day they may be used as a convenient, compact source of electricity in homes (see Impact 9.2 at the end of the chapter). Electric eels and electric catfish are biological versions of fuel cells in which the fuel is food and the cells are adaptations of muscle cells. Electrolytic cells include the arrangement used to electrolyse water into hydrogen and oxygen and to obtain aluminium from its oxide in the **Hall–Hérault process.** Electrolysis is the only commercially viable means for the production of fluorine. The electron transfer processes that occur in respiration and photosynthesis can be modelled by electrochemical cells in which electrons are transferred between proteins.

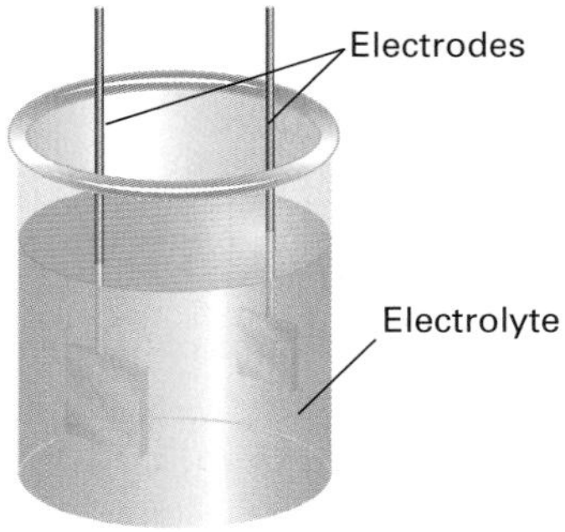

Fig. 9.6 The arrangement for an electrochemical cell in which the two electrodes share a common electrolyte.

9.3 Half-reactions and electrodes

A redox reaction is the outcome of the loss of electrons, and perhaps atoms, from one species and their gain by another species. It will be familiar from introductory chemistry that we identify the loss of electrons (oxidation) by noting whether an element

The chemist's toolkit 9.3 Oxidation numbers

An **oxidation number**, N_{ox}, is a formal measure of the extent to which an atom can be considered to have gained or lost electrons when it is part of a compound. The oxidation number of an uncombined form of the element, such as oxygen as O_2 and O_3, is zero. For monatomic cations, the oxidation number is the same as the charge number of the ion. Thus, the oxidation number of magnesium present as Mg^{2+} is +2 and that of chlorine present as Cl^- is −1. The oxidation number of elements present in polyatomic species is calculated formally by regarding ownership of electrons having passed to the more electronegative atom. Thus, if oxygen is present in a covalent compound, it is regarded as having acquired two electrons, to be present as O^{2-}, and therefore to have oxidation number −2. The oxidation numbers of the elements other than oxygen are then calculated by arranging for the sum of oxidation numbers of all the atoms in the species to be equal to the charge (including 0) of the species. Thus, SO_4^{2-} is regarded (but only for this purpose) as being $S^{+6}(O^{2-})_4$ and therefore as being a species in which sulfur is present with oxidation number +6, denoted S(+6). Oxygen has oxidation number −2 in all its compounds, except those with fluorine. Oxidation numbers are often denoted by roman numerals, as in S(VI) or Fe(III) for Fe^{3+}.

We, but not everyone, distinguish oxidation number from **oxidation state**. The oxidation state is the physical state in which an atom of that oxidation number is regarded as being in. Thus, oxygen is in the −2 oxidation state when its oxidation number is −2; sulfur is in the oxidation state +6 in the ion SO_4^{2-}.

has undergone an increase in oxidation number (see The chemist's toolkit 9.3 for a review of oxidation numbers). We identify the gain of electrons (reduction) by noting whether an element has undergone a decrease in oxidation number. The requirement to break and form covalent bonds in some redox reactions, as in the conversion of PCl_3 to PCl_5 or of NO_2^- to NO_3^-, is one of the reasons why redox reactions often achieve equilibrium quite slowly, often much more slowly than acid–base proton transfer reactions.

Brief illustration 9.3 Oxidizing and reducing agents

To identify the species that have undergone oxidation and reduction in the reaction CuS(s) + O_2(g) → Cu(s) + SO_2(g) we note the following oxidation numbers (in red)

$$\overset{+2\ -2}{\mathrm{CuS(s)}} + \overset{0}{\mathrm{O_2(g)}} \rightarrow \overset{0}{\mathrm{Cu(s)}} + \overset{+4\ -2}{\mathrm{SO_2(g)}}$$

We see that Cu(+2) is reduced to Cu(0), S(−2) is oxidized to S(+4), and O(0) reduced to O(−2).

Self-test 9.4

Identify which elements undergo oxidation and which reduction in the reaction 2 H_2(g) + O_2(g) → 2 H_2O(l).

Answer: H is oxidized, O is reduced

Any redox reaction may be expressed as the difference of two reduction **half-reactions**. Two examples of half reactions are

$$\text{Reduction of } Cu^{2+}\text{:}\quad Cu^{2+}(aq) + 2\,e^- \rightarrow Cu(s)$$

$$\text{Reduction of } Zn^{2+}\text{:}\quad Zn^{2+}(aq) + 2\,e^- \rightarrow Zn(s)$$

$$\text{Difference:}\quad Cu^{2+}(aq) + Zn(s) \rightarrow Cu(s) + Zn^{2+}(aq) \quad (A)$$

A half-reaction in which atom transfer accompanies electron transfer is

$$\text{Reduction of } MnO_4^-\text{:}\ MnO_4^-(aq) + 8\,H^+(aq) + 5\,e^- \rightarrow Mn^{2+}(aq) + 4\,H_2O(l) \quad (B)$$

where oxygen atoms are lost from MnO_4^-(aq) and form H_2O(l). In the discussion of redox reactions, the hydrogen ion is commonly denoted simply H^+(aq) rather than treated as a hydronium ion, H_3O^+(aq), as proton transfer is less of an issue and the chemical equations are simplified.

Half-reactions are *conceptual*. Redox reactions normally proceed by a much more complex mechanism in which the electron is never free. The electrons in these conceptual reactions are regarded as being 'in transit' and are not ascribed a state. The oxidized and reduced species in a half-reaction form a **redox couple**, denoted Ox/Red. Thus, the redox couples mentioned so far are Cu^{2+}/Cu, Zn^{2+}/Zn, and MnO_4^-, H^+/Mn^{2+}, H_2O. In general, we adopt the notation

$$\text{Couple: Ox/Red}\quad \text{Half-reaction: Ox} + \nu\,e^- \rightarrow \text{Red}$$

with ν the number of electrons regarded as transferred.

Example 9.2

Expressing a reaction in terms of half-reactions

Express the oxidation of NADH (nicotinamide adenine dinucleotide, **1**), which participates in the chain of oxidations that constitutes respiration, to NAD^+ (**2**) by oxygen, when the latter is reduced to H_2O_2, in aqueous solution as the difference of two reduction half-reactions. The overall reaction is NADH(aq) + O_2(g) + H^+(aq) → NAD^+(aq) + H_2O_2(aq).

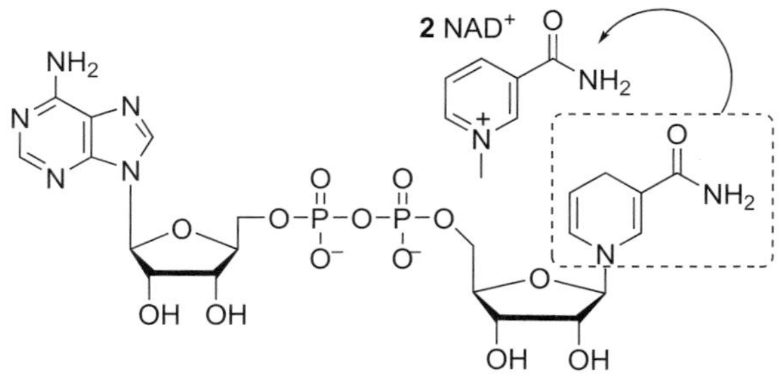

1 Nicotinamide adenine dinucleotide, reduced form (NADH)

Strategy To express a reaction as the difference of two reduction half-reactions, identify one reactant species that undergoes reduction, its corresponding reduction product, and write the half-reaction for this process. To find the second half-reaction, subtract the first half-reaction from the overall reaction and rearrange the species so that all the stoichiometric coefficients are positive and the equation is written as a reduction.

Solution Oxygen is reduced to H_2O_2, so one half-reaction is

$$O_2(g) + 2\,H^+(aq) + 2\,e^- \rightarrow H_2O_2(aq)$$

Subtraction of this half-reaction from the overall equation gives

$$NADH(aq) - H^+(aq) - 2\,e^- \rightarrow NAD^+(aq)$$

Addition of $H^+(aq) + 2\,e^-$ to both sides gives

$$NADH(aq) \rightarrow NAD^+(aq) + H^+(aq) + 2\,e^-$$

This is an oxidation half-reaction. Reverse it to find the corresponding reduction half-reaction:

$$NAD^+(aq) + H^+(aq) + 2\,e^- \rightarrow NADH(aq)$$

Self-test 9.5

Express the formation of H_2O from H_2 and O_2 in acidic solution as the difference of two reduction half-reactions.

Answer: $4\,H^+(aq) + 4\,e^- \rightarrow 2\,H_2(g)$, $O_2(g) + 4\,H^+(aq) + 4\,e^- \rightarrow 2\,H_2O(l)$

A chemical reaction need not be a redox reaction for it to be expressed in terms of reduction half-reactions. For instance, the expansion of a gas

$$H_2(g, p_i) \rightarrow H_2(g, p_f)$$

can be expressed as the difference of two reductions:

$$2\,H^+(aq) + 2\,e^- \rightarrow H_2(g, p_f)$$
$$2\,H^+(aq) + 2\,e^- \rightarrow H_2(g, p_i)$$

The two couples are both H^+/H_2 with the gas at a different pressure in each case. Similarly, the dissolution of the sparingly soluble salt silver chloride

$$AgCl(s) \rightarrow Ag^+(aq) + Cl^-(aq)$$

can be expressed as the difference of the following two reduction half-reactions:

$$AgCl(s) + e^- \rightarrow Ag(s) + Cl^-(aq)$$
$$Ag^+(aq) + e^- \rightarrow Ag(s)$$

We saw in Chapter 7 that a natural way to express the composition of a system is in terms of the reaction quotient, Q. The quotient for a half-reaction is defined like the quotient for the overall reaction, but with the electrons ignored. Thus, for the half-reaction of the $NAD^+/NADH$ couple in Example 9.2 we would write

$$NAD^+(aq) + H^+(aq) + 2\,e^- \rightarrow NADH(aq)$$

$$Q = \frac{a_{NADH}}{a_{NAD^+}a_{H^+}} \approx \frac{[NADH]}{[NAD^+][H^+]}$$

In elementary work, and provided the solution is very dilute, the activities are interpreted as the numerical values of the molar concentrations (see Table 6.2). The replacement of activities by molar concentrations is very hazardous for ionic solutions, as we have seen, so wherever possible we delay taking that final step.

9.4 Reactions at electrodes

In an electrochemical cell, the **anode** is where oxidation takes place and the **cathode** is where reduction takes place. As the reaction proceeds in a galvanic cell, the electrons released at the anode travel through the external circuit (Fig. 9.8). They re-enter the cell at the cathode, where they bring about reduction.

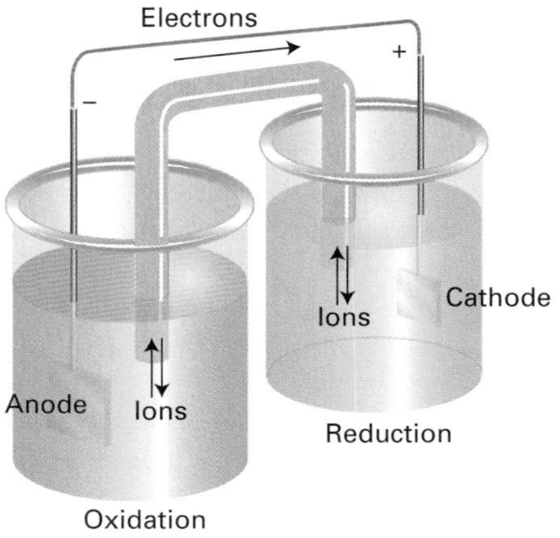

Fig. 9.8 The flow of electrons in the external circuit is from the anode of a galvanic cell, where they have been lost in the oxidation reaction, to the cathode, where they are used in the reduction reaction. Electrical neutrality is preserved in the electrolytes by the flow of cations and anions in opposite directions through the salt bridge.

Because negatively charged electrons tend to travel to regions of higher (more positive) potential, this flow of current in the external circuit, from anode to cathode, corresponds to the cathode having a higher potential than the anode. In an electrolytic cell, the anode is also the location of oxidation (by definition). Now, though, electrons must be withdrawn from the species in the anode compartment, so the anode must be connected to the positive terminal of an external supply. Similarly, electrons must pass from the cathode to the species undergoing reduction, so the cathode must be connected to the negative terminal of a supply (Fig. 9.9).

In a **gas electrode** (Fig. 9.10), a gas is in equilibrium with a solution of its ions in the presence of an inert metal. The inert metal, which is often platinum, acts as a source or sink of electrons but takes no other part in the reaction except perhaps acting as a catalyst. One important example is the *hydrogen electrode*, in which hydrogen is bubbled through an aqueous solution of hydrogen ions and the redox couple is H^+/H_2. This electrode is denoted $Pt(s)|H_2(g)|H^+(aq)$. The vertical lines denote junctions between phases. In this electrode, the junctions are between the platinum and the gas and between the gas and the liquid containing its ions.

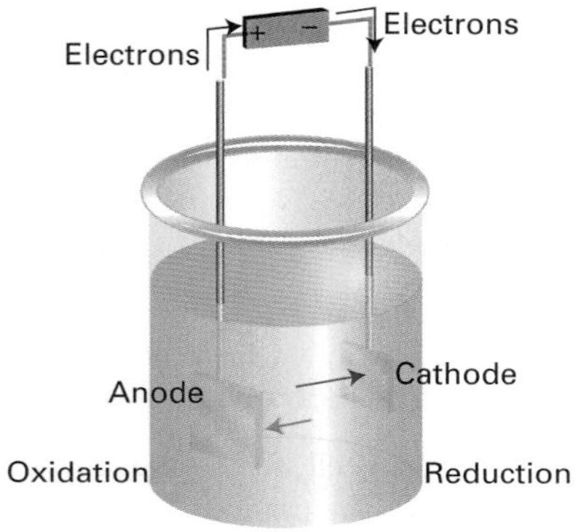

Fig. 9.9 The flow of electrons and ions in an electrolytic cell. An external supply forces electrons into the cathode, where they are used to bring about a reduction, and withdraws them from the anode, which results in an oxidation reaction at that electrode. Cations migrate towards the negatively charged cathode and anions migrate towards the positively charged anode. An electrolytic cell usually consists of a single compartment, but a number of industrial versions have two compartments.

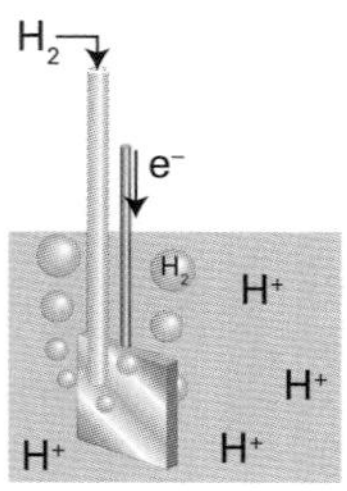

Fig. 9.10 The schematic structure of a hydrogen electrode, which is like other gas electrodes. Hydrogen is bubbled over a black (that is, finely divided) platinum surface that is in contact with a solution containing hydrogen ions. The platinum, as well as acting as a source or sink for electrons speeds the electrode reaction because hydrogen attaches to (adsorbs on) the surface as atoms.

Example 9.3

Writing the half-reaction for a gas electrode

Write the half-reaction and the reaction quotient for the reduction of oxygen to water in acidic solution.

Strategy Write the chemical equation for the half-reaction. Then express the reaction quotient in terms of the activities and the corresponding stoichiometric coefficients, with products in the numerator and reactants in the denominator. Pure (and nearly pure) solids and liquids do not appear in Q; nor does the electron. The activity of a gas is set equal to the numerical value of its partial pressure in bar (more formally: $a_J = p_J/p^{\ominus}$).

Solution The equation for the reduction of O_2 in acidic solution is

$$O_2(g) + 4\,H^+(aq) + 4\,e^- \rightarrow 2\,H_2O(l)$$

The reaction quotient for the half-reaction is therefore

$$Q = \frac{\overbrace{a_{H_2O}^2}^{1}}{\underbrace{a_{O_2}}_{p(O_2)}\,a_{H^+}^4} = \frac{1}{p(O_2)a_{H^+}^4}$$

Note the very strong dependence of Q on the hydrogen ion activity.

Self-test 9.6

Write the half-reaction and the reaction quotient for a chlorine gas electrode.

Answer: $Cl_2(g) + 2\,e^- \rightarrow 2\,Cl^-(aq)$, $Q = a_{Cl^-}^2/p(Cl_2)$

A **metal–insoluble salt electrode** consists of a metal M covered by a porous layer of insoluble salt MX, the whole being immersed in a solution containing X^- ions (Fig. 9.11). The electrode is denoted $M|MX|X^-$, where the vertical line denotes a boundary across which electron transfer takes place. An example is the silver–silver chloride electrode, $Ag(s)|AgCl(s)|Cl^-(aq)$, for which the reduction half-reaction is

$$AgCl(s) + e^- \rightarrow Ag(s) + Cl^-(aq)$$

Self-test 9.13

Given the standard potentials $E^\ominus(Fe^{3+},Fe) = -0.04$ V and $E^\ominus(Fe^{2+},Fe) = -0.44$ V, calculate $E^\ominus(Fe^{3+},Fe^{2+})$.

Answer: +0.76 V

Once we have measured $\Delta_r G^\ominus$ we can use thermodynamic relations to determine other properties. For instance, as we show in the following Derivation, the standard entropy of the cell reaction can be obtained from the change in the cell potential with temperature:

$$\Delta_r S^\ominus = \frac{\nu F\{E^\ominus_{cell} - E^{\ominus\prime}_{cell}\}}{T - T'} \qquad \text{Standard entropy of reaction} \qquad (9.19)$$

Derivation 9.4

The reaction entropy from the cell potential

The definition of the Gibbs energy is $G = H - TS$. This formula applies to all substances involved in a reaction, so at a given temperature $\Delta_r G^\ominus(T) = \Delta_r H^\ominus - T\Delta_r S^\ominus$. If we can ignore the weak temperature dependence of $\Delta_r H^\ominus$ and $\Delta_r S^\ominus$, at a temperature T' we can write $\Delta_r G^\ominus(T') = \Delta_r H^\ominus - T'\Delta_r S^\ominus$. Therefore,

$$\Delta_r G^\ominus(T') - \Delta_r G^\ominus(T) = -(T' - T)\Delta_r S^\ominus$$

Substitution of $\Delta_r G^\ominus = -\nu F E^\ominus_{cell}$ then gives

$$-\nu F E^{\ominus\prime}_{cell} + \nu F E^\ominus_{cell} = -(T' - T)\Delta_r S^\ominus$$

which becomes eqn 9.19.

We see from eqn 9.19 that the standard cell potential increases with temperature if the standard reaction entropy is positive, and that the slope of a plot of potential against temperature is proportional to the reaction entropy (Fig. 9.16). An implication is that if the cell reaction produces a lot of gas, then its potential will increase with temperature. The opposite is true for a reaction that consumes gas.

Finally, we can combine the results obtained so far by using $G = H - TS$ in the form $H = G + TS$ to obtain the standard reaction enthalpy:

$$\Delta_r H^\ominus = \Delta_r G^\ominus + T\Delta_r S^\ominus \qquad (9.20)$$

with $\Delta_r G^\ominus$ determined from the cell potential and $\Delta_r S^\ominus$ from its temperature variation. Thus, we now have a non-calorimetric method of measuring a reaction enthalpy.

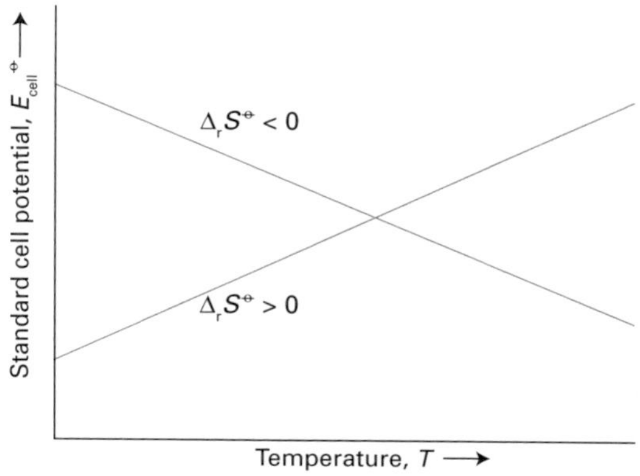

Fig. 9.16 The variation of the standard potential of a cell with temperature depends on the standard entropy of the cell reaction.

Example 9.8

Using the temperature dependence of the cell potential

The standard potential of the cell

$Pt(s)|H_2(g)|HCl(aq)|Hg_2Cl_2(s)|Hg(l)$

was found to be +0.2699 V at 293 K and +0.2669 V at 303 K. Evaluate the standard Gibbs energy, enthalpy, and entropy at 298 K of the reaction

$Hg_2Cl_2(s) + H_2(g) \rightarrow 2\,Hg(l) + 2\,HCl(aq)$

Strategy We find the standard reaction Gibbs energy from the standard cell potential by using eqn 9.16 and making a linear interpolation between the two temperatures (in this case, we take the mean $E^\ominus_{cell}$ because 298 K lies midway between 293 K and 303 K). The standard reaction entropy is obtained by substituting the data into eqn 9.19. Then the standard reaction enthalpy is obtained by combining these two quantities by using eqn 9.20. Use 1 C V = 1 J.

Solution Because the mean standard cell potential is +0.2684 V and $\nu = 2$ for the reaction,

$$\Delta_r G^\ominus = -\nu F E^\ominus_{cell} = -2 \times (9.6485 \times 10^4\ \mathrm{C\ mol^{-1}}) \times (0.2684\ \mathrm{V})$$
$$= -51.79\ \mathrm{kJ\ mol^{-1}}$$

Then, from eqn 9.19, the standard reaction entropy is

$$\Delta_r S^\ominus = \frac{\overbrace{2}^{\nu} \times \overbrace{(9.6485 \times 10^4\ \mathrm{C\,mol^{-1}})}^{F} \times \left(\overbrace{0.2699\ \mathrm{V}}^{E^\ominus_{cell}} - \overbrace{0.2669\ \mathrm{V}}^{E^{\ominus\prime}_{cell}}\right)}{\underbrace{293\ \mathrm{K}}_{T} - \underbrace{303\ \mathrm{K}}_{T'}}$$
$$= -57.9\ \mathrm{J\ K^{-1}\ mol^{-1}}$$

For the next stage of the calculation it is convenient to write the last value as -0.0579 kJ K^{-1} mol^{-1}. Then, from eqn 9.20 we find

$$\Delta_r H^\ominus = (-51.79\ \mathrm{kJ\ mol^{-1}}) + (298\ \mathrm{K}) \times (-0.0579\ \mathrm{kJ\ K^{-1}\ mol^{-1}})$$
$$= -69\ \mathrm{kJ\ mol^{-1}}$$

One difficulty with this procedure lies in the accurate measurement of small temperature variations of cell potential. Nevertheless, it is another example of the striking ability of thermodynamics to relate the apparently unrelated, in this case to relate electrical measurements to thermal properties.

Self-test 9.14

Predict the standard potential of the *Harned cell*

Pt(s)|H_2(g)|HCl(aq)|AgCl(s)|Ag(s)

at 303 K from tables of thermodynamic data for 298 K.

Answer: +0.2168 V

Impact on technology 9.2

Fuel cells

A fuel cell operates like a conventional galvanic cell with the exception that the reactants are supplied from outside rather than forming an integral part of its construction. A fundamental and important example of a fuel cell is the *hydrogen/oxygen cell*, such as the ones used in the Apollo Moon missions. One of the electrolytes used is concentrated aqueous potassium hydroxide maintained at 200 °C and 20–40 atm; the electrodes may be porous nickel in the form of sheets of compressed powder. The cathode reaction is the reduction

$$O_2(g) + 2\,H_2O(l) + 4e^- \rightarrow 4\,OH^-(aq) \qquad E^\ominus = +0.40\ V$$

and the anode reaction is the oxidation

$$H_2(g) + 2\,OH^-(aq) \rightarrow 2\,H_2O(l) + 2\,e^-$$

For the corresponding reduction, $E^\ominus = -0.83$ V. Because the overall reaction

$$2\,H_2(g) + O_2(g) \rightarrow 2\,H_2O(l) \qquad E^\ominus_{cell} = +1.23\ V$$

is exothermic as well as spontaneous, it is less favourable thermodynamically at 200 °C than at 25 °C, so the cell potential is lower at the higher temperature. However, the increased pressure compensates for the increased temperature, and at 200 °C and 40 atm $E_{cell} \approx +1.2$ V.

A property that determines the efficiency of an electrode is the **current density**, the electric current flowing through a region of an electrode divided by the area of the region. One advantage of the hydrogen/oxygen system is the large **exchange-current density**, the magnitude of the equal but opposite current densities when the electrode is at equilibrium, of the hydrogen reaction. Unfortunately, the oxygen reaction has an exchange-current density of only about 0.1 nA cm^{-2}, which limits the current available from the cell. One way round the difficulty is to use a catalytic surface with a large surface area. One type of highly developed fuel cell has phosphoric acid as the electrolyte and operates with hydrogen and air at about 200 °C; the hydrogen is obtained from a reforming reaction on natural gas

$$\text{Anode: } 2\,H_2(g) \rightarrow 4\,H^+(aq) + 4\,e^-$$

$$\text{Cathode: } O_2(g) + 4\,H^+(aq) + 4\,e^- \rightarrow 2\,H_2O(l)$$

This fuel cell has shown promise for combined heat and power systems (CHP systems). In such systems, the waste heat is used to heat buildings or to do work. Efficiency in a CHP plant can reach 80 per cent. The power output of batteries of such cells has reached the order of 10 MW. Although hydrogen gas is an attractive fuel, it has disadvantages for mobile applications: it is difficult to store and dangerous to handle. One possibility for portable fuel cells is to store the hydrogen in carbon nanotubes. It has been shown that carbon nanofibres in herringbone patterns can store huge amounts of hydrogen and result in energy densities twice that of gasoline.

Cells with molten carbonate electrolytes at about 600 °C can make use of natural gas directly. Until these materials have been developed, one attractive fuel is methanol, which is easy to handle and is rich in hydrogen atoms:

$$\text{Anode: } CH_3OH(l) + 6\,OH^-(aq) \rightarrow 5\,H_2O(l) + CO_2(g) + 6\,e^-$$

$$\text{Cathode: } O_2(g) + 4\,e^- + 2\,H_2O(l) \rightarrow 4\,OH^-(aq)$$

One disadvantage of methanol, however, is the phenomenon of **electro-osmotic drag** in which protons moving through the polymer electrolyte membrane separating the anode and cathode carry water and methanol with them into the cathode compartment where the potential is sufficient to oxidize CH_3OH to CO_2, so reducing the efficiency of the cell. Solid ionic conducting oxide cells operate at about 1000 °C and can use hydrocarbons directly as fuel.

A biofuel cell is like a conventional fuel cell but in place of a platinum catalyst it uses enzymes or even whole organisms. The electricity will be extracted through organic molecules that can support the transfer of electrons. One application will be as the power source for medical implants, such as pacemakers, perhaps using the glucose present in the bloodstream as the fuel.

Checklist of key concepts

☐ 1 Deviations from ideal behaviour in ionic solutions are ascribed to the interaction of an ion with its ionic atmosphere.

☐ 2 The Debye–Hückel limiting law relates the mean activity of ions in a solution to the ionic strength.

☐ 3 The molar conductivity of a strong electrolyte follows the Kohlrausch law.

☐ 4 The rate at which an ion migrates through solution is determined by its mobility, which depends on its charge, its hydrodynamic radius, and the viscosity of the solution.

☐ 5 Protons migrate by the Grotthus mechanism, Fig. 9.4.

☐ 6 A galvanic cell is an electrochemical cell in which a spontaneous chemical reaction produces a potential difference.

☐ 7 An electrolytic cell is an electrochemical cell in which an external source of current is used to drive a non-spontaneous chemical reaction.

☐ 8 A redox reaction is expressed as the difference of two reduction half-reactions.

☐ 9 A cathode is the site of reduction; an anode is the site of oxidation.

☐ 10 The cell potential is the potential difference it produces when operating reversibly.

☐ 11 The Nernst equation relates the cell potential to the composition of the reaction mixture.

☐ 12 The standard potential of a couple is the standard cell potential in which it forms the right-hand electrode and a hydrogen electrode is on the left.

☐ 13 The pH of a solution is determined by measuring the potential of a glass electrode.

☐ 14 A couple with a low standard potential has a thermodynamic tendency (in the sense $K > 1$) to reduce a couple with a high standard potential.

☐ 15 The entropy and enthalpy of a cell reaction are measured from the temperature dependence of the cell potential.

Road map of key equations

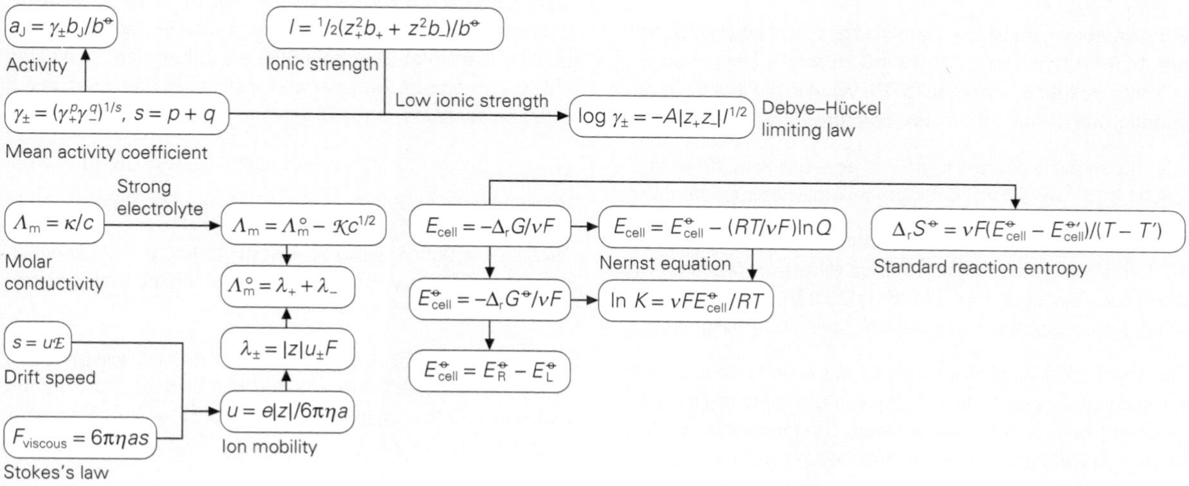

Questions and exercises

Discussion questions

9.1 Describe the general features of the Debye–Hückel theory of electrolyte solutions. Which approximations limit its reliability to very low concentrations?

9.2 Describe the mechanism of proton conduction in water. Could a similar mechanism apply to proton conduction in liquid ammonia?

9.3 Distinguish between galvanic, electrolytic, and fuel cells. Explain why salt bridges are routinely used in electrochemical cell measurements.

9.4 Discuss how the electrochemical series can be used to determine if a redox reaction is spontaneous.

9.5 Describe an electrochemical method for the determination of thermodynamic properties of a chemical reaction.

Exercises

9.1 Calculate the ionic strength of a solution that is 0.15 mol kg^{-1} in KCl(aq) and 0.30 mol kg^{-1} in $CuSO_4$(aq).

9.2 Calculate the masses of (a) $Ca(NO_3)_2$ and, separately, (b) NaCl to add to a 0.150 mol kg^{-1} solution of KNO_3(aq) containing 500 g of solvent to raise its ionic strength to 0.250.

9.3 Express the mean activity coefficient of the ions in a solution of MgF_2 in terms of the activity coefficients of the individual ions.

9.4 Estimate the mean ionic activity coefficient and activity of a solution that is 0.015 mol kg^{-1} MgF_2(aq) and 0.025 mol kg^{-1} NaCl(aq).

9.5 The mean activity coefficients of HBr in three dilute aqueous solutions at 25 °C are 0.930 (at 5.0 mmol kg^{-1}), 0.907 (at 10.0 mmol kg^{-1}), and 0.879 (at 20.0 mmol kg^{-1}). Estimate the value of *B* in the extended Debye–Hückel law.

9.6 The limiting molar conductivities of KCl, KNO_3, and $AgNO_3$ are 14.99 mS $m^2 mol^{-1}$, 14.50 mS $m^2 mol^{-1}$, and 13.34 mS $m^2 mol^{-1}$, respectively (all at 25 °C). What is the limiting molar conductivity of AgCl at this temperature?

9.7 The mobility of a chloride ion in aqueous solution at 25 °C is 7.91×10^{-8} $m^2 s^{-1} V^{-1}$. Calculate its molar ionic conductivity.

9.8 The mobility of a Rb^+ ion in aqueous solution is 7.92×10^{-8} $m^2 s^{-1} V^{-1}$ at 25 °C. The potential difference between two electrodes placed in the solution is 35.0 V. If the electrodes are 8.00 mm apart, what is the drift speed of the Rb^+ ion?

9.9 The resistances of a series of aqueous NaCl solutions, formed by successive dilution of a sample, were measured in a cell with cell constant (the constant *C* in the relation $\kappa = C/R$) equal to 0.2063 cm^{-1}. The following values were found:

$c/(mol\ dm^{-3})$	0.00050	0.0010	0.0050	0.010	0.020	0.050
R/Ω	3314	1669	342.1	174.1	89.08	37.14

(a) Verify that the molar conductivity follows the Kohlrausch law and find the limiting molar conductivity. (b) Determine the coefficient $\mathcal{K}$. (c) Use the value of $\mathcal{K}$ (which should depend only on the nature, not the identity of the ions) and the information that $\lambda(Na^+) = 5.01$ mS $m^2 mol^{-1}$ and $\lambda(I^-) = 7.68$ mS $m^2 mol^{-1}$ to predict (i) the molar conductivity, (ii) the conductivity, (iii) the resistance it would show in the cell, of 0.010 mol dm^{-3} NaI(aq) at 25 °C.

9.10 After correction for the water conductivity, the conductivity of a saturated aqueous solution of AgCl at 25 °C was found to be 0.1887 mS m^{-1}. What is the solubility of silver chloride at this temperature?

9.11 The molar conductivity of 0.020 M HCOOH(aq) is 3.83 mS $m^2 mol^{-1}$. What is the value of pK_a for formic acid?

9.12 The mobility of an ion depends on its charge and if a large molecule, such as a protein, can be contrived to have zero net charge, then it does not respond to an electric field. This 'isoelectric point' can be reached by varying the pH of the medium. The speed with which bovine serum albumin (BSA) moves through water under the influence of an electric field was monitored at several values of pH, and the data are listed below. What is the isoelectric point of the protein?

pH	4.20	4.56	5.20	5.65	6.30	7.00
Velocity/(μm s^{-1})	0.50	0.18	−0.25	−0.65	−0.90	−1.25

Hint: Use a plot of speed against pH to find the pH at which the speed is zero, which is the pH at which the molecule has zero net charge.

9.13 Express the oxidation of cysteine ($HSCH_2CH(NH_2)COOH$) to cystine ($HOOCCH(NH_2)CH_2SSCH_2CH(NH_2)COOH$) as the difference of two half-reactions, one of which is $O_2(g) + 4 H^+(aq) + 4 e^- \rightarrow 2 H_2O(l)$.

9.14 From the biological standard half-cell potentials $E^\oplus(O_2,H^+,H_2O) = +0.82$ V and $E^\oplus(NADH^+,H^+,NADH) = -0.32$ V, calculate the standard potential arising from the reaction in which NADH is oxidized to NAD^+ and the corresponding biological standard reaction Gibbs energy.

9.15 Consider a hydrogen electrode in HBr(aq) at 25 °C operating at 1.45 bar. Estimate the change in the electrode potential when the solution is changed from 5.0 mmol dm^{-3} to 15.0 mmol dm^{-3}.

9.16 Devise a cell in which the cell reaction is $Mn(s) + Cl_2(g) \rightarrow MnCl_2(aq)$. Give the half-reactions for the electrodes and from the standard cell potential of +2.54 V deduce the standard potential of the Mn^{2+}/Mn couple.

9.17 Write the cell reactions, electrode half-reactions, and Nernst equations for the following cells:

(a) $Ag(s)|AgNO_3(aq,m_L)||AgNO_3(aq,m_R)|Ag(s)$

(b) $Pt(s)|H_2(g,p_L)|HCl(aq)|H_2(g,p_L)|Pt(s)$

(c) $Pt(s)|K_3[Fe(CN)_6](aq),K_4[Fe(CN)_6](aq)||Mn^{2+}(aq),H^+(aq)|MnO_2(s)|Pt(s)$

(d) $Pt(s)|Cl_2(g)|HCl(aq)||HBr(aq)|Br_2(l)|Pt(s)$

(e) $Pt(s)|Fe^{3+}(aq),Fe^{2+}(aq)||Sn^{4+}(aq),Sn^{2+}(aq)|Pt(s)$

(f) $Fe(s)|Fe^{2+}(aq)||Mn^{2+}(aq),H^+(aq)|MnO_2(s)|Pt(s)$

9.18 Use the standard potentials of the electrodes to calculate the standard potentials of the cells in Exercise 9.19.

9.19 Devise cells in which the following are the reactions. In each case state the value for ν to use in the Nernst equation.

(a) $Fe(s) + PbSO_4(aq) \rightarrow FeSO_4(aq) + Pb(s)$

(b) $Hg_2Cl_2(s) + H_2(g) \rightarrow 2\ HCl(aq) + 2\ Hg(l)$

(c) $2\ H_2(g) + O_2(g) \rightarrow 2\ H_2O(l)$

(d) $H_2(g) + O_2(g) \rightarrow H_2O_2(aq)$

(e) $H_2(g) + I_2(g) \rightarrow 2\ HI(aq)$

(f) $2\ CuCl(aq) \rightarrow Cu(s) + CuCl_2(aq)$

9.20 Use the standard potentials of the electrodes to calculate the standard potentials of the cells devised in Exercise 9.19.

9.21 A fuel cell develops an electric potential from the chemical reaction between reagents supplied from an outside source. What is the potential of a cell fuelled by (a) hydrogen and oxygen, (b) the complete oxidation of benzene at 1.0 bar and 298 K?

9.22 A fuel cell is constructed in which both electrodes make use of the oxidation of methane. The left-hand electrode makes use of the complete oxidation of methane to carbon dioxide and water; the right-hand electrode makes use of the partial oxidation of methane to carbon monoxide and water. (a) Which electrode is the cathode? (b) What is the cell potential at 25 °C when all gases are at 1 bar?

9.23 State what you would expect to happen to the cell potential when the following changes are made to the corresponding cells in Exercise 9.18. Confirm your prediction by using the Nernst equation in each case.

(a) The molar concentration of silver nitrate in the left-hand compartment is increased.

(b) The pressure of hydrogen in the left-hand compartment is increased.

(c) The pH of the right-hand compartment is decreased.

(d) The concentration of HCl is increased.

(e) Some iron(III) chloride is added to both compartments.

(f) Acid is added to both compartments.

9.24 State what you would expect to happen to the cell potential when the following changes are made to the corresponding cells devised in Exercise 9.29. Confirm your prediction by using the Nernst equation in each case.

(a) The molar concentration of $FeSO_4$ is increased.

(b) Some nitric acid is added to both cell compartments.

(c) The pressure of oxygen is increased.

(d) The pressure of hydrogen is increased.

(e) Some (i) hydrochloric acid, (ii) hydroiodic acid is added to both compartments.

(f) Hydrochloric acid is added to both compartments.

9.25 (a) Calculate the standard potential of the cell $Hg(l)|HgCl_2(aq)||TlNO_3(aq)|Tl(s)$ at 25 °C. (b) Calculate the cell potential when the molar concentration of the Hg^{2+} ion is 0.230 mol dm^{-3} and that of the Tl^+ ion is 0.720 mol dm^{-3}.

9.26 Calculate the standard Gibbs energies at 25 °C of the following reactions from the standard potential data in the *Data section*.

(a) $Ca(s) + 2\ H_2O(l) \rightarrow Ca(OH)_2(aq) + H_2(g)$

(b) $2\ Ca(s) + 4\ H_2O(l) \rightarrow 2\ Ca(OH)_2(aq) + 2\ H_2(g)$

(c) $Fe(s) + 2\ H_2O(l) \rightarrow Fe(OH)_2(aq) + H_2(g)$

(d) $Na_2S_2O_8(aq) + 2\ NaI(aq) \rightarrow I_2(s) + 2\ Na_2SO_4(aq)$

(e) $Na_2S_2O_8(aq) + 2\ KI(aq) \rightarrow I_2(s) + Na_2SO_4(aq) + K_2SO_4(aq)$

(f) $Pb(s) + Na_2CO_3(aq) \rightarrow PbCO_3(aq) + 2\ Na(s)$

9.27 Calculate the biological standard Gibbs energies of reactions of the following reactions and half-reactions:

(a) $2\ NADH(aq) + O_2(g) + 2\ H^+(aq) \rightarrow 2\ NAD^+(aq) + 2\ H_2O(l)$ $E^{\oplus} = +1.14$ V

(b) $Malate(aq) + NAD^+(aq) \rightarrow oxaloacetate(aq) + NADH(aq) + H^+(aq)$ $E^{\oplus} = -0.154$ V

(c) $O_2(g) + 4H^+(aq) + 4\ e^- \rightarrow 2\ H_2O(l)$ $E^{\oplus} = +0.81$ V

9.28 Tabulated thermodynamic data can be used to predict the standard potential of a cell even if it cannot be measured directly. The standard Gibbs energy of the reaction $K_2CrO_4(aq) + 2\ Ag(s) + 2\ FeCl_3(aq) \rightarrow Ag_2CrO_4(s) + 2\ FeCl_2(aq) + 2\ KCl(aq)$ is −62.5 kJ mol^{-1} at 298 K. (a) Calculate the standard potential of the corresponding galvanic cell and (b) the standard potential of the $Ag_2CrO_4/Ag,CrO_4^{2-}$ couple.

9.29 Estimate the potential at 25 °C of the cell

$Ag(s)|AgCl(s)|KCl(aq,\ 0.025\ mol\ kg^{-1})||AgNO_3(aq,\ 0.010\ mol\ kg^{-1})|Ag(s)$

9.30 (a) Use the information in the *Data section* to calculate the standard potential of the cell $Ag(s)|AgNO_3(aq)||Cu(NO_3)_2(aq)|Cu(s)$ and the standard Gibbs energy and enthalpy of the cell reaction at 25 °C. (b) Estimate the value of $\Delta_r G^{\ominus}$ at 35 °C.

9.31 (a) Calculate the standard potential of the cell Pt(s)|cystine(aq), cysteine(aq)||H^+(aq)|O_2(g)|Pt(s) and the standard Gibbs energy and enthalpy of the cell reaction at 25 °C. (b) Estimate the value of $\Delta_r G^\ominus$ at 35 °C. Use $E^\ominus = -0.34$ V for the cysteine/cystine couple.

9.32 Calculate the equilibrium constants of the following reactions at 25 °C from standard potential data:

(a) $Sn(s) + Sn^{4+}(aq) \rightleftharpoons 2\ Sn^{2+}(aq)$

(b) $Sn(s) + 2\ AgBr(s) \rightleftharpoons SnBr_2(aq) + 2\ Ag(s)$

(c) $Fe(s) + Hg(NO_3)_2(aq) \rightleftharpoons Hg(l) + Fe(NO_3)_2(aq)$

(d) $Cd(s) + CuSO_4(aq) \rightleftharpoons Cu(s) + CdSO_4(aq)$

(e) $Cu^{2+}(aq) + Cu(s) \rightleftharpoons 2\ Cu^+(aq)$

(f) $3\ Au^{2+}(aq) \rightleftharpoons Au(s) + 2\ Au^{3+}(aq)$

9.33 The dichromate ion in acidic solution is a common oxidizing agent for organic compounds. Derive an expression for the potential of an electrode for which the half-reaction is the reduction of $Cr_2O_7^{2-}$ ions to Cr^{3+} ions in acidic solution.

9.34 The permanganate ion is a common oxidizing agent. What is the standard potential of the MnO_4^-,H^+/Mn^{2+} couple at (a) pH = 6.00, (b) general pH?

9.35 The biological standard potential of the couple pyruvic acid/lactic acid is −0.19 V at 25 °C. What is the thermodynamic standard potential of the couple? Pyruvic acid is $CH_3COCOOH$ and lactic acid is $CH_3CH(OH)COOH$.

9.36 One ecologically important equilibrium is that between carbonate and hydrogencarbonate (bicarbonate) ions in natural water. (a) The standard Gibbs energies of formation of CO_3^{2-}(aq) and HCO_3^-(aq) are −527.81 kJ mol^{-1} and −586.77 kJ mol^{-1}, respectively. What is the standard potential of the HCO_3^-/CO_3^{2-},H_2 couple? (b) Calculate the standard potential of a cell in which the cell reaction is $Na_2CO_3(aq) + H_2O(l) \rightarrow NaHCO_3(aq) + NaOH(aq)$. (c) Write the Nernst equation for the cell, and (d) predict and calculate the change in potential when the pH is change to 7.0. (e) Calculate the value of pK_a for HCO_3^-(aq).

9.37 (a) Can mercury produce zinc metal from aqueous zinc sulfate under standard conditions? (b) Can chlorine gas oxidize water to oxygen gas under standard conditions in basic solution?

9.38 For a hydrogen/oxygen fuel cell, with an overall four-electron cell reaction $2\ H_2(g) + O_2(g) \rightarrow 2\ H_2O(l)$, the standard cell potential is +1.2335 V at 293 K and +1.2251 V at 303 K. Calculate the standard reaction enthalpy and entropy within this temperature range.

Projects

9.39 Consider the Harned cell, Pt(s)|H_2(g, 1 bar)|HCl(aq, *b*)|AgCl(s)|Ag(s). Show that the standard potential of the silver–silver chloride electrode may be determined by plotting $E-(RT/F)\ln b$ against $b^{1/2}$. *Hint:* Express the cell potential in terms of activities, and use the Debye–Hückel law to estimate the mean activity coefficient. (b) Use the procedure you devised in part (a) and the following data at 25 °C to determine the standard potential of the silver–silver chloride electrode

$b/(10^{-3}\ b^\ominus)$	3.215	5.619	9.138	25.63
E/V	0.52053	0.49257	0.46860	0.41824

9.40 The standard potentials of proteins are not commonly measured by the methods described in this chapter because proteins often lose their native structure and their function when they react on the surfaces of electrodes. In an alternative method, the oxidized protein is allowed to react with an appropriate electron donor in solution. The standard potential of the protein is then determined from the Nernst equation, the equilibrium concentrations of all species in solution, and the known standard potential of the electron donor. We illustrate this method with the protein cytochrome *c*. (a) The one-electron reaction between cytochrome *c*, cyt, and 2,6-dichloroindophenol, D, can be written as

$$cyt_{ox} + D_{red} \rightleftharpoons cyt_{red} + D_{ox}$$

Consider E_{cyt} and E_D to be the standard potentials of cytochrome *c* and D, respectively. Show that, at equilibrium (eq), a plot of $\ln([D_{ox}]_{eq}/[D_{red}]_{eq})$ against $\ln([cyt_{ox}]_{eq}/[cyt_{red}]_{eq})$ is linear with slope of 1 and *y*-intercept $F(E_{cyt} - E_D)/RT$, where equilibrium activities are replaced by the numerical values of equilibrium molar concentrations. (b) The following data were obtained for the reaction between oxidized cytochrome *c* and reduced D at pH = 6.5 buffer and 298 K. The ratios $[D_{ox}]_{eq}/[D_{red}]_{eq}$ and $[cyt_{ox}]_{eq}/[cyt_{red}]_{eq}$ were adjusted by adding known volumes of a solution of sodium ascorbate, a reducing agent, to a solution containing oxidized cytochrome *c* and reduced D. From the data and the standard potential of D of 0.237 V, determine the standard potential of cytochrome c at pH = 6.5 and 298 K.

$[D_{ox}]_{eq}/[D_{red}]_{eq}$	0.00279	0.00843	0.0257	0.0497	0.0748	0.238	0.534
$[cyt_{ox}]_{eq}/[cyt_{red}]_{eq}$	0.0106	0.00230	0.0894	0.197	0.335	0.809	1.39

10

Chemical kinetics: the rates of reactions

The branch of physical chemistry called **chemical kinetics** is concerned with the rates of chemical reactions. Chemical kinetics deals with how rapidly reactants are consumed and products formed, how reaction rates respond to changes in the conditions or the presence of a catalyst, and the identification of the steps by which a reaction takes place.

One reason for studying the rates of reactions is the practical importance of being able to predict how quickly a reaction mixture approaches equilibrium. The rate might depend on variables under our control, such as the pressure, the temperature, and the presence of a catalyst, and we might be able to optimize it by the appropriate choice of conditions. Another reason is that the study of reaction rates leads to an understanding of the **mechanism** of a reaction, its analysis into a sequence of elementary steps. For example, we might discover that the reaction of hydrogen and bromine to form hydrogen bromide proceeds by the dissociation of a Br_2 molecule, the attack of a Br atom on an H_2 molecule, and several subsequent steps. By analysing the rate of a biochemical reaction we may discover how an enzyme, a biological catalyst, acts. **Enzyme kinetics**, the study of the effect of enzymes on the rates of reactions, is also an important window on how these macromolecules work.

We need to cope with a wide variety of different rates and a process that appears to be slow may be the outcome of many faster steps. That is particularly true in the chemical reactions that underlie life. Photobiological processes like those responsible for photosynthesis and the slow growth of a plant may take place in about 1 ps. The binding of a neurotransmitter can have an effect after about 1 μs. Once a gene has been activated, a protein may emerge in about 100 s; but even that timescale incorporates many others, including the wriggling of a newly formed polypeptide chain into its working conformation, each step of which may take about 1 ps. On

Empirical chemical kinetics 236

10.1 Spectrophotometry 236

10.2 Experimental techniques 237

Reaction rates 238

10.3 The definition of rate 238

10.4 Rate laws and rate constants 239

10.5 Reaction order 240

10.6 The determination of the rate law 241

10.7 Integrated rate laws 243

10.8 Half-lives and time constants 247

The temperature dependence of reaction rates 248

10.9 The Arrhenius parameters 249

10.10 Collision theory 251

10.11 Transition-state theory 253

CHECKLIST OF KEY CONCEPTS 256

ROAD MAP OF KEY EQUATIONS 257

QUESTIONS AND EXERCISES 257

a grander view, some of the equations of chemical kinetics are applicable to the behaviour of whole populations of organisms; such societies change on timescales of 10^7–10^9 s.

Empirical chemical kinetics

The first step in the investigation of the rate and mechanism of a reaction is the determination of the overall stoichiometry of the reaction and the identification of any side reactions. The next step is to determine how the concentrations of the reactants and products change with time after the reaction has been initiated. Because the rates of chemical reactions are sensitive to temperature, the temperature of the reaction mixture must be held constant throughout the course of the reaction, for otherwise the observed rate would be a meaningless average of the rates for different temperatures.

The method used to monitor the concentrations of reactants and products and their variation with time depends on the substances involved and the rapidity with which their concentrations change (Table 10.1). We shall see that **spectrophotometry**, the measurement of the absorption of light by a material, is used widely to monitor concentration. If a reaction changes the number or type of ions present in a solution, then concentrations may be followed by monitoring the conductivity of the solution. Reactions that change the concentration of hydrogen ions may be studied by monitoring the pH of the solution with a glass electrode. Other methods of monitoring the composition include the detection of light emission, titration, mass spectrometry, gas chromatography, and magnetic resonance (both EPR and NMR, Chapter 21). Polarimetry, the observation of the optical activity of a reaction mixture, is occasionally applicable.

Table 10.1
Kinetic techniques for fast reactions

Technique	Range of timescales/s
Flash photolysis	$>10^{-15}$
Fluorescence decay[a]	10^{-10}–10^{-6}
Ultrasonic absorption	10^{-10}–10^{-4}
EPR[b]	10^{-9}–10^{-4}
Electric field jump[c]	10^{-7}–1
Temperature jump[c]	10^{-6}–1
Phosphorescence decay[a]	10^{-6}–10
NMR[b]	10^{-5}–1
Pressure jump[c]	$>10^{-5}$
Stopped flow	$>10^{-3}$

[a] Fluorescence and phosphorescence are modes of emission of radiation from a material; see Chapter 20.

[b] EPR is electron paramagnetic resonance (or electron spin resonance, ESR); NMR is nuclear magnetic resonance; see Chapter 21.

[c] These techniques are discussed in Chapter 11.

10.1 Spectrophotometry

The key result for using the intensity of absorption of radiation at a particular wavelength to determine the concentration [J] of the absorbing species is the empirical **Beer–Lambert law** (Fig. 10.1):

$$A = \log \frac{I_0}{I} = \varepsilon[\mathrm{J}]L \qquad \text{Beer–Lambert law} \qquad (10.1)$$

(Note: common logarithms, to the base 10.) This expression is discussed further in Chapter 20. For now, we need to know only how to use it in the interpretation of experimental results. In eqn 10.1, A is the **absorbance** (a dimensionless quantity), I_0 and I are the incident and transmitted intensities, respectively, and L is the length of the sample. The quantity ε (epsilon) is called the **molar absorption coefficient** (formerly, and still widely, the 'extinction coefficient'; its dimensions are those of (concentration)$^{-1}$ (length)$^{-1}$ and is typically reported in $\mathrm{dm^3\,mol^{-1}\,cm^{-1}}$): it depends on the wavelength of the incident radiation and is greatest where the absorption is most intense. In a typical spectrophotometer, the absorbance is plotted as a function of wavelength, so A may be determined directly from the data at a given wavelength.

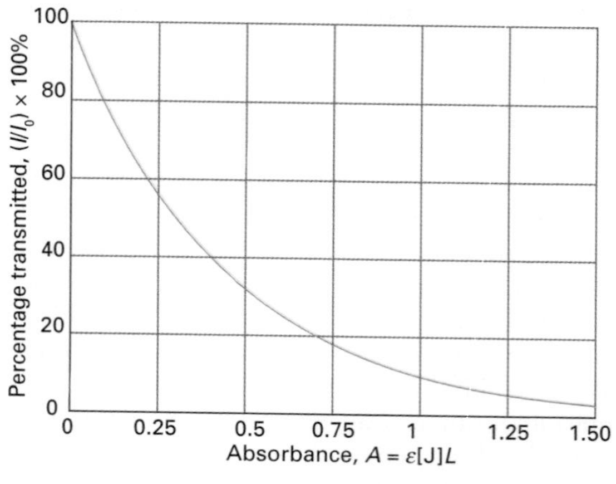

Fig. 10.1 The intensity of light transmitted by an absorbing sample decreases exponentially with the path length through the sample. (The absorbance is proportional to the path length when the concentration is uniform.)

Brief illustration 10.1 The Beer–Lambert law

The molar absorption coefficient of benzene in a certain solvent at a wavelength of 256 nm is $1.6 \times 10^2\ \text{dm}^3\ \text{mol}^{-1}\ \text{cm}^{-1}$. In an experiment examining the rate at which benzene reacts in a non-absorbing solvent the absorbance was measured as $A = 0.80$ in a cell of length $L = 1.0\ \text{mm} = 0.10\ \text{cm}$. The concentration of benzene is calculated from eqn 10.1 rewritten as $[\text{J}] = A/\varepsilon L$:

$$[\text{J}] = \frac{\overbrace{0.80}^{A}}{\underbrace{(1.6\times10^2\ \text{dm}^3\ \text{mol}^{-1}\ \text{cm}^{-1})}_{\varepsilon}\times\underbrace{(0.10\ \text{cm})}_{L}}$$

$$= \frac{0.80}{(1.6\times10^2)\times0.10}\ \text{mol dm}^{-3}$$

$$= 5.0\times10^{-2}\ \text{mol dm}^{-3}$$

10.2 Experimental techniques

In a **real-time analysis**, the composition of a system is analysed while the reaction is in progress by direct spectroscopic observation of the reaction mixture. In the **flow method**, the reactants are mixed as they flow together in a chamber (Fig. 10.2). The reaction continues as the thoroughly mixed solutions flow through a capillary outlet tube at about 10 m s^{-1}, and different points along the tube correspond to different times after the start of the reaction. Spectrophotometric determination of the composition at different positions along the tube is equivalent to the determination of the composition of the reaction mixture at different times after mixing. This technique was originally developed in connection with the study of the rate at which oxygen combined with haemoglobin. Its disadvantage is that a large volume of reactant solution is necessary, because the mixture must flow continuously through the apparatus. This

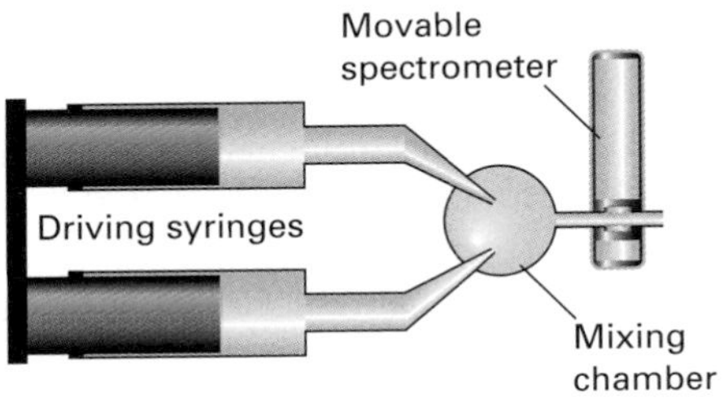

Fig. 10.2 The arrangement used in the flow technique for studying reaction rates. The reactants are squirted into the mixing chamber at a steady rate from the syringes or by using peristaltic pumps (pumps that squeeze the fluid through flexible tubes, like in our intestines). The location of the spectrometer corresponds to different times after initiation.

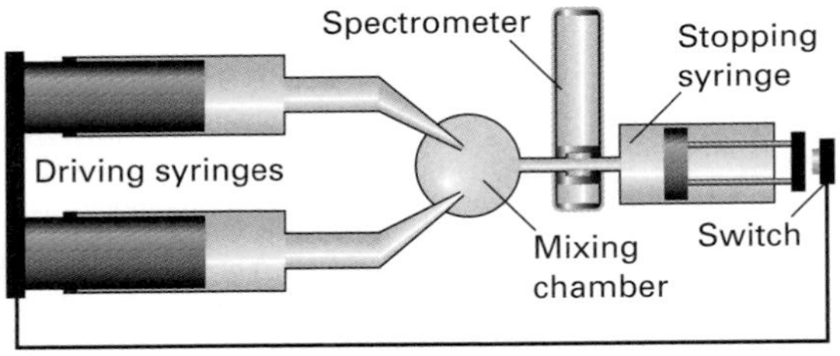

Fig. 10.3 In the stopped-flow technique the reagents are driven quickly into the mixing chamber and then the time dependence of the concentrations is monitored. The switch ensures that the flow stops after a certain volume of reagents has been injected.

disadvantage is particularly important for reactions that take place very rapidly, because the flow must be rapid if it is to spread the reaction over an appreciable length of tube.

The **stopped-flow technique** avoids this disadvantage (Fig. 10.3). The two solutions are mixed very rapidly (in less than 1 ms) by injecting them into a mixing chamber designed to ensure that the flow is turbulent and that complete mixing occurs very quickly. Behind the reaction chamber there is an observation cell fitted with a plunger that moves back as the liquids flood in, but which comes up against a stop after a certain volume has been admitted. The filling of that chamber corresponds to the sudden creation of an initial sample of the reaction mixture. The reaction then continues in the thoroughly mixed solution and is monitored spectrophotometrically. Because only a small, single charge of the reaction chamber is prepared, the technique is much more economical than the flow method. The suitability of the stopped-flow technique to the study of small samples means that it is appropriate for biochemical reactions, and it has been widely used to study the kinetics of enzyme action. Modern techniques of monitoring composition spectrophotometrically can span repetitively a wavelength range of 300 nm at 1 ms intervals.

Very fast reactions can be studied by **flash photolysis**, in which the sample is exposed to a brief flash of light that initiates the reaction, and then the contents of the reaction chamber are monitored spectrophotometrically. Lasers can be used to generate nanosecond flashes routinely, picosecond flashes quite readily, and flashes as brief as a few femtoseconds (1 fs = 10^{-15} s) and even attoseconds (1 as = 10^{-18} s) in special arrangements. Fast reactions are also studied by **pulse radiolysis** in which the flash of electromagnetic radiation is replaced by a short burst of high-velocity electrons.

In contrast to real-time analysis, **quenching methods** are based on stopping, or quenching, the reaction after it has been allowed to proceed for a certain time and the composition is analysed at leisure. The quenching (of the entire mixture or of a sample drawn from it) can be achieved either by cooling suddenly, by adding the mixture to a large volume of solvent, or by rapid neutralization of an acid reagent. This method is suitable only for reactions that are slow enough for there to be little reaction during the time it takes to quench the mixture.

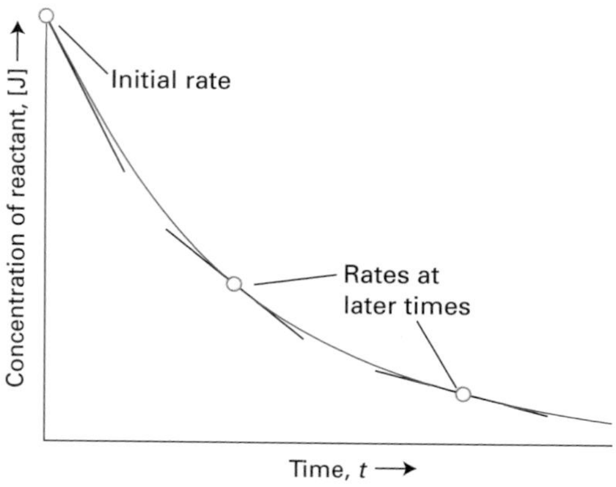

Fig. 10.4 The rate of a chemical reaction is the slope of the tangent to the curve showing the variation of concentration of a species with time. This graph is a plot of the concentration of a reactant, which is consumed as the reaction progresses. The rate of consumption decreases in the course of the reaction as the concentration of reactant decreases.

Reaction rates

The raw data from experiments to measure reaction rates are quantities (such as the absorbance of a sample) that are proportional to the concentrations or partial pressures of reactants and products at a series of times after the reaction is initiated. Ideally, information on any intermediates should also be obtained, but often they cannot be studied because their existence is so fleeting or their concentration so low. More information about the reaction can be extracted if data are obtained at a series of different temperatures. The next few sections look at these observations in more detail.

10.3 The definition of rate

The rate of a reaction taking place in a container of fixed volume is defined in terms of the rate of change of the concentration of a designated species:

$$\text{Rate} = \frac{\overbrace{|\Delta[\text{J}]|}^{\substack{\text{Change in [J]} \\ |\ldots| \text{ means: ignore negative signs}}}}{\underbrace{\Delta t}_{\text{Time interval of interest}}} \qquad \text{Definition} \quad \text{Average rate} \quad (10.2)$$

where $\Delta[\text{J}]$ is the change in the molar concentration of the species J that occurs during the time interval Δt. We have put the change in concentration between modulus signs (|. . .|) to ensure that all rates are positive: if J is a reactant, its concentration will decrease and $\Delta[\text{J}]$ will be negative, but $|\Delta[\text{J}]|$ is positive.

Because the rates at which reactants are consumed and products are formed change in the course of a reaction, it is necessary to consider the **instantaneous rate** of the reaction, its rate at a specific instant. The instantaneous rate of consumption of a reactant is the slope of a graph of its molar concentration plotted against the time, with the slope evaluated as the tangent to the graph at the instant of interest (Fig. 10.4) and reported as a positive quantity. The instantaneous rate of formation of a product is also the slope of the tangent to the graph of its molar concentration plotted, and also reported as a positive quantity. The steeper the slope in either case, the greater the rate of the reaction. With the concentration measured in moles per cubic decimetre and the time in seconds, the reaction rate is reported in moles per cubic decimetre per second (mol dm^{-3} s^{-1}). From now on, we shall denote the instantaneous rate v (for 'velocity').

A note on good practice The rate of a reaction is best expressed by using the language of calculus. In that language, the slope of the tangent to a plot of [J] against time curve at any instant is expressed as the magnitude of the derivative, $|d[\text{J}]/dt|$, where the interval Δt (and consequently the change in concentration $\Delta[\text{J}]$) in eqn 10.2 have been allowed to become infinitesimal. Therefore, the precise definition of reaction rate is $v = |d[\text{J}]|/dt$.

In general, the various reactants in a given reaction are consumed at different rates, and the various products are also formed at different rates. However, these rates are related by the stoichiometry of the reaction.

● **Brief illustration 10.2** The rate of reaction and the reaction stoichiometry

In the decomposition of urea, $(NH_2)_2CO$, in acidic solution

$$(NH_2)_2CO(aq) + 2\ H_2O(l) \rightarrow 2\ NH_4^+(aq) + CO_3^{2-}(aq)$$

the rate of formation of NH_4^+ is twice the rate of disappearance of $(NH_2)_2CO$, because, for 1 mol $(NH_2)_2CO$ consumed, 2 mol NH_4^+ is formed. Once we know the rate of formation or consumption of one substance, we can use the reaction stoichiometry to deduce the rates of formation or consumption of the other participants in the reaction. In this example, for instance,

Rate of formation of NH_4^+ = 2 × rate of consumption of $(NH_2)_2CO$

There is a complication: if the reactants form a slowly decaying intermediate (we see examples later), then the products do not form at the same rate as the reactants turn into the intermediate. In such cases, we have to be very careful about the interpretation of the measured rate of reaction. This complication can be turned to advantage: the observation that the consumption and formation rates are not related by the reaction stoichiometry is a good sign that a long-lived intermediate is involved in the reaction.

One consequence of using a relation like eqn 10.2 is that we have to be careful to specify exactly what species we mean when we report a reaction rate. For this reason, the most sophisticated definition of a **unique rate** of a reaction is in terms of the stoichiometric numbers, ν_J, that appear in the chemical equation. **Stoichiometric numbers** are the stoichiometric coefficients but written as positive for products and as negative for reactants. Then

$$v = \frac{1}{\nu_J}\frac{\Delta[J]}{\Delta t} \qquad \text{Definition} \quad \text{Unique rate} \quad (10.3)$$

Note that we have removed the modulus sign from $\Delta[J]$. The rate is always positive because, whenever $\Delta[J]/\Delta t$ is negative (signalling consumption of a reactant), so is the stoichiometric number (which is negative for a reactant).

Brief illustration 10.3 The unique rate of reaction

If the rate of formation of NH_3 in the reaction $N_2(g) + 3\,H_2(g) \rightarrow 2\,NH_3(g)$ was reported as 1.2 mmol dm^{-3} s^{-1} under a certain set of conditions. For this reaction, $\nu_{NH_3} = +2$, $\nu_{N_2} = -1$, and $\nu_{H_2} = -3$. Then the unique rate of the reaction, written in terms of the change in concentration of ammonia, is

$$v = \frac{1}{\nu_{NH_3}}\frac{\Delta[NH_3]}{\Delta t} = \tfrac{1}{2} \times (1.2\ \text{mmol dm}^{-3}\ \text{s}^{-1})$$
$$= 0.6\ \text{mmol dm}^{-3}\ \text{s}^{-1}$$

We could also write the unique rate as

$$v = -\frac{\Delta[N_2]}{\Delta t} = -\frac{1}{3}\frac{\Delta[H_2]}{\Delta t}$$

and predict, for example, that the rate of change of $[H_2]$ is

$$\frac{\Delta[H_2]}{\Delta t} = -3v = -1.8\ \text{mmol dm}^{-3}\ \text{s}^{-1}$$

In this case, we could also report the rate of consumption (a positive number) of H_2 as 1.8 mmol dm^{-3} s^{-1}.

10.4 Rate laws and rate constants

An empirical observation of the greatest importance is that *the rate of reaction is often found to be proportional to the molar concentrations of the reactants raised to a simple power*. For example, it may be found that the rate is directly proportional to the concentrations of the reactants A and B, so

$$v = k_r[A][B] \qquad (10.4)$$

The coefficient k_r, which is characteristic of the reaction being studied, is called the **rate constant** (or 'rate coefficient'). The rate constant is independent of the concentrations of the species taking part in the reaction but depends on the temperature. An *experimentally determined* equation of this kind is called the 'rate law' of the reaction. More formally, a **rate law** is an equation that expresses the rate of reaction in terms of the molar concentrations (or partial pressures) of the species in the overall reaction (including, possibly, the products that might be present).

The units of k_r are always such as to convert the product of concentrations into a rate expressed as a change in concentration divided by time. For example, if the rate law is the one shown in eqn 10.4, with concentrations expressed in mol dm^{-3}, then the units of k_r will be dm^3 mol^{-1} s^{-1} because

$$\overbrace{\text{dm}^3\ \text{mol}^{-1}\ \text{s}^{-1}}^{\text{units of } k_r} \times \overbrace{\text{mol dm}^{-3}}^{\text{units of [A]}} \times \overbrace{\text{mol dm}^{-3}}^{\text{units of [B]}} = \overbrace{\text{mol dm}^{-3}\ \text{s}^{-1}}^{\text{units of } v}$$

In gas-phase studies, including studies of the processes taking place in the atmosphere, concentrations are commonly expressed in molecules cm^{-3}, so the rate constant for the reaction above would be expressed in cm^3 molecule^{-1} s^{-1}. We can use the approach just illustrated to determine the units of the rate constant for rate laws of any form. For example, the rate constant for a reaction with rate law of the form $v = k_r[A]$ is commonly expressed in s^{-1}.

Example 10.1

Expressing the rate constant in different units

The rate constant for the reaction $O(g) + O_3(g) \rightarrow 2\ O_2(g)$ is 8.0×10^{-15} cm^3 molecule^{-1} s^{-1} at 298 K. Express this rate constant in dm^3 mol^{-1} s^{-1}.

Strategy Make use of the following relations:

$$1\ \text{cm} = 10^{-2}\ \text{m} = 10^{-2} \times 10\ \text{dm} = 10^{-1}\ \text{dm} = \frac{1\ \text{dm}}{10}$$

$$1\ \text{mol} = 6.022 \times 10^{23}\ \text{molecules},$$

$$\text{so } 1\ \text{molecule} = \frac{1\ \text{mol}}{6.022 \times 10^{23}}$$

Solution It follows from the relations given above that

$$k_r = 8.0 \times 10^{-15}\ \text{cm}^3\ \text{molecule}^{-1}\ \text{s}^{-1}$$

$$= 8.0 \times 10^{-15} \overbrace{\left(\frac{1\ \text{dm}}{10}\right)^3}^{\text{cm}^3} \overbrace{\left(\frac{1\ \text{mol}}{6.022 \times 10^{23}}\right)^{-1}}^{\text{molecule}^{-1}} \text{s}^{-1}$$

$$= \frac{8.0 \times 10^{-15} \times 6.022 \times 10^{23}}{10^3}\ \text{dm}^3\ \text{mol}^{-1}\ \text{s}^{-1}$$

$$= 4.8 \times 10^6\ \text{dm}^3\ \text{mol}^{-1}\ \text{s}^{-1}$$

Note that, as should be expected (but is a good point to check), the rate per mole of molecules is numerically much greater (by a factor of 6.022×10^{23}) than the rate per molecule.

Self-test 10.1

A reaction has a rate law of the form $k_r[A]^2[B]$. What are the units of the rate constant k_r if the reaction rate was measured in mol dm^{-3} s^{-1}?

Answer: dm^6 mol^{-2} s^{-1}

Once we know the rate law and the rate constant of the reaction, we can predict the rate of the reaction for any given composition of the reaction mixture. We shall also see that we can use a rate law to predict the concentrations of the reactants and products at any time after the start of the reaction. Furthermore, an observed rate law is also an important guide to the mechanism of the reaction, for any proposed mechanism must be consistent with it.

10.5 Reaction order

A rate law provides a basis for the classification of reactions according to their kinetics. The advantage of having such a classification is that reactions belonging to the same class have similar kinetic behaviour—their rates and the concentrations of the reactants and products vary with composition in a similar way. The classification of reactions is based on their **order,** the power to which the concentration of a species is raised in the rate law. For example:

First-order in A: $$v = k_r[A] \qquad (10.5a)$$

First-order in A, first-order in B: $$v = k_r[A][B] \qquad (10.5b)$$

Second-order in A: $$v = k_r[A]^2 \qquad (10.5c)$$

The **overall order** of a reaction with a rate law of the form $v = k_r[A]^a[B]^b[C]^c \ldots$ is the sum, $a + b + c + \cdots$, of the orders of all the components. The rate laws in eqns 10.5b and 10.5c both correspond to reactions that are *second-order* overall.

Brief illustration 10.4 Reaction orders

An example of the type of reaction in eqn 10.5b is the reformation of a DNA double helix after the double helix has been separated into two strands by raising the temperature or the pH:

Strand + complementary strand → double helix

$v = k_r$ [strand][complementary strand]

This reaction is first-order in each strand and second-order overall. An example of a reaction like eqn 10.5c is the reduction of nitrogen dioxide by carbon monoxide,

$$NO_2(g) + CO(g) \rightarrow NO(g) + CO_2(g) \qquad v = k_r[NO_2]^2$$

which is second-order in NO_2 and, because no other species occurs in the rate law, second-order overall. The rate of this reaction is independent of the concentration of CO provided that some CO is present. This independence of concentration is expressed by saying that the reaction is *zeroth-order* in CO, because a concentration raised to the power zero is 1 ($[CO]^0 = 1$, just as $x^0 = 1$ in algebra).

A reaction need not have an integral order, and many gas-phase reactions do not. For example, if a reaction is found to have the rate law

$$v = k_r[A]^{1/2}[B] \qquad (10.6)$$

then it is *half-order* in A, first-order in B, and three-halves order overall.

If a rate law is not of the form $v = k_r[A]^a[B]^b[C]^c \ldots$ then the reaction does not have an overall order. Thus, the experimentally determined rate law for the gas-phase reaction $H_2(g) + Br_2(g) \rightarrow 2\ HBr(g)$ is

$$v = \frac{k_a[H_2][Br_2]^{3/2}}{[Br_2] + k_b[HBr]} \qquad (10.7)$$

Although the reaction is first-order in H_2, it has an indefinite order with respect to both Br_2 and HBr and an indefinite order overall. Similarly, a typical rate law for the action of an enzyme E on a substrate S is (see Chapter 11)

$$v = \frac{k_r[E][S]}{[S]+K_M} \tag{10.8}$$

where K_M is a constant. This rate law is first-order in the enzyme but does not have a specific order with respect to the substrate.

Under certain circumstances a complicated rate law without an overall order may simplify into a law with a definite order. For example, if the substrate concentration in the enzyme catalysed reaction is so low that $[S] \ll K_M$, then we can ignore [S] in the denominator of eqn 10.8, which simplifies to

$$v = \frac{k_r}{K_M}[E][S] \tag{10.9}$$

which is first-order in S, first-order in E, and second-order overall.

It is very important to note that *a rate law is established experimentally, and cannot in general be inferred from the chemical equation for the reaction.* The reaction of hydrogen and bromine, for example, has a very simple stoichiometry,

$$H_2(g) + Br_2(g) \rightarrow 2\ HBr(g)$$

but its rate law (eqn 10.7) is very complicated. In some cases, however, the rate law does happen to reflect the reaction stoichiometry. This is the case with the reaction of hydrogen and iodine, which has the same stoichiometry as the reaction of hydrogen with bromine but a much simpler rate law:

$$H_2(g) + I_2(g) \rightarrow 2\ HI(g) \qquad v = k_r[H_2][I_2]$$

10.6 The determination of the rate law

There are many ways to investigate chemical reactions and extract information about their rate laws. Here we highlight two useful and common strategies.

(a) The isolation method

The determination of a rate law is simplified by the **isolation method**, in which all the reactants except one are present in large excess. We can find the dependence of the rate on each of the reactants by isolating each of them in turn—by having all the other substances present in large excess—and piecing together a picture of the overall rate law. For instance, we might use CH_3I in solution at a concentration of 0.2 mol dm^{-3} and an attacking nucleophile at only 0.01 mol dm^{-3}.

If a reactant B is in large excess it is a good approximation to take its concentration as constant throughout the reaction. Then, although the true rate law might be $v = k_r[A][B]^2$ we can approximate [B] by its initial value $[B]_0$ (from which it hardly changes in the course of the reaction) and write

$$v = k_{r,\text{eff}}[A], \text{ with } k_{r,\text{eff}} = k_r[B]_0^2$$

A pseudofirst-order reaction, B in excess (10.10a)

Because the true rate law has been forced into first-order form by assuming a constant B concentration, the effective rate law is classified as **pseudofirst-order** and $k_{r,\text{eff}}$ is called the **effective rate constant** for a given, fixed concentration of B. If, instead, the concentration of A is in large excess, and hence effectively constant, then the rate law simplifies to

$$v = k_{r,\text{eff}}[B]^2, \text{ now with } k_{r,\text{eff}} = k_r[A]_0$$

A pseudofirst-order reaction, A in excess (10.10b)

This **pseudosecond-order rate law** is also much easier to analyse and identify than the complete law. Note that the order of the reaction and the form of the effective rate constant change according to whether A or B is in excess. In a similar manner, a reaction may even appear to be zeroth order.

Brief illustration 10.5 Pseudo orders

The oxidation of ethanol to acetaldehyde (ethanal) by NAD^+ in the liver in the presence of the enzyme liver alcohol dehydrogenase

$$CH_3CH_2OH(aq) + NAD^+(aq) + H_2O(l) \rightarrow CH_3CHO(aq) + NADH(aq) + H_3O^+(aq)$$

is zeroth-order overall as the ethanol is in excess and the concentration of the NAD^+ is maintained at a constant level by normal metabolic processes. Many reactions in aqueous solution that are reported as first- or second-order are actually pseudofirst- or pseudosecond-order: the solvent water participates in the reaction but it is in such large excess that its concentration remains constant.

(b) The method of initial rates

In the method of **initial rates**, which is often used in conjunction with the isolation method, the instantaneous rate is measured at the beginning of the reaction for several different initial concentrations of reactants. For example, suppose the rate law for a reaction with A isolated is

$$v = k_{r,\text{eff}}[A]^a$$

Then the initial rate of the reaction, v_0, is given by the initial concentration of A:

$$v_0 = k_{r,\text{eff}}[A]_0^a$$

Initial rate of an *a*th-order reaction (10.11)

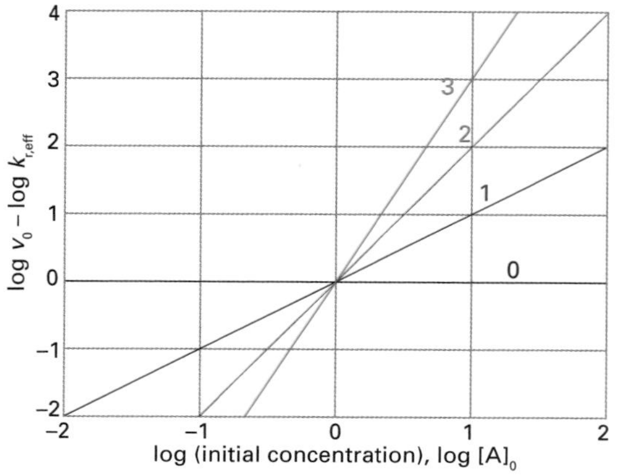

Fig. 10.5 The plot of log v_0 (and, as shown here, of log v_0 − log $k_{r,eff}$) against log$[A]_0$ gives straight lines with slopes equal to the order of the reaction.

Taking (common) logarithms gives

$$\log v_0 = \log(k_{r,eff}[A]_0^a) \overset{\log xy = \log x + \log y}{=} \log k_{r,eff} + \log[A]_0^a$$

$$\overset{\log x^a = a\log x}{=} \log k_{r,eff} + a\log[A]_0 \qquad (10.12)$$

This equation has the form of the equation for a straight line:

$$\underbrace{\log v_0}_{y} = \underbrace{\log k_{r,eff}}_{\text{intercept}} + \underbrace{a\log[A]_0}_{\text{slope} \times x}$$

It follows that, for a series of initial concentrations, a plot of the logarithms of the initial rates against the logarithms of the initial concentrations of A should be a straight line, and that the slope of the graph will be a, the order of the reaction with respect to A (Fig. 10.5).

Example 10.2

Using the method of initial rates

The recombination of iodine atoms in the gas phase in the presence of argon (which removes the energy released by the formation of an I—I bond, and so prevents the immediate dissociation of a newly formed I_2 molecule) was investigated and the order of the reaction was determined by the method of initial rates. The initial rates of reaction of 2 I(g) + Ar(g) → I_2(g) + Ar(g) were as follows:

	$[I]_0/(10^{-5}$ mol dm$^{-3})$			
	1.0	2.0	4.0	6.0
v_0/(mol dm^{-3} s^{-1})				
(a)	8.70×10^{-4}	3.48×10^{-3}	1.39×10^{-2}	3.13×10^{-2}
(b)	4.35×10^{-3}	1.74×10^{-2}	6.96×10^{-2}	1.57×10^{-1}
(c)	8.69×10^{-3}	3.47×10^{-2}	1.38×10^{-1}	3.13×10^{-1}

The Ar concentrations are (a) 1.0×10^{-3} mol dm^{-3}, (b) 5.0×10^{-3} mol dm^{-3}, and (c) 1.0×10^{-2} mol dm^{-3}. Find the orders of reaction with respect to I and Ar and the rate constant.

Strategy For constant $[Ar]_0$, the initial rate law has the form $v_0 = k_{r,eff}[I]_0^a$, with $k_{r,eff} = k_r[Ar]_0^b$, so

$$\log v_0 = \log k_{r,eff} + a \log[I]_0$$

We need to make a plot of log v_0 against log$[I]_0$ for a given $[Ar]_0$ and find the order from the slope and the value of $k_{r,eff}$ from the intercept at log$[I]_0 = 0$. Then, because

$$\log k_{r,eff} = \log k_r + b \log[Ar]_0$$

plot log $k_{r,eff}$ against log$[Ar]_0$ to find log k_r from the intercept and b from the slope.

Solution The data give the following points for the graph:

log($[I]_0$/mol dm^{-3})		−5.00	−4.70	−4.40	−4.22
log(v_0/mol dm^{-3} s^{-1})	(a)	−2.971	−2.458	−1.857	−1.504
	(b)	−2.362	−1.760	−1.157	−0.804
	(c)	−1.971	−1.460	−0.860	−0.504

The graph of the data is shown in Fig. 10.6. The slopes of the lines are 2, so the reaction is second-order with respect to I. The effective rate constants $k_{r,eff}$ are as follows:

$[Ar]_0$/(mol dm^{-3})	1.0×10^{-3}	5.0×10^{-3}	1.0×10^{-2}
log($[Ar]_0$/mol dm^{-3})	−3.00	−2.30	−2.00
log($k_{r,eff}$/mol^{-1} dm^3 s^{-1})	6.94	7.64	7.93

Figure 10.7 is the plot of log $k_{r,eff}$ against log$[Ar]_0$. The slope is 1, so $b = 1$ and the reaction is first order with respect to Ar. The intercept at log$[Ar]_0 = 0$ is log $k_{r,eff} = 9.94$, so $k_r = 8.7 \times 10^9$ mol^{-2} dm^6 s^{-1}. The overall (initial) rate law is therefore

$$v = k_r[I]_0^2[Ar]_0$$

A note on good practice When taking the common logarithm of a number of the form $x.xx \times 10^n$, there are *four* significant figures in the answer: the figure before the decimal point is simply the power of 10. Conversely, when taking the common antilogarithm of $y.yyy$, there are *three* significant figures in the answer. Strictly, the logarithms are of the quantity divided by its units.

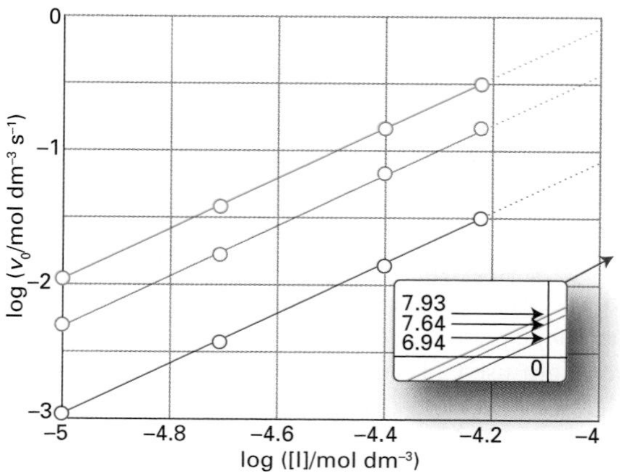

Fig. 10.6 The plots of the data in Example 10.2 for finding the order with respect to I.

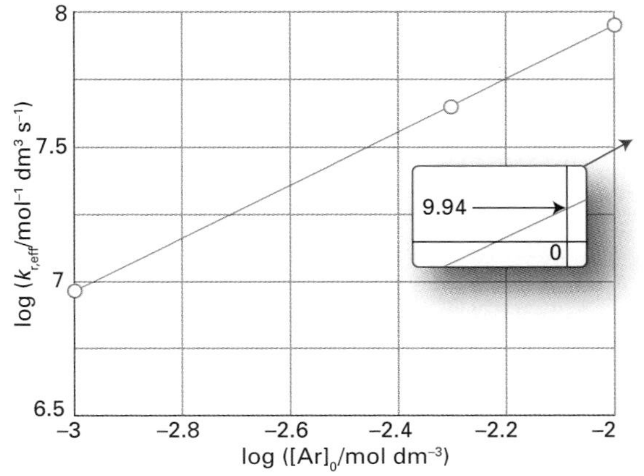

Fig. 10.7 The plots of the data in Example 10.2 for finding the order with respect to Ar.

Self-test 10.2

The initial rate of a certain reaction depended on concentration of a substance J as follows:

$[J]_0/(10^{-3}\ \text{mol dm}^{-3})$	5.0	10.2	17	30
$v_0/(10^{-7}\ \text{mol dm}^{-3}\ \text{s}^{-1})$	3.6	9.6	4	130

Find the order of the reaction with respect to J and the rate constant.

Answer: 2, 1.6×10^{-2} mol⁻¹ dm³ s⁻¹

The method of initial rates might not reveal the entire rate law, for in a complex reaction the products themselves might affect the rate. That is the case for the synthesis of HBr, for eqn 10.7 shows that the rate law depends on the concentration of HBr, none of which is present initially.

10.7 Integrated rate laws

A rate law tells us the rate of the reaction at a given instant (when the reaction mixture has a particular composition). That is rather like being given the speed of a car at each point of its journey. For a car journey, we may want to know the distance that a car has travelled at a certain time given its varying speed. Similarly, for a chemical reaction, we may want to know the composition of the reaction mixture at a given time given the varying rate of the reaction. An **integrated rate law** is an expression that gives the concentration of a species as a function of the time.

Integrated rate laws have two principal uses. One is to predict the concentration of a species at any time after the start of the reaction. Another is to help find the rate constant and order of the reaction. Indeed, although we have introduced rate laws through a discussion of the determination of reaction rates, these rates are rarely measured directly because slopes are so difficult to determine accurately. Almost all experimental work in chemical kinetics deals with integrated rate laws; their great advantage being that they are expressed in terms of the experimental observables of concentration and time. Computers can be used to find the integrated form of even the most complex rate laws numerically and in some cases can be used to obtain closed, algebraic expressions. However, in a number of simple cases solutions can be by elementary techniques and prove to be very useful.

For a chemical reaction and first-order rate law of the form

$$\text{A} \rightarrow \text{products,}$$

$$\text{Rate of consumption of A} = k_r[\text{A}] \qquad (10.13)$$

we show in the following Derivation that the integrated rate law is

$$\ln\frac{[\text{A}]}{[\text{A}]_0} = -k_r t \qquad \text{First-order integrated rate law} \qquad (10.14\text{a})$$

where $[\text{A}]_0$ is the initial concentration of A. Another form of this expression is

$$[\text{A}] = [\text{A}]_0 e^{-k_r t} \qquad \text{Exponential decay} \qquad (10.14\text{b})$$

Equation 10.14b has the form of an **exponential decay** (Fig. 10.8). A common feature of all first-order reactions, therefore, is that *the concentration of the reactant decays exponentially with time*.

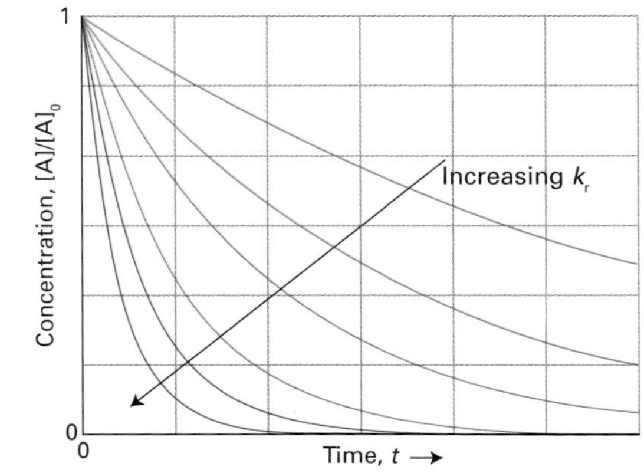

Fig. 10.8 The exponential decay of the reactant in a first-order reaction. The greater the rate constant, the more rapid is the decay.

Derivation 10.1

First-order integrated rate law

Our first step is to express the rate of consumption of a reactant A mathematically. As remarked in Section 10.3, the rate of a reaction is $|d[A]|/dt$. Because A is a reactant, the change d[A] is negative (the concentration of A decreases with time), so $-d[A]$ is positive. We can therefore interpret the rate as $-d[A]/dt$. It follows that a first-order rate equation has the form

$$-\frac{d[A]}{dt} = k_r[A]$$

This expression is an example of a 'differential equation' (see The chemist's toolkit 10.1). Because the terms d[A] and dt may be manipulated like any algebraic quantity, we rearrange the differential equation into

$$\frac{d[A]}{[A]} = -k_r dt$$

and then integrate both sides. Integration from $t = 0$, when the concentration of A is $[A]_0$, to the time of interest, t, when the molar concentration of A is [A], is written as

$$\int_{[A]_0}^{[A]} \frac{d[A]}{[A]} = -k_r \overbrace{\int_0^t dt}^{t}$$

To develop the term on the left, we use the standard integral (The chemist's toolkit 2.1)

$$\int \frac{dx}{x} = \ln x + \text{constant}$$

and the relation $\ln(x/y) = \ln x - \ln y$ (The chemist's toolkit 2.2) to obtain

$$\int_{[A]_0}^{[A]} \frac{d[A]}{[A]} = \ln\frac{[A]}{[A]_0}$$

Equation 10.14a follows after combining the results for the left- and right-hand terms.

The chemist's toolkit 10.1 Ordinary differential equations

An *ordinary differential equation* is a relation between derivatives of a function of one variable and the function itself, as in

$$\text{(A) } a\frac{dy}{dx} + bx + c = 0 \qquad \text{(B) } a\frac{d^2y}{dx^2} + b\frac{dy}{dx} + cx + d = 0$$

The coefficients a, b, etc., may be constants or functions of x. The *order* of the equation is the order of the highest derivative that occurs in it, so (A) is a first-order equation and (B) is a second-order equation. 'Solving' a differential equation is the process of determining the function, in this case $y(x)$, that satisfies it.

In many cases it is found that various constants appear in the solution, such as $y(x)$ + constant. These constants are determined by imposing various *boundary conditions* on the solutions, values that the solution must have at specified points. A second-order differential equation requires two boundary conditions, a first-order equation requires one. For time-dependent solutions, the 'boundary condition' is termed an *initial condition*, and is typically the value that the solution must have at $t = 0$.

Equation 10.14b lets us predict the concentration of A at any time after the start of the reaction. Equation 10.14a shows that if we plot $\ln([A]/[A]_0)$ against t, then we get a straight line if the reaction is first-order. If the experimental data do not give a straight line when plotted in this way, then the reaction is not first-order. If the line is straight, then it follows from eqn 10.14a that its slope is $-k_r$, so we can also determine the rate constant from the graph. Some rate constants determined in this way are given in Table 10.2.

Table 10.2
Kinetic data for first-order reactions

Reaction	Phase	θ/°C	k_r/s^{-1}*	$t_{1/2}$
$2\ N_2O_5$ $\rightarrow 4\ NO_2 + O_2$	g	25	3.38×10^{-5}	2.85 h
$2\ N_2O_5$ $\rightarrow 4\ NO_2 + O_2$	Br_2(l)	25	4.27×10^{-5}	2.25 h
C_2H_6 $\rightarrow 2\ CH_3$	g	700	5.46×10^{-4}	21.2 min
Cyclopropane $\rightarrow$ propene	g	500	6.17×10^{-4}	17.2 min

*The rate constant is for the rate of formation or consumption of the species in bold type. The rate laws for the other species may be obtained from the reaction stoichiometry.

Example 10.3

Analysing a first-order reaction

The variation in the partial pressure of azomethane with time was followed at 600 K, with the results given below. Confirm that the decomposition

$CH_3N_2CH_3(g) \rightarrow CH_3CH_3(g) + N_2(g)$

is first-order in azomethane, and find the rate constant at 600 K.

t/s	0	1000	2000	3000	4000
p/Pa	10.9	7.63	5.32	3.71	2.59

Strategy To confirm that a reaction is first-order, plot $\ln([A]/[A]_0)$ against time and expect a straight line. Because the partial pressure of a gas is proportional to its concentration, an equivalent procedure is to plot $\ln(p/p_0)$ against t. If a straight line is obtained, its slope can be identified with $-k_r$. Follow the guidelines for plotting graphs given in The chemist's toolkit 1.1.

Solution Draw up the following table:

t/s	0	1000	2000	3000	4000
$\ln(p/p_0)$	0	−0.357	−0.717	−1.078	−1.437

Figure 10.9 shows the plot of $\ln(p/p_0)$ against t. The plot is straight, confirming a first-order reaction, and its slope is -3.6×10^{-4}. Therefore, $k_r = 3.6 \times 10^{-4}\ s^{-1}$.

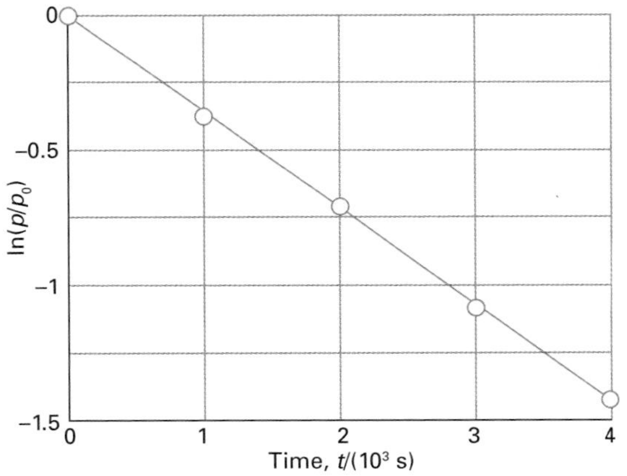

Fig. 10.9 The determination of the rate constant of a first-order reaction. A straight line is obtained when ln[A] (or ln p, where p is the partial pressure of the species of interest) is plotted against t; the slope is $-k_r$. The data are from Example 10.3.

Self-test 10.3

The concentration of N_2O_5 in liquid bromine varied with time as follows:

t/s	0	200	400	600	1000
$[N_2O_5]$/(mol dm^{-3})	0.110	0.073	0.048	0.032	0.014

Confirm that the reaction is first-order in N_2O_5 and determine the rate constant.

Answer: $2.1 \times 10^{-3}\ s^{-1}$

Now we need to see how the concentration varies with time for a reaction and second-order rate law if the form

$$A \rightarrow \text{products},$$
$$\text{Rate of consumption of A} = k_r[A]^2 \qquad (10.1)$$

As before, we suppose that the concentration of A at $t=0$ is $[A]_0$ and, as shown in the following Derivation, find that

$$\frac{1}{[A]_0} - \frac{1}{[A]} = -k_r t \qquad \text{Second-order integrated rate law} \qquad (10.16a)$$

which may also be written

$$[A] = \frac{[A]_0}{1 + k_r t[A]_0} \qquad (10.16b)$$

Derivation 10.2

Second-order integrated rate law

As before, the rate of consumption of the reactant A is $-d[A]/dt$, so the differential equation for the rate law is

$$-\frac{d[A]}{dt} = k_r[A]^2$$

To solve this equation, we rearrange it into

$$\frac{d[A]}{[A]^2} = -k_r dt$$

and integrate it between $t=0$, when the concentration of A is $[A]_0$, and the time of interest t, when the concentration of A is [A]:

$$\int_{[A]_0}^{[A]} \frac{d[A]}{[A]^2} = -k_r \int_0^t dt$$

The term on the right is $-k_r t$. We evaluate the integral on the left by using the standard form

$$\int \frac{dx}{x^2} = -\frac{1}{x} + \text{constant}$$

which implies that

$$\int_a^b \frac{dx}{x^2} = \left\{-\frac{1}{x} + \text{constant}\right\}\Bigg|_b - \left\{-\frac{1}{x} + \text{constant}\right\}\Bigg|_a = -\frac{1}{b} + \frac{1}{a}$$

and so obtain eqn 10.16a.

Equation 10.16a shows that to test for a second-order reaction we should plot $1/[A]$ against t and expect a straight line. If the line is straight, then the reaction is second-order in A and the slope of the line is equal to the rate constant (Fig. 10.10). Some rate constants determined in this way are given in

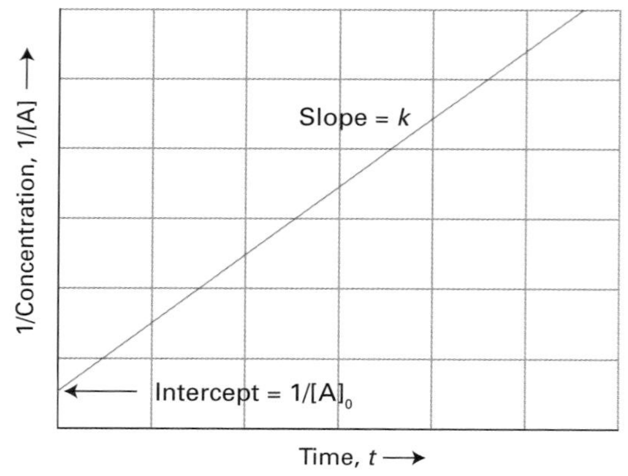

Fig. 10.10 The determination of the rate constant of a second-order reaction. A straight line is obtained when 1/[A] (or 1/*p*, where *p* is the partial pressure of the species of interest) is plotted against *t*; the slope is k_r.

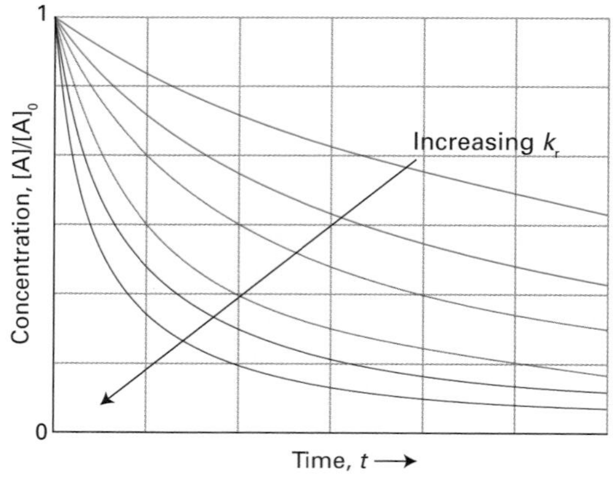

Fig. 10.11 The variation with time of the concentration of a reactant in a second-order reaction.

Table 10.3

Kinetic data for second-order reactions

Reaction	Phase	θ/°C	k_r/(dm^3 mol^{-1} s^{-1})*
2 NOBr → 2 NO + **Br$_2$**	g	10	0.80
2 NO$_2$ → 2 NO + **O$_2$**	g	300	0.54
H$_2$ + I$_2$ → 2 HI	g	400	2.42×10^{-2}
D$_2$ + HCl → DH + DCl	g	600	0.141
2 I → I$_2$	g	23	7×10^{9}
	hexane	50	1.8×10^{10}
CH$_3$Cl + CH$_3$O$^-$	CH$_3$OH(l)	20	2.29×10^{-6}
CH$_3$Br + CH$_3$O$^-$	CH$_3$OH(l)	20	9.23×10^{-6}
H$^+$ + OH$^-$ → H$_2$O	water	25	1.5×10^{11}

*The rate constant is for the rate of formation or consumption of the species in bold type. The rate laws for the other species may be obtained from the reaction stoichiometry.

Table 10.3. Equation 10.16b enables us to predict the concentration of A at any time after the start of the reaction (Fig. 10.11).

From plots of [A] against *t*, we see that the concentration of A approaches zero more slowly in a second-order reaction than in a first-order reaction with the same initial rate (Fig. 10.12). That is, reactants that decay by a second-order process die away more slowly at low concentrations than would be expected if the decay were first-order. A point of interest in this connection is that pollutants commonly disappear by second-order processes, so it takes a very long time for them to decline to acceptable levels.

Table 10.4 summarizes the integrated rate laws for a variety of simple reaction types.

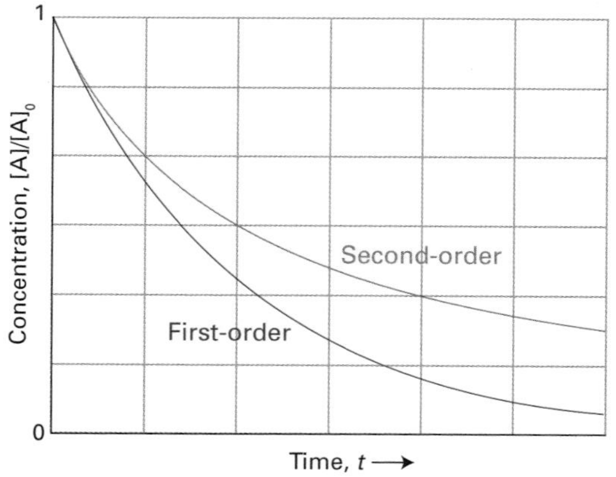

Fig. 10.12 Although the initial decay of a second-order reaction may be rapid, later the concentration approaches zero more slowly than in a first-order reaction with the same initial rate.

Table 10.4
Integrated rate laws

Order	Reaction type	Rate law	Integrated rate law
0	A → P	$v = k_r$	$[P] = k_r t$ for $k_r t \leq [A]_0$
1	A → P	$v = k_r[A]$	$[P] = [A]_0(1 - e^{-k_r t})$
2	A → P	$v = k_r[A]^2$	$[P] = \dfrac{k_r t[A]_0^2}{1 + k_r t[A]_0}$
	A + B → P	$v = k_r[A][B]$	$[P] = \dfrac{[A]_0[B]_0(1 - e^{([B]_0-[A]_0)k_r t})}{[A]_0 - [B]_0 e^{([B]_0-[A]_0)k_r t}}$

10.8 Half-lives and time constants

A useful indication of the rate of a first-order chemical reaction is the **half-life**, $t_{1/2}$, of a reactant, which is the time it takes for the concentration of the species to fall to half its initial value. We can find the half-life of a species A that decays in a first-order reaction (eqn 10.13) by substituting $[A] = \frac{1}{2}[A]_0$ and $t = t_{1/2}$ into eqn 10.14a:

$$k_r t_{1/2} = -\ln\frac{\frac{1}{2}[A]_0}{[A]_0} \overset{\text{cancel terms}}{=} -\ln\frac{1}{2} \overset{\ln(1/x)=-\ln x}{=} \ln 2$$

It follows that

$$t_{1/2} = \frac{\ln 2}{k_r} \quad \text{Half-life of a first-order reaction} \quad (10.17)$$

Brief illustration 10.6 The half-life of a first-order reaction

Because the rate constant for the first-order reaction

$N_2O_5(g) \rightarrow 2\ NO_2(g) + \frac{1}{2}\ O_2(g)$

Rate of consumption of $N_2O_5 = k_r[N_2O_5]$

is equal to $6.76 \times 10^{-5}\ s^{-1}$ at 25 °C, the half-life of N_2O_5 is

$$t_{1/2} = \frac{\ln 2}{\underbrace{6.76 \times 10^{-5}\ s^{-1}}_{k_r}} = 1.03 \times 10^4\ s$$

or 2.85 h. Hence, the concentration of N_2O_5 falls to half its initial value in 2.85 h, and then to half that concentration again in a further 2.85 h, and so on (Fig. 10.13). This procedure is commonly used in reverse: the half-life is measured and then eqn 10.17 is used to determine k_r from $k_r = (\ln 2)/t_{1/2}$.

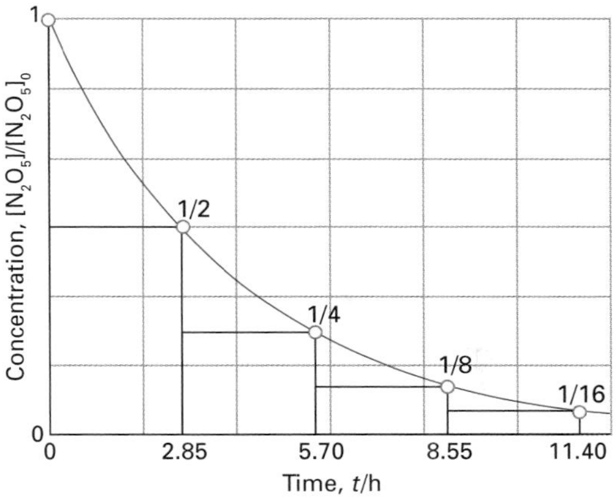

Fig. 10.13 The molar concentration of N_2O_5 after a succession of half-lives.

Self-test 10.4

The rate constant for the first-order isomerization of cyclopropane to propene is $6.71 \times 10^4\ s^{-1}$ at 25 °C. What is the half-life of cyclopropane?

Answer: 1490 s

The main point to note about eqn 10.17 is that *for a first-order reaction, the half-life of a reactant is independent of its concentration.* It follows that if the concentration of A at some arbitrary stage of the reaction is [A], then the concentration will fall to $\frac{1}{2}[A]$ after an interval of $(\ln 2)/k_r$ whatever the actual value of [A] (Fig. 10.14). Some half-lives are given in Table 10.2.

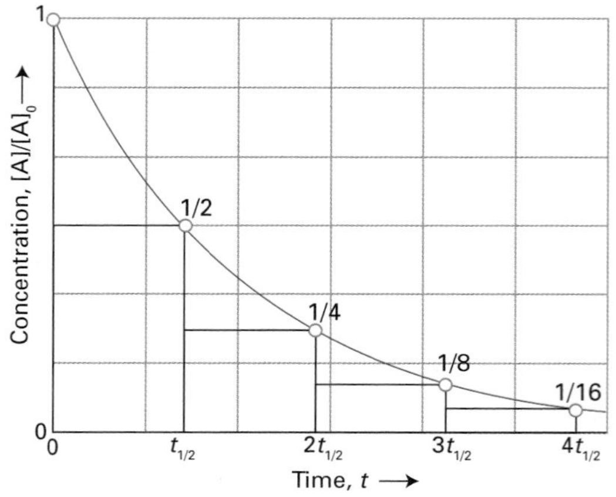

Fig. 10.14 In each successive period of duration $t_{1/2}$, the concentration of a reactant in a first-order reaction decays to half its value at the start of that period. After *n* such periods, the concentration is $(1/2)^n$ of its initial concentration.

● **Brief illustration 10.7** The half-life and the progress of a reaction

In acidic solution, the disaccharide sucrose (cane sugar) is converted to a mixture of the monosaccharides glucose and fructose in a pseudofirst-order reaction. Under certain conditions of pH, the half-life of sucrose is 28.4 min. To calculate how long it takes for the concentration of a sample to fall from 8.0 mmol dm^{-3} to 1.0 mmol dm^{-3} we note that

Molar concentration/(mmol dm^{-3}):

$$8.0 \xrightarrow{28.4\ \text{min}} 4.0 \xrightarrow{28.4\ \text{min}} 2.0 \xrightarrow{28.4\ \text{min}} 1.0$$

The total time required is 3 × 28.4 min = 85.2 min.

Self-test 10.5

The half-life of a substrate in a certain enzyme-catalysed first-order reaction is 138 s. How long is required for the concentration of substrate to fall from 1.28 mmol dm^{-3} to 0.040 mmol dm^{-3}?

Answer: 690 s

We can use the half-life of a substance to recognize first-order reactions. All we need do is inspect a set of data of composition against time. If we see that the initial concentration falls to half its value in a certain time, and that another concentration falls to half its value in the same time, then we can infer that the reaction is first-order. The first-order character can then be confirmed by plotting ln[A] against t and obtaining a straight line, as indicated earlier.

Another indication of the rate of a first-order reaction is the **time constant**, τ, the time required for the concentration of a reactant to fall to 1/e of its initial value. From eqn 10.14a it follows that

$$k_r\tau \overbrace{=}^{\text{Set }[A]=[A]_0/e} -\ln\left(\frac{[A]_0/e}{[A]_0}\right) \overbrace{=}^{\text{cancel terms}} -\ln\frac{1}{e} = 1$$

Hence, the time constant is the reciprocal of the rate constant:

$$\tau = \frac{1}{k_r} \qquad \text{Time constant of a first-order decay} \qquad (10.18)$$

It follows from eqn 10.16a, by substituting $t = t_{1/2}$ and $[A] = \tfrac{1}{2}[A]_0$, that the half-life of a species A that is consumed in a second-order reaction is

$$t_{1/2} = \frac{1}{k_r[A]_0} \qquad \text{Half-life for a second-order reaction} \qquad (10.19)$$

Therefore, unlike a first-order reaction, the half-life of a substance in a second-order reaction varies with the initial concentration.

● **Brief illustration 10.8** The half-life of a second-order reaction

Consider a second-order reaction 2 A → P with rate constant $k_r = 5.0 \times 10^4$ dm^3 mol^{-1} s^{-1}. The half-life of A when $[A]_0 = 2.50 \times 10^{-2}$ mol dm^{-3} is

$$t_{1/2} = \frac{1}{\underbrace{(5.0\times10^4\,\text{dm}^3\,\text{mol}^{-1}\,\text{s}^{-1})}_{k_r}\times\underbrace{(2.50\times10^{-2}\,\text{mol dm}^{-3})}_{[A]_0}} = 8.0\times10^{-4}\ \text{s}$$

but when $[A]_0$ is lowered to 1.00×10^{-3} mol dm^{-3} the half-life increases to

$$t_{1/2} = \frac{1}{\underbrace{(5.0\times10^4\,\text{dm}^3\,\text{mol}^{-1}\,\text{s}^{-1})}_{k_r}\times\underbrace{(1.00\times10^{-3}\,\text{mol dm}^{-3})}_{[A]_0}} = 2.0\times10^{-2}\ \text{s}$$

A practical consequence of this dependence of the half-life on concentration is that species that decay by second-order reactions (which includes some environmentally harmful substances) may persist in low concentrations for long periods because their half-lives are long when their concentrations are low.

The temperature dependence of reaction rates

The rates of most chemical reactions increase as the temperature is raised. Many organic reactions in solution lie somewhere in the range spanned by the hydrolysis of methyl ethanoate (for which the rate constant at 35 °C is 1.8 times that at 25 °C) and the hydrolysis of sucrose (for which the factor is 4.1). Enzyme-catalysed reactions may show a more complex temperature dependence because raising the temperature may provoke conformational changes that lower the effectiveness of the enzyme. Indeed, one of the reasons why we fight infection with a fever is to upset the balance of reaction rates in the infecting organism, and hence destroy it, by the increase in temperature. There is a fine line, though, between killing an invader and killing the invaded!

10.9 The Arrhenius parameters

As data on reaction rates were accumulated towards the end of the nineteenth century, the Swedish chemist Svante Arrhenius noted that almost all of them showed a similar dependence on the temperature. In particular, he noted that a graph of ln k_r, where k_r is the rate constant for the reaction, against $1/T$, where T is the (absolute) temperature at which k_r is measured, gives a straight line with a slope that is characteristic of the reaction (Fig. 10.15). The mathematical expression of this conclusion is that the rate constant varies with temperature as

$$\ln k_r = \text{intercept} + \text{slope} \times \frac{1}{T} \qquad \text{Empirical temperature dependence} \qquad (10.20a)$$

This expression is normally written as the **Arrhenius equation**

$$\ln k_r = \ln A - \frac{E_a}{RT} \qquad \text{Arrhenius equation} \qquad (10.20b)$$

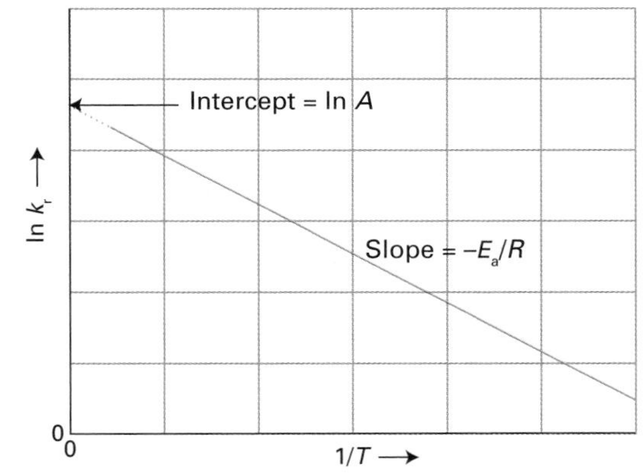

Fig. 10.15 The general form of an Arrhenius plot of ln k_r against $1/T$. The slope is equal to $-E_a/R$ and the intercept at $1/T = 0$ is equal to ln A.

An alternative form of eqn 10.20b is obtained by using $\ln k_r - \ln A = \ln(k_r/A)$ and taking natural antilogarithms, e^x, of both sides:

$$k_r = Ae^{-E_a/RT} \qquad \text{Alternative form} \qquad \text{Arrhenius equation} \qquad (10.20c)$$

The parameter A (which has the same units as k_r) is called the **pre-exponential factor**, and E_a (which is a molar energy and normally expressed as kilojoules per mole) is called the **activation energy**. Collectively, A and E_a are called the **Arrhenius parameters** of the reaction (Table 10.5).

A practical point to note from eqn 10.20 and illustrated in Fig. 10.16 is that a high activation energy corresponds to a reaction rate that is very sensitive to temperature (the Arrhenius plot has a steep slope). Conversely, a small activation energy indicates a reaction rate that varies only slightly with temperature (the slope is shallow). A reaction with zero activation energy, such as for some radical recombination reactions in the gas phase, has a rate that is largely independent of temperature.

Table 10.5
Arrhenius parameters

	A/s^{-1}	$E_a/(kJ\ mol^{-1})$
First-order reactions		
Cyclopropene → propane	1.58×10^{15}	272
$CH_3NC \rightarrow CH_3CN$	3.98×10^{13}	160
cis-CHD=CHD → *trans*-CHD=CHD	3.16×10^{12}	256
Cyclobutane → 2 C_2H_4	3.98×10^{15}	261
$2\ N_2O_5 \rightarrow 4\ NO_2 + O_2$	4.94×10^{13}	103
$N_2O \rightarrow N_2 + O$	7.94×10^{11}	250
Second-order, gas phase		
$O + N_2 \rightarrow NO + H$	1×10^{11}	315
$OH + H_2 \rightarrow H_2O + H$	8.0×10^{10}	42
$Cl + H_2 \rightarrow HCl + H$	8×10^{10}	23
$CH_3 + CH_3 \rightarrow C_2H_6$	2×10^{10}	0
$NO + Cl_2 \rightarrow NOCl + Cl$	4×10^{9}	85
Second-order, solution		
$NaC_2H_5O + CH_3I$ in ethanol	2.42×10^{11}	81.6
$C_2H_5Br + OH^-$ in water	4.30×10^{11}	89.5
$CH_3I + S_2O_3^{2-}$ in water	2.19×10^{12}	78.7
Sucrose + H_2O in acidic water	1.50×10^{15}	107.9

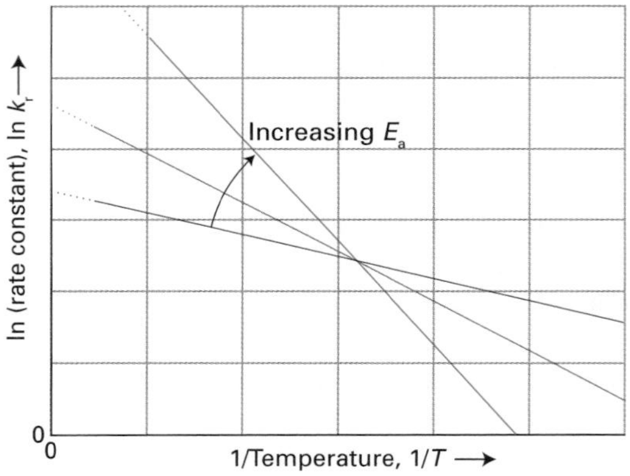

Fig. 10.16 These three Arrhenius plots correspond to three different activation energies. Note that the fact that the plot corresponding to the higher activation energy indicates that the rate of that reaction is more sensitive to temperature.

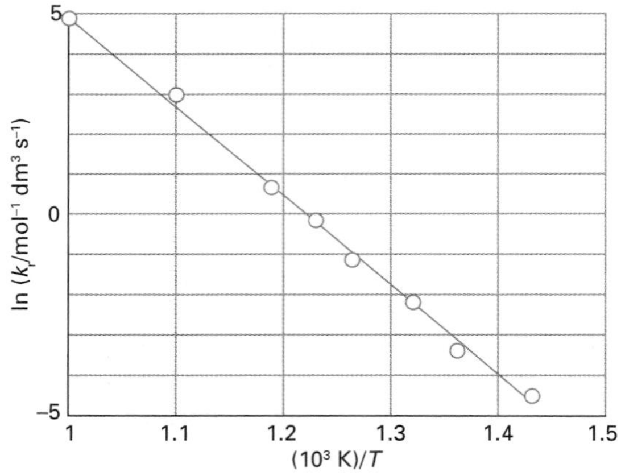

Fig. 10.17 The Arrhenius plot for the decomposition of CH_3CHO, and the best (least-squares) straight line fitted to the data points. The data are from Example 10.4.

Example 10.4

Determining the Arrhenius parameters

The rate of the second-order decomposition of acetaldehyde (ethanal, CH_3CHO) was measured over the temperature range 700–1000 K, and the rate constants are reported below. Find E_a and A.

T/K	700	730	760	790
k_r/(dm^3 mol^{-1} s^{-1})	0.011	0.035	0.105	0.343
T/K	810	840	910	1000
k_r/(dm^3 mol^{-1} s^{-1})	0.789	2.17	20.0	145

Strategy According to eqn 10.20b, the data can be analysed by plotting ln(k_r/dm^3 mol^{-1} s^{-1}) against 1/(T/K), or more conveniently (10^3 K)/T, and getting a straight line. Obtain the activation energy from the dimensionless slope by writing $-E_a/R$ = slope/units, where in this case 'units' = 1/(10^3 K), so $E_a = -$slope $\times R \times 10^3$ K. The intercept at $1/T = 0$ is ln(A/dm^3 mol^{-1} s^{-1}). Use a least-squares procedure to determine the plot parameters.

Solution We draw up the following table:

(10^3 K)/T	1.43	1.37	1.32	1.27
ln(k_r/dm^3 mol^{-1} s^{-1})	−4.51	−3.35	−2.25	−1.07
(10^3 K)/T	1.23	1.19	1.10	1.00
ln(k_r/dm^3 mol^{-1} s^{-1})	−0.24	0.77	3.00	4.98

Now plot ln k_r against 1/T (Fig. 10.17). The least-squares fit results in a line with slope $E_a/R = -22.7$ and intercept ln(A/dm^3 mol^{-1} s^{-1}) = 27.7. Therefore,

$$E_a = \text{slope} \times R = 22.7 \times (8.3145 \text{ J K}^{-1} \text{ mol}^{-1}) \times 10^3 \text{ K} = 189 \text{ kJ mol}^{-1}$$

$$A = e^{\text{intercept}} \times \text{dm}^3 \text{ mol}^{-1} \text{ s}^{-1} = e^{27.7} \times \text{dm}^3 \text{ mol}^{-1} \text{ s}^{-1} = 1.1 \times 10^{12} \text{ dm}^3 \text{ mol}^{-1} \text{ s}^{-1}$$

Note that A has the same units as k_r.

Self-test 10.6

Determine A and E_a from the following data:

T/K	300	350	400
k_r/(mol^{-1} dm^3 s^{-1})	7.9 × 10^6	3.0 × 10^7	7.9 × 10^7
T/K	450	500	
k_r/(mol^{-1} dm^3 s^{-1})	1.7 × 10^8	3.2 × 10^8	

Answer: 8 × 10^{10} mol^{-1} dm^3 s^{-1}, 23 kJ mol^{-1}

Once the activation energy of a reaction is known, it is a simple matter to predict the value of a rate constant $k_r(T')$ at a temperature T' from its value $k_r(T)$ at another temperature T. To do so, we write

$$\ln k_{r,2} = \ln A - \frac{E_a}{RT_2}$$

and then subtract eqn 10.20b (with T identified as T_1 and k_r as $k_{r,1}$), so obtaining

$$\ln k_{r,2} - \ln k_{r,1} = -\frac{E_a}{RT_2} + \frac{E_a}{RT_1}$$

We can rearrange this expression to

$$\ln \frac{k_{r,2}}{k_{r,1}} = \frac{E_a}{R}\left(\frac{1}{T_1} - \frac{1}{T_2}\right) \qquad (10.21)$$

Temperature dependence of the rate constant

Brief illustration 10.9 The Arrhenius equation

For a reaction with an activation energy of 50 kJ mol^{-1}, an increase in the temperature from 25 °C to 37 °C (body temperature) corresponds to

$$\ln \frac{k_{r,2}}{k_{r,1}} = \overbrace{\frac{50\times10^3\ \text{J mol}^{-1}}{\underbrace{8.3145\ \text{J K}^{-1}\,\text{mol}^{-1}}_{R}}}^{E_a}\left(\overbrace{\frac{1}{298\ \text{K}}}^{1/T_1} - \overbrace{\frac{1}{310\ \text{K}}}^{1/T_2}\right)$$

$$\overset{\text{cancel terms}}{=} \frac{50\times10^3}{8.3145}\left(\frac{1}{298} - \frac{1}{310}\right)$$

(The right-hand side evaluates to 0.781 . . . , but we take the next step before evaluating the answer.) By taking natural antilogarithms (that is, by forming e^x), $k_{r,2} = 2.18k_{r,1}$. This result corresponds to slightly more than a doubling of the rate constant.

Self-test 10.7

The activation energy of one of the reactions in a biochemical process is 87 kJ mol^{-1}. What is the change in rate constant when the temperature falls from 37 °C to 15 °C?

Answer: $k_r(15\ °C) = 0.076k_r(37\ °C)$

10.10 Collision theory

The origin of the Arrhenius parameters can be understood most simply by considering a class of gas-phase reactions in which reaction occurs when two molecules meet. That is, in the terminology to be introduced in Section 11.4, we are considering bimolecular gas-phase reactions. In this **collision theory** of reaction rates it is supposed that reaction occurs only if two molecules collide with a certain minimum kinetic energy along their line of approach (Fig. 10.18). In collision theory, a reaction resembles the collision of two defective billiard balls: the balls bounce apart if they collide with only a small energy, but might smash each other into fragments (products) if they collide with more than a certain minimum kinetic energy. This model of a reaction is a reasonable first approximation to the types of process that take place in planetary atmospheres and govern their compositions and temperature profiles.

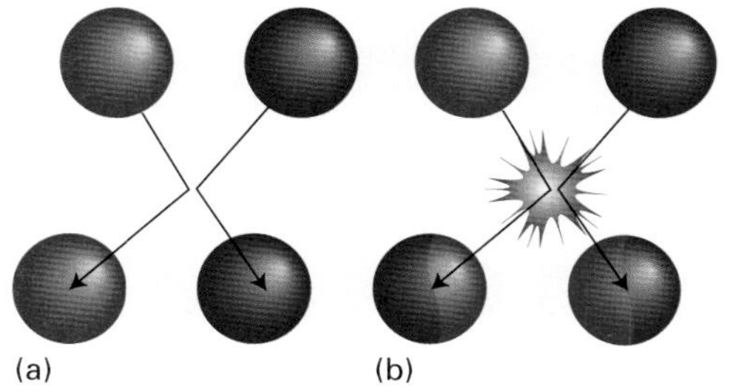

Fig. 10.18 In the collision theory of gas phase chemical reactions, reaction occurs when two molecules collide, but only if the collision is sufficiently vigorous. (a) An insufficiently vigorous collision: the reactant molecules collide but bounce apart unchanged. (b) A sufficiently vigorous collision results in a reaction.

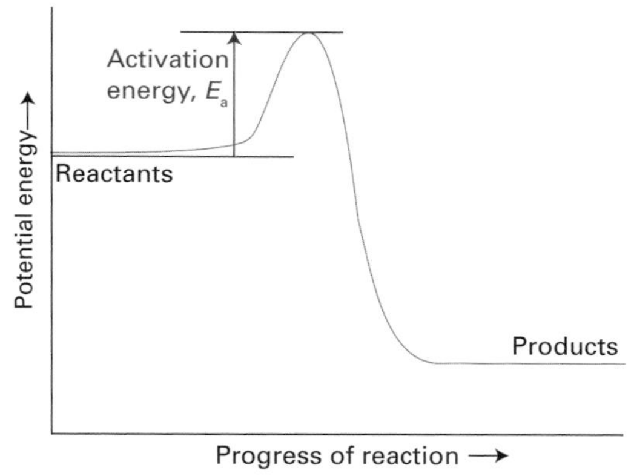

Fig. 10.19 A reaction profile. The graph depicts schematically the changing potential energy of two species that approach, collide, and then go on to form products. The activation energy is the height of the barrier above the potential energy of the reactants.

A **reaction profile** in collision theory is a graph showing the variation in potential energy as one reactant molecule approaches another and the products then separate (Fig. 10.19). On the left, the horizontal line represents the potential energy of the two reactant molecules that are far apart from one another. The potential energy rises from this value only when the separation of the molecules is so small that they are in contact, when it rises as bonds bend and start to break. The potential energy reaches a peak when the two molecules are highly distorted. Then it starts to decrease as new bonds are formed. At separations to the right of the maximum, the potential energy rapidly falls to a low value as the product molecules separate. For the reaction to be successful, the reactant molecules must approach with sufficient kinetic energy along their line of approach to carry them over the **activation barrier**, the peak in the reaction profile. As we shall see, we can identify the height of the activation barrier with the activation energy of the reaction.

With the reaction profile in mind, it is quite easy to establish that collision theory accounts for Arrhenius behaviour. Thus, **collision frequency**, the rate of collisions between species A and B, is proportional to both their concentrations: if the concentration of B is doubled, then the rate at which A molecules collide with B molecules is doubled, and if the concentration of A is doubled, then the rate at which B molecules collide with A molecules is also doubled. It follows that the collision frequency of A and B molecules is directly proportional to the concentrations of A and B, and we can write

$$\text{Collision frequency} \propto [\text{A}][\text{B}]$$

Next, we need to multiply the collision frequency by a factor f that represents the fraction of collisions that occur with at least a kinetic energy E_a along the line of approach (Fig. 10.20), for only these collisions will lead to the formation of products. Molecules that approach with less than a kinetic energy E_a will behave like a ball that rolls toward the activation barrier, fails to surmount it, and rolls back. We saw in Section 1.6 that only small fractions of molecules in the gas phase have very high speeds and that the fraction with very high speeds increases sharply as the temperature is raised. Because the kinetic energy increases as the square of the speed (for a body of mass m moving at a speed v, the kinetic energy is $E_k = \frac{1}{2}mv^2$), we expect that, at higher temperatures, a larger fraction of molecules will have a speed and kinetic energy that exceed the minimum values required for collisions that lead to formation of products (Fig. 10.21). The fraction of collisions that occur with at least a kinetic energy E_a can be calculated from general arguments developed in Chapter 22 concerning the probability that a molecule has a specified energy. The result is

$$f = e^{-E_a/RT} \qquad (10.22)$$

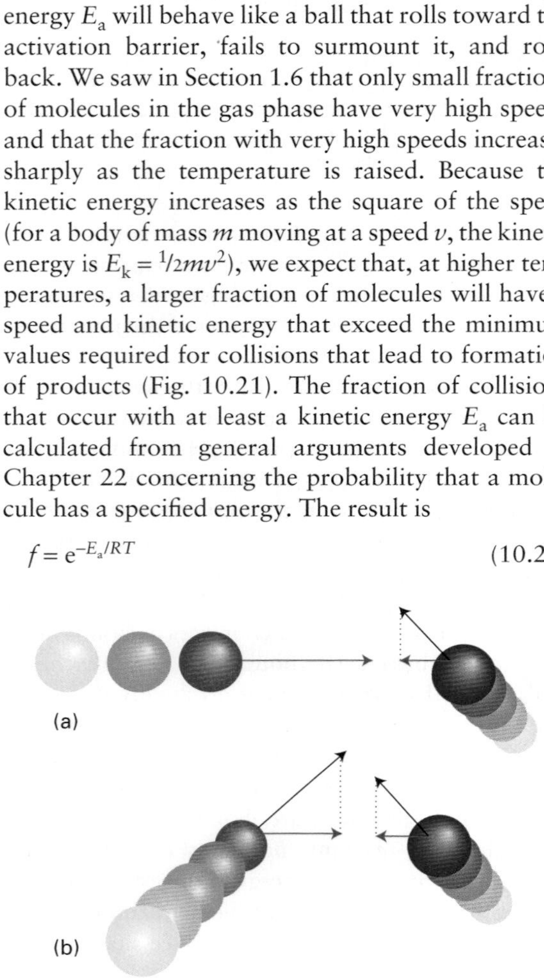

Fig. 10.20 (a) The criterion for a successful collision is that the two reactant species should collide with a kinetic energy along their line of approach that exceeds a certain minimum value E_a that is characteristic of the reaction. (b) The two molecules might also have components of velocity (and an associated kinetic energy) in other directions (for example, the two molecules depicted here might be moving up the page as well as towards each other); but only the energy associated with their mutual approach can be used to overcome the activation energy.

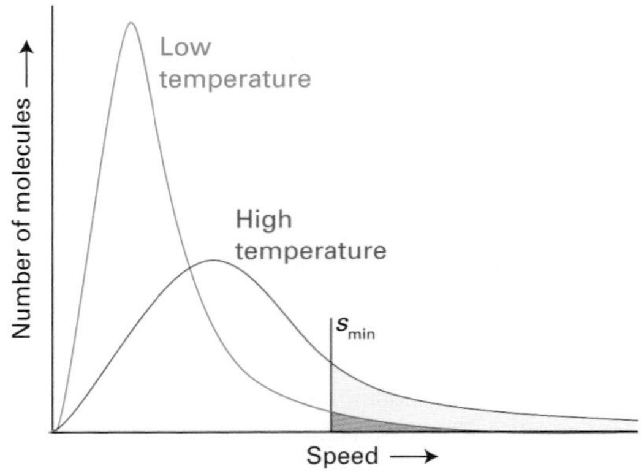

Fig. 10.21 According to the Maxwell distribution of speeds (Section 1.6), as the temperature increases, so does the fraction of gas phase molecules with a speed that exceeds a minimum value s_{min}. Because the kinetic energy is proportional to the square of the speed, it follows that more molecules can collide with a minimum kinetic energy E_a (the activation energy) at higher temperatures.

This fraction increases from 0 when $T = 0$ to 1 when T is infinite.

Brief illustration 10.10 The fraction of energetic collisions

If the activation energy is 50 kJ mol^{-1}, the fraction of collisions that have sufficient energy for reaction at 25 °C is

$$f = e^{-\overbrace{(5.0\times10^4\,\text{J mol}^{-1})}^{E_a}/\overbrace{(8.3145\,\text{J K}^{-1}\,\text{mol}^{-1})}^{R}\times\overbrace{(298\,\text{K})}^{T}} = 1.7\times10^{-9}$$

Self-test 10.8

What is the fraction of collisions that have sufficient energy for the same reaction at 500 °C?

Answer: 4.2×10^{-4}

At this stage we can conclude that the rate of reaction, which is proportional to the collision frequency multiplied by the fraction of successful collisions, is

$$v \propto [\text{A}][\text{B}]e^{-E_a/RT}$$

If we compare this expression with a second-order rate law,

$$v = k_r[\text{A}][\text{B}]$$

it follows that

$$k_r \propto e^{-E_a/RT}$$

This expression has exactly the Arrhenius form (eqn 10.20) if we identify the constant of proportionality with A. Collision theory therefore suggests the following interpretations:

- The *pre-exponential factor*, A, is the constant of proportionality between the concentrations of the reactants and the rate at which the reactant molecules collide.
- The *activation energy*, E_a, is the minimum kinetic energy required for a collision to result in reaction.

The value of A can be calculated from the kinetic theory of gases (Chapter 1):

$$A = \sigma\left(\frac{8kT}{\pi\mu}\right)^{1/2} N_A \qquad \mu = \frac{m_A m_B}{m_A + m_B} \qquad (10.23)$$

The pre-exponential factor from the kinetic theory of gases

where m_A and m_B are the masses of the molecules A and B and σ is the collision cross-section (Section 1.8). However, it is often found that the experimental value of A is smaller than that calculated from the kinetic theory. One possible explanation is that not only must the molecules collide with sufficient kinetic energy, but they must also come together in a specific relative orientation (Fig. 10.22). It follows that the reaction rate is proportional to the probability that the encounter occurs in the correct relative orientation. The pre-exponential factor A should therefore include a **steric factor**, P, which usually lies between 0 (no relative orientations lead to reaction) and 1 (all relative orientations lead to reaction).

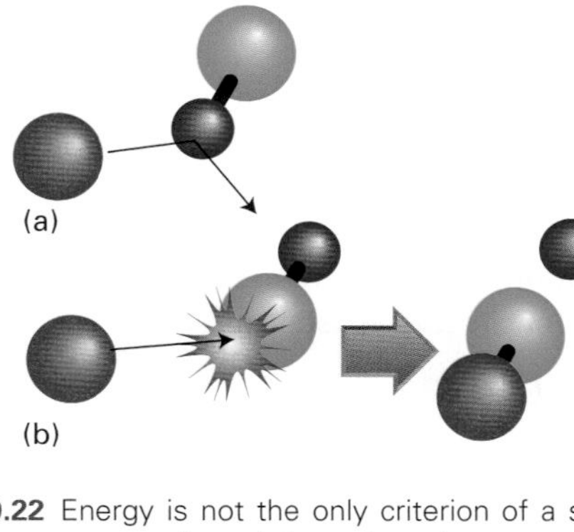

Fig. 10.22 Energy is not the only criterion of a successful reactive encounter, for relative orientation may also play a role. (a) In this collision, the reactants approach in an inappropriate relative orientation and no reaction occurs even though their energy is sufficient. (b) In this encounter, both the energy and the orientation are suitable for reaction.

Brief illustration 10.11 The steric factor

For the reactive collision,

$$NOCl + NOCl \rightarrow NO + NO + Cl_2$$

in which two NOCl molecules collide and break apart into two NO molecules and a Cl_2 molecule, $P \approx 0.16$. For the hydrogen addition reaction

$$H_2 + H_2C{=}CH_2 \rightarrow H_3C{-}CH_3$$

in which a hydrogen molecule attaches directly to an ethene molecule to form an ethane molecule, P is only 1.7×10^{-6}, which suggests that the reaction has very stringent orientational requirements.

Some reactions have $P > 1$. Such a value may seem absurd, because it appears to suggest that the reaction occurs more often than the molecules meet! An example of a reaction of this kind is

$$K + Br_2 \rightarrow KBr + Br$$

in which a K atom plucks a Br atom out of a Br_2 molecule; for this reaction the experimental value of P is 4.10. In this reaction, the distance of approach at which reaction can occur seems to be considerably larger than the distance needed for deflection of the path of the approaching molecules in a non-reactive collision! To explain this surprising conclusion, it has been proposed that the reaction proceeds by a 'harpoon mechanism'. This brilliant name is based on a model of the reaction which pictures the K atom as approaching the Br_2 molecules, and when the two are close enough an electron (the harpoon) flips across to the Br_2 molecule. In place of two neutral particles there are now two ions, and so there is a Coulombic attraction between them: this attraction is the line on the harpoon. Under its influence the ions move together (the line is wound in), the reaction takes place, and KBr and Br emerge. The harpoon extends the cross-section for the reactive encounter and we would greatly underestimate the reaction rate if we used for the collision cross-section the value for simple mechanical contact between K and Br_2.

10.11 Transition-state theory

There is a more sophisticated theory of reaction rates that can be applied to reactions taking place in solution as well as in the gas phase. In the **transition-state theory** (also called the 'activated complex theory') of reactions, it is supposed that as two reactants approach, their potential energy rises and reaches a maximum, as illustrated by the reaction profile in

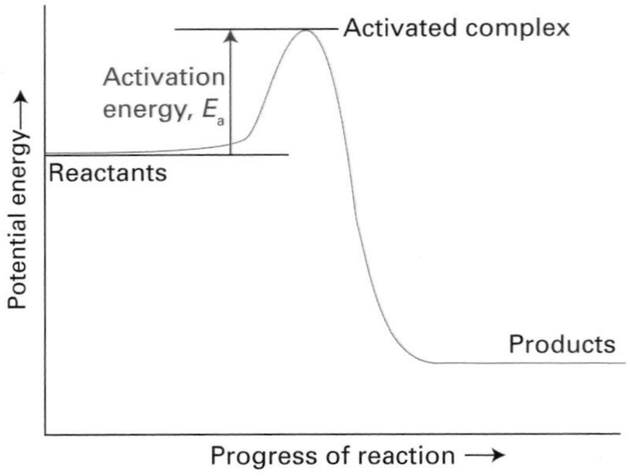

Fig. 10.23 The same type of graph as in Fig. 10.19 represents the reaction profile that is considered in activated complex theory. The activation energy is the potential energy of the activated complex relative to that of the reactants.

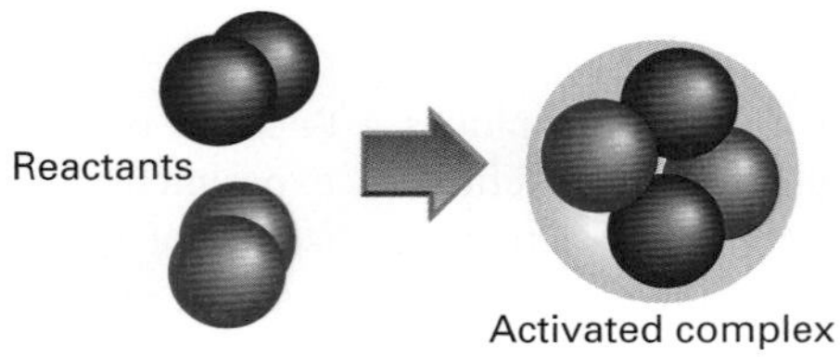

Fig. 10.24 The activated complex is depicted here by a relatively loose cluster of atoms that may undergo rearrangement into products. In an actual reaction, only some atoms—those at the actual reaction site—might be significantly loosened in the complex, the bonding of the others remaining almost unchanged.

Fig. 10.23. This maximum corresponds to the formation of an **activated complex**, a cluster of atoms that is poised to pass on to products or to collapse back into the reactants from which it was formed (Fig. 10.24). An activated complex is not a reaction intermediate that can be isolated and studied like ordinary molecules. The concept of an activated complex is applicable to reactions in solutions as well as to the gas phase, because we can think of the activated complex as perhaps involving any solvent molecules that may be present.

Initially only the reactants A and B are present. As the reaction event proceeds, A and B come into contact, distort, and begin to exchange or discard atoms. The potential energy rises to a maximum, and the cluster of atoms that corresponds to the region close to the maximum is the activated complex. The potential energy falls as the atoms rearrange in the cluster, and reaches a value characteristic of the products. The climax of the reaction is at the peak of the potential energy. Here, two reactant molecules have come to such a degree of closeness and distortion that a small further distortion will send them in the direction of products. This crucial configuration is called the **transition state** of the reaction. Although some molecules entering the transition state might revert to reactants, if they pass through this configuration it is probable that products will emerge from the encounter.

The **reaction coordinate** is an indication of the stage reached in this process. On the left are the undistorted, widely separated reactants. On the right are the products. Somewhere in the middle is the stage of the reaction corresponding to the formation of the activated complex. The principal goal of transition state theory is to write an expression for the rate constant by tracking the history of the activated complex from its formation by encounters between the reactants to its decay into product.

(a) Outline of the theory

Here we outline the steps involved in the calculation of the rate constant, with an eye toward gaining insight into the molecular events that optimize the rate constant.

The activated complex $C^{\ddagger}$ is formed from the reactants A and B and it is supposed—without much justification—that there is an equilibrium, with equilibrium constant $K^{\ddagger}$, between the concentrations of A, B, and $C^{\ddagger}$:

$$A + B \rightleftharpoons C^{\ddagger} \qquad K^{\ddagger} = \frac{[C^{\ddagger}]c^{\ominus}}{[A][B]}$$

where $c^{\ominus} = 1$ mol dm^{-3} (Section 6.1). At the transition state, motion along the reaction coordinate corresponds to some complicated collective vibration-like motion of all the atoms in the complex (and the motion of the solvent molecules if they are involved too). However, it is possible that not every motion along the reaction coordinate takes the complex through the transition state and to the product P. By taking into account the equilibrium between A, B, and $C^{\ddagger}$ and the rate of successful passage of $C^{\ddagger}$ through the transition state, it is possible to derive the **Eyring equation** for the rate constant:

$$k_r = \kappa \times \frac{kT}{h} \times \frac{K^{\ddagger}}{c^{\ominus}} \qquad \text{Eyring equation} \quad (10.24)$$

where $k = 1.381 \times 10^{-23}$ J K^{-1} is Boltzmann's constant and $h = 6.626 \times 10^{-34}$ J s is Planck's constant (both constants are introduced in Foundations). The factor κ (kappa) is the **transmission coefficient**, which takes into account the fact that the activated complex does not always pass through to the transition state. In the absence of information to the contrary, κ is assumed to be about 1.

A note on good practice Be very careful to distinguish the Boltzmann constant k from the symbol for a rate constant, k_r. In some expositions, you will see Boltzmann's constant denoted k_B to emphasize its significance (and sometimes, confusingly, the rate constant is denoted k).

The term kT/h in eqn 10.24 (which has the dimensions of a frequency, as kT is an energy, and division by Planck's constant turns an energy into a frequency; with kT in joules, kT/h has the units s^{-1}) arises from consideration of the motions of atoms that lead to the decay of $C^{\ddagger}$ into products, as specific bonds are broken and formed. It follows that one way in which an increase in temperature enhances the rate is by causing more vigorous motion in the activated complex, facilitating the rearrangement of atoms and the formation of new bonds.

Calculation of the equilibrium constant $K^{\ddagger}$ is very difficult, except in certain simple model cases. For example, if we suppose that the reactants are two atoms and that the activated complex is a diatomic molecule of bond length R_{bond}, then k_r turns out to be the same as for collision theory provided we interpret the collision cross-section in eqn 10.23 as πR_{bond}^2. It is more useful to express the Eyring equation in terms of thermodynamic parameters and to discuss reactions in terms of their empirical values. Thus, we saw in Section 7.3 that an equilibrium constant may be expressed in terms of the standard reaction Gibbs energy ($-RT \ln K = \Delta_r G^{\ominus}$). In this context, the Gibbs energy is called the **activation Gibbs energy**, and written $\Delta^{\ddagger}G$. It follows that

$$\Delta^{\ddagger}G = -RT \ln K^{\ddagger} \quad \text{and} \quad K^{\ddagger} = e^{-\Delta^{\ddagger}G/RT}$$

Therefore, by writing

$$\Delta^{\ddagger}G = \Delta^{\ddagger}H - T\Delta^{\ddagger}S \qquad (10.25)$$

we conclude that (with $\kappa = 1$)

$$k_r = \frac{kT}{h} e^{-\overbrace{(\Delta^{\ddagger}H - T\Delta^{\ddagger}S)}^{\Delta^{\ddagger}G}/RT} \overset{e^{x+y}=e^x e^y}{=} \left(\frac{kT}{h} e^{\Delta^{\ddagger}S/R}\right) e^{-\Delta^{\ddagger}H/RT}$$

The Eyring equation in terms of thermodynamic parameters (10.26)

This expression has the form of the Arrhenius expression, eqn 10.20, if we identify the **enthalpy of activation**, $\Delta^{\ddagger}H$, with the activation energy and the term in parentheses, which depends on the **entropy of activation**, $\Delta^{\ddagger}S$, with the pre-exponential factor.

The advantage of transition state theory over collision theory is that it is applicable to reactions in solution as well as in the gas phase. It also gives some clue to the calculation of the steric factor P, for the orientation requirements are carried in the entropy of activation. Thus, if there are strict orientation requirements (for example, in the approach of a substrate molecule to an enzyme), then the entropy of activation will be strongly negative (representing a decrease in disorder when the activated complex forms), and the pre-exponential factor will be small. In practice, it is occasionally possible to estimate the sign and magnitude of the entropy of activation and hence to estimate the rate constant. The general importance of transition state theory is that it shows that even a complex series of events—not only a collisional encounter in the gas phase—displays Arrhenius-like behaviour, and that the concept of activation energy is applicable.

Brief illustration 10.12 The entropy of activation

Consider a particular reaction in water, for which it is proposed that two ions of opposite charge come together to form an electrically neutral activated complex. The contribution of the solvent to the entropy of activation is likely to be positive because H_2O is less organized around the neutral species than around the initial charged reactants.

(b) Observation of the activated complex

Until recently, activated complexes were not observed directly, for they have a very fleeting existence and often survive for only a few picoseconds. However, the development of femtosecond pulsed lasers (1 fs = 10^{-15} s) and their application to chemistry in the form of **femtochemistry** has made it possible to make observations on species that have such short lifetimes that in a number of respects they resemble activated complexes. Further developments have even brought attosecond investigations (1 as = 10^{-18} s) within reach.

In a typical experiment, energy from a femtosecond pulse is used to dissociate a molecule, and then a second femtosecond pulse is fired at an interval after the pulse. The frequency of the second pulse is set at an absorption of one of the free fragmentation products, so its absorption is a measure of the abundance of the dissociation product. For example, when ICN is dissociated by the first pulse, the emergence of CN can be monitored by watching the growth of the free CN absorption. In this way it has been found that the CN signal remains zero until the fragments have separated by about 600 pm, which takes about 205 fs.

Some sense of the progress that has been made in the study of the intimate mechanism of chemical reactions can be obtained by considering the decay of the ion pair Na^+I^-. Absorption of energy from the femtosecond laser by the ionic species leads to redistribution of electrons and forms a state that corresponds to a covalently bonded NaI molecule. The probe pulse examines the system at an absorption frequency either of the free Na atom or at a frequency at which the atom absorbs when it is a part of the complex. The latter frequency depends on the Na–I distance, so an absorption is obtained each time the vibration of the complex returns it to that separation.

A typical set of results is shown in Fig. 10.25. The bound Na absorption intensity shows up as a series of pulses that recur in about 1 ps, showing that the complex vibrates with about that period. The decline in intensity shows the rate at which the complex can dissociate as the two atoms swing away from each other. The free Na absorption also grows in an oscillating manner, showing the periodicity of the vibration of the complex, each swing of which gives it a chance to dissociate. The precise period of the oscillation in NaI is 1.25 ps. The complex survives for about ten oscillations. In contrast, although the oscillation frequency of NaBr is similar, it barely survives one oscillation.

Femtochemistry techniques have also been used to examine analogues of the activated complex involved in bimolecular reactions. As an example, consider the weakly bound complex (also called a 'van der Waals molecule'), IH...OCO. The HI bond can be dissociated by a femtosecond pulse, and the H atom is ejected towards the O atom of the neighbouring CO_2 molecule to form HOCO. Hence, the complex is a source of a species that resembles the activated complex of the reaction

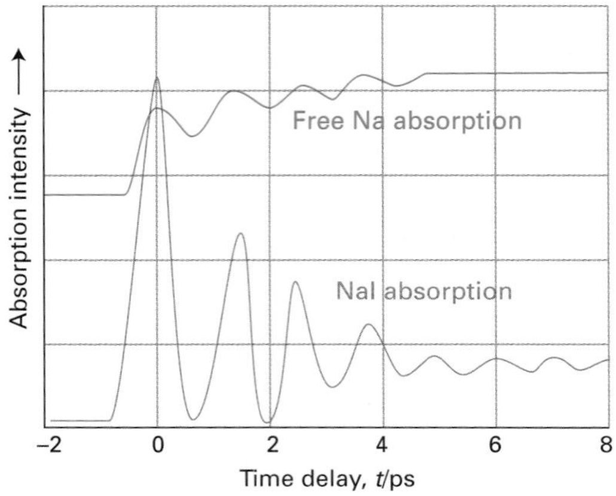

Fig. 10.25 The absorption spectra of the species NaI and Na immediately after a femtosecond flash. The oscillations show how the species incipiently form then reform their precursors before finally forming products. (Adapted from A. H. Zewail, *Science*, **242**, 1645 (1988).)

$$H + CO_2 \rightarrow [HOCO]^{\ddagger} \rightarrow HO + CO$$

The probe pulse is tuned to the OH radical, which enables the evolution of $[HOCO]^{\ddagger}$ to be studied in real time. Femtosecond techniques have also been used to study more complex reactions, such as the Diels–Alder reaction, nucleophilic substitution reactions, and pericyclic addition and cleavage reactions.

Checklist of key concepts

- ☐ 1 The rates of chemical reactions are measured by using techniques that monitor the concentrations of species present in the reaction mixture (Table 10.1).
- ☐ 2 Spectrophotometry is the measurement of the absorption of light by a material.
- ☐ 3 The Beer–Lambert law relates the absorbance of a sample to the concentration of an absorbing species.
- ☐ 4 Techniques include real-time and quenching procedures, flow and stopped-flow techniques, and flash photolysis.
- ☐ 5 The instantaneous rate of a reaction is the slope of the tangent to the graph of concentration against time (expressed as a positive quantity).
- ☐ 6 A rate law is an expression for the reaction rate in terms of the concentrations of the species that occur in the overall chemical reaction.
- ☐ 7 For a rate law of the form $v = k_r[A]^a[B]^b \ldots$, the order with respect to A is a and the overall order is $a + b + \cdots$.
- ☐ 8 An integrated rate law is an expression for the variation of the concentration of a reactant or product with time.

☐ 9 The half-life of a first-order reaction is the time it takes for the concentration of a species to fall to half its initial value.

☐ 10 The temperature dependence of the rate constant of a reaction typically follows the Arrhenius law.

☐ 11 The larger the activation energy, the more sensitive the rate constant is to the temperature.

☐ 12 In collision theory, it is supposed that the rate is proportional to the collision frequency, a steric factor, and the fraction of collision that occur with at least the kinetic energy E_a along their lines of centres.

☐ 13 In transition state theory, it is supposed that an activated complex is in equilibrium with the reactants, and that the rate at which that complex forms products depends on the rate at which it passes through a transition state. The result is the Eyring equation.

Road map of key equations

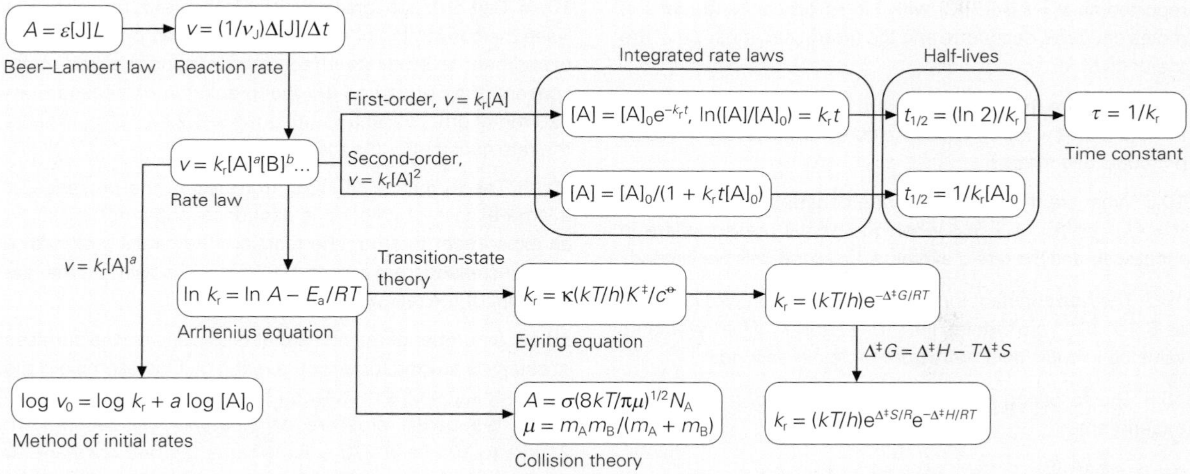

Questions and exercises

Discussion questions

10.1 Consult literature sources and list the observed ranges of timescales during which the following processes occur: proton-transfer reactions, electron-transfer events between complex ions in solution, harpoon reactions, and collisions in liquids.

10.2 What information can be extracted from the determination of the rate of a chemical reaction under different conditions?

10.3 Describe the main features, including advantages and disadvantages, of the following experimental methods for determining the rate law of a reaction: the isolation method, the method of initial rates, and fitting data to integrated rate law expressions.

10.4 Distinguish between zeroth-order, first-order, second-order, and pseudofirst-order reactions; under what conditions can the apparent order of a reaction change?

10.5 Define the terms in, and limit the generality of, the expression $\ln k_r = \ln A - E_a/RT$; why might there be deviations from the Arrhenius expression?

10.6 State, explain, and justify the steady-state approximation.

10.7 Describe the formulation of the Eyring equation; in what sense is it superior to the collision theory of reaction rates?

Exercises

10.1 The molar absorption coefficient of cytochrome P450, an enzyme involved in the breakdown of harmful substances in the liver and small intestine, at 522 nm is 291 $dm^3\ mol^{-1}\ cm^{-1}$. When light of that wavelength passes through a cell of length 6.5 mm containing a solution of the solute, 39.8 per cent of the light was absorbed. What is the molar concentration of the solute?

10.2 The rate of formation of C in the reaction $2\,A + B \rightarrow 4\,C + 3\,D$ is 3.2 mol $dm^{-3}\ s^{-1}$. State the rates of formation and consumption of A, B, and D.

10.3 Equation 10.3 defines the unique rate of a reaction. Write expressions for v for each of the species in the reaction $2\,A + B \rightarrow 4\,C + 3\,D$. What is the value of v given the information in Exercise 10.2?

10.4 The rate law for the reaction in Exercise 10.2 was reported as $v = k_r[A][B][C]$ with the molar concentrations in moles per cubic decimetre and the time in seconds. What are the units of k_r?

10.5 What are the units of the rate constants in the rate law $v = k_{r1}[A][B]/(1 + k_{r2}[B])$ when the concentrations are in moles per cubic decimetre?

10.6 What are the units of the rate constants in the rate law $v = k_{r1}p_Bp_A^{3/2}/(p_A + k_{r2}p_B)$ when the partial pressures are in kilopascals and the rate is expressed in kilopascals per second?

10.7 The rate constant for a gas-phase reaction was reported as 6.2×10^{-14} $cm^3\ molecule^{-1}\ s^{-1}$ at 298 K. What would its value be in cubic decimetres per mole per second?

10.8 The following rate law was established in a series of experiments:

$$v = \frac{k_{r1}[A][B]}{k_{r2} + k_{r3}[B]^{1/2}}$$

Identify the conditions under which the reaction can be classified by its order.

10.9 The following initial-rate data were obtained on the rate of binding of glucose with the enzyme hexokinase (obtained from yeast) present at a concentration of 1.34 mmol dm^{-3}. What is (a) the order of reaction with respect to glucose, (b) the rate constant?

$[C_6H_{12}O_6]/(mmol\ dm^{-3})$	1.00	1.54	3.12	4.02
$v_0/(mol\ dm^{-3}\ s^{-1})$	5.0	7.6	15.5	20.0

10.10 The following data were obtained on the initial rates of a reaction of a d-metal complex in aqueous solution. What is (a) the order of reaction with respect to the complex and the reactant Y, (b) the rate constant? For the experiments (a), [Y] = 2.7 mmol dm^{-3} and for experiments (b) [Y] = 6.1 mmol dm^{-3}.

$[complex]/(mmol\ dm^{-3})$		8.01	9.22	12.11
$v/(mol\ dm^{-3}\ s^{-1})$	(a)	125	144	190
	(b)	640	730	960

10.11 In a study of the alcohol dehydrogenase catalysed oxidation of ethanol, the molar concentration of ethanol decreased in a first-order reaction from 220 mmol dm^{-3} to 56.0 mmol dm^{-3} in 1.22×10^4 s. What is the rate constant of the reaction?

10.12 The elimination of carbon dioxide from pyruvate ions by a decarboxylase enzyme was monitored by measuring the partial pressure of the gas as it was formed. In one experiment, the partial pressure increased from zero to 100 Pa in 422 s in a first-order reaction. What is the rate constant of the reaction?

10.13 In the study of a second-order gas-phase reaction, it was found that the molar concentration of a reactant fell from 220 mmol mol^{-1} to 56.0 mmol mol^{-1} in 1.22×10^4 s. What is the rate constant of the reaction?

10.14 Carbonic anhydrase is a zinc-based enzyme that catalyses the conversion of carbon dioxide to carbonic acid. In an experiment to study its effect, it was found that the molar concentration of carbon dioxide in solution decreased from 220 mmol dm^{-3} to 56.0 mmol dm^{-3} in 1.22×10^4 s. What is the rate constant of the reaction?

10.15 The formation of NOCl from NO in the presence of a large excess of chlorine is pseudosecond-order in NO. In an experiment to study the reaction, the partial pressure of NOCl increased from zero to 100 Pa in 522 s. What is the rate constant of the reaction?

10.16 A number of reactions that take place on the surfaces of catalysts are zero order in the reactant. One example is the decomposition of ammonia on hot tungsten. In one experiment, the partial pressure of ammonia decreased from 21 kPa to 10 kPa in 770 s. (a) What is the rate constant for the zero-order reaction? (b) How long will it take for all the ammonia to disappear?

10.17 The following kinetic data (r_0 is the initial rate) were obtained for the reaction $2\,ICl(g) + H_2(g) \rightarrow I_2(g) + 2\,HCl(g)$:

Experiment	$[ICl]_0/(mmol\ dm^{-3})$	$[H_2]_0/(mmol\ dm^{-3})$	$v_0/(mol\ dm^{-3}\ s^{-1})$
1	1.5	1.5	3.7×10^{-7}
2	3.0	1.5	7.4×10^{-7}
3	3.0	4.5	22×10^{-7}
4	4.7	2.7	?

(a) Write the rate law for the reaction. (b) From the data, determine the value of the rate constant. (c) Use the data to predict the reaction rate for Experiment 4.

10.18 The variation in the partial pressure p of mercury dimethyl with time was followed at 800 K, with the results given below. Confirm that the decomposition $Hg(CH_3)_2(g) \rightarrow Hg(g) + 2\,CH_3(g)$ is first order in $Hg(CH_3)_2$ and find the rate constant at this temperature.

t/s	0	1.0	2.0	3.0	4.0
p/kPa	15.1	11.8	9.21	7.2	5.6

10.19 The following data were collected for the reaction $2\ HI(g) \rightarrow H_2(g) + I_2(g)$ at 580 K:

t/s	0	1000	2000	3000	4000
[HI]/(mol dm^{-3})	1.00	0.112	0.061	0.041	0.031

(a) Plot the data in an appropriate fashion to determine the order of the reaction. (b) From the graph, determine the rate constant.

10.20 The following data were collected for the reaction $H_2(g) + I_2(g) \rightarrow 2\ HI(g)$ at 780 K:

t/s	0	1	2	3	4
[HI]/(mol dm^{-3})	1	0.43	0.27	0.2	0.16

(a) Plot the data in an appropriate fashion to determine the order of the reaction. (b) From the graph, determine the rate constant.

10.21 Laser flash photolysis is often used to measure the binding rate of CO to haem proteins, such as myoglobin (Mb), because CO dissociates from the bound state relatively easily upon absorption of energy from an intense and narrow pulse of light. The reaction is usually run under pseudofirst-order conditions. For a reaction in which $[\mathrm{Mb}]_0$ = 10 mmol dm^{-3}, [CO] = 400 mmol dm^{-3}, and the rate constant is 5.8×10^5 dm^3 mol^{-1} s^{-1}, plot a curve of [Mb] against time. The observed reaction is Mb + CO → MbCO.

10.22 The integrated rate law of a second-order reaction of the form 3 A → B is $[\mathrm{A}] = [\mathrm{A}]_0/(1 + k_r t[\mathrm{A}]_0)$. How does the concentration of B change with time?

10.23 The composition of a liquid-phase reaction 2 A → B was followed spectrophotometrically with the following results:

t/min	0	10	20	30	40	∞
[B]/(mol dm^{-3})	0	0.372	0.426	0.448	0.460	0.500

Determine the order of the reaction and its rate constant (written in the form of eqn 10.5c).

10.24 Example 10.3 provided data on a first-order gas-phase reaction. How does the total pressure of the sample change with time?

10.25 The half-life of pyruvic acid in the presence of an aminotransferase enzyme (which converts it to alanine) was found to be 221 s. How long will it take for the concentration of pyruvic acid to fall to $^1/_{64}$ of its initial value in this first-order reaction?

10.26 The half-life for the (first-order) radioactive decay of ^{14}C is 5730 a (1 a is the SI unit 1 annum, for 1 year; the nuclide emits β particles, high-energy electrons, with an energy of 0.16 MeV). An archaeological sample contained wood that had only 69 per cent of the ^{14}C found in living trees. What is its age?

10.27 One of the hazards of nuclear explosions is the generation of ^{90}Sr and its subsequent incorporation in place of calcium in bones. This nuclide emits β particles of energy 0.55 MeV, and has a half-life of 28.1 a (1 a is the SI unit annum, for 1 year). Suppose 1.00 μg was absorbed by a newly born child. How much will remain after (a) 19 a, (b) 75 a if none is lost metabolically?

10.28 The second-order rate constant for the reaction $CH_3COOC_2H_5(aq) + OH^-(aq) \rightarrow CH_3CO_2^-(aq) + CH_3CH_2OH(aq)$ is 0.11 dm^3 mol^{-1} s^{-1}. What is the concentration of ester after (a) 15 s, (b) 15 min when ethyl acetate is added to sodium hydroxide so that the initial concentrations are [NaOH] = 0.055 mol dm^{-3} and $[CH_3COOC_2H_5]$ = 0.150 mol dm^{-3}?

10.29 A reaction 2 A → P has a second-order rate law with k_r = 1.44 dm^3 mol^{-1} s^{-1}. Calculate the time required for the concentration of A to change from 0.460 mol dm^{-3} to 0.046 mol dm^{-3}.

10.30 The rate constant for the first-order decomposition of N_2O_5 in the reaction $2\ N_2O_5(g) \rightarrow 4\ NO_2(g) + O_2(g)$ with $r = k_r[N_2O_5]$ is $k_r = 3.38 \times 10^{-5}$ s^{-1} at 25 °C. What is the half-life of N_2O_5? What will be the total pressure, initially 78.4 kPa for the pure N_2O_5 vapour, (a) 5.0 s, (b) 5.0 min after initiation of the reaction?

10.31 The half-life of a first-order reaction was found to be 439 s; what is the time constant for the reaction?

10.32 The Arrhenius parameters for the reaction $C_4H_8(g) \rightarrow 2\ C_2H_4(g)$, where C_4H_8 is cyclobutane, are log(A/s^{-1}) = 15.6 and E_a = 261 kJ mol^{-1}. What is the half-life of cyclobutane at (a) 20 °C, (b) 500 °C?

10.33 A rate constant is 2.78×10^{-4} dm^3 mol^{-1} s^{-1} at 19 °C and 3.38×10^{-3} dm^3 mol^{-1} s^{-1} at 37 °C. Evaluate the Arrhenius parameters of the reaction.

10.34 The activation energy for the decomposition of benzene diazonium chloride is 99.1 kJ mol^{-1}. At what temperature will the rate be 10 per cent greater than its rate at 25 °C?

10.35 Which reaction responds more strongly to changes of temperature, one with an activation energy of 52 kJ mol^{-1} or one with an activation energy of 25 kJ mol^{-1}?

10.36 The rate constant of a reaction increases by a factor of 1.41 when the temperature is increased from 20 °C to 27 °C. What is the activation energy of the reaction?

10.37 Make an appropriate Arrhenius plot of the following data for the conversion of cyclopropane to propene and calculate the activation energy for the reaction.

T/K	750	800	850	900
k_r/s^{-1}	1.8×10^{-4}	2.7×10^{-3}	3.0×10^{-2}	0.26

10.38 Food rots about 40 times more rapidly at 25 °C than when it is stored at 4 °C. Estimate the overall activation energy for the processes responsible for its decomposition.

10.39 Suppose that the rate constant of a reaction *decreases* by a factor of 1.23 when the temperature is increased from

20 °C to 27 °C. How should you report the activation energy of the reaction?

10.40 The enzyme urease catalyses the reaction in which urea is hydrolysed to ammonia and carbon dioxide. The half-life of urea in the pseudofirst-order reaction for a certain amount of urease doubles when the temperature is lowered from 20 °C to 10 °C and the Michaelis constant is largely unchanged. What is the activation energy of the reaction?

10.41 What proportion of the collisions between NO_2 molecules have enough energy to result in reaction when the temperature is (a) 20 °C, (b) 200 °C given that the activation energy for the reaction $2\ NO_2(g) \rightarrow 2\ NO(g) + O_2(g)$ is 111 kJ mol^{-1}? Use collision theory to estimate the pre-exponential factor for the reaction. The experimental value is 2×10^9 dm^3 mol^{-1} s^{-1}. Suggest a reason for any discrepancy.

10.42 The collision cross-section for methyl radicals is 2.62 nm^2. Use collision theory to estimate the Arrhenius pre-exponential factor for the reaction $CH_3(g) + CH_3(g) \rightarrow C_2H_6(g)$ at 298 K. Why might the measured value be rather lower than that calculated?

10.43 Suppose an electronegative reactant needs to come to within 500 pm of a reactant with low ionization energy before an electron can flip across from one to the other (as in the harpoon mechanism). Estimate the reaction cross-section.

10.44 Estimate the activation Gibbs energy for the decomposition of urea in the reaction $CO(NH_2)(aq) + 2\ H_2O(l) \rightarrow 2\ NH_4^+(aq) + CO_3^{2-}(aq)$ for which the pseudofirst-order rate constant is 1.2×10^{-7} s^{-1} at 60 °C and 4.6×10^{-7} s^{-1} at 70 °C.

10.45 Calculate the entropy of activation of the reaction in Exercise 10.44 at the two temperatures.

Projects

The symbol ‡ indicates that calculus is required.

10.46‡ Here we explore integrated rate laws in more detail. (a) Establish the integrated form of a third-order rate law of the form $v = k_r[A]^3$. What would be appropriate to plot to confirm that a reaction is third order? (b) Establish the integrated form of a second-order rate law of the form $v = k_r[A][B]$ for a reaction A + B → products (i) with different initial concentrations of A and B, (ii) with the same concentrations of the two reactants. Note that when the concentration of A falls to $[A]_0 - x$, the concentration of B falls to $[B]_0 - x$. Use these relations to show that the rate law may be written as

$$\frac{dx}{dt} = k_r([A]_0 - x)([B]_0 - x)$$

Hint: To make progress with integration of the rate law, use the form:

$$\int \frac{dx}{(a-x)(b-x)} = \frac{1}{b-a}\left(\ln\frac{1}{a-x} - \ln\frac{1}{b-x}\right) + \text{constant}$$

10.47 Prebiotic reactions are reactions that might have occurred under the conditions prevalent on the Earth before the first living creatures emerged and that can lead to analogues of molecules necessary for life as we now know it. To qualify, a reaction must proceed with a favourable rate and have a reasonable value for the equilibrium constant. An example of a prebiotic reaction is the formation of 5-hydroxymethyluracil (HMU) from uracil and formaldehyde (HCHO). Amino acid analogues can be formed from HMU under prebiotic conditions by reaction with various nucleophiles, such as H_2S, HCN, indole, imidazole, etc. For the synthesis of HMU at pH = 7, the temperature dependence of the rate constant is given by

$$\log k_r/(\text{dm}^3\ \text{mol}^{-1}\ \text{s}^{-1}) = 11.75 - 5488/(T/\text{K})$$

and the temperature dependence of the equilibrium constant is given by

$$\log K = -1.36 + 1794/(T/\text{K})$$

(a) Calculate the rate constants and equilibrium constants over a range of temperatures corresponding to possible prebiotic conditions, such as 0–50 °C, and plot them against temperature. (b) Calculate the activation energy and the standard reaction Gibbs energy and enthalpy at 25 °C. (c) Prebiotic conditions are not likely to be standard conditions. Speculate about how the actual values of the reaction Gibbs energy and enthalpy might differ from the standard values. Do you expect that the reaction would still be favourable?

11

Chemical kinetics: accounting for the rate laws

Even quite simple rate laws can give rise to complicated behaviour. The fact that the heart maintains a steady pulse throughout a lifetime, but may break into fibrillation during a heart attack, is one sign of that complexity. On a less personal scale, reaction intermediates come and go, and all reactions approach equilibrium. However, the complexity of the behaviour of reaction rates means that the study of reaction rates can give deep insight into the way that reactions actually take place. As remarked in Chapter 10, a rate law is a window on to the mechanism, the sequence of elementary molecular events that lead from the reactants to the products, of the reaction it summarizes.

Reaction schemes 261

11.1 The approach to equilibrium 261

11.2 Relaxation methods 263

11.3 Consecutive reactions 265

Reaction mechanisms 266

11.4 Elementary reactions 266

11.5 The formulation of rate laws 267

11.6 The steady-state approximation 268

11.7 The rate-determining step 269

11.8 Kinetic control 270

11.9 Unimolecular reactions 270

Reactions in solution 272

11.10 Activation control and diffusion control 272

11.11 Diffusion 273

Homogeneous catalysis 276

11.12 Acid and base catalysis 276

11.13 Enzymes 277

Chain reactions 280

11.14 The structure of chain reactions 280

11.15 The rate laws of chain reactions 281

FURTHER INFORMATION 11.1 282

CHECKLIST OF KEY CONCEPTS 283

ROAD MAP OF KEY EQUATIONS 283

QUESTIONS AND EXERCISES 284

Reaction schemes

So far, we have considered very simple rate laws, in which reactants are consumed or products formed. However, all reactions actually proceed towards a state of equilibrium in which the reverse reaction becomes increasingly important. Moreover, many reactions proceed to products through a series of intermediates. In industry, one of the intermediates may be of crucial importance and the ultimate products may represent waste.

Throughout this chapter we write k_r for the rate constant of a general forward reaction and k_r' for the rate constant of the corresponding reverse reaction. When there are several steps a, b, . . . in a mechanism, we write the forward and reverse rate constants k_a, k_b, . . . and k_a', k_b', . . . , respectively.

11.1 The approach to equilibrium

All forward reactions are accompanied by their reverse reactions. Close to the start of a reaction, when little or no product is present, the rate of the reverse reaction is negligible. However, as the concentration of products increases, the rate at which they decompose into reactants becomes greater. At equilibrium, the reverse rate matches the forward rate and the reactants and products are present in abundances given by the equilibrium constant for the reaction.

We can analyse this behaviour by thinking of a very simple reaction of the form

Forward: A → B Rate of formation of B = k_r[A]

Reverse: B → A Rate of decomposition of B = k_r'[B]

For instance, we could envisage this scheme as the interconversion of coiled (A) and uncoiled (B) DNA molecules. The *net* rate of formation of B, the difference of its rates of formation and decomposition, is

$$\text{Net rate of formation of B} = k_r[\text{A}] - k_r'[\text{B}]$$

When the reaction has reached equilibrium the concentrations of A and B are $[\text{A}]_{eq}$ and $[\text{B}]_{eq}$ and there is no net formation of either substance. It follows that

$$k_r[\text{A}]_{eq} = k_r'[\text{B}]_{eq}$$

and therefore, provided activities can be approximated by molar concentrations, that the equilibrium constant for the reaction is related to the rate constants by

$$K = \frac{[\text{B}]_{eq}}{[\text{A}]_{eq}} = \frac{k_r}{k_r'} \qquad \text{A} \rightleftharpoons \text{B} \qquad (11.1\text{a})$$

Relation of equilibrium constant to rate constants.

If the forward rate constant is much larger than the reverse rate constant, then $K \gg 1$. If the opposite is true, then $K \ll 1$. This result is a crucial connection between the kinetics of a reaction and its equilibrium properties. It is also very useful in practice, for we may be able to measure the equilibrium constant and one of the rate constants, and can then calculate the missing rate constant from eqn 11.1a. Alternatively, we can use the relation to calculate the equilibrium constant from kinetic measurements. This relation is valid even if the forward and reverse reactions have different orders. In that case we need to be careful with units. For instance, if the reaction A + B → C is second-order forward and first-order in reverse, the condition for equilibrium is $k_r[\text{A}]_{eq}[\text{B}]_{eq} = k_r'[\text{C}]_{eq}$ and the dimensionless equilibrium constant in full dress is

$$K = \frac{[\text{C}]_{eq}/c^\ominus}{([\text{A}]_{eq}/c^\ominus)([\text{B}]_{eq}/c^\ominus)} = \frac{[\text{C}]_{eq}c^\ominus}{[\text{A}]_{eq}[\text{B}]_{eq}} = \frac{k_r}{k_r'} \times c^\ominus \qquad \text{A}+\text{B} \rightleftharpoons \text{C} \qquad (11.1\text{b})$$

The presence of $c^\ominus = 1$ mol dm^{-3} in the last term ensures that the ratio of a second-order to first-order rate constants, with their different units, is turned into a dimensionless quantity.

Brief illustration 11.1 Relation between rate and equilibrium constants

The rates of the forward and reverse reactions for the dimerization of an antibacterial agent were found to be 8.1×10^8 dm^3 mol^{-1} s^{-1} (second-order) and 2.0×10^6 s^{-1} (first-order), respectively. The equilibrium constant for the dimerization is therefore

$$K = \frac{\overbrace{8.1 \times 10^8 \text{ dm}^3 \text{ mol}^{-1} \text{ s}^{-1}}^{k_r}}{\underbrace{2.0 \times 10^6 \text{ s}^{-1}}_{k_r'}} \times \overbrace{1 \text{ mol dm}^{-3}}^{c^\ominus} = 4.0 \times 10^2$$

Self-test 11.1

The equilibrium constant for the isomerization reaction $CH_3CN \rightarrow CH_3NC$ is 3.6×10^{-11} at 500 K. The rate constant for the forward reaction is 7.68×10^{-4} s^{-1} at this temperature. Calculate the value of the rate constant for the backward reaction.

Answer: 21×10^6 s^{-1}

Equation 11.1 also gives us insight into the temperature dependence of equilibrium constants. First, we suppose that both the forward and reverse reactions show Arrhenius behaviour (Section 10.9). As we see from Fig. 11.1, for an exothermic reaction the activation energy of the forward reaction is smaller than that of the reverse reaction. Therefore, the forward rate constant increases less sharply with temperature than the reverse reaction does. Consequently, when we increase the temperature of a system at equilibrium, k_r' increases more steeply than k_r does, and the ratio k_r/k_r', and therefore K, decreases. This

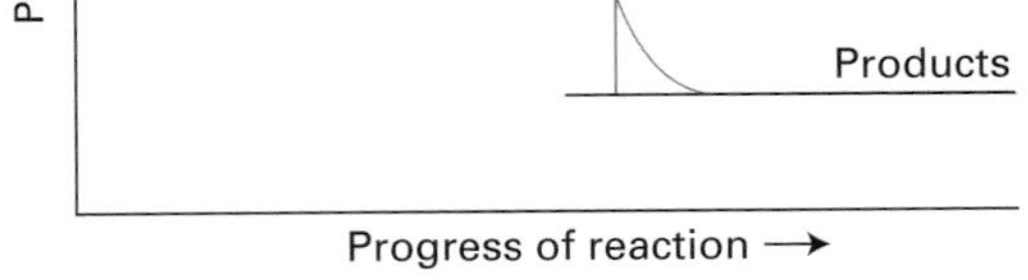

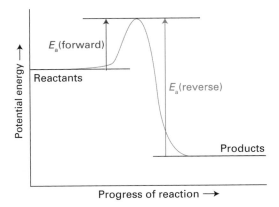

Fig. 11.1 The reaction profile for an exothermic reaction. The activation energy is greater for the reverse reaction than for the forward reaction, so the rate of the forward reaction increases less sharply with temperature. As a result, the equilibrium constant shifts in favour of the reactants as the temperature is raised.

is exactly the conclusion we drew from the van 't Hoff equation (eqn 7.15), which was based on thermodynamic arguments.

Equation 11.1 tells us the ratio of concentrations after a long time has passed and the reaction has reached equilibrium. To find the concentrations at an intermediate stage, we need the integrated rate equation. If no B is present initially, we show in the following Derivation that

$$[\mathrm{A}] = \frac{(k_r' + k_r e^{-(k_r+k_r')t})[\mathrm{A}]_0}{k_r + k_r'} \qquad \text{Approach to equilibrium} \qquad (11.2a)$$

$$[\mathrm{B}] = \frac{k_r(1 - e^{-(k_r+k_r')t})[\mathrm{A}]_0}{k_r + k_r'} \qquad (11.2b)$$

where $[\mathrm{A}]_0$ is the initial concentration of A.

Derivation 11.1

The approach to equilibrium

The concentration of A is reduced by the forward reaction (at a rate $k_r[\mathrm{A}]$) but it is increased by the reverse reaction (at a rate $k_r'[\mathrm{B}]$). Therefore, the net rate of change is

$$\frac{d[\mathrm{A}]}{dt} = -k_r[\mathrm{A}] + k_r'[\mathrm{B}]$$

If the initial concentration of A is $[\mathrm{A}]_0$, and no B is present initially, then at all times $[\mathrm{A}] + [\mathrm{B}] = [\mathrm{A}]_0$. Therefore, by replacing [B] with $[\mathrm{A}]_0 - [\mathrm{A}]$, we obtain

$$\frac{d[\mathrm{A}]}{dt} = -k_r[\mathrm{A}] + k_r'([\mathrm{A}]_0 - [\mathrm{A}]) = -(k_r + k_r')[\mathrm{A}] + k_r'[\mathrm{A}]_0$$

The solution of this differential equation is eqn 11.2a, as may be verified by differentiation of that expression:

$$\frac{d[\mathrm{A}]}{dt} = \frac{d}{dt}\frac{(k_r' + k_r e^{-(k_r+k_r')t})[\mathrm{A}]_0}{k_r + k_r'}$$

$$\overset{\text{Rearrange}}{=} \underbrace{-(k_r' + k_r e^{-(k_r+k_r')t})[\mathrm{A}]_0}_{-(k_r+k_r')[\mathrm{A}]} + k_r'[\mathrm{A}]_0$$

To obtain eqn 11.2b, we use eqn 11.2a and $[\mathrm{B}] = [\mathrm{A}]_0 - [\mathrm{A}]$ again.

As we see in Fig. 11.2, the concentrations start from their initial values and move gradually towards their final equilibrium values as t approaches infinity. We find the latter by setting t equal to infinity in eqn 11.2 and using $e^{-x} = 0$ at $x = \infty$:

$$[\mathrm{A}]_{eq} = \frac{k_r'[\mathrm{A}]_0}{k_r + k_r'} \qquad [\mathrm{B}]_{eq} = \frac{k_r[\mathrm{A}]_0}{k_r + k_r'} \qquad (11.3)$$

As may be verified, the ratio of these two expressions is the equilibrium constant in eqn 11.1 (that is, $[\mathrm{B}]_{eq}/[\mathrm{A}]_{eq} = k_r/k_r'$).

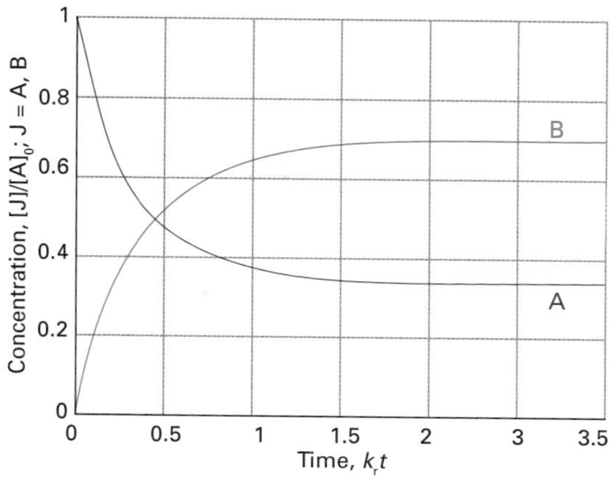

Fig. 11.2 The approach to equilibrium of a reaction that is first-order in both directions. Here we have taken $k_r = 2k_r'$. Note how, at equilibrium, the ratio of concentrations is 2:1, corresponding to $K = 2$.

11.2 Relaxation methods

The term **relaxation** denotes the return of a system to equilibrium. It is used in chemical kinetics to indicate that an externally applied influence has shifted the equilibrium position of a reaction, normally abruptly, and that the reaction is adjusting to the equilibrium composition characteristic of the new conditions (Fig. 11.3). We shall consider the response of reaction rates to a **temperature jump**, a sudden change in temperature. We know from Section 7.8 that the equilibrium composition of a reaction depends on the temperature (provided $\Delta_r H^\ominus$ is nonzero), so a shift in temperature acts as a perturbation on the system. One way of achieving a temperature jump

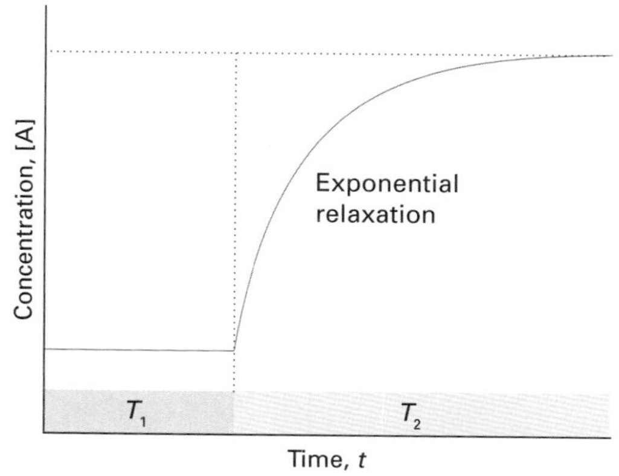

Fig. 11.3 The relaxation to the new equilibrium composition when a reaction initially at equilibrium at a temperature T_1 is subjected to a sudden change of temperature, which takes it to T_2.

is to discharge a capacitor through a sample made conducting by the addition of ions, but laser or microwave discharges can also be used. Temperature jumps of between 5 and 10 K can be achieved in about 1 μs with electrical discharges. The high energy output of pulsed lasers is sufficient to generate temperature jumps of between 10 and 30 K within nanoseconds in aqueous samples, making the technique suitable for the study of the events involved in protein folding. Equilibria for which there is a change in volume between reactants and products are also sensitive to pressure, and **pressure-jump techniques** may then also be used.

We show in the following Derivation that when a sudden temperature increase is applied to a simple $A \rightleftharpoons B$ equilibrium that is first-order in each direction, the composition relaxes exponentially to the new equilibrium composition:

$$x = x_0 e^{-t/\tau} \qquad \tau = \frac{1}{k_r + k_r'} \qquad A \rightleftharpoons B \quad \text{Relaxation} \quad (11.4)$$

where x is the departure from equilibrium at the new temperature, x_0 is the departure from equilibrium immediately after the temperature jump, and τ (tau) is the **relaxation time.**

Derivation 11.2

Relaxation to equilibrium

We need to keep track of the fact that rate constants depend on temperature. At the initial temperature, when the rate constants are $k_{r,\text{initial}}$ and $k'_{r,\text{initial}}$, the net rate of change of [A] is

$$\frac{d[A]}{dt} = \overbrace{-k_{r,\text{initial}}[A]}^{\text{Loss of A}} + \overbrace{k'_{r,\text{initial}}[B]}^{\text{Formation of A from B}}$$

At equilibrium under these conditions, we write the concentrations as $[A]_{\text{eq,initial}}$ and $[B]_{\text{eq,initial}}$ and because d[A]/dt is then zero,

$$k_{r,\text{initial}}[A]_{\text{eq,initial}} = k'_{r,\text{initial}}[B]_{\text{eq,initial}}$$

When the temperature is increased suddenly, the rate constants change to k_r and k_r' but the concentrations of A and B remain for an instant at their old equilibrium values. As the system is no longer at equilibrium, it readjusts to the new equilibrium concentrations, which are now given by

$$k_r[A]_{\text{eq,final}} = k_r'[B]_{\text{eq,final}}$$

and it does so at a rate that depends on the new rate constants.

We write the deviation of [A] from its new equilibrium value as x, so $[A] = x + [A]_{\text{eq,final}}$ and $[B] = [B]_{\text{eq,final}} - x$. The concentration of A then changes as follows:

$$\begin{aligned}\frac{d[A]}{dt} &= -k_r[A] + k_r'[B] \\ &= -k_r(x + [A]_{\text{eq,final}}) + k_r'(-x + [B]_{\text{eq,final}}) \\ &= -k_r x - k_r' x \overbrace{- k_r[A]_{\text{eq,final}} + k_r'[B]_{\text{eq,final}}}^{0} \\ &= -(k_r + k_r')x\end{aligned}$$

From $[A] = x + [A]_{\text{eq,final}}$ it follows that d[A]/dt = dx/dt, because $[A]_{\text{eq,final}}$ is a constant, and therefore that

$$\frac{dx}{dt} = -(k_r + k_r')x$$

To solve this equation we divide both sides by x and multiply by dt:

$$\frac{dx}{x} = -(k_r + k_r')dt$$

Now integrate both sides. When $t = 0$, $x = x_0$, its initial value, so the integrated equation has the form given in The chemist's toolkit 2.1:

$$\overbrace{\int_{x_0}^{x} \frac{dx}{x}}^{\ln(x/x_0)} = -(k_r + k_r')\overbrace{\int_0^t dt}^{t}$$

The integrated equation is therefore

$$\ln \frac{x}{x_0} = -(k_r + k_r')t$$

When antilogarithms are taken of both sides, the result is eqn 11.4.

Equation 11.4 shows that the concentrations of A and B relax into the new equilibrium at a rate determined by the *sum* of the two new rate constants. Because the equilibrium constant under the new conditions is $K \approx k_r/k_r'$, its value may be combined with the relaxation time measurement to find the individual k_r and k_r'.

Brief illustration 11.2 Relaxation

The forward and reverse rate constants for a first-order reaction (in both directions) were found to be $5.0 \times 10^4\ \text{s}^{-1}$ and $2.0 \times 10^3\ \text{s}^{-1}$, respectively. The relaxation time for return to equilibrium after a thermal pulse is therefore

$$\begin{aligned}\tau &= \frac{1}{(5.0\times10^4 + 2.0\times10^3)\ \text{s}^{-1}} = \frac{1}{5.2\times10^4\ \text{s}^{-1}} \\ &= \frac{1}{5.2\times10^4\ \text{s}^{-1}} \\ &= 1.9\times10^{-5}\ \text{s}\end{aligned}$$

The relaxation time is therefore 19 μs.

Self-test 11.2

The relaxation time for another first-order reaction was measured as 1.3 ms. Determine the rate constant for the reverse reaction given that the rate constant for the forward reaction is 320 s^{-1}.

Answer: 450 s^{-1}

11.3 Consecutive reactions

It is commonly the case that a reactant produces an intermediate, which subsequently decays into a product. Radioactive decay is often of this type, with one nuclide decaying into another, and then that nuclide decaying into a third:

$$^{239}\text{U} \xrightarrow{2.35\ \text{min}} {}^{239}\text{Np} \xrightarrow{2.35\ \text{d}} {}^{239}\text{Pu}$$

The times are half-lives. Biochemical processes are often elaborate versions of this simple model. For instance, the restriction enzyme EcoRI catalyses the cleavage of DNA and brings about the sequence of reactions

Supercoiled DNA → open-circle DNA → linear DNA

To illustrate the kinds of considerations involved, we suppose that a reaction takes place in two steps. First, an intermediate I (the open-circle DNA, for instance) is formed from the reactant A (the supercoiled DNA) in a first-order reaction. Then I decays in a first-order reaction to form the product P (the linear DNA):

$$\text{A} \rightarrow \text{I} \quad \text{Rate of formation of I} = k_a[\text{A}]$$

$$\text{I} \rightarrow \text{P} \quad \text{Rate of formation of P} = k_b[\text{I}]$$

For simplicity, we are ignoring the reverse reactions, which is valid if they are very slow. The first of these rate laws implies that

$$[\text{A}] = [\text{A}]_0 e^{-k_a t} \qquad (11.5\text{a})$$

The net rate of formation of I is the difference between its rate of formation and its rate of consumption, so we can write

$$\text{Net rate of formation of I} = k_a[\text{A}] - k_b[\text{I}]$$

with [A] given by eqn 11.5a. This equation, which in mathematical form is $d[\text{I}]/dt = k_a[\text{A}] - k_b[\text{I}]$ is a standard form given in The chemist's toolkit 11.1, with the following solution:

$$[\text{I}] = \frac{k_a}{k_b - k_a}(e^{-k_a t} - e^{-k_b t})[\text{A}]_0 \qquad (11.5\text{b})$$

Finally, because $[\text{A}] + [\text{I}] + [\text{P}] = [\text{A}]_0$ at all stages of the reaction, the concentration of P is $[\text{P}] = [\text{A}]_0 - [\text{A}] - [\text{I}]$, and therefore, after substituting the expressions for [A] and [I],

$$[\text{P}] = \left\{1 + \frac{k_a e^{-k_b t} - k_b e^{-k_a t}}{k_b - k_a}\right\}[\text{A}]_0 \qquad (11.5\text{c})$$

These solutions are illustrated in Fig. 11.4. We see that the intermediate grows in concentration initially, then decays as A is exhausted. Meanwhile, the concentration of P rises smoothly to its final value. As shown in the following Derivation, the intermediate reaches its maximum concentration at

$$t = \frac{1}{k_a - k_b}\ln\frac{k_a}{k_b} \qquad \text{Time for [I] at maximum} \qquad (11.6)$$

This is the optimum time for a manufacturer trying to make the intermediate in a batch process to extract it.

The chemist's toolkit 11.1 Differential equations for kinetics

Ordinary differential equations are described in The chemist's toolkit 10.1. We need two types of differential equations in this chapter. One has the form

$$\frac{dy}{dx} + ay = b$$

where a and b are constants. We encounter an equation of this form in Derivation 11.1. The solution is

$$y(x) = ce^{-ax} + b/a$$

where c is a constant. To verify this solution you should use the relation

$$\frac{d}{dx}e^{\pm ax} = \pm ae^{-ax}$$

The second type has the more complicated form

$$\frac{dy}{dx} + a(x)y = f(x)$$

This equation has the solution

$$y(x) = e^{-F(x)}\int e^{F(x)} f(x)dx + ce^{-F(x)}$$

where c is a constant and

$$F(x) = \int a(x)dx$$

We encounter an equation of this type in Section 11.3, where a is a constant (so $F(x) = ax$) and where $f(x) = be^{-b'x}$ with b and b' constants. In that special case, the solution is

$$y = \frac{be^{-b'x}}{a - b'} + ce^{-ax}$$

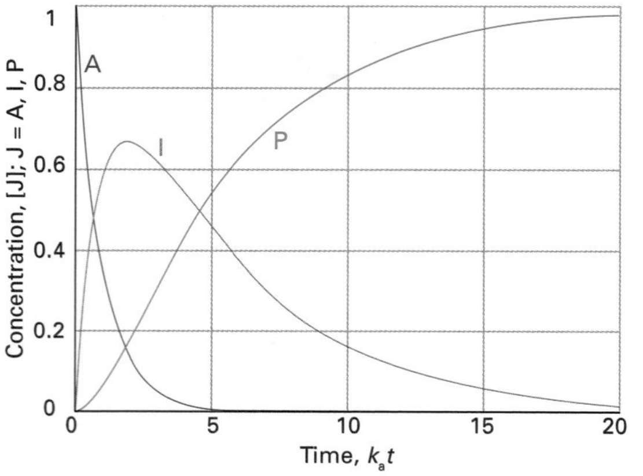

Fig. 11.4 The concentrations of the substances involved in a consecutive reaction of the form $A \rightarrow I \rightarrow P$, where I is an intermediate and P a product. We have used $k_a = 5k_b$. Note how, at each time, the sum of the three concentrations is a constant.

Brief illustration 11.3 The time for the intermediate to be a maximum

Suppose that in a certain batch process $k_a = 0.120\ h^{-1}$ and $k_b = 0.012\ h^{-1}$, then the intermediate is at a maximum at

$$t = \frac{1}{(0.120 - 0.012)\ h^{-1}} \ln \frac{0.120}{0.012} = \frac{h}{0.108} \ln \frac{0.120}{0.012} = 21\ h$$

$$= 21\ h$$

Derivation 11.3

The time to reach maximum concentration

To find the time corresponding to the maximum concentration of intermediate, we differentiate eqn 11.5b and look for the time at which $d[I]/dt = 0$. First, because $de^{at}/dt = a\,e^{at}$, we obtain

$$\frac{d[I]}{dt} = \frac{d}{dt}\frac{k_a}{k_b - k_a}(e^{-k_a t} - e^{-k_b t})[A]_0 = \frac{k_a[A]_0}{k_b - k_a}\frac{d}{dt}(e^{-k_a t} - e^{-k_b t})$$

$$= \frac{k_a[A]_0}{k_b - k_a}(-k_a e^{-k_a t} + k_b e^{-k_b t}) = 0$$

This equation is satisfied if

$$k_a e^{-k_a t} = k_b e^{-k_b t}$$

Because $e^x e^y = e^{x+y}$, this relation becomes

$$\frac{k_a}{k_b} = e^{k_a t - k_b t}$$

Taking logarithms of both sides leads to eqn 11.6.

Reaction mechanisms

We have seen how two simple types of reaction—approach to equilibrium and consecutive reactions—result in a characteristic dependence of the concentration on the time. We can suspect that other variations with time will act as the signatures of other reaction mechanisms.

11.4 Elementary reactions

Many reactions occur in a series of steps called **elementary reactions**, each of which involves only one or two molecules. We shall denote an elementary reaction by writing its chemical equation without displaying the physical state of the species, as in

$$H + Br_2 \rightarrow HBr + Br$$

We have already used this convention without comment in some of the reactions discussed in Chapter 10. This equation signifies that a specific H atom attacks a specific Br_2 molecule to produce a molecule of HBr and a Br atom. Ordinary chemical equations summarize the overall stoichiometry of the reaction and do not imply any specific mechanism.

The **molecularity** of an elementary reaction is the number of molecules coming together to react. In a **unimolecular reaction** a single molecule shakes itself apart or shakes its atoms into a new arrangement (Fig. 11.5). An example is the isomerization of cyclopropane into propene. The radioactive decay of nuclei (for example, the emission of a β particle from the nucleus of a tritium atom, which is used in mechanistic studies to follow the course of particular groups of atoms) is 'unimolecular' in the sense that a single nucleus shakes itself apart. In a **bimolecular reaction,** two molecules collide and exchange energy, atoms, or groups of atoms, or undergo some other kind of change, as in the reaction between H and F_2 or between H and Br_2 (Fig. 11.6).

It is important to distinguish molecularity from order: the *order* of a reaction is an empirical quantity, and is obtained by inspection of the experimentally determined rate law; the *molecularity* of a reaction refers to an individual elementary reaction that has been

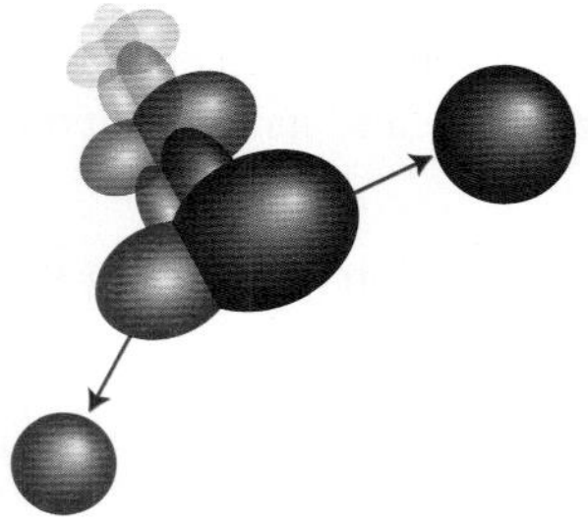

Fig. 11.5 In a unimolecular elementary reaction, an energetically excited species decomposes into products: it simply shakes itself apart.

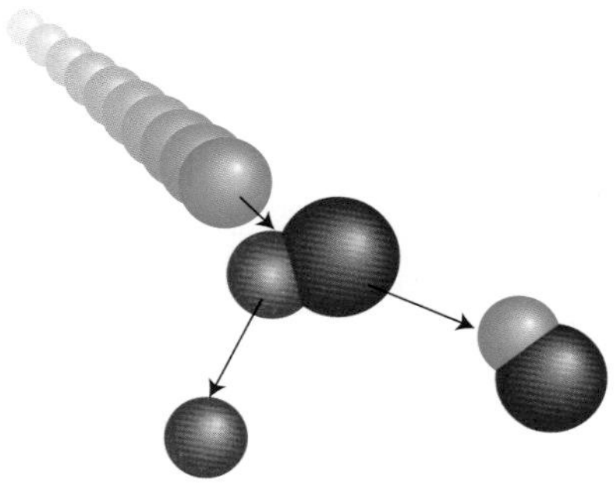

Fig. 11.6 In a bimolecular elementary reaction, two species are involved in the process.

postulated as a step in a proposed mechanism. Many substitution reactions in organic chemistry (for instance, S_N2 nucleophilic substitutions) are bimolecular and involve an activated complex that is formed from two reactant species. Enzyme-catalysed reactions (Section 11.13) can be regarded, to a good approximation, as bimolecular in the sense that they depend on the encounter of a substrate molecule and an enzyme molecule.

We can write down the rate law of an elementary reaction from its chemical equation. First, consider a unimolecular reaction. In a given interval, ten times as many A molecules decay when there are initially 1000 A molecules as when there are only 100 A molecules present. Therefore the rate of decomposition of A is proportional to its concentration and we can conclude that *a unimolecular reaction is first-order*:

$$\mathrm{A} \rightarrow \text{products} \quad v = k_r[\mathrm{A}] \qquad \text{Unimolecular reaction} \qquad (11.7)$$

The rate of a bimolecular reaction is proportional to the rate at which the reactants meet, which in turn is proportional to both their concentrations. Therefore, the rate of the reaction is proportional to the product of the two concentrations and *an elementary bimolecular reaction is second-order overall*:

$$\mathrm{A} + \mathrm{B} \rightarrow \text{products} \quad v = k_r[\mathrm{A}][\mathrm{B}] \qquad \text{Bimolecular reaction} \qquad (11.8)$$

We must now explore how to string simple steps together into a mechanism and how to arrive at the corresponding overall rate law. For the present we emphasize that if the reaction is an elementary bimolecular process, then it has second-order kinetics; however, if the kinetics are second-order, then the reaction could be bimolecular but might be complex.

11.5 The formulation of rate laws

Suppose we propose that a particular reaction is the outcome of a sequence of elementary steps. How do we arrive at the rate law implied by the mechanism? We introduce the technique by considering the rate law for the gas-phase oxidation of nitric oxide (nitrogen monoxide, NO):

$$2\,\mathrm{NO(g)} + \mathrm{O_2(g)} \rightarrow 2\,\mathrm{NO_2(g)}$$

Nitric oxide is a very important component of polluted atmospheres. It is formed in the hot exhausts of vehicles and the jet engines of aircraft and its oxidation is a step in the formation of acid rain. The compound is also a neurotransmitter involved in the physiological changes taking place during sexual arousal. Experimentally, the reaction is found to be third-order overall:

$$v = k_r[\mathrm{NO}]^2[\mathrm{O_2}]$$

One explanation of the observed reaction order might be that the reaction is a single termolecular (three-molecule) elementary step involving the simultaneous collision of two NO molecules and one O_2 molecule. However, such collisions occur very infrequently and so termolecular processes can usually be ignored. Because a termolecular mechanism is so slow another mechanism usually dominates.

The following mechanism has been proposed:

Step 1. Two NO molecules combine to form a dimer:

$$\mathrm{NO} + \mathrm{NO} \rightarrow \mathrm{N_2O_2}$$

Rate of formation of $N_2O_2 = k_a[\mathrm{NO}]^2$

This step is plausible, because NO is an odd-electron species, a radical, and two radicals can pair their electrons and form a covalent bond when they meet. That the N_2O_2 dimer is also known in the solid makes the suggestion plausible: it is often a good strategy to decide whether a proposed intermediate is the analogue of a known compound.

Step 2. The N_2O_2 dimer decomposes into NO molecules:

$$\mathrm{N_2O_2} \rightarrow \mathrm{NO} + \mathrm{NO}$$

Rate of decomposition of $N_2O_2 = k_a'[\mathrm{N_2O_2}]$

This step, the reverse of Step 1, is a unimolecular decay: the dimer shakes itself apart. We adopt the convention in which the rate constant of a reverse reaction is marked with a prime (as in k_a for the forward reaction and k_a' for its reverse).

Step 3. Alternatively, an O_2 molecule collides with the dimer and results in the formation of NO_2:

$$\mathrm{N_2O_2} + \mathrm{O_2} \rightarrow \mathrm{NO_2} + \mathrm{NO_2}$$

Rate of consumption of $N_2O_2 = k_b[\mathrm{N_2O_2}][\mathrm{O_2}]$

Now we proceed to derive the rate law on the basis of this proposed mechanism. The rate of formation of product comes directly from Step 3:

$$\text{Rate of formation of } NO_2 = 2k_b[N_2O_2][O_2]$$

The 2 appears in the rate law because two NO_2 molecules are formed in each reaction event, so the concentration of NO_2 increases at twice the rate that the concentration of N_2O_2 decays. However, this expression is not an acceptable overall rate law because it is expressed in terms of the concentration of the intermediate N_2O_2: an acceptable rate law for an overall reaction is expressed solely in terms of the species that appear in the overall reaction. Therefore, we need to find an expression for the concentration of N_2O_2. To do so, we consider the net rate of formation of the intermediate, the difference between its rates of formation and decay. Because N_2O_2 is formed by Step 1 but decays by Steps 2 and 3, its net rate of formation is

$$\text{Net rate of formation of } N_2O_2 \\ = k_a[NO]^2 - k_a'[N_2O_2] - k_b[N_2O_2][O_2]$$

Notice that formation terms occur with a positive sign and decay terms occur with a negative sign because they reduce the net rate of formation.

If we could solve this equation for the concentration of N_2O_2 in terms of the concentrations of NO and O_2, we could substitute the result into the preceding expression and obtain the overall rate law. However, this involves solving a very difficult differential equation, and will give an enormously complex expression. In fact, even in this relatively simple case, we can obtain only a numerical solution using a computer. To make progress towards obtaining a simple formula, we must make an approximation.

11.6 The steady-state approximation

It is common at this stage of formulating a rate law to introduce the **steady-state approximation**, in which we suppose that *the concentrations of all intermediates remain constant and small throughout the reaction* (except right at the beginning and right at the end when no intermediates are present). An **intermediate** is any species that does not appear in the overall reaction but which has been invoked in the mechanism. For our mechanism the intermediate is N_2O_2, so we write

$$\text{Net rate of formation of } N_2O_2 = 0$$

which implies that

$$k_a[NO]^2 - k_a'[N_2O_2] - k_b[N_2O_2][O_2] = 0$$

We can rearrange this equation first to

$$k_a[NO]^2 - (k_a + k_b[O_2])[N_2O_2] = 0$$

and then into an expression for the concentration of N_2O_2:

$$[N_2O_2] = \frac{k_a[NO]^2}{k_a' + k_b[O_2]}$$

It follows that the rate of formation of NO_2 is

$$\text{Rate of formation of } NO_2 = 2k_b[N_2O_2][O_2] \\ = \frac{2k_ak_b[NO]^2[O_2]}{k_a' + k_b[O_2]} \qquad (11.9)$$

At this stage, the rate law is more complex than the observed law, but the numerator resembles it. The two expressions become identical if we suppose that the rate of decomposition of the dimer is much greater than its rate of reaction with oxygen, for then $k_a'[N_2O_2] \gg k_b[N_2O_2][O_2]$, or, after cancelling the $[N_2O_2]$, $k_a' \gg k_b[O_2]$. When this condition is satisfied, we can approximate the denominator in the overall rate law by k_a' alone and conclude that

$$\text{Rate of formation of } NO_2 \\ = \frac{2k_ak_b[NO]^2[O_2]}{\underbrace{k_a' + k_b[O_2]}_{\approx k_a' \text{ if } k_a' \gg k_b[O_2]}} = \frac{2k_ak_b}{k_a'}[NO]^2[O_2] \qquad (11.10)$$

This expression has the observed overall third-order form. Moreover, we can identify the observed rate constant as the following combination of rate constants for the elementary reactions:

$$k_r = \frac{2k_ak_b}{k_a'} \qquad (11.11)$$

Example 11.1

Using the steady-state assumption

An alternative mechanism that may apply when the concentration of O_2 is high and that of NO is low is one in which the first step is $NO + O_2 \rightarrow NO\cdots O_2$, where the dotted line indicates a loosely bound cluster, and its reverse, followed by $NO\cdots O_2 + NO \rightarrow NO_2 + NO_2$. Confirm that this mechanism also leads to the observed rate law when the concentration of NO is low.

Strategy Set up the rate laws for the individual elementary reactions, identify the reaction intermediate, and set its net rate of formation equal to zero. Then solve the equations for the rate of formation of product. Finally, impose the conditions and simplify the resulting rate law.

Answer The elementary reactions and their rate laws are

(a) $NO + O_2 \rightarrow NO\cdots O_2$
Rate of formation of $NO\cdots O_2 = k_a[NO][O_2]$

(a′) $NO\cdots O_2 \rightarrow NO + O_2$
Rate of decomposition of $NO\cdots O_2 = k_a'[NO\cdots O_2]$

(b) $NO\cdots O_2 + NO \rightarrow NO_2 + NO_2$
Rate of loss of $NO\cdots O_2 = k_b[NO\cdots O_2][NO]$
Rate of formation of $NO_2 = 2k_b[NO\cdots O_2][NO]$

The intermediate is $NO\cdots O_2$; its net rate of formation (its rate of formation less the sum of the rates at which it is lost, schematically a – a′ – b) is

$$\text{Net rate of formation of } NO\cdots O_2 = k_a[NO][O_2] - k_a'[NO\cdots O_2] - k_b[NO\cdots O_2][NO] = 0$$

or

$$k_a[NO][O_2] - (k_a' + k_b[NO])[NO\cdots O_2] = 0$$

Therefore, the steady-state concentration of the intermediate is

$$[NO\ldots O_2] = \frac{k_a[NO][O_2]}{k_a' + k_b[NO]}$$

The rate law for the formation of product, from (b), is therefore

$$\text{Rate of formation of } NO_2 = 2k_b \times \frac{k_a[NO][O_2]}{k_a' + k_b[NO]} \times [NO] = \frac{2k_ak_b[NO]^2[O_2]}{k_a' + k_b[NO]}$$

When the concentration of NO is low in the sense $k_b[NO] \ll k_a'$, the second term in the denominator may be neglected, and we find

$$\text{Rate of formation of } NO_2 = \overbrace{\frac{2k_ak_b}{k_a'}}^{k_r}[NO]^2[O_2] = k_r[NO]^2[O_2]$$

This rate law is also in accord with the one observed.

Self-test 11.3

The proposed reaction mechanism for renaturation of a double helix from its strands A and B is A + B ⇄ unstable helix (fast, k_a and k_a') followed by unstable helix → stable double helix (slow, k_b). Deduce the rate law for the formation of the double helix making a steady-state approximation.

Answer: $v = k_ak_b[A][B]/(k_a' + k_b)$

One feature to note is that although each of the rate constants in eqn 11.11 increases with temperature, that might not be true of k_r itself. Thus, if the rate constant k_a' increases more rapidly than the product k_ak_b increases, then k_r will decrease with increasing temperature and the reaction will go more slowly as the temperature is raised. The physical reason is that the dimer N_2O_2 shakes itself apart so quickly at the higher temperature that its reaction with O_2 is less able to take place, and products are formed more slowly. Mathematically, we would say that the composite reaction had a 'negative activation energy'. For instance, suppose that each rate constant in eqn 11.11 exhibits an Arrhenius temperature dependence. The overall rate constant would be

$$k_r = \frac{2\overbrace{(A_a e^{-E_{a,a}/RT})}^{k_a}\overbrace{(A_b e^{-E_{a,b}/RT})}^{k_b}}{\underbrace{A_a' e^{-E_{a,a'}/RT}}_{k_a'}}$$

$$= \frac{2A_aA_b}{A_a'}e^{-(E_{a,a}+E_{a,b}-E_{a,a'})/RT} = Ae^{-E_a/RT}$$

$$\text{with } A = \frac{2A_aA_b}{A_a'} \text{ and } E_a = E_{a,a} + E_{a,b} - E_{a,a'}$$

In this case we see that $E_a < 0$ if $E_{a,a'} > E_{a,a} + E_{a,b}$. We have to be very cautious about making predictions about the effect of temperature on reactions that are the outcome of several steps.

11.7 The rate-determining step

The oxidation of nitrogen monoxide introduces another important concept. Let's suppose that Step 3 is very fast, so k_a' may be neglected relative to $k_b[O_2]$ in eqn 11.9. One way to achieve this condition is to increase the concentration of O_2 in the reaction mixture. Then the rate law simplifies to

$$\text{Rate of formation of } NO_2 = \frac{2k_ak_b[NO]^2[O_2]}{\underbrace{k_a' + k_b[O_2]}_{\approx k_b[O_2] \text{ if } k_b[O_2] \gg k_a'}} = \frac{2k_ak_b[NO]^2[O_2]}{k_b[O_2]}$$

$$\overset{\text{Cancel } k_b \text{ and } [O_2]}{=} 2k_a[NO]^2 \qquad (11.12)$$

Now the reaction is second-order in NO and the concentration of O_2 does not appear in the rate law. The physical explanation is that the rate of reaction of N_2O_2 is so great on account of the high concentration of O_2 in the system, that N_2O_2 reacts as soon as it is formed. Therefore, under these conditions, the rate of formation of NO_2 is determined by the rate at which N_2O_2 is formed. This step is an example of a **rate-determining step**, the slowest step in a reaction mechanism, which controls the rate of the overall reaction.

The rate-determining step is not just the slowest step: it must be slow *and* be a crucial gateway for the

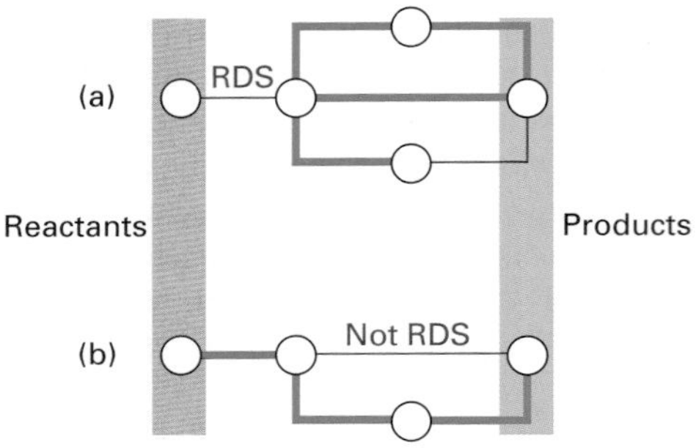

Fig. 11.7 The rate-determining step is the slowest step of a reaction *and* acts as a bottleneck. In this schematic diagram, fast reactions are represented by heavy lines (freeways) and slow reactions by thin lines (country lanes). Circles represent substances. (a) The first step is rate-determining. (b) Although the second step is the slowest, it is not rate-determining because it does not act as a bottleneck (there is a faster route that circumvents it).

formation of products. If a faster reaction can also lead to products, then the slowest step is irrelevant because the slow reaction can then be side-stepped (Fig. 11.7). A rate-determining step is like a slow ferry crossing between two fast highways: the overall rate at which traffic can reach its destination is determined by the rate at which it can make the ferry crossing. If a bridge is built that circumvents the ferry, the ferry remains the slowest step but it is no longer rate-determining.

The rate law of a reaction that has a rate-determining step can often be written down almost by inspection. If the first step in a mechanism is rate-determining, then the rate of the overall reaction is equal to the rate of the first step because all subsequent steps are so fast that once the first intermediate is formed it results immediately in the formation of products. Figure 11.8 shows the reaction profile for a mechanism of this kind in which the slowest step is the one with the highest activation energy. Once over the initial barrier, the intermediates cascade into products.

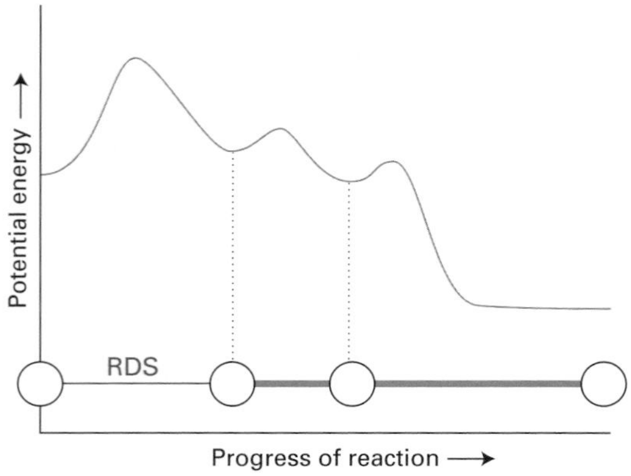

Fig. 11.8 The reaction profile for a mechanism in which the first step is rate-determining.

11.8 Kinetic control

In some cases reactants can give rise to a variety of products, as in nitrations of mono-substituted benzene, when various proportions of the ortho-, meta-, and para-substituted products are obtained, depending on the directing power of the original substituent. Suppose two products, P_1 and P_2, are produced by the following competing reactions:

$$A + B \rightarrow P_1 \quad \text{Rate of formation of } P_1 = k_{r,1}[A][B]$$
$$A + B \rightarrow P_2 \quad \text{Rate of formation of } P_2 = k_{r,2}[A][B]$$

The relative proportion in which the two products have been produced at a given stage of the reaction (before it has reached equilibrium) is given by the ratio of the two rates, and therefore to the two rate constants:

$$\frac{[P_2]}{[P_1]} = \frac{k_{r,2}}{k_{r,1}} \qquad \text{Kinetic control} \qquad (11.13)$$

This ratio represents the **kinetic control** over the proportions of products, and is a common feature of the reactions encountered in organic chemistry where reactants are chosen that facilitate pathways favouring the formation of a desired product. If a reaction is allowed to reach equilibrium, then the proportion of products is determined by thermodynamic rather than kinetic considerations and the ratio of concentrations is controlled by considerations of the standard Gibbs energies of all the reactants and products.

11.9 Unimolecular reactions

A number of gas-phase reactions follow first-order kinetics, as in the isomerization of cyclopropane mentioned earlier, in which the strained triangular molecule bursts open into an acyclic alkene:

$$cyclo\text{-}C_3H_6 \rightarrow CH_3CH{=}CH_2 \quad v = k_r[cyclo\text{-}C_3H_6]$$

The problem with reactions like this is that the reactant molecule presumably acquires the energy it needs to react by collisions with other molecules. Collisions, though, are simple *bimolecular* events, so how can they result in a first-order rate law? First-order gas-phase reactions are widely called 'unimolecular reactions' because (as we shall see) the rate-determining step is an elementary unimolecular reaction in which the reactant molecule changes into the product. This term must be used with caution, however, because the composite mechanism has bimolecular as well as unimolecular steps.

The first successful explanation of unimolecular reactions is ascribed to Frederick Lindemann in 1921. The **Lindemann mechanism** (or *Lindemann–Hinshelwood mechanism*) is as follows:

Step 1. A reactant molecule A becomes energetically excited (denoted A*) by collision with another A molecule:

$$\mathrm{A+A \rightarrow A^{*}+A} \quad \text{Rate of formation of A}^{*} = k_a[\mathrm{A}]^2$$

Step 2. The energized molecule might lose its excess energy by collision with another molecule:

$$\mathrm{A^{*}+A \rightarrow A+A} \quad \text{Rate of deactivation of A}^{*} = k_a'[\mathrm{A}^{*}][\mathrm{A}]$$

Step 3. Alternatively, the excited molecule might shake itself apart and form products P. That is, it might undergo the unimolecular decay

$$\mathrm{A^{*} \rightarrow P} \quad \text{Rate of formation of P} = k_b[\mathrm{A}^{*}]$$
$$\text{Rate of consumption of A}^{*} = k_b[\mathrm{A}^{*}]$$

If the unimolecular step, Step 3, is slow enough to be the rate-determining step, then the overall reaction will have first-order kinetics, as observed. We can demonstrate this conclusion explicitly by applying the steady-state approximation to the net rate of formation of the intermediate A* and, as shown in the following Derivation, find that

$$\text{Rate of formation of P} = k_r[\mathrm{A}], \quad \text{with } k_r = \frac{k_a k_b}{k_a'} \tag{11.14}$$

This rate law is first-order, as we set out to show.

Derivation 11.4

The Lindemann mechanism

First, we write down the expression for the net rate of formation of A*, and set this rate equal to zero:

$$\text{Net rate of formation of A}^{*} = \underbrace{k_a[\mathrm{A}]^2}_{\text{Formation in Step 1}} \underbrace{- k_a'[\mathrm{A}^{*}][\mathrm{A}]}_{\text{Loss in Step 2}} \underbrace{- k_b[\mathrm{A}^{*}]}_{\text{Loss in Step 3}} = 0$$

and therefore

$$k_a[\mathrm{A}]^2 - (k_a'[\mathrm{A}] + k_b)[\mathrm{A}^{*}] = 0$$

The solution of this equation is

$$[\mathrm{A}^{*}] = \frac{k_a[\mathrm{A}]^2}{k_b + k_a'[\mathrm{A}]}$$

It follows that the rate law for the formation of products is

$$\text{Rate of formation of P} = k_b[\mathrm{A}^{*}] = \frac{k_a k_b[\mathrm{A}]^2}{k_b + k_a'[\mathrm{A}]}$$

At this stage the rate law is not first-order in A. However, we can suppose that the rate of deactivation of A* by (A*,A) collisions is much greater than the rate of unimolecular decay of A* to products. That is, we suppose that the unimolecular decay of A* is the rate-determining step. Then $k_a'[\mathrm{A}^{*}][\mathrm{A}] \gg k_b[\mathrm{A}^{*}]$, which corresponds to $k_a'[\mathrm{A}] \gg k_b$. If that is the case, we can neglect k_b in the denominator of the rate law and obtain eqn 11.14.

Example 11.2

Devising a rate law

Suppose that an inert gas M is present and dominates the excitation of A and de-excitation (quenching) of A*. Devise the rate law for the formation of products.

Strategy Write the elementary reactions that are likely to take place, and the corresponding rate laws for the unimolecular and bimolecular processes. Identify the reaction intermediate, and set its net rate of formation to zero. Solve the resulting equations for the concentration of the intermediate, and substitute that expression into the rate law for the formation of product. At that stage, assume that the processes involving M overwhelm all other competing processes and simplify the rate law accordingly.

Answer The elementary reactions and their rate laws are

(a)	$\mathrm{A+A \rightarrow A^{*}+A}$	Rate of formation of A* = $k_a[\mathrm{A}]^2$
(a′)	$\mathrm{A^{*}+A \rightarrow A+A}$	Rate of removal of A* = $k_a'[\mathrm{A}][\mathrm{A}^{*}]$
(b)	$\mathrm{A+M \rightarrow A^{*}+M}$	Rate of formation of A* = $k_b[\mathrm{A}][\mathrm{M}]$
(b′)	$\mathrm{A^{*}+M \rightarrow A+M}$	Rate of removal of A* = $k_b'[\mathrm{M}][\mathrm{A}^{*}]$
(c)	$\mathrm{A^{*} \rightarrow P}$	Rate of removal of A* = $k_c[\mathrm{A}^{*}]$ Rate of formation of P = $k_c[\mathrm{A}^{*}]$

The intermediate is A*, and its net rate of formation (the sum of the rates of formation less the sum of the rates of removal; or a + b − a′ − b′ − c in terms of the labels given to the elementary reactions) is

$$k_a[\mathrm{A}]^2 + k_b[\mathrm{A}][\mathrm{M}] - k_a'[\mathrm{A}][\mathrm{A}^{*}] - k_b'[\mathrm{M}][\mathrm{A}^{*}] - k_c[\mathrm{A}^{*}] = 0$$

which rearranges into

$$k_a[\mathrm{A}]^2 + k_b[\mathrm{A}][\mathrm{M}] - (k_a'[\mathrm{A}] + k_b'[\mathrm{M}] + k_c)[\mathrm{A}^{*}] = 0$$

The steady-state concentration of A* is therefore

$$[\mathrm{A}^{*}] = \frac{k_a[\mathrm{A}]^2 + k_b[\mathrm{A}][\mathrm{M}]}{k_a' + k_b'[\mathrm{M}] + k_c}$$

The rate for formation of product (c) is therefore

$$\text{Rate of formation of P} = k_c[\mathrm{A}^{*}] = k_c \times \frac{k_a[\mathrm{A}]^2 + k_b[\mathrm{A}][\mathrm{M}]}{k_a' + k_b'[\mathrm{M}] + k_c}$$

At this point we suppose that $k_b[\mathrm{A}][\mathrm{M}] \gg k_a[\mathrm{A}]^2$ and $k_b'[\mathrm{M}] \gg k_a' + k_c$, whereupon

$$\text{Rate of formation of P} \approx k_c \times \frac{k_b[\mathrm{A}][\mathrm{M}]}{k_b'[\mathrm{M}]} = \frac{k_b k_c}{k_b'}[\mathrm{A}]$$

and the reaction is effectively first-order in A.

Self-test 11.4

Now suppose that P itself is the most effective quencher of A* (with no M present). Devise the rate law.

Answer: Rate of formation of P ≈ $(k_a k_c / k_b)[A]^2/[P]$

Reactions in solution

We now turn specifically to reactions in solution, where the reactant molecules do not fly freely through a gaseous medium and collide with each other, but wriggle past their closely packed neighbours as gaps open up in the structure.

11.10 Activation control and diffusion control

The concept of the rate-determining step plays an important role for reactions in solution where it leads to the distinction between 'diffusion control' and 'activation control'. To develop this point, let's suppose that a reaction between two solute molecules A and B occurs by the following mechanism. First, A and B drift into each other's vicinity by the process of **diffusion**, or migration through the system in a series of small steps in random directions, and form an **encounter pair**, AB, at a rate proportional to each of their concentrations:

$$A + B \rightarrow AB \quad \text{Rate of formation of AB} = k_{r,d}[A][B]$$

The 'd' subscript reminds us that this process is diffusional. The encounter pair may persist for some time as a result of the **cage effect**, the trapping of A and B near each other by their inability to escape rapidly through the surrounding solvent molecules. However, the encounter pair can break up when A and B have the opportunity to diffuse apart, so we must allow for the following process:

$$AB \rightarrow A + B \quad \text{Rate of loss of AB} = k'_{r,d}[AB]$$

We suppose that this process is first-order in AB. Competing with this process is the reaction between A and B while they exist as an encounter pair. This process depends on their ability to acquire sufficient energy to react. That energy might come from the jostling of the thermal motion of the solvent molecules. We assume that this step is first-order in AB, but if the solvent molecules are involved it is more accurate to regard it as pseudofirst-order with the solvent molecules in great and constant excess. In any event, we can suppose that the reaction is

$$AB \rightarrow \text{products} \quad \text{Rate of reactive loss of AB} = k_{r,a}[AB]$$

The 'a' subscript reminds us that this process is activated in the sense that it depends on the acquisition by AB of at least a minimum energy.

Now we use the steady-state approximation to set up the rate law for the formation of products. As shown in the following Derivation, we find

$$\text{Rate of formation of products} = k_r[A][B]$$

$$k_r = \frac{k_{r,a}k_{r,d}}{k_{r,a} + k'_{r,d}} \qquad (11.15)$$

Derivation 11.5

Diffusion control

The net rate of formation of AB is

$$\text{Net rate of formation of AB} = k_{r,d}[A][B] - k'_{r,d}[AB] - k_{r,a}[AB]$$

In a steady state, this rate is zero, so we can write

$$k_{r,d}[A][B] - k'_{r,d}[AB] - k_{r,a}[AB] = 0$$

which we can rearrange to find [AB]:

$$[AB] = \frac{k_{r,d}[A][B]}{k_{r,a} + k'_{r,d}}$$

The rate of formation of products (which is the same as the rate of reactive loss of AB) is therefore

$$\text{Rate of formation of products} = k_{r,a}[AB] = \frac{k_{r,a}k_{r,d}[A][B]}{k_{r,a} + k'_{r,d}}$$

which is eqn 11.15.

Now we distinguish two limits. Suppose the rate of reaction is much faster than the rate at which the encounter pair breaks up. In this case, $k_{r,a} \gg k'_{r,d}$ and we can neglect $k'_{r,d}$ in the denominator of the expression for k_r in eqn 11.15:

$$\text{Rate of formation of products} = \frac{k_{r,a}k_{r,d}[A][B]}{k_{r,a}}$$

The $k_{r,a}$ in the numerator and denominator then cancel and we are left with

$$\text{Rate of formation of products} = k_{r,d}[A][B]$$

Diffusion-controlled limit (11.16)

In this **diffusion-controlled limit**, the rate of the reaction is controlled by the rate at which the reactants diffuse together (as expressed by $k_{r,d}$), for once they have encountered the reaction is so fast that they will certainly go on to form products rather than diffuse apart before reacting. Alternatively, we may suppose that the rate at which the encounter pair accumulated enough energy to react is so low that it is highly likely that the pair will break up. In this case, we can set $k_{r,a} \ll k'_{r,d}$ in the expression for k_r, and obtain

$$\text{Rate of formation of products} = \frac{k_{r,a}k_{r,d}}{k'_{r,d}}[A][B]$$

Activation-controlled limit (11.17)

In this **activation-controlled limit**, the reaction rate depends on the rate at which energy accumulates in the encounter pair (as expressed by $k_{r,a}$).

A lesson to learn from this analysis is that the concept of the rate-determining stage is rather subtle. Thus, in the diffusion-controlled limit, the condition for the encounter rate to be rate-determining is not that it is the slowest step, but that the reaction rate of the encounter pair is much greater than the rate at which the pair breaks up. In the activation-controlled limit, the condition for the rate of energy accumulation to be rate-determining is likewise a competition between the rate of reaction of the pair and the rate at which it breaks up, and all three rate-constants control the overall rate. The best way to analyse competing rates is to do as we have done here: to set up the overall rate law, and then to analyse how it simplifies as we allow particular elementary processes to dominate others.

11.11 Diffusion

Diffusion plays such a central role in the processes involved in reactions in solution that we need to examine it more closely. The picture to hold in mind is that a molecule in a liquid is surrounded by other molecules and can move only a fraction of a diameter, perhaps because its neighbours move aside momentarily, before colliding. Molecular motion in liquids, the diffusion of a reactant for instance, is a series of short steps, with incessantly changing directions, like people in an aimless, milling crowd. We can think of the motion of the molecule as a series of short jumps in random directions, a so-called **random walk**.

If there is an initial concentration gradient in the liquid—for instance, a solution may have a high concentration of solute in one region, such as when a reactant is first injected into a region—then the rate at which the molecules spread out is proportional to the concentration gradient, and we write

$$\text{Rate of diffusion} \propto \text{concentration gradient}$$

To express this relation mathematically, we introduce the **flux**, J, which is the number of particles passing through an imaginary window in a given time interval, divided by the area of the window and the duration of the interval:

$$J = \frac{\text{number of particles passing through window}}{\text{area of window} \times \text{time interval}}$$

Definition Flux (11.18a)

The statement about rate being proportional to gradient is then written

$$J = -D \times \text{concentration gradient}$$

Fick's first law (11.18b)

where D is a constant of proportionality. Equation 11.18b is a verbal interpretation of the equation

$$J = -D\frac{d\mathcal{N}}{dx}$$

Fick's first law (11.18c)

where $\mathcal{N}$ is the number density (the number of solute molecules in a given volume divided by the volume) and $d\mathcal{N}/dx$ is the gradient of that density. The number density is proportional to the molar concentration, because $\mathcal{N} = N/V = nN_A/V = cN_A$, where N_A is Avogadro's constant. Equation 11.18b is called **Fick's first law** of diffusion (see Further information 11.1 for a derivation). The coefficient D, which has the dimensions of area divided by time (with units $m^2\,s^{-1}$), is called the **diffusion coefficient**: if D is large, molecules diffuse rapidly. Some values are given in Table 11.1. The negative sign in eqn 11.18b simply means that if the concentration gradient is negative (down from left to right, Fig. 11.9), then the flux is positive (flowing from left to right). To get the number of molecules passing through a given window in a given time interval, we multiply the flux by the area of the window and the time interval. If the concentration in eqn 11.18b is a molar concentration, then the flux is expressed in moles rather than numbers of molecules and $\mathcal{N}$ is replaced by c.

Table 11.1
Diffusion coefficients at 25 °C, $D/(10^{-9}\,m^2\,s^{-1})$

Ar in tetrachloromethane	3.63
$C_{12}H_{22}O_{11}$ (sucrose) in water	0.522
CH_3OH in water	1.58
H_2O in water	2.26
NH_2CH_2COOH in water	0.673
O_2 in tetrachloromethane	3.82

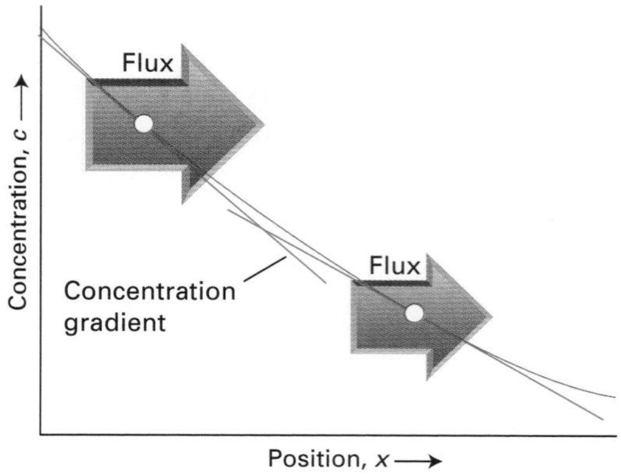

Fig. 11.9 The flux of solute particles is proportional to the concentration gradient. Here we see a solution in which the concentration falls from left to right (as depicted by the curve). The gradient is negative (down from left to right) and the flux is positive (towards the right). The greatest flux is found where the gradient is steepest (on the left).

Brief illustration 11.4 Flux

Suppose that in a region of an unstirred aqueous solution of sucrose the molar concentration gradient is $-0.010\ \text{mol dm}^{-3}\ \text{cm}^{-1}$. Then, using the value for the diffusion coefficient from Table 11.1, the flux arising from this gradient is

$$J = -\overbrace{(0.522\times10^{-9}\ \text{m}^2\ \text{s}^{-1})}^{D}\times\overbrace{(-0.010\ \text{mol dm}^{-3}\ \text{cm}^{-1})}^{\text{d}c/\text{d}x}$$

$$= 5.22\times1.0\times10^{-11}\frac{\text{m}^2\ \text{s}^{-1}\,\text{mol}}{\text{dm}^3\ \text{cm}}$$

$$= 5.22\times1.0\times10^{-11}\frac{\text{m}^2\ \text{s}^{-1}\,\text{mol}}{(10^3\ \text{m}^3)\times(10^{-2}\ \text{m})}$$

$$= 5.2\times10^{-6}\ \text{mol m}^{-2}\ \text{s}^{-1}$$

The amount of sucrose molecules passing through a 10-cm square window in 10 minutes is therefore

$$n = JA\Delta t = (5.2\times10^{-6}\ \text{mol m}^{-2}\ \text{s}^{-1})\times(1.0\times10^{-2}\ \text{m})^2 \times(10\times60\ \text{s})$$

$$= 3.1\times10^{-7}\ \text{mol}$$

The diffusion of molecules may be assisted—and normally greatly dominated—by bulk motion of the fluid as a whole (as when a wind blows in the atmosphere and currents flow in lakes). This motion is called **convection**. Because diffusion is so slow, we speed up the spread of solute molecules by inducing convection by stirring a fluid, turning on an extractor fan, or relying on natural phenomena such as winds and storms.

One of the most important equations in the physical chemistry of fluids is the **diffusion equation**, which enables us to predict the rate at which the concentration of a solute changes in a non-uniform solution. In essence, the diffusion equation expresses the fact that wrinkles in the concentration tend to disperse. The formal (but still verbal) statement of the diffusion equation, which is also known as **Fick's second law** of diffusion, is:

Rate of change of concentration in a region
= D × (curvature of the concentration in the region)

Fick's second law (11.19a)

The 'curvature' is a measure of the wrinkliness of the concentration (see below). The derivation of this expression is given in Further information 11.1, which shows how to derive this law from Fick's first law. The concentrations on the left and right of this equation may be either number density ($\mathcal{N}$, as molecules m^{-3}, for instance) or molar concentration (c). For the record, the mathematical form of the diffusion equation is

$$\frac{\text{d}\mathcal{N}}{\text{d}t} = D\frac{\text{d}^2\mathcal{N}}{\text{d}x^2} \qquad (11.19\text{b})$$

We are interpreting the second derivative $\text{d}^2\mathcal{N}/\text{d}x^2$ as a measure of the curvature of the concentration c.[1]

The diffusion equation tells us the following:

- If the concentration is uniform or its profile has a constant slope, there is no net change in concentration in the region.

In this case the rate of influx through one wall of the region is equal to the rate of efflux through the opposite wall. Only if the slope of the concentration varies through region—only if the concentration is wrinkled—is there a change in concentration. Then:

- Where the curvature is positive (a dip, Fig. 11.10), the change in concentration is positive: the dip tends to fill.
- Where the curvature is negative (a heap), the change in concentration is negative: the heap tends to spread.

The diffusion coefficient increases with temperature (the molecule becomes more mobile) because an increase in temperature enables a molecule to escape more easily from the attractive forces exerted by its neighbours. If we suppose that the rate ($1/\tau$) of the

[1] Because the concentration is a function of both time and location, the derivatives are in fact *partial* derivatives and you will normally see it written $\partial\mathcal{N}/\partial t = D\partial^2\mathcal{N}/\partial x^2$ and more fully as $(\partial\mathcal{N}/\partial t)_x = D(\partial^2\mathcal{N}/\partial x^2)_t$.

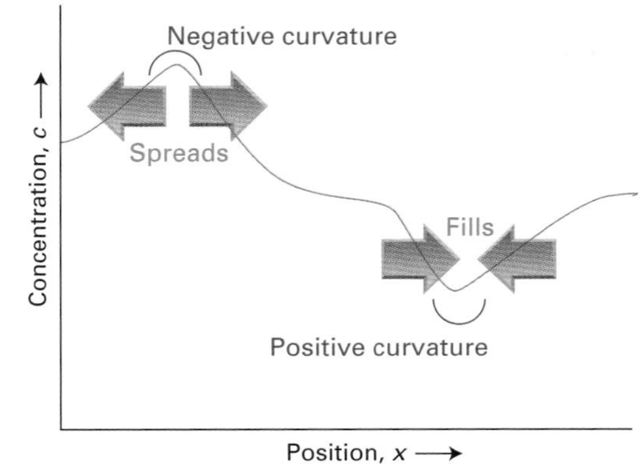

Fig. 11.10 Nature abhors a wrinkle. The diffusion equation tells us that peaks in a distribution (regions of negative curvature) spread and troughs (regions of positive curvature) fill in.

random walk follows an Arrhenius temperature dependence with an activation energy E_a, then the diffusion coefficient will follow the relation

$$D(T) = D_0 e^{-E_a/RT} \qquad (11.20)$$

The rate at which particles diffuse through a liquid is related to the viscosity, and we should expect a high diffusion coefficient to be found for fluids that have a low viscosity. That is, we can suspect that $\eta \propto 1/D$, where η is the coefficient of viscosity. In fact, the **Einstein relation** states that

$$D = \frac{kT}{6\pi\eta a} \qquad \text{Einstein relation} \qquad (11.21)$$

where a is the radius of the molecule. It follows that

$$\eta(T) = \eta_0 e^{E_a/RT} \qquad (11.22)$$

Note the change in sign of the exponent: viscosity *decreases* as the temperature is raised. We are supposing that the strong temperature dependence of the exponential term dominates the weak linear dependence on T in the numerator of eqn 11.21. The temperature dependence described by eqn 11.22 is observed, at least over reasonably small temperature ranges (Fig. 11.11). The forces acting between the molecules govern the magnitude of E_a, but the problem of calculating it is immensely difficult and still largely unsolved.

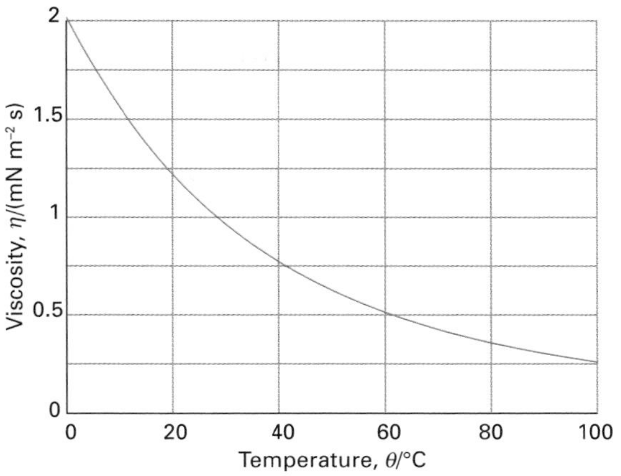

Fig. 11.11 The experimental temperature dependence of the viscosity of water. As the temperature is increased, more molecules are able to escape from the potential wells provided by their neighbours, so the liquid becomes more fluid.

Example 11.3

The activation energy for viscosity

Estimate the activation energy for the viscosity of water from the graph in Fig. 11.11, by using the viscosities at 40 °C (for greater precision, tables give this value as 0.785 mN m^{-2} s) and 80 °C (0.416 mN m^{-2} s).

Strategy Use an equation like eqn 11.22 to formulate an expression for the logarithm of the ratio of the two viscosities, and then solve for the activation energy E_a.

Answer On taking the logarithm of both sides of eqn 11.22 and using the rules from The chemist's toolkit 2.2, we obtain

$$\ln \eta(T) \overbrace{=}^{\ln xy = \ln x + \ln y} \ln \eta_0 + \ln(e^{E_a/RT}) \overbrace{=}^{\ln e^x = x} \ln \eta_0 + \frac{E_a}{RT}$$

Therefore, the difference between two such expressions at temperatures T_1 and T_2 is

$$\ln \eta(T_2) - \ln \eta(T_1) = \left(\ln \eta_0 + \frac{E_a}{RT_2}\right) - \left(\ln \eta_0 + \frac{E_a}{RT_1}\right) = \frac{E_a}{RT_2} - \frac{E_a}{RT_1} = \frac{E_a}{R}\left(\frac{1}{T_2} - \frac{1}{T_1}\right)$$

It follows that

$$E_a = \frac{\ln \eta(T_2) - \ln \eta(T_1)}{\frac{1}{R}\left(\frac{1}{T_2} - \frac{1}{T_1}\right)} \overbrace{=}^{\ln x - \ln y = \ln(x/y)} \frac{R}{1/T_2 - 1/T_1} \ln \frac{\eta(T_2)}{\eta(T_1)}$$

At this point we insert the data from the graph, noting that 40 °C corresponds to 313 K and 80 °C corresponds to 353 K. With $\eta(313\text{ K}) = 0.785$ mN m^{-2} s and $\eta(353\text{ K}) = 0.416$ mN m^{-2} s we find

$$E_a = \frac{8.3145 \text{ J K}^{-1}\text{ mol}^{-1}}{(1/353\text{ K}) - (1/313\text{ K})} \ln \frac{0.416 \text{ mN m}^{-1}\text{ s}}{0.785 \text{ mN m}^{-1}\text{ s}}$$

$$= \frac{8.3145 \text{ J mol}^{-1}}{(1/353) - (1/313)} \ln \frac{0.416}{0.785}$$

$$= 1.46 \times 10^4 \text{ J mol}^{-1}$$

or 14.6 kJ mol^{-1}.

Self-test 11.5

Does the activation energy depend on the temperature? Evaluate it using the viscosities at 80 °C and 100 °C (0.416 mN m^{-2} and 0.329 mN m^{-2}).

Answer: 12.8 kJ mol^{-1}

We can now bring this material back into the context of reaction kinetics. A detailed analysis of the rates of diffusion of molecules in liquids shows that the rate constant $k_{r,d}$ is related to the **coefficient of viscosity**, η (eta), of the medium by[2]

$$k_{r,d} = \frac{8RT}{3\eta} \qquad (11.23)$$

We see that the higher the viscosity, the smaller the diffusional rate constant, and therefore the slower the rate of a diffusion-controlled reaction.

Brief illustration 11.5 The diffusion-controlled rate constant

For a diffusion-controlled reaction in water, for which $\eta = 8.9\times10^{-4}$ kg m^{-1} s^{-1} at 25 °C, we find

$$k_{r,d} = \frac{8\times(8.3145\ \text{J K}^{-1}\ \text{mol}^{-1})\times(298\ \text{K})}{3\times(8.9\times10^{-4}\ \text{kg m}^{-1}\ \text{s}^{-1})}$$

$$= \frac{8\times8.3145\times298}{3\times8.9\times10^{-4}} \frac{\overbrace{\text{J}}^{\text{kg m}^2\ \text{s}^{-2}}\ \ \text{mol}^{-1}}{\text{kg m}^{-1}\ \text{s}^{-1}}$$

$$= 7.4\times10^{6}\frac{\text{kg m}^2\ \text{s}^{-2}\ \text{mol}^{-1}}{\text{kg m}^{-1}\ \text{s}^{-1}}$$

$$= 7.4\times10^{6}\ \text{m}^3\ \text{s}^{-1}\ \text{mol}^{-1}$$

Because 1 m^3 = 10^3 dm^3, this result can be written $k_{r,d} = 7.4\times10^{9}$ dm^3 mol^{-1} s^{-1}, which is a useful approximate estimate to keep in mind for such reactions.

Homogeneous catalysis

A **catalyst** is a substance that accelerates a reaction but undergoes no net chemical change. The catalyst lowers the activation energy of the reaction by providing an alternative path that avoids the slow, rate-determining step of the uncatalysed reaction (Fig. 11.12). Catalysts can be very effective; for instance, the activation energy for the decomposition of hydrogen peroxide in solution is 76 kJ mol^{-1}, and the reaction is slow at room temperature. When a little iodide ion is added, the activation energy falls to 57 kJ mol^{-1} and the rate constant increases by a factor of 2000. **Enzymes**, which are biological catalysts, are very selective and can have a dramatic effect on the reactions they control. For example, the enzyme catalase reduces the activation energy for the decomposition of hydrogen peroxide to 8 kJ mol^{-1}, corresponding to an acceleration of the reaction by a factor of 10^{15} at 298 K.

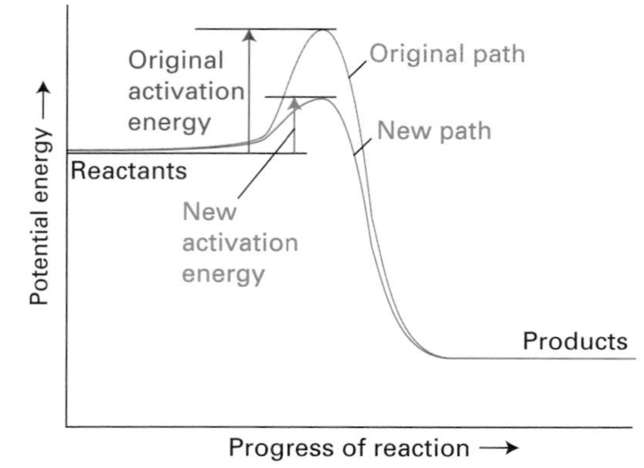

Fig. 11.12 A catalyst acts by providing a new reaction pathway between reactants and products, with a lower activation energy than the original pathway.

A **homogeneous catalyst** is a catalyst in the same phase as the reaction mixture. For example, the decomposition of hydrogen peroxide in aqueous solution is catalysed by bromide ion or catalase. A **heterogeneous catalyst** is a catalyst in a different phase from the reaction mixture. For example, the hydrogenation of ethene to ethane, a gas-phase reaction, is accelerated in the presence of a solid catalyst such as palladium, platinum, or nickel. The metal provides a surface upon which the reactants bind; this binding facilitates encounters between reactants and increases the rate of the reaction. We examine heterogeneous catalysis in Chapter 18 and consider only homogeneous catalysis here.

11.12 Acid and base catalysis

In **acid catalysis** the crucial step is the transfer of a proton to the substrate:

$$X + HA \rightarrow HX^+ + A^- \qquad HX^+ \rightarrow \text{products}$$

Acid catalysis is the primary process in keto–enol tautomerism:

$$CH_3COCH_2CH_3 \xrightarrow{H^+} CH_3C(OH){=}CHCH_3$$

In **base catalysis**, a proton is transferred from the substrate to a base,

$$HX + B \rightarrow X^- + HB^+$$

[2] For a derivation, see our *Physical Chemistry* (2010).

An example is the hydrolysis of esters:

$$CH_3COOCH_2CH_3 + H_2O \xrightarrow{OH^-} CH_3COOH \text{ (as } CH_3CO_2^-) + CH_3CH_2OH$$

11.13 Enzymes

One of the earliest descriptions of the action of enzymes is the **Michaelis–Menten mechanism.** The proposed mechanism, with all species in an aqueous environment, is as follows:

Step 1 The bimolecular formation of a combination, ES, of the enzyme E and the substrate S:

$E + S \rightarrow ES$ Rate of formation of ES $= k_a[E][S]$

Step 2 The unimolecular decomposition of the complex:

$ES \rightarrow E + S$ Rate of decomposition of ES $= k_a'[ES]$

Step 3 The unimolecular formation of products P and the release of the enzyme from its combination with the substrate:

$ES \rightarrow P + E$ Rate of formation of P $= k_b[ES]$

Rate of consumption of ES $= k_b[ES]$

As shown in the following Derivation, the rate law for the rate of formation of product in terms of the concentrations of enzyme and substrate turns out to be

Rate of formation of P $= k_r[E]_0$,

$$\text{with } k_r = \frac{k_b[S]}{[S]+K_M} \qquad \text{Michaelis–Menten rate law} \qquad (11.24)$$

The **Michaelis constant,** K_M (which has the dimensions of a concentration), is

$$K_M = \frac{k_a' + k_b}{k_a} \qquad \text{Michaelis constant} \qquad (11.25)$$

and $[E]_0$ is the total concentration of enzyme (both bound and unbound).

Derivation 11.6

The Michaelis–Menten rate law

The product is formed (irreversibly) in Step 3, so we begin by writing

Rate of formation of P $= k_b[ES]$

To calculate the concentration [ES] we set up an expression for the net rate of formation of ES allowing for its formation in Step 1 and its removal in Steps 2 and 3. Then we use the steady-state approximation and set that net rate equal to zero:

Net rate of formation of ES $= k_a[E][S] - k_a'[ES] - k_b[ES] = 0$

and therefore

$$k_a[E][S] - (k_a' + k_b)[ES] = 0$$

It follows that

$$[ES] = \frac{k_a[E][S]}{k_a' + k_b}$$

However, [E] and [S] are the molar concentrations of the *free* enzyme and *free* substrate. If $[E]_0$ is the total concentration of enzyme, then $[E] + [ES] = [E]_0$ and we can replace [E] in this expression by $[E]_0 - [ES]$. Therefore,

$$[ES] = \frac{k_a([E]_0 - [ES])[S]}{k_a' + k_b}$$

Multiplication by $k_a' + k_b$ gives first

$$k_a'[ES] + k_b[ES] = k_a[E]_0[S] - k_a[ES][S]$$

and then

$$(k_a' + k_b + k_a[S])[ES] = k_a[E]_0[S]$$

Division by k_a turns this expression into

$$\left(\overbrace{\frac{k_a' + k_b}{k_a}}^{K_M} + [S]\right)[ES] = [E]_0[S]$$

We recognize the first term inside the parentheses as K_M, so this expression rearranges first to

$$(K_M + [S])[ES] = [E]_0[S]$$

and then to

$$[ES] = \frac{k_a[E]_0[S]}{[S] + K_M}$$

It follows from the first equation in this derivation that the rate of formation of product is given by eqn 11.24.

According to eqn 11.24, the rate of enzymolysis is first-order in the added enzyme concentration, but the effective rate constant k_r depends on the concentration of substrate. We can infer from eqn 11.24 that:

- When $[S] \ll K_M$, the effective rate constant is equal to $k_b[S]/K_M$. Therefore, the rate increases linearly with [S] at low concentrations.
- When $[S] \gg K_M$, the effective rate constant is equal to k_b, and the rate law in eqn 11.24 reduces to

$$\text{Rate of formation of P} = k_b[E]_0 \qquad (11.26)$$

When $[S] \gg K_M$, the rate is independent of the concentration of S because there is so much substrate present that it remains at effectively the same concentration even though products are being formed. Under these conditions, the rate of formation of

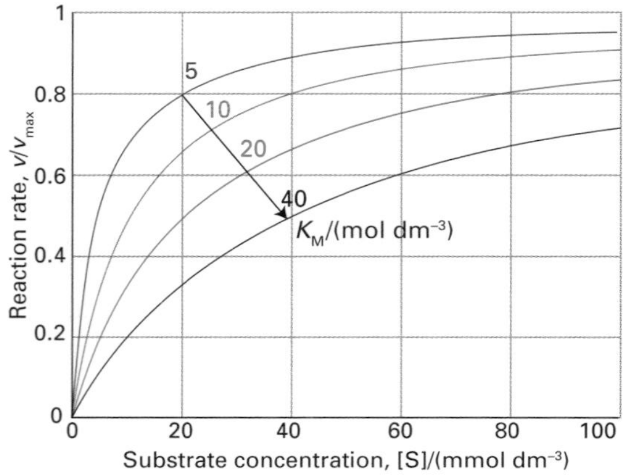

Fig. 11.13 The variation of the rate of an enzyme-catalysed reaction with concentration of the substrate according to the Michaelis–Menten model. When [S] ≪ K_M, the rate is proportional to [S]; when [S] ≫ K_M, the rate is independent of [S].

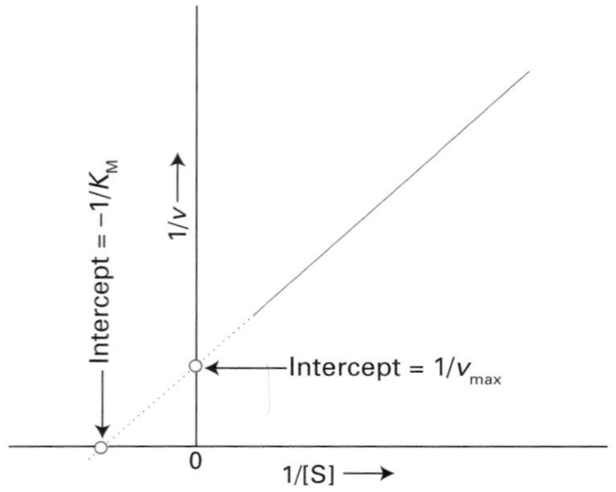

Fig. 11.14 A Lineweaver–Burk plot is used to analyse kinetic data on enzyme-catalysed reactions. The reciprocal of the rate of formation of products ($1/v$) is plotted against the reciprocal of the substrate concentration (1/[S]). All the data points (which typically lie in the full region of the line) correspond to the same overall enzyme concentration, $[E]_0$. The intercept of the extrapolated (dotted) straight line with the horizontal axis is used to obtain the Michaelis constant, K_M. The intercept with the vertical axis, is used to determine $v_{max} = k_b[E]_0$, and hence k_b. The slope may also be used, for it is equal to K_M/v_{max}.

product is a maximum, and $k_b[E]_0$ is called the **maximum velocity**, v_{max}, of the enzymolysis:

$$v_{max} = k_b[E]_0 \qquad \text{Definition} \quad \text{Maximum velocity} \qquad (11.27)$$

The rate-determining step is Step 3, because there is ample ES present (because S is so abundant), and the rate is determined by the rate at which ES reacts to form the product.

It follows from eqns 11.24 and 11.27 that the reaction rate v at a general substrate composition is related to the maximum velocity by

$$v = k_b\frac{[S][E]_0}{[S]+K_M} = \frac{[S]}{[S]+K_M}v_{max} \qquad (11.28)$$

This relation is illustrated in Fig. 11.13. The graph gives a clue to the significance of K_M, for we see (and as eqn 11.28 shows), the rate of enzymolysis reaches $\frac{1}{2}v_{max}$ when [S] = K_M; broadly speaking, therefore, K_M is a measure of the substrate concentration at and above which the enzyme is effective.

Equation 11.28 is the basis of the analysis of enzyme kinetic data by using a **Lineweaver–Burk plot**, a graph of $1/v$ (the reciprocal of the reaction rate) against 1/[S] (the reciprocal of the substrate concentration). If we take the reciprocal of both sides of eqn 11.28 it becomes

$$\frac{1}{v} = \left(\frac{[S]+K_M}{[S]}\right)\frac{1}{v_{max}} = \left(1+\frac{K_M}{[S]}\right)\frac{1}{v_{max}}$$

$$= \frac{1}{v_{max}} + \left(\frac{K_M}{v_{max}}\right)\frac{1}{[S]} \qquad (11.29a)$$

Because this expression has the form $y = b + mx$,

$$\overbrace{\frac{1}{v}}^{y} = \overbrace{\frac{1}{v_{max}}}^{b} + \overbrace{\left(\frac{K_M}{v_{max}}\right)}^{m}\overbrace{\frac{1}{[S]}}^{x} \qquad \text{Lineweaver–Burk plot} \qquad (11.29b)$$

with $y = 1/v$ and $x = 1/[S]$, we should obtain a straight line when we plot $1/v$ against 1/[S] (recall The chemist's toolkit 1.1). The slope of the straight line is K_M/v_{max} and the extrapolated intercept at 1/[S] = 0 is equal to $1/v_{max}$ (Fig. 11.14). Therefore, the intercept can be used to find v_{max}, and then that value combined with the slope to find the value of K_M. Alternatively, note that the extrapolated intercept with the horizontal axis (where $1/v$ = 0) occurs at $1/[S] = -1/K_M$.

We can calculate further parameters from those derived from a Lineweaver–Burk plot and which allow us to compare the catalytic properties of different enzymes. The **turnover frequency**, or **catalytic constant**, of an enzyme, k_{cat}, is the number of catalytic cycles (turnovers) performed by the active site in a given interval divided by the duration of the interval:

$$k_{cat} = \frac{\text{number of catalytic cycles}}{\text{time interval}} \qquad \text{Definition} \quad \text{Turnover frequency} \qquad (11.30)$$

The time constant for the first-order decomposition of ES is $1/k_b$. So the number of catalytic events in an interval Δt is $\Delta t/(1/k_b) = k_b\Delta t$. The frequency of such events is therefore this number divided by the interval Δt, which is k_b itself. That is, $k_{cat} = k_b$ (it has a more complicated form for more complex reaction schemes). It follows from the identification of k_{cat} with k_b and from eqn 11.27 that

$$k_{cat} = k_b = \frac{v_{max}}{[E]_0} \qquad (11.31)$$

An alternative interpretation of k_{cat}, therefore, is a measure of the effectiveness of an active site, as it is the maximum rate of enzymolysis that can be achieved divided by the concentration of active sites.

The **catalytic efficiency**, η (eta), of an enzyme is the ratio

$$\eta = \frac{k_{cat}}{K_M} \qquad \text{Definition Catalytic efficiency} \qquad (11.32)$$

This relation is inspired by recognizing that k_{cat} is a measure of the effectiveness of an enzyme and K_M is a measure of the concentration at which the enzyme becomes effective. The higher the value of η, the more efficient is the enzyme. We can think of the catalytic efficiency as the effective rate constant of the enzymatic reaction. From $K_M = (k'_a + k_b)/k_a$ and eqn 11.32, it follows that

$$\eta = \frac{\overbrace{k_b}^{k_{cat}}}{\underbrace{(k'_a + k_b)/K_a}_{K_M}} = \frac{k_a k_b}{k'_a + k_b} \qquad (11.33)$$

The efficiency reaches its maximum value of k_a when $k_b \gg k'_a$. Because k_a is the rate constant for the formation of a complex from two species that are diffusing freely in solution, the maximum efficiency is related to the maximum rate of diffusion of E and S in solution, as we saw in Section 11.11. In this limit rate constants are about 10^8–10^9 $dm^3\,mol^{-1}\,s^{-1}$ for molecules as large as enzymes at room temperature. The enzyme catalase has $\eta = 4.0 \times 10^8$ $dm^3\,mol^{-1}\,s^{-1}$ and is said to have attained 'catalytic perfection', in the sense that the rate of the reaction it catalyses is controlled only by diffusion: it acts as soon as a substrate makes contact.

Example 11.4

Determining the catalytic efficiency of an enzyme

The enzyme carbonic anhydrase catalyses the hydration of CO_2 in red blood cells to give bicarbonate (hydrogencarbonate) ion:

$$CO_2(g) + H_2O(l) \rightarrow HCO_3^-(aq) + H^+(aq)$$

The following data were obtained for the reaction at pH = 7.1, 273.5 K, and an enzyme concentration of 2.3 nmol dm^{-3}:

$[CO_2]/(mmol\ dm^{-3})$	1.25	2.5
$v/(mmol\ dm^{-3}\ s^{-1})$	2.78×10^{-2}	5.00×10^{-2}
$[CO_2]/(mmol\ dm^{-3})$	5	20
$v/(mmol\ dm^{-3}\ s^{-1})$	8.33×10^{-2}	1.67×10^{-1}

Determine the catalytic efficiency of carbonic anhydrase at 273.5 K.

Strategy We construct a Lineweaver–Burk plot by drawing up a table of 1/[S] and $1/v$. The intercept at 1/[S] = 0 is v_{max} and the slope of the line through the points is K_M/v_{max}, so K_M is found from the slope divided by the intercept. From eqn 11.31 and the enzyme concentration, we calculate k_{cat} and the catalytic efficiency from eqn 11.32.

Answer We draw up the following table:

$1/([CO_2]/(mmol\ dm^{-3}))$	0.8	0.4	0.2	0.05
$1/(v/(mmol\ dm^{-3}\ s^{-1}))$	36	20	12	5.99

The data are plotted in Fig. 11.15. A least-squares analysis gives an intercept at 4.00 and a slope of 40.0. It follows that

$$v_{max}/(mmol\ dm^{-3}\ s^{-1}) = \frac{1}{\text{intercept}} = \frac{1}{4.00} = 0.250$$

and

$$K_M/(mmol\ dm^{-3}) = \frac{\text{slope}}{\text{intercept}} = \frac{40.0}{4.00} = 10.0$$

It follows that

$$k_{cat} = \frac{v_{max}}{[E]_0} = \frac{2.5\times10^{-4}\ mol\ dm^{-3}\ s^{-1}}{2.3\times10^{-9}\ mol\ dm^{-3}} = 1.1\times10^5\ s^{-1}$$

and

$$\eta = \frac{k_{cat}}{K_M} = \frac{1.1\times10^5\ s^{-1}}{1.0\times10^{-2}\ mol\ dm^{-3}} = 1.1\times10^7\ dm^3\ mol^{-1}\ s^{-1}$$

A note on good practice The slope and the intercept are unitless: see The chemist's toolkit 1.1.

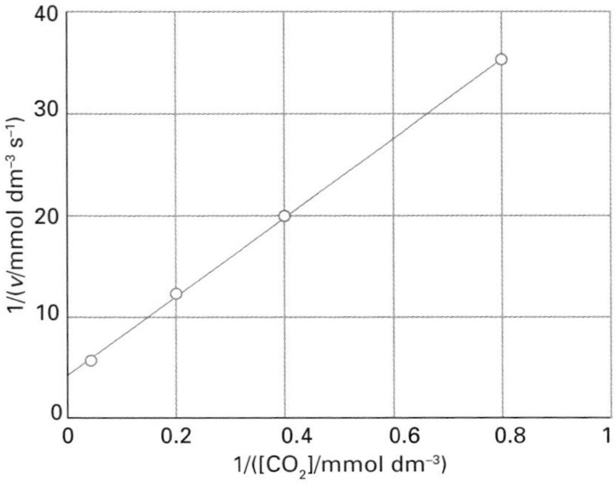

Fig. 11.15 The Lineweaver–Burke plot based on the data in Example 11.4.

Self-test 11.6

The enzyme α-chymotrypsin is secreted in the pancreas of mammals and cleaves peptide bonds made between certain amino acids. Several solutions containing the small peptide *N*-glutaryl-L-phenylalanine-*p*-nitroanilide at different concentrations were prepared and the same small amount of α-chymotrypsin was added to each one. The following data were obtained on the initial rates of the formation of product:

[S]/(mmol dm^{-3})	0.334	0.450	0.667	1.00	1.33	1.67
v/(mmol dm^{-3} s^{-1})	0.152	0.201	0.269	0.417	0.505	0.667

Determine the maximum velocity and the Michaelis constant for the reaction.

Answer: v_{max} = 2.80 mmol dm^{-3} s^{-1}, K_M = 5.89 mmol dm^{-3}

The action of an enzyme may be partially suppressed by the presence of a foreign substance, which is called an **inhibitor.** An inhibitor may be a poison that has been administered to the organism, or it may be a substance that is naturally present in a cell and involved in its regulatory mechanism. In **competitive inhibition** the inhibitor competes for the active site and reduces the ability of the enzyme to bind the substrate (Fig. 11.16). In **non-competitive inhibition** the inhibitor attaches to another part of the enzyme molecule, thereby distorting it and reducing its ability to bind the substrate (Fig. 11.17).

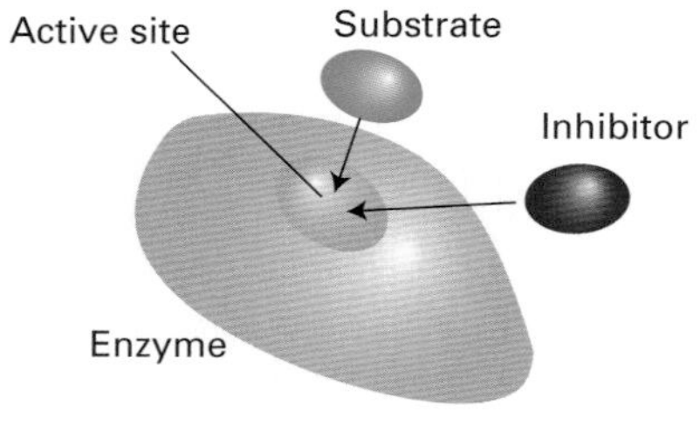

Fig. 11.16 In competitive inhibition, both the substrate and the inhibitor compete for the active site, and reaction ensues only if the substrate is successful in attaching there.

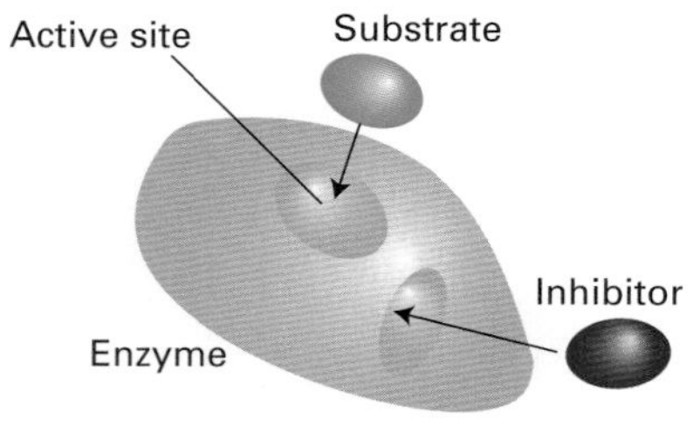

Fig. 11.17 In one version of non-competitive inhibition, the substrate and the inhibitor attach to distant sites of the enzyme molecule, and a complex in which they are both attached (IES) does not lead to the formation of product.

Chain reactions

Many gas-phase reactions and liquid-phase polymerization reactions are **chain reactions,** reactions in which an intermediate produced in one step generates a reactive intermediate in a subsequent step, then that intermediate generates another reactive intermediate, and so on.

11.14 The structure of chain reactions

The intermediates responsible for the propagation of a chain reaction are called **chain carriers.** In a **radical chain reaction** the chain carriers are radicals. Ions may also propagate chains, and in nuclear fission the chain carriers are neutrons. We focus on radicals as chain carriers, and when we judge it helpful mark them with a dot (as in $\cdot CH_3$) to signify the unpaired electron.

A radical chain reaction typically involves some but not necessarily all of the following steps:

- **Initiation,** in which radicals are formed from non-radical molecules.
- **Propagation,** in which a radical attacks a molecule and forms a new radical.
- **Branching,** in which the attack by a radical produces two radicals.
- **Retardation,** in which a product molecule is destroyed by a radical.
- **Inhibition,** in which radicals are removed other than by termination, perhaps by reaction with the walls of the reaction vessel or an added agent.
- **Termination,** in which two radicals combine to form a molecule and so end the chain.

Brief illustration 11.6 Chain reactions

In an initiation step, $\cdot CH_3$ radicals are formed by the dissociation of CH_3CH_3 molecules either as a result of vigorous intermolecular collisions in a thermolysis reaction or as a result of absorption of a photon in a photolysis reaction. The radicals attack other reactant molecules in the propagation steps, as in $\cdot CH_3 + CH_3CH_3 \rightarrow CH_4 + \cdot CH_2CH_3$. A termination step might be the combination of two radicals, as in $CH_3CH_2\cdot + \cdot CH_2CH_3 \rightarrow CH_3CH_2CH_2CH_3$. Inhibition would occur if a $\cdot CH_2CH_3$ radical were to react with the wall of the reaction vessel and be removed from the gas phase. If the product were $CH_3CH_2CH_2CH_3$, then a retardation step might be the atom-extraction event $CH_3CH_2CH_2CH_3 + \cdot CH_2CH_3 \rightarrow \cdot CH_2CH_2CH_2CH_3 + CH_3CH_3$.

Self-test 11.7

Identify the type of step in the process $\cdot O\cdot + H_2O \rightarrow HO\cdot + HO\cdot$.

Answer: Branching

The NO molecule has an unpaired electron and is a very efficient chain inhibitor. The observation that a gas-phase reaction is quenched when NO is introduced is a good indication that a radical chain mechanism is in operation.

11.15 The rate laws of chain reactions

A chain reaction often leads to a complicated rate law (but not always). As a first example, consider the thermal reaction of H_2 with Br_2. The overall reaction and the observed rate law are

$$H_2(g) + Br_2(g) \rightarrow 2\,HBr(g)$$

$$\text{Rate of formation of HBr} = \frac{k_{r1}[H_2][Br_2]^{3/2}}{[Br_2] + k_{r2}[HBr]} \qquad (11.34)$$

The complexity of the rate law suggests that a complicated mechanism is involved. The following radical chain mechanism has been proposed:

Step 1 Initiation: $Br_2 \rightarrow Br\cdot + Br\cdot$

Rate of consumption of $Br_2 = k_a[Br_2]$

Step 2 Propagation:

$Br\cdot + H_2 \rightarrow HBr + H\cdot \quad v = k_b[Br][H_2]$

$H\cdot + Br_2 \rightarrow HBr + Br\cdot \quad v = k_c[H][Br_2]$

In this and the following steps, 'rate' v means either the rate of formation of one of the products or the rate of consumption of one of the reactants.

Step 3 Retardation:

$H\cdot + HBr \rightarrow H_2 + Br\cdot \quad v = k_d[H][HBr]$

Step 4 Termination:

$Br\cdot + \cdot Br + M \rightarrow Br_2 + M \quad v = k_e[Br]^2$

The 'third body', M, a molecule of an inert gas, removes the energy of recombination; the constant concentration of M has been absorbed into the rate constant k_d. Other possible termination steps include the recombination of H atoms to form H_2 and the combination of H and Br atoms; however, it turns out that only Br atom recombination is important.

As we show in the following Derivation, this mechanism result in a rate law of the observed form with the empirical rate constant related to the individual rate constants as follows:

$$k_{r1} = 2k_b\left(\frac{k_a}{k_c}\right)^{1/2} \qquad k_{r2} = \frac{k_d}{k_c} \qquad (11.35)$$

We can conclude that the proposed mechanism is at least consistent with the observed rate law. Additional support for the mechanism would come from the detection of the proposed intermediates (by spectroscopy), and the measurement of individual rate constants for the elementary steps and confirming that they correctly reproduced the observed composite rate constants.

Derivation 11.7

The rate law of a chain reaction

The experimental rate law is expressed in terms of the rate of formation of product, HBr, so we start by writing an expression for its net rate of formation. Because HBr is formed in Step 2 (by both reactions) and consumed in Step 3,

Net rate of formation of HBr
$= k_b[Br][H_2] + k_c[H][Br_2] - k_d[H][HBr]$

To make progress, we need the concentrations of the intermediates Br and H. Therefore, we set up the expressions for their net rate of formation and apply the steady-state assumption to both:

Net rate of formation of H
$= k_b[Br][H_2] - k_c[H][Br_2] - k_d[H][HBr] = 0$

Net rate of formation of Br
$= 2k_a[Br_2] - k_b[Br][H_2] + k_c[H][Br_2] + k_d[H][HBr] - 2k_e[Br]^2 = 0$

The steady-state concentrations of the intermediates are found by solving these two equations and are

$$[Br] = \left(\frac{k_a[Br_2]}{k_e}\right)^{1/2} \qquad [H] = \frac{k_b(k_a/k_c)^{1/2}[H_2][Br_2]^{3/2}}{k_c[Br_2] + k_d[HBr]}$$

When we substitute these concentrations into the equation for the rate of formation of HBr we obtain

$$\text{Rate of formation of HBr} = \frac{\overbrace{2k_b(k_a/k_c)^{1/2}}^{k_{r1}}[H_2][Br_2]^{3/2}}{[Br_2] + \underbrace{(k_d/k_c)}_{k_{r2}}[HBr]}$$

This equation has the same form as the empirical rate law, and we can identify the two empirical rate constants, as shown in eqn 11.35.

Further information 11.1

Fick's laws of diffusion

1. Fick's first law of diffusion

Consider the arrangement in Fig. 11.18. Let's suppose that in an interval Δt the number of molecules passing through the window of area A from the left is proportional to the number in the slab of thickness l and area A, and therefore volume lA, just to the left of the window where the average (number) concentration is $\mathcal{N}(x - {}^1\!/\!_2 l)$ and to the length of the interval Δt:

$$\text{Number coming from left} \propto \mathcal{N}(x - {}^1\!/\!_2 l)lA\Delta t$$

Likewise, the number coming from the right in the same interval is

$$\text{Number coming from right} \propto \mathcal{N}(x + {}^1\!/\!_2 l)lA\Delta t$$

The net flux is therefore the difference in these numbers divided by the area and the time interval:

$$J \propto \frac{\mathcal{N}(x - \frac{1}{2}l)lA\Delta t - \mathcal{N}(x + \frac{1}{2}l)lA\Delta t}{A\Delta t}$$

$$\overset{\text{Cancel } A\Delta t}{\propto} \{\mathcal{N}(x - \tfrac{1}{2}l) - \mathcal{N}(x + \tfrac{1}{2}l)\}l$$

We now express the two concentrations in terms of the concentration at the window itself, $\mathcal{N}(x)$, and the concentration gradient, $\Delta\mathcal{N}/\Delta x$, as follows:

$$\mathcal{N}(x + \tfrac{1}{2}l) = \mathcal{N}(x) + \tfrac{1}{2}l\frac{\Delta\mathcal{N}}{\Delta x} \qquad \mathcal{N}(x - \tfrac{1}{2}l) = \mathcal{N}(x) - \tfrac{1}{2}l\frac{\Delta\mathcal{N}}{\Delta x}$$

from which it follows that

$$J \propto \left\{\left(\mathcal{N}(x) - \tfrac{1}{2}l\frac{\Delta\mathcal{N}}{\Delta x}\right) - \left(\mathcal{N}(x) + \tfrac{1}{2}l\frac{\Delta\mathcal{N}}{\Delta x}\right)\right\}l \propto -l^2\frac{\Delta\mathcal{N}}{\Delta x}$$

On writing the constant of proportionality as D (and absorbing l^2 into it), we obtain eqn 11.18.

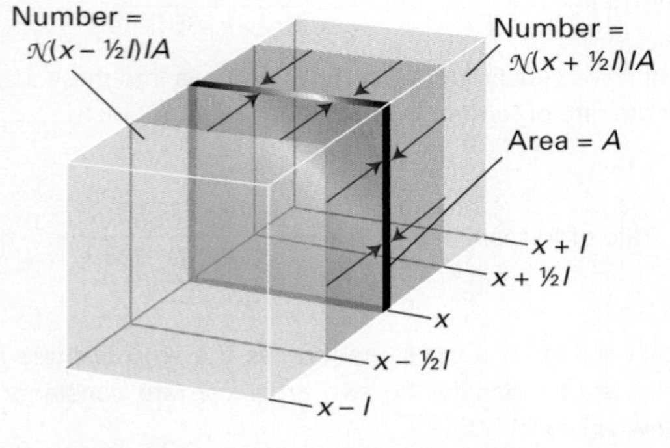

Fig. 11.18 The calculation of the rate of diffusion considers the net flux of molecules through a plane of area A as a result of arrivals from on average a distance ${}^1\!/\!_2 l$ in each direction.

2. Fick's second law

Consider the arrangement in Fig. 11.19. The number of solute particles passing through the window of area A located at x in an interval Δt is $J(x)A\Delta t$, where $J(x)$ is the flux at the location x. The number of particles passing out of the region through a window of area A at a short distance away, at $x + \Delta x$, is $J(x + \Delta x)A\Delta t$, where $J(x + \Delta x)$ is the flux at the location of this window. The flux in and the flux out will be different if the concentration gradients are different at the two windows. The net change in the number of solute particles in the region between the two windows is

$$\begin{aligned}\text{Net change in number} &= J(x)A\Delta t - J(x + \Delta x)A\Delta t\\ &= \{J(x) - J(x + \Delta x)\}A\Delta t\end{aligned}$$

Now we express the flux at $x + \Delta x$ in terms of the flux at x and the gradient of the flux, $\Delta J/\Delta x$:

$$J(x + \Delta x) = J(x) + \frac{\Delta J}{\Delta x} \times \Delta x$$

It follows that

$$\begin{aligned}\text{Net change in number} &= \left\{J(x) - \left(J(x) + \frac{\Delta J}{\Delta x} \times \Delta x\right)\right\} \times A\Delta t\\ &= \frac{\Delta J}{\Delta x} \times \Delta x \times A\Delta t\end{aligned}$$

The change in concentration inside the region between the two windows is the net change in number divided by the volume of the region (which is $A\Delta x$), and the net rate of change is obtained by dividing that change in concentration by the time interval Δt. Therefore, on dividing by both $A\Delta x$ and Δt we obtain

$$\text{Rate of change of concentration} = -\frac{\Delta J}{\Delta x}$$

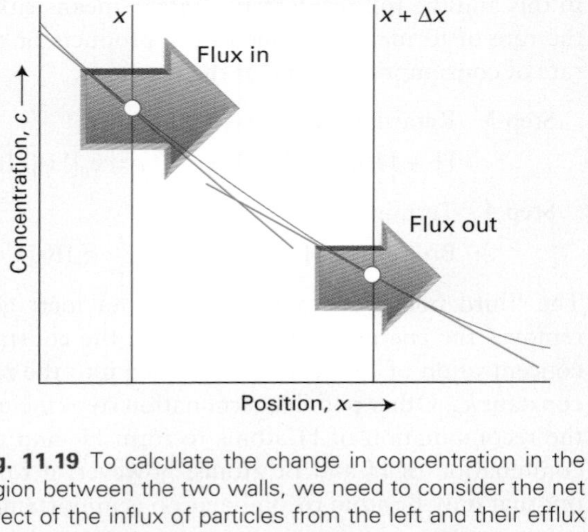

Fig. 11.19 To calculate the change in concentration in the region between the two walls, we need to consider the net effect of the influx of particles from the left and their efflux towards the right. Only if the slope of the concentrations is different at the two walls will there be a net change.

Finally, we express the flux by using Fick's first law:

$$\text{Rate of change of concentration} = -\frac{\Delta(-D \times \Delta\mathcal{N}/\Delta x)}{\Delta x} = D\frac{\Delta^2\mathcal{N}}{\Delta x^2}$$

The 'gradient of the gradient' of the concentration is what we have called the 'curvature' of the concentration, and thus we obtain eqn 11.19, where the Δ have been interpreted, more formally, as the infinitesimal change d.

Checklist of key concepts

- ☐ 1 In relaxation methods of kinetic analysis, the equilibrium position of a reaction is first shifted suddenly and then allowed to readjust the equilibrium composition characteristic of the new conditions.
- ☐ 2 The molecularity of an elementary reaction is the number of molecules coming together to react.
- ☐ 3 An elementary unimolecular reaction has first-order kinetics; an elementary bimolecular reaction has second-order kinetics.
- ☐ 4 In the steady-state approximation, it is assumed that the concentrations of all reaction intermediates remain constant and small throughout the reaction.
- ☐ 5 The rate-determining step is the slowest step in a reaction mechanism that controls the rate of the overall reaction.
- ☐ 6 Provided a reaction has not reached equilibrium, the products of competing reactions are controlled by kinetics.
- ☐ 7 The Lindemann mechanism of 'unimolecular' reactions is a theory that accounts for the first-order kinetics of gas-phase reactions.
- ☐ 8 A reaction in solution may be diffusion-controlled or activation-controlled.
- ☐ 9 Catalysts are substances that accelerate reactions but undergo no net chemical change.
- ☐ 10 A homogeneous catalyst is a catalyst in the same phase as the reaction mixture.
- ☐ 11 Enzymes are homogeneous, biological catalysts.
- ☐ 12 The Michaelis–Menten mechanism of enzyme kinetics accounts for the dependence of rate on the concentration of the substrate.
- ☐ 13 In a chain reaction, an intermediate (the chain carrier) produced in one step generates a reactive intermediate in a subsequent step.
- ☐ 14 The steps in a chain reaction may include initiation, propagation, inhibition, retardation, branching, and termination.

Road map of key equations

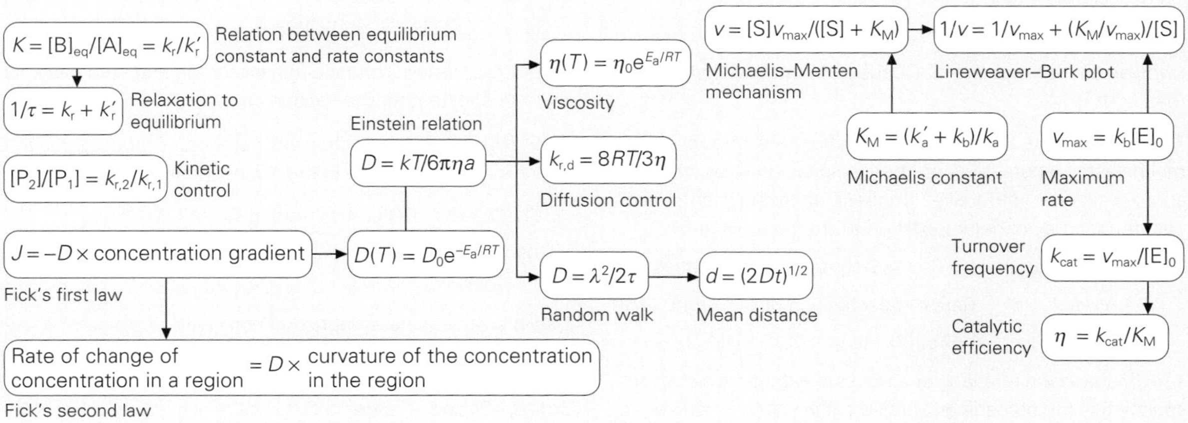

Questions and exercises

Discussion questions

11.1 Sketch, without carrying out the calculation, the variation of concentration with time for the approach to equilibrium when both forward and reverse reactions are second-order. How does your graph differ from that in Fig. 11.2?

11.2 Assess the validity of the following statement: The rate-determining step is the slowest step in a reaction mechanism.

11.3 Specify the pre-equilibrium and steady-state approximations and explain why they might lead to different conclusions.

11.4 Distinguish between kinetic and thermodynamic control of a reaction. Suggest criteria for expecting one rather than the other.

11.5 Why may some gas-phase reactions show first-order kinetics?

11.6 Discuss the features, applications, and limitations of the Michaelis–Menten mechanism of enzyme action.

Exercises

11.1 The equilibrium constant for the attachment of a substrate to the active site of an enzyme was measured as 200. In a separate experiment, the rate constant for the second-order attachment was found to be $1.5 \times 10^8\ dm^3\ mol^{-1}\ s^{-1}$. What is the rate constant for the loss of the unreacted substrate from the active site?

11.2 The equilibrium $NH_3(aq) + H_2O(l) \rightleftharpoons NH_4^+(aq) + OH^-(aq)$ at 25 °C is subjected to a temperature jump which slightly increases the concentration of $NH_4^+(aq)$ and $OH^-(aq)$. The measured relaxation time is 7.61 ns. The equilibrium constant for the system is 1.78×10^{-5} at 25 °C, and the equilibrium concentration of $NH_3(aq)$ is 0.15 mol dm^{-3}. Calculate the rate constants for the forward and reverse steps.

11.3 Two radioactive nuclides decay by successive first-order processes:

$$X \xrightarrow{22.5\ d} Y \xrightarrow{33.0\ d} Z$$

The times are half-lives in days. Suppose that Y is an isotope that is required for medical applications. At what time after X is first formed will Y be most abundant?

11.4 The reaction mechanism

$$A_2 \rightleftharpoons A + A \quad \text{(fast)}$$

$$A + B \rightarrow P \quad \text{(slow)}$$

involves an intermediate A. Deduce the rate law for the formation of P.

11.5 Deduce the rate law for a reaction with the following mechanism, where M is an inert species, and identify any approximations you make. Suggest an experimental procedure that may either support or refute the mechanism.

$A + M \rightarrow A^* + M$	Rate of formation of $A^* = k_a[A][M]$
$A^* + M \rightarrow A + M$	Rate of deactivation of $A^* = k_a'[A^*][M]$
$A^* \rightarrow P$	Rate of formation of $P = k_b[A^*]$

11.6 Deduce the rate law for a reaction with the mechanism specified in the preceding exercise but in which A, as well as M, can also participate in the activation of A and the deactivation of A*. Suggest an experimental procedure that may either support or refute the mechanism.

11.7 The conversion of hypochlorite ions, ClO^-, to chlorate ions, ClO_3^-, in aqueous solution proceeds by a two-step mechanism:

(1) $ClO^- + ClO^- \rightarrow ClO_2^- + Cl^-$

(2) $ClO_2^- + ClO^- \rightarrow ClO_3^- + Cl^-$

The rate of formation of chlorate ions is found to depend upon the square of the concentration of hypochlorite ions. (a) Write an equation for the overall reaction. (b) Identify the rate-determining step in the mechanism.

11.8 The mechanism for the reaction between 2-chloroethanol, CH_2ClCH_2OH, and hydroxide ions in aqueous solution to form ethylene oxide, $(CH_2CH_2)O$, is thought to consist of two steps

(1) $CH_2ClCH_2OH + OH^- \rightleftharpoons CH_2ClCH_2O^- + H_2O$ (fast)

(2) $CH_2ClCH_2O^- \rightarrow (CH_2CH_2)O + Cl^-$ (slow)

Show that, for this mechanism, the rate of formation of ethylene oxide is

$$\text{Rate} = k_2K[CH_2ClCH_2OH][OH^-]$$

where K is the equilibrium constant for the first step and k_2 is the rate constant for the second step.

11.9 The following mechanism has been proposed for the decomposition of ozone in the atmosphere:

(1) $O_3 \rightarrow O_2 + O$ and its reverse (k_1, k_1')

(2) $O + O_3 \rightarrow O_2 + O_2$ (k_2; the reverse reaction is negligibly slow)

Use the steady-state approximation, with O treated as the intermediate, to find an expression for the rate of decomposition of O_3. Show that if step 2 is slow, then the rate is second-order in O_3 and −1 order in O_2.

11.10 Consider the following mechanism for formation of a double helix from its strands A and B:

A + B $\rightleftharpoons$ unstable helix (fast)

unstable helix $\rightarrow$ stable double helix (slow)

Derive the rate equation for the formation of the double helix and express the rate constant of the reaction in terms of the rate constants of the individual steps. What would be your conclusion if the pre-equilibrium assumption was replaced by the steady-state approximation?

11.11 Two products are formed in reactions in which there is kinetic control of the ratio of products. The activation energy for the reaction leading to Product 1 is greater than that leading to Product 2. Will the ratio of product concentrations $[P_1]/[P_2]$ increase or decrease if the temperature is raised?

11.12 The effective rate constant for a gaseous reaction which has a Lindemann mechanism is 2.50×10^{-4} s^{-1} at 1.30 kPa and 2.10×10^{-5} s^{-1} at 12 Pa. Calculate the rate constant for the activation step in the mechanism.

11.13 Calculate the magnitude of the diffusion-controlled rate constant at 298 K for a species in (a) water, (b) pentane. The viscosities are 1.00×10^{-3} kg m^{-1} s^{-1}, and 2.2×10^{-4} kg m^{-1} s^{-1}, respectively.

11.14 What is (a) the flux of nutrient molecules down a concentration gradient of 0.10 mol dm^{-3} m^{-1}, (b) the amount of molecules (in moles) passing through an area of 5.0 mm^2 in 1.0 min? Take for the diffusion coefficient the value for sucrose in water (5.22×10^{-10} m^2 s^{-1}).

11.15 How long does it take a sucrose molecule in water at 25 °C to diffuse along a single dimension by (a) 10 mm, (b) 10 cm, (c) 10 m from its starting point?

11.16 The mobility of species through fluids is of the greatest importance for nutritional processes. (a) Estimate the diffusion coefficient for a molecule that leaps along a single dimension by 150 pm each 1.8 ps. (b) What would be the diffusion coefficient if the molecule travelled only half as far on each step?

11.17 Is diffusion important in lakes? How long would it take a small pollutant molecule about the size of H_2O to diffuse across a lake of width 100 m?

11.18 Pollutants spread through the environment by convection (winds and currents) and by diffusion. How many steps must a molecule take to be likely to be found 1000 step lengths away from its origin if it undergoes a one-dimensional random walk?

11.19 The viscosity of water at 20 °C is 1.0019 mN s m^{-2} and at 30 °C it is 0.7982 mN s m^{-2}. What is the activation energy for the motion of water molecules?

11.20 Calculate the ratio of rates of catalysed to non-catalysed reactions at 37 °C given that the Gibbs energy of activation for a particular reaction is reduced from 150 kJ mol^{-1} to 15 kJ mol^{-1}.

11.21 The reaction $2\,H_2O_2(aq) \rightarrow 2\,H_2O(l) + O_2(g)$ is catalysed by Br^- ions. If the mechanism is

$H_2O_2 + Br^- \rightarrow H_2O + BrO^-$ (slow)

$BrO^- + H_2O_2 \rightarrow H_2O + O_2 + Br^-$ (fast)

give the predicted order of the reaction with respect to the various participants.

11.22 Consider the acid-catalysed reaction

$HA + H^+ \rightleftharpoons HAH^+$ (fast)

$HAH^+ + B \rightarrow BH^+ + AH$ (slow)

Deduce the rate law and show that it can be made independent of the specific term $[H^+]$.

11.23 The condensation reaction of acetone, $(CH_3)_2CO$ (propanone), in aqueous solution is catalysed by bases, B, which react reversibly with acetone to form the carbanion $C_3H_5O^-$. The carbanion then reacts with a molecule of acetone to give the product. A simplified version of the mechanism is

(1) $AH + B \rightarrow BH^+ + A^-$

(2) $A^- + BH^+ \rightarrow AH + B$

(3) $A^- + HA \rightarrow$ product

where AH stands for acetone and A^- its carbanion. Use the steady-state approximation to find the concentration of the carbanion and derive the rate equation for the formation of the product.

11.24 As remarked in Derivation 11.6, Michaelis and Menten derived their rate law by assuming a rapid pre-equilibrium of E, S, and ES. Derive the rate law in this manner, and identify the conditions under which it becomes the same as that based on the steady-state approximation (eqn 11.24).

11.25 The enzyme-catalysed conversion of a substrate at 25 °C has a Michaelis constant of 0.045 mol dm^{-3}. The rate of the reaction is 1.15 mmol dm^{-3} s^{-1} when the substrate concentration is 0.110 mol dm^{-3}. What is the maximum velocity of this reaction?

11.26 The enzyme-catalysed conversion of a substrate at 25 °C has a Michaelis constant of 0.015 mol dm^{-3}, and a maximum velocity of 4.25×10^{-4} mol dm^{-3} s^{-1} when the enzyme concentration is 3.60×10^{-9} mol dm^{-3}. Calculate k_{cat} and the catalytic efficiency η. Is the enzyme 'catalytically perfect'?

11.27 The following results were obtained for the action of an ATPase on ATP at 20 °C, when the concentration of the ATPase is 20 nmol dm^{-3}:

[ATP]/(μmol dm^{-3})	0.60	0.80	1.4	2.0	3.0
v/(μmol dm^{-3} s^{-1})	0.81	0.97	1.30	1.47	1.69

Determine the Michaelis–Menten constant, the maximum velocity of the reaction, and the maximum turnover number of the enzyme.

11.28 There are different ways to represent and analyse data for enzyme catalysed reactions. For example, in the *Eadie–Hofstee plot*, $v/[S]_0$ is plotted against v. Alternatively, in the

Hanes plot, $v/[S]_0$ is plotted against $[S]_0$. (a) Use the simple Michaelis–Menten mechanism to derive relations between $v/[S]_0$ and v and between $v/[S]_0$ and $[S]_0$. (b) Discuss how the values of K_M and v_{max} are obtained from analysis of the Eadie–Hofstee and Hanes plots. (c) Determine the Michaelis constant and the maximum velocity of the reaction from Exercise 11.27 by using Eadie–Hofstee and Hanes plots to analyse the data.

11.29 Consider the following chain mechanism:

(1) $AH \rightarrow A\cdot + H\cdot$

(2) $A\cdot \rightarrow B\cdot + C$

(3) $AH + B\cdot \rightarrow A\cdot + D$

(4) $A\cdot + B\cdot \rightarrow P$

Identify the initiation, propagation, and termination steps, and use the steady-state approximation to deduce that the decomposition of AH is first-order in AH.

11.30 Consider the following mechanism for the thermal decomposition of R_2:

(1) $R_2 \rightarrow R + R$

(2) $R + R_2 \rightarrow P_B + R'$

(3) $R' \rightarrow P_A + R$

(4) $R + R \rightarrow P_A + P_B$

where R_2, P_A, and P_B are stable hydrocarbons and R and R′ are radicals. Find the dependence of the rate of decomposition of R_2 on the concentration of R_2.

11.31 (a) Refer to Derivation 11.7 and deduce the expressions for the steady-state concentrations [Br] and [H] from the expressions for the net rates of formation of H and Br. (b) What are the orders of the reaction (with respect to each species) when the concentration of HBr is (i) very low, (ii) very high. Suggest an interpretation in each case.

Projects

The symbol ‡ indicates that calculus is required.

11.32 Prepare a report on the application of the experimental strategies described in Chapter 10 to the study of enzyme-catalysed reactions. Devote some attention to the following topics: (a) the determination of reaction rates over a long timescale, (b) the determination of the rate constants and equilibrium constant of binding of substrate to an enzyme, and (c) the characterization of intermediates in a catalytic cycle. Your report should be similar in content and extent to one of the Impacts found throughout this text.

11.33‡ Here we explore more quantitatively the kinetic analysis of a reaction approaching equilibrium. (a) Confirm (by differentiation) that the expressions in eqn 11.2 are the correct solutions of the rate laws for approach to equilibrium. (b) Find the solutions of the same rate laws that led to eqn 11.2, but for some B present initially. Go on to confirm that the solutions you find reduce to those in eqn 11.2 when $[B]_0 = 0$.

11.34‡ Complete the kinetic analysis of consecutive reactions by confirming that the three expressions in eqn 11.5 are correct solutions of the rate laws for consecutive first-order reactions.

11.35 Proteins are polymers that attain well-defined three-dimensional structures both in solution and in biological cells. They are polypeptides formed from different amino acids strung together by the peptide link, —CONH—. Hydrogen bonds between amino acids of a polypeptide give rise to stable helical or sheet structures, which may collapse into a random coil when certain conditions are changed. Consider a mechanism for the helix–coil transition of a polypeptide chain in which initiation occurs in the middle of the chain:

hhhh . . . ⇄ *hchh* . . .

hchh . . . ⇄ *cccc* . . .

where *h* and *c* denote amino acid residues belonging to a helical and coil region, respectively. Both steps can be relatively slow, so neither step may be rate-determining. (a) Set up the rate equations for this alternative mechanism. (b) Apply the steady-state approximation and show that, under these circumstances, the mechanism is equivalent to *hhhh* . . . ⇄ *cccc* . . . (c) Use your knowledge of experimental techniques and your results from the previous exercise to support or refute the following statement: It is very difficult to obtain experimental evidence for intermediates in protein folding by performing simple rate measurements and one must resort to special time-resolved or trapping techniques to detect intermediates directly.

12

Quantum theory

The emergence of quantum theory 287

12.1 Atomic and molecular spectra: discrete energies 288

12.2 The photoelectric effect: light as particles 289

12.3 Electron diffraction: electrons as waves 290

The dynamics of microscopic systems 292

12.4 The Schrödinger equation 292

12.5 The Born interpretation 293

12.6 The uncertainty principle 295

Applications of quantum mechanics 297

12.7 Translation 298

12.8 Rotational motion 303

12.9 Vibration: the harmonic oscillator 307

FURTHER INFORMATION 12.1 310

CHECKLIST OF KEY CONCEPTS 311

ROAD MAP OF KEY EQUATIONS 311

QUESTIONS AND EXERCISES 312

The phenomena of chemistry cannot be understood thoroughly without a firm understanding of the principal concepts of quantum mechanics, the most fundamental description of matter that we currently possess. The same is true of virtually all the spectroscopic techniques that are now so central to investigations of composition and structure. Present-day techniques for studying chemical reactions have progressed to the point where the information is so detailed that quantum mechanics has to be used in its interpretation. And, of course, the very currency of chemistry—the electronic structures of atoms and molecules—cannot be discussed without making use of quantum mechanical concepts.

The role—indeed, the existence—of quantum mechanics was appreciated only during the twentieth century. Until then it was thought that the motion of atomic and subatomic particles could be expressed in terms of the laws of classical mechanics introduced in the seventeenth century by Isaac Newton (see Foundations), for these laws were very successful at explaining the motion of planets and everyday objects such as pendulums and projectiles. However, towards the end of the nineteenth century experimental evidence accumulated showing that classical mechanics failed when it was applied to very small particles, such as individual atoms, nuclei, and electrons, and when the transfers of energy were very small. It took until 1926 to identify the appropriate concepts and equations for describing them.

The emergence of quantum theory

Quantum theory emerged from a series of observations made during the late nineteenth century. As far as we are concerned, there are three crucially important experiments. One shows—contrary to what had been supposed for two centuries—that energy can be transferred between systems only in discrete amounts. Another showed that electromagnetic radiation (light), which had long been considered to be a wave, in fact behaved like a stream of particles. A third showed that electrons, which since their discovery in 1897 had been supposed to be particles, in fact behaved like waves. In this section we review these three experiments and establish the properties that a valid system of mechanics must accommodate.

12.1 Atomic and molecular spectra: discrete energies

A **spectrum** is a display of the frequencies or wavelengths (which are related by $\lambda = c/\nu$) of electromagnetic radiation that are absorbed or emitted by an atom or molecule. Figure 12.1 shows a typical atomic emission spectrum and Fig. 12.2 shows a typical molecular absorption spectrum. The obvious feature of both is that *radiation is absorbed or emitted at a series of discrete frequencies*. The emission of light at

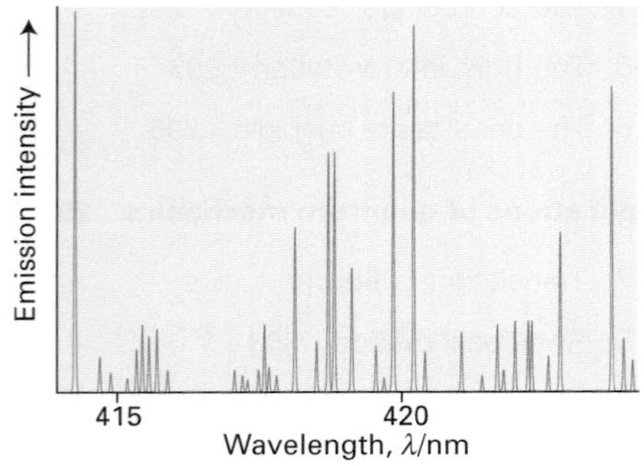

Fig. 12.1 A region of the spectrum of radiation emitted by excited iron atoms consists of radiation at a series of discrete wavelengths (or frequencies).

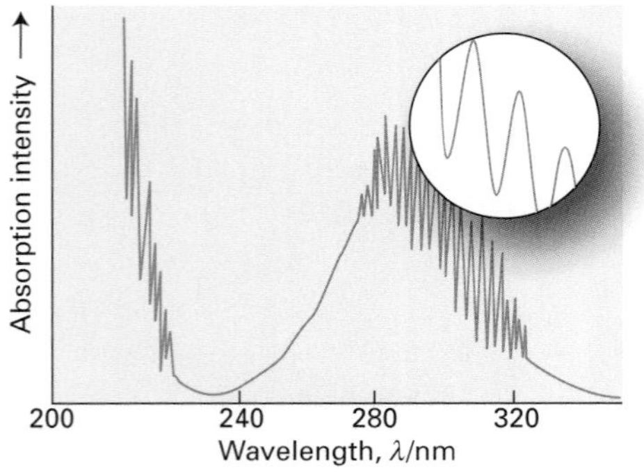

Fig. 12.2 When a molecule changes its state, it does so by absorbing radiation at definite frequencies. This spectrum is part of that due to sulfur dioxide (SO_2) molecules. This observation suggests that molecules can possess only discrete energies, not a continuously variable energy.

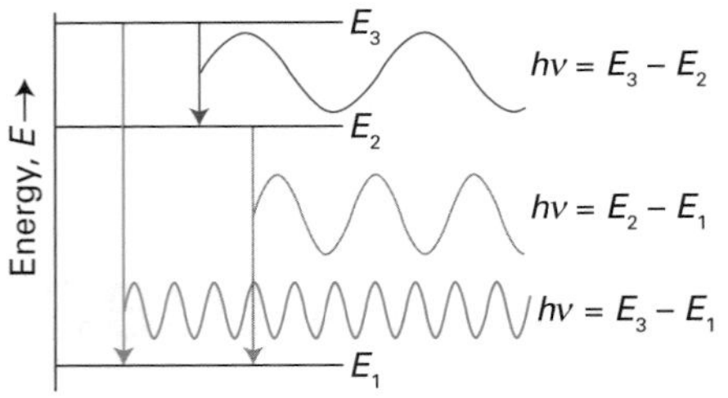

Fig. 12.3 Spectral lines can be accounted for if we assume that a molecule emits a photon as it changes between discrete energy levels. High-frequency radiation is emitted when the two states involved in the transition are widely separated in energy; low-frequency radiation is emitted when the two states are close in energy.

discrete frequencies can be understood if we suppose that:

- The energy of the atoms or molecules is confined to discrete values, for then energy can be discarded or absorbed only in packets as the atom or molecule jumps between its allowed states (Fig. 12.3).
- The frequency of the radiation is related to the energy difference between the initial and final states.

The simplest assumption is the **Bohr frequency condition**, that the frequency ν (nu) is directly proportional to the difference in energy ΔE, and that we can write

$$\Delta E = h\nu \qquad \text{Bohr frequency condition} \qquad (12.1)$$

where h is the constant of proportionality. The additional evidence that we describe below confirms this simple relation and gives the value $h = 6.626 \times 10^{-34}$ J s. This constant is now known as **Planck's constant,** for it arose in a context that had been suggested by the German physicist Max Planck (see Section 12.2).

Brief illustration 12.1 The Bohr frequency condition

The bright yellow light emitted by sodium atoms in some street lamps has wavelength 590 nm. Wavelength and frequency are related by $\nu = c/\lambda$, so the light is emitted when an atom loses an energy $\Delta E = hc/\lambda$. In this case,

$$\Delta E = \frac{\overbrace{(6.626\times10^{-34}\ \text{J s})}^{h}\times\overbrace{(2.998\times10^{8}\ \text{m s}^{-1})}^{c}}{\underbrace{5.9\times10^{-7}\ \text{m}}_{\lambda}}$$

$$= 3.4\times10^{-19}\ \text{J}$$

or 0.34 aJ. This energy difference can be expressed in a variety of ways:

- Multiplication by Avogadro's constant results in an energy separation per mole of atoms, of 200 kJ mol^{-1}, comparable to the energy of a weak chemical bond.
- A very useful conventional unit is the electronvolt (eV), with 1 eV corresponding to the kinetic energy gained by an electron when it is accelerated through a potential difference of 1 V: 1 eV = 1.602×10^{-19} J. Therefore the calculated value of ΔE corresponds to 2.1 eV. The ionization energies of atoms are typically several electronvolts.

Self-test 12.1

Neon lamps emit red radiation of wavelength 736 nm. What is the energy separation of the levels in joules, kilojoules per mole, and electronvolts responsible for the emission?

Answer: 2.70×10^{-19} J, 163 kJ mol^{-1}, 1.69 eV

At this point we can conclude that one feature of nature that any system of mechanics must accommodate is that the internal modes of atoms and molecules can possess only certain energies; that is, the energy of these modes is **quantized**.

12.2 The photoelectric effect: light as particles

By the middle of the nineteenth century, the generally acceptable view was that electromagnetic radiation is a wave (see Foundations). There was a great deal of compelling information that supported this view, specifically that light underwent **diffraction**, the interference between waves caused by an object in their path, and which results in a series of bright and dark fringes where the waves are detected.

A new view of electromagnetic radiation began to emerge in 1900 when the German physicist Max Planck discovered that the energy of an electromagnetic oscillator is limited to discrete values and cannot be varied arbitrarily. This proposal is quite contrary to the viewpoint of classical physics, in which all possible energies are allowed. In particular, Planck found that he could account for the form of the radiation emitted by a hot body only if he assumed that the permitted energies of an electromagnetic oscillator of frequency ν are integer multiples of $h\nu$:

$$E = nh\nu \quad n = 0, 1, 2, \ldots \qquad (12.2)$$

Quantization of energy in electromagnetic oscillators

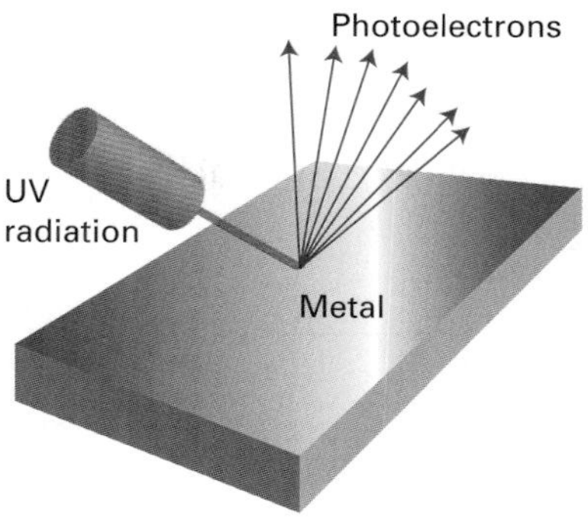

Fig. 12.4 The experimental arrangement to demonstrate the photoelectric effect. A beam of ultraviolet radiation is used to irradiate a patch of the surface of a metal, and electrons are ejected from the surface if the frequency of the radiation is above a threshold value that depends on the metal.

where h is Planck's constant. This conclusion inspired Albert Einstein to conceive of radiation as consisting of a stream of particles, each particle having an energy $h\nu$. When there is only one such particle present, the energy of the radiation is $h\nu$, when there are two particles of that frequency, their total energy is $2h\nu$, and so on. These particles of electromagnetic radiation are now called **photons.** According to the photon picture of radiation, an intense beam of monochromatic (single-frequency) radiation consists of a dense stream of identical photons; a weak beam of radiation of the same frequency consists of a relatively small number of the same type of photons.

Evidence that confirms the view that radiation can be interpreted as a stream of particles comes from the **photoelectric effect**, the ejection of electrons from metals when they are exposed to ultraviolet radiation (Fig. 12.4). The characteristics of the photoelectric effect are as follows:

1. No electrons are ejected, regardless of the intensity of the radiation, unless the frequency exceeds a threshold value characteristic of the metal.
2. The kinetic energy of the ejected electrons varies linearly with the frequency of the incident radiation but is independent of its intensity.
3. Even at low light intensities, electrons are ejected immediately if the frequency is above the threshold value.

A note on good practice We say that y *varies linearly* with x if the relation between them is $y = b + mx$; we say that y is *proportional* to x if the relation is $y = mx$.

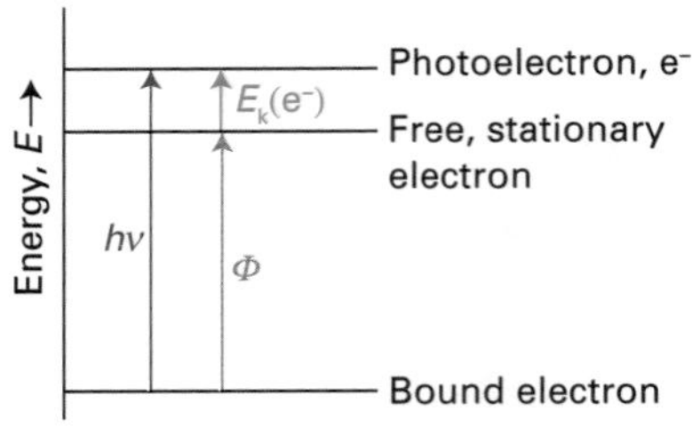

Fig. 12.5 In the photoelectric effect, an incoming photon brings a definite quantity of energy, $h\nu$. It collides with an electron close to the surface of the metal target, and transfers its energy to it. The difference between the work function, Φ, and the energy $h\nu$ appears as the kinetic energy of the ejected electron.

These observations strongly suggest an interpretation of the photoelectric effect in which an electron is ejected in a collision with a particle-like projectile, provided the projectile carries enough energy to expel the electron from the metal. If we suppose that the projectile is a photon of energy $h\nu$, where ν is the frequency of the radiation, then the conservation of energy requires that the kinetic energy, E_k, of the electron (which is equal to $\frac{1}{2}m_e v^2$, when the speed of the electron is v) should be equal to the energy supplied by the photon less the energy Φ (uppercase phi) required to remove the electron from the metal (Fig. 12.5):

$$E_k = h\nu - \Phi \qquad \text{The photoelectric effect} \qquad (12.3)$$

The quantity Φ is called the **work function** of the metal, the analogue of the ionization energy of an atom.

When $h\nu < \Phi$, photoejection (the ejection of electrons by light) cannot occur because the photon supplies insufficient energy to expel the electron: this conclusion is consistent with observation 1. Equation 12.3 predicts that the kinetic energy of an ejected electron should increase linearly with the frequency, in agreement with observation 2. When a photon collides with an electron, it gives up all its energy, so we should expect electrons to appear as soon as the collisions begin, provided the photons carry sufficient energy: this conclusion agrees with observation 3.

Thus, the photoelectric effect is strong evidence for the particle-like nature of light and the existence of photons. Moreover, it provides a route to the determination of h, for a plot of E_k against ν is a straight line of slope h.

Example 12.1

Calculating a threshold for photoejection

A photon of radiation with λ = 305 nm ejects an electron from a metal with E_k = 1.77 eV. Calculate the threshold for photoejection, the frequency (or wavelength) of the radiation able to remove the electron from the metal but not give it any excess energy.

Strategy Begin by calculating the work function of the metal by using $\nu = c/\lambda$ and eqn 12.3, rearranged as $\Phi = h\nu - E_k$. The threshold for photoejection corresponds to radiation of frequency $\nu_{min} = \Phi/h$ and wavelength $\lambda_{max} = c/\nu_{min}$.

Solution From the expression for the work function $\Phi = h\nu - E_k$ the minimum frequency for photoejection is

$$\nu_{min} = \frac{\Phi}{h} = \frac{h\nu - E_k}{h} \overset{\nu=c/\lambda}{=} \frac{c}{\lambda} - \frac{E_k}{h}$$

The wavelength of the incident photon is $\lambda = 305\text{ nm} = 3.05 \times 10^{-7}$ m and kinetic energy of the electron is

$$E_k = 1.77\text{ eV} \times (1.602 \times 10^{-19}\text{ J eV}^{-1}) = 2.83 \times 10^{-19}\text{ J}$$

so it follows that

$$\nu_{min} = \frac{\overbrace{2.998\times10^{8}\text{ m s}^{-1}}^{c}}{\underbrace{3.05\times10^{-7}\text{ m}}_{\lambda}} - \frac{\overbrace{2.83\times10^{-19}\text{ J}}^{E_k}}{\underbrace{6.626\times10^{-34}\text{ J s}}_{h}}$$

$$= 5.56 \times 10^{14}\text{ s}^{-1}$$

The maximum wavelength is therefore

$$\lambda_{max} = \frac{\overbrace{2.998\times10^{8}\text{ m s}^{-1}}^{c}}{\underbrace{5.56\times10^{14}\text{ s}^{-1}}_{\nu_{min}}} = 5.40\times10^{-7}\text{ m}$$

or 540 nm.

Self-test 12.2

When ultraviolet radiation of wavelength 165 nm strikes a certain metal surface, electrons are ejected with a speed of 1.24 Mm s^{-1} (1 Mm = 10^6 m). Calculate the speed of electrons ejected by radiation of wavelength 265 nm.

Answer: 735 km s^{-1}

12.3 Electron diffraction: electrons as waves

The photoelectric effect shows that light has certain properties of particles. Although contrary to the long-established wave theory of light, a similar view had been held before, but discarded. No significant scientist, however, had taken the view that matter is

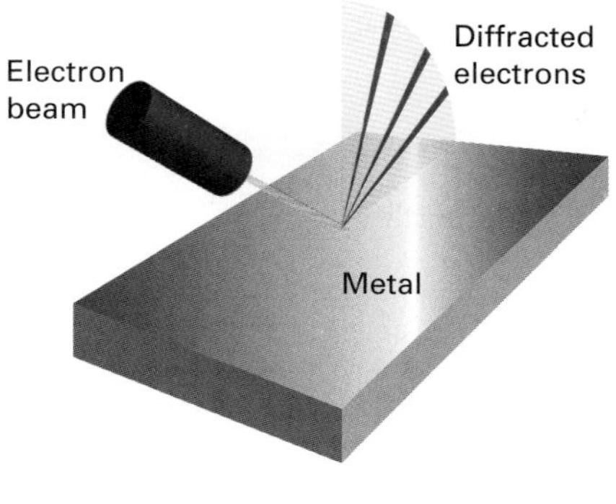

Fig. 12.6 In the Davisson–Germer experiment, a beam of electrons was directed on a single crystal of nickel, and the scattered electrons showed a variation in intensity with angle that corresponded to the pattern that would be expected if the electrons had a wave character and were diffracted by the layers of atoms in the solid.

wavelike. Nevertheless, experiments carried out in the early 1920s forced people to question even that conclusion. The crucial experiment was performed by the American physicists Clinton Davisson and Lester Germer, who observed the diffraction of electrons by a crystal (Fig. 12.6).

There was an understandable confusion—which continues to this day—about how to combine both aspects of matter into a single description. Some progress was made by Louis de Broglie when, in 1924, he suggested that any particle travelling with a linear momentum, $p = mv$, should have (in some sense) a wavelength λ given by what we now call the **de Broglie relation:**

$$\lambda = \frac{h}{p} \qquad \text{de Broglie relation} \qquad (12.4)$$

The wave corresponding to this wavelength, what de Broglie called a 'matter wave', has the mathematical form $\sin(2\pi x/\lambda)$. The de Broglie relation implies that the wavelength of a 'matter wave' should decrease as the particle's speed increases (Fig. 12.7). Equation 12.4 was confirmed by the Davisson–Germer experiment, for the wavelength it predicts for the electrons they used in their experiment agrees with the details of the diffraction pattern they observed.

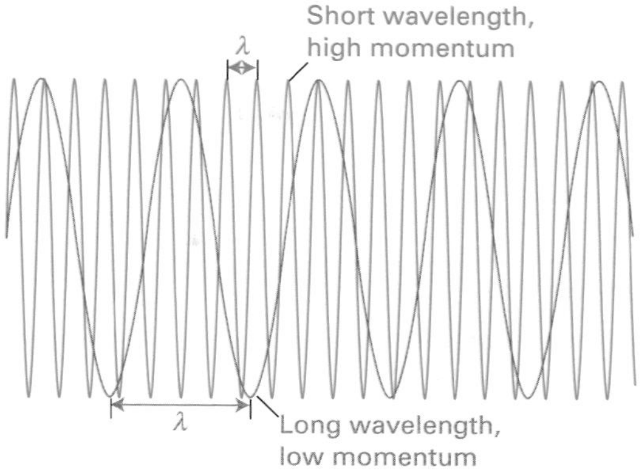

Fig. 12.7 According to the de Broglie relation, a particle with low momentum has a long wavelength whereas a particle with high momentum has a short wavelength. A high momentum can result either from a high mass or from a high velocity (because $p = mv$). Macroscopic objects have such large masses that, even if they are travelling very slowly, their wavelengths are undetectably short.

Example 12.2

Estimating the de Broglie wavelength

Estimate the wavelength of electrons that have been accelerated from rest through a potential difference of 1.00 kV.

Strategy We need to establish a string of relations: from the potential difference we can deduce the kinetic energy acquired by the accelerated electron; then we need to find the electron's linear momentum from its kinetic energy; finally, we use that linear momentum in the de Broglie relation to calculate the wavelength.

Solution The kinetic energy acquired by an electron of charge $-e$ accelerated from rest by falling through a potential difference $\Delta\phi$ is

$$E_k = e\Delta\phi$$

Because $E_k = \frac{1}{2}m_e v^2$ and $p = m_e v$ the linear momentum is related to the kinetic energy by $p = (2m_e E_k)^{1/2}$ and therefore

$$p = (2m_e e\Delta\phi)^{1/2}$$

This is the expression we use in the de Broglie relation, which becomes

$$\lambda = \frac{h}{(2m_e e\Delta\phi)^{1/2}}$$

At this stage, all we need do is to substitute the data and use the relations 1 C V = 1 J and 1 J = 1 kg m^2 s^{-2}:

$$\lambda = \frac{\overbrace{6.626\times10^{-34}\,\text{J s}}^{h}}{\left\{2\times\underbrace{(9.109\times10^{-31}\,\text{kg})}_{m_e}\times\underbrace{(1.602\times10^{-19}\,\text{C})}_{e}\times\underbrace{(1.00\times10^{3}\,\text{V})}_{\Delta\phi}\right\}^{1/2}}$$

$$= \frac{6.626\times10^{-34}}{\{2\times(9.109\times10^{-31})\times(1.602\times10^{-19})\times(1.00\times10^{3})\}^{1/2}}$$

$$\times\frac{\overbrace{\text{J s}}^{\text{kg m}^2\,\text{s}^{-1}}}{\underbrace{(\text{kg C V})^{1/2}}_{\text{kg m s}^{-1}}} = 3.88\times10^{-11}\text{ m}$$

The wavelength of 38.8 pm is comparable to typical bond lengths in molecules (about 100 pm). Electrons accelerated in this way are used in the technique of **electron diffraction**, in which the diffraction pattern generated by interference when a beam of electrons passes through a sample is interpreted in terms of the locations of the atoms.

Self-test 12.3

Calculate the wavelength of an electron in a 10 MeV particle accelerator (1 MeV = 10^6 eV).

Answer: 0.39 pm

The Davisson–Germer experiment, which has since been repeated with other particles (including molecular hydrogen and C_{60}), shows clearly that 'particles' have wavelike properties. We have also seen that 'waves' have particle-like properties. Thus we are brought to the heart of modern physics. When examined on an atomic scale, the concepts of particle and wave melt together, particles taking on the characteristics of waves, and waves the characteristics of particles. This joint wave–particle character of matter and radiation is called **wave–particle duality.** It will be central to all that follows.

The dynamics of microscopic systems

How can we accommodate the fact that atoms and molecules exist with only certain energies, waves exhibit the properties of particles, and particles exhibit the properties of waves?

We shall take the de Broglie relation as our starting point, and abandon the classical concept of particles moving along 'trajectories', precise paths at definite speeds. From now on, we adopt the quantum mechanical view that *a particle is spread through space like a wave.* To describe this distribution, we introduce the concept of a **wavefunction,** ψ (psi), in place of the precise path, and then set up a scheme for calculating and interpreting ψ. A 'wavefunction' is the modern term for de Broglie's 'matter wave'. To a very crude first approximation, we can visualize a wavefunction as a blurred version of a path (Fig. 12.8); however, we refine this picture considerably in the following sections. The formal definition of a wavefunction is that it is *a mathematical function that contains all the dynamical information about the state of a system.*

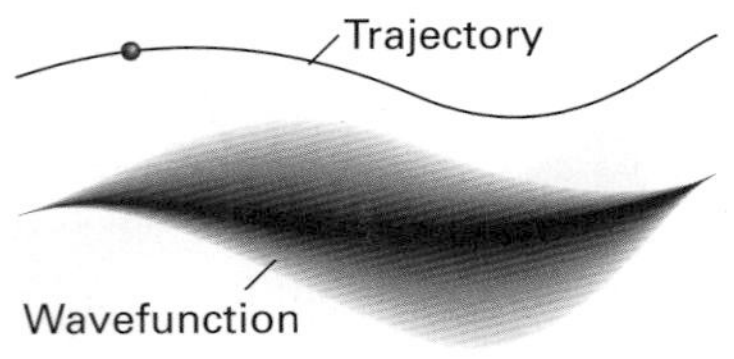

Fig. 12.8 According to classical mechanics, a particle may have a well-defined trajectory, with a precisely specified position and momentum at each instant (as represented by the precise path in the diagram). According to quantum mechanics, a particle cannot have a precise trajectory; instead, there is only a probability that it may be found at a specific location at any instant. The wavefunction that determines its probability distribution is a kind of blurred version of the trajectory. Here, the wavefunction is represented by areas of shading: the darker the area, the greater the probability of finding the particle there.

12.4 The Schrödinger equation

In 1926, the Austrian physicist Erwin Schrödinger proposed an equation for calculating wavefunctions. The **Schrödinger equation,** specifically the *time-independent* Schrödinger equation, for a single particle of mass m moving with energy E in one dimension is

$$-\frac{\hbar^2}{2m}\frac{d^2\psi}{dx^2} + V(x)\psi = E\psi \qquad \text{The Schrödinger equation} \qquad (12.5a)$$

The Shrödinger equation is a differential equation, and in this case a second-order ordinary differential equation (see The chemist's toolkit 10.1). In eqn 12.5a $V(x)$ is the potential energy; $\hbar$ (which is read h-bar) is a convenient modification of Planck's constant:

$$\hbar = \frac{h}{2\pi} = 1.055 \times 10^{-34}\ \text{J s}$$

The term proportional to $d^2\psi/dx^2$ is closely related to the kinetic energy (so that its sum with V is the total energy, E). Mathematically, it can be interpreted as the way of expressing the curvature of the wavefunction at each point. Thus, if the wavefunction is sharply curved, then $d^2\psi/dx^2$ is large; if it is only slightly curved, then $d^2\psi/dx^2$ is small. We shall develop this interpretation later: just keep it in mind for now.

You will often see eqn 12.5 written in the very compact form

$$\hat{H}\psi = E\psi \qquad \text{Alternative form} \qquad \text{The Schrödinger equation} \qquad (12.5b)$$

where '$\hat{H}\psi$' stands for everything on the left of eqn 12.5a. The quantity $\hat{H}$ is called the **hamiltonian** of the system after the mathematician William Hamilton who had formulated a version of classical mechanics that used the concept. It is written with a ^ to signify that it is an 'operator', a series of mathematical operations that acts in a particular way on ψ rather than just multiplying it (as E multiplies ψ in $E\psi$); see the following Derivation. You should be aware that a lot of quantum mechanics is formulated in terms of various operators but we shall encounter them only very rarely in this text.[1]

For a justification of the form of the Schrödinger equation, see the following Derivation. The fact that the Schrödinger equation is a differential equation should not cause too much consternation for in most cases we shall simply quote the solutions and not go into the details of how they are found. The rare cases where we need to see the explicit forms of its solution will involve very simple functions.

Brief illustration 12.2 Simple wavefunctions

Three simple but important cases, but not putting in various constants in order to emphasize the form of the functions and to show that they are familiar, are as follows:

- The wavefunction for a freely moving particle is sin x, exactly as for de Broglie's matter wave.
- The wavefunction for a particle free to oscillate to-and-fro near a point is e^{-x^2}, where x is the displacement from the point.
- The wavefunction for an electron in the lowest energy state of a hydrogen atom is e^{-r}, where r is the distance from the nucleus.

Derivation 12.1

A justification of the Schrödinger equation

The form of the Schrödinger equation can be justified to a certain extent by showing that it implies the de Broglie relation for a freely moving particle. By free motion we mean motion in a region where the potential energy is zero ($V = 0$ everywhere). Then, eqn 12.5a simplifies to

$$-\frac{\hbar^2}{2m}\frac{d^2\psi}{dx^2} = E\psi \quad \text{Freely moving particle} \quad \text{The Schrödinger equation} \qquad (12.6)$$

A solution of this equation is

$$\psi = \sin(kx) \qquad k = \frac{(2mE)^{1/2}}{\hbar}$$

as you should verify by substitution of the solution into both sides of the equation.

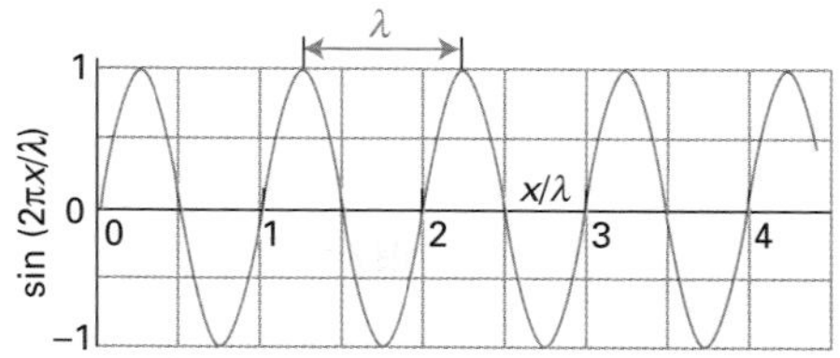

Fig. 12.9 The wavelength of a harmonic wave of the form $\sin(2\pi x/\lambda)$. The amplitude of the wave is the maximum height above the centre line.

The function $\sin(kx)$ is a wave of wavelength $\lambda = 2\pi/k$, as we can see by comparing $\sin(kx)$ with $\sin(2\pi x/\lambda)$, the standard form of a harmonic wave with wavelength λ (Fig. 12.9). Next, we note that the energy of the particle is entirely kinetic (because $V = 0$ everywhere), so the total energy of the particle is just its kinetic energy:

$$E = E_k = \frac{p^2}{2m}$$

Because E is related to k by $E = k^2\hbar^2/2m$, it follows from a comparison of the two equations that $p = k\hbar$. Therefore, the linear momentum is related to the wavelength of the wavefunction by

$$p = \overbrace{\frac{2\pi}{\lambda}}^{k} \times \overbrace{\frac{h}{2\pi}}^{\hbar} = \frac{h}{\lambda}$$

which is the de Broglie relation. We see, in the case of a freely moving particle, that the Schrödinger equation has led to an experimentally verified conclusion.

12.5 The Born interpretation

Before going any further, it will be helpful to understand the physical significance of a wavefunction. The interpretation that is widely used is based on a suggestion made by the German physicist Max Born. He made use of an analogy with the wave theory of light, in which the square of the amplitude of an electromagnetic wave is interpreted as its intensity and therefore (in quantum terms) as the number of photons present. He argued that, by analogy, the square of a wavefunction gives an indication of the probability of finding a particle in a particular region of space. To be precise, the **Born interpretation** asserts that:

> The probability of finding a particle in a small region of space of volume δV is proportional to $\psi^2\delta V$, where ψ is the value of the wavefunction in the region.

In other words, ψ^2 is a **probability density**. As for other kinds of density, such as mass density (ordinary 'density'), we get the probability itself by multiplying

[1] See, for instance, our *Physical Chemistry* (2010).

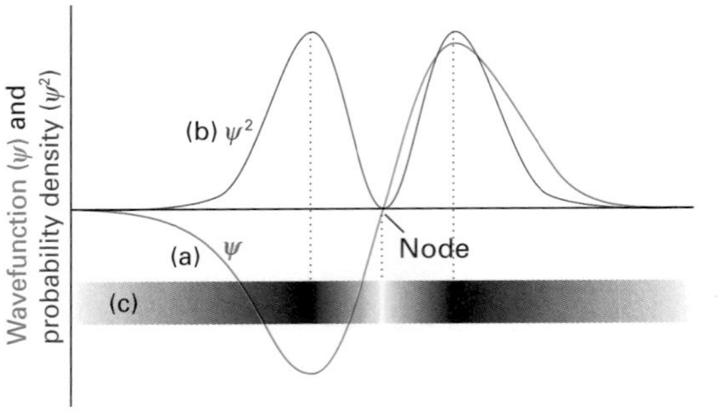

Fig. 12.10 (a) A wavefunction does not have a direct physical interpretation. However, (b) its square (its square modulus if it is complex) tells us the probability of finding a particle at each point. The probability density implied by the wavefunction shown here is depicted by the density of shading in (c).

the probability density ψ^2 by the volume δV of the region of interest.[2]

A note on good practice The symbol δ is used to indicate a small change in a parameter, as in x changing to $x + \delta x$. The symbol Δ is used to indicate a finite (measurable) difference between two quantities, as in $\Delta X = X_{\text{final}} - X_{\text{initial}}$. The sequence of symbls is Δ for measurable, δ for small, and d for infinitesimal.

For a small 'inspection volume' δV of given size, the Born interpretation implies that wherever ψ^2 is large, there is a high probability of finding the particle. Wherever ψ^2 is small, there is only a small chance of finding the particle. The density of shading in Fig. 12.10 represents this **probabilistic interpretation**, an interpretation that accepts that we can make predictions only about the probability of finding a particle somewhere. This interpretation is in contrast to classical physics, which claims to be able to predict precisely that a particle will be at a given point on its path at a given instant.

Example 12.3

Interpreting a wavefunction

The wavefunction of an electron in the lowest energy state of a hydrogen atom is proportional to e^{-r/a_0}, with $a_0 = 52.9$ pm and r the distance from the nucleus (Fig. 12.11). Calculate the relative probabilities of finding the electron inside a small cubic volume located at (a) the nucleus, (b) a distance a_0 from the nucleus.

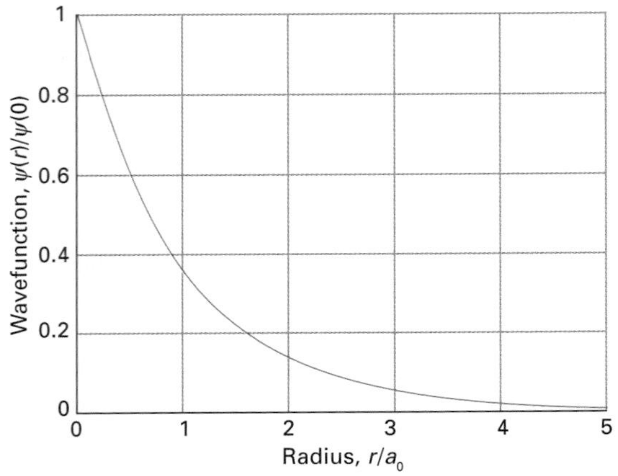

Fig. 12.11 The wavefunction for an electron in the ground state of a hydrogen atom is an exponentially decaying function of the form e^{-r/a_0}, where a_0 is the Bohr radius.

Strategy The probability is proportional to $\psi^2\delta V$ evaluated at the specified location. The volume of interest is so small (even on the scale of the atom) that we can ignore the variation of ψ within it and write

$$\text{Probability} \propto \psi^2\delta V$$

with ψ evaluated at the point in question.

Solution (a) At the nucleus, $r = 0$, so there $\psi^2 \propto 1.0$ (because $e^0 = 1$) and the probability is proportional to $1.0 \times \delta V$. (b) At a distance $r = a_0$ in an arbitrary direction, $\psi^2 \propto e^{-2} \times \delta V = 0.14 \times \delta V$. Therefore, the ratio of probabilities is 1.0/0.14 = 7.1. It is more probable (by a factor of 7.1) that the electron will be found at the nucleus than in the same tiny volume located at a distance a_0 from the nucleus.

Self-test 12.4

The wavefunction for the lowest energy state in the ion He^+ is proportional to e^{-2r/a_0}. Repeat the calculation for this ion. Any comment?

Answer: 55; a more compact wavefunction on account of the higher nuclear charge

There is more information embedded in ψ than the probability that a particle will be found at a location. We saw a hint of that in the discussion of eqn 12.5 when we identified the first term as an indication of the relation between the kinetic energy of the particle and the curvature of the wavefunction: if the wavefunction is sharply curved, then the particle it describes has a high kinetic energy; if the wavefunction has only a low curvature, then the particle has only a low kinetic energy. This interpretation is consistent with the de Broglie relation, for a short

[2] We are supposing throughout that ψ is a real function (that is, one that does not depend on i, the square root of −1). In general, ψ is complex (has both real and imaginary components); in such cases ψ^2 is replaced by $\psi^*\psi$, where ψ^* is the complex conjugate of ψ. We do not consider complex functions in this book. For the role, properties, and interpretation of complex wavefunctions, see our *Physical Chemistry* (2010).

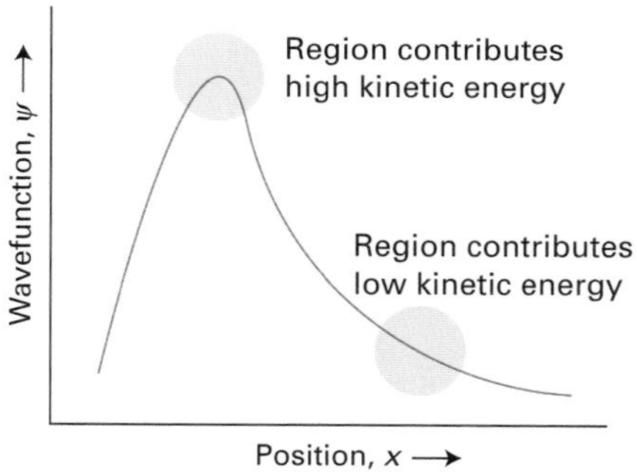

Fig. 12.12 The observed kinetic energy of a particle is the average of contributions from the entire space covered by the wavefunction. Sharply curved regions contribute a high kinetic energy to the average; slightly curved regions contribute only a small kinetic energy.

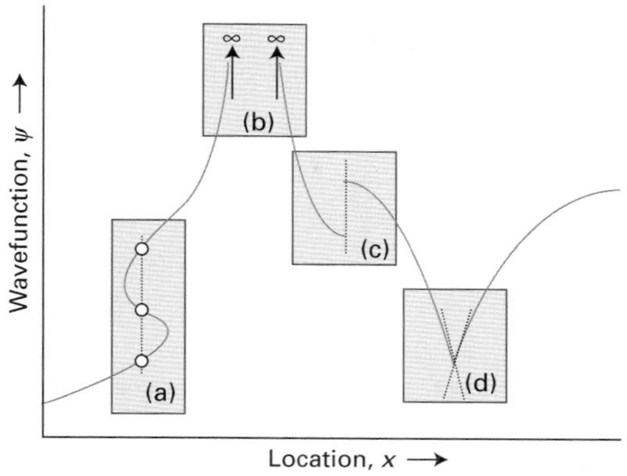

Fig. 12.13 This wavefunction is unacceptable because (a) it is not single-valued, (b) it is infinite over a finite range, (c) it is not continuous, (d) its slope is not continuous.

wavelength corresponds to both a sharply curved wavefunction and a high linear momentum and therefore a high kinetic energy (Fig. 12.12). For more complicated wavefunctions, the curvature changes from point to point, and the total contribution to the kinetic energy is an average over the entire region of space.

The central point to remember is that *the wavefunction contains all the dynamical information about the particle it describes*. By 'dynamical' we mean all aspects of the particle's motion. Its amplitude at any point tells us the probability density of the particle at that point and other details of its shape tells us all that it is possible to know about other aspects of its motion, such as its momentum and its kinetic energy.

The Born interpretation has a further important implication: it helps us identify the conditions that a wavefunction must satisfy for it to be acceptable:

- It must be single valued (that is, have only a single value at each point): there cannot be more than one probability density at each point.
- It cannot become infinite over a finite region of space: the total probability of finding a particle in a region cannot exceed 1.

These conditions turn out to be satisfied if the wavefunction takes on particular values at various points, such as at a nucleus, at the edge of a region, or at infinity. That is, the wavefunction must satisfy certain **boundary conditions**, values that the wavefunction must adopt at certain positions. We shall see plenty of examples later. Two further conditions stem from the Schrödinger equation itself, which could not be written unless:

- The wavefunction is continuous everywhere.
- It has a continuous slope everywhere.

These last two conditions mean that the 'curvature' term, the first term in eqn 12.5, is well defined everywhere. All four conditions are summarized in Fig. 12.13.

These requirements have a profound implication. One feature of the solution of any given Schrödinger equation, a feature common to all differential equations, is that an infinite number of possible solutions are allowed mathematically. For instance, if $\sin x$ is a solution of the equation, then so too is $a \sin(bx)$, where a and b are arbitrary constants, with each solution corresponding to a particular value of E. However, it turns out that only some of these solutions fulfill the requirements stated above. Suddenly, we are at the heart of quantum mechanics: *the fact that only some solutions are acceptable, together with the fact that each solution corresponds to a characteristic value of E, implies that only certain values of the energy are acceptable*. That is, *when the Schrödinger equation is solved subject to the boundary conditions that the solutions must satisfy, we find that the energy of the system is quantized* (Fig. 12.14).

12.6 The uncertainty principle

We have seen that, according to the de Broglie relation, a wave of constant wavelength, the wavefunction $\sin(2\pi x/\lambda)$, corresponds to a particle with a definite linear momentum $p = h/\lambda$. However, a wave does not have a definite location at a single point in space, so we cannot speak of the precise position of the particle if it has a definite momentum. Indeed,

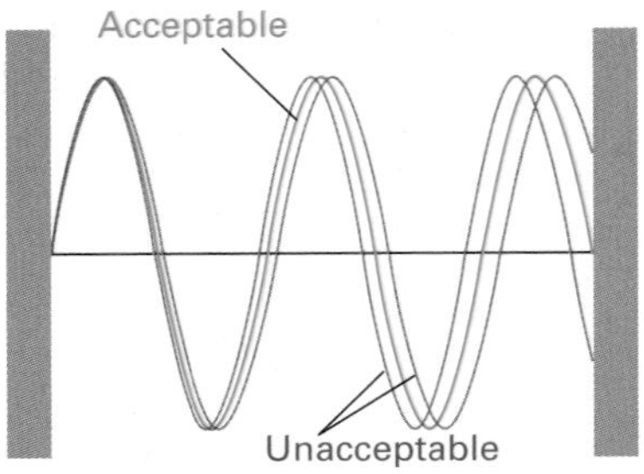

Fig. 12.14 Although an infinite number of solutions of the Schrödinger equation exist, not all of them are physically acceptable. Acceptable wavefunctions have to satisfy certain boundary conditions, which vary from system to system. In the example shown here, where the particle is confined between two impenetrable walls, the only acceptable wavefunctions are those that fit between the walls (like the vibrations of a stretched string). Because each wavefunction corresponds to a characteristic energy, and the boundary conditions rule out many solutions, only certain energies are permissible.

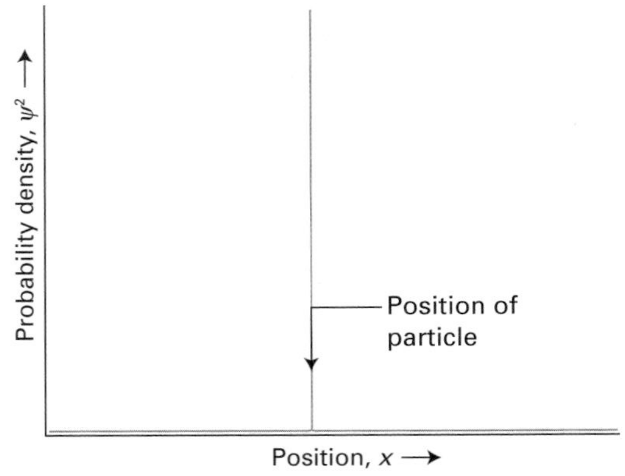

Fig. 12.15 The wavefunction for a particle with a well-defined position is a sharply spiked function that has zero amplitude everywhere except at the particle's position.

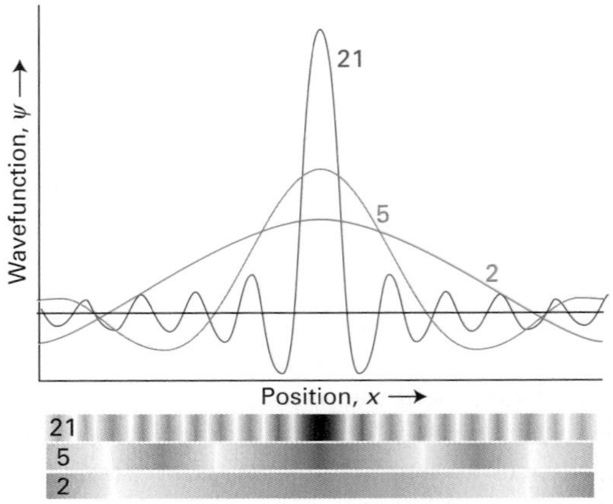

Fig. 12.16 The wavefunction for a particle with an ill-defined location can be regarded as the sum (superposition) of several wavefunctions of different wavelengths that interfere constructively in one place but destructively elsewhere. As more waves are used in the superposition, the location becomes more precise at the expense of uncertainty in the particle's momentum. An infinite number of waves are needed to construct the wavefunction of a perfectly localized particle. The numbers against each curve are the number of sine waves used in the superpositions.

because a sine wave spreads throughout the whole of space we cannot say anything about the location of the particle: because the wave spreads everywhere, the particle may be found anywhere in the whole of space. This statement is one half of the **uncertainty principle** proposed by Werner Heisenberg in 1927, in one of the most celebrated results of quantum mechanics:

> It is impossible to specify simultaneously, with arbitrary precision, both the momentum and the position of a particle.

More precisely, this is the *position–momentum uncertainty principle*.

Before discussing the principle further, we must establish the other half: that if we know the position of a particle exactly, then we can say nothing about its momentum. If the particle is at a definite location, then its wavefunction must be nonzero there and zero everywhere else (Fig. 12.15). We can simulate such a wavefunction by forming a **superposition** of many wavefunctions; that is, by adding together the amplitudes of a large number of sine functions (Fig. 12.16). This procedure is successful because the amplitudes of the waves add together at one location to give a nonzero total amplitude, but cancel everywhere else. In other words, we can create a sharply localized wavefunction by adding together wavefunctions corresponding to many different wavelengths, and therefore, by the de Broglie relation, of many different linear momenta.

The superposition of a few sine functions gives a broad, ill-defined wavefunction. As the number of functions increases, the wavefunction becomes sharper because of the more complete interference between the positive and negative regions of the components. When an infinite number of components are used, the wavefunction is a sharp, infinitely narrow spike like that in Fig. 12.15, which corresponds to perfect localization of the particle. Now the particle is perfectly localized, but at the expense of discarding all information about its momentum.

The quantitative version of the **position–momentum uncertainty relation** is

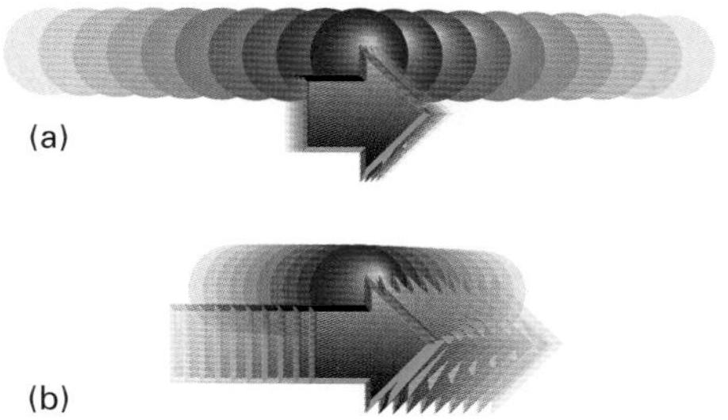

Fig. 12.17 A representation of the content of the uncertainty principle. The range of locations of a particle is shown by the spheres, and the range of momenta by the arrows. In (a), the position is quite uncertain, and the range of momenta is small. In (b), the location is much better defined, and now the momentum of the particle is quite uncertain.

$$\Delta p \Delta x \geq \tfrac{1}{2}\hbar \qquad \text{Position–momentum uncertainty relation} \qquad (12.7)$$

The quantity Δp is the 'uncertainty' in the linear momentum and Δx is the uncertainty in position (which is proportional to the width of the peak in Fig. 12.16). Equation 12.7 expresses quantitatively the fact that the more closely the location of a particle is specified (the smaller the value of Δx), then the greater the uncertainty in its momentum (the larger the value of Δp) parallel to that coordinate, and vice versa (Fig. 12.17). When $\Delta x = 0$, which is the case when the location of the particle is known exactly, Δp must be infinite, implying complete ignorance about its linear momentum. Likewise, if $\Delta p = 0$ (certainty about the linear momentum), $\Delta x = \infty$ (total ignorance about where the particle will be found). The position–momentum uncertainty principle applies to location and momentum *along the same axis*. It does not limit our ability to specify location on one axis and momentum along a perpendicular axis.

Brief illustration 12.3 The uncertainty principle

Suppose that the speed of a certain projectile of mass 1.0 g is known to within $\Delta v = 1.0\ \mu\text{m s}^{-1}$. From $\Delta p\Delta x \geq \tfrac{1}{2}\hbar$, the uncertainty in its position along its line of flight is

$$\Delta x \geq \frac{\hbar}{2\Delta p} = \frac{\hbar}{2m\Delta v}$$

$$= \frac{\overbrace{1.055\times10^{-34}\ \text{J s}}^{h}}{\underbrace{2\times\underbrace{(1.0\times10^{-3}\,\text{kg})}_{m}\times\underbrace{(1.0\times10^{-6}\,\text{m s}^{-1})}_{\Delta v}}_{\Delta p}}$$

$$\overset{\text{Use J = kg m}^2\,\text{s}^{-2}}{=} \quad 5.3\times10^{-26}\ \text{m}$$

This degree of uncertainty is completely negligible for all practical purposes, which is why the need for quantum mechanics was not recognized for over 200 years after Newton had proposed his system of mechanics and why in daily life we are completely unaware of the restrictions it implies. However, when the mass is that of an electron, the same uncertainty in speed implies an uncertainty in position far larger than the diameter of an atom, so the concept of a trajectory—the simultaneous possession of a precise position and momentum—is untenable.

Self-test 12.5

Estimate the minimum uncertainty in the speed of an electron in a hydrogen atom (taking its diameter as 100 pm).

Answer: 580 km s^{-1}

The uncertainty principle captures one of the principal differences between classical and quantum mechanics. Classical mechanics supposed, falsely as we now know, that the position and momentum of a particle can be specified simultaneously with arbitrary precision. However, quantum mechanics shows that position and momentum are **complementary**, that is, not simultaneously specifiable. Quantum mechanics requires us to make a choice: we can specify position at the expense of momentum, or momentum at the expense of position.[3]

The uncertainty principle has profound implications for the description of electrons in atoms and molecules and therefore for chemistry as a whole. When the nuclear model of the atom was first proposed it was supposed that the motion of an electron around the nucleus could be described by classical mechanics and that it would move in some kind of orbit. But to specify an orbit, we need to specify the position and momentum of the electron at each point of its path. The possibility of doing so is ruled out by the uncertainty principle. The properties of electrons in atoms, and therefore the foundations of chemistry, have had to be formulated (as we shall see) in a completely different way.

Applications of quantum mechanics

To prepare for applying quantum mechanics to chemistry we need to understand three basic types of

[3] See our *Quanta, Matter, and Change* (2009) for additional examples of complementarity.

motion: translation (motion through space), rotation, and vibration. It turns out that the wavefunctions for free translational and rotational motion in a plane can be constructed directly from the de Broglie relation, without solving the Schrödinger equation itself, and we shall take that simple route. That is not possible for rotation in three dimensions and vibrational motion where the motion is more complicated, so there we shall have to use the Schrödinger equation to find the wavefunctions.

12.7 Translation

The simplest type of motion is translation in one dimension. We have already seen how classical mechanics handles translational motion in one dimension (Foundations 0.4–0.6). The crucial difference between that treatment and quantum mechanics is that when applying the theory to the motion of a particle confined between two infinitely high walls, the appropriate boundary conditions imply that only certain wavefunctions and their corresponding energies are acceptable. That is, the translational energy is quantized. When the walls are of finite height, the solutions of the Schrödinger equation reveal surprising features of a particle, especially its ability to penetrate into and through regions where classical physics would forbid it to be found.

(a) Motion in one dimension

First, we consider the translational motion of a 'particle in a box', a particle of mass m that can travel in a straight line in one dimension (along the x-axis) but is confined between two walls separated by a distance L. The potential energy, V, of the particle is zero inside the box but rises abruptly to infinity at the walls (Fig. 12.18). The particle might be a bead free to slide along a horizontal wire between two stops. Although this problem is very elementary, there has been a resurgence of research interest in it now that nanometre-scale structures are used to trap electrons in cavities resembling square wells.

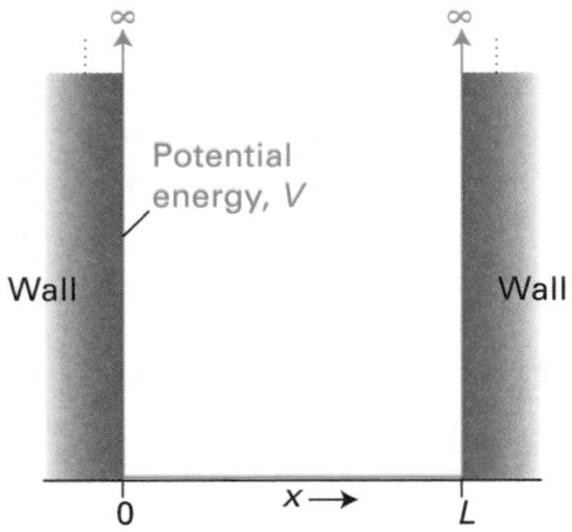

Fig. 12.18 A particle in a one-dimensional region with impenetrable walls at either end. Its potential energy is zero between $x = 0$ and $x = L$ and rises abruptly to infinity as soon as the particle touches either wall.

The boundary conditions for this system are the requirement that each acceptable wavefunction of the particle must fit inside the box exactly, like the vibrations of a violin string (as in Fig. 12.14). More precisely, the boundary conditions stem from the requirement that the wavefunction is continuous everywhere: because the wavefunction is zero outside the box, it must therefore be zero at $x = 0$ and at $x = L$.

It follows that the wavelength, λ, of the permitted wavefunctions must be one of the values

$$\lambda = 2L, L, \tfrac{2}{3}L, \ldots \quad \text{or} \quad \lambda = \frac{2L}{n}, \text{ with } n = 1, 2, 3, \ldots$$

The value $n = 0$ is ruled out because it would result in a line of constant, zero amplitude. Each wavefunction is a sine wave with one of these wavelengths; therefore, because a sine wave of wavelength λ has the form $\sin(2\pi x/\lambda)$, the permitted wavefunctions are

$$\psi_n = N \sin\frac{n\pi x}{L} \qquad n = 1, 2, \ldots \qquad \text{Wavefunctions} \quad \text{Particle in a one-dimensional box} \qquad (12.8)$$

The constant N is called the **normalization constant**. It is chosen so that the total probability of finding the particle inside the box is 1, and as we show in the following Derivation, has the value $N = (2/L)^{1/2}$. The integer n is an example of a **quantum number**, an integer (in some cases a half-integer) that specifies the state of the system.

Derivation 12.2

The normalization constant

According to the Born interpretation, the probability of finding a particle in the infinitesimal region of length dx at the point x given that its normalized wavefunction has the value ψ at that point, is equal to $\psi^2 dx$. Therefore, the total probability of finding the particle between $x = 0$ and $x = L$ is the sum (integral) of all the probabilities of its being in each infinitesimal region. That total probability is 1 (the particle is certainly in the range somewhere), so we know that

$$\underbrace{\int_0^L \psi^2 dx}_{\text{Total probability of finding the particle between 0 and L}} = 1$$

Substitution of the form of the wavefunction, $\psi_n = N\sin(n\pi x/L)$, turns this expression into

$$N^2\int_0^L \left(\sin\frac{n\pi x}{L}\right)^2 dx = 1$$

Our task is to find N. To do so, we use the standard integral

$$\int (\sin ax)^2 dx = \tfrac{1}{2}x - \frac{\sin 2ax}{4a} + \text{constant}$$

and write, after recognizing that $a = n\pi/L$,

$$\int_0^L \left(\sin\frac{n\pi x}{L}\right)^2 dx = \left(\tfrac{1}{2}L - \frac{\overbrace{\sin 2n\pi}^{0}}{4n\pi/L} + \text{constant}\right) - \left(\tfrac{1}{2}\times 0 - \frac{\overbrace{\sin 0}^{0}}{4n\pi/L} + \text{constant}\right) = \tfrac{1}{2}L$$

Therefore,

$$N^2 \times \tfrac{1}{2}L = 1$$

and hence $N = (2/L)^{1/2}$. Note that, in this case but not in general, the same normalization factor applies to all the wavefunctions regardless of the value of n.

It is now a simple matter to find the permitted energy levels because the only contribution to the energy is the kinetic energy of the particle: the potential energy is zero everywhere inside the box, and the particle is never outside the box. First, we note that it follows from the de Broglie relation that the only acceptable values of the linear momentum are

$$p = \frac{h}{\lambda} \overset{\lambda=2L/n}{=} \frac{nh}{2L} \qquad n = 1, 2, \ldots$$

Then, because the kinetic energy of a particle of momentum p and mass m is $E_k = p^2/2m$ and the potential energy is zero, it follows that the permitted energies of the particle are

$$E_n = \frac{n^2h^2}{8mL^2} \qquad n = 1, 2, \ldots \qquad \text{Quantized energies} \quad \text{Particle in a one-dimensional box} \qquad (12.9)$$

As we see in eqns 12.8 and 12.9, the wavefunctions and energies of a particle in a box are labelled with the quantum number n. As well as acting as a label, a quantum number specifies certain physical properties of the system: in the present example, n specifies the energy of the particle through eqn 12.9.

The permitted energies of the particle are shown in Fig. 12.19 together with the shapes of the wavefunctions for $n = 1$ to 6. All the wavefunctions except the one of lowest energy ($n = 1$) possess points called **nodes** where the function passes through zero. Passing *through* zero is an essential part of the definition: just becoming zero is not sufficient. The points at the edges of the box where $\psi = 0$ are not nodes, because the wavefunction does not pass through zero there. The number of nodes in the wavefunctions shown in the illustration increases from 0 (for $n = 1$) to 5 (for $n = 6$), and is $n - 1$ for a particle in a box in general. It is a general feature of quantum mechanics that the wavefunction corresponding to the state of lowest energy has no nodes, and as the number of nodes in the wavefunctions increase, the energy increases too.

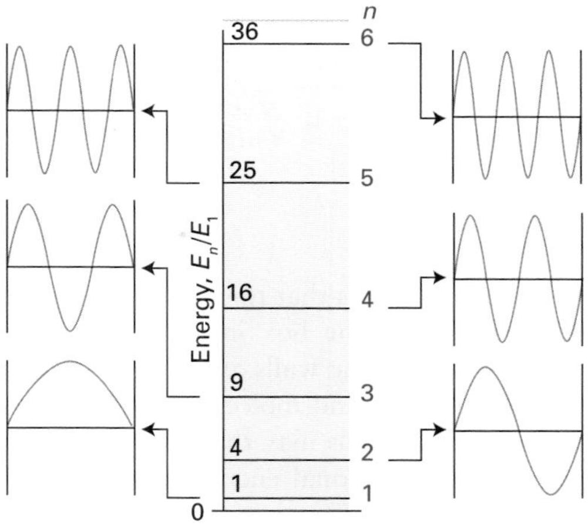

Fig. 12.19 The allowed energy levels and the corresponding (sine wave) wave functions for a particle in a box. Note that the energy levels increase as n^2, and so their spacing increases as n increases. Each wavefunction is a standing wave, and successive functions possess one more half wave and a correspondingly shorter wavelength.

The solutions of a particle in a box introduce another important general feature of quantum mechanics. Because the quantum number n cannot be zero (for this system), the lowest energy that the particle may possess is not zero, as would be allowed by classical mechanics, but $h^2/8mL^2$ (the energy when $n = 1$). This lowest, irremovable energy is called the **zero-point energy**. The existence of a zero-point energy is consistent with the uncertainty principle. If a particle is confined to a finite region, its location is not completely indefinite; consequently its momentum cannot be specified precisely as zero, and therefore its kinetic energy cannot be precisely zero either. The zero-point energy is not a special, mysterious kind of energy. It is simply the last remnant of energy that a particle cannot give up. For a particle in a box it can interpreted as the energy arising from a ceaseless fluctuating motion of the particle between the two confining walls of the box.

The energy difference between adjacent levels is

$$\Delta E = E_{n+1} - E_n = \overbrace{(n+1)^2}^{n^2+2n+1}\frac{h^2}{8mL^2} - n^2\frac{h^2}{8mL^2}$$

$$= (2n+1)\frac{h^2}{8mL^2} \qquad (12.10)$$

This expression shows that the difference decreases as the length L of the box increases, and that it becomes zero when the walls are infinitely far apart (Fig. 12.20). Atoms and molecules free to move in laboratory-sized vessels may therefore be treated as though their translational energy is not quantized, because L is so large. The expression also shows that the separation decreases as the mass of the particle increases. Particles of macroscopic mass (like balls and planets, and even minute specks of dust) behave as though their translational motion is unquantized. Both the following conclusions are true in general:

- The greater the extent of the confining region, the less important are the effects of quantization. Quantization is very important for highly confining regions.
- The greater the mass of the particle, the less important are the effects of quantization. Quantization is very important for particles of very small mass.

This chapter opened with the remark that the correct description of nature must account for the observation of transitions at discrete frequencies. This is exactly what is predicted for a system that can be modelled as a particle in a box, for it follows that when a particle makes a transition from a state with quantum number n_{initial} to one with quantum number n_{final}, the change in energy is

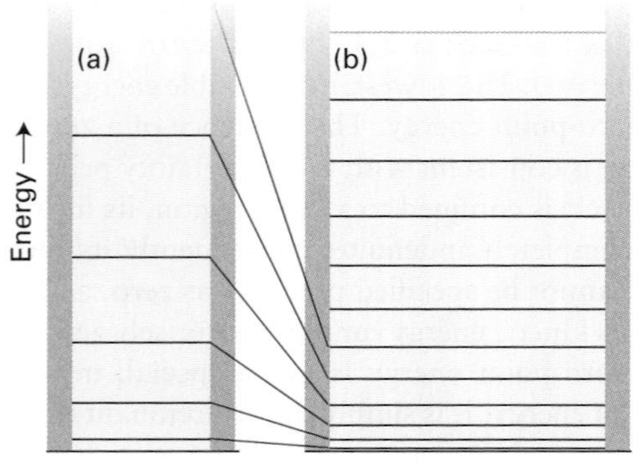

Fig. 12.20 (a) A narrow box has widely spaced energy levels; (b) a wide box has closely spaced energy levels. (In each case, the separations depend on the mass of the particle too.)

$$\Delta E = E_{\text{final}} - E_{\text{initial}} = (n_{\text{final}}^2 - n_{\text{initial}}^2)\frac{h^2}{8mL^2} \qquad (12.11)$$

Because the two quantum numbers can take only integer values, only certain energy changes are allowed, and therefore, through $\nu = \Delta E/h$, only certain frequencies will appear in the spectrum of transitions.

Example 12.4

Estimating an absorption wavelength

β-Carotene (**1**) is a linear polyene in which 10 single and 11 double bonds alternate along a chain of 22 carbon atoms. If we take each CC bond length to be about 140 pm, then the length L of the molecular box in β-carotene is $L = 2.94$ nm. Estimate the wavelength of the light absorbed by this molecule from its ground state to the next-higher excited state.

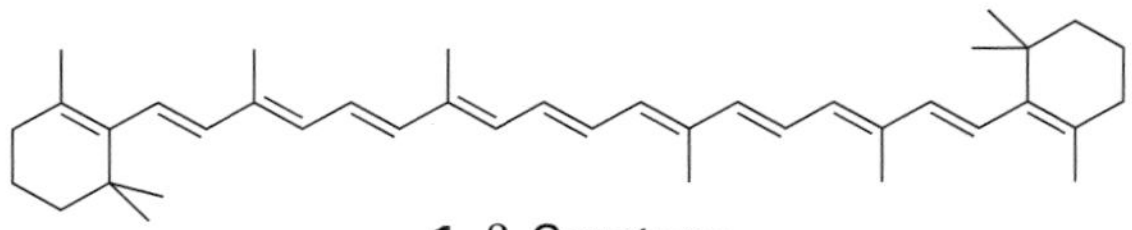

1 β-Carotene

Strategy For reasons that will be familiar from introductory chemistry (and which will be discussed in more detail in Chapter 14), each C atom contributes one p electron to the π orbitals. Use eqn 12.11 to calculate the energy separation between the highest occupied and the lowest unoccupied levels, and convert that energy to a wavelength by using the Bohr frequency condition (eqn 12.1).

Solution There are 22 C atoms in the conjugated chain; each contributes one p electron to the levels, so each level up to $n = 11$ is occupied by two electrons. The separation in energy between the ground state and the state in which one electron is promoted from $n_{\text{initial}} = 11$ to $n_{\text{final}} = 12$ is

$$\Delta E = E_{12} - E_{11}$$

$$= (\underbrace{12^2}_{n_{\text{final}}^2} - \underbrace{11^2}_{n_{\text{initial}}^2})\frac{\overbrace{(6.626\times10^{-34}\ \text{J s})^2}^{h^2}}{8\times\underbrace{(9.109\times10^{-31}\ \text{kg})}_{m_e}\times\underbrace{(2.94\times10^{-9}\ \text{m})^2}_{L^2}}$$

$$= 1.60\times10^{-19}\ \text{J}$$

It follows from the Bohr frequency condition ($\Delta E = h\nu$) that the frequency of radiation required to cause this transition is

$$\nu = \frac{\Delta E}{h} = \frac{1.60\times10^{-19}\ \text{J}}{6.626\times10^{-34}\ \text{J s}} = 2.41\times10^{14}\ \text{s}^{-1}$$

or 241 THz (1 THz = 10^{12} Hz), corresponding to a wavelength $\lambda = 1240$ nm. The experimental value is 603 THz ($\lambda = 497$ nm), corresponding to radiation in the visible range of the electromagnetic spectrum. Considering the crudeness of the model we have adopted here, we should be encouraged that the computed and observed frequencies agree to within a factor of 2.5.

A note on good practice The ability to make such quick 'back-of-the-envelope' estimates of orders of magnitude of physical properties should be a part of every scientist's toolkit.

Self-test 12.6

Estimate a typical nuclear excitation energy in electronvolts ($1\ \text{eV} = 1.602 \times 10^{-19}$ J; $1\ \text{GeV} = 10^9$ eV) by calculating the first excitation energy of a proton confined to a one-dimensional box with a length equal to the diameter of a nucleus (approximately 1×10^{-15} m, or 1 fm).

Answer: 0.6 GeV

(b) Tunnelling

If the potential energy, V, of a particle does not rise to infinity when it is in the walls of the container, and $E < V$ (so that the total energy is less than the potential energy and classically the particle cannot escape from the container), the wavefunction does not decay abruptly to zero. The wavefunction oscillates inside the box (eqn 12.6), decays exponentially inside the region representing the wall, and oscillates again on the other side of the wall outside the box (Fig. 12.21). Hence, if the walls are so thin and the mass of particle so small that the exponential decay of the wavefunction has not brought it to zero by the time it emerges on the right, the particle might be found on the outside of a container even though according to classical mechanics it has insufficient energy to escape. Such leakage by penetration into or through classically forbidden zones is called **tunnelling**.

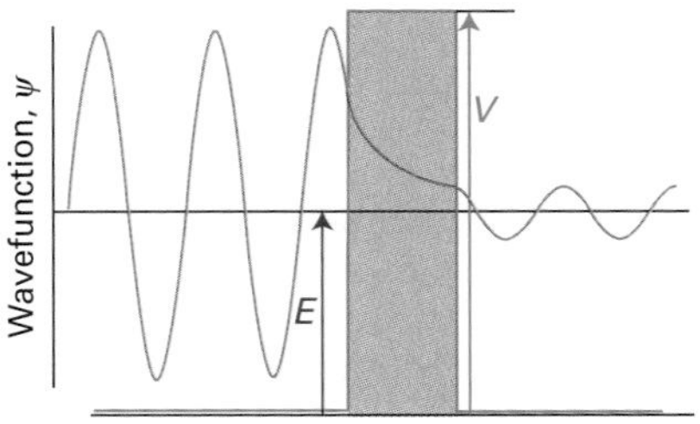

Fig. 12.21 A particle incident on a barrier from the left has an oscillating wavefunction, but inside the barrier there are no oscillations (for $E < V$). If the barrier is not too thick, the wavefunction is nonzero at its opposite face, and so oscillation begins again there.

The Schrödinger equation can be used to determine the probability of tunnelling, the **transmission probability**, T, of a particle incident on a finite barrier. When the barrier is high (in the sense that $V/E \gg 1$) and wide (in the sense that the wavefunction loses much of its amplitude inside the barrier), we may write[4]

$$T \approx 16\varepsilon(1-\varepsilon)e^{-2\kappa L} \qquad \kappa = \frac{\{2m(V-E)\}^{1/2}}{\hbar} \qquad \text{High and wide one-dimensional barrier} \quad \text{Transmission probability} \qquad (12.12)$$

where $\varepsilon = E/V$ and L is the thickness of the barrier. The transmission probability decreases exponentially with L and with $m^{1/2}$. It follows that particles of low mass are more able to tunnel through barriers than heavy ones (Fig. 12.21). Hence, tunnelling is very important for electrons, moderately important for protons, and negligible for most other heavier particles.

Brief illustration 12.4 Tunnelling in proton and hydrogen transfer reactions

The very rapid equilibration of proton transfer reactions (Chapter 8) is also a manifestation of the ability of protons to tunnel through barriers and transfer quickly from an acid to a base. The process may be visualized as a proton passing *through* an activation barrier rather than having to acquire enough energy to travel over it (Fig. 12.22). Quantum mechanical tunnelling can be the dominant process in reactions involving hydrogen atom or proton transfer when the temperature is so

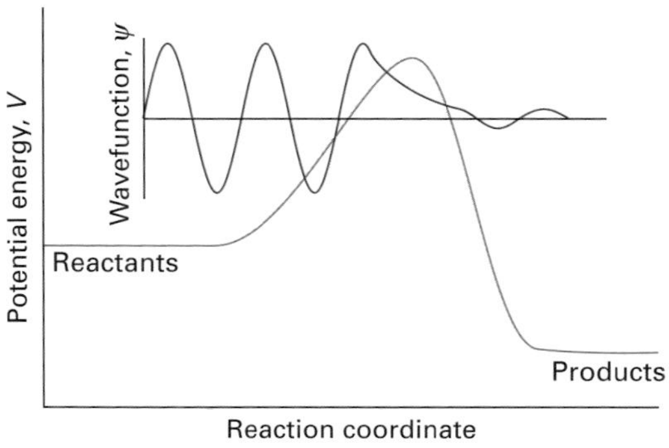

Fig. 12.22 A proton can tunnel through the activation energy barrier that separates reactants from products, so the effective height of the barrier is reduced and the rate of the proton transfer reaction increases. The effect is represented by drawing the wavefunction of the proton near the barrier. Proton tunnelling is important only at low temperatures, when most of the reactants are trapped on the left of the barrier.

4 For details of the calculation, see our *Physical Chemistry* (2010).

low that very few reactant molecules can overcome the activation energy barrier. One indication that a proton transfer is taking place by tunnelling is that an Arrhenius plot (Section 10.9) deviates from a straight line at low temperatures and the rate is higher than would be expected by extrapolation from room temperature.

(c) Motion in two dimensions

Once we have dealt with translation in one dimension it is quite easy to step into higher dimension. In doing so, we encounter two very important features of quantum mechanics that will occur many times in what follows. One feature is the simplification of the Schrödinger equation by the technique known as 'separation of variables'; the other is the existence of 'degeneracy'.

The arrangement we shall consider is like a particle confined to the floor of a rectangular box (Fig. 12.23). The box is of side L_X in the x-direction and L_Y in the y-direction. The wavefunction varies from place to place across the floor of the box, so it is a function of both the x- and y-coordinates; we write it $\psi(x,y)$. We show in Further information 12.1 that for this problem, according to the **separation of variables procedure**, the wavefunction can be expressed as a product of wavefunctions for each direction:

$$\psi(x,y) = X(x)Y(y) \tag{12.13}$$

with each wavefunction satisfying its 'own' Schrödinger equation like that in eqn 12.5, and that the solutions are

$$\psi_{n_X,n_Y}(x,y) = X_{n_X}(x)Y_{n_Y}(y)$$

$$= \overbrace{\left(\frac{2}{L_X}\right)^{1/2} \sin\left(\frac{n_X\pi x}{L_X}\right)}^{X_{n_X}(x)} \overbrace{\left(\frac{2}{L_Y}\right)^{1/2} \sin\left(\frac{n_Y\pi y}{L_Y}\right)}^{Y_{n_Y}(y)}$$

$$= \left(\frac{4}{L_XL_Y}\right)^{1/2} \sin\left(\frac{n_X\pi x}{L_X}\right) \sin\left(\frac{n_Y\pi y}{L_Y}\right)$$

Wavefunctions — Particle in a two-dimensional box (12.14a)

with energies

$$E_{n_X,n_Y} = E_{n_X} + E_{n_Y}$$

$$= \frac{n_X^2h^2}{8mL_X^2} + \frac{n_Y^2h^2}{8mL_Y^2}$$

$$= \left(\frac{n_X^2}{L_X^2} + \frac{n_Y^2}{L_Y^2}\right)\frac{h^2}{8m}$$

Energies — Particle in a two-dimensional box (12.14b)

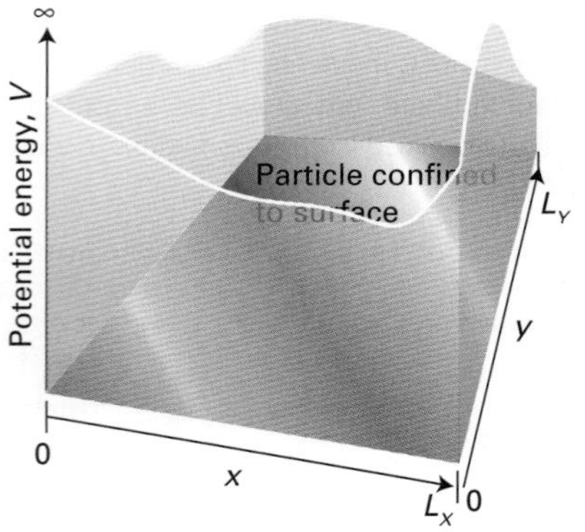

Fig. 12.23 A two-dimensional square well. The particle is confined to a rectangular plane bounded by impenetrable walls. As soon as the particle touches a wall, its potential energy rises to infinity.

There are two quantum numbers (n_X and n_Y), each allowed the values 1, 2, . . . independently. The separation of variables procedure is very important and occurs (sometimes without its use being acknowledged) throughout chemistry, for it underlies the fact that energies of independent systems are additive and that their wavefunctions are products of simpler component wavefunctions. We shall encounter it several times in later chapters.

Brief illustration 12.5 The zero-point energy of a particle in a two-dimensional box

An electron trapped in a cavity of dimensions L_X = 1.0 nm and L_Y = 2.0 nm can be described by a particle-in-a-box wavefunction. The zero-point energy of the electron is given by eqn 12.14b with $n_X = 1$ and $n_Y = 1$:

$$E_{1,1} = \left\{\frac{\overbrace{1^2}^{n_X^2}}{\underbrace{(1.0\times10^{-9}\,\text{m})^2}_{L_X^2}} + \frac{\overbrace{1^2}^{n_Y^2}}{\underbrace{(2.0\times10^{-9}\,\text{m})^2}_{L_Y^2}}\right\} \frac{\overbrace{(6.626\times10^{-34}\,\text{J s})^2}^{h^2}}{8\times\underbrace{(9.109\times10^{-31}\,\text{kg})}_{m_e}} = 7.5\times10^{-20}\,\text{J}$$

This energy, which could be reported as 75 zJ (1 zJ = 10^{-21} J), corresponds to 0.47 eV.

Self-test 12.7

Calculate the energy separation between the levels $n_X = n_Y = 2$ and $n_X = n_Y = 1$ of an electron trapped in a square cavity with sides of length 1.0 nm.

Answer: 3.6×10^{-19} J, 2.2 eV

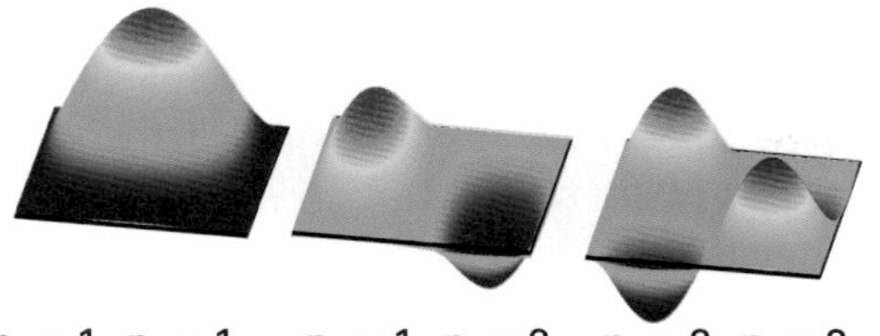

Fig. 12.24 Three wavefunctions of a particle confined to a rectangular surface.

Figure 12.24 shows some wavefunctions for the two-dimensional case: in one dimension the wavefunctions are like the vibrations of a violin string clamped at each end; in two dimensions the functions are like the vibrations of a rectangular sheet clamped at its edges.

A specially interesting case arises when the rectangular region is square with $L_X = L_Y = L$. The allowed energies are then

$$E_{n_X,n_Y} = (n_X^2 + n_Y^2)\frac{h^2}{8mL^2} \qquad (12.15a)$$

This expression is interesting because it shows that different wavefunctions may correspond to the same energy. For example, the wavefunctions with $n_X = 1$, $n_Y = 2$ and $n_X = 2$, $n_Y = 1$ are different:

$$\begin{aligned}\psi_{1,2}(x,y) &= \frac{2}{L}\sin\left(\frac{\pi x}{L}\right)\sin\left(\frac{2\pi y}{L}\right)\\ \psi_{2,1}(x,y) &= \frac{2}{L}\sin\left(\frac{2\pi x}{L}\right)\sin\left(\frac{\pi y}{L}\right)\end{aligned} \qquad (12.15b)$$

but both have the energy $5h^2/8mL^2$. Different states with the same energy are said to be **degenerate**. Degeneracy is always associated with an aspect of symmetry. In this case, it is easy to understand, because the confining region is square, and can be rotated through 90°, which takes the $n_X = 1$, $n_Y = 2$ wavefunction into the $n_X = 2$, $n_Y = 1$ wavefunction. In other cases the symmetry might be harder to identify, but it is always there.

The separation of variables will appear again when we discuss rotational motion and the structures of atoms. Degeneracy is very important in atoms, and is a feature that underlies the structure of the periodic table.

12.8 Rotational motion

Rotational motion is important in chemistry for a number of reasons. First, electrons circulate around nuclei in atoms, and an understanding of their orbital rotational behaviour is essential for understanding the structure of the periodic table and the properties it summarizes. Also, molecules rotate in the gas phase, and transitions between their allowed rotational states give rise to a variety of spectroscopic methods for determining their shapes and the lengths of their bonds. In fact, angular momenta, the momenta associated with rotational motion (see Foundations), are related to all manner of directional effects in chemistry and physics, including the shapes of electron distributions in atoms and hence the directions along which atoms can form chemical bonds.

(a) Rotation in two dimensions

The discussion of translational motion focussed on linear momentum, p. When we turn to rotational motion we have to focus instead on the analogous **angular momentum,** $\mathcal{J}$ (Foundations 0.4). The angular momentum of a particle that is travelling on a circular path of radius r about the z axis, and therefore confined to the xy-plane, is defined as

$$\mathcal{J}_z = pr \qquad \text{Magnitude of the angular momentum of a particle moving on a circular path} \qquad (12.16)$$

where p is its linear momentum ($p = mv$) at any instant. A particle that is travelling at high speed in a circle has a higher angular momentum than a particle of the same mass travelling more slowly. An object with a high angular momentum (like a flywheel) requires a strong braking force (more precisely, a strong 'torque') to bring it to a standstill.

To see what quantum mechanics tells us about rotational motion, we consider a particle of mass m moving in a horizontal circular path of radius r. The energy of the particle is entirely kinetic because the potential energy is constant and can be set equal to zero everywhere. We can therefore write $E = p^2/2m$. By using eqn 12.16 in the form $p = \mathcal{J}_z/r$, we can express this energy in terms of the angular momentum as

$$E = \frac{p^2}{2m} \overset{p=\mathcal{J}_z/r}{=} \frac{\mathcal{J}_z^2}{2\underbrace{mr^2}_{I}} \qquad \text{Kinetic energy} \quad \text{Particle on a ring} \qquad (12.17)$$

As we saw in Foundations 0.4, The quantity mr^2 is the **moment of inertia** of the particle about the z-axis, and denoted I (Fig. 12.25). It follows that the energy of the particle is

$$E = \frac{\mathcal{J}_z^2}{2I} \qquad \text{Kinetic energy in terms of the moment of inertia} \quad \text{Particle on a ring} \qquad (12.18)$$

Now we use the de Broglie relation ($\lambda = h/p$) to see that the energy of rotation is quantized. To do so, we

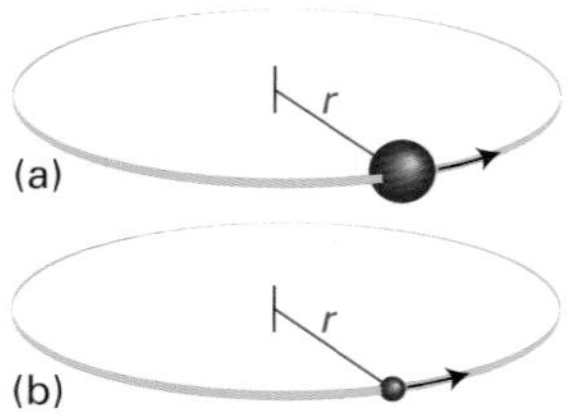

Fig. 12.25 A particle travelling on a circular path has a moment of inertia I that is given by mr^2. (a) This heavy particle has a large moment of inertia about the central point; (b) this light particle is travelling on a path of the same radius, but it has a smaller moment of inertia. The moment of inertia plays a role in circular motion that is the analogue of the mass for linear motion: a particle with a high moment of inertia is difficult to accelerate into a given state of rotation, and requires a strong braking force to stop its rotation.

express the angular momentum in terms of the wavelength of the particle:

$$\mathcal{J}_z = pr \overbrace{=}^{p=h/\lambda} \frac{hr}{\lambda} \qquad \text{Angular momentum in terms of the moment of inertia} \quad \text{Particle on a ring} \qquad (12.19)$$

Suppose for the moment that λ can take an arbitrary value. In that case, the amplitude of the wavefunction depends on the angle, as shown in Fig. 12.26. When the angle increases beyond 2π (that is, beyond 360°), the wavefunction continues to change on its next circuit. For an arbitrary wavelength it gives rise to a different amplitude at each point and the wavefunction will not be single valued (a requirement for acceptable wavefunctions, Section 12.5). Thus, this arbitrary wave is not acceptable. An acceptable solution is obtained if the wavefunction reproduces itself on successive circuits in the sense that the wavefunction at $\phi = 2\pi$ (after a complete revolution) must be the same as the wavefunction at $\phi = 0$: we say that the wavefunction must satisfy **cyclic boundary conditions**. Specifically, the acceptable wavefunctions that match after each circuit have wavelengths that are given by the expression

$$\lambda = \frac{\overbrace{2\pi r}^{\text{Circumference of ring}}}{n} \qquad n = 0, 1, \ldots$$

where the value $n = 0$, which gives an infinite wavelength, corresponds to a uniform nonzero amplitude. It follows that the permitted energies are

$$E_n \overbrace{=}^{E=\mathcal{J}_z^2/2I \text{ and } \mathcal{J}_z=hr/\lambda} \frac{(hr/\lambda)^2}{2I} \overbrace{=}^{\lambda=2\pi r/n} \frac{(nh/2\pi)^2}{2I} \overbrace{=}^{h/2\pi=\hbar} \frac{n^2\hbar^2}{2I}$$

with $n = 0, \pm1, \pm2, \ldots$.

In the discussion of rotational motion it is conventional—for reasons that will become clear—to denote the quantum number by m_l in place of n. Therefore, the final expression for the energy levels is

$$E_{m_l} = \frac{m_l^2\hbar^2}{2I} \qquad m_l = 0, \pm1, \ldots \qquad \text{Quantized energies} \quad \text{Particle on a ring} \qquad (12.20)$$

These energy levels are drawn in Fig. 12.27. The occurrence of m_l^2 in the expression for the energy means that two states of motion with opposite values of m_l, such as those with $m_l = +1$ and $m_l = -1$, correspond to the same energy. This degeneracy arises from the fact that the energy is independent of the direction of travel. The state with $m_l = 0$ is nondegenerate. A further point is that the particle does

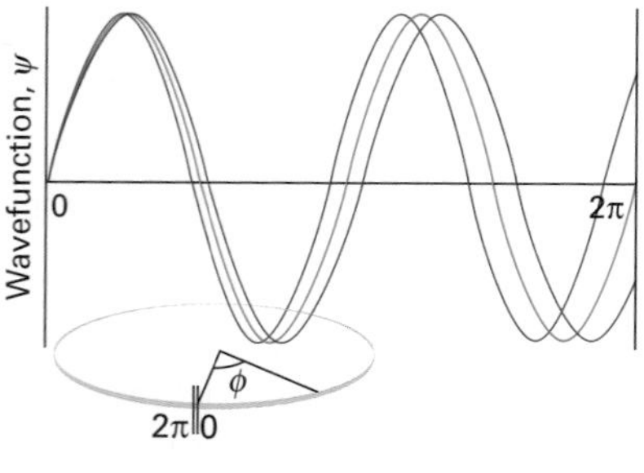

Fig. 12.26 Two solutions of the Schrödinger equation for a particle on a ring. The circumference has been opened out into a straight line; the points at $\phi = 0$ and 2π are identical. The solutions in red are unacceptable because they have different values after each circuit and so interfere destructively with themselves. The solution in green is acceptable because it reproduces itself on successive circuits.

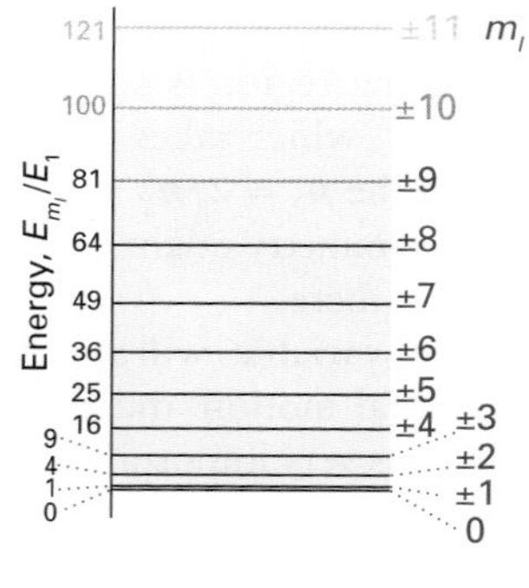

Fig. 12.27 The energy levels of a particle that can move on a circular path. Classical physics allowed the particle to travel with any energy (as represented by the continuous tinted band); quantum mechanics, however, allows only discrete energies. Each energy level, other than the one with $m_l = 0$, is doubly degenerate, because the particle may rotate either clockwise or anticlockwise with the same energy.

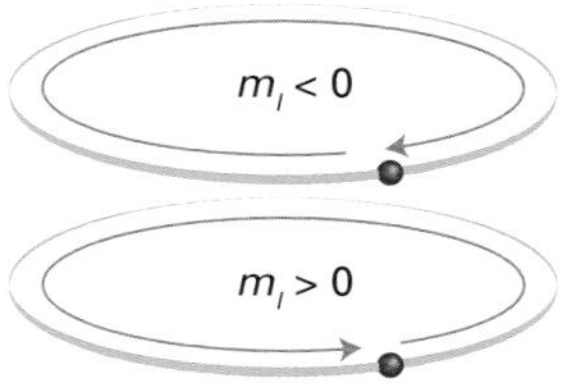

Fig. 12.28 The significance of the sign of m_l. When $m_l < 0$, the particle travels in an anticlockwise direction as viewed from below; then $m_l > 0$, the motion is clockwise.

not have a zero-point energy: m_l may take the value 0, and $E_0 = 0$.

An important additional conclusion is that *the angular momentum of the particle is quantized*. We can use the relation between angular momentum and linear momentum ($\mathcal{J}_z = pr$), and between linear momentum and the allowed wavelengths of the particle ($\lambda = 2\pi r/m_l$), to conclude that the angular momentum of a particle around the z-axis is confined to the values

$$\mathcal{J}_z = pr \overbrace{=}^{p=h/\lambda} \frac{hr}{\lambda} \overbrace{=}^{\lambda=2\pi r/m_l} \frac{hr}{2\pi r/m_l} \overbrace{=}^{\text{cancel terms and rearrange}} m_l \times \frac{h}{2\pi}$$

That is, the angular momentum of the particle around the axis is confined to the values

$$\mathcal{J}_z = m_l\hbar \qquad \text{z-Component of the angular momentum} \quad \text{Particle on a ring} \qquad (12.21)$$

with $m_l = 0, \pm1, \pm2, \ldots$. Positive values of m_l correspond to clockwise rotation (as seen from below) and negative values correspond to counter clockwise rotation (Fig. 12.28). The quantized motion can be thought of in terms of the rotation of a bicycle wheel that can rotate only with a discrete series of angular momenta, so that as the wheel is accelerated, the angular momentum jerks from the values 0 (when the wheel is stationary) to $\hbar, 2\hbar, \ldots$, but can have no intermediate value.

Brief illustration 12.6 The electronic structure of benzene

Consider the π electrons of an aromatic molecule, such as benzene. As a first approximation, we may treat the group as a circular ring of radius 140 pm, with six electrons in the conjugated system moving along the perimeter of the ring. As in Example 12.4, we assume that only one electron from each carbon atom is allowed to move freely around the ring and that in the ground state of the molecule, each level is occupied by two electrons. Therefore, only the $m_l = 0, +1$, and -1 levels are occupied (with the last two states being degenerate). From eqn 12.20, the energy separation between the $m_l = \pm1$ and the $m_l = \pm2$ levels is

$$\Delta E = E_{\pm2} - E_{\pm1}$$

$$= \left(\overbrace{4}^{m_{l,\text{final}}^2} - \overbrace{1}^{m_{l,\text{initial}}^2}\right) \frac{\overbrace{(1.055\times10^{-34}\,\text{J})^2}^{\hbar^2}}{2\times\underbrace{(9.109\times10^{31}\,\text{kg})\times(1.40\times10^{-10}\,\text{m})^2}_{I=m_er^2}}$$

$$= 9.33\times10^{-19}\,\text{J}$$

This energy separation, of 5.82 eV, corresponds to an absorption frequency of 1.41 PHz (1 PHz = 10^{15} Hz) and a wavelength of 213 nm, which is in close agreement with the experimental value of 260 nm for a transition of this kind.

(b) Rotation in three dimensions

Rotational motion in three dimensions includes the motion of electrons around nuclei in atoms. Consequently, understanding rotational motion in three dimensions is crucial to understanding the electronic structures of atoms. Gas-phase molecules also rotate freely in three dimensions and by studying their allowed energies (using the spectroscopic techniques described in Chapter 19) we can infer bond lengths, bond angles, and dipole moments.

Just as the location of a city on the surface of the Earth is specified by giving its latitude and longitude, the location of a particle free to move at a constant distance from a point is specified by two angles, the **colatitude** θ (theta) and the **azimuth** ϕ (phi) (Fig. 12.29). The wavefunction for the particle is therefore a function of both angles and is written $\psi(\theta,\phi)$. It turns out that this wavefunction factorizes by the separation of variables procedure into the product of a function of θ and a function of ϕ, and that the latter are exactly the same as those we have already found for a particle on a ring. In other words, motion of a particle over the surface of a sphere is like the motion of the particle over a stack of rings, with the additional freedom to migrate between rings.

There are two sets of cyclic boundary conditions that limit the selection of solutions of the Schrödinger

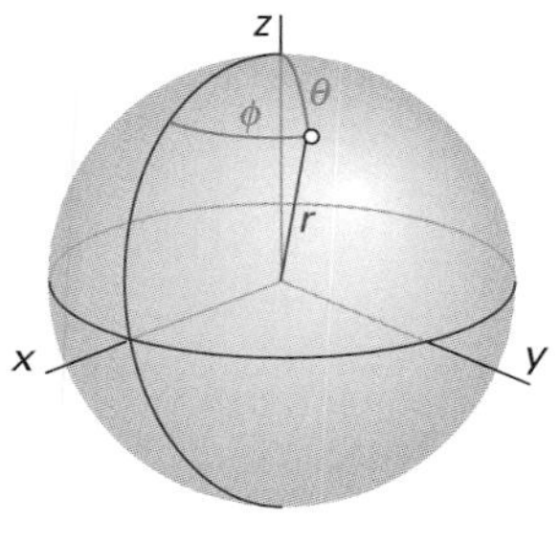

Fig. 12.29 The spherical polar coordinates r (the radius), θ (the colatitude), and ϕ (the azimuth).

equation. One is that the wavefunctions must match as we travel round the equator (just like the particle on a ring); as we have seen, that boundary condition introduces the quantum number m_l. The other condition is that the wavefunction must match as we travel over the poles. This constraint introduces a second quantum number, which is called the **orbital angular momentum quantum number** and denoted l. We shall not go into the details of the solution, but just quote the results. It turns out that the quantum numbers are allowed the following values:

$$l = 0, 1, 2, \ldots \qquad m_l = l, l-1, \ldots, -l$$

Note that there are $2l + 1$ values of m_l for a given value of l. The energy of the particle is given by the expression

$$E = l(l+1)\frac{\hbar^2}{2I} \qquad l = 0, 1, 2, \ldots \qquad \text{Quantized energies} \quad \text{Particle on a sphere} \quad (12.22)$$

where r is the radius of the surface of the sphere on which the particle moves. Note that, for reasons that will become clear in a moment, the energy depends on l and is independent of the value of m_l. The wavefunctions appear in a number of applications, and are called **spherical harmonics**. They are commonly denoted $Y_{l,m_l}(\theta,\phi)$ and can be imagined as wavelike distortions of a spherical shell (Fig. 12.30).

We can draw a very important additional conclusion by comparing the expression for the energy in eqn 12.22 with the classical expression for the energy:

Classical	Quantum mechanical
$E = \dfrac{\mathcal{J}^2}{2mr^2}$	$E_l = \dfrac{l(l+1)\hbar^2}{2mr^2}$

where $\mathcal{J}$ is the magnitude of the angular momentum of the particle. We can conclude that the magnitude of the angular momentum is quantized and limited to the values

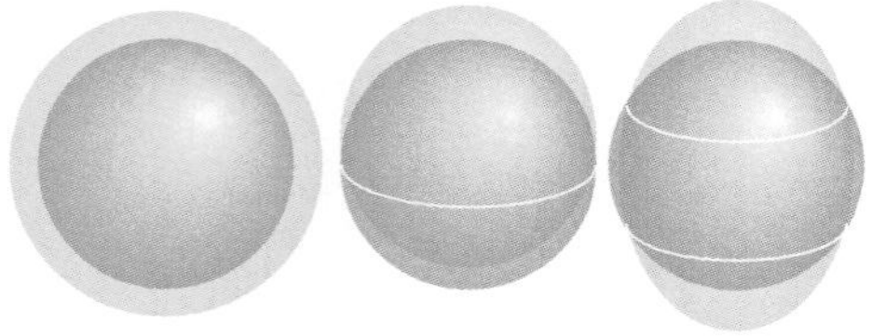

Fig. 12.30 The wavefunctions of a particle on a sphere can be imagined as having the shapes that the surface would have when the sphere is distorted. Three of these 'spherical harmonics' are shown here.

$$\mathcal{J} = \{l(l+1)\}^{1/2}\hbar \qquad l = 0, 1, 2 \ldots \qquad \text{Magnitude of the angular momentum} \quad \text{Particle on a sphere} \quad (12.23)$$

Thus, the allowed values of the magnitude of the angular momentum are $0, 2^{1/2}\hbar, 6^{1/2}\hbar, \ldots$. As we remarked in Foundations, the angular momentum is a vector (The chemist's toolkit 12.1). For rotation in three dimensions, the angular momentum has three components: $\mathcal{J}_x$, $\mathcal{J}_y$, and $\mathcal{J}_z$. We have already seen that m_l tells us the value, as $m_l\hbar$, of the angular momentum around the z-axis (the polar axis of a sphere). In summary:

- The orbital angular momentum quantum number l can have the non-negative integral values 0, 1, 2, . . . ; it tells us (through eqn 12.23) the magnitude of the orbital angular momentum of the particle.
- The magnetic quantum number m_l is limited to the $2l+1$ values $l, l-1, \ldots, -l$; it tells us, through $m_l\hbar$, the z-component of the orbital angular momentum.

The chemist's toolkit 12.1 Vectors

A vector quantity has both magnitude and direction. The vector $\boldsymbol{v}$ shown in Sketch 12.1 has components on the x, y, and z axes with magnitudes v_x, v_y, and v_z, respectively. The direction of each of the components is denoted with a plus sign or minus sign. For example, if $v_x = -1.0$, the x-component of the vector $\boldsymbol{v}$ has a magnitude of 1.0 and points in the $-x$ direction. The magnitude of the vector is denoted v or $|\boldsymbol{v}|$ and is given by

$$v = (v_x^2 + v_y^2 + v_z^2)^{1/2}$$

Thus, a vector with components $v_x = -1.0$, $v_y = +2.5$, and $v_z = +1.1$ has magnitude 2.9 and would be represented by an arrow of 2.9 units. Operations involving vectors are not as straightforward as those involving numbers. We describe the operations we need for this text in The chemist's toolkit 13.1.

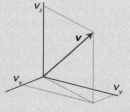

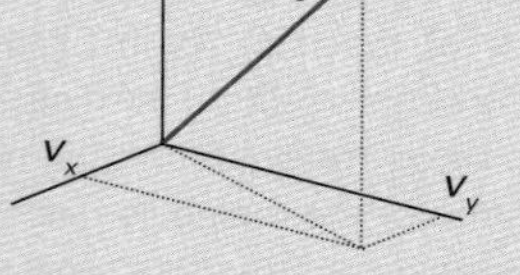

Sketch 12.1 The magnitude of a vector in terms of its components.

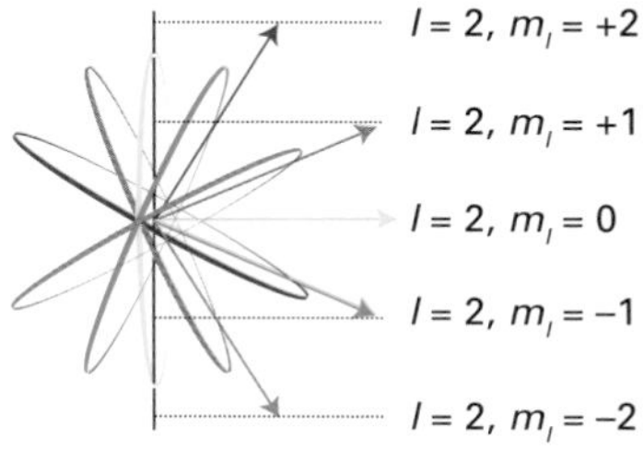

Fig. 12.31 The significance of the quantum numbers l and m_l shown for $l = 2$: l determines the magnitude of the angular momentum (as represented by the length of the arrow), and m_l the component of that angular momentum about the z-axis.

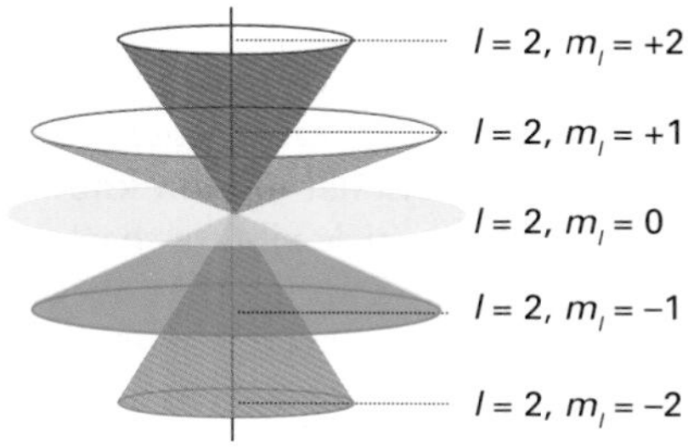

Fig. 12.32 The vector model of angular momentum acknowledges that nothing can be said about the x- and y-components of angular momentum if the z-component is known, by representing the states of angular momentum by cones.

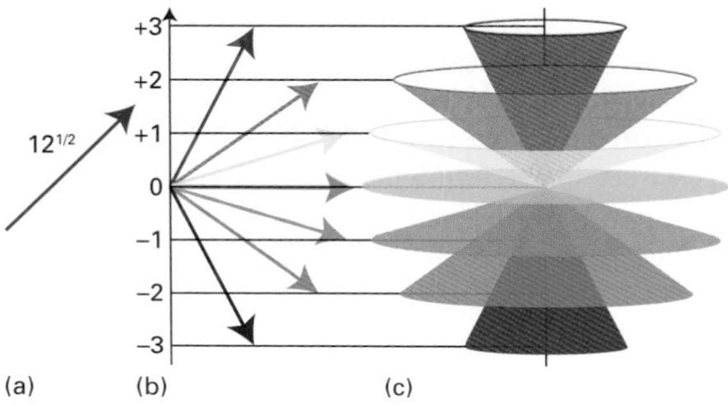

Fig. 12.33 (a) The angular momentum of a particle with $l = 3$. (b) The seven orientations of the angular momentum vector. (c) The cones representing possible but unspecified orientations around the z-axis.

Several features now fall into place. First, we can now see why m_l is confined to a *range* of values that depend on l: the angular momentum around a single axis (as expressed by m_l) cannot exceed the magnitude of the angular momentum (as expressed by l). Second, for a given magnitude to correspond to different values of the angular momentum around the z-axis, the angular momentum must lie at different angles (Fig. 12.31). The value of m_l therefore indicates the angle to the z-axis of the motion of the particle. Providing the particle has a given amount of angular momentum, its kinetic energy (its only source of energy) is independent of the orientation of its path: hence, the energy is independent of m_l, as asserted above.

What can we say about the component of angular momentum about the x- and y-axes? Almost nothing. We know that these components cannot exceed the magnitude of the angular momentum, but there is no quantum number that tells us their precise values. In fact, $\mathcal{J}_x$, $\mathcal{J}_y$, and $\mathcal{J}_z$, the three components of angular momentum, are complementary observables in the sense described in Section 12.6 in connection with the uncertainty principle, and if one is known exactly (the value of $\mathcal{J}_z$, for instance, as $m_l\hbar$), then the values of the other two cannot be specified. For this reason, the angular momentum is often represented as lying anywhere on a cone with a given z-component (indicating the value of m_l) and side (indicating the value of $\{l(l+1)\}^{1/2}$, but with indefinite projection on the x- and y-axes (Fig. 12.32). This **vector model** of angular momentum is intended to be only a representation of the quantum mechanical aspects of angular momentum, expressing the fact that the magnitude is well defined, one component is well defined, and the two other components are indeterminate.

Brief illustration 12.7 The vector model of angular momentum

Suppose that a particle is in a state with $l = 3$. We would know that the magnitude of its angular momentum is $12^{1/2}\hbar$, so the angular momentum itself is represented by an arrow of length $12^{1/2} = 3.46\ldots$ units in Fig. 12.33a. The angular momentum could have any of seven orientations with z-components $m_l\hbar$, with $m_l = +3, +2, +1, 0, -1, -2$, or -3, so it is represented by the arrow lying in the orientations that give projections of length m_l units on the z-axis (Fig. 12.33b). Then we invoke the uncertainty principle, which denies knowledge of the x- and y-components, and so replace the arrows by cones representing possible but unspecified orientations around the z-axis (Fig. 12.33c).

12.9 Vibration: the harmonic oscillator

One very important type of motion of a molecule is the vibration of its atoms—bonds stretching, compressing, and bending. A molecule is not just a frozen, static array of atoms: all of them are in constant motion relative to one another. As we shall see, the discussion in this section is the foundation for a discussion of the technique of vibrational spectroscopy in Chapter 19; it is also used in the discussion of various aspects of chemical kinetics.

In the type of vibrational motion known as **harmonic oscillation**, a particle vibrates backwards and forwards restrained by a spring that obeys **Hooke's law** of force. Hooke's law states that the restoring force is proportional to the displacement, x:

$$\text{Restoring force} = -k_f x \qquad \text{Hooke's law} \qquad (12.24a)$$

The constant of proportionality k_f is called the **force constant:** a stiff spring has a high force constant (the restoring force is strong even for a small displacement) and a weak spring has a low force constant. The SI units of k_f are newtons per metre (N m^{-1}). The negative sign in eqn 12.24a is included because a displacement to the right (to positive x) corresponds to a force directed to the left (towards negative x). As we see in the following Derivation, the potential energy of a particle subjected to this force increases as the square of the displacement, and specifically

$$V(x) = \tfrac{1}{2}k_f x^2 \qquad \text{Potential energy} \quad \text{Harmonic oscillator} \qquad (12.24b)$$

The variation of V with x is shown in Fig. 12.34: it has the shape of a parabola (a curve of the form $y = ax^2$), and we say that a particle undergoing harmonic motion has a 'parabolic potential energy'.

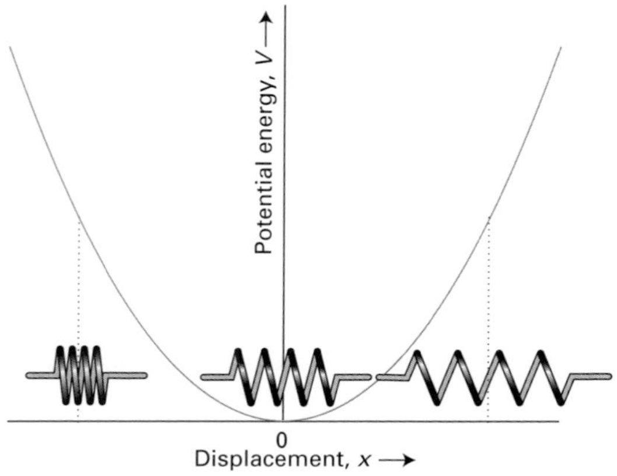

Fig. 12.34 The parabolic potential energy characteristic of a harmonic oscillator. Positive displacements correspond to extension of the spring; negative displacements correspond to compression of the spring.

Derivation 12.3

Potential energy of a harmonic oscillator

Force is the negative slope of the potential energy: $F = -\mathrm{d}V/\mathrm{d}x$. Because the infinitesimal quantities may be treated as any other quantity in algebraic manipulations, we rearrange the expression into $\mathrm{d}V = -F\mathrm{d}x$ and then integrate both sides from $x = 0$, where the potential energy is $V(0)$, to x, where the potential energy is $V(x)$:

$$V(x) - V(0) = -\int_0^x F(x)\mathrm{d}x$$

Now substitute $F(x) = -k_f x$:

$$V(x) - V(0) = -\int_0^x \overbrace{(-k_f x)}^{F(x)}\mathrm{d}x = k_f \overbrace{\int_0^x x\mathrm{d}x}^{\frac{1}{2}x^2} = \tfrac{1}{2}k_f x^2$$

We are free to choose $V(0) = 0$, which then gives eqn 12.24b.

Unlike the earlier cases we considered, the potential energy varies with position in the regions where the particle may be found, so we have to use $V(x)$ in the Schrödinger equation. Then we have to select the solutions that satisfy the boundary equations, which in this case means that they must fit into the parabola representing the potential energy. More precisely, the wavefunctions must all go to zero for large displacements in either direction from $x = 0$: they do not have to go abruptly to zero at the edges of the parabola.

The solutions turn out to be very simple. For instance, the energies of the solutions that satisfy the boundary conditions are

$$E_v = (v + \tfrac{1}{2})h\nu \qquad v = 0, 1, 2, \ldots \qquad \nu = \frac{1}{2\pi}\left(\frac{k_f}{m}\right)^{1/2} \qquad \text{Quantized energies} \quad \text{Harmonic oscillator} \qquad (12.25)$$

where m is the mass of the particle and v is the **vibrational quantum number**. Be very careful to distinguish the quantum number v (italic vee) from the frequency ν (Greek nu). These energies form a uniform ladder of values separated by $h\nu$ (Fig. 12.35). The quantity ν is a frequency (in cycles per second, or hertz, Hz), and is in fact the frequency that a classical

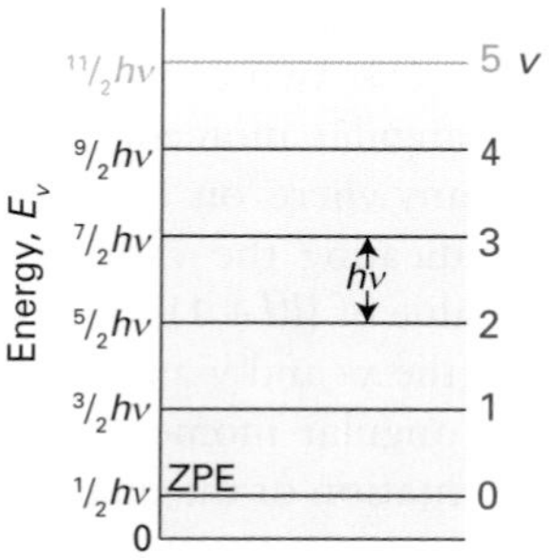

Fig. 12.35 The array of energy levels of a harmonic oscillator (the levels continue upwards to infinity). The separation depends on the mass and the force constant. Note the zero-point energy.

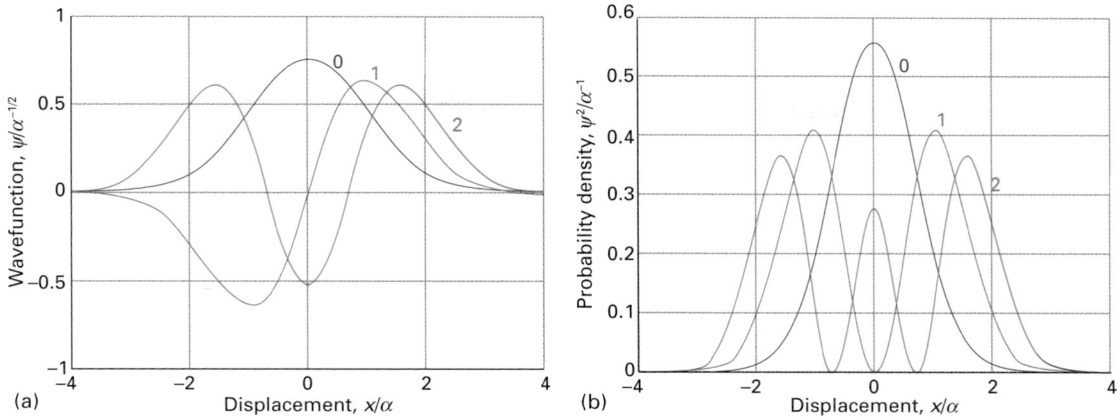

Fig. 12.36 (a) The wavefunctions and (b) the probability densities of the first three states of a harmonic oscillator. Note how the probability of finding the oscillator at large displacements increases as the state of excitation increases. The wavefunctions and displacements are expressed in terms of the parameter $\alpha = (\hbar^2/mk_f)^{1/4}$.

oscillator of mass m and force constant k_f would be calculated to have. In quantum mechanics, though, ν tells us (through $h\nu$) the separation of any pair of adjacent energy levels. The separation is large for stiff springs and low masses.

Brief illustration 12.8 The vibrational frequency of a chemical bond

The force constant for an H—Cl bond is 516 N m^{-1}, where the newton (N) is the SI unit of force (1 N = 1 kg m s^{-2}). If we suppose that, because the chlorine atom is relatively very heavy, only the hydrogen atom moves, we take m as the mass of the H atom (1.67 × 10^{-27} kg for ^{1}H). We find

$$\nu = \frac{1}{2\pi}\left(\frac{k_f}{m}\right)^{1/2} = \frac{1}{2\pi}\left(\frac{516\ \overbrace{\mathrm{N}}^{\mathrm{kg\,m\,s^{-2}}}\ \mathrm{m^{-1}}}{1.67\times10^{-27}\ \mathrm{kg}}\right)^{1/2} \underbrace{=}_{\text{cancel terms and use } 1\,\mathrm{s^{-1}}=1\,\mathrm{Hz}} 8.85\times10^{13}\ \mathrm{Hz}$$

The separation between adjacent levels is h times this frequency, or 5.86 × 10^{-20} J (58.6 zJ), corresponding to 35.3 kJ mol^{-1}.

Self-test 12.8

The vibrational frequency of an H—I bond is 6.95 × 10^{13} s^{-1}. Estimate the force constant for the bond on the basis that it is only the H atoms that moves.

Answer: 318 N m^{-1}

Figure 12.36 shows the shapes of the first few wavefunctions of a harmonic oscillator. The ground-state wavefunction (corresponding to $v = 0$ and having the zero-point energy $\frac{1}{2}h\nu$) is a bell-shaped curve, a curve of the form e^{-x^2} (a Gaussian function; see The chemist's toolkit 1.2), with no nodes. This shape shows that the particle is most likely to be found at $x = 0$ (zero displacement), but may be found at greater displacements with decreasing probability. The first excited wavefunction has a node at $x = 0$ and positive and negative peaks on either side. Therefore, in this state, the particle will be found most probably with the 'spring' stretched or compressed to the same amount. However, the wavefunctions extend beyond the limits of motion of a classical oscillator (Fig. 12.37), which is another example of quantum mechanical tunnelling. We explore the consequences of these characteristics in later chapters, especially Chapter 19.

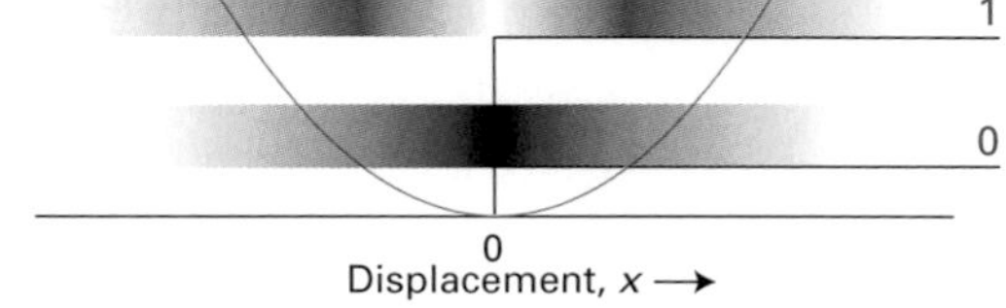

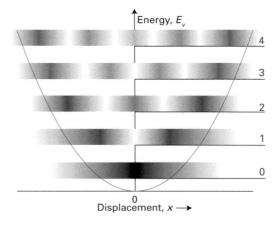

Fig. 12.37 A schematic illustration of the probability density for finding a harmonic oscillator at a given displacement. Classically, the oscillator cannot be found at displacements at which its total energy is less than its potential energy (because the kinetic energy cannot be negative). A quantum oscillator, though, may tunnel into regions that are classically forbidden.

Further information 12.1

The separation of variables procedure

The wavefunction for a particle in a two-dimensional box depends on two variables, x and y. Therefore, and as we saw in Section 2.8, $\psi(x,y)$ has two partial derivatives, $\partial\psi(x,y)/\partial x$ with respect to x (calculated by holding y constant), and $\partial\psi(x,y)/\partial y$ with respect to y (calculated by holding x constant). In such cases, the Schrödinger equation is written as a partial differential equation (The chemist's toolkit 12.2):

$$-\frac{\hbar^2}{2m}\frac{\partial^2\psi(x,y)}{\partial x^2}-\frac{\hbar^2}{2m}\frac{\partial^2\psi(x,y)}{\partial y^2}=E\psi(x,y)$$

For simplicity, we can write this expression as

$$\hat{H}_X\psi(x,y)+\hat{H}_Y\psi(x,y)=E\psi(x,y)$$

The chemist's toolkit 12.2 Partial differential equations

We encountered ordinary differential equations in The chemist's toolkit 10.1. If the unknown function f depends on more than one variable, as in

$$a\left(\frac{\partial^2 f}{\partial x^2}\right)_y+b\left(\frac{\partial^2 f}{\partial y^2}\right)_x=cf$$

then the equation is called a *partial differential equation*. Here, f is a function of x and y, which we could write explicitly as $f(x,y)$, and the factors a, b, c may be either constants or functions of either or both variables. Note the change in symbol from d to ∂ to signify a *partial derivative*, which is obtained by differentiating a function with respect to one variable while holding the other variable, which is denoted by a suffix, constant (Section 2.8).

The only partial differential equations that we need to know about are those that can be separated into two or more ordinary differential equations by the technique known as **separation of variables**. To discover if a differential equation with solution $f(x,y)$ can be solved by this method we suppose that the full solution can be factored into functions that depend only on x or only on y, and write $f(x,y) = X(x)Y(y)$. The method is illustrated in Further information 12.1.

where $\hat{H}_X$ affects—mathematicians say 'operates on'—only functions of x and $\hat{H}_Y$ operates only on functions of y. Thus, generalizing slightly from Derivation 12.1, $\hat{H}_X$ just means 'take the second derivative with respect to x' and $\hat{H}_Y$ means the same for y. To see if $\psi(x,y) = X(x)Y(y)$ is indeed a solution, we substitute this product on both sides of the last equation,

$$\hat{H}_X X(x)Y(y)+\hat{H}_Y X(x)Y(y)=EX(x)Y(y)$$

and note that $\hat{H}_X$ acts on only $X(x)$, with $Y(y)$ being treated as a constant, and $\hat{H}_Y$ acts on only $Y(y)$, with $X(x)$ being treated as a constant. Therefore, $\hat{H}_X X(x)Y(y) = Y(y)\hat{H}_X X(x)$ and $\hat{H}_Y X(x)Y(y) = X(x)\hat{H}_Y Y(y)$, and the preceding equation becomes

$$Y(y)\hat{H}_X X(x)+X(x)\hat{H}_Y Y(y)=EX(x)Y(y)$$

When we divide both sides by $X(x)Y(y)$, we obtain

$$\frac{1}{X(x)}\hat{H}_X X(x)+\frac{1}{Y(y)}\hat{H}_Y Y(y)=E$$

Now we come to the crucial part of the argument. The first term on the left depends only on x and the second term depends only on y. Therefore, if x changes, only the first term can change. But its sum with the unchanging second term is the constant E. Therefore, the first term cannot in fact change when x changes. That is, the first term is equal to a constant, which we write E_X. The same argument applies to the second term when y is changed; so it too is equal to a constant, which we write E_Y, and the sum of these two constants is E. That is, we have shown that

$$\frac{1}{X(x)}\hat{H}_X X(x)=E_X \qquad \frac{1}{Y(y)}\hat{H}_Y Y(y)=E_Y$$

with $E_X + E_Y = E$. These two equations are easily turned into

$$\hat{H}_X X(x)=E_X X(x) \qquad \hat{H}_Y Y(y)=E_Y Y(y)$$

which we should recognize as the Schrödinger equations for one-dimensional motion, one along the x-axis and the other along the y-axis. Thus, the variables have been separated, and because the boundary conditions are essentially the same for each axis (the only difference being the actual values of the lengths L_X and L_Y), the individual wavefunctions are essentially the same as those found for the one-dimensional case.

Checklist of key concepts

□ **1** Atomic and molecular spectra show that the energies of atoms and molecules are quantized.

□ **2** The photoelectric effect is the ejection of electrons when radiation of greater than a threshold frequency is incident on a metal.

□ **3** The wave-like character of electrons was demonstrated by the Davisson–Germer diffraction experiment.

□ **4** The joint wave–particle character of matter and radiation is called wave–particle duality.

□ **5** A wavefunction, ψ, contains all the dynamical information about a system and is found by solving the appropriate Schrödinger equation subject to the constraints on the solutions known as boundary conditions.

□ **6** According to the Born interpretation, the probability of finding a particle in a small region of space of volume δV is proportional to $\psi^2 \delta V$, where ψ is the value of the wavefunction in the region.

□ **7** According to the Heisenberg uncertainty principle, it is impossible to specify simultaneously, with arbitrary precision, both the momentum and the position of a particle.

□ **8** The energy levels of a particle of mass m in a box of length L are quantized and the wavefunctions are sine functions (see the following *Road map*).

□ **9** The zero-point energy is the lowest permissible energy of a system.

□ **10** Different states with the same energy are said to be degenerate.

□ **11** Because wavefunctions do not, in general, decay abruptly to zero, particles may tunnel into classically forbidden regions.

□ **12** The angular momentum and the kinetic energy of a particle free to move on a circular ring are quantized; the quantum number is denoted m_l.

□ **13** A particle on a ring and on a sphere must satisfy cyclic boundary conditions (the wavefunctions must repeat on successive cycles).

□ **14** The angular momentum and the kinetic energy of a particle on a sphere are quantized with values determined by the quantum numbers l and m_l (see the following *Road map*); the wavefunctions are the spherical harmonics.

□ **15** A particle undergoes harmonic motion if it is subjected to a Hooke's-law restoring force (a force proportional to the displacement).

□ **16** The energy levels of a harmonic oscillator are equally spaced and specified by the quantum number $v = 0, 1, 2, \ldots$.

Road map of key equations

Bohr frequency condition

$$\Delta E = h\nu$$

$$\Delta p_x \Delta x \geq \tfrac{1}{2}\hbar \quad \text{Uncertainty principle}$$

2D box
$$E_{n_X n_Y} = (n_X^2/L_X^2 + n_Y^2/L_Y^2)h^2/8m$$
$$\psi_{n_X n_Y} = (4/L_X L_Y)^{1/2}\sin(\pi n_X x/L_X)\sin(\pi n_Y y/L_Y)$$

Particle in 1D box
$$E_n = n^2 h^2/8mL^2 \quad \psi_n = (2/L)^{1/2}\sin(\pi n x/L)$$

Transitions

Tunnelling
$$T = 16\varepsilon(1 - \varepsilon)\mathrm{e}^{-2\kappa L}$$
$$\kappa = \{2m(V - E)\}^{1/2}/\hbar, \ \varepsilon = E/V$$

Transmission probability, $E < V$

$$E = h\nu \qquad \hat{H}\psi = E\psi$$

Particle on a ring
$$E_{m_l} = m_l^2 \hbar^2/2I \quad \mathcal{J}_z = m_l \hbar$$

Photon energy / Schrödinger equation / Free particle

Particle on a sphere
$$E_{l,m_l} = l(l + 1)\hbar^2/2I \quad \mathcal{J} = \{l(l + 1)\}^{1/2}\hbar, \ \mathcal{J}_z = m_l \hbar$$

$$E_k = h\nu - \Phi \qquad \lambda = h/p$$

Photoelectric effect / de Broglie relation

Harmonic oscillator
$$E_v = (v + \tfrac{1}{2})h\nu, \ \nu = (1/2\pi)(k_f/m)^{1/2}$$

$$V = \tfrac{1}{2}k_f x^2$$

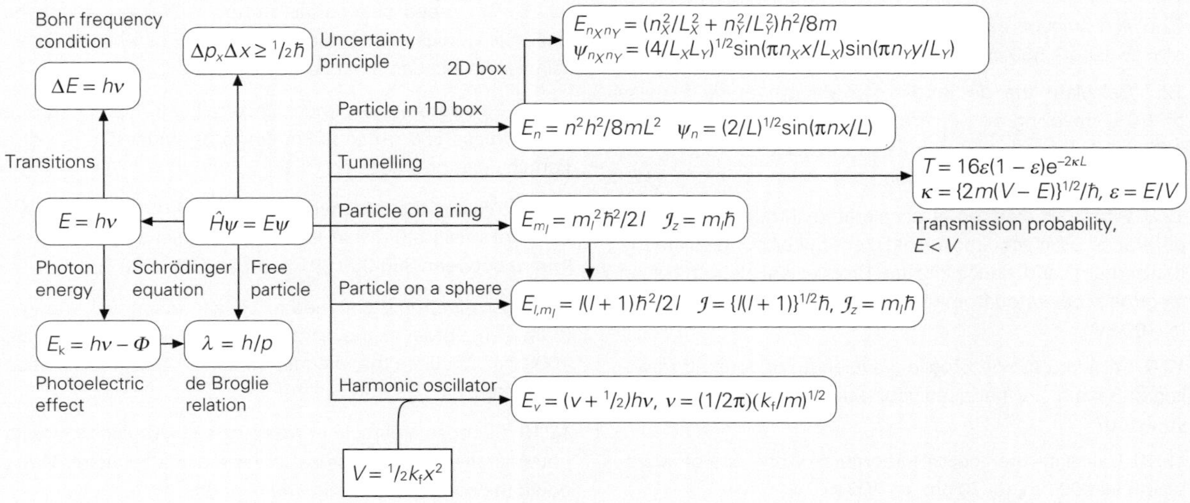

Questions and exercises

Discussion questions

12.1 Summarize the evidence that led to the introduction of quantum theory.

12.2 Discuss the physical origin of quantization energy for a particle confined to moving inside a one-dimensional box or on a ring.

12.3 Define, justify, and provide examples of zero-point energy.

12.4 Describe and justify the Born interpretation of the wavefunction.

12.5 What are the implications of the uncertainty principle?

12.6 Discuss the physical origins of quantum mechanical tunnelling. How does tunnelling appear in chemistry?

Exercises

12.1 The wavelength of the bright red line in the spectrum of atomic hydrogen is 652 nm. What is the energy of the photon generated in the transition?

12.2 What is the wavenumber of the radiation emitted when a hydrogen atom makes a transition corresponding to a change in energy of 10.20 eV?

12.3 Calculate the size of the quantum involved in the excitation of (a) an electronic motion of frequency 1.0×10^{15} Hz, (b) a molecular vibration of period 20 fs, (c) a pendulum of period 0.50 s. Express the results in joules and in kilojoules per mole.

12.4 The work function for metallic caesium is 2.14 eV. Calculate the kinetic energy and the speed of the electrons ejected by light of wavelength (a) 750 nm, (b) 250 nm.

12.5 Use the following data on the kinetic energy of photoelectrons ejected by radiation of different wavelengths from a metal to determine the value of Planck's constant and the work function of the metal.

λ/nm	300	350	400	450
E_k/eV	1.613	1.022	0.579	0.235

12.6 A diffraction experiment requires the use of electrons of wavelength 550 pm. Calculate the speed of the electrons.

12.7 Calculate the de Broglie wavelength of (a) a mass of 1.0 g travelling at 1.0 m s^{-1}, (b) the same, travelling at 1.00×10^5 km s^{-1}, (c) a He atom travelling at 1000 m s^{-1} (a typical speed at room temperature).

12.8 When an electron is accelerated through a potential difference $\Delta\phi$ it acquires a kinetic energy $e\Delta\phi$. Calculate the momentum, and hence the de Broglie wavelength, of an electron accelerated from rest through (a) 1.00 V, (b) 1.00 kV, (c) 100 kV.

12.9 Calculate the de Broglie wavelength of yourself travelling at 8 km h^{-1}. What does your wavelength become when you stop?

12.10 Calculate the linear momentum of photons of wavelength (a) 600 nm, (b) 70 pm, (c) 200 m.

12.11 How fast would a particle of mass 1.0 g need to travel to have the same linear momentum as a photon of radiation of wavelength 300 nm?

12.12 Suppose that you designed a spacecraft to work by photon pressure. The sail was a completely absorbing fabric of area 1.0 km^2 and you directed a red laser beam of wavelength 650 nm on to it from a base on the Moon. What is (a) the force, (b) the pressure exerted by the radiation on the sail? (c) Suppose the mass of the spacecraft was 1.0 kg. Given that, after a period of acceleration from standstill, speed = (force/mass) × time, how long would it take for the craft to accelerate to a speed of 1.0 m s^{-1}?

12.13 A certain wavefunction is zero everywhere except between $x = 0$ and $x = L$, where it has the constant value A. Normalize the wavefunction.

12.14 Suppose a particle has a wavefunction $\psi(x) = Ne^{-ax^2}$. Sketch the form of this wavefunction. Where is the particle most likely to be found? At what values of x is the probability of finding the particle reduced by 50 per cent from its maximum value?

12.15 The speed of a certain proton is 350 km s^{-1}. If the uncertainty in its momentum is 0.0100 per cent, what uncertainty in its location must be tolerated?

12.16 Calculate the minimum uncertainty in the speed of a ball of mass 500 g that is known to be within 5.0 μm of a certain point on a bat.

12.17 What is the minimum uncertainty in the position of a bullet of mass 5.0 g that is known to have a speed somewhere between 350.000 001 m s^{-1} and 350.000 000 m s^{-1}?

12.18 An electron is confined to a linear region with a length of the same order as the diameter of an atom (take that to be 100 pm). Calculate the minimum uncertainties in its position and speed.

12.19 Suppose a particle of mass m is in a region where its potential energy varies as ax^4, where a is a constant. Write down the corresponding Schrödinger equation.

12.20 Calculate the probability that an electron will be found (a) between $x = 0.1$ and 0.2 nm, (b) between 4.9 and 5.2 nm in a box of length L = 10 nm when its wavefunction is $\psi = (2/L)^{1/2}\sin(2\pi x/L)$. Treat the wavefunction as a constant in the small region of interest and interpret δV as δx in this one-dimensional system.

12.21 A hydrogen atom, treated as a point mass, is confined to a one-dimensional square well of side 1.0 nm. How much energy does it have to give up to fall from the level with $n = 2$ to the lowest energy level?

12.22 Consider a rectangular area of sides L and $2L$. Are there any degenerate states? If there are, identify the two lowest.

12.23 The pores in zeolite catalysts are so small that quantum mechanical effects on the distribution of atoms and molecules within them can be significant. Calculate the location in a box of length L at which the probability of a particle being found is 50 per cent of its maximum probability when $n = 1$.

12.24 The blue solution formed when an alkali metal dissolves in liquid ammonia consists of the metal cations and electrons trapped in a cavity formed by ammonia molecules. (a) Calculate the spacing between the levels with $n = 4$ and $n = 5$ of an electron in a one-dimensional box of length 5.0 nm. (b) What is the wavelength of the radiation emitted when the electron makes a transition between the two levels?

12.25 As indicated in the text, a particle in a box is a crude model of the distribution and energy of electrons in conjugated polyenes, such as carotene and related molecules. Carotene itself is a molecule in which 22 single and double bonds alternate (11 of each) along a chain of carbon atoms. Take each C—C bond length to be about 140 pm and suppose that the first possible upward transition is from $n = 11$ to $n = 12$. Estimate the wavelength of this transition.

12.26 Suppose a particle has zero potential energy for $x < 0$, a constant value V, for $0 \le x \le L$, and then zero for $x > L$. Sketch the potential. Now suppose that wavefunction is a sine wave on the left of the barrier, declines exponentially inside the barrier, and then becomes a sine wave on the right, being continuous everywhere and having a continuous slope. Sketch the wavefunction on your sketch of the potential energy.

12.27 An electron is incident on a potential-energy barrier of width L and height $V = \hbar^2/2m_eL^2$. Estimate the probability that the electron will tunnel through the barrier if it has an energy (a) $V/10$ and (b) $V/100$?

12.28 Treat a rotating HI molecule as a stationary I atom around which an H atom circulates in a plane at a distance of 161 pm. Calculate (a) the moment of inertia of the molecule, (b) the greatest wavelength of the radiation that can excite the molecule into rotation.

12.29 The moment of inertia of an H_2O molecule about an axis bisecting the HOH angle is 1.91×10^{-47} kg m^2. Its minimum angular momentum about that axis (other than zero) is $\hbar$. In classical terms, how many revolutions per second do the H atoms make about the axis when in that state?

12.30 What is the minimum energy needed to excite the rotation of an H_2O molecule about the axis described in the preceding exercise?

12.31 The moment of inertia of CH_4 can be calculated from the expression $I = \frac{8}{3}m_HR^2$ where R is the C—H bond length (take $R = 109$ pm). Calculate the minimum rotational energy (other than zero) of the molecule and the degeneracy of that rotational state.

12.32 A bee of mass 1 g lands on the end of a horizontal twig, which starts to oscillate up and down with a period of 1 s. Treat the twig as a massless spring, and estimate its force constant.

12.33 Treat a vibrating HI molecule as a stationary I atom with the H atom oscillating towards and away from the I atom. Given the force constant of the HI bond is 314 N m^{-1}, calculate (a) the vibrational frequency of the molecule, (b) the wavelength required to excite the molecule into vibration.

12.34 By what factor will the vibrational frequency of HI change when H is replaced by deuterium?

Projects

The symbol ‡ indicates that calculus is required.

12.35‡ Here we use calculus to carry out more accurate calculations of probabilities. (a) Repeat Exercise 12.20, but allow for the variation of the wavefunction in the region of interest. What are the percentage errors in the procedure used in Exercise 12.20? What is the probability of finding a particle of mass m in (a) the left-hand one-third, (b) the central one-third, (c) the right-hand one-third of a box of length L when it is in the state with $n = 1$? *Hint*: You will need to integrate $\psi^2 dx$ between the limits of interest. The indefinite integral you require is given in Derivation 12.2.

12.36‡ Now we explore the quantum-mechanical harmonic oscillator in more quantitative detail. (a) The ground-state wavefunction of a harmonic oscillator is proportional to $e^{-ax^2/2}$, where a depends on the mass and force constant. (i) Normalize this wavefunction; you will need the integral $\int_{-\infty}^{\infty} e^{-ax^2}dx = (\pi/a)^{1/2}$. (ii) At what displacement is the oscillator most likely to be found in its ground state? Recall that the maximum (or minimum) of a function $f(x)$ occurs at the value of x for which $df/dx = 0$. (b) Repeat part (a) for the first excited state of a harmonic oscillator, for which the wavefunction is proportional to $xe^{-ax^2/2}$.

12.37 The solutions of the Schrödinger equation for a harmonic oscillator also apply to diatomic molecules. The only complication is that both atoms joined by the bond move, so the 'mass' of the oscillator has to be interpreted carefully. Detailed calculation shows that for two atoms of masses m_A and m_B joined by a bond of force constant k_f, the energy levels are given by eqn 12.25 but with m replaced by the 'effective mass' $\mu = m_A m_B/(m_A + m_B)$. Consider the vibration of carbon monoxide, a poison that prevents the transport and storage of O_2. The bond in a $^{12}C^{16}O$ molecule has a force constant of 1860 N m^{-1}. (a) Calculate the vibrational frequency, ν, of the molecule. (b) In infrared spectroscopy it is common to convert the vibrational frequency of a molecule to its vibrational wavenumber, $\tilde{\nu}$, given by $\tilde{\nu} = \nu/c$. What is the vibrational wavenumber of a $^{12}C^{16}O$ molecule? (c) Assuming that isotopic substitution does not affect the force constant of the C≡O bond, calculate the vibrational wavenumbers of the following molecules: $^{12}C^{16}O$, $^{13}C^{16}O$, $^{12}C^{18}O$, $^{13}C^{18}O$.

13

Quantum chemistry: atomic structure

Chapter 12 provided enough background for us to be able to move on to the discussion of the atomic structure. Atomic structure—the description of the arrangement of electrons in atoms—is an essential part of chemistry because it is the basis for understanding molecular and solid structures and all the physical and chemical properties of elements and their compounds.

A **hydrogenic atom** is a one-electron atom or ion of general atomic number Z. Hydrogenic atoms include H, He^+, Li^{2+}, C^{5+}, and even U^{91+}. Such very highly ionized atoms may be found in the outer regions of stars. A **many-electron atom** is an atom or ion that has more than one electron. Many-electron atoms include all neutral atoms other than H. For instance, helium, with its two electrons, is a many-electron atom in this sense. Hydrogenic atoms, and H in particular, are important because the Schrödinger equation can be solved for them and their structures can be discussed exactly. They provide a set of concepts that are used to describe the structures of many-electron atoms and (as we shall see in the next chapter) the structures of molecules too.

Hydrogenic atoms 315

13.1 The spectra of hydrogenic atoms 316
13.2 The permitted energies of hydrogenic atoms 316
13.3 Quantum numbers 318
13.4 The wavefunctions: s orbitals 321
13.5 The wavefunctions: p and d orbitals 324
13.6 Electron spin 325
13.7 Spectral transitions and selection rules 326

The structures of many-electron atoms 327

13.8 The orbital approximation 327
13.9 The Pauli principle 328
13.10 Penetration and shielding 328
13.11 The building-up principle 329
13.12 The occupation of d orbitals 330
13.13 The configurations of cations and anions 331
13.14 Self-consistent field orbitals 331

Periodic trends in atomic properties 332

13.15 Atomic radius 332
13.16 Ionization energy and electron affinity 333

The spectra of complex atoms 335

13.17 Term symbols 335
13.18 Spin–orbit coupling 338
13.19 Selection rules 338

FURTHER INFORMATION 13.1 339
CHECKLIST OF KEY CONCEPTS 340
ROAD MAP OF KEY EQUATIONS 341
QUESTIONS AND EXERCISES 341

Hydrogenic atoms

Energetically excited atoms are produced when an electric discharge is passed through a gas or vapour or when an element is exposed to a hot flame. These atoms emit electromagnetic radiation of discrete frequencies as they discard energy and return to the **ground state**, their state of lowest energy (Fig. 13.1). The record of frequencies (ν, typically in hertz, Hz), wavenumbers ($\tilde{\nu} = \nu/c$, typically in reciprocal centimetres, cm^{-1}), or wavelengths ($\lambda = c/\nu$, typically in nanometres, nm), of the radiation emitted is called the **emission spectrum** of the atom. In its earliest form, the radiation was detected photographically as a series of lines (the focussed image of the slit that the

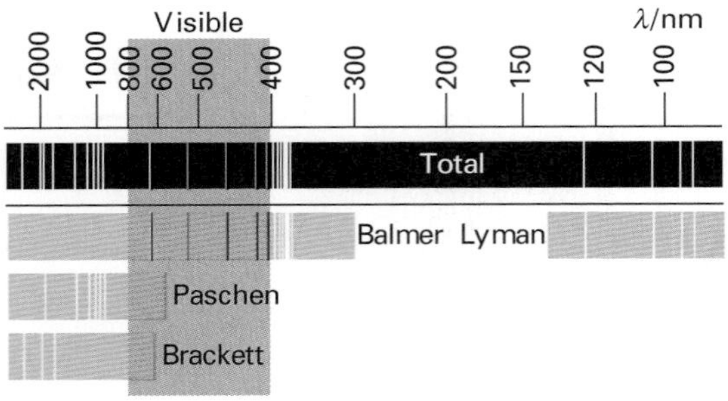

Fig. 13.1 The spectrum of atomic hydrogen. The spectrum is shown at the top, and is analysed into overlapping series below. The Balmer series lies largely in the visible region.

light was sampled through), and the components of radiation present in a spectrum are still widely referred to as spectroscopic 'lines'.

13.1 The spectra of hydrogenic atoms

The first important contribution to understanding the spectrum of atomic hydrogen, which is observed when an electric discharge is passed through hydrogen gas, was made by the Swiss schoolteacher Johann Balmer. In 1885 he pointed out that (in modern terms) the wavenumbers of the light in the visible region of the electromagnetic spectrum fit the expression

$$\tilde{\nu} \propto \frac{1}{2^2} - \frac{1}{n^2}$$

with $n = 3, 4, \ldots$. The lines described by this formula are now called the **Balmer series** of the spectrum. Later, another set of lines was discovered in the ultraviolet region of the spectrum, and is called the **Lyman series.** Yet another set was discovered in the infrared region when detectors became available for that region, and is called the **Paschen series.** With this additional information available, the Swedish spectroscopist Johannes Rydberg noted (in 1890) that all the lines are described by the expression

$$\tilde{\nu} = R_H\left(\frac{1}{n_1^2} - \frac{1}{n_2^2}\right) \qquad \text{Rydberg formula} \qquad (13.1)$$

with $n_1 = 1, 2, \ldots$, $n_2 = n_1 + 1, n_1 + 2, \ldots$, and $R_H = 109\,677\ \text{cm}^{-1}$. The constant R_H is now called the **Rydberg constant** for hydrogen. The first five series of lines then correspond to n_1 taking the values 1 (Lyman), 2 (Balmer), 3 (Paschen), 4 (Brackett), and 5 (Pfund).

As we saw in Section 12.1, the existence of discrete spectroscopic lines strongly suggests that the energy of atoms is quantized and that when an atom changes its energy by ΔE, this difference is carried away as a photon of frequency ν related to ΔE by the **Bohr frequency condition** (eqn 12.1):

$$\Delta E = h\nu \qquad \text{Bohr frequency condition} \qquad (13.2)$$

In terms of the wavenumber $\tilde{\nu}$ of the radiation the Bohr frequency condition is $\Delta E = hc\tilde{\nu}$. It follows that we can expect to observe discrete lines if an electron in an atom can exist only in certain energy states and electromagnetic radiation induces transitions between them.

13.2 The permitted energies of hydrogenic atoms

The quantum mechanical description of the structure of a hydrogenic atom is based on Rutherford's **nuclear model**, in which the atom is pictured as consisting of an electron outside a central nucleus of charge Ze. To derive the details of the structure of this type of atom, we have to set up and solve the Schrödinger equation in which the potential energy, V, is the Coulomb potential energy for the interaction between the nucleus of charge $+Ze$ and the electron of charge $-e$. We saw in The chemist's toolkit 9.1 that the Coulombic potential energy of a charge Q_1 at a distance r from another charge Q_2 is

$$V(r) = \frac{Q_1 Q_2}{4\pi\varepsilon_0 r} \qquad \text{Coulomb potential energy} \qquad (13.3a)$$

where $\varepsilon_0 = 8.854 \times 10^{-12}\ \text{J}^{-1}\ \text{C}^2\ \text{m}^{-1}$ is the vacuum permittivity. On setting $Q_1 = +Ze$ and $Q_2 = -e$ this expression becomes

$$V(r) = -\frac{Ze^2}{4\pi\varepsilon_0 r} \qquad \text{Electron in a hydrogenic atom} \quad \text{Coulomb potential energy} \qquad (13.3b)$$

The negative sign indicates that the potential energy falls (becomes more negative) as the distance between the nucleus and the electron decreases.

We need to identify the appropriate conditions that the wavefunctions must satisfy in order to be acceptable. For the hydrogen atom, these conditions are that the wavefunction must not become infinite anywhere and that it must repeat itself (just like the particle on the surface of a sphere) on circling the nucleus either over the poles or round the equator. We should expect that, with three conditions to satisfy, three quantum numbers will emerge.

With a lot of work, the Schrödinger equation with this potential energy and these conditions can be solved. As usual, the need to satisfy conditions leads to the conclusion that the electron can have only certain energies, which is qualitatively in accord with the spectroscopic evidence. Schrödinger himself found that for a hydrogenic atom of atomic number Z with a nucleus of mass m_N, the allowed energy levels are given by the expression

$$E_n = -\frac{hcR_N Z^2}{n^2} \qquad \text{Hydrogenic atom} \quad \text{Energy levels} \qquad (13.4a)$$

where

$$R_N = \frac{\mu e^4}{8\varepsilon_0^2 h^3 c} \quad \mu = \frac{m_e m_N}{m_e + m_N} \qquad \text{Rydberg constant, reduced mass} \qquad (13.4b)$$

and $n = 1, 2, \ldots$. The quantity μ is called the **reduced mass** of the atom, which is a kind of effective mass that takes into account the fact that although most of the motion within an atom is that of the electron, the heavy nucleus is not completely stationary. The constant R_N is numerically identical to the experimental Rydberg constant R_H when m_N is set equal to the mass of the proton. Because $m_N \gg m_e$, even for hydrogen, in all but the most precise calculations, we can set $\mu = m_e$. Schrödinger must have been thrilled to find that when he calculated R_H, the value he obtained was in almost exact agreement with the experimental value.

Here we shall focus on eqn 13.4a, and unpack its significance. We shall examine three aspects of it: the role of n, the significance of the negative sign, and the appearance in the equation of Z^2.

The quantum number n is called the **principal quantum number**. We use it to calculate the energy of the electron in the atom by substituting its value into eqn 13.4a. The resulting energy levels are depicted in Fig. 13.2. Note how they are widely separated at low values of n, but then converge as n increases. At

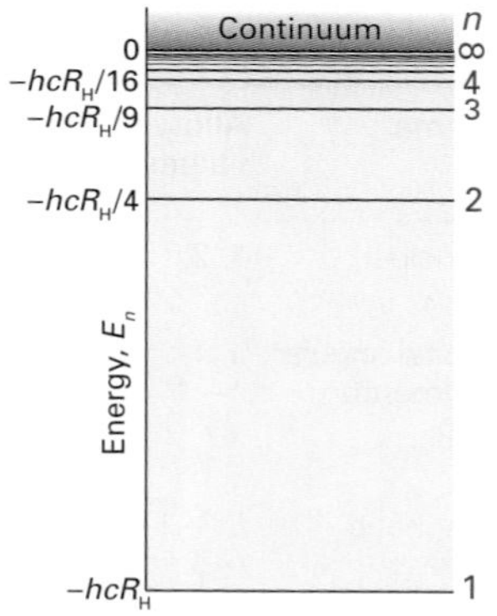

Fig. 13.2 The energy levels of the hydrogen atom. The energies are relative to a proton and an infinitely distant, stationary electron.

low values of n the electron is confined close to the nucleus by the attraction of opposite charges and the energy levels are widely spaced like those of a particle in a narrow box. At high values of n, when the electron has such a high energy that it can travel out to large distances, the energy levels are close together, like those of a particle in a large box.

Now for the sign in eqn 13.4a. All the energies are negative, which signifies that an electron in an atom has a lower energy than when it is free. The zero of energy (which occurs at $n = \infty$) corresponds to the infinitely widely separated (so that the Coulomb potential energy is zero) and stationary (so that the kinetic energy is zero) electron and nucleus. The state of lowest, most negative, energy, the ground state of the atom, is the one with $n = 1$ (the smallest permitted value of n and hence the most negative value of the energy). The energy of this state is $E_1 = -hcR_N Z^2$: the negative sign means that the ground state lies $hcR_N Z^2$ *below* the energy of the infinitely separated and stationary electron and nucleus. The first excited state of the atom, the state with $n = 2$, lies at $E_2 = -\frac{1}{4}hcR_N Z^2$. This energy level is $\frac{3}{4}hcR_N Z^2$ above the ground state.

These results allow us to explain the empirical expression for the spectroscopic lines observed in the emission spectrum of atomic hydrogen (for which $R_N = R_H$ and $Z = 1$). In a transition, an electron jumps from an energy level with one quantum number (n_2) to a level with a lower energy (with quantum number n_1). As a result, the energy discarded is

$$\Delta E = \left(-\frac{hcR_H}{n_2^2}\right) - \left(-\frac{hcR_H}{n_1^2}\right) = hcR_H\left(\frac{1}{n_1^2} - \frac{1}{n_2^2}\right)$$

This energy is carried away by a photon of energy $hc\tilde{\nu}$. By equating this energy to ΔE, we immediately obtain eqn 13.1.

Now consider the significance of Z^2 in eqn 13.4a. The fact that the energy levels are proportional to Z^2 stems from two effects. First, an electron at a given distance from a nucleus of charge Ze has a potential energy that is Z times more negative than an electron at the same distance from a proton (for which $Z = 1$). However, the electron is drawn in to the vicinity of the nucleus by the greater nuclear charge, so it is more likely to be found closer to the nucleus of charge Z than the proton. This effect is also proportional to Z, so overall the energy of an electron can be expected to be proportional to the square of Z, one factor representing the Z times greater strength of the nuclear field and the second factor representing the fact that the electron is Z times more likely to be found closer to the nucleus.

Brief illustration 13.1 Transitions in other hydrogenic atoms

The shortest wavelength transition in the Paschen series in hydrogen occurs at 821 nm. According to eqn 13.4a, the allowed energies of one-electron atoms and ions are proportional to Z^2, so we can expect the frequencies of the transitions between the same two levels to scale as Z^2 too. Because wavelength is inversely proportional to frequency, it follows that the wavelength of the transition should scale as $1/Z^2$. Therefore, the same transition in Li^{2+} (for which $Z=3$) can be expected to lie at $\frac{1}{9}\times 821$ nm = 91.2 nm.

Self-test 13.1

A transition in the Brackett series of Li^{2+} is observed at a wavelength of 450 nm. What is the quantum number of the upper energy level?

Answer: 5

The minimum energy needed to remove an electron completely from an atom is called the **ionization energy**, I. For a hydrogen atom, the ionization energy is the energy required to raise the electron from the ground state (with $n = 1$ and energy $E_1 = -hcR_H$) to the state corresponding to complete removal of the electron (the state with $n = \infty$ and zero energy). Therefore, the energy that must be supplied is $I = hcR_H = 2.180 \times 10^{-18}$ J, which corresponds to 1312 kJ mol^{-1} or 13.59 eV.

Brief illustration 13.2 Ionization energies of other hydrogenic atoms

As in Brief illustration 13.1, we note from eqn 13.4a that the allowed energies are proportional to Z^2, and therefore conclude that ionization energies of one-electron species should also scale as Z^2. Because the ionization energy of H is 13.59 eV, we can predict that the ionization energy of He^+ (for which $Z=2$) should be $I_{He^+} = 4I_H = 54.36$ eV. That is in fact its experimentally determined value.

13.3 Quantum numbers

The wavefunction of the electron in a hydrogenic atom is called an **atomic orbital.** The name is intended to express something less definite than the 'orbit' of classical mechanics. An electron that is described by a particular wavefunction is said to 'occupy' that orbital. So, in the ground state of the atom, the electron occupies the orbital of lowest energy (that with $n = 1$).

We have remarked that there are three mathematical conditions on the orbitals: that the wavefunctions must decay to zero as it extends to infinity, that they must match as we encircle the equator, and that they must match as we encircle the poles. Each condition gives rise to a quantum number, so each orbital is specified by three quantum numbers that act as a kind of 'address' of the electron in the atom. We can suspect that the values allowed to the three quantum numbers are linked because, as we saw in the discussion of a particle on a sphere, to get the right shape on a polar journey we also have to note how the wavefunction changes shape as we travel round the equator. It turns out that the relations between the allowed values are very simple.

We saw in Chapter 12 that in certain cases a wavefunction can be separated into factors that depend on different coordinates and that the Schrödinger equation separates into simpler versions for each variable. As may be expected for a system like a hydrogen atom, by using the separation of variables procedure, its Schrödinger equation separates into one equation for the electron moving around the nucleus (the analogue of the particle on a sphere treated in Section 12.8) and an equation for the radial dependence. The wavefunction correspondingly factorizes, and is written

$$\psi_{n,l,m_l}(r,\theta,\phi) = \overbrace{R_{n,l}(r)}^{\text{Radial wavefunction}} \times \overbrace{Y_{l,m_l}(\theta,\phi)}^{\text{Angular wavefunction}} \qquad \text{Hydrogenic orbital} \quad (13.5)$$

The factor $R_{n,l}(r)$ is called the **radial wavefunction.** The factor $Y_{l,m_l}(\theta,\phi)$ is called the **angular wavefunction** and is exactly the wavefunction we found for a particle on a sphere. As can be seen from this expression, the wavefunction is specified by three quantum numbers, all of which we have already met in different guises (Section 13.2 and Chapter 12):

Quantum number	Name	Allowed values	Determines
n	principal	1, 2, . . . ,	Energy, through $E_n = -hcR_N Z^2/n^2$
l	orbital angular momentum	0, 1, . . . , $n-1$	Orbital angular momentum, through $\mathcal{J} = \{l(l+1)\}^{1/2}\hbar$
m_l	magnetic	$l, l-1, l-2, \ldots, -l$	z-Component of orbital angular momentum, through $\mathcal{J}_z = m_l\hbar$

Note that the radial wavefunction $R_{nl}(r)$ depends only on n and l, so all wavefunctions of a given n and l have the same radial shape regardless of the value of m_l. Similarly, the angular wavefunction $Y_{l,m_l}(\theta,\phi)$ depends only on l and m_l, so all wavefunctions of a given l and m_l have the same angular shape regardless of the value of n. The explicit expressions for some of the orbitals are shown in Table 13.1; we use the notation that is described immediately below.

Table 13.1
*Hydrogenic wavefunctions**

Orbital	Radial wavefunction	Angular wavefunction
1s	$2\left(\frac{Z}{a_0}\right)^{3/2} e^{-Zr/a_0}$	$\frac{1}{2\pi^{1/2}}$
2s	$\frac{1}{8^{1/2}}\left(\frac{Z}{a_0}\right)^{3/2}\left(2-\frac{Zr}{a_0}\right)e^{-Zr/2a_0}$	$\frac{1}{2\pi^{1/2}}$
$2p_x$	$\frac{1}{24^{1/2}}\left(\frac{Z}{a_0}\right)^{3/2}\left(\frac{Zr}{a_0}\right)e^{-Zr/2a_0}$	$\frac{1}{2}\left(\frac{3}{\pi}\right)^{1/2}\sin\theta\cos\phi$
$2p_y$		$\frac{1}{2}\left(\frac{3}{\pi}\right)^{1/2}\sin\theta\sin\phi$
$2p_z$		$\frac{1}{2}\left(\frac{3}{\pi}\right)^{1/2}\cos\theta$
3s	$\frac{2}{243^{1/2}}\left(\frac{Z}{a_0}\right)^{3/2}\left(3-\frac{2Zr}{a_0}+\frac{2Z^2r^2}{9a_0^2}\right)e^{-Zr/3a_0}$	$\frac{1}{2\pi^{1/2}}$
$3p_x$	$\frac{2}{486^{1/2}}\left(\frac{Z}{a_0}\right)^{3/2}\left(2-\frac{Zr}{3a_0}\right)e^{-Zr/3a_0}$	$\frac{1}{2}\left(\frac{3}{\pi}\right)^{1/2}\sin\theta\cos\phi$
$3p_y$		$\frac{1}{2}\left(\frac{3}{\pi}\right)^{1/2}\sin\theta\sin\phi$
$3p_z$		$\frac{1}{2}\left(\frac{3}{\pi}\right)^{1/2}\cos\theta$
$3d_{xy}$	$\frac{1}{2430^{1/2}}\left(\frac{Z}{a_0}\right)^{3/2}\left(\frac{2Zr}{3a_0}\right)^2 e^{-Zr/3a_0}$	$\frac{1}{4}\left(\frac{15}{\pi}\right)^{1/2}\sin^2\theta\sin 2\phi$
$3d_{yz}$		$\frac{1}{2}\left(\frac{15}{\pi}\right)^{1/2}\cos\theta\sin\theta\sin\phi$
$3d_{zx}$		$\frac{1}{2}\left(\frac{15}{\pi}\right)^{1/2}\cos\theta\sin\theta\cos\phi$
$3d_{x^2-y^2}$		$\frac{1}{4}\left(\frac{15}{\pi}\right)^{1/2}\sin^2\theta\cos 2\phi$
$3d_{z^2}$		$\frac{1}{4}\left(\frac{5}{\pi}\right)^{1/2}(3\cos^2\theta-1)$

* $a_0 = 4\pi\varepsilon_0\hbar^2/m_e e^2$, the Bohr radius. All orbitals of a given subshell have the same radial wavefunction.

Brief illustration 13.3 Quantum numbers and orbitals

It follows from the restrictions on the values of the quantum numbers that there is only one orbital with $n = 1$, because when $n = 1$ the only value that l can have is 0, and that in turn implies that m_l can have only the value 0. Likewise, there are four orbitals with $n = 2$, because l can take the values 0 and 1, and in the latter case m_l can have the three values +1, 0, and −1. In general, there are n^2 orbitals with a given value of n.

A note on good practice Always give the sign of m_l, even when it is positive. So, write $m_l = +1$, not $m_l = 1$.

Although we need all three quantum numbers to specify a given orbital, eqn 13.4 reveals that for hydrogenic atoms—and, as we shall see, *only* for hydrogenic atoms—the energy depends only on the principal quantum number, n. Therefore, in hydrogenic atoms, and only in hydrogenic atoms, *all orbitals of the same value of n but different values of l and* m_l *have the same energy*. Recall from Section 12.7 that when we have more than one wavefunction corresponding to the same energy, we say that the wavefunctions are 'degenerate'; so, now we can say that in hydrogenic atoms all orbitals with the same value of n are degenerate. This degeneracy arises because of the symmetry of the centrosymmetric Coulomb potential. We might expect, and shall indeed discover later in this chapter, that this degeneracy is partly removed for atoms with more than one electron because the high level of symmetry is broken.

The degeneracy of all orbitals with the same value of n (remember from Brief illustration 13.3 that there are n^2 of them) and, as we shall see, their similar mean radii, is the basis for saying that they all belong to the same **shell** of the atom. It is common to refer to successive shells by letters:

n	1	2	3	4 . . .
	K	L	M	N . . .

Thus, all four orbitals of the shell with $n = 2$ form the L shell of the atom.

Orbitals with the same value of n but different values of l belong to different **subshells** of a given shell. These subshells are denoted by the letters s, p, . . . using the following correspondence:[1]

l	0	1	2	3 . . .
	s	p	d	f . . .

[1] The letters have only a historical significance: they originally stood for *sharp*, *principal*, *diffuse*, and *fundamental*, referring to the characteristics of certain spectral lines.

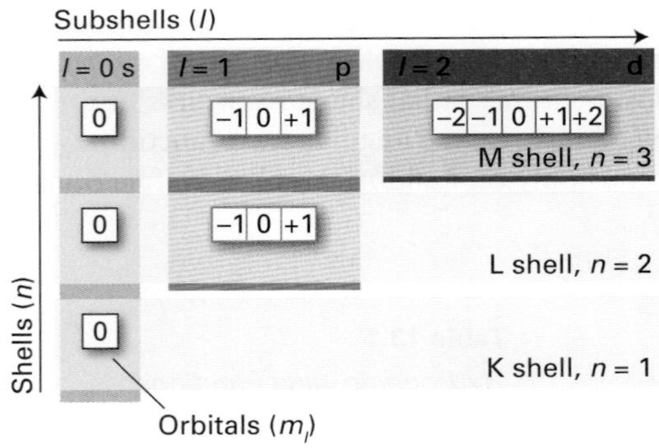

Fig. 13.3 The structures of atoms are described in terms of shells of electrons that are labelled by the principal quantum number n, and a series of n subshells of these shells, with each subshell of a shell being labelled by the quantum number l. Each subshell consists of $2l + 1$ orbitals.

Only these four types of subshell are important in practice. For the shell with $n = 1$, there is only one subshell, the one with $l = 0$. For the shell with $n = 2$ (which allows $l = 0, 1$), there are two subshells, namely the 2s subshell (with $l = 0$) and the 2p subshell (with $l = 1$). The general pattern of the first three shells and their subshells is shown in Fig. 13.3. In a hydrogenic atom, all the subshells of a given shell correspond to the same energy (because, as we have seen, the energy depends on n and not on l).

We have seen that if the orbital angular momentum quantum number is l, then m_l can take the $2l + 1$ values $m_l = 0, \pm1, \ldots, \pm l$. Therefore, each subshell contains $2l + 1$ individual orbitals (corresponding to the $2l + 1$ values of m_l for each value of l). It follows that in any given subshell, the number of orbitals is

s	p	d	f
1	3	5	7

An orbital with $l = 0$ (and necessarily $m_l = 0$) is called an **s orbital.** A p subshell ($l = 1$) consists of three **p orbitals** (corresponding to $m_l = +1, 0, -1$). An electron that occupies an *s* orbital is called an **s electron.** Similarly, we can speak of p, d, . . . electrons according to the orbitals they occupy.

Brief illustration 13.4 The composition of a shell

As remarked in Brief illustration 13.3, there are n^2 orbitals in a shell with principal quantum number n, so there are 25 orbitals in the shell with $n = 5$. The orbital angular quantum number l can take the values 0,1,2,3, and 4, with $2l + 1$ orbitals in each subshell. We conclude that the shell consists of one s, three p, five d, seven f, and nine g orbitals.

13.4 The wavefunctions: s orbitals

The mathematical form of a 1s orbital (the wavefunction with $n = 1$, $l = 0$, and $m_l = 0$) for a hydrogen atom is

$$\psi_{1s} = \overbrace{2\left(\frac{1}{a_0^3}\right)^{1/2} e^{-r/a_0}}^{R_{1,0}} \times \overbrace{\frac{1}{2\pi^{1/2}}}^{Y_{0,0}} = \frac{1}{(\pi a_0^3)^{1/2}} e^{-r/a_0} \quad \text{H 1s orbital} \quad (13.6a)$$

where

$$a_0 = \frac{4\pi\varepsilon_0\hbar^2}{m_e e^2} \quad \text{Definition} \quad \text{Bohr radius} \quad (13.6b)$$

In this case the angular wavefunction, $Y_{0,0} = 1/2\pi^{1/2}$, is a constant, independent of the angles θ and ϕ. You should recall that in Example 12.3 we anticipated that a wavefunction for an electron in a hydrogen atom is proportional to e^{-r}: eqn 13.6a is its precise form. The constant a_0 is called the **Bohr radius** (because it occurred in Bohr's calculation of the properties of the hydrogen atom) and has the value 52.9177 pm. The wavefunction in eqn 13.6 is normalized to 1 (Section 12.7), so the probability of finding the electron in a small volume of magnitude δV at a given point is *equal* to $\psi^2\delta V$, with ψ evaluated at a point in the region of interest. We are supposing that the volume δV is so small that the wavefunction does not vary inside it.

The general form of the wavefunction can be understood by considering the contributions of the potential and kinetic energies to the total energy of the atom. The closer the electron is to the nucleus on average, the lower its average potential energy. This dependence suggests that the lowest potential energy should be obtained with a sharply peaked wavefunction that has a large amplitude at the nucleus and is almost zero everywhere else (Fig. 13.4). However, this shape implies a high kinetic energy, because such a wavefunction has a very high average curvature. The electron would have very low kinetic energy if its wavefunction had only a very low average curvature. However, such a wavefunction spreads to great distances from the nucleus and the average potential energy of the electron will be correspondingly high. The actual wavefunction is a compromise between these two extremes: the wavefunction spreads away from the nucleus (so the potential energy is not as low as in the first example, but nor is it very high) and has a reasonably low average curvature (so the kinetic energy is not very low, but nor is it as high as in the first example).

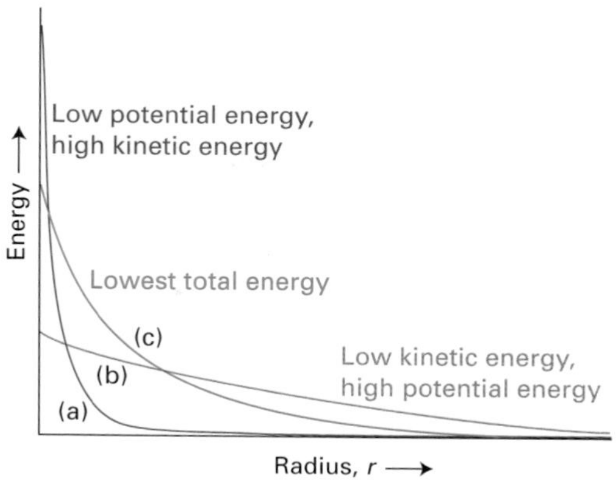

Fig. 13.4 The balance of kinetic and potential energies that accounts for the structure of the ground state of hydrogen (and similar atoms). (a) The sharply curved but localized orbital has high mean kinetic energy, but low mean potential energy; (b) the mean kinetic energy is low, but the potential energy is not very favourable; (c) the compromise of moderate kinetic energy and moderately favourable potential energy.

A 1s orbital depends only on the radius, r, of the point of interest and is independent of angle (the latitude and longitude of the point). Therefore, the orbital has the same amplitude at all points at the same distance from the nucleus regardless of direction. Because the probability of finding an electron is proportional to the square of the wavefunction, we now know that the electron will be found with the same probability in any direction (for a given distance from the nucleus). We summarize this angular independence by saying that a 1s orbital is **spherically symmetrical**. Because the same factor Y occurs in all orbitals with $l = 0$, all s orbitals have the same spherical symmetry.

The wavefunction in eqn 13.6 decays exponentially towards zero from a maximum value at the nucleus (Fig. 13.5). It follows that *the most probable point at*

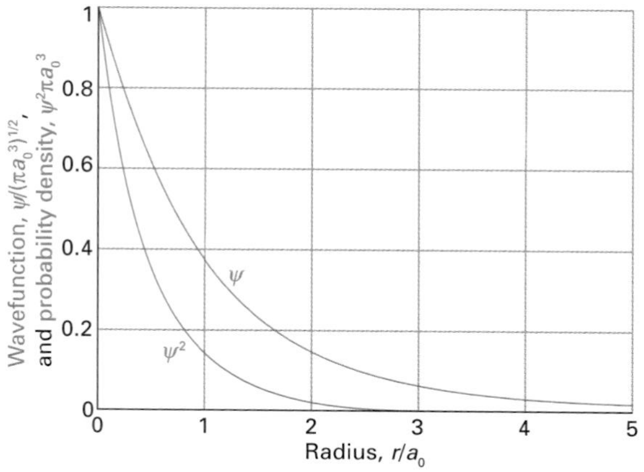

Fig. 13.5 The radial dependence of the wavefunction of a 1s orbital ($n = 1$, $l = 0$) and the corresponding probability density. The quantity a_0 is the Bohr radius (52.9 pm).

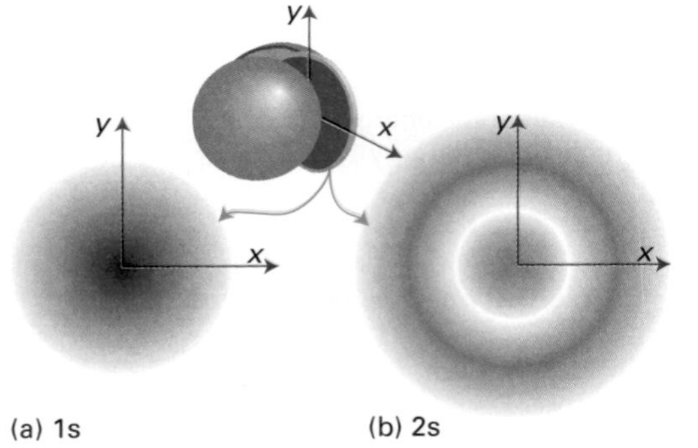

Fig. 13.6 Representations of the first two hydrogenic s orbitals, (a) 1s, (b) 2s, in terms of the electron densities in a slice through the centre of the atom (as represented by the density of shading) shown at the origin of the two green arrows.

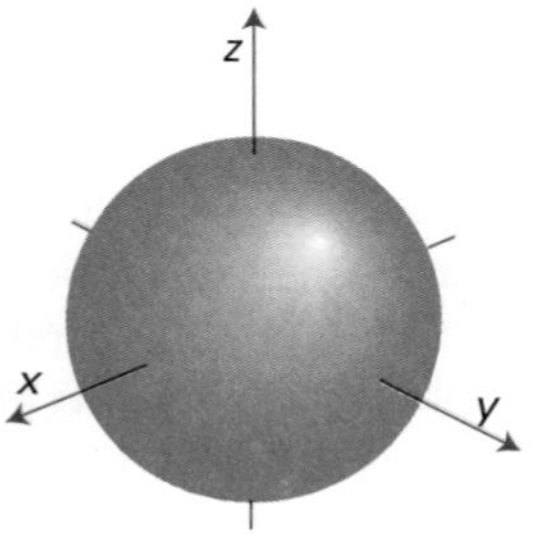

Fig. 13.7 The boundary surface of an s orbital within which there is a high probability of finding the electron.

which the electron will be found is at the nucleus itself. A method of depicting the probability of finding the electron at each point in space is to represent ψ^2 by the density of shading in a diagram (Fig. 13.6). A simpler procedure is to show only the **boundary surface**, the shape that captures about 90 per cent of the electron probability. For the 1s orbital, the boundary surface is a sphere centred on the nucleus (Fig. 13.7).

Brief illustration 13.5 Probability distribution

We can calculate the probability of finding the electron in a volume of 1.0 pm^3 centred on the nucleus in a hydrogen atom by setting $r = 0$ in the expression for ψ, using $e^0 = 1$, and taking $\delta V = 1.0$ pm^3. The value of ψ at the nucleus is $1/(\pi a_0^3)^{1/2}$. Therefore, $\psi^2 = 1/\pi a_0^3$ at the nucleus, and we can write

$$\text{Probability} = \frac{1}{\pi a_0^3} \times \delta V = \frac{1}{\pi \times (52.9\ \text{pm})^3} \times (1.0\ \text{pm}^3)$$

$$= \frac{1.0}{\pi \times 52.9^3} = 2.2 \times 10^{-5}$$

This result means that the electron will be found in the volume on one observation in 455 000.

Self-test 13.2

Repeat the calculation for finding the electron in the same volume located at the Bohr radius.

Answer: 3.0×10^{-7}, 1 in 3 300 000 observations

We often need to know the probability that an electron will be found at a given distance from a nucleus regardless of its angular position (Fig. 13.8). We can calculate this probability by combining the wavefunction in eqn 13.5 with the Born interpretation and, as shown in the following Derivation, find that, for an s orbital, the answer can be expressed as

$$\text{Probability} = P(r)\delta r \quad \text{with } P(r) = 4\pi r^2 \psi^2 \qquad \text{s orbitals} \quad \text{Radial distribution function} \qquad (13.7a)$$

The function P is called the **radial distribution function**. The more general form, which also applies to orbitals that depend on angle, is

$$P(r) = r^2 R(r)^2 \qquad \text{General form} \quad \text{Radial distribution function} \qquad (13.7b)$$

where $R(r)$ is the radial wavefunction.

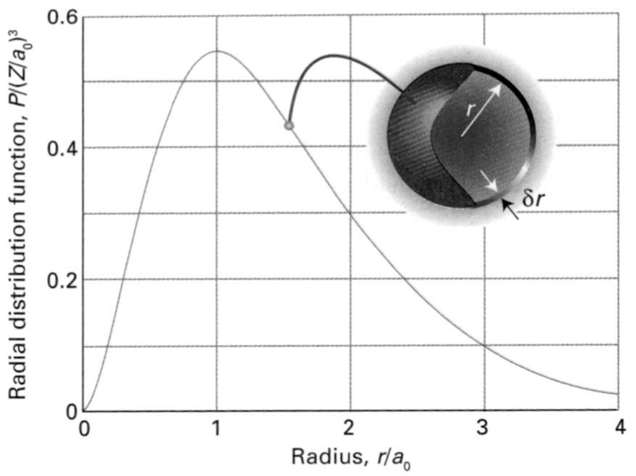

Fig. 13.8 The radial distribution function gives the probability that the electron will be found anywhere in a shell of radius r and thickness δr regardless of angle. The graph shows the output from an imaginary shell-like detector of variable radius and fixed thickness δr.

Example 13.1

Using the radial distribution function

Calculate the probability that the electron will be found anywhere between a shell of radius a_0 and a shell of radius 1.0 pm greater when it occupies a 1s orbital of a hydrogenic atom of atomic number Z, and apply the result to $Z = 1$.

Strategy Take the form of the orbital (for general Z) from Table 13.1 and substitute it into the expression for P in eqn 13.7a. Treat the thickness of the shell as so small that the wavefunction is uniform inside it.

Solution The hydrogenic 1s wavefunction is

$$\psi_{1s} = \overbrace{2\left(\frac{Z}{a_0}\right)^{3/2} e^{-Zr/a_0}}^{R_{1,0}} \times \overbrace{\frac{1}{2\pi^{1/2}}}^{Y_{0,0}} = \frac{1}{\pi^{1/2}}\left(\frac{Z}{a_0}\right)^{3/2} e^{-Zr/a_0}$$

So the radial distribution function for this orbital is

$$P(r) = 4\pi r^2\left(\frac{1}{\pi^{1/2}}\left(\frac{Z}{a_0}\right)^{3/2} e^{-Zr/a_0}\right)^2 = 4\left(\frac{Z}{a_0}\right)^3 r^2 e^{-2Zr/a_0}$$

Now we substitute $r = a_0$ to find the radial distribution function at that distance:

$$P(a_0) = 4\left(\frac{Z}{a_0}\right)^3 a_0^2 e^{-2Za_0/a_0} = \frac{4Z^3}{a_0} e^{-2Z}$$

For hydrogen, with $\delta r = 1.0$ pm and $Z = 1$,

$$\text{Probability} = P(a_0)\delta r = \frac{4}{a_0} e^{-2} \times \delta r$$

$$= \frac{4}{\underbrace{52.92 \text{ pm}}_{a_0}} \times e^{-2} \times \overbrace{(1.0 \text{ pm})}^{\delta r} = 0.010$$

or about 1 inspection in 100.

Self-test 13.3

Repeat the analysis for a 2s orbital.

Answer: 8.7×10^{-4}, 1 inspection in 1100

Derivation 13.1

The radial distribution function

Consider two spherical shells centred on the nucleus, one of radius r and the other of radius $r + \delta r$. The probability of finding the electron at a radius r regardless of its direction is equal to the probability of finding it between these two spherical surfaces. The volume of the region of space between the surfaces is equal to the surface area of the inner shell, $4\pi r^2$, multiplied by the thickness, δr, of the region, and is therefore $4\pi r^2\delta r$. According to the Born interpretation, the probability of finding an electron inside a small volume of magnitude δV is given, for a normalized wavefunction that is constant throughout the region, by the value of $\psi^2\delta V$. An s orbital has the same value at all angles at a given distance from the nucleus, so it is constant throughout the shell (provided δr is very small). Therefore, interpreting δV as the volume of the shell, we obtain

$$\text{Probability} = \psi^2 \times (4\pi r^2\delta r)$$

as in eqn 13.7a. The result we have derived applies only to s orbitals.

The radial distribution function tells us the probability of finding an electron at a distance r from the nucleus regardless of its direction. Because r^2 increases from 0 as r increases but ψ^2 decreases towards 0 exponentially, P starts at 0, goes through a maximum, and declines to 0 again. The location of the maximum marks the most probable *radius* (not point) at which the electron will be found. For a 1s orbital of hydrogen, the maximum occurs at a_0, the Bohr radius. An analogy that might help to fix the significance of the radial distribution function for an electron is the corresponding distribution for the population of the Earth regarded as a perfect sphere. The radial distribution function is zero at the centre of the Earth and for the next 6400 km (to the surface of the planet), when it peaks sharply and then rapidly decays again to zero. It remains virtually zero for all radii more than about 10 km above the surface. Almost all the population will be found very close to $r = 6400$ km, and it is not relevant that people are dispersed non-uniformly over a very wide range of latitudes and longitudes. The small probabilities of finding people above and below 6400 km anywhere in the world correspond to the population that happens to be down mines or living in places as high as Denver or Tibet at the time.

A 2s orbital (an orbital with $n = 2$, $l = 0$, and $m_l = 0$) is also spherical, so its boundary surface is a sphere. Because a 2s orbital spreads further out from the nucleus than a 1s orbital—because the electron it describes has more energy to climb away from the nucleus—its boundary surface is a sphere of larger radius. The orbital also differs from a 1s orbital in its radial dependence (Fig. 13.9), for although the wavefunction has a nonzero value at the nucleus (like all s orbitals), it passes through zero before commencing its exponential decay towards zero at large distances. We summarize the fact that the wavefunction passes through zero everywhere at a certain radius by saying that the orbital has a **radial node.** A 3s orbital has two radial nodes, a 4s orbital has three radial nodes. In general, an ns orbital has $n - 1$ radial nodes.

A general feature of orbitals is that their mean radii increase with n, for more radial nodes have to be fitted into the wavefunction, with the result that it spreads out to greater radii. All orbitals of the same principal quantum number have similar mean radii, which reinforces the notion of the shell structure of the atom. Mean radii decrease with increasing Z, because the increased nuclear charge attracts the electron more strongly and it is confined more closely to the nucleus.

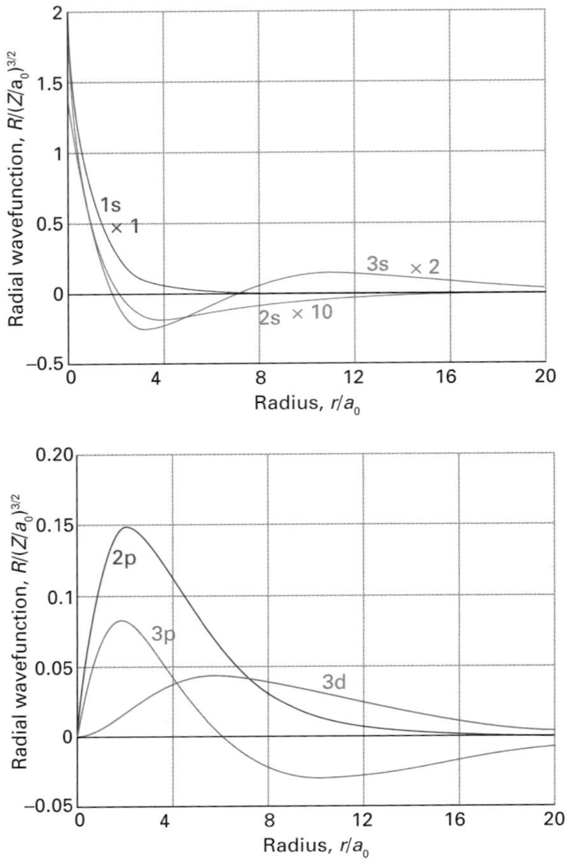

Fig. 13.9 The radial wavefunctions of some hydrogenic s, p, and d orbitals. Note that the s orbitals have a nonzero and finite value at the nucleus. The vertical scales are different in each case.

13.5 The wavefunctions: p and d orbitals

All p orbitals (orbitals with $l = 1$) have a double-lobed appearance like that shown in Fig. 13.10. The two lobes are separated by a **nodal plane** that cuts through the nucleus and arises from the angular wavefunction $Y_{l,m_l}(\theta,\phi)$. There is zero probability density for an electron on this plane. Here, for instance, is the explicit form of the $2p_z$ orbital:

$$\psi = \overbrace{\frac{1}{2}\left(\frac{1}{6a_0^3}\right)^{1/2}\frac{r}{a_0}e^{-r/2a_0}}^{R_{2,1}} \times \overbrace{\left(\frac{3}{4\pi}\right)^{1/2}\cos\theta}^{Y_{1,0}}$$

$$= \left(\frac{1}{32\pi a_0^5}\right)^{1/2} r\cos\theta\, e^{-r/2a_0}$$

Note that because ψ has a factor r, it is zero at the nucleus, so there is zero probability density of the electron at the nucleus. That is not a radial node because the wavefunction does not pass *through* zero there because r does not extend to negative values. However, there is a nodal plane through the nucleus: the orbital is also zero everywhere on the plane with $\cos\theta = 0$, corresponding to $\theta = 90°$ and the wavefunction changes sign on passing through this plane. The p_x and p_y orbitals are similar, but have nodal planes perpendicular to this one.

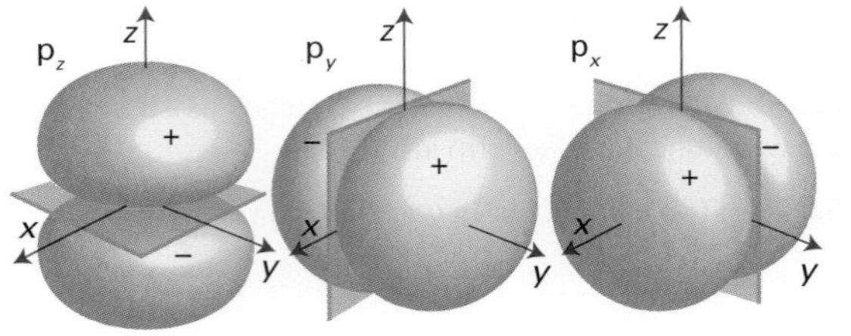

Fig. 13.10 The boundary surfaces of p orbitals. A nodal plane passes through the nucleus and separates the two lobes of each orbital.

The exclusion of the electron from the nucleus is a common feature of all atomic orbitals except s orbitals. To understand its origin, we need to note that the value of the quantum number l tells us the magnitude of the angular momentum of the electron around the nucleus (in classical terms, how rapidly it is circulating around the nucleus). For an s orbital, the orbital angular momentum is zero (because $l = 0$), and in classical terms the electron does not circulate around the nucleus. Because $l = 1$ for a p orbital, the magnitude of the angular momentum of a p electron is $2^{1/2}\hbar$. As a result, a p electron—in classical terms—is flung away from the nucleus by the centrifugal force arising from its motion, but an s electron is not. The same centrifugal effect appears in all orbitals with angular momentum (those for which $l > 0$), such as d orbitals and f orbitals, and all such orbitals have nodal planes that cut through the nucleus.

Each p subshell consists of three orbitals ($m_l = +1, 0, -1$). The three orbitals are normally represented by their boundary surfaces, as depicted in Fig. 13.10. The p_x orbital has a symmetrical double-lobed shape directed along the x-axis, and similarly the p_y and p_z orbitals are directed along the y- and z-axes, respectively. As n increases, the p orbitals become bigger (for the same reason as s orbitals) and have $n - 2$ radial nodes. However, their boundary surfaces retain the double-lobed shape shown in the illustration.

Each d subshell ($l = 2$) consists of five orbitals ($m_l = +2, +1, 0, -1, -2$). These five orbitals are normally represented by the boundary surfaces shown in Fig. 13.11 and labelled as shown there.

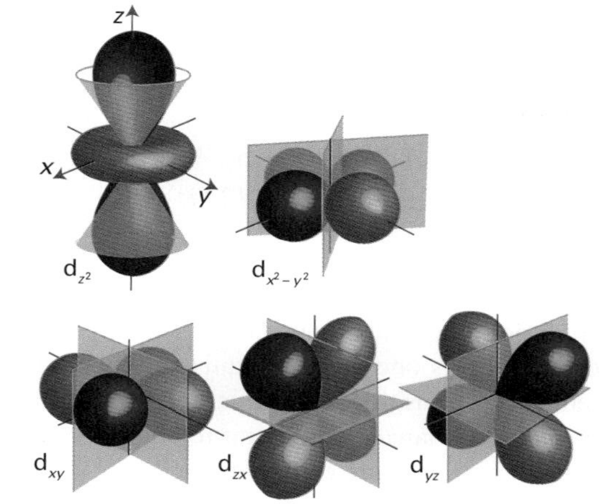

Fig. 13.11 The boundary surfaces of d orbitals. Two nodal planes in each orbital intersect at the nucleus and separate the four lobes of each orbital. (For a d_{z^2} orbital the planes are replaced by conical surfaces.) The light and dark tones denote regions of opposite sign of the wavefunction.

The quantum number m_l indicates, through the expression $m_l\hbar$, the component of the electron's orbital angular momentum around an arbitrary axis passing through the nucleus. As explained in Section 12.8, positive values of m_l correspond to clockwise motion seen from below and negative values correspond to anticlockwise motion. An s electron has $m_l = 0$, and has no orbital angular momentum about any axis. A p electron can circulate clockwise about an axis as seen from below ($m_l = +1$). Of its total orbital angular momentum of $2^{1/2}\hbar = 1.414\hbar$, an amount $\hbar$ is due to motion around the selected axis (the rest is due to motion around the other two axes). A p electron can also circulate anticlockwise as seen from below ($m_l = -1$), or not at all ($m_l = 0$) about that selected axis. An electron in the d subshell can circulate with five different amounts of orbital angular momentum about an arbitrary axis ($+2\hbar, +\hbar, 0, -\hbar, -2\hbar$).

Except for orbitals with $m_l = 0$, there is not a one-to-one correspondence between the value of m_l and the orbitals shown in the illustrations: we cannot say, for instance, that a p_x orbital has $m_l = +1$. For technical reasons, the orbitals we draw are combinations of orbitals with opposite values of m_l (p_x, for instance, is the sum—a superposition—of the orbitals with $m_l = +1$ and -1).

13.6 Electron spin

To complete the description of the state of a hydrogenic atom, we need to introduce one more concept, that of electron spin. The **spin** of an electron is an

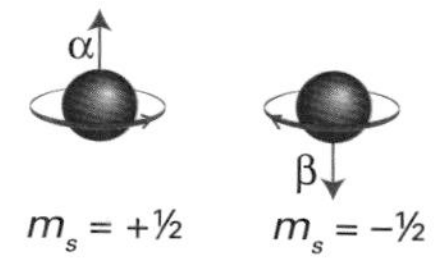

Fig. 13.12 A classical representation of the two allowed spin states of an electron. The magnitude of the spin angular momentum is $(3^{1/2}/2)\hbar$ in each case, but the directions of spin are opposite.

intrinsic angular momentum that every electron possesses and which cannot be changed or eliminated (just like its mass or its charge). The name 'spin' is evocative of a ball spinning on its axis, and (so long as it is treated with caution) this classical interpretation can be used to help to visualize the motion. However, in fact spin is a purely quantum mechanical phenomenon and has no classical counterpart, so the analogy must be used with care.

We shall make use of two properties of electron spin (Fig. 13.12):

1. Electron spin is described by a **spin quantum number**, s (the analogue of l for orbital angular momentum), with s fixed at the single (positive) value of $\frac{1}{2}$ for all electrons at all times.
2. The spin can be clockwise or anticlockwise; these two states are distinguished by the **spin magnetic quantum number**, m_s, which can take the values $+\frac{1}{2}$ or $-\frac{1}{2}$ but no other values.

An electron with $m_s = +\frac{1}{2}$ is called an **α electron** and commonly denoted α or $\uparrow$; an electron with $m_s = -\frac{1}{2}$ is called a **β electron** and denoted β or $\downarrow$.

A note on good practice The quantum number s is equal to $\frac{1}{2}$ for electrons. You will occasionally see its value written incorrectly as $s = +\frac{1}{2}$ or $s = -\frac{1}{2}$. For the projection, use m_s.

The existence of the electron spin was demonstrated by an experiment performed by Otto Stern and Walther Gerlach in 1921, who shot a beam of silver atoms through a strong, inhomogeneous magnetic field (Fig. 13.13). A silver atom has 47 electrons, and (as will be familiar from introductory chemistry and will be reviewed later in this chapter) 23 of the spins are $\uparrow$ and 23 spins are $\downarrow$; the one remaining spin may be either $\uparrow$ or $\downarrow$. Because the spin angular momenta of the $\uparrow$ and $\downarrow$ electrons cancel each other, the atom behaves as if it had the spin of a single electron. The idea behind the Stern–Gerlach experiment was that a rotating, charged body—in this case an electron—behaves like a magnet and interacts with the applied field. Because the magnetic field pushes or pulls the electron according to the orientation of

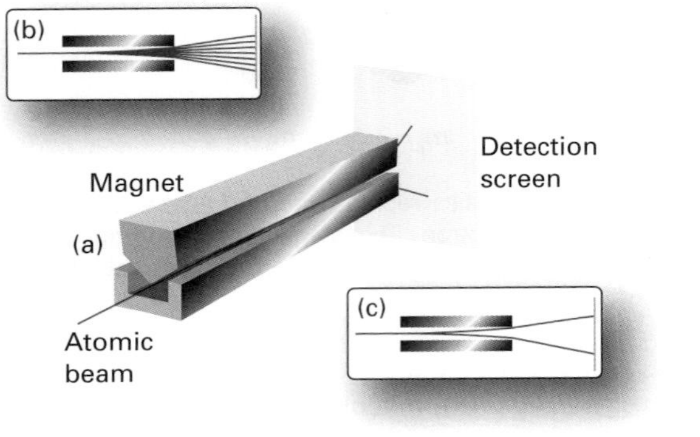

Fig. 13.13 (a) The experimental arrangement for the Stern–Gerlach experiment: the magnet is the source of an inhomogeneous field. (b) The classically expected result, when the orientations of the electron spins can take all angles. (c) The observed outcome using silver atoms, when the electron spins can adopt only two orientations (↑ and ↓).

the electron's spin, the initial beam of atoms should split into two beams, one corresponding to atoms with ↑ spin and the other to atoms with ↓ spin. This result was observed.

Other fundamental particles also have characteristic spins. For example, protons and neutrons are **spin-½ particles** (that is, for them $s = \frac{1}{2}$) so always spin with a single, irremovable angular momentum. Because the masses of a proton and a neutron are so much greater than the mass of an electron, yet they all have the same spin angular momentum, the classical picture of proton and neutron spin would be of particles spinning much more slowly than an electron. Some elementary particles have $s = 1$ and therefore have a higher intrinsic angular momentum than an electron. For our purposes the most important **spin-1 particle** is the photon.

It is a very deep feature of nature, that the fundamental particles from which matter is built have half–integral spin (such as electrons and quarks, all of which have $s = \frac{1}{2}$). The particles that transmit forces between these particles, so binding them together into entities like nuclei, atoms, and planets, all have integral spin (such as $s = 1$ for the photon, which transmits the electromagnetic interaction between charged particles). Fundamental particles with half-integral spin are called **fermions**; those with integral spin are called **bosons.** Matter therefore consists of fermions bound together by the exchange of bosons.

13.7 Spectral transitions and selection rules

We can think of the sudden change in the distribution of the electron as it changes its spatial distribution from one orbital to another orbital as jolting the electromagnetic field into oscillation, and that oscillation corresponds to the generation of a photon of light. It turns out, however, that not all transitions between all available orbitals are possible. For example, it is not possible for an electron in a 3d orbital to make a transition to a 1s orbital. Transitions are classified as either **allowed,** if they can contribute to the spectrum, or **forbidden,** if they cannot. The allowed or forbidden character of a transition can be traced to the role of the photon spin, which we mentioned above. When a photon, with its one unit of angular momentum, is generated in a transition, the angular momentum of the electron must change by one unit to compensate for the angular momentum carried away by the photon. That is, the angular momentum must be conserved—neither created nor destroyed—just as linear momentum is conserved in collisions. Thus, an electron in a d orbital (with $l = 2$) cannot make a transition into an s orbital (with $l = 0$) because the photon cannot carry away enough angular momentum. Similarly, an s electron cannot make a transition to another s orbital, because then there is no change in the electron's angular momentum to make up for the angular momentum carried away by the photon.

A **selection rule** is a statement about which spectroscopic transitions are allowed. They are derived (for atoms) by identifying the transitions that conserve angular momentum when a photon is emitted or absorbed. The selection rules for hydrogenic atoms are

$$\Delta l = \pm 1 \qquad \Delta m_l = 0, \pm 1$$

The principal quantum number n can change by any amount consistent with the Δl for the transition because it does not relate directly to the angular momentum.

Brief illustration 13.6 Selection rules for hydrogenic atoms

To identify the orbitals to which an electron in a 4d orbital of a hydrogenic atom may make spectroscopic transitions we apply the selection rules, principally the rule concerning l. Because $l = 2$, the final orbital must have $l = 1$ or 3. Thus, an electron may make a transition from a 4d orbital to any np orbital (subject to $\Delta m_l = 0, \pm 1$) and to any nf orbital (subject to the same rule). However, it cannot undergo a transition to any other orbital, so a transition to any ns orbital or another nd orbital is forbidden.

Self-test 13.4

To what orbitals may a 4s electron make spectroscopic transitions?

Answer: np orbitals only

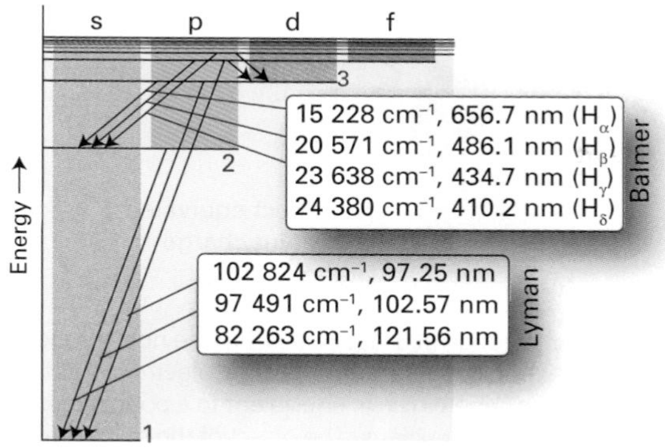

Fig. 13.14 A Grotrian diagram that summarizes the appearance and analysis of the spectrum of atomic hydrogen.

Selection rules enable us to construct a **Grotrian diagram** (Fig. 13.14), which is a diagram that summarizes the energies of the states and the allowed transitions between them. The thickness of a transition line in the diagram is sometimes used to indicate in a general way its relative intensity in the spectrum.

The structures of many-electron atoms

The Schrödinger equation for a many-electron atom is highly complicated because all the electrons interact with one another. Even for a He atom, with its two electrons, no mathematical expression for the orbitals and energies can be given and we are forced to make approximations. Modern computational techniques, though, are able to refine the approximations we are about to make, and permit highly accurate numerical calculations of energies and wavefunctions.

13.8 The orbital approximation

We show in the following Derivation that it is a general rule in quantum mechanics that the wavefunction for several non-interacting particles is the product of the wavefunctions for each particle. This rule justifies the **orbital approximation**, in which we suppose that a reasonable first approximation to the exact wavefunction is obtained by letting each electron occupy its 'own' orbital, and writing

$$\psi = \psi(1)\psi(2)\ldots \qquad (13.8)$$

where $\psi(1)$ is the wavefunction of electron 1, $\psi(2)$ that of electron 2, and so on.

Derivation 13.2

Many-electron wavefunctions

Consider a two-electron system. If the electrons do not interact with one another, the total hamiltonian that appears in the Schrödinger equation is the sum of contributions from each electron, and the equation itself, using notation like that in eqn 12.5b, is

$$\{\hat{H}(1) + \hat{H}(2)\}\psi(1,2) = E\psi(1,2)$$

We need to verify that $\psi(1,2) = \psi(1)\psi(2)$ is a solution, where each individual wavefunction is a solution of its 'own' Schrödinger equation:

$$\hat{H}(1)\psi(1) = E(1)\psi(1) \qquad \hat{H}(2)\psi(2) = E(2)\psi(2)$$

To do so, we substitute $\psi(1,2) = \psi(1)\psi(2)$ into the full equation, then let $\hat{H}(1)$ operate on $\psi(1)$ and $\hat{H}(2)$ operate on $\psi(2)$:

$$\begin{aligned}\{\hat{H}(1) + \hat{H}(2)\}\psi(1)\psi(2) &= \hat{H}(1)\psi(1)\psi(2) + \psi(1)\hat{H}(2)\psi(2)\\ &= E(1)\psi(1)\psi(2) + \psi(1)E(2)\psi(2)\\ &= \{E(1) + E(2)\}\psi(1)\psi(2)\end{aligned}$$

This expression has the form of the original Schrödinger equation, so $\psi(1,2) = \psi(1)\psi(2)$ is indeed a solution, and we can identify the total energy as $E = E(1) + E(2)$. Note that this argument fails if the electrons interact with one another, because then there is an additional term in the hamiltonian and the variables cannot be separated. For electrons in the same atom, therefore, writing $\psi(1,2) = \psi(1)\psi(2)$ is an approximation.

We can think of the individual orbitals as resembling the hydrogenic orbitals, but with nuclear charges that are modified by the presence of all the other electrons in the atom. This description is only approximate, but it is a useful model for discussing the properties of atoms, and is the starting point for more sophisticated descriptions of atomic structure.

Brief illustration 13.7 A two-electron wavefunction

If both electrons occupy the same 1s orbital for an atom with atomic number Z, the wavefunction for each electron in He ($Z = 2$) is $\psi = (8/\pi a_0^3)^{1/2}e^{-2r/a_0}$. If electron 1 is at a radius r_1 and electron 2 is at a radius r_2 (and at any angle), then the overall wavefunction for the two-electron atom is

$$\psi = \psi(1)\psi(2) = \overbrace{\left(\frac{8}{\pi a_0^3}\right)^{1/2} e^{-2r_1/a_0}}^{\psi(1)} \times \overbrace{\left(\frac{8}{\pi a_0^3}\right)^{1/2} e^{-2r_2/a_0}}^{\psi(2)}$$

$$\overset{e^x e^y = e^{x+y}}{=} \frac{8}{\pi a_0^3} e^{-2(r_1+r_2)/a_0}$$

The orbital approximation allows us to express the electronic structure of an atom by reporting its **configuration**, a statement of the orbitals that are occupied (usually, but not necessarily, in its ground state). For example, because the ground state of a hydrogen atom consists of a single electron in a 1s orbital, we report its configuration as $1s^1$ (read 'one s one'). A helium atom has two electrons. We can imagine forming the atom by adding the electrons in succession to the orbitals of the bare nucleus (of charge $2e$). The first electron occupies a hydrogenic 1s orbital, but because $Z = 2$, the orbital is more compact than in H itself. The second electron joins the first in the same 1s orbital, and so the electron configuration of the ground state of He is $1s^2$ (read 'one s two').

13.9 The Pauli principle

Lithium, with $Z = 3$, has three electrons. Two of its electrons occupy a 1s orbital drawn even more closely than in He around the more highly charged nucleus. The third electron, however, does not join the first two in the 1s orbital because a $1s^3$ configuration is forbidden by a fundamental feature of nature summarized by the **Pauli exclusion principle:**

> No more than two electrons may occupy any given orbital, and if two electrons do occupy one orbital, then their spins must be paired.

Electrons with **paired spins**, denoted ↑↓, have zero net spin angular momentum because the spin angular momentum of one electron is cancelled by the spin of the other. The exclusion principle is the key to understanding the structures of complex atoms, to chemical periodicity, and to molecular structure. It was proposed by the Austrian Wolfgang Pauli in 1924 when he was trying to account for the absence of some lines in the spectrum of helium. In Further information 13.1 we see that the exclusion principle is a consequence of a very deep statement about wavefunctions.

Lithium's third electron cannot enter the 1s orbital because that orbital is already full: we say the K shell ($n = 1$) is **complete** and that the two electrons form a **closed shell**. Because a similar closed shell occurs in the He atom, we denote it [He]. The third electron is excluded from the K shell and must occupy the next available orbital, which is one with $n = 2$ and hence belonging to the L shell. However, we now have to decide whether the next available orbital is the 2s orbital or a 2p orbital, and therefore whether the lowest energy configuration of the atom is $[He]2s^1$ or $[He]2p^1$.

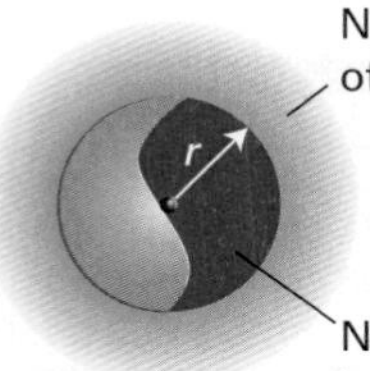

Fig. 13.15 An electron at a distance r from the nucleus experiences a Coulombic repulsion from all the electrons within a sphere of radius r and which is equivalent to a point negative charge located on the nucleus. The effect of the point charge is to reduce the apparent nuclear charge of the nucleus from Ze to $Z_{eff}e$.

13.10 Penetration and shielding

Unlike in hydrogenic atoms, in many-electron atoms the 2s and 2p orbitals (and, in general, all the subshells of a given shell) are not degenerate. For reasons we shall now explain, s electrons generally lie lower in energy than p electrons of a given shell, and p electrons lie lower than d electrons.

An electron in a many-electron atom experiences a Coulombic repulsion from all the other electrons present. When the electron is at a distance r from the nucleus, the repulsion it experiences from the other electrons can be modelled by a point negative charge located on the nucleus and having a magnitude equal to the charge of the electrons within a sphere of radius r (Fig. 13.15). The effect of the point negative charge is to lower the full charge of the nucleus from Ze to $Z_{eff}e$, the **effective nuclear charge.** To express the fact that an electron experiences a nuclear charge that has been modified by the other electrons present, we say that the electron experiences a **shielded nuclear charge.** The electrons do not actually 'block' the full Coulombic attraction of the nucleus: the effective charge is simply a way of expressing the net outcome of the nuclear attraction and the electronic repulsions in terms of a single equivalent charge at the centre of the atom.

> **A note on good practice** Commonly, Z_{eff} itself is referred to as the 'effective nuclear charge', although strictly that quantity is $Z_{eff}e$.

The effective nuclear charges experienced by s and p electrons are different because the electrons have different wavefunctions and therefore different distributions around the nucleus (Fig. 13.16). An s electron has a greater **penetration** through inner shells than a p electron of the same shell in the sense that an s electron is more likely to be found close to the nucleus

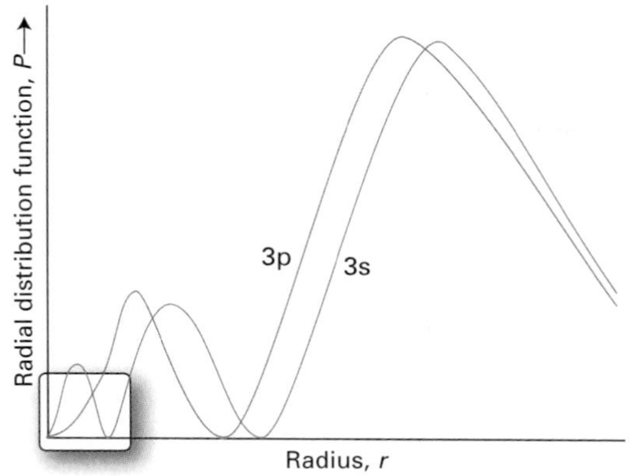

Fig. 13.16 An electron in an s orbital (here a 3s orbital) is more likely to be found close to the nucleus than an electron in a p orbital of the same shell.

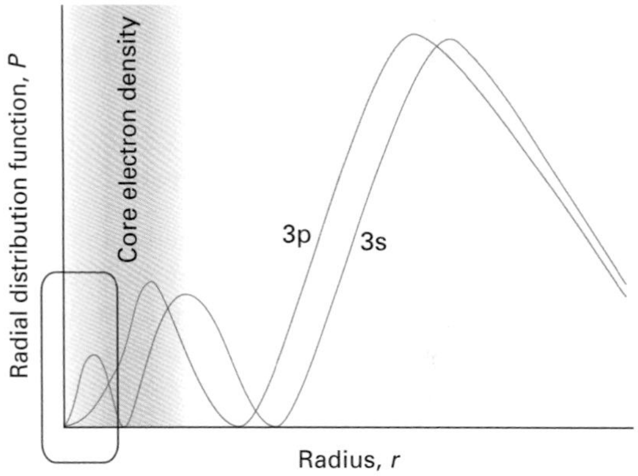

Fig. 13.17 The radial distribution function of an *n*s orbital (here, $n = 3$) shows that the electron that occupies it penetrates through the core electron density more than an electron in an *n*p orbital (see the highlighted region) with the result that it experiences a less shielded nuclear charge.

than a p electron of the same shell (Fig. 13.17). As a result of this greater penetration, an s electron experiences less shielding than a p electron of the same shell and therefore experiences a larger $Z_{eff}e$. Consequently, by the combined effects of penetration and shielding, an s electron is more tightly bound than a p electron of the same shell. Similarly, a d electron penetrates less than a p electron of the same shell, and it therefore experiences more shielding and an even smaller $Z_{eff}e$.

The consequence of penetration and shielding is that, in general, the energies of orbitals in the same shell of a many-electron atom lie in the order s < p < d < f. The individual orbitals of a given subshell (such as the three p orbitals of the p subshell) remain degenerate because they all have the same radial characteristics and so experience the same effective nuclear charge.

We can now complete the Li story. Because the shell with $n = 2$ has two non-degenerate subshells, with the 2s orbital lower in energy than the three 2p orbitals, the third electron occupies the 2s orbital. This arrangement results in the ground-state configuration $1s^22s^1$, or $[He]2s^1$. It follows that we can think of the structure of the atom as consisting of a central nucleus surrounded by a complete helium-like shell of two 1s electrons, and around that a more diffuse 2s electron. The electrons in the outermost shell of an atom in its ground state are called the **valence electrons** because they are largely responsible for the chemical bonds that the atom forms (and, as we shall see, the extent to which an atom can form bonds is called its 'valence'). Thus, the valence electron in Li is a 2s electron, and lithium's other two electrons belong to its core, where they take little part in bond formation.

13.11 The building-up principle

The extension of the procedure used for H, He, and Li to other atoms is called the **building-up principle.** The building-up principle, which is still widely called the *Aufbau principle* (from the German word for building up), specifies an order of occupation of atomic orbitals that reproduces the experimentally determined ground-state configurations of neutral atoms.

We imagine the bare nucleus of atomic number Z, and then feed into the available orbitals Z electrons one after the other. The first two rules of the building-up principle are:

- The order of occupation of orbitals is
  1s 2s 2p 3s 3p 4s 3d 4p 5s 4d 5p 6s 5d 4f 6p . . .
- According to the Pauli exclusion principle, each orbital may accommodate up to two electrons.

The order of occupation is approximately the order of energies of the individual orbitals, because in general the lower the energy of the orbital, the lower the total energy of the atom as a whole when that orbital is occupied. An s subshell is complete as soon as two electrons are present in it. Each of the three p orbitals of a shell can accommodate two electrons, so a p subshell is complete as soon as six electrons are present in it. A d subshell, which consists of five orbitals, can accommodate up to ten electrons.

Brief illustration 13.8 An electron configuration

Consider a carbon atom. Because $Z = 6$ for carbon, there are six electrons to accommodate. Two enter and fill the 1*s* orbital, two enter and fill the 2s orbital, leaving two electrons to occupy the orbitals of the 2p subshell. Hence its ground configuration is $1s^22s^22p^2$, or more succinctly $[He]2s^22p^2$, with [He] the helium-like $1s^2$ core.

However, it is possible to be more specific than the conclusion in Brief illustration 13.8 that the ground-state electron configuration is $[He]2s^22p^2$. On electrostatic grounds, we can expect the last two electrons to occupy different 2p orbitals, for they will then be farther apart on average and repel each other less than if they were in the same orbital. Thus, one electron can be thought of as occupying the $2p_x$ orbital and the other the $2p_y$ orbital, and the lowest energy configuration of the atom is $[He]2s^22p_x^12p_y^1$. The same rule applies whenever degenerate orbitals of a subshell are available for occupation. Therefore, another rule of the building-up principle is:

- Electrons occupy different orbitals of a given subshell before doubly occupying any one of them.

It follows that a nitrogen atom ($Z = 7$) has the configuration $[He]2s^22p_x^12p_y^12p_z^1$. Only when we get to oxygen ($Z = 8$) is a 2p orbital doubly occupied, giving the configuration $[He]2s^22p_x^22p_y^12p_z^1$.

An additional point arises when electrons occupy degenerate orbitals (such as the three 2p orbitals) singly, as they do in C, N, and O, for there is then no requirement that their spins should be paired. We need to know whether the lowest energy is achieved when the electron spins are the same (both ↑, for instance, denoted ↑↑, if there are two electrons in question, as in C) or when they are paired (↑↓). This question is resolved by **Hund's rule:**

- In its ground state, an atom adopts a configuration with the greatest number of unpaired electrons.

The explanation of Hund's rule is complicated, but it reflects the quantum mechanical property of **spin correlation**, that electrons in different orbitals with parallel spins have a quantum-mechanical tendency to stay well apart (a tendency that has nothing to do with their charge: even two 'uncharged electrons' would behave in the same way). Their mutual avoidance allows the atom to shrink slightly, so the electron–nucleus interaction is improved when the spins are parallel. We can now conclude that in the ground state of a C atom, the two 2p electrons have the same spin, that all three 2p electrons in an N atom have the same spin, and that the two electrons that singly occupy different 2p orbitals in an O atom have the same spin (the two in the $2p_x$ orbital are necessarily paired).

Neon, with $Z = 10$, has the configuration $[He]2s^22p^6$, which completes the L ($n = 2$) shell. This closed-shell configuration is denoted [Ne], and acts as a core for subsequent elements. The next electron must enter the 3s orbital and begin a new shell, and so a Na atom, with $Z = 11$, has the configuration $[Ne]3s^1$. Like lithium with the configuration $[He]2s^1$, sodium has a single s electron outside a complete core.

This analysis has brought us to the origin of chemical periodicity. The L shell is completed by eight electrons, and so the element with $Z = 3$ (Li) should have similar properties to the element with $Z = 11$ (Na). Likewise, Be ($Z = 4$) should be similar to Mg ($Z = 12$), and so on up to the noble gases He ($Z = 2$), Ne ($Z = 10$), and Ar ($Z = 18$).

13.12 The occupation of d orbitals

Argon has complete 3s and 3p subshells, and as the 3d orbitals are high in energy, the atom effectively has a closed-shell configuration. Indeed, the 4s orbitals are so lowered in energy by their ability to penetrate close to the nucleus that the next electron (for potassium) occupies a 4s orbital rather than a 3d orbital and the K atom resembles a Na atom. The same is true of a Ca atom, which has the configuration $[Ar]4s^2$, resembling that of its partner in the same group, Mg, which is $[Ne]3s^2$.

Ten electrons can be accommodated in the five 3d orbitals, which accounts for the electron configurations of scandium to zinc. The building-up principle has less clear-cut predictions about the ground-state configurations of these elements and a simple analysis no longer works. Calculations show that for these atoms the energies of the 3d orbitals are always lower than the energy of the 4s orbital. However, spectroscopic results show that Sc has the configuration $[Ar]3d^14s^2$, instead of $[Ar]3d^3$ or $[Ar]3d^24s^1$. To understand this observation, we have to consider the nature of electron–electron repulsions in 3d and 4s orbitals. The most probable distance of a 3d electron from the nucleus is less than that for a 4s electron, so two 3d electrons repel each other more strongly than two 4s electrons. As a result, Sc has the configuration $[Ar]3d^14s^2$ rather than the two alternatives, for then the strong electron–electron repulsions in the 3d

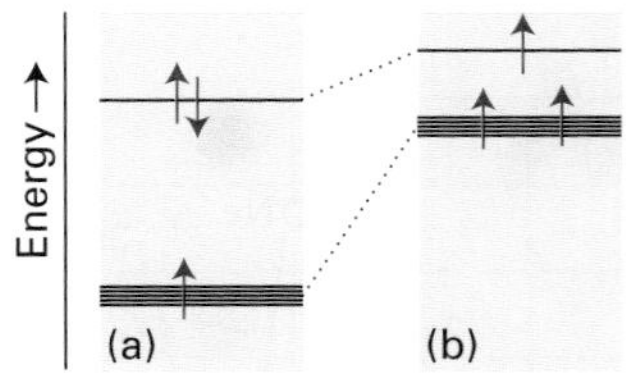

Fig. 13.18 Strong electron–electron repulsions in the 3d orbitals are minimized in the ground state of a scandium atom if (a) the atom has the configuration [Ar]3d^{1}4s^2 instead of (b) [Ar]3d^{2}4s^1. The total energy of the atom is lower when it has the configuration [Ar]3d^{1}4s^2 despite the cost of populating the high-energy 4s orbital.

orbitals are minimized. The total energy of the atom is least despite the cost of allowing electrons to populate the high energy 4s orbital (Fig. 13.18). The effect just described is generally, but not always, true for scandium to zinc, so their electron configurations are of the form [Ar]3d^{N}4s^2, where $N = 1$ for scandium and $N = 10$ for zinc.

At gallium, the energy of the 3d orbitals has fallen so far below those of the 4s and 4p orbitals that they (the full 3d orbitals) can be largely ignored, and the building-up principle can be used in the same way as in preceding periods. Now the 4s and 4p subshells constitute the valence shell, and the period terminates with krypton. Because 18 electrons have intervened since argon, this period is the first **long period** of the periodic table. The existence of the **d block** (the 'transition metals') reflects the stepwise occupation of the 3d orbitals, and the subtle shades of energy differences along this series gives rise to the rich complexity of inorganic (and bioinorganic) d-metal chemistry. A similar intrusion of the f orbitals in Periods 6 and 7 accounts for the existence of the **f block** of the periodic table (the lanthanoids and actinoids).

13.13 The configurations of cations and anions

The configurations of cations of elements in the s, p, and d blocks of the periodic table are derived by removing electrons from the ground-state configuration of the neutral atom in a specific order. First, we remove any valence p electrons, then the valence s electrons, and then as many d electrons as are necessary to achieve the stated charge. The configurations of anions are derived by continuing the building-up procedure and adding electrons to the neutral atom until the configuration of the next noble gas has been reached.

Brief illustration 13.9 The electron configuration of an ion

Because the configuration of Fe is [Ar]3d^{6}4s^2, an Fe^{3+} cation has the configuration [Ar]3d^5. The configuration of an O^{2-} ion is achieved by adding two electrons to [He]2s^{2}2p^4, giving [He]2s^{2}2p^6, the configuration of Ne.

Self-test 13.5

Predict the ground-state electron configurations of (a) a Cu^{2+} ion and (b) an S^{2-} ion.

Answer: (a) [Ar]3d^9, (b) [Ne]3s^{2}3p^6

13.14 Self-consistent field orbitals

The treatment we have given to the electronic configuration of many-electron species is only approximate because it is hopeless to expect to find exact solutions of a Schrödinger equation that takes into account the interaction of all the electrons with one another. However, computational techniques are available that give very detailed and reliable approximate solutions for the wavefunctions and energies. The techniques were originally introduced by D.R. Hartree (before computers were available) and then modified by V. Fock to take into account the Pauli principle correctly. In broad outline, the **Hartree–Fock self-consistent field** (HF-SCF) procedure is as follows.

Imagine that we have a rough idea of the structure of the atom. In the Ne atom, for instance, the orbital approximation suggests the configuration 1s^{2}2s^{2}2p^6 with the orbitals approximated by hydrogenic atomic orbitals. Now consider one of the 2p electrons. A Schrödinger equation can be written for this electron by ascribing to it a potential energy due to the nuclear attraction and the repulsion from the other electrons. Although the equation is for the 2p orbital, it depends on the wavefunctions of all the other occupied orbitals in the atom. To solve the equation, we guess an approximate form of the wavefunctions of all the orbitals except 2p and then solve the Schrödinger equation for the 2p orbital. The procedure is then repeated for the 1s and 2s orbitals. This sequence of calculations gives the form of the 2p, 2s, and 1s orbitals, and in general they will differ from the set used initially to start the calculation. These improved orbitals can be used in another cycle of calculation, and a second improved set of orbitals and a better energy are obtained. The recycling continues until the orbitals and energies obtained are insignificantly

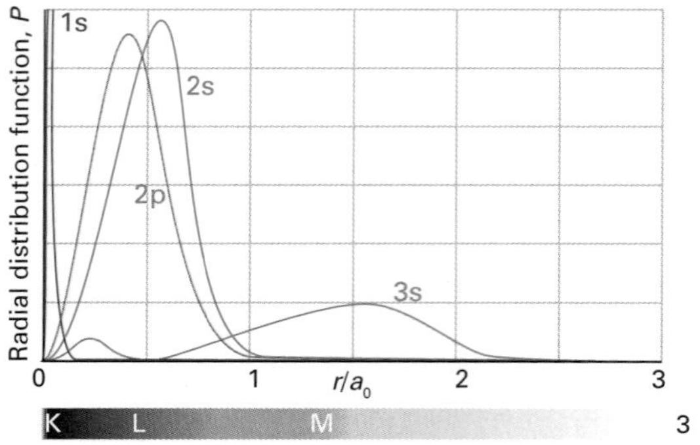

Fig. 13.19 The radial distribution functions for the orbitals of Na based on SCF calculations. Note the shell-like structure, with the 3s orbital outside the inner K and L shells.

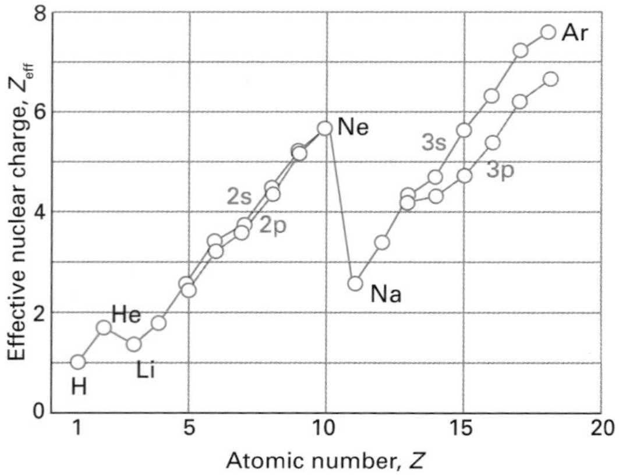

Fig. 13.20 The variation of the effective atomic number (Z_{eff}) with actual atomic number for the elements of the first three periods. The value of Z_{eff} depends on the identity of the orbital occupied by the electron: we show the values only for the valence electrons. The effective nuclear charge is $Z_{\text{eff}}e$.

different from those used at the start of the current cycle. The solutions are then self-consistent and accepted as solutions of the problem.

Figure 13.19 shows plots of some of the HF-SCF radial distribution functions for sodium. They show the grouping of electron density into shells, as was anticipated by the early chemists, and the differences of penetration as discussed above. These SCF calculations therefore support the qualitative discussions that are used to explain chemical periodicity. They also considerably extend that discussion by providing detailed wavefunctions and precise energies.

Periodic trends in atomic properties

The periodic recurrence of analogous ground-state electron configurations as the atomic number increases accounts for the periodic variation in the properties of atoms. Here we concentrate on two aspects of atomic periodicity: atomic radius and ionization energy. Both can be correlated with the effective nuclear charge, and Fig. 13.20 shows how this quantity varies through the first three periods.

13.15 Atomic radius

The **atomic radius** of an element is half the distance between the centres of neighbouring atoms in a solid (such as Cu) or, for non-metals, in a homonuclear molecule (such as H_2 or S_8). It is of great significance in chemistry, for the size of an atom is one of the most important controls on the number of chemical bonds the atom can form. Moreover, the size and shape of a molecule depend on the sizes of the atoms of which it is composed, and molecular shape and size are crucial aspects of a molecule's biological function. Atomic radius also has an important technological aspect, because the similarity of the atomic radii of the d-block elements is the main reason why they can be blended together to form so many different alloys, particularly varieties of steel.

In general, atomic radii decrease from left to right across a period and increase down each group (Table 13.2 and Fig. 13.21). The decrease across a period can be traced to the increase in nuclear charge, which draws the electrons in closer to the nucleus. The increase in nuclear charge is partly cancelled by the increase in the number of electrons, but because electrons are spread over a region of space, one electron does not fully shield one nuclear charge, so the increase in nuclear charge dominates. The increase in

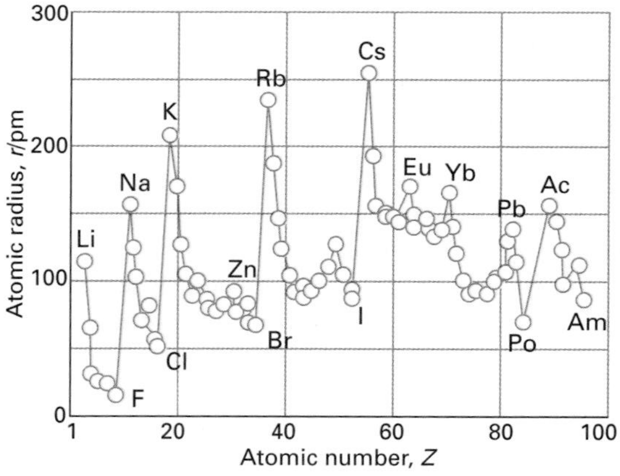

Fig. 13.21 The variation of atomic radius through the periodic table. Note the contraction of radius following the lanthanoids in Period 6 (following Yb, ytterbium).

Table 13.2
Atomic radii of main-group elements, r/pm

Li	Be	B	C	N	O	F
157	112	88	77	74	66	64
Na	Mg	Al	Si	P	S	Cl
191	160	143	118	110	104	99
K	Ca	Ga	Ge	As	Se	Br
235	197	153	122	121	117	114
Rb	Sr	In	Sn	Sb	Te	I
250	215	167	158	141	137	133
Cs	Ba	Tl	Pb	Bi	Po	
272	224	171	175	182	167	

atomic radius down a group (despite the increase in nuclear charge) is explained by the fact that the valence shells of successive periods correspond to higher principal quantum numbers. That is, successive periods correspond to the start and then completion of successive (and more distant) shells of the atom that surround each other like the successive layers of an onion. The need to occupy a more distant shell leads to a larger atom despite the increased nuclear charge.

A modification of the increase down a group is encountered in Period 6, for the radii of the atoms late in the d block and in the following regions of the p block are not as large as would be expected by simple extrapolation down the group. The reason can be traced to the fact that in Period 6 the f orbitals are in the process of being occupied. An f electron is a very inefficient shielder of nuclear charge (for reasons connected with its radial extension), and as the atomic number increases from La to Yb, there is a considerable contraction in radius. By the time the d block resumes (at hafnium, Hf), the poorly shielded but considerably increased nuclear charge has drawn in the surrounding electrons, and the atoms are compact. They are so compact, that the metals in this region of the periodic table (iridium to lead) are very dense. The reduction in radius below that expected by extrapolation from preceding periods is called the **lanthanide contraction**.

13.16 Ionization energy and electron affinity

The minimum energy necessary to remove an electron from a many-electron atom is its **first ionization energy**, I_1. The **second ionization energy**, I_2, is the minimum energy needed to remove a second electron (from the singly charged cation):

$$\begin{aligned} &X(g) \rightarrow X^+(g) + e^-(g) \\ &X^+(g) \rightarrow X^{2+}(g) + e^-(g) \\ &I_1 = E(X^+) - E(X) \\ &I_2 = E(X^{2+}) - E(X^+) \end{aligned} \quad \text{Ionization energy} \quad (13.9)$$

The variation of the first ionization energy through the periodic table is shown in Fig. 13.22 and some numerical values are given in Table 13.3. The ionization energy of an element plays a central role in determining the ability of its atoms to participate in bond formation (for bond formation, as we shall see in Chapter 14, is a consequence of the relocation of electrons from one atom to another). After atomic radius, it is the most important property for determining an element's chemical characteristics.

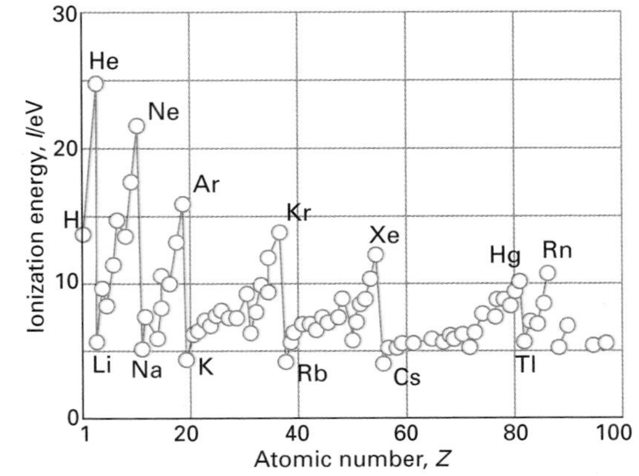

Fig. 13.22 The periodic variation of the first ionization energies of the elements.

Table 13.3

First ionization energies of main-group elements, I/eV *

H 13.60							He **24.59**
Li 5.32	Be 9.32	B 8.30	C 11.26	N 14.53	O 13.62	F 17.42	Ne 21.56
Na 5.14	Mg 7.65	Al 5.98	Si 8.15	P 10.49	S 10.36	Cl 12.97	Ar 15.76
K 4.34	Ca 6.11	Ga 6.00	Ge 7.90	As 9.81	Se 9.75	Br 11.81	Kr 14.00
Rb 4.18	Sr 5.70	In 5.79	Sn 7.34	Sb 8.64	Te 9.01	I 10.45	Xe 12.13
Cs 3.89	Ba 5.21	Tl 6.11	Pb 7.42	Bi 7.29	Po 8.42	At 9.64	Rn 10.78

* 1 eV = 96.485 kJ mol^{-1}. See also Table 3.4.

The following trends are noteworthy:

- Lithium has a low first ionization energy: its outermost electron is well-shielded from the weakly charged nucleus by the core ($Z_{\text{eff}} = 1.3$ compared with $Z = 3$) and it is easily removed.
- Beryllium has a higher nuclear charge than lithium, and its outermost electron (one of the two 2s electrons) is more difficult to remove: its ionization energy is larger.
- The ionization energy decreases between beryllium and boron because in the latter the outermost electron occupies a 2p orbital and is less strongly bound than if it had been a 2s electron.
- The ionization energy increases between boron and carbon because the latter's outermost electron is also 2p and the nuclear charge has increased.
- Nitrogen has a still higher ionization energy because of the further increase in nuclear charge.

There is now a kink in the curve because the ionization energy of oxygen is lower than would be expected by simple extrapolation.

- At oxygen a 2p orbital must become doubly occupied, and the electron–electron repulsions are increased above what would be expected by simple extrapolation along the row. (The kink is less pronounced in the next row, between phosphorus and sulfur, because their orbitals are more diffuse.)
- The values for oxygen, fluorine, and neon fall roughly on the same line, the increase of their ionization energies reflecting the increasing attraction of the nucleus for the outermost electrons.
- The outermost electron in sodium is 3s. It is far from the nucleus, and the latter's charge is shielded by the compact, complete neon-like core. As a result, the ionization energy of sodium is substantially lower than that of neon.
- The periodic cycle starts again along this row, and the variation of the ionization energy can be traced to similar reasons.

The **electron affinity**, E_{ea}, is the difference in energy between a neutral atom and its anion. It is the energy *released* in the process

$$\mathrm{X(g)} + \mathrm{e^-(g)} \rightarrow \mathrm{X^-(g)} \qquad E_{\text{ea}} = E(\mathrm{X}) - E(\mathrm{X^-}) \qquad \text{Electron affinity} \quad (13.10a)$$

The electron affinity is positive if the anion has a lower energy than the neutral atom. Care should be taken to distinguish the electron affinity from the electron-gain enthalpy (Section 3.2): they have very similar numerical values but differ in sign:

$$\mathrm{X(g)} + \mathrm{e^-(g)} \rightarrow \mathrm{X^-(g)} \qquad \Delta_{\text{eg}}H^{\ominus} = H^{\ominus}_{\text{m}}(\mathrm{X^-}) - H^{\ominus}_{\text{m}}(\mathrm{X}) \qquad \text{Electron-gain enthalpy} \quad (13.10b)$$

Electron affinities (Table 13.4) vary much less systematically through the periodic table than ionization energies. However, the following general observations are important:

- Broadly speaking the highest electron affinities are found close to fluorine. In the halogens, the

Table 13.4
*Electron affinities of main-group elements, E_{ea}/eV**

H +0.75							He $<0^{\dagger}$
Li +0.62	Be −0.19	B +0.28	C +1.26	N −0.07	O +1.46	F +3.40	Ne $-0.30^{\dagger}$
Na +0.55	Mg −0.22	Al +0.46	Si +1.38	P +0.46	S +2.08	Cl +3.62	Ar $-0.36^{\dagger}$
K +0.50	Ca −1.99	Ga +0.3	Ge +1.20	As +0.81	Se +2.02	Br +3.37	Kr $-0.40^{\dagger}$
Rb +0.49	Sr +1.51	In +0.3	Sn +1.20	Sb +1.05	Te +1.97	I +3.06	Xe $-0.42^{\dagger}$
Cs +0.47	Ba −0.48	Tl +0.2	Pb +0.36	Bi +0.95	Po +1.90	At +2.80	Rn $-0.42^{\dagger}$

* 1 eV = 96.485 kJ mol^{-1}. See also Table 3.3.
† Calculated.

incoming electron enters the valence shell and experiences a strong attraction from the nucleus.

- The electron affinities of the noble gases are negative—which means that the anion has a higher energy than the neutral atom—because the incoming electron occupies an orbital outside the closed valence shell. It is then far from the nucleus and repelled by the electrons of the closed shells.
- The first electron affinity of oxygen is positive for the same reason as for the halogens. However, the second electron affinity (for the formation of O^{2-} from O^-) is strongly negative because although the incoming electron enters the valence shell, it experiences a strong repulsion from the net negative charge of the O^- ion.

The spectra of complex atoms

The spectra of many-electron atoms can be very complicated, yet that complexity contains a great deal of detailed information about the interactions between electrons. Here we consider the notation used to specify the states of atoms. Chemists need to know how to designate the states of atoms when they are describing photochemical events in the atmosphere and the chemical composition of stars (see Impact 13.1). We also describe an interaction that has important consequences in molecular spectroscopy and magnetism.

13.17 Term symbols

For historical reasons, the energy level of an atom is called a **term** and the notation used to specify the term is called a **term symbol**. A term symbol looks like $^3D_2$, with each component (the 3, the D, and the 2) telling us something about the angular momentum of the electrons in the atom. The scheme we shall use to arrive at the term symbols of atoms is called **Russell–Saunders coupling** and is based on the notion that the orbital and spin angular momenta of each electron couple together, and then all these resultants couple together to give an overall total angular momentum. The calculations focus on the valence electrons, for core electrons do not contribute to the overall angular momentum of an atom.

The letter (D, for instance) tells us the total orbital angular momentum of the electrons in the atom. To find it, we work out the **total orbital angular momentum quantum number**, L, in the manner described below, and then we use the following code:

L	0	1	2	3...
	S	P	D	F...

Note that the code is the same as for orbitals but uses uppercase Roman letters. To find L we identify the orbital angular momentum quantum numbers (l_1 and l_2 for instance) of the electrons in the valence shell of the atom, and then form the following series:

$$L = l_1 + l_2, l_1 + l_2 - 1, \ldots, |l_1 - l_2|$$

This and the analogous series introduced later for other types of angular momenta are called **Clebsch–Gordan series.** The modulus signs (|. . .|) simply mean that the series terminates at a positive value. The highest total orbital angular momentum occurs when the two electrons are orbiting in the same direction (in classical terms, like the planets round the Sun); the lowest occurs when they are orbiting in opposite directions.

● Brief illustration 13.10 Atomic term symbols

Suppose we are considering the excited-state configuration of carbon [He]$2s^2 2p^1 3p^1$ in which a 2p electron has been promoted to a 3p orbital (though not by electromagnetic radiation; see Section 13.19). We concentrate on the p electrons because the s electrons have no orbital angular momentum. For each electron $l = 1$ (that is, $l_1 = 1$ and $l_2 = 1$ for the two electrons we are considering). It follows that

$$L = 1 + 1, 1 + 1 - 1, \ldots |1 - 1| = 2, 1, 0$$

This result, which is shown pictorially in Fig. 13.23, uses the rules for addition and subtraction of vectors (The chemist's toolkit 13.1). It follows that the configuration gives rise to D, P, and S terms, corresponding to the three allowed values of the total orbital angular momentum.

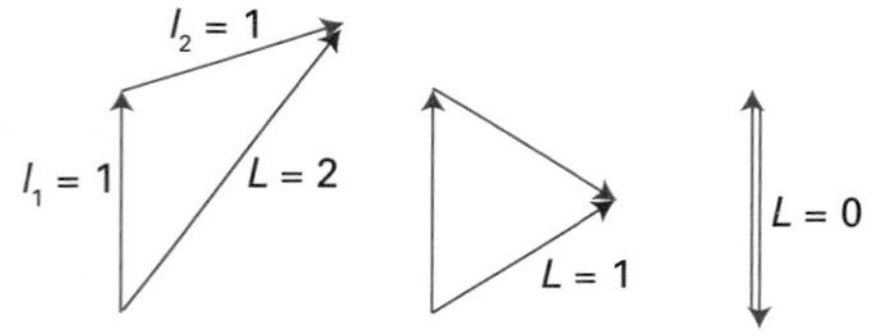

Fig. 13.23 A depiction of the rules for coupling two angular momenta into a resultant. In this case, $l_1 = l_2 = 1$ to give resultants with $L = 2$, 1, and 0. The lengths of the vectors are proportional to $\{l(l+1)\}^{1/2}$ in each case.

Self-test 13.6

Determine the possible terms that may arise from an excited calcium atom with an electronic configuration [Ar]$3d^1 4d^1$.

Answer: G, F, D, P, S

The chemist's toolkit 13.1 Addition and subtraction of vectors

Consider two vectors $\boldsymbol{v}_1$ and $\boldsymbol{v}_2$ making an angle θ:

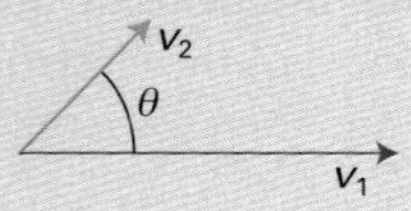

The first step in the addition of $\boldsymbol{v}_2$ to $\boldsymbol{v}_1$ consists of joining the tail of $\boldsymbol{v}_2$ to the head of $\boldsymbol{v}_1$:

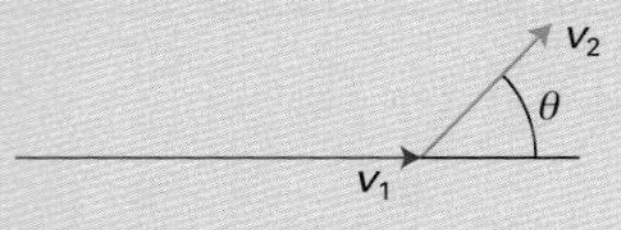

In the second step, we draw a vector $\boldsymbol{v}_{res}$, the **resultant vector**, originating from the tail of $\boldsymbol{v}_1$ to the head of $\boldsymbol{v}_2$:

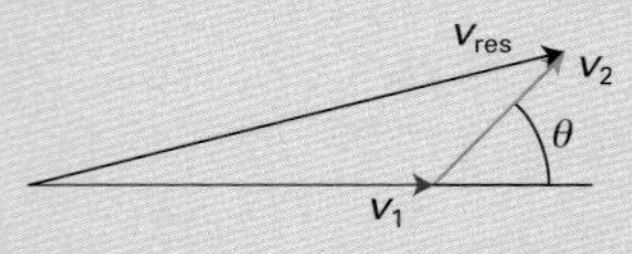

The magnitude v_{res} of the resultant vector is given by:

$$v_{res} = (v_1^2 + v_2^2 + 2v_1 v_2 \cos\theta)^{1/2}$$

where v_1 and v_2 are the magnitudes of the vectors $\boldsymbol{v}_1$ and $\boldsymbol{v}_2$, respectively.

The subtraction of vectors follows the same principles outlined above for addition by noting that subtraction of $\boldsymbol{v}_2$ from $\boldsymbol{v}_1$ amounts to addition of $-\boldsymbol{v}_2$ to $\boldsymbol{v}_1$:

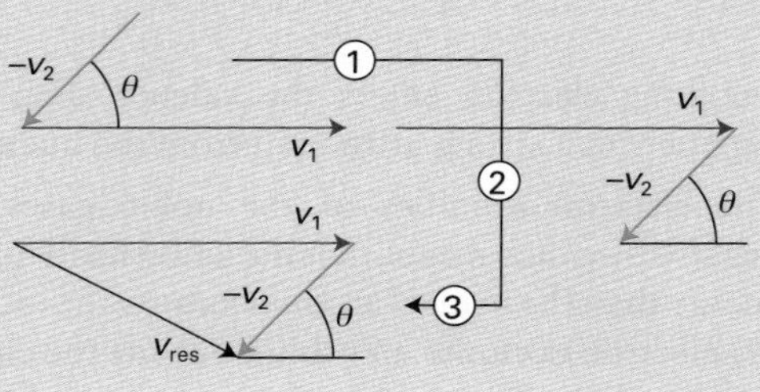

Example 13.2

Identifying the terms of a three-electron shell

Identify the terms that arise from the configuration p^3.

Strategy When there are more than two electrons to couple together, we use two Clebsch–Gordan series in succession: first couple two electrons, and then couple the third to each combined state, and so on.

Solution Coupling two electrons with orbital angular momentum quantum numbers $l_1 = l_2 = 1$ gives a minimum value of $|1 - 1| = 0$. Therefore, using L' to denote the total orbital angular momentum quantum number for these two electrons only, we obtain

$$L' = 1 + 1, 1 + 1 - 1, \ldots, 0 = 2, 1, 0$$

Now go on to calculate the values of L, the total orbital angular momentum quantum number for the three-electron system.

Couple l_3 with $L' = 2$, to give $L = 3, 2, 1$

Couple l_3 with $L' = 1$, to give $L = 2, 1, 0$

Couple l_3 with $L' = 0$, to give $L = 1$

The overall result is

$L = 3, 2, 2, 1, 1, 1, 0$

giving one F, two D, three P, and one S term.

Self-test 13.7

Identify the terms that can arise from the configuration d^2p^1.

Answer: H, 2G, 3F, 3D, 3P, S

Next, we consider the total **spin angular momentum quantum number**, S. This quantum number is obtained in the same way as L, by adding together the individual spin angular momentum quantum numbers:

$$S = s_1 + s_2, s_1 + s_2 - 1, \ldots, |s_1 - s_2|$$

For electrons, $s = \tfrac{1}{2}$, so for two electrons

$$S = \tfrac{1}{2} + \tfrac{1}{2}, \tfrac{1}{2} + \tfrac{1}{2} - 1, \ldots, |\tfrac{1}{2} - \tfrac{1}{2}| = 1, 0$$

Then the value of S is represented in the term symbol by writing the **multiplicity** of the term, the value of $2S + 1$, as a left superscript. The higher the multiplicity of a term, the more electrons there are in the atom that are spinning in the same direction. In more advanced treatments, it turns out that certain combinations of spin and orbital angular momentum are forbidden by the Pauli principle.

Brief illustration 13.11 The multiplicities of terms

For the excited configuration of carbon, $[He]2s^22p^13p^1$, that we are considering, the two p electrons each have $s = \tfrac{1}{2}$, so $S = 1,0$. The corresponding multiplicities are $2 \times 1 + 1 = 3$ (a 'triplet' term) and $2 \times 0 + 1 = 1$ (a 'singlet' term). The corresponding term symbols are

Triplet terms: ^3D, ^3P, ^3S Singlet terms: ^1D, ^1P, ^1S

A note on good practice Except in casual conversation, the name 'state' should not be used in place of 'term'. As we shall see, in general a term consists of a number of different states.

Finally, we come to the right subscript. This label is the **total angular momentum quantum number,** J, the total angular momentum being the sum of the orbital angular momentum and the spin angular momentum. We find J (a positive number) by forming the series

$$J = L + S, L + S - 1, \ldots, |L - S|$$

If there are many electrons having spins in the same direction as their orbital motion, then J is large. If the spins are aligned against the orbital motion, then J is small. Each value of J corresponds to a particular **level** of a term.

Brief illustration 13.12 The levels of a term

The levels that occur in the ^3D term are found by setting $L = 2$ and $S = 1$; then

$$J = 2 + 1, 2 + 1 - 1, \ldots, |2 - 1| = 3, 2, 1$$

That is, the levels of the ^3D term are $^3D_3$, $^3D_2$, and $^3D_1$ (note that there are three levels for this triplet term). In the $^3D_3$ level, not only are the two p electrons orbiting in the same sense, the two spins are spinning in the same direction as each other, and the total spin is in the same direction as the orbital angular momentum. In $^3D_1$ the total spin is aligned oppositely to the total orbital momentum and the overall total angular momentum is relatively low.

A note on good practice 'Levels' are still not 'states'. Each level with a quantum number J consists of $2J + 1$ individual states distinguished by the quantum number M_J.

Self-Test 13.8

What terms and levels can arise from the configuration ... $4p^13d^1$?

Answer: $^1F_3$, $^1D_2$, $^1P_1$, $^3F_{4,3,2}$, $^3D_{3,2,1}$, $^3P_{2,1,0}$

The terms of a configuration in general have different energies because they correspond to the occupation of different orbitals and to different numbers of electrons with parallel spins. Typically, Hund's rule enables us to identify the term of lowest energy as the one with the greatest number of parallel spins (Section 13.11), for parallel spins tend to stay apart and that allows the atom to shrink slightly. In other words,

- The term with the greatest multiplicity lies lowest in energy.

In the excited configuration of carbon in Brief illustration 13.11, the triplet terms lie lower than the singlet terms, so one of the terms ^3D, ^3P, ^3S lies lowest. It is also commonly found, having sorted the terms by multiplicity, that

- The term with the greatest orbital angular momentum lies lowest in energy.

Classically, we can think of the term with the greatest orbital angular momentum as having electrons circulating in the same direction, like cars on a traffic circle, and therefore being able to stay far apart. We therefore predict that the ^3D term will lie lowest of all the terms arising from the $[He]2s^22p^13p^1$ configuration. To find which of the three *levels* of this term lies lowest, we need another concept.

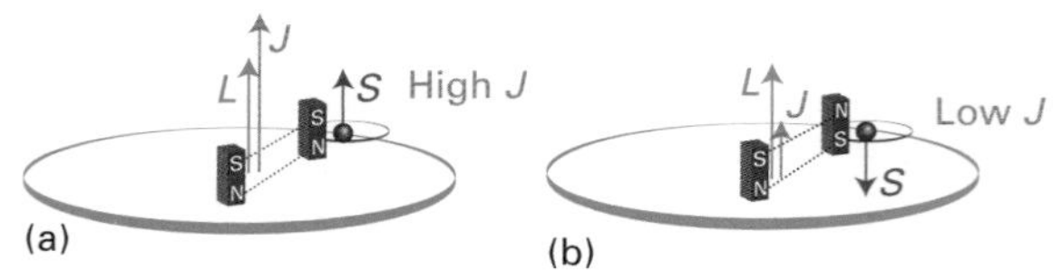

Fig. 13.24 The magnetic interaction responsible for spin–orbit coupling. (a) A high total angular momentum corresponds to a parallel arrangement of magnetic moments (represented by the bar magnets), and hence a high energy. (b) A low total angular momentum corresponds to an antiparallel arrangement of magnetic moments, and hence a low energy. Note that the difference in energy is not due *directly* to the differences in total angular momentum: the total simply tells us the relative orientations of the two magnetic moments.

13.18 Spin–orbit coupling

An electron is a charged particle, so its orbital angular momentum gives rise to a magnetic field, just as an electric current in a loop gives rise to a magnetic field in an electromagnet. That is, an electron with orbital angular momentum acts like a tiny bar magnet. An electron also has a spin angular momentum, and this intrinsic 'spinning motion' means that it also acts as a tiny bar magnet. The magnet arising from the spin interacts with the magnet arising from the orbital motion and gives rise to the interaction called **spin–orbit coupling.**

The two magnets have a higher energy when they are parallel than when they are antiparallel (Fig. 13.24). Therefore, because the relative orientation of the magnets reflects the relative orientation of the orbital and spin angular momenta, the energy of the atom depends on the total angular momentum quantum number J (because its value also reflects the relative orientation of the two kinds of momentum). A low energy is obtained when the angular momenta, and therefore the bar magnets, are antiparallel to each other. That arrangement of angular momenta corresponds to a low value of J. Therefore, we can predict that the level with lowest value of J will lie lowest in energy. In our current example, we predict that the lowest level of the ^{3}D term is ^{3}D$_1$. A more general statement, which applies to many-electron systems, is as follows:

> For atoms with shells that are less than half full, the level with lowest J lies lowest in energy; for atoms with shells that are more than half full, the level with highest J lies lowest.

The strength of the spin–orbit coupling increases sharply with atomic number. In Period 2 atoms it gives rise to splittings between levels of the order of 10^2 cm^{-1}, but in Period 3 the difference approaches 10^3 cm^{-1}. We can understand this increase by thinking about the source of the orbital magnetic field. To do so, imagine that we are riding on the electron as it orbits the nucleus. From our viewpoint, the nucleus appears to orbit around us (rather as the pre-Copernicans thought the Sun revolved around the Earth). If the nucleus has a high atomic number it will have a high charge, we shall be at the centre of a strong electric current, and we experience a strong magnetic field. If the nucleus has a low atomic number, we experience a feeble magnetic field arising from the low current that encircles us.

Spin–orbit coupling has important consequences in photochemistry and in particular for the existence of the property of 'phosphorescence', which we discuss in Chapter 20.

13.19 Selection rules

Now that we have described the energy levels of complex atoms, we can decide which spectroscopic transitions are allowed or forbidden. We have seen that spectroscopic selection rules arise from the conservation of angular momentum during a transition and from the fact that a photon has a spin of 1. They can therefore be expressed in terms of the term symbols, because the latter carry information about angular momentum. A detailed analysis leads to the following rules:

$$\Delta S = 0 \qquad \Delta L = 0, \pm 1 \qquad \Delta l = \pm 1$$
$$\Delta J = 0, \pm 1, \text{ but } J = 0 \leftrightarrow J = 0 \text{ is forbidden}$$

The rule about ΔS (no change of overall spin) stems from the fact that the light does not affect the spin directly. The rules about ΔL and Δl express the fact that the orbital angular momentum of an individual electron must change (so $\Delta l = \pm 1$), but whether or not this results in an overall change of orbital momentum depends on the coupling of angular momenta. These selection rules apply strictly for relatively light atoms, those near the top of the periodic table. As the atomic number increases, the rules progressively fail on account of significant spin–orbit coupling. So, for example, transitions between singlet and triplet states are allowed in heavy atoms.

Impact on astronomy 13.1

The spectroscopy of stars

The bulk of stellar material consists of neutral and ionized forms of hydrogen and helium atoms, with helium being the product of 'hydrogen burning' by nuclear fusion. However, nuclear fusion also makes heavier elements. It is generally accepted that the outer layers of stars are composed of lighter elements, such as H, He, C, N, O, and Ne in both neutral and ionized forms. Heavier elements, including neutral and ionized forms of Si, Mg, Ca, S, and Ar, are found closer to the stellar core. The core itself contains the heaviest elements and ^{56}Fe is particularly abundant because it is very stable. All of these elements are in the gas phase on account of the very high temperatures in stellar interiors. For example, the temperature is estimated to be 3.6 MK (1 MK = 10^6 K) halfway to the centre of the Sun.

Astronomers use spectroscopic techniques to determine the chemical composition of stars because each element, and indeed each isotope of an element, has a characteristic spectral signature that is transmitted through space by the star's light. To understand the spectra of stars, we must first know why they shine. Nuclear reactions in the dense stellar interior generate radiation that travels to less dense outer layers. Absorption and re-emission of photons by the atoms and ions in the interior give rise to a quasi-continuum of radiation energy that is emitted into space by a thin layer of gas called the *photosphere*. To a good approximation, the distribution of energy emitted from a star's photosphere resembles that for a very hot object. For example, the energy distribution of our Sun's photosphere is like that of an object heated to a temperature of 5800 K. Superimposed on the radiation continuum are sharp absorption and emission lines from neutral atoms and ions present in the photosphere. Analysis of stellar radiation with a spectrometer mounted on to a telescope, such as the Hubble Space Telescope, yields the chemical composition of the star's photosphere by comparison with known spectra of the elements. The data can also reveal the presence of small molecules, such as CN, C_2, TiO, and ZrO, in certain 'cold' stars, which are stars with relatively low effective temperatures.

The two outermost layers of a star are the *chromosphere*, a region just above the photosphere, and the *corona*, a region above the chromosphere that can be seen (with proper care) during eclipses. The photosphere, chromosphere, and corona comprise a star's 'atmosphere'. Our Sun's chromosphere is much less dense than its photosphere and its temperature is much higher, rising to about 10 kK (1 kK = 10^3 K). The reasons for this increase in temperature are not fully understood. The temperature of our Sun's corona is very high, rising up to 1.5 MK, so emission is strong from the X-ray to the radiofrequency region of the spectrum. The spectrum of the Sun's corona is dominated by emission lines from electronically excited species, such as neutral atoms and a number of highly ionized species. The most intense emission lines in the visible range are from the Fe^{13+} ion at 530.3 nm, the Fe^{9+} ion at 637.4 nm, and the Ca^{4+} ion at 569.4 nm.

Because our telescopes detect only light from the outer layers of stars, their overall chemical composition must be inferred from theoretical work on their interiors and from spectral analysis of their atmospheres. Data on our Sun indicate that it is 92 per cent hydrogen and 7.8 per cent helium. The remaining 0.2 per cent are due to heavier elements, among which C, N, O, Ne, and Fe are the most abundant. More advanced analysis of spectra also permit the determination of other properties of stars, such as their relative speeds and their effective temperatures.

Further information 13.1

The Pauli principle

The Pauli exclusion principle is a special case of a general statement called the *Pauli principle*:

> When the labels of any two identical fermions are exchanged, the total wavefunction changes sign. When the labels of any two identical bosons are exchanged, the total wavefunction retains the same sign.

As remarked in the text, a *fermion* is a particle with half integral spin (such as electrons, protons, and neutrons); a *boson* is a particle with integral spin (such as photons, which have spin 1). The Pauli *exclusion* principle applies only to fermions. By 'total wavefunction' is meant the entire wavefunction, including the spin of the particles.

Consider the wavefunction for two electrons $\Psi(1,2)$. The Pauli principle implies that it is a fact of nature that the wavefunction must change sign if we interchange the labels 1 and 2 wherever they occur in the function: $\Psi(2,1) = -\Psi(1,2)$.

Suppose the two electrons in an atom occupy an orbital ψ, then in the orbital approximation the overall wavefunction is $\psi(1)\psi(2)$. To apply the Pauli principle, we must deal with the total wavefunction, the wavefunction including spin. There are four possibilities for two spins:

$$\alpha(1)\alpha(2) \qquad \alpha(1)\beta(2) \qquad \beta(1)\alpha(2) \qquad \beta(1)\beta(2)$$

Let's consider two of these possibilities: the state $\alpha(1)\alpha(2)$ corresponds to parallel spins whereas (for technical reasons related to the cancellation of each spin's angular momentum by the other) the combination $\alpha(1)\beta(2) - \beta(1)\alpha(2)$ corresponds to paired spins. The total wavefunction of the system is one of the following:

Parallel spins: $\psi(1)\psi(2)\alpha(1)\alpha(2)$

Paired spins: $\psi(1)\psi(2)\{\alpha(1)\beta(2) - \beta(1)\alpha(2)\}$

The Pauli principle, however, asserts that, for a wavefunction to be acceptable (for electrons), it must change sign when the electrons are exchanged. In each case, exchanging the labels 1 and 2 converts the factor $\psi(1)\psi(2)$ into $\psi(2)\psi(1)$, which is the same, because the order of multiplying the functions does not change the value of the product. The same is true of $\alpha(1)\alpha(2)$. Therefore, the first combination is not allowed, because it does not change sign. The second combination, however, changes to

$$\psi(2)\psi(1)\{\alpha(1)\beta(2) - \beta(1)\alpha(2)\} = -\psi(1)\psi(2)\{\alpha(1)\beta(2) - \beta(1)\alpha(2)\}$$

This combination does change sign (it is 'antisymmetric'), and is therefore acceptable.

Now we see that the only possible state of two electrons in the same orbital allowed by the Pauli principle is the one that has paired spins. This is the content of the Pauli exclusion principle. The exclusion principle is irrelevant when the orbitals occupied by the electrons are different, and both electrons may then have (but need not have) the same spin state. Nevertheless, even then the overall wavefunction must still be antisymmetric overall, and must still satisfy the Pauli principle itself.

Checklist of key concepts

☐ 1 Hydrogenic atoms are atoms with a single electron.

☐ 2 The wavefunctions of hydrogenic atoms are labelled with three quantum numbers: the principal quantum number $n = 1, 2, \ldots$, the orbital angular momentum quantum number $l = 0, 1, \ldots, n-1$, and the magnetic quantum number $m_l = l, l-1, \ldots, -l$.

☐ 3 s Orbitals are spherically symmetrical and have nonzero amplitude at the nucleus.

☐ 4 A radial distribution function, $P(r)$, is the probability density for finding an electron at a radius r; the probability of finding the electron between r and $r + \delta r$ is $P(r)\delta r$.

☐ 5 The magnitude of the orbital angular momentum of an electron is $\{l(l+1)\}^{1/2}\hbar$ and the component of angular momentum about an axis is $m_l\hbar$.

☐ 6 An electron possesses an intrinsic angular momentum, its spin, which is described by the quantum numbers $s = \frac{1}{2}$ and $m_s = \pm\frac{1}{2}$.

☐ 7 A selection rule is a statement about which spectroscopic transitions are allowed.

☐ 8 In the orbital approximation, each electron in a many-electron atom is supposed to occupy its own orbital.

☐ 9 The Pauli exclusion principle states that no more than two electrons may occupy any given orbital and, if two electrons do occupy one orbital, then their spins must be paired.

☐ 10 In a many-electron atom, the orbitals of a given shell lie in the order s < p < d < f as a result of the effects of penetration and shielding.

☐ 11 Atomic radii decrease from left to right across a period and increase down a group.

☐ 12 Ionization energies increase from left to right across a period and decrease down a group.

☐ 13 Electron affinities are highest towards the top right of the periodic table (near fluorine).

☐ 14 A term symbol has the form $^{2S+1}\{L\}_J$, where $2S+1$ is the multiplicity and $\{L\}$ is a letter denoting the value of L, the total orbital angular momentum quantum number.

☐ 15 For a given configuration (and most reliably for the ground-state configuration) the term with the greatest multiplicity lies lowest in energy, that with the highest value of L lies lowest, and for atoms with shells that are less than half full, the level with the lowest J lies lowest.

☐ 16 Different levels of a term have different energies on account of spin–orbit coupling, and the strength of spin–orbit coupling increases sharply with increasing atomic number.

Road map of key equations

Hydrogenic atoms

$$J = j_1 + j_2, j_1 + j_2 - 1, ..., |j_1 - j_2|$$

$$j = l \text{ or } s$$

Clebsch–Gordan series

$$E_n = -hcR_N Z^2/n^2$$
$$n = 1, 2, ...$$

Energy levels

$$\Delta l = \pm 1, \Delta m_l = 0, \pm 1$$

Selection rules

$$\Delta S = 0, \Delta L = 0, \pm 1, \Delta l = \pm 1$$
$$\Delta J = 0, \pm 1 \text{ but } J = 0 \longleftrightarrow\!\!\!\!| \longrightarrow J = 0$$

Many-electron atoms

$$\psi_{n,l,m_l}(r,\theta,\phi) = R_{n,l}(r) Y_{l,m_l}(\theta,\phi)$$

Wavefunctions

$$P(r) = r^2 R(r)^2$$

Radial distribution function

s orbitals

$$P(r) = 4\pi r^2 \psi^2$$

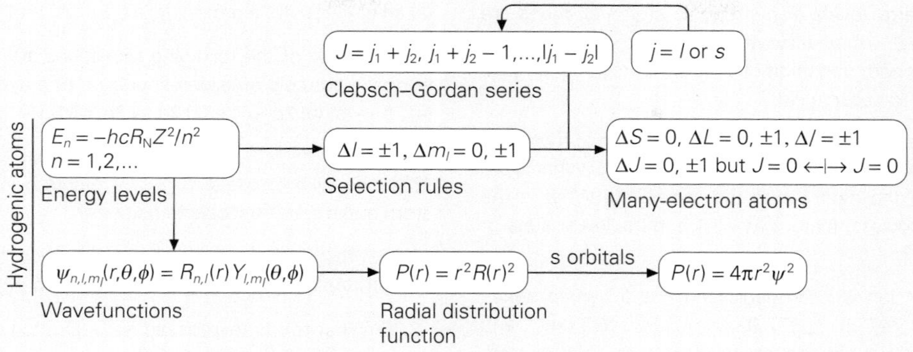

Questions and exercises

Discussion questions

13.1 List and describe the significance of the quantum numbers needed to specify the internal state of a hydrogenic atom.

13.2 Explain the significance of (a) a boundary surface and (b) the radial distribution function for hydrogenic orbitals.

13.3 Describe the orbital approximation for the wavefunction of a many-electron atom. What are the limitations of the approximation?

13.4 Discuss the relationship between the location of a many-electron atom in the periodic table and its electron configuration.

13.5 Describe the self-consistent field procedure for calculating the form of the orbitals and the energies of many-electron atoms.

13.6 Describe and account for the variation of first ionization energies along Period 2 of the periodic table. Would you expect the same variation in Period 3?

13.7 Explain the origin of spin–orbit coupling and how it affects the appearance of a spectrum.

13.8 Specify and account for the selections rules for spectroscopic transitions in (a) hydrogenic atoms and (b) many-electron atoms.

Exercises

13.1 Calculate the wavelength of the line with $n = 6$ in the Balmer series of the spectrum of atomic hydrogen.

13.2 The frequency of one of the lines in the Paschen series of the spectrum of atomic hydrogen is 2.7415×10^{15} Hz. Identify the principal quantum number of the upper state in the transition.

13.3 One of the terms of the H atom is at 27 414 cm^{-1}. What is (a) the wavenumber, (b) the energy of the term with which it combines to produce light of wavelength 486.1 nm?

13.4 The Rydberg constant, eqn 13.4b, depends on the mass of the nucleus. What is the difference in wavenumbers of the 3p → 1s transition in hydrogen and deuterium?

13.5 What transition in He$^+$ has the same frequency (disregarding mass differences) as the 2p → 1s transition in H?

13.6 Hydrogen is the most abundant element in all stars. However, neither absorption nor emission lines due to neutral hydrogen are found in the spectra of stars with effective temperatures higher than 25 000 K. Account for this observation.

13.7 The distribution of isotopes of an element may yield clues about the nuclear reactions that occur in the interior of a star. Show that it is possible to use spectroscopy to confirm the presence of both ^{4}He$^+$ and ^{3}He$^+$ in a star by calculating the wavenumbers of the $n = 3 \rightarrow n = 2$ and of the $n = 2 \rightarrow n = 1$ transitions for each isotope.

13.8 Predict the ionization energy of Li^{2+} given that the ionization energy of He$^+$ is 54.36 eV.

13.9 The Humphreys series is another group of lines in the spectrum of atomic hydrogen. It begins at 12 368 nm and has been traced to 3281.4 nm. (a) What are the transitions involved? (b) What are the wavelengths of the intermediate transitions?

13.10 The energy levels of hydrogenic atoms and ions are proportional to Z^2. At what wavelength would you expect the longest wavelength transition of the Humphreys series (see Exercise 13.9) to occur in He^+?

13.11 A series of lines in the spectrum of atomic hydrogen lies at 656.46, 486.27, 434.17, and 410.29 nm. (a) What is the wavelength of the next line in the series? (b) What is the ionization energy of the atom when it is in the lower state of the transitions?

13.12 The Li^{2+} ion is hydrogenic and has a Lyman series of lines at 740 747 cm^{-1}, 877 924 cm^{-1}, 925 933 cm^{-1}, and beyond. (a) Show that the energy levels are of the form $-hcR_{Li}/n^2$ and find the value of R_{Li} for this ion. (b) Go on to predict the wavenumbers of the two longest wavelength transitions of the Balmer series of the ion and (c) find the ionization energy of the ion.

13.13 How is the ionization energy of a singly charged anion related to the electron affinity of the parent atom?

13.14 When ultraviolet radiation of wavelength 58.4 nm from a helium lamp is directed on to a sample of krypton, electrons are ejected with a speed of 1.59×10^6 m s^{-1}. Calculate the ionization energy of krypton.

13.15 How many orbitals are present in the N shell of an atom?

13.16 What is the orbital angular momentum (as multiples of $\hbar$) of an electron in the orbitals (a) 1s, (b) 3s, (c) 3d, (d) 2p, (e) 3p? Give the numbers of angular and radial nodes in each case.

13.17 State the orbital degeneracy of the levels in the hydrogen atom that have energy (a) $-hcR_H$, (b) $-\frac{1}{9}hcR_H$, and (c) $-\frac{1}{49}hcR_H$.

13.18 How many electrons can occupy subshells with the following values of l: (a) 0, (b) 3, (c) 5?

13.19 At what radius does the probability of finding an electron in a small volume located at a point in the ground state of an H atom fall to 30 per cent of its maximum value?

13.20 At what radius in the H atom does the radial distribution function of the ground state have (a) 30 per cent, (b) 5 per cent of its maximum value?

13.21 What is the probability of finding the electron in a volume of 6.5 pm^3 centred on the nucleus in (a) a hydrogen atom, (b) a He^+ ion?

13.22 Locate the radial nodes in (a) the 3s orbital, (b) the 4s orbital of an H atom.

13.23 What is the probability of finding an electron anywhere in one lobe of a p orbital given that it occupies the orbital?

13.24 The wavefunction of one of the d orbitals is proportional to $\sin\theta \cos\theta$. At what angles does it have nodal planes?

13.25 For an atomic orbital with quantum numbers n and l, (a) how many radial, (b) angular, and (c) nodes are there?

13.26 Classify the following as either a fermion or a boson: (a) an electron, (b) a proton, (c) a neutron, and (d) a photon.

13.27 Which of the following transitions are allowed in the normal electronic emission spectrum of a hydrogenic atom: (a) 2s $\rightarrow$ 1s, (b) 2p $\rightarrow$ 1s, (c) 3d $\rightarrow$ 2p, (d) 5d $\rightarrow$ 2s, (e) 5p $\rightarrow$ 3s, (f) 6f $\rightarrow$ 4p?

13.28 To what orbitals may a 5f electron in a hydrogenic atom make spectroscopic transitions?

13.29 Why is the electronic configuration of the yttrium atom $[Kr]4d^15s^2$ and that of the silver atom $[Kr]4d^{10}5s^1$?

13.30 What terms (expressed as S, D, etc.) can arise from the $[He]2s^22p^13d^1$ excited configuration of carbon?

13.31 The Clebsch–Gordan series may be used successively to determine the possible values for the total angular momentum of several electrons. What are the total spin angular momenta that can arise from four electrons?

13.32 What levels can the following terms possess: (a) ^1S, (b) ^3F, (c) ^5S, (d) ^5P?

13.33 The ground configuration of a Ti^{2+} ion is $[Ar]3d^2$. (a) What is the term of lowest energy and which level of that term lies lowest? (b) How many states belong to that lowest level?

13.34 What is the ground-state configuration of a Sc^{2+} ion? What levels arise from this configuration? Which of these levels lies lowest in energy?

13.35 The lowest term of both the Al and Cl atoms is ^2P, which gives rise to $^2P_{1/2}$ and $^2P_{3/2}$ levels. Predict which level lies lower in energy for each atom.

13.36 One important function of atomic and ionic radius is in regulating the uptake of oxygen by haemoglobin, for the change in ionic radius that accompanies the conversion of Fe(II) to Fe(III) when O_2 attaches triggers a conformational change in the protein. Which do you expect to be larger: Fe^{2+} or Fe^{3+}? Why?

13.37 The ionization energy of potassium is 4.34 eV and the electron affinity of Br is 3.36 eV. What is the change in energy for the reaction $K(g) + Br(g) \rightarrow K^+(g) + Br^-(g)$? Express your answer in electronvolts and kilojoules per mole.

13.38 Which of the following transitions between terms are allowed in the normal electronic emission spectrum of a many-electron atom: (a) $^3D_2 \rightarrow {}^3P_1$, (b) $^3P_2 \rightarrow {}^1S_0$, (c) $^3F_4 \rightarrow {}^3D_3$?

13.39 Two transitions in the $[Ne]3p^1 \rightarrow [Ne]3s^1$ emission spectrum of atomic sodium are observed at 589.0 nm and 589.6 nm. What is the difference in energy, expressed in electronvolts, between the $^2P_{3/2}$ and $^2P_{1/2}$ levels of the upper term?

Projects

The symbol ‡ indicates that calculus is required.

13.40‡ Here we explore hydrogenic wavefunctions in more quantitative detail. (a) Find, by identifying the maximum in the radial distribution function of a hydrogenic 1s electron, the most probable distance of an electron from the nucleus in a hydrogen atom in its ground state. (b) The (normalized) wavefunction for a 2s orbital in hydrogen is

$$\psi = \left(\frac{1}{32\pi a_0^3}\right)^{1/2}\left(2 - \frac{r}{a_0}\right)e^{-r/2a_0}$$

Calculate the probability of finding an electron that is described by this wavefunction in a volume of 1.0 pm^3 (i) centred on the nucleus, (ii) at the Bohr radius, (iii) at twice the Bohr radius. (c) Construct an expression for the radial distribution function of a hydrogenic 2s electron (see part (b) for the form of the orbital), and plot the function against *r*. What is the most probable radius at which the electron will be found? (d) For a more accurate determination of the most probable radius at which an electron will be found in an H2s orbital, differentiate the radial distribution function to find where it is a maximum.

13.41 Thallium, a neurotoxin, is the heaviest member of Group 13 of the periodic table and is found most usually in the +1 oxidation state. Aluminium, which causes anaemia and dementia, is also a member of the group but its chemical properties are dominated by the +3 oxidation state. Examine this issue by plotting the first, second, and third ionization energies for the Group 13 elements against atomic number. Explain the trends you observe. For data, see the links to databases of atomic properties provided in the text's website.

13.42 The spectrum of a star is used to measure its *radial velocity* with respect to the Sun, the component of the star's velocity vector that is parallel to a vector connecting the star's centre to the centre of the Sun. The measurement relies on the Doppler effect, in which radiation is shifted in frequency when the source is moving towards or away from the observer. When a star emitting electromagnetic radiation of frequency ν moves with a speed s relative to an observer, the observer detects radiation of frequency $\nu_{\text{receding}} = \nu f$ or $\nu_{\text{approaching}} = \nu/f$, where $f = \{(1 - s/c)/(1 + s/c)\}^{1/2}$ and c is the speed of light. (a) Three lines in the spectrum of atomic iron of the star HDE 271 182, which belongs to the Large Magellanic Cloud, occur at 438.882 nm, 441.000 nm, and 442.020 nm. The same lines occur at 438.392 nm, 440.510 nm, and 441.510 nm in the spectrum of an Earth-bound iron arc. Determine whether HDE 271 182 is receding from or approaching the Earth and estimate the star's radial speed with respect to the Earth. (b) What additional information would you need to calculate the radial velocity of HDE 271 182 with respect to the Sun?

14

Quantum chemistry: the chemical bond

Introductory concepts 345

14.1 The classification of bonds 345

14.2 Potential energy curves 345

Valence bond theory 346

14.3 Diatomic molecules 346

14.4 Polyatomic molecules 348

14.5 Promotion and hybridization 349

14.6 Resonance 352

14.7 The language of valence bonding 353

Molecular orbitals 353

14.8 Linear combinations of atomic orbitals 353

14.9 Bonding and antibonding orbitals 355

14.10 The structures of homonuclear diatomic molecules 356

14.11 The structures of heteronuclear diatomic molecules 363

14.12 The structures of polyatomic molecules 365

14.13 The Hückel method 366

Computational chemistry 369

14.14 Techniques 369

14.15 Graphical output 370

14.16 Applications 371

CHECKLIST OF KEY CONCEPTS 372

ROAD MAP OF KEY EQUATIONS 373

QUESTIONS AND EXERCISES 373

The **chemical bond**, a link between atoms, is central to all aspects of chemistry. Reactions make them and break them, and the structures of solids and individual molecules depend on them. The physical properties of individual molecules and of bulk samples of matter also stem in large part from the shifts in electron density that take place when atoms form bonds. The theory of the origin of the numbers, strengths, and three-dimensional arrangements of chemical bonds between atoms is called **valence theory**. (The name comes from a Latin word for strength.)

Valence theory is an attempt to explain the properties of molecules ranging from the smallest to the largest. For instance, it explains why N_2 is so inert that it dilutes the aggressive oxidizing power of atmospheric oxygen. At the other end of the scale, valence theory deals with the structural origins of the function of protein molecules and the molecular biology of DNA. The description of chemical bonding has become highly developed through the use of computers, and it is now possible to compute details of the electron distribution in molecules of almost any complexity. However, much can also be achieved in terms of a simple qualitative understanding of bond formation, and that is the initial focus of this chapter.

There are two major approaches to the calculation of molecular structure, **valence bond theory** (VB theory) and **molecular orbital theory** (MO theory). Almost all modern computational work makes use of MO theory, and we concentrate on it in this chapter. Valence bond theory, though, has left its imprint on the language of chemistry, and it is important to know the significance of terms that chemists use every day. The structure of this chapter is therefore as follows. First, we set out a few concepts common to all levels of description. Then we present the concepts of VB theory that continue to be used in chemistry (such as hybridization and resonance). Next, we present the basic ideas of MO theory, and finally we see how computational techniques pervade all current discussions of molecular structure.

Introductory concepts

Certain ideas of valence theory will be well known from introductory chemistry. This section reviews this background.

14.1 The classification of bonds

We distinguish between two types of bond:

An **ionic bond** is formed by the transfer of electrons from one atom to another and the consequent attraction between the ions so formed.

A **covalent bond** is formed when two atoms share a pair of electrons.

The character of a covalent bond, on which we concentrate in this chapter, was identified by G.N. Lewis in 1916, before quantum mechanics was fully developed. We shall assume that Lewis's ideas are familiar. In this chapter we develop the modern theory of chemical bond formation in terms of the quantum mechanical properties of electrons and set Lewis's ideas in a modern context. We shall see that ionic and covalent bonds are two extremes of a single type of bond. However, because there are certain aspects of ionic solids that require special attention, we treat them separately in Chapter 17.

Lewis's original theory was unable to account for the shapes adopted by molecules. The most elementary but qualitatively quite successful explanation of the shapes adopted by molecules is the **valence-shell electron pair repulsion model** (VSEPR model) in which we suppose that the shape of a molecule is determined by the repulsions between electron pairs in the valence shell. This model is fully discussed in introductory chemistry texts, but we give a brief review of it in The chemist's toolkits 14.1 and 14.2. Once again, the purpose of this chapter is to extend these elementary arguments and to indicate some of the contributions that quantum theory has made to understanding why a molecule adopts its characteristic shape.

14.2 Potential energy curves

All theories of molecular structure adopt the **Born–Oppenheimer approximation**. In this approximation, it is supposed that the nuclei, being so much heavier than an electron, move relatively slowly and

The chemist's toolkit 14.1 The Lewis theory of covalent bonding

In his original formulation of a theory of the covalent bond, which we assume to be familiar from introductory chemistry courses, G.N. Lewis proposed that each bond consisted of one electron pair. Each atom in a molecule shared electrons until it had acquired an octet characteristic of a noble gas atom near it in the periodic table. (Hydrogen is an exception: it acquires a duplet of electrons.) Thus, to write down a Lewis structure:

1. Arrange the atoms as they are found in the molecule.
2. Add one electron pair (represented by dots, :) between each bonded atom.
3. Use the remaining electron pairs to complete the octets of all the atoms present either by forming lone pairs or by forming multiple bonds.
4. Replace bonding electron pairs by bond lines (–) but leave lone pairs as dots (:).

A Lewis structure does not (except in very simple cases), portray the actual geometrical structure of the molecule; it is a topological map of the arrangement of bonds.

The chemist's toolkit 14.2 The VSEPR model

The basic assumption of the **valence shell electron pair repulsion** (VSEPR) model is that *the valence-shell electron pairs of the central atom adopt positions that maximize their separations*. Thus, if the atom has four electron pairs in its valence shell, then the pairs adopt a tetrahedral arrangement around the atom; if the atom has five pairs, then the arrangement is trigonal bipyramidal. The arrangements adopted by electron pairs are:

Number of groups	Arrangement
2	Linear
3	Trigonal planar
4	Tetrahedral
5	Trigonal bipyramidal
6	Octahedral
7	Pentagonal bipyramidal

Once the basic shape of the arrangement of electron pairs has been identified, the pairs are identified as bonding or nonbonding and the shape of the molecule is classified by noting the arrangement of the atoms around the central atom.

The next stage in the application of the VSEPR model is to accommodate the greater repelling effect of lone pairs compared with that of bonding pairs. That is, *bonding pairs tend to move away from lone pairs even though that might reduce their separation from other bonding pairs.* To take into account multiple bonds, each set of two or three electron pairs is treated as a single region of high electron density.

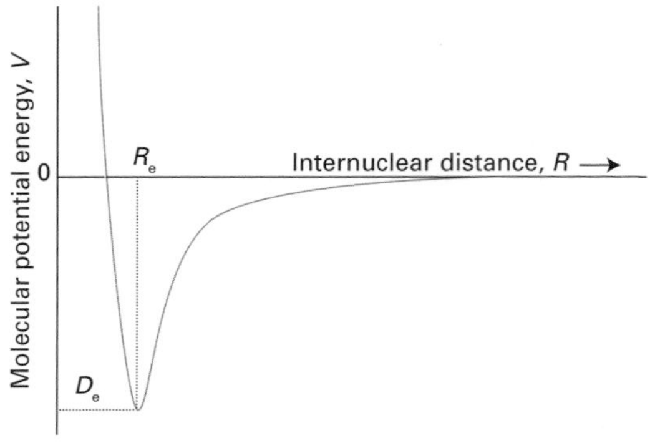

Fig. 14.1 A molecular potential energy curve. The equilibrium bond length R_e corresponds to the energy minimum D_e.

may be treated as stationary while the electrons move around them. We can therefore think of the nuclei as being fixed at arbitrary locations, and then solve the Schrödinger equation for the electrons alone. The approximation is quite good for molecules in their electronic ground states, for calculations suggest that (in classical terms) the nuclei in H_2 move through only about 1 pm while the electron speeds through 1000 pm.

By invoking the Born–Oppenheimer approximation, we can select an internuclear separation in a diatomic molecule and solve the Schrödinger equation for the electrons for that nuclear separation. Then we can choose a different separation and repeat the calculation, and so on. In this way we can explore how the energy of the molecule varies with bond length and obtain a **molecular potential energy curve,** a graph showing how the molecular energy depends on the internuclear separation (Fig. 14.1). The graph is called a *potential* energy curve because the nuclei are stationary and contribute no kinetic energy. Once the curve has been calculated, we can identify the **equilibrium bond length,** R_e, the internuclear separation at the minimum of the curve, and D_e, the depth of the minimum below the energy of the infinitely widely separated atoms. In Chapter 19 we shall also see that the narrowness of the potential well is an indication of the stiffness of the bond. Similar considerations apply to polyatomic molecules, where bond angles may be varied as well as bond lengths.

Valence bond theory

In valence bond theory, a bond is regarded as forming when an electron in an atomic orbital on one atom pairs its spin with that of an electron in an atomic orbital on another atom (Fig. 14.2). To understand why this pairing leads to bonding, we have to examine the wavefunction for the two electrons that form the bond.

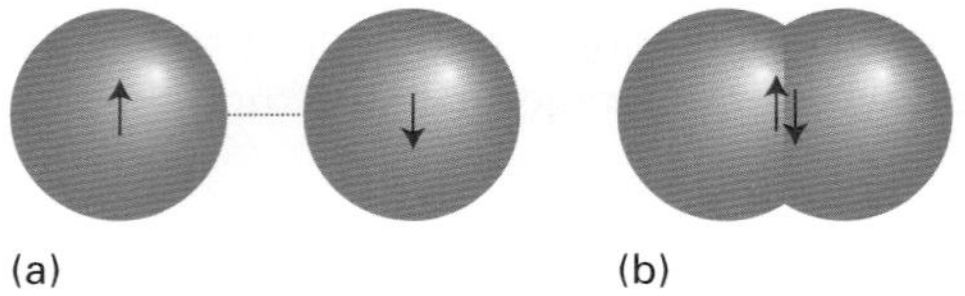

Fig. 14.2 In the valence bond theory, a σ bond is formed when two electrons in orbitals on neighbouring atoms, as in (a), pair and the orbitals merge to form a cylindrical electron cloud, as in (b).

14.3 Diatomic molecules

We begin by considering the simplest possible chemical bond, the one in molecular hydrogen, H—H. When the two ground-state H atoms are far apart, we can be confident that electron 1 is in the 1s orbital of atom A, which we denote $\psi_A(1)$, and electron 2 is in the 1s orbital of atom B, which we denote $\psi_B(2)$. We saw in Section 12.7 that it is a general rule in quantum mechanics that the wavefunction for several non-interacting particles is the product of the wavefunctions for each particle (this is the separation of variables argument), so providing we can ignore the interactions between the electrons we can write $\psi(1,2) = \psi_A(1)\psi_B(2)$.

When the two atoms are at their bonding distance, it may still be true that electron 1 is on A and electron 2 is on B. However, an equally likely arrangement is for electron 1 to escape from A and be found on B and for electron 2 to be on A. In this case the wavefunction is $\psi(1,2) = \psi_A(2)\psi_B(1)$. Whenever two outcomes are equally likely, the rules of quantum mechanics tell us to add together, formally to **superimpose,** the two corresponding wavefunctions. Therefore, the (unnormalized) wavefunction for the two electrons in a hydrogen molecule is

$$\psi_{H-H}(1,2) = \psi_A(1)\psi_B(2) + \psi_A(2)\psi_B(1) \qquad (14.1)$$

This expression is the VB wavefunction for the bond in molecular hydrogen. It expresses the impossibility of keeping track of either electron, and therefore that their distributions blend together. The wavefunction is only an approximation, because when the two atoms are close together it is not true that the electrons do not interact. However, this approximate wavefunction is a reasonable starting point for all discussions of the VB theory of bonding.

We show in the following Derivation that for technical reasons related to the Pauli exclusion principle, the wavefunction in eqn 14.1 can exist only if the two electrons it describes have opposite spins. It follows that the merging of orbitals that gives rise to a bond is accompanied by the pairing of the two electrons that contribute to it. Bonds do not form *because* electrons tend to pair: bonds are *allowed* to form by the electrons pairing their spins.

Derivation 14.1

The Pauli principle and bond formation

The VB wavefunction in eqn 14.1 does not change sign when the labels 1 and 2 are interchanged. To formulate a wavefunction that obeys the Pauli principle (Further information 13.1) and does change sign when the labels 1 and 2 are interchanged, we must combine this symmetric spatial wavefunction $\psi(1)\psi(2)$ with the antisymmetric spin function $\alpha(1)\beta(2) - \beta(1)\alpha(2)$ and write

$$\psi_{A-B}(2,1) = \{\psi_A(1)\psi_B(2) + \psi_A(2)\psi_B(1)\} \times \{\alpha(1)\beta(2) - \beta(1)\alpha(2)\}$$

This is the only permitted combination of space and spin functions. Because $\alpha(1)\beta(2) - \beta(1)\alpha(2)$ represents a spin-paired state of the two electrons, we see that the Pauli principle requires the two electrons in the bond to be paired (↑↓).

Because ψ_{H-H} is built from the merging of H1s orbitals, we can expect the overall distribution of the electrons in the molecule to be sausage-shaped (as in Fig. 14.2). A VB wavefunction with cylindrical symmetry around the internuclear axis is called a **σ bond**. The bond is so called because, when viewed along the internuclear axis it resembles a pair of electrons in an s orbital (and σ, sigma, is the Greek equivalent of s). All VB wavefunctions are constructed in a similar way, by using the atomic orbitals available on the participating atoms. In general, therefore, the (unnormalized) VB wavefunction for an A—B bond is

$$\psi_{A-B}(1,2) = \psi_A(1)\psi_B(2) + \psi_A(2)\psi_B(1) \qquad \text{A valence bond wavefunction} \quad (14.2)$$

To calculate the energy of a molecule for a series of internuclear separations R, we substitute the VB wavefunction into the Schrödinger equation for the molecule and carry out the necessary mathematical manipulations to calculate the corresponding values of the energy. When this energy is plotted against R, we get a curve like that shown in Fig. 14.1. As R decreases from infinity, the energy falls below that of two separated H atoms as each electron becomes free

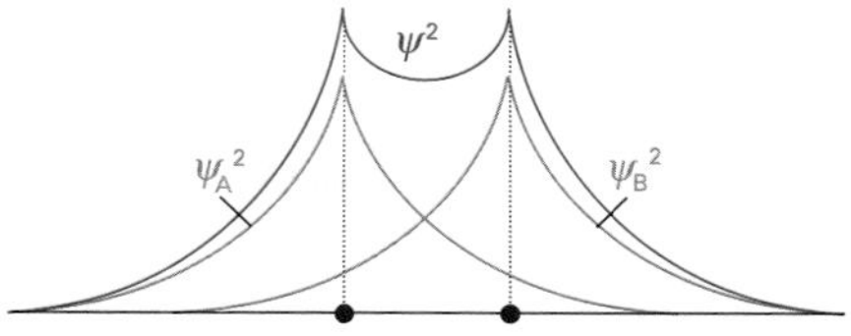

Fig. 14.3 The electron density in H_2 according to the valence bond model of the chemical bond and the electron densities corresponding to the contributing atomic orbitals. The nuclei are denoted by large dots on the horizontal line. Note the accumulation of electron density in the internuclear region.

to migrate to the other atom. This decrease in energy is the outcome of several effects:

- As the two atoms approach each other, there is an accumulation of electron density between the two nuclei (Fig. 14.3). The electrons attract the two nuclei, and the potential energy is lowered.
- This accumulation between the nuclei is at the expense of removing electron density from close to the nuclei, which contributes an increase in potential energy.
- The freedom of the electrons to migrate between the atoms is like the transfer of an electron from a small box to a bigger box, which (as we saw in the discussion of a particle in a box) results in a lowering of their kinetic energy.

In H_2 the last is the dominant effect, but the relative importance of changes in potential and kinetic energy is still unclear in more complex molecules.

The overall decrease in energy due to the redistribution of electrons is counteracted by an increase in energy from the Coulombic repulsion between the two positively charged nuclei of charges Z_Ae and Z_Be, which has the form

$$V_{\text{nuc-nuc}}(R) = \frac{Z_A Z_B e^2}{4\pi\varepsilon_0 R} \qquad \text{Coulombic repulsion between two nuclei} \quad (14.3)$$

(For H_2, $Z_A = Z_B = 1$.) This positive contribution to the energy becomes large as R becomes small (and the decrease in electronic kinetic energy becomes less significant as the 'big box' is no longer much bigger than the initial two 'little boxes'). As a result, the total energy curve passes through a minimum and then climbs to a strongly positive value as the two nuclei are pressed together.

A similar description is used for molecules built from atoms that contribute more than one electron to the bonding. For example, to construct the VB description of N_2, we consider the valence-electron

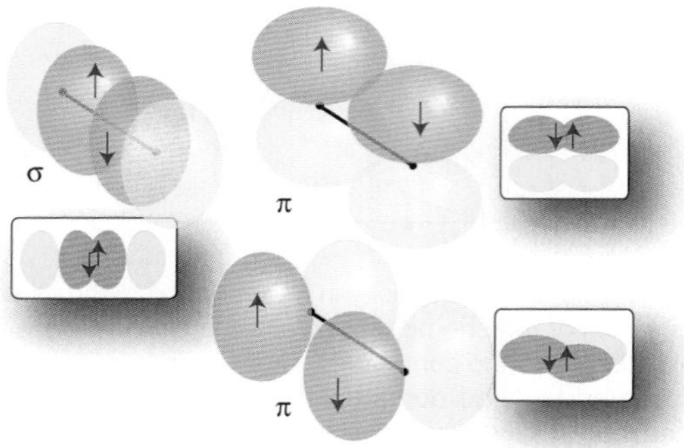

Fig. 14.4 The bonds in N_2 are built by allowing the electrons in the N2p orbitals to pair. However, only one orbital on each atom can form a σ bond: the orbitals perpendicular to the axis form π-bonds.

configuration of each atom, which is $2s^22p_x^12p_y^12p_z^1$. It is conventional to take the z-axis to be the internuclear axis, so we can imagine each atom as having a $2p_z$ orbital pointing towards a $2p_z$ orbital on the other atom, with the $2p_x$ and $2p_y$ orbitals perpendicular to the axis (Fig. 14.4). According to the building-up principle, each of these p orbitals is occupied by one electron, so we can think of bonds as being formed by the merging of matching orbitals on neighbouring atoms and the pairing of the electrons that occupy them. We get a cylindrically symmetric σ bond from the merging of the two $2p_z$ orbitals and the pairing of the electrons they contain. However, the remaining p orbitals cannot merge to give σ bonds because they do not have cylindrical symmetry around the internuclear axis. Instead, the $2p_x$ orbitals merge and the two electrons pair to form a **π bond.** A π bond is so called because, viewed along the internuclear axis, it resembles a pair of electrons in a p orbital (and π is the Greek equivalent of p). Similarly, the $2p_y$ orbitals merge and their electrons pair to form another π bond. In general, a π bond arises from the merging of two p orbitals that approach side-by-side and the pairing of the electrons that they contain.

Brief illustration 14.1 The bonding pattern in N_2 and O_2

It follows from the discussion above that the overall bonding pattern in N_2 is a σ bond plus two π bonds, which is consistent with the Lewis structure :N≡N: in which the atoms are linked by a triple bond. Similarly, the bonding pattern in O_2 is predicted to be one σ(O2p$_z$,O2p$_z$) bond and one π(O2p$_x$,O2p$_x$) bond. However, see Section 14.10 for the continuing story about O_2.

14.4 Polyatomic molecules

Each σ bond in a polyatomic molecule is formed by the merging of orbitals with cylindrical symmetry about the relevant internuclear axis and the pairing of the spins of the electrons they contain. Likewise, π bonds are formed by pairing electrons that occupy atomic orbitals of the appropriate symmetry (broadly speaking, of the appropriate shape).

Brief illustration 14.2 A valence bond description of H_2O

The valence electron configuration of an O atom is $2s^22p_x^22p_y^12p_z^1$. The two unpaired electrons in the O2p orbitals can each pair with an electron in a H1s orbital, and each combination results in the formation of a σ bond (each bond has cylindrical symmetry about the respective O—H internuclear distance). Because the $2p_y$ and $2p_z$ orbitals lie at 90° to each other, the two σ bonds they form also lie at 90° to each other (Fig. 14.5). We predict, therefore, that H_2O should be an angular molecule, which it is. However, the model predicts a bond angle of 90°, whereas the actual bond angle is 104°.

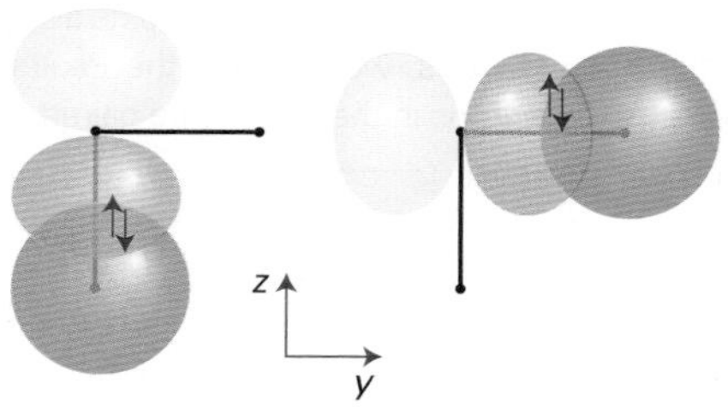

Fig. 14.5 The bonding in an H_2O molecule can be pictured in terms of the pairing of an electron belonging to one H atom with an electron in an O2p orbital; the other bond is formed likewise, but using a perpendicular O2p orbital. The predicted bond angle is 90°, which is in poor agreement with the experimental bond angle (104°).

Self-test 14.1

Give a VB description of NH_3, and predict the bond angle of the molecule on the basis of this description.

Answer: three σ(N2p,H1s) bonds; 90°; the experimental bond angle is 107°.

While broadly correct, VB theory seems to have two deficiencies. One is the poor estimate it provides for the bond angle in H_2O and other molecules, such as NH_3. Indeed, the theory appears to make worse predictions than the qualitative VSEPR model, which predicts HOH and HNH bond angles of slightly less than 109° in H_2O and NH_3, respectively. The second major deficiency is the apparent inability of VB theory to account for the number of bonds that atoms can form, and in particular the tetravalence of carbon. To appreciate the latter problem, we note that the ground-state valence configuration of a carbon atom is $2s^2 2p_x^1 2p_y^1$, which suggests that it should be capable of forming only two bonds, not four.

14.5 Promotion and hybridization

Two modifications solve all these problems. We have assumed that the atoms are in their lowest energy configurations, as predicted by the building-up principle. Now we allow a valence electron to be **promoted** from a full atomic orbital to an empty atomic orbital as a bond is formed: that results in two unpaired electrons instead of two paired electrons, and each unpaired electron can participate in bond formation. In carbon, for example, the promotion of a 2s electron to a 2p orbital leads to the configuration $2s^1 2p_x^1 2p_y^1 2p_z^1$, with four unpaired electrons in separate orbitals. These electrons may pair with four electrons in orbitals provided by four other atoms (such as four H1s orbitals if the molecule is CH_4), and as a result the atom can form four σ bonds. Promotion is worthwhile if the energy it requires can be more than recovered in the greater strength or number of bonds that can be formed.

We can now see why tetravalent carbon is so common. The promotion energy of carbon is small because the promoted electron leaves a doubly occupied 2s orbital and enters a vacant 2p orbital, hence significantly relieving the electron–electron repulsion it experiences in the former. Furthermore, the energy required for promotion is more than recovered by the atom's ability to form four bonds in place of the two bonds of the unpromoted atom.

Promotion, however, appears to imply the presence of three σ bonds of one type (in CH_4, from the merging of H1s and C2p orbitals) and a fourth σ bond of a distinctly different type (formed from the merging of H1s and C2s). It is well known, however, that all four bonds in methane are exactly equivalent both in terms of their chemical properties and their physical properties (their lengths, strengths, and stiffnesses).

This problem is overcome in VB theory by drawing on another technical feature of quantum mechanics which allows the same electron distribution to be described in different ways. In this case, we can describe the electron distribution in the promoted atom either as arising from four electrons in one s and three p orbitals, or as arising from four electrons in four different *mixtures* of these orbitals. Mixtures (more formally, linear combinations) of atomic orbitals on the same atom are called **hybrid orbitals**. These wavefunctions interfere destructively or constructively in different regions and give rise to four new shapes (we saw in Section 12.2 that interference is a characteristic of waves). The specific linear combinations that give rise to four equivalent hybrid orbitals are

$$h_1 = s + p_x + p_y + p_z \qquad h_2 = s - p_x - p_y + p_z$$
$$h_3 = s - p_x + p_y - p_z \qquad h_4 = s + p_x - p_y - p_z$$

sp³ hybrid orbitals (14.4)

(In general, a 'linear combination' of two functions f and g is $c_1 f + c_2 g$, where c_1 and c_2 are numerical coefficients, so a linear combination is a more general term than 'sum'. In a sum, $c_1 = c_2 = 1$.)

As a result of the constructive and destructive interference between the positive and negative regions of the component orbitals, each hybrid orbital has a large lobe pointing towards one corner of a regular tetrahedron (Fig. 14.6). Because each hybrid is built from one s orbital and three p orbitals, it is called an **sp³ hybrid orbital**.

It is now easy to see how the valence bond description of the methane molecule leads to a tetrahedral molecule containing four equivalent C—H bonds. It is energetically favourable (in the end, after bonding has been taken into account) for the carbon atom to undergo promotion. The promoted configuration has a distribution of electrons that is equivalent to one electron occupying each of four tetrahedral hybrid orbitals. Each hybrid orbital of the promoted atom contains a single unpaired electron; a hydrogen 1s electron can pair with each one, giving rise to a σ

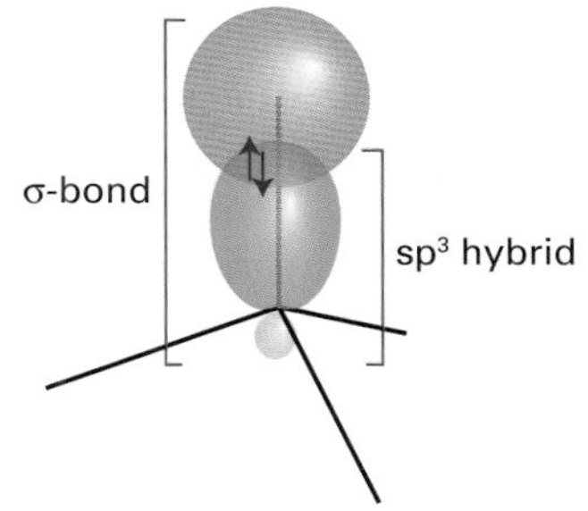

Fig. 14.6 The 2s and three 2p orbitals of a carbon atom hybridize, and the resulting hybrid orbitals point towards the corners of a regular tetrahedron. Each σ bond is formed by the pairing of an electron in an H1s orbital with an electron in one of the hybrid orbitals. The resulting molecule is regular tetrahedral.

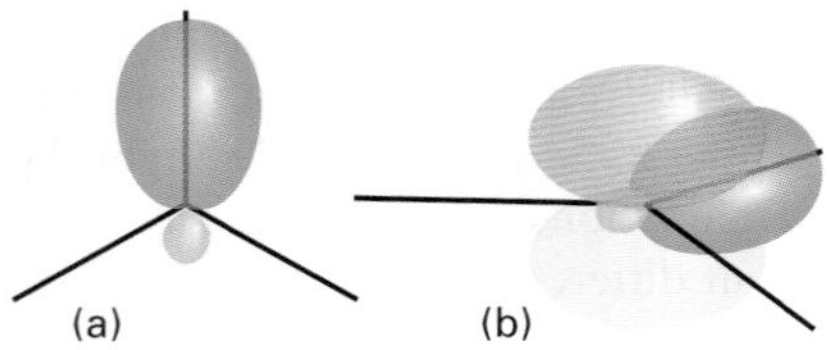

Fig. 14.7 (a) Trigonal planar hybridization is obtained when an s and two p orbitals are hybridized. The three lobes lie in a plane and make an angle of 120° to each other. (b) The remaining p orbital in the valence shell of an sp^2-hybridized atom lies perpendicular to the plane of the three hybrids.

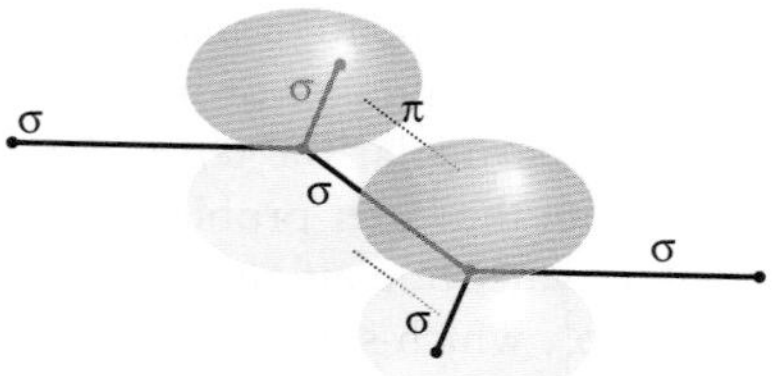

Fig. 14.8 The valence bond description of the structure of a carbon–carbon double bond, as in ethene. The electrons in the two sp^2 hybrids that point towards each other pair and form a σ bond. Electrons in the two p orbitals that are perpendicular to the plane of the hybrids pair, and form a π bond. The electrons in the remaining hybrid orbitals are used to form bonds to other atoms (in ethene itself, to H atoms).

bond pointing in a tetrahedral direction. Because each sp^3 hybrid orbital has the same composition, all four σ bonds are identical apart from their orientation in space.

Hybridization is also used in the VB description of alkenes. An ethene molecule is planar, with HCH and HCC bond angles close to 120°. To reproduce this σ-bonding structure, we think of each C atom as being promoted to a $2s^12p_x^12p_y^12p_z^1$ configuration. However, instead of using all four orbitals to form hybrids, we form **sp^2 hybrid orbitals** by allowing the s orbital and two of the p orbitals to interfere. As shown in Fig. 14.7, the three hybrid orbitals

$$\begin{aligned} h_1 &= s + 2^{1/2}\,p_x \\ h_2 &= s + (3/2)^{1/2}p_x - (1/2)^{1/2}p_y \\ h_3 &= s - (3/2)^{1/2}p_x - (1/2)^{1/2}p_y \end{aligned} \qquad \text{sp}^2\text{ hybrid orbitals} \qquad (14.5)$$

lie in a plane and point towards the corners of an equilateral triangle. The third 2p orbital ($2p_z$) is not included in the hybridization, and its axis is perpendicular to the plane in which the hybrids lie. The coefficients $2^{1/2}$, etc in the hybrids have been chosen to give the correct directional properties of the hybrids. The *squares* of the coefficients give the proportion of each atomic orbital in the hybrid. All three hybrids have s and p orbitals in the ratio 1:2, as indicated by the designation sp^2.

The sp^2-hybridized C atoms each form three σ bonds with either the h_1 hybrid of the other C atom or with the H1s orbitals. The σ framework therefore consists of bonds at 120° to each other. Moreover, provided the two CH_2 groups lie in the same plane, the two electrons in the unhybridized $C2p_z$ orbitals can pair and form a π bond (Fig. 14.8). The formation of this π bond locks the framework into the planar arrangement, for any rotation of one CH_2 group relative to the other leads to a weakening of the π bond (and consequently an increase in energy of the molecule).

Now consider a linear ethyne (acetylene) molecule, H—C≡C—H. The carbon atoms are **sp-hybridized**, and the σ bonds are built from hybrid atomic orbitals of the form

$$h_1 = s + p_z \qquad h_2 = s - p_z \qquad \text{sp hybrid orbitals} \qquad (14.6)$$

Note that the s and p orbitals contribute in equal proportions. The two hybrids lie along the z-axis. The electrons in them pair either with an electron in the corresponding hybrid orbital on the other C atom or with an electron in the H1s orbitals. Electrons in the two remaining p orbitals on each atom, which are perpendicular to the molecular axis, pair to form two perpendicular π bonds (as in Fig. 14.9).

Other hybridization schemes, particularly those involving d orbitals, are often invoked to account for (or at least be consistent with) other molecular geometries (Table 14.1).

An important point to note is that

The hybridization of N atomic orbitals always results in the formation of N hybrid orbitals.

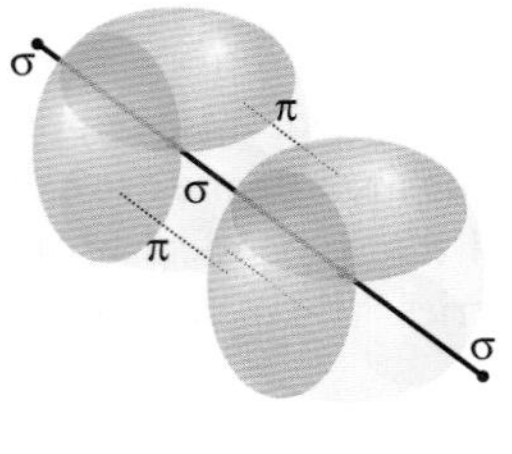

Fig. 14.9 The electronic structure of ethyne (acetylene). The electrons in the two sp hybrids on each atom pair to form σ bonds either with the other C atom or with an H atom. The remaining two unhybridized 2p orbitals on each atom are perpendicular to the axis: the electrons in corresponding orbitals on each atom pair to form two π bonds. The overall electron distribution is cylindrical.

Table 14.1
Hybrid orbitals

Number	Shape	Hybridization*
2	Linear	sp
3	Trigonal planar	sp^2
4	Tetrahedral	sp^3
5	Trigonal bipyramidal	sp^3d
6	Tetrahedral	sp^3d^2

*Other combinations are possible.

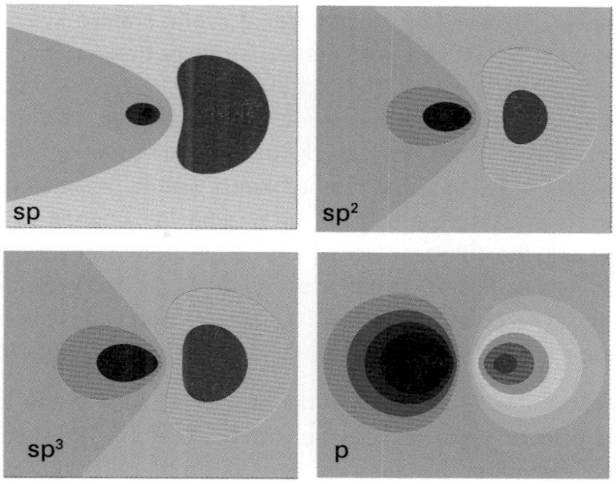

Fig. 14.11 Contour plots showing the amplitudes of sp^n hybrid orbitals. To construct these plots, we have used hydrogenic 2s and 2p orbitals.

● **Brief illustration 14.3** Bonding in polyatomic molecules

There are five equivalent P—Cl bonds in PCl_5, with the P atom at the centre of the molecule. All bonds are σ bonds formed from sp^3d hybrid orbitals on the central P atom. In SF_6, the six S—F bonds are σ bonds formed from sp^3d^2 hybrid orbitals on the central S atom, pointing towards the corners of a regular octahedron. This octahedral hybridization scheme is sometimes invoked to account for the structure of octahedral molecules, such as SF_6.

The 'pure' schemes in Table 14.1 are not the only possibilities: it is possible to form hybrid orbitals with intermediate proportions of atomic orbitals. For example, as more p-orbital character is included in an sp-hybridization scheme, the hybridization changes towards sp^2 and the angle between the hybrids changes continuously from 180° for pure sp hybridization to 120° for pure sp^2 hybridization. If the proportion of p character continues to be increased (by reducing the proportion of s orbital), then the hybrids eventually become pure p orbitals at an angle of 90° to each other (Fig. 14.10). Figure 14.11 shows contour plots of hybrid orbitals as the ratio of 2p character to 2s character increases.

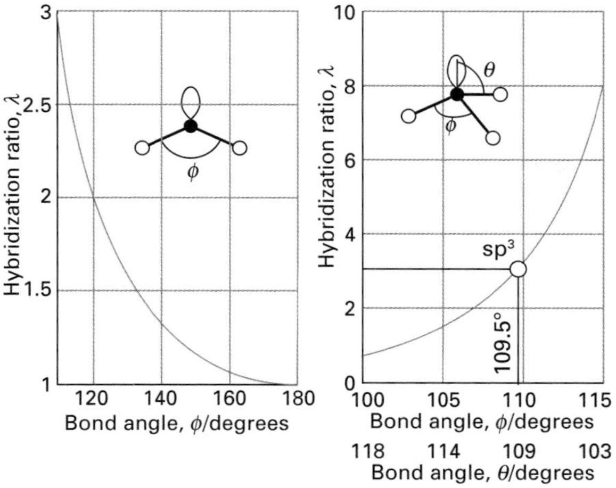

Fig. 14.10 The variation of hybridization with bond angle in (a) angular, (b) trigonal pyramidal molecules. The vertical axis gives the ratio of p to s character, so high values indicate mostly p character.

● **Brief illustration 14.4** A better valence bond description of H_2O

Now we can solve the problems from Brief illustration 14.2 and account for the structure of H_2O, with its bond angle of 104°. Each O—H σ bond is formed from an O atom hybrid orbital with a composition that lies between pure p (which would lead to a bond angle of 90°) and pure sp^2 (which would lead to a bond angle of 120°). The actual bond angle and hybridization adopted are found by calculating the energy of the molecule as the bond angle is varied, and looking for the angle at which the energy is a minimum.

Example 14.1

Bonding in the amide group

Use VB theory to describe the CO, CN, and NH bonds of the amide group based on the structure shown in (**1**).

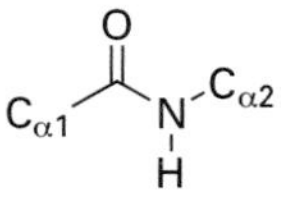

1

Strategy To calculate the number of hybrid orbitals, we note that each orbital can hold either one or two electrons. If it contains one electron, the orbital is ready to make a σ bond with an orbital on another atom. If it contains a pair of electrons, then it does not participate in bonding but acts as a lone pair. It follows that the number of hybrid orbitals on an atom is equal to the sum of the number of σ bonds to the atom and the number of lone pairs on the atom. Unhybridized p orbitals can participate in π bonds, as described in Section 14.3. As noted in Section 14.10, a double bond consists of a σ and a π bond.

Solution The O atom is sp^2-hybridized because it has two lone pairs and makes a σ bond with the C atom. The C atom is sp^2-hybridized because it makes three σ bonds: one with the O atom, one with the $C_{\alpha 1}$ atom, and one with the N atom. The N atom is sp^3-hybridized because it has one lone pair and makes three σ bonds: one with the H atom, one with the C atom, and one with the $C_{\alpha 2}$ atom.

We can infer that the CO group has a σ bond between Csp^2 and Osp^2 hybrid orbitals and a π bond between unhybridized $C2p_z$ and $O2p_z$ orbitals (where again we have taken the *z*-axis to be perpendicular to the plane containing the hybrid orbitals). The CN group has a σ bond between Csp^2 and Nsp^3 hybrid orbitals. Finally, the NH group has a σ bond between a Nsp^3 hybrid orbital and a H1s atomic orbital.

Self-test 14.2

Estimate the values of the $C_{\alpha 1}$CN and CN$C_{\alpha 2}$ bond angles for the structure shown in (**1**).

Answer: 120°, <109°

14.6 Resonance

Another term introduced into chemistry by VB theory is **resonance**, the superposition of the wavefunctions representing different electron distributions in the same nuclear framework. To understand what this means, consider the VB description of a purely covalently bonded HCl molecule, which could be written

$$\psi_{H-Cl}(1,2) = \psi_H(1)\psi_{Cl}(2) + \psi_H(2)\psi_{Cl}(1)$$

We have supposed that the bond is formed by the spin pairing of electrons in the H1s orbital, ψ_H, and the $Cl2p_z$ orbital, ψ_{Cl}. However, there is something wrong with this description: it allows electron 1 to be on the H atom when electron 2 is on the Cl atom, and vice versa, but it does not allow for unequal sharing of electron density between the atoms. On physical grounds, we should expect the purely covalent character of HCl to be an incomplete description of the molecule: because the Cl atom has higher ionization energy and electron affinity than the H atom, we can expect the ionic form H^+Cl^- to play a role. The wavefunction for this ionic structure, in which both electrons are in the $Cl2p_z$ orbital, is

$$\psi_{H^+Cl^-}(1,2) = \psi_{Cl}(1)\psi_{Cl}(2)$$

However, this wavefunction alone is unrealistic, because HCl is not an ionic species. A better description of the wavefunction for the molecule is as a superposition of the covalent and ionic descriptions, and we write (with a slightly simplified notation)

$$\psi_{HCl} = \psi_{H-Cl} + \lambda\psi_{H^+Cl^-}$$

with λ (lambda) some numerical coefficient. In general, we write

$$\psi = \psi_{\text{covalent}} + \lambda\psi_{\text{ionic}} \qquad (14.7)$$

where ψ_{covalent} is the wavefunction for the purely covalent form of the bond and ψ_{ionic} is the wavefunction for the ionic form of the bond. According to the general rules of quantum mechanics, in which probabilities are related to squares of wavefunctions, we interpret the square of λ as the relative proportion of the ionic contribution. If λ^2 is very small, the covalent description is dominant. If λ^2 is very large, the ionic description is dominant.

We find the numerical value of λ by using the **variation theorem**. First, we write down a plausible wavefunction, a **trial wavefunction**, for the molecule, such as the wavefunction in eqn 14.7, where λ is a variable parameter. The variation theorem then states that:

> The energy of a trial wavefunction is never less than the true energy.

The theorem implies that if we vary λ until we achieve the lowest energy, then the wavefunction with that value of λ is the best available of that particular kind.

The approach summarized by eqn 14.7, in which we express a wavefunction as the superposition of wavefunctions corresponding to a variety of structures *with the nuclei in the same locations*, is called **resonance**. In this case, where one structure is pure covalent and the other pure ionic, it is called **ionic–covalent resonance**. The interpretation of the wavefunction, which is called a **resonance hybrid**, is that if we were to inspect the molecule, then the probability that it would be found with an ionic structure is proportional to λ^2.

Brief illustration 14.5 Resonance hybrids

Consider a bond described by eqn 14.7. We might find that the lowest energy is reached when $\lambda = 0.1$, so the best description of the bond in the molecule is a resonance structure described by the wavefunction $\psi = \psi_{\text{covalent}} + 0.1\psi_{\text{ionic}}$. This wavefunction implies that the probabilities of finding the molecule in its covalent and ionic forms are in the ratio 100:1 (because $0.1^2 = 0.01$).

One of the most famous examples of resonance is in the VB description of benzene, where the wavefunction of the molecule is written as a superposition of the wavefunctions of the two covalent Kekulé structures (**2**) and (**3**):

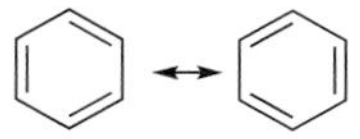

2 **3**

$$\psi = \psi_{\text{Kek}_2} + \psi_{\text{Kek}_3} \tag{14.8}$$

The two contributing structures have identical energies, so they contribute equally to the superposition. The effect of resonance (which is represented by a double-headed arrow) in this case is to distribute double-bond character around the ring and to make the lengths and strengths of all the carbon–carbon bonds identical. The wavefunction is improved by allowing resonance because it allows for a more accurate description of the location of the electrons, and in particular the distribution can adjust into a state of lower energy. This lowering is called the **resonance stabilization** of the molecule and, in the context of VB theory, is largely responsible for the unusual stability of aromatic rings. Resonance always lowers the energy, and the lowering is greatest when the contributing structures have similar energies. The wavefunction of benzene is improved still further, and the calculated energy of the molecule is lowered further still, if we allow ionic–covalent resonance too, by allowing a small admixture of structures such as that shown in (**4**).

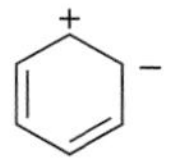

4

Resonance is not a flickering between the contributing states: it is a blending of their characteristics, much as a mule is a blend of a horse and a donkey. It is only a mathematical device for achieving a closer approximation to the true wavefunction of the molecule than that represented by any single contributing structure alone.

14.7 The language of valence bonding

It might be helpful at this point to summarize the concepts that VB theory has introduced into chemistry and which still survive even though MO theory is the dominant computational mode:

1. *The names of bond types*: σ and π bonds are formed by spin pairing of electrons on adjacent atoms.
2. *Promotion:* valence electrons may be promoted to empty orbitals if overall that results in a lowering of energy.
3. *Hybridization:* atomic orbitals may be hybridized to match the observed geometry of a molecule.
4. *Resonance:* the superposition of individual structures. Resonance distributes multiple-bond character over the molecule and lowers the overall energy.

Molecular orbitals

In molecular orbital theory, electrons are treated as spreading throughout the entire molecule: every electron contributes to the strength of every bond. This theory has been more fully developed than valence bond theory and provides the language that is widely used in modern discussions of bonding in small inorganic molecules, d-metal complexes, and solids. To introduce it, we follow the same strategy as in Chapter 13, where the one-electron hydrogen atom was taken as the fundamental species for discussing atomic structure, and then developed into a description of many-electron atoms. In this section we use the simplest molecule of all, the one-electron hydrogen molecule-ion, H_2^+, to introduce the essential features of bonding, and then use it as a guide to the structures of more complex systems.

14.8 Linear combinations of atomic orbitals

A **molecular orbital** is a one-electron wavefunction that spreads throughout the molecule. The mathematical forms of such orbitals are highly complicated, even for such a simple species as H_2^+, and they are unknown in general. All modern work builds approximations to the true molecular orbital by formulating models based on linear combinations of the atomic orbitals on the atoms in the molecule.

First, we recall the general principle of quantum mechanics—which we used earlier to construct VB wavefunctions—that if there are several possible outcomes, then we superimpose—add together—the wavefunctions that represent those outcomes. In H_2^+, there are two possible outcomes: because an electron spreads throughout the molecule, it may be found either in an atomic orbital centred on A, ψ_A, or in an orbital centred on B, ψ_B. Therefore, we write

$$\psi = c_A\psi_A + c_B\psi_B \quad \text{An LCAO} \quad (14.9a)$$

where c_A and c_B are numerical coefficients. A wavefunction constructed in this way is called a **linear combination of atomic orbitals** (LCAO) and the corresponding molecular orbital is called an LCAO-MO. The squares of the coefficients tell us the relative proportions of the atomic orbitals contributing to the molecular orbital. In a homonuclear diatomic molecule an electron can be found with equal probability in orbital A or orbital B, so the *squares* of the coefficients must be equal, which implies that $c_B = \pm c_A$. The two possible (unnormalized) wavefunctions are therefore

$$\psi = \psi_A \pm \psi_B \quad \text{An LCAO for a diatomic molecule} \quad (14.9b)$$

First, we consider the LCAO with the plus sign, $\psi = \psi_A + \psi_B$, as this molecular orbital will turn out to have the lower energy of the two. The form of this orbital is shown in Fig. 14.12. It is called a **σ orbital** because it resembles an s orbital when viewed along the axis. More precisely, it is so called because an electron that occupies a σ orbital has zero orbital angular momentum around the internuclear axis, just as an s electron has zero orbital angular momentum around an axis passing through the nucleus.

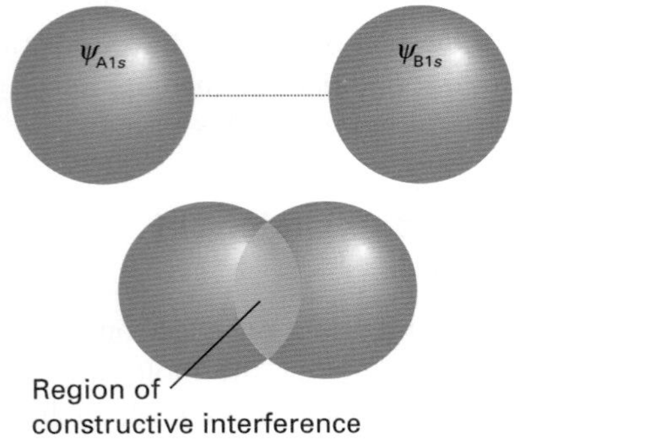

Fig. 14.12 The formation of a bonding molecular orbital (a σ orbital). (a) Two H1s orbitals come together. (b) The atomic orbitals overlap, interfere constructively, and give rise to an enhanced amplitude in the internuclear region. The resulting orbital has cylindrical symmetry about the internuclear axis. When it is occupied by two paired electrons, to give the configuration σ^2, we have a σ bond.

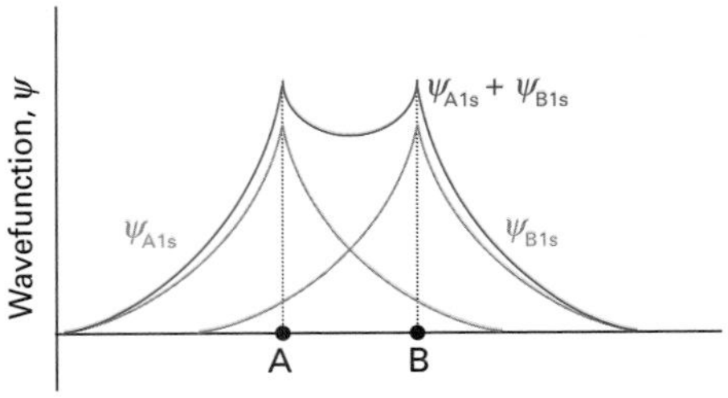

Fig. 14.13 The bonding molecular orbital wavefunction along the internuclear axis. Note that there is an enhancement of amplitude between the nuclei, so there is an increased probability of finding the bonding electrons in that region.

Because, as we shall see, it is the σ orbital of lowest energy, it is labelled 1σ. An electron that occupies a σ orbital is called a **σ electron**. In the ground state of the H_2^+ ion, there is a single 1σ electron, so we report the ground-state configuration of H_2^+ as $1\sigma^1$.

We can see the origin of the lowering of energy that is responsible for the formation of the bond by examining the LCAO-MO. The two atomic orbitals are like waves centred on adjacent nuclei. In the internuclear region, the amplitudes interfere constructively and the wavefunction has an enhanced amplitude there (Fig. 14.13). The three contributions that we listed for bonding in VB theory (Section 14.3) apply here too: there is an accumulation of electron density between the two nuclei, a removal of electron density from close to the nuclei, and a lowering of kinetic energy as a result of the electron spreading over both nuclei.

The accumulation of probability density in the internuclear region is measured by the **overlap integral**, *S*. As we show in the following Derivation, when $S = 1$, there is perfect overlap between two atomic orbitals; when $S = 0$, there is no overlap at all (Fig. 14.14). Broadly speaking, the greater the overlap integral, the stronger is the bonding effect of electrons in the molecular orbital they form.

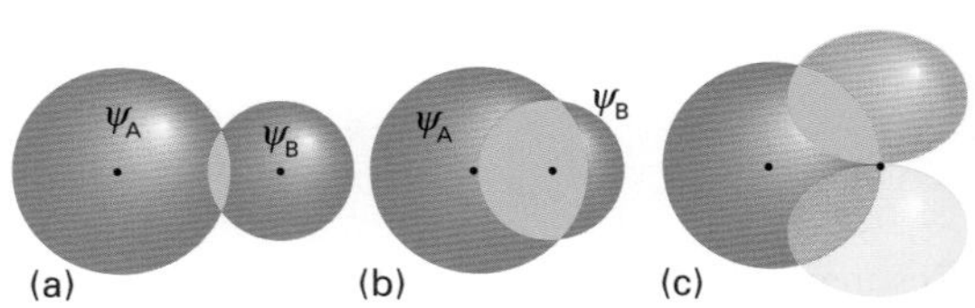

Fig. 14.14 A schematic representation of the contributions to the overlap integral. (a) $S \approx 0$ because the orbitals are far apart and their product is always small. (b) *S* is large (but less than 1) because the product $\psi_A\psi_B$ is large over a substantial region. (c) $S = 0$ because the positive region of overlap is exactly cancelled by the negative region.

Derivation 14.2

Overlap integrals

An overlap integral is calculated by dividing space up into a large number of small regions, multiplying together the values of ψ_A and ψ_B in each region, then adding together (integrating) the resulting products for all the regions. Formally, we express this rule by writing

$$S_{AB} = \int \psi_A \psi_B d\tau \qquad \text{An overlap integral} \quad (14.10)$$

where $d\tau$ (dee tau) is an infinitesimal volume element (for instance, $d\tau = dxdydz$ in three-dimensional Cartesian coordinates). If ψ_B is small wherever ψ_A is large, and vice versa (such as when two hydrogen nuclei are far apart), the products are all small and the integral is also small: this corresponds to a small value of S_{AB}. At typical bonding distances, ψ_A and ψ_B are both large in the internuclear region, so their products there are large, and the integral is also large: this corresponds to a value of S_{AB} approaching 1 (typically, about 0.4). If the two nuclei are coincident, the two atomic orbitals have identical values everywhere, and the integral of their products gives $S_{AB} = 1$.

It is possible, but not easy, to evaluate the overlap integral for hydrogenic orbitals, and for two 1s orbitals on hydrogen nuclei separated by a distance R, the result is

$$S_{HH} = \left\{1 + \frac{R}{a_0} + \frac{R^2}{3a_0^2}\right\} e^{-R/a_0} \qquad (14.11)$$

This function is plotted in Fig. 14.15. The exponential factor guarantees that the overlap integral goes to zero at large separations.

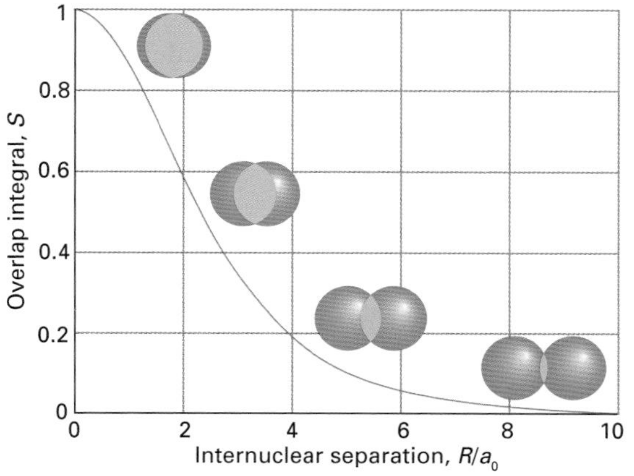

Fig. 14.15 The variation of the overlap integral with internuclear distance for two H1s orbitals.

14.9 Bonding and antibonding orbitals

A 1σ orbital is an example of a **bonding orbital**, a molecular orbital which, if occupied, contributes to the strength of a bond between two atoms. As in VB theory, we can substitute the wavefunction $\psi = \psi_A + \psi_B$ into the Schrödinger equation for the molecule-ion with the nuclei at a fixed separation R and solve the equation for the energy. The molecular potential energy curve obtained by plotting the energy against R is very similar to the one drawn in Fig. 14.1. The energy of the molecule falls as R is decreased from large values because the electron is increasingly likely to be found in the internuclear region as the two atomic orbitals interfere more effectively. However, at small separations, there is too little space between the nuclei for significant accumulation of electron density there. In addition, the nucleus–nucleus repulsion $V_{nuc,nuc}$ (given in eqn 14.3) becomes large and the kinetic energy of the electron is not lowered by very much. As a result, after an initial decrease, at small internuclear separations the potential energy curve passes through a minimum and then rises sharply to high values. Calculations on H_2^+ give the equilibrium bond length as 130 pm and the bond dissociation energy as 171 kJ mol^{-1}; the experimental values are 106 pm and 250 kJ mol^{-1}, so this simple LCAO-MO description of the molecule, while inaccurate, is not absurdly wrong.

Now consider the alternative LCAO, the one with a minus sign: $\psi = \psi_A - \psi_B$. Because this wavefunction is also cylindrically symmetrical around the internuclear axis it is also a σ orbital, which we denote 1σ* (Fig. 14.16). When substituted into the Schrödinger

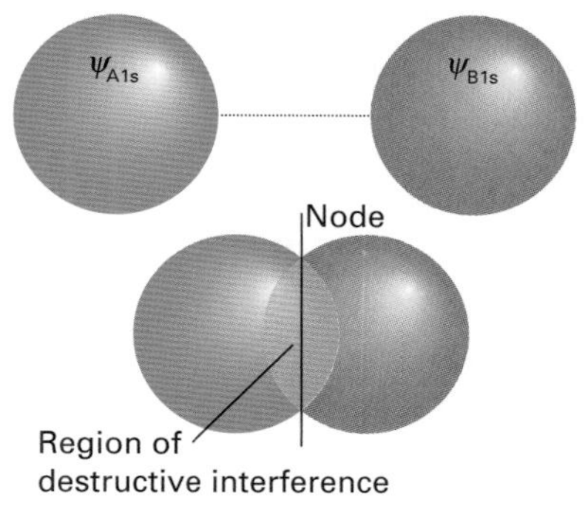

Fig. 14.16 The formation of an antibonding molecular orbital (a σ* orbital). (a) Two H1s orbitals come together. (b) The atomic orbitals overlap with opposite signs (as depicted by different colours), interfere destructively, and give rise to a decreased amplitude in the internuclear region. There is a nodal plane exactly halfway between the nuclei, on which any electrons that occupy the orbital will not be found.

equation, we find that it has a higher energy than the bonding 1σ orbital and, indeed, it has a higher energy than either of the two atomic orbitals.

We can trace the origin of the high energy of the 1σ* orbital to the existence of a **nodal plane**, a plane on which the wavefunction passes through zero. This plane lies halfway between the nuclei and cuts through the internuclear axis. The two atomic orbitals cancel on this plane as a result of their destructive interference, because they have opposite signs. In drawings like that in Figs. 14.12 and 14.16, we represent overlap of orbitals with the same sign (as in the formation of 1σ) by shading of the same tint; the overlap of orbitals of opposite sign (as in the formation of 1σ*) is represented by one orbital of a light tint and another orbital of a dark tint.

The 1σ* orbital is an example of an **antibonding orbital**, an orbital that, if occupied, decreases the strength of a bond between two atoms. The antibonding character of the 1σ* orbital is partly a result of the exclusion of the electron from the internuclear region and its relocation outside the bonding region where it helps to pull the nuclei apart rather than pulling them together (Fig. 14.17). An antibonding orbital is often slightly more strongly antibonding than the corresponding bonding orbital is bonding: although the 'gluing' effect of a bonding electron and the 'anti-gluing' effect of an antibonding electron are similar, the nuclei repel each other in both cases, and this repulsion pushes both levels up in energy.

We need to be aware of a few points regarding notation. For homonuclear diatomic molecules, it is helpful to identify the **inversion symmetry** of a molecular orbital, especially when discussing electronic transitions (Chapter 20). By 'inversion symmetry' is meant the behaviour of a wavefunction when it is inverted through the centre (more formally, the centre of inversion) of the molecule. Thus, if we consider any point of the σ bonding orbital, and then project it through the centre of the molecule and out an equal distance on the other side, then we arrive at an identical value of the wavefunction (Fig. 14.18). This so-called ***gerade*** **symmetry** (from the German word for 'even') is denoted by a subscript g, as in σ_g. On the other hand, the same procedure applied to the antibonding σ* orbital results in the same size but opposite sign of the wavefunction. This ***ungerade*** **symmetry** ('odd symmetry') is denoted by a subscript u, as in σ_u. This inversion symmetry classification (or 'parity') is not applicable to heteronuclear diatomic molecules (like CO) as they do not have a centre of inversion.

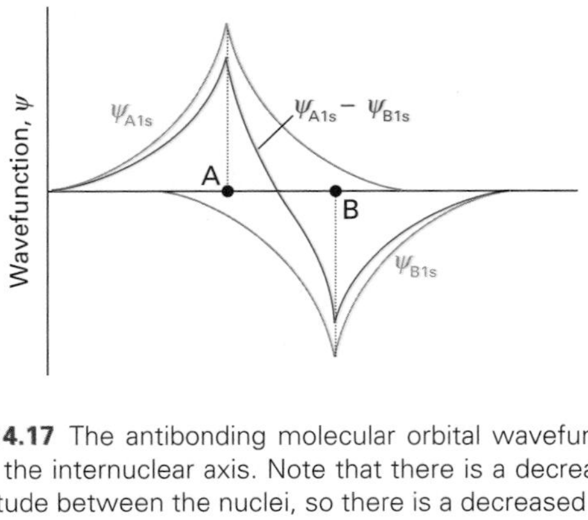

Fig. 14.17 The antibonding molecular orbital wavefunction along the internuclear axis. Note that there is a decrease in amplitude between the nuclei, so there is a decreased probability of finding the bonding electrons in that region.

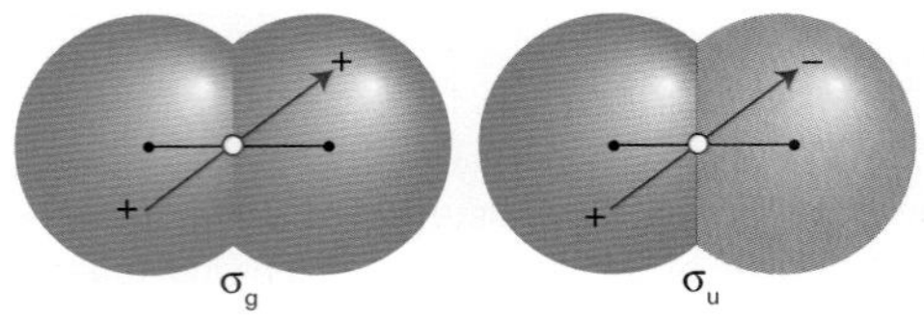

Fig. 14.18 The *gerade/ungerade* character of σ bonding and antibonding orbitals.

14.10 The structures of homonuclear diatomic molecules

In Chapter 13 we used the hydrogenic atomic orbitals and the building-up principle to deduce the ground electronic configurations of many-electron atoms. Here we use the same procedure for many-electron diatomic molecules (such as H_2 with two electrons and even Br_2 with 70), but using the H_2^+ molecular orbitals as a basis. The general procedure is as follows:

1. Construct molecular orbitals by forming linear combinations of all suitable valence atomic orbitals supplied by the atoms (the meaning of 'suitable' will be explained shortly); *N* atomic orbitals result in *N* molecular orbitals.
2. Accommodate the valence electrons supplied by the atoms so as to achieve the lowest overall energy subject to the constraint of the Pauli exclusion principle, that no more than two electrons may occupy a single orbital (and then must be paired).
3. If more than one molecular orbital of the same energy is available, add the electrons to each individual orbital before doubly occupying any one orbital (because that minimizes electron–electron repulsions).

4. Take note of Hund's rule (Section 13.11), that if electrons occupy different degenerate orbitals, then they do so with parallel spins.

The following sections show how these rules are used in practice.

(a) Hydrogen and helium molecules

The first step in the discussion of H_2, the simplest many-electron diatomic molecule, is to build the molecular orbitals. Because each H atom of H_2 contributes a 1s orbital (as in H_2^+), we can form the 1σ (more precisely, $1\sigma_g$) and $1\sigma^*$ (that is, $1\sigma_u$) bonding and antibonding orbitals from them, as we have seen already. At the equilibrium internuclear separation these orbitals have the energies represented by the horizontal lines in Fig. 14.19.

There are two electrons to accommodate (one from each atom). Both can enter the $1\sigma_g$ orbital by pairing their spins (Fig. 14.20). The ground-state configuration is therefore $1\sigma_g^2$, and the atoms are joined by a bond consisting of an electron pair in a bonding σ orbital. These two electrons bind the two nuclei together more strongly and closely than the single electron in H_2^+, and the bond length is reduced from 106 pm to 74 pm. A pair of electrons in a σ orbital is called a **σ bond**, and is very similar to the σ bond of VB theory. The two differ in certain details of the electron distribution between the two atoms joined by the bond, but both have an accumulation of density between the nuclei.

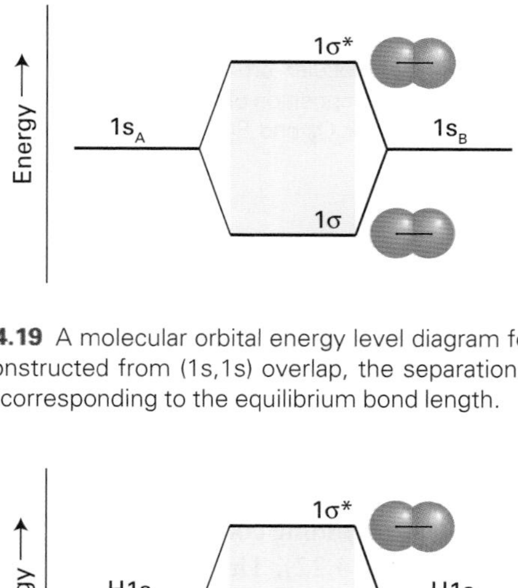

Fig. 14.19 A molecular orbital energy level diagram for orbitals constructed from (1s,1s) overlap, the separation of the levels corresponding to the equilibrium bond length.

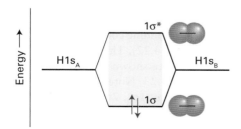

Fig. 14.20 The ground electronic configuration of H_2 is obtained by accommodating the two electrons in the lowest available orbital (the bonding orbital).

We can conclude that *the importance of an electron pair in bonding stems from the fact that two is the maximum number of electrons that can enter each bonding molecular orbital.* Electrons do not 'want' to pair: they pair, as we show in the following brief Derivation, because the Pauli exclusion principle implies that:

- only if electrons pair their spins can they both occupy a bonding orbital
- no more than two electrons can occupy any given orbital.

Derivation 14.3

Electron pairing in MO theory

The spatial wavefunction for two electrons in a bonding molecular orbital ψ such as the bonding orbital in eqn 14.9 is $\psi(1)\psi(2)$. This two-electron wavefunction is obviously symmetric under interchange of the electron labels. To satisfy the Pauli principle, it must be multiplied by the antisymmetric spin state, $\alpha(1)\beta(2) - \beta(1)\alpha(2)$, to give the overall antisymmetric combination

$$\psi(1,2) = \psi(1)\psi(2) \times \{\alpha(1)\beta(2) - \beta(1)\alpha(2)\}$$

Because $\alpha(1)\beta(2) - \beta(1)\alpha(2)$ corresponds to paired electron spins (Further information 13.1), we see that two electrons can occupy the same molecular orbital (in this case, the bonding orbital) only if their spins are paired.

A similar argument shows why helium is a monatomic gas. Consider a hypothetical He_2 molecule. Each He atom contributes a 1s orbital to the linear combination used to form the molecular orbitals, and so we can construct $1\sigma_g$ and $1\sigma_u$ molecular orbitals. They differ in detail from those in H_2 because the He1s orbitals are more compact, but the general shape is the same, and for qualitative discussions we can use the same molecular orbital energy level diagram as for H_2. Because each atom provides two electrons, there are four electrons to accommodate. Two can enter the $1\sigma_g$ orbital, but then it is full (by the Pauli exclusion principle). The next two electrons must enter the antibonding $1\sigma_u$ orbital (Fig. 14.21). The ground electronic configuration of He_2 is therefore $1\sigma_g^2 1\sigma_u^2$. Because an antibonding orbital is slightly more antibonding than a bonding orbital is bonding, the He_2 molecule has a higher energy than the separated atoms and is unstable. Hence, two

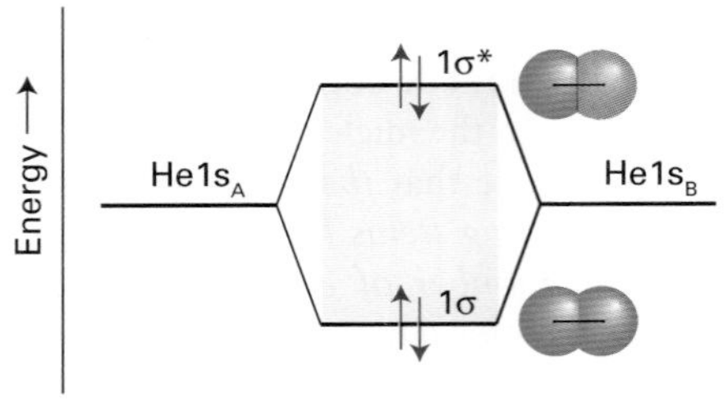

Fig. 14.21 The ground electronic configuration of the four-electron molecule He_2 has two bonding electrons and two antibonding electrons. It has a higher energy than the separated atoms, and so He_2 is unstable relative to two He atoms.

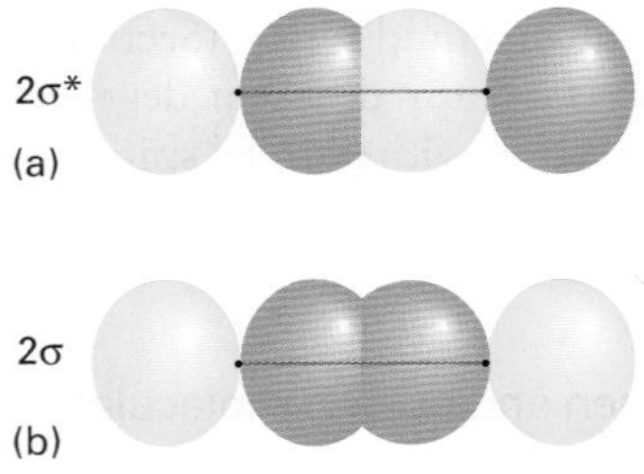

Fig. 14.22 (a) The interference leading to the formation of a σ bonding orbital and (b) the corresponding antibonding orbital when two p orbitals overlap along an internuclear axis.

ground-state He atoms do not form bonds to each other, and helium is a monatomic gas.

Example 14.2

Judging the stability of diatomic molecules

Decide whether Li_2 is likely to exist on the assumption that only the valence s orbitals contribute to its molecular orbitals.

Strategy Decide what molecular orbitals can be formed from the available valence orbitals, rank them in order of energy, then feed in the electrons supplied by the valence orbitals of the atoms. Judge whether there is a net bonding or net antibonding effect between the atoms.

Solution Each molecular orbital is built from 2s atomic orbitals, which give one bonding and one antibonding combination ($1\sigma_g$ and $1\sigma_u$, respectively). Each Li atom supplies one valence electron; the two electrons fill the $1\sigma_g$ orbital, to give the configuration $1\sigma_g^2$, which is bonding.

Self-test 14.3

Is LiH likely to exist if the Li atom uses only its 2s orbital for bonding?

Answer: Yes, $\sigma(\text{Li2s,H1s})^2$

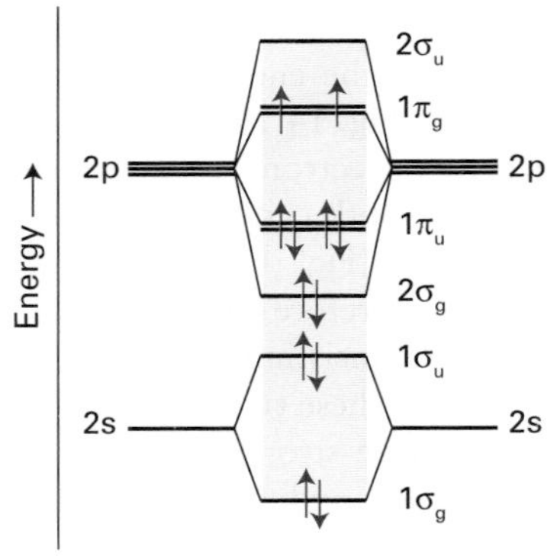

Fig. 14.23 A typical molecular orbital energy level diagram for Period 2 homonuclear diatomic molecules. The valence atomic orbitals are drawn in the columns on the left and the right; the molecular orbitals are shown in the middle. Note that the π orbitals form doubly degenerate pairs. The sloping lines joining the molecular orbitals to the atomic orbitals show the principal composition of the molecular orbitals. This diagram is suitable for O_2 and F_2; the configuration of O_2 is shown.

(b) π bonds

We shall now see how the concepts we have introduced apply to other homonuclear diatomic molecules, such as N_2 and Cl_2, and diatomic ions such as O_2^{2-}. In line with the building-up procedure, we first consider the molecular orbitals that may be formed from the valence orbitals and do not (at this stage) trouble about how many electrons are available.

In Period 2, the valence orbitals are 2s and 2p. Suppose first that we consider these two types of orbital separately. Then the 2s orbitals on each atom overlap to form bonding and antibonding combinations that we denote $1\sigma_g$ and $1\sigma_u$, respectively. Likewise, the two $2p_z$ orbitals (by convention, the internuclear axis is the z-axis) have cylindrical symmetry around the internuclear axis. They may therefore participate in σ-orbital formation to give the bonding and antibonding combinations $2\sigma_g$ and $2\sigma_u$, respectively (Fig. 14.22). The resulting energy levels of the σ orbitals are shown in the MO energy level diagram in Fig. 14.23. Note that we number the σ_g orbitals in sequence ($1\sigma_g$, $2\sigma_g$, . . .) and the σ_u orbitals likewise.

Strictly, we should not consider the 2s and $2p_z$ orbitals separately, because both of them can contribute to the formation of σ molecular orbitals. Therefore, in a more advanced treatment, we should

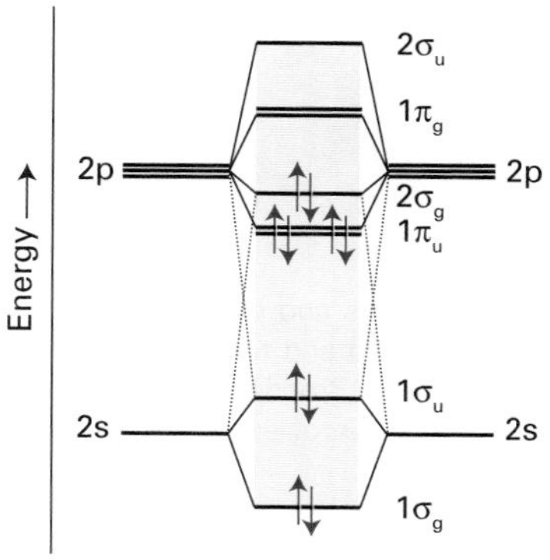

Fig. 14.24 A typical molecular orbital energy level diagram for Period 2 homonuclear diatomic molecules up to and including N_2.

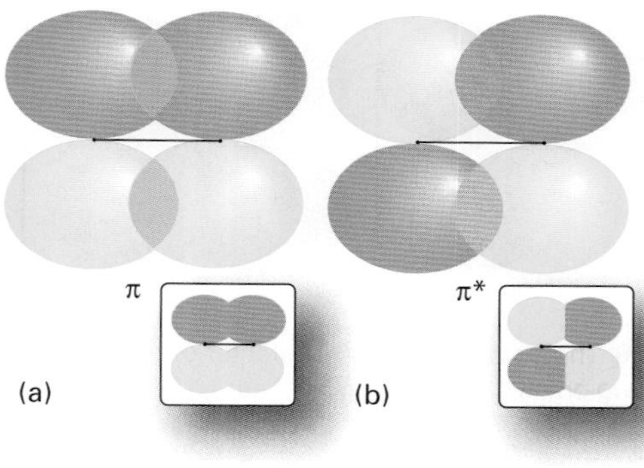

Fig. 14.25 (a) The interference leading to the formation of a π bonding orbital and (b) the corresponding antibonding orbital.

combine all four orbitals together to form four σ molecular orbitals, each one of the form

$$\psi = c_1\psi_{A2s} + c_2\psi_{B2s} + c_3\psi_{A2p_z} + c_4\psi_{B2p_z}$$

We find the four coefficients, which represent the different contributions that each atomic orbital makes to the overall molecular orbital, by using the variation theorem. However, in practice, the two lowest energy combinations of this kind are very similar to the combination $1\sigma_g$ and $1\sigma_u$ of 2s orbitals that we have described, and the two highest energy combinations are very similar to the $2\sigma_g$ and $2\sigma_u$ combinations of $2p_z$ orbitals. In each case there are small differences: the $1\sigma_g$ orbital, for instance, is contaminated by some $2p_z$ character and the $2\sigma_g$ orbital is contaminated by some 2s character, and their energies will be slightly shifted from where they would be if we considered only the 'pure' combinations. Nevertheless, the changes are not great, and we can continue to think of $1\sigma_g$ and $1\sigma_u$ as being one bonding and antibonding pair, and of $2\sigma_g$ and $2\sigma_u$ as being another pair. The four orbitals are shown in the centre column of Fig. 14.24. There is no guarantee that $1\sigma_u$ and $2\sigma_g$ will be in the exact location shown in the illustration and the locations shown in Fig. 14.23 are found in some molecules (see below).

There is one further point in this connection. As soon as we allow all four atomic orbitals to contribute to an LCAO it is no longer clear—except by appealing to the form of the simple pairwise LCAOs that each one resembles—whether a particular combination is bonding or antibonding: all we can say is that the four linear combinations have successively increasing energies. However, the parity classification is unaffected, and the orbitals can still be classified as g or u; in homonuclear diatomic molecules, inversion symmetry is a more fundamental classification scheme than bonding and antibonding.

Now consider the $2p_x$ and $2p_y$ orbitals of each atom, which are perpendicular to the internuclear axis and may overlap side-by-side. This overlap may be constructive or destructive and results in a bonding and an antibonding **π orbital**, which initially we label $1\pi_u$ and $1\pi^*$, respectively. The notation π is the analogue of p in atoms, for when viewed along the axis of the molecule, a π orbital looks like a p orbital (Fig. 14.25). More precisely, an electron in a π orbital has one unit of orbital angular momentum about the internuclear axis. The two $2p_x$ orbitals overlap to give a bonding and an antibonding π orbital, as do the two $2p_y$ orbitals too. The two bonding combinations have the same energy; likewise, the two antibonding combinations have the same energy. Hence, each π energy level is doubly degenerate and consists of two distinct orbitals. Typically (but not universally) the bonding effect of electrons in a π orbital is less than for a σ orbital in the same molecule because the electron density it represents does not lie between the nuclei so completely. Likewise, the antibonding effect of electrons in a π* orbital is typically less than when they occupy a σ* orbital in the same molecule. Two electrons in a π orbital constitute a **π bond**: such a bond resembles a π bond of valence bond theory, but the details of the electron distribution are slightly different.

The inversion-symmetry classification also applies to π orbitals. As we see from Fig. 14.26, a bonding π orbital changes sign on inversion through the centre of the molecule, and is therefore classified as u. On the other hand, the antibonding π* orbital does not change sign on inversion, and is therefore g. The bonding and antibonding combinations will

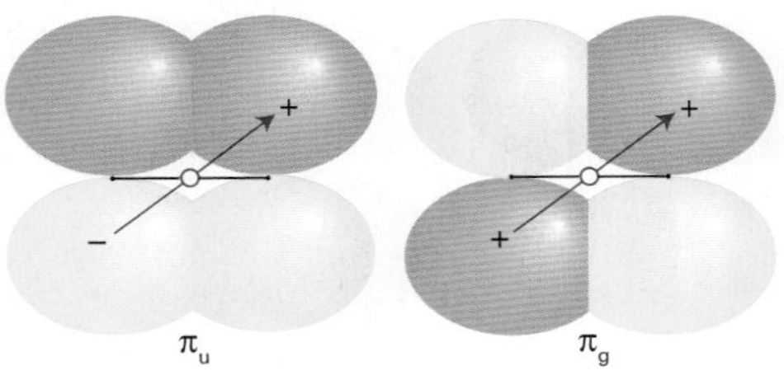

Fig. 14.26 The *gerade/ungerade* character of π bonding and antibonding orbitals.

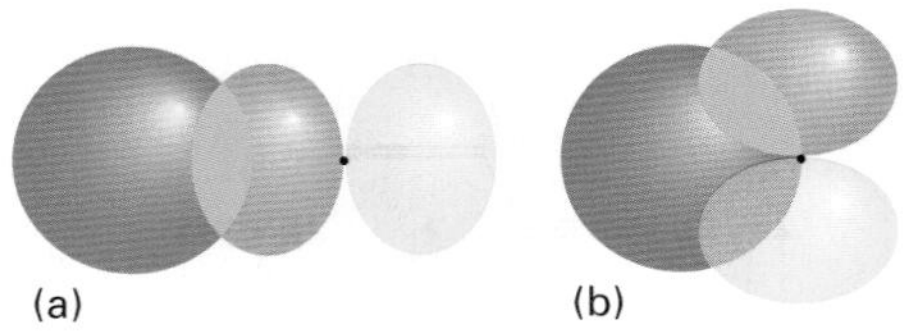

Fig. 14.28 Overlapping s and p orbitals. (a) End-on overlap leads to nonzero overlap and to the formation of an axially symmetric σ orbital. (b) Broadside overlap leads to no net accumulation or reduction of electron density and does not contribute to bonding.

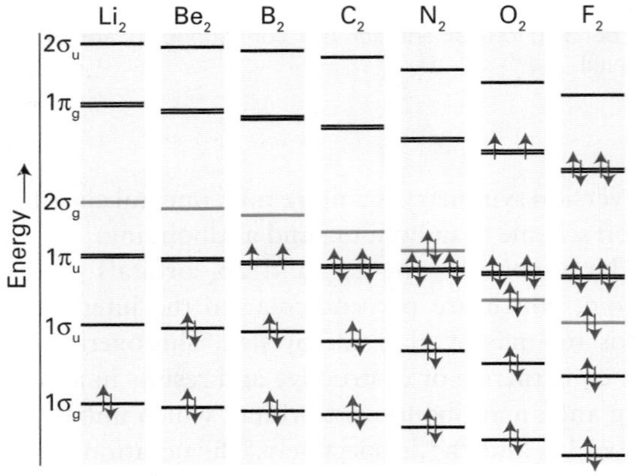

Fig. 14.27 The variation of the orbital energies of Period 2 homonuclear diatomic molecules. Only the valence-shell orbitals are shown.

henceforth be denoted $1\pi_u$ and $1\pi_g$. The relative order of the σ and π orbitals in a molecule cannot be predicted without detailed calculation and varies with the energy separation between the 2s and 2p orbitals of the atoms; in some molecules the order shown in Fig. 14.23 applies, whereas others have the order shown in Fig. 14.24. The change in order can be seen in Fig. 14.27, which shows the calculated energy levels for the Period 2 homonuclear diatomic molecules. A useful rule is that, for neutral molecules, the order shown in Fig. 14.23 is valid for O_2 and F_2, whereas the order shown in Fig. 14.24 is valid for the preceding elements of the period.

(c) Symmetry and overlap

One central feature of molecular orbital theory can now be addressed. We have seen that s and p_z orbitals may contribute to the formation of σ orbitals, and that p_x and p_y orbitals may contribute to π orbitals. However, we never have to consider orbitals formed by the overlap of s and p_x orbitals (or p_y orbitals). When building molecular orbitals, *we need consider linear combinations only of atomic orbitals of the same symmetry with respect to the internuclear axis*. Because an s orbital has cylindrical symmetry around the internuclear axis, but a p_x orbital does not, the two atomic orbitals cannot contribute to the same molecular orbital. The reason for this distinction based on symmetry can be understood by considering the interference between an s orbital and a p_x orbital (Fig. 14.28): although there is constructive interference between the two orbitals on one side of the axis, there is an exactly compensating amount of destructive interference on the other side of the axis, and the net bonding or antibonding effect is zero.

Consistent with this interpretation, the overlap of a 1s orbital on one atom and a $2p_x$ orbital on another atom (with z the internuclear axis) is zero. In terms of the discussion in Derivation 14.2, we see in Fig. 14.28 that at some point the product $\psi_A\psi_B$ may be large. However, there is a matching point in the lower half of the figure point where $\psi_A\psi_B$ has exactly the same magnitude but an opposite sign. When the integral is evaluated, these two contributions are added together and cancel. For every point in the upper half of the diagram, there is a point in the lower half that cancels it, so $S = 0$. Therefore, for symmetry reasons, there is no net overlap between the s and p orbitals in this arrangement.

(d) Criteria for formation of molecular orbitals

We now have the criteria for selecting atomic orbitals from which molecular orbitals are to be built:

1. Use all available valence orbitals from both atoms.
2. Classify the atomic orbitals as having σ and π symmetry with respect to the internuclear axis, and build σ and π orbitals from all atomic orbitals of a given symmetry.
3. From N_σ atomic orbitals of σ symmetry, N_σσ orbitals can be built with progressively higher

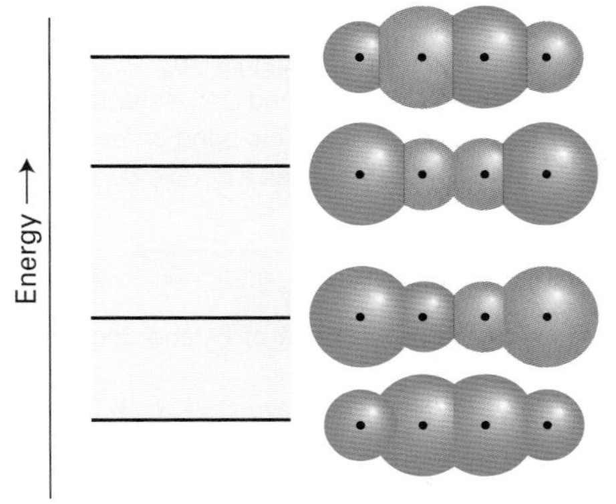

Fig. 14.29 A schematic representation of the four molecular orbitals that can be formed from four s orbitals in a chain of four atoms. The lowest energy combination (the bottom diagram) is formed from atomic orbitals with the same sign, and there are no internuclear nodes. The next-higher orbital has one node (at the centre of the molecule). The next-higher orbital has two internuclear nodes, and the uppermost, highest energy orbital, has three internuclear nodes, one between each neighbouring pair of atoms, and is fully antibonding. The sizes of the spheres reflect the contributions of each atom to the molecular orbital; the shading represents different signs.

energy from strongly bonding to strongly antibonding.

4. From N_π atomic orbitals of π symmetry, N_π π orbitals can be built with progressively higher energy from strongly bonding to strongly antibonding. The π orbitals occur in doubly degenerate pairs.

As a general rule, the energy of each type of orbital (σ or π) increases with the number of internuclear nodes. The lowest energy orbital of a given species has no internuclear nodes and the highest energy orbital has a nodal plane between each pair of adjacent atoms (Fig. 14.29).

Example 14.3

Assessing the contribution of d orbitals

To get a sense of how molecular orbitals can be built from d orbitals, show how they can contribute to formation of σ and π orbitals in diatomic molecules.

Strategy We need to assess the symmetry of d orbitals with respect to the internuclear *z*-axis: orbitals of the same symmetry can contribute to a given molecular orbital.

Solution A d_{z^2} orbital has cylindrical symmetry around *z* and so can contribute to σ orbitals. The d_{zx} and d_{yz} orbitals have π symmetry with respect to the axis (Fig. 14.30), so they can contribute to π orbitals.

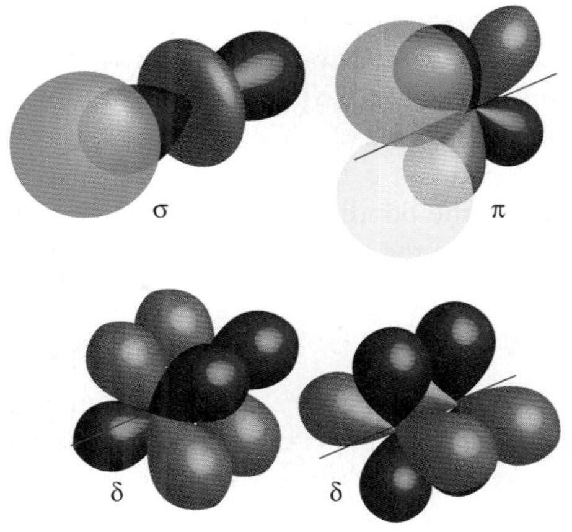

Fig. 14.30 The types of molecular orbital to which d orbitals can contribute. The σ and π combinations can be formed with s, p, and d orbitals of the appropriate symmetry, but the δ orbitals can be formed only by the d orbitals of the two atoms.

Self-test 14.4

Sketch the 'δ orbitals' (orbitals that resemble four-lobed d orbitals when viewed along the internuclear axis) that may be formed by the remaining two d orbitals (and which contribute to bonding in some d-metal cluster compounds). Give their inversion-symmetry classification.

Answer: see Fig. 14.30: bonding are g, antibonding are u

(e) Period 2 homonuclear diatomic molecules

Figures 14.23 and 14.24 show the general layout of the valence-shell atomic orbitals of Period 2 atoms on the left and right. The lines in the middle are an indication of the energies of the molecular orbitals that can be formed by overlap of atomic orbitals. From the eight valence-shell orbitals (four from each atom), we can form eight molecular orbitals: four are σ orbitals and four, in two pairs, are doubly degenerate π orbitals. With the orbitals established, we derive the ground-state electron configurations of the molecules by adding the appropriate number of electrons to the orbitals and following the building-up rules. Charged species (such as the peroxide ion, O_2^{2-}, and C_2^+) need either more or fewer electrons (for anions and cations, respectively) than the neutral molecules.

We illustrate the procedure with N_2, which has ten valence electrons; for this molecule we use Fig. 14.24. The first two electrons pair, enter, and fill the $1\sigma_g$ orbital. The next two electrons enter and fill the $1\sigma_u$ orbital. Six electrons remain. There are two $1\pi_u$ orbitals, so four electrons can be accommodated in them. The two remaining electrons enter the $2\sigma_g$ orbital. The ground-state configuration of N_2 is therefore $1\sigma_g^2 1\sigma_u^2 1\pi_u^4 2\sigma_g^2$. This configuration is also depicted in Fig. 14.24.

The strength of a bond in a molecule is the net outcome of the bonding and antibonding effects of the electrons in the orbitals. The **bond order**, *b*, in a diatomic molecule is defined as

$$b = \tfrac{1}{2}(N - N^*) \qquad \text{Bond order} \quad (14.12)$$

where N is the number of electrons in bonding orbitals and N^* is the number of electrons in antibonding orbitals (as judged by their resemblance to the simple pairwise LCAO). Each electron pair in a bonding orbital increases the bond order by 1 and each pair in an antibonding orbital decreases it by 1.

Brief illustration 14.6 Bond order

For H_2, $b = 1$, corresponding to a single bond between the two atoms: this bond order is consistent with the Lewis structure H—H for the molecule. In He_2, which has equal numbers of bonding and antibonding electrons (with $N = 2$ and $N^* = 2$), the bond order is $b = 0$, and there is no bond. In N_2, $1\sigma_g$, $2\sigma_g$, and $1\pi_u$ are bonding orbitals, and $N = 2 + 2 + 4 = 8$; however, $1\sigma_u$ (the antibonding partner of $1\sigma_g$) is antibonding, so $N^* = 2$ and the bond order of N_2 is $b = {}^1/_2(8 - 2) = 3$. This value is consistent with the Lewis structure :N≡N:, in which there is a triple bond between the two atoms.

The bond order is a useful parameter for discussing the characteristics of bonds, because it correlates with bond length, and the greater the bond order between atoms of a given pair of atoms, the shorter the bond. The bond order also correlates with bond strength, and the greater the bond order, the greater the strength. The high bond order of N_2 is consistent with its high dissociation energy (942 kJ mol^{-1}).

Example 14.4

Writing the electron configuration of a diatomic molecule

Write the ground-state electron configuration of O_2 and calculate the bond order.

Strategy Decide which MO energy level diagram to use (Fig. 14.23 or Fig. 14.24). Count the valence electrons and accommodate them by using the building-up principle.

Solution Figure 14.23 is appropriate for oxygen. There are 12 valence electrons to accommodate. The first 10 electrons recreate the N_2 configuration (with a reversal of the order of the $2\sigma_g$ and $1\pi_u$ orbitals); the remaining two electrons must occupy the $1\pi_g$ orbitals. The configuration and bond order are therefore $1\sigma_g^2 1\sigma_u^2 2\sigma_g^2 1\pi_u^4 1\pi_g^2$. This configuration is also depicted in Fig. 14.23. Because $1\sigma_g$, $2\sigma_g$, and $1\pi_u$ are regarded as bonding and $1\sigma_u$ and $1\pi_g$ as antibonding, the bond order is $b = \tfrac{1}{2}(8 - 4) = 2$. This bond order accords with the classical view that oxygen has a double bond.

Self-test 14.5

Write the electron configuration of F_2 and deduce its bond order.

Answer: $1\sigma_g^2 1\sigma_u^2 2\sigma_g^2 1\pi_u^4 1\pi_g^4$, $b = 1$

We see from Example 14.4 that the electron configuration of O_2 is $1\sigma_g^2 1\sigma_u^2 2\sigma_g^2 1\pi_u^4 1\pi_g^2$. According to the building-up principle, the two $1\pi_g$ electrons in O_2 will occupy different orbitals. One enters the $1\pi_g$ orbital formed by overlap of $2p_x$. The other enters its degenerate partner, the $1\pi_g$ orbital formed from overlap of the $2p_y$ orbitals. Because the two electrons occupy different orbitals, by Hund's rule they will have parallel spins (↑↑). Consequently, an O_2 molecule is sometimes regarded as a biradical, a species with two unpaired electrons. (A true biradical has two electron spins with random relative orientations; in O_2 the two spins are parallel.) Molecular orbital theory therefore suggests—correctly—that O_2 is a reactive component of the Earth's atmosphere; its most important biological role is as an oxidizing agent. By contrast N_2, the major component of the air we breathe, is so unreactive that nitrogen fixation, the reduction of atmospheric N_2 to NH_3 by certain microorganisms, is among the most thermodynamically demanding of biological processes in the sense that it requires a great deal of energy derived from metabolic processes.

The electronic configuration of O_2 also suggests that it will be magnetic because the magnetic fields generated by the two unpaired spins do not cancel. Specifically, O_2 is predicted to be a **paramagnetic** substance, a substance that is drawn into a magnetic field. Most substances (those with paired electron spins) are **diamagnetic**, and are pushed out of a magnetic field. That O_2 is in fact a paramagnetic gas is a striking confirmation of the superiority of the molecular orbital description of the molecule over the Lewis and VB descriptions (which require all the electrons to be paired). The property of paramagnetism is utilized to monitor the oxygen content of incubators by measuring the magnetism of the gases they contain.

An F_2 molecule has two more electrons than an O_2 molecule, so its configuration is $1\sigma_g^2 1\sigma_u^2 2\sigma_g^2 1\pi_u^4 1\pi_g^4$

and its bond order is 1. We conclude that F_2 is a singly bonded molecule, in agreement with its Lewis structure. The low bond order is consistent with the low dissociation energy of F_2 (154 kJ mol^{-1}). A hypothetical Ne_2 molecule would have two further electrons: its configuration would be $1\sigma_g^2 1\sigma_u^{*2} 2\sigma_g^2 1\pi_u^4 1\pi_g^{*4} 2\pi_u^{*2}$ and its bond order 0. The bond order of zero, which implies that two neon atoms do not bond together, is consistent with the monatomic character of neon.

Example 14.5

Judging the relative bond strengths of molecules and ions

The superoxide ion, O_2^-, plays an important role in the ageing processes that take place in organisms. Judge whether O_2^- is likely to have a larger or smaller dissociation energy than O_2.

Strategy Because a species with the larger bond order is likely to have the larger dissociation energy, we should compare their electronic configurations, and assess their bond orders.

Solution From Fig. 14.23,

O_2 $\quad 1\sigma_g^2 1\sigma_u^{*2} 2\sigma_g^2 1\pi_u^4 1\pi_g^{*2}$ $\quad b = 2$

O_2^- $\quad 1\sigma_g^2 1\sigma_u^{*2} 2\sigma_g^2 1\pi_u^4 1\pi_g^{*3}$ $\quad b = 1.5$

Because the anion has the smaller bond order, we expect it to have the smaller dissociation energy.

Self-test 14.6

Which can be expected to have the higher dissociation energy, F_2 or F_2^+?

Answer: F_2^+

14.11 The structures of heteronuclear diatomic molecules

A **heteronuclear diatomic molecule** is a diatomic molecule formed from atoms of two different elements; two examples are CO and HCl. The electron distribution in the covalent bond between the atoms is not symmetrical between the atoms because it is energetically favourable for a bonding electron pair to be found closer to one atom rather than the other. This imbalance results in a **polar bond**, which is a covalent bond in which the electron pair is shared unequally by the two atoms. The **electronegativity**, χ (chi), of an element is the power of its atoms to draw electrons to itself when it is part of a compound, so we can expect the polarity of a bond to depend on the relative electronegativities of the elements.

Linus Pauling formulated a numerical scale of electronegativity based on considerations of bond dissociation energies, E(A—B):

$$|\chi_A - \chi_B| = 0.102 \times (\Delta E/\text{kJ mol}^{-1})^{1/2} \quad \text{Pauling electronegativity scale} \quad (14.13a)$$

with

$$\Delta E = E(\text{A—B}) - \tfrac{1}{2}\{E(\text{A—A}) + E(\text{B—B})\} \quad (14.13b)$$

Table 14.2 lists values for the main-group elements. Robert Mulliken proposed an alternative definition in terms of the ionization energy, I, and the electron affinity, E_{ea}, of the element expressed in electronvolts:

$$\chi = \tfrac{1}{2}(I + E_{ea})/\text{eV} \quad \text{Mulliken electronegativity scale} \quad (14.14)$$

This relation is plausible, because an atom that has a high electronegativity is likely to be one that has a high ionization energy (so that it is unlikely to lose electrons to another atom in the molecule) and a high electron affinity (so that it is energetically favourable for an electron to move towards it). The Mulliken electronegativities are broadly in line with the Pauling electronegativities. Electronegativities show a periodicity, and the elements with the highest electronegativities are those close to fluorine in the periodic table.

The location of the bonding electron pair close to one atom in a heteronuclear molecule results in that atom having a net negative charge, which is called a **partial negative charge** and denoted δ−. There is a compensating **partial positive charge**, δ+, on the other atom. In a typical heteronuclear diatomic molecule, the more electronegative element has the

Table 14.2
*Electronegativities of the main-group elements**

H 2.20						
Li 0.98	Be 1.57	B 2.04	C 2.55	N 3.04	O 3.44	F 3.98
Na 0.93	Mg 1.31	Al 1.61	Si 1.90	P 2.19	S 2.58	Cl 3.16
K 0.82	Ca 1.00	Ga 1.81	Ge 2.01	As 2.18	Se 2.55	Br 2.96
Rb 0.82	Sr 0.95	In 1.78	Sn 1.96	Sb 2.05	Te 2.1	I 2.66
Cs 0.79	Ba 0.89	Tl 1.8	Pb 1.8	Bi 1.9	Po 2.0	

*Pauling values.

partial negative charge and the less electronegative element has the partial positive charge.

Molecular orbital theory takes polar bonds into its stride. A polar bond consists of two electrons in an orbital of the form

$$\psi = c_A\psi_A + c_B\psi_B \quad \text{A general LCAO} \quad (14.15)$$

with c_B^2 no longer equal to c_A^2. If $c_B^2 > c_A^2$, the electrons have a greater probability of being found on B than on A and the molecule is polar in the sense $^{\delta+}$A—B$^{\delta-}$. A nonpolar bond, a covalent bond in which the electron pair is shared equally between the two atoms and there are zero partial charges on each atom, has $c_A^2 = c_B^2$. A pure ionic bond, in which one atom has obtained virtually sole possession of the electron pair (as in Cs^+F^-, to a first approximation), has one coefficient zero (so that A^+B^- would have $c_A^2 = 0$ and $c_B^2 = 1$).

A general feature of molecular orbitals between dissimilar atoms is that the atomic orbital with the lower energy (that belonging to the more electronegative atom) makes the larger contribution to the lowest energy molecular orbital. The opposite is true of the highest (most antibonding) orbital, for which the principal contribution comes from the atomic orbital with higher energy (the less electronegative atom):

Bonding orbitals: for $\chi_A > \chi_B$, $c_A^2 > c_B^2$
Antibonding orbitals: for $\chi_A > \chi_B$, $c_A^2 < c_B^2$

Figure 14.31 shows a schematic representation of this point.

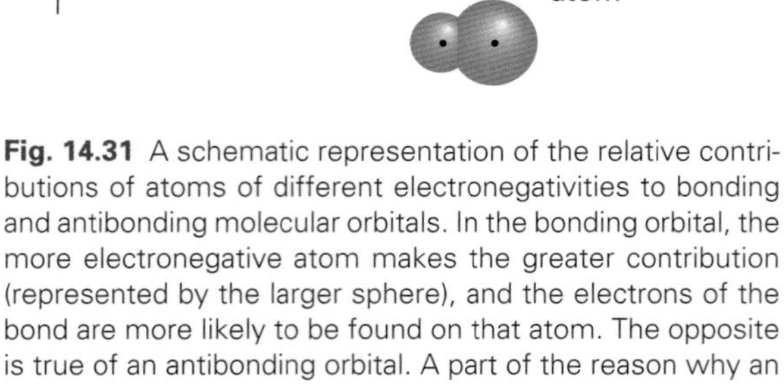

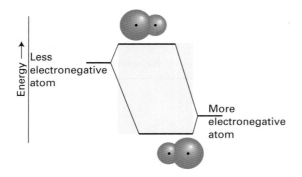

Fig. 14.31 A schematic representation of the relative contributions of atoms of different electronegativities to bonding and antibonding molecular orbitals. In the bonding orbital, the more electronegative atom makes the greater contribution (represented by the larger sphere), and the electrons of the bond are more likely to be found on that atom. The opposite is true of an antibonding orbital. A part of the reason why an antibonding orbital is of high energy is that the electrons that occupy it are likely to be found on the less electronegative atom.

Brief illustration 14.7 The molecular orbitals of HF

The features of polar bonds can be illustrated by considering HF. The general form of the molecular orbitals of HF is $\psi = c_H\psi_H + c_F\psi_F$, where ψ_H is an H1s orbital and ψ_F is an F2p$_z$ orbital. The energies of the orbitals may be estimated from ionization energies and electron affinities. Thus, the extreme cases of an atom X in a molecule are X^+ if it has lost control of the electron it supplied, X if it is sharing the electron pair equally with its bonded partner, and X^- if it has gained control of both electrons in the bond. If X^+ is taken as defining the energy 0, then X lies at $-I(X)$ and X^- lies at $-\{I(X) + E_{ea}(X)\}$, where I is the ionization energy and E_{ea} the electron affinity. The actual energy of the orbital lies at an intermediate value, and in the absence of further information, we estimate it as halfway down to the lowest of these values, namely $-\frac{1}{2}\{I(X) + E_{ea}(X)\}$. Applying this estimate to H1s and F2p$_z$ gives energies of −7.2 eV and −10.4 eV, respectively (Fig. 14.32). It follows that the bonding σ orbital in HF is mainly F2p$_z$ and the antibonding σ orbital is mainly H1s orbital in character. The two electrons in the bonding orbital are most likely to be found in the F2p$_z$ orbital, so there is a partial negative charge on the F atom and a partial positive charge on the H atom.

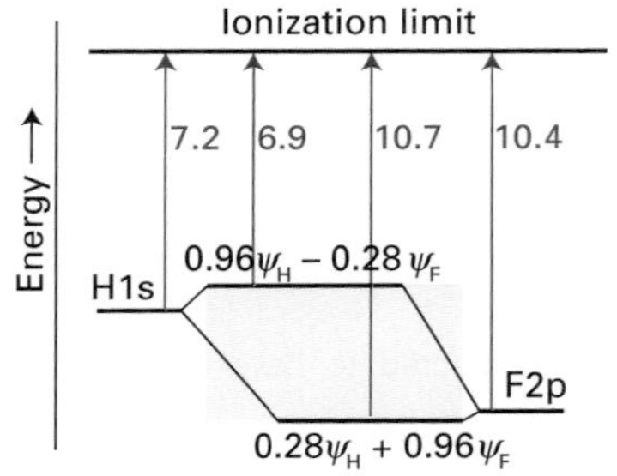

Fig. 14.32 The atomic orbital energy levels of H and F atoms and the molecular orbitals they form. Energies are in electronvolts.

A systematic way of finding the coefficients c_A and c_B in the linear combinations is to use the variation theorem and to look for the values of the coefficients that result in the lowest energy (Section 14.6)

Brief illustration 14.8 The variation theorem

When the variation theorem is applied to an H_2 molecule, the calculated energy is lowest when the two H1s orbitals contribute equally to a bonding orbital. However, when we apply the principle to HF, the lowest energy is obtained for the orbital $\psi = 0.28\psi_H + 0.96\psi_F$. We see that indeed the F2p$_z$ orbital does make the greater contribution to the bonding σ orbital. The probability of finding a σ electron in HF in a F2p$_z$ orbital is

$$c_F^2 \times 100 \text{ per cent} = (0.96)^2 \times 100 \text{ per cent} = 92 \text{ per cent}$$

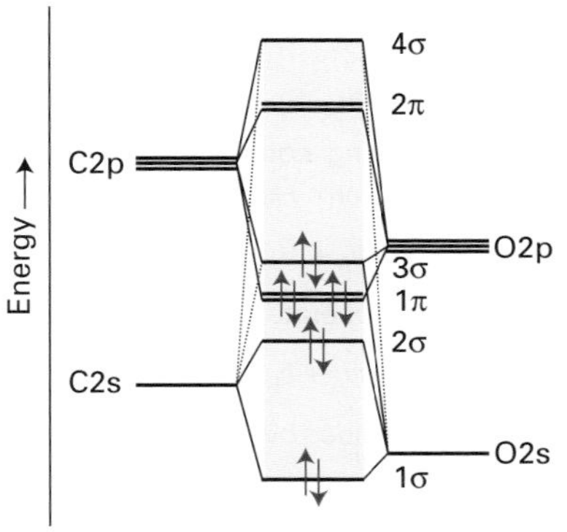

Fig. 14.33 The molecular orbital energy level diagram for CO.

An even lower energy is obtained—with a lot more calculation—if even more orbitals are included in the linear combination (such as F2s and F3p_z orbitals) but the principal lowering of energy is achieved from atomic orbitals of similar energies.

Figure 14.33 shows the bonding scheme in CO and illustrates a number of points we have made. The ground configuration is $1\sigma^2 2\sigma^2 1\pi^4 3\sigma^2$. (The g,u designation is inapplicable because the molecule is heteronuclear and the σ orbitals are simply numbered in sequence, 1σ, 2σ, . . . , and the π orbitals likewise.) The lowest energy orbitals are predominantly of O character as that is the more electronegative element. The **highest occupied molecular orbital** (HOMO) is 3σ, which is a largely nonbonding orbital centred on C, so the two electrons that occupy it can be regarded as a lone pair on the C atom. The **lowest unoccupied molecular orbital** (LUMO) is 2π, which is largely a doubly degenerate orbital of 2p character on carbon.

This combination of a lone pair orbital on C and a pair of empty π orbitals also largely on C is at the root of the importance of carbon monoxide in d-block chemistry, because it enables it to form an extensive series of carbonyl complexes by a combination of electron donation from the 3σ orbital and electron acceptance into the 2π orbitals. The HOMO and the LUMO jointly form the **frontier orbitals** of the molecule, and are of great importance for assessing its reactions.

14.12 The structures of polyatomic molecules

The bonds in polyatomic molecules are built in the same way as in diatomic molecules, the only difference being that we use more atomic orbitals to construct the molecular orbitals, and these molecular orbitals spread over the entire molecule, not just the adjacent atoms of the bond. In general, a molecular orbital is a linear combination of all the atomic orbitals of all the atoms in the molecule. In H_2O, for instance, the atomic orbitals are the two H1s orbitals, the O2s orbital, and the three O2p orbitals (if we consider only the valence shell). From these six atomic orbitals we can construct six molecular orbitals that spread over all three atoms. The molecular orbitals differ in energy. The lowest energy, most strongly bonding orbital has the least number of nodes between adjacent atoms. The highest energy, most strongly antibonding orbital has the greatest numbers of nodes between neighbouring atoms.

According to MO theory, the bonding influence of a single electron pair is distributed over all the atoms, and each electron pair (the maximum number of electrons that can occupy any single molecular orbital) helps to bind all the atoms together. In the LCAO approximation, each molecular orbital is modelled as a sum of atomic orbitals, with atomic orbitals contributed by all the atoms in the molecule. Thus, a typical molecular orbital in H_2O constructed from H1s orbitals (denoted $\psi_{\mathrm{H1s(A)}}$ and $\psi_{\mathrm{H1s(B)}}$) and O2s and O2p_z orbitals (denoted ψ_{O2s} and $\psi_{\mathrm{O2p_z}}$) will have the composition

$$\psi = c_1\psi_{\mathrm{H1s(A)}} + c_2\psi_{\mathrm{O2s}} + c_3\psi_{\mathrm{O2p_z}} + c_4\psi_{\mathrm{H1s(B)}} \tag{14.16}$$

Because four atomic orbitals are being used to form the LCAO, there will be four possible molecular orbitals of this kind: the lowest energy (most bonding) orbital will have no internuclear nodes and the highest energy (most antibonding) orbital will have a node between each pair of neighbouring nuclei (Fig. 14.34).

A final point is that the σ,π classification of molecular orbitals is not strictly applicable to nonlinear

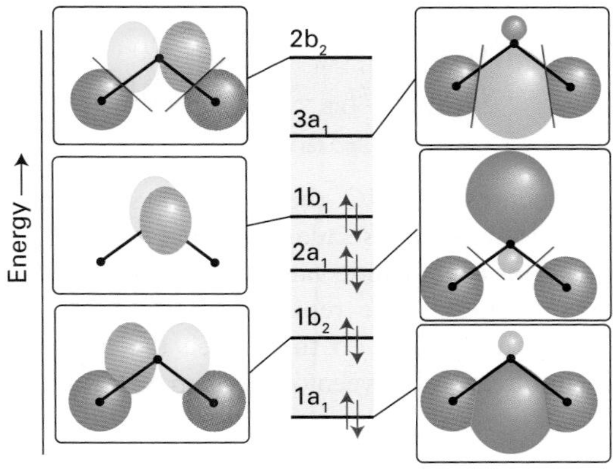

Fig. 14.34 Schematic form of the molecular orbitals of H_2O.

polyatomic molecules. Instead, a classification scheme based on the actual symmetry of the molecule is used, and you will see symbols such as a_1, b_1, and so on instead.[1] However, the σ,π classification is relevant *locally*, in the sense that we can speak of the σ bond between the O and an H atom in H_2O.

14.13 The Hückel method

An important example of the application of MO theory is to the orbitals that may be formed from the p orbitals perpendicular to a molecular plane, such as that of the phenyl ring of the amino acid phenylalanine. A computational scheme was proposed by Erich Hückel and provides a simple way of establishing the molecular orbitals of π-electron systems, especially hydrocarbons such as ethene, benzene, and their derivatives. A common procedure in elementary applications is to treat the σ-bonding framework using the language of VB theory, and to treat the π-electron system separately by MO theory. We use that approach here.

(a) Ethene

Each carbon atom in ethene, $CH_2{=}CH_2$, is regarded as sp^2-hybridized and forming C—C and C—H σ-bonds at 120° to each other by spin-pairing and either (Csp^2,Csp^2)- or (Csp^2,H1s)-orbital overlap (note the VB language). The unhybridized $C2p_z$ orbitals perpendicular to the σ-framework (ψ_A and ψ_B), each of which is occupied by a single electron, are then used to construct molecular orbitals (Fig. 14.35):

$$\psi = c_A\psi_A + c_B\psi_B \qquad (14.17)$$

We show in the following Derivation that to find the energies and coefficients of the two molecular orbitals that can be formed from these two atomic orbitals, we need to solve the following simultaneous equations:

$$(H_{AA} - ES_{AA})c_A + (H_{AB} - ES_{AB})c_B = 0$$
$$(H_{BA} - ES_{BA})c_A + (H_{BB} - ES_{BB})c_B = 0$$

Secular equations for ethene (14.18)

In the context of MO theory, these simultaneous equations are called **secular equations.** The H_{JK} are expressions that include various contributions to the energy, including the repulsion between electrons and their attractions to the nuclei; the S_{JK} are the overlap integrals between orbitals on atoms J and K.

[1] The symmetry classification of orbitals in polyatomic molecules is described in our *Physical Chemistry* (2010).

Derivation 14.4

The secular equations

We begin by substituting eqn 14.17 into the Schrödinger equation written in the form $\hat{H}\psi = E\psi$:

$$\hat{H}(c_A\psi_A + c_B\psi_B) = \hat{H}c_A\psi_A + \hat{H}c_B\psi_B$$

Because the operators operate only on the ψ, we may write

$$c_A\hat{H}\psi_A + c_B\hat{H}\psi_B = c_AE\psi_A + c_BE\psi_B$$

Then we multiply both sides by ψ_A

$$c_A\psi_A\hat{H}\psi_A + c_B\psi_A\hat{H}\psi_B = c_A\psi_AE\psi_A + c_B\psi_AE\psi_B$$

and integrate over all space (with the term $d\tau$ denoting an infinitesimal volume element in three dimensions; in Cartesian coordinates, $d\tau = dxdydz$):

$$c_A\overbrace{\int\psi_A\hat{H}\psi_A d\tau}^{H_{AA}} + c_B\overbrace{\int\psi_A\hat{H}\psi_B d\tau}^{H_{AB}} = c_AE\overbrace{\int\psi_A\psi_A d\tau}^{1} + c_BE\overbrace{\int\psi_A\psi_B d\tau}^{S_{AB}}$$

where E is a constant, so we have been able to take it outside the integral. It follows that

$$c_AH_{AA} + c_BH_{AB} = c_AES_{AA} + c_BES_{AB}$$

which is easy to rearrange into the first of eqn 14.18. If instead of multiplying through by ψ_A we multiply by ψ_B, we obtain the second of eqn 14.18.

To simplify the solution of the secular equations Hückel introduced the following drastic approximations:

- All H_{JJ} are set equal to a single quantity α called the **Coulomb integral.**
- All H_{JK} are set equal to zero unless atoms J and K are adjacent, when it is set equal to a single quantity β (a negative quantity) called the **resonance integral.**
- All S_{JJ} are set equal to 1 and all S_{JK} are set equal to 0 whether or not J and K are adjacent.

With these 'Hückel approximations' the secular equations become

$$(\alpha - E)c_A + \beta c_B = 0$$
$$\beta c_A + (\alpha - E)c_B = 0$$

Hückel approximation for ethene (14.19a)

As set out in The chemist's toolkit 14.3, these two simultaneous equations have a solution only if the **secular determinant** vanishes:

$$\begin{vmatrix} \alpha - E & \beta \\ \beta & \alpha - E \end{vmatrix} = (\alpha - E)^2 - \beta^2 = 0$$

Hückel secular determinant for ethene (14.19b)

This condition is satisfied if

$$E = \alpha \pm \beta \qquad \text{Hückel energies for ethene} \quad (14.19c)$$

When each value is substituted into eqn 14.19a, we find:

For $E = \alpha + \beta$ $\quad c_A = c_B$, so $\psi = c_A(\psi_A + \psi_B)$

For $E = \alpha - \beta$ $\quad c_A = -c_B$, so $\psi = c_A(\psi_A - \psi_B)$

(Remember that $\beta < 0$, so $E = \alpha + \beta$ is the lower energy of the two.) These energies and orbitals are represented in Fig. 14.35: they will be recognized as the bonding and antibonding combinations of the C2p$_z$ atomic orbitals. The value of the one unknown, c_A, is found by ensuring that each orbital is normalized, but we do not need its explicit value.

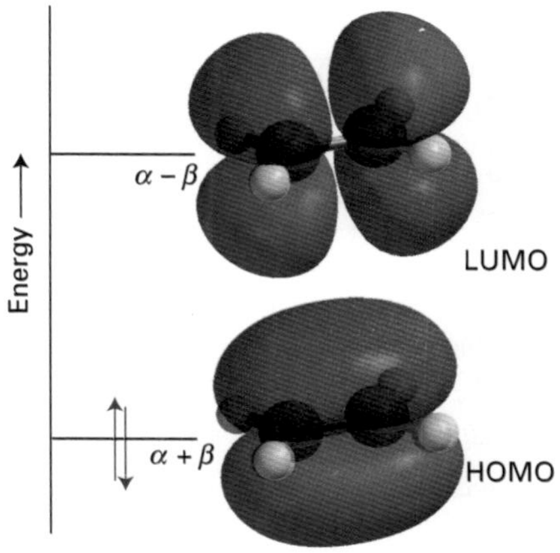

Fig. 14.35 The bonding and antibonding π molecular orbitals of ethene and their energies.

The chemist's toolkit 14.3 Simultaneous equations

Two simultaneous equations of the form

$$ax + by = 0$$
$$cx + dy = 0$$

have solutions only if the 'determinant' of the coefficients is equal to zero. In this case we write

$$\begin{vmatrix} a & b \\ c & d \end{vmatrix} = 0$$

where the term on the left is the determinant and has the following meaning:

$$\begin{vmatrix} a & b \\ c & d \end{vmatrix} = ad - bc$$

Three simultaneous equations of the form

$$ax + by + cz = 0$$
$$dx + ey + fz = 0$$
$$gx + hy + iz = 0$$

have a solution only if

$$\begin{vmatrix} a & b & c \\ d & e & f \\ g & h & i \end{vmatrix} = 0$$

This 3 × 3 determinant expands as follows:

$$\begin{vmatrix} a & b & c \\ d & e & f \\ g & h & i \end{vmatrix} = a\begin{vmatrix} e & f \\ h & i \end{vmatrix} - b\begin{vmatrix} d & f \\ g & i \end{vmatrix} + c\begin{vmatrix} d & e \\ g & h \end{vmatrix}$$

Note the alternation in signs for successive columns. The 2 × 2 determinants then expand like the one above.

Because there are two electrons to be accommodated, both enter the lower energy orbital and give a contribution of $2\alpha + 2\beta$ to the energy of the molecule. We can also infer that the energy needed to excite a π electron to the antibonding combination is $2|\beta|$. A typical value of β in hydrocarbons is about −2.4 eV, or −230 kJ mol^{-1}.

Example 14.6

Calculating the Hückel energies for butadiene

The Hückel equations for butadiene (**5**) are:

$$(\alpha - E)c_A + \beta c_B = 0$$
$$\beta c_A + (\alpha - E)c_B + \beta c_C = 0$$
$$\beta c_B + (\alpha - E)c_C + \beta c_D = 0$$
$$\beta c_C + (\alpha - E)c_D = 0$$

Write down and solve the secular determinant for the energies of the π orbitals.

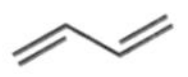

5 Butadiene

Strategy Because there are four equations and four c_i, the secular determinant in this case has four rows and four columns. From left to right, the elements of the first row of the determinant are $\alpha - E$, β, 0, and 0; that is, the terms multiplying c_A, c_B, c_C, and c_D, respectively, in the top equation. The remaining rows are constructed similarly by inspection of the remaining equations. It is possible to solve this determinant manually by extension of the method described in The chemist's toolkit 14.3. However, it is far easier and faster to use mathematical software.

Solution The secular determinant is

$$\begin{vmatrix} \alpha-E & \beta & 0 & 0 \\ \beta & \alpha-E & \beta & 0 \\ 0 & \beta & \alpha-E & \beta \\ 0 & 0 & \beta & \alpha-E \end{vmatrix}$$

Mathematical software routines for solving this determinant first set it to 0. Then the software looks for the four values of E that satisfy the expression 'determinant = 0'. In this case, the outcome is

$$E = \alpha \pm 1.62\beta,\ \alpha \pm 0.62\beta$$

To solve the equation by hand, first expand the determinant, and find

$$(\alpha - E)^4 - 3(\alpha - E)^2\beta^2 + \beta^4 = 0$$

With $x = (\alpha - E)^2$, this is a quadratic equation:

$$x^2 - 3\beta^2 x + \beta^4 = 0$$

with the solutions (The chemist's toolkit 7.1) $x = \frac{1}{2}(3 \pm 5^{1/2})\beta^2$, which results in the four values of E calculated with mathematical software. Because α and β are both negative quantities, the order of increasing energy, from most bonding to most antibonding, is $\alpha + 1.62\beta$, $\alpha - 0.62\beta$, $\alpha - 0.62\beta$, $\alpha - 1.62\beta$.

Self-test 14.7

The π-electron binding energy, E_π, is the total energy due to the electrons occupying p orbitals. What is E_π of butadiene in its ground state?

Answer: $4\alpha + 4.48\beta$

(b) Benzene

Exactly the same procedure can be used for benzene, C_6H_6. Each C atom is regarded as sp^2-hybridized (note the VB language again) and forms a planar hexagonal framework of σ bonds (Fig. 14.36). There is an unhybridized $C2p_z$ orbital on each atom perpendicular to the ring from which we form molecular orbitals. From these six atomic orbitals we construct six molecular orbitals of the form

$$\psi = c_A\psi_A + c_B\psi_B + c_C\psi_C + c_D\psi_D + c_E\psi_E + c_F\psi_F \qquad (14.20)$$

Then we set up the six simultaneous equations for the coefficients and the corresponding 6 × 6 secular determinant and apply the Hückel approximations. Full-frontal attack on it to determine the six values of E is rather tedious, especially as there are procedures that make use of symmetry that greatly simplifies the solution. The energies and the corresponding (unnormalized) molecular orbitals obtained are as follows (Fig. 14.37):

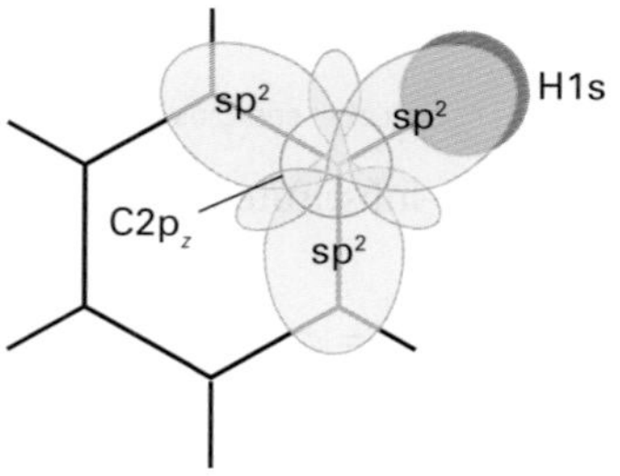

Fig. 14.36 The orbitals used to construct the molecular orbitals of benzene.

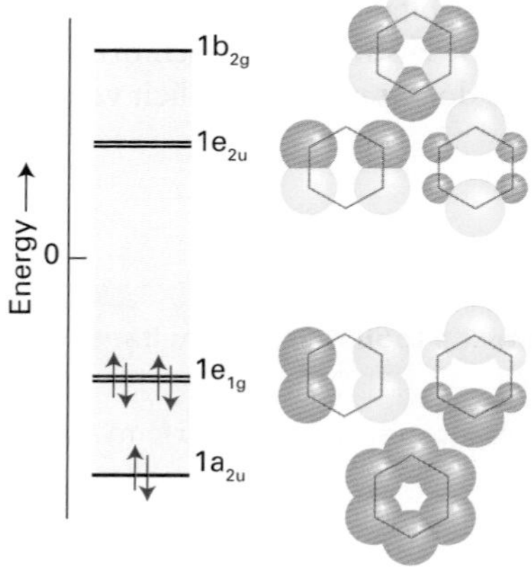

Fig. 14.37 The π orbitals of benzene and their energies. The lowest-energy orbital is fully bonding between neighbouring atoms, but the uppermost orbital is fully antibonding. The two pairs of doubly degenerate molecular orbitals have an intermediate number of internuclear nodes. As usual, different colours represent different signs of the wavefunction.

Energy	Orbital
Highest (most antibonding)	
$\alpha - 2\beta$	$\psi = \psi_A - \psi_B + \psi_C - \psi_D + \psi_E - \psi_F$
$\alpha - \beta$	$\psi = 2^{1/2}\psi_A - \psi_B - \psi_C + 2^{1/2}\psi_D - \psi_E - \psi_F$
$\alpha - \beta$	$\psi = \psi_B - \psi_C + \psi_E - \psi_F$
$\alpha + \beta$	$\psi = 2^{1/2}\psi_A + \psi_B + \psi_C - 2^{1/2}\psi_D - \psi_E - \psi_F$
$\alpha + \beta$	$\psi = \psi_B + \psi_C - \psi_E - \psi_F$
$\alpha + 2\beta$	$\psi = \psi_A + \psi_B + \psi_C + \psi_D + \psi_E + \psi_F$
Lowest (most bonding)	

Note that the lowest energy, most bonding orbital has no internuclear nodes. It is strongly bonding because the constructive interference between neighbouring p orbitals results in a good accumulation of electron density between the nuclei (but slightly off the internuclear axis, as in the π bonds of diatomic molecules). In the most antibonding orbital the alternation of signs in the linear combination results in destructive interference between neighbours, and the molecular orbital has a nodal plane between each pair of neighbours, as shown in the illustration. The four intermediate orbitals form two doubly degenerate pairs, one net bonding and the other net antibonding.

There are six electrons to be accommodated (one is supplied by each C atom), and they occupy the lowest three orbitals in Fig. 14.37. The resulting electron distribution is like a double doughnut. It is an important feature of the configuration that the only molecular orbitals occupied have a net bonding character, for this is one contribution to the stability (in the sense of low energy) of the benzene molecule.

A feature of the molecular orbital description of benzene is that each molecular orbital spreads either all around or partially around the C_6 ring. That is, π bonding is **delocalized**, and each electron pair helps to bind together several or all of the C atoms. The delocalization of bonding influence is a primary feature of molecular orbital theory that we shall use time and again when discussing conjugated systems. The **delocalization energy**, E_{deloc}, is the additional lowering of energy of the molecule due to the spreading of the p electrons throughout the molecule instead of being localized in discrete bonding regions.

Brief illustration 14.9 The delocalization energy of benzene

If the six π electrons in benzene occupied three localized ethane-like orbitals, then their energy would be $3 \times (2\alpha + 2\beta) = 6\alpha + 6\beta$. However, their energy in benzene is $2(\alpha+2\beta) + 4(\alpha+\beta) = 6\alpha + 8\beta$. The delocalization energy is the difference of these two energies:

$$E_{deloc} = (6\alpha + 8\beta) - (6\alpha + 6\beta) = 2\beta$$

or about $-460\ kJ\ mol^{-1}$.

Self-test 14.8

Calculate the delocalization energy of butadiene.

Answer: 0.48β

Computational chemistry

Computational chemistry is now a standard part of chemical research. One major application is in pharmaceutical chemistry, where the likely pharmacological activity of a molecule can be assessed computationally from its shape and electron density distribution before expensive and ethically controversial *in vivo* trials are started. Commercial software is now widely available for calculating the electronic structures of molecules and displaying the results graphically. All such calculations work within the Born–Oppenheimer approximation and express the molecular orbitals as linear combinations of atomic orbitals.

14.14 Techniques

There are two principal approaches to solving the Schrödinger equation for many-electron polyatomic molecules. In the **semi-empirical methods**, certain expressions that occur in the Schrödinger equation are set equal to parameters that have been chosen to lead to the best fit to experimental quantities, such as enthalpies of formation. Semi-empirical methods are applicable to a wide range of molecules with a virtually limitless number of atoms, and are widely popular. In the more fundamental ***ab initio* methods**, an attempt is made to calculate structures from first principles, using only the atomic numbers of the atoms present. Such an approach is intrinsically more reliable than a semi-empirical procedure but is much more demanding computationally.

Both types of procedure typically adopt a **self-consistent field** (SCF) procedure, in which an initial guess about the composition of the LCAO is successively refined until the solution remains unchanged in a cycle of calculation. First, we guess the values of the coefficients in the LCAO used to build the molecular orbitals—and solve the Schrödinger equation for the coefficients of one LCAO on the basis of that guess for the coefficients of all the other occupied orbitals. Now we have a first approximation to the coefficients of one LCAO. We then repeat the procedure for all the other occupied molecular orbitals. At the end of that stage we have a new set of LCAO coefficients that differ from our first guess, and we also have an estimate of the energy of the molecule. We use that refined set of coefficients to repeat the calculation and calculate a new set of coefficients and a new energy. In general, these will differ from the new starting point. However, there comes a stage when repetition of the calculation leaves the coefficients and energy unchanged. The orbitals are now said to be 'self-consistent', and are accepted as a description of the molecule.

The severe approximations of the Hückel method have been removed over the years in a succession of better approximations. Each has given rise to an acronym, such as CNDO ('complete neglect of differential overlap'), INDO ('intermediate neglect of differential overlap'), MINDO ('modified neglect of differential overlap'), and AM1 ('Austin Model 1', version 2 of MINDO). Software for all these procedures is now readily available, and reasonably sophisticated calculations can now be run even on hand-held computers. A semi-empirical technique that has gained considerable ground in recent years to become one of the most widely used techniques for the calculation of molecular structure is **density functional theory** (DFT). Its advantages include less demanding computational effort, less computer time, and—in some cases, particularly d-metal complexes—better agreement with experimental values than is obtained from other procedures.

The *ab initio* methods also simplify the calculations, but they do so by setting up the problem in a different manner, avoiding the need to estimate parameters by appeal to experimental data. In these methods, sophisticated techniques are used to solve the Schrödinger equation numerically. The difficulty with this procedure it the enormous time it takes to carry out the detailed calculation. That time can be reduced by replacing the hydrogenic atomic orbitals used to form the LCAO by a **gaussian-type orbital** (GTO) in which the exponential function e^{-r} characteristic of actual orbitals is replaced by a sum of gaussian functions of the form e^{-r^2}.

14.15 Graphical output

One of the most significant developments in computational chemistry has been the introduction of graphical representations of molecular orbitals and electron densities. The raw output of a molecular structure calculation is a list of the coefficients of the atomic orbitals in each molecular orbital and the energies of these orbitals. The graphical representation of a molecular orbital uses stylized shapes to represent the basis set, and then scales their size to indicate the value of the coefficient in the LCAO. Different signs of the wavefunctions are represented by different colours (Fig. 14.38).

Once the coefficients are known, we can build up a representation of the electron density in the molecule by noting which orbitals are occupied and then forming the squares of those orbitals. The total electron density at any point is then the sum of the squares of the wavefunctions evaluated at that point. The outcome is commonly represented by an **isodensity surface**, a surface of constant total electron density (Fig. 14.39). There are several styles of representing an isodensity surface, as a solid form, as a transparent form with a ball-and-stick representation of the molecule within, or as a mesh.

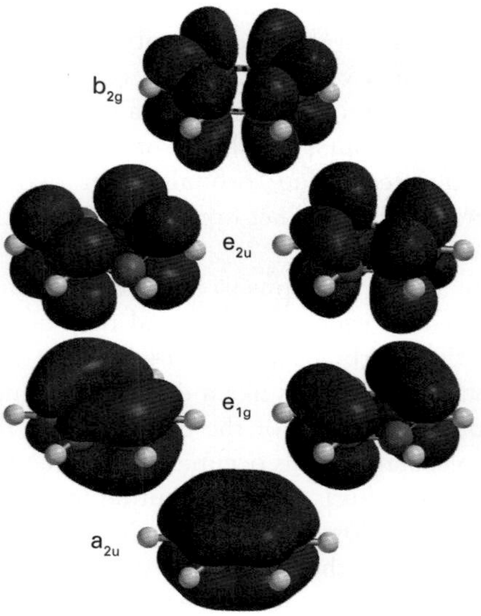

Fig. 14.38 The output of a computation of the π orbitals of benzene: opposite signs of the wavefunctions are represented by different colours. Compare these molecular orbitals with the more diagrammatic representation in Fig. 14.37.

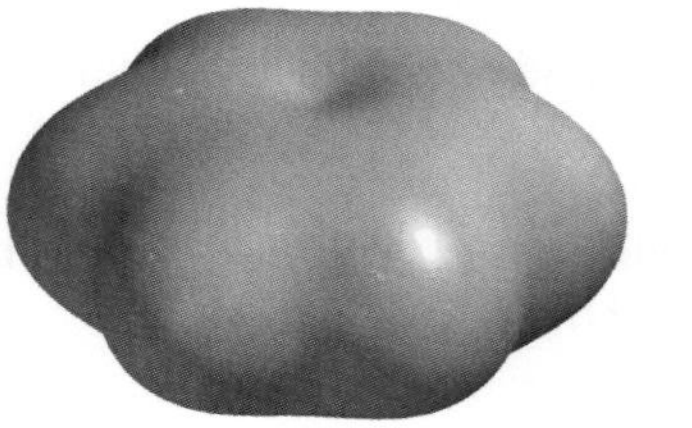

Fig. 14.39 The isodensity surface of benzene obtained by using the same software as in Fig. 14.38.

One of the most important aspects of a molecule other than its geometrical shape is the distribution of electric potential over its surface. A common procedure begins with calculation of the net potential at each point on an isodensity surface. The result is an **electrostatic potential surface** (an 'elpot surface') in which net positive potential is shown in one colour and net negative potential is shown in another, with intermediate gradations of colour (Fig. 14.40).

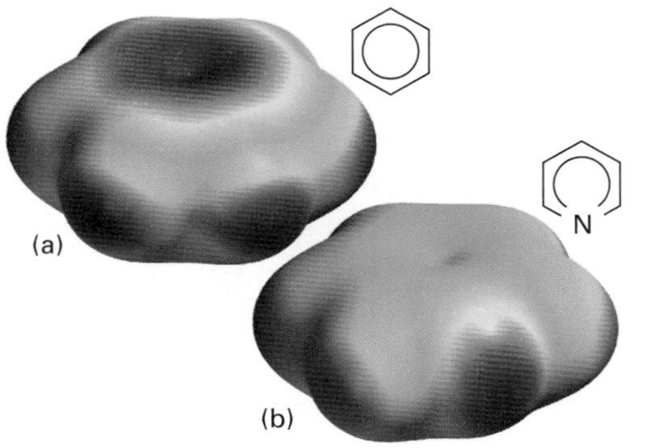

Fig. 14.40 The electrostatic potential surfaces of (a) benzene and (b) pyridine. Note the negative potential on the nitrogen atom of pyridine at the expense of the other atoms.

14.16 Applications

One goal of computational chemistry—at least when applied to large molecules—is to gain insight into trends in molecular properties without necessarily striving for ultimate accuracy. For example, consider the prediction of standard enthalpies of formation of the equatorial (**6**) and axial (**7**) conformations of methylcyclohexane. If we proceed as in Example 3.3 (in which mean bond energies are used) we obtain the same enthalpy of formation for both conformers. However, it has been observed experimentally that molecules in these two conformations have different standard enthalpies of formation as a result of the greater steric repulsion when the methyl group is in an axial position than when it is equatorial.

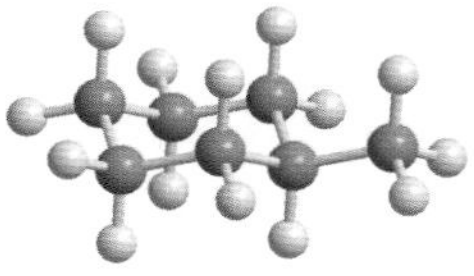

6 Equatorial

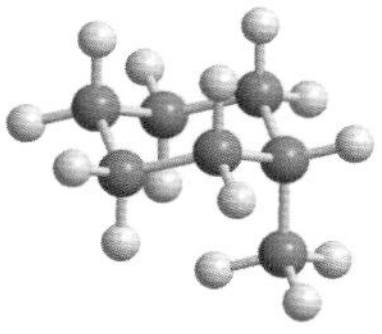

7 Axial

Computational chemistry is now widely used to estimate standard enthalpies of formation of molecules with complex three-dimensional structures, and can distinguish between different conformations of the same molecule. However, good agreement between calculated and experimental values is relatively rare. Computational methods almost always predict correctly which conformation of a molecule is most stable but do not always predict the correct numerical values of the difference in enthalpies of formation.

Brief illustration 14.10 Standard enthalpies of formation from computational methods

Each software package has its own procedures; the general approach, though, is the same in most cases: the structure of the molecule is specified and the nature of the calculation selected. When the procedure is applied to the two isomers of methylcyclohexane, a typical value for the standard enthalpy of formation of equatorial isomer in the gas phase is -153 kJ mol^{-1} whereas that for the axial isomer is -139 kJ mol^{-1}, a difference of 14 kJ mol^{-1}. The experimental difference is 7.5 kJ mol^{-1}.

A calculation performed in the absence of solvent molecules estimates the properties of the molecule of interest in the gas phase. Computational methods are available that model the effect of the solvent on the enthalpy of formation of the solute. Again, the numerical results are only estimates and the primary purpose of the calculation is to predict whether interactions with the solvent increase or decrease the enthalpy of formation. As an example, consider the amino acid glycine, which can exist in a neutral or zwitterionic form, H_2NHCH_2COOH and $^+H_3NCH_2CO_2^-$, respectively, in which in the latter the amino group is protonated and the carboxyl group is deprotonated. Molecular modelling shows that in the gas phase the neutral form has a lower enthalpy of formation than the zwitterionic form. However, in water the opposite is true on account of the strong interactions between the polar solvent and the charges on the zwitter ion.

Molecular orbital calculations may also be used to predict trends in electrochemical properties, such as standard potentials (Chapter 9). Several experimental and computational studies of aromatic hydrocarbons indicate that decreasing the energy of the LUMO enhances the ability of a molecule to accept an electron into the LUMO, with an attendant increase in the value of the molecule's standard potential.

We remarked in Chapter 12 that a molecule can absorb or emit a photon of energy hc/λ, resulting in a transition between two quantized molecular energy levels. The transition of lowest energy (and longest wavelength) occurs between the HOMO and LUMO. Also in Chapter 12 we used the particle in a box model to estimate the transition wavelengths for linear polyenes, remarking that the model was crude and offered only qualitative insight. A better model makes use of computational chemistry to correlate the HOMO–LUMO energy gap with the wavelength of absorption. For example, consider the linear polyenes shown in Table 14.3, all of which absorb in the ultraviolet region of the spectrum. The table also shows that, as expected, the wavelength of the lowest-energy electronic transition decreases as the energy separation between the HOMO and LUMO increases. We also see that the smallest HOMO–LUMO gap and longest wavelength of absorption correspond to octatetraene, the longest polyene in the group. It follows that the wavelength of the transition increases with increasing number of conjugated double bonds in linear polyenes. Extrapolation of the trend suggests that a sufficiently long linear polyene should absorb light in the visible region of the electromagnetic spectrum. This is indeed the case for β-carotene (structure **1** in Section 12.7), which absorbs light with $\lambda \approx 450$ nm. The ability of β-carotene to absorb visible light is part of the strategy employed by plants to harvest solar energy for use in photosynthesis (see Impact 20.2).

There are several ways in which molecular orbital calculations lend insight into reactivity. For example, electrostatic potential surfaces may be used to reveal an electron-poor region of a molecule, a region that is susceptible to association with or chemical attack by an electron-rich region of another molecule. Such considerations are important for assessing the pharmacological activity of potential drugs. Computational chemistry may also be used to model species that may be too unstable or short-lived to be studied experimentally. For this reason, it is often used to study the transition state, with an eye toward describing factors that increase the reaction rate.

Table 14.3

Summary of ab initio *calculations and spectroscopic data for four linear polyenes*

	$\Delta E_{\text{HOMO–LUMO}}$/eV*	$\lambda_{\text{transition}}$/nm
	18.1	163
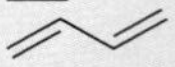	14.5	217
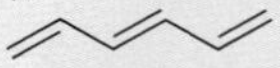	12.7	252
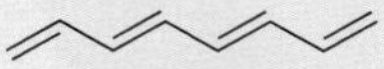	11.6	304

*1 eV = 1.602×10^{-19} J

Checklist of key concepts

- ☐ **1** An ionic bond is formed by transfer of electrons from one atom to another and the attraction between the ions.
- ☐ **2** A covalent bond is formed when two atoms share a pair of electrons.
- ☐ **3** In the Born–Oppenheimer approximation, nuclei are stationary while the electrons move around them.
- ☐ **4** In valence bond theory (VB theory), a bond is regarded as forming when an electron in an atomic orbital on one atom pairs its spin with that of an electron in an atomic orbital on another atom.
- ☐ **5** A valence bond wavefunction with cylindrical symmetry around the internuclear axis is a σ bond.
- ☐ **6** A π bond arises from the merging of two p orbitals that approach side-by-side and the pairing of electrons that they contain.
- ☐ **7** Hybrid orbitals are mixtures or atomic orbitals on the same atom. In VB theory, hybridization is invoked to be consistent with molecular geometries.
- ☐ **8** Resonance is the superposition of the wavefunctions representing different electron distributions in the same nuclear framework.
- ☐ **9** In molecular orbital theory (MO theory), electrons are treated as spreading throughout the entire molecule.
- ☐ **10** A bonding orbital is a molecular orbital that, if occupied, contributes to the strength of a bond between two atoms.
- ☐ **11** An antibonding orbital is a molecular orbital that, if occupied, decreases the strength of a bond between two atoms.
- ☐ **12** The building-up principle suggests procedures for constructing the electron configuration of molecules on the basis of their molecular orbital energy level diagram.
- ☐ **13** When constructing molecular orbitals, we need to consider only combinations of atomic orbitals of similar energies and of the same symmetry around the internuclear axis.

☐ **14** The electronegativity of an element is the power of its atoms to draw electrons to itself when it is part of a compound.

☐ **15** In a bond between dissimilar atoms, the atomic orbital belonging to the more electronegative atom makes the larger contribution to the molecular orbital with the lowest energy. For the molecular orbital with the highest energy, the principal contribution comes from the atomic orbital belonging to the less electronegative atom.

☐ **16** Hückel theory is a simple treatment of the molecular orbitals of π-electron systems. In hydrocarbons the technique consists of forming linear combinations of unhybridized C2p orbitals.

☐ **17** In the self-consistent field procedure, an initial guess about the composition of the molecular orbitals is successively refined until the solution remains unchanged in a cycle of calculations.

☐ **18** In semi-empirical methods for the determination of electronic structure, the Schrödinger equation is written in terms of parameters chosen to agree with selected experimental quantities.

☐ **19** In *ab initio* methods an attempt is made to calculate structures from first principles.

Road map of key equations

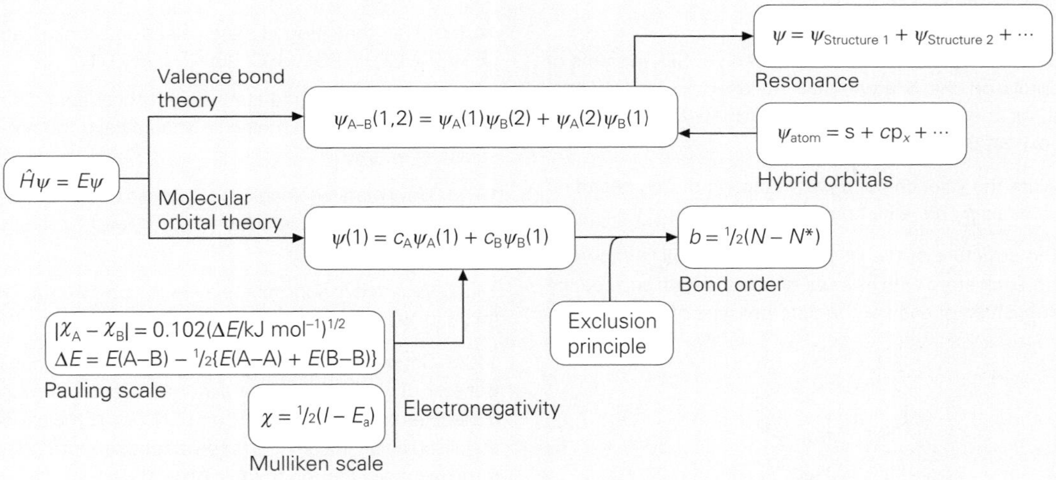

Questions and exercises

Discussion questions

14.1 Compare the approximations built into valence bond theory and molecular orbital theory.

14.2 Discuss the steps involved in the construction of sp^3, sp^2, and sp hybrid orbitals.

14.3 Describe how molecular orbital theory accommodates all the conventional types of bonding.

14.4 Why is the electron pair such a central concept in theories of the chemical bond?

14.5 Distinguish between the Pauling and Mulliken electronegativity scales.

14.6 Why is orbital overlap a guide to assessing the strengths of chemical bonds? Why, sometimes, is it not?

14.7 Identify and justify the approximations used in the Hückel theory of hydrocarbons.

14.8 Explain the differences between semi-empirical and *ab initio* methods of electronic structure determination.

Exercises

14.1 The Morse potential-energy function

$$V(R) = D_e\{1 - e^{-a(R-R_e)}\}^2$$

may be used to describe how the potential energy, V, of a diatomic molecule varies with internuclear separation, R. Sketch the function, taking $D_e = 50$ kJ mol^{-1}, $R_e = 0.30$ nm and $a = 0.18$ nm^{-1} as typical values of the parameters. Hence explain the significance of these three parameters.

14.2 Calculate the energy of Coulombic repulsion between two hydrogen nuclei at their separation in H_2 (74.1 pm). Express your answer in kilojoules per mole. The result is the energy that must be overcome by the attraction from the electrons that form the bond.

14.3 Write the valence bond wavefunction of the bond in a C—H group of a molecule.

14.4 Write the valence bond wavefunction of a P_2 molecule. Why is P_4 a more stable form of molecular phosphorus?

14.5 Write down the valence bond wavefunction for a nitrogen molecule.

14.6 Give the valence bond wavefunction of SO_2 in terms of the orbitals on the S atom that are used to form bonds. Predict, on the basis of valence bond theory, whether you would expect SO_2 to be linear or bent.

14.7 Write the valence bond wavefunction of CH_4 based on hybrid orbitals h on the carbon atom.

14.8 The structure of the visual pigment retinal is shown in (8). Label each atom with its state of hybridization and specify the composition of each of the different type of bond.

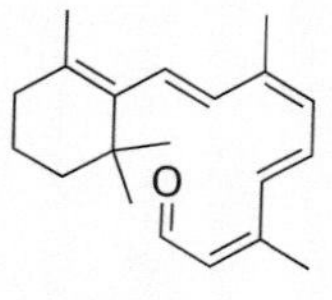

8 Retinal

14.9 Write down three non-ionic contributions to the resonance structure of naphthalene, $C_{10}H_8$.

14.10 A normalized valence bond wavefunction turned out to have the form $\psi = 0.889\psi_{cov} + 0.458\psi_{ion}$. What is the chance that, in 1000 inspections of the molecule, both electrons of the bond will be found on one atom?

14.11 Benzene is commonly regarded as a resonance hybrid of the two Kekulé structures, but other possible structures can also contribute. Draw three other structures in which there are only covalent π bonds (allowing for bonding between some non-adjacent C atoms) and two structures in which there is one ionic bond. Why may these structures be ignored in simple descriptions of the molecule?

14.12 Give the (g,u) parities of the first four levels of a particle-in-a-box wavefunction.

14.13 Give the parities of the wavefunctions for the first four levels of a harmonic oscillator. (b) How may the parity be expressed in terms of the quantum number v?

14.14 Suppose that the π-electron molecular orbitals of naphthalene can be represented by the wavefunctions of a particle in a two-dimensional rectangular box. What are the parities of the occupied orbitals?

14.15 Draw diagrams to show the various orientations in which a p orbital and a d orbital on adjacent atoms may form bonding and antibonding molecular orbitals.

14.16 Try to anticipate the form that the bonding and antibonding 'ϕ orbitals' that could be constructed from two neighbouring f orbitals. What are their parities?

14.17 Put the following species in order of increasing bond length: F_2^-, F_2, F_2^+. Identify the bond order of each of the species.

14.18 Arrange the species O_2^+, O^2, O_2^-, O_2^{2-} in order of increasing bond length and identify the bond order of each.

14.19 Give the ground-state electron configurations of (a) H_2^-, (b) Li^2, (c) Be^2, (d) C^2, (e) N^2, and (f) O_2.

14.20 From the ground-state electron configurations of B_2 and C_2, predict which molecule should have the greater dissociation energy.

14.21 How many molecular orbitals can be constructed from a diatomic molecule in which s, p, d, and f orbitals are all important for bonding?

14.22 Where it is appropriate, give the parity of (a) 2π* in F_2, (b) 2σ in NO, (c) 1δ in Tl_2, (d) 2δ* in Fe_2.

14.23 Some chemical reactions proceed by the initial loss or transfer of an electron to a diatomic species. Which of the molecules N_2, NO, O_2, C_2, F_2, and CN would you expect to be stabilized by (a) the addition of an electron to form AB^-, (b) the removal of an electron to form AB^+?

14.24 Two important diatomic molecules for the welfare of humanity are NO and N_2: the former is both a pollutant and a neurotransmitter, and the latter is the ultimate source of the nitrogen of proteins and other biomolecules. Use the electron configurations of NO and N_2 to predict which is likely to have the shorter bond length.

14.25 Three biologically important diatomic species, either because they promote or inhibit life, are (a) CO, (b) NO, and (c) CN^-. The first binds to haemoglobin, the second is a neurotransmitter, and the third interrupts the respiratory electron-transfer chain. Their biochemical action is a reflection of their orbital structure. Deduce their ground-state electron configurations. For heteronuclear diatomic molecules, a good first approximation is that the energy level diagram is much the same as for homonuclear diatomic molecules.

14.26 The existence of compounds of the noble gases was once a great surprise and stimulated a great deal of theoretical work. Sketch the molecular orbital energy level diagram for XeF and deduce its ground-state electron configurations. Is XeF likely to have a shorter bond length than XeF^+?

14.27 Construct the molecular-orbital energy level diagrams of (a) ethene and (b) ethyne on the basis that the molecules are formed from the appropriately hybridized CH_2 or CH fragments.

14.28 Predict the polarities of the bonds (a) P—H, (b) B—H.

14.29 State the parities of the six π orbitals of benzene (see Fig. 14.37).

14.30 Many of the colours of vegetation are due to electronic transitions in conjugated π-electron systems. In the *free-electron molecular orbital* (FEMO) theory, the electrons in a conjugated molecule are treated as independent particles in a box of length L. (a) Sketch the form of the two occupied orbitals in butadiene predicted by this model and predict the minimum excitation energy of the molecule. (b) In many cases, an extra half bond-length is often added at each end of the box. The tetraene CH_2=CHCH=CHCH=CHCH=CH_2 can therefore be treated as a box of length $8R$, where R = 140 pm. Calculate the minimum excitation energy of the molecule and sketch the HOMO and LUMO.

14.31 The FEMO theory (Exercise 14.30) of conjugated molecules is rather crude and better results are obtained with simple Hückel theory. (a) For a linear conjugated polyene with each of N carbon atoms contributing an electron in a 2p orbital, the energies E_k of the resulting π molecular orbitals are given by:

$$E_k = \alpha + 2\beta\cos\frac{k\pi}{N+1} \qquad k = 1, 2, 3, \ldots, N$$

Use this expression to determine a reasonable empirical estimate of the resonance integral β for the series consisting of ethene, butadiene, hexatriene, and octatetraene given that ultraviolet absorptions from the HOMO, which is a bonding π orbital, to the LUMO, which is an antibonding π* orbital, occur at 61 500, 46 080, 39 750, and 32 900 cm^{-1}, respectively. (b) Calculate the π-electron delocalization energy, $E_{deloc} = E_\pi - n(\alpha+\beta)$, of octatetraene, where E_π is the total π-electron binding energy and n is the total number of π-electrons.

14.32 For monocyclic conjugated polyenes (such as cyclobutadiene and benzene) with each of N carbon atoms contributing an electron in a 2p orbital, simple Hückel theory gives the following expression for the energies E_k of the resulting π molecular orbitals

$$E_k = \alpha + 2\beta\cos\frac{2k\pi}{N} \qquad k = 0, \pm1, \pm2, \pm3, \ldots \pm N/2 \text{ (even } N)$$

$$k = 0, \pm1, \pm2, \pm3, \ldots \pm(N-1)/2 \text{ (odd } N)$$

(a) Calculate the energies of the π molecular orbitals of benzene and cyclooctaene. Comment on the presence or absence of degenerate energy levels. (b) Calculate and compare the delocalization energies of benzene and hexatriene using the expression from Exercise 14.31. What do you conclude from your results? (c) Calculate and compare the delocalization energies of cyclooctaene and octatetraene. Are your conclusions for this pair of molecules the same as for the pair of molecules investigated in part (b)?

14.33 Predict the electronic configurations of (a) the benzene anion, (b) the benzene cation. Estimate the π-bond energy in each case.

Projects

The symbol ‡ indicates that calculus is required.

14.34‡ Here we explore hybrid orbitals in more quantitative detail. Mathematical functions are said to be orthogonal if the integral of their product is zero. (a) Show that the orbitals $h_1 = s + p_x + p_y + p_z$ and $h_2 = s - p_x - p_y + p_z$ are orthogonal. Each atomic orbital individually normalized to 1. Also, note that: (i) s and p orbitals are orthogonal and, (ii) p orbitals with perpendicular orientations are orthogonal. (b) Show that the sp^2 hybrid orbital $(s + 2^{1/2}p)/3^{1/2}$ is normalized to 1 if the s and p orbitals are each normalized to 1. (c) Find another sp^2 hybrid orbital that is orthogonal to the hybrid orbital in par (b).

14.35‡ Show, if overlap is ignored, (a) that any molecular orbital expressed as a linear combination of two atomic orbitals may be written in the form $\psi = \psi_A \cos\theta + \psi_B \sin\theta$, where θ is a parameter that varies between 0 and π/2, and (b) that if ψ_A and ψ_B are orthogonal and normalized to 1, then ψ is also normalized to 1. (c) To what values of θ do the bonding and antibonding orbitals in a homonuclear diatomic molecule correspond?

14.36‡ Now we explore orbital overlap and overlap integrals in detail. (a) Without doing a calculation, sketch how the overlap between an H1s orbital and a 2p orbital can be expected to depend on their separation. (b) The overlap integral between an H1s orbital and a H2p orbital on nuclei separated by a distance R is $S = (R/a_0)\{1 + (R/a_0) + \frac{1}{3}(R/a_0)^2\}e^{-R/a_0}$. Plot this function, and find the separation for which the overlap is a maximum. (c) Suppose that a molecular orbital has the form $N(0.245A + 0.644B)$. Find a linear combination of the orbitals A and B that does not overlap with (that is, is orthogonal to) this combination. (d) Normalize the wavefunction $\psi = \psi_{cov} + \lambda\psi_{ion}$ in terms of the parameter λ and the overlap integral S between the covalent and ionic wavefunctions.

14.37 Use computational chemistry software to explore the bonding in pyridine, C_6H_5N. (a) Use the various procedures available in your chosen software or on the advice of your instructor to determine the shapes and energies of the highest occupied and lowest unoccupied molecular orbitals. (b) Hence determine the wavelength of the LUMO–HOMO transition. Compare the values that you obtain with the various methods with the observed visible spectrum of pyridine, which shows a maximum absorption at 352 nm. (c) Calculate the enthalpy of formation of pyridine in the gas phase by using the different methods. Are the calculated values consistent with the experimentally determined value of 140.2 kJ mol^{-1} at 298 K?

15

Molecular interactions

van der Waals interactions 376

15.1 Interactions between partial charges 377

15.2 Electric dipole moments 377

15.3 Interactions between dipoles 380

15.4 Induced dipole moments 382

15.5 Dispersion interactions 384

The total interaction 385

15.6 Hydrogen bonding 385

15.7 The hydrophobic effect 386

15.8 Modelling the total interaction 387

Molecules in motion 390

CHECKLIST OF KEY CONCEPTS 390

ROAD MAP OF KEY EQUATIONS 391

QUESTIONS AND EXERCISES 391

Atoms and molecules with complete valence shells are still able to interact with one another even though all their valences are satisfied. They attract one another over the range of several atomic diameters and they repel one another when pressed together. These residual interactions are highly important. They account, for instance, for the condensation of gases to liquids and the structures of molecular solids. All organic liquids and solids, ranging from small molecules like benzene to virtually infinite cellulose and the polymers from which fabrics are made, are bound together by the cohesive interactions we explore in this chapter. These interactions are also responsible for the structural organization of biological macromolecules, for they pin molecular building blocks—such as polypeptides, polynucleotides, and lipids—together in the arrangement essential to their proper physiological function.

In this chapter we present the basic theory of molecular interactions and then explore how they play a role in the properties of liquids. In the following chapter we explore how the same interactions contribute to the properties of macromolecules and molecular aggregates. A great deal of this chapter is based on the Coulomb law (The chemist's toolkit 9.1 and Foundations 0.9).

van der Waals interactions

The interactions between or within molecules (for example, within macromolecules) include the attractive and repulsive interactions involving partial electric charges and electron clouds of polar and nonpolar molecules or functional groups, and the repulsive interactions that prevent the complete collapse of matter to densities as high as those characteristic of atomic nuclei. The repulsive interactions arise from the exclusion of electrons from regions of space where the orbitals of closed-shell species overlap. These interactions are called **van der Waals interactions**; the term excludes interactions that result in the formation of covalent or ionic bonds. We shall see that the potential energy arising from an attractive van der Waals interaction is commonly proportional to the inverse sixth power of the separation between molecules or functional groups. The intermolecular *force* depends inversely on one higher power of the separation, so a van der Waals interaction for which the potential energy is proportional to the inverse sixth power of the separation corresponds to a force that is proportional to the inverse seventh power. To confirm this conclusion, we use the relation from Derivation 12.3 that $F = -\mathrm{d}V/\mathrm{d}r$, where $V(r)$ is the potential energy, to write

$$\text{For } V(r) = -\frac{C}{r^6}, \qquad (15.1)$$

$$F(r) = -\frac{\mathrm{d}V(r)}{\mathrm{d}r} = -\frac{\mathrm{d}}{\mathrm{d}r}\left(-\frac{C}{r^6}\right) \overset{\mathrm{d}x^n/\mathrm{d}x = nx^{n-1},\ n=-6}{=} -\frac{6C}{r^7}$$

15.1 Interactions between partial charges

Atoms in molecules in general have partial charges. Table 15.1 gives the partial charges typically found on the atoms in peptides. When partial charges were separated by a vacuum, they attract or repel each other in accord with Coulomb's law (with the potential energy E_p denoted V, as is conventional in this context) and we write

$$V(r) = \frac{Q_1 Q_2}{4\pi\varepsilon_0 r} \qquad \text{Interaction between partial charges} \qquad (15.2a)$$

where Q_1 and Q_2 are the partial charges and r is their separation. However, we need to take into account the possibility that other parts of the molecule, or other molecules, lie between the charges, and decrease

Table 15.1
Partial charges in polypeptides

Atom	Partial charge/e
C(=O)	+0.45
C(—CO)	+0.06
H(—C)	+0.02
H(—N)	+0.18
H(—O)	+0.42
N	−0.36
O	−0.38

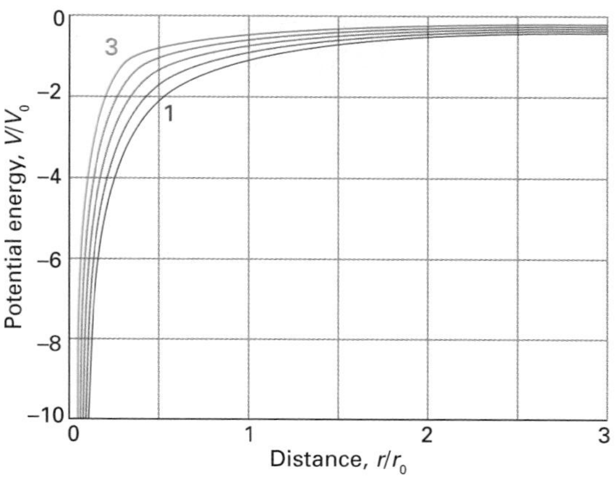

Fig. 15.1 The Coulomb potential for two charges and its dependence on their separation. The curves correspond to different relative permittivities (1 for a vacuum, 3 for a fluid).

the strength of the interaction. The simplest procedure for taking into account these very complicated effects is to treat the medium as a uniform substance and to write

$$V(r) = \frac{Q_1 Q_2}{4\pi\varepsilon r} \qquad (15.2b)$$

where ε is the permittivity of the medium (see The chemist's toolkit 9.1): a high permittivity means that the medium reduces the strength of the interaction between the two charges. As explained in the toolkit, the permittivity is usually expressed as a multiple of the vacuum permittivity by writing $\varepsilon = \varepsilon_\mathrm{r}\varepsilon_0$, where the dimensionless quantity ε_r is the relative permittivity. The effect of the medium can be very large: for water at 25 °C, $\varepsilon_\mathrm{r} = 78$, so the potential energy of two charges separated by bulk water is reduced by nearly two orders of magnitude compared to the value it would have if the charges were separated by a vacuum (Fig. 15.1). The problem is made worse in calculations on polypeptides and nucleic acids by the fact that two partial charges may have water and a biopolymer chain lying between them. Various models have been proposed to take this effect into account, the simplest being to set $\varepsilon_\mathrm{r} = 3.5$ and to hope for the best.

15.2 Electric dipole moments

When the molecules or groups that we are considering are widely separated, it turns out to be simpler to express the principal features of their interaction in terms of the dipole moments associated with the charge distributions rather than with each individual partial charge. At its simplest, an **electric dipole** consists of two charges Q and $-Q$ separated by a distance l. The product Ql is called the **electric dipole moment**, μ. We represent dipole moments by an arrow with a length proportional to μ and pointing from the negative charge to the positive charge (**1**). (Be careful with this convention: for historical reasons the opposite convention is still widely used.)

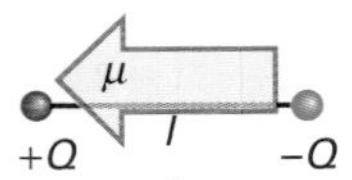

1

Because a dipole moment is the product of a charge (in coulombs, C) and a length (in metres, m), the SI unit of dipole moment is the coulomb metre (C m). However, it is often much more convenient to report a dipole moment in **debye**, D, where

$$1\ \text{D} = 3.335\ 64 \times 10^{-30}\ \text{C m} \qquad \text{Definition of debye}$$

because then experimental values for molecules are close to 1 D (Table 15.2). The unit is named after Peter Debye, the Dutch pioneer of the study of dipole moments of molecules. The dipole moment of charges e and $-e$ separated by 100 pm, which is the order of the length of a typical chemical bond, is 1.6×10^{-29} C m, corresponding to 4.8 D. Dipole moments of small molecules are typically smaller than that, at about 1 D, confirming that the charge separation in simple molecules is only partial.

A **polar molecule** has a permanent electric dipole moment arising from the partial charges on its atoms (Section 14.11). A **nonpolar molecule** has no permanent electric dipole moment. All heteronuclear diatomic molecules are polar because the difference in electronegativities of their two atoms results in nonzero partial charges. Typical dipole moments are 1.08 D for HCl and 0.42 D for HI (Table 15.2). A very approximate relation between the dipole moment and the difference $\Delta\chi$ of Pauling electronegativities (Table 14.2) χ_A and χ_B of two atoms A and B, is

$$\mu/\text{D} \approx \chi_A - \chi_B = \Delta\chi \qquad \text{Dipole moment and electronegativity difference} \qquad (15.3)$$

Table 15.2

Dipole moments, mean polarizabilities, and polarizability volumes

	μ/D	$\alpha/(10^{-40}\ \text{J}^{-1}\ \text{C}^2\ \text{m}^2)$	$\alpha'/(10^{-30}\ \text{m}^3)$
Ar	0	1.85	1.66
CCl_4	0	11.7	10.3
C_6H_6	0	11.6	10.4
H_2	0	0.911	0.819
H_2O	1.85	1.65	1.48
NH_3	1.47	2.47	2.22
HCl	1.08	2.93	2.63
HBr	0.80	4.01	3.61
HI	0.42	6.06	5.45

Brief illustration 15.1 Dipole moment and electronegativity

The electronegativities of hydrogen and bromine are 2.1 and 2.8, respectively. The difference is 0.7, so we predict an electric dipole moment of about 0.7 D for HBr. The experimental value is 0.80 D.

Self-test 15.1

What is the dipole moment of a C—H fragment in an organic molecule; which atom lies at the negative end?

Answer: μ = 0.4 D; C

Because it attracts the electrons more strongly, the more electronegative atom is usually the negative end of the dipole. However, there are exceptions, particularly when antibonding orbitals are occupied. Thus, the dipole moment of CO is small (0.12 D) but the negative end of the dipole is on the C atom even though the O atom is more electronegative. This apparent paradox is resolved as soon as we realize that antibonding orbitals are occupied in CO (see Fig. 14.33). Because electrons in antibonding orbitals tend to be found closer to the less electronegative atom, they contribute a negative partial charge to that atom. If this contribution is larger than the opposite contribution from the electrons in bonding orbitals, then the net effect will be a small negative partial charge on the *less* electronegative atom.

Molecular symmetry is of the greatest importance in deciding whether a polyatomic molecule is polar or not. Indeed, molecular symmetry is more important than the question of whether or not the atoms in the molecule belong to the same element. Homonuclear polyatomic molecules may be polar if they have low symmetry and the atoms are in inequivalent positions. For instance, the angular molecule ozone, O_3 (**2**), is homonuclear; however, it is polar because the central O atom is different from the outer two (it is bonded to two atoms, they are bonded only to one); moreover, the dipole moments associated with each bond make an angle to each other and do not cancel. Heteronuclear polyatomic molecules may be nonpolar if they have high symmetry, because individual bond dipoles may then cancel. The heteronuclear linear triatomic molecule CO_2, for example, is nonpolar because, although there are partial charges on all three atoms, the dipole moment associated with the OC bond points in the opposite direction to the dipole moment associated with the CO bond, and the two cancel (**3**).

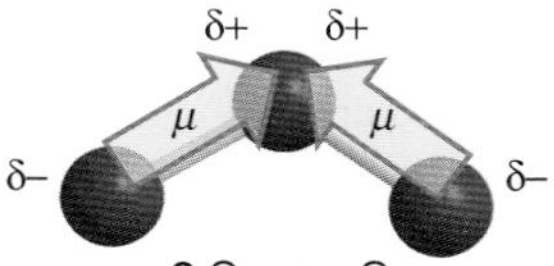

2 Ozone, O_3

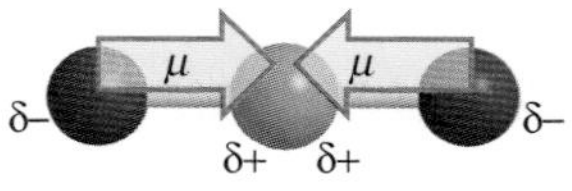

3 Carbon dioxide, CO_2

Example 15.1

Assessing the polarity of a molecule

Is a ClF_3 molecule polar or nonpolar?

Strategy First judge whether the bonds are polar: if they are not, the molecule will be nonpolar. (Electronegativity differences are a cautious but unreliable guide because atoms of the same element that lie in inequivalent locations might have different partial charges.) Then judge the shape of the molecule by using the VSEPR theory (The chemist's toolkit 14.2) and judge whether the bond dipoles, if any, cancel. If they do (as in a tetrahedral molecule), the molecule is nonpolar. If they do not cancel, then the molecule is polar.

Solution The electronegativities of Cl and F are 3.0 and 4.0, respectively; so each Cl—F fragment is polar with a dipole moment of about 1 D. To apply VSEPR theory, we note that the Lewis structure is shown in (**4**), with five regions of high electron density (three bonds, two lone pairs), so these regions are likely to be arranged as a trigonal bipyramid (**5**). The repulsion between the two lone pairs is least if they occupy the equatorial positions and move apart slightly. The three F atoms lie at the remaining positions, with the result that the molecule is T-shaped (**6**). The three bond dipole moments do not cancel in this arrangement, so the molecule is polar.

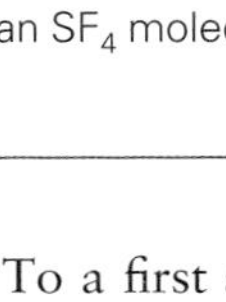

4

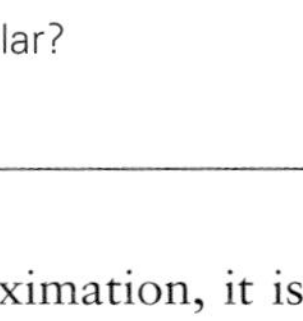

5

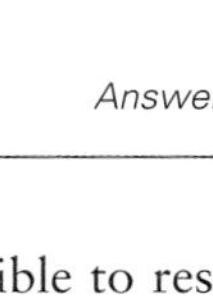

6

Self-test 15.2

Is an SF_4 molecule polar?

Answer: Yes

To a first approximation, it is possible to resolve the dipole moment of a polyatomic molecule into contributions from various groups of atoms in the molecule and the directions in which these individual contributions lie (Fig. 15.2). Thus, 1,4-dichlorobenzene

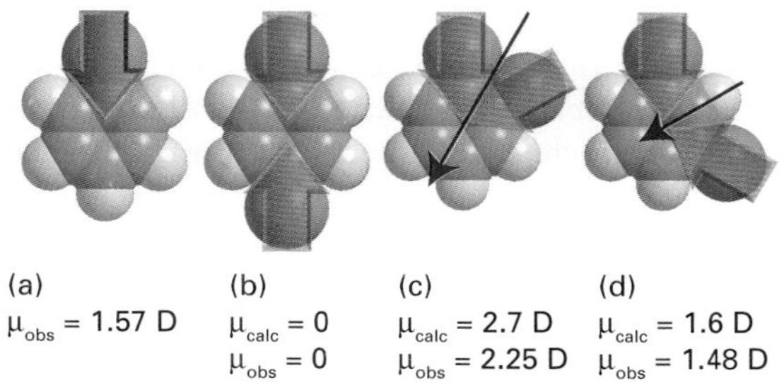

Fig. 15.2 The dipole moments of the dichlorobenzene isomers can be obtained approximately by vector addition of two chlorobenzene dipole moments (1.57 D).

is nonpolar by symmetry on account of the cancellation of two equal but opposing C—Cl moments (exactly as in carbon dioxide). 1,2-Dichlorobenzene, however, has a dipole moment which is approximately the resultant of two chlorobenzene dipole moments arranged at 60° to each other. This technique of 'vector addition' (see The chemist's toolkit 13.1) can be applied with fair success to other series of related molecules, and the resultant μ_{res} of two dipole moments μ_1 and μ_2 that make an angle θ to each other (**7**) is approximately (see The chemist's toolkit 13.1)

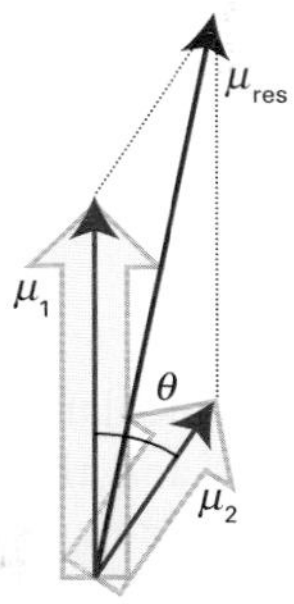

7

$$\mu_{res} \approx (\mu_1^2 + \mu_2^2 + 2\mu_1\mu_2 \cos\theta)^{1/2} \qquad (15.4)$$

This relation is not exact because the dipole moments are not strictly additive.

Brief illustration 15.2 The resultant dipole

To estimate the ratio of the electric dipole moments of *ortho* (1,2-) and *meta* (1,3-) similarly disubstituted benzenes we note that for the *ortho* isomer $\theta = 60°$ and for the *meta* isomer $\theta = 120°$. The C—R group dipole moments are the same, so we can use eqn 15.4 in the form $\mu_{res} = 2^{1/2}\mu(1 + \cos\theta)^{1/2}$ in each case. Then the ratio is

$$\frac{\mu(ortho)}{\mu(meta)} = \frac{2^{1/2}\mu(1+\cos 60°)^{1/2}}{2^{1/2}\mu(1+\cos 120°)^{1/2}} = 1.7$$

Self-test 15.3

The O—H bond dipole moment is approximately 1.4 D. Estimate the dipole moment of an H_2O_2 molecule, in which the O—H bonds lie at 90° to each other and the O—O bond.

Answer: 2.0 D

A better approach to the calculation of dipole moments is to take into account the locations and magnitudes of the partial charges on all the atoms. These partial charges are included in the output of many molecular structure software packages. The programs calculate the dipole moments of the molecules by noting that an electric dipole moment is actually a vector, $\boldsymbol{\mu}$, with three components, μ_x, μ_y, and μ_z (8). The direction of $\boldsymbol{\mu}$ shows the orientation of the dipole in the molecule and the length of the vector is the magnitude, μ, of the dipole moment. In common with all vectors, the magnitude is related to the three components by

$$\mu = (\mu_x^2 + \mu_y^2 + \mu_z^2)^{1/2} \qquad (15.5a)$$

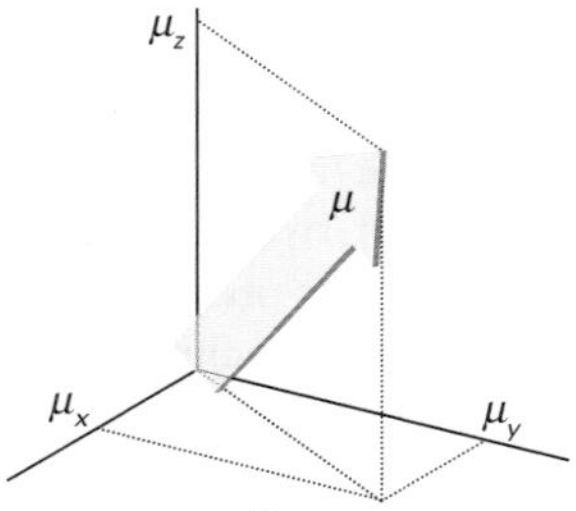

8

To calculate μ we need to calculate the three components and then substitute them into this expression. To calculate the x-component, for instance, we need to know the partial charge on each atom and the atom's x-coordinate relative to a point in the molecule and form the sum

$$\mu_x = \sum_J Q_J x_J \qquad (15.5b)$$

Here Q_J is the partial charge of atom J, x_J is the x-coordinate of atom J, and the sum is over all the atoms in the molecule. Analogous expressions are used for the y- and z-components. For an electrically neutral molecule, the origin of the coordinates is arbitrary, so it is best chosen to simplify the measurements.

Example 15.2

Calculating a molecular dipole moment

Estimate the electric dipole moment of the peptide group using the partial charges (as multiples of e) in Table 15.1 and the locations of the atoms shown in (9).

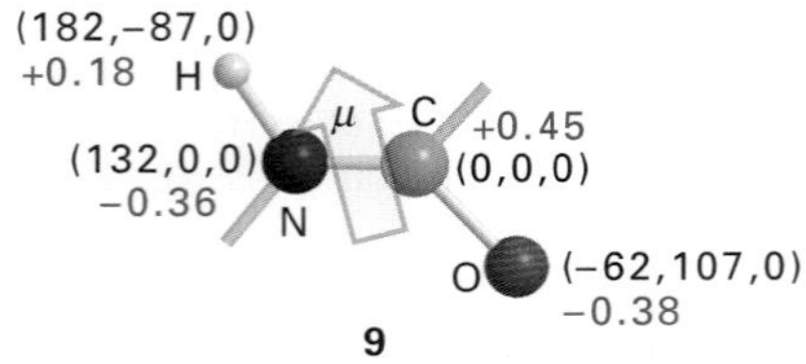

9

Strategy We use eqn 15.5b to calculate each of the components of the dipole moment and then eqn 15.5a to assemble the three components into the magnitude of the dipole moment. Note that the partial charges are multiples of the fundamental charge, $e = 1.609 \times 10^{-19}$ C.

Solution The expression for μ_x is

$$\begin{aligned}\mu_x &= (-0.36e) \times (132\ \text{pm}) + (0.45e) \times (0\ \text{pm}) \\ &\quad + (0.18e) \times (182\ \text{pm}) + (-0.38e) \times (-62\ \text{pm}) \\ &= 8.8e\ \text{pm} \\ &= 8.8 \times (1.609 \times 10^{-19}\ \text{C}) \times (10^{-12}\ \text{m}) = 1.4 \times 10^{-30}\ \text{C m}\end{aligned}$$

corresponding to $\mu_x = +0.42$ D. The expression for μ_y is

$$\begin{aligned}\mu_y &= (-0.36e) \times (0\ \text{pm}) + (0.45e) \times (0\ \text{pm}) \\ &\quad + (0.18e) \times (-87\ \text{pm}) + (-0.38e) \times (107\ \text{pm}) \\ &= -56e\ \text{pm} = -9.1 \times 10^{-30}\ \text{C m}\end{aligned}$$

It follows that $\mu_y = -2.7$ D. Therefore, because $\mu_z = 0$,

$$\mu = \{(0.42\ \text{D})^2 + (-2.7\ \text{D})^2\}^{1/2} = 2.7\ \text{D}$$

We can find the orientation of the dipole moment by arranging an arrow of length 2.7 units of length to have x, y, and z components of 0.42, −2.7, and 0 units; the orientation is superimposed on (9).

Self-test 15.4

Calculate the electric dipole moment of formaldehyde, using the information in (10).

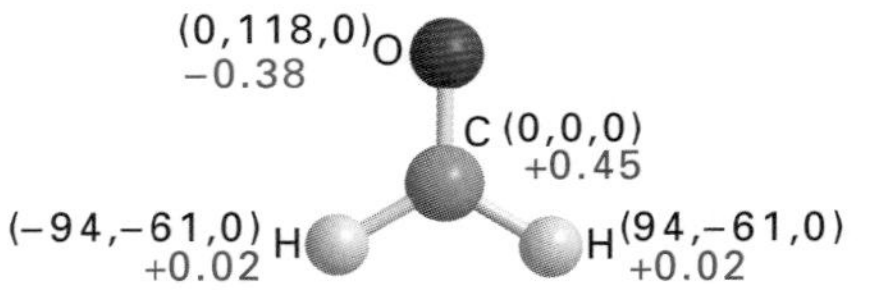

10

Answer: −7.6 D

15.3 Interactions between dipoles

The potential energy of a dipole μ_1 in the presence of a charge Q_2 is calculated by taking into account the interaction of the charge with the two partial charges of the dipole, one resulting in a repulsion and the other an attraction. As shown in the following Derivation, the result for the arrangement shown in (11) is

$$V(r) = -\frac{\mu_1 Q_2}{4\pi\varepsilon_0 r^2} \qquad \text{Charge–dipole interaction} \qquad (15.6a)$$

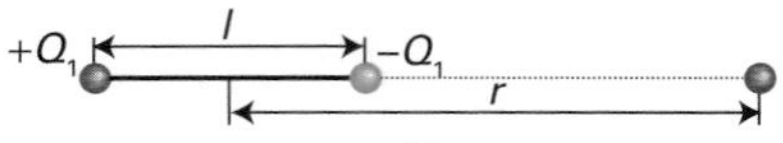

11

Derivation 15.1

The interaction of a charge with a dipole

When the charge and dipole are collinear, as in (**11**), the potential energy is

$$V(r) = \overbrace{\frac{Q_1Q_2}{4\pi\varepsilon_0(r+\frac{1}{2}l)}}^{\text{Repulsion between } Q_1 \text{ and } +Q_2} - \overbrace{\frac{Q_1Q_2}{4\pi\varepsilon_0(r-\frac{1}{2}l)}}^{\text{Attraction between } Q_1 \text{ and } -Q_2}$$

$$= \frac{Q_1Q_2}{4\pi\varepsilon_0 r(1+\frac{l}{2r})} - \frac{Q_1Q_2}{4\pi\varepsilon_0 r(1-\frac{l}{2r})}$$

Next, we suppose that the separation of charges in the dipole is much smaller than the distance of the charge Q_2 in the sense that $l/2r \ll 1$. Then we can use (see The chemist's toolkit 6.1)

$$\frac{1}{1+x} \approx 1-x \qquad \frac{1}{1-x} \approx 1+x$$

to write

$$V(r) \approx \frac{Q_1Q_2}{4\pi\varepsilon_0 r}\left\{\left(1-\frac{l}{2r}\right)-\left(1+\frac{l}{2r}\right)\right\} = -\frac{Q_1Q_2 l}{4\pi\varepsilon_0 r^2}$$

Now we recognize that $Q_1 l = \mu_1$, the dipole moment of molecule 1, and obtain eqn 15.6a.

A similar calculation for the more general orientation shown in (**12**) gives

$$V(r) = -\frac{\mu_1 Q_2 \cos\theta}{4\pi\varepsilon_0 r^2} \qquad \text{Charge–dipole interaction} \qquad (15.6b)$$

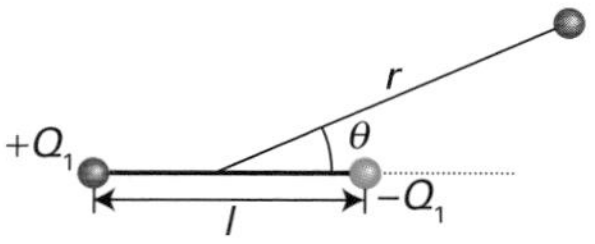

12

If Q_2 is positive, the energy is lowest when $\theta = 0$ (and $\cos\theta = 1$), because then the partial negative charge of the dipole lies closer than the partial positive charge to the point charge and the attraction outweighs the repulsion. This interaction energy decreases more rapidly with distance than that between two point charges (as $1/r^2$ rather than $1/r$) because, from the viewpoint of the single charge, the partial charges of the point dipole seem to merge and cancel as the distance r increases.

We can calculate the interaction energy between two dipoles μ_1 and μ_2 in the orientation shown in (**13**) in a similar way, by taking into account all four charges of the two dipoles. The outcome is[1]

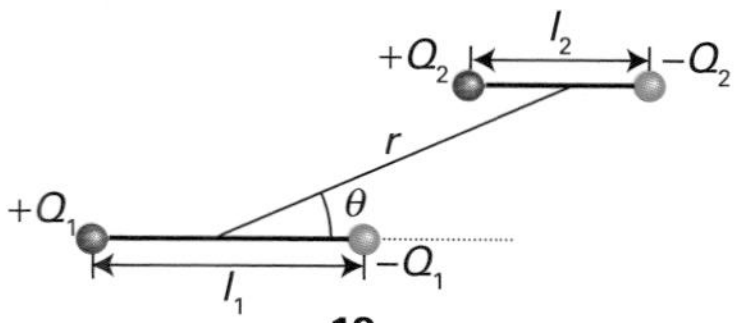

13

$$V(r,\theta) = -\frac{\mu_1\mu_2(1-3\cos^2\theta)}{4\pi\varepsilon_0 r^3} \qquad \text{Fixed orientations} \quad \text{Dipole–dipole interaction} \qquad (15.7)$$

Let's interpret this expression, and refer to Fig. 15.3, which shows the angular dependence of the potential energy:

- The potential energy decreases even more rapidly than in eqn 15.6 (as $1/r^3$ instead of $1/r^2$) because the charges of *both* dipoles seem to merge as the separation of the dipoles increases.
- The angular factor takes into account how the like or opposite charges come closer to one another as the relative orientation of the dipoles is changed.
- The energy is lowest when $\theta = 0$ or 180° (when $1 - 3\cos^2\theta = -2$), because opposite partial charges then lie closer together than like partial charges.
- The potential energy is negative (attractive) in some orientations when $\theta < 54.7°$ (the angle at which $1 - 3\cos^2\theta = 0$, corresponding to $\cos\theta = 1/3^{1/2}$) because opposite charges are closer than like charges.
- The potential energy is positive (repulsive) when $\theta > 54.7°$ because then like charges are closer than unlike charges.

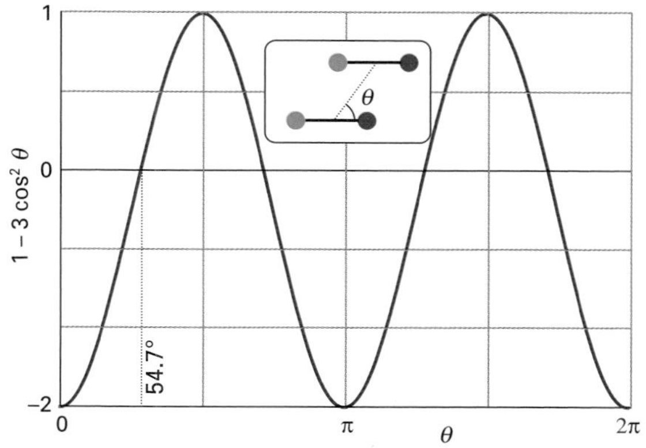

Fig. 15.3 The angular dependence of the potential energy of two parallel electric dipoles.

[1] For a derivation of eqn 15.7, see our *Physical Chemistry* (2010).

- The potential energy is zero on the lines at 54.7° and 180 − 54.7 = 123.3° because at those angles the two attractions and the two repulsions cancel (**14**).

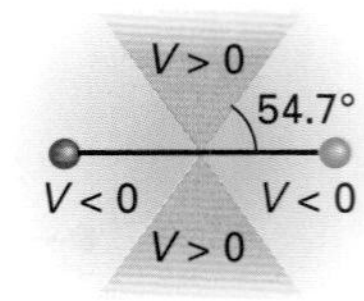

14

Brief illustration 15.3 The dipole–dipole interaction

To calculate the molar potential energy of the dipolar interaction between two peptide links separated by 3.0 nm in different regions of a polypeptide chain with $\theta = 180°$, we take $\mu_1 = \mu_2 = 2.7$ D, corresponding to 9.1×10^{-30} C m, and find

$$V(3.0\text{ nm},180°) = \frac{\overbrace{(9.1\times10^{-30}\text{ C m})^2}^{\mu_1\mu_2=\mu^2} \times \overbrace{(-2)}^{(1-3\cos^2\theta),\ \theta=180°}}{\underbrace{4\pi\times(8.854\times10^{-12}\text{ J}^{-1}\text{ C}^2\text{ m}^{-1})}_{4\pi\varepsilon_0}\times\underbrace{(3.0\times10^{-9}\text{ m})^3}_{r^3}}$$

$$= \frac{(9.1\times10^{-30})^2\times(-2)}{4\pi\times(8.854\times10^{-12})\times(3.0\times10^{-9})^3}\,\frac{\text{C}^2\text{ m}^2}{\text{J}^{-1}\text{ C}^2\text{ m}^{-1}\text{ m}^3}$$

$$= -5.5\times10^{-23}\text{ J}$$

This value corresponds (after multiplication by Avogadro's constant) to −34 J mol^{-1}.

A note on good practice We reiterate the importance of including the units at every stage of the calculation, in part because the correct cancellation helps to monitor whether the calculation has been set up and carried out correctly.

The average potential energy of interaction between polar molecules that are freely rotating in a fluid (a gas or liquid) is zero because the attractions and repulsions cancel. However, because the potential energy of a dipole near another dipole depends on their relative orientations, the molecules exert forces on each other and therefore do not in fact rotate completely freely, even in a gas. As a result, the lower energy orientations are marginally favoured, so there is a nonzero interaction between rotating polar molecules (Fig. 15.4). The detailed calculation of the average interaction energy is quite complicated, but the final answer is very simple:

$$V(r) = -\frac{\mu_1^2\mu_2^2}{3(4\pi\varepsilon_0)^2kTr^6} \quad \text{Rotating molecules} \quad \text{Dipole–dipole interaction} \qquad (15.8)$$

where k is Boltzmann's constant. As before, let's 'read' this expression:

- The interaction between dipoles is an example of a van der Waals interaction that varies as the inverse sixth power of the distance.

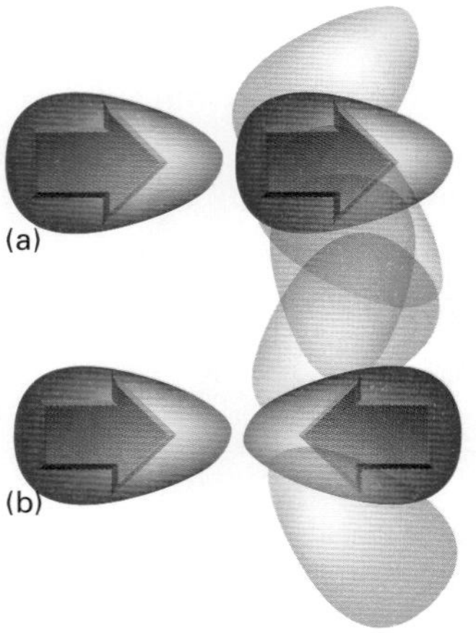

Fig. 15.4 A dipole–dipole interaction. When a pair of molecules can adopt all relative orientations with equal probability, the favourable orientations (a) and the unfavourable ones (b) cancel, and the average interaction is zero. In an actual fluid, the interactions in (a) slightly predominate.

- The inverse temperature dependence ($V \propto 1/T$) reflects the way that at higher temperatures the greater thermal motion overcomes the mutual orientating effects of the dipoles.

Brief illustration 15.4 The dipolar interaction of rotating molecules

At 25 °C the average interaction energy for pairs of molecules with $\mu = 1.0$ D when the separation is 0.30 nm is

$$V(0.30\text{ nm}) = \frac{\overbrace{(3.3\times10^{-30}\text{ C m})^2}^{\mu_1^2}\,\overbrace{(3.3\times10^{-30}\text{ C m})^2}^{\mu_2^2}}{3\underbrace{(4\pi\times8.85\times10^{-12}\text{ J}^{-1}\text{ C}^2\text{ m}^{-1})^2}_{(4\pi\varepsilon_0)^2}\times\underbrace{(1.381\times10^{-23}\text{ J K}^{-1})}_{k}}$$

$$\times\frac{1}{\underbrace{(298\text{ K})}_{T}\times\underbrace{(3.0\times10^{-10}\text{ m})^6}_{r^6}}$$

$$= \frac{3.3^4\times10^{-120}}{3(4\pi\times8.85)^2\times298\times3.0^6\times10^{-107}}\,\frac{\text{C}^4\text{ m}^4}{\text{J}^{-1}\text{ C}^4\text{ m}^4}$$

$$= -1.5\times10^{-21}\text{ J}$$

or about −0.9 kJ mol^{-1}. This energy should be compared with the average molar kinetic energy of $\frac{3}{2}RT$ = 3.7 kJ mol^{-1} at the same temperature (Foundations 0.8): the two are not very dissimilar, but they are both much less than the energies involved in the making and breaking of chemical bonds.

15.4 Induced dipole moments

A nonpolar molecule may acquire a temporary **induced dipole moment**, μ^*, as a result of the influence of an electric field generated by a nearby ion or polar molecule. The field distorts the electron distribution of the molecule, and gives rise to an electric dipole. The molecule is said to be *polarizable*. The magnitude of the induced dipole moment is proportional to the strength of the electric field, $\mathcal{E}$, and we write

$$\mu^* = \alpha\mathcal{E} \qquad \text{Induced dipole moment} \quad (15.9)$$

The proportionality constant α is the **polarizability** of the molecule. It is important to understand the following features of the polarizability:

- The larger the polarizability of the molecule, the greater is the distortion caused by a given strength of electric field.
- If the molecule has few electrons (such as N_2), they are tightly controlled by the nuclear charges and the polarizability of the molecule is low. If the molecule contains large atoms with electrons some distance from the nucleus (such as I_2), the nuclear control is less and the polarizability of the molecule is greater.
- The polarizability is low when the ionization energy of the molecule is high: the more tightly the electrons are bound, the more difficult it is to distort the electron distribution around the nuclei.
- The polarizability depends on the orientation of the molecule with respect to the field unless the molecule is tetrahedral (such as CCl_4), octahedral (such as SF_6), or icosahedral (such as C_{60}). Atoms, tetrahedral, octahedral, and icosahedral molecules have isotropic (orientation independent) polarizabilities; all other molecules have anisotropic (orientation dependent) polarizabilities.

The polarizabilities reported in Table 15.2 are given as **polarizability volumes**, α', which are often easier to use than the polarizabilities themselves:

$$\alpha' = \frac{\alpha}{4\pi\varepsilon_0} \qquad \text{Definition} \quad \text{Polarizability volume} \quad (15.10)$$

The polarizability volume has the dimensions of volume (hence its name) and is comparable in magnitude to the volume of the molecule.

Brief illustration 15.5 The induced dipole

To assess the strength of electric field required to induce an electric dipole moment of 1.0 μD in a molecule of polarizability volume 1.0×10^{-29} m^3 (like CCl_4) we write eqn 15.9 as $\mathcal{E} = \mu^*/\alpha$ and substitute the data. We need to use the fact that 1 D = 3.336×10^{-30} C m and convert the polarizability volume to polarizability by using eqn 15.10 in the form $\alpha = 4\pi\varepsilon_0\alpha'$. Then

$$\mathcal{E} = \frac{\mu^*}{\alpha}$$

$$= \frac{\overbrace{3.336\times10^{-30}\times1.0\times10^{-6}\ \text{C m}}^{1.0\ \mu\text{D}}}{4\pi\times\underbrace{(8.854\times10^{-12}\ \text{J}^{-1}\ \text{C}^2\ \text{m}^{-1})}_{\varepsilon_0}\times\underbrace{(1.0\times10^{-29}\ \text{m}^3)}_{\alpha'}}$$

$$= \frac{3.3\times10^{-36}}{4\pi\times(8.854\times10^{-12})\times(1.0\times10^{-29})}\frac{\text{C m}}{\text{J}^{-1}\ \text{C}^2\ \text{m}^2}$$

$$= 9.3\times10^3\ \overbrace{\text{J C}^{-1}}^{\text{V}}\ \text{m}^{-1} = 9.3\ \text{kV m}^{-1}$$

A polar molecule with dipole moment μ_1 can induce a dipole moment in a polarizable molecule (which may itself be either polar or nonpolar) because the partial charges of the polar molecule give rise to an electric field that distorts the second molecule. That induced dipole interacts with the permanent dipole of the first molecule, and the two are attracted together (Fig. 15.5). The formula for the **dipole–induced dipole interaction energy** is

$$V(r) = -\frac{\mu_1^2\alpha_2'}{4\pi\varepsilon_0 r^6} \qquad \text{Dipole–induced dipole interaction} \quad (15.11)$$

where α_2' is the polarizability volume of molecule 2. The negative sign, which denotes a lowering of energy from zero as the molecules approach, shows that the interaction is attractive. The interaction between a dipole and an induced dipole is another example of a van der Waals interaction that varies as the inverse sixth power of the distance.

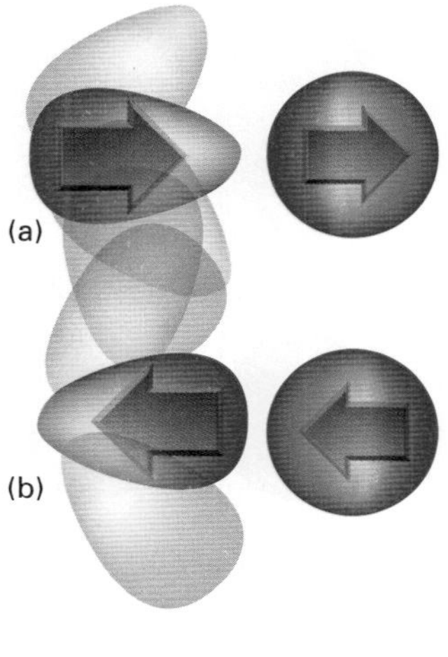

Fig. 15.5 A dipole–induced dipole interaction. The induced dipole (yellow arrows) follows the changing orientation of the permanent dipole (purple arrow).

Brief illustration 15.6 The dipole–induced dipole interaction energy

For a molecule with $\mu = 1.0$ D (such as HCl) near a molecule of polarizability volume $\alpha_2' = 1.0 \times 10^{-29}$ m^3 (such as benzene, Table 15.2) the average interaction energy when the separation is 0.30 nm is

$$V(0.30\text{ nm}) = -\frac{\overbrace{(3.3\times10^{-30}\text{ C m})^2}^{\mu_1^2}\times\overbrace{(1.0\times10^{-29}\text{ m}^3)}^{\alpha_2'}}{\underbrace{(4\pi\times8.85\times10^{-12}\text{ J}^{-1}\text{ C}^2\text{ m}^{-1})}_{4\pi\varepsilon_0}\times\underbrace{(3.0\times10^{-10}\text{ m})^6}_{r^6}}$$

$$= -\frac{3.3^2\times1.0\times10^{-89}}{4\pi\times8.85\times3.0^6\times10^{-72}}\frac{\text{C}^2\text{ m}^5}{\text{J}^{-1}\text{ C}^2\text{ m}^5}$$

$$= -1.3\times10^{-21}\text{ J}$$

After multiplication by Avogadro's constant, this energy corresponds to about −0.8 kJ mol^{-1}.

15.5 Dispersion interactions

Finally, we consider the interactions between species that have neither a net charge nor a permanent electric dipole moment (such as two Xe atoms in a gas or two nonpolar groups on the peptide residues of a protein). Despite their absence of partial charges, we know that uncharged, nonpolar species can interact because they form condensed phases, such as benzene, liquid hydrogen, and liquid xenon.

The **dispersion interaction**, or **London interaction**, between nonpolar species arises from the transient dipoles that they possess as a result of fluctuations in the electron density distribution (Fig. 15.6). Suppose, for instance, that the electrons in one molecule flicker into an arrangement that results in partial positive and negative charges and thus gives it an instantaneous dipole moment μ_1. While it exists, this dipole can polarize the other molecule and induce in it an instantaneous dipole moment μ_2. The two dipoles attract each other and the potential energy of the pair is lowered. Although the first molecule will go on to change the size and direction of its dipole (perhaps within 10^{-16} s), the second will follow it; that is, the two dipoles are *correlated* in direction like two meshing gears, with a positive partial charge on one molecule appearing close to a negative partial charge on the other molecule and vice versa. Because of this correlation of the relative positions of the partial charges, and their resulting attractive interaction, the attraction between the two instantaneous dipoles does not average to zero. Instead, it gives rise to a net attractive interaction. Polar molecules interact by a dispersion interaction as well as by dipole–dipole interactions, with the dispersion interaction often dominant.

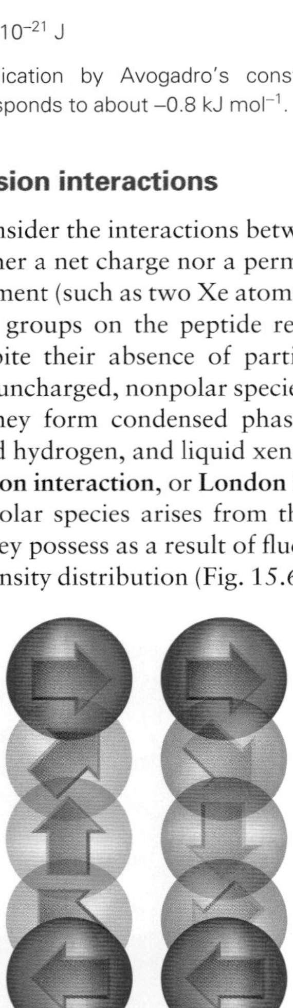

Fig. 15.6 In the dispersion interaction, an instantaneous dipole on one molecule induces a dipole on another molecule, and the two dipoles then interact to lower the energy. The directions of the two instantaneous dipoles are correlated, and, although they occur in different orientations at different instants, the interaction does not average to zero.

The calculation of the dispersion interaction is quite involved, but a reasonable approximation to the interaction energy is the **London formula:**

$$V(r) = -\frac{2}{3}\times\frac{\alpha_1'\alpha_2'}{r^6}\times\frac{I_1 I_2}{I_1+I_2} \qquad \text{London formula} \qquad (15.12)$$

where I_1 and I_2 are the ionization energies of the two molecules. The strength of the dispersion interaction depends on the polarizability of the first molecule because the magnitude of the instantaneous dipole moment μ_1 depends on the looseness of the control that the nuclear charge has over the outer electrons. If that control is loose, the electron distribution can undergo relatively large fluctuations. Moreover, if the control is loose, then the electron distribution can also respond strongly to applied electric fields and hence have a high polarizability. It follows that a high polarizability is a sign of large fluctuations in local charge density. The strength also depends on the polarizability of the second molecule, for that polarizability determines how readily a dipole can be induced in molecule 2 by molecule 1. We therefore expect $V \propto \alpha_1\alpha_2$, as in eqn 15.12.

Brief illustration 15.7 The strength of a dispersion interaction

For two molecules with polarizability volume 1.0×10^{-29} m^3 (as for benzene) with $I = 9.2$ eV (corresponding to 1.5×10^{-18} J) separated by 0.30 nm, noting that $I_1I_2/(I_1 + I_2) = \frac{1}{2}I$ when the ionization energies are the same,

$$V(0.30\text{ nm}) = -\frac{2}{3}\times\frac{\overbrace{(1.0\times10^{-29}\text{ m}^3)^2}^{\alpha_1'\alpha_2'=\alpha^2}}{\underbrace{(3.0\times10^{-10}\text{ m})^6}_{r^6}}\times\frac{1}{2}\overbrace{(1.5\times10^{-18}\text{ J})}^{I}$$

$$= -\frac{1.0^2\times1.5\times10^{-76}\text{ J m}^6}{3\times3.0^6\times10^{-60}\quad\text{m}^6} = -6.9\times10^{-20}\text{ J}$$

or about -41 kJ mol^{-1}.

As usual, we should interpret mathematical expressions and from eqn 15.12 we see that:

- The potential energy of interaction increases with decreasing ionization energies.

This conclusion may be puzzling at first sight, for the product of the ionization energies appears in the numerator of the right-hand side of eqn 15.12. However, broadly speaking, the polarizability is inversely proportional to the ionization energy (Section 15.4), so it follows that $\alpha_1'\alpha_2' \propto (I_1I_2)^{-1}$ and, for a constant separation r, $V \propto (I_1 + I_2)^{-1}$: the potential energy is inversely proportional to the sum of the ionization energies.

- The potential energy of interaction is proportional to the inverse sixth power of the separation.

We have seen this result for the other interactions considered thus far in this chapter. It is consistent with our previous statement that the potential energy of an attractive van der Waals interaction is commonly proportional to r^{-6}.

The total interaction

So far we have discussed attractive interactions that vary as the inverse sixth power of the separation. However, there are several other types of interaction, both attractive and repulsive, some of which dominate the interactions we have explored when they are present.

15.6 Hydrogen bonding

The strongest intermolecular interaction arises from the formation of a **hydrogen bond**, in which a hydrogen atom lies between two strongly electronegative atoms and binds them together. The bond is normally denoted X—H···Y, with X and Y being nitrogen, oxygen, or fluorine. Unlike the other interactions we have considered, hydrogen bonding is not universal but is restricted to molecules that contain these atoms. A common hydrogen bond is formed between O—H groups and O atoms, as in liquid water and ice. The distance dependence of the hydrogen bond is quite different from the other interactions we have considered, and is best regarded as a 'contact' interaction, which turns on when the X—H group is in direct contact with the Y atom.

The most elementary description of the formation of a hydrogen bond is that it is the result of a Coulombic interaction between the partly exposed positive charge of a proton bound to an electron-withdrawing X atom (in the fragment X—H) and the negative charge of a lone pair on the second atom Y, as in $^{\delta-}$X—H$^{\delta+}\cdots$:Y$^{\delta-}$.

Brief illustration 15.8 The hydrogen bond

In Exercise 15.22, you are invited to use the electrostatic model to calculate the dependence of the molar potential energy of interaction on the OOH angle, denoted θ in (**15**), and the results are plotted in Fig. 15.7. We see that at $\theta = 0$, when the OHO atoms lie in a straight line, the potential energy is -19 kJ mol^{-1}. Note how sharply the energy depends on angle: it is negative only within $\pm12°$ of linearity.

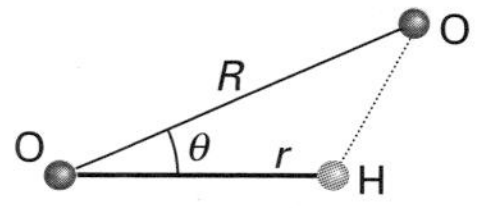

15

Molecular orbital theory provides an alternative description that is more in line with the concept of delocalized bonding and the ability of an electron pair to bind more than one pair of atoms (Section 14.13). Thus, if the X—H bond is regarded as formed from the overlap of an orbital on X, ψ_X, and a hydrogen 1s orbital, ψ_H, and the lone pair on Y occupies an orbital on Y, ψ_Y, then when the two molecules are

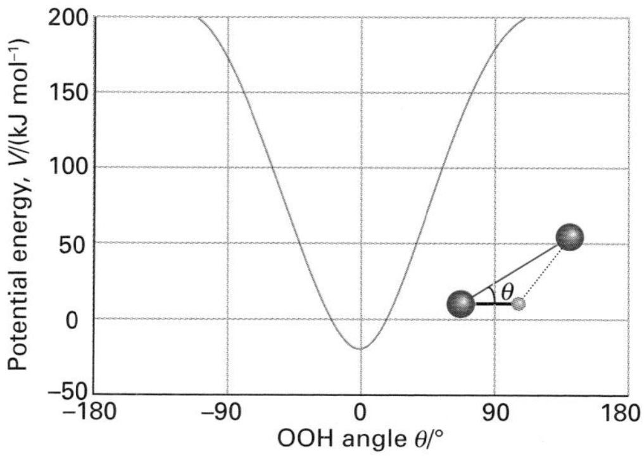

Fig. 15.7 The variation of the energy of interaction (on the electrostatic model) of a hydrogen bond as the angle between the O—H and :O groups is changed.

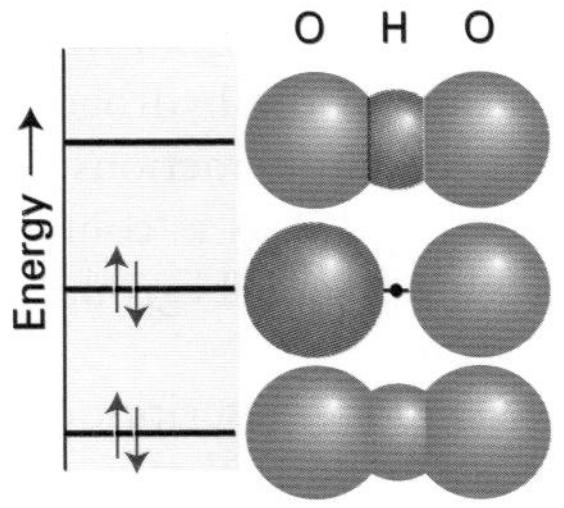

Fig. 15.8 A schematic portrayal of the molecular orbitals that can be formed from an X, H, and Y orbital and which gives rise to an X—H···Y hydrogen bond. The lowest energy combination is fully bonding, the next nonbonding, and the uppermost is antibonding. The antibonding orbital is not occupied by the electrons provided by the X—H bond and the :Y lone pair, so the configuration shown may result in a net lowering of energy in certain cases (namely when the X and Y atoms are N, O, or F).

close together, we can build three molecular orbitals from the three atomic orbitals:

$$\psi = c_1\psi_X + c_2\psi_H + c_3\psi_Y$$

One of the molecular orbitals is bonding, one almost nonbonding, and the third antibonding (Fig. 15.8). These three orbitals need to accommodate four electrons (two from the original X—H bond and two from the lone pair of Y), so two enter the bonding orbital and two enter the nonbonding orbital. Because the antibonding orbital remains empty, the net effect—depending on the precise location of the almost nonbonding orbital—may be a lowering of energy.

Experimental evidence and theoretical arguments have been presented in favour of both the electrostatic and molecular orbital models. Recent experiments suggest that the hydrogen bonds in ice have significant covalent character and are best described by a molecular orbital treatment. However, this interpretation of experimental results has been challenged by theoretical studies, which favour the electrostatic model. The matter has not yet been resolved.

Hydrogen bond formation, which has a typical strength of the order of 20 kJ mol^{-1}, dominates the van der Waals interactions when it can occur. It accounts for the rigidity of molecular solids such as sucrose and ice, the low vapour pressure, high viscosity, and surface tension of liquids such as water, the secondary structure of proteins (the formation of helices and sheets of polypeptide chains), the structure of DNA and hence the transmission of genetic information, and the attachment of drugs to receptor sites in proteins. Hydrogen bonding also contributes to the solubility in water of species such as ammonia and compounds containing hydroxyl groups and to the hydration of anions. In this last case, even ions such as Cl^- and HS^- can participate in hydrogen bond formation with water, for their charge enables them to interact with the hydroxylic protons of H_2O.

Table 15.3 summarizes the strengths and distance dependence of the attractive interactions that we have considered so far.

15.7 The hydrophobic effect

There is one further type of interaction that we need to consider: it is an *apparent* force that influences the shape of a macromolecule and which is mediated by the solvent, water. First, we need to understand why hydrocarbon molecules do not dissolve appreciably in water. Experiments indicate that the transfer of a hydrocarbon molecule from a nonpolar solvent into water is often exothermic ($\Delta H < 0$). Therefore, the fact that dissolving is not spontaneous must mean that entropy change is negative ($\Delta S < 0$). Substances characterized by a positive Gibbs energy of transfer

Table 15.3
Potential energy of molecular interactions

Interaction type	Distance dependence of potential energy	Typical energy/ (kJ mol^{-1})	Comment
Ion–ion	$1/r$	250	Only between ions
Ion–dipole	$1/r^2$	15	
Dipole–dipole	$1/r^3$	2	Between stationary polar molecules
	$1/r^6$	0.3	Between rotating polar molecules
London (dispersion)	$1/r^6$	2	Between all types of molecules and ions
Hydrogen bonding		20	The interaction is for X—H···Y and occurs on contact for X, Y = N, O, or F.

from a nonpolar to a polar solvent are classified as **hydrophobic**; if the Gibbs energy is negative, then the substance is **hydrophilic**. For example, the process CH_4(in CCl_4) → CH_4(aq) has $\Delta G = +12$ kJ mol^{-1}, $\Delta H = -10$ kJ mol^{-1}, and $\Delta S = -75$ J K^{-1} mol^{-1} at 298 K. Thus, CH_4 is characterized as hydrophobic.

Brief illustration 15.9 Hydrophobic and hydrophilic substances

When methylbenzene is transferred from water to aqueous solutions of sodium chloride, the Gibbs energy of transfer becomes more positive, suggesting that methylbenzene is less soluble in the salt solution because it has difficulty in breaking up the solvation of the Na^+ and Cl^- ions. The opposite effect is observed when toluene is transferred to aqueous solutions of guanidinium chloride, $(NH_2)_2C{=}NH_2^+Cl^-$: the Gibbs energy of transfer becomes more negative as the salt concentration is increased, indicating that methylbenzene is more soluble in the presence of this salt, presumably on account of favorable interactions between the two species.

The origin of the decrease in entropy that prevents hydrocarbons from dissolving in water is the formation of a solvent cage around the hydrophobic molecule (Fig. 15.9). The formation of this cage decreases the entropy of the system because the water molecules must adopt a less disordered arrangement than in the bulk liquid. However, when many solute molecules cluster together, fewer (though larger) cages are required and more solvent molecules are free to move. The net effect of formation of large clusters of hydrophobic molecules is then a decrease in the organization of the solvent and therefore a net increase in entropy of the system. This increase in entropy of the solvent is large enough to render spontaneous the association of hydrophobic molecules in a polar solvent.

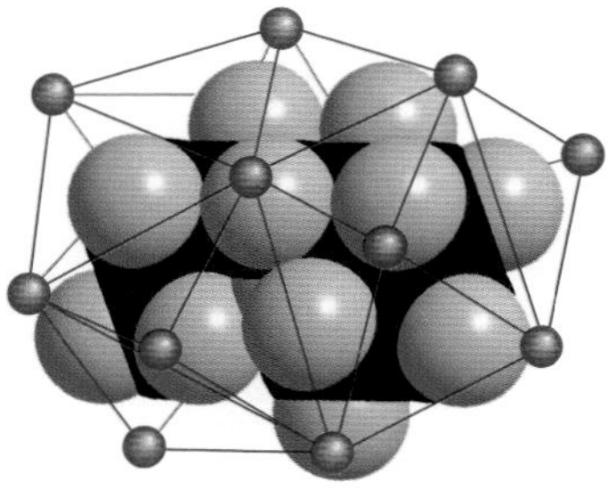

Fig. 15.9 When a hydrocarbon molecule is surrounded by water, the water molecules form a cage called a *clathrate*. As a result of this acquisition of structure, the entropy of the water decreases, so the dispersal of the hydrocarbon into water is entropy-opposed; the coalescence of the hydrocarbon into a single large blob is entropy-favoured.

The increase in entropy that results from the decrease in structural demands on the solvent is the origin of the **hydrophobic effect**, which tends to encourage the clustering together of hydrophobic groups in micelles and biopolymers. Thus, the presence of hydrophobic groups in polypeptides results in an increase in structure of the surrounding water and a decrease in entropy. The entropy can increase if the hydrophobic groups are twisted into the interior of the molecule, which liberates the water molecules and results in an increase in their disorder. The hydrophobic interaction is an example of an ordering process, a kind of virtual force, that is mediated by a tendency toward greater disorder of the solvent.

15.8 Modelling the total interaction

The total attractive interaction energy between rotating molecules that cannot participate in hydrogen bonding is the sum of the contributions from the dipole–dipole, dipole–induced dipole, and dispersion interactions. (In more advanced treatments, there may be other 'multipolar' contributions.[2]) Only the dispersion interaction contributes if both molecules are nonpolar. All three interactions vary as the inverse sixth power of the separation, so we may write the total attractive contribution to the van der Waals interaction energy as

$$V(r) = -\frac{C}{r^6} \qquad \text{van der Waals attraction} \qquad (15.13)$$

where C is a coefficient that depends on the identity of the molecules and the type of interaction between them.

Repulsive terms become important and begin to dominate the attractive forces when molecules are squeezed together (Fig. 15.10), for instance, during the impact of a collision, under the force exerted by a weight pressing on a substance, or simply as a result of the attractive forces drawing the molecules together. These repulsive interactions arise in large measure from the Pauli exclusion principle, which forbids pairs of electrons being in the same region of space. The repulsions increase steeply with decreasing separation in a way that can be deduced only by very extensive, complicated molecular structure calculations. In many cases, however, progress can be made by using a greatly simplified representation of

[2] See our *Physical Chemistry* (2010) for more information.

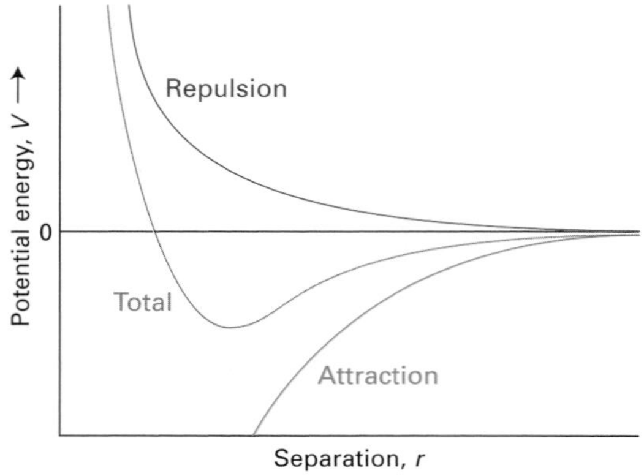

Fig. 15.10 The general form of an intermolecular potential energy curve (the graph of the potential energy of two closed-shell species as the distance between them is changed). The attractive (negative) contribution has a long range, but the repulsive (positive) interaction increases more sharply once the molecules come into contact. The overall potential energy is shown by the orange line.

the potential energy, where the details are ignored and the general features expressed by a few adjustable parameters.

One such approximation is the **hard-sphere potential energy**, in which it is assumed that the potential energy rises abruptly to infinity as soon as the particles come within a separation σ (Fig. 15.11):

$$V(r) = \begin{cases} \infty & \text{for } r \le \sigma \\ 0 & \text{for } r > \sigma \end{cases} \qquad \text{Hard-sphere potential energy} \qquad (15.14)$$

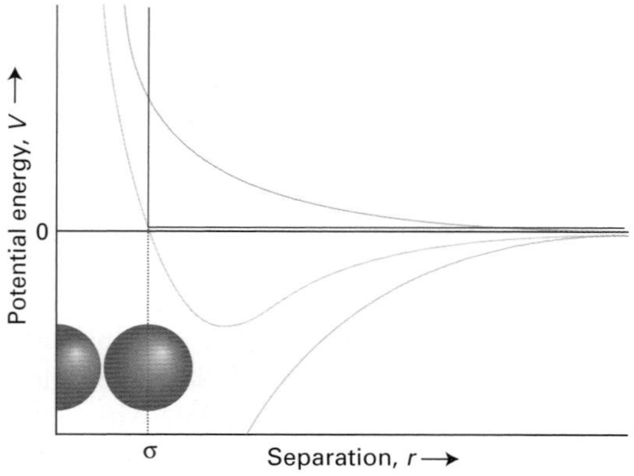

Fig. 15.11 The true intermolecular potential can be modelled in a variety of ways. One of the simplest is this *hard-sphere potential energy*, in which there is no potential energy of interaction until the two molecules are separated by a distance σ when the potential energy rises abruptly to infinity as the impenetrable hard spheres repel each other.

This very simple potential energy is surprisingly useful for assessing a number of properties.

Another widely used approximation is to express the short-range repulsive potential energy as inversely proportional to a high power of r:

$$V(r) = +\frac{C^*}{r^n} \qquad \text{Repulsive contribution} \qquad (15.15)$$

where C^* is another constant (the star signifies repulsion). Typically, n is set equal to 12, in which case the repulsion dominates the $1/r^6$ attractions strongly at short separations because then $C^*/r^{12} \gg C/r^6$. The sum of the repulsive interaction with $n = 12$ and the attractive interaction given by eqn 15.13 is called the **Lennard-Jones (12,6) potential energy**. It is normally written in the form

$$V(r) = 4\varepsilon\left\{\overbrace{\left(\frac{\sigma}{r}\right)^{12}}^{\text{Repulsion}} - \overbrace{\left(\frac{\sigma}{r}\right)^{6}}^{\text{Attraction}}\right\} \qquad \text{Lennard-Jones (12,6) potential energy} \qquad (15.16)$$

and is drawn in Fig. 15.12. The two parameters are now ε (epsilon), the depth of the well (don't confuse this with the electric permittivity), and σ, the separation at which $V = 0$. Some typical values are listed in Table 15.4. The well minimum occurs at $r = 2^{1/6}\sigma$.

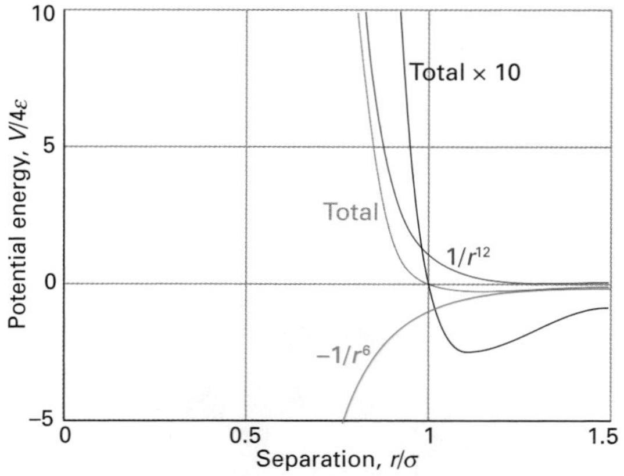

Fig. 15.12 The Lennard-Jones potential energy is another approximation to the true intermolecular potential energy curves. It models the attractive component by a contribution that is proportional to $1/r^6$, and the repulsive component by a contribution that is proportional to $1/r^{12}$. Specifically, these choices result in the Lennard-Jones (12,6) potential energy. Although there are good theoretical reasons for the former, there is plenty of evidence to show that $1/r^{12}$ is only a very poor approximation to the repulsive part of the curve.

Table 15.4
Lennard-Jones parameters for the (12,6) potential

	$\varepsilon/(\text{kJ mol}^{-1})$	σ/pm
Ar	128	342
Br_2	536	427
C_6H_6	454	527
Cl_2	368	412
H_2	34	297
He	11	258
Xe	236	406

Brief illustration 15.10 The Lennard-Jones potential energy

The Lennard-Jones potential energies for pairs of atoms of the noble gases argon and xenon as a function of separation are shown in Fig. 15.13. We see that the location of the minimum moves to greater separations on going down the Group in the periodic table, as would be expected for these increasingly large atoms. The depth of the minimum increases too, because the polarizabilities of the atoms increase with the number of electrons.

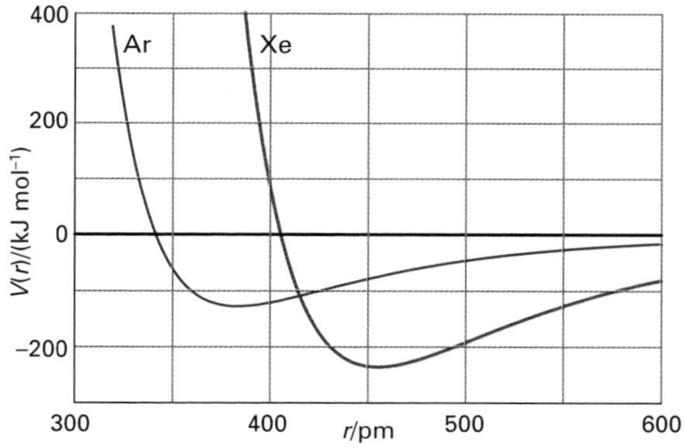

Fig. 15.13 The Lennard-Jones potential energy of pairs of noble gas atoms as a function of their separation.

Although the (12,6) potential energy has been used in many calculations, there is plenty of evidence to show that $1/r^{12}$ is a very poor representation of the repulsive potential energy, and that the exponential form $e^{-r/\sigma}$ is superior. An exponential function is more faithful to the exponential decay of atomic wavefunctions at large distances, and hence to the distance-dependence of the overlap that is responsible for repulsion. However, a disadvantage of the exponential form is that it is slower to compute, which is important when considering the interactions between the large numbers of atoms in liquids and macromolecules.

Impact on medicine 15.1

Molecular recognition and drug design

Molecular interactions are responsible for the assembly of many biological structures. Hydrogen bonding and hydrophobic interactions are primarily responsible for the three-dimensional structures of biopolymers, such as proteins, nucleic acids, and cell membranes. The binding of a ligand, or *guest*, to a biopolymer, or *host*, is also governed by molecular interactions. Examples of biological *host–guest complexes* include enzyme–substrate complexes, antigen–antibody complexes, and drug–receptor complexes. In all these cases, a site on the guest contains functional groups that can interact with complementary functional groups of the host. For example, a hydrogen bond donor group of the guest must be positioned near a hydrogen bond acceptor group of the host for tight binding to occur. It is generally true that many specific intermolecular contacts must be made in a biological host–guest complex and, as a result, a guest binds only chemically similar hosts. The strict rules governing molecular recognition of a guest by a host control every biological process, from metabolism to immunological response, and provide important clues for the design of effective drugs for the treatment of disease.

Interactions between nonpolar groups can be important in the binding of a guest to a host. For example, many active sites of enzymes have hydrophobic pockets that bind nonpolar groups of a substrate. In addition to dispersion, repulsive, and hydrophobic interactions, so-called *π-stacking interactions* are also possible, in which the planar π systems of aromatic macrocycles lie one on top of the other, in a nearly parallel orientation. Such interactions are responsible for the stacking of hydrogen-bonded base pairs in DNA, as shown in Fig. 15.14. Some drugs with planar π systems, shown as a rectangle in the illustration, are effective because they insert between base pairs through π-stacking interactions, causing the helix to unwind slightly and altering the function of DNA.

Coulombic interactions can be important in the interior of a biopolymer host, where the relative permittivity can be much lower than that of the aqueous exterior. For example, at physiological pH, amino acid side chains containing carboxylic acid or amine groups are negatively and positively charged, respectively, and can attract each other. Dipole–dipole interactions

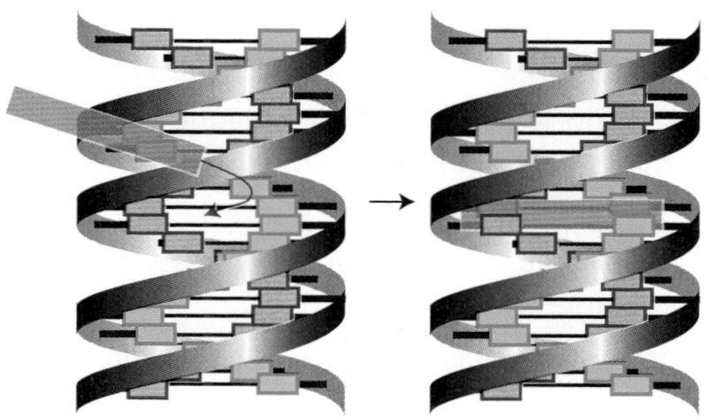

Fig. 15.14 Some drugs with planar π systems, shown by a blue rectangle, intercalate between the base pairs of DNA.

are also possible because many of the building blocks of biopolymers are polar, including the peptide link, —CONH— (Example 15.2). However, hydrogen bonding interactions are by far the most common in biological host–guest complexes. Many effective drugs on the market bind tightly and inhibit the action of enzymes that are associated with the progress of a disease. In many cases, a successful inhibitor will be able to form the same hydrogen bonds with the binding site that the normal substrate of the enzyme can form, except that the drug is chemically inert toward the enzyme. This strategy has been used in the design of drugs for the treatment of acquired immunodeficiency syndrome (AIDS), caused by the human immunodeficiency virus (HIV) (see Project 15.28).

Molecules in motion

The intermolecular interactions we have described govern a variety of properties, including the shapes that complicated molecules adopt and the motion of molecules in liquids. We deal with the structural aspects of these interactions in the next chapter. In this section, we consider how to take the interactions into account to describe molecular motion.

In a **molecular dynamics** simulation, the molecule is set in motion by heating it to a specified temperature and the possible trajectories of all atoms under the influence of the intermolecular forces are calculated from Newton's laws of motion. The equations of motion are solved numerically, allowing the molecules to adjust their locations and velocities in femtosecond steps (1 fs = 10^{-15} s). The calculation is repeated for tens of thousands of such steps.

The same technique can be used to examine the internal motion of macromolecules, such as the proteins considered in Chapter 16, and software packages are available that calculate the trajectories of a large number of atoms in three dimensions. The trajectories correspond to the conformations that the molecule can sample at the temperature of the simulation. At very low temperatures, the neighbouring components of the molecule are trapped in wells like that in Fig. 15.12, the atomic motion is restricted, and only a few conformations are possible. At high temperatures, more potential energy barriers can be overcome and more conformations are possible.

In the **Monte Carlo method**, the atoms of a macromolecule or the molecules of a liquid are moved through small but otherwise random distances, and the change in potential energy is calculated. If the potential energy is not greater than before the change, then the new arrangement is accepted. However, if the potential energy is greater than before the change, it is necessary to use a criterion for rejecting or accepting it. To establish this criterion, we use the Boltzmann distribution (Foundations 0.11), which states that at equilibrium at a temperature T the ratio of populations of two states that differ in energy by ΔE is $e^{-\Delta E/kT}$, where k is Boltzmann's constant. In the Monte Carlo method, the exponential factor is calculated for the new atomic arrangement and compared with a random number between 0 and 1; if the factor is larger than the random number, the new arrangement is accepted; if the factor is not larger, then the new arrangement is rejected and another one is generated instead.

Checklist of key concepts

☐ 1 van der Waals interactions between or within molecules are the attractive and repulsive nonbonding interactions.

☐ 2 A polar molecule is a molecule with a permanent electric dipole moment; the magnitude of a dipole moment is the product of the partial charge and the separation.

☐ 3 Dipole moments are approximately additive (as vectors).

☐ 4 The polarizability is a measure of the ability of an electric field to induce a dipole moment in a molecule.

☐ 5 A hydrogen bond is an interaction of the form X—H···Y, where X and Y are N, O, or F.

☐ 6 The hydrophobic interaction is an ordering process mediated by a tendency toward greater disorder of the solvent: it causes hydrophobic groups to cluster together.

☐ 7 The Lennard-Jones (12,6) potential is a model of the total intermolecular potential energy.

☐ 8 A molecular dynamics calculation uses Newton's laws of motion to calculate the motion of molecules in a fluid (and the motion of atoms in macromolecules).

☐ 9 A Monte Carlo simulation uses a selection criterion for accepting or rejecting a new arrangement of atoms or molecules.

Road map of key equations

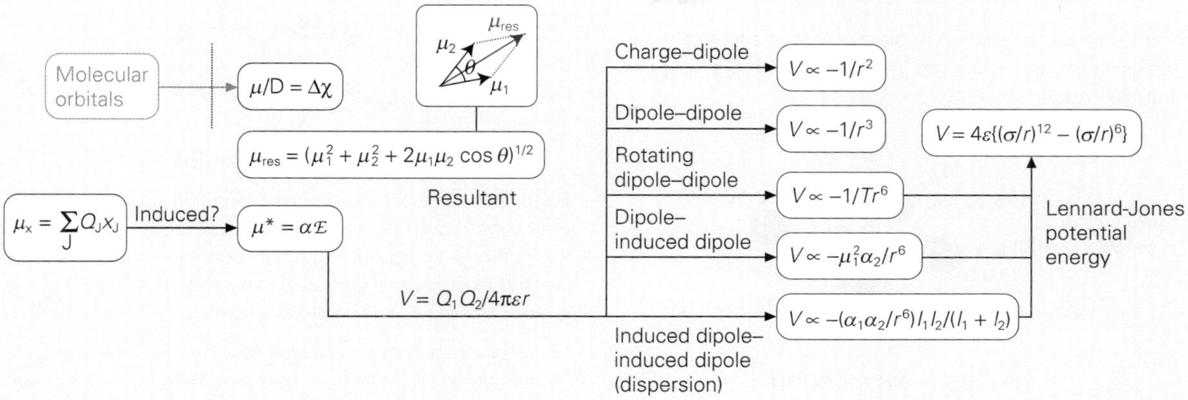

Questions and exercises

Discussion questions

15.1 Explain how the permanent dipole moment and the polarizability of a molecule arise and explain how they depend on the structure of the molecule.

15.2 Account for the theoretical conclusion that many attractive interactions between molecules vary with their separation as $1/r^6$.

15.3 Describe how van der Waals interactions depend on the structure of the molecules.

15.4 Explain why, for the noble gases, the values for the Lennard-Jones parameters ε and σ increase down Group 18.

15.5 Describe the formation of a hydrogen bond in terms of (a) electrostatic interactions and (b) molecular orbitals. How would you identify the better model?

15.6 Account for the hydrophobic effect and discuss its manifestations.

15.7 Outline the procedures used to calculate the motion of molecules in fluids and atoms in molecules.

Exercises

15.1 Calculate the molar potential energy of interaction between singly charged positive and negative ions that are separated by a distance of 50 nm in water.

15.2 Estimate the dipole moment of an HF molecule from the electronegativities of the elements and express the answer in debye and coulomb-metres.

15.3 Use the VSEPR model to judge whether PCl_5 is polar.

15.4 The electric dipole moment of toluene (methylbenzene) is 0.40 D. Estimate the dipole moments of the three xylenes (dimethylbenzenes). Which value can you be sure about?

15.5 Calculate the resultant of two dipoles of magnitude 1.20 D and 0.60 D that make an angle 107° to each other.

15.6 From the information in Exercise 15.4, estimate the dipole moments of (a) 1,2,3-trimethylbenzene, (b) 1,2,4-trimethylbenzene, and (c) 1,3,5-trimethylbenzene. Which value can you be sure about?

15.7 At low temperatures a substituted 1,2-dichloroethane molecule can adopt the three conformations (**16**), (**17**), and (**18**) with different probabilities. Suppose that the dipole moment of each bond is 1.50 D. Calculate the mean dipole moment of the molecule when (a) all three conformations are equally likely, (b) only conformation (**17**) occurs, (c) the three conformations occur with probabilities in the ratio 2:1:1 and (d) 1:2:2.

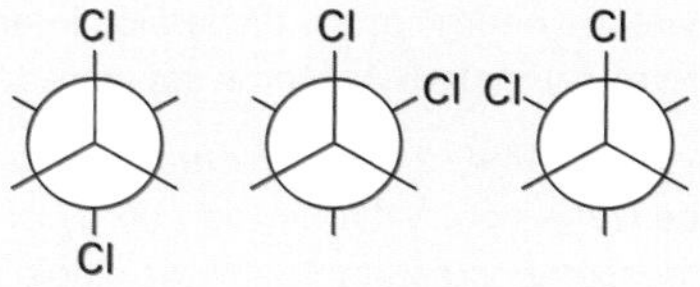

16 **17** **18**

15.8 Calculate the magnitude and direction of the dipole moment of the following arrangement of charges in the *xy*-plane: $3e$ at (0,0), $-e$ at (0.32 nm, 0), and $-2e$ at an angle of 20° from the *x*-axis and a distance of 0.23 nm from the origin.

15.9 Calculate the electric dipole moment of a glycine molecule using the partial charges in Table 15.1 and the locations of the atoms shown in (**19**).

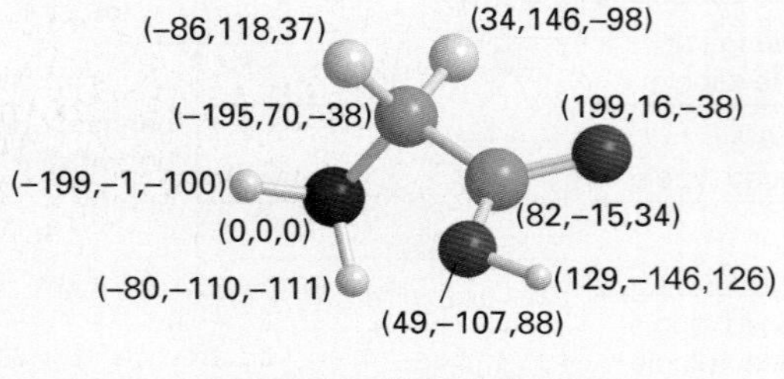

19 Glycine, NH_2CH_2COOH

15.10 (a) Plot the magnitude of the electric dipole moment of hydrogen peroxide as the H—OO—H (azimuthal) angle ϕ changes. Use the dimensions shown in (**20**). (b) Devise a way for depicting how the angle as well as the magnitude changes.

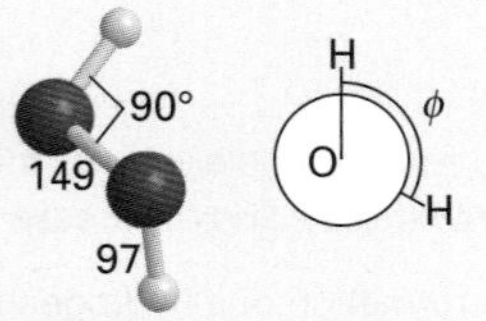

20 Hydrogen peroxide, H_2O_2

15.11 Calculate the molar energy required to reverse the direction of a water molecule located (a) 150 pm, (b) 350 pm from a Li^+ ion. Take the dipole moment of water as 1.85 D.

15.12 Show, by following the procedure in Derivation 15.1, that eqn 15.7 describes the potential energy of two electric dipole moments in the orientation shown in diagram (**13**) of the text.

15.13 What is the contribution to the total molar energy of (a) the kinetic energy, (b) the potential energy of interaction between hydrogen chloride molecules in a gas at 298 K when 0.50 mol of molecules is confined to 1.0 dm^3? Is the kinetic theory of gases justifiable in this case?

15.14 (a) What are the units of the polarizability α? (b) Show that the units of polarizability volume are cubic metres (m^3).

15.15 The magnitude of the electric field at a distance r from a point charge Q is equal to $Q/4\pi\varepsilon_0 r^2$. How close to a water molecule (of polarizability volume 1.48×10^{-30} m^3) must a proton approach before the dipole moment it induces is equal to the permanent dipole moment of the molecule (1.85 D)?

15.16 Estimate the energy of the dispersion interaction (use the London formula) for two Ar atoms separated by 1.0 nm.

15.17 Phenylanine (Phe, **21**) is a naturally occurring amino acid with a benzene ring. What is the energy of interaction between its benzene ring and the electric dipole moment of a neighbouring peptide group? Take the distance between the groups as 4.0 nm and treat the benzene ring as benzene itself. Take the dipole moment of the peptide group as 2.7 D.

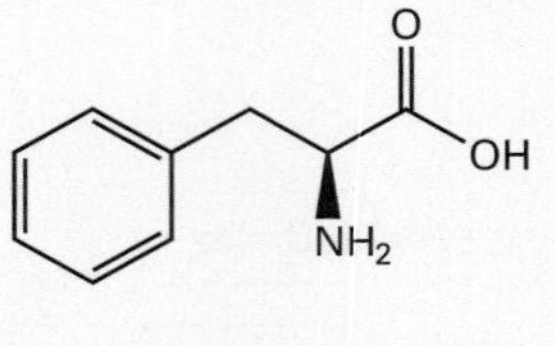

21 Phenylalanine

15.18 Now consider the London interaction between the benzene rings of two Phe residues (see Exercise 15.17). Estimate the potential energy of attraction between two such rings (treated as benzene molecules) separated by 4.0 nm. For the ionization energy, use $I = 5.0$ eV, where 1 eV $= 1.602 \times 10^{-19}$ J.

15.19 In a region of the oxygen-storage protein myoglobin, the OH group of a tyrosine residue is linked by a hydrogen bond to the N atom of a histidine residue in the geometry shown in (**22**). Use the partial charges in Table 15.1 to estimate the potential energy of this interaction.

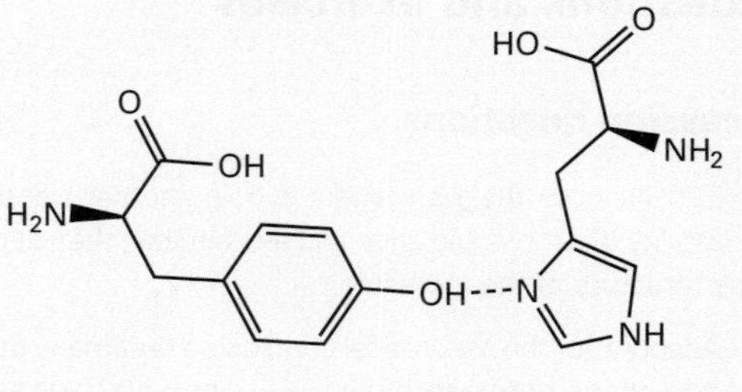

22

15.20 Acetic acid vapour contains a proportion of planar, hydrogen-bonded dimers (**23**). The apparent dipole moment of molecules in pure gaseous acetic acid increases with increasing temperature. Suggest an interpretation of the latter observation.

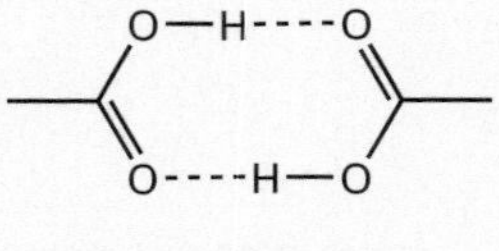

23

15.21 The coordinates of the atoms of an acetic acid dimer are set out in more detail in (**24**). Consider only the Coulombic interactions between the partial charges indicated by the dashed lines and their symmetry-related equivalents. At what distance R does the attraction become attractive?

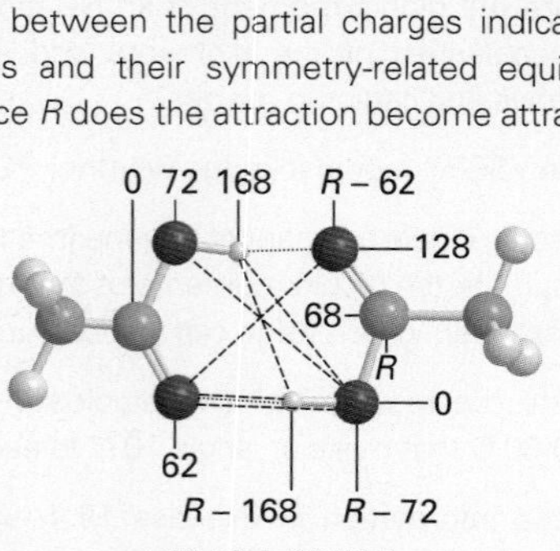

24 $(CH_3COOH)_2$

15.22 Consider the arrangement shown in (**15**) in the text for a system consisting of an O—H group and an O atom, and then use the electrostatic model of the hydrogen bond to calculate the dependence of the molar potential energy of interaction on the angle θ. Set the partial charges on H and O to $0.45e$ and $-0.83e$, respectively, and take $R = 200$ pm and $r = 95.7$ pm.

15.23 Using the parameters for the Lennard-Jones potential energy in Table 15.4, calculate the separation at which the potential energy of interaction between two bromine molecules is lowest.

15.24 The Lennard-Jones potential-energy function is sometimes expressed in the form $V(r) = A/r^{12} - B/r^6$. For tetrachloromethane, CCl_4, $A = 7.31 \times 10^{13}$ J pm^{12} and $B = 1.24 \times 10^{-3}$ J pm^6. What is the depth of the well and the separation at which the potential energy is lowest?

15.25 The potential energy of a CH_3 group in ethane as it is rotated around the C—C bond can be written $V = \frac{1}{2}V_0(1 + \cos 3\phi)$, where ϕ is the azimuthal angle (**25**) and $V_0 = 11.6$ kJ mol^{-1}. (a) What is the change in potential energy between the *trans* and fully eclipsed conformations? (b) Show that for small variations in angle, the torsional (twisting) motion around the C—C bond can be expected to be that of a harmonic oscillator. (d) Estimate the vibrational frequency of this torsional oscillation.

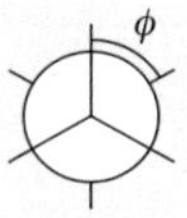

25

Projects

The symbol ‡ indicates that calculus is required.

15.26‡ Here we explore London interactions in more detail. Given that force is the negative slope of the potential calculate the distance dependence of the force acting between two non-bonded groups of atoms in a polymeric chain that have a London dispersion interaction with each other. (a) What is the separation at which the force is zero? (b) First calculate the slope by considering the potential energy at R and $R + \delta R$, with $\delta R \ll R$, and evaluating $\{V(R+\delta R) - V(R)\}/\delta R$. You should use the expansions $(1 + x)^{-1} \approx 1 - x + \cdots$, $(1 \pm x + \cdots)^6 = 1 \pm 6x + \cdots$ and $(1 \pm x + \cdots)^{12} = 1 \pm 12x + \cdots$. At the end of the calculation, let δR become vanishingly small. (c) Then repeat part (b) noting that $F(R) = -dV/dR$, and differentiating the expression for V.

15.27‡ Now we explore an alternative to the Lennard-Jones potential. (a) Suppose you distrusted the Lennard-Jones (12,6) potential for assessing a particular polymer conformation, and replaced the repulsive term by an exponential function of the form $e^{-r/\sigma}$. Sketch the form of the potential energy and locate the distance at which it is a minimum. (b) Use calculus to identify the distance at which the exponential-6 potential described in part (a) is a minimum.

15.28 For mature HIV particles to form in cells of the host organism, several large proteins coded for by the viral genetic material must be cleaved by a protease enzyme. The drug Crixivan (**26**) is a competitive inhibitor of HIV protease and has several molecular features that optimize binding to the active site of the enzyme. Consult the literature and prepare a brief report summarizing molecular interactions between Crixivan and HIV protease that are thought to be responsible for the drug's efficacy.

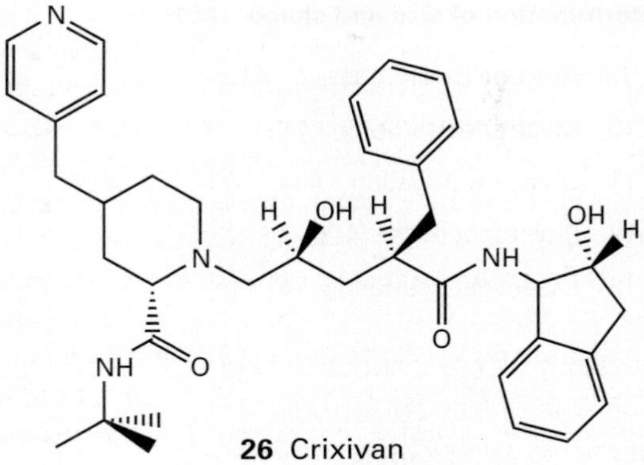

26 Crixivan

16

Macromolecules and aggregates

Biological and synthetic macromolecules 395

16.1 Models of structure 395

16.2 Mechanical properties of polymers 401

16.3 Electrical properties of polymers 403

Mesophases and disperse systems 403

16.4 Liquid crystals 403

16.5 Classification of disperse systems 404

16.6 Surface, structure, and stability 405

16.7 The electric double layer 408

16.8 Liquid surfaces and surfactants 409

Determination of size and shape 411

16.9 Average molar masses 412

16.10 Mass spectrometry 413

16.11 Ultracentrifugation 414

16.12 Electrophoresis 415

16.13 Laser light scattering 415

CHECKLIST OF KEY CONCEPTS 416

ROAD MAP OF KEY EQUATIONS 417

QUESTIONS AND EXERCISES 417

Macromolecules are very large molecules assembled from smaller molecules biosynthetically in organisms, by chemists in the laboratory, or in an industrial reactor. Naturally occurring macromolecules include polysaccharides such as cellulose, polypeptides such as protein enzymes, and polynucleotides such as deoxyribonucleic acid (DNA). Synthetic macromolecules include **polymers** such as nylon and polystyrene that are manufactured by stringing together and in some cases cross-linking smaller units known as **monomers** (Fig. 16.1).

Natural macromolecules differ in certain respects from synthetic macromolecules, particularly in their composition and the resulting structure, but the two share a number of common properties. We concentrate on these common properties in the first part of the chapter. In the second part we explore the grouping of small molecules into large particles in a process called 'self-assembly', which gives rise to aggregates. One example is the assembly of haemoglobin from four myoglobin-like polypeptides. A similar type of aggregation leads to a variety of **disperse phases**, which include colloids. The properties

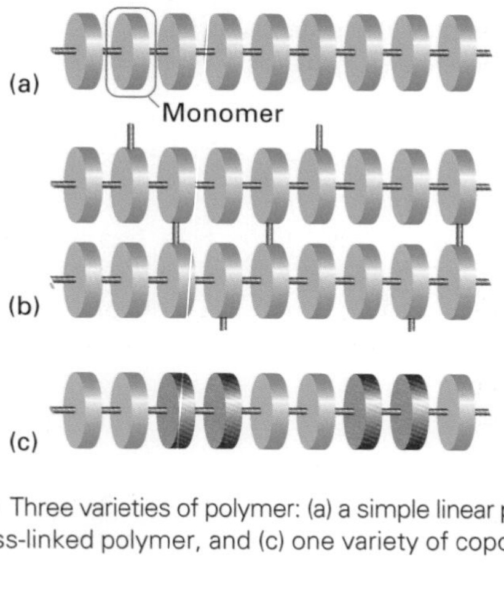

Fig. 16.1 Three varieties of polymer: (a) a simple linear polymer, (b) a cross-linked polymer, and (c) one variety of copolymer.

of these disperse phases resemble to a certain extent the properties of solutions of macromolecules, and we describe their common attributes. The third part of the chapter focusses on techniques for the determination of the size and shape of macromolecules and disperse phases.

Biological and synthetic macromolecules

Macromolecules provide an interesting and important illustration of how the interactions described in Chapter 15 determine the shape of a molecule and its properties. The overall shape of a polypeptide, for instance, is maintained by a variety of molecular interactions, including van der Waals interactions, hydrogen bonding, and the hydrophobic effect.

16.1 Models of structure

The concept of the 'structure' of a macromolecule takes on different meanings at the different levels at which we think about the arrangement of the chain or network of monomers. The **primary structure** of a macromolecule is the sequence of small molecular residues making up the polymer. The residues may form either a chain, as in polyethene, or a more complex network in which cross-links connect different chains, as in cross-linked polyacrylamide. In a synthetic polymer, virtually all the residues are identical and it is sufficient to name the monomer used in the synthesis. Thus, the repeating unit of polyethene and its derivatives is $-CHXCH_2-$, and the primary structure of the chain is specified by denoting it as $-(CHXCH_2)_n-$.

The concept of primary structure ceases to be trivial in the case of synthetic copolymers and biological macromolecules, for in general these substances are chains formed from different molecules. For example, proteins are **polypeptides** formed from different amino acids (about twenty occur naturally) strung together by the **peptide link**, $-CONH-$. The determination of the primary structure is then a highly complex problem of chemical analysis called **sequencing**. The **degradation** of a polymer is a disruption of its primary structure, when the chain breaks into shorter components.

The term **conformation** refers to the spatial arrangement of the different parts of a chain, and one conformation can be changed into another by rotating one part of a chain around a bond. The

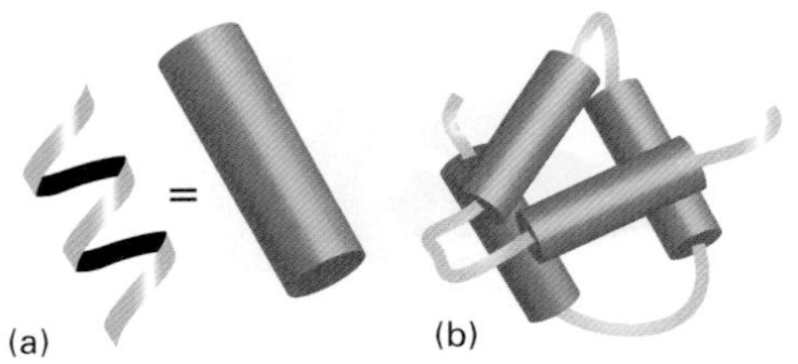

Fig. 16.2 (a) A polymer adopts a highly organized helical conformation, an example of a secondary structure. The helix is represented as a cylinder. (b) Several helical segments connected by short random coils pack together, providing an example of tertiary structure.

conformation of a macromolecule is manifested at three levels of structure. The **secondary structure** of a macromolecule is the (often local) spatial arrangement of a chain. The secondary structure of a molecule of polyethene in a good solvent is typically a random coil; in the absence of a solvent polyethene forms crystals consisting of stacked sheets with a hairpin-like bend about every 100 monomer units, presumably because for that number of monomers the intermolecular (in this case *intra*molecular) potential energy is sufficient to overcome thermal disordering. The secondary structure of a protein is a highly organized arrangement determined largely by hydrogen bonds, and taking the form of random coils, helices (Fig. 16.2a), or sheets in various segments of the molecule.

The **tertiary structure** is the overall three-dimensional structure of a macromolecule, so gives a higher level of conformation. For instance, the hypothetical protein shown in Fig. 16.2b has helical regions connected by short random-coil sections. The helices interact to form a compact tertiary structure. Denaturation may also occur at this level.

The **quaternary structure** of a macromolecule is the manner in which large molecules are formed by the aggregation of others. Figure 16.3 shows how four molecular subunits, each with a specific tertiary structure, aggregate together. Quaternary structure can be very important in biology. For example, the oxygen-transport protein haemoglobin consists of four subunits that work together to take up and release O_2.

(a) Random coils

The most likely secondary structure of a chain of identical units not capable of forming hydrogen bonds or any other type of specific bond is a **random coil**. Polyethene is a simple example. The random coil model is a helpful starting point for estimating the

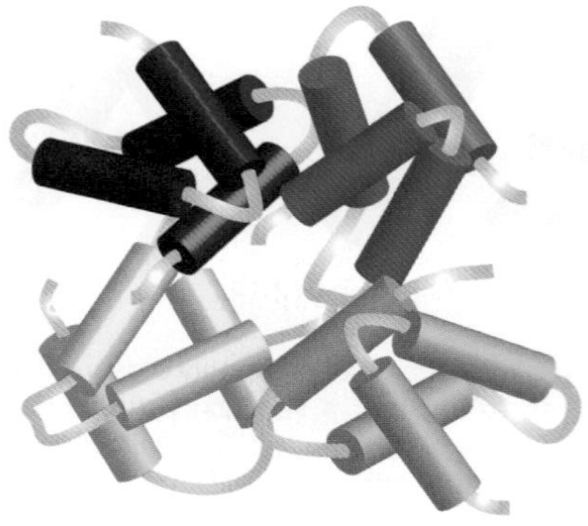

Fig. 16.3 Several subunits with specific tertiary structures pack together, providing an example of quaternary structure.

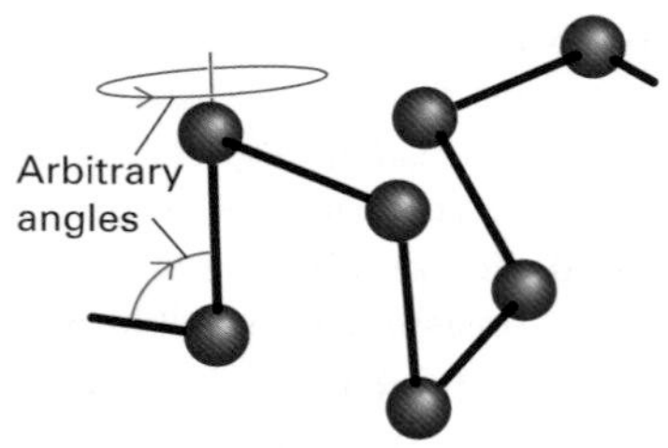

Fig. 16.4 A freely jointed chain is like a three-dimensional random walk, each step being in an arbitrary direction but of the same length.

orders of magnitude of the properties of polymers and denatured proteins in solution.[1]

The simplest model of a random coil is a **freely jointed chain**, in which any bond is free to make any angle with respect to the preceding one (Fig. 16.4). We assume that the residues occupy zero volume, so different parts of the chain can occupy the same region of space. The model is obviously an over-simplification because a bond is actually constrained to a cone of angles around a direction defined by its neighbour and has bulk. In a hypothetical one-dimensional freely jointed chain all the residues lie in a straight line, and the angle between neighbours is either 0° or 180°. The residues in a three-dimensional freely jointed chain are not restricted to lie in a line or a plane.

The **contour length**, R_c, of a polymer is the length of the molecule measured along its backbone from monomer to monomer:

$$R_c = Nl \quad \text{Random coil} \quad \text{Contour length} \qquad (16.1a)$$

The contour length is proportional to the number of monomers, N, in the polymer and the length l occupied by each monomer unit. The radius of the random coil such a molecule forms, however, is proportional only to the square root of N because the coil consists of steps (neighbouring bonds) that might double back on themselves as the chain grows. Specifically, the **root-mean-square separation,** R_{rms}, the square root of the mean value of the square of the end-to-end separation, is a measure of the average separation of the two ends of a random coil:

$$R_{rms} = N^{1/2}l \quad \text{Random coil} \quad \text{Root-mean-square separation} \qquad (16.1b)$$

Consequently, the volume of the coil increases as $N^{3/2}$. The **radius of gyration,** R_g, of a random coil is the radius of a thin shell (think of a table-tennis ball) that has the same mass as the molecule and the same moment of inertia. The radius of gyration of a table-tennis ball is the same as its actual radius; that of a solid sphere of radius r is $R_g = (2/5)^{1/2}r$. For a one-dimensional and three-dimensional random coil, respectively,

$$R_g = N^{1/2}l \quad \text{One-dimensional random coil} \quad \text{Radius of gyration} \qquad (16.1c)$$

$$R_g = \left(\frac{N}{6}\right)^{1/2} l \quad \text{Three-dimensional random coil} \quad \text{Radius of gyration} \qquad (16.1d)$$

The term 'coil' is used even in the one-dimensional case, when it simply means a linear structure with the bonds folded back on themselves.

Brief illustration 16.1 Measures of size of a random coil

Consider a polyethene chain with $M = 112$ kg mol^{-1}, corresponding to $N = 4000$. Because $l = 154$ pm for a C—C bond, we find (by using 10^3 pm = 1 nm)

From eqn 16.1a: $R_c = 4000 \times 154$ pm = 616 nm

From eqn 16.1b: $R_{rms} = (4000)^{1/2} \times 154$ pm = 9.74 nm

From eqn 16.1d: $R_g = \left(\frac{4000}{6}\right)^{1/2} \times 154$ pm = 3.98 nm

Self-test 16.1

The contour length of a different sample of polyethene is 17 μm. Assuming a random-coil model, calculate the root-mean-square separation and radius of gyration. What is the molar mass?

Answer: 51 nm, 21 nm, 3.1×10^3 kg mol^{-1}

The random-coil model ignores the role of the solvent: a poor solvent will tend to cause the coil to tighten so that solute–solvent contacts are minimized; a good solvent does the opposite. Therefore,

[1] For the derivation of the expressions in this section, see our *Physical Chemistry* (2010).

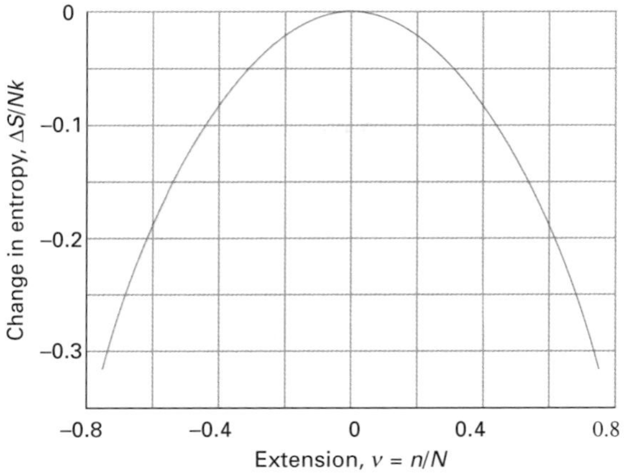

Fig. 16.5 The change in molar entropy of a one-dimensional perfect elastomer as its extension changes; $v = 1$ corresponds to complete extension; $v = 0$, the conformation of highest entropy, corresponds to the random coil.

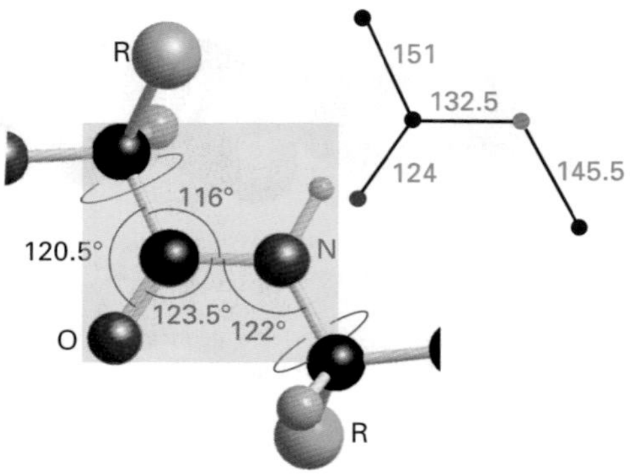

Fig. 16.6 The dimensions that characterize the peptide link. The C—NH—CO—C atoms define a plane (the C—N bond has partial double-bond character), but there is rotational freedom around the C—CO and N—C bonds.

calculations based on this model are better regarded as lower bounds to the dimensions for a polymer in a good solvent and as an upper bound for a polymer in a poor solvent. The model is most reliable for a polymer in a bulk solid sample, where the coil is likely to have its natural dimensions.

A random coil is the least structured conformation of a polymer chain in the sense that it can be achieved in the greatest possible number of ways (in contrast, for instance, to the straight chain conformation, which can be achieved in only one way) and corresponds to the state of greatest entropy. Any stretching of the coil introduces order and reduces the entropy. Conversely, the formation of a random coil from a more extended form is a spontaneous process (provided enthalpy contributions do not interfere). The change in **conformational entropy**, the entropy arising from the arrangement of bonds, when a one-dimensional coil containing N bonds of length l is stretched or compressed by nl is

$$\Delta S = -\tfrac{1}{2}kN \ln\{(1+v)^{1+v}(1-v)^{1-v}\} \quad v = n/N$$

Random coil Conformational entropy (16.2)

where k is Boltzmann's constant. This function is plotted in Fig. 16.5, and we see that minimum extension—fully coiled ($n = 0$)—corresponds to maximum entropy. This spontaneous tendency to form a coil is responsible for the tendency of rubber (or at least, an ideal rubber with no intermolecular interactions) to spring back into shape after being stretched.

(b) Polypeptides and polynucleotides

Polypeptides are almost at the opposite end of the scale of structure from random coils, for they can become highly ordered: they need to be, for in biology structure is almost synonymous with function. We can rationalize the secondary structures of proteins in large part in terms of the hydrogen bonds between the —NH— and —CO— groups of the peptide links (Fig. 16.6). These bonds lead to two principal structures. One, which is stabilized by hydrogen bonding between peptide links of the same chain, is the **α-helix**. The other, which is stabilized by hydrogen bonding links to different chains or more distant parts of the same chain, is the **β-sheet** (or *β-pleated sheet*).

The α-helix is illustrated in Fig. 16.7. Each turn of the helix contains 3.6 amino acid residues, so there are 18 residues in 5 turns of the helix. The pitch of a single turn (the lateral movement corresponding to one complete rotation) is 544 pm. The N—H···O bonds lie parallel to the axis and link every fifth group (so residue i is linked to residues $i - 4$ and $i + 4$). There is freedom for the helix to be arranged as either a right- or a left-handed screw, but the overwhelming majority of natural polypeptides are right-handed on account of the preponderance of the L-configuration of the naturally occurring amino acids. It turns out, in agreement with experience, that a right-handed α-helix of L-amino acids has a marginally lower energy than a left-handed helix of the same acids. A β-sheet is formed by hydrogen bonding between two extended polypeptide chains. Some of the side chains lie above the sheet and some lie below it. Two types of structures can be distinguished from the pattern of hydrogen bonding between the constituent chains: (a) in an antiparallel β-sheet (Fig. 16.8), the N—H···O atoms of the hydrogen bonds form a straight line; (b) in a parallel β-sheet

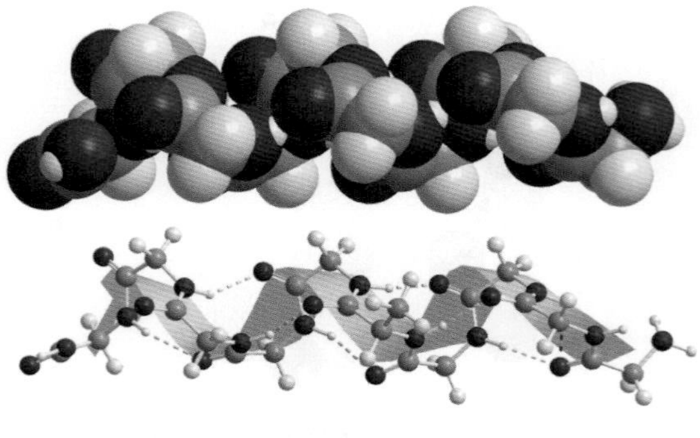

Fig. 16.7 The polypeptide α helix, with poly-L-glycine as an example. There are 3.6 residues per turn, and a translation along the helix of 150 pm per residue, giving a pitch of 540 pm. The diameter (ignoring side chains) is about 600 pm.

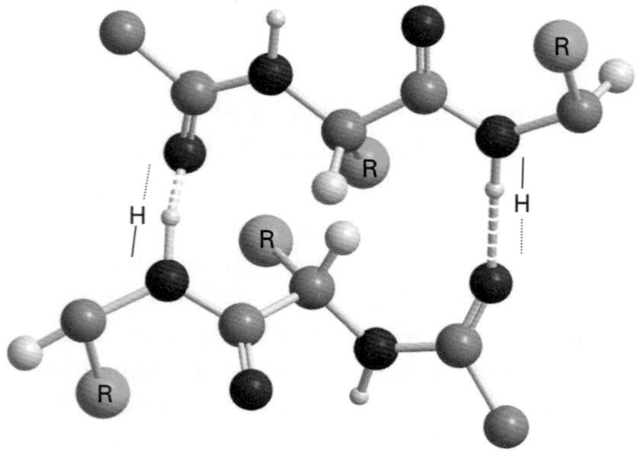

Fig. 16.8 An antiparallel β sheet in which the N—H—O atoms of the hydrogen bonds form a nearly straight line.

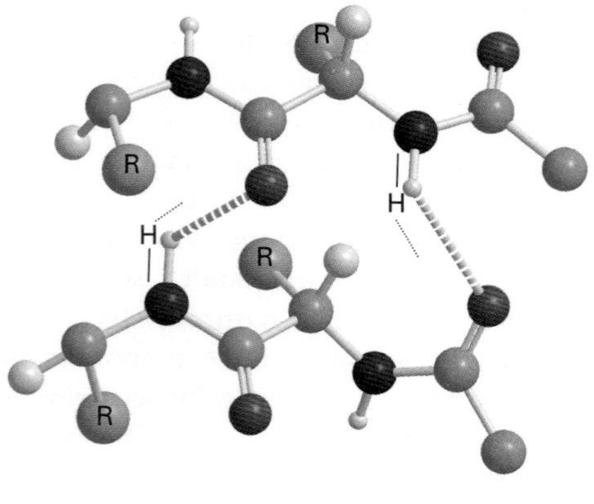

Fig. 16.9 A parallel β sheet in which the N—H—O atoms of the hydrogen bonds are not as well aligned as in the antiparallel version.

(Fig. 16.9), the N—H···O atoms of the hydrogen bonds are not perfectly aligned.

Helical and sheet-like polypeptide chains are folded into a tertiary structure if there are other bonding influences between the residues of the chain

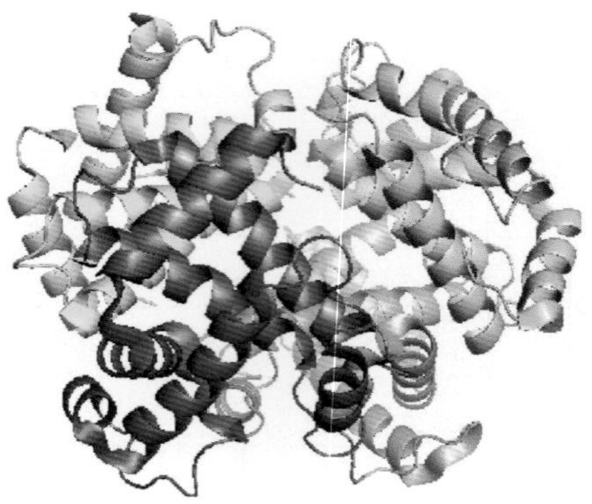

Fig. 16.10 A haemoglobin molecule consists of four myoglobin-like units. An O_2 molecule attaches to the iron atom in the haem group.

that are strong enough to overcome the interactions responsible for the secondary structure. The folding influences include —S—S— **disulfide links**, van der Waals interactions, hydrophobic interactions, ionic interactions (which depend on the pH), and strong hydrogen bonds (such as O—H···O).

Proteins with $M > 50$ kg mol^{-1} are often found to be aggregates of two or more polypeptide chains. The possibility of such quaternary structure often confuses the determination of their molar masses because different techniques might give values differing by factors of 2 or more. Haemoglobin, which consists of four myoglobin-like chains (Fig. 16.10), is an example of a quaternary structure. Myoglobin is an oxygen-storage protein. The subtle differences that arise when four such molecules coalesce to form haemoglobin result in the latter being an oxygen transport protein, able to load and unload O_2 cooperatively (see Impact 7.2).

Deoxyribonucleic acid (DNA) and ribonucleic acid (RNA), which are key components of the mechanism of storage and transfer of genetic information in biological cells, are **polynucleotides**. The backbones of these molecules consist of alternating sugar and phosphate groups, and one of the bases adenine (A), cytosine (C), guanine (G), and thymine (T, found in DNA only), and uracil (U, found in RNA only) is attached to each sugar. In B-DNA, the most common form of DNA in biological cells, two polynucleotide chains held together by A—T and C—G base pairs (**1** and **2**) wind around each other to form a right-handed double helix (Fig. 16.11). The structure is stabilized further by the π-stacking interactions mentioned in Impact 15.1. In contrast, RNA exists primarily as single chains that can fold into complex structures by formation of A—U and G—C base pairs.

Fig. 16.11 DNA double helix, in which two polynucleotide chains are linked together by hydrogen bonds between adenine (A) and thymine (T) and between cytosine (C) and guanine (G).

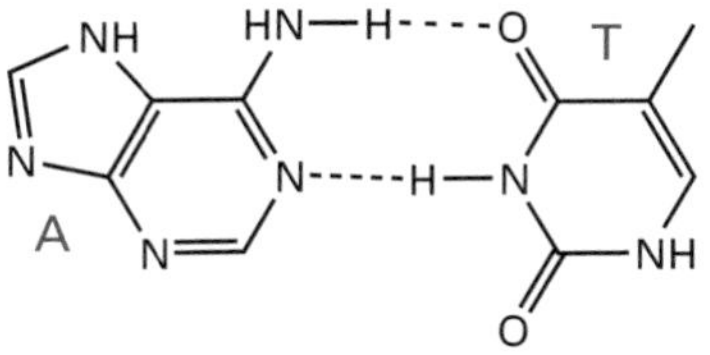

1 A–T base pair

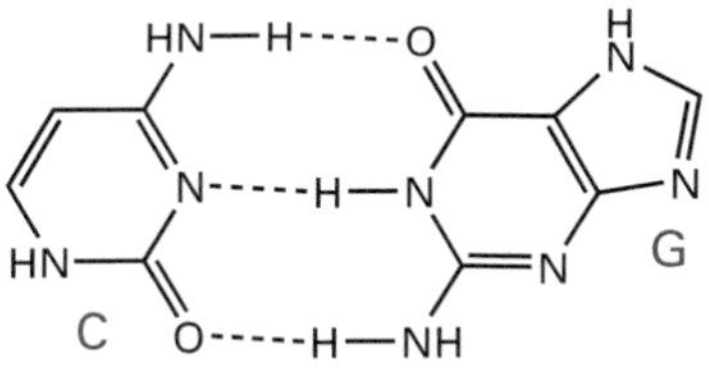

2 C–G base pair

Biopolymer **denaturation**, or loss of structure, can be caused by several means, and different aspects of structure may be affected. Denaturation at the secondary level is brought about by agents that destroy hydrogen bonds. Thermal motion may be sufficient, in which case denaturation is a kind of intramolecular melting. When eggs are cooked the albumin is denatured irreversibly, and the protein collapses into a structure resembling a random coil. The **helix–coil transition** of polypeptides is sharp, like ordinary melting, because it is a cooperative process in the sense that when one hydrogen bond has been broken it is easier to break its neighbours, and then even easier to break theirs, and so on. The disruption cascades down the helix, and the transition occurs sharply. Denaturation may also be brought about chemically. For instance, a solvent that forms stronger hydrogen bonds than those within the helix will compete successfully for the NH and CO groups. Acids and bases can cause denaturation by protonation or deprotonation of various groups.

In contemporary physical chemistry and molecular biophysics, a great deal of work is being done on the rationalization and prediction of the structures of biomolecules such as the polypeptides and nucleic acids described here, using the interactions described in Chapter 15 (Impact 16.1).

Impact on biochemistry 16.1

The prediction of protein structure

A polypeptide chain adopts a conformation corresponding to a minimum Gibbs energy, which depends on the **conformational energy**, the energy of interaction between different parts of the chain, and the energy of interaction between the chain and surrounding solvent molecules. In the aqueous environment of biological cells, the outer surface of a protein molecule is covered by a mobile sheath of water molecules, and its interior contains pockets of water molecules. These water molecules play an important role in determining the conformation that the chain adopts through hydrophobic interactions and hydrogen bonding to amino acids in the chain.

The simplest calculations of the conformational energy of a polypeptide chain ignore entropy and solvent effects and concentrate on the total potential energy of all the interactions between non-bonded atoms. For example, as remarked in the text, these calculations predict that a right-handed α-helix of L-amino acids is marginally more stable than a left-handed helix of the same amino acids.

To calculate the energy of a conformation, we need to make use of many of the molecular interactions described in Chapter 15, and also of some additional interactions:

1. *Bond stretching.* Bonds are not rigid, and it may be advantageous for some bonds to stretch and others to be compressed slightly as parts of the chain press against one another. If we liken the bond to a spring, then the potential energy takes the form (see Section 12.9):

$$V_{\text{stretch}}(R) = \tfrac{1}{2}k_{\text{f,stretch}}(R - R_e)^2$$

where R_e is the equilibrium bond length and $k_{\text{f,stretch}}$ is the force constant, a measure of the stiffness of the bond in question.

2. *Bond bending.* An O—C—H bond angle (or some other angle) may open out or close in slightly to enable the molecule as a whole to fit together better. If the equilibrium bond angle is θ_e, we write

$$V_{\text{bend}}(\theta) = \tfrac{1}{2}k_{\text{f,bend}}(\theta - \theta_e)^2$$

where $k_{\text{f,bend}}$ is the force constant, a measure of how difficult it is to change the bond angle.

Brief illustration 16.2 The energy of bond angle bending

Theoretical studies have estimated that the lumiflavin isoalloazine ring system (**3**) has an energy minimum at the bending angle of 15°, but that it requires only 1.41×10^{-20} J or 8.50 kJ mol^{-1} to increase the angle

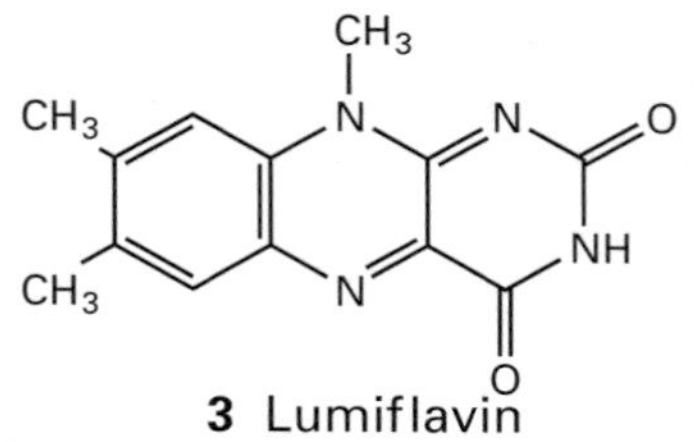

3 Lumiflavin

to 30°. The force constant for lumiflavin bending is therefore

$$k_{f,\text{bend}} = \frac{2V_{\text{bend}}(\theta)}{(\theta-\theta_e)^2} = \frac{2\times\overbrace{(1.41\times10^{-20}\text{ J})}^{V_{\text{bend}}(30°)}}{(30°-15°)^2}$$

$$= 1.3\times10^{-22}\text{ J deg}^{-2}$$

corresponding to 75 J deg^{-2} mol^{-1}.

Self-test 16.2

It takes 0.90 aJ to bend a C—C—C bond by 2.0° from its equilibrium bond angle. What is the force constant for this bending motion?

Answer: 4.5×10^{-19} J deg^{-2}, 270 kJ deg^{-2} mol^{-1}

3. *Bond torsion.* There is a barrier to internal rotation of one bond relative to another (just like the barrier to internal rotation in ethane). Because the planar peptide link is fairly rigid, the geometry of a polypeptide chain can be specified by the two angles that two neighbouring planar peptide links make to each other. Figure 16.12 shows the two angles ϕ and ψ commonly used to specify this relative orientation. The sign convention is that a positive angle means that the front atom must be rotated clockwise to bring it in front of the rear atom. Specifically:

   Right-handed α-helix, all $\phi = -57°$ and all $\psi = -47°$

   Left-handed α helix, all $\phi = 57°$ and all $\psi = 47°$

   Antiparallel β-sheet, $\phi = -139°$, $\psi = 113°$.

   The torsional contribution to the total potential energy is

   $$V_{\text{torsion}}(\phi,\psi) = A(1+\cos 3\phi) + B(1+\cos 3\psi)$$

   in which A and B are constants of the order of 1 kJ mol^{-1}. Because only two angles are needed to specify the conformation of a helix, and they range from −180° to +180°, the torsional potential energy of the entire molecule can be represented on a **Ramachandran plot**, a contour diagram in which one axis represents ϕ and the other represents ψ (Fig. 16.13).

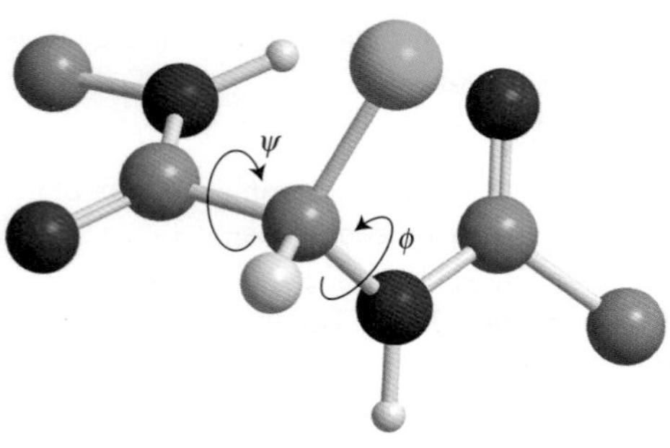

Fig. 16.12 The definition of the torsional angles ψ and ϕ between two peptide units.

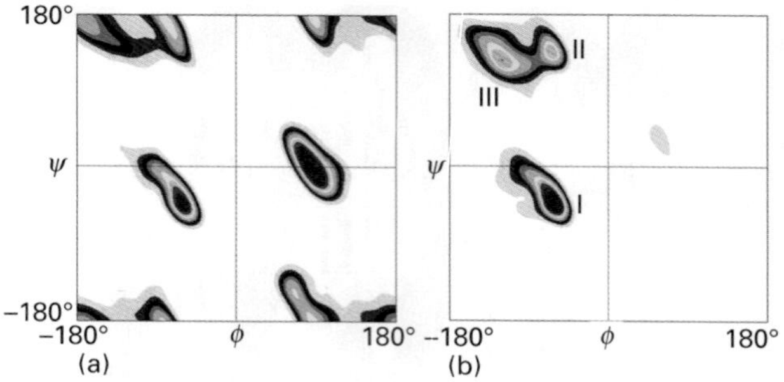

Fig. 16.13 Contour plots of potential energy against the angles ψ and ϕ, also known as a Ramachandran diagram, for (a) a glycyl residue of a polypeptide chain and (b) an alanyl residue. (Hovmoller, et al., *Acta Cryst.* **D58**, 768 (2002).)

4. *Interaction between partial charges.* If the partial charges Q_i and Q_j on the atoms i and j are known, a Coulombic contribution of the form $1/r$ can be included:

   $$V_{\text{Coulomb}}(r) = \frac{Q_iQ_j}{4\pi\varepsilon r}$$

   where ε is the permittivity of the medium in which the charges are embedded. Partial charges of $-0.28e$ and $+0.28e$ are assigned to N and H, respectively, and $-0.39e$ and $+0.39e$ to O and C, respectively. The interaction between partial charges does away with the need to take dipole–dipole interactions into account, for they are taken care of by dealing with each partial charge explicitly.

5. *Dispersive and repulsive interactions.* The interaction energy of two atoms separated by a distance r (which we know once ϕ and ψ are specified) can be given by the Lennard-Jones (12,6) form (Section 15.8):

   $$V_{\text{LJ}}(r) = \frac{C^*}{r^{12}} - \frac{C}{r^6}$$

6. *Hydrogen bonding.* In some models of structure, the interaction between partial charges is judged to take into account the effect of hydrogen bonding. In other models, hydrogen bonding is added as another interaction of the form

   $$V_{\text{H-bonding}}(r) = \frac{D^*}{r^{12}} - \frac{D}{r^{10}}$$

The total potential energy of a given conformation (ϕ,ψ) is calculated by summing the contributions given by the preceding equations for all bond angles (including torsional angles) and pairs of atoms in the molecule. The procedure is known as a **molecular mechanics** simulation and is automated in commercially available molecular modelling software. For large molecules, plots of potential energy versus bond distance or bond angle often show several local minima and a global minimum (Fig. 16.14). The software packages

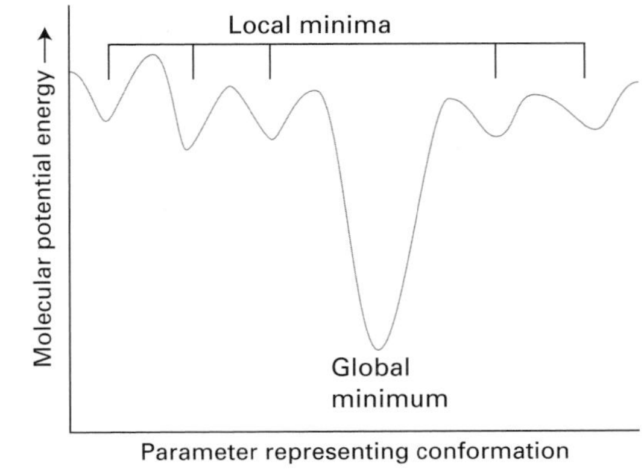

Fig. 16.14 For large molecules, a plot of potential energy against the molecular geometry often shows several local minima and a global minimum.

include schemes for modifying the locations of the atoms and searching for these minima systematically.

Figure 16.13 shows the potential-energy contours for the helical form of polypeptide chains formed from the non-chiral amino acid glycine (R = H) and the chiral amino acid L-alanine (R = CH_3). The contours have been computed by summing all the contributions described above for each choice of angles, and then plotting contours of equal potential energy. The glycine map is symmetrical, with global minima of equal depth at $\phi = -80°$, $\psi = -60°$ and at $\phi = +80°$, $\psi = 0°$. In contrast, the map for L-alanine is unsymmetrical, and there are three distinct low-energy conformations (marked I, II, III). The minima of regions I and II lie close to the angles typical of right- and left-handed α helices, but the former has a lower minimum, which is consistent with the formation of right-handed helices from the naturally occurring L-amino acids.

The structure corresponding to the global minimum of a molecular mechanics simulation is a snapshot of the molecule at $T = 0$ because only the potential energy is included in the calculation; contributions to the total energy from kinetic energy are excluded. In a **molecular dynamics** simulation, the molecule is set in motion by heating it to a specified temperature (Chapter 15). The possible trajectories of all atoms under the influence of the intermolecular potentials are then calculated by integration of Newton's second law of motion. These trajectories correspond to the conformations that the molecule can sample at the temperature of the simulation. Therefore, molecular dynamics calculations are useful tools for the study of the flexibility of polymers.

16.2 Mechanical properties of polymers

Synthetic polymers are classified broadly as *elastomers*, *fibres*, and *plastics*, depending on their **crystallinity**, the degree of three-dimensional long-range order in the solid state.

An **elastomer** is a flexible polymer that can expand or contract easily upon application of an external force. Elastomers are polymers with numerous crosslinks that pull them back into their original shape when a stress is removed. The weak directional constraint on silicon–oxygen bonds is responsible for the high elasticity of silicones. A **perfect elastomer,** a polymer in which the internal energy is independent of the extension of the random coil, can be modelled as a freely jointed chain.

We saw in Section 16.1 that the contraction of an extended chain to a random coil is spontaneous in the sense that it corresponds to an increase in entropy; the entropy change of the surroundings is zero because no energy is released or absorbed when the coil forms. The conformational entropy can be used to deduce that the restoring force, $\mathcal{F}$, of a one-dimensional perfect elastomer at a temperature T is[2]

$$\mathcal{F} = \frac{kT}{2l}\ln\left(\frac{1+\nu}{1-\nu}\right) \quad \text{One-dimensional random coil} \quad \text{Restoring force} \quad (16.3a)$$

$$\nu = n/N$$

where N is the total number of bonds of length l and the polymer is stretched or compressed by nl (k is Boltzmann's constant). This function is plotted in Fig. 16.15. At low extensions, $\nu \ll 1$ and we can use the techniques of approximation in The chemist's toolkit 6.1:

$$\mathcal{F} \overbrace{=}^{\ln(x/y)=\ln x-\ln y} \frac{kT}{2l}\{\ln(1+\nu)-\ln(1-\nu)\}$$

$$\overbrace{=}^{\ln(1+x)=x-\frac{1}{2}x^2+\cdots,\ \ln(1-x)=-x-\frac{1}{2}x^2+\cdots} \frac{kT}{2l}\{(\nu-\tfrac{1}{2}\nu^2+\cdots)-(-\nu-\tfrac{1}{2}\nu^2+\cdots)\}$$

$$\overbrace{=}^{\text{cancel terms}} \frac{kT}{2l}\{2\nu+\cdots\} \approx \frac{\nu kT}{l}$$

That is,

$$\mathcal{F} \approx \frac{\overbrace{\nu}^{n/N} kT}{l} = \frac{nkT}{Nl} \quad \text{Approximate form} \quad \text{Restoring force} \quad (16.3b)$$

and the restoring force is proportional to the displacement (which is proportional to n). Therefore, the sample obeys Hooke's law (in the notation of Chapter 12, $\mathcal{F} = -k_f x$; eqn 12.24a) and for small displacements the whole coil shakes with simple harmonic motion.

[2] For the derivation of this expression and its small-extension form, see our *Physical Chemistry* (2010).

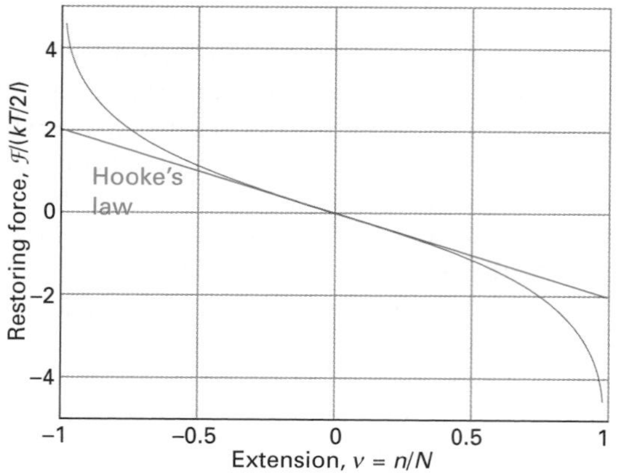

Fig. 16.15 The restoring force, $\mathcal{F}$, of a one-dimensional perfect elastomer. For small strains, $\mathcal{F}$ is linearly proportional to the extension, corresponding to Hooke's law.

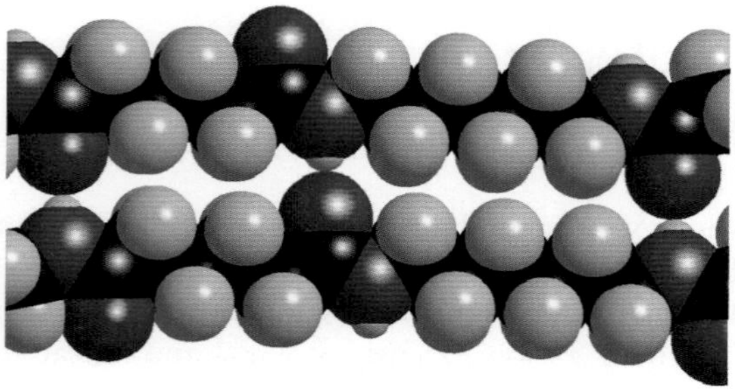

Fig. 16.16 A fragment of two nylon-66 polymer chains showing the pattern of hydrogen bonds that are responsible for the cohesion between the chains.

A **fibre** is a polymeric material with such a low degree of branching that the molecules can be made to lie parallel to one another and acquire strength from the interactions between them. One example is nylon-66 (Fig. 16.16). In contrast to elastomers, fibres need to have a resistance to stretching, which requires the chains to be nearly fully extended and for there to be strong interactions between them. Hydrogen bonding between chains, as in nylon, is one way to achieve this resistance, and side chains are undesirable as they hinder the formation of ordered microcrystalline regions. Under certain conditions, nylon-66 can be prepared in a state of high crystallinity, in which hydrogen bonding between the peptide links of neighbouring chains results in an ordered array.

A **plastic** is a polymer that can attain only a limited degree of crystallinity and as a result is neither as strong as a fibre nor as resilient as an elastomer. Certain materials, such as nylon-66, can be prepared either as a fibre or as a plastic. A sample of plastic nylon-66 may be visualized as consisting of crystalline hydrogen-bonded regions of varying size interspersed amongst amorphous, random coil regions. A single type of polymer may exhibit more than one characteristic, for to display fibrous character, the polymers need to be aligned; if the chains are not aligned, then the substance may be plastic. That is the case with nylon, poly(vinyl chloride), and the siloxanes.

The crystallinity of synthetic polymers can be destroyed by thermal motion at sufficiently high temperatures. This loss of crystallinity may be thought of as a kind of intramolecular melting from a crystalline solid to a more fluid-like random coil. Polymer melting also occurs at a specific **melting temperature,** T_m, which increases with the strength and number of intermolecular interactions in the material.

Brief illustration 16.3 Melting temperatures of synthetic and biological polymers

Polyethene, which has chains that interact only weakly in the solid, has T_m = 414 K, and nylon-66 fibres, in which there are strong hydrogen bonds between chains, has T_m = 530 K. High melting temperatures are desirable in most practical applications involving fibres and plastics. The 'melting' of biopolymers from an ordered structure, such as a helix or sheet, to a flexible random coil also occurs at a specific temperature which increases with the strength and number of intermolecular interactions in the material. The melting temperature, and therefore the thermal stability, of DNA increases with the number of G—C base pairs in the sequence because each G—C base pair has three hydrogen bonds whereas each A—T base pair has only two. More energy is required to unravel a double helix that, on average, has more hydrogen bonding interactions per base pair.

All synthetic polymers undergo a transition from a state of high to low chain mobility when they are cooled through the **glass transition temperature,** T_g. The transition is commonly detected by using differential scanning calorimetry (DSC, Impact 3.1). To visualize the glass transition, we consider what happens to an elastomer as we lower its temperature. There is sufficient energy available at normal temperatures for limited bond rotation to occur and the flexible chains writhe about. At lower temperatures, the amplitudes of the writhing motion decrease until a specific temperature, T_g, is reached at which motion is frozen almost completely and the sample forms a glass. Glass transition temperatures well below 300 K are desirable in elastomers that are to be used at normal temperatures.

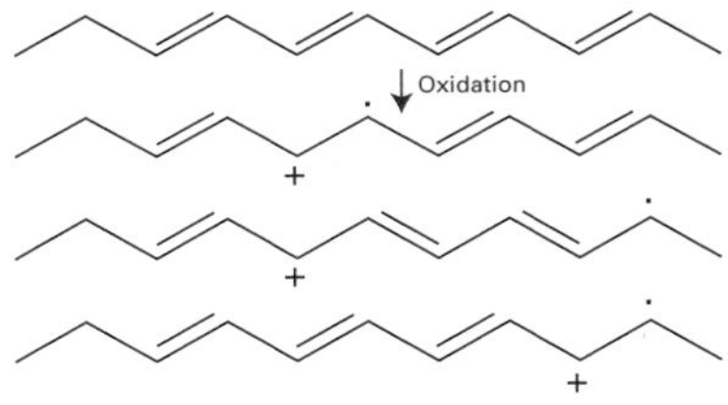

Fig. 16.17 The mechanism of migration of a partially localized cation radical, or polaron, in polyacetylene.

16.3 Electrical properties of polymers

Most of the macromolecules and self-assembled structures considered in this chapter are insulators, or very poor electrical conductors. However, a variety of newly developed macromolecular materials have electrical conductivities that rival those of silicon-based semiconductors and even metallic conductors (Chapter 17).

In **conducting polymers** extensively conjugated double bonds facilitate electron conduction along the polymer chain. One example of a conducting polymer is polyacetylene (polyethyne, Fig. 16.17). Whereas the delocalized π bonds do suggest that electrons can move up and down the chain, the electrical conductivity of polyacetylene increases significantly when it is partially oxidized by I_2 and other strong oxidants. The product is a **polaron**, a partially localized cation radical that travels through the chain, as shown in Fig. 16.17. Oxidation of the polymer by one more equivalent forms either **bipolarons**, a di-cation that moves as a unit through the chain, or **solitons**, two separate cation radicals that move independently. Polarons and solitons contribute to the mechanism of charge conduction in polyacetylene.

Conducting polymers are slightly better electrical conductors than silicon semiconductors but are far worse than metallic conductors. They are currently used in a number of devices, such as electrodes in batteries, electrolytic capacitors, and sensors. Recent studies of photon emission by conducting polymers may lead to new technologies for light-emitting diodes and flat-panel displays. Conducting polymers also show promise as molecular wires that can be incorporated into nanometre-sized electronic devices.

Mesophases and disperse systems

A **mesophase** is a bulk phase that is intermediate in character between a solid and a liquid. The most important type of mesophase is a **liquid crystal**, which is a substance having liquid-like imperfect long-range order in some directions but some aspects of crystal-like short-range order in other directions. Liquid crystals can be used as models of biological membranes and studied to gain insight into the process of transport of molecules into and out of cells. They are also of considerable technological importance for their use in displays on electronic equipment. A **disperse system** is a dispersion of small particles of one material in another. The small particles are commonly called **colloids**. In this context, 'small' means something less than about 1 μm in diameter (about twice the wavelength of visible light). In general, they are aggregates of numerous atoms or molecules, but are too small to be seen with an ordinary optical microscope. They pass through most filter papers, but can be detected by light-scattering, sedimentation, and osmosis.

16.4 Liquid crystals

There are three important types of liquid crystal; they differ in the type of long-range order that they retain. One type of retained long-range order gives rise to a **smectic phase** (from the Greek word for soapy), in which the molecules align themselves in layers (Fig. 16.18). Other materials, and some smectic liquid crystals at higher temperatures, lack the layered structure but retain a nearly parallel alignment (Fig. 16.19): this mesophase is the **nematic phase** (from the Greek for thread). The strongly anisotropic optical properties of nematic liquid crystals, and their response to electric fields, are the basis of their use as data displays. In the **cholesteric phase**, which is so called because some derivatives of cholesterol form them, the molecules lie in sheets at angles

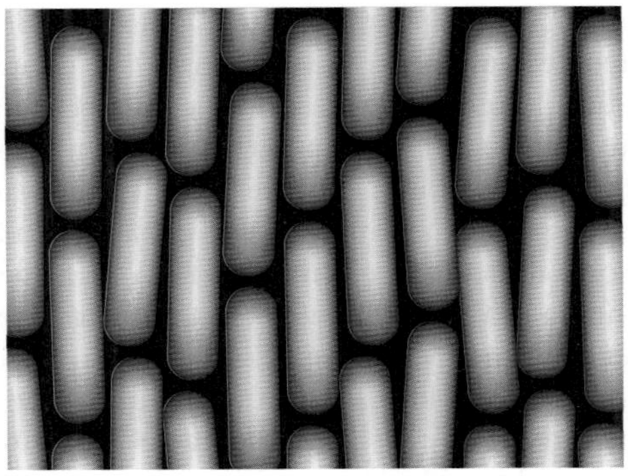

Fig. 16.18 The arrangement of molecules in the smectic phase of a liquid crystal.

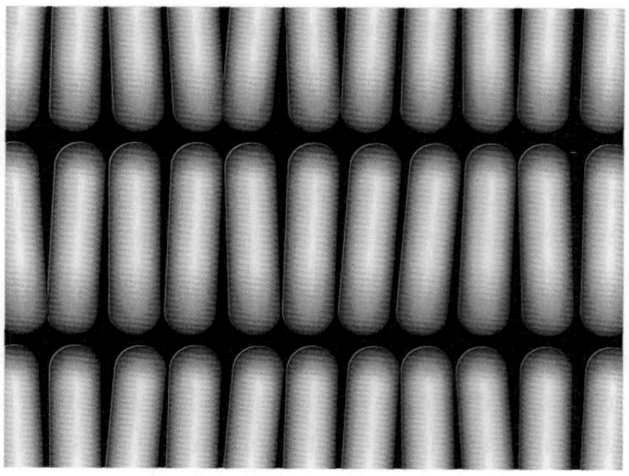

Fig. 16.19 The arrangement of molecules in the nematic phase of a liquid crystal.

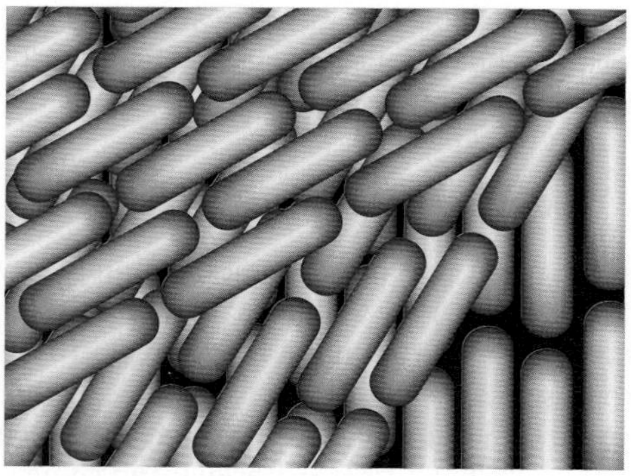

Fig. 16.20 The arrangement of molecules in the cholesteric phase of a liquid crystal. Three layers are shown: the relative orientation of these layers is repeated in successive layers to give a helical structure.

that change slightly between neighbouring sheets (Fig. 16.20), so forming helical structures. The pitch of the helix varies with temperature. As a result, the colours of cholesteric liquid crystals, which are due to diffraction and hence vary with the pitch, depend on the temperature. They are used for detecting temperature distributions in living material, including human patients, and have even been incorporated into fabrics. Liquid crystals are models for the membranes of biological cells (see Impact 16.2).

Although there are many liquid crystalline materials, some difficulty is often experienced in achieving a technologically useful temperature range for the existence of the mesophase. To overcome this difficulty, mixtures can be used. An example of the type of phase diagram that is then obtained is shown in Fig. 16.21. As can be seen, the mesophase exists over a wider range of temperatures than either liquid crystalline material alone.

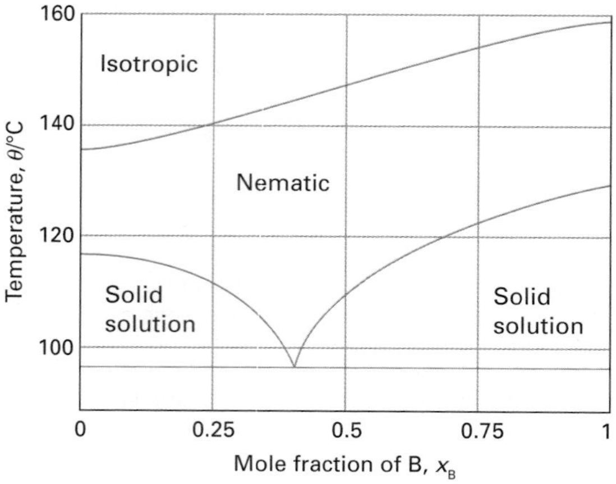

Fig. 16.21 The phase diagram at 1 atm of a binary system of two liquid crystalline materials, 4,4-dimethoxyazoxybenzene (A) and 4,4-diethoxyazoxybenzene (B).

16.5 Classification of disperse systems

The name given to a disperse system depends on the nature of the substances involved. A **sol** is a dispersion of a solid in a liquid (such as clusters of gold atoms in water) or of a solid in a solid (such as ruby glass, which is a gold-in-glass sol, and achieves its colour by scattering). An **aerosol** is a dispersion of a liquid in a gas (like fog and many sprays) and of a solid in a gas (such as smoke): the particles are often large enough to be seen with a microscope. An **emulsion** is a dispersion of a liquid in a liquid (such as milk and some paints). A **gel** is a system in which at least one component has a low rigidity (such as a cross-linked polymer or a lipid bilayer) and at least one component has a high mobility (for example, the solvent).

The preparation of aerosols can be as simple as sneezing (which produces an aerosol). Laboratory and commercial methods make use of several techniques. Material (for example, quartz) may be ground in the presence of the dispersion medium. Passing a heavy electric current through a cell may lead to the crumbling of an electrode into colloidal particles; arcing between electrodes immersed in the support medium also produces a colloid. Chemical precipitation sometimes results in a colloid. A precipitate (for example, silver iodide) already formed may be converted to a colloid by the addition of **peptizing agent**, a substance that disperses a colloid. An example of a peptizing agent is potassium iodide, which provides ions that adhere to the colloidal particles and cause them to repel one another. Clays may be peptized by alkalis, the OH^- ion being the active agent.

Emulsions are normally prepared by shaking the two components together, although some kind of **emulsifying agent** has to be used in order to stabilize the product. This emulsifier may be a soap (the salt of a long-chain fatty acid), a surfactant, or a lyophilic sol that forms a protective film around the dispersed phase. In milk, which is an emulsion of fats in water, the emulsifying agent is casein, a protein containing phosphate groups. That casein is not completely successful in stabilizing milk is apparent from the formation of cream: the dispersed fats coalesce into oily droplets which float to the surface. This separation may be prevented by ensuring that the emulsion is dispersed very finely initially: violent agitation with ultrasonics or extrusion through a very fine mesh brings this about, the product being 'homogenized' milk.

Aerosols are formed when a spray of liquid is torn apart by a jet of gas. The dispersal is aided if a charge is applied to the liquid, for then the electrostatic repulsions blast the jet apart into droplets. This procedure may also be used to produce emulsions, for the charged liquid phase may be squirted into another liquid.

Disperse systems are often purified by **dialysis**, a technique based on osmosis (Section 6.8) and in which a membrane (for instance, cellulose) is selected that is permeable to solvent and ions but not to the bigger colloid particles. The aim is to remove much (but not all, for reasons explained later) of the ionic material that may have accompanied their formation. Dialysis is very slow, and is normally accelerated by applying an electric field and making use of the charge carried by many colloids; the technique is then called **electrodialysis**.

16.6 Surface, structure, and stability

The principal feature of colloids is the very great surface area of the dispersed phase in comparison with the same amount of ordinary material. For example, a cube of side 1 cm has a surface area of 6 cm^2. When it is dispersed as 10^{18} little 10 nm cubes the total surface area is 6×10^6 cm^2 (about the size of a tennis court). This dramatic increase in area means that surface effects are of dominating importance in the chemistry of disperse systems.

As a result of their great surface area, many colloids are thermodynamically unstable with respect to the bulk: that is, many colloids have a thermodynamic tendency to reduce their surface area (like a liquid). Their apparent stability must therefore be a consequence of the kinetics of collapse: such disperse systems are kinetically non-labile (that is, the activation energy for collapse is high), not thermodynamically stable. At first sight, though, even the kinetic argument seems to fail: colloidal particles attract one another over large distances by the dispersion interaction, so there is a long-range force tending to collapse them down into a single blob.

Several factors oppose the long-range dispersion attraction. There may be a protective film at the surface of the colloid particles that stabilizes the interface and cannot be penetrated when two particles touch. For example, the surface atoms of a platinum sol in water react chemically, becoming coordinated with $-(OH)_3H_3$, and this layer encases the particle like a shell. A fat can be emulsified by a soap because the long hydrocarbon tails penetrate the oil droplet but the $-CO_2^-$ groups (or other hydrophilic groups in detergents) surround the surface, form hydrogen bonds with water, and give rise to a shell of negative charge that repels a possible approach from another similarly charged particle.

By a **surfactant** we mean a species that accumulates at the interface of two phases or substances (one of which may be air) and modifies the properties of the surface. An effective surfactant accumulates at the interface between the phases and does not dissolve well in either of the bulk phases. A typical surfactant consists of a long hydrocarbon tail that dissolves in hydrocarbon and other nonpolar materials, and a hydrophilic **head group** that dissolves in a polar solvent (typically water). Typical head groups include the ionic species $-CO_2^-$ and $-SO_3^-$; typical non-ionic species include $-(OC_2H_4)_6OH$ and $-(OC_2H_4)_8OH$. A surfactant is an **amphiphilic** substance, meaning that it has both hydrophobic and hydrophilic regions (the *amphi-* part of the name is from the Greek word for both). Soaps, for example, consist of the alkali metal salts of long-chain carboxylic acids, and the surfactant in detergents is typically a long chain benzenesulfonic acid ($R-C_6H_4SO_3H$) or its salt. The mode of action of a surfactant in a detergent, and of soap, is to dissolve in both the aqueous phase and the hydrocarbon phase where their surfaces are in contact, and hence to solubilize the hydrocarbon phase so that it can be washed away (Fig. 16.22).

Surfactant molecules can group together as **micelles**, colloid-sized clusters of molecules, even in the absence of grease droplets, for their hydrophobic tails tend to congregate, and their hydrophilic heads provide protection (Fig. 16.23). Micelles form only above when the concentration of surfactant is equal to or greater than a value called the **critical micelle concentration** (CMC). Surfactants form micelles only when

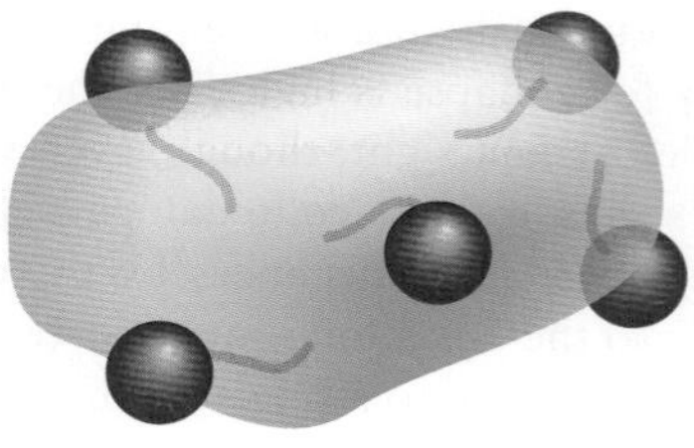

Fig. 16.22 A surfactant molecule in a detergent or soap acts by sinking its hydrophobic hydrocarbon tail into the grease, so leaving its hydrophilic head groups on the surface of the grease where they can interact attractively with the surrounding water.

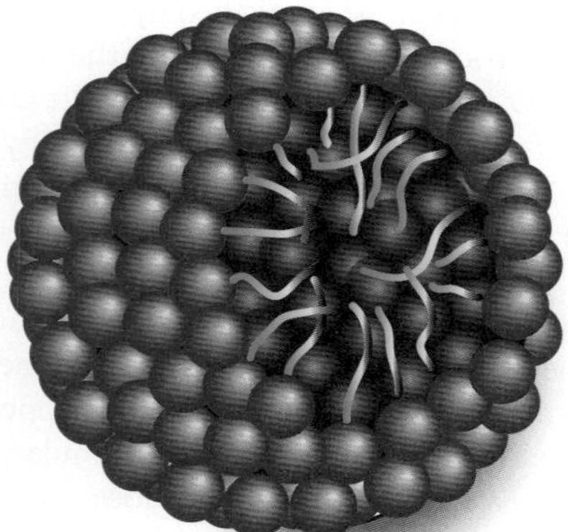

Fig. 16.23 A representation of a spherical micelle. The hydrophilic groups are represented by spheres and the hydrophobic hydrocarbon chains are represented by the stalks. The latter are mobile.

the temperature is above a critical value called the **Krafft temperature,** T_K.

The thermodynamics of micelle formation account for the existence of a critical temperature. Experiments show that the enthalpy of formation of micelles in aqueous systems is probably positive (that is, that they are endothermic) with $\Delta H \approx 1\text{–}2$ kJ per mole of surfactant molecules, due in large part to repulsions between the head groups of the surfactants. That micelles do form above the CMC indicates that the entropy change accompanying their formation must be positive in order for the Gibbs energy accompanying the formation process, $\Delta G = \Delta H - T\Delta S$, to be negative, and measurements suggest a value of about $+140\ \mathrm{J\,K^{-1}\,mol^{-1}}$ at room temperature. That the entropy change is positive even though the molecules are clustering together shows that there must be a contribution to the entropy from the solvent: the surrounding solvent molecules no longer have to solvate individual surfactant molecules and so become less ordered, as in the hydrophobic effect (Section 15.7). The role of entropy is magnified by the temperature (the factor T in $T\Delta S$), and ΔG may become negative and micelle formation spontaneous when the temperature is high enough.

The self-assembly of a micelle has the characteristics of a cooperative process in which the addition of a surfactant molecule to a cluster that is forming becomes more probable the larger the size of the aggregate, so after a slow start there is a cascade of formation of micelles. If we suppose that the dominant micelle M_N consists of N monomers M, then the dominant equilibrium we have to consider is

$$N\,\mathrm{M} \rightleftharpoons \mathrm{M}_N \qquad K = \frac{[\mathrm{M}_N]}{[\mathrm{M}]^N} \tag{16.4a}$$

We have assumed, probably dangerously on account of the large sizes of monomers, that the solution is ideal and that activities can be replaced by molar concentrations. The total concentration of surfactant is $[\mathrm{M}]_{\text{total}} = [\mathrm{M}] + N[\mathrm{M}_N]$ because each micelle consists of N monomer molecules. Therefore,

$$K = \frac{[\mathrm{M}_N]}{\left([\mathrm{M}]_{\text{total}} - N[\mathrm{M}_N]\right)^N} \tag{16.4b}$$

Brief illustration 16.4 The fraction of surfactant molecules in micelles

Equation 16.4b can be solved numerically for the micelle concentration as a function of the total surfactant concentration and some results for $K = 1$ are shown in Fig. 16.24. We see that for large N, there is a reasonably sharp transition in the relative concentrations of surfactant molecules that are present in micelles, which corresponds to the existence of a CMC.

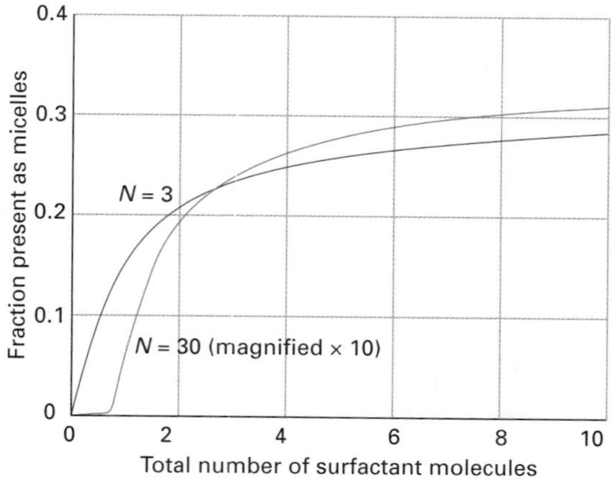

Fig. 16.24 The micelle concentration as a function of the total surfactant concentration and $K = 1$.

Ionic species tend to be disrupted by the Coulomb repulsions between head groups and are normally limited to groups of between 10 and 100 molecules. Non-ionic surfactants may cluster together in swarms of 1000 or more, and as the temperature is raised these large aggregates separate into a distinct phase at a temperature known as the **cloud point**. The shapes of the individual micelles vary with concentration. Although spherical micelles do occur, they are more commonly flattened spheres close to the CMC, and rod-like at higher concentrations. The interior of a micelle is like a droplet of oil, and magnetic resonance spectroscopy shows that the hydrocarbon tails are mobile, but slightly more restricted than in the bulk.

Micelles are important in industry and biology on account of their solubilizing function: matter can be transported by water after it has been dissolved in their hydrocarbon interiors. For this reason, micellar systems are used as detergents and drug carriers, and for organic synthesis, froth flotation, and petroleum recovery. They can be perceived as a part of a family of similar structures formed when amphiphilic substances are present in water (Fig. 16.25). A **monolayer** forms at the air–water interface, with the hydrophilic head groups facing the water. Micelles are like monolayers that enclose a region. A **bilayer vesicle** is like a double-micelle, with an inward pointing inner surface of molecules surrounded by an outward pointing outer layer. The 'flat' version of a bilayer vesicle is the analogue of a cell membrane.

Some micelles at concentrations well above the CMC form extended parallel sheets, called *lamellar micelles*, two molecules thick. The individual molecules lie perpendicular to the sheets, with hydrophilic groups on the outside in aqueous solution and on the inside in nonpolar media. Such lamellar micelles show a close resemblance to biological membranes, and are often a useful model on which to base investigations of biological structures.

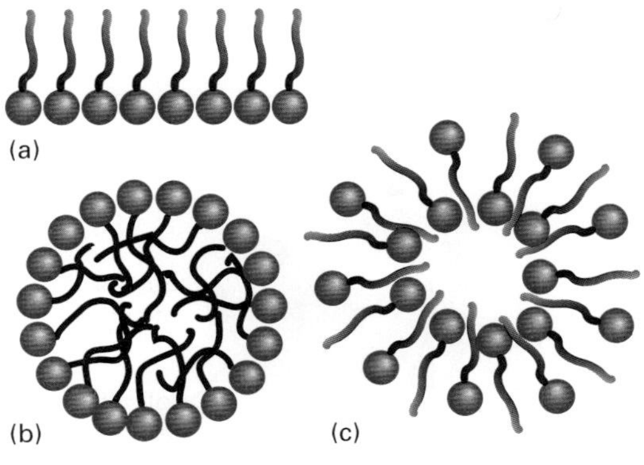

Fig. 16.25 Amphiphilic molecules form a variety of related structures in water: (a) a monolayer, (b) a spherical micelle, (c) a bilayer vesicle.

Impact on biochemistry 16.2

Biological membranes

Although lamellar micelles are convenient models of cell membranes, actual membranes are highly sophisticated structures. The basic structural element of a membrane is a phospholipid, such as phosphatidyl choline (**4**), which contains long hydrocarbon chains (typically in the range C_{14}–C_{24}) and a variety of polar groups, such as $-CH_2CH_2N(CH_3)_3^+$. The hydrophobic chains stack together to form an extensive bilayer about 5 nm across. The lipid molecules form layers instead of spherical micelles because the hydrocarbon chains are too bulky to allow packing into nearly spherical clusters.

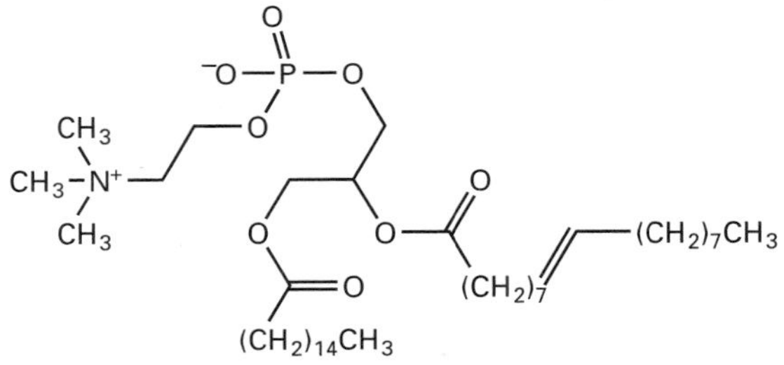

4 Phosphatidyl choline

A bilayer is a highly mobile structure. Not only are the hydrocarbon chains ceaselessly twisting and turning in the region between the polar groups, but the phospholipid and other molecules inserted into the bilayer migrate over the surface. It is better to think of the membrane as a viscous fluid rather than a permanent structure, with a viscosity about a hundred times that of water. In common with diffusional behaviour in general (Section 11.11), the average distance a phospholipid molecule diffuses is proportional to the square root of the time. Typically, a phospholipid molecule migrates through about 1 μm (the diameter of a cell) in about 1 min.

Peripheral proteins are proteins attached to the bilayer. **Integral proteins** are proteins immersed in the mobile but viscous bilayer. These proteins may span the depth of the bilayer and consist of tightly packed α helices or, in some cases, β sheets containing hydrophobic residues that sit comfortably within the hydrocarbon region of the bilayer. There are two views of the motion of integral proteins in the bilayer. In the **fluid mosaic model**, the proteins are mobile, but their diffusion coefficients are much smaller than those of the lipids. In the **lipid raft model**, a number of lipid and cholesterol molecules form ordered structures, or 'rafts', that envelope proteins and help carry them to specific parts of the cell.

The mobility of the bilayer enables it to flow round a molecule close to the outer surface, to engulf it, and incorporate it into the cell by the process of **endocytosis**. Alternatively, material from the cell interior wrapped in cell membrane may coalesce with the cell membrane itself, which then withdraws and ejects the material in the process of **exocytosis**. The function of the proteins embedded in the bilayer, though, is to act as devices for transporting matter into and out of the cell in a more subtle manner. By providing hydrophilic channels through an otherwise alien hydrophobic environment, some proteins act as ion channels and ion pumps (Impact 9.1).

All lipid bilayers undergo a transition from a state of high to low chain mobility at a temperature that depends on the structure of the lipid. There is sufficient energy available at normal temperatures for limited bond rotation to occur and the flexible chains writhe about. However, the membrane still has a great deal of order in the sense that the bilayer structure does not come apart and the system is best described as a liquid crystal (Fig. 16.26a). At lower temperatures, the amplitudes of the writhing motion decrease until a specific temperature is reached at which motion is largely frozen. The membrane is then said to exist as a gel (Fig. 16.26b). Biological membranes exist as liquid crystals at physiological temperatures.

Interspersed among the phospholipids of biological membranes are sterols, such as cholesterol (**5**), which is largely hydrophobic but does contain a hydrophilic —OH group. Sterols, which are present in different proportions in different types of cells, prevent the hydrophobic chains of lipids from 'freezing' into a gel and, by disrupting the packing of the chains, spread the melting point of the membrane over a range of temperatures.

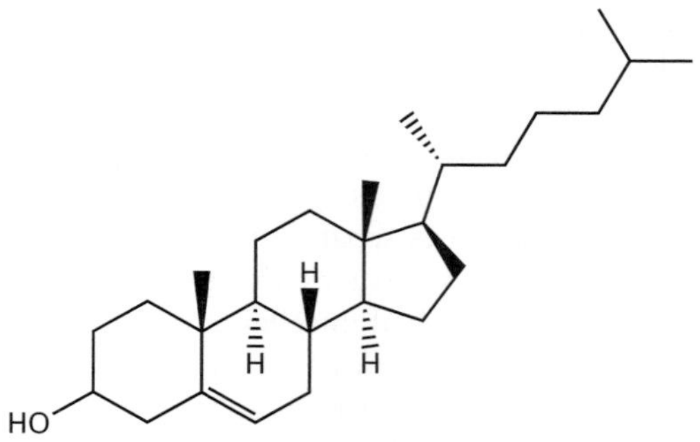

5 Cholesterol

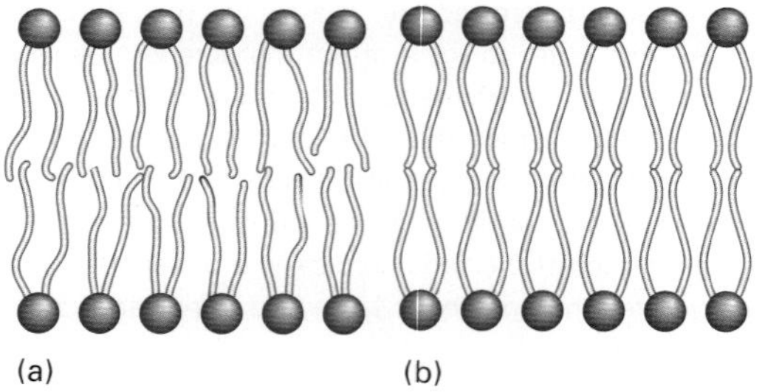

Fig. 16.26 A depiction of the variation with temperature of the flexibility of hydrocarbon chains in a lipid bilayer. (a) At physiological temperature, the bilayer exists as a liquid crystal, in which some order exists but the chains writhe. (b) At a specific temperature, the chains are largely frozen and the bilayer is said to exist as a gel.

16.7 The electric double layer

Apart from the physical stabilization of disperse systems, a major source of kinetic non-lability is the existence of an electric charge on the surfaces of the colloidal particles. On account of this charge, ions of opposite charge tend to cluster nearby.

Two regions of charge must be distinguished. First, there is a fairly immobile layer of ions that stick tightly to the surface of the colloidal particle, and which may include water molecules (if that is the support medium). The radius of the sphere that captures this rigid layer is called the **radius of shear**, and is the major factor determining the mobility of the particles (Fig. 16.27). The electric potential at the radius of shear relative to its value in the distant, bulk medium is called the **electrokinetic potential**, ζ (zeta). The charged unit attracts an oppositely charged ionic atmosphere. The inner shell of charge and the outer atmosphere jointly constitute the **electric double layer**.

At high concentrations of ions of high charge number, the atmosphere is dense and the potential falls to its bulk value within a short distance. In this

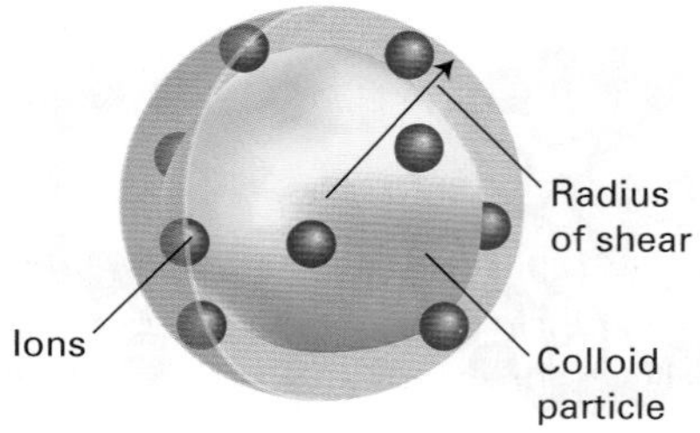

Fig. 16.27 The definition of the radius of shear for a colloidal particle. The spheres are ions attached to the surface of the particle.

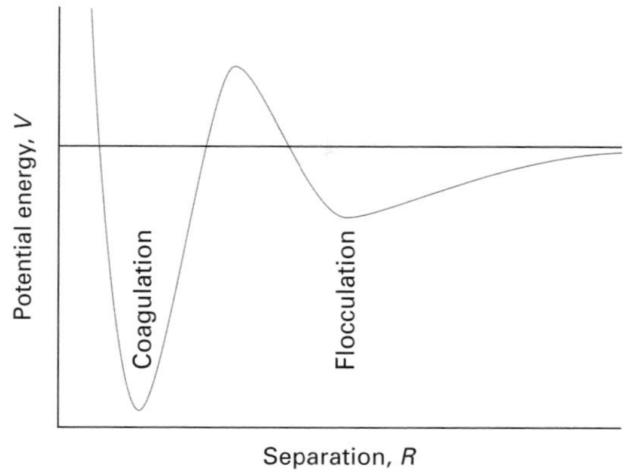

Fig. 16.28 The potential energy of interaction of two colloidal particles varies with distance as shown here. The shallow outer well represents the van der Waals interactions between the particles and accounts for flocculation; the deep inner well represents the merging—the coagulation—of the particles.

case there is little electrostatic repulsion to hinder the close approach of two colloid particles. As a result, **flocculation**, the aggregation of the colloidal particles, occurs as a consequence of the van der Waals forces (Fig. 16.28). Flocculation is often reversible, and should be distinguished from **coagulation**, which is the irreversible collapse of the colloid into a bulk phase. When river water containing colloidal clay flows into the sea, the brine induces coagulation and is a major cause of silting in estuaries.

Metal oxide and sulfide sols have charges that depend on the pH; sulfur and the noble metals tend to be negatively charged. Naturally occurring macromolecules also acquire a charge when dispersed in water, and an important feature of proteins and other natural macromolecules is that their overall charge depends on the pH of the medium. For instance, in acid environments protons attach to basic groups and the net charge of the macromolecule is positive; in basic media the net charge is negative as a result of proton loss. At the **isoelectric point**, the pH is such that there is no net charge on the macromolecule.

The primary role of the electric double layer is to render the colloid kinetically non-labile (that is, to ensure that it survives for long periods despite being thermodynamically unstable). Colliding colloidal particles break through the double layer and coalesce only if the collision is sufficiently energetic to disrupt the layers of ions and solvating molecules, or if thermal motion has stirred away the surface accumulation of charge. This kind of disruption of the double layer may occur at high temperatures, which is one reason why sols precipitate when they are heated. The protective role of the double layer is the reason why it is important not to remove all the ions (other than those needed to ensure overall electrical neutrality) when a colloid is being purified by dialysis, and why proteins coagulate most readily at their isoelectric point.

The presence of charge on colloidal particles and natural macromolecules also permits us to control their motion, as in dialysis and electrophoresis. Apart from its application to the determination of molar mass, electrophoresis has several analytical and technological applications. One analytical application is to the separation of different macromolecules, as discussed in Section 16.12. Technical applications include silent ink-jet printers, the painting of objects by airborne charged paint droplets, and electrophoretic rubber forming by deposition of charged rubber molecules on anodes formed into the shape of the desired product (for example, surgical gloves).

16.8 Liquid surfaces and surfactants

Liquid surfaces are mobile interfaces where solutes might gather and influence its properties. The smooth surface of stationary liquids is due to the imbalance of forces, for whereas a molecule in the interior of a bulk sample experiences attractions from all directions, those at the surface experience only inward forces. A molecule at an air–liquid surface has a higher potential energy than one in the bulk because it interacts with fewer neighbours, so work must be done to bring a molecule from the bulk into the surface layer. The work required to increase the area of surface by $\Delta\sigma$ is proportional to that increase and we write $w = \gamma\Delta\sigma$, where the constant of proportionality γ is called the **surface tension.** For $\gamma\Delta\sigma$ to be expressed in joules, γ must be in newtons per metre, N m^{-1} (because then 1 N m^{-1} × 1 m^2 = 1 N m = 1 J). Some values of the surface tension are given in Table 16.1. Broadly speaking, surface tensions are high when

Table 16.1
Surface tensions of liquids at 293 K

	γ/(mN m^{-1})
Benzene	28.88
Mercury	472
Methanol	22.6
Water	72.75

Note that 1 N m^{-1} = 1 J m^{-2}.

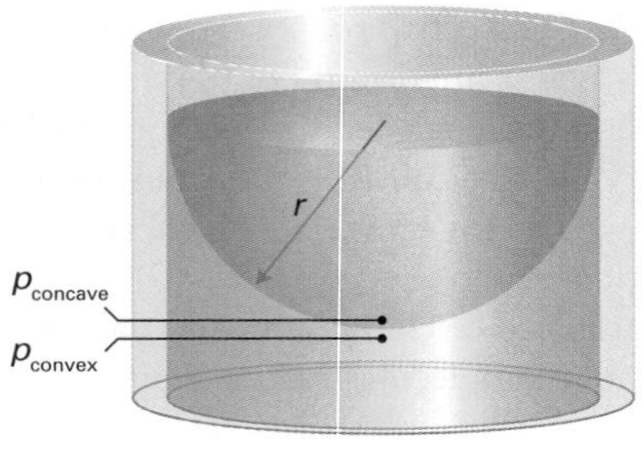

Fig. 16.29 The pressure just outside a curved surface (on the 'convex' side) is lower than that just inside the surface (on the 'concave' side); the difference is greater the greater the surface tension of the liquid.

there are strong forces acting between the molecules or atoms, as in water and mercury. Surface tensions typically decrease as the temperature is raised and vanish at the boiling point.

We saw in Chapter 4 that non-expansion work (work that does not involve expansion against an external pressure) can be identified with a change in Gibbs energy, ΔG, so we can write

$$\Delta G = \gamma \Delta \sigma \qquad (16.5)$$

This is the link between surface properties and thermodynamics.

One consequence of the Gibbs energy change expressed in eqn 16.5 is that the pressure is different on either side of a curved liquid surface (as in a droplet or a cavity in a liquid). As we show in the following Derivation, the pressure on either side of a spherical surface of radius r is given by the **Laplace equation:**

$$p_{\text{concave}} = p_{\text{convex}} + \frac{2\gamma}{r} \qquad \text{Laplace equation} \qquad (16.6)$$

This equation tells us that the pressure just outside a curved surface (on the convex side, Fig. 16.29) is lower than that just inside the surface and that the difference is greater the greater the surface tension of the liquid.

Derivation 16.1

The Laplace equation

If the pressure outside a spherical cavity is p_{concave}, the total force (which is pressure × area) acting on the wall of the cavity of area $4\pi r^2$ is $4\pi r^2 p_{\text{concave}}$. The force tending to compress the cavity is the sum of effects due to the pressure inside the cavity—on the convex side of the surface—and the surface tension. The former gives rise to a force $4\pi r^2 p_{\text{convex}}$. The force due to the surface tension is calculated as follows.

The change in surface area when the radius of the cavity increases by a small amount δr from r to $r + \delta r$ is

$$\delta\sigma = 4\pi(r+\delta r)^2 - 4\pi r^2 = 4\pi(r^2 + 2r\delta r + \delta r^2) - 4\pi r^2$$

$$\approx 8\pi r\delta r + 4\pi \overbrace{(\delta r)^2}^{\text{Negligibly small}}$$

The change in Gibbs energy, from eqn 16.5, is therefore $8\pi\gamma r\delta r$. At constant pressure and temperature, a difference in Gibbs energy is equal to the non-expansion work associated with the change (in this case, the expansion is not against an external atmospheric pressure, so it is actually the additional 'non-expansion' work in the sense of Chapter 4), so the work done when the cavity expands by δr is $w = 8\pi\gamma r\delta r$. From Foundations 0.9, we know that the magnitude of the work done is equal to the product of force and distance, so in this case the magnitude of the opposing force must be $8\pi\gamma r$. The total opposing force is therefore $4\pi r^2 p_{\text{convex}} + 8\pi\gamma r$. When the inward and outward forces are balanced,

$$4\pi r^2 p_{\text{concave}} = 4\pi r^2 p_{\text{convex}} + 8\pi\gamma r$$

This relation can now be rearranged into eqn 16.6 by dividing both sides by $4\pi r^2$.

The difference in pressure across a curved interface has a number of consequences. One is that it gives rise to **capillary action,** in which a liquid climbs up the interior of a narrow tube. As can be seen from Fig. 16.30, the pressure just below the meniscus of a liquid in a narrow tube is less, by $2\gamma/r$, than the atmospheric pressure so the liquid is pushed up the tube until the hydrostatic pressure, the pressure due to the column of liquid, cancels the reduction in pressure due to the curvature. The hydrostatic pressure of a column of liquid of height h is equal to ρgh,

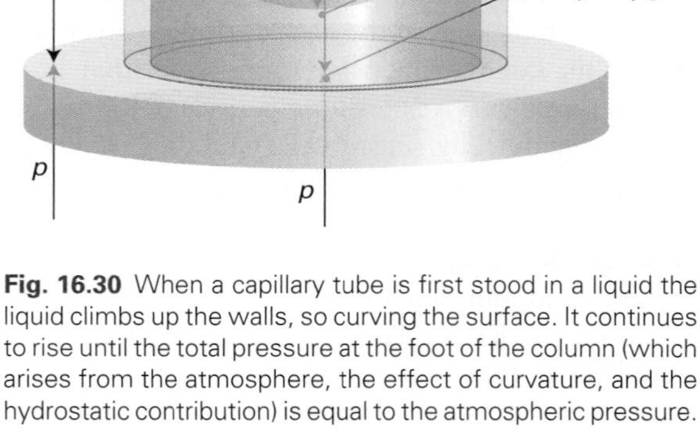

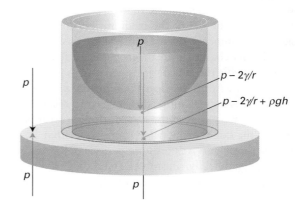

Fig. 16.30 When a capillary tube is first stood in a liquid the liquid climbs up the walls, so curving the surface. It continues to rise until the total pressure at the foot of the column (which arises from the atmosphere, the effect of curvature, and the hydrostatic contribution) is equal to the atmospheric pressure.

where ρ is the mass density of the liquid and g is the acceleration of free fall. That is, the liquid rises to a height at which $\rho gh = 2\gamma/r$, and therefore

$$h = \frac{2\gamma}{\rho g r} \qquad \text{Capillary action} \quad (16.7)$$

This expression gives a simple method for estimating the surface tension of a liquid (by rearranging it into $\gamma = \frac{1}{2}\rho grh$).

Brief illustration 16.5 Capillary action

If water at 25 °C rises to a height of 7.36 cm in a capillary tube of internal radius 0.20 mm, the surface tension is

$$\gamma = \tfrac{1}{2}(997.1\ \text{kg m}^{-3}) \times (9.81\ \text{m s}^{-2}) \times (7.36 \times 10^{-2}\ \text{m}) \times (2.0 \times 10^{-4}\ \text{m})$$
$$= 7.2 \times 10^{-2}\ \text{kg s}^{-2} = 7.2 \times 10^{-2}\ \text{N m}^{-1}$$

This value could be reported as 72 mN m^{-1}.

The surface tension of a liquid changes markedly if a surfactant is present. Amphiphilic molecules accumulate at the water–air surface with their hydrophobic tails exposed to the air to minimize interaction with the water. Their accumulation at the surface relative to the bulk is reported as the **surface excess**, Γ (uppercase gamma). In a simple case where no surfactant appears in the vapour above the surface, this quantity is measured by noting the total amount of surfactant in a sample of the liquid, n_{total}, and subtracting from that total the amount known to be in the bulk solution, n_{solution}, from measurement of its concentration. Then

$$\Gamma = \frac{n_{\text{total}} - n_{\text{solution}}}{\sigma} \qquad \text{Surface excess} \quad (16.8)$$

where σ is the area of the surface. Below the critical micelle concentration the slope of a plot of surface tension against the logarithm of the concentration is equal to $-RT\Gamma$, so Γ can be determined.[3] Above the CMC the surface tension is independent of the concentration of surfactant, so the CMC can be determined graphically (Fig. 16.31).

Pure liquids do not form foams: the Gibbs energy increases when a surface is formed, so there is always a spontaneous tendency for a cavity in a liquid to collapse. A bubble in boiling water will rise to the surface and break when it arrives. If a surfactant is present, however, there is a smaller pressure difference between its interior and the surroundings (because the surface tension is lower) and the surface

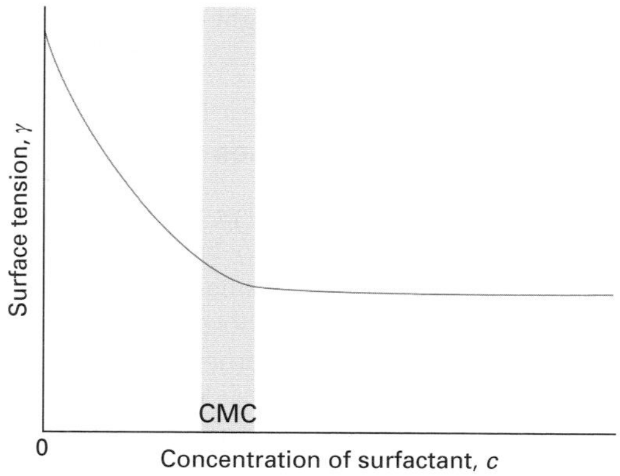

Fig. 16.31 The variation of surface tension with concentration of surfactant.

is stabilized by the surface excess of surfactants. A bubble in a surfactant solution will rise to the surface after it has been formed and will survive at the surface. They will be joined by others, and a foam will develop. The structure of that foam is itself a highly interesting mathematical problem, for the originally spherical bubbles deform into polyhedra that minimize the total surface area. The most common polyhedra are predicted mathematically to have 13.4 sides, and indeed it is observed that most have 14 sides, with the second most abundant having 12.

Determination of size and shape

X-ray diffraction, a technique discussed in detail in Chapter 17, can reveal the position of almost every atom other than hydrogen even in very large molecules. However, there are several reasons why other techniques must also be used. In the first place, the sample might be a mixture of molecules with different chain lengths and extent of cross-linking, in which case sharp X-ray images are not obtained. Even if all the molecules in the sample are identical, it might prove impossible to obtain a single crystal. Furthermore, although X-ray information about proteins and DNA has shown how immensely interesting and motivating the data can be, it is incomplete. For instance, what can be said about the shape of the molecule in its natural environment, a biological cell? What can be said about the response of its shape to changes in its environment?

X-ray and NMR techniques are so important that they are treated separately in Chapters 17 and 21, respectively. Here we concentrate on techniques that are used when they are not available or appropriate. First, though, we need to clarify what is meant by the

[3] See our *Physical Chemistry* (2010) for the procedure.

molar mass of a sample that might contain molecules of a range of molar masses.

16.9 Average molar masses

Many proteins (and specifically protein enzymes) are **monodisperse**, meaning that they have a single, definite molar mass. There may be small variations, such as one amino acid replacing another, depending on the source of the sample. A synthetic polymer, however, is **polydisperse**, in the sense that a sample is a mixture of molecules with various chain lengths and molar masses. The various techniques that are used to measure molar masses result in different types of mean values of polydisperse systems. The **number-average molar mass,** $\bar{M}_n$, is the value obtained by multiplying each molar mass by the numerical fraction (N_i/N) of molecules of that mass present in the sample:

$$\bar{M}_n = \frac{1}{N}\sum_i N_i M_i \quad \text{Definition} \quad \text{Number-average molar mass} \qquad (16.9a)$$

Here N_i (with $i = 1, 2, \ldots$) is the number of molecules with molar mass M_i and there are N molecules in all. By dividing the terms in the numerator and denominator by Avogadro's constant N_A and writing $N_i/N_A = n_i$ and $N/N_A = n$, we can express this equation in terms of the amounts (in moles) rather than the actual numbers:

$$\bar{M}_n = \frac{1}{n}\sum_i n_i M_i \quad \text{Alternative form} \quad \text{Number-average molar mass} \qquad (16.9b)$$

The **weight-average molar mass,** $\bar{M}_w$, is the average calculated by multiplying the molar masses of the molecules by the mass fraction (m_i/m) of each one present in the sample:

$$\bar{M}_w = \frac{1}{m}\sum_i m_i M_i \quad \text{Definition} \quad \text{Weight-average molar mass} \qquad (16.9c)$$

In this expression, m_i is the total mass of molecules of molar mass M_i and m is the total mass of the sample. In general, these two averages are different and the ratio $\bar{M}_w/\bar{M}_n$ is called the **heterogeneity index** (or 'polydispersity index'). In the determination of protein molar masses we expect the various averages to be the same because unless there has been degradation the sample is monodisperse. A synthetic polymer normally spans a range of molar masses and the different averages yield different values. Typical synthetic materials have $\bar{M}_w/\bar{M}_n \approx 4$. The term 'monodisperse' is conventionally applied to synthetic polymers in which this index is less than 1.1; commercial polyethene samples might be much more heterogeneous, with an index of close to 30. One consequence of a narrow molar mass distribution for synthetic polymers is often a higher degree of crystallinity in the solid and therefore higher density and melting point. The spread of values is controlled by the choice of catalyst and reaction conditions.

Example 16.1

Determining the heterogeneity index of a polymer sample

Determine the heterogeneity index of a sample of poly(vinyl chloride) from the following data:

Molar mass interval/(kg mol^{-1})	Average molar mass within interval/(kg mol^{-1})	Mass of sample within interval/g
5–10	7.5	9.6
10–15	12.5	8.7
15–20	17.5	8.9
20–25	22.5	5.6
25–30	27.5	3.1
30–35	32.5	1.7

Strategy Begin by calculating the number-average and weight-average molar masses from eqns 16.9b and 16.9c, respectively. To do so, multiply the molar mass within each interval by the number and mass fractions, respectively, of the molecule in each interval. Obtain the amount (in moles) in each interval by dividing the mass of the sample in each interval by the average molar mass for that interval and then use eqn 16.9b. Finally, use the average molar masses to calculate the heterogeneity index of the sample as the ratio $\bar{M}_w/\bar{M}_n$.

Solution The amounts in each interval are as follows:

Interval	5–10	10–15	15–20	20–25	25–30	30–35
Molar mass/(kg mol^{-1})	7.5	12.5	17.5	22.5	27.5	32.5
Amount/mmol	1.30	0.70	0.51	0.25	0.11	0.052
					Total amount/mmol:	2.92

The number-average molar mass is therefore

$$\bar{M}_n/(\text{kg mol}^{-1}) = \frac{1}{2.92}(1.3\times 7.5 + 0.70\times 12.5 + 0.51\times 17.5 + 0.25\times 22.5 + 0.11\times 27.5 + 0.052\times 32.5) = 13$$

The weight-average molar mass is calculated directly from the data by first noting that adding the masses in each interval gives the total mass of the sample, 37.6 g. It follows that

$$\bar{M}_w/(\text{kg mol}^{-1}) = \frac{1}{37.6}(9.6\times 7.5 + 8.7\times 12.5 + 8.9\times 17.5 + 5.6\times 22.5 + 3.1\times 27.5 + 1.7\times 32.5) = 16$$

The heterogeneity index is $\bar{M}_w/\bar{M}_n = 1.2$.

Self-test 16.3

The *Z-average molar mass* is defined as

$$\bar{M}_Z = \frac{\sum_i N_i M_i^3}{\sum_i N_i M_i^2}$$

Evaluate the *Z*-average molar mass of the sample described in Example 16.1.

Answer: 19 kg mol^{-1}

Number-average molar masses may be determined by measuring the osmotic pressure of polymer solutions (Section 6.8). The upper limit for the reliability of membrane osmometry is about 1000 kg mol^{-1}. A major problem for macromolecules of relatively low molar mass (less than about 10 kg mol^{-1}), however, is their ability to percolate through the membrane. One consequence of this partial permeability is that membrane osmometry tends to overestimate the average molar mass of a polydisperse mixture. Techniques for the determination of molar mass and polydispersity that are not limited in this way include mass spectrometry, laser light scattering, ultracentrifugation, electrophoresis, and chromatography.

A note on good practice The masses of macromolecules are often reported in daltons (Da), where 1 Da = m_u (with $m_u = 1.661 \times 10^{-27}$ kg). Note that 1 Da is a measure of *molecular* mass not of *molar* mass. We might say that the mass (not the molar mass) of a certain macromolecule is 100 kDa (that is, its mass is $100 \times 10^3 \times m_u$); we could also say that its molar mass is 100 kg mol^{-1}; we should not say (even though it is common practice) that its molar mass is 100 kDa.

16.10 Mass spectrometry

Mass spectrometry is among the most accurate techniques for the determination of molar masses. The procedure consists of ionizing the sample in the gas phase and then measuring the mass-to-charge number ratio (*m*/*z*) of all ions. Macromolecules present a challenge because it is difficult to produce gaseous ions of large species without fragmentation. However, **matrix-assisted laser desorption/ionization** (MALDI) has overcome this problem. In this technique, the macromolecule is embedded in a solid matrix composed of an organic material and inorganic salts, such as sodium chloride or silver trifluoroethanoate, $AgCF_3CO_2$. This sample is then irradiated with a pulsed laser. The laser energy, which is absorbed by the matrix, ejects electronically excited matrix ions, cations, and neutral macromolecules, thus creating a dense gas plume above the sample surface. The macromolecule is ionized by collisions and complexation with small cations, such as H^+, Na^+, and Ag^+, and the masses of the resulting ions are determined in a mass spectrometer.

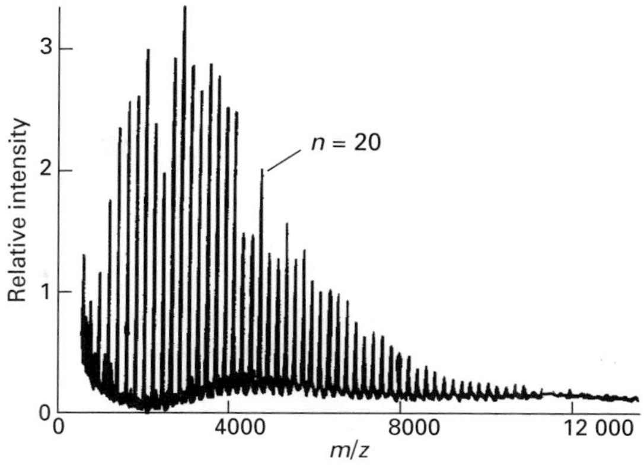

Fig. 16.32 MALDI spectrum of a sample of poly(butylene adipate) with $\bar{M}_n$ = 4525 g mol^{-1}. Adapted from Mudiman et al., *J. Chem. Educ.*, **74**, 1288 (1997).

Figure 16.32 shows the MALDI mass spectrum of a polydisperse sample of poly(oxybuteneoxyadipoyl), commonly called poly(butylene adipate) (**6**), obtained with NaCl in the matrix. The MALDI technique produces mostly singly charged molecular ions that are not fragmented. Therefore, the multiple peaks in the spectrum arise from polymers of different lengths (different '*N*-mers', where *N* is the number of repeating units), with the intensity of each peak being proportional to the abundance of each *N*-mer in the sample. Values of $\bar{M}_n$, $\bar{M}_w$, and the heterogeneity index can be calculated from the data. It is also possible to use the mass spectrum to verify the structure of a polymer, as shown in the following example.

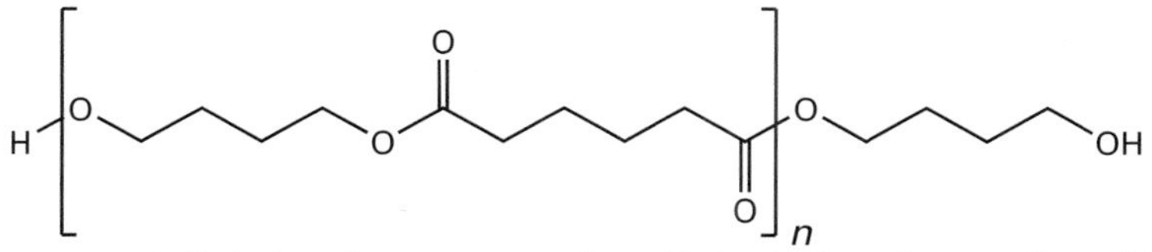

6 Poly(oxybuteneoxyadipoyl), 'poly(butylene adipate)'

Example 16.2

Interpreting the mass spectrum of a polymer

The mass spectrum in Fig. 16.32 consists of peaks spaced by 200 g mol^{-1}. The peak at 4113 g mol^{-1} corresponds to a polymer with N = 20 repeating units. The matrix used contained NaCl. From these data, verify that the sample consists of polymers with the general structure given by (**6**).

Strategy Because each peak corresponds to a different value of N, the molar mass difference, ΔM, between peaks corresponds to the molar mass, M, of the repeating unit (the group inside the brackets in **6**). Furthermore, the molar mass of the terminal groups (the groups outside the brackets in **6**) may be obtained from the molar mass of any peak, by using

$$M(\text{terminal groups}) = M(N\text{-mer}) - N\Delta M - M(\text{cation})$$

where the last term corresponds to the molar mass of the cation that attaches to the macromolecule during ionization.

Solution The value of ΔM is consistent with the molar mass of the repeating unit shown in (**6**), which is 200 g mol^{-1}. The molar mass of the terminal group is calculated by noting that Na^+ is the cation in the matrix:

$$\begin{aligned} M(\text{terminal group}) &= 4113\ \text{g mol}^{-1} - 20(200\ \text{g mol}^{-1}) \\ &\quad - 23\ \text{g mol}^{-1} \\ &= 90\ \text{g mol}^{-1} \end{aligned}$$

The result is consistent with the molar mass of the —$O(CH_2)_4OH$ terminal group (89 g mol^{-1}) plus the molar mass of the —H terminal group (1 g mol^{-1}).

Self-test 16.4

What would be the molar mass of the N = 20 polymer if silver trifluoroethanoate were used instead of NaCl in the preparation of the matrix?

Answer: 4.2 kg mol^{-1}

16.11 Ultracentrifugation

In a gravitational field, heavy particles settle towards the foot of a column of solution by the process called **sedimentation**. The rate of sedimentation depends on the strength of the field and on the masses and shapes of the particles. Spherical molecules (and compact molecules in general) sediment faster than rod-like or extended molecules. For example, DNA helices sediment much faster when they are collapsed into a random coil, so sedimentation rates can be used to study denaturation (the loss of structure). Sedimentation is normally very slow, but it can be accelerated by **ultracentrifugation**, a technique that replaces the gravitational field with a centrifugal field. The effect is achieved in an ultracentrifuge, which is essentially a cylinder that can be rotated at high speed about its axis with a sample in a cell near its periphery (Fig. 16.33). Modern ultracentrifuges can produce accelerations equivalent to about 10^5 that of gravity ('10^5g'). Initially the sample is uniform, but the solute molecules move towards the outer edge of the cell at a rate that can be interpreted in terms of the number-average molar mass. In an alternative 'equilibrium' version of the technique, the weight-average molar mass can be obtained from the ratio of concentrations c of the macromolecules at two different radii in a centrifuge operating at angular frequency ω (in radians per second):

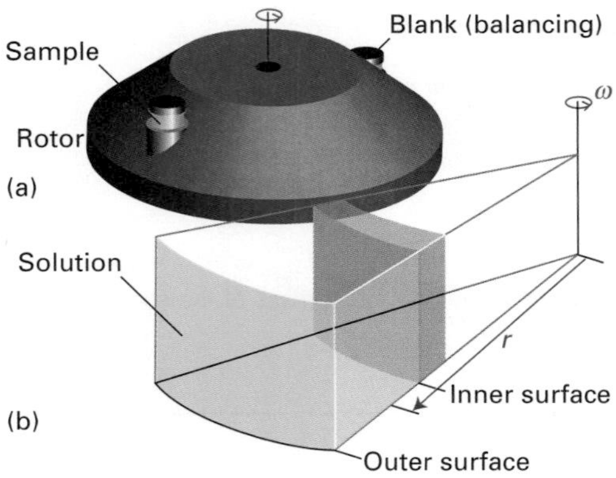

Fig. 16.33 (a) An ultracentrifuge head. The sample on one side is balanced by a blank diametrically opposite. (b) Detail of the sample cavity: the 'top' surface is the inner surface, and the centrifugal force causes sedimentation towards the outer surface; a particle at a radius r experiences a force of magnitude $mr\omega^2$.

$$\bar{M}_w = \frac{2RT}{(r_2^2 - r_1^2)b\omega^2}\ln\frac{c_2}{c_1} \qquad (16.10)$$

Here b is a factor that takes into account the buoyancy of the medium. The centrifuge is run more slowly in this technique than in the sedimentation rate method to avoid having all the solute pressed in a thin film against the bottom of the cell. At these slower speeds, several days may be needed for equilibrium to be reached.

Example 16.3

Determining molar mass by ultracentrifugation

The data from an equilibrium ultracentrifugation experiment performed at 295 K on an aqueous solution of a protein show that a graph of ln c against $(r/\text{cm})^2$ is a straight line with a slope of 0.959. The rotational rate of the centrifuge was 50 000 rotations per minute and b = 0.55. Calculate the weight-average molar mass of the protein.

Strategy We need to reinterpret eqn 16.10 in terms of the slope of a plot of ln c against r^2. To do so, we rewrite eqn 16.10 as

$$\bar{M}_w = \frac{2RT}{(r_2^2 - r_1^2)b\omega^2}\ln\frac{c_2}{c_1} \overbrace{=}^{\ln(x/y)=\ln x - \ln y} \frac{2RT}{(r_2^2 - r_1^2)b\omega^2}(\ln c_2 - \ln c_1)$$

$$\overbrace{=}^{\text{rearrangement}} \frac{2RT}{b\omega^2}\times\overbrace{\frac{\ln c_2 - \ln c_1}{r_2^2 - r_1^2}}^{\text{slope of } \ln c \text{ against } r^2}$$

If a plot of ln c against r^2 is linear, then the ratio (ln c_2 − ln c_1)/($r_2^2 - r_1^2$) has the form of the slope of the line. In practice, ln(c/g cm^{-3}) is plotted against (r/cm)^2 to give a dimensionless slope. It follows that

$$\bar{M}_w = \frac{2RT}{b\omega^2}\times(\text{slope}\times\text{cm}^{-2}) \qquad (16.11)$$

and we can use the data provided to calculate the weight-average molar mass $\bar{M}_w$. Each full revolution of the rotor corresponds to an angular change of 2π radians, so, to obtain the angular frequency ω, we multiply the rotation rate in cycles per second by 2π.

Solution The angular frequency is

$$\omega = 2\pi\times 50\,000\ \text{min}^{-1}\times\frac{1\ \text{min}}{60\ \text{s}} = 2\pi\times\frac{50\,000}{60}\text{s}^{-1}$$

It follows from eqn 16.11 with 1 cm^{-2} = 10^4 m^{-2} and the slope 0.959 that the weight-average molar mass is

$$\bar{M}_w = \frac{2\times\overbrace{(8.3145\ \overbrace{\text{J}}^{\text{kg m}^2\text{ s}^{-2}}\ \text{K}^{-1}\ \text{mol}^{-1})}^{R}\times\overbrace{(295\ \text{K})}^{T}\times(\overbrace{0.959}^{\text{slope}}\times 10^4\ \text{m}^{-2})}{\underbrace{0.55}_{b}\times\underbrace{\left(2\pi\times\frac{50\,000}{60}\ \text{s}^{-1}\right)^2}_{\omega}}$$

$$= 3.1\ \text{kg mol}^{-1}$$

Self-test 16.5

The data from a sedimentation equilibrium experiment performed at 293 K on a macromolecular solute in aqueous solution show that a graph of ln(c/g cm^{-3}) against (r/cm)^2 is a straight line with a slope of 0.821. The rotation rate of the centrifuge was 450 Hz (1 Hz = 1 s^{-1}) and b = 0.60. Calculate the weight-average molar mass of the solute.

Answer: 8.3 kg mol^{-1}

16.12 Electrophoresis

Many macromolecules, such as DNA, are charged and move in response to an electric field. This motion is called **electrophoresis**. Electrophoretic mobility is a result of a constant drift speed reached by an ion when the electrical driving force is matched by the frictional drag force. Electrophoresis is a very valuable tool in the separation of biopolymers from complex mixtures, such as those resulting from fractionation of biological cells. In **gel electrophoresis**, migration takes place through a gel slab. In **capillary electrophoresis**, the sample is dispersed in a medium (such as methylcellulose) and held in a thin glass or plastic tube with diameters ranging from 20 to 100 μm. The small size of the apparatus makes it easy to dissipate heat when large electric fields are applied. Excellent separations may be effected in minutes rather than hours. Each polymer fraction emerging from the capillary can be characterized further by other techniques, such as MALDI.

16.13 Laser light scattering

Light scattering measurements of polymer size are based on the observation that large particles scatter light very efficiently. A familiar example is the light scattered by specks of dust in a sunbeam. Analysis of the intensity of light scattered by a sample at different angles relative to the incident radiation from a monochromatic laser beam yields the size and molar mass of a polymer, large aggregate (such as a colloid; see Section 16.6), or biological system ranging in size from a protein to a virus.

Dynamic light scattering is used to investigate the diffusion of polymers in solution. Consider two polymer molecules being irradiated by a laser beam. Suppose that at one instant the scattered waves from these particles interfere constructively at the detector, leading to a large signal. However, as the molecules move through the solution, the scattered waves may interfere destructively at a later instant and result in no signal. When this behaviour is extended to a very large number of molecules in solution, it results in fluctuations in light intensity that can be analysed to reveal the molar mass and diffusion coefficient of the polymer.

Checklist of key concepts

☐ 1 Macromolecules are very large molecules assembled from smaller molecules.

☐ 2 Synthetic polymers are manufactured by stringing together and in some cases cross-linking smaller units known as monomers.

☐ 3 The conformation of a macromolecule is the spatial arrangement of the different parts of a chain.

☐ 4 The primary structure of a polymer is the sequence of its monomer units.

☐ 5 The least structured model of a macromolecule is as a random coil.

☐ 6 The secondary structure of a protein is the spatial arrangement of the polypeptide chain and includes the α-helix and β-sheet.

☐ 7 Helical and sheet-like polypeptide chains are folded into a tertiary structure by bonding influences between the residues of the chain.

☐ 8 Some macromolecules have a quaternary structure as aggregates of two or more polypeptide chains.

☐ 9 Protein denaturation is loss of structure; a helix–coil transition is a cooperative process.

☐ 10 Synthetic polymers are classified as elastomers, fibres, or plastics.

☐ 11 A perfect elastomer is a polymer in which the internal energy is independent of the extension of the random coil; for small extensions, a random coil model obeys a Hooke's-law restoring force.

☐ 12 Synthetic polymers undergo a transition from a state of high to low chain mobility at the glass transition temperature, T_g.

☐ 13 A mesophase is a bulk phase that is intermediate in character between a solid and a liquid.

☐ 14 A disperse system is a dispersion of small particles of one material in another.

☐ 15 Liquid crystals are classified as smectic, nematic, or cholesteric.

☐ 16 A surfactant is a species that accumulates at the interface of two phases or substances and modifies the properties of the surface.

☐ 17 The radius of shear is the radius of the sphere that captures the rigid layer of charge attached to a colloid particle.

☐ 18 The electrokinetic potential is the electric potential at the radius of shear relative to its value in the distant, bulk medium.

☐ 19 The inner shell of charge and the outer atmosphere jointly constitute the electric double layer.

☐ 20 Many colloid particles are thermodynamically unstable but kinetically non-labile.

☐ 21 Surface tension is a measure of the work needed to produce a liquid surface.

☐ 22 The pressure on the convex side of a curved surface is lower than that on the concave side; the difference gives rise to capillary action.

☐ 23 The accumulation of a surfactant at a surface lowers the surface tension.

☐ 24 Many proteins (specifically protein enzymes) are monodisperse; a synthetic polymer is polydisperse.

☐ 25 Techniques for the determination of the mean molar masses of macromolecules include osmometry, mass spectrometry (as MALDI), sedimentation rates and equilibria, gel and capillary electrophoresis, and laser light scattering.

Road map of key equations

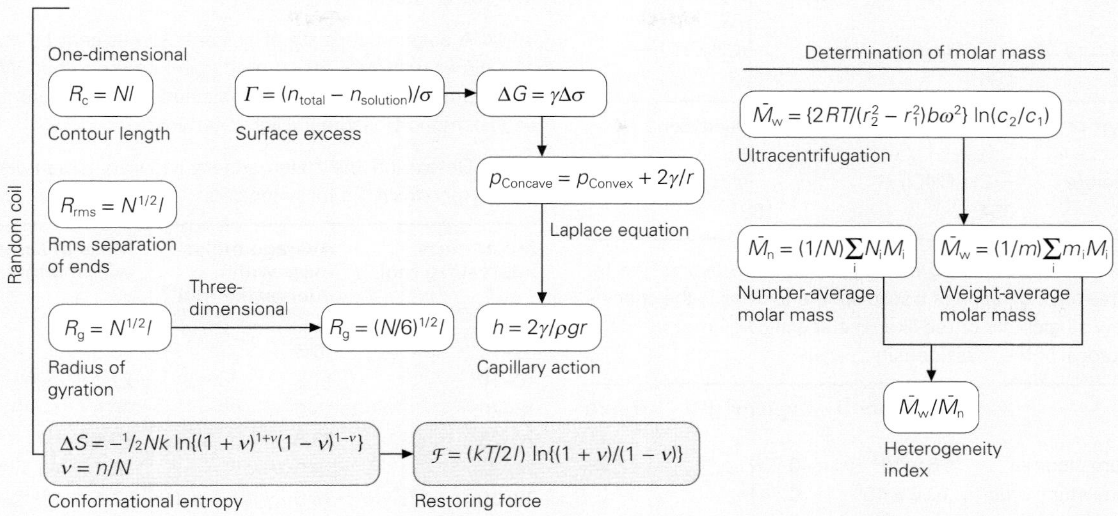

The blue boxes indicate expressions valid for perfect elastomers.

Questions and exercises

Discussion questions

16.1 Distinguish between number-average and weight-average molar masses. Why might they differ?

16.2 Distinguish between contour length, root-mean-square separation, and radius of gyration of a random coil.

16.3 What are the limitations of the random coil model of a polymer?

16.4 Describe the methods available for the determination of the molar masses of macromolecules and polymers.

16.5 Why does a perfect elastomer act like a coiled spring?

16.6 What molecular interactions contribute to the formation, thermal stability, and mechanical strength of polymeric material?

16.7 Explain the physical origins of surface activity by surfactant molecules.

16.8 Explain the formation and importance of the electric double layer in the context of disperse systems.

Exercises

16.1 A polymer chain consists of 800 segments, each 1.10 nm long. If the chain were ideally flexible, what would be (a) the contour length, (b) the root-mean-square separation of the ends of the chain?

16.2 Calculate the contour length and the root-mean-square separation of the ends of the chain for polyethene with a molar mass of 250 kg mol^{-1}.

16.3 The radius of gyration of a long chain molecule is found to be 7.3 nm. The chain consists of C—C links. Assume the chain is randomly coiled and estimate the number of links in the chain.

16.4 What is the change in conformational entropy when a random coil is stretched from fully coiled to 10 per cent (that is, $\nu = 0.1$ in eqn 16.2).

16.5 An elastomeric chain of polybutadiene, $-(CH_2CHCHCH_2)_n-$, consisting of n = 4000 units, each of length 150 pm, is extended by 5.0 per cent of its overall length. What is the magnitude of the restoring force at 25 °C? Use Hooke's law to calculate the force constant of the elastomer and hence the frequency with which the chain vibrates.

16.6 The following table lists the glass transition temperatures, T_g, of several polymers. Discuss the reasons why the structure of the monomer unit has an effect on the value of T_g.

Polymer	Poly(oxymethylene)	Polyethene
Structure	$-(OCH_2)_n-$	$-(CH_2CH_2)_n-$
T_g/K	198	253
Polymer	**Poly(vinyl chloride)**	**Polystyrene**
Structure	$-(CH_2CHCl)_n-$	$-(CH_2CH(C_6H_5))_n-$
T_g/K	354	381

16.7 Use the following information and the expression for the radius of gyration of a solid sphere to classify the species below as globular or rod-like. The specific volume, v_s, is the reciprocal of the mass density.

	M/(g mol^{-1})	v_s/(cm^3 g^{-1})	R_g/nm
Serum albumin	66×10^3	0.752	2.98
Bushy stunt virus	10.6×10^6	0.741	12.0
DNA	4×10^6	0.556	117.0

16.8 Determine the work that must be done to double the volume of a spherical bubble (treated as a spherical cavity) of air of initial radius 5.0 mm in water at 298 K. The surface tension of water at this temperature is 72 mN m^{-1}.

16.9 To what height would you expect ethanol ($\gamma = 22.39$ mN m^{-1} at 298 K, $\rho = 789$ kg m^{-3}) to climb by capillary action in a tube of internal radius 0.10 mm?

16.10 In an experiment to determine the surface tension of methanol ($\rho = 791$ kg m^{-3} at 298 K) it was found that it rose to a height of 5.8 cm in a tube of internal diameter 0.20 mm. What is the surface tension of methanol at 298 K?

16.11 Use the Laplace equation to calculate the difference in pressure on either side of a curved surface of water ($\gamma = 72$ mN m^{-1} at 298 K) of radius (a) 0.10 mm, (b) 1.0 mm.

16.12 Calculate the surface excess of solute given the following data:

Molar concentration in bulk solution as prepared: 0.100 mol dm^{-3}

Molar concentration in bulk solution as determined: 0.981 mol dm^{-3}

Total volume of solution: 100 cm^3

Radius of beaker containing the solution: 2.50 cm

16.13 Calculate the number-average molar mass and the weight-average molar mass of a mixture of equal amounts of two polymers, one having $M = 82$ kg mol^{-1} and the other $M = 108$ kg mol^{-1}.

16.14 A solution consists of solvent, 30 per cent by mass of a dimer with $M = 30$ kg mol^{-1}, and its monomer. What average molar mass would be obtained from measurement of (a) osmotic pressure, (b) light scattering?

16.15 Determine the heterogeneity index of a sample of polystyrene from the following data:

Molar mass interval/(kg mol^{-1})	Average molar mass within interval/(kg mol^{-1})	Mass of sample within interval/g
5–10	6.5	16.0
10–15	11.5	27.1
15–20	19.5	29.5
20–25	23.5	13.4
25–30	28.5	8.7
30–35	35.5	3.5

16.16 Polystyrene is a synthetic polymer of composition $-(CH_2CH(C_6H_5))_n-$. A batch of polydisperse polystyrene was prepared by initiating the polymerization with *t*-butyl radicals. As a result, the *t*-butyl group is expected to be attached covalently to the end of the final products. A sample from this batch was embedded in an organic matrix containing silver trifluoroethanoate and the resulting MALDI-TOF spectrum consisted of a large number of peaks separated by 104 g mol^{-1}, with the most intense peak at 25 578 g mol^{-1}. Comment on the purity of this sample and determine the number of $-CH_2CH(C_6H_5)-$ units in the species that gives rise to the most intense peak in the spectrum.

16.17 The data from a sedimentation equilibrium experiment performed at 300 K on a macromolecular solute in aqueous solution show that a graph of ln c against $(r/\text{cm})^2$ is a straight line with a slope of 659. The rotation rate of the centrifuge was 55 000 r.p.m. The specific volume of the solute is $v_s = 0.61$ cm^3 g^{-1}. Calculate the molar mass of the solute. *Hint*: Use eqn 16.10; you need to know that the buoyancy correction is $b = 1 - \rho v_s$; take $\rho = 0.996$ g cm^{-3}.

Projects

The symbol ‡ indicates that calculus is required.

16.18‡ The probability that the ends of a three-dimensional random coil of N links each of length l will be found in the range R to $R + dR$ is $f(R)dR$, where $f(R) = 4\pi(a/\pi^{1/2})R^2e^{-a^2R^2}$ with $a = (\frac{3}{2}Nl^2)^{1/2}$. Use this expression to deduce expressions for (a) the root-mean-square separation of the ends of the chain, (b) the mean separation of the ends, and (c) their most probable separation. Evaluate these three quantities for a fully flexible chain with $N = 5000$ and $l = 154$ pm.

16.19 Construct a two-dimensional random walk by using a random number generating routine with mathematical software or electronic spreadsheet. Construct a walk of 50 and 100 steps. If there are many people working on the problem, investigate the mean and most probable separations in the plots by direct measurement. Do they vary as $N^{1/2}$?

16.20‡ Here we explore elastomers in quantitative detail. (a) Estimate the force required to expand a random coil (a perfect elastomer) consisting of 1000 links by 10 per cent of its fully coiled state at 300 K. (b) The restoring force acting when a random coil is extended by dx is related to the conformational entropy by $F = -T\mathrm{d}S/\mathrm{d}x$. Use this expression to deduce eqn 16.3a and eqn 16.3b.

16.21 Equation 16.4b is surprisingly tricky to solve. Convince yourself of that by taking the very simple case of $N = 2$ and $K = 1$, and find an expression for $[M_2]$. *Hint:* Use the fact that $[M_2] < [M]_{total}$ to eliminate one of the roots of the quadratic equation. Now extend your approach to solving eqn 16.4b by using mathematical software, increasing N systematically until the transition becomes sharp. Take $K = 1$ initially, but once you have established the procedure, explore the consequences of changing K.

17

Metallic, ionic, and covalent solids

Bonding in solids 420

17.1 The band theory of solids 421
17.2 The occupation of bands 422
17.3 The optical properties of junctions 424
17.4 Superconductivity 424
17.5 The ionic model of bonding 425
17.6 Lattice enthalpy 425
17.7 The origin of lattice enthalpy 427
17.8 Covalent networks 429
17.9 Magnetic properties of solids 430

Crystal structure 432

17.10 Unit cells 432
17.11 The identification of crystal planes 433
17.12 The determination of structure 435
17.13 Bragg's law 437
17.14 Experimental techniques 438
17.15 Metal crystals 440
17.16 Ionic crystals 442
17.17 Molecular crystals 443

CHECKLIST OF KEY CONCEPTS 445
ROAD MAP OF KEY EQUATIONS 446
QUESTIONS AND EXERCISES 446

Modern chemistry is closely concerned with the properties of solids. Apart from their intrinsic usefulness for construction, modern solids have made possible the semiconductor revolution and recent advances in ceramics have given rise to the hope that we may now be on the verge of a superconductor revolution. Advances in our understanding of electron mobility in solids are also useful in biology, where electron transport is responsible for many biochemical processes, particularly photosynthesis and respiration.

The principal technique for investigating the arrangements of atoms in condensed phases, primarily crystalline solids, is X-ray diffraction, but nuclear magnetic resonance (NMR, Chapter 21) is now also making significant contributions. Information from X-ray diffraction and NMR is the basis of much of molecular biology, so the material presented here is the foundation for our discussion of biomolecular structures in Chapter 16. In each case, the observed crystal structure is Nature's solution to the problem of condensing objects of various shapes into an aggregate of minimum energy and, for temperatures above zero, of minimum Gibbs energy.

Bonding in solids

The bonding within a solid may be of various kinds. Simplest of all (in principle) is the bonding in an elemental **metallic solid**, in which electrons are delocalized over arrays of identical cations and bind the whole together into a rigid but malleable structure. Because the delocalized electrons can accommodate bonding patterns with very little directional character, the crystal structures of metals are determined largely by the geometrical problem of packing spherical atoms into a dense, orderly array. In an **ionic solid**, the ions (in general, of different radii, and not always spherical) are held together by their Coulombic interaction, and pack together to give an electrically neutral structure. In a **covalent solid** (or *network solid*), covalent bonds in a definite spatial orientation link the atoms in a network extending through the crystal. The stereochemical demands of valence now override the geometrical problem of packing spheres together, and elaborate and extensive structures may be formed. Important examples of covalent solids are diamond and graphite (Section 17.8). **Molecular solids**, which are the subject of the overwhelming majority of modern structural determinations, consist of discrete molecules attracted to one another by the interactions described in Chapter 15.

Some solids—notably the metals—conduct electricity because they have mobile electrons. These **electronic conductors** are classified on the basis of the variation of their electrical conductivity with temperature (Fig. 17.1):

- A **metallic conductor** is an electronic conductor with a conductivity that *decreases* as the temperature is raised.
- A **semiconductor** is an electronic conductor with a conductivity that *increases* as the temperature is raised.

Metallic conductors include the metallic elements, their alloys, and graphite (parallel to the graphene planes). Some organic solids are metallic conductors. Semiconductors include silicon, diamond, and gallium arsenide. A semiconductor generally has a lower conductivity than that typical of metals, but the magnitude of the conductivity is not relevant to the distinction. It is conventional to classify substances with very low electrical conductivities, such as most ionic solids, as **insulators**. We shall use this term, but it is one of convenience rather than one of fundamental significance. **Superconductors** are substances that conduct electricity with zero resistance. The mechanism of superconductivity in metals at very low (liquid helium) temperatures is well understood: that of the potentially more useful **high-temperature superconductors** (HTSCs), which are ceramic mixed oxides such as $YBa_2Cu_3O_7$, is still unresolved.

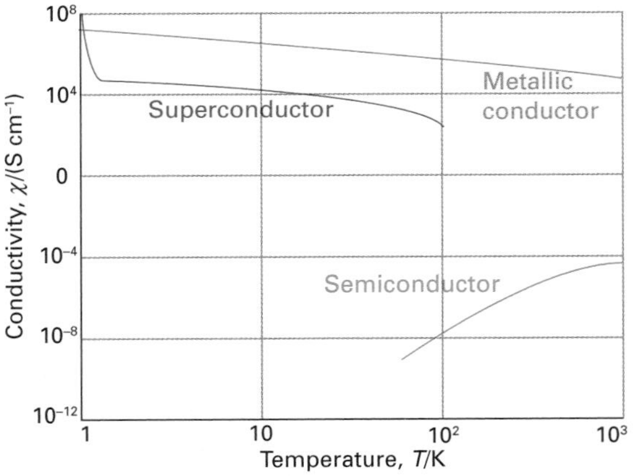

Fig. 17.1 The typical variation with temperature of the electrical conductivities of different classes of electronic conductor.

17.1 The band theory of solids

Metallic and ionic solids can both be treated by molecular orbital theory. The advantage of that approach is that we can then see both types of solid as two extremes of a single kind. In each case, the electrons responsible for the bonding are delocalized throughout the solid (like in a benzene molecule, but on a much bigger scale). In an elemental metal, the electrons can be found on all the atoms with equal probability, which matches the primitive picture of a metal as consisting of cations embedded in a nearly uniform electron 'sea'. In an ionic solid the wavefunctions occupied by the delocalized electrons are almost entirely concentrated on the anions, so the Cl atoms in NaCl, for instance, are present as Cl^- ions and the Na atoms, which have low valence electron density, are present as Na^+ ions.

To set up the molecular orbital theory of solids we shall consider initially a single, infinitely long line of identical atoms, each one having an s orbital available for forming molecular orbitals (as in sodium). One atom of the solid contributes an s orbital with a certain energy (Fig. 17.2). When a second atom is

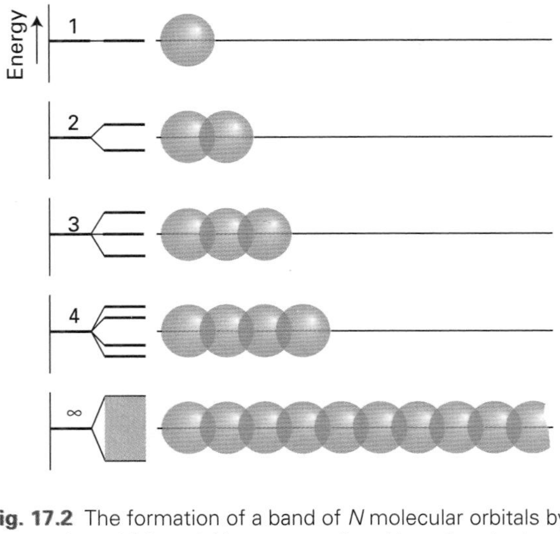

Fig. 17.2 The formation of a band of *N* molecular orbitals by successive addition of *N* atoms to a line. Note that the band remains of finite width, and although it looks continuous when *N* is large, it consists of *N* different orbitals.

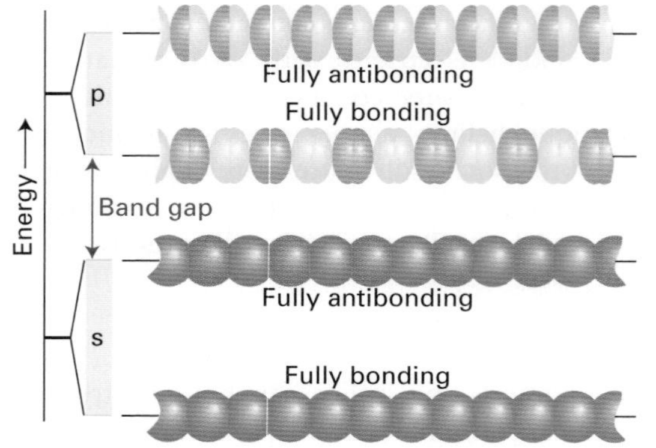

Fig. 17.3 The overlap of s orbitals gives rise to an s band and the overlap of p orbitals gives rise to a p band. In this case the s and p orbitals of the atoms are so widely spaced that there is a band gap. In many cases the separation is less, and the bands overlap.

Derivation 17.1

The width of a band

The energy of the level with $k = 1$ is

$$E_1 = \alpha + 2\beta\cos\left(\frac{\pi}{N+1}\right)$$

As N becomes infinite, the cosine term becomes cos 0, which is equal to 1. Therefore, in this limit, $E_1 = \alpha + 2\beta$. When k has its maximum value of N,

$$E_N = \alpha + 2\beta\cos\left(\frac{N\pi}{N+1}\right)$$

As N approaches infinity, we can ignore the 1 in the denominator, and the cosine term becomes cos π, which is equal to −1. Therefore, in this limit, $E_N = \alpha - 2\beta$. The difference between the upper and lower energies of the band is therefore $4|\beta|$. We use modulus signs (|. . .|, the instruction to ignore the negative sign) around β because β itself is negative but the width of the band, like any width, is a positive quantity.

brought up it forms a bonding and antibonding orbital. The orbital of the third atom overlaps its nearest neighbour (and only slightly the next-nearest), and three molecular orbitals are formed from these three atomic orbitals. The fourth atom leads to the formation of a fourth molecular orbital. At this stage we can begin to see that the general effect of bringing up successive atoms is to spread the range of energies covered by the molecular orbitals, and also to fill in the range of energies with more and more orbitals (one more for each additional atom). When N atoms have been added to the line, there are N molecular orbitals covering a band of finite width. The lowest-energy orbital of this band is fully bonding and the highest-energy orbital is fully antibonding between adjacent atoms (Fig. 17.3). In the Hückel approximation (Section 14.13), the energies of the orbitals are given by

$$E_k = \alpha + 2\beta\cos\left(\frac{k\pi}{N+1}\right) \quad k = 1, 2, \ldots, N \qquad \text{Metal orbital energies} \qquad (17.1)$$

where α is approximately equal to (the negative of) the ionization energy of the atom and β is a negative quantity that represents the lowering of energy due to interaction between neighbouring atoms. As N becomes infinite, the separation between neighbouring levels, $E_{k+1} - E_k$ goes to zero but, as shown in the following Derivation, the width of the band, $E_N - E_1$, becomes $4|\beta|$, a finite quantity.

A band formed from overlap of s orbitals is called an **s band.** If the atoms have p orbitals available, then the same procedure leads to a **p band** (as in the upper half of Fig. 17.3, with different values of α and β in eqn 17.1). The p band in the illustration has supposed σ overlap along the chain; a p band may also arise from π overlap between neighbours. If the atomic p orbitals lie higher in energy than the s orbitals, then the p band lies higher than the s band, and there may be a **band gap**, a range of energies for which no molecular orbitals exist. If the separation of the atomic orbitals is not large, the two types of band might overlap.

17.2 The occupation of bands

Now consider the electronic structure of a solid formed from atoms each of which is able to contribute one valence orbital and one electron (for example, the alkali metals). There are N atomic orbitals and therefore N molecular orbitals squashed into a band of finite width. There are N electrons to accommodate; they form pairs that occupy the lowest $\frac{1}{2}N$ molecular orbitals (Fig. 17.4). The highest occupied molecular orbital is called the **Fermi level.** However, unlike in the discrete molecules we considered in Chapter 14, there are empty orbitals just above and very close in energy to the Fermi level, so it requires hardly any energy to excite the uppermost electrons. Some of the electrons are therefore very mobile and give rise to electrical conductivity. An unfilled band of orbitals is called a **conduction band.**

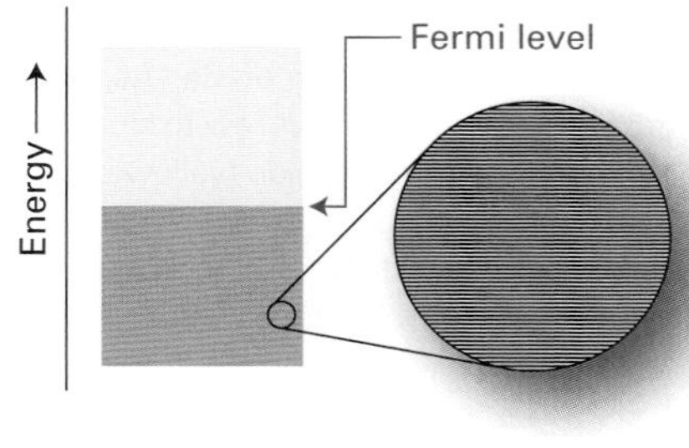

Fig. 17.4 When N electrons occupy a band of N orbitals, it is only half full and the electrons near the Fermi level (the top of the filled levels) are mobile.

As we have remarked, metallic conductivity is characterized by a decrease in electrical conductivity with increasing temperature. This behaviour is accommodated in the present model because an increase in temperature causes more vigorous thermal motion of the atoms, with the result that there are more collisions between the moving electrons and the atoms. That is, at high temperatures the electrons are scattered out of their paths through the solid and are less efficient at transporting charge.

When each atom provides one valence orbital but two electrons, the $2N$ electrons fill the N orbitals of the s band. The Fermi level now lies at the top of the band and there is a gap before the next band begins (Fig. 17.5a). A filled band is called a **valence band.** It might be suspected that such elements, which include members of Group 2, will be insulators. However, p orbitals also form bands that in some cases (as in Group 2) overlap the s bands. The bands available to the electrons are then not full and the elements are metallic conductors.

When there is a gap between an s band and a p band the element and the former is full, the substance is not a metallic conductor. However, as the temperature is increased, electrons can populate the empty orbitals of the upper band (Fig. 17.5b): we saw in

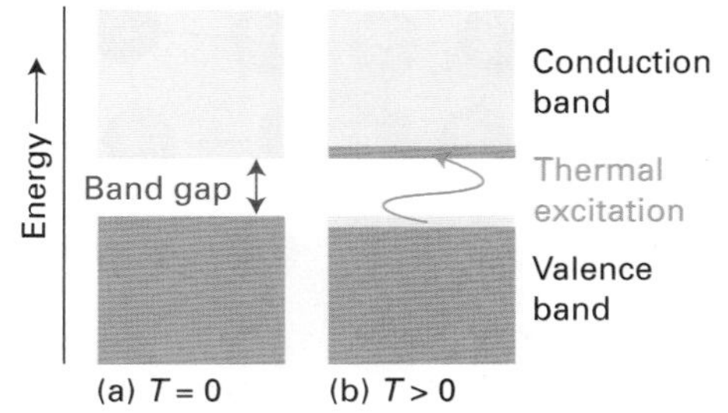

Fig. 17.5 (a) When $2N$ electrons are present, the band is full and the material is an insulator at $T = 0$. (b) At temperatures above $T = 0$, electrons populate the levels of the conduction band at the expense of the valence band, and the solid is a semiconductor.

Foundations that the Boltzmann distribution spreads into higher energy levels as the temperature is raised. They are now mobile, and the solid has become an electronic conductor. In fact, it is a semiconductor, because the electrical conductivity depends on the number of electrons that are promoted across the gap and that number increases, and the electrical conductivity increases accordingly, as the temperature is raised. This condition is observed for elements, such as Si and Ge, where bands formed from valence electrons do not overlap, so leaving a gap.

If the gap is large, very few electrons will be excited across it at ordinary temperatures and the conductivity will remain close to zero, giving an insulator. Thus, the conventional distinction between an insulator and a semiconductor is related to the size of the band gap and is not absolute like the distinction between a metal (incomplete bands at $T = 0$) and a semiconductor (full bands at $T = 0$).

Another method of increasing the number of charge carriers and enhancing the semiconductivity of a solid is to implant foreign atoms into an otherwise pure material. If these **dopants** can trap electrons (as indium or gallium atoms can in silicon, because In and Ga atoms have one fewer valence electron than Si), they withdraw electrons from the filled band, leaving holes which allow the remaining electrons to move (Fig. 17.6a). This doping procedure gives rise to **p-type semiconductivity**, the p indicating that the holes are positive relative to the electrons in the band. Alternatively, a dopant might carry excess electrons (for example, phosphorus atoms introduced into germanium), and these additional electrons occupy otherwise empty bands, giving **n-type semiconductivity**

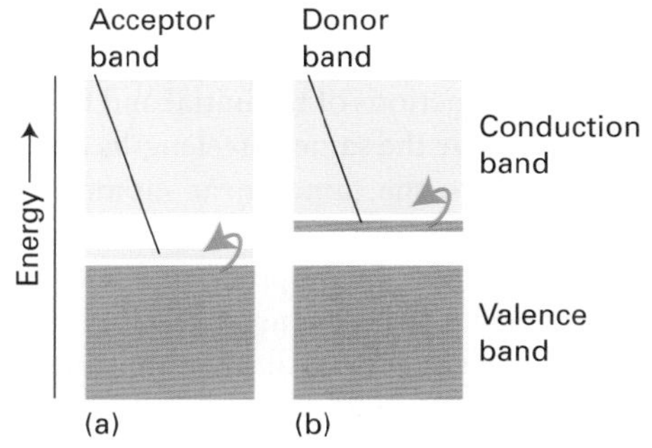

Fig. 17.6 (a) A dopant with fewer electrons than its host can form a narrow band that accepts electrons from the valence band. The holes in the valence band are mobile, and the substance is a *p-type semiconductor*. (b) A dopant with more electrons than its host forms a narrow band that can supply electrons to the conduction band. The electrons it supplies are mobile, and the substance is an *n-type semiconductor*.

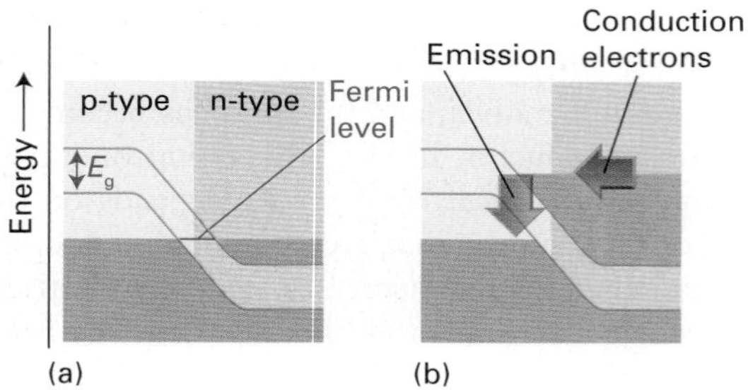

Fig. 17.7 The structure of a diode junction (a) without bias, (b) with bias (that is, with a potential difference applied).

(Fig. 17.6b), where n denotes the negative charge of the carriers.

17.3 The optical properties of junctions

The band structure of a **p–n junction**, the interface of the two types of semiconductor, is shown in Fig. 17.7. When electrons are supplied through an external circuit to the n side of the junction, the electrons in the conduction band of that semiconductor fall into the holes in the valence band of the p-type semiconductor.

As the electrons fall from the upper band into the lower, they release energy. In some solids the wavelengths of the wavefunctions in the upper and lower states are different, which means that the linear momenta (through the de Broglie relation, $p = h/\lambda$) of the electron in the initial and final states are different. As a result, the transition can occur only if the electron transfers linear momentum to the lattice: the device becomes warm as the atoms are stimulated to vibrate. This is the case for silicon semiconductors, and is one reason why computers need efficient cooling systems.

In some materials, most notably gallium arsenide, GaAs, the wavefunctions of the initial and final states of the electron have the same wavelengths and therefore correspond to the same linear momentum. As a result, transitions can occur without the lattice needing to participate by mopping up the difference in linear momenta. The energy difference is therefore emitted as light. Practical **light-emitting diodes** of this kind are widely used in electronic displays. Gallium arsenide itself emits infrared light, but the width of the band gap is increased by incorporating phosphorus.

Brief illustration 17.1 Light-emitting diodes

A material of composition approximately $GaAs_{0.6}P_{0.4}$ emits light in the red region of the spectrum, and diodes emitting orange and amber light can also be made with different proportions of Ga, As, and P. The spectral region ranging from yellow to blue can be covered by using gallium phosphide (yellow or green light) and gallium nitride (green or blue light). With some modification, these materials can also be used in the fabrication of *diode lasers*, as we shall see in Chapter 20.

17.4 Superconductivity

Following the discovery by the Dutch physicist Heike Kamerlingh Onnes in 1911 that mercury is a superconductor below the **critical temperature**, T_c, of 4.2 K, the boiling point of liquid helium, physicists and chemists made slow but steady progress in the discovery of superconductors with higher critical temperatures. Metals, such as tungsten, mercury, and lead, tend to have critical temperatures below about 10 K. Intermetallic compounds, such as Nb_3X (X = Sn, Al, or Ge), and alloys, such as Nb/Ti and Nb/Zr, have critical temperatures between 10 K and 23 K. In 1986, however, an entirely new range of high-temperature superconductors (HTSC) was discovered with critical temperatures well above 77 K, the boiling point of the inexpensive refrigerant liquid nitrogen. For example, $HgBa_2Ca_2Cu_2O_8$ has T_c = 153 K.

The central concept of superconductivity is the existence of a **Cooper pair**, a pair of electrons that exists on account of the indirect electron–electron interactions mediated by the nuclei of the atoms in the lattice. Thus, if one electron is in a particular region of a solid, the nuclei there move toward it and give rise to a distorted local structure (Fig. 17.8).

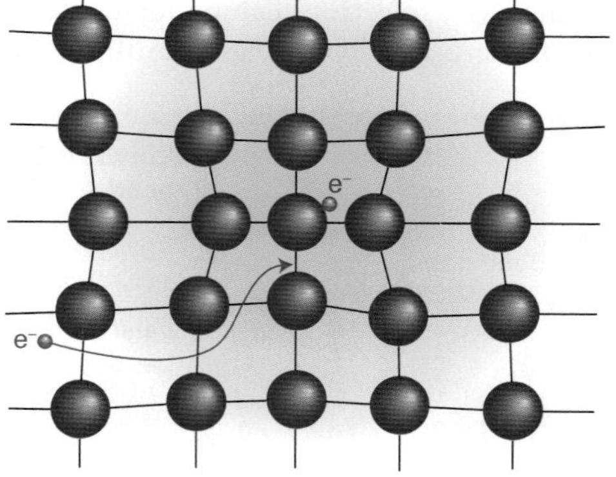

Fig. 17.8 The formation of a Cooper pair. One electron distorts the crystal lattice and the second electron has a lower energy if it goes to that region. These electron–lattice interactions effectively bind the two electrons into a pair.

Because that local distortion is rich in positive charge, it is favourable for a second electron to join the first. Hence, there is a virtual attraction between the two electrons, and they move together as a pair. A Cooper pair undergoes less scattering than an individual electron as it travels through the solid because the distortion caused by one electron can attract back the other electron should it be scattered out of its path in a collision. Because the Cooper pair is stable against scattering, it can carry charge freely through the solid, and hence give rise to superconductivity. The local distortion is disrupted by thermal motion of the ions in the solid, so the virtual attraction occurs only at very low temperatures.

The Cooper pairs responsible for low-temperature superconductivity are likely to be important in HTSCs, but the mechanism for pairing is hotly debated.

Brief illustration 17.2 High-temperature superconductors

One of the most widely studied superconductors is $YBa_2Cu_3O_7$, which consists of layers of square-pyramidal CuO_5 units and almost flat sheets of square-planar CuO_4 units (Fig. 17.9). It is believed that movement of electrons along the linked CuO_4 units accounts for superconductivity, whereas the linked CuO_5 units act as 'charge reservoirs' that maintain an appropriate number of electrons in the superconducting layers.

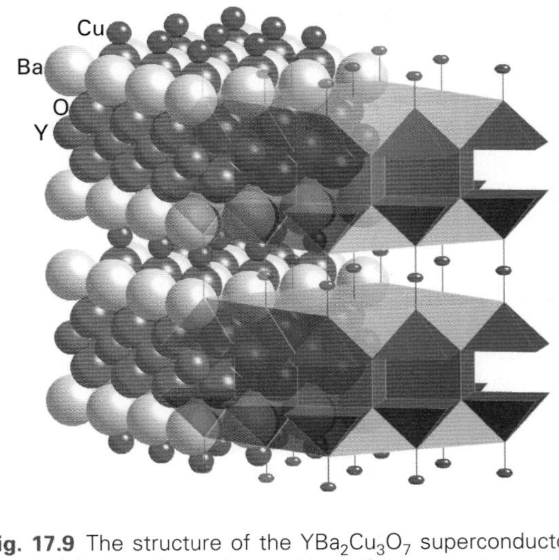

Fig. 17.9 The structure of the $YBa_2Cu_3O_7$ superconductor. The polyhedra show the position of oxygen atoms and indicate that the metal ions are in square-planar and square-pyramidal coordination environments.

17.5 The ionic model of bonding

Suppose we have a line of atoms with different electronegativities, such as a one-dimensional array of sodium and chlorine atoms rather than the identical atoms treated so far. Each sodium atom contributes an s orbital and one electron. Each chlorine atom contributes a p orbital and its one electron.

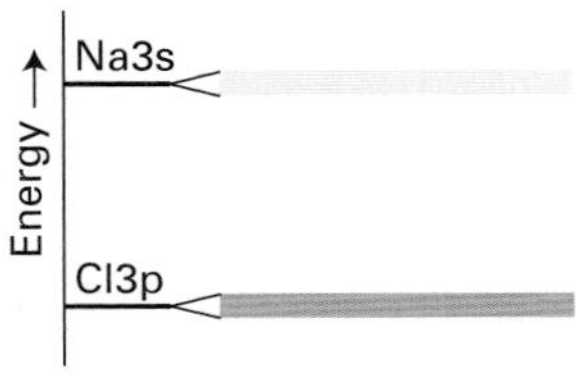

Fig. 17.10 The bands formed from two elements of widely different electronegativity (such as sodium and chlorine): they are widely separated and narrow. If each atom provides one electron, the lower band is full and the substance is an insulator.

We use the s and p orbitals to build molecular orbitals that spread throughout the solid. Now, though, there is a crucial difference. The orbitals on the two types of atom have markedly different energies, so (just as in the construction of molecular orbitals for diatomic molecules, Section 14.11) we consider them separately. The Cl3p orbitals interact to form one band and the higher energy Na3s orbitals interact to form another band. However, because the sodium atoms have very little overlap with one another (they are separated by a chlorine atom), the Na3s band is very narrow; so is the Cl3p band, for a similar reason. As a result, there is a big gap between two narrow bands (Fig. 17.10).

Now consider the occupation of the bands. If there are N sodium atoms and N chlorine atoms, there will be $2N$ electrons to accommodate (one from each Na atom and one from each Cl atom). These electrons occupy and fill the lower Cl3p band. As a result of the big band gap, the substance is an insulator. Moreover, because only the Cl3p band is occupied, the electron density is almost entirely on the chlorine atoms. In other words, we can treat the solid as composed of Na^+ cations and Cl^- anions, just as in an elementary picture of ionic bonding.

Now that we know where the electron density is largely located, we can adopt a much simpler model of the solid. Instead of expressing the structure in terms of molecular orbitals, we treat it as a collection of cations and anions. This simplification is the basis of the **ionic model** of bonding.

17.6 Lattice enthalpy

The strength of a covalent bond is measured by its dissociation energy, the energy needed to separate the two atoms joined by the bond. For thermodynamic applications we express this energy in terms of the bond enthalpy (Section 3.2). The strength of an ionic bond is measured similarly, but now we have to take into account the energy required to separate *all* the ions of a solid sample from one another and, for thermodynamic applications, express this energy as a change in enthalpy. The **lattice enthalpy**, $\Delta H_L^\ominus$, is the standard enthalpy change accompanying the separation of the species that compose the solid (such as ions if the solid is ionic, and molecules if the solid is molecular) per mole of formula units. For example, the lattice enthalpy of an ionic solid such as sodium chloride is the standard molar enthalpy change accompanying the process:

$$\text{NaCl(s)} \rightarrow \text{Na}^+\text{(g)} + \text{Cl}^-\text{(g)} \quad \Delta H_L^\ominus = 786 \text{ kJ mol}^{-1}$$

Because the lattice enthalpy is invariably a positive quantity, it is normally reported without its + sign. The lattice enthalpy of a molecular solid, such as ice, is the standard molar enthalpy of sublimation; the lattice enthalpy of a metal is its enthalpy of atomization.

Lattice enthalpies of solids are determined from other experimental data by using a **Born–Haber cycle**, which is a cycle (a closed path) of steps that includes lattice formation as one stage. The value of the lattice enthalpy—the only unknown in a well-chosen cycle—is found from the requirement that the sum of the enthalpy changes measured at a single temperature round a complete cycle is zero (because enthalpy is a state property). A typical cycle for an ionic compound has the form shown in Fig. 17.11.

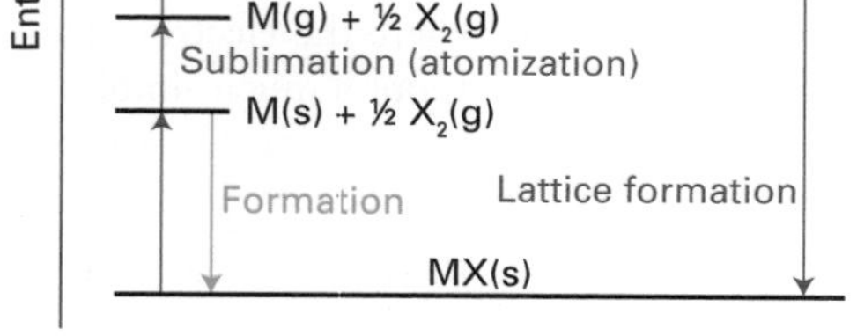

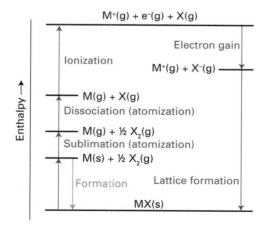

Fig. 17.11 The Born–Haber cycle for the determination of one of the unknown enthalpies, most commonly the lattice enthalpy. Upward pointing arrows denote positive changes in enthalpy; downward pointing arrows denote negative enthalpy changes. All the steps in the cycle correspond to the same temperature.

Table 17.1
Lattice enthalpies, $\Delta H_L^\ominus$/(kJ mol^{-1})

LiF	1037	LiCl	852	LiBr	815	LiI	761
NaF	926	NaCl	786	NaBr	752	NaI	705
KF	821	KCl	717	KBr	689	KI	649
MgO	3850	CaO	3461	SrO	3283	BaO	3114
MgS	3406	CaS	3119	SrS	2974	BaS	2832
Al_2O_3	15 900						

The following example illustrates how the cycle is used and Table 17.1 gives characteristic values.

Example 17.1

Using a Born–Haber cycle to determine a lattice enthalpy

Calculate the lattice enthalpy of KCl(s) using a Born–Haber cycle and the following data, which are all for 25 °C.

Process	$\Delta H^\ominus$/(kJ mol^{-1})
Sublimation of K(s)	+89
Ionization of K(g)	+418
Dissociation of Cl_2(g)	+244
Electron attachment to Cl(g)	−349
Formation of KCl(s)	−437

Strategy First, draw the cycle, showing the atomization of the elements, their ionization, and the formation of the solid lattice; then complete the cycle (for the step *solid compound → original elements*) by using the enthalpy of formation. The sum of enthalpy changes round the cycle is zero, so include the numerical data and set the sum of all the terms equal to zero; then solve the equation for the one unknown (the lattice enthalpy).

Solution Figure 17.12 shows the cycle required. The first step is the sublimation (atomization) of solid potassium:

	$\Delta H^\ominus$/(kJ mol^{-1})	
K(s) → K(g)	+89	(the enthalpy of sublimation or atomization of potassium)

Chlorine atoms are formed by dissociation of Cl_2:

$\frac{1}{2}$ Cl_2(g) → Cl(g)	+122	(half the bond enthalpy of Cl—Cl)

Now, potassium ions are formed by ionization of the gas-phase atoms:

K(g) → K^+(g) + e^-(g)	+418	(the ionization enthalpy of potassium)

and chloride ions are formed from the chlorine atoms:

Cl(g) + e^-(g) → Cl^-(g)	−349	(the electron-gain enthalpy of chlorine)

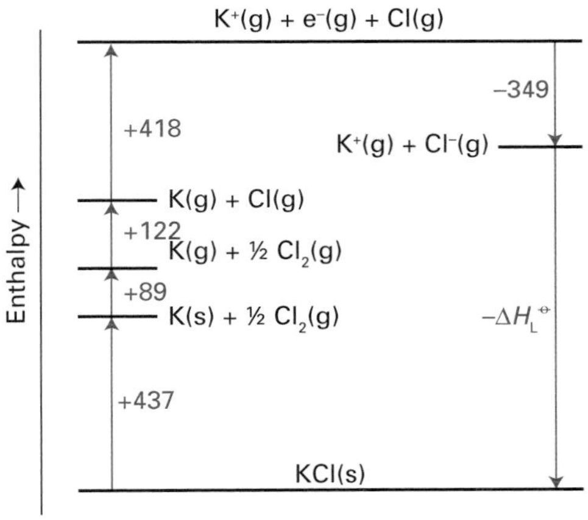

Fig. 17.12 The Born–Haber cycle for the calculation of the lattice enthalpy of potassium chloride. The sum of the enthalpy changes around the cycle is zero. The numerical values are in kilojoules per mole.

The solid is now formed:

$K^+(g) + Cl^-(g) \rightarrow KCl(s)$	$-\Delta H_L^\ominus$ (the enthalpy change when the lattice *forms* is the negative of the lattice enthalpy)

and the cycle is completed by decomposing KCl(s) into its elements:

$KCl(s) \rightarrow K(s) + \tfrac{1}{2}\,Cl_2(g)$	+437 (the negative of the enthalpy of formation of KCl)

The sum of the enthalpy changes is $-\Delta H_L^\ominus + 717$ kJ mol^{-1}; however, the sum must be equal to zero, so $\Delta H_L^\ominus = 717$ kJ mol^{-1}.

Self-test 17.1

Calculate the lattice enthalpy of magnesium bromide from the following data and the information in the *Data section.*

Process	$\Delta H^\ominus$/(kJ mol^{-1})
Sublimation of Mg(s)	+148
Ionization of Mg(g) to Mg^{2+}(g)	+2187
Dissociation of Br_2(g)	+193
Electron attachment to Br(g)	−325

Answer: 2402 kJ mol^{-1}

17.7 The origin of lattice enthalpy

Our next task is to account for the values of lattice enthalpies. The dominant attractive interaction in an ionic lattice is the Coulombic interaction between ions, which is far stronger than any other attractive interaction, so we concentrate on that.

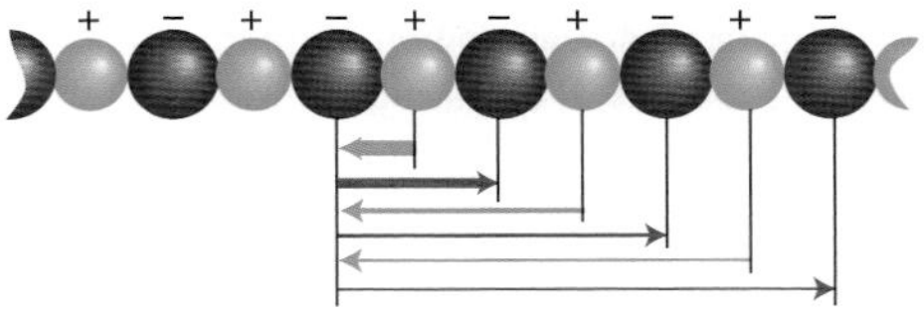

Fig. 17.13 There are alternating positive and negative contributions to the potential energy of a crystal lattice on account of the repulsions between ions of like charge and attractions of ions of opposite charge. The total potential energy is negative, but the sum might converge quite slowly.

The starting point is the Coulombic potential energy (The chemist's toolkit 9.1 and Foundations 0.9) for the interaction of two ions of charge numbers z_1 and z_2 (with cations having positive charge numbers and anions negative charge numbers) with centres separated by a distance r_{12}:

$$V_{12} = \frac{(z_1 e)(z_2 e)}{4\pi\varepsilon_0 r_{12}} \qquad (17.2)$$

where ε_0 is the vacuum permittivity. To calculate the total potential energy of all the ions in a crystal, we have to sum this expression over all the ions present. Nearest neighbours (which have opposite signs) attract and contribute a large negative term, second-nearest neighbours (which have the same sign) repel and contribute a slightly weaker positive term, and so on (Fig. 17.13). The overall result, however, is that there is a net attraction between the cations and anions and a favourable (negative) contribution to the energy of the solid. For instance, as shown in the following Derivation, for a uniformly spaced line of alternating cations and anions for which $z_1 = +z$ and $z_2 = -z$, with d the distance between the centres of adjacent ions,

$$V = -\frac{z^2 e^2}{4\pi\varepsilon_0 d} \times 2\ln 2 \qquad (17.3)$$

Derivation 17.2

The lattice energy of a one-dimensional crystal

Consider a line of alternating cations and anions extending in an infinite direction to the left and right of the ion of interest. The Coulombic potential energy of interaction of the ions on the right is the sum of the following terms:

$$V = \frac{1}{4\pi\varepsilon_0} \times \Bigg(\underbrace{-\frac{z^2e^2}{d}}_{\text{Attraction}} \underbrace{+\frac{z^2e^2}{2d}}_{\text{Repulsion}} \underbrace{-\frac{z^2e^2}{3d}}_{\text{Attraction}} \underbrace{+\frac{z^2e^2}{4d}}_{\text{Repulsion}} + \cdots \Bigg)$$

$$= -\frac{z^2e^2}{4\pi\varepsilon_0 d}\left(1 - \tfrac{1}{2} + \tfrac{1}{3} - \tfrac{1}{4} + \cdots\right)$$

The series in blue is well known to mathematicians (see The chemist's toolkit 6.1) as having the value ln 2:

$$1-\tfrac{1}{2}+\tfrac{1}{3}-\tfrac{1}{4}+\cdots=\ln 2$$

Therefore, we can conclude that

$$V=-\frac{z^2e^2}{4\pi\varepsilon_0 d}\times\ln 2$$

The interaction of the ion of interest with the ions to its left is the same, so the total potential energy of interaction is twice this expression for V, which is eqn 17.3.

When the calculation is repeated for more realistic, three-dimensional arrays of ions it is also found that the potential energy depends on the charge numbers of the ions and the value of a single parameter d, which may be taken as the distance between the centres of nearest neighbours:

$$V=\frac{e^2}{4\pi\varepsilon_0}\times\frac{z_1z_2}{d}\times A \qquad \text{Total Coulombic interaction} \quad (17.4)$$

Here A is a number called the **Madelung constant.** The value of the Madelung constant for a single line of ions is $2\ln 2 = 1.386\ldots$, as we have already seen; Table 17.2 gives the computed values of the constant for a variety of lattices with structures that we describe later in the chapter. Because the charge number of cations is positive and that of anions is negative, the product z_1z_2 is negative. Therefore, V is also negative, which corresponds to a lowering in potential energy relative to the gas of widely separated ions.

So far, we have considered only the Coulombic interaction between ions. However, regardless of their signs, the ions repel each other when they are pressed together and their wavefunctions overlap. These additional repulsions work against the net Coulombic attraction between ions, so they raise the energy of the solid. When their effect is taken into account,[1] it turns out that the lattice enthalpy is given by the **Born–Mayer equation:**

$$\Delta H_L^{\ominus}=\frac{|z_1z_2|}{d}\times\frac{N_Ae^2}{4\pi\varepsilon_0}\times\left(1-\frac{d^*}{d}\right)\times A \qquad \text{Born–Mayer equation} \quad (17.5)$$

where d^* is an empirical parameter that is often taken as 34.5 pm (simply because that value is found to give reasonable agreement with experiment). The modulus signs $(|\ldots|)$ mean that we should remove any minus sign from the product of z_1 and z_2, which results in a positive value for the lattice enthalpy. The important features of this expression are:

- Because $\Delta H_L^{\ominus}\propto|z_1z_2|$, the lattice enthalpy increases with increasing charge number of the ions.
- Because $\Delta H_L^{\ominus}\propto 1/d$, the lattice enthalpy increases with decreasing ionic radius.

The second conclusion follows from the fact that the smaller the ionic radii, the smaller the value of d. These features are in accord with the variation in the experimental values in Table 17.1.

Table 17.2
Madelung constants

Structural type	A
Caesium chloride	1.763
Fluorite	2.519
Rock salt	1.748
Rutile	2.408

Brief illustration 17.3 The Born–Mayer equation

To estimate the lattice enthalpy of MgO, which has a rock-salt structure ($A = 1.748$), we use $d = r(Mg^{2+}) + r(O^{2-}) = 72 + 140\text{ pm} = 212\text{ pm}$ from Table 17.6 later in this chapter. We also use (for future reference too)

$$\frac{N_Ae^2}{4\pi\varepsilon_0}=\frac{(6.022\ldots\times10^{23}\text{ mol}^{-1})\times(1.602\ldots\times10^{-19}\text{ C})^2}{4\pi\times(8.854\ldots\times10^{-12}\text{ J}^{-1}\text{ C}^2\text{ m}^{-1})}$$
$$=1.389\,354\ldots\times10^{-4}\text{ J mmol}^{-1}$$

and obtain

$$\Delta H_L^{\ominus}=\frac{4}{2.12\times10^{-10}\text{ m}}\times(1.389\,354\ldots\times10^{-4}\text{ J mmol}^{-1})\times\left(1-\frac{34.5\text{ pm}}{212\text{ pm}}\right)\times1.748$$
$$=3840\text{ kJ mol}^{-1}$$

to three significant figures. The experimental value is 3850 kJ mol^{-1}, suggesting that the ionic model is reliable for this compound.

Self-test 17.2

Which can be expected to have the greater lattice enthalpy, magnesium oxide or strontium oxide?

Answer: MgO

[1] See our *Physical Chemistry* (2010) for a derivation.

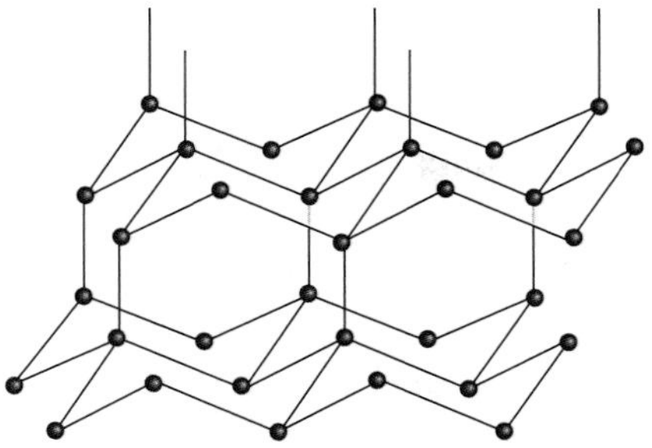

Fig. 17.14 A fragment of the structure of diamond. Each carbon atom is tetrahedrally bonded to four neighbours. This framework-like structure results in a rigid crystal with a high thermal conductivity.

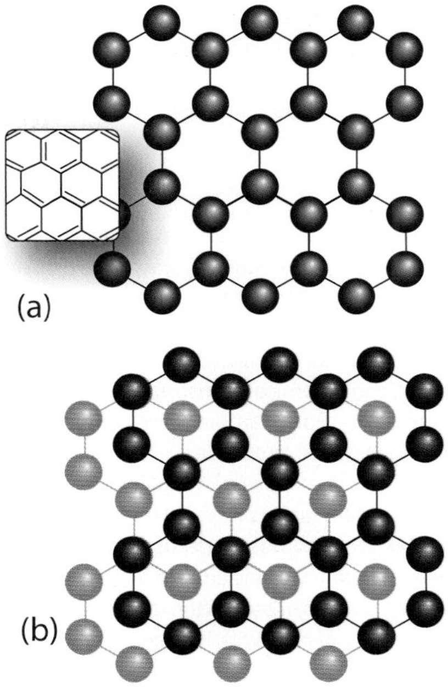

Fig. 17.15 Graphite consists of flat planes of hexagons of carbon atoms lying above one another. (a) The arrangement of carbon atoms in a sheet; (b) the relative arrangement of neighbouring sheets. When impurities are present, the planes can slide over one another easily. Graphite conducts well within the planes but less well perpendicular to the planes.

17.8 Covalent networks

We have already noted that covalent bonds in a definite spatial orientation link the atoms in covalent network solids. Covalent solids are typically hard and often unreactive. Examples include silicon, red phosphorus, boron nitride, and—very importantly—diamond and graphite, which we discuss in detail.

Diamond and graphite are two allotropes of carbon. In diamond each sp^3-hybridized carbon is bonded tetrahedrally to its four neighbours (Fig. 17.14). The network of strong C—C bonds is repeated throughout the crystal and, as a result, diamond is the hardest known substance.

A note on good practice Allotropes are distinct forms of an element that differ in the way that atoms are linked. Whereas the term allotrope is applied only to elements (and includes different molecular species, such as O_2 and O_3), the term *polymorph* applies to the different *solid* structures that an element or compound may adopt, such as the different phases of iron (which are also allotropes) or of calcium carbonate (which are polymorphs but not allotropes).

In graphite, σ bonds between sp^2-hybridized carbon atoms form hexagonal rings which, when repeated throughout a plane, give rise to sheets (Fig. 17.15). Because the sheets can slide against each other when impurities are present, graphite is used widely as a lubricant. It cannot be so used in space, because the impurities out-gas and the layers become immobile.

The electrical properties of diamond and graphite are determined by differences in the bonding patterns in these solids. Graphite is an electronic conductor because electrons are free to move through bands formed by the overlap of partially filled, unhybridized p orbitals that are perpendicular to the hexagonal sheets. This band model explains the experimental observation that graphite conducts electricity well within the sheets but less well between them. The electrical conductivity of these graphene (graphite-like) sheets of carbon atoms is now being considered in the design of nanometer-sized electronic devices. We see from Fig. 17.14 that delocalized π networks are not possible in diamond, which—in contrast to graphite—is an insulator (more precisely, a large-band-gap semiconductor).

Impact on technology 17.1

Nanowires

A great deal of research effort is now being expended in the fabrication of nanometre-sized assemblies of atoms and molecules that can be used as tiny building blocks in a variety of technological applications. The future economic impact of **nanotechnology**, the aggregate of applications of devices built from nanometre-sized components, could be very significant. For example, increased demand for very small digital electronic devices has driven the design of ever smaller and more powerful microprocessors. However, there is an upper limit on the density of electronic circuits that can be incorporated into silicon-based chips with current fabrication technologies. As the ability to process data increases with the number of circuits in a chip, it follows that soon chips and the devices that use them will have to become bigger if processing power is to increase indefinitely. One way to circumvent this problem is to fabricate devices from nanometre-sized components. Another advantage of making nanometre-sized electronic devices, or *nanodevices*, is the possibility of using quantum mechanical effects. For example, electron tunnelling between two conducting regions separated by a thin insulating region can increase the speed of electron conduction and, consequently, the data processing speed in a digital nanoprocessor.

The study of nanodevices can also advance our basic understanding of chemical reactions. Nanometre-sized chemical reactors can serve as laboratories for the study of chemical reactions in constrained environments. Some of these reactions could comprise the foundation for the construction of nanometre-sized chemical sensors, with potential applications in medicine. For example, nanodevices with carefully designed biochemical properties could replace viruses and bacteria as the active species in vaccines.

A number of techniques have already been developed for the fabrication of nanometre-sized structures. The synthesis of *nanowires*, nanometre-sized atomic assemblies that conduct electricity, is a major step in the fabrication of nanodevices. An important type of nanowire is based on *carbon nanotubes*, thin cylinders of carbon atoms that are both mechanically strong and highly conducting. In recent years, methods for selective synthesis of nanotubes have been developed and they consist of different ways to condense a carbon plasma either in the presence or absence of a catalyst. The simplest structural motif is called a *single-walled nanotube* (SWNT) and is shown in Fig. 17.16. In a SWNT, sp^2-hybridized carbon atoms form hexagonal rings reminiscent of the structure of the carbon sheets found in graphite. The tubes have diameters of between 1 and 2 nm and lengths of several micrometres. The features shown in the illustration have been confirmed by direct observation with scanning tunnelling microscopy. A *multi-walled nanotube* (MWNT) consists of several concentric SWNTs and its diameter varies between 0.4 and 25 nm.

The origin of electrical conductivity in carbon nanotubes is the delocalization of π electrons that occupy unhydridized p orbitals, just as in graphite (Section 17.8). Recent studies have shown a correlation between structure and conductivity in SWNTs. The illustration shows a SWNT that is a semiconductor. If the hexagons are rotated by 60°, the resulting SWNT is a metallic conductor.

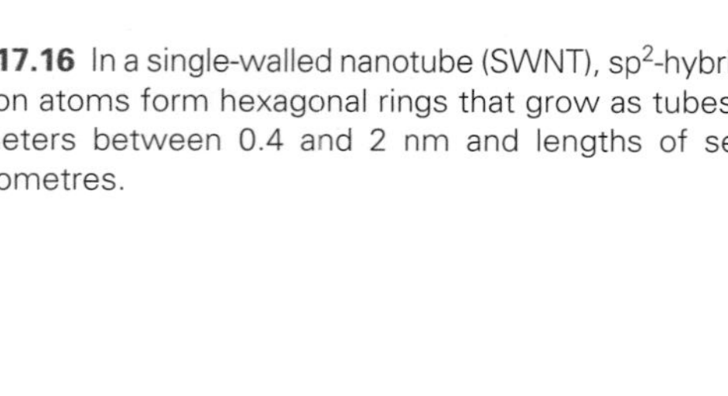

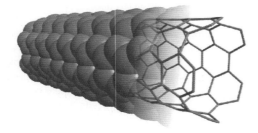

Fig. 17.16 In a single-walled nanotube (SWNT), sp^2-hybridized carbon atoms form hexagonal rings that grow as tubes with diameters between 0.4 and 2 nm and lengths of several micrometres.

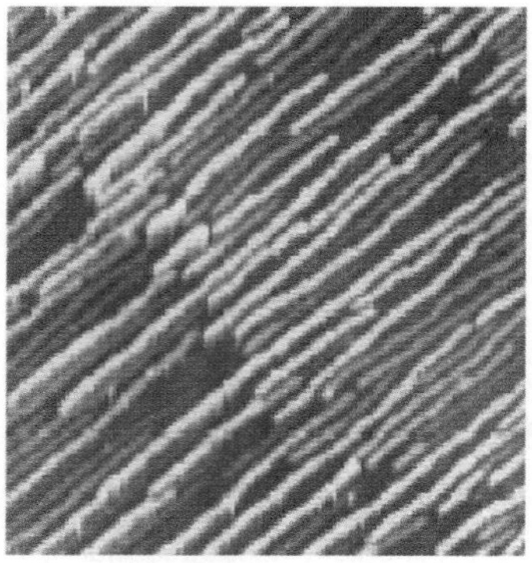

Fig. 17.17 Germanium nanowires fabricated on to a silicon surface by molecular beam epitaxy and imaged by atomic force microscopy. Reproduced with permission from T. Ogino et al., *Acc. Chem. Res.* **32**, 447 (1999).

Silicon nanowires can be made by focussing a pulsed laser beam on to a solid target composed of silicon and iron. The laser ejects Fe and Si atoms from the surface of the target, forming a vapour phase that can condense into liquid $FeSi_n$ nanoclusters at sufficiently low temperatures. The phase diagram for this complex mixture shows that solid silicon and liquid $FeSi_n$ coexist at temperatures higher than 1473 K. Hence, it is possible to precipitate solid silicon from the mixture if the experimental conditions are controlled to maintain the $FeSi_n$ nanoclusters in a liquid state that is saturated with silicon. It is observed that the silicon precipitate consists of nanowires with diameters of about 10 nm and lengths greater than 1 μm.

Nanowires are also fabricated by *molecular beam epitaxy* (MBE), in which gaseous atoms or molecules are sprayed on to a crystalline surface in an evacuated chamber. Through careful control of the chamber temperature and of the spraying process, it is possible to create nanometre-sized assemblies with specific shapes. For example, Fig. 17.17 shows an image of germanium nanowires on a silicon surface. The wires are about 2 nm high, 10–32 nm wide, and 10–600 nm long. It is also possible to deposit *quantum dots*, nanometre-sized boxes or spheres of atoms, on a surface. Semiconducting quantum dots could be important building blocks of nanometre-sized lasers.

Direct manipulation of atoms on a surface also leads to the formation of nanowires. The Coulomb attraction between an atom and the tip of a scanning tunnelling microscope (STM, Section 18.2) can be exploited to move atoms along a surface, arranging them into patterns, such as wires.

17.9 Magnetic properties of solids

The magnetic properties of solids are determined by interactions between the spins of its electrons. Some materials are magnetic and others may become magnetized when placed in an external magnetic field. A bulk sample exposed to a magnetic field of strength $\mathcal{H}$ acquires a **magnetization**, $\mathcal{M}$, which is proportional to $\mathcal{H}$:

$$\mathcal{M} = \chi\mathcal{H} \quad \text{Definition of } \chi \quad \text{Magnetization} \quad (17.6)$$

where χ (chi) is the dimensionless **volume magnetic susceptibility** (Table 17.3). We can think of the magnetization as contributing to the density of lines of force in the material (Fig. 17.18). Materials for which χ is negative are called **diamagnetic** and tend to move out of a magnetic field; the density of lines of force within them is lower than in a vacuum. Those for which χ is positive are called **paramagnetic**; they tend to move into a magnetic field and the density of lines of force within them is greater than in a vacuum.

Diamagnetism arises from the effect of the magnetic field on the electrons of molecules. Specifically, an applied magnetic field induces the circulation of electronic currents which give rise to a magnetic field that usually opposes the applied field and reduces the density of lines of force. The great majority of molecules with no unpaired electron spins are diamagnetic. In these cases, the induced electron currents occur within the orbitals of the molecule that are occupied in its ground state.

The most common kind of paramagnetism arises from unpaired electron spins, which behave like tiny bar magnets that tend to line up with the applied field. The more that can line up in this way, the greater the lowering of the energy and the sample tends to move into the applied field. Many compounds of the d-block elements are paramagnetic because they have various numbers of unpaired d electrons. Molecules, specifically radicals, with unpaired electrons are paramagnetic. Examples include the brown gas nitrogen dioxide (NO_2) and the peroxyl radical (HO_2), which plays a role in atmospheric chemistry. In a few cases the induced field augments the applied field and increases the density of lines of force within the material even though there are no unpaired electrons present. In these cases, the induced electron currents arise from migration of electrons through unoccupied orbitals, so this kind of paramagnetism occurs only if the excited states are low in energy (as in some d- and f-block complexes).

Table 17.3

*Magnetic susceptibilities at 298 K**

	$\chi/10^{-6}$	$\chi_m/(10^{-5}\ cm^3\ mol^{-1})$
Al(s)	+22	+2.2
Cu(s)	−9.6	−6.8
$CuSO_4\cdot 5H_2O(s)$	+176	+1930
$H_2O(l)$	−9.06	−160
$MnSO_4\cdot 4H_2O(s)$	+2640	+2790
NaCl(s)	−13.9	−38
S(s)	−12.9	−2.0

* χ is the dimensionless magnetic susceptibility; χ_m is the molar magnetic susceptibility. The two are related by $\chi_m = \chi V_m$, where V_m is the molar volume of the substance.

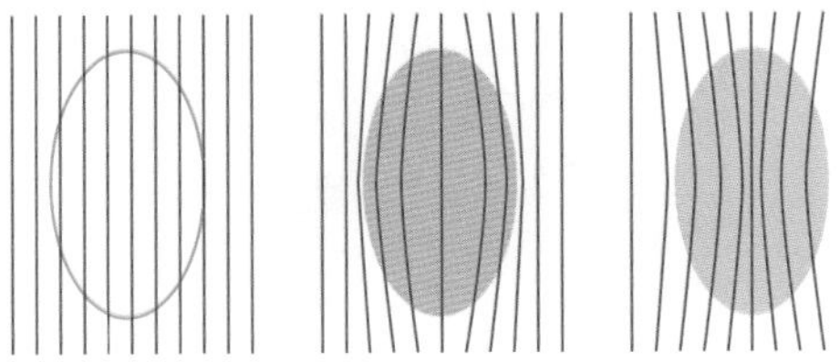

Fig. 17.18 (a) In a vacuum, the strength of a magnetic field can be represented by the density of lines of force; (b) in a diamagnetic material, the lines of force are reduced; (c) in a paramagnetic material, the lines of force are increased.

Brief illustration 17.4 Magnetic character

Solid magnesium is a metal in which the two valence electrons of each Mg atom are donated to a band of orbitals constructed from 3s orbitals. From N atomic orbitals we can construct N molecular orbitals spreading through the metal. Each atom supplies two electrons, so there are $2N$ electrons to accommodate. These occupy and fill the N molecular orbitals. There are no unpaired electrons, so the metal is diamagnetic. An O_2 molecule has the electronic structure described in Section 14.10, where we see that two electrons occupy separate antibonding π orbitals with parallel spins. We conclude that oxygen is a paramagnetic gas.

Self-test 17.3

Repeat the analysis for Zn(s) and NO(g).

Answer: Zn diamagnetic, NO paramagnetic

At low temperatures, some paramagnetic solids make a transition to a phase in which large regions, called **domains**, of electron spins align with parallel orientations. This cooperative alignment gives rise to a very strong magnetization—in some cases millions of times greater—and is called **ferromagnetism** (Fig. 17.19). In other cases, the cooperative effect leads to alternating spin orientations: the spins are locked into a low-magnetization arrangement to

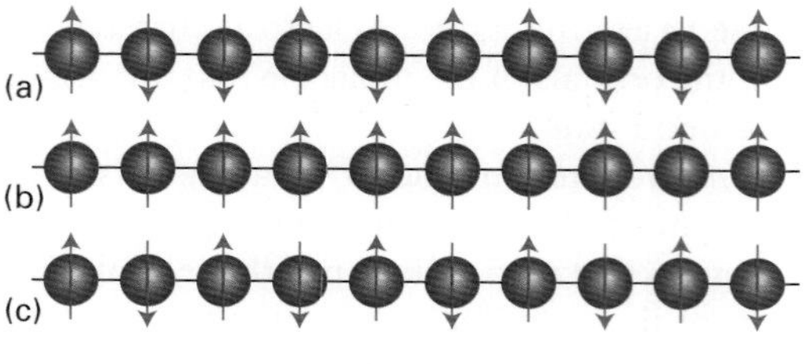

Fig. 17.19 (a) In a paramagnetic material, the electron spins are aligned at random in the absence of an applied magnetic field; (b) in a ferromagnetic material, the electron spins are locked into a parallel alignment over large domains; (c) in an antiferromagnetic material, the electron spins are locked into an antiparallel arrangement. The latter two arrangements survive even in the absence of an applied field.

give an **antiferromagnetic phase** which has a zero magnetization because the contributions from different spins cancel. The transition to the ferromagnetic phase occurs at the **Curie temperature,** T_C, and the transition to the antiferromagnetic occurs at the **Néel temperature,** T_N.

Superconductors have unique magnetic properties. Some superconductors, classed as Type I, show abrupt loss of superconductivity when an applied magnetic field exceeds a critical value $\mathcal{H}_c$ characteristic of the material. Type I superconductors are also completely diamagnetic—the lines of force are completely excluded—below $\mathcal{H}_c$. This exclusion of a magnetic field in a material is known as the **Meissner effect,** which can be demonstrated by the levitation of a superconductor above a magnet. Type II superconductors, which include the HTSCs, show a gradual loss of superconductivity and diamagnetism with increasing magnetic field.

Crystal structure

Now we turn to the structures adopted by atoms and ions when they stack together to give a crystalline solid. The structures of crystals are of considerable practical importance, for they have implications for geology, materials, technologically advanced materials such as semiconductors and high-temperature superconductors, and in biology. The first, and often very demanding, step in an X-ray structural analysis of biological macromolecules is to form crystals in which the large molecules lie in orderly ranks. On the other hand, the crystallization of a virus particle would take it out of circulation, and one of the strategies adopted by viruses for avoiding this kind of entombment makes unconscious use of the geometry of crystal packing.

17.10 Unit cells

The pattern that atoms, ions, or molecules adopt in a crystal is expressed in terms of an array of points making up the **lattice** that identify the locations of the individual species (Fig. 17.20). A **unit cell** of a crystal is the small three-dimensional figure obtained by joining typically eight of these points, and which may be used to construct the entire crystal lattice by purely translational displacements, much as a wall may be constructed from bricks (Fig. 17.21). An infinite number of different unit cells can describe the same structure, but it is conventional to choose the cell with the greatest symmetry and the smallest dimensions.

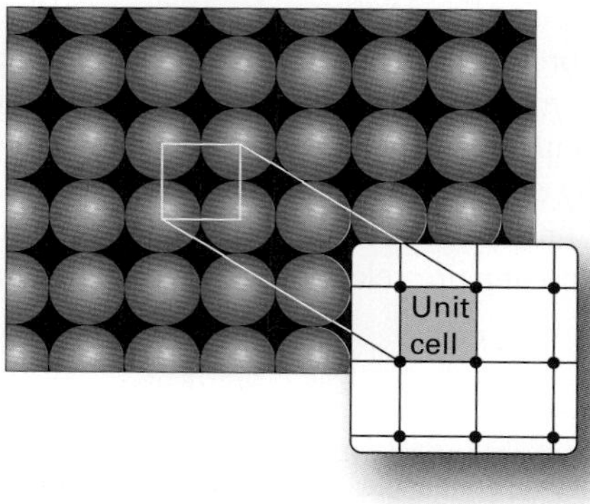

Fig. 17.20 A crystal consists of a uniform array of atoms, molecules, or ions, as represented by these spheres. In many cases, the components of the crystal are far from spherical, but this diagram illustrates the general idea. The location of each atom, molecule, or ion can be represented by a single point; here (for convenience only), the locations are denoted by a point at the centre of the sphere. The unit cell, which is shaded in the inset, is the smallest block from which the entire array of points can be constructed without rotating or otherwise modifying the block.

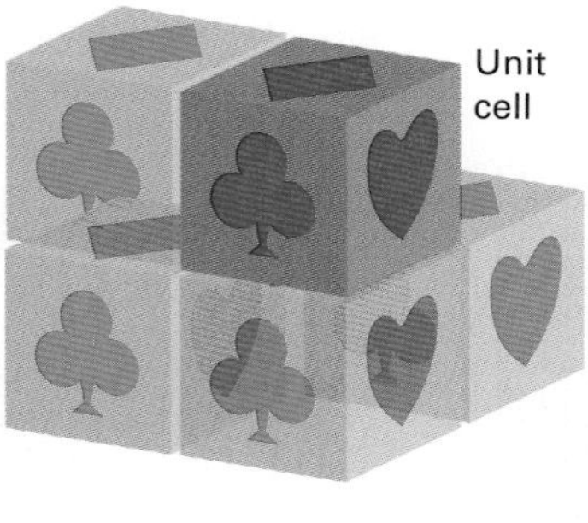

Fig. 17.21 A unit cell, here shown in three dimensions, is like a brick used to construct a wall. Once again, only pure translations are allowed in the construction of the crystal. (Some bonding patterns for actual walls use rotations of bricks, so for these patterns a single brick is not a unit cell.)

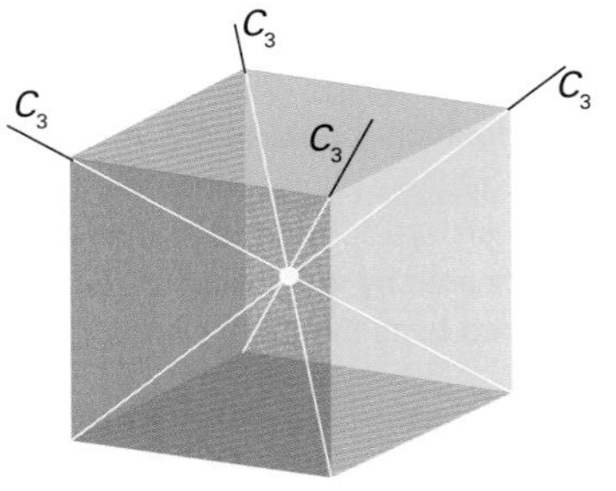

Fig. 17.22 A unit cell belonging to the cubic system has four threefold axes (denoted C_3) arranged tetrahedrally.

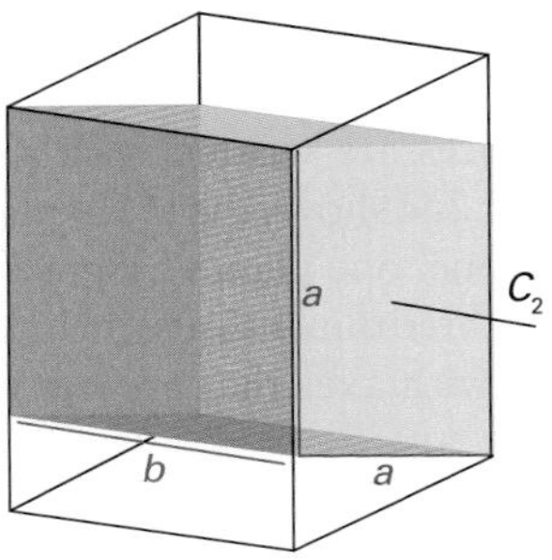

Fig. 17.23 A unit cell belonging to the monoclinic system has one twofold (denoted C_2) axis (parallel to b).

Unit cells are classified into one of seven **crystal systems** according to the symmetry they possess under rotations about different axes. The *cubic system*, for example, has four threefold axes (Fig. 17.22). A threefold axis is an axis of a rotation that restores the unit cell to the same appearance three times during a complete revolution, after rotations through 120°, 240°, and 360°. The four axes of a cube make the tetrahedral angle to each other. The *monoclinic system* has one twofold axis (Fig. 17.23). A twofold axis is an axis of a rotation that leaves the cell apparently unchanged twice during a complete revolution, after rotations through 180° and 360°. The **essential symmetries**, the properties that must be present for the unit cell to belong to a particular system, are listed in Table 17.4.

Table 17.4
The essential symmetries of the seven crystal systems

The systems	Essential symmetries
Triclinic	None
Monoclinic	One twofold axis
Orthorhombic	Three perpendicular twofold axes
Rhombohedral	One threefold axis
Tetragonal	One fourfold axis
Hexagonal	One sixfold axis
Cubic	Four threefold axes in a tetrahedral arrangement

A unit cell need not have perpendicular faces and may have lattice points other than at its corners, so each crystal system can occur in a number of different varieties. For example, in some cases points may occur on the faces and in the body of the cell without destroying the cell's essential symmetry. These various possibilities give rise to fourteen distinct types of unit cell, which are called the **Bravais lattices** (Fig. 17.24).

17.11 The identification of crystal planes

To specify a unit cell fully, we also need to know its size, such as the lengths of its sides. There is a useful relation between the spacing of the planes passing

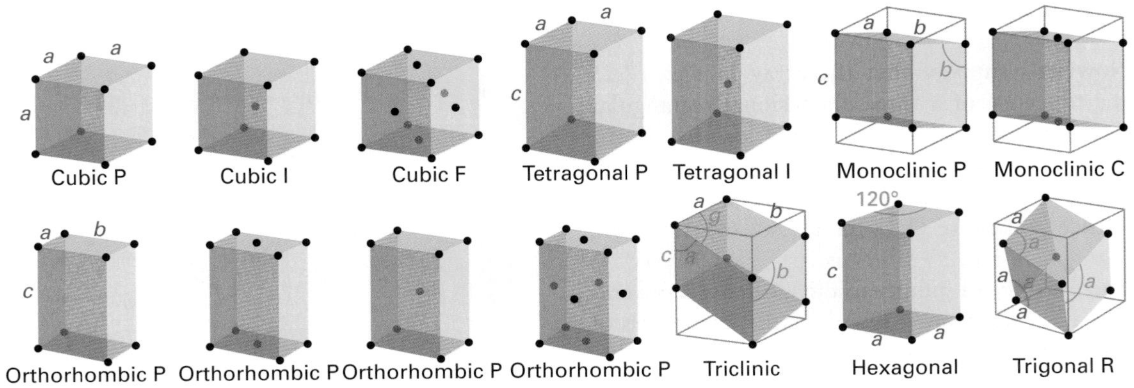

Fig. 17.24 The fourteen Bravais lattices. The letter P denotes a primitive unit cell, I a body-centred unit cell, F a face-centred unit cell, and C (or A or B) a cell with lattice points on two opposite faces.

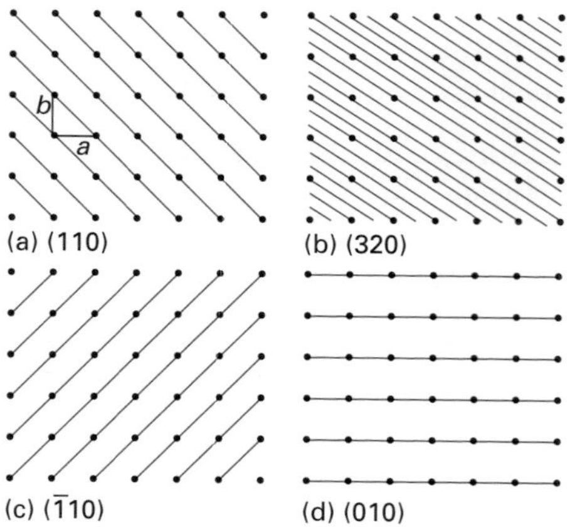

Fig. 17.25 Some of the planes that can be drawn through the points of the space lattice and their corresponding Miller indices (*hkl*). The origin of the coordinate system used for labelling the planes coincides with the position of the lattice point in the lower left-hand corner of each lattice.

through the lattice points, which (as we shall see) we can measure, and the lengths we need to know.

Because two-dimensional arrays of points are easier to visualize than three-dimensional arrays, we shall introduce the concepts we need by referring to two dimensions initially and then extend the conclusions to three dimensions. In particular, consider the two-dimensional lattice formed from an orthogonal (rectangular) unit cell of sides a and b (Fig. 17.25). We can distinguish the four sets of planes shown in the illustration by the distances at which they intersect the axes. One way of labelling the planes would therefore be to denote each set by the smallest intersection distances. For example, we could denote the four sets in the illustration as $(1a,1b)$, $(2a,3b)$, $(-1a,1b)$, and $(\infty a,1b)$. If, however, we agreed always to quote distances along the axes as multiples of the lengths of the unit cell, then we could omit the a and b and label the planes more simply as (1,1), (2,3), (−1,1), and (∞,1).

Now let's suppose that the array in Fig. 17.25 is the top view of a three-dimensional rectangular lattice in which the unit cell has a length c in the z direction. All four sets of planes intersect the z-axis at infinity, so the full labels of the sets of planes of lattice points are (1,1,∞), (2,3,∞), (−1,1,∞), and (∞,1,∞).

The presence of infinity in the labels is inconvenient. We can eliminate it by taking the reciprocals of the numbers in the labels; this step also turns out to have further advantages, as we shall see. The resulting **Miller indices**, (*hkl*), are the reciprocals of the numbers in the parentheses with fractions cleared.

Brief illustration 17.5 Miller indices

The (1,1,∞) planes in Fig. 17.25 are the (110) planes in the Miller notation (because 1/1 = 1 and 1/∞ = 0). Similarly, the (2,3,∞) planes become first ($\frac{1}{2}$,$\frac{1}{3}$,0) when reciprocals are formed, and then (3,2,0) when fractions are cleared by multiplication through by 6, so they are referred to as the (320) planes.

Self-test 17.4

A representative member of a set of planes in a crystal intersects the axes at $3a$, $2b$, and $2c$; what are the Miller indices of the planes?

Answer: (233)

Negative indices are written with a bar over the number: Fig. 17.25c shows the ($\bar{1}$10) planes, which is read 'bar-one, one, one planes'. Figure 17.26 shows some planes in three dimensions, including an example of a lattice with axes that are not mutually perpendicular.

It is helpful to keep in mind the fact, as illustrated in Fig. 17.25, that the smaller the value of h in the Miller index (*hkl*), the more nearly parallel the plane is to the a axis. The same is true of k and the b axis and l and the c axis. When $h = 0$, the planes intersect the a axis at infinity, so the (0*kl*) planes are parallel to the a axis. Similarly, the (*h*0*l*) planes are parallel to b and the (*hk*0) planes are parallel to c.

The Miller indices are very useful for calculating the separation of planes. For instance, we show in the following Derivation that they can be used to derive the following very simple expression for the separation, d, of the (*hkl*) planes in an orthogonal lattice:

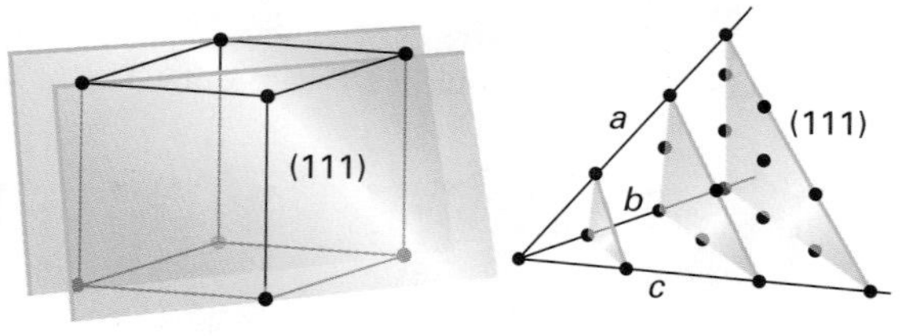

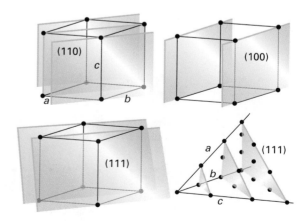

Fig. 17.26 Some representative planes in three dimensions and their Miller indices. Note that a 0 indicates that a plane is parallel to the corresponding axis, and that the indexing may also be used for unit cells with non-orthogonal axes.

$$\frac{1}{d^2} = \frac{h^2}{a^2} + \frac{k^2}{b^2} + \frac{l^2}{c^2} \qquad \text{Orthogonal lattice} \quad \text{Plane separation} \qquad (17.7)$$

Derivation 17.3

The separation of lattice planes

Consider the ($hk0$) planes of a rectangular lattice built from an orthorhombic unit cell of sides of lengths a and b (Fig. 17.27). We can write the following trigonometric expressions for the angle ϕ shown in the illustration:

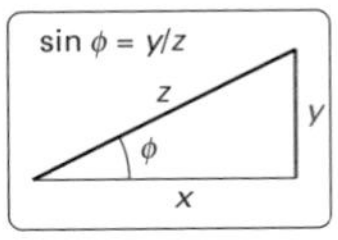

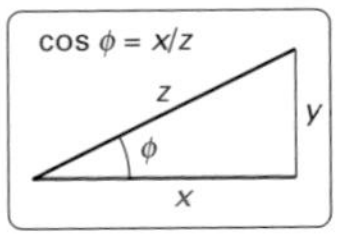

$$\sin\phi = \frac{d}{(a/h)} = \frac{hd}{a} \qquad \cos\phi = \frac{d}{(b/k)} = \frac{kd}{b}$$

Because the lattice planes intersect the a axis h times and the b axis k times, the length of each hypotenuse is calculated by dividing a by h and b by k. Then, because $\sin^2\phi + \cos^2\phi = 1$, we obtain

$$\left(\frac{hd}{a}\right)^2 + \left(\frac{kd}{b}\right)^2 = 1$$

which we can rearrange, by dividing both sides by d^2, into

$$\frac{1}{d^2} = \frac{h^2}{a^2} + \frac{k^2}{b^2}$$

In three dimensions, this expression generalizes to eqn 17.7.

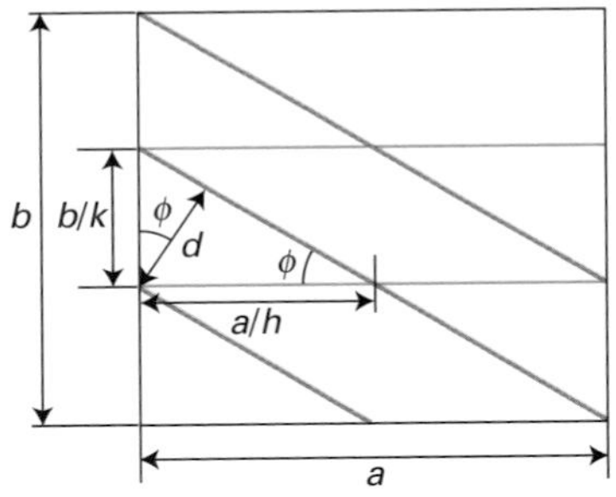

Fig. 17.27 The geometrical construction used to relate the separation of planes to the dimensions of the unit cell.

Example 17.2

Using the Miller indices

Calculate the separation of (a) the (123) planes and (b) the (246) planes of an orthorhombic cell with a = 0.82 nm, b = 0.94 nm, and c = 0.75 nm.

Strategy For the first part, we simply substitute the information into eqn 17.7. For the second part, instead of repeating the calculation, we should examine how d in eqn 17.7 changes when all three Miller indices are multiplied by 2 (or by a more general factor, n).

Solution Substituting the data into eqn 17.7 gives

$$\frac{1}{d^2} = \frac{1^2}{(0.82\ \text{nm})^2} + \frac{2^2}{(0.94\ \text{nm})^2} + \frac{3^2}{(0.75\ \text{nm})^2} = \frac{22}{\text{nm}^2}$$

It follows that d = 0.21 nm. When the indices are all increased by a factor of 2, the separation becomes

$$\frac{1}{d^2} = \frac{(2\times1)^2}{(0.82\ \text{nm})^2} + \frac{(2\times2)^2}{(0.94\ \text{nm})^2} + \frac{(2\times3)^2}{(0.75\ \text{nm})^2} = 2^2 \times \frac{22}{\text{nm}^2}$$

So, for these planes d = 0.11 nm. In general, increasing the indices uniformly by a factor n decreases the separation of the planes by n.

Self-test 17.5

Calculate the separation of the (133) and (399) planes in the same lattice.

Answer: 0.19 nm, 0.063 nm

17.12 The determination of structure

One of the most important techniques for the determination of the structures of crystals is **X-ray diffraction.** In its simplest form, the technique is used to identify the lattice type and the separation of the planes of lattice points (and hence the distance between the centres of atoms and ions). In its most sophisticated version, X-ray diffraction provides detailed information about the location of all the atoms in a molecule, even those as complicated as proteins. Special techniques are also available for the study of structural changes that accompany chemical reactions. The current considerable success of modern molecular biology has stemmed from X-ray diffraction techniques that have grown in sensitivity and scope as computing techniques have become more powerful and X-ray sources more intense. Here we concentrate on the principles of the technique and illustrate how it may be used to determine the spacing of atoms in a crystal.

A characteristic property of waves is that they **interfere** with one another, which means that they result in a greater amplitude where their displacements add and a smaller amplitude where their displacements subtract (Fig. 17.28). The former is called 'constructive interference' and the latter 'destructive interference'. Because the intensity of electromagnetic radiation is proportional to the square of the amplitude of the waves, the regions of constructive and destructive interference show up as regions of enhanced and diminished intensities. The phenomenon of **diffraction** is the interference caused by an object in the path of waves, and the pattern of varying intensity that results is called the **diffraction**

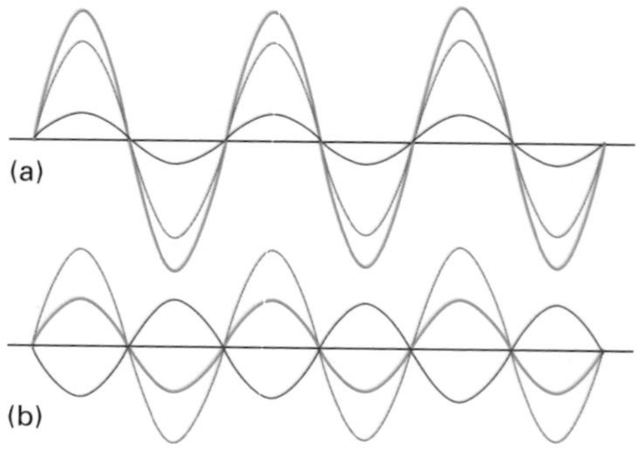

Fig. 17.28 When two waves (drawn as red and orange lines) are in the same region of space they interfere. Depending on their relative phase, they may interfere (a) constructively, to give an enhanced amplitude, or (b) destructively, to give a smaller amplitude.

pattern. Diffraction occurs when the dimensions of the diffracting object are comparable to the wavelength of the radiation. Sound waves, with wavelengths of the order of 1 m, are diffracted by macroscopic objects. Light waves, with wavelengths of the order of 500 nm, are diffracted by narrow slits.

X-rays have wavelengths comparable to bond lengths in molecules and the spacing of atoms in crystals (about 100 pm), so they are diffracted by them. By analysing the diffraction pattern, it is possible to draw up a detailed picture of the location of atoms. Electrons moving at about 2×10^4 km s^{-1} (after acceleration through about 4 kV) have wavelengths of about 20 pm (recall Example 12.2), and may also be diffracted by molecules. Neutrons generated in a nuclear reactor, and then slowed to thermal velocities (that is, bouncing off the nuclei of atoms until their kinetic energy has become the same as that of the targets), have similar wavelengths and may also be used for diffraction studies.

Example 17.3

Assessing the wavelength of thermal neutrons

Estimate the wavelength of neutrons that have been slowed by collisions with atoms in the surroundings at 300 K.

Strategy We need to combine the de Broglie relation $\lambda = h/p$ with an estimate of the mean linear momentum from the mean energy of the particles, $E_k = \frac{1}{2}mv^2 = p^2/2m$. The mean kinetic energy can be estimated from the equipartition theorem (Foundations 0.12), which for translational motion in three dimensions is $E_k = \frac{3}{2}kT$.

Solution From $E_k = p^2/2m$, $p = (2mE_k)^{1/2}$, and from $E_k = \frac{3}{2}kT$, $p = (3mkT)^{1/2}$. It then follows from the de Broglie relation that

$$\lambda = \frac{h}{p} = \frac{h}{(3mkT)^{1/2}}$$

We now insert the data, using for m the mass of a neutron, $m_n = 1.675 \times 10^{-27}$ kg:

$$\lambda = \frac{\overbrace{6.626\times10^{-34}\ \mathrm{J\ s}}^{h}}{\left\{3\times\underbrace{(1.675\times10^{-27}\ \mathrm{kg})}_{m_n}\times\underbrace{(1.381\times10^{-23}\ \mathrm{J\ K^{-1}})}_{k}\times\underbrace{(300\ \mathrm{K})}_{T}\right\}^{1/2}}$$

$$= \frac{6.626\times10^{-34}}{(3\times1.675\times1.381\times10^{-50}\times300)^{1/2}}\ \underbrace{\frac{\overbrace{\mathrm{J\ s}}^{\mathrm{kg\ m^2\ s^{-1}}}}{\left(\underbrace{\mathrm{J\ kg}}_{\mathrm{kg^2\ m^2\ s^{-2}}}\right)^{1/2}}}_{\mathrm{kg\ m\ s^{-1}}}$$

$= 1.45 \times 10^{-10}$ m

For the cancellation of units, we have used 1 J = 1 kg m^2 s^{-2}. The wavelength is 145 pm.

Self-test 17.6

You are investigating the use of protons for a diffraction experiment. What is the wavelength of a proton accelerated from rest through 100 kV? *Hint:* Refer to Example 12.2.

Answer: 905 pm

The short-wavelength electromagnetic radiation we call X-rays is produced by bombarding a metal with high-energy electrons. The electrons decelerate as they plunge into the metal and generate radiation with a continuous range of wavelengths. This radiation is called **bremsstrahlung** (*Bremse* is German for brake, *Strahlung* for ray). Superimposed on the continuum are a few high-intensity, sharp peaks. These peaks arise from the interaction of the incoming electrons with the electrons in the inner shells of the atoms: the collision expels an electron (Fig. 17.29), and an electron of higher energy drops into the vacancy, emitting the excess energy as an X-ray photon. An

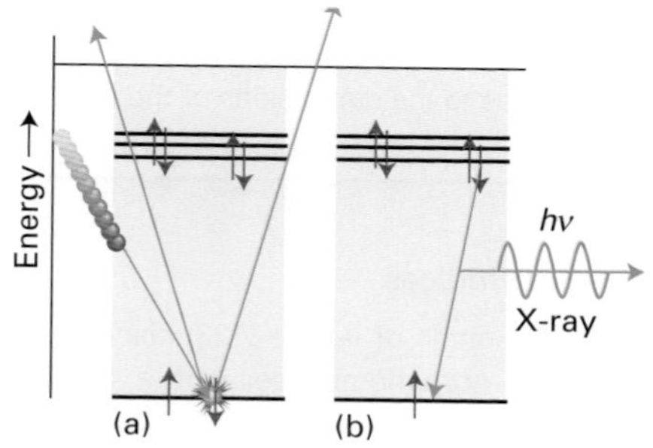

Fig. 17.29 The formation of X-rays. (a) When a metal is subjected to a high-energy electron beam, an electron in an inner shell of an atom is ejected. (b) When an electron falls into the vacated orbital from an orbital of much higher energy, the excess energy is released as an X-ray photon.

example of the process is the expulsion of an electron from the K shell (the shell with $n = 1$) of a copper atom, followed by the transition of an outer electron into the vacancy. The energy so released gives rise to copper's 'K_α radiation' of wavelength 154 pm. Currently, however, there is a major shift in emphasis to using **synchrotron radiation** as a source of high-intensity, monochromatic X-rays. Synchrotron radiation is produced when electrons move at high speed in a circle, their constantly changing direction corresponds to acceleration, and accelerated charges generate electromagnetic radiation. The high speeds achieved in the particle accelerators known as *synchrotrons* result in the production of very high-frequency radiation. The principal drawback is that synchrotron sources are costly and must be built as national facilities.

In 1923, the German physicist Max von Laue suggested that X-rays might be diffracted when passed through a crystal, for the wavelengths of X-rays are comparable to the separation of atoms, and diffraction occurs when the wavelength of radiation is comparable to the dimensions of a target. Laue's suggestion was confirmed almost immediately by Walter Friedrich and Paul Knipping, and then developed by William and Lawrence Bragg, who later jointly received the Nobel Prize. It has grown since then into a technique of extraordinary power.

17.13 Bragg's law

The earliest approach to the analysis of X-ray diffraction patterns treated a plane of atoms as a semi-transparent mirror and modelled the crystal as stacks of reflecting planes of separation d (Fig. 17.30). The model makes it easy to calculate the angle the crystal must make to the incoming beam of X-rays for constructive interference to occur. It has also given rise to the name **reflection** to denote an intense spot arising from constructive interference.

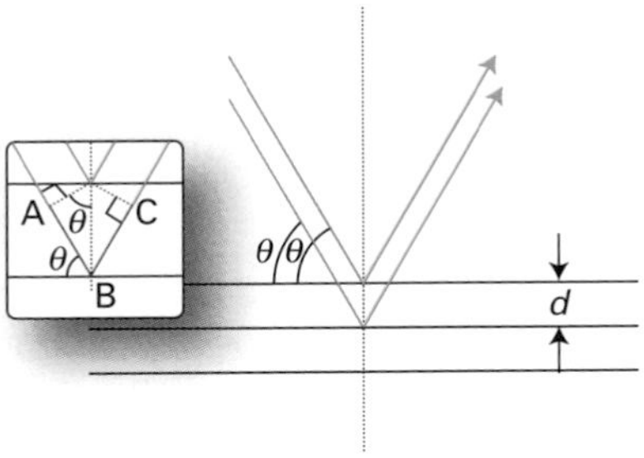

Fig. 17.30 The derivation of Bragg's law treats each lattice plane as reflecting the incident radiation. The path lengths differ by AB + BC, which depends on the angle θ. Constructive interference (a 'reflection') occurs when AB + BC is equal to an integral number of wavelengths.

The path-length difference of the two rays shown in the illustration is

$$AB + BC = 2d \sin \theta$$

where 2θ is the **glancing angle.** When the path-length difference is equal to one wavelength ($AB + BC = \lambda$), the reflected waves interfere constructively. It follows that a reflection should be observed when the glancing angle satisfies **Bragg's law:**

$$\lambda = 2d \sin \theta \qquad \text{Bragg's law} \qquad (17.8)$$

The primary use of Bragg's law is to determine the spacing between the layers of atoms, for once the angle θ corresponding to a reflection has been determined, d may readily be calculated.

A note on good practice You will often see Bragg's law in the form $n\lambda = 2d \sin \theta$, where n, an integer, is the 'order' of the diffraction. The modern tendency is to omit n and to ascribe the diffraction to planes of separation d/n instead (recall the discussion in Example 17.2).

Example 17.4

Using Bragg's law

A reflection from the (111) planes of a cubic crystal was observed at a glancing angle of 11.2° when Cu K_α X-rays of wavelength 154 pm were used. What is the length of the side of the unit cell?

Strategy We can find the separation, d, of the lattice planes from eqn 17.8 and the data. Then we find the length of the side of the unit cell by using eqn 17.7. Because the unit cell is cubic, $a = b = c$, so eqn 17.7 simplifies to

$$\frac{1}{d^2} = \frac{h^2}{a^2} + \frac{k^2}{a^2} + \frac{l^2}{a^2} = \frac{h^2 + k^2 + l^2}{a^2}$$

which rearranges to $a^2 = d^2 \times (h^2 + k^2 + l^2)$ and therefore to

$$a = d \times (h^2 + k^2 + l^2)^{1/2}$$

Solution According to Bragg's law, the separation of the (111) planes responsible for the diffraction is

$$d = \frac{\lambda}{2 \sin \theta} = \frac{154 \text{ pm}}{2 \sin 11.2°}$$

It then follows that, with $h = k = l = 1$,

$$a = \overbrace{\frac{154 \text{ pm}}{2 \sin 11.2°}}^{d} \times \overbrace{3^{1/2}}^{(h^2+k^2+l^2)^{1/2}} = 687 \text{ pm}$$

Self-test 17.7

Calculate the angle at which the same lattice will give a reflection from the (123) planes.

Answer: 24.8°

17.14 Experimental techniques

Laue's original method consisted of passing a beam of X-rays of a wide range of wavelengths into a single crystal and recording the diffraction pattern photographically. The idea behind the approach was that a crystal might not be suitably orientated to act as a diffraction grating for a single wavelength, but whatever its orientation the Bragg law would be satisfied for at least one of the wavelengths when a range of wavelengths is present in the beam.

An alternative technique was developed by Peter Debye and Paul Scherrer and independently by Albert Hull. They used monochromatic (single frequency) X-rays and a powdered sample. When the sample is a powder, we can be sure that some of the randomly distributed crystallites will be orientated so as to satisfy Bragg's law. For example, some of them will be orientated so that their (111) planes, of spacing d, give rise to a reflection at a particular angle and others will be orientated so that their (230) planes give rise to a reflection at a different angle. Each set of (hkl) planes gives rise to reflections at a different angle. In the modern version of the technique, which uses a **powder diffractometer**, the sample is spread on a flat plate and the diffraction pattern is monitored electronically. The major application is for qualitative analysis because the diffraction pattern is a kind of fingerprint and may be recognizable by reference to a library of patterns (Fig. 17.31). The technique is also used for the characterization of substances that cannot be crystallized or for the initial determination of the dimensions and symmetries of unit cells.

Modern X-ray diffraction, which utilizes an **X-ray diffractometer** (Fig. 17.32), is now a highly sophisticated technique. By far the most detailed information comes from developments of the techniques pioneered by the Braggs, in which a single crystal is employed as the diffracting object and a monochromatic beam of X-rays is used to generate the diffraction pattern. The single crystal, which may be only a fraction of a millimetre in length, is rotated relative to the beam and the diffraction pattern is monitored and recorded electronically for each crystal orientation. The primary data are therefore a set of intensities arising from the Miller planes (hkl), with each set of planes giving a reflection of intensity I_{hkl}. For our purposes, we focus on the $(h00)$ planes and write the intensities I_h.

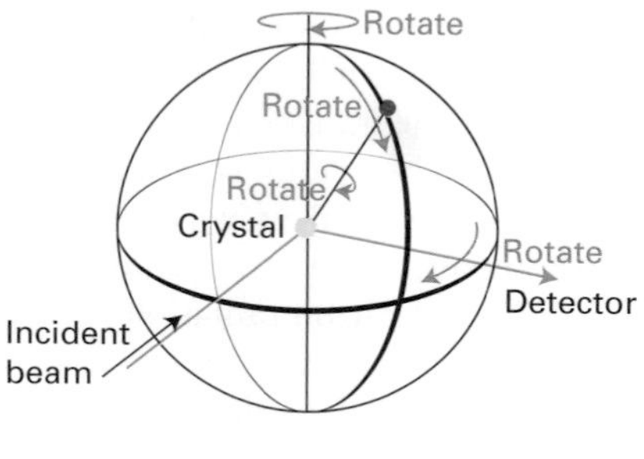

Fig. 17.32 A four-circle diffractometer. The settings of the orientations of the components are controlled by computer; each reflection is monitored in turn, and their intensities are recorded.

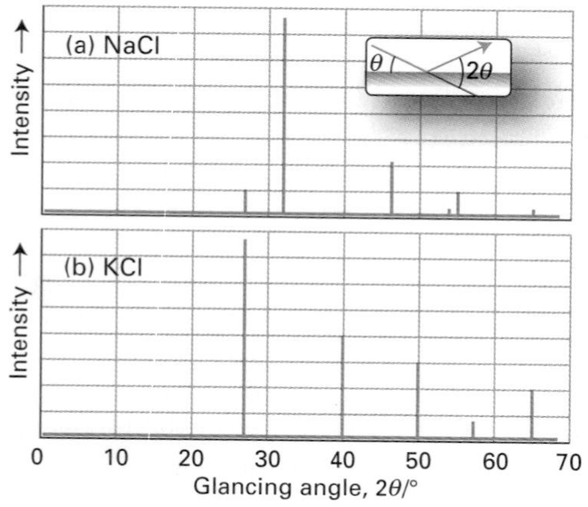

Fig. 17.31 Typical X-ray powder diffraction patterns (for (a) sodium chloride, (b) potassium chloride) that can be used to identify the material and determine the size of its unit cell.

To derive the structure of the crystal from the intensities we need to convert them to the *amplitude* of the wave responsible for the signal. Because the intensity of electromagnetic radiation is given by the square of the amplitude, we need to form the **structure factors** $F_h = I_h^{1/2}$. Here is the first difficulty: we do not know the sign to take. For instance, if $I_h = 4$, then F_h can be either +2 or −2. This ambiguity is the **phase problem** of X-ray diffraction. However, once we have the structure factors, we can calculate the electron density $\rho(x)$ by forming the following sum:

$$\rho(x) = \frac{1}{V}\left\{F_0 + 2\sum_{h=1}^{\infty} F_h \cos(2h\pi x)\right\}$$

Electron density reconstruction (17.9)

where V is the volume of the unit cell. This expression is called a **Fourier synthesis** of the electron density: we show how it is used in the following example. A Fourier synthesis seeks to reconstruct the varying electron density in a unit cell by superimposing a lot of cosine waves of different wavelengths. Low values of the index h give the major features of the structure (they correspond to long-wavelength cosine terms) and high values of h give the fine detail (short wavelength cosine terms). Clearly, if we do not know the sign of F_h, we do not know whether the corresponding term in the sum is positive or negative and we get different electron densities, and hence crystal structures, for different choices of sign.

Example 17.5

Constructing the electron density

The following intensities were obtained in a diffraction experiment on an organic solid:

h	0	1	2	3	4	5	6	7	8	9
I_h	256	100	5	1	50	100	8	10	5	10
h	10	11	12	13	14	15				
I_h	40	25	9	4	4	9				

Construct the electron density along the x direction.

Strategy Begin by finding the structure factors from the corresponding values of I_h by using $F_h = I_h^{1/2}$. Then use eqn 17.9 to plot the electron density as $V\rho(x)$ against x. However, because F_h can be either positive or negative, you will need to make guesses about the signs of F_h, generate different plots for different guesses, and assess the plausibility of each guess.

Solution To find the structure factors, we take square roots of the intensities:

h	0	1	2	3	4	5	6	7	8	9
F_h	±16.0	±10	±2.2	±1	±7.1	±10	±2.8	±3.2	±2.2	±3.2
h	10	11	12	13	14	15				
F_h	±6.3	±5	±3	±2	±2	±3				

Suppose we guess that the signs alternate + − + − . . . ; then according to eqn 17.9 the electron density is

$$V\rho(x) = 16 - 20\cos(2\pi x) + 4.4\cos(4\pi x) - \ldots - 6\cos(30\pi x)$$

This function is shown in Fig. 17.33 as the green line, and the locations of several types of atom are easy to identify as peaks in the electron density. If we guess + signs up to $h = 5$ and − signs thereafter, the electron density is

$$V\rho(x) = 16 + 20\cos(2\pi x) + 4.4\cos(4\pi x) + \ldots - 6\cos(30\pi x)$$

This density is shown in Fig. 17.33 as the red line. This structure has more regions of illegal negative electron density, so is less plausible than the structure obtained from the first choice of phases.

Self-test 17.8

In an X-ray investigation, the following structure factors were determined. Construct the electron density along the corresponding direction.

h	0	1	2	3	4	5	6	7	8	9
F_h	10	−10	8	−8	6	−6	4	−4	2	−2

Answer: Fig. 17.34

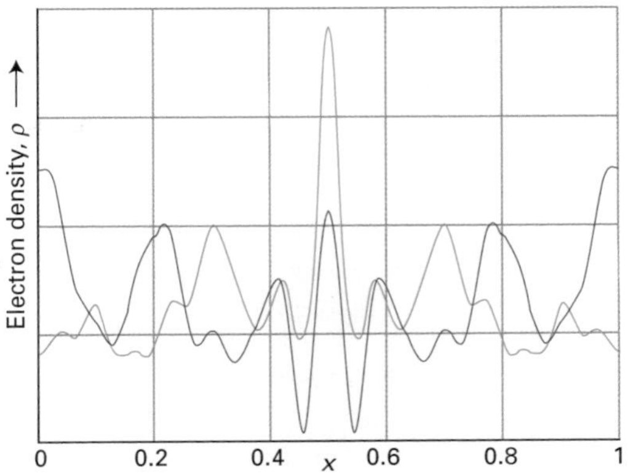

Fig. 17.33 The Fourier synthesis of the electron density of a one-dimensional crystal, using the data in Example 17.5. Green: using alternating signs for the structure factors; red: using positive signs for h up to 5, then negative signs.

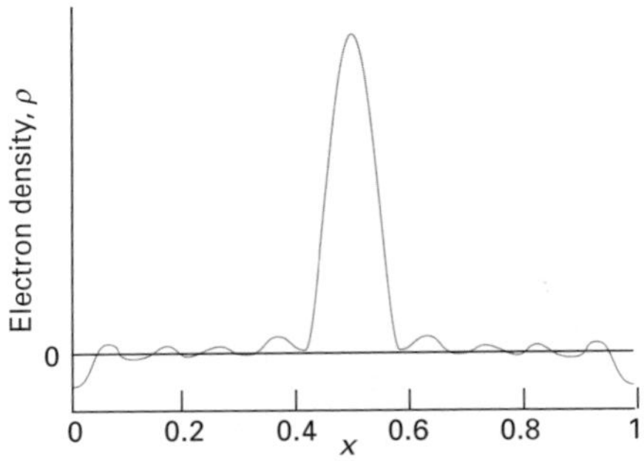

Fig. 17.34 The electron density calculated from data given in Self-test 17.8.

The phase problem can be overcome to some extent by the method of **isomorphous replacement**, in which heavy atoms are introduced into the crystal. The technique relies on the fact that the scattering of X-rays is caused by the oscillations an incoming electromagnetic wave generates in the electrons of atoms, and heavy atoms give rise to stronger scattering than light atoms. So heavy atoms dominate the diffraction pattern and greatly simplify its interpretation. The phase problem can also be resolved by judging whether the calculated structure is chemically plausible, whether the electron density is positive throughout, and by using more refined mathematical techniques. Huge numbers of crystal structures have been determined in this way. In the following sections we review how some of them can be rationalized. For metals and monatomic ions we can model the atoms and ions as hard spheres, and consider how such spheres can be stacked together in a regular, electrically neutral array.

17.15 Metal crystals

Most metallic elements crystallize in one of three simple forms, two of which can be explained in terms of stacking spheres to give the closest possible packing. In such **close-packed structures** the spheres representing the atoms are packed together with least waste of space and each sphere has the greatest possible number of nearest neighbours.

We can form a close-packed layer of identical spheres, one with maximum utilization of space, as shown in Fig. 17.35a. Then we can form a second close-packed layer by placing spheres in the depressions of the first layer (Fig. 17.35b). The third layer may be added in either of two ways, both of which result in the same degree of close packing. In one, the spheres are placed so that they reproduce the first layer (Fig. 17.35c), to give an ABA pattern of layers. Alternatively, the spheres may be placed over the gaps in the first layer (Fig. 17.35d), so giving an ABC pattern.

Two types of structures are formed if the two stacking patterns are repeated. The spheres are **hexagonally close-packed** (hcp) if the ABA pattern is repeated to give the sequence of layers ABABAB. . . . The name reflects the symmetry of the unit cell (Fig. 17.36). Metals with hcp structures include beryllium, cadmium, cobalt, manganese, titanium, and zinc. Solid helium (which forms only under pressure) also adopts this arrangement of atoms. Alternatively, the spheres are **cubic close-packed** (ccp) if the ABC pattern is repeated to give the sequence of layers ABCABC. . . . Here too, the name reflects the symmetry of the unit cell (Fig. 17.37). Metals with this structure include

(a) (b) (c) (d)

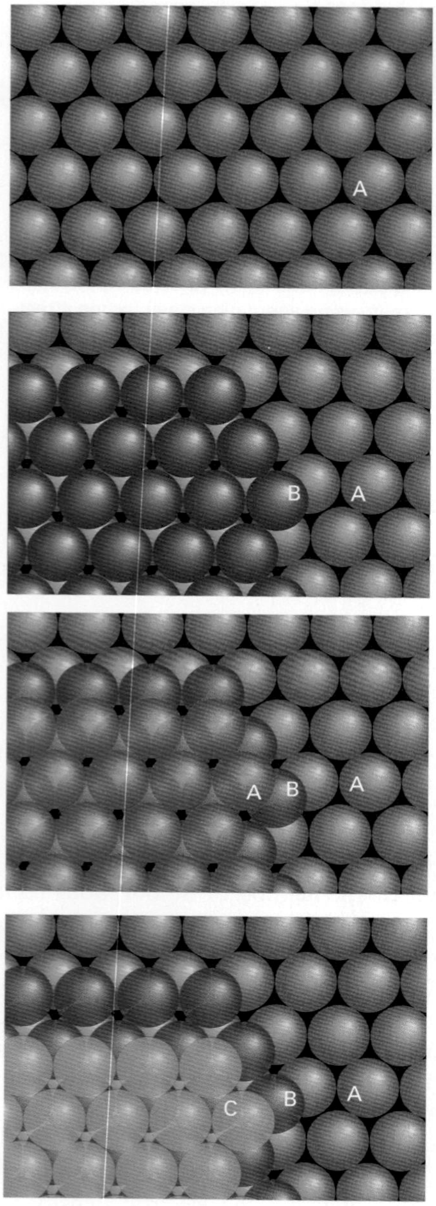

Fig. 17.35 The close-packing of identical spheres. (a) The first layer of close-packed spheres. (b) The second layer of close-packed spheres occupies the dips of the first layer. The two layers are the AB component of the structure. (c) The third layer of close-packed spheres might occupy the dips lying directly above the spheres in the first layer, resulting in an ABA structure. (d) Alternatively, the third layer might lie in the dips that are not above the spheres in the first layer, resulting in an ABC structure.

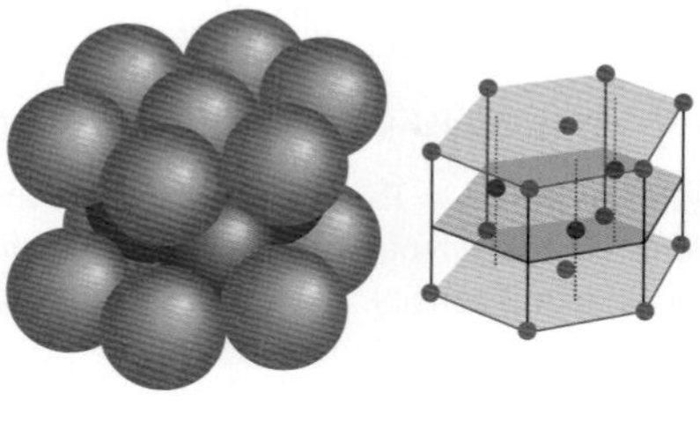

Fig. 17.36 A hexagonal close-packed structure. The colours of the spheres (denoting the three layers of atoms) are the same as in Fig. 17.35.

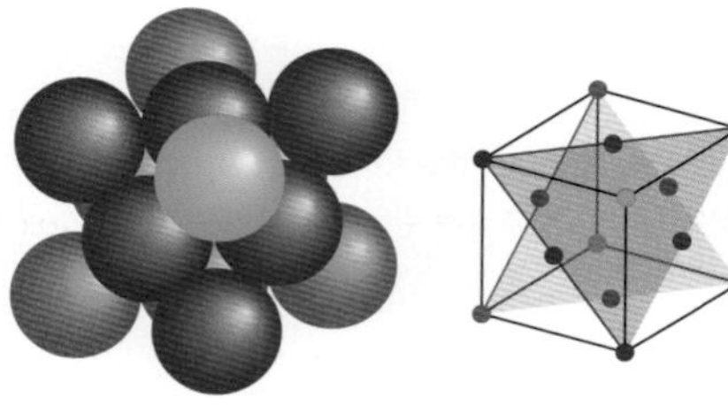

Fig. 17.37 A cubic close-packed structure. The colours of the spheres are the same as in Fig. 17.35.

silver, aluminium, gold, calcium, copper, nickel, lead, and platinum. The noble gases other than helium also adopt a ccp structure in the solid state.

The compactness of the ccp and hcp structures is indicated by their **coordination number**, the number of atoms immediately surrounding any selected atom, which is 12 in both cases. Another measure of their compactness is the **packing fraction**, the fraction of space occupied by the spheres, which is 0.740, as we show in the following Derivation. That is, in a close-packed solid of identical hard spheres, 74.0 per cent of the available space is occupied and only 26.0 per cent of the total volume is empty space.

Derivation 17.4

The packing fraction

Consider the cubic close-packed unit cell in Fig. 17.38. There are six half-spheres within the cell, corresponding to 3 spheres, and eight octants of a sphere, corresponding to one further sphere, for a total of 4 spheres. The volume of each sphere is $\frac{4}{3}\pi r^3$, so the total volume occupied by the spheres is $^{16}/_{3}\pi r^3$. The diagonal of a face is of length $4r$, so by Pythagoras' theorem, the length, a, of the side of the cube is such that $a^2 + a^2 = (4r)^2$, so $a = 8^{1/2}r$. The volume of the cubic unit cell is therefore $a^3 = 8^{3/2}r^3$. It follows that the fraction of the cube occupied by spheres is

$$\frac{\text{Total volume of spheres}}{\text{Volume of cube}} = \frac{(16/3)\pi r^3}{8^{3/2}r^3} = \frac{2\pi}{3\times 8^{1/2}} = 0.7405$$

This fraction corresponds to 74.05 per cent occupancy. An hcp unit cell must have the same packing fraction as it is equally close-packed.

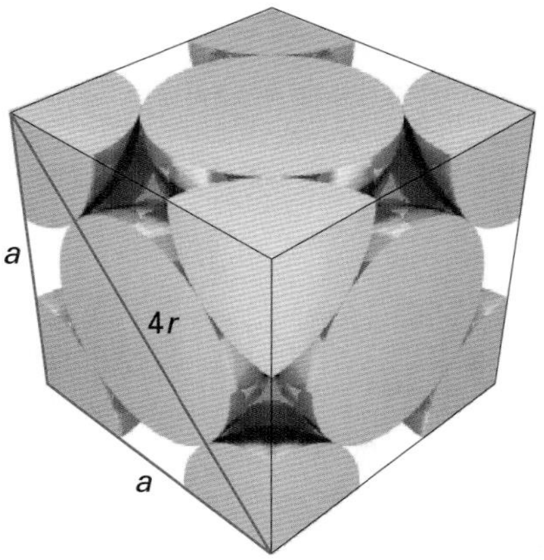

Fig. 17.38 The dimensions of a ccp unit cell used to calculate the packing fraction.

The fact that many metals are close-packed accounts for one of their common characteristics, their high density. However, there is a difference between ccp and hcp metals. In cubic close packing, the faces of the cubes extend throughout the solid, and give rise to a **slip plane**. Careful analysis of the ccp structure shows that there are eight slip planes in various orientations whereas an hcp structure has only one set of slip planes. When the metal is under stress, the layers of atoms may slip past one another along a slip plane. Because a ccp metal has more slip planes than an hcp metal, a ccp metal is more malleable than an hcp metal. Thus, copper, which is ccp, is highly malleable, but zinc, which is hcp, is more brittle. It must be born in mind, however, that metals in real use are not single crystals: they are polycrystalline, with numerous grain-like regions and defects that permeate the structure. Much of metallurgy is associated with the control of the density of grains and grain boundaries.

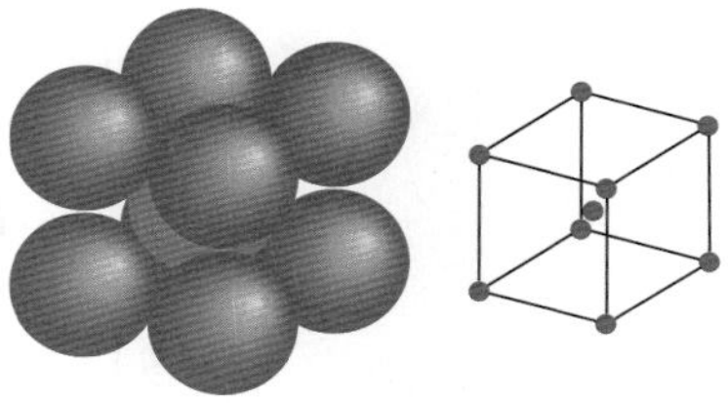

Fig. 17.39 A body-centred cubic unit cell. The spheres on the corners touch the central sphere but the packing pattern leaves more empty space than in the two close-packed structures.

A number of common metals adopt structures that are not close-packed, which suggests that directional covalent bonding between neighbouring atoms is beginning to influence the structure and impose a specific geometrical arrangement. One such arrangement results in a **body-centred cubic** (bcc) lattice, with one sphere at the centre of a cube formed by eight others (Fig. 17.39). The bcc structure is adopted by a number of common metals, including barium, caesium, chromium, iron, potassium, and tungsten. The coordination number of a bcc lattice is 8 and its packing fraction is only 0.68, showing that only about two-thirds of the available space is occupied.

Example 17.6

Assessing a packing fraction

What is the coordination number and the packing fraction of a primitive cubic lattice in which there is a lattice point at each corner of a cube?

Strategy Refer to a diagram of the unit cell in Fig. 17.40 and assess the coordination number by examining the number of nearest neighbours of one of the lattice points at the corner of the cell and imagining the presence of neighbouring cells. For the packing fraction, evaluate the volume of the cell and

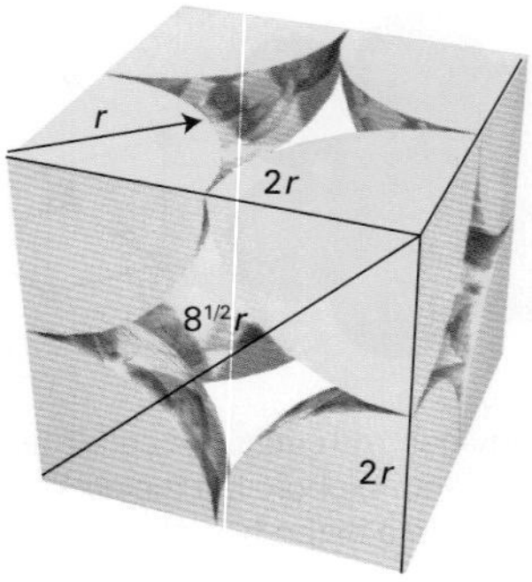

Fig. 17.40 The dimensions of a primitive cubic unit cell used to calculate the packing fraction.

the total volume of the spheres located at each corner, assuming that the spheres are in contact. For the packing fraction, evaluate the ratio of these two volumes.

Solution Each corner point has six nearest neighbours, so the coordination number is 6. The total volume of the cell of side a is a^3. The radius of each sphere is $\frac{1}{2}a$. There are 8 spheres, but each one contributes $^1/_8$ of its volume to the interior of the unit cell, so there is effectively one complete sphere in the cell. The volume of that sphere is $\frac{4}{3}\pi(\frac{1}{2}a)^3 = \frac{1}{6}\pi a^3$. The packing fraction is therefore $\frac{1}{6}\pi a^3/a^3 = \frac{1}{6}\pi$, or 0.52.

Self-test 17.9

Evaluate the packing fraction of a stack of close-packed cylinders. *Hint:* The area of a triangle is $^1/_2$ base × height.

Answer: $\pi/2(3)^{1/2} = 0.91$

17.16 Ionic crystals

To model the structures of ionic crystals by stacks of spheres we must allow for the fact that the two or more types of ion present in the compound have different radii (typically with the cations smaller than the anions) and different charges.

The coordination number of an ion in an ionic crystal is the number of nearest neighbours of opposite charge. Even if, by chance, the ions have the same size, the problem of ensuring that the unit cells are electrically neutral makes it impossible to achieve 12-coordinate close-packed structures, which is one reason why ionic solids are generally less dense than metals. The closest packing that can be achieved is the 8-coordination of the **caesium-chloride structure** in which each cation is surrounded by eight anions and each anion is surrounded by eight cations (Fig. 17.41). In the caesium-chloride structure, an ion of one charge occupies the centre of a cubic unit cell with eight ions of opposite charge at its corners. This structure is adopted by caesium chloride itself and by calcium sulfide, caesium cyanide (with some distortion), and one type of brass (CuZn).

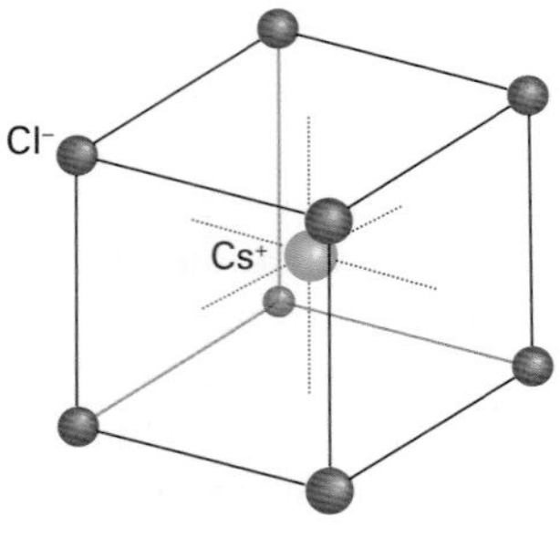

Fig. 17.41 The caesium-chloride structure consists of two interpenetrating simple cubic lattices, one of cations and the other of anions, so that each cube of ions of one kind has a counter-ion at its centre. This illustration shows a single unit cell with a Cs^+ ion at the centre. By imagining eight of these unit cells stacked together to form a bigger cube, it should be possible to imagine an alternative form of the unit cell with Cs^+ at the corners and a Cl^- ion at the centre.

When the radii of the ions differ by more than in caesium chloride, even 8-coordinate packing cannot be achieved. One common structure adopted is the 6-coordinated **rock-salt structure** typified by sodium chloride (rock salt is a mineral form of sodium chloride) in which each cation is surrounded by six anions and each anion is surrounded by six cations (Fig. 17.42). The rock-salt structure is the structure of sodium chloride itself and of several other compounds of formula MX, including potassium bromide, silver chloride, and magnesium oxide.

The switch from the caesium-chloride structure to the rock-salt structure (in a number of examples) can be correlated with the **radius ratio**

$$\gamma = \frac{r_{\text{smaller}}}{r_{\text{larger}}} \qquad \text{Definition} \quad \text{Radius ratio} \qquad (17.10)$$

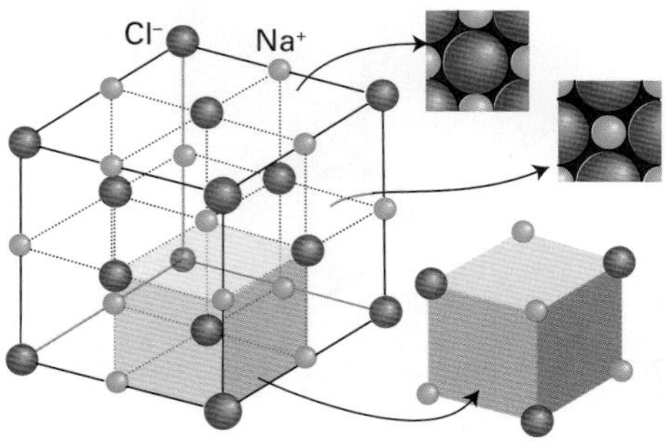

Fig. 17.42 The rock-salt (NaCl) structure consists of two mutually interpenetrating slightly expanded face-centred cubic lattices. The additional diagrams in this illustration show various details of the structure.

Table 17.5
Radius ratio and crystal type

Radius ratio		Coordination	Crystal type
$\gamma > 3^{1/2} - 1 = 0.732$		(8,8)	caesium chloride
$2^{1/2} - 1 = 0.414 < \gamma < 0.732$	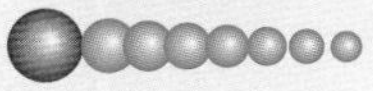	(6,6)	rock salt
$\gamma < 0.414$	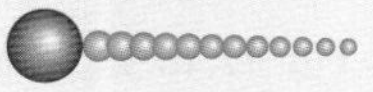	(4,4)	sphalerite, zinc blende

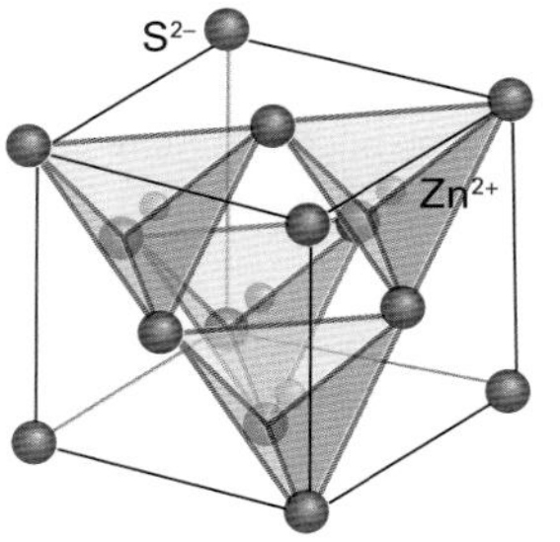

Fig. 17.43 The sphalerite (zinc-blende, ZnS) structure. This structure is typical of ions that have markedly different radii and equal but opposite charges.

Table 17.6
Ionic radii, r/pm

Li^+ 159	Be^{2+} 27	B^{3+} 12	N^{3-} 171	O^{2-} 140	F^- 133
Na^+ 102	Mg^{2+} 72	Al^{3+} 53	P^{3-} 212	S^{2-} 184	Cl^- 181
K^+ 138	Ca^{2+} 100	Ga^{3+} 62	As^{3-} 222	Se^{2-} 198	Br^- 196
Rb^+ 149	Sr^{2+} 116				
Cs^+ 170	Ba^{2+} 136				

The two radii are those of the smaller and larger ions in the crystal. The **radius-ratio rule**, which is derived by analysing the geometrical problem of stacking together spheres of different radii, then suggests the structural types shown in Table 17.5. Sphalerite (zinc blende), which is mentioned in the table, is a form of zinc sulfide, ZnS (Fig. 17.43). The radius-ratio rule is moderately well supported by observation. The deviation of a structure from the prediction is often taken to be an indication of a shift from ionic towards covalent bonding.

The **ionic radii** used to calculate γ, and wherever else it is important to know the sizes of ions, are derived from the distance between the centres of adjacent ions in a crystal. However, in a diffraction experiment we measure the distance between the centres of ions. It is necessary to apportion that total distance by defining the radius of one ion and reporting all others on that basis. One scale that is widely used is based on the value 140 pm for the radius of the O^{2-} ion (Table 17.6). Other scales are also available (such as one based on F^- for discussing halides), and it is essential not to mix values from different scales. Because ionic radii are so arbitrary, predictions based on them (such as those made by using the radius-ratio rule) must be viewed cautiously.

Brief illustration 17.6 The radius-ratio rule

The ionic radii of magnesium (Mg^{2+}) and oxygen (O^{2-}) are 72 pm and 140 pm, respectively. The radius ratio for MgO is therefore

$$\gamma = \frac{72\ \text{pm}}{140\ \text{pm}} = 0.51$$

According to Table 17.5, this ratio suggests a rock-salt structure, as in fact is found.

Self-test 17.10

Is sodium iodide likely to have a rock-salt or caesium-chloride structure?

Answer: rock salt

17.17 Molecular crystals

X-ray diffraction studies of solids reveal a huge amount of information, including interatomic distances, bond angles, stereochemistry, and vibrational parameters. **Molecular solids**, which are the subject of the overwhelming majority of modern structural

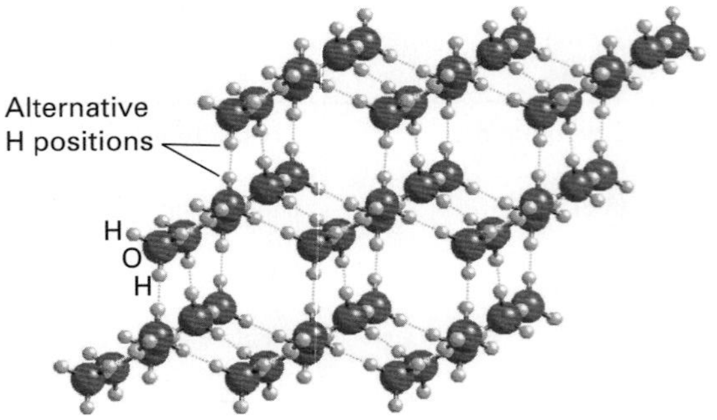

Fig. 17.44 A fragment of the crystal structure of ice (ice-I). Each O atom is at the centre of a tetrahedron of four O atoms at a distance of 276 pm. The central O atom is attached by two short O—H bonds to two H atoms and by two long hydrogen bonds to the H atoms of two of the neighbouring molecules. Overall, the structure consists of planes of hexagonal puckered rings of H_2O molecules (like the chair form of cyclohexane). The two H atoms between each O atom show the former's two alternative locations.

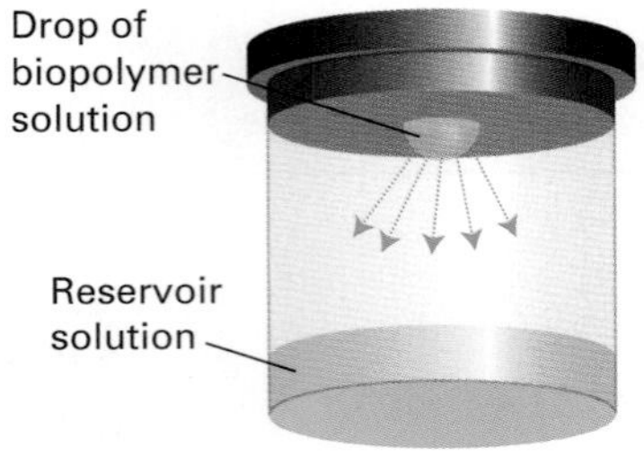

Fig. 17.45 In a common implementation of the vapour diffusion method of biopolymer crystallization, a single drop of biopolymer solution hangs above a reservoir solution that is very concentrated in a non-volatile solute. Solvent evaporates from the more dilute drop until the vapour pressure of water in the closed container reaches a constant equilibrium value. In the course of evaporation (denoted by the downward arrows), the biopolymer solution becomes more concentrated and, at some point, crystals may form.

determinations, are held together by van der Waals interactions and hydrogen bonds (Chapter 15). The observed crystal structure is Nature's solution to the problem of condensing objects of various shapes into an aggregate of minimum energy (actually, for $T > 0$, of minimum Gibbs energy). The prediction of the structure is difficult, especially when the molecules are large, but software specifically designed to explore interaction energies can now make reasonably reliable predictions. The problem is made more complicated by the role of hydrogen bonds, which in some cases dominate the crystal structure, as in ice (Fig. 17.44), but in others (for example, in phenol) distort a structure that is determined largely by the van der Waals interactions.

Impact on biochemistry 17.2

X-ray crystallography of biological macromolecules

X-ray crystallography is the deployment of X-ray diffraction techniques for the determination of the location of all the atoms in molecules as complicated as biopolymers. The success of modern biochemistry in explaining such processes as DNA replication, protein biosynthesis, and enzyme catalysis is a direct result of developments in preparatory, instrumental, and computational procedures that have led to the determination of large numbers of structures of biological macromolecules by X-ray crystallography. Most work is now done not on fibres but on crystals, in which the large molecules lie in orderly ranks. A technique that works well for charged proteins consists of adding large amounts of a salt, such as $(NH_4)_2SO_4$, to a buffer solution containing the biopolymer. The increase in the ionic strength of the solution decreases the solubility of the protein to such an extent that the protein precipitates, sometimes as crystals that are amenable to analysis by X-ray diffraction. A common strategy for inducing crystallization involves the gradual removal of solvent from a biopolymer solution by *vapour diffusion*. In one implementation of the method, a single drop of biopolymer solution hangs above an aqueous solution (the reservoir), as shown in Fig. 17.45. If the reservoir solution is more concentrated in a non-volatile solute (for example, a salt) than is the biopolymer solution, then solvent will evaporate slowly from the drop. At the same time, the concentration of biopolymer in the drop increases gradually until crystals begin to form.

Special techniques are used to crystallize hydrophobic proteins, such as those spanning the bilayer of a cell membrane. In such cases, surfactant molecules, which like phospholipids contain polar head groups and hydrophobic tails, are used to encase the protein molecules and make them soluble in aqueous buffer solutions. Vapour diffusion may then be used to induce crystallization.

After suitable crystals are obtained, X-ray diffraction data are collected and analysed as described in the text. The three-dimensional structures of a very large number of biological polymers have been determined in this way. However, the techniques discussed so far give only static pictures and are not useful in studies of dynamics and reactivity. This limitation stems from the fact that the Bragg rotation method requires stable crystals that do not change structure during the lengthy data acquisition times required. However, special time-resolved X-ray diffraction techniques have become available in recent years and it is now possible to make exquisitely detailed measurements of atomic motions during chemical and biochemical reactions.

Time-resolved X-ray diffraction techniques make use of synchrotron sources, which can emit intense polychromatic pulses of X-ray radiation with pulse widths varying from 100 ps to 200 ps (1 ps = 10^{-12} s). Instead of the Bragg method, the Laue method is used because many reflections can be collected simultaneously, rotation of the sample is not required, and data acquisition times are short. However, good diffraction data cannot be obtained from a single X-ray pulse and reflections from several pulses must be averaged together. In practice, this averaging dictates the time resolution of the experiment, which is commonly tens of microseconds or less.

The progress of a reaction may be studied either by real-time analysis of the evolving system or by trapping intermediates by chemical or physical means. Regardless of the strategy, all the molecules in the crystal must be made to react at the same time, so special reaction initiation schemes are required. One way to initiate a reaction is to allow a solution containing one of the reactants to diffuse into a crystal containing the other reactant. This method is simple, but limited to relatively long reaction times, as diffusion of solutions into crystals large enough for crystallographic measurements is of the order of seconds to minutes.

Checklist of key concepts

☐ **1** Solids are classified as metallic, ionic, covalent, or molecular.

☐ **2** Electronic conductors are classified as metallic conductors or semiconductors according to the temperature dependence of their conductivities.

☐ **3** A superconductor is an electronic conductor with zero resistance.

☐ **4** According to the band theory, electrons occupy molecular orbitals formed from the overlap of atomic orbitals.

☐ **5** Full bands are called valence bands and empty bands are called conduction bands.

☐ **6** Semiconductors are classified as p-type or n-type according to whether conduction is due to holes in the valence band or electrons in the conduction band.

☐ **7** The lattice enthalpy is the change in enthalpy (per mole of formula units) accompanying the complete separation of the components of the solid.

☐ **8** A material is diamagnetic if it tends to move out of a magnetic field, and paramagnetic if it tends to move into a magnetic field.

☐ **9** Ferromagnetism is the cooperative alignment of electron spins in a material and gives rise to strong magnetization.

☐ **10** Antiferromagnetism results from alternating spin orientations in a material and leads to weak magnetization.

☐ **11** Type I superconductors show abrupt loss of superconductivity when an applied magnetic field exceeds a critical value $\mathcal{H}_c$ characteristic of the material. They are also completely diamagnetic below $\mathcal{H}_c$.

☐ **12** Type II superconductors show a gradual loss of superconductivity and diamagnetism with increasing magnetic field.

☐ **13** Unit cells are classified into seven crystal systems according to their rotational symmetries.

☐ **14** A unit cell is the small three-dimensional figure that may be used to construct the entire crystal lattice by purely translational displacements.

☐ **15** A Bravais lattice is one of 14 types of unit cell shown in Fig. 17.24.

☐ **16** Lattice planes are specified by a set of Miller indices (*hkl*).

☐ **17** Many elemental metals have close-packed structures with coordination number 12.

☐ **18** Close-packed structures may be either cubic (ccp) or hexagonal (hcp).

☐ **19** Representative ionic structures include the caesium-chloride, rock-salt, and zinc-blende structures.

☐ **20** The radius-ratio rule may be used cautiously to predict which of these three structures is likely.

Road map of key equations

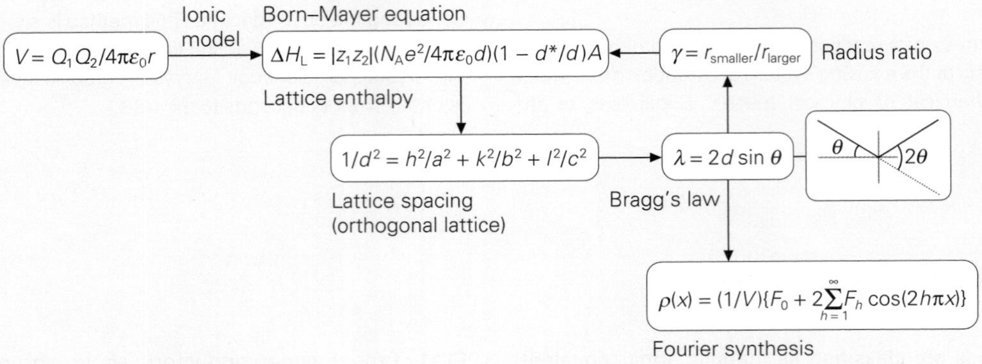

Questions and exercises

Discussion questions

17.1 Explain how metallic conductors, semiconductors, and insulators are identified and explain their properties in terms of band theory. Why is graphite an electronic conductor and diamond an insulator?

17.2 Explain how planes of lattice points are labelled.

17.3 Describe the consequences of the phase problem in determining structure factors and how the problem is overcome.

17.4 Describe the structures of elemental metallic solids in terms of the packing of hard spheres.

17.5 Describe the caesium-chloride and rock-salt structures. How does the radius-ratio rule help in the classification of a structure into each type?

17.6 Describe the different types of magnetism that materials can display and account for their origins.

Exercises

17.1 The electrical resistance of a sample increased from 100 Ω to 120 Ω when the temperature was changed from 0 °C to 100 °C. Is the substance a metallic conductor or a semiconductor?

17.2 The energy levels of N atoms in the Hückel approximation are given by (eqn 17.1)

$$E_k = \alpha + 2\beta\cos\frac{k\pi}{N+1} \qquad k = 1, 2, \ldots, N$$

If the atoms are arranged in a ring, the energy levels are given by

$$E_k = \alpha + 2\beta\cos\frac{2k\pi}{N} \qquad k = 0, \pm1, \pm2, \ldots, \pm\tfrac{1}{2}N$$

(for N even). Discuss the consequences, if any, of joining the ends of an initially straight length of material.

17.3 Classify as n-type or p-type a semiconductor formed by doping (a) germanium with phosphorus, (b) germanium with indium.

17.4 Type I superconductors show abrupt loss of superconductivity when an applied magnetic field exceeds a critical value $\mathcal{H}_c$ that depends on temperature and T_c as

$$\mathcal{H}_c(T) = \mathcal{H}_c(0)\left\{1 - \frac{T^2}{T_c^2}\right\}$$

where $\mathcal{H}_c(0)$ is the value of $\mathcal{H}_c$ as $T \to 0$. Lead has $T_c = 7.19$ K and $\mathcal{H}_c = 63.9$ kA m^{-1}. At what temperature does lead become superconducting in a magnetic field of 20.0 kA m^{-1}?

17.5 Describe the bonding in calcium oxide, CaO, in terms of bands composed of Ca and O atomic orbitals. How does this model justify the ionic model of this compound?

17.6 Calculate the lattice enthalpy of CaO from the following data:

	$\Delta H/(\text{kJ mol}^{-1})$
Sublimation of Ca(s)	+178
Ionization of Ca(g) to Ca^{2+}(g)	+1735
Dissociation of O_2(g)	+249
Electron attachment to O(g)	−141
Electron attachment to O^-(g)	+844
Formation of CaO(s) from Ca(s) and O_2(g)	−635

17.7 Calculate the lattice enthalpy of $MgBr_2$ from the following data:

	$\Delta H/(\text{kJ mol}^{-1})$
Sublimation of Mg(s)	+148
Ionization of Mg(g) to Mg^{2+}(g)	+2187
Vaporization of Br_2(l)	+31
Dissociation of Br_2(g)	+193
Electron attachment to Br(g)	−331
Formation of $MgBr_2$(s) from Mg(s) and Br_2(l)	−524

17.8 Estimate the ratio of the lattice enthalpies of SrO and CaO from the Born–Mayer equation by using the ionic radii in Table 17.6.

17.9 Calculate the potential energy of an ion at the centre of a diffuse 'spherical crystal' in which concentric spheres of ions of opposite charge surround the ion and the numbers of ions on the spherical surfaces fall away rapidly with distance. Let successive spheres lie at radii d, $2d$, . . . , and the number of ions (all of the same charge) on each successive sphere is inversely proportional to the radius of the sphere. You will need the following sum:

$$1-\frac{1}{2^2}+\frac{1}{3^2}-\frac{1}{4^2}+\cdots=\frac{\pi^2}{12}$$

17.10 The tip of a scanning tunnelling microscope can be used to move atoms on a surface. The movement of atoms and ions depends on their ability to leave one position and stick to another, and therefore on the energy changes that occur. As an illustration, consider a two-dimensional square lattice of univalent positive and negative ions separated by 200 pm, and consider a cation on top of this array. Calculate, by direct summation, its Coulombic interaction when it is in an empty lattice point directly above an anion.

17.11 The compound Rb_3TlF_6 has a tetragonal unit cell with dimensions $a = 651$ pm and $c = 934$ pm. Calculate the volume of the unit cell and the density of the solid.

17.12 The orthorhombic unit cell of $NiSO_4$ has the dimensions $a = 634$ pm, $b = 784$ pm, and $c = 516$ pm, and the density of the solid is estimated as 3.9 g cm^{-3}. Determine the number of formula units per unit cell and calculate a more precise value of the density.

17.13 The unit cells of $SbCl_3$ are orthorhombic with dimensions $a = 812$ pm, $b = 947$ pm, and $c = 637$ pm. Calculate the spacing of (a) the (321) planes, (b) the (642) planes.

17.14 Draw a set of points as a rectangular array based on unit cells of side a and b, and mark the planes with Miller indices (10), (01), (11), (12), (23), (41), $(4\bar{1})$.

17.15 Repeat Exercise 17.14 for an array of points in which the a and b axes make 60° to each other.

17.16 In a certain unit cell, planes cut through the crystal axes at ($2a$, $3b$, c), (a, b, c), ($6a$, $3b$, $3c$), ($2a$, $-3b$, $-3c$). Identify the Miller indices of the planes.

17.17 Draw an orthorhombic unit cell and mark on it the (100), (010), (001), (011), (101), and (101) planes.

17.18 Draw a triclinic unit cell and mark on it the (100), (010), (001), (011), (101), and (101) planes.

17.19 Calculate the separations of the planes (111), (211), and (100) in a crystal in which the cubic unit cell has sides of length 572 pm.

17.20 Calculate the separations of the planes (123) and (236) in an orthorhombic crystal in which the unit cell has sides of lengths 784, 633, and 454 pm.

17.21 The glancing angle of a Bragg reflection from a set of crystal planes separated by 97.3 pm is 19.85°. Calculate the wavelength of the X-rays.

17.22 Reflections from the (110) planes of barium titanate, $BaTiO_3$, which has a tetragonal crystalline structure, were observed at glancing angles of 9.13° with Co K_α radiation of wavelength 179 pm. What is the value of the unit cell parameter a?

17.23 Construct the electron density along the x-axis of a crystal given the following structure factors:

h	0	1	2	3	4	5	6	7	8	9
F_h	+30.0	+8.2	+6.5	+4.1	+5.5	−2.4	+5.4	+3.2	−1.0	+1.1

h	10	11	12	13	14	15
F_h	+6.5	+5.2	−4.3	−1.2	+0.1	+2.1

17.24 The separation of (100) planes of lithium metal is 350 pm and its density is 0.53 g cm^{-3}. Is the structure of lithium face-centred cubic or body-centred cubic?

17.25 Copper crystallizes in a face-centred cubic structure with unit cells of side 361 pm. (a) Predict the appearance of the powder diffraction pattern using 154 pm radiation. (b) Calculate the density of copper on the basis of this information.

17.26 Calculate the packing fraction of a stack of cylinders.

17.27 Calculate the packing fraction of a cubic close-packed structure.

17.28 Suppose a virus can be regarded as a sphere and that it stacks together in a hexagonal close-packed arrangement. If the density of the virus is the same as that of water (1.00 g cm^{-3}), what is the density of the solid?

17.29 How many (a) nearest neighbours, (b) next-nearest neighbours are there in a body-centred cubic structure? What are their distances if the side of the cube is 600 nm?

17.30 How many (a) nearest neighbours, (b) next-nearest neighbours are there in a cubic close-packed structure? What are their distances if the side of the cube is 600 nm?

17.31 The thermal and mechanical processing of materials is an important step in ensuring that they have the appropriate physical properties for their intended application. Suppose a metallic element underwent a phase transition in which its crystal structure changed from cubic close-packed to body-centred cubic. (a) Would it become more or less dense? (b) By what factor would its density change?

17.32 Use the radius-ratio rule to predict the kind of crystal structure expected for magnesium oxide.

Projects

The symbol ‡ indicates that calculus is required.

17.33‡ Here we explore the band theory of solids in more detail. (a) Use eqn 17.1 to find an expression for the separation between neighbouring levels in a band of N atoms and show that the separation goes to zero as N increases to infinity. (b) Calculate the density of states for a long line of atoms, where the density of states is the quantity $\rho(k)$ in the expression $dE = \rho(k)dk$ and draw a graph of $\rho(k)$. Where is the density of states greatest? *Hint:* Use eqn 17.1 and form dE/dk. (c) The treatment in parts (a) and (b) applies only to one-dimensional solids. In three dimensions, the variation of density of states is more like that shown in Fig. 17.46. Account for the fact that in a three-dimensional solid the greatest density of states is near the centre of the band and the lowest density is at the edges.

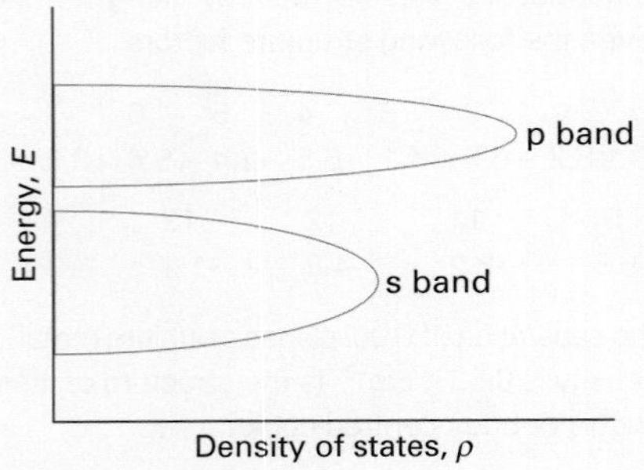

Fig. 17.46 Typical densities of states in a solid.

17.34 A photoactive yellow protein is involved in the 'negative phototactic response', or movement away from light, of the bacterium *Ectothiorhodospira halophila*. Within 1 ns after absorption of a photon with λ = 446 nm, a protein-bound phenolate ion (**1**) undergoes *trans–cis* isomerization to form the intermediate shown in the second illustration. What follows is a series of rearrangements that include the ejection of the chromophore from its binding site deep in the protein, its return to the site, and reformation of the *trans* conformation. Consult the current literature and prepare a brief report on the use of time-resolved X-ray diffraction techniques in the description of structural changes that follow electronic excitation of the chromophore with a laser pulse.

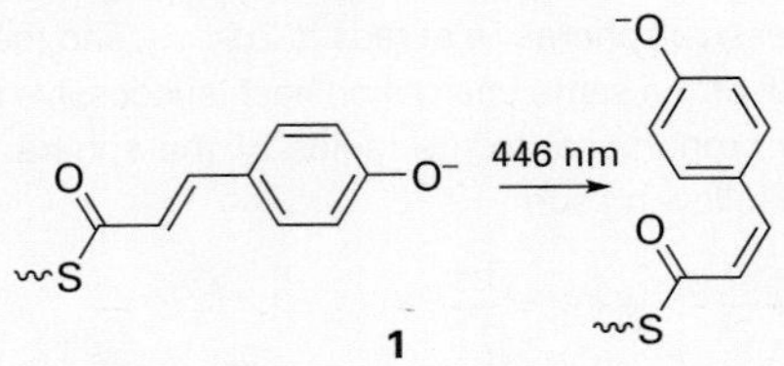

1

17.35 A transistor is a semiconducting device that is commonly used either as a switch or an amplifier of electrical signals. Prepare a brief report on the design of a nanometre-sized transistor that uses a carbon nanotube as a component. A useful starting point is the work summarized by Tans, et al. (*Nature* **393**, 49 (1998)).

18

Solid surfaces

Processes at solid surfaces govern the viability of industry constructively, as in catalysis, and the permanence of its products destructively, as in corrosion. Chemical reactions at solid surfaces may differ sharply from reactions in the bulk, for reaction pathways of much lower activation energy may be provided by the surface, and hence result in catalysis. The concept of a solid surface has been extended in recent years with the availability of microporous materials as catalysts, where the 'surface' includes the interior of the porous solid.

Although we start the chapter with a discussion of clean surfaces, you should not lose sight of the fact that for chemists the important aspects of a surface are the attachment of substances to it and the reactions that take place there. Also of interest are surfaces immersed in solvents and in gases at high pressure, when the concept of a 'clean' surface loses much of its meaning. Moreover, the structure and even the elemental composition at the surface may be entirely different from that of the underlying bulk material, as in the presence of an oxide layer on aluminium. Because the reactions that take place at a surface typically involve only a few surface layers of atoms, the reactivity of a surface may be determined solely by this different composition and have little to do with the composition of the bulk.

Reactions at surfaces include the processes that lie at the heart of electrochemistry. Therefore, in the final part of the chapter we revisit the topics treated in Chapter 9, but focus on the dynamics of electrode processes rather than the equilibrium properties treated there.

The growth and structure of surfaces 449

18.1 Surface growth 450

18.2 Surface composition and structure 450

The extent of adsorption 455

18.3 Physisorption and chemisorption 455

18.4 Adsorption isotherms 456

18.5 The rates of surface processes 461

Catalytic activity at surfaces 463

18.6 Unimolecular reactions 463

18.7 The Langmuir–Hinshelwood mechanism 464

18.8 The Eley–Rideal mechanism 464

Processes at electrodes 467

18.9 The electrode–solution interface 467

18.10 The rate of electron transfer 468

18.11 Voltammetry 470

18.12 Electrolysis 472

CHECKLIST OF KEY CONCEPTS 473

ROAD MAP OF KEY EQUATIONS 474

QUESTIONS AND EXERCISES 474

The growth and structure of surfaces

The attachment of molecules to a surface is called **adsorption**. The substance that adsorbs is the **adsorbate** and the underlying material that we are concerned with in this section is the **adsorbent** or **substrate**. The reverse of adsorption is **desorption**.

18.1 Surface growth

A simple picture of a perfect crystal surface is as a tray of oranges in a grocery store (Fig. 18.1). A gas molecule that collides with the surface can be imagined as a table-tennis ball bouncing erratically over the oranges. The molecule loses energy as it bounces under the influence of intermolecular forces, but it is likely to escape from the surface before it has lost so much kinetic energy that it has become trapped. The same is true, to some extent, of an ionic crystal in contact with a solution. There is little energy advantage for an ion in solution to discard some of its solvating molecules and stick at an exposed position on a flat surface.

The picture changes when the surface has defects, for then there are ridges of incomplete layers of atoms or ions. A typical type of surface defect is a **step** between two otherwise flat layers of atoms called **terraces** (Fig. 18.2). A step defect might itself have defects, including kinks. When an atom settles on a terrace it bounces across it under the influence of the intermolecular potential, and might come to a step or a corner formed by a kink. Instead of interacting with a single terrace atom, the molecule now interacts with several,

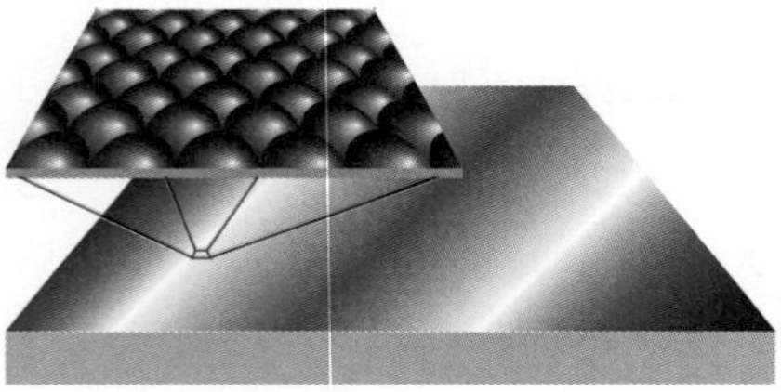

Fig. 18.1 A schematic diagram of the flat surface of a solid. This primitive model is largely supported by scanning tunnelling microscope images.

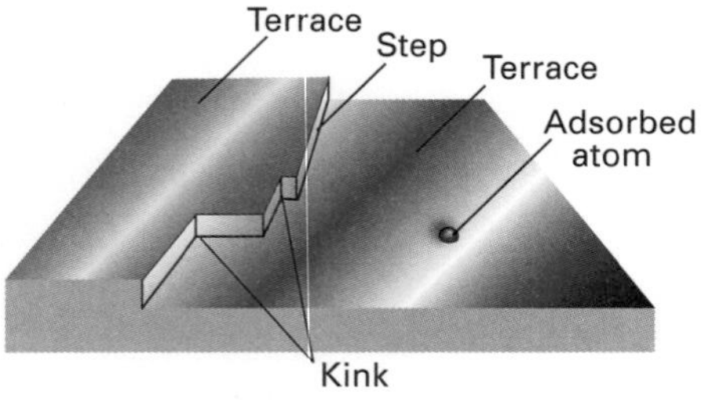

Fig. 18.2 Some of the kinds of defects that may occur on otherwise perfect terraces. Defects play an important role in surface growth and catalysis.

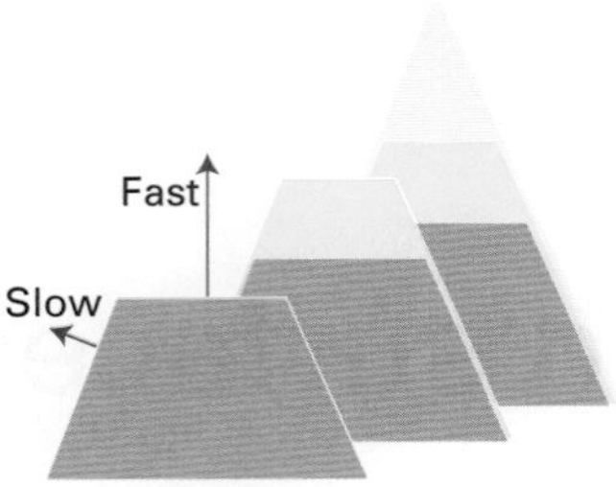

Fig. 18.3 The slower-growing faces of a crystal dominate its final external appearance. Three successive stages of the growth are shown.

and the interaction may be strong enough to trap it. Likewise, when ions deposit from solution, the loss of the solvation interaction is offset by a strong Coulombic interaction between the arriving ions and several ions at the surface defect.

The rapidity of growth depends on the crystal plane concerned and—perhaps surprisingly—the slowest growing faces dominate the appearance of the crystal. This feature is explained in Fig. 18.3, where we see that although the horizontal face grows forward most rapidly, it grows itself out of existence and the more slowly growing faces survive.

18.2 Surface composition and structure

Under normal conditions, a surface exposed to a gas is constantly bombarded with molecules and a freshly prepared surface is covered very quickly. Just how quickly can be estimated by using the kinetic theory of gases and an expression for the rate at which the surface is bombarded with atoms or molecules. In the laboratory we need special techniques, described below, for studying the composition and structure of surfaces.

(a) The collision flux

The **collision flux**, Z_W, is the number of hits on a region of a surface during an interval divided by the area of the region and the duration of the interval:[1]

$$Z_W = \frac{p}{(2\pi mkT)^{1/2}} \qquad \text{Collision flux} \quad (18.1a)$$

where m is the mass of the molecules. If we write $m = M/N_A$, where M is the molar mass of the gas, eqn 18.1 becomes

$$Z_W = \frac{(N_A/2\pi k)^{1/2}p}{(TM)^{1/2}} \qquad \text{Alternative form} \quad \text{Collision flux} \quad (18.1b)$$

[1] For the origin of this equation, see our *Physical Chemistry* (2010).

After inserting numerical values for the constants and selecting units for the variables, the practical form of this expression is

$$Z_W = \frac{Z_0(p/\text{Pa})}{\{(T/\text{K})(M/(\text{g mol}^{-1}))\}^{1/2}},$$
with $Z_0 = 2.63 \times 10^{24}\ \text{m}^{-2}\ \text{s}^{-1}$

Alternative form Collision flux (18.1c)

As usual, let's extract physical significance from the mathematical expression. At first glance, eqn 18.1 might suggest that the collision flux decreases as the temperature rises, even though molecules are moving faster! In fact, in a container of constant volume the pressure is proportional to the temperature, so the overall temperature dependence goes as $T/T^{1/2}$; that is, $Z_W \propto T^{1/2}$, and the flux increases with temperature in proportion to the speed of the molecules.

Brief illustration 18.1 The collision flux

For air, with $M \approx 29$ g mol^{-1}, at $p = 1$ atm $= 1.013\ 25 \times 10^5$ Pa and $T = 298$ K, we obtain

$$Z_W = \frac{\overbrace{(2.63\times10^{24}\ \text{m}^{-2}\ \text{s}^{-1})}^{Z_0}\times\overbrace{(1.013\ 25\times10^5)}^{p/p_0}}{(\underbrace{298}_{T/\text{K}} \times \underbrace{29}_{M/\text{g mol}^{-1}})^{1/2}}$$
$$= 2.9\times10^{27}\ \text{m}^{-2}\ \text{s}^{-1}$$

Because 1 m^2 of metal surface consists of about 10^{19} atoms, each atom is struck about 10^8 times each second. Even if only a few collisions leave a molecule adsorbed to the surface, the time for which a freshly prepared surface remains clean is very short.

Self-test 18.1

Calculate the collision flux with a surface of a vessel containing propane at 25 °C when the pressure is 100 Pa.

Answer: $Z_W = 2.30 \times 10^{24}\ \text{m}^{-2}\ \text{s}^{-1}$

(b) Experimental techniques

The obvious way to retain the cleanliness of the surface is to reduce the pressure. When it is reduced to 0.1 mPa (as in a simple vacuum system) the collision flux falls to about $10^{18}\ \text{m}^{-2}\ \text{s}^{-1}$, corresponding to one hit per surface atom in each 0.1 s. Even that is too brief in most experiments, and in **ultra-high vacuum** (UHV) techniques pressures as low as 0.1 μPa (when $Z_W = 10^{15}\ \text{m}^{-2}\ \text{s}^{-1}$) are reached on a routine basis and 1 nPa (when $Z_W = 10^{13}\ \text{m}^{-2}\ \text{s}^{-1}$) are reached with special care. These collision fluxes correspond to each surface atom being hit once every 10^5 to 10^6 s, or about once a day.

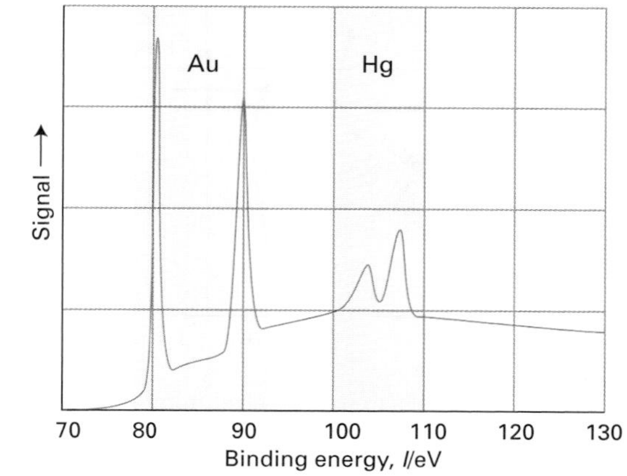

Fig. 18.4 The X-ray photoelectron emission spectrum of a sample of gold contaminated with a surface layer of mercury. (M.W. Roberts and C.S. McKee, *Chemistry of the metal–gas interface*, Oxford (1978).)

The chemical composition of a surface can be determined by a variety of ionization techniques. The same techniques can be used to detect any remaining contamination after cleaning and to detect layers of material adsorbed later in the experiment. One technique that may be used is **photoemission spectroscopy,** a derivative of the photoelectric effect, in which X-rays (for XPS) or hard (short-wavelength) ultraviolet (for UPS) ionizing radiation is used to eject electrons from adsorbed species. The kinetic energies of the electrons ejected from their orbitals are measured and the pattern of energies is used to identify the material present (Fig. 18.4). UPS, which examines electrons ejected from valence shells, is also used to establish the bonding characteristics and the details of electronic structures of substances on the surface. Its usefulness is its ability to reveal which orbitals of the adsorbate are involved in the bond to the substrate. For instance, the principal difference between the photoemission results on free benzene and benzene adsorbed on palladium is in the energies of the π electrons. This difference is interpreted as meaning that the C_6H_6 molecules lie parallel to the surface and are attached to it by their π orbitals. In contrast, pyridine (C_6H_5N) stands almost perpendicular to the surface, and is attached by a σ bond formed by the nitrogen lone pair.

A very important technique, which is widely used in the microelectronics industry, is **Auger electron spectroscopy** (AES). The **Auger effect** (pronounced oh-zhey) is the emission of a second electron after high-energy radiation has expelled another electron. The first electron to depart leaves a hole in a low-lying orbital, and an upper electron falls into it. The energy

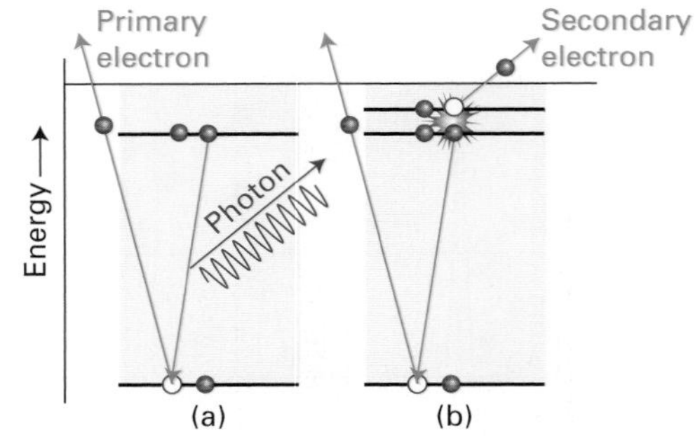

Fig. 18.5 When an electron is expelled from a solid (a) an electron of higher energy may fall into the vacated orbital and emit an X-ray photon to produce X-ray fluorescence. Alternatively (b) the electron falling into the orbital may give up its energy to another electron, which is ejected in the Auger effect.

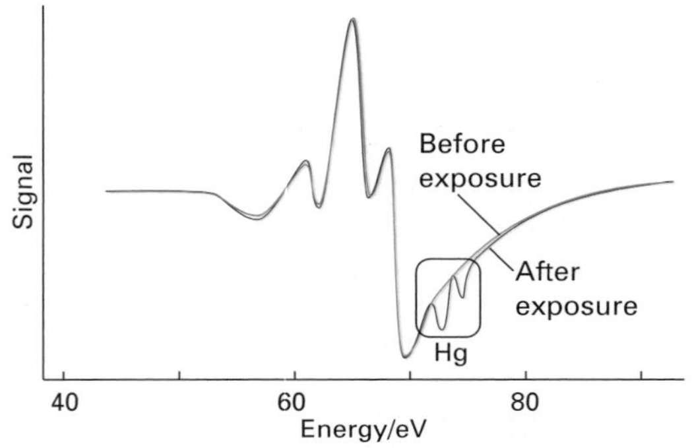

Fig. 18.6 An Auger spectrum of the same sample used for Fig. 18.4 taken before and after deposition of mercury. (M.W. Roberts and C.S. McKee, *Chemistry of the metal–gas interface*, Oxford (1978).)

released in this transition may result either in the generation of radiation, which is called **X-ray fluorescence** (Fig. 18.5a), or in the ejection of another electron (Fig. 18.5b). The latter is the secondary electron of the Auger effect. The energies of the secondary electrons are characteristic of the material present, so the Auger effect takes a 'fingerprint' of the sample (Fig. 18.6). In practice, the Auger spectrum is normally obtained by irradiating the sample with an electron beam rather than electromagnetic radiation. In **scanning Auger electron microscopy** (SAM), the finely focussed electron beam is scanned over the surface and a map of composition is compiled; the resolution can reach to below about 50 nm.

One of the most informative techniques for determining the arrangement of the atoms close to and adsorbed on the surface is **low-energy electron diffraction** (LEED). This technique is like X-ray diffraction but uses the wave character of electrons. The use of low-energy electrons (with energies in the range 10–200 eV, corresponding to wavelengths in the range 100–400 pm) ensures that the diffraction is caused only by atoms on and close to the surface. The experimental arrangement is shown in Fig. 18.7, and typical LEED patterns, obtained by photographing the fluorescent screen through the viewing port, are shown in Fig. 18.8.

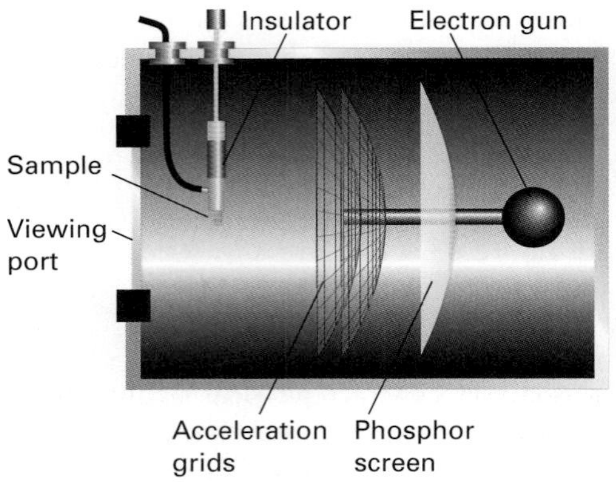

Fig. 18.7 A schematic diagram of the apparatus used for a LEED experiment. The electrons diffracted by the surface layers are detected by the fluorescence they cause on the phosphor screen.

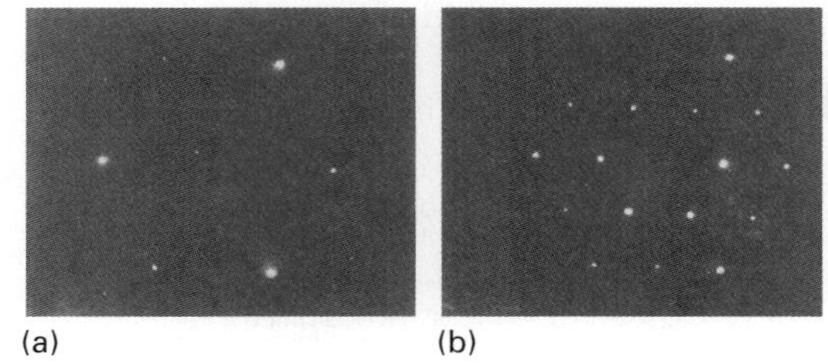

Fig. 18.8 LEED photographs of (a) a clean platinum surface and (b) after its exposure to propyne, $CH_3C{\equiv}CH$. (Photographs provided by Professor G.A. Somorjai.)

Example 18.1

Interpreting a LEED pattern

The LEED pattern from a clean unreconstructed (110) face of palladium is shown in (a) below. The reconstructed surface gives a LEED pattern shown as (b). What can be inferred about the structure of the reconstructed surface?

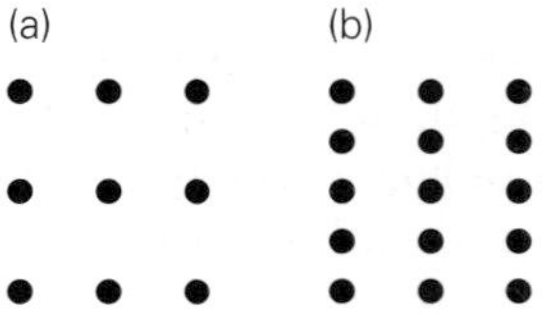

Strategy Recall from Bragg's law (Section 17.13), $\lambda = 2d\sin\theta$, that for a given wavelength, the smaller the separation d of the layers, the greater the scattering angle (so that $2d\sin\theta$

remains constant and equal to the wavelength). In terms of the LEED pattern, the farther apart the atoms responsible for the pattern, the closer the spots appear in the pattern. Twice the separation between the atoms corresponds to half the separation between the spots, and vice versa. Therefore, inspect the two patterns and identify how the new pattern relates to the old.

Answer The horizontal separation of the spots is unchanged, which indicates that the atoms remain in the same position in that dimension when reconstruction occurs. However, the vertical spacing is halved, which suggests that the atoms are twice as far apart in that direction as they are in the unreconstructed surface.

Self-test 18.2

Sketch the LEED pattern for a surface that was reconstructed from that shown in (a) above by tripling the vertical separation of the atoms.

Answer: 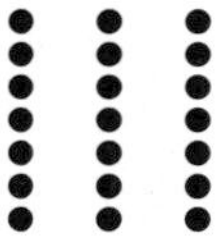

LEED experiments show that the surface of a crystal rarely has exactly the same form as a hypothetical slice through the bulk. As a general rule, it is found that metal surfaces are often simply truncations of the bulk lattice but the distance between the top layer of atoms and the one below is contracted by around 5 per cent. Semiconductors generally have surfaces reconstructed to a depth of several layers. Reconstruction occurs in ionic solids. For example, in lithium fluoride the Li^+ and F^- ions close to the surface are found to lie on slightly different planes. An actual example of the detail that can now be obtained from refined LEED techniques is shown in Fig. 18.9 for $CH_3C—$ adsorbed on a (111) plane of rhodium.

The presence of terraces, steps, and kinks in a surface shows up in LEED patterns, and their surface density (the number of defects in a region divided by the area of the region) can be estimated. Three examples of how steps and kinks affect the pattern are shown in Fig. 18.10. The samples used were obtained by cleaving a crystal at different angles to a plane of atoms. Only terraces are produced when the cut is parallel to the plane and the density of steps increases as the angle of the cut increases. The observation of additional structure in the LEED patterns, rather than blurring, shows that the steps are arrayed regularly.

Terraces, steps, kinks, and dislocations on a surface may be observed by **scanning tunnelling microscopy** (STM) and **atomic force microscopy** (AFM), two

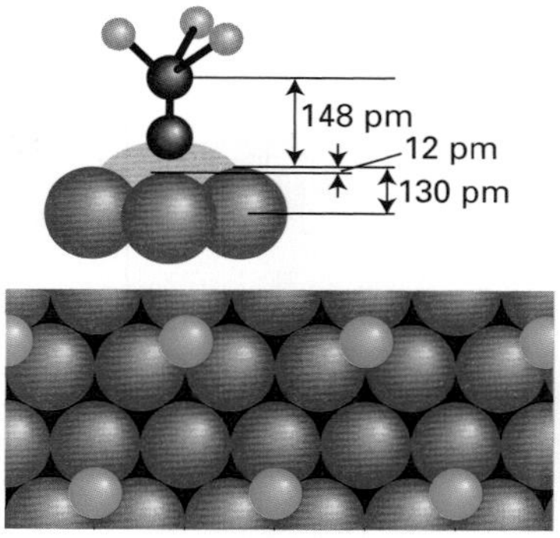

Fig. 18.9 The structure of a surface close to the point of attachment of $CH_3C—$ to the (111) surface of rhodium at 300 K and the changes in positions of the metal atoms that accompany chemisorption.

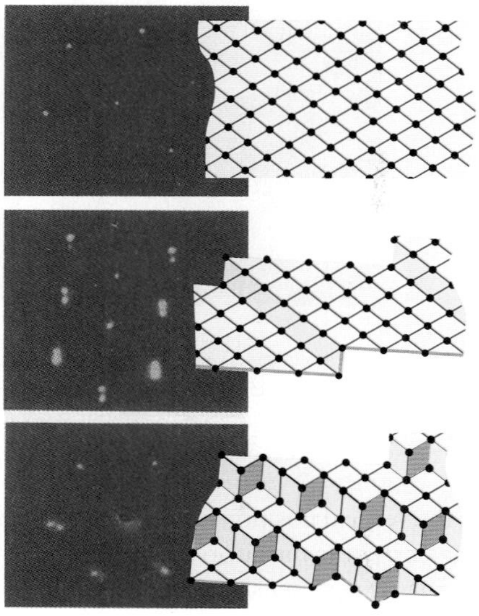

(a)

(b)

(c)

Fig. 18.10 LEED patterns may be used to assess the defect density of a surface. The photographs correspond to a platinum surface with (a) low defect density, (b) regular steps separated by about four atoms, and (c) regular steps with kinks. (Photographs provided by Professor G.A. Somorjai.)

techniques that have revolutionized the study of surfaces. In STM, a platinum–rhodium or tungsten needle is scanned across the surface of a conducting solid. When the tip of the needle is brought very close to the surface, electrons tunnel across the intervening space (Fig. 18.11). In the *constant-current mode* of operation, the stylus moves up and down corresponding to the form of the surface, and the topography of the surface, including any adsorbates, can be mapped on an atomic scale. The vertical motion of the stylus is achieved by fixing it to a piezoelectric cylinder, which contracts or expands according to the potential difference it experiences. In the *constant-z*

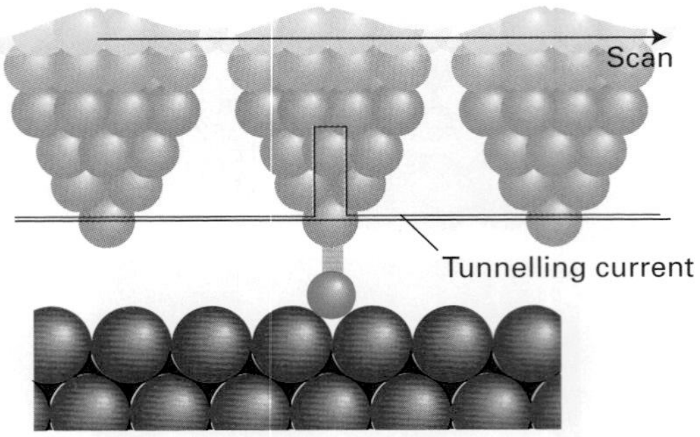

Fig. 18.11 A scanning tunnelling microscope makes use of the current of electrons that tunnel between the surface and the tip. That current is very sensitive to the distance of the tip above the surface.

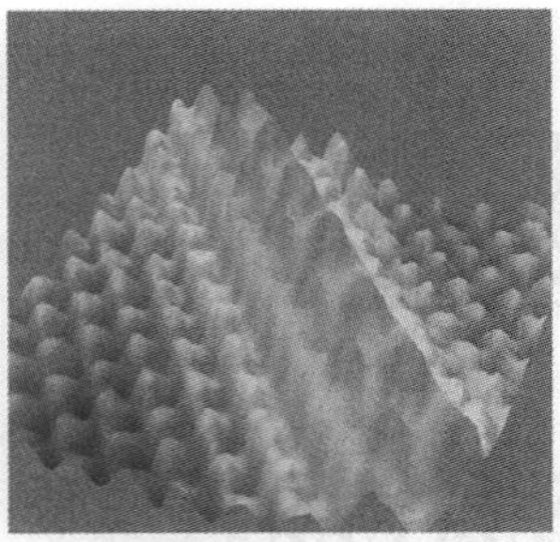

Fig. 18.12 An STM image of caesium atoms on a gallium arsenide surface.

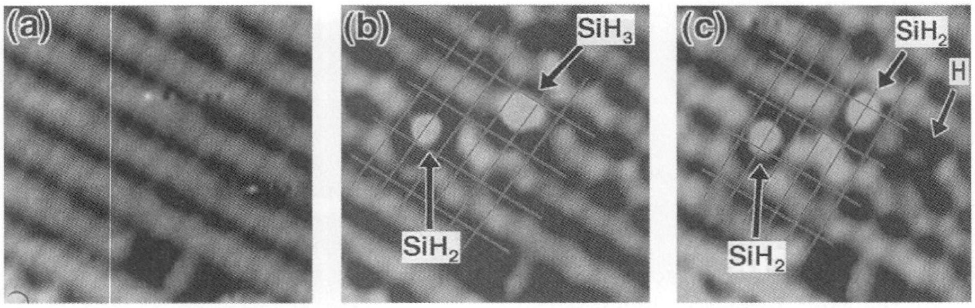

Fig. 18.13 Visualization by STM of the reaction $SiH_3 \rightarrow SiH_2 + H$ on a 4.7 nm × 4.7 nm area of a Si(001) surface. (a) The Si(001) surface before exposure to Si_2H_6(g). (b) Adsorbed Si_2H_6 dissociates into SiH_2(surface), on the left of the image, and SiH_3(surface), on the right. (c) After 8 min, SiH_3(surface) dissociates to SiH_2(surface) and H(surface). Reproduced with permission from Y. Wang, M.J. Bronikowski, and R.J. Hamers, *Surface Science* **64**, 311 (1994).

mode, the vertical position of the stylus is held constant and the current is monitored. Because the tunnelling probability is very sensitive to the size of the gap, the microscope can detect tiny, atom-scale variations in the height of the surface. An example of the kind of image obtained with a clean surface is shown in Fig. 18.12, where the cliff is only one atom high. Figure 18.13 shows the dissociation of SiH_3 adsorbed on to a Si(001) surface into adsorbed SiH_2 units and H atoms. The tip of the scanning tunnelling microscope can also be used to manipulate adsorbed atoms on a surface, making possible the fabrication of complex and yet very tiny structures, such as nanometre-sized electronic devices.

In AFM, a sharpened stylus attached to a beam is scanned across the surface. The force exerted by the surface and any adsorbate pushes or pulls on the stylus and deflects the beam (Fig. 18.14). The deflection is monitored by using a laser beam. Because no current is needed between the sample and the probe, the technique can be applied to non-conducting surfaces too. A spectacular demonstration of the power of AFM is given in Fig. 18.15, which shows individual DNA molecules on a solid surface.

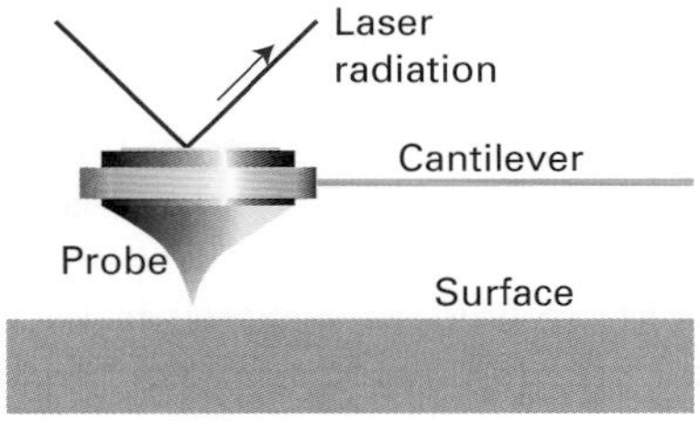

Fig. 18.14 In atomic force microscopy, a laser beam is used to monitor the tiny changes in position of a probe as it is attracted to or repelled from atoms on a surface.

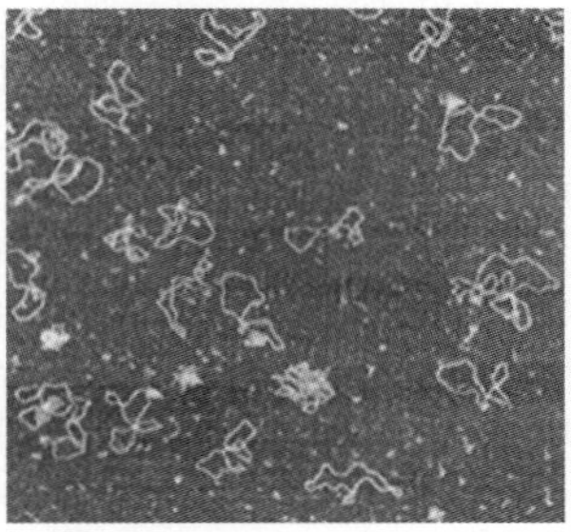

Fig. 18.15 An AFM image of bacterial DNA plasmids on a mica surface. (Courtesy of Veeco Instruments.)

The extent of adsorption

The extent of surface coverage is normally expressed as the **fractional coverage**, θ (theta):

$$\theta = \frac{\text{number of adsorption sites occupied}}{\text{number of adsorption sites available}}$$

Definition Fractional coverage (18.2)

The fractional coverage can be inferred from the volume of adsorbate adsorbed by $\theta = V/V_\infty$, where V_∞ is the volume of adsorbate corresponding to complete monolayer coverage (a single layer of adsorbed molecules). In each case, the volume in the definition of θ is that of the free gas measured under standard conditions of temperature and pressure, not the volume the adsorbed gas occupies when attached to the surface. The **rate of adsorption** is the rate of change of surface coverage and is measured by observing the change of fractional coverage with time.

Brief illustration 18.2 The fractional coverage

For the adsorption of CO on charcoal at 273 K, V_∞ = 111 cm^3, a value corrected to 1 atm. When the partial pressure of CO is 80.0 kPa, the value of V (also corrected to 1 atm) is 41.6 cm^3, so it follows that

$$\theta = \frac{V}{V_\infty} = \frac{41.6\ \text{cm}^3}{111\ \text{cm}^3} = 0.375$$

corresponding to 37.5 per cent of the surface being covered.

Among the principal techniques for measuring the rate of desorption are flow methods, in which the sample itself acts as a pump because adsorption removes molecules from the gas. One commonly used technique is therefore to monitor the rates of flow of gas into and out of the system: the difference is the rate of gas uptake by the sample. In **flash desorption** the sample is suddenly heated (electrically) and the resulting rise of pressure is interpreted in terms of the amount of adsorbate originally on the sample. The interpretation may be confused by the desorption of a compound (for example, WO_3 from oxygen on tungsten). **Surface plasmon resonance** (SPR) is a technique in which the kinetics and thermodynamics of surface processes, particularly of biological systems, are monitored by detecting the effect of adsorption and desorption on optical properties of a gold substrate. **Gravimetry**, in which the sample is weighed on a microbalance during the experiment, can also be used. A common instrument for gravimetric measurements is the **quartz crystal microbalance** (QCM), in which the mass of a sample adsorbed on the surface of a quartz crystal is related to changes in the latter's mechanical properties. The key principle behind the operation of a QCM is the ability of a quartz crystal to vibrate at a characteristic frequency when an oscillating electric field is applied. The vibrational frequency decreases when material is spread over the surface of the crystal and the change in frequency is proportional to the mass of material. Masses as small as a few nanograms (1 ng = 10^{-9} g) can be measured reliably in this way.

18.3 Physisorption and chemisorption

Molecules and atoms can attach to surfaces in two ways, although there is no clear frontier between the two types of adsorption. In **physisorption** (an abbreviation of 'physical adsorption'), there is a van der Waals interaction between the adsorbate and the substrate (for example, a dispersion or a dipolar interaction, Section 15.3, of the kind responsible for the condensation of vapours to liquids). The energy released when a molecule is physisorbed is of the same order of magnitude as the enthalpy of condensation. Such small energies can be absorbed as vibrations of the lattice and dissipated as thermal motion, and a molecule bouncing across the surface will gradually lose its energy and finally adsorb to it in the process called **accommodation**. The enthalpy of physisorption can be measured by monitoring the rise in temperature of a sample of known heat capacity, and typical values are in the region of −20 kJ mol^{-1} (Table 18.1). This small enthalpy change is insufficient to lead to bond breaking, so a physisorbed molecule retains its identity but might be distorted. Enthalpies of physisorption may also be measured by observing the temperature dependence of the parameters that occur in the adsorption isotherm (see Section 18.4).

In **chemisorption** (an abbreviation of 'chemical adsorption'), the molecules (or atoms) adsorb to the surface by forming a chemical (usually covalent) bond and tend to find sites that maximize their coordination number with the substrate. The enthalpy of chemisorption is much more negative than that for

Table 18.1
Maximum observed enthalpies of physisorption

Adsorbate	$\Delta_{ad}H^\ominus/(\text{kJ mol}^{-1})$
CH_4	−21
H_2	−84
H_2O	−59
N_2	−21

Table 18.2

Enthalpies of chemisorption, $\Delta_{ad}H^{\ominus}/(kJ\ mol^{-1})$

Adsorbate	Adsorbent (substrate)		
	Cr	Fe	Ni
C_2H_4	−427	−285	−243
CO		−192	
H_2	−188	−134	
NH_3		−188	−155

physisorption, and typical values are in the region of −200 kJ mol^{-1} (Table 18.2). The distance between the surface and the closest adsorbate atom is also typically shorter for chemisorption than for physisorption. A chemisorbed molecule may be torn apart at the demand of the unsatisfied valencies of the surface atoms and the existence of molecular fragments on the surface as a result of chemisorption is one reason why solid surfaces catalyse many reactions.

A type of chemisorption that has received much attention recently is the formation of a **self-assembled monolayer** (SAM), which is an ordered molecular aggregate that forms a single layer of organic material on a surface. To understand the formation of a SAM, consider the result of exposing molecules such as alkyl thiols, RSH, where R represents an alkyl chain, to a gold surface. The thiols chemisorb to the surface, forming RS—Au(I) adducts. If we represent the atoms close to the adsorption site as Au_n, then the attachment can be written as

$$RSH + Au_n \rightarrow RS\text{—}Au(I)\cdot Au_{n-1} + \tfrac{1}{2} H_2(g)$$

If R has a sufficiently long chain, van der Waals interactions between the adsorbed RS— units lead to the formation of a highly ordered monolayer on the surface (Fig. 18.16).

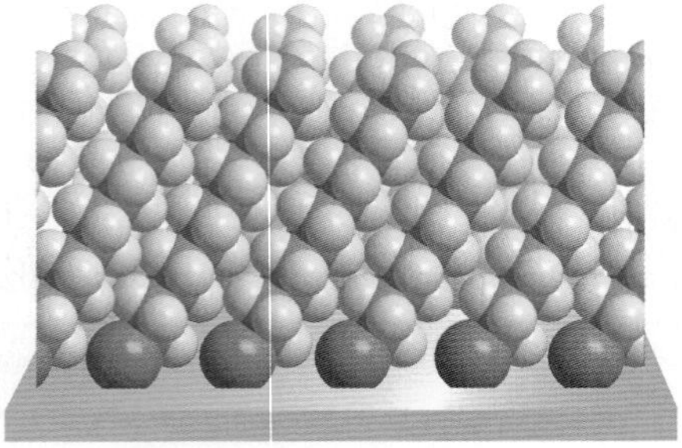

Fig. 18.16 Self-assembled monolayers of alkylthiols formed on to a gold surface by chemisorption of the thiol groups and aggregation of the alkyl chains.

18.4 Adsorption isotherms

The free gas A and the adsorbed gas are in a dynamic equilibrium of the form

$$A(g) + M(surface) \rightleftharpoons AM(surface)$$

and the fractional coverage of the surface depends on the pressure of the overlying gas. The enthalpy change associated with the forward reaction (per mole of adsorbed species) is the **enthalpy of adsorption, $\Delta_{ads}H$.** The variation of θ with pressure at a chosen temperature is called the **adsorption isotherm.**

The simplest physically plausible adsorption isotherm is based on three assumptions:

1. Adsorption cannot proceed beyond monolayer coverage.
2. All sites are equivalent and the surface is uniform (that is, the surface is perfectly flat on a microscopic scale).
3. There are no interactions between adsorbed molecules, so the ability of a molecule to adsorb at a given site is independent of the occupation of neighbouring sites.

(a) The Langmuir isotherm

Assumptions 2 and 3 imply, respectively, that the enthalpy of adsorption is the same for all sites and is independent of the extent of surface coverage. We show in the following Derivation that the relation between the fractional coverage θ and the partial pressure of A, p, that results from these three assumptions is the **Langmuir isotherm**:

$$\theta = \frac{\alpha p}{1+\alpha p} \qquad \alpha = \frac{k_a}{k_d} \qquad \text{Langmuir isotherm} \qquad (18.3)$$

where k_a and k_d are, respectively, the rate constants for adsorption and desorption. Note that α, being the ratio of forward and reverse rate constants, is a type of equilibrium constant (Chapter 7), but is not a true equilibrium constant because it has units. This expression is plotted for various values of α (which has the dimensions of 1/pressure) in Fig. 18.17. We see that:

- As the partial pressure of A increases, the fractional coverage increases towards 1. Half the surface is covered when $p = 1/\alpha$.
- At low pressures (in the sense that $\alpha p \ll 1$), the denominator can be replaced by 1, and $\theta = \alpha p$. Under these conditions, the surface coverage increases linearly with pressure.
- At high pressure (in the sense that $\alpha p \gg 1$), the 1 in the denominator can be neglected, the αp cancel, and $\theta = 1$. Now the surface is saturated.

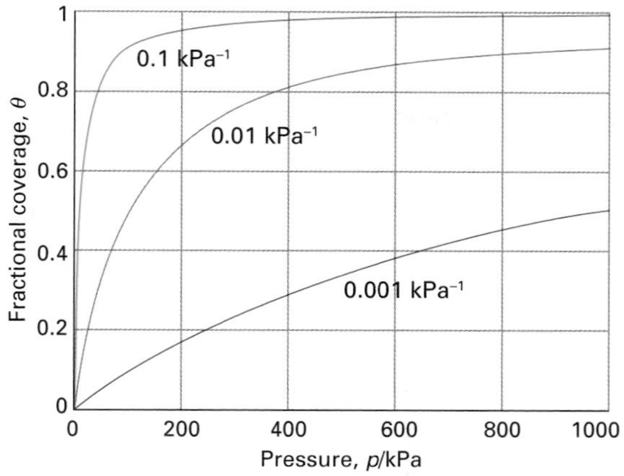

Fig. 18.17 The Langmuir isotherm for non-dissociative adsorption for different values of α.

Derivation 18.1

The Langmuir isotherm

To obtain the Langmuir isotherm, we suppose that the rate at which A adsorbs to the surface is proportional to the partial pressure (because the rate at which molecules strike the surface is proportional to the pressure), and to the number of sites that are not occupied at the time, which is $(1-\theta)N$:

$$\text{Rate of adsorption} = k_a N(1-\theta)p$$

where k_a is the adsorption rate constant. The rate at which the adsorbed molecules leave the surface is proportional to the number currently on the surface ($N\theta$):

$$\text{Rate of desorption} = k_d N\theta$$

where k_d is the desorption rate constant. At equilibrium, the two rates are equal, so we can write

$$k_a N(1-\theta)p = k_d N\theta$$

The Ns cancel and, by using $\alpha = k_a/k_d$, we obtain

$$\alpha p(1-\theta) = \theta$$

which rearranges into eqn 18.3.

Example 18.2

Using the Langmuir isotherm

The data given below are for the adsorption of CO on charcoal at 273 K. Confirm that they fit the Langmuir isotherm, and find the constant K and the volume corresponding to complete coverage. In each case V has been corrected to 1 (more precisely, to 101.325 kPa).

p/kPa	13.3	26.7	40.0	53.3	66.7	80.0	93.3
V/cm^3	10.2	18.6	25.5	31.5	36.9	41.6	46.1

Strategy From eqn 18.3,

$$\frac{1}{\theta} = \frac{1+\alpha p}{\alpha p} = \frac{1}{\alpha p} + 1$$

Then, by substituting $\theta = V/V_\infty$, where V_∞ is the volume corresponding to complete coverage (as measured at 273 K and 1 atm)

$$\frac{V_\infty}{V} = \frac{1}{\alpha p} + 1$$

Division of both sides by V_∞ and multiplication by p then gives

$$\underbrace{\frac{p}{V}}_{y} = \underbrace{\frac{1}{\alpha V_\infty}}_{\text{intercept}} + \underbrace{\frac{1}{V_\infty} \times p}_{\text{slope} \times x}$$

Hence, a plot of p/V against p should give a straight line of slope $1/V_\infty$ and intercept $1/\alpha V_\infty$, and the ratio slope:intercept gives α:

$$\frac{\overbrace{\text{slope}}^{1/V_\infty}}{\underbrace{\text{intercept}}_{1/\alpha V_\infty}} = \frac{1}{V_\infty} \times \alpha V_\infty = \alpha$$

Answer The data for the plot are as follows:

p/kPa	13.3	26.7	40.0	53.3	66.7	80.0	93.3
(p/kPa)/(V/cm^3)	1.30	1.44	1.57	1.69	1.81	1.92	2.02

The points are plotted in Fig. 18.18. The (least-squares) slope is 0.00900, so $V_\infty = 111$ cm^3. The intercept at $p = 0$ is 1.20, so

$$\alpha = \frac{\overbrace{0.00900}^{\text{Slope}} \quad \overbrace{\text{cm}^{-3}}^{\text{Units of }(p/V)/p}}{\underbrace{1.20}_{\text{Intercept}} \quad \underbrace{\text{kPa cm}^{-3}}_{\text{Units of } p/V}} = 7.51 \times 10^{-3}\ \text{kPa}^{-1}$$

To attach units to the dimensionless intercept and slope, we have used the procedure outlined in The chemist's toolkit

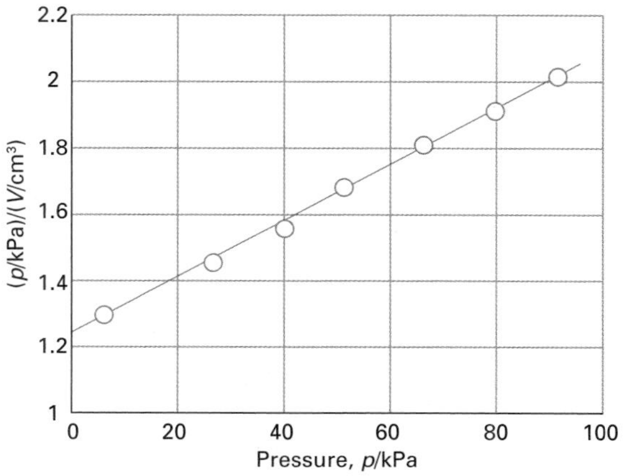

Fig. 18.18 The plot of the data in Example 18.2. As illustrated here, the Langmuir isotherm predicts that a straight line should be obtained when p/V is plotted against p.

1.1: the slope is that of p/V plotted against p, so it is given the units of $(p/V)/p$, and the intercept is on the vertical axis, the value of p/V, so it is given the units of p/V.

Self-test 18.3

Repeat the calculation for the following data:

p/kPa	13.3	26.7	40.0	53.3	66.7	80.0	93.3
V/cm^3	10.3	19.3	27.3	34.1	40.0	45.5	48.0

Answer: 128 cm^3, 6.70 × 10^{-3} kPa^{-1}

(b) The isosteric enthalpy of adsorption

A further point is that, as remarked in connection with eqn 18.3, because α is essentially an equilibrium constant, then its temperature dependence is given by the van 't Hoff equation (Section 7.8):

$$\ln\alpha = \ln\alpha' - \frac{\Delta_{ads}H^\ominus}{R}\left(\frac{1}{T}-\frac{1}{T'}\right) \quad \text{Isosteric enthalpy of adsorption} \quad (18.4)$$

It follows that if we plot ln α against $1/T$, then the slope of the graph is equal to $-\Delta_{ads}H^\ominus/R$, where $\Delta_{ads}H^\ominus$ is the standard enthalpy of adsorption. However, because this quantity might vary with the extent of surface coverage either because the adsorbate molecules interact with each other or because adsorption occurs at a sequence of different sites, care must be taken to measure α at the same value of the fractional coverage. The resulting value of $-\Delta_{ads}H^\ominus$ is called the **isosteric enthalpy of adsorption.** The variation of $-\Delta_{ads}H^\ominus$ with θ allows us to explore the validity of the assumptions on which the Langmuir isotherm is based.

Example 18.3

The isosteric enthalpy of adsorption

The pressure of nitrogen gas in equilibrium with a layer of nitrogen adsorbed on rutile (TiO_2) with a fractional coverage of $\theta = 0.10$ varied with temperature as follows:

T/K	220	240	260	280	300
p/kPa	2.8	7.7	17.0	38.0	68.0

Determine the isosteric enthalpy of adsorption at $\theta = 0.10$.

Strategy First, find the relation between K in the Langmuir isotherm and p for a given fractional coverage. Then convert the van 't Hoff equation to an equation relating p and T in place of α and T, and plot the data appropriately.

Answer We rearrange eqn 18.3 into

$$\alpha = \frac{\theta}{1-\theta}\times\frac{1}{p}$$

and then, on taking logarithms,

$$\ln\alpha = \underbrace{\ln\left(\frac{\theta}{1-\theta}\right)}_{\text{constant because }\theta\text{ is fixed}} + \ln\frac{1}{p} \overbrace{=}^{\ln(1/x)=-\ln x} \text{constant} - \ln p$$

The van 't Hoff equation then becomes

$$\overbrace{\text{constant} - \ln p}^{\ln\alpha} = \overbrace{\text{constant} - \ln p'}^{\ln\alpha'} - \frac{\Delta_{ads}H^\ominus}{R}\left(\frac{1}{T}-\frac{1}{T'}\right)$$

After cancellation of the constants and changing the signs on both sides, we can rearrange this equation into

$$\overbrace{\ln p}^{y} = \overbrace{\ln p' - \frac{\Delta_{ads}H^\ominus}{RT'}}^{\text{Intercept}} + \overbrace{\frac{\Delta_{ads}H^\ominus}{R}}^{\text{Slope}} \times \overbrace{\frac{1}{T}}^{\times x}$$

Therefore, a plot of ln p against $1/T$ should be a straight line of slope $\Delta_{ads}H^\ominus/R$. We draw up the following table:

(10^3 K)/T	4.55	4.17	3.85	3.57	3.33
ln(p/kPa)	1.03	2.04	2.83	3.64	4.22

The points are plotted in Fig. 18.19. The (least-squares) slope of the straight line is −0.381, so

$$\frac{\Delta_{ads}H^\ominus}{R} = \overbrace{-0.381\times10^3}^{\text{Slope}} \overbrace{\text{K}}^{\text{Units of }\ln p/(1/T)}$$

Therefore,

$$\Delta_{ads}H^\ominus = (-0.381\times10^3\text{ K})\times(8.3145\text{ J K}^{-1}\text{ mol}^{-1}) = -3.17\times10^3\text{ J mol}^{-1}$$

or −3.17 kJ mol^{-1}.

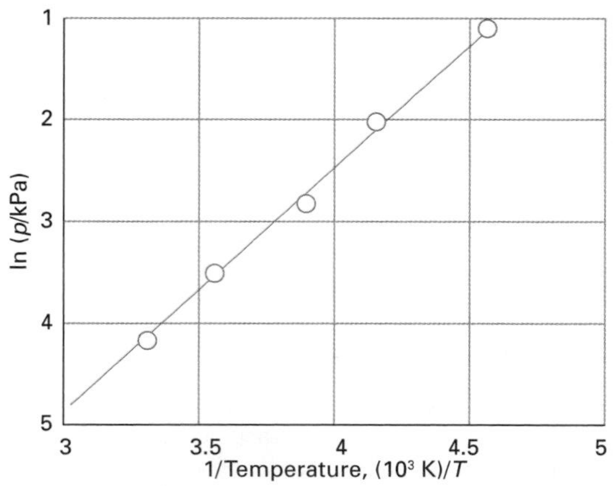

Fig. 18.19 The isosteric enthalpy of adsorption can be obtained from the slope of the plot of ln p against $1/T$, where p is the pressure needed to achieve the specified surface coverage. The data used are from Example 18.3.

Self-test 18.4

The data below show the pressures of CO needed for the volume of adsorption (corrected to 1.00 atm and 273 K) to be 10.0 cm^3. Calculate the adsorption enthalpy at this surface coverage.

T/K	200	210	220	230	240	250
p/kPa	4.00	4.95	6.03	7.20	8.47	9.85

Answer: −7.52 kJ mol^{-1}

Now suppose the substrate dissociates on adsorption, as in

$$A_2(g) + M(\text{surface}) \rightleftharpoons A\text{–}M(\text{surface}) + A\text{–}M(\text{surface})$$

We show in the following Derivation that the resulting isotherm is

$$\theta = \frac{(\alpha p)^{1/2}}{1+(\alpha p)^{1/2}} \qquad \text{Langmuir isotherm for adsorption with dissociation} \qquad (18.5)$$

The surface coverage now depends on the square root of the pressure in place of the pressure itself (Fig. 18.20).

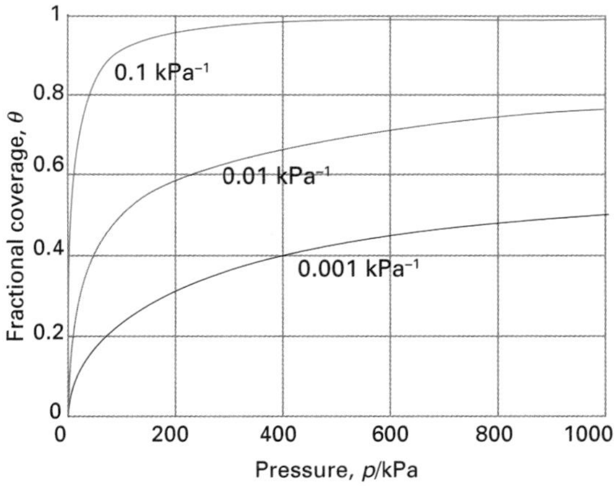

Fig. 18.20 The Langmuir isotherm for dissociative adsorption, $A_2(g) \rightarrow 2\,A(\text{surface})$, for different values of α.

Derivation 18.2

The effect of substrate dissociation on the Langmuir isotherm

When the substrate dissociates on adsorption, the rate of adsorption is proportional to the pressure and to the probability that *both* atoms will find sites, which is proportional to the square of the number of vacant sites:

$$\text{Rate of adsorption} = k_a p\{N(1-\theta)\}^2$$

The rate of desorption is proportional to the frequency of encounters of atoms on the surface, and is therefore second-order in the number of atoms present:

$$\text{Rate of desorption} = k_d(N\theta)^2$$

The condition for no net change (equal rates of adsorption and desorption) is

$$k_a p\{N(1-\theta)\}^2 = k_d(N\theta)^2$$

After using $\alpha = k_a/k_d$, cancelling the Ns, to get

$$\alpha p(1-\theta)^2 = \theta^2$$

and taking the square root of both sides of the expression, we obtain

$$(\alpha p)^{1/2}(1-\theta) = \theta, \text{ or } \quad (\alpha p)^{1/2} = \{1 + (\alpha p)^{1/2}\}\theta$$

which rearranges into eqn 18.5.

A second modification of the Langmuir isotherm that we need to consider deals with **co-adsorption**, in which a mixture of two gases A and B compete for the same sites on the surface. It is left as an exercise (Exercise 18.13) for you to show that if A and B both follow Langmuir isotherms, and adsorb without dissociation, then

$$\theta_A = \frac{\alpha_A p_A}{1+\alpha_A p_A + \alpha_B p_B} \qquad \theta_B = \frac{\alpha_B p_B}{1+\alpha_A p_A + \alpha_B p_B} \qquad \text{Langmuir isotherm with co-adsorption} \qquad (18.6)$$

where α_J (with J = A or B) is the ratio of adsorption and desorption rate constants for species J, p_J is its partial pressure in the gas phase, and θ_J is the fraction of total sites occupied by J. Co-adsorption of this kind is important in catalysis and we use these isotherms later.

(c) The BET isotherm

If the initial adsorbed layer can act as a substrate for further (for example, physical) adsorption, then instead of the isotherm levelling off to some saturated value at high pressures, it can be expected to rise indefinitely as more and more molecules condense on to the surface, just like water vapour can condense indefinitely on to the surface of liquid water. The most widely used isotherm dealing with multilayer adsorption was derived by Stephen Brunauer, Paul Emmett, and Edward Teller, and is called the **BET isotherm:**

$$\frac{V}{V_{mon}} = \frac{cz}{(1-z)\{1-(1-c)z\}} \qquad \text{with } z = \frac{p}{p^*} \qquad \text{BET isotherm} \qquad (18.7)$$

In this expression, p^* is the vapour pressure above a layer of adsorbate that is more than one molecule thick and can therefore be taken to be the vapour pressure of the bulk liquid, V_{mon} is the volume corresponding to monolayer coverage, and $c = \alpha_0/\alpha_1$, where α_0 is the equilibrium constant for adsorption on to the substrate and α_1 the equilibrium constant for physisorption on to the overlaying layers already present (Fig. 18.21). Provided the entropy of chemisorption and physisorption are the same,

$$c = e^{(\Delta_{des}H^{\ominus} - \Delta_{vap}H^{\ominus})/RT} \tag{18.8}$$

where $\Delta_{des}H^{\ominus}$ is the standard enthalpy of desorption from the substrate and $\Delta_{vap}H^{\ominus}$ is the standard enthalpy of vaporization of the liquid adsorbate.

Figure 18.22 illustrates the shapes of BET isotherms. At low pressures, the dominant effect is monolayer adsorption, so we can expect the BET isotherm to resemble the Langmuir isotherm. Indeed, when $p \ll p^*$, so $z \ll 1$, we can write

$$\frac{V}{V_{mon}} = \frac{cz}{\underbrace{(1-z)}_{\approx 1}\underbrace{(1-z+cz)}_{\approx 1}} \approx \frac{cz}{1+cz} \tag{18.9}$$

which has the form of the Langmuir isotherm. As the pressure is increased, though, multilayer coverage becomes important and the extent of coverage rises without limit. A BET isotherm is not accurate at all pressures, but it is widely used in industry to determine the surface areas of solids.

When $c \gg 1$ and $cz \gg 1$, which is the case when the enthalpy of desorption from the substrates is very high, the BET isotherm takes the form

$$\frac{V}{V_{mon}} = \frac{cz}{(1-z)\underbrace{(1-z+cz)}_{\approx cz}} \approx \frac{cz}{(1-z)cz} = \frac{1}{1-z} \tag{18.10}$$

This expression is applicable to unreactive gases on polar surfaces, for which $c \approx 10^2$.

The BET isotherm is reasonably reliable in the range $0.8 < \theta < 2$, and so provides a reasonably reliable technique for measuring the surface area of solids (corresponding to $\theta = 1$). The adsorbate is then typically nitrogen gas.

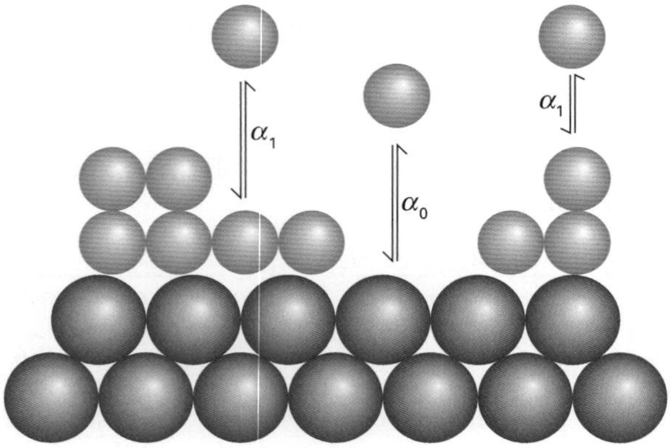

Fig. 18.21 Adsorption on to the substrate occurs with an equilibrium constant α_0, and physisorption on to the overlaying layers already present occurs with an equilibrium constant α_1.

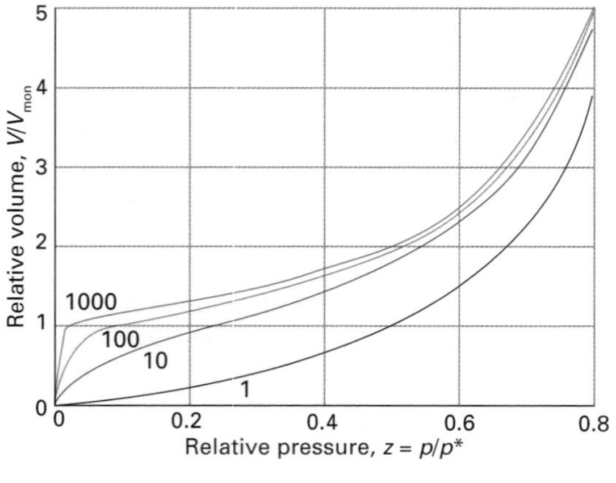

Fig. 18.22 Plots of the BET isotherm for different values of c. The value of V/V_{mon} rises indefinitely because the adsorbate may condense on the covered substrate surface.

Example 18.4

Using the BET isotherm to determine the area of a surface

The amount of N_2 adsorbed on 0.30 g of silica at 77 K (the normal boiling point of nitrogen) was determined by measuring the volume adsorbed and then using the perfect gas law to calculate the amount. The following values were obtained:

p/Torr	100	200	300	400
n/mmol	0.90	1.10	1.40	1.90

Determine the value of c and the number of adsorption sites on the sample.

Strategy To use the BET isotherm, we first take the reciprocal of both sides of eqn 18.7:

$$\frac{V_{mon}}{V} = \frac{(1-z)\{1-(1-c)z\}}{cz} = \frac{(1-z)-(1-z)(1-c)z}{cz} \overset{(A-B)/C=A/C-B/C}{=} \frac{1-z}{cz} - \frac{(1-c)(1-z)}{c}$$

The ratio V_{mon}/V can be set equal to n_{mon}/n, where n is the amount of adsorbate molecules adsorbed and n_{mon} is the amount of adsorbate molecules in a monolayer, to give

$$\frac{n_{mon}}{n} = \frac{1-z}{cz} - \frac{(1-c)(1-z)}{c}$$

To get an expression of the form y = intercept + slope × x we multiply both sides by $z/n_{mon}(1-z)$, to obtain

$$\frac{zn_{mon}}{n_{mon}(1-z)n} = \frac{z}{n_{mon}(1-z)}\left(\frac{1-z}{cz}\right) - \frac{z(1-c)(1-z)}{n_{mon}(1-z)c}$$

which, after cancelling the various terms in blue, becomes

$$\overbrace{\frac{z}{(1-z)n}}^{y} = \overbrace{\frac{1}{cn_{mon}}}^{\text{Intercept}} - \overbrace{\frac{(1-c)}{cn_{mon}}}^{\text{Slope}} \times \overbrace{z}^{x} \qquad (18.11)$$

Use $z = p/p^*$, with p^* = 760 Torr (because the vapour pressure of a substance at its normal boiling point is 1 atm). This expression is the equation of a straight line when the left-hand side is plotted against z, with an intercept at $1/cn_{mon}$ and a slope $(c-1)/cn_{mon}$. From the intercept and slope the values of c and n_{mon} can be determined.

Answer Draw up the following table:

p/Torr	100	200	300	400
$z = p/p^*$	0.132	0.263	0.395	0.526
$z/\{(1-z)(n/\text{mmol})\}$	0.17	0.32	0.47	0.58

Figure 18.23 shows a plot of the data. The intercept is at 0.039, so $1/\{c(n_{mon}/\text{mmol})\} = 0.039$ and therefore cn_{mon} = 15 mmol, The slope is 1.1, so $(c-1)/\{c(n_{mon}/\text{mmol})\} = 1.1$, and therefore

$$c - 1 = 1.1 \times 15 = 16$$

and hence c = 17 and n_{mon} = (15 mmol)/16 = 0.94 mmol. This amount corresponds to

$$N_{mon} = n_{mon}N_A = (9.4 \times 10^{-4}\ \text{mol}) \times (6.022 \times 10^{23}\ \text{mol}^{-1}) = 5.7 \times 10^{20}$$

adsorption sites.

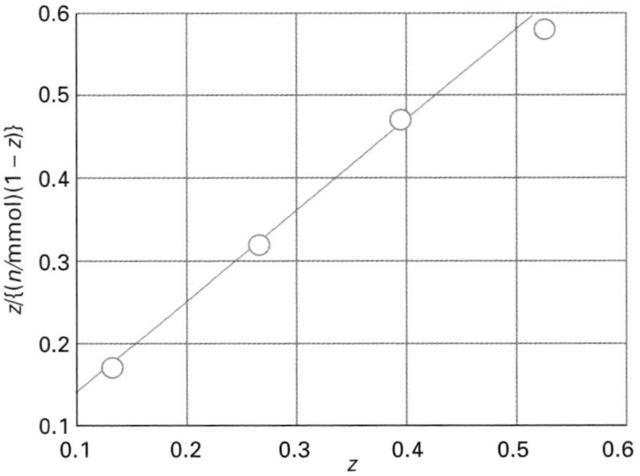

Fig. 18.23 The plot of the data in Example 18.4.

Self-test 18.5

Repeat the analysis using the following data for a different sample of silica:

p/Torr	100	150	200	250	300	350
n/mmol	1.28	1.55	1.79	2.05	2.33	2.67

Answer: n_{mon} = 1.54 mmol, $N_{mon} = 9.27 \times 10^{20}$

18.5 The rates of surface processes

Figure 18.24 shows how the potential energy of a molecule varies with its distance above the adsorption site. As the molecule approaches the surface its potential energy decreases as it becomes physisorbed into the **precursor state** for chemisorption. Dissociation into fragments often takes place as a molecule moves into its chemisorbed state, and, after an initial increase of energy as the bonds stretch, there is a sharp decrease as the adsorbate–substrate bonds reach their full strength. Even if the molecule does not fragment, there is likely to be an initial increase of potential energy as the bonds adjust when the molecule approaches the surface.

In most cases, therefore, we can expect there to be a potential-energy barrier separating the precursor and chemisorbed states. This barrier, though, might be low and might not rise above the energy of a distant, stationary molecule (as in Fig. 18.24a). In this case, chemisorption is not an activated process because every molecule that approaches the surface can reach it regardless of its kinetic energy, and chemisorption can be expected to be rapid. Many gas adsorptions

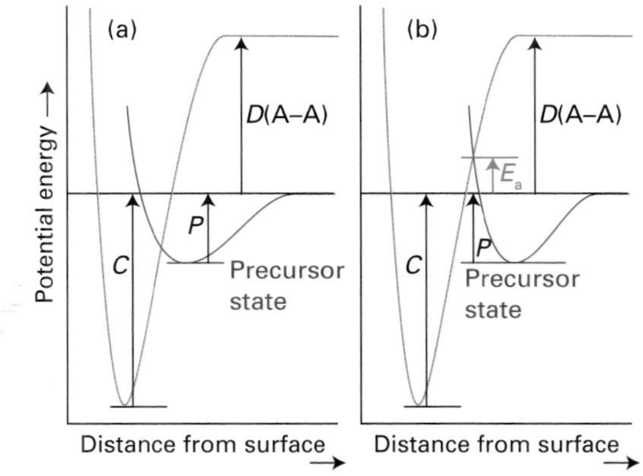

Fig. 18.24 The potential energy profiles for the dissociative chemisorption of an A_2 molecule. In each case, P is the enthalpy of (non-dissociative) physisorption and C that for chemisorption (at T = 0). The relative locations of the curves determine whether the chemisorption is (a) not activated or (b) activated.

on clean metals appear to be non-activated. In some cases the barrier rises above the zero axis (as in Fig. 18.24b); such chemisorption is activated and slower than the non-activated kind because only a fraction of the molecules approaching the surface can reach it and become chemisorbed. An example is the adsorption of H_2 on copper, which has an activation energy in the region of 20–40 kJ mol^{-1}.

One point that emerges from this discussion is that rates are not good criteria for distinguishing between physisorption and chemisorption. Chemisorption can be fast if the activation energy is small or zero; but it may be slow if the activation energy is large. Physisorption is usually fast, but it can appear to be slow if adsorption is taking place on a porous medium.

(a) The sticking probability

The rate at which a surface is covered by adsorbate depends on the ability of the substrate to dissipate the energy of the incoming molecule as thermal motion as it crashes on to the surface. If the energy is not dissipated quickly, the molecule migrates over the surface until a vibration expels it into the overlying gas or it reaches an edge. The proportion of collisions with the surface that successfully lead to adsorption is called the **sticking probability**, s:

$$s = \frac{\text{rate of adsorption of particles by the surface}}{\text{rate of collision of particles with the surface}}$$

Definition Sticking probability (18.12)

The denominator can be calculated from kinetic theory (by using eqn 18.1), and the numerator can be measured by observing the rate of change of pressure. Values of s vary widely. For example, at room temperature CO has s in the range 0.1–1.0 for several d-metal surfaces, suggesting that almost every collision sticks, but for N_2 on rhenium $s < 10^{-2}$, indicating that more than a hundred collisions are needed before one molecule sticks successfully.

Desorption is always an activated process because the molecules have to be lifted from the foot of a potential well. A physisorbed molecule vibrates in its shallow potential well, and might shake itself off the surface after a short time. The temperature dependence of the first-order rate of departure can be expected to be Arrhenius-like,

$$k_d = Ae^{-E_d/RT}$$

Activated desorption (18.13)

where A is a pre-exponential factor (obtained from the intercept of an Arrhenius plot, Section 10.9, at $1/T = 0$) and the activation energy for desorption, E_d, is likely to be comparable to the enthalpy of physisorption. In the discussion of half-lives of first-order reactions (Section 10.8), which desorption is, we saw that $t_{1/2} = (\ln 2)/k$; so the half-life for remaining on the surface has a temperature dependence given by

$$t_{1/2} = \frac{\ln 2}{k_d} = \tau_0 e^{E_d/RT} \qquad \tau_0 = \frac{\ln 2}{A}$$

Residence half-life (18.14)

Note the positive sign in the exponent: the half-life *decreases* as the temperature is raised.

Brief illustration 18.3 Residence half-lives

If we suppose that $1/\tau_0$ is approximately the same as the vibrational frequency of the weak particle–surface bond (about 10^{12} Hz) and $E_d \approx 25$ kJ mol^{-1}, then residence half-lives of around 10 ns are predicted at room temperature. Lifetimes close to 1 s are obtained only by lowering the temperature to about 100 K. For chemisorption, with E_d = 100 kJ mol^{-1} and guessing that $\tau_0 = 10^{-14}$ s (because the adsorbate–substrate bond is quite stiff), we expect a residence half-life of about 3×10^3 s (about an hour) at room temperature, decreasing to 1 s at about 350 K.

Self-test 18.6

What is the half-life for an atom on a surface at 800 K if its desorption activation energy is 200 kJ mol^{-1}? Take $\tau_0 = 0.10$ ps.

Answer: $t_{1/2} = 1.3$ s

(b) Experimental techniques

One way to measure the desorption activation energy is to monitor the rate of increase in pressure when the sample is maintained at a series of temperatures and then to attempt to make an Arrhenius plot. A more sophisticated technique is **temperature programmed desorption** (TPD) or **thermal desorption spectroscopy** (TDS). The basic observation is a surge in desorption rate (as monitored by a mass spectrometer) when the temperature is raised linearly to the temperature at which desorption occurs rapidly; but once the desorption has occurred there is no more adsorbate to escape from the surface, so the desorption flux falls again as the temperature continues to rise. The TPD spectrum, the plot of desorption flux against temperature, therefore shows a peak, the location of which depends on the desorption activation energy. There are three maxima in the example shown in

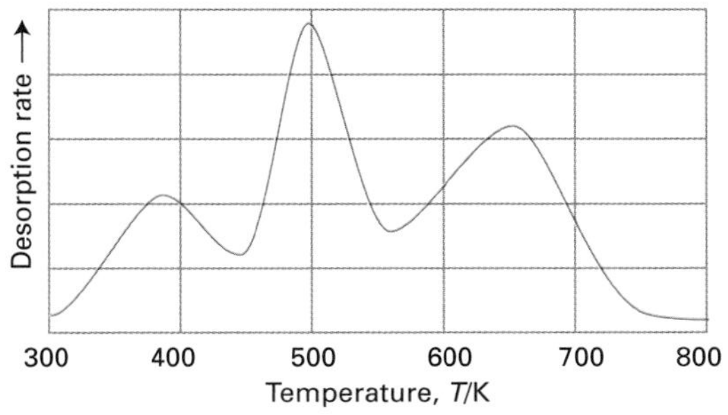

Fig. 18.25 The thermal desorption spectrum of H_2 on the (100) face of tungsten. The three peaks indicate the presence of three sites with different adsorption enthalpies and therefore different desorption activation energies. (P.W. Tamm and L.D. Schmidt, *J. Chem. Phys.*, 51, 5352 (1969).)

Fig. 18.25, indicating the presence of three adsorption sites with different activation energies.

In many cases only a single desorption activation energy (and a single peak in the TPD spectrum) is observed. When several peaks are observed they might correspond to adsorption on different crystal planes or to multilayer adsorption.

Brief illustration 18.4 Temperature programmed desorption

Cadmium atoms on tungsten show two desorption activation energies, one of 18 kJ mol^{-1} and the other of 90 kJ mol^{-1}. The explanation is that the more tightly bound Cd atoms are attached directly to the substrate, and the less strongly bound are in a layer (or layers) above the primary layer. Another example of a system showing two desorption activation energies is CO on tungsten, the values being 120 kJ mol^{-1} and 300 kJ mol^{-1}. The explanation is believed to be the existence of two types of metal–adsorbate binding site, one involving a simple M—CO bond, the other adsorption with dissociation into individually adsorbed C and O atoms.

Catalytic activity at surfaces

We saw in Chapter 11 that a catalyst acts by providing an alternative reaction path with a lower activation energy. A catalyst does not disturb the final equilibrium composition of the system, only the rate at which that equilibrium is approached. In this section we shall consider **heterogeneous catalysis**, in which the catalyst and the reagents are in different phases. A common example is a solid introduced as a heterogeneous catalyst into a gas-phase reaction. Many industrial processes make use of heterogeneous catalysts, which include platinum, rhodium, zeolites (Section 18.7), and various metal oxides, but increasingly attention is turning to homogeneous catalysts, partly because they are easier to cool. However, their use typically requires additional separation steps, and such catalysts are generally immobilized on a support, in which case they become heterogeneous. In general, heterogeneous catalysts are highly selective and to find an appropriate catalyst each reaction must be investigated individually. Computational procedures are beginning to be a fruitful source of prediction of catalytic activity.

A metal acts as a heterogeneous catalyst for certain gas-phase reactions by providing a surface to which a reactant can attach by chemisorption. For example, hydrogen molecules may attach as atoms to a nickel surface and these atoms react much more readily with another species (such as an alkene) than the original molecules. The chemisorption step therefore results in a new reaction pathway with a lower activation energy than in the absence of the catalyst. Note that chemisorption is normally required for catalytic activity: physisorption might precede chemisorption but is not itself sufficient.

Heterogeneous catalysis normally depends on at least one reactant being adsorbed (usually chemisorbed) and modified to a form in which it readily undergoes reaction. Often this modification takes the form of a fragmentation of the reactant molecules. The **catalyst ensemble** is the minimum arrangement of atoms at the surface active site that can be used to model the action of the catalyst. It may be determined, for instance, by diluting the active metal with a chemically inert metal and observing the catalytic activity of the resulting alloy. In this way it has been found, for instance, that as many as 12 neighbouring Ni atoms are needed for the cleavage of the C—C bond in the conversion of ethane to methane.

18.6 Unimolecular reactions

The rate law of a surface-catalysed unimolecular reaction, such as the decomposition of a substance on a surface, can be written in terms of an adsorption isotherm if the rate is supposed to be proportional to the surface coverage. For example, if θ is given by the Langmuir isotherm (eqn 18.3, $\theta = \alpha p/(1+\alpha p)$), we would write.

$$\text{Rate} = k_r\theta = \frac{k_r \alpha p}{1+\alpha p} \tag{18.15}$$

where p is the pressure of the adsorbing substance.

Brief illustration 18.5 Surface-catalysed unimolecular decomposition

Consider the decomposition of phosphine (PH_3) on tungsten, which is first-order at low pressures. We can use eqn 18.15 to account for this observation. When the pressure is so low that $\alpha p \ll 1$, we can neglect αp in the denominator of eqn 18.15 and obtain

$$\text{Rate} = \frac{k_r \alpha p}{\underbrace{1+\alpha p}_{\approx 1}} \approx k_r \alpha p$$

The decomposition is predicted to be first-order, as observed experimentally.

Self-test 18.7

Write a rate law for the decomposition of PH_3 on tungsten at high pressures.

Answer: Rate = k_r; the reaction is zeroth-order at high pressures

18.7 The Langmuir–Hinshelwood mechanism

In the **Langmuir–Hinshelwood mechanism** of surface-catalysed reactions, the reaction takes place by encounters between molecular fragments and atoms adsorbed on the surface. We therefore expect the rate law to be overall second-order in the extent of surface coverage:

$$\text{A} + \text{B} \rightarrow \text{P} \qquad \text{Rate} = k_r \theta_A \theta_B$$

Insertion of the appropriate isotherms for A and B then gives the reaction rate in terms of the partial pressures of the reactants. For example, if A and B follow the co-adsorption isotherms given in eqn 18.6, then the rate law can be expected to be

$$\text{Rate} = k_r \times \overbrace{\frac{\alpha_A p_A}{(1+\alpha_A p_A + \alpha_B p_B)}}^{\text{Langmuir isotherm for A}} \times \overbrace{\frac{\alpha_B p_B}{(1+\alpha_A p_A + \alpha_B p_B)}}^{\text{Langmuir isotherm for B}}$$

$$= \frac{k_r \alpha_A \alpha_B p_A p_B}{(1+\alpha_A p_A + \alpha_B p_B)^2} \qquad (18.16)$$

The parameters α in the isotherms and the rate constant k_r are all temperature dependent, so the overall temperature dependence of the rate may be strongly non-Arrhenius, in the sense that the reaction rate is unlikely to be proportional to $e^{-E_a/RT}$. The Langmuir–Hinshelwood mechanism is dominant for the catalytic oxidation of CO to CO_2 on the (111) surface of platinum.

18.8 The Eley–Rideal mechanism

In the **Eley–Rideal mechanism** of a surface-catalysed reaction, a gas-phase molecule collides with another molecule already adsorbed on the surface. We can therefore expect the rate of formation of product to be proportional to the partial pressure, p_B, of the non-adsorbed gas B and the extent of surface coverage, θ_A, of the adsorbed gas A. It follows that the rate law should be

$$\text{A} + \text{B} \rightarrow \text{P} \qquad \text{Rate} = k_r p_B \theta_A$$

The rate constant, k_r, might be much larger than for the uncatalysed gas-phase reaction because the reaction on the surface has a low activation energy and the adsorption itself is often not activated. If we know the adsorption isotherm for A, we can express the rate law in terms of its partial pressure, p_A. For example, if the adsorption of A follows a Langmuir isotherm, $\theta_A = \alpha_A p_A/(1 + \alpha_A p_A)$, in the pressure range of interest, then the rate law would be

$$\text{Rate} = \frac{k_r \alpha p_A p_B}{1+\alpha p_A} \qquad (18.17)$$

If A were a diatomic molecule that adsorbed as atoms, then we would substitute the isotherm given in eqn 18.5 instead.

Brief illustration 18.6 The Eley–Rideal mechanism

According to eqn 18.17, when the partial pressure of A is high (in the sense $\alpha p_A \gg 1$) there is almost complete surface coverage, and

$$\text{Rate} = \frac{k_r \alpha p_A p_B}{\underbrace{1+\alpha p_A}_{\approx \alpha p_A}} \approx \frac{k_r \alpha p_A p_B}{\alpha p_A} \overset{\text{Cancel } \alpha p_A}{\approx} k_r p_B$$

The rate is equal to $k_r p_B$. Now the rate-determining step is the collision of B with the adsorbed fragments. When the pressure of A is low ($\alpha p_A \ll 1$), perhaps because of its reaction, the rate is equal to $k_r \alpha p_A p_B$; now the extent of surface coverage is important in the determination of the rate.

Self-test 18.8

Rewrite eqn 18.17 for cases where A is a diatomic molecule that adsorbs as atoms.

Answer: Rate = $k_r p_B (\alpha p_A)^{1/2}/(1 + (\alpha p_A)^{1/2})$

Almost all thermal surface-catalysed reactions are thought to take place by the Langmuir–Hinshelwood mechanism, but a number of reactions with an Eley–

Rideal mechanism have also been identified from molecular-beam investigations. For example, the reaction between H(g) and D(ad) to form HD(g) is thought to be by an Eley–Rideal mechanism involving the direct collision and pick-up of the adsorbed D atom by the incident H atom. However, the two mechanisms should really be thought of as ideal limits, and all reactions lie somewhere between the two and show features of both.

Impact on technology 18.1

Examples of heterogeneous catalysis

Almost the whole of modern chemical industry depends on the development, selection, and application of catalysts (Table 18.3). All we can hope to do in this section is to give a brief indication of some of the problems involved. Other than the ones we consider, these problems include the danger of the catalyst being poisoned by by-products or impurities, and economic considerations relating to cost, regeneration, and lifetime.

The activity of a catalyst depends on the strength of chemisorption as indicated by the 'volcano' curve in Fig. 18.26 (which is so called on account of its general shape; but note that the vertical axis is logarithmic, so the high activities are very much higher than the low activities). To be active, the catalyst should be extensively covered by adsorbate, which is the case if chemisorption is strong. On the other hand, if the strength of the substrate–adsorbate bond becomes too great, then the activity declines either because the other reactant molecules cannot react with the adsorbate or because the adsorbate molecules are immobilized on the surface. This pattern of behaviour suggests that the activity of a catalyst should initially increase with strength of adsorption (as measured, for instance, by the enthalpy of adsorption) and then decline, and that the most active catalysts should be those lying near the summit of the volcano. The most active metals are those lying close to the middle of the d block.

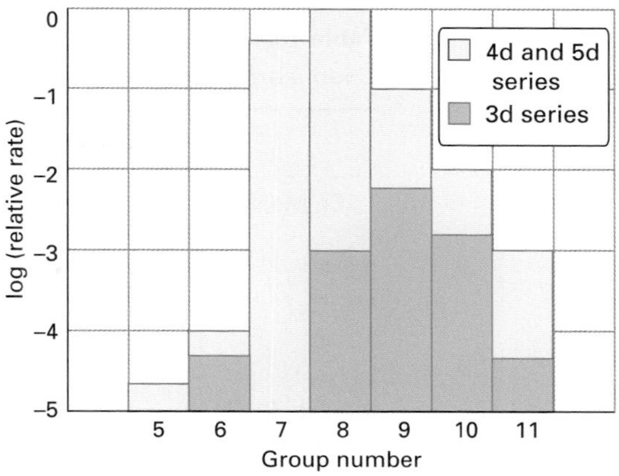

Fig. 18.26 A 'volcano curve' of catalytic activity arises because, although the reactants must adsorb reasonably strongly, they must not adsorb so strongly that they are immobilized. The green and yellow rectangles correspond to the 3d and (4d, 5d) series of metals. The group numbers relate to the periodic table (see inside back cover).

Because heterogeneous catalysis is a surface phenomenon, it is essential to achieve high surface areas. Thus, solid catalysts may be finely divided or structures with internal channels and cavities (as in zeolites; see below). Inactive catalyst supports are used to stabilize catalytic nanoparticles dispersed over them.

Many metals are suitable for adsorbing gases, and the general order of adsorption strengths decreases along the series O_2, C_2H_2, C_2H_4, CO, H_2, CO_2, N_2. Some of these molecules adsorb dissociatively (for example, H_2). Elements from the d block, such as iron, vanadium, and chromium, show a strong activity towards all these gases, but manganese and copper are unable to adsorb N_2 and CO_2. Metals towards the left of the periodic table (for example, magnesium and lithium) can adsorb (and, in fact, react with) only the most active gas (O_2). These trends are summarized in Table 18.4.

Table 18.3
Properties of catalysts

Catalyst	Function	Examples
Metals	Hydrogenation Dehydrogenation	Fe, Ni, Pt
Semiconducting oxides and sulfides	Oxidation Desulfurization	NiO, ZnO, MgO, Bi_2O_3/MoO_3, MoS_2
Insulating oxides	Dehydration	Al_2O_3, SiO_2, MgO
Acids	Polymerization Isomerization Cracking Alkylation	H_3PO_4, H_2SO_4, SiO_3/Al_2O_3, zeolites

Table 18.4
*Chemisorption abilities**

	O_2	C_2H_2	C_2H_4	CO	H_2	CO_2	N_2
Ti, Cr, Mo, Fe	+	+	+	+	+	+	+
Ni, Co	+	+	+	+	+	+	−
Pd, Pt	+	+	+	+	+	−	−
Mn, Cu	+	+	+	+	±	−	−
Al, Au	+	+	+	−	−	−	−
Li, Na, K	+	+	−	−	−	−	−
Mg, Ag, Zn, Pb	+	−	−	−	−	−	−

* +, Strong chemisorption; ±, chemisorption; −, no chemisorption.

As an example of catalytic action, consider the hydrogenation of alkenes. The alkene (**1**) adsorbs by forming two bonds with the surface (**2**), and on the same surface there may be adsorbed H atoms. When an encounter occurs, one of the alkene–surface bonds is broken (forming **3** or **4**) and later an encounter with a second H atom releases the fully hydrogenated hydrocarbon, which is the thermodynamically more stable species. The evidence for a two-stage reaction is the appearance of different isomeric alkenes in the mixture. The formation of isomers comes about because while the hydrocarbon chain is waving about over the surface of the metal, an atom in the chain might chemisorb again to form (**5**) and then desorb to (**6**), an isomer of the original molecule. The new alkene would not be formed if the two hydrogen atoms attached simultaneously.

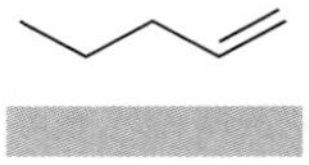

1

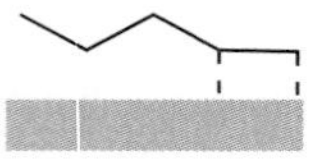

2

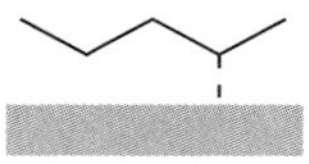

3

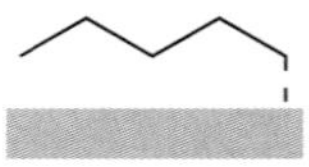

4

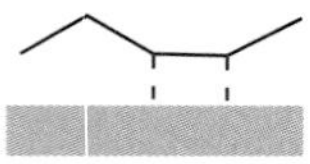

5

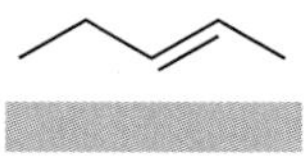

6

Catalytic oxidation is widely used in industry and in pollution control. Although in some cases it is desirable to achieve complete oxidation (as in the production of nitric acid from ammonia), in others partial oxidation is the aim. For example, the complete oxidation of propene to carbon dioxide and water is wasteful, but its partial oxidation to propenal (acrolein, $CH_2{=}CHCHO$) is the start of important industrial processes. Likewise, the controlled oxidations of ethene to ethanol, ethanal (acetaldehyde), and (in the presence of chlorine) to chloroethene (vinyl chloride, for the manufacture of PVC) are the initial stages of very important chemical industries.

Some of these reactions are catalysed by d-metal oxides of various kinds. The physical chemistry of oxide surfaces is very complex, as can be appreciated by considering what happens during the oxidation of propene to propenal on bismuth molybdate. The first stage is the adsorption of the propene molecule with loss of a hydrogen to form the propenyl (allyl) radical, $CH_2{=}CHCH_2\cdot$. An O atom in the surface can now transfer to this radical, leading to the formation of propenal and its desorption from the surface. The H atom also escapes with a surface O atom, and goes on to form H_2O, which leaves the surface. The surface is left with vacancies and metal ions in lower oxidation states. These vacancies are attacked by O_2 molecules in the overlying gas, which then chemisorb as O_2^- ions, so reforming the catalyst. This sequence of events, which is called the **Mars van Krevelen mechanism**, involves great upheavals of the surface, and some materials break up under the stress.

Many of the small organic molecules used in the preparation of all kinds of chemical products come from petroleum. These small building blocks of polymers, and petrochemicals in general, are usually cut from the long-chain hydrocarbons drawn from the Earth as petroleum. The catalytically induced fragmentation of the long-chain hydrocarbons is called **cracking**, and is often brought about on silica–alumina catalysts. These catalysts act by forming unstable carbocations, which dissociate and rearrange to more highly branched isomers. These branched isomers burn more smoothly and efficiently in internal combustion engines, and are used to produce higher octane fuels.

Catalytic **reforming** uses a dual-function catalyst, such as a dispersion of platinum and acidic alumina. The platinum provides the metal function, and brings about dehydrogenation and hydrogenation. The alumina provides the acidic function, being able to form carbocations from alkenes. The sequence of events in catalytic reforming shows up very clearly the complications that must be unravelled if a reaction as important as this is to be understood and improved. The first step is the attachment of the long-chain hydrocarbon by chemisorption to the platinum. In this process first one and then a second H atom is lost, and an alkene is formed. The alkene migrates to a Brønsted acid site, where it accepts a proton and attaches to the surface as a carbocation. This carbocation can undergo several different reactions. It can break into two, isomerize into a more highly branched form, or undergo

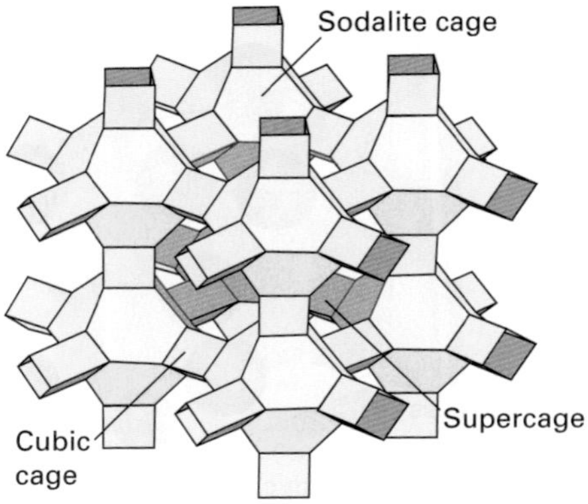

Fig. 18.27 A framework representation of the general layout of the Si, Al, and O atoms in a zeolite material. Each vertex corresponds to a Si or Al atom and each edge corresponds to the approximate location of an O atom. Note the large central pore, which can hold cations, water molecules, or other small molecules.

varieties of ring closure. Then the adsorbed molecule loses a proton, escapes from the surface, and migrates (possibly through the gas) as an alkene to a metal part of the catalyst where it is hydrogenated. We end up with a rich selection of smaller molecules which can be withdrawn, fractionated, and then used as raw materials for other products.

The concept of a solid surface has been extended in recent years with the availability of **microporous materials**, in which the surface effectively extends deep inside the solid. Zeolites are microporous aluminosilicates with the general formula $\{[M^{n+}]_{x/n}\cdot[H_2O]_m\}\{[AlO_2]_x[SiO_2]_y\}^{x-}$, where M^{n+} cations and H_2O molecules bind inside the cavities, or pores, of the Al—O—Si framework (Fig. 18.27). Small neutral molecules, such as CO_2, NH_3, and hydrocarbons (including aromatic compounds), can also adsorb to the internal surfaces and we shall see that this partially accounts for the utility of zeolites as catalysts.

Some zeolites for which $M = H^+$ are very strong acids and catalyse a variety of reactions that are of particular importance to the petrochemical industry. Examples include the dehydration of methanol to form hydrocarbons such as gasoline and other fuels:

$$x\,CH_3OH \xrightarrow{\text{zeolite}} (CH_2)_x + x\,H_2O$$

and the isomerization of 1,3-dimethylbenzene (*m*-xylene) to 1,4-dimethylbenzene (*p*-xylene). The catalytically important form of these acidic zeolites may be either a Brønsted acid (**7**) or a Lewis acid (**8**). Like enzymes, a zeolite catalyst with a specific composition and structure is very selective toward certain reactants and products because only molecules of certain sizes can enter and exit the pores in which catalysis occurs. It is also possible that zeolites derive their selectivity from the ability to bind and to stabilize only transition states that fit properly in the pores. The analysis of the mechanism of zeolite catalysis is greatly facilitated by computer simulation of microporous systems, which shows how molecules fit in the pores, migrate through the connecting tunnels, and react at the appropriate active sites.

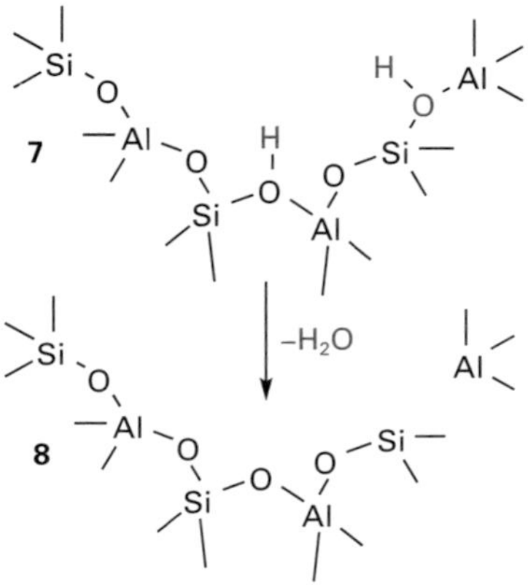

Processes at electrodes

A very special kind of surface is that of an electrode in contact with an electrolyte. Studies of processes on electrode surfaces are of enormous importance in electrochemistry where they give information about the rate of electron transfer between the electrode and electroactive species in solution, and are essential to the improvement of the performance of batteries and fuel cells (Impact 9.2). Detailed knowledge of the factors that determine the rate of electron transfer leads to a better understanding of power production in batteries and of electron conduction in metals, semiconductors, and nanometre-sized electronic devices. Indeed, the economic consequences of electrode processes are almost incalculable. Most of the modern methods of generating electricity are inefficient, and the development of fuel cells could enhance our production and deployment of energy, not least by the reduction of the generation of polluting nitrogen oxides. Today we produce energy inefficiently to produce goods that then decay by corrosion. Each step of this wasteful sequence could be improved by discovering more about the kinetics of electrochemical processes. Similarly, the techniques of organic and inorganic electrosynthesis, where an electrode is an active component of an industrial process, depend on intimate understanding of the kinetics of the processes taking place at electrodes.

18.9 The electrode–solution interface

Whereas most of the preceding discussion focussed on the gas–solid interface, we now have to turn our attention to a metallic conductor immersed in an aqueous solution of ions. The most primitive model of the boundary between the solid and liquid phases is an **electrical double layer**, which consists of a sheet of positive charge at the surface of the electrode and a sheet of negative charge next to it in the solution (or vice versa). This arrangement creates an electrical potential difference, called the **Galvani potential difference**, between the bulk of the electrode and the bulk of the solution. For simplicity in the following, we shall identify the Galvani potential difference with what in Chapter 9 we called the electrode potential.

We can construct a more detailed picture of the interface by speculating about the arrangement of ions and electric dipoles in the solution. In the **Helmholtz layer model** of the interface the solvated ions arrange themselves along the surface of the electrode but are held away from it by their hydration spheres (Fig. 18.28). The location of the sheet of ionic charge, which is called the **outer Helmholtz plane** (OHP), is identified as the plane running through the solvated ions. In this simple model, the electrical potential changes linearly within the layer bounded by the electrode surface on one side and the OHP on the other. In a refinement of this model, ions that have discarded their solvating molecules and have become attached to the electrode surface by chemical bonds are regarded as forming the **inner Helmholtz plane** (IHP). The Helmholtz layer model ignores the disrupting effect of thermal motion, which tends to break up and disperse the rigid outer plane of charge. In the **Gouy–Chapman model** of the **diffuse double layer**, the disordering effect of thermal motion is taken into account in much the same way as the Debye–Hückel model describes the ionic atmosphere of an ion (Section 9.1) with the latter's single central ion replaced by an infinite, plane electrode (Fig. 18.29).

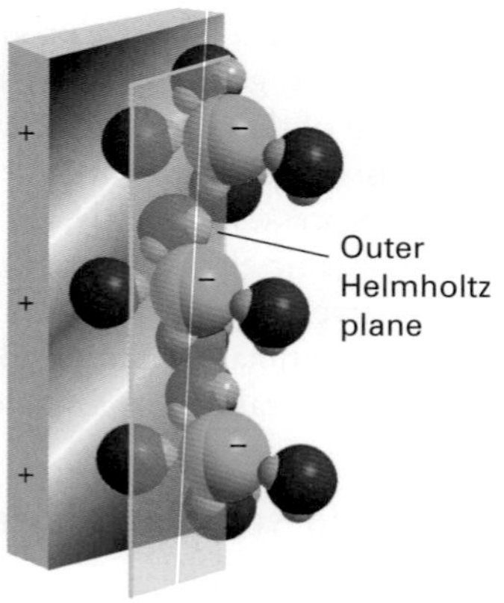

Fig. 18.28 A simple model of the electrode–solution interface treats it as two rigid planes of charge. One plane, the outer Helmholtz plane (OHP), is due to the ions with their solvating molecules and the other plane is that of the electrode itself.

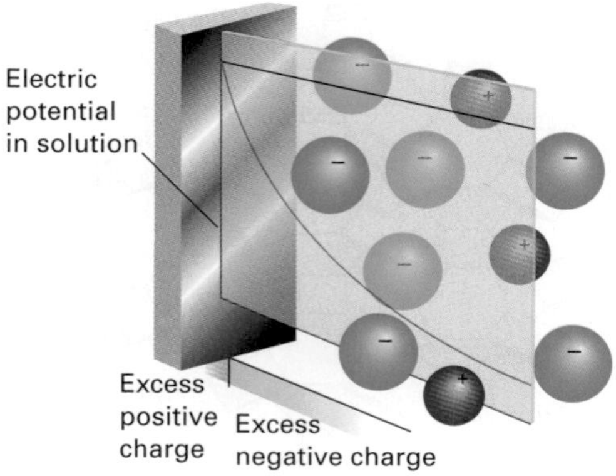

Fig. 18.29 The Gouy–Chapman model of the electrical double layer treats the outer region as an atmosphere of countercharge, similar to the Debye–Hückel theory of ion atmospheres. The plot of electrical potential against distance from the electrode surface shows the meaning of the diffuse double layer (see text for details).

18.10 The rate of electron transfer

We shall consider a reaction at the electrode in which an ion is reduced by the transfer of a single electron in the rate-determining step. The last phrase is important: in the deposition of cadmium, for instance, only one electron is transferred in the rate-determining step even though overall the deposition involves the transfer of two electrons. The quantity we focus on is the **current density**, j, the electric current flowing through a region of an electrode divided by the area of the region. An analysis of the effect of the Galvani potential difference at the electrode on the current density using a version of transition-state theory (Section 10.11) leads to the **Butler–Volmer equation:**[2]

$$j = j_0\{e^{(1-\alpha)f\eta} - e^{-\alpha f\eta}\} \quad \text{Butler–Volmer equation} \quad (18.18)$$

We have written $f = F/RT$, where F is Faraday's constant (Section 9.7; at 298 K, $f = 38.9\ V^{-1}$). Let's go over the other parameters in this equation:

- The quantity η (eta) is the **overpotential:**

$$\eta = E' - E \quad \text{Overpotential} \quad (18.19)$$

where E is the electrode potential at equilibrium, when there is no net flow of current, and E' is the electrode potential when a current is being drawn from the cell.

[2] For a derivation of this equation, see our *Physical Chemistry* (2010).

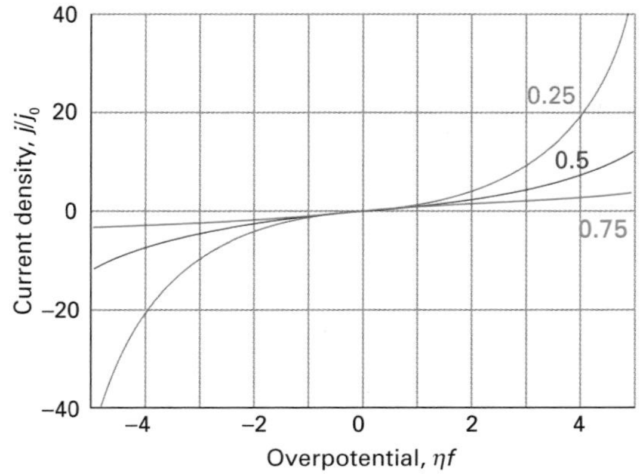

Fig. 18.30 The dependence of the current density on the overpotential for different values of the transfer coefficient.

- The quantity α is the **transfer coefficient**, and is an indication of whether the transition state between the reduced and oxidized forms of the electroactive species in solution is reactant-like ($\alpha = 0$) or product-like ($\alpha = 1$): typical values are close to 0.5.
- The quantity j_0 is the **exchange-current density**, the magnitude of the equal but opposite current densities when the electrode is at equilibrium. As usual in chemistry, equilibrium is dynamic, so even though there may be no net flow of current at an electrode, there are matching inward and outward flows of electrons.

Figure 18.30 shows how eqn 18.18 predicts the current density to depend on the overpotential for different values of the transfer coefficient.

When the overpotential is so small that $f\eta \ll 1$ (in practice, η less than about 0.01 V) the exponentials in eqn 18.18 can be expanded by using $e^x = 1 + x + \cdots$ and $e^{-x} = 1 - x + \cdots$ (The chemist's toolkit 6.1) to give

$$j = j_0\{\overbrace{1+(1-\alpha)f\eta+\cdots}^{e^{(1-\alpha)f\eta}} - \overbrace{(1-\alpha f\eta+\cdots)}^{e^{-\alpha f\eta}}\} \approx j_0 f\eta \qquad (18.20)$$

This equation shows that the current density is proportional to the overpotential, so at low overpotentials the interface behaves like a conductor that obeys Ohm's law, which states that the current is proportional to the potential difference.

When the overpotential is large and positive (in practice, $\eta \geq 0.12$ V), the second exponential in eqn 18.18 is much smaller than the first, and may be neglected. For instance, if $\eta = 0.2$ V and $\alpha = 0.5$, $e^{-\alpha f\eta} = 0.02$ whereas $e^{(1-\alpha)f\eta} = 49$. Then (ignoring signs, which indicate the direction of the current)

$$j = j_0 e^{(1-\alpha)f\eta}$$

By taking logarithms of both sides (and using the rules for logarithms outlined in The chemist's toolkit 2.2, $\ln xy = \ln x + \ln y$, $\ln e^x = x$) we obtain

$$\ln j = \ln j_0 + (1-\alpha)f\eta \qquad (18.21a)$$

The plot of the logarithm of the current density against the overpotential is called a **Tafel plot.** the slope, which is equal to $(1-\alpha)f$, gives the value of α and the intercept at $\eta = 0$ gives the exchange-current density. If instead the overpotential is large but negative (in practice, $\eta \leq -0.12$ V), the first exponential in eqn 18.18 may be neglected. Then

$$j = j_0 e^{-\alpha f\eta}$$

so, after taking logarithms as before,

$$\ln j = \ln j_0 - \alpha f\eta \qquad (18.21b)$$

In this case the slope of the Tafel plot is $-\alpha f$.

Example 18.5

Interpreting a Tafel plot

The data below refer to the anodic current through a platinum electrode of area 2.0 cm^2 in contact with an Fe^{3+},Fe^{2+} aqueous solution at 298 K. Calculate the exchange-current density and the transfer coefficient for the electrode process.

η/mV	50	100	150	200	250
I/mA	8.8	25.0	58.0	131	298

Strategy The anodic process is the oxidation $Fe^{2+}(aq) \rightarrow Fe^{3+}(aq) + e^-$. To analyse the data, we make a Tafel plot (of $\ln j$ against η) using the anodic form (eqn 18.21a). The intercept at $\eta = 0$ is $\ln j_0$ and the slope is $(1-\alpha)f$.

Answer Draw up the following table:

η/mV	50	100	150	200	250
j/(mA cm^{-2})	4.4	12.5	29.0	65.5	149
ln(j/(mA cm^{-2}))	1.48	2.53	3.37	4.18	5.00

The points are plotted in Fig. 18.31. The high overpotential region gives a straight line of intercept 0.88 and slope 0.0165. From the former it follows that $\ln(j_0/(\text{mA cm}^{-2})) = 0.88$, so $j_0 = 2.4$ mA cm^{-2}. From the latter,

$$\overbrace{(1-\alpha)f = 0.0165}^{\text{Slope}} \quad \overbrace{\text{mV}^{-1}}^{\text{Units of } \overbrace{\ln j}^{1} / \overbrace{\eta}^{\text{mV}}}$$

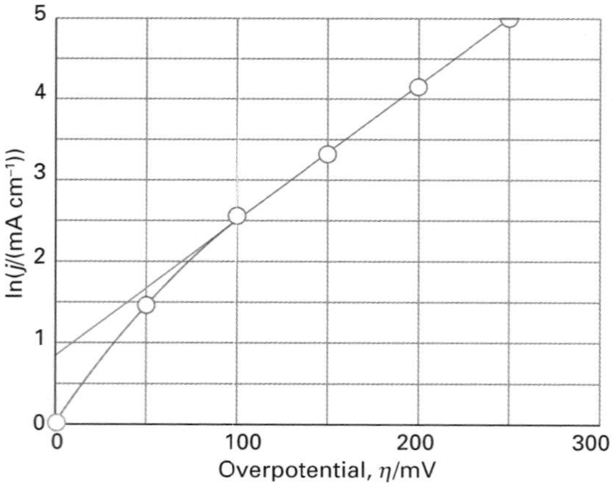

Fig. 18.31 A Tafel plot is used to measure the exchange-current density (given by the extrapolated intercept at $\eta = 0$) and the transfer coefficient (from the slope). The data are from Example 18.5.

so

$$\alpha = 1 - \frac{\overbrace{(0.0165\ \text{mV}^{-1})}^{16.5\ \text{V}^{-1}}}{\underbrace{f}_{38.9\ \text{V}^{-1}}} = 1 - 0.42\ldots = 0.58$$

Note that the Tafel plot is nonlinear for $\eta < 100$ mV; in this region $\alpha f\eta = 2.3$ and the approximation that $\alpha f\eta \gg 1$ fails.

Self-test 18.9

Repeat the analysis using the following cathodic current data:

η/mV	−50	−100	−150	−200	−250	−300
I/mA	0.3	1.5	6.4	27.6	118.6	510

Answer: $\alpha = 0.75$, $j_0 = 0.041$ mA cm^{-2}

Some experimental values for the Butler–Volmer parameters are given in Table 18.5. From them we can see that exchange-current densities vary over a very wide range. For example, the N_2/N_3^- couple on platinum has $j_0 = 10^{-76}$ A cm^{-2}, whereas the H^+/H_2 couple on platinum has $j_0 = 8 \times 10^{-4}$ A cm^{-2}, a difference of 73 orders of magnitude. Exchange currents are generally large when the redox process involves no bond breaking (as in the $[Fe(CN)_6]^{3-}/[Fe(CN)_6]^{4-}$ couple) or if only weak bonds are broken (as in Cl_2/Cl^-). They are generally small when more than one electron needs to be transferred, or when multiple or strong bonds are broken, as in the N_2/N_3^- couple and in redox reactions of organic compounds.

Table 18.5
Exchange current densities and transfer coefficients at 298 K

Reaction	Electrode	j_0/(A cm^{-2})	α
$2\,H^+ + 2\,e^- \rightarrow H_2$	Pt	7.9×10^{-4}	
	Ni	6.3×10^{-6}	0.58
	Pb	5.0×10^{-12}	
$Fe^{3+} + e^- \rightarrow Fe^{2+}$	Pt	2.5×10^{-3}	0.58

Electrodes with potentials that change only slightly when a current passes through them are classified as **nonpolarizable**. Those with strongly current-dependent potentials are classified as **polarizable**. From the linearized equation (eqn 18.21) it is clear that the criterion for low polarizability is high exchange-current density (so η may be small even though j is large). The calomel and H_2|Pt electrodes are both highly nonpolarizable, which is one reason why they are so extensively used as reference electrodes in electrochemistry.

18.11 Voltammetry

One of the assumptions in the derivation of the Butler–Volmer equation is the negligible conversion of the electroactive species at low current densities, resulting in uniformity of concentration near the electrode. This assumption fails at high current densities because the consumption of electroactive species close to the electrode results in a concentration gradient. The diffusion of the species towards the electrode from the bulk is slow and may become rate-determining; a larger overpotential is then needed to produce a given current. This effect is called **concentration polarization**. Concentration polarization is important in the interpretation of **voltammetry**, the study of the current through an electrode as a function of the applied potential difference.

The kind of output from **linear-sweep voltammetry** is illustrated in Fig. 18.32. Initially, the absolute value of the potential is low, and the current is due to the migration of ions in the solution. However, as the potential approaches the reduction potential of the reducible solute, the current grows. Soon after the potential exceeds the reduction potential the current rises and reaches a maximum value. This maximum current is proportional to the molar concentration of the species, so that concentration can be determined

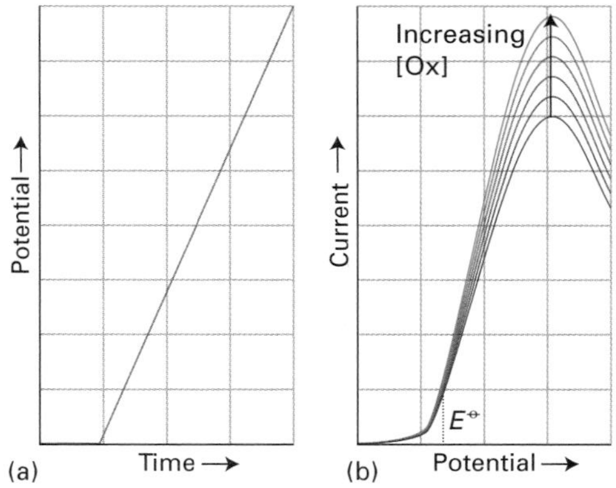

Fig. 18.32 (a) The change of potential with time and (b) the resulting current/potential curve in a voltammetry experiment. The peak value of the current density is proportional to the concentration of electroactive species (for instance, [Ox]) in solution.

from the peak height after subtraction of an extrapolated baseline.

In **cyclic voltammetry** the potential is applied with a triangular waveform (linearly up, then linearly down) and the current is monitored. A typical cyclic voltammogram is shown in Fig. 18.33. The shape of the curve is initially like that of a linear-sweep experiment, but after reversal of the sweep there is a rapid change in current on account of the high concentration of oxidizable species close to the electrode that was generated on the reductive sweep. When the potential is close to the value required to oxidize the reduced species, there is a substantial current until all the oxidation is complete, and the current returns to zero. Cyclic voltammetry data are obtained at scan rates of about 50 mV s^{-1}, so a scan over a range of 2 V takes about 80 s.

When the reduction reaction at the electrode can be reversed, as in the case of the $[Fe(CN)_6]^{3-}/[Fe(CN)_6]^{4-}$ couple, the cyclic voltammogram is broadly symmetric about the standard potential of the couple (as in Fig. 18.33). The scan is initiated with $[Fe(CN)_6]^{3-}$ present in solution, and as the potential approaches $E^{\ominus}$ for the couple, the $[Fe(CN)_6]^{3-}$ near the electrode is reduced and current begins to flow. As the potential continues to change, the current begins to decline again because all the $[Fe(CN)_6]^{3-}$ near the electrode has been reduced and the current reaches its limiting value. The potential is now returned linearly to its initial value, and the reverse series of events occurs with the $[Fe(CN)_6]^{4-}$ produced during the forward scan now undergoing oxidation. The peak of current lies on the other side of $E^{\ominus}$, so the species present and its standard potential can be identified, as indicated in the illustration, by noting the locations of the two peaks.

The overall shape of the curve gives details of the kinetics of the electrode process and the change in shape as the rate of change of potential is altered gives information on the rates of the processes involved. For example, the matching peak on the return phase of the potential sweep may be missing, which indicates that the oxidation (or reduction) is irreversible. The appearance of the curve may also depend on the timescale of the sweep, for if the sweep is too fast some processes might not have time to occur. This style of analysis is illustrated in the following example.

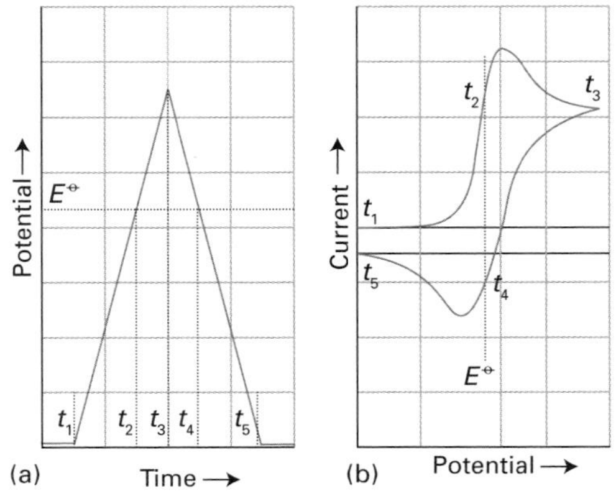

Fig. 18.33 (a) The change of potential with time and (b) the resulting current/potential curve in a cyclic voltammetry experiment.

Example 18.6

Analysing a cyclic voltammetry experiment

The electroreduction of *p*-bromonitrobenzene in liquid ammonia is believed to occur by the following mechanism:

$$BrC_6H_4NO_2 + e^- \rightarrow BrC_6H_4NO_2^-$$
$$BrC_6H_4NO_2^- \rightarrow \cdot C_6H_4NO_2 + Br^-$$
$$\cdot C_6H_4NO_2 + e^- \rightarrow C_6H_4NO_2^-$$
$$C_6H_4NO_2^- + H^+ \rightarrow C_6H_5NO_2$$

Suggest the likely form of the cyclic voltammogram expected on the basis of this mechanism.

Strategy Decide which steps are likely to be reversible on the timescale of the potential sweep: such processes will give symmetrical voltammograms. Irreversible processes will give unsymmetrical shapes as reduction (or oxidation) might

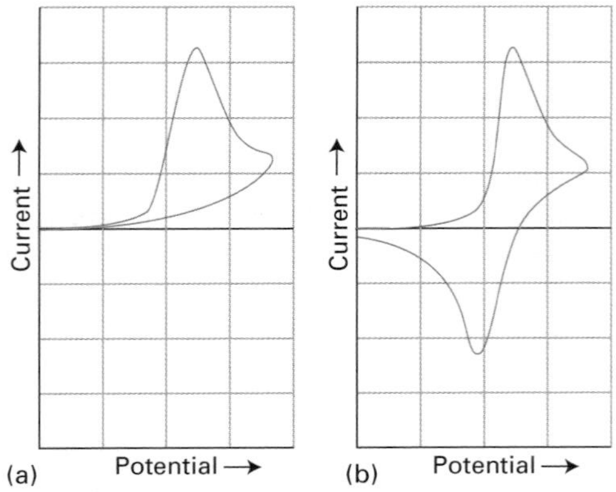

Fig. 18.34 (a) When a non-reversible step in a reaction mechanism has time to occur, the cyclic voltammogram may not show the reverse oxidation or reduction peak. (b) However, if the rate of sweep is increased, the return step may be caused to occur before the irreversible step has had time to intervene, and a typical 'reversible' voltammogram is obtained.

not occur. However, at fast sweep rates, an intermediate might not have time to react, and a reversible shape will be observed.

Answer At slow sweep rates, the second reaction has time to occur, and a curve typical of a two-electron reduction will be observed, but there will be no oxidation peak on the second half of the cycle because the product, $C_6H_5NO_2$, cannot be oxidized (Fig. 18.34a). At fast sweep rates, the second reaction does not have time to take place before oxidation of the $BrC_6H_4NO_2^-$ intermediate starts to occur during the reverse scan, so the voltammogram will be typical of a reversible one-electron reduction (Fig. 18.34b).

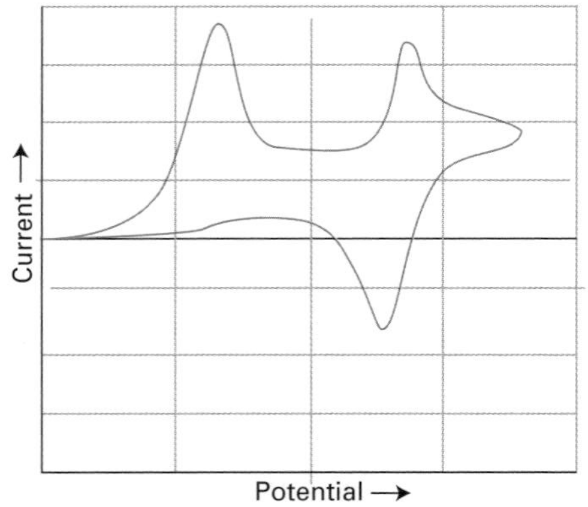

Fig. 18.35 The cyclic voltammogram referred to in Self-test 18.10.

Self-test 18.10

Suggest an interpretation of the cyclic voltammogram shown in Fig. 18.35. The electroactive material is ClC_6H_4CN in acid solution; after reduction to $ClC_6H_4CN^-$, the radical anion may form C_6H_5CN irreversibly

Answer: $ClC_6H_4CN + e^- \rightleftharpoons ClC_6H_4CN^-$, $ClC_6H_4CN^- + H^+ + e^- \rightarrow C_6H_5CN + Cl^-$, $C_6H_5CN + e^- \rightleftharpoons C_6H_5CN^-$

18.12 Electrolysis

To induce current to flow through an electrolytic cell and bring about a non-spontaneous cell reaction, the applied potential difference must exceed the cell potential by at least the **cell overpotential.** The cell overpotential is the sum of the overpotentials at the two electrodes and the ohmic drop (IR_s, where R_s is the internal resistance of the cell) due to the current through the electrolyte. The additional potential needed to achieve a detectable rate of reaction may need to be large when the exchange-current density at the electrodes is small.

The rate of gas evolution or metal deposition during electrolysis can be estimated from the Butler–Volmer equation and tables of exchange-current densities. The exchange-current density depends strongly on the nature of the electrode surface, and changes in the course of the electrodeposition of one metal on another. A very crude criterion is that significant evolution or deposition occurs only if the overpotential exceeds about 0.6 V.

A glance at Table 18.5 shows the wide range of exchange-current densities for a metal/hydrogen electrode. The most sluggish exchange currents occur for lead and mercury, and the value of 1 pA cm^{-2} corresponds to a monolayer of atoms being replaced in about 5 a (a is the SI symbol for annum, year). For such systems, a high overpotential is needed to induce significant hydrogen evolution. In contrast, the value for platinum (1 mA cm^{-2}) corresponds to a monolayer being replaced in 0.1 s, so gas evolution occurs for a much lower overpotential.

Checklist of key concepts

☐ 1 Adsorption is the attachment of molecules to a surface; the reverse of adsorption is desorption.

☐ 2 The substance that adsorbs is the adsorbate and the underlying material is the adsorbent or substrate.

☐ 3 Techniques for studying surface composition and structure include scanning tunnelling microscopy (STM), atomic force microscopy (AFM), photoemission spectroscopy, Auger electron spectroscopy (AES), and low-energy electron diffraction (LEED).

☐ 4 The fractional coverage, θ, is the ratio of the number of occupied sites to the number of available sites.

☐ 5 Techniques for studying the rates of surface processes include flash desorption, surface plasmon resonance (SPR), and gravimetry by using a quartz crystal microbalance (QCM).

☐ 6 Physisorption is adsorption by a van der Waals interaction; chemisorption is adsorption by formation of a chemical (usually covalent) bond.

☐ 7 The isosteric enthalpy of adsorption is determined from a plot of $\ln \alpha$ against $1/T$.

☐ 8 The sticking probability, s, is the proportion of collisions with the surface that successfully lead to adsorption.

☐ 9 Desorption is an activated process; the desorption activation energy is measured by temperature programmed desorption (TPD) or thermal desorption spectroscopy (TDS).

☐ 10 In the Langmuir–Hinshelwood mechanism of surface-catalysed reactions, the reaction takes place by encounters between molecular fragments and atoms adsorbed on the surface.

☐ 11 In the Eley–Rideal mechanism of a surface-catalysed reaction, a gas-phase molecule collides with another molecule already adsorbed on the surface.

☐ 12 An electrical double layer consists of a sheet of positive charge at the surface of the electrode and a sheet of negative charge next to it in the solution (or vice versa).

☐ 13 The Galvani potential difference is the potential difference between the bulk of the metal electrode and the bulk of the solution.

☐ 14 Models of the double layer include the Helmholtz layer model and the Gouy–Chapman model.

☐ 15 A Tafel plot is a plot of the logarithm of the current density against the overpotential: the slope gives the value of α and the intercept at $\eta = 0$ gives the exchange-current density.

☐ 16 Voltammetry is the study of the current through an electrode as a function of the applied potential difference.

☐ 17 To induce current to flow through an electrolytic cell, the applied potential difference must exceed the cell potential by at least the cell overpotential.

Road map of key equations

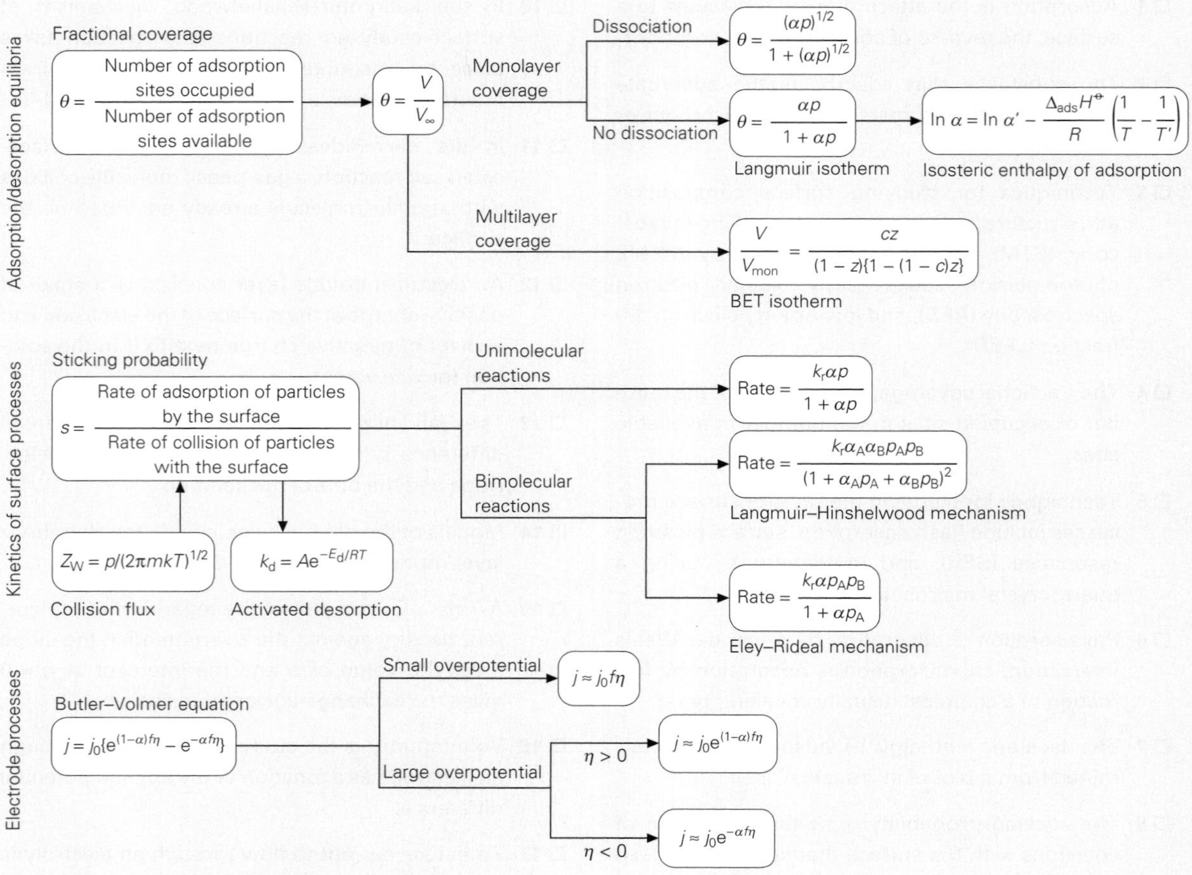

Questions and exercises

Discussion questions

18.1 Summarize the techniques available for characterizing the composition and structure of a surface.

18.2 Explain and justify the assumptions that are used to derive the Langmuir isotherm.

18.3 Demonstrate that the BET isotherm describes multilayer adsorption, showing that it behaves in the manner that you would expect on physical grounds as the various parameters are changed.

18.4 Describe the essential features of the Langmuir–Hinshelwood and Eley–Rideal mechanisms for surface-catalysed reactions. How can they be tested experimentally?

18.5 Outline the steps in the Mars van Krevelen mechanism of a catalysed reaction. How could it be tested experimentally?

18.6 Describe the various models of the structure of the electrode–electrolyte interface.

18.7 Discuss the technique of cyclic voltammetry and account for the characteristic shape of a cyclic voltammogram, such as those shown in Figs. 18.33 and 18.34.

Exercises

18.1 Calculate the frequency of molecular collisions per square centimetre of surface in a vessel containing (a) hydrogen, (b) propane at 25 °C when the pressure is (i) 100 Pa, (ii) 0.10 μTorr.

18.2 What pressure of argon gas is required to produce a collision rate of $8.5 \times 10^{20}\ s^{-1}$ at 450 K on a circular surface of diameter 2.5 mm?

18.3 Calculate the average rate at which He atoms strike a Cu atom in a surface formed by exposing a (100) plane in metallic copper to helium gas at 100 K and a pressure of 25 Pa. Crystals of copper are face-centred cubic with a cell edge of 361 pm.

18.4 In an adsorption experiment, the temperature of the apparatus of constant volume and containing a fixed amount of gaseous adsorbate is increased from 300 K to 400 K. By what factor does the collision flux increase?

18.5 A monolayer of CO molecules is adsorbed on the surface of 1.00 g of an Fe/Al_2O_3 catalyst at 77 K, the boiling point of liquid nitrogen. Upon warming, the carbon monoxide occupies 4.25 cm^3 at 0 °C and 1.00 bar. What is the surface area of the catalyst?

18.6 The adsorption of a gas is described by the Langmuir isotherm with $\alpha = 1.85\ kPa^{-1}$ at 25 °C. Calculate the pressure at which the fractional surface coverage is (a) 0.10, (b) 0.90.

18.7 Derive a version of the Langmuir isotherm starting from eqn 18.1 for the rate at which molecules strike the surface.

18.8 The data below are for the chemisorption of hydrogen on copper powder at 25 °C. Confirm that they fit the Langmuir isotherm at low coverages. Then find the value of α for the adsorption equilibrium and the adsorption volume corresponding to complete coverage.

p/Pa	25	129	253	540	1000	1593
V/cm^3	0.042	0.163	0.221	0.321	0.411	0.471

18.9 The values of α for the adsorption of CO on charcoal are $1.0 \times 10^{-3}\ Torr^{-1}$ at 273 K and $2.7 \times 10^{-3}\ Torr^{-1}$ at 250 K. Estimate the enthalpy of adsorption.

18.10 The data below show the pressures of CO needed for the volume of adsorption (corrected to 1.00 atm and 273 K) to be 10.0 cm^3 using the same sample as in Example 18.2. Calculate the adsorption enthalpy at this surface.

T/K	200	210	220	230	240	250
p/kPa	4.32	5.59	7.07	8.80	10.67	12.80

18.11 Suppose you wanted to achieve a certain surface coverage of an adsorbate that dissociates. Determine from eqn 18.5 how p depends on θ.

18.12 Suppose that an ozone molecule dissociates into three oxygen atoms when it adsorbs to a surface. Deduce the corresponding isotherm.

18.13 Confirm that the adsorption isotherms for two reactants A and B that compete for the same sites on a surface are given by eqn 18.6.

18.14 The data for the adsorption of ammonia on barium fluoride at 0 °C, when $p^* = 429.6$ kPa, are reported below. Confirm that they fit a BET isotherm and find values of c and V_{mon}.

p/kPa	14.0	37.6	65.6	79.2	82.7	100.7	106.4
V/cm^3	11.1	13.5	14.9	16.0	15.5	17.3	16.5

18.15 The enthalpy of adsorption of ammonia on a nickel surface is found to be $-155\ kJ\ mol^{-1}$. Estimate the mean lifetime of an NH_3 molecule on the surface at 600 K.

18.16 The average time for which an oxygen atom remains adsorbed to a tungsten surface is 0.36 s at 2548 K and 3.49 s at 2362 K. (a) Find the activation energy for desorption. (b) What is the pre-exponential factor for these tightly chemisorbed atoms?

18.17 In an experiment on the adsorption of oxygen on tungsten it was found that the same volume of oxygen was desorbed in 27 min at 1856 K and 2.0 min at 1978 K. What is the activation energy of desorption? How long would it take for the same amount to desorb at (a) 298 K, (b) 3000 K?

18.18 Ammonia at 10.0 Pa and 210 K was adsorbed on a surface of area 10 cm^2 at the rate of 0.33 mmol s^{-1}. What is the sticking probability?

18.19 Hydrogen iodide is very strongly adsorbed on gold but only slightly adsorbed on platinum. Assume the adsorption follows the Langmuir isotherm and predict the order of the HI decomposition reaction on each of the two metal surfaces.

18.20 According to the Langmuir–Hinshelwood mechanism of surface-catalysed reactions, the rate of reaction between A and B depends on the rate at which the adsorbed species meet. (a) Write the rate law for the reaction according to this mechanism. (b) Find the limiting form of the rate law when the partial pressures of the reactants are low. (c) Could this mechanism ever account for zero-order kinetics?

18.21 The transfer coefficient of a certain electrode in contact with M^{2+} and M^{3+} in aqueous solution at 25 °C is 0.48. The current density is found to be 17.0 mA cm^{-2} when the overpotential is 115 mV. What is the overpotential required for a current density of 38 mA cm^{-2}?

18.22 Determine the exchange-current density from the information given in Exercise 18.21.

18.23 A typical exchange-current density, that for H^+ discharge at platinum, is 0.79 mA cm^{-2} at 25 °C. What is the current density at an electrode when its overpotential is (a) 10 mV, (b) 100 mV, (c) −5.0 V? Take $\alpha = 0.5$.

18.24 How many electrons or protons are transported through the double layer in each second when the $Pt,H_2|H^+$, $Pt|Fe^{3+},Fe^{2+}$, and Pb, $H_2|H^+$ electrodes are at equilibrium at 25 °C? Take the area as 1.0 cm^2 in each case. Estimate the number of times each second a single atom on the surface takes part in a electron transfer event, assuming an electrode atom occupies about $(280\ pm)^2$ of the surface.

18.25 In an experiment on the $Pt|H_2|H^+$ electrode in dilute H_2SO_4 the following current densities were observed at 25 °C. Evaluate α and j_0 for the electrode.

η/mV	50	100	150	200	250
j/(mA cm^{-2})	2.66	8.91	29.9	100	335

How would the current density at this electrode depend on the overpotential of the same set of magnitudes but of opposite sign?

18.26 The following current–voltage data are for an indium anode relative to a standard hydrogen electrode at 293 K:

$-E$/V	0.388	0.365	0.350	0.335
j/(A m^{-2})	0	0.590	1.438	3.507

Use the data to calculate the transfer coefficient and the exchange-current density. What is the cathodic current density when the potential is 0.365 V?

18.27 The following data are for the overpotential for H_2 evolution with a mercury electrode in dilute aqueous solutions of H_2SO_4 at 25 °C. Determine the exchange-current density and transfer coefficient, α.

η/V	0.60	0.65	0.73	0.79	0.84	0.89	0.93	0.96
j/(mA m^{-2})	2.9	6.3	28	100	250	630	1650	3300

Explain any deviations from the result expected from the Tafel equation.

18.28 The illustrations below are four different examples of voltammograms. Identify the processes occurring in each system. In each case the vertical axis is the current and the horizontal axis is the (negative) electrode potential.

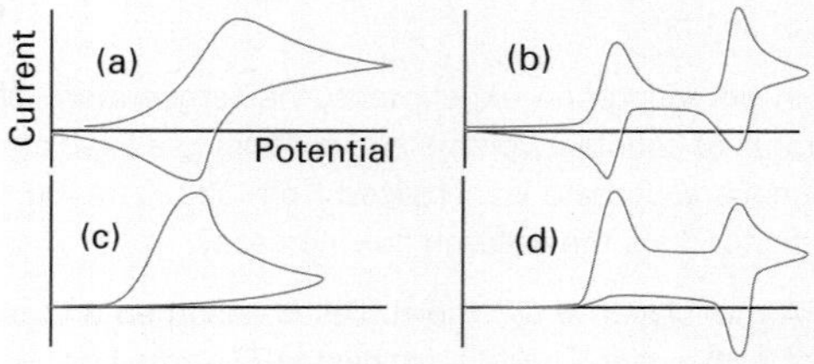

Projects

The symbol ‡ indicates that calculus is necessary.

18.29‡ Here we explore atomic force microscopy (AFM) quantitatively. (a) We saw in Foundations that the potential energy of interaction between two charges Q_1 and Q_2 separated by a distance r is $V = Q_1Q_2/4\pi\varepsilon_0 r$. To get an idea of the magnitudes of forces measured by AFM, calculate the force acting between two electrons separated by 0.50 nm. By what factor does the force drop if the distance between the electrons increases to 0.60 nm? *Hint*: The relation between force and potential energy is $F = -dV/dr$. (b) Suppose that the interaction probed by an AFM experiment can be expressed as a Lennard-Jones potential (Section 15.8): how does the force vary with distance?

18.30‡ The differential form of the van 't Hoff equation for the temperature dependence of equilibrium constants is $d(\ln K)/dT = \Delta_r H^{\ominus}/RT^2$. Find the corresponding expression for the temperature dependence of the pressure corresponding to a given fractional coverage on the basis of the Langmuir isotherm.

18.31 Here we explore further the design and operation of fuel cells (Impact 9.2): (a) Calculate the thermodynamic limit to the cell potential of fuel cells operating on (i) hydrogen and oxygen, (ii) methane and air. Use the Gibbs energy information in the *Data section*, and take the species to be in their standard states at 25 °C. (b) The reaction $2\ H^+ + 2\ e^- \rightarrow H_2$ is important for the operation of hydrogen/oxygen fuel cells. Use the data in Table 18.5 for the exchange-current density and transfer coefficient for the reaction $2\ H^+ + 2\ e^- \rightarrow H_2$ on nickel at 25 °C to determine what current density would be needed to obtain an overpotential of 0.20 V as calculated from (i) the Butler–Volmer equation and (ii) the Tafel equation. Is the validity of the Tafel approximation affected at higher overpotentials (of 0.4 V and more)?

19

Spectroscopy: molecular rotations and vibrations

Rotational spectroscopy 478

19.1 The rotational energy levels of molecules 478

19.2 Forbidden and allowed rotational states 482

19.3 Populations at thermal equilibrium 483

19.4 Rotational transitions: microwave spectroscopy 484

19.5 Linewidths 486

19.6 Rotational Raman spectra 488

Vibrational spectroscopy 489

19.7 The vibrations of molecules 489

19.8 Vibrational transitions 490

19.9 Anharmonicity 491

19.10 Vibrational Raman spectra of diatomic molecules 492

19.11 The vibrations of polyatomic molecules 492

19.12 Vibration–rotation spectra 495

19.13 Vibrational Raman spectra of polyatomic molecules 496

CHECKLIST OF KEY CONCEPTS 498

ROAD MAP OF KEY EQUATIONS 499

QUESTIONS AND EXERCISES 499

Spectroscopy is the analysis of the electromagnetic radiation emitted, absorbed, or scattered by molecules. We saw in Chapter 13 that photons act as messengers from inside atoms, and that atomic spectra can be used to obtain detailed information about electronic structure. Photons of radiation ranging from radio waves to the ultraviolet also bring information to us about molecules. The difference between molecular and atomic spectroscopy, however, is that the energy of a molecule can change not only as a result of electronic transitions but also because it can make transitions between its rotational and vibrational states. Molecular spectra are more complicated than atomic spectra but they contain more information, including electronic energy levels, bond lengths, bond angles, and bond strength. Molecular spectroscopy is also used to analyse materials and to monitor changing concentrations in kinetic studies (Section 10.1).

As in the discussion of atomic spectra, the frequency of a photon emitted or absorbed is given by the Bohr frequency condition (Section 13.1):

$$h\nu = |E_1 - E_2| \qquad \text{Bohr frequency condition} \qquad (19.1)$$

Here E_1 and E_2 are the energies of the two states between which the transition occurs and h is Planck's constant. This relation is often expressed in terms of the wavelength, λ (lambda), of the radiation by using the relation (as explained in Foundations 0.13)

$$\lambda = \frac{c}{\nu} \qquad \text{Wavelength} \qquad (19.2a)$$

where c is the speed of light, or in terms of the wavenumber, $\tilde{\nu}$ (nu tilde):

$$\tilde{\nu} = \frac{1}{\lambda} = \frac{\nu}{c} \qquad \text{Wavenumber} \qquad (19.2b)$$

The units of wavenumber are almost always chosen as reciprocal centimetres (cm^{-1}), so we can picture the wavenumber of radiation as the number of complete wavelengths per centimetre. In Foundations, Fig. 0.8 summarizes the various regions of the electromagnetic spectrum.

A note on good practice You will often hear people speak of 'a frequency as so many wavenumbers'. This usage is doubly wrong. First, *frequency* and *wavenumber* are two distinct physical observables with different units, and should be distinguished. Second, 'wavenumber' is not a unit, it is an observable with the dimensions of 1/length and commonly reported in reciprocal centimetres (cm^{-1}).

In this chapter, which explores rotational and vibrational spectroscopy, we first establish the allowed rotational and vibrational energies of molecules and then discuss the transitions between the various states.

Rotational spectroscopy

Very little energy is needed to change the state of rotation of a molecule, and the electromagnetic radiation emitted or absorbed lies in the microwave region, with wavelengths of the order of 0.1–1 cm and frequencies close to 10 GHz. The rotational spectroscopy of gas-phase samples is therefore also known as **microwave spectroscopy**. To achieve sufficient absorption, the path lengths of gaseous samples must be very long, of the order of metres. Long path lengths are achieved by multiple passage of the beam between two parallel mirrors at each end of the sample cavity. A *klystron* (which is also used in radar installations and microwave ovens) or, more commonly now, a semiconductor device known as a *Gunn diode*, is used to generate microwaves. A microwave detector is typically a *crystal diode* consisting of a tungsten tip in contact with a semiconductor, such as germanium, silicon, or gallium arsenide. The intensity of the radiation arriving at the detector is usually modulated, because alternating signals are easier to amplify than a steady signal. In most cases the beam is chopped by a rotating shutter. Gaseous samples are essential for rotational (microwave) spectroscopy, for in that phase molecules rotate freely.

19.1 The rotational energy levels of molecules

To a first approximation, the rotational states of molecules are based on a model system called a **rigid rotor**, a body that is not distorted by the stress of rotation. The simplest type of rigid rotor is called a **linear rotor**, and corresponds to a linear molecule, such as HCl, CO_2, or HC≡CH, that is supposed

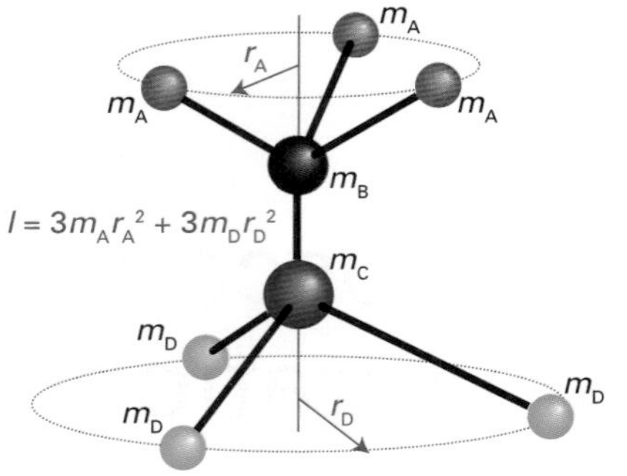

Fig. 19.1 The definition of moment of inertia. In this molecule there are three identical atoms attached to the B atom and three different but mutually identical atoms attached to the C atom. In this example, the centre of mass lies on an axis passing through the B and C atom, and the perpendicular distances are measured from this axis.

not to be able to bend or stretch under the stress of rotation. As shown in the following Derivation, the energies of a linear rotor are found to be

$$E_J = hBJ(J+1) \quad J = 0, 1, 2, \ldots \qquad \text{Linear rotor} \quad \text{Rotational energy levels} \qquad (19.3)$$

where J is the **rotational quantum number**. The constant B (a frequency, with the units hertz, Hz, with 1 Hz denoting 1 cycle per second) is called the **rotational constant** of the molecule, and is defined as

$$B = \frac{\hbar}{4\pi I} \qquad \text{Rotational constant} \qquad (19.4)$$

where I is the **moment of inertia** of the molecule (this property was introduced in Foundations 0.5). The moment of inertia of a molecule is the mass of each atom multiplied by the square of its distance from the axis of rotation (Fig. 19.1):

$$I = \sum_i m_i r_i^2 \qquad \text{Moment of inertia} \qquad (19.5)$$

A note on good practice To calculate the moment of inertia precisely, we need to specify the nuclide. Also, the mass to use is the actual atomic mass, not the element's molar mass. Nuclide masses are reported as multiples of the atomic mass constant (a constant, not a unit), so we write, for instance, $16.00m_u$ not $16.00\ m_u$.

Derivation 19.1

Energy levels of a linear rotor

The starting point for this derivation draws on the concepts of energy and angular momentum introduced in Foundations 0.5 and 0.9. A linear rotor can rotate about two perpendicular axes fixed in the molecule with the *x*- and *y*-axes perpendicular to the molecular axis and the z-axis lying along it. The moment of inertia around the x- and y-axes is I (the same in each case) and is zero about the z-axis (because all the atoms lie on that axis); so the total rotational kinetic energy, and therefore the total energy, is the sum of two contributions:

$$E = \tfrac{1}{2}I\omega_x^2 + \tfrac{1}{2}I\omega_y^2$$

In terms of the angular momentum $\mathcal{J}_q = I\omega_q$ around each perpendicular axis,

$$E = \frac{\mathcal{J}_x^2}{2I} + \frac{\mathcal{J}_y^2}{2I} = \frac{\mathcal{J}^2}{2I}$$

where $\mathcal{J}$ is the total angular momentum (there is no z-component for a linear rotor).

Now make the transition from classical to quantum mechanics. According to quantum mechanics, the square of the magnitude of angular momentum is $J(J+1)\hbar^2$, with the quantum number $J = 0, 1, 2, \ldots$ (Section 12.8). It follows that the quantum mechanical expression for the energy of a linear rotor is

$$E = J(J+1)\frac{\hbar^2}{2I} \overbrace{=}^{\hbar=h/2\pi} J(J+1)\frac{\hbar}{2I}\times\frac{h}{2\pi} = h\overbrace{\frac{\hbar}{4\pi I}}^{B}J(J+1)$$

This expression is the same as eqn 19.3 when we recognize that $B = \hbar/4\pi I$.

The moment of inertia plays a role in rotation analogous to the role played by mass in translation. A body with a high moment of inertia (like that of a flywheel or a heavy molecule) undergoes only a small rotational acceleration when a twisting force (a torque) is applied, but a body with a small moment of inertia undergoes a large acceleration when subjected to the same torque. Table 19.1 gives the expressions for the moments of inertia of various types of molecules in terms of the masses of their atoms and their bond lengths and bond angles.

Example 19.1

Evaluating a rotational constant

Evaluate the rotational constant of a $^1H^{35}Cl$ molecule.

Strategy Begin by evaluating the moment of inertia of the molecule by using the appropriate expression from Table 19.1. Then convert the moment of inertia to the rotational constant by using eqn 19.4.

Solution The masses of the two atoms are $1.008m_u$ and $34.969m_u$, for ^1H and ^{35}Cl, respectively; the equilibrium bond length is 127.4 pm. From Table 19.1 the value of μ is

$$\mu = \frac{\overbrace{m_A}^{m(^1\mathrm{H})}\overbrace{m_B}^{m(^{35}\mathrm{Cl})}}{\underbrace{m}_{m(^1\mathrm{H})+m(^{35}\mathrm{Cl})}} = \frac{(1.008m_u)\times(34.969m_u)}{1.008m_u + 34.969m_u}$$

$$= \frac{1.008\times 34.969}{1.008+34.969}m_u = 0.9798m_u$$

Therefore, the moment of inertia is

$$I = \mu R^2 = \overbrace{0.9798\times\overbrace{(1.660\,54\times10^{-27}\ \mathrm{kg})}^{m_u}}^{\mu}\times\overbrace{(1.274\times10^{-10}\ \mathrm{m})^2}^{R^2}$$

$$= 2.6407\times10^{-47}\ \mathrm{kg\ m^2}$$

It then follows from eqn 19.4 that the rotational constant of $^1H^{35}Cl$ is

$$B = \frac{\hbar}{4\pi I} = \frac{\overbrace{1.054\,57\times10^{-34}\ \mathrm{J\ s}}^{\hbar}}{\underbrace{4\pi\times(2.6407\times10^{-47}\ \mathrm{kg\ m^2})}_{I}}$$

$$= 3.1779\times10^{11}\frac{\overbrace{\mathrm{J}}^{1\ \mathrm{kg\ m^2\ s^{-2}}}\ \mathrm{s}}{\mathrm{kg\ m^2}} = 3.1779\times10^{11}\frac{\mathrm{kg\ m^2\ s^{-2}\ s}}{\mathrm{kg\ m^2}}$$

$$= 3.1779\times10^{11}\ \overbrace{\mathrm{s^{-1}}}^{\mathrm{Hz}}$$

This value corresponds to 3.1779×10^{11} Hz (or 0.317 79 THz). Expressed as a wavenumber, when it is denoted $\tilde{B}$ ('B tilde') with $\tilde{B} = B/c$, it is 66.604 cm^{-1}.

Self-test 19.1

Evaluate the rotational constant of $^2H^{35}Cl$ ($m(^2H) = 2.0141m_u$).

Answer: 0.163 50 THz

Figure 19.2 shows the energy levels predicted by eqn 19.3; note that the separation of neighbouring levels increases with J. Note also that, because J may be 0 (Section 12.8), the lowest possible energy is $E_0 = 0$: there is no zero-point rotational energy for molecules.

Molecules are not really *rigid* rotors: they distort under the stress of rotation. As their bond lengths increase, their energy levels become slightly closer together. This effect is taken into account by supposing that eqn 19.3 can be modified to

$$E_J = hBJ(J+1) - hDJ^2(J+1)^2 \qquad \text{Linear rotor} \quad \text{Centrifugal distortion} \qquad (19.6)$$

The parameter D is the **centrifugal distortion constant.** It is large when the bond is easily stretched,

Table 19.1
*Moments of inertia**

1. Diatomic molecules

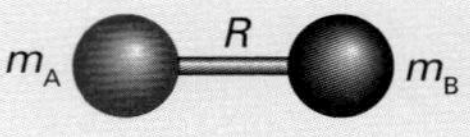

$$I = \mu R^2 \qquad \mu = \frac{m_A m_B}{m}$$

2. Triatomic linear rotors

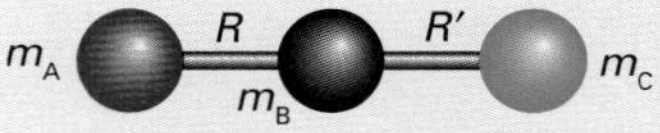

$$I = m_A R^2 + m_C R'^2 - \frac{(m_A R - m_C R')^2}{m}$$

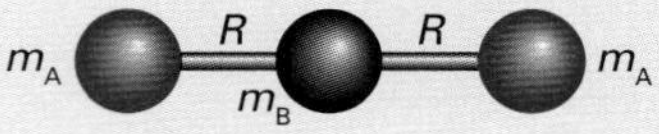

$$I = 2m_A R^2$$

3. Symmetric rotors

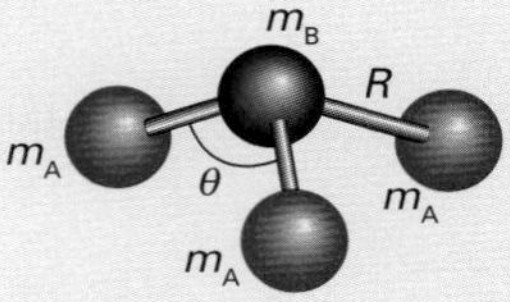

$$I_{\parallel} = 2m_A(1 - \cos\theta)R^2$$

$$I_{\perp} = m_A(1 - \cos\theta)R^2 + \frac{m_A}{m}(m_B + m_C)(1 + 2\cos\theta)R^2$$

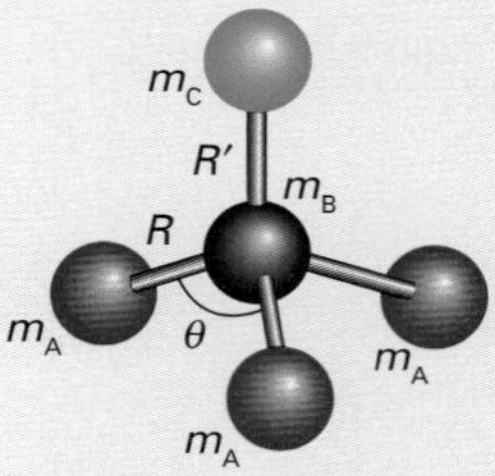

$$I_{\parallel} = 2m_A(1 - \cos\theta)R^2$$

$$I_{\perp} = m_A(1 - \cos\theta)R^2 + \frac{m_A}{m}(m_B + m_C)(1 + 2\cos\theta)R^2$$

$$+ \frac{m_C}{m}(3m_A + m_B)R'^2 + 6m_A\{{}^1\!/_3(1 + 2\cos\theta)\}^{1/2}RR'$$

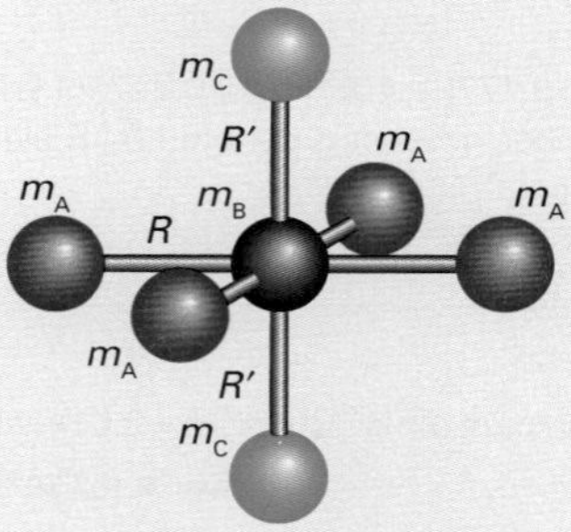

$$I_{\parallel} = 4m_A R^2$$

$$I_{\perp} = 2m_A R^2 + 2m_C R'^2$$

4. Spherical rotor

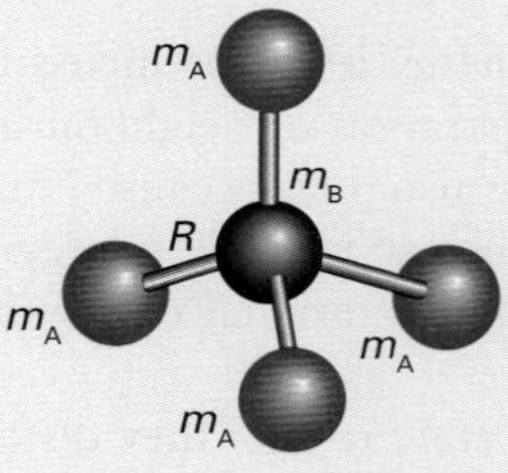

$$I = {}^8\!/_3 m_A R^2$$

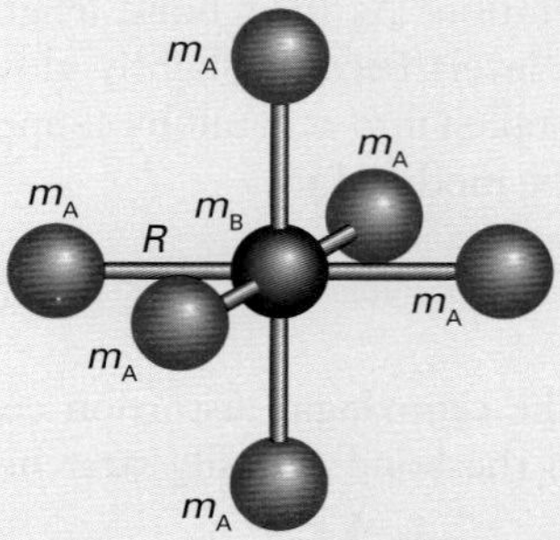

$$I = 4m_A R^2$$

* In each case, *m* is the total mass of the molecule.

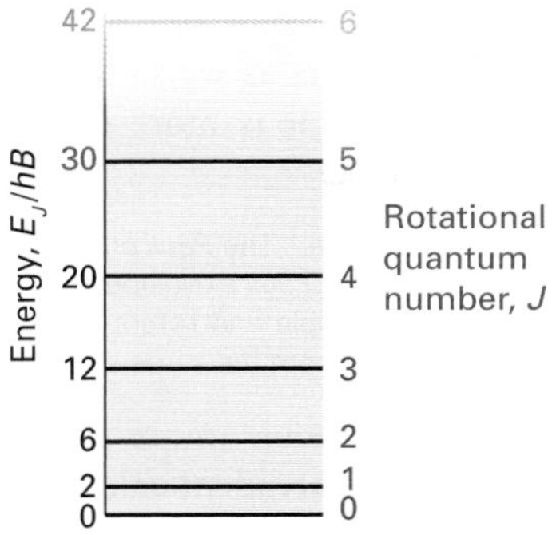

Fig. 19.2 The energy levels of a linear rigid rotor as multiples of hB.

and so its magnitude is related to the force constants of bonds, a measure of their rigidity (Section 19.7). Later, when we consider molecular vibration, we shall see that B depends on the vibrational state of the molecule, because the moment of inertia of the molecule changes as it vibrates.

A number of nonlinear molecules, all of which can rotate around three axes, can be modelled as a **symmetric rotor**, a rigid rotor in which the moments of inertia about two axes are the same but different from a third (and all three are nonzero). The formal criterion of a molecule being a symmetric rotor is that it has an axis of threefold or higher symmetry. An example of a symmetric rotor is ammonia, NH_3, and another is phosphorus pentachloride, PCl_5 (Fig. 19.3). As shown in the following Derivation, the rotational energy levels of a symmetric rotor are

$$E_{J,K} = hBJ(J+1) + h(A-B)K^2$$
$$J = 0, 1, 2, \ldots \qquad \text{Symmetric rotor} \qquad (19.7)$$
$$K = J, J-1, \ldots, -J$$

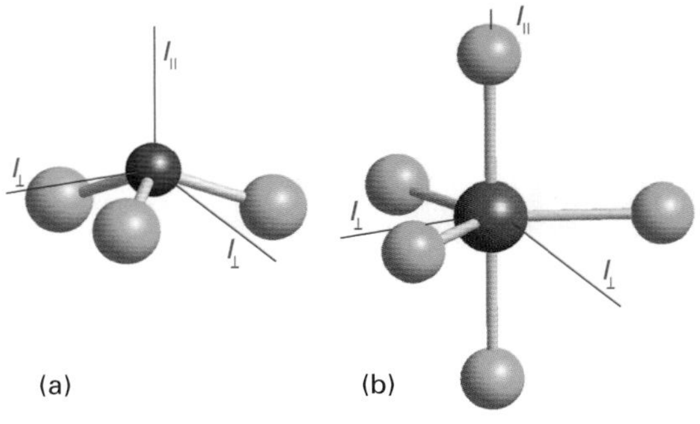

Fig. 19.3 The two different moments of inertia of (a) a trigonal pyramidal molecule and (b) a trigonal bipyramidal molecule.

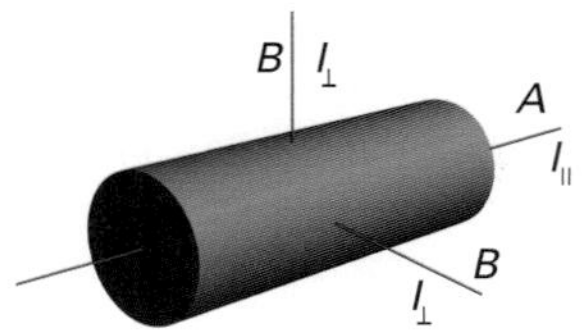

Fig. 19.4 The two rotational constants of a symmetric rotor, which are inversely proportional to the moments of inertia parallel and perpendicular to the axis of the molecule.

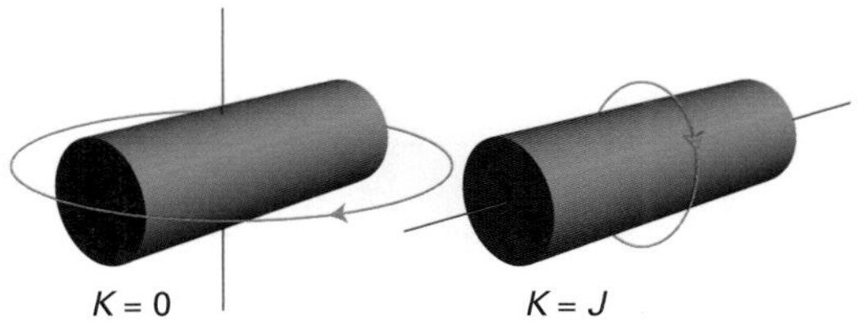

Fig. 19.5 When $K = 0$ for a symmetric rotor, the entire motion of the molecule is around an axis perpendicular to the symmetry axis of the rotor. When the value of $|K|$ is close to J, almost all the motion is around the symmetry axis.

The rotational constants A and B are inversely proportional to the moments of inertia parallel and perpendicular to the axis of the molecule (Fig. 19.4):

$$A = \frac{\hbar}{4\pi I_\parallel} \qquad B = \frac{\hbar}{4\pi I_\perp} \qquad \text{Rotational constants} \qquad (19.8)$$

The quantum number K tells us, through $K\hbar$, the component of angular momentum around the molecular axis (Fig. 19.5). When $K = 0$, the molecule is rotating end-over-end and not at all around its own axis. When $K = \pm J$ (the greatest values in its range), the molecule is rotating mainly about its axis. Intermediate values of K correspond to a combination of the two modes of rotation.

Derivation 19.2

Energy levels of a symmetric rotor

The starting point for this derivation is a generalization of the expression for the total energy of a linear rotor to

$$E = \frac{\mathcal{J}_x^2}{2I_\perp} + \frac{\mathcal{J}_y^2}{2I_\perp} + \frac{\mathcal{J}_z^2}{2I_\parallel}$$

It is convenient to write this expression in terms of the magnitude of the angular momentum $\mathcal{J}^2 = \mathcal{J}_x^2 + \mathcal{J}_y^2 + \mathcal{J}_z^2$:

$$E = \frac{\mathcal{J}_x^2}{2I_\perp} + \frac{\mathcal{J}_y^2}{2I_\perp} + \frac{\mathcal{J}_z^2}{2I_\parallel} + \overbrace{\frac{\mathcal{J}_z^2}{2I_\perp} - \frac{\mathcal{J}_z^2}{2I_\perp}}^{0} = \frac{\mathcal{J}^2}{2I_\perp} + \left(\frac{1}{2I_\parallel} - \frac{1}{2I_\perp}\right)\mathcal{J}_z^2$$

As in the previous Derivation, at this point we make the transition from classical to quantum mechanics. According to quantum mechanics, the square of the magnitude of angular momentum is $J(J+1)\hbar^2$, with $J = 0, 1, 2, \ldots$ and any component (such as $\mathcal{J}_z$) is limited to the values $K\hbar$ with $K = J, J-1, \ldots, -J$. By convention, the quantum number K plays the role of M_J for the component on an internally defined axis while M_J is preserved for the projection of the angular momentum on to an externally defined z-axis. It follows that the quantum mechanical expression for the energy of a symmetric rotor is

$$E = \overbrace{\frac{J(J+1)\hbar^2}{2I_\perp}}^{\mathcal{J}^2} + \left(\frac{1}{2I_\parallel} - \frac{1}{2I_\perp}\right)\overbrace{K^2\hbar^2}^{\mathcal{J}_z^2}$$

Finally, with A and B defined as in eqn 19.8, we obtain eqn 19.7.

A special case of a symmetric rotor is a **spherical rotor**, a rigid body with three equal moments of inertia (like a sphere). Tetrahedral, octahedral, and icosahedral molecules (CH_4, SF_6, and C_{60}, for instance) are spherical rotors. Their energy levels are very simple: when $I_\parallel = I_\perp$, the rotational constants A and B are equal and eqn 19.7 simplifies to eqn 19.3.

19.2 Forbidden and allowed rotational states

Not all the rotational states with $J = 0, 1, 2, \ldots$ of symmetrical molecules, like H_2 and CO_2, are permitted. The elimination of certain states is a consequence of the Pauli exclusion principle which, as we saw in Chapter 13, also forbids the occurrence of certain atomic states (such as those with three electrons in one orbital, or two electrons with the same spin in the same orbital). The restrictions on the permitted rotational states due to the Pauli principle can be traced to the effect of nuclear spin called **nuclear statistics.**

To understand how the Pauli exclusion principle excludes certain rotational states, we need to express the principle in a more general way than in Further information 13.1, which referred only to electrons. The most general form of the Pauli principle states:

> When any two indistinguishable fermions are interchanged, the wavefunction must change sign; when any two indistinguishable bosons are interchanged, the wavefunction must remain the same.

(Bosons are particles with integral spin quantum number including 0; fermions are particles with half-integral spin quantum number; Section 13.6.) Thus, if A and B are indistinguishable particles, then

For fermions: $\psi(B,A) = -\psi(A,B)$

For bosons: $\psi(B,A) = \psi(A,B)$

The 'fermion' part of this principle implies the Pauli *exclusion* principle, as we saw in Chapter 13. However, in this form it is more general and has wider implications.

A note on good practice The *Pauli principle* is the general statement given here. The *Pauli exclusion principle* is a consequence of the Pauli principle and refers to the exclusion of more than two electrons from the same state.

Consider a CO_2 molecule (more precisely, a CO_2 in which both O atoms are identical, as in $^{16}OC^{16}O$), which we denote O_ACO_B. When the molecule rotates through 180°, it becomes O_BCO_A, with the two O atoms interchanged. The nuclear spin quantum number of oxygen-16 is zero, so it is a boson, and therefore the wavefunction must remain unchanged by this interchange. However, when *any* molecule is rotated through 180°, its wavefunction changes by a factor of $(-1)^J$. To see why that is so, we have drawn the first few wavefunctions for a particle travelling on a ring in Fig. 19.6, and we see that a rotation of 180° leaves wavefunctions with $J=0, 2, \ldots$ unchanged but changes the sign of those with $J = 1, 3, \ldots$. The only way for the two requirements (that the wavefunction does not change sign and the fact that it changes by a factor of $(-1)^J$) to be consistent is for J to be restricted to even values. That is, a CO_2 molecule can exist only in the rotational states with $J = 0, 2, 4, \ldots$.

The analysis of the implications of nuclear statistics is more complex for molecules in which the nuclei have nonzero spin (which includes H_2, with its spin-½ nuclei) because the permitted rotational states depend on the relative orientation of the nuclear spins. However, the results can be expressed quite simply:

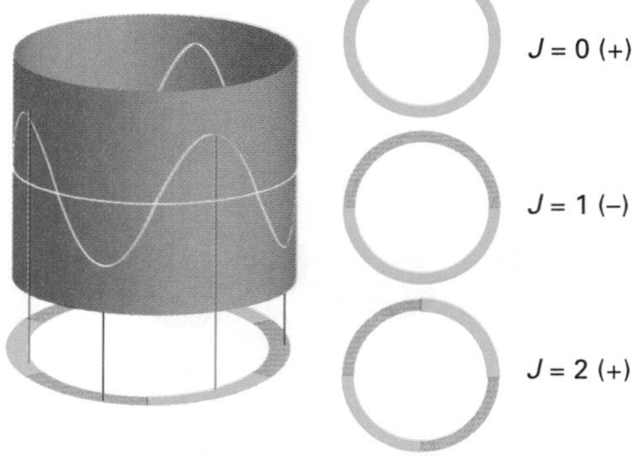

Fig. 19.6 The phases of the wavefunctions of a particle on a ring for the first few states: note that the parity of the wavefunction (its behaviour under inversion through the centre of the ring) is even, odd, even, ...

$$\frac{\text{Number of ways of achieving odd } J}{\text{Number of ways of achieving even } J} \quad \text{Nuclear statistics} \quad (19.9)$$

$$= \begin{cases} (I+1)/I \text{ for half integral spin nuclei} \\ I/(I+1) \text{ for integral spin nuclei} \end{cases}$$

where I is the nuclear spin quantum number.

Brief Illustration 19.1 Nuclear spin statistics

For H_2, with its spin-½ nuclei,

$$\frac{\text{Number of ways of achieving odd } J}{\text{Number of ways of achieving even } J} = \frac{I+1}{I}$$

$$= \frac{1/2+1}{1/2} = 3$$

There are therefore three times as many ways of achieving rotational levels with odd J than with even J. The levels with even J correspond to molecules with parallel nuclear spins, which are called *ortho*-hydrogen; levels with odd J correspond to paired nuclear spins and are called *para*-hydrogen. Different relative nuclear spin orientations change into one another only very slowly, so an H_2 molecule with parallel nuclear spins remains distinct from one with paired nuclear spins for long periods. The two forms of hydrogen can be separated by physical techniques, and stored.

Self-test 19.2

Determine the ratio of the number of ways of achieving rotational levels with odd and even J for D_2 where D is deuterium, ^{2}H, for which $I = 1$.

Answer: 1:2

19.3 Populations at thermal equilibrium

Because the rotational states of molecules are close together in energy, we can expect many states to be occupied at ordinary temperatures. However, we have to take into account the degeneracy of the rotational levels, because although a given *state* may have a low population, there may be many states of the same energy, and the total population of an *energy level* may be quite large.

To avoid the additional complication of nuclear spin statistics, we consider only linear molecules that do not possess a centre of symmetry, such as HCl and OCS; there is then no restriction on which states are allowed by the Pauli principle. The angular momentum of the molecule may have $2J+1$ different orientations with respect to an external axis, each designated by the value of the quantum number

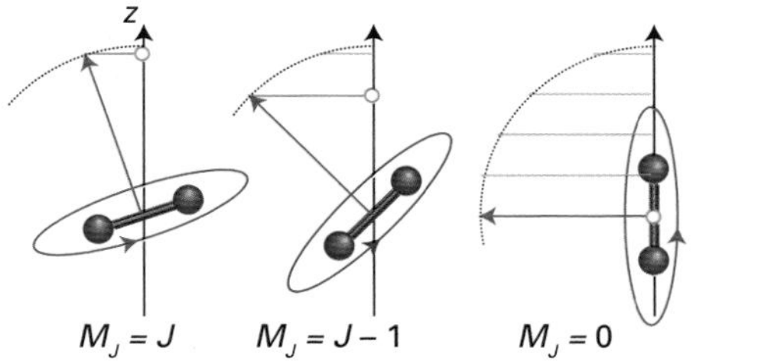

Fig. 19.7 The significance of the quantum number M_J (in this case, for $J = 4$): it indicates the orientation of the molecular rotational angular momentum with respect to an external axis.

$M_J = J, J-1, \ldots, -J$ (Fig. 19.7, just as in atoms, where there are $2l+1$ orientations of the orbital angular momentum, one corresponding to each permitted value of m_l). The energy of the molecule is independent of its plane of rotation, so all $2J+1$ states have the same energy. There are therefore $(2J+1)$ states of the same energy for each J rotational level. As explained in Foundations 0.11, each *state* (not level) has a population that is proportional to the Boltzmann factor, $e^{-E_J/kT}$, in this case with $E_J = hBJ(J+1)$. To apply the Boltzmann distribution to levels rather than states, we must multiply by the number of states in each level, $g_J = 2J+1$. The total population, P_J, of a given level consisting of $2J+1$ individual states, relative to that of the lowest level with $J = 0$ and its one state ($M_J = 0$), is thus

$$\frac{P_J}{P_0} = (2J+1)e^{-hBJ(J+1)/kT} \quad \text{Linear rotor} \quad \text{Boltzmann population} \quad (19.10)$$

Figure 19.8 shows how this population varies with J. As shown in the following Derivation, it passes through a maximum at an integer value of J close to

$$J_{\max} = \left(\frac{kT}{2hB}\right)^{1/2} - \frac{1}{2} \quad \text{Linear rotor} \quad \text{Maximally populated level} \quad (19.11)$$

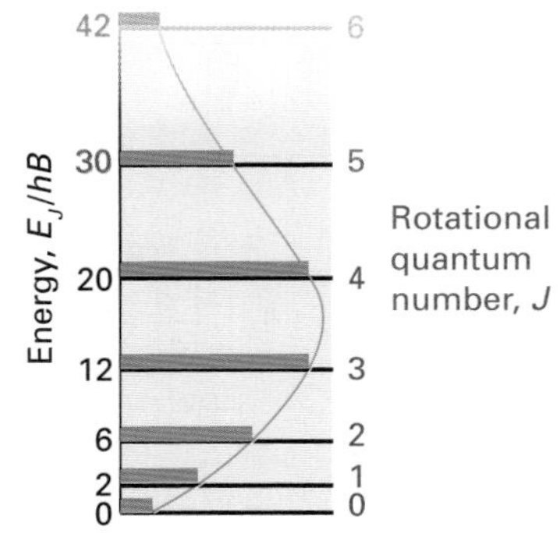

Fig. 19.8 The thermal equilibrium relative populations of the rotational energy levels of a linear rotor.

For a typical linear molecule (for example, OCS, with $B = 6$ GHz) at room temperature, $J_{max} = 22$. Broadly speaking, then, the absorption spectrum of the molecule should show a similar distribution of intensities.

Brief illustration 19.2 The most populated level

In Example 19.1 we established that $B = 3.1779 \times 10^{11}$ Hz for $^1H^{35}Cl$. Therefore, the most populated rotational energy level at 298 K is at an integer value of J close to

$$J_{max} = \left\{\frac{\overbrace{(1.381\times10^{-23}\,\mathrm{J\,K^{-1}})}^{k}\times\overbrace{(298\,\mathrm{K})}^{T}}{2\times\underbrace{(6.626\times10^{-34}\,\mathrm{J\,s})}_{h}\times\underbrace{(3.1779\times10^{11}\,\mathrm{Hz})}_{B}}\right\}^{1/2} - \frac{1}{2}$$

$$= 2.6$$

The closest integer is 3, so the level with $J = 3$ (with its 7 states) is the most populated at 298 K. The molecule has a low moment of inertia as virtually the whole of the rotational motion is that of the H atom around a nearly stationary Cl atom, so its energy levels are widely separated and only a few are thermally accessible at low temperatures.

Derivation 19.3

The most populated level

Here we need to find the value of J for which P_J is a maximum. To proceed, we recall from The chemist's toolkit 1.3 that to find the value of x corresponding to the extremum (maximum or minimum) of any function $f(x)$, we differentiate the function, set the result equal to zero, and solve the resulting equation for x. Applying this procedure to eqn 19.10 with J treated, at this stage, as a continuous variable, we obtain

$$\frac{\mathrm{d}}{\mathrm{d}J}\overbrace{(2J+1)}^{f}\overbrace{\mathrm{e}^{-hBJ(J+1)/kT}}^{g}$$

$$\overset{d(fg)/dx=(df/dx)g+f(dg/dx)}{=} \overbrace{\underbrace{\left\{\frac{\mathrm{d}}{\mathrm{d}J}(2J+1)\right\}}_{2}\mathrm{e}^{-hBJ(J+1)/kT}}^{(df/dx)g}$$

$$+\overbrace{(2J+1)\underbrace{\left\{\frac{\mathrm{d}}{\mathrm{d}J}\mathrm{e}^{-hBJ(J+1)/kT}\right\}}_{-\{hB(2J+1)/kT\}\mathrm{e}^{-hBJ(J+1)/kT}}}^{f(dg/dx)}$$

$$= 2\mathrm{e}^{-hBJ(J+1)/kT} + (2J+1)\left\{-\frac{hB(2J+1)}{kT}\mathrm{e}^{-hBJ(J+1)/kT}\right\}$$

$$= \left\{2 - \frac{hB(2J+1)^2}{kT}\right\}\mathrm{e}^{-hBJ(J+1)/kT}$$

This expression is equal to zero when the term multiplying the (blue) exponential function is zero. Therefore, after setting $J = J_{max}$, we need to solve

$$2 - \frac{hB(2J_{max}+1)^2}{kT} = 0$$

which gives eqn 19.11. This derivation has treated J as a continuous variable, whereas in fact it is confined to integer values. Therefore, the actual value of J corresponding to the maximum population is interpreted as the integer lying closest to the calculated J_{max}.

19.4 Rotational transitions: microwave spectroscopy

Whether or not a transition can be driven by or drive the oscillations of the surrounding electromagnetic field depends on a quantity called the **transition dipole moment.** This quantity is a measure of the dipole moment associated with the shift of electric charge that accompanies a transition (Fig. 19.9).[1] The intensity of the transition is proportional to the square of the associated transition dipole moment. A large transition dipole moment indicates that the transition gives a strong 'thump' to the electromagnetic field, and conversely that the electromagnetic field interacts strongly with the molecule. A **selection rule** is a statement about when a transition dipole may be nonzero. There are two parts to a selection rule:

- A **gross selection rule** specifies the general features a molecule must have if it is to have a spectrum of a given kind.

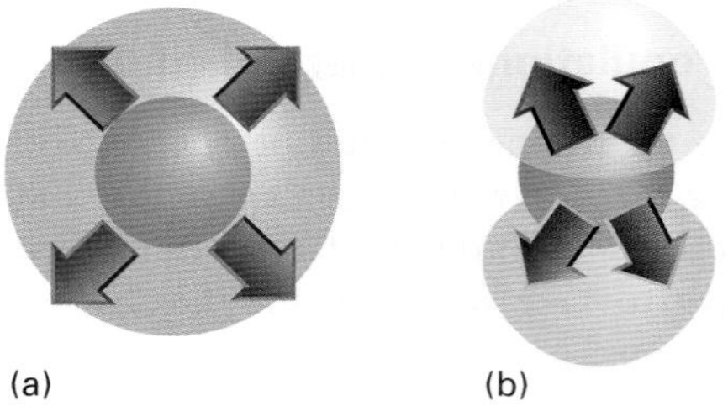

Fig. 19.9 The transition moment is a measure of the magnitude of the shift in charge during a transition. (a) A spherical redistribution of charge as in this transition has no associated dipole moment, and does not give rise to electromagnetic radiation. (b) This redistribution of charge has an associated dipole moment.

[1] To be (nearly) precise, the transition dipole moment for the transition *initial state* → *final state* is the value of the integral $\int\psi_{final}\mu\psi_{initial}\mathrm{d}\tau$; see our *Physical Chemistry* (2010) for more information.

- A **specific selection rule** specifies what changes in quantum numbers may occur.

A transition that is permitted by a specific selection rule is classified as **allowed.** Transitions that are disallowed by a specific selection rule are called **forbidden.** Forbidden transitions sometimes occur weakly because the selection rule is based on an approximation that turns out to be slightly invalid.

The gross selection rule for rotational transitions is that *the molecule must be polar*. The classical basis of this rule is that a stationary observer watching a rotating polar molecule sees its partial charges moving backwards and forwards and their motion shakes the electromagnetic field into oscillation (Fig. 19.10). Because the molecule must be polar, it follows that tetrahedral (CH_4, for instance), octahedral (SF_6), symmetric linear (CO_2), and homonuclear diatomic (H_2) molecules do not have rotational spectra. On the other hand, heteronuclear diatomic (HCl) and less symmetrical polar polyatomic molecules (NH_3) are polar and do have rotational spectra. We say that polar molecules are **rotationally active** whereas nonpolar molecules are **rotationally inactive.**

The specific selection rules for rotational transitions are

$$\Delta J = \pm 1 \qquad \Delta K = 0 \qquad \text{Rotational selection rules} \qquad (19.12)$$

The first of these selection rules can be traced, like the rule $\Delta l = \pm 1$ for atoms (Section 13.7), to the conservation of angular momentum when a photon is absorbed or created. A photon is a spin-1 particle, and when one is absorbed or created the angular momentum of the molecule must change by a compensating amount. Because J is a measure of the angular momentum of the molecule, J can change only by ± 1 (for pure rotational transitions, $\Delta J = +1$ corresponds to absorption, $\Delta J = -1$ to emission). The second selection rule ($\Delta K = 0$; that is, the quantum number K may not change) can be traced to the fact that the dipole moment of a polar molecule does not move when a molecule rotates around its symmetry axis (think of NH_3 rotating around its threefold axis). As a result, there can be no acceleration or deceleration of the rotation of the molecule about that axis by the absorption or emission of electromagnetic radiation.

When a rigid, unsymmetrical linear molecule changes its rotational quantum number from J to $J+1$ in an absorption, the change in rotational energy of the molecule is

$$\begin{aligned}\Delta E = E_{J+1} - E_J &\overset{E_J = hBJ(J+1)}{=} hB(J+1)(J+2) - hBJ(J+1)\\ &= hB\{J^2 + 3J + 2 - (J^2 + J)\}\\ &= 2hB(J+1)\end{aligned}$$

The same expression applies to a symmetric rotor because K does not change in a transition. The frequency of the radiation absorbed in a transition starting from the level J is therefore

$$\nu_J = 2B(J+1) \qquad \text{Rigid rotor} \quad \text{Rotational transition frequencies} \qquad (19.13a)$$

and the absorption lines occur at $2B, 4B, 6B, \ldots$. The intensity distribution will be like that in Fig. 19.11 with a maximum intensity at $\nu_{J_{max}}$, with J_{max} given by eqn 19.11. A rotational spectrum of a polar linear molecule (HCl) and of a polar symmetric rotor (NH_3) therefore consists of a series of lines at frequencies separated by $2B$.

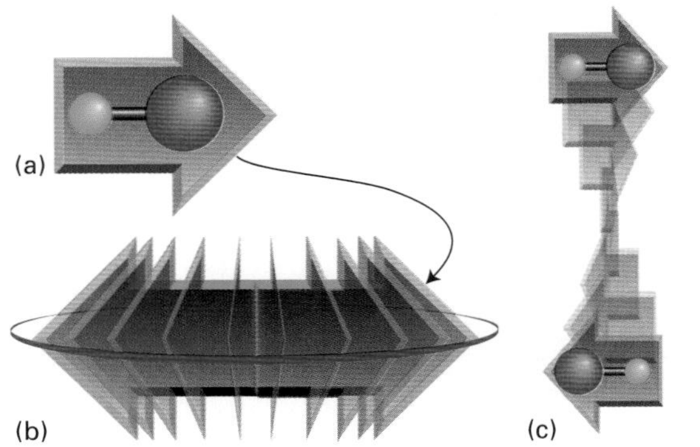

Fig. 19.10 To an external observer, (a) a rotating polar molecule has (b) an electric dipole (the arrow) that (c) appears to oscillate. This oscillating dipole can interact with the electromagnetic field.

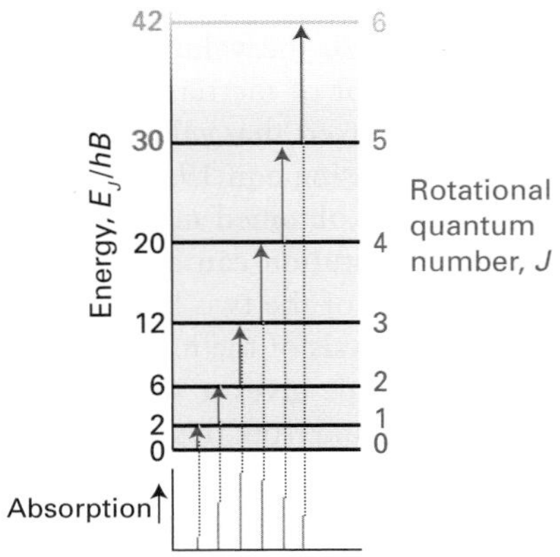

Fig. 19.11 The allowed rotational transitions (shown as absorptions) for a linear molecule.

Brief illustration 19.3 Rotational transition frequencies

In Example 19.1 we calculated the rotational constant of $^1H^{35}Cl$ as 3.1779×10^{11} Hz (or 317.79 GHz); therefore the rotational spectrum of this molecule will consist of a series of lines spaced by 635.6 GHz and therefore at the frequencies 635.6 GHz, 1271.2 GHz, 1.906.8 GHz, ... The wavelength of the first of these lines is 0.472 mm.

Self-test 19.3

What is the frequency and wavelength of the $J = 1 \leftarrow 0$ transition in the $^2H^{35}Cl$ molecule? The mass of ^2H is $2.014m_u$. Before commencing the calculation, decide whether the frequency should be higher or lower than for $^1H^{35}Cl$.

Answer: 327.0 GHz, 0.9167 mm

If centrifugal distortion is significant, we use eqn 19.6 in the same way, and find

$$\nu_J = 2B(J+1) - 4D(J+1)^3 \qquad \text{Non-rigid rotor} \quad \text{Rotational transition frequencies} \qquad (19.13b)$$

Now, because the second term subtracts from the first ever larger amounts as J increases, the lines converge as J increases. To find a way to determine B and D experimentally, we divide both sides by $J + 1$, to obtain

$$\overbrace{\frac{\nu_J}{J+1}}^{y} = \overbrace{2B}^{\text{Intercept}} \; \overbrace{-4D}^{+\,\text{Slope}} \times \overbrace{(J+1)^2}^{\times\, x} \qquad (19.14)$$

Therefore, by plotting the $\nu_J/(J + 1)$ against $(J + 1)^2$, we should get a straight line with intercept $2B$ and slope $-4D$ (see Exercise 19.17).

Once the separation between adjacent lines in a rotational spectrum of a molecule has been measured and converted it to B, the value of B can be used to obtain the moment of inertia $I_\perp$. For a diatomic molecule, we can convert that value to a value of the bond length, R, by using eqn 19.4. Highly accurate bond lengths can be obtained in this way. In some cases, isotopic substitution can help. A classic case is the determination of the two bond lengths in the molecule OCS. Analysis of the microwave spectrum of this linear molecule gives a single quantity, the rotational constant, and from this single quantity it is not possible to infer the two different bond lengths. However, by recording the absorption of the two isotopologues (molecules of different isotopic composition) $^{16}O^{12}C^{33}S$ and $^{16}O^{12}C^{34}S$ and assuming that isotopic substitution leaves the bond lengths unchanged, we get two pieces of information, the moment of inertia of each isotopomer, and it is now possible to infer the two bond lengths (see Exercise 19.18).

19.5 Linewidths

Spectral lines are not infinitely narrow. An important broadening process in gaseous samples is the **Doppler effect**, in which radiation is shifted in frequency when the source is moving towards or away from the observer (Fig. 19.12). Molecules reach high speeds in all directions in a gas, and a stationary observer, the spectrometer, detects the corresponding Doppler-shifted range of frequencies. Some molecules approach the observer, some move away; some move quickly, others slowly. The detected spectroscopic 'line' is the absorption or emission profile arising from all the resulting Doppler shifts. The profile reflects the Maxwell distribution of molecular speeds (Section 1.6) towards or away from the observer, and the outcome is that we observe a bell-shaped Gaussian curve (a curve of the form e^{-x^2}, Fig. 19.13, and The chemist's toolkit 1.2). When the temperature is T and the molar mass of the molecule is M, the width of the line at half its maximum height (the 'width at half-height') is

$$\delta\nu = \frac{2\nu}{c}\left(\frac{2RT\ln 2}{M}\right)^{1/2} \qquad \text{Width at half-height} \qquad (19.15)$$

which is best remembered as $\delta\nu \propto (T/M)^{1/2}$. The Doppler width increases with temperature because

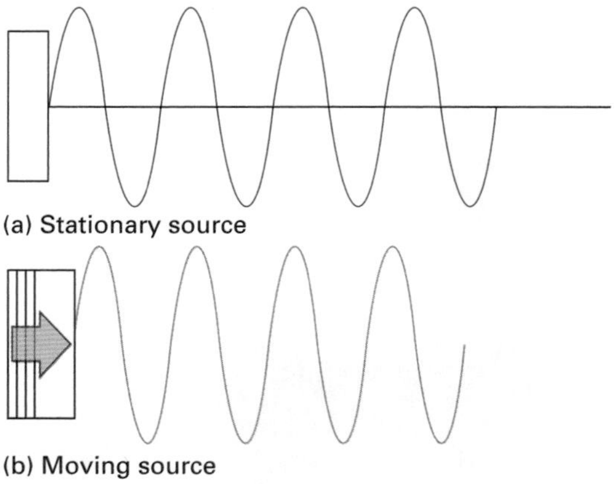

Fig. 19.12 The Doppler effect. (a) The radiation emitted by a stationary source. (b) When the same source moves towards the observer, the radiation appears to be shifted to higher frequencies. Similarly, a receding source shifts the radiation to lower frequencies.

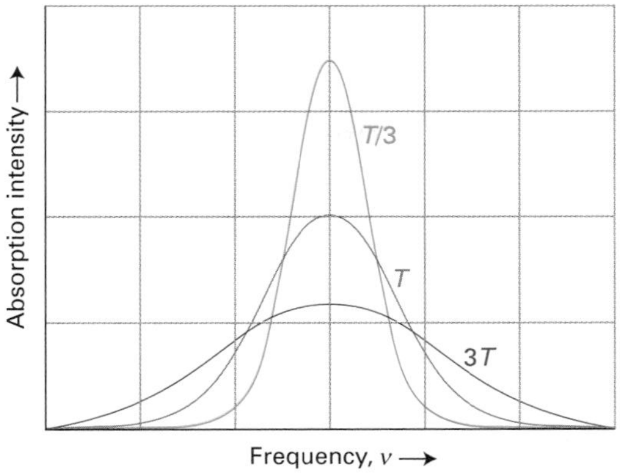

Fig. 19.13 The shape of a Doppler-broadened spectral line reflects the Maxwell distribution of speeds in the sample at the temperature of the experiment. Notice that the line broadens as the temperature is increased. The width at half-height is given by eqn 19.15.

the molecules acquire a wider range of speeds (recall Fig. 1.9). Therefore, to obtain spectra of maximum sharpness, it is best to work with cold gaseous samples.

Brief illustration 19.4 Doppler linewidth

The Doppler width of the $J = 1 \leftarrow 0$ transition of $^1H^{35}Cl$ (of molar mass 35.973 g mol^{-1}, corresponding to 3.5973×10^{-2} kg mol^{-1}) at 298 K is

$$\delta\nu = \frac{\overbrace{2\times(6.356\times10^{11}\ \mathrm{s^{-1}})}^{\nu}}{\underbrace{2.998\times10^{8}\ \mathrm{m\ s^{-1}}}_{c}} \times\left(\frac{2\times\overbrace{(8.3145\ \mathrm{J\ K^{-1}\ mol^{-1}})}^{R}\times\overbrace{(298\ \mathrm{K})}^{T}\times\ln 2}{\underbrace{3.5973\times10^{-2}\ \mathrm{kg\ mol^{-1}}}_{M}}\right)^{1/2}$$

$$= \frac{2\times6.356\times10^{11}}{2.998\times10^{8}} \times\left(\frac{2\times8.3145\times298\times\ln 2}{3.5973\times10^{-2}}\right)^{1/2}\overbrace{\frac{1}{\mathrm{m}}\left(\frac{\overbrace{\mathrm{kg\ m^2\ s^{-2}}}^{\mathrm{J}}\ \mathrm{mol^{-1}}}{\mathrm{kg\ mol^{-1}}}\right)^{1/2}}^{\mathrm{m\ s^{-1}}}$$

$$= 1.310 \times 10^{6}\ \mathrm{s^{-1}}$$

This value corresponds to a width of 1.310 MHz.

Self-test 19.4

Determine the Doppler width of the $J = 4 \leftarrow 3$ transition of $^{12}C^{16}O$, which is observed at 461.0 MHz, at 400 K.

Answer: 1.248 kHz

Another source of line broadening is the finite lifetime of the states involved in the transition. When the Schrödinger equation is solved for a system that is changing with time, it is found that the states of the system do not have precisely defined energies. If the time-constant for the decay of a state is τ (tau), which is called the **lifetime** of the state, then its energy levels are blurred by δE (corresponding to a frequency range of $\delta\nu = \delta E/h$), with

$$\delta E \approx \frac{\hbar}{\tau} \quad \text{or} \quad \delta\nu \approx \frac{1}{2\pi\tau} \qquad \text{Lifetime broadening} \qquad (19.16)$$

and where the decay of the state is assumed to be exponential and proportional to $e^{-t/\tau}$. We see that the shorter the lifetime of a state, the less well defined is its energy. The energy spread inherent to the states of systems that have finite lifetimes is called **lifetime broadening.** Only if τ is infinite can the energy of a state be specified exactly (with $\delta E = 0$). However, no excited state has an infinite lifetime; therefore, all states are subject to some lifetime broadening, and the shorter the lifetimes of the states involved in a transition, the broader the spectral lines.

Brief illustration 19.5 Lifetime broadening

For a transition from a state with lifetime 50 ps the broadening is

$$\delta\nu \approx \frac{1}{2\pi\times(5.0\times10^{-11}\ \mathrm{s})} = 3.2\times10^{9}\ \mathrm{s^{-1}}$$

This width corresponds to 3.2 GHz.

Self-test 19.5

The widths of lines in the spectrum of a short-lived excited state of NO_2 are due to lifetime broadening. Calculate the lifetime of the state that gives rise to a line with a lifetime-broadened width of 47 kHz.

Answer: 3.4 μs

A note on good practice Lifetime broadening is sometimes referred to as 'uncertainty broadening' because eqn 19.16 can be written as $\tau\delta E \approx \hbar$, which resembles the form of a Heisenberg uncertainty principle for energy and time. However, there are technical reasons for not regarding this expression as a true uncertainty principle (essentially that there is no operator for time in quantum mechanics) and the term 'lifetime broadening' is to be preferred.

Two processes are principally responsible for the finite lifetimes of excited states, and hence for the widths of transitions to or from them. The dominant one is **collisional deactivation**, which arises from collisions between molecules or with the walls of the container. If the collisional lifetime is τ_{col}, then the resulting collisional linewidth is $\delta E_{col} \approx \hbar/\tau_{col}$. In gases, the collisional lifetime can be lengthened, and the broadening—which in this case is also called *pressure broadening*—minimized, by working at low pressures. The second contribution is **spontaneous emission** (see Further information 20.2), the emission of radiation when an excited state collapses into a lower state. The rate of spontaneous emission depends on details of the wavefunctions of the excited and lower states. Because the rate of spontaneous emission cannot be changed (without changing the molecule), it is a natural limit to the lifetime of an excited state. The resulting lifetime broadening is the **natural linewidth** of the transition.

The natural linewidth of a transition cannot be changed by modifying the temperature or pressure. Natural linewidths depend strongly on the transition frequency ν (they increase as ν^3), so low-frequency transitions (such as the microwave transitions of rotational spectroscopy) have very small natural linewidths; for such transitions, collisional and Doppler line-broadening processes are dominant.

19.6 Rotational Raman spectra

In **Raman spectroscopy**, molecular energy levels are explored by examining the frequencies present in the radiation scattered by molecules. In a typical experiment, a monochromatic incident laser beam is passed through the sample and the radiation scattered from the front face of the sample is monitored. About 1 in 10^7 of the incident photons collide with the molecules, give up some of their energy, and emerge with a lower energy and therefore lower frequency. These scattered photons constitute the lower-frequency **Stokes radiation** from the sample. Other incident photons may collect energy from the molecules (if they are already excited), and emerge as higher-frequency **anti-Stokes radiation.** The component of radiation scattered into the forward direction without change of frequency is called **Rayleigh radiation.**

Lasers are used as the radiation sources in Raman spectrometers for two reasons. First, the shifts in frequency of the scattered radiation from the incident radiation are quite small, so highly monochromatic radiation from a laser is required if the shifts are to be observed. Second, the intensity of scattered radiation is low, so intense incident beams, such as those from a laser, are needed.

The gross selection rule for rotational Raman spectra is that the polarizability of *the molecule must be anisotropic*. We saw in Section 15.4 that the

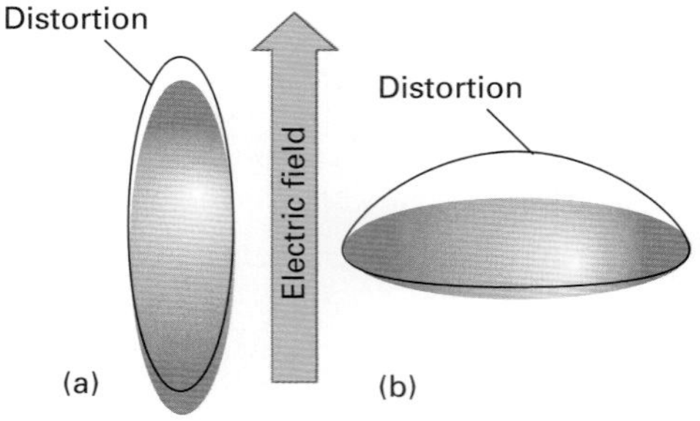

Fig. 19.14 The anisotropy of polarizability is depicted here by the different distortion induced by an electric field when the molecule is aligned (a) parallel to and (b) perpendicular to the field.

polarizability of a molecule is a measure of the extent to which an applied electric field can induce an electric dipole moment ($\mu^* = \alpha\mathcal{E}$). The *anisotropy* of this polarizability is its variation with the orientation of the molecule (Fig. 19.14). Tetrahedral (CH_4), octahedral (SF_6), and icosahedral (C_{60}) molecules, like all spherical rotors, have the same polarizability regardless of their orientations, so these molecules are **rotationally Raman inactive:** they do not have rotational Raman spectra. All other molecules, including homonuclear diatomic molecules such as H_2, are **rotationally Raman active.**

The specific selection rules for the rotational Raman transitions of linear molecules (the only ones we consider) are

$$\Delta J = +2 \text{ (Stokes lines)} \qquad \Delta J = -2 \text{ (anti-Stokes lines)} \qquad \text{Rotational Raman selection rules} \qquad (19.17)$$

It follows that the change in energy when a rigid rotor makes the transition $J \to J+2$ is

$$\begin{aligned}\Delta E = E_{J+2} - E_J &\overset{E_J = hBJ(J+1)}{=} hB(J+2)(J+3) - hBJ(J+1)\\ &= hB\{J^2 + 5J + 6 - (J^2 + J)\}\\ &= 2hB(2J+3)\end{aligned}$$

Therefore, the shift in frequency for the transition $J \to J+2$ is

$$\Delta\nu = 2B(2J+3) \qquad \text{Raman shift} \qquad (19.18)$$

It follows that when a photon scatters from molecules in the rotational states $J = 0, 1, 2, \ldots,$ and transfers some of its energy to the molecule, the frequency of the photon is decreased by $6B$, $10B$, $14B, \ldots$ from the frequency of the incident radiation. If the photon acquires energy during the collision, then a similar argument shows that the anti-Stokes lines occur with frequencies $6B$, $10B$,

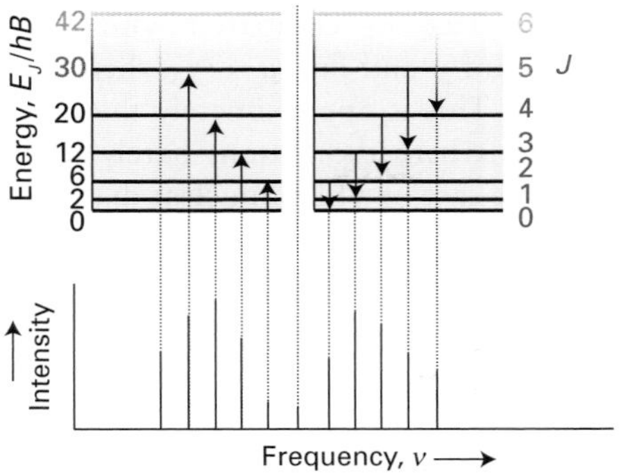

Fig. 19.15 The transitions responsible for the Stokes and anti-Stokes lines of a rotational Raman spectrum of a linear molecule.

$14B, \ldots$ higher than the incident radiation (Fig. 19.15). It follows that from a measurement of the separation of the Raman lines, we can determine the value of B and hence calculate the bond length. Because homonuclear diatomic species are rotationally Raman active, this technique can be applied to them as well as to heteronuclear species.

There is an important qualification of these remarks for symmetrical molecules, such as H_2 and $C^{16}O_2$. We saw in Section 19.2 that nuclear statistics either rules out certain states or leads to an alternation of populations. We saw, for instance, that $C^{16}O_2$ can exist only in states with even values of J. As a result, its rotational Raman spectrum consists of lines at $6B$, $14B$, $22B, \ldots$ and separated by $8B$ because the lines starting from odd values of J are missing. For molecules with nonzero nuclear spin, all the Raman lines are present but they show an alternation of intensities: for H_2, the odd-J lines are three times more intense than the even-J lines, whereas for D_2 and N_2, even-J lines are twice as intense as the odd-J lines.

Vibrational spectroscopy

All molecules are capable of vibrating, and complicated molecules may do so in a large number of different modes. Even a benzene molecule, with 12 atoms, can vibrate in 30 different modes, some of which involve the periodic swelling and shrinking of the ring and others its buckling into various distorted shapes. A molecule as big as a protein can vibrate in thousands of different ways, twisting, stretching, and buckling in different regions and in different manners. Vibrations can be excited by the absorption of electromagnetic radiation. Observing the frequencies at which this absorption occurs gives very valuable information about the identity of the molecule and provides quantitative information about the flexibility of its bonds.

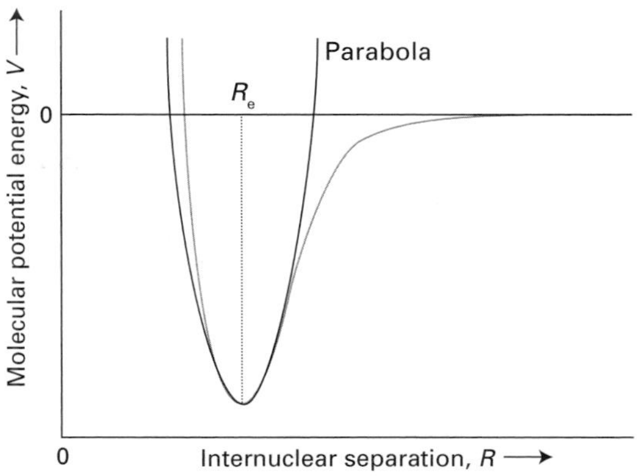

Fig. 19.16 A molecular potential energy curve can be approximated by a parabola near the bottom of the well. A parabolic potential results in harmonic oscillation. At high vibrational excitation energies the parabolic approximation is poor.

19.7 The vibrations of molecules

We base our discussion on Fig. 19.16, which shows a typical potential energy curve (it is a reproduction of Fig. 14.1) of a diatomic molecule as its bond is lengthened by pulling one atom away from the other or pressing it into the other. In regions close to the equilibrium bond length R_e (at the minimum of the curve) the potential energy can be approximated by a parabola (a curve of the form $y \propto x^2$), and we write

$$V(x) = \tfrac{1}{2}k_f x^2 \qquad \text{Parabolic potential approximation} \qquad (19.19)$$

where $x = R - R_e$ is the displacement from equilibrium and k_f is the force constant of the bond (units: newton per metre, N m^{-1}), as in the discussion of vibrations in Section 12.9. The steeper the walls of the potential (the stiffer the bond), the greater is the force constant.

The potential energy in eqn 19.19 has the same form as that for the harmonic oscillator (Section 12.9), so we can use the solutions of the Schrödinger equation given there. The only complication is that both atoms joined by the bond move, so the 'mass' of the oscillator has to be interpreted carefully. Detailed calculation shows that for two atoms of masses m_A and m_B joined by a bond of force constant k_f, the energy levels are

$$E_v = (v + \tfrac{1}{2})h\nu \quad v = 0, 1, 2, \ldots \qquad \text{Harmonic approximation} \quad \text{Energy levels} \qquad (19.20a)$$

where

$$\nu = \frac{1}{2\pi}\left(\frac{k_f}{\mu}\right)^{1/2} \quad \text{Vibrational frequency} \quad (19.20b)$$

and

$$\mu = \frac{m_A m_B}{m_A + m_B} \quad \text{Diatomic molecule} \quad \text{Effective mass} \quad (19.20c)$$

The **effective mass** of the molecule, μ, is a measure of the quantity of matter moved during the vibration. The effective masses of polyatomic molecules are complicated combinations of the atomic masses with each atomic mass contributing in a manner reflecting how much it moves. Vibrational transitions are commonly expressed as a wavenumber (in reciprocal centimetres), so it is often convenient to write eqn 19.20a as

$$E_v = (v + \tfrac{1}{2})hc\tilde{\nu} \qquad \tilde{\nu} = \nu/c \qquad (19.20d)$$

Figure 19.17 (a repeat of Fig. 12.35) illustrates these energy levels: we see that they form a uniform ladder of separation $hc\tilde{\nu}$ between neighbours.

A note on good practice The *effective mass* is widely called the *reduced mass*. However, that is only because the effective vibrational mass of a diatomic molecule happens to be given by the same expression as its reduced mass, a quantity that occurs in the separation of the internal motion of a molecule from its overall translation. For polyatomic molecules the effective mass is not the same as the reduced mass, and depends on the vibrational mode. It is better to distinguish the two from the outset.

At first sight it might be puzzling that the effective mass appears rather than the total mass of the two atoms. However, the presence of μ is physically plausible. If atom A were as heavy as a brick wall, it would not move at all during the vibration and the vibrational frequency would be determined by the lighter, mobile atom. Indeed, if A were a brick wall,

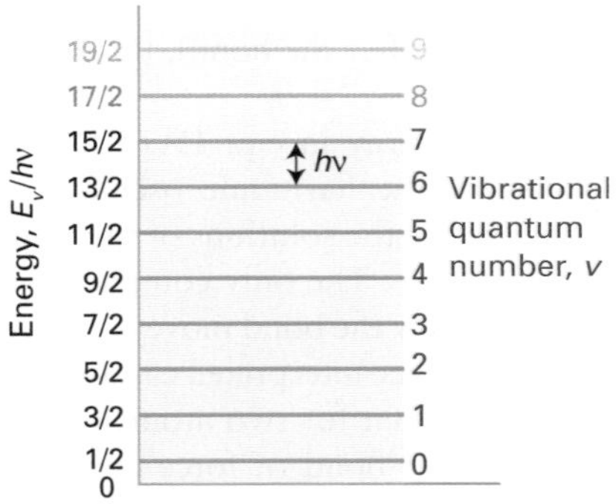

Fig. 19.17 The energy levels of a harmonic oscillator. The quantum number v ranges from 0 to infinity, and the permitted energy levels form a uniform ladder with spacing $h\nu$.

we could neglect m_B compared with m_A in the denominator of μ and find $\mu \approx m_B$, the mass of the lighter atom. This is approximately the case in HI, for example, where the I atom barely moves and $\mu \approx m_H$. In the case of a homonuclear diatomic molecule, for which $m_A = m_B = m$, the effective mass is half the mass of one atom: $\mu = \frac{1}{2}m$.

Brief illustration 19.6 The vibrational frequency

A $^1H^{35}Cl$ molecule has a force constant of 516 N m^{-1}, a reasonably typical value. Its effective mass (which we calculated as its 'reduced mass' in Example 19.1) is $0.9798m_u$. Therefore, its vibrational frequency is

$$\nu = \frac{1}{2\pi}\left(\frac{\overbrace{516\ \overset{\text{kg m s}^{-1}}{\widehat{\text{N}}}\ \text{m}^{-1}}^{k_f}}{0.9798\times\underbrace{(1.660\,54\times10^{-27}\ \text{kg})}_{m_u}}\right)^{1/2}$$

$$= 8.96\times10^{13}\ \text{s}^{-1}$$

For the unit cancellation, we have used 1 N = 1 kg m s^{-2}. The frequency corresponds to 89.6 THz. You might recall that in Brief illustration 12.8 we calculated the frequency on the basis that the Cl atom was stationary and obtained 88.5 THz. The corresponding wavenumber in the present case is

$$\tilde{\nu} = \frac{\nu}{c} = \frac{8.96\times10^{13}\ \text{s}^{-1}}{2.998\times10^{8}\ \text{m s}^{-1}} = 2.99\times10^{5}\ \text{m}^{-1}$$

This value corresponds to 2.99×10^3 cm^{-1}.

19.8 Vibrational transitions

Because typical vibrational frequencies are of the order of 10^{13}–10^{14} Hz, transitions can be induced with radiation of this frequency, which corresponds to infrared radiation. That is, vibrational transitions are observed by **infrared spectroscopy**. As we have remarked, in infrared spectroscopy, transitions are normally expressed in terms of their wavenumbers and lie typically in the range 300–3000 cm^{-1}.

The gross selection rule for vibrational spectra is that *the electric dipole moment of the molecule must change during the vibration*. The basis of this rule is that the molecule can shake the electromagnetic field into oscillation only if it has an electric dipole moment that oscillates as the molecule vibrates (Fig. 19.18). The molecule need not have a permanent dipole: the rule requires only a *change* in dipole moment, possibly from zero. The stretching motion of a homonuclear diatomic molecule does not change its electric dipole moment from zero, so

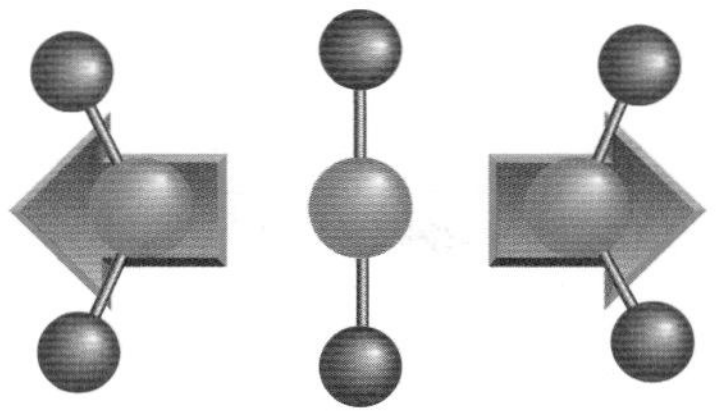

Fig. 19.18 The oscillation of a molecule, even if it is nonpolar, may result in an oscillating dipole that can interact with the electromagnetic field. Here we see a representation of a bending mode of CO_2.

the vibrations of such molecules neither absorb nor generate radiation. We say that homonuclear diatomic molecules are **infrared inactive**, because their dipole moments remain zero however long the bond. Heteronuclear diatomic molecules, which have a dipole moment that changes as the bond lengthens and contracts, are **infrared active**.

Example 19.2

Using the gross selection rule

State which of the following molecules are infrared active: N_2, CO_2, OCS, H_2O, $CH_2{=}CH_2$, C_6H_6.

Strategy Molecules that are infrared active (that is, have vibrational spectra) have dipole moments that change during the course of a vibration. Therefore, judge whether a distortion of the molecule can change its dipole moment (including changing it from zero).

Solution All the molecules except N_2 possess at least one vibrational mode that results in a change of dipole moment, so all except N_2 are infrared active. It should be noted that not all the modes of complicated molecules are infrared active. For example, a vibration of CO_2 in which the O—C—O bonds stretch and contract symmetrically is inactive because it leaves the dipole moment unchanged (at zero). A bending motion of the molecule, however, is active and can absorb radiation.

Self-test 19.6

Repeat the question for H_2, NO, and N_2O.

Answer: NO and N_2O

The specific selection rule for vibrational transitions is

$$\Delta v = \pm 1 \qquad \text{Vibrational selection rule} \qquad (19.21)$$

The change in energy for the transition from a state with quantum number v to one with quantum number $v + 1$ is

$$\Delta E = E_{v+1} - E_v \overset{E_v=(v+\frac{1}{2})h\tilde{\nu}}{=} (v+\tfrac{3}{2})hc\tilde{\nu} - (v+\tfrac{1}{2})hc\tilde{\nu} = hc\tilde{\nu} \qquad (19.22)$$

It follows that absorption occurs when the incident radiation provides photons with this energy, and therefore when the incident radiation has a wavenumber given by eqn 19.20d. Molecules with stiff bonds (large k_f) joining atoms with low masses (small μ) have high vibrational wavenumbers. Bending modes are usually less stiff than stretching modes, so bends typically occur at lower wavenumbers than stretches in a spectrum.

At room temperature, almost all the molecules are in their vibrational ground states initially (the state with $v = 0$). Therefore, the most important spectral transition is from $v = 0$ to $v = 1$.

Brief illustration 19.7 Vibrational transitions

It follows from the calculation of $\tilde{\nu}$ for HCl (in Brief illustration 19.6), that $\tilde{\nu} = 2992\ \text{cm}^{-1}$, so the infrared spectrum of the molecule will be an absorption at that wavenumber. The corresponding frequency and wavelength are 89.6 THz and 3.35 μm, respectively.

Self-test 19.7

The force constant of the bond in the CO group of a peptide link is approximately 1.2 kN m^{-1}. At what wavenumber would you expect it to absorb? *Hint:* For the effective mass, treat the group as a $^{12}C^{16}O$ molecule.

Answer: at approximately 1720 cm^{-1}

19.9 Anharmonicity

The vibrational terms in eqn 19.20 are only approximate because they are based on a parabolic approximation to the actual potential energy curve. A parabola cannot be correct at all extensions because it does not allow a molecule to dissociate. At high vibrational excitations the swing of the atoms (more precisely, the spread of vibrational wavefunction) allows the molecule to explore regions of the potential energy curve where the parabolic approximation is poor. The motion then becomes **anharmonic**, in the sense that the restoring force is no longer proportional to the displacement. Because the actual curve is less confining than a parabola, we can anticipate that the energy levels become less widely spaced at high excitation, just as the energy levels of a particle in a box get closer together as the length of the box is increased.

The convergence of levels at high vibrational quantum numbers is expressed by replacing eqn 19.20 by

$$E_v = (v + \tfrac{1}{2})hc\tilde{\nu} - (v + \tfrac{1}{2})^2 hc\tilde{\nu}x_e + \cdots$$

Anharmonicity correction (19.23)

where x_e is the (dimensionless) **anharmonicity constant.** Anharmonicity also accounts for the appearance of additional weak absorption lines called **overtones** corresponding to the transitions with $\Delta v = +2, +3, \ldots$. These overtones appear because the usual selection rule is derived from the properties of harmonic oscillator wavefunctions, which are only approximately valid when anharmonicity is present. Overtones in a vibrational spectrum can appear in the near infrared region and **overtone spectroscopy** is a technique used by analytical chemists in the characterization of food.

Table 19.2
Properties of diatomic molecules

	$\tilde{\nu}$/cm^{-1}	R_e/pm	k_f/(N m^{-1})	D/(kJ mol^{-1})
$^1H_2^+$	2333	106	160	256
$^1H_2$	4401	74	575	432
$^2H_2$	3118	74	577	440
$^1H^{19}F$	4138	92	955	564
$^1H^{35}Cl$	2991	127	516	428
$^1H^{81}Br$	264	141	412	363
$^1H^{127}I$	2308	161	314	295
$^{14}N_2$	235S	110	2294	942
$^{16}O_2$	158	121	1177	494
$^{19}F_2$	892	142	445	154
$^{35}Cl_2$	560	199	323	239

19.10 Vibrational Raman spectra of diatomic molecules

In **vibrational Raman spectroscopy** the incident photon leaves some of its energy in the vibrational modes of the molecule it strikes, or collects additional energy from a vibration that has already been excited.

The gross selection rule for vibrational Raman transitions is that *the molecular polarizability must change as the molecule vibrates*. The polarizability plays a role in vibrational Raman spectroscopy because the molecule must be squeezed and stretched by the incident radiation in order that a vibrational excitation may occur during the photon–molecule collision. Both homonuclear and heteronuclear diatomic molecules swell and contract during a vibration, and the control of the nuclei over the electrons, and hence the molecular polarizability, changes too. Both types of diatomic molecule are therefore vibrationally Raman active.

The specific selection rule for vibrational Raman transitions is the same as for infrared transitions:

$$\Delta v = \pm 1$$

Raman selection rule (19.24)

The photons that are scattered with a lower wavenumber than that of the incident light, the Stokes lines, are those for which $\Delta v = +1$. The Stokes lines are more intense than the anti-Stokes lines (for which $\Delta v = -1$), because very few molecules are in an excited vibrational state initially.

The information available from vibrational Raman spectra adds to that from infrared spectroscopy because homonuclear diatomic molecules can also be studied. The spectra can be interpreted in terms of the force constants, dissociation energies, and bond lengths, and some of the information obtained is included in Table 19.2. It is used, for example, in the calculation of equilibrium constants by the techniques of statistical thermodynamics.

19.11 The vibrations of polyatomic molecules

How many modes of vibration are there for a polyatomic molecule? We can answer this question by thinking about how each atom may change its location, and we show in the following Derivation that for a molecule built from N atoms the number of vibrational modes, N_{vib}, is

Nonlinear molecules: $N_{vib} = 3N - 6$

Linear molecules: $N_{vib} = 3N - 5$

Brief illustration 19.8 The number of vibrational modes

A water molecule, H_2O, is triatomic and nonlinear, and has three modes of vibration. Naphthalene, $C_{10}H_8$, has 48 distinct modes of vibration. Any diatomic molecule ($N = 2$) has one vibrational mode; carbon dioxide ($N = 3$) has four vibrational modes.

Self-test 19.8

How many normal modes of vibration are there in (a) ethyne (HC≡CH) and (b) a protein molecule of 4000 atoms?

Answer: (a) 7, (b) 11 994

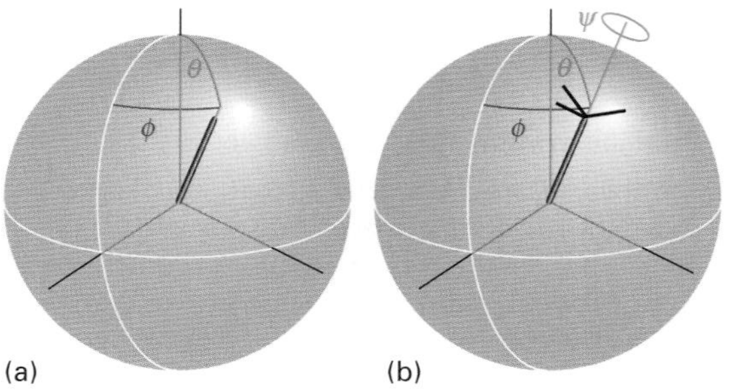

Fig. 19.19 (a) The orientation of a linear molecule requires the specification of two angles (the latitude and longitude of its axis). (b) The orientation of a nonlinear molecule requires the specification of three angles (the latitude and longitude of its axis and the angle of twist—the azimuthal angle—around that axis).

Derivation 19.4

The number of normal modes

Each atom may move along any of three perpendicular axes. Therefore, the total number of such displacements in a molecule consisting of N atoms is $3N$. Three of these displacements correspond to movement of the centre of mass of the molecule, so these three displacements correspond to the translational motion of the molecule as a whole. The remaining $3N-3$ displacements are 'internal' modes of the molecule that leave its centre of mass unchanged. Three angles are needed to specify the orientation of a nonlinear molecule in space (Fig. 19.19). Therefore three of the $3N-3$ internal displacements leave all bond angles and bond lengths unchanged but change the orientation of the molecule as a whole. These three displacements are therefore rotations. That leaves $3N-6$ displacements which change neither the centre of mass of the molecule nor the orientation of the molecule in space. These $3N-6$ displacements are the vibrational modes. A similar calculation for a linear molecule, which requires only two angles to specify its orientation in space, gives $3N-5$ as the number of vibrational modes.

The description of the vibrational motion of a polyatomic molecule is much simpler if we consider combinations of the stretching and bending motions of individual bonds. For example, although we could describe two of the four vibrations of a CO_2 molecule as individual carbon–oxygen bond stretches, ν_L and ν_R in Fig. 19.20, the description of the motion is much simpler if we use two combinations of these vibrations. One problem with dealing with individual bond stretches is that they are not independent: if one bond is stimulated to vibrate, the motion of the shared C atom quickly stimulates the other bond to vibrate. One combination is ν_1 in Fig. 19.21: this combination is the **symmetric stretch**. The other combination is ν_3, the **antisymmetric stretch**, in which the two O atoms always move in the same directions and opposite to the C atom. The two modes are independent in the sense that if one is excited, then its motion does not excite the other. They are two of the four 'normal modes' of the molecule, its independent, collective vibrational displacements. The two other (degenerate) normal modes are the **bending modes**, ν_2. In general, a **normal mode** is an independent, synchronous motion of atoms or groups of atoms that may be excited without leading to the excitation of any other normal mode.

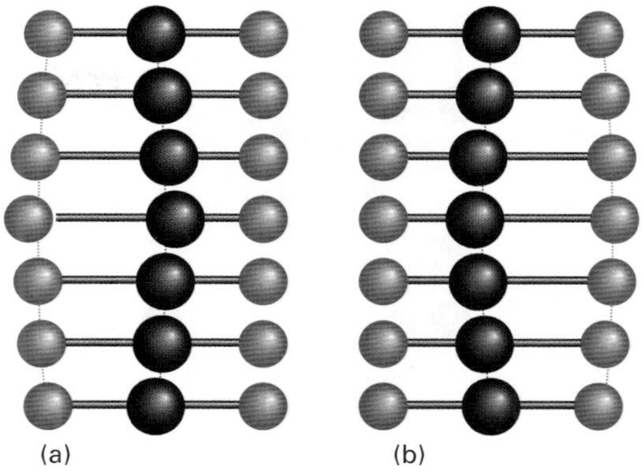

Fig. 19.20 The stretching vibrations of a CO_2 molecule can be represented in a number of ways. In this representation, (a) one O═C bond vibrates and the remaining O atom is stationary, and (b) the C═O bond vibrates while the other O atom is stationary. Because the stationary atom is linked to the C atom, it does not remain stationary for long. That is, if one vibration begins, it rapidly stimulates the other to occur.

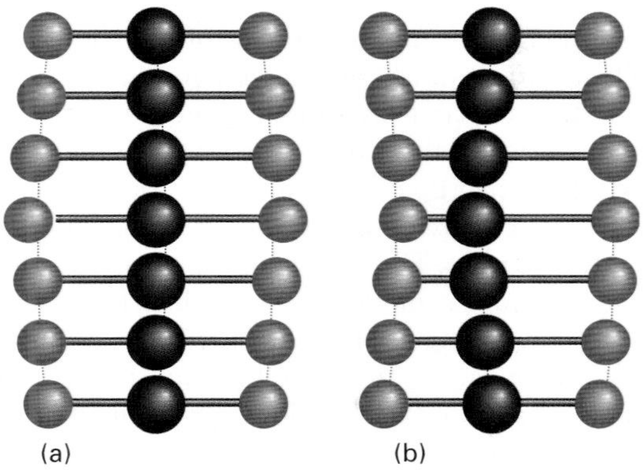

Fig. 19.21 Alternatively, linear combinations of the two modes can be taken to give these two normal modes of the molecule. The mode in (a) is the symmetric stretch and that in (b) is the antisymmetric stretch. The two modes are independent, and if either of them is stimulated, the other remains unexcited. Normal modes greatly simplify the description of the vibrations of the molecule.

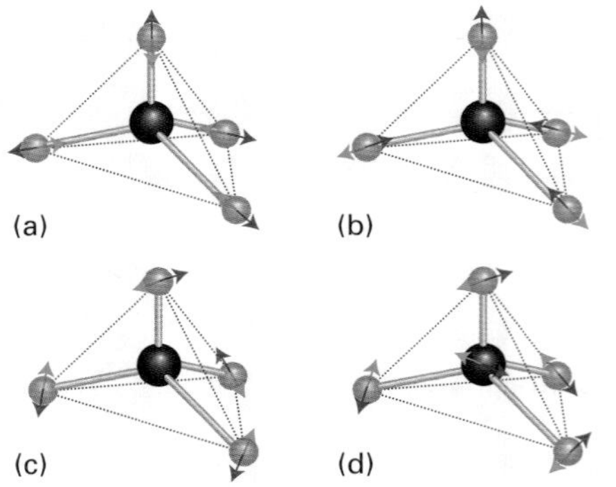

Fig. 19.22 Some of the normal modes of vibration of CH_4. An arrow indicates the direction of motion of an atom during the vibration.

The four normal modes of CO_2, and the $3N-6$ (or $3N-5$) normal modes of polyatomic molecules in general (for instance, those of methane, Fig. 19.22), are the key to the description of molecular vibrations. Each normal mode behaves like an independent harmonic oscillator and the energies of the vibrational levels are given by the same expression as in eqn 19.20, but with an effective mass that depends on the extent to which each of the atoms contributes to the vibration. Atoms that do not move, such as the C atom in the symmetric stretch of CO_2, do not contribute to the effective mass. The force constant also depends in a complicated way on the extent to which bonds bend and stretch during a vibration. Typically, a normal mode that is largely a bending motion has a lower force constant (and hence a lower frequency) than a normal mode that is largely a stretching motion.

The gross selection rule for the infrared activity of a normal mode is that *the motion corresponding to a normal mode must give rise to a changing dipole moment*. Deciding whether this is so can sometimes be done by inspection. For example, the symmetric stretch of CO_2 leaves the dipole moment unchanged (at zero), so this mode is infrared inactive and makes no contribution to the molecule's infrared spectrum. The antisymmetric stretch, however, changes the dipole moment because the molecule becomes unsymmetrical as it vibrates, so this mode is infrared active. The fact that the mode does absorb infrared radiation enables carbon dioxide to act as a 'greenhouse gas' by absorbing infrared radiation emitted from the surface of the Earth. Because the dipole moment change is parallel to the molecular axis in the antisymmetric stretching mode, the transitions arising from this mode are classified as **parallel bands** in the spectrum. Both bending modes are also infrared active: they are accompanied by a changing dipole perpendicular to the molecular axis, so transitions involving them lead to a **perpendicular band** in the spectrum.

Brief illustration 19.9 Infrared activity

The infrared spectrum of dinitrogen oxide (nitrous oxide, N_2O) differs from that of carbon dioxide in a variety of ways despite both being linear triatomic molecules. First, the corresponding vibrational modes have different frequencies on account of different atomic masses and force constants. In CO_2, the symmetric stretch is not infrared active, so only three modes (the asymmetric stretch and the two degenerate bending modes) are active. In contrast, all four vibrational modes of N_2O are infrared active.

Some of the normal modes of organic molecules can be regarded as motions of individual functional groups. Others cannot be regarded as localized in this way and are better regarded as collective motions of the molecule as a whole. The latter are generally of relatively low frequency, and occur at wavenumbers below about 1500 cm^{-1} in the spectrum. The resulting whole-molecule region of the absorption spectrum is called the **fingerprint region** of the spectrum, for it is characteristic of the molecule. The matching of the fingerprint region with a spectrum of a known compound in a library of infrared spectra is a very powerful way of confirming the presence of a particular substance.

The characteristic vibrations of functional groups that occur outside the fingerprint region are very useful for the identification of an unknown compound. Most of these vibrations can be regarded as stretching modes, for the lower frequency bending modes usually occur in the fingerprint region and so are less readily identified. The characteristic wavenumbers of some functional groups are listed in Table 19.3.

Table 19.3
Typical vibrational wavenumbers

Vibration type	$\tilde{\nu}/cm^{-1}$
C—H	2850–2960
C—H	1340–1465
C—C stretch, bend	700–1250
C=C stretch	1620–1680
C≡C stretch	2100–2260
O—H stretch	3590–3650
C=O stretch	1640–1780
C≡N stretch	2215–2275
N—H stretch	3200–3500
Hydrogen bonds	3200–3570

Example 19.3

Interpreting an infrared spectrum

The infrared spectrum of an organic compound is shown in Fig. 19.23. Suggest an identification.

Strategy Some of the features at wavenumbers above 1500 cm^{-1} can be identified by comparison with the data in Table 19.3.

Solution (a) C—H stretch of a benzene ring, indicating a substituted benzene; (b) carboxylic acid O—H stretch, indicating a carboxylic acid; (c) the strong absorption of a conjugated C≡C group, indicating a substituted alkyne; (d) this strong absorption is also characteristic of a carboxylic acid that is conjugated to a carbon–carbon multiple bond; (e) a characteristic vibration of a benzene ring, confirming the deduction drawn from (a); (f) a characteristic absorption of a nitro group (—NO_2) connected to a multiply bonded carbon–carbon system, suggesting a nitro-substituted benzene. The molecule contains as components a benzene ring, an aromatic carbon–carbon bond, a —COOH group, and a —NO_2 group. The molecule is in fact O_2N—C_6H_4—C≡C—COOH. A more detailed analysis and comparison of the fingerprint region shows it to be the 1,4-isomer.

Self-test 19.9

Suggest an identification of the organic compound responsible for the spectrum shown in Fig. 19.24. *Hint:* The molecular formula of the compound is C_3H_5ClO.

Answer: CH_2=CClCH$_2$OH

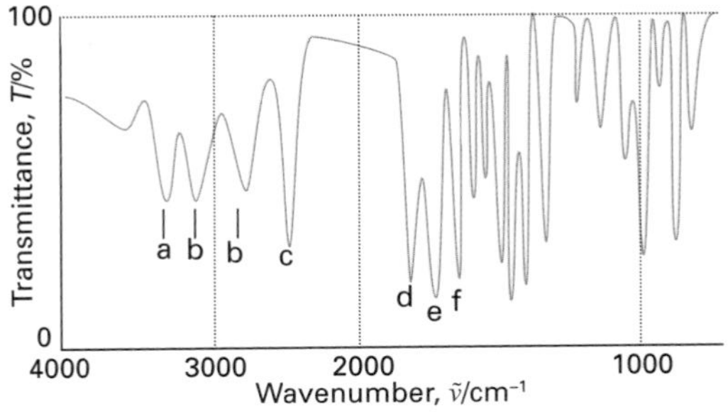

Fig. 19.23 A typical infrared absorption spectrum taken by forming a sample into a disk with potassium bromide. As explained in the example, the substance can be identified as $O_2NC_6H_4$—C≡C—COOH.

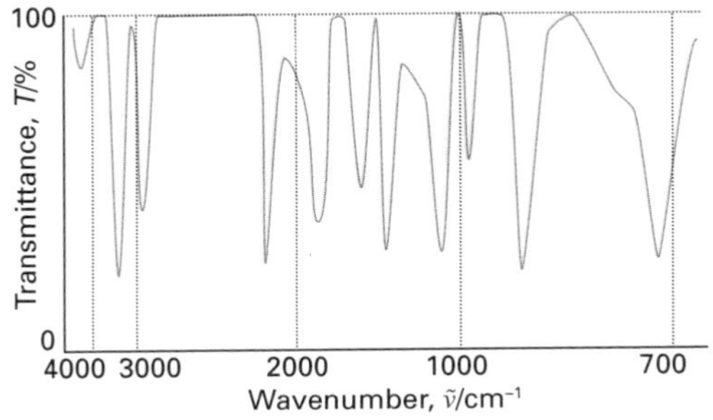

Fig. 19.24 The spectrum considered in Self-test 19.9.

19.12 Vibration–rotation spectra

The vibrational spectra of gas-phase molecules are more complicated than this discussion implies, because the excitation of a vibration also results in the excitation of rotation. The effect is rather like what happens when ice skaters throw out or draw in their arms: they rotate more slowly or more rapidly. The effect on the spectrum is to break the single line resulting from a vibrational transition into a multitude of lines with separations between neighbours that depend on the rotational constant of the molecule.

To establish the so-called 'band structure' of a vibrational transition, we begin by writing the expressions for the vibrational and rotational levels. For a rigid linear molecule (the only type we consider), we combine eqns 19.3 and 19.20 and write

$$E_{v,J} = (v + \tfrac{1}{2})h\nu + hBJ(J+1)$$

Vibration–rotation energies (19.25)

(For this part of the discussion it is simpler to express vibrational transitions as frequencies rather than wavenumbers, but the conversion between them is straightforward.) As mentioned in Section 19.1, B depends on the vibrational state and therefore on v, but we shall ignore that complication in this discussion. Next, we apply the selection rules. Provided the molecule is polar, or at least acquires a dipole moment in a vibrational transition (as when CO_2 bends or undergoes an asymmetric stretch), the rotational quantum number may change by ±1 or (in some cases, see below) 0. The absorptions then fall into three groups called **branches** of the spectrum.

P branch, transitions with $\Delta J = -1$: $\nu_J = \nu - 2BJ$

Q branch, transitions with $\Delta J = 0$: $\nu_J = \nu$

R branch, transitions with $\Delta J = +1$: $\nu_J = \nu + 2B(J+1)$

The Q branch is not always allowed. For example, it is observed in the spectrum of NO, but not in the spectrum of HCl: the difference can be traced to the fact that NO, with an electron in a π orbital, has electronic angular momentum around its internuclear axis but HCl does not.[2]

Figure 19.25 shows the resulting appearance of the branches of a typical spectrum. The separation between the lines in the P and R branches of a vibrational transition is $2B$. Therefore, the bond length can be deduced without needing to take a pure rotational microwave spectrum. However, the latter is more precise.

[2] For more information, see our *Physical Chemistry* (2010).

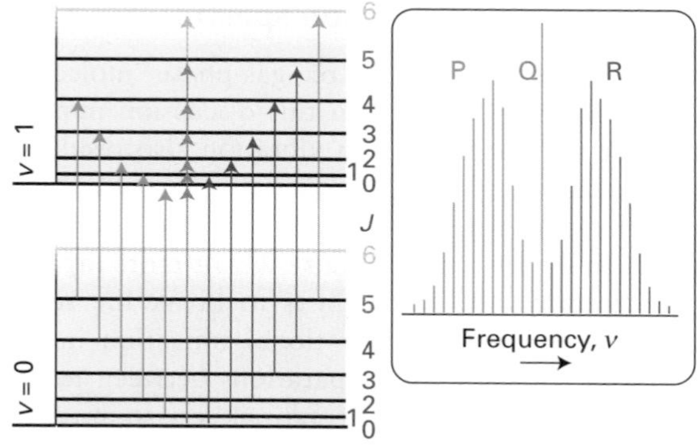

Fig. 19.25 The formation of P, Q, and R branches in a vibration–rotation spectrum. The intensities reflect the populations of the initial rotational levels.

19.13 Vibrational Raman spectra of polyatomic molecules

The gross selection rule for the vibrational Raman spectrum of a polyatomic molecule is that *the normal mode of vibration is accompanied by a changing polarizability*. However, it is often quite difficult to judge by inspection when this is so. The symmetric stretch of CO_2, for example, alternately swells and contracts the molecule: this motion changes its polarizability, so the mode is Raman active. The other modes of CO_2 are Raman inactive because the polarizability does not change as the atoms move collectively. A very simple explanation (which is not reliable in all cases) is that the polarizability of a molecule depends on its size, and whereas the symmetric stretch changes the size of the molecule neither the antisymmetric stretch nor the bending modes do—at least, to a first approximation.

In some cases it is possible to make use of a very general rule about the infrared and Raman activity of vibrational modes:

> The **exclusion rule** states that if the molecule has a centre of inversion, then no modes can be both infrared and Raman active.

(A mode may be inactive in both.) A molecule has a centre of inversion if it looks unchanged when each atom is projected through a single point and out an equal distance on the other side (Fig. 19.26). Because we can often judge intuitively when a mode changes the molecular dipole moment, we can use this rule to identify modes that are not Raman active. The rule applies to CO_2 but to neither H_2O nor CH_4 because they have no centre of symmetry. Thus, both the antisymmetric stretch and the bending modes of CO_2 are infrared active, so we know at once that they are Raman inactive, as we asserted above.

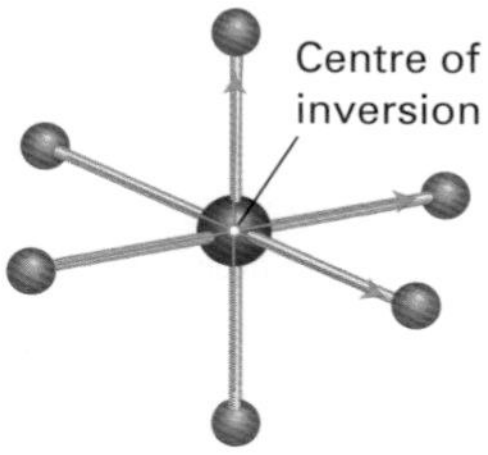

Fig. 19.26 In an inversion operation, we consider every point in a molecule, and project them all through the centre of the molecule out to an equal distance on the other side.

Brief illustration 19.10 Raman activity

One vibrational mode of benzene is a 'breathing mode' in which the entire ring alternately expands and contracts symmetrically (Fig. 19.27). As it does so, the polarizability of the molecule changes because the electron distribution can be modified by an electric field differently when the molecule is compressed or extended. As a result, this mode is Raman active. It is not infrared active because the molecular dipole moment remains unchanged (at zero).

Self-test 19.10

Predict whether the symmetric stretching mode of ethene, C_2H_4, in which all of the C—H bonds vibrate in phase, is infrared active, Raman active, or both.

Answer: Raman active only

A modification of the basic Raman effect involves using incident radiation that nearly coincides with the frequency of an electronic transition of the sample (Fig. 19.28). The technique is then called **resonance**

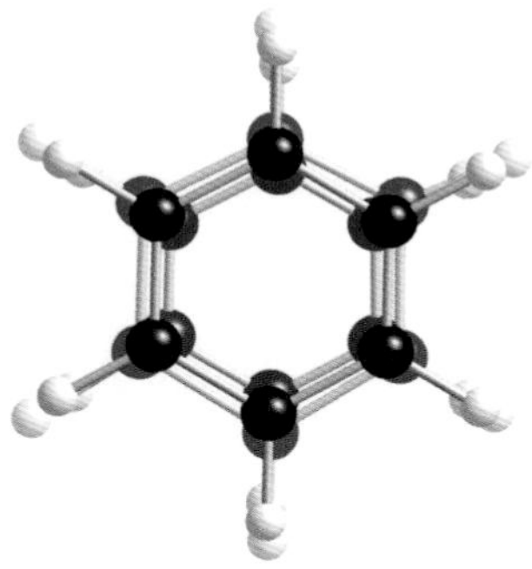

Fig. 19.27 The symmetrical breathing vibrational mode of a benzene molecule.

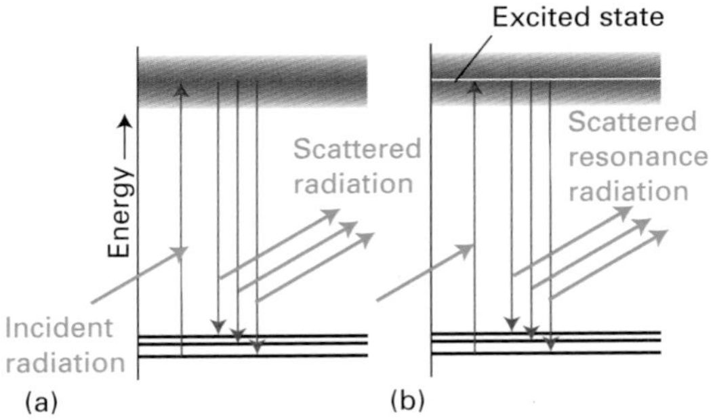

Fig. 19.28 (a) In Raman spectroscopy, an incident photon is scattered from a molecule with either an increase in frequency (if the radiation collects energy from the molecule) or—as shown here—with a lower frequency if it loses energy to the molecule. The process can be regarded as taking place by an excitation of the molecule to a wide range of states (represented by the shaded band), and the subsequent return of the molecule to a lower state; the net energy change is then carried away by the photon. (b) In the *resonance Raman effect*, the incident radiation has a frequency corresponding to an actual electronic excitation of the molecule. A photon is emitted when the excited state returns to a state close to the ground state.

Raman spectroscopy. It is characterized by a much greater intensity in the scattered radiation. Furthermore, because it is often the case that only a few vibrational modes contribute to the more intense scattering, the spectrum is greatly simplified. Resonance Raman spectroscopy is used to study biological molecules that absorb strongly in the ultraviolet and visible regions of the spectrum. Examples include the haem co-factors in haemoglobin and the cytochromes and the pigments β-carotene and chlorophyll, which capture solar energy during plant photosynthesis.

Impact on the environment 19.1

Climate change

Solar energy strikes the top of the Earth's atmosphere at a rate of 343 W m^{-2}. About 30 per cent of this energy is reflected back into space by the Earth or the atmosphere. The Earth–atmosphere system absorbs the remaining energy and re-emits it into space as so-called 'black-body radiation', radiation characteristic of a hot body, with most of the intensity being carried by infrared radiation in the range 200–2500 cm^{-1} (4–50 μm). The Earth's average temperature is maintained by an energy balance between solar radiation absorbed by the Earth and black-body radiation emitted by the Earth.

The trapping of infrared radiation by certain gases in the atmosphere is known as the *greenhouse effect*, so called because it warms the Earth as if the planet were enclosed in a huge greenhouse. The result is that the natural greenhouse effect raises the average surface temperature well above the freezing point of water and creates an environment in which life is possible. The major constituents to the Earth's atmosphere, O_2 and N_2, do not contribute to the greenhouse effect because homonuclear diatomic molecules cannot absorb infrared radiation. However, the minor atmospheric gases water vapour and CO_2 do absorb infrared radiation and hence are responsible for the greenhouse effect (Fig. 19.29). Water vapour absorbs strongly in the ranges 1300–1900 cm^{-1} (5.3–7.7 μm) and 3550–3900 cm^{-1} (2.6–2.8 μm), whereas CO_2 shows strong absorption in the ranges 500–725 cm^{-1} (14–20 μm) and 2250–2400 cm^{-1} (4.2–4.4 μm).

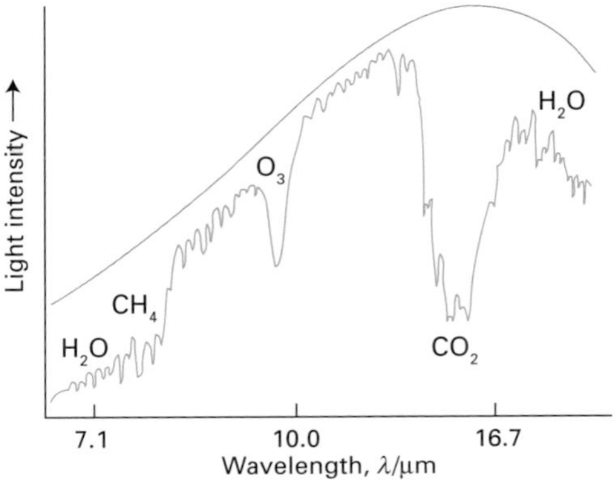

Fig. 19.29 The intensity of infrared radiation that would be lost from Earth in the absence of greenhouse gases is shown by the smooth line. The jagged line is the intensity of the radiation actually emitted. The maximum wavelength of radiation absorbed by each greenhouse gas is indicated.

Increases in the levels of greenhouse gases, which also include methane, dinitrogen oxide, ozone, and certain chlorofluorocarbons, as a result of human activity have the potential to enhance the natural greenhouse effect, leading to significant warming of the planet. This problem is referred to as *global warming*, which we now explore in some detail.

The concentration of water vapour in the atmosphere has remained steady over time, but concentrations of some other greenhouse gases are rising. From about the year 1000 until about 1750, the CO_2 concentration remained fairly stable, but, since then, it has increased by 28 per cent. The concentration of methane, CH_4, has more than doubled during this time and is now at its highest level for 160 000 years (160 ka; 1 a is the SI unit denoting 1 year). Studies of air pockets in ice cores taken from Antarctica show that increases in the concentration of both atmospheric CO_2 and CH_4 over the past 160 ka correlate well with increases in the global surface temperature.

Human activities are primarily responsible for the rising concentrations of atmospheric CO_2 and CH_4. Most of the atmospheric CO_2 comes from the burning of hydrocarbon fuels, which began on a large scale with the Industrial Revolution in the middle of the nineteenth century. The additional methane comes mainly from the petroleum industry and from agriculture.

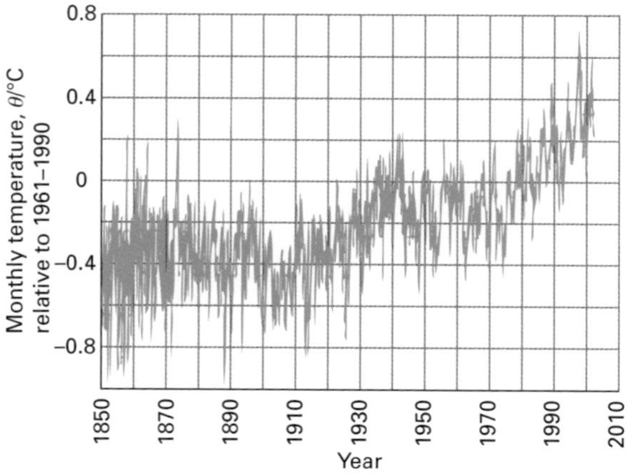

Fig. 19.30 The progressive change in the average surface temperature of the Earth, based on the IPCC report in 2007.

The temperature of the surface of the Earth has increased by about 0.8 °C since the middle of the nineteenth century (Fig. 19.30). In 2007, the Intergovernmental Panel on Climate Change (IPCC) estimated that our continued reliance on hydrocarbon fuels, coupled to current trends in population growth, could result in an additional increase of 1–3 °C in the temperature of the Earth by 2100, relative to the surface temperature in 2000. Furthermore, the rate of temperature change is likely to be greater than at any time in the last 10 ka. To place a temperature rise of 3 °C in perspective, note that the average temperature of the Earth during the last ice age was only 6 °C colder than at present. Just as cooling the planet (for example, during an ice age) can lead to detrimental effects on ecosystems, so too can a dramatic warming. One example of a significant change in the environment caused by a temperature increase of 3 °C is a rise in sea level by about 0.5 m, which is sufficient to alter weather patterns and submerge currently coastal ecosystems.

Computer projections for the next 200 years predict further increases in atmospheric CO_2 levels and suggest that, to maintain CO_2 at its current concentration, we would have to reduce hydrocarbon fuel consumption immediately. Clearly, in order to reverse global warming trends, we need to develop alternatives to fossil fuels, such as hydrogen (which can be used in fuel cells, Impact 9.2) and solar energy technologies.

Checklist of key concepts

☐ 1 The populations of rotational energy levels are given by the Boltzmann distribution in connection with noting the degeneracy of each level.

☐ 2 The intensity of a transition is proportional to the square of the transition dipole moment.

☐ 3 A selection rule is a statement about when the transition dipole may be nonzero.

☐ 4 A gross selection rule specifies the general features a molecule must have if it is to have a spectrum of a given kind.

☐ 5 A specific selection rule is a statement about which changes in quantum number may occur in a transition.

☐ 6 The gross selection rule for rotational transitions is that the molecule must be polar. The specific selection rules are in the following *Road map*.

☐ 7 The Pauli principle states for fermions $\psi(\mathrm{B,A}) = -\psi(\mathrm{A,B})$ and for bosons $\psi(\mathrm{B,A}) = \psi(\mathrm{A,B})$.

☐ 8 The consequences of the Pauli principle for rotational states are called nuclear statistics.

☐ 9 The rotational spectrum of a polar linear molecule and of a polar symmetric rotor consists of a series of lines at frequencies separated by $2B$.

☐ 10 One contribution to the linewidth is the Doppler effect; another contribution is lifetime broadening.

☐ 11 In a Raman spectrum, lines shifted to lower frequency than the incident radiation are called Stokes lines and lines shifted to higher frequency are called anti-Stokes lines.

☐ 12 The gross selection rule for rotational Raman spectra is that the polarizability of the molecule must be anisotropic. The specific selection rules are in the following *Road map*.

☐ 13 The gross selection rule for vibrational spectra is that the electric dipole moment of the molecule must change during the vibration. The specific selection rules are in the following *Road map*.

☐ 14 The number of vibrational modes of nonlinear molecules is $3N - 6$; for linear molecules the number is $3N - 5$.

☐ 15 Rotational transitions accompany vibrational transitions and split the spectrum into a P branch ($\Delta J = -1$), a Q branch ($\Delta J = 0$), and an R branch ($\Delta J = +1$).

☐ 16 A Q branch is observed only when the molecule possesses angular momentum around its axis.

☐ 17 The gross selection rule for the vibrational Raman spectrum of a polyatomic molecule is that the normal mode of vibration is accompanied by a changing polarizability.

☐ 18 The exclusion rule states that if the molecule has a centre of inversion, then no modes can be both infrared and Raman active.

☐ 19 In resonance Raman spectroscopy, radiation that nearly coincides with the frequency of an electronic transition is used to excite the sample and the result is a much greater intensity in the scattered radiation.

Road map of key equations

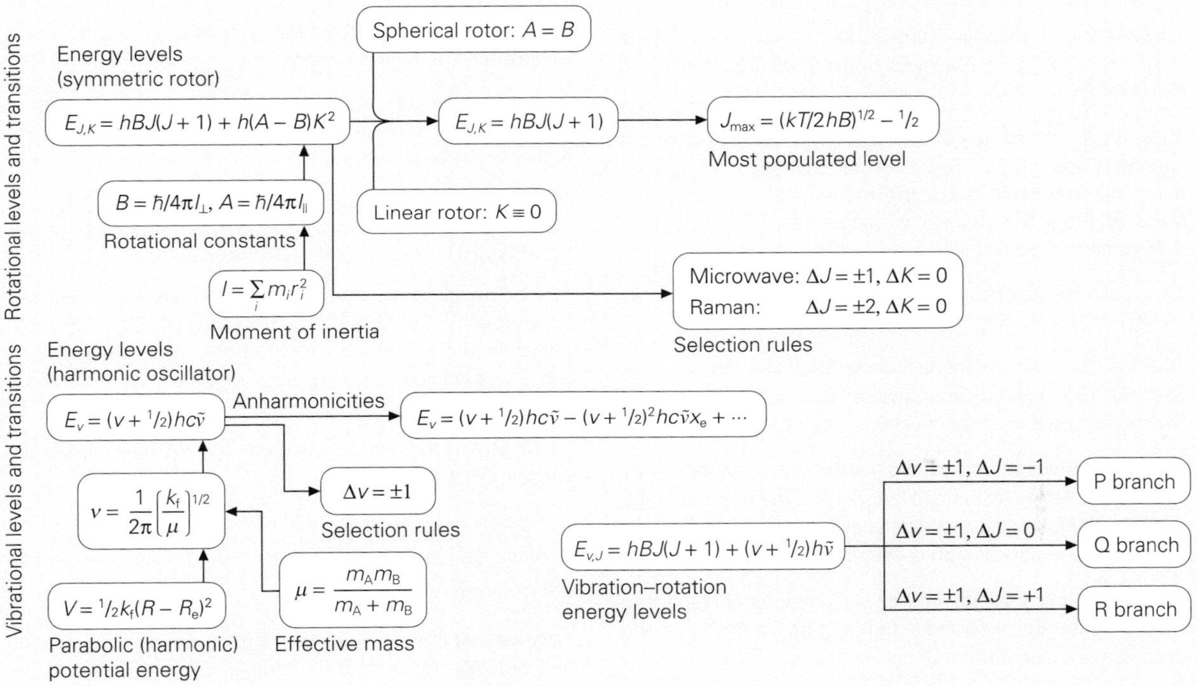

Questions and exercises

Discussion questions

19.1 Discuss the physical origins of the gross selection rules for microwave spectroscopy and rotational Raman spectroscopy.

19.2 Describe the physical origins of linewidths in the absorption and emission spectra of gases, liquids, and solids. How may they be reduced?

19.3 Consider a diatomic molecule that is highly susceptible to centrifugal distortion in its ground vibrational state. Do you expect excitation to high rotational energy levels to change the equilibrium bond length of this molecule? Justify your answer.

19.4 Why does the vibrational state of a diatomic molecule affect its rotational constant? Is there an effect even if the potential is strictly parabolic?

19.5 Account physically for the fact that a linear polyatomic molecule composed of N atoms has one more vibrational mode than a nonlinear molecule of N atoms.

19.6 (a) Discuss the physical origins of the gross selection rules for infrared spectroscopy and vibrational Raman spectroscopy. (b) Suppose that you wish to characterize the normal modes of benzene in the gas phase. Why is it important to obtain both infrared absorption and Raman spectra of your sample?

19.7 Suggest a reason why the replacement of ^{12}C by ^{13}C in CO_2 affects some of its vibrational frequencies but not all.

19.8 Account for the appearance of P, Q, and R branches in the vibration–rotation spectrum of a diatomic molecule.

Exercises

For these exercises, use $m(^1H) = 1.0078m_u$, $m(^2H) = 2.0140m_u$, $m(^{12}C) = 12.0000m_u$, $m(^{13}C) = 13.0034m_u$, $m(^{16}O) = 15.9949m_u$, $m(^{19}F) = 18.9984m_u$, $m(^{32}S) = 31.9721m_u$, $m(^{34}S) = 33.9679m_u$, $m(^{35}Cl) = 34.9688m_u$, $m(^{127}I) = 126.9045m_u$.

19.1 Express a wavelength of 442 nm as (a) a frequency, (b) a wavenumber.

19.2 What is (a) the wavenumber, (b) the wavelength of the radiation used by an FM radio transmitter broadcasting at 88.0 MHz?

19.3 The kinetic energy of a bicycle wheel rotating once per second is about 0.2 J. To what rotational quantum number does that correspond? For the moment of inertia, let the mass of the wheel (which is concentrated in its rim) be 0.75 kg and its radius be 70 cm.

19.4 Calculate the moment of inertia of (a) $^1H_2$, (b) $^2H_2$, (c) $^{12}C^{16}O_2$, (d) $^{13}C^{16}O_2$.

19.5 Calculate the rotational constants of the molecules in Exercise 19.4; express your answer as a frequency in hertz (Hz) and as a wavenumber in cm^{-1}.

19.6 (a) Express the moment of inertia of an octahedral AB_6 molecule in terms of its bond lengths and the masses of the B atoms. (b) Calculate the rotational constant of $^{32}S^{19}F_6$, for which the S—F bond length is 158 pm.

19.7 Derive expressions for the two moments of inertia of a square-planar AB_4 molecule in terms of its bond lengths and the masses of the B atoms.

19.8 Suppose you were seeking the presence of (planar) SO_3 molecules in the microwave spectra of interstellar gas clouds. (a) You would need to know the rotational constants A and B. Calculate these parameters for $^{32}S^{16}O_3$, for which the S—O bond length is 143 pm. (b) Could you use microwave spectroscopy to distinguish the relative abundances of $^{32}S^{16}O_3$ and $^{33}S^{16}O_3$?

19.9 Which of the following molecules can have a pure rotational spectrum: (a) HCl, (b) N_2O, (c) O_3, (d) SF_4, (e) XeF_4?

19.10 Which of the molecules in Exercise 19.9 can have a rotational Raman spectrum?

19.11 A rotating methane molecule is described by the quantum numbers J, M_J, and K. How many rotational states have an energy equal to $hBJ(J+1)$ with $J=8$?

19.12 Suppose the methane molecule in Exercise 19.11 is replaced by chloromethane. How many rotational states now have an energy equal to $hBJ(J+1)$ with $J=8$?

19.13 The rotational constant of $^1H^{35}Cl$ is 318.0 GHz. What is the separation of the line in its pure rotational spectrum (a) in gigahertz, (b) in reciprocal centimetres?

19.14 The rotational constant of $^{127}I^{35}Cl$ is 0.1142 cm^{-1}. Calculate the I—Cl bond length.

19.15 Suppose that hydrogen is replaced by deuterium in $^1H^{35}Cl$. Would you expect the $J=1 \leftarrow 0$ transition to move to higher or lower wavenumber?

19.16 The microwave spectrum of $^1H^{127}I$ consists of a series of lines separated by 384 GHz. Compute its bond length. What would be the separation of the lines in $^2H^{127}I$?

19.17 The following wavenumbers are observed in the rotational spectrum of OCS: 1.217 1054 cm^{-1}, 1.622 8005 cm^{-1}, 2.028 4883 cm^{-1}, and 2.434 1708 cm^{-1}. Use the graphical procedure implied by eqn 19.14 to infer the values of B and D for this molecule.

19.18 The microwave spectrum of $^{16}O^{12}CS$ gave absorption lines (in GHz) as follows:

J	1	2	3	4
^{32}S	24.325 92	36.488 82	48.651 64	60.814 08
^{34}S	23.732 33			47.462 40

Assume that the bond lengths are unchanged by substitution and calculate the CO and CS bond lengths in OCS. *Hint:* The moment of inertia of a linear molecule of the form ABC is

$$I = m_A R_{AB}^2 + m_C R_{BC}^2 - \frac{(m_A R_{AB} - m_C R_{BC})^2}{m_A + m_B + m_C}$$

where r_{AB} and r_{BC} are the A—B and B—C bond lengths, respectively.

19.19 What is the Doppler-shifted wavelength of a red (660 nm) traffic light approached at 65 m.p.h.? At what speed would it appear green (520 nm)?

19.20 A spectral line of $^{48}Ti^{8+}$ in a distant star was found to be shifted from 654.2 nm to 706.5 nm and to be broadened to 61.8 pm. What is the speed of recession and the surface temperature of the star?

19.21 Estimate the lifetime of a state that gives rise to a line of width (a) 0.10 cm^{-1}, (b) 1.0 cm^{-1}, (c) 1.0 GHz.

19.22 A molecule in a liquid undergoes about 1.0×10^{13} collisions in each second. Suppose that (a) every collision is effective in deactivating the molecule vibrationally and (b) that one collision in 200 is effective. Calculate the width (in cm^{-1}) of vibrational transitions in the molecule.

19.23 The wavenumber of the incident radiation in a Raman spectrometer is 20 623 cm^{-1}. What is the wavenumber of the scattered Stokes radiation for the $J = 4 \leftarrow 2$ transition of $^{16}O_2$?

19.24 The rotational constant of $^{12}C^{16}O_2$ (from Raman spectroscopy) is 11.70 GHz. What is the CO bond length in the molecule?

19.25 Suppose the C=O group in a peptide bond can be regarded as isolated from the rest of the molecule. Given the force constant of the bond in a carbonyl group is 908 N m^{-1}, calculate the vibrational frequency of (a) $^{12}C{=}^{16}O$, (b) $^{13}C{=}^{16}O$.

19.26 The wavenumber of the fundamental vibrational transition of Cl_2 is 565 cm^{-1}. Calculate the force constant of the bond.

19.27 The hydrogen halides have the following fundamental vibrational wavenumbers:

	HF	HCl	HBr	HI
$\tilde{\nu}/cm^{-1}$	4141.3	2988.9	2649.7	2309.5

Calculate the force constants of the hydrogen–halogen bonds.

19.28 From the data in Exercise 19.27, predict the fundamental vibrational wavenumbers of the deuterium halides.

19.29 Which of the following molecules may show infrared absorption spectra: (a) H_2, (b) HCl, (c) CO_2, (d) H_2O, (e) CH_3CH_3, (f) CH_4, (g) CH_3Cl, (h) N_2?

19.30 In the infrared spectrum of CO a strong vibrational transition is observed centred at 2143.29 cm^{-1} with a weaker transition at 4259.66 cm^{-1}. What is the vibrational wavenumber and anharmonicity constant for CO?

19.31 How many normal modes of vibration are there for (a) NO_2, (b) N_2O, (c) cyclohexane, (d) hexane?

19.32 Infrared absorption by $^1H^{81}Br$ gives rise to an R branch from $v = 0$. What is the wavenumber of the line originating from the rotational state with $J = 2$?

19.33 Vibration–rotation transitions in the infrared spectrum of $^1H^{19}F$ were observed at wavenumbers of 2886.50, 2908.51, 2930.43, 2974.55, 2996.57, 3018.58 cm^{-1}. Assign the transitions by reference to Fig. 19.25. Hence determine the bond length of $^1H^{19}F$.

19.34 Consider the vibrational mode that corresponds to the uniform expansion of the benzene ring. Is it (a) Raman, (b) infrared active?

19.35 Suppose that three conformations are proposed for the nonlinear molecule H_2O_2 (**1**, **2**, and **3**). The infrared absorption spectrum of gaseous H_2O_2 has bands at 870, 1370, 2869, and 3417 cm^{-1}. The Raman spectrum of the same sample has bands at 877, 1408, 1435, and 3407 cm^{-1}. All bands correspond to fundamental vibrational wavenumbers and you may assume that: (i) the 870 and 877 cm^{-1} bands arise from the same normal mode, and (ii) the 3417 and 3407 cm^{-1} bands arise from the same normal mode. (a) If H_2O_2 were linear, how many normal modes of vibration would it have? (b) Determine which of the proposed conformations is inconsistent with the spectroscopic data. Explain your reasoning.

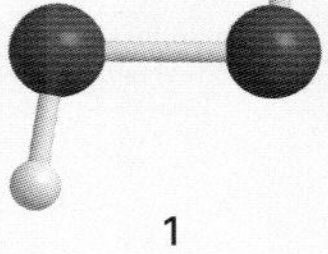
1

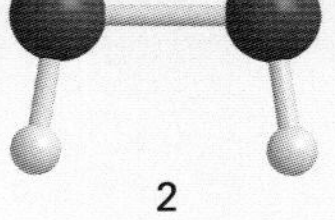
2

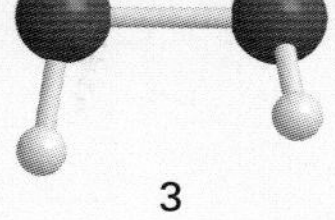
3

Projects

The symbol ‡ indicates that calculus is required.

19.36‡ The most populated rotational energy level of a linear rotor is given in eqn 19.11. What is the most populated rotational level of a spherical rotor, given that its degeneracy is $(2J + 1)^2$?

19.37 The protein haemerythrin (Her) is responsible for binding and carrying O_2 in some invertebrates. Each protein molecule has two Fe^{2+} ions that are in very close proximity and work together to bind one molecule of O_2. The Fe_2O_2 group of oxygenated haemerythrin is coloured and has an electronic absorption band at 500 nm. (a) The resonance Raman spectrum of oxygenated haemerythrin obtained with laser excitation at 500 nm has a band at 844 cm^{-1} that has been attributed to the O—O stretching mode of bound $^{16}O_2$. Why is resonance Raman spectroscopy and not infrared spectroscopy the method of choice for the study of the binding of O_2 to haemerythrin? (b) Proof that the 844 cm^{-1} band in the resonance Raman spectrum of oxygenated haemerythrin arises from a bound O_2 species may be obtained by conducting experiments on samples of haemerythrin that have been mixed with $^{18}O_2$, instead of $^{16}O_2$. Predict the fundamental vibrational wavenumber of the ^{18}O—^{18}O stretching mode in a sample of haemerythrin that has been treated with $^{18}O_2$. (c) The fundamental vibrational wavenumbers for the O—O stretching modes of O_2, O_2^- (superoxide anion), and O_2^{2-} (peroxide anion) are 1555, 1107, and 878 cm^{-1}, respectively. (i) Explain this trend in terms of the electronic structures of O_2, O_2^-, and O_2^{2-}. (ii) What are the bond orders of O_2, O_2^-, and O_2^{2-}? (d) Based on the data given in part (c), which of the following species best describes the Fe_2O_2 group of haemerythrin: $Fe_2^{2+}O_2$, $Fe^{2+}Fe^{3+}O_2^-$, or $Fe_2^{3+}O_2^{2-}$? Explain your reasoning. (e) The resonance Raman spectrum of haemerythrin mixed with $^{16}O^{18}O$ has two bands that can be attributed to the O—O stretching mode of bound oxygen. Discuss how this observation may be used to exclude one or more of the four proposed schemes (**4–7**) for binding of O_2 to the Fe_2 site of haemerythrin.

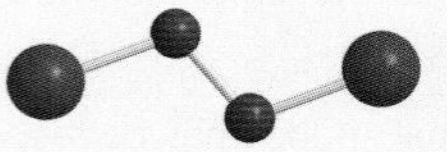

4

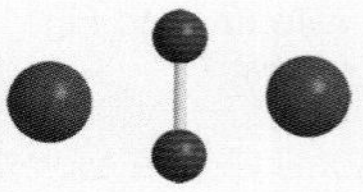

5

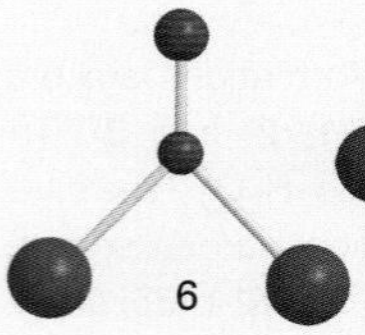

6

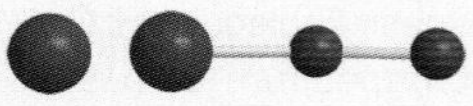

7

19.38 We saw in Impact on the environment 19.1 that water, carbon dioxide, and methane are able to absorb some of the Earth's infrared emissions whereas nitrogen and oxygen cannot. The computational methods discussed in Section 14.14 can be also be used to simulate vibrational spectra, and from the results of the calculation it is possible to determine the correspondence between a vibrational frequency and the atomic displacements that give rise to a normal mode. (a) Using molecular modelling software and the computational method of your instructor's choice, investigate and depict pictorially the vibrational normal modes of CH_4, CO_2, and H_2O in the gas phase. (b) Which vibrational modes of CH_4, CO_2, and H_2O are responsible for absorption of infrared radiation?

20

Spectroscopy: electronic transitions

Ultraviolet and visible spectra 503

20.1 Practical considerations 505

20.2 Absorption intensities 505

20.3 The Franck–Condon principle 507

20.4 Specific types of transitions 508

Radiative and non-radiative decay 510

20.5 Fluorescence 511

20.6 Phosphorescence 511

20.7 Quenching 512

20.8 Lasers 518

Photoelectron spectroscopy 522

FURTHER INFORMATION 20.1 524

FURTHER INFORMATION 20.2 524

CHECKLIST OF KEY CONCEPTS 525

ROAD MAP OF KEY EQUATIONS 526

QUESTIONS AND EXERCISES 526

The energy needed to change the occupation of orbitals in a molecule is of the order of several electronvolts (an energy difference of 1 eV is equivalent to radiation of wavenumber 8066 cm^{-1}). Consequently, the photons emitted or absorbed when such changes occur lie in the visible and ultraviolet regions of the spectrum, which spread from about 14 000 cm^{-1} for red light to 21 000 cm^{-1} for blue light, and on to 50 000 cm^{-1} for ultraviolet radiation (Table 20.1).

Many of the colours of the objects in the world around us, including the green of vegetation, the colours of flowers and of synthetic dyes, and the colours of pigments and minerals, stem from transitions in which an electron makes a transition from one orbital of a molecule or ion into another. The change in the distribution of probability density of an electron that takes place when chlorophyll absorbs red and blue light (leaving green to be reflected) is the primary energy harvesting step by which our planet captures energy from the Sun and uses it to drive the non-spontaneous reactions of photosynthesis. In some cases the relocation of an electron may be so extensive that it results in the breaking of a bond and the dissociation of the molecule: such processes give rise to the numerous reactions of photochemistry, including the reactions that sustain or damage the atmosphere.

Ultraviolet and visible spectra

White light is a mixture of light of all different colours. The removal, by absorption, of any one of these colours from white light results in the complementary colour being observed. For instance, the absorption of red light from white light by an object results in that object appearing green, the complementary colour of red. Conversely, the absorption of green results in the object appearing red. The pairs of complementary colours are neatly summarized by

Table 20.1
Colour, frequency, and energy of light

Colour	λ/nm	$\nu/(10^{14}$ Hz)	$\tilde{\nu}/(10^4$ cm^{-1})	E/eV	E/(kJ mol^{-1})
Infrared	1000	3.00	1.00	1.24	120
Red	700	4.28	1.43	1.77	171
Orange	620	4.84	1.61	2.00	193
Yellow	580	5.17	1.72	2.14	206
Green	530	5.66	1.89	2.34	226
Blue	470	6.38	2.13	2.64	254
Violet	420	7.14	2.38	2.95	285
Near ultraviolet	300	10.0	3.3	4.15	400
Far ultraviolet	200	15.0	5.00	6.20	598

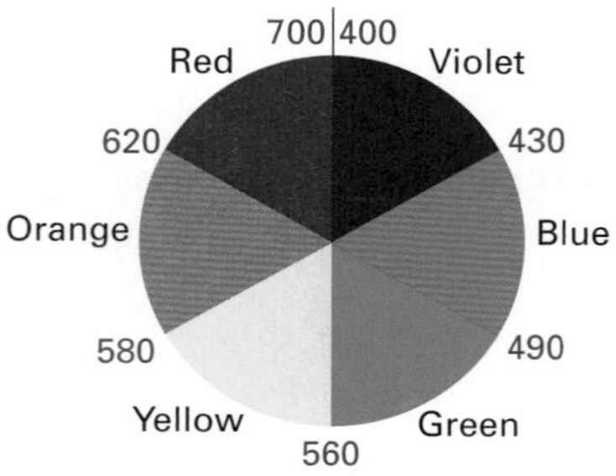

Fig. 20.1 An artist's colour wheel: complementary colours are opposite one another on a diameter. The numbers correspond to wavelengths of light in nanometres (nm).

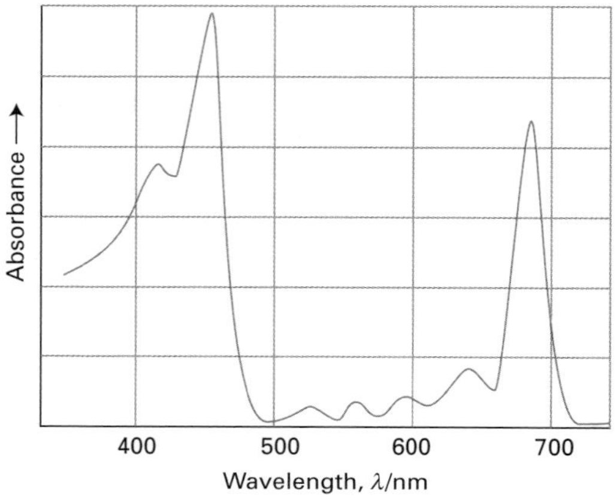

Fig. 20.2 The absorption spectrum of chlorophyll in the visible region. Note that it absorbs in the red and blue regions, and that green light is not absorbed.

the artist's colour wheel, shown in Fig. 20.1, where complementary colours lie opposite one another along a diameter.

It should be stressed, however, that the perception of colour is a very subtle phenomenon. Although an object may appear green because it absorbs red light, it may also appear green because it absorbs all colours from the incident light *except* green. This is the origin of the colour of vegetation, because chlorophyll absorbs in two regions of the spectrum, leaving green to be reflected (Fig. 20.2). Moreover, an absorption band may be very broad, and although it may be a maximum at one particular wavelength, it may have a long tail that spreads into other regions (Fig. 20.3). In such cases, it is very difficult to predict the perceived colour from the location of the absorption maximum.

In Chapter 19 we discussed the general principles that determine the extent of absorption of electromagnetic radiation by a sample and contribute to the linewidths of absorption spectra. Here we apply

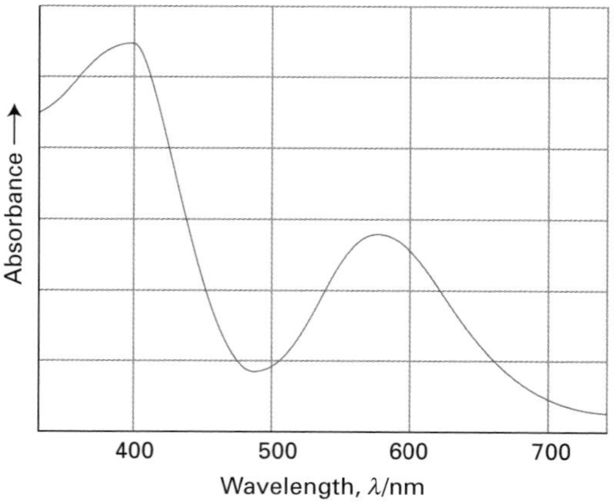

Fig. 20.3 An electronic absorption of a species in solution is typically very broad and consists of several broad bands.

those principles to electronic transitions in the ultraviolet and visible regions of the electromagnetic spectrum.

20.1 Practical considerations

For the visible region of the spectrum, common light sources in absorption spectrometers include light-emitting diodes (Section 17.3) or tungsten–iodine lamps. A discharge through deuterium gas or xenon in quartz is still widely used for the near ultraviolet.

The simplest dispersing element is a glass or quartz prism, but modern instruments use a diffraction grating. For work in the visible region of the spectrum, the device consists of a glass or ceramic plate into which fine grooves have been cut about 1000 nm apart (a spacing comparable to the wavelength of visible light) and covered with a reflective aluminium coating. The grating causes interference between waves reflected from its surface, and constructive interference occurs at specific angles that depend on the frequency of the radiation being used. Thus, each wavelength of light is directed into a specific direction (Fig. 20.4). In a monochromator, a narrow exit slit allows only a narrow range of wavelengths to reach the detector. Turning the grating around an axis perpendicular to the incident and diffracted beams allows different wavelengths to be analysed; in this way, the absorption spectrum is built up one narrow wavelength range at a time.

Detectors may consist of a single radiation sensing element or of several small elements arranged in one or two-dimensional arrays. A common detector is a photodiode, a solid-state device that conducts electricity when struck by photons because light-induced electron transfer reactions in the detector material create mobile charge carriers (negatively charged electrons and positively charged 'holes'). Silicon is sensitive in the visible region. A charge-coupled device (CCD) is a two-dimensional array of several million photodiode detectors. With a CCD, a wide range of wavelengths that emerge from a polychromator are detected simultaneously, thus eliminating the need to measure light intensity one narrow wavelength range at a time. CCD detectors are used widely to monitor absorption, emission, and Raman scattering.

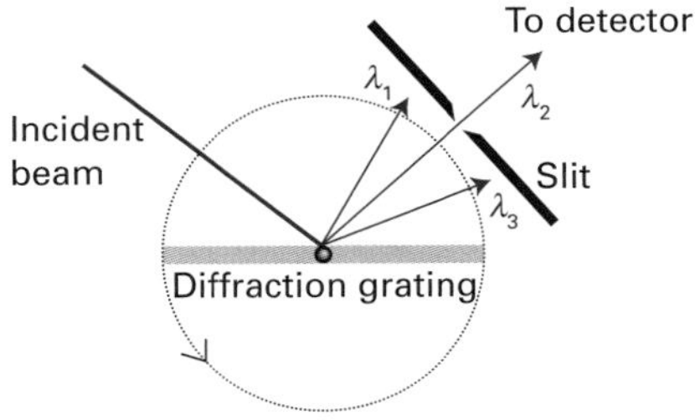

Fig. 20.4 A beam of light is dispersed by a diffraction grating into three component wavelengths λ_1, λ_2, and λ_3. In the configuration shown, only radiation with λ_2 passes through a narrow slit and reaches the detector. Rotating the diffraction grating allows λ_1 and λ_3 to reach the detector.

20.2 Absorption intensities

The intensity of absorption of radiation at a particular wavelength is related to the molar concentration [J] of the absorbing species by the Beer–Lambert law (Section 10.1):

$$I = I_0 10^{-\varepsilon[\mathrm{J}]L} \qquad \text{Beer–Lambert law} \qquad (20.1)$$

I_0 and I are the incident and transmitted intensities, respectively, L is the length of the sample, and ε (epsilon) is the **molar absorption coefficient** (formerly and still widely the 'extinction coefficient'), with dimensions of 1/(concentration × length). Typical values of ε for strong transitions are of the order of 10^4–10^5 dm^3 mol^{-1} cm^{-1}, indicating that in a solution of molar concentration 0.01 mol dm^{-3} the intensity of light (of frequency corresponding to the maximum absorption) falls to 10 per cent of its initial value after passing through about 0.1 mm of solution (Fig. 20.5). The Beer–Lambert law is an empirical result, but its form can be justified by considering the passage of light through a uniform absorbing medium (Further information 20.1).

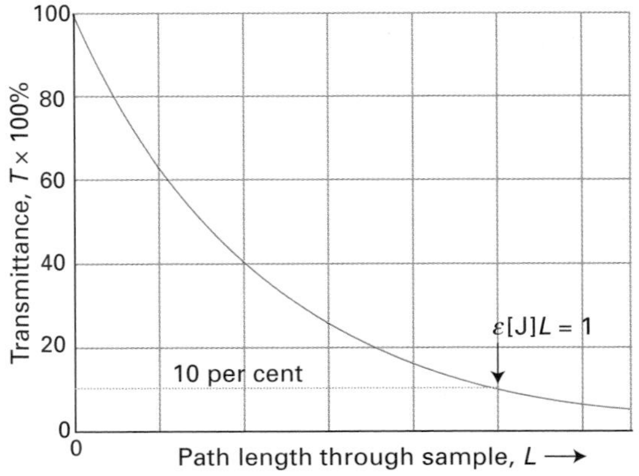

Fig. 20.5 The intensity of light transmitted by an absorbing sample decreases exponentially with the path length through the sample.

The **absorbance** $A = \varepsilon[\mathrm{J}]L$ of a sample is measured by using the incident and final intensities of a light beam:

$$A = \log\frac{I_0}{I} \quad \text{Definition Absorbance} \quad (20.2)$$

(The logarithm is a common logarithm, to the base 10.) It is common to report the absorption of radiation in terms of the **transmittance**, T, of a sample at a given frequency, where

$$T = \frac{I}{I_0} \quad \text{Definition Transmittance} \quad (20.3)$$

Thus, $A = -\log T$. The Beer–Lambert law then takes either of the two following forms:

$$A = \varepsilon L[\mathrm{J}],\ T = 10^{-\varepsilon[\mathrm{J}]L} \quad \text{Alternative forms Beer–Lambert law} \quad (20.4)$$

Brief illustration 20.1 The Beer–Lambert law

The transmittance through an absorbing sample changes from $T_1 = 10^{-\varepsilon[\mathrm{J}]L_1}$ to $T_2 = 10^{-\varepsilon[\mathrm{J}]L_2}$ when the path length through the sample is changed from L_1 to L_2. If the path length is doubled, then $L_2 = 2L_1$ and

$$T_2 = 10^{-2\varepsilon[\mathrm{J}]L_1} \overset{e^{ax}=(e^x)^a}{=} (10^{-\varepsilon[\mathrm{J}]L_1})^2 = T_1^2$$

If the transmittance is 0.1 for a path length of 1 cm (corresponding to a 90 per cent reduction in intensity), then it would be $(0.1)^2 = 0.01$ for a path of double the length (corresponding to a 99 per cent reduction in intensity overall).

The concentration of an absorbing species can be determined as explained in Section 10.1, by using eqn 20.4 in the form $[\mathrm{J}] = A/\varepsilon L$. Measurements at two wavelengths can be used to find the individual concentrations of two components A and B in a mixture. For this analysis, we write the total absorbance at a given wavelength as

$$A = A_\mathrm{A} + A_\mathrm{B} = \varepsilon_\mathrm{A}[\mathrm{A}]L + \varepsilon_\mathrm{B}[\mathrm{B}]L = (\varepsilon_\mathrm{A}[\mathrm{A}] + \varepsilon_\mathrm{B}[\mathrm{B}])L$$

Then, for two measurements of the total absorbance at wavelengths λ_1 and λ_2 at which the molar absorption coefficients are ε_1 and ε_2 (Fig. 20.6), we have

$$A_1 = (\varepsilon_{\mathrm{A}1}[\mathrm{A}] + \varepsilon_{\mathrm{B}1}[\mathrm{B}])L \qquad A_2 = (\varepsilon_{\mathrm{A}2}[\mathrm{A}] + \varepsilon_{\mathrm{B}2}[\mathrm{B}])L$$

As shown in the following Derivation, these two simultaneous equations can be solved for the two unknowns, the molar concentrations of A and B:

$$[\mathrm{A}] = \frac{\varepsilon_{\mathrm{B}2}A_1 - \varepsilon_{\mathrm{B}1}A_2}{(\varepsilon_{\mathrm{A}1}\varepsilon_{\mathrm{B}2} - \varepsilon_{\mathrm{A}2}\varepsilon_{\mathrm{B}1})L} \quad (20.5\mathrm{a})$$

$$[\mathrm{B}] = \frac{\varepsilon_{\mathrm{A}1}A_2 - \varepsilon_{\mathrm{A}2}A_1}{(\varepsilon_{\mathrm{A}1}\varepsilon_{\mathrm{B}2} - \varepsilon_{\mathrm{A}2}\varepsilon_{\mathrm{B}1})L} \quad (20.5\mathrm{b})$$

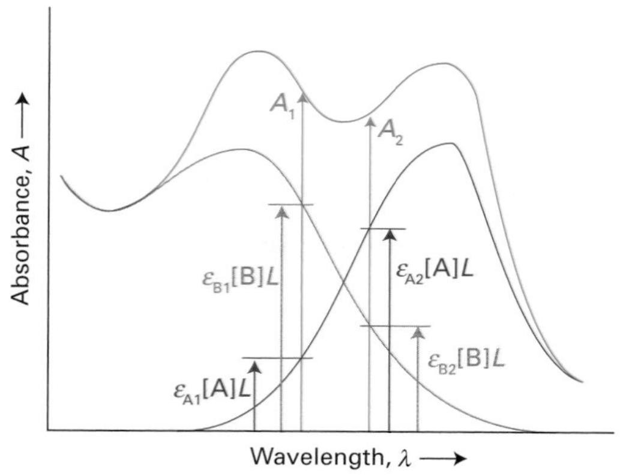

Fig. 20.6 The concentrations of two absorbing species in a mixture can be determined from their molar absorption coefficients and the measurement of their absorbances at two different wavelengths lying within their joint absorption region.

Derivation 20.1

Determining concentrations in a mixture

The two equations to solve for [A] and [B] are

$$\varepsilon_{\mathrm{A}1}[\mathrm{A}]L + \varepsilon_{\mathrm{B}1}[\mathrm{B}]L = A_1 \qquad \varepsilon_{\mathrm{A}2}[\mathrm{A}]L + \varepsilon_{\mathrm{B}2}[\mathrm{B}]L = A_2$$

To match the two second terms, multiply the first by $\varepsilon_{\mathrm{B}2}$ and the second by $\varepsilon_{\mathrm{B}1}$, to obtain

$$\varepsilon_{\mathrm{B}2}\varepsilon_{\mathrm{A}1}[\mathrm{A}]L + \varepsilon_{\mathrm{B}2}\varepsilon_{\mathrm{B}1}[\mathrm{B}]L = \varepsilon_{\mathrm{B}2}A_1$$
$$\varepsilon_{\mathrm{B}1}\varepsilon_{\mathrm{A}2}[\mathrm{A}]L + \varepsilon_{\mathrm{B}1}\varepsilon_{\mathrm{B}2}[\mathrm{B}]L = \varepsilon_{\mathrm{B}1}A_2$$

When the second is subtracted from the first, we obtain

$$\varepsilon_{\mathrm{B}2}\varepsilon_{\mathrm{A}1}[\mathrm{A}]L - \varepsilon_{\mathrm{B}1}\varepsilon_{\mathrm{A}2}[\mathrm{A}]L = \varepsilon_{\mathrm{B}2}A_1 - \varepsilon_{\mathrm{B}1}A_2$$

which rearranges into eqn 20.5a. To obtain eqn 20.5b, repeat the process by multiplying the first equation by $\varepsilon_{\mathrm{A}2}$ and the second by $\varepsilon_{\mathrm{A}1}$ so that the [A] terms cancel when the two equations are subtracted.

There may be a wavelength at which the molar absorption coefficients of the two species are equal; we write this common value as ε_iso. The total absorbance of the mixture at this wavelength is

$$A_\mathrm{iso} = (\varepsilon_\mathrm{iso}[\mathrm{A}] + \varepsilon_\mathrm{iso}[\mathrm{B}])L = \varepsilon_\mathrm{iso}([\mathrm{A}] + [\mathrm{B}])L \quad (20.6)$$

Even if A and B are interconverted in a reaction of the form A → B or its reverse, then because their total concentration remains constant, so does A_iso. As a result, it is possible to observe one or more **isosbestic points** (the name 'isosbestic' comes from the Greek words for 'the same' and 'extinguish'), which are invariant points in the absorption spectrum (Fig. 20.7). It is very unlikely that three or more species would have the same molar absorption coefficients at

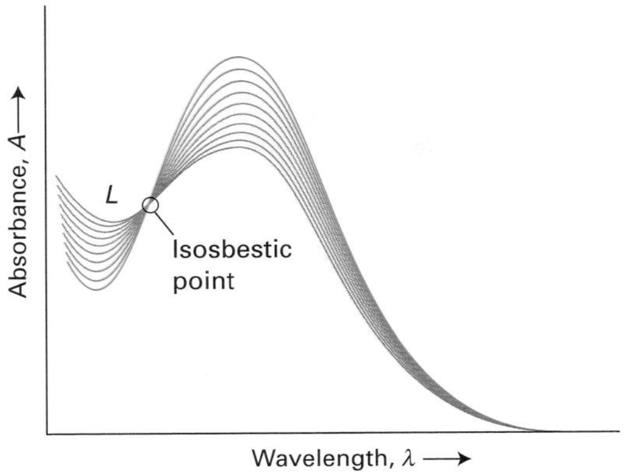

Fig. 20.7 One or more isosbestic points are formed when there are two interrelated absorbing species in solution. The curves correspond to different stages of the reaction A → B.

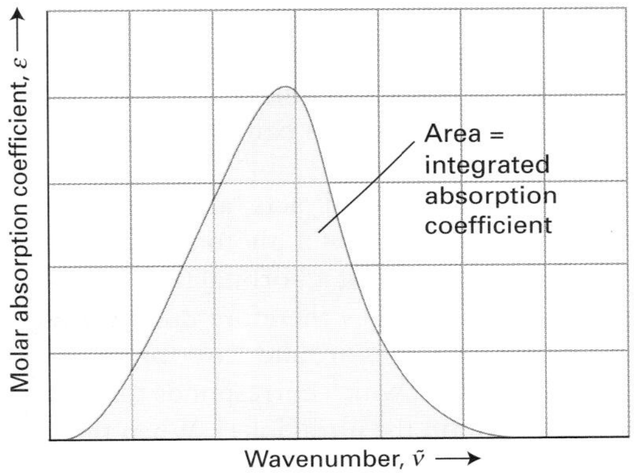

Fig. 20.8 The integrated absorption coefficient of a transition is the area under a plot of the molar absorption coefficient against the wavenumber of the incident radiation.

a single wavelength. Therefore, the observation of an isosbestic point, or at least not more than one such point, is compelling evidence that a solution consists of only two solutes in equilibrium with each other with no intermediates.

The molar absorption coefficient depends on the wavenumber (or, equivalently, the frequency and wavelength) of the incident radiation and is greatest where the absorption is most intense. The maximum value of the molar absorption coefficient, ε_{max}, is an indication of the intensity of a transition. However, because absorption bands generally spread over a range of wavenumbers, the absorption at a single wavenumber might not give a true indication of the intensity. The latter is best reported as the **integrated absorption coefficient,** $\mathcal{A}$, the area under the plot of the molar absorption coefficient against wavenumber (Fig. 20.8).

20.3 The Franck–Condon principle

Whenever an electronic transition takes place it is accompanied by the excitation of vibrations of the molecule. In the electronic ground state of a molecule, the nuclei take up locations in response to the Coulombic forces acting on them. These forces arise from the electrons and the other nuclei. After an electronic transition, when electron density has migrated to a different part of the molecule, the nuclei are subjected to different forces and the molecule may respond by bursting into vibration. As a result, some of the energy used to redistribute an electron is in fact used to stimulate the vibrations of the absorbing molecules. Therefore, instead of a single, sharp, and purely electronic absorption line being observed, the absorption spectrum consists of many lines. This **vibrational structure** of an electronic transition can be resolved if the sample is gaseous, but in a liquid or solid the lines usually merge together and result in a broad, almost featureless band (Fig. 20.9).

The vibrational structure of a band is explained by the **Franck–Condon principle:**

> Because nuclei are so much more massive than electrons, an electronic transition takes place faster than the nuclei can respond.

In an electronic transition, electron density is lost rapidly from some regions of the molecule and is built up rapidly in others. As a result, the initially stationary nuclei suddenly experience a new force field. They respond by beginning to vibrate, and (in classical terms) swing backwards and forwards from their original separation, which they maintained during the rapid electronic excitation. The initial,

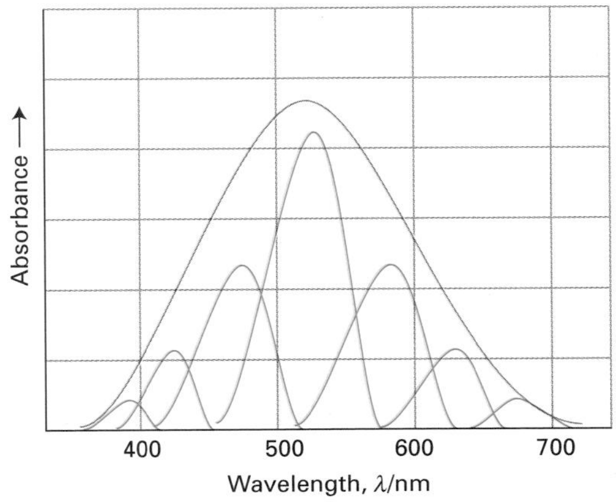

Fig. 20.9 An electronic absorption band consists of many superimposed bands which merge together to give a single broad band with unresolved vibrational structure.

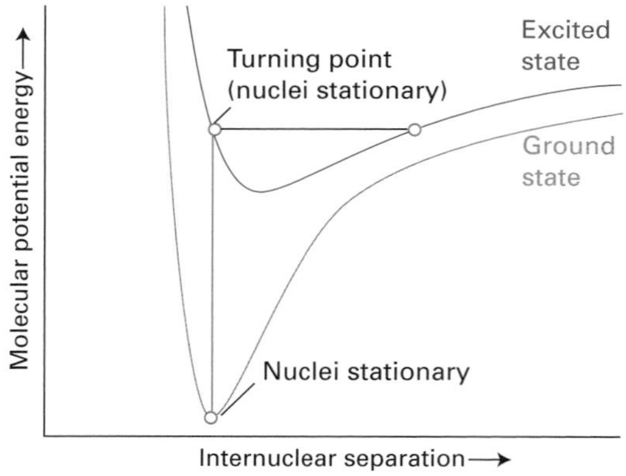

Fig. 20.10 According to the Franck–Condon principle, the most intense electronic transition is from the ground vibrational state to the vibrational state that lies vertically above it in the upper electronic state. Transitions to other vibrational levels also occur, but with lower intensity.

stationary, equilibrium separation of the nuclei in the initial electronic state therefore becomes the new, stationary, **turning point**, one of the end points of a nuclear swing, in the final electronic state (Fig. 20.10).

To predict the most likely final vibrational state we draw a vertical line from the minimum of the lower curve (the starting point for the transition) up to the point at which the line intersects the curve representing the upper electronic state (the turning point of the newly stimulated vibration). This procedure gives rise to the name **vertical transition** for a transition in accord with the Franck–Condon principle. In practice, the electronically excited molecule may be formed in one of several excited vibrational states all with turning points nearly vertically above the minimum of the lower curve, so the absorption occurs at several different frequencies. As remarked above, in a condensed medium, the individual transitions merge together to give a broad, largely featureless band of absorption.

20.4 Specific types of transitions

The absorption of a photon can often be traced to the excitation of an electron that is localized on a small group of atoms. For example, an absorption at about 290 nm is normally observed when a carbonyl group is present. Groups with characteristic optical absorptions are called **chromophores** (from the Greek for 'colour bringer'), and their presence often accounts for the colours of many substances.

The transition responsible for absorption in carbonyl compounds can be traced to the lone pairs of electrons on the O atom. One of these electrons may

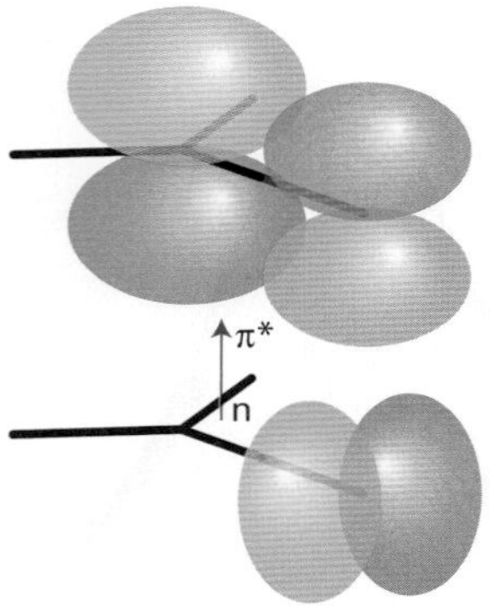

Fig. 20.11 A carbonyl group acts as a chromophore primarily on account of the excitation of a nonbonding O lone-pair electron to an antibonding CO π* orbital.

be excited into an empty π* orbital of the carbonyl group (Fig. 20.11), which gives rise to an **n-to-π* transition**, where n denotes a nonbonding orbital, an orbital that is neither bonding nor antibonding, such as that occupied by a lone pair. Typical absorption energies are about 4 eV.

A C=C double bond acts as a chromophore because the absorption of a photon excites a π electron into an antibonding π* orbital (Fig. 20.12). The chromophore activity is therefore due to a **π-to-π* transition.** Its energy is around 7 eV for an unconjugated double bond, which corresponds to an absorption at 180 nm (in the ultraviolet). When the double bond is part of a conjugated chain, the energies of the molecular orbitals lie closer together and the transition shifts into the visible region of the spectrum (see Section 14.16).

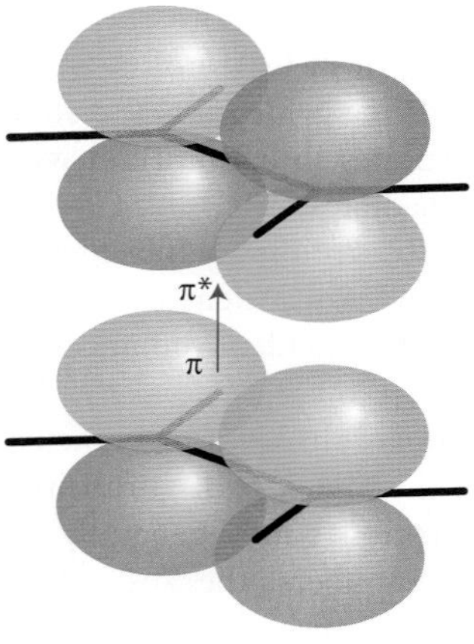

Fig. 20.12 A carbon–carbon double bond acts as a chromophore. One of its important transitions is the π-to-π* transition illustrated here, in which an electron is promoted from a π orbital to the corresponding antibonding orbital.

Brief illustration 20.2 π-to-π* Transitions

Many of the reds and yellows of vegetation are due to π-to-π* transitions. For example, the carotenes that are present in green leaves (but are concealed by the intense absorption of the chlorophyll until the latter decays) collect some of the solar radiation incident on the leaf by a π-to-π* transition in their long conjugated hydrocarbon chains. A similar type of absorption is responsible for the primary process of vision (Impact 20.1).

A d-metal complex may absorb light as a result of the transfer of an electron from the ligands into the d orbitals of the central atom, or vice versa. In such **charge-transfer transitions** the electron moves through a considerable distance, which means that the redistribution of charge as measured by the transition dipole moment may be large and the absorption correspondingly intense. Indeed, the most intense electronic transitions responsible for the colours of many d-metal complexes are charge-transfer transitions. In the permanganate ion, MnO_4^-, the charge redistribution that accompanies the migration of an electron from the O atoms to the central Mn atom results in a charge-transfer transition in the range 420–700 nm and accounts for the intense purple colour of the ion.

Impact on biochemistry 20.1

Vision

The eye is an exquisite photochemical organ that acts as a transducer, converting radiant energy into electrical signals that travel along neurons. Here we concentrate on the events taking place in the human eye, but similar processes occur in all animals. Indeed, a single type of protein, rhodopsin, is the primary receptor for light throughout the animal kingdom, which indicates that vision emerged very early in evolutionary history, no doubt because of its enormous value for survival.

Photons enter the eye through the cornea, pass through the ocular fluid that fills the eye, and falls on the retina. The ocular fluid is principally water, and passage of light through this medium is largely responsible for the *chromatic aberration* of the eye, the blurring of the image as a result of different frequencies being brought to slightly different focuses. The chromatic aberration is reduced to some extent by the tinted region called the *macular pigment* that covers part of the retina. The pigments in this region are the carotene-like xanthophylls (**1**), which remove some of the blue light and hence help to sharpen the image. They also protect the photoreceptor molecules from too great a flux of potentially dangerous high-energy photons. The xanthophylls have delocalized electrons that spread along the chain of conjugated double bonds, and the π-to-π* transition lies in the visible.

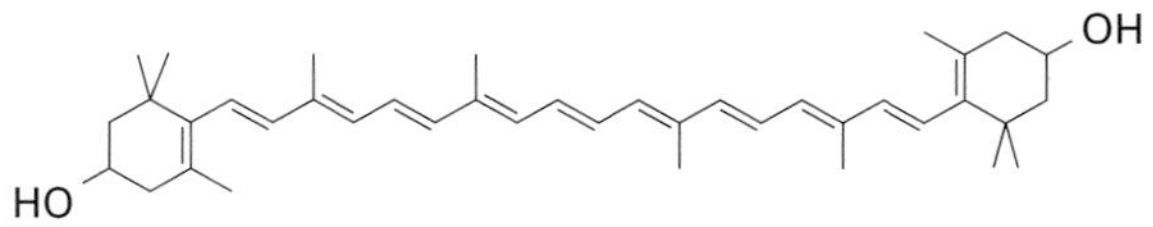

1 A xanthophyll

About 57 per cent of the photons that enter the eye reach the retina; the rest are scattered or absorbed by the ocular fluid. Here the primary act of vision takes place, in which the chromophore of a rhodopsin molecule absorbs a photon in another π-to-π* transition. A rhodopsin molecule consists of an opsin protein molecule to which is attached an 11-*cis*-retinal molecule (**2**). The latter resembles half a carotene molecule, showing Nature's economy in its use of available materials. The attachment is by the formation of a Schiff's base, utilizing the —CHO group of the chromophore. The free 11-*cis*-retinal molecule absorbs in the ultraviolet, but attachment to the opsin protein molecule shifts the absorption into the visible region. The rhodopsin molecules are situated in the membranes of special cells (the 'rods' and the 'cones') that cover the retina. The opsin molecule is anchored into the cell membrane by two hydrophobic groups and largely surrounds the chromophore (Fig. 20.13).

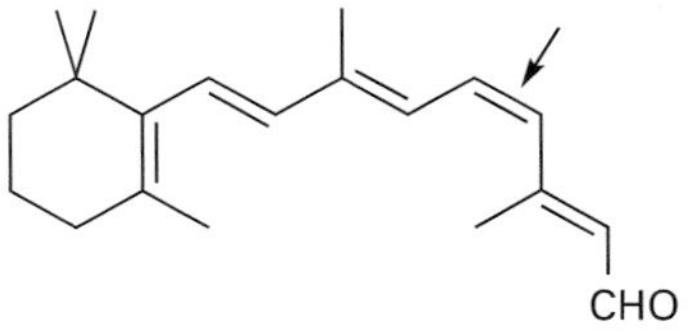

2 11-*cis*-Retinal

Immediately after the absorption of a photon, the 11-*cis*-retinal molecule undergoes photoisomerization into all-*trans*-retinal (**3**). Photoisomerization takes about 200 fs and about 67 pigment molecules isomerize for every 100 photons that are absorbed. The process is able to occur because the π-to-π* excitation of an electron loosens one of the π bonds (the one indicated by the arrow in the diagram), its torsional rigidity is lost, and one part of the molecule swings round into its new position. At that point, the molecule returns to its ground

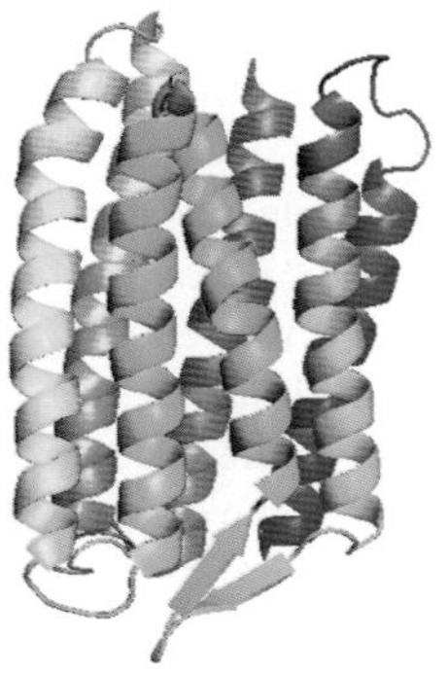

Fig. 20.13 The structure of the rhodopsin molecule.

state, but is now trapped in its new conformation. The straightened tail of all-*trans*-retinal results in the molecule taking up more space than 11-*cis*-retinal did, so the molecule presses against the coils of the opsin molecule that surrounds it. Thus, in about 0.25–0.50 ms from the initial absorption event, the rhodopsin molecule is activated.

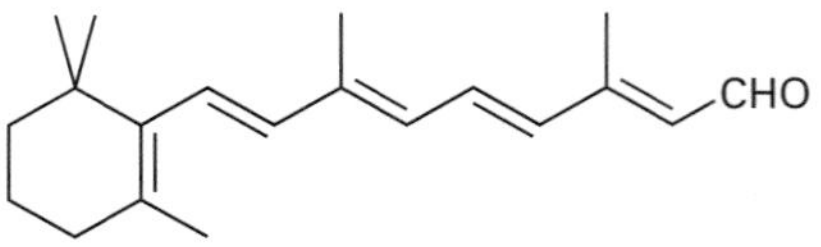

3 All-*trans*-retinal

Now a sequence of biochemical events—the *biochemical cascade*—converts the altered configuration of the rhodopsin molecule into a pulse of electric potential that travels through the optical nerve into the optical cortex, where it is interpreted as a signal and incorporated into the web of events we call 'vision'. At the same time, the resting state of the rhodopsin molecule is restored by a series of non-radiative chemical events powered by ATP. The process involves the escape of all-*trans*-retinal as all-*trans*-retinol (in which —CHO has been reduced to —CH_2OH) from the opsin molecule by a process catalysed by the enzyme rhodopsin kinase and the attachment of another protein molecule, arrestin. The free all-*trans*-retinol molecule now undergoes enzyme-catalysed isomerization into 11-*cis*-retinol followed by dehydrogenation to form 11-*cis*-retinal, which is then delivered back into an opsin molecule. At this point, the cycle of excitation, photoisomerization, and regeneration is ready to begin again.

Radiative and non-radiative decay

In most cases, the excitation energy of a molecule that has absorbed a photon is degraded by the process called **non-radiative decay** into the disordered thermal motion of its surroundings. In other cases, its electrons may undergo a redistribution that results in it undergoing an **internal conversion** (IC), a radiationless conversion to another state of the same multiplicity. However, one process by which an electronically excited molecule can discard its excess energy is by **radiative decay**, in which an electron relaxes back into a lower energy orbital and in the process generates a photon. As a result, and if the emitted radiation is in the visible region of the spectrum, an observer sees the sample glowing.

There are two principal modes of radiative decay, fluorescence and phosphorescence (Fig. 20.14). In **fluorescence**, the spontaneously emitted radiation ceases very soon (within nanoseconds) after the exciting radiation is extinguished. In **phosphorescence**, the spontaneous emission may persist for long periods—even hours, but characteristically seconds or fractions of seconds. The difference suggests that fluorescence is an immediate conversion of absorbed light into re-emitted radiant energy and that phosphorescence involves the storage of energy in a reservoir from which it slowly leaks.

Absorption might result in **dissociation**, or fragmentation (Fig. 20.15). The onset of dissociation can be detected in an absorption spectrum by seeing that the vibrational structure of a band terminates at a certain energy. Absorption occurs in a continuous band above this **dissociation limit**, the highest frequency before the onset of continuous absorption, because the final state is unquantized translational

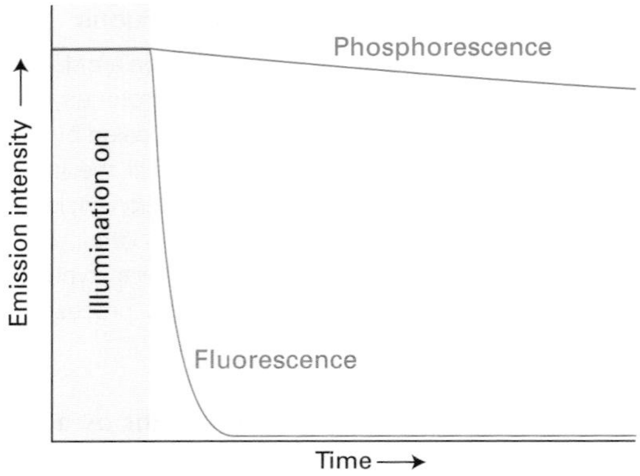

Fig. 20.14 The empirical (observation-based) distinction between fluorescence and phosphorescence is that the former is extinguished very quickly after the exciting source is removed, whereas the latter continues with relatively slowly diminishing intensity.

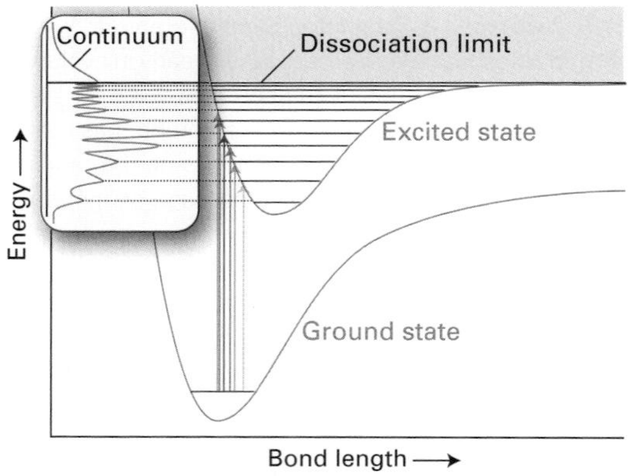

Fig. 20.15 When absorption occurs to unbound states of the upper electronic state, the molecule dissociates and the absorption is a continuum. Below the dissociation limit the electronic spectrum has a normal vibrational structure.

motion of the fragments. Locating the dissociation limit is a valuable way of determining the bond dissociation energy. Dissociation may also occur if the electronic transition takes place directly to a purely repulsive state, one that shows no minimum corresponding to bonding. Internal conversion from a bound state to a repulsive state might also result in dissociation. Because it occurs at wavenumbers below those needed for dissociation of the initial excited state, this process is called **predissociation.**

20.5 Fluorescence

Figure 20.16 is a simple example of a **Jablonski diagram**, a schematic portrayal of molecular electronic and vibrational energy levels, which shows the sequence of steps involved in fluorescence. The initial absorption takes the molecule to an excited electronic state, and if the absorption spectrum were monitored it would look like the one shown in Fig. 20.17a. The excited molecule is subjected to collisions with the surrounding molecules, and as it gives up energy it steps down the ladder of vibrational levels. The surrounding molecules, however, might be unable to accept the larger energy needed to lower the molecule to the ground electronic state. The excited state might therefore survive long enough to generate a photon and emit the remaining excess energy as radiation. The downward electronic transition is **vertical**, which means in accord with the Franck–Condon principle, and the fluorescence spectrum has a vibrational structure characteristic of the lower electronic state (Fig. 20.17b).

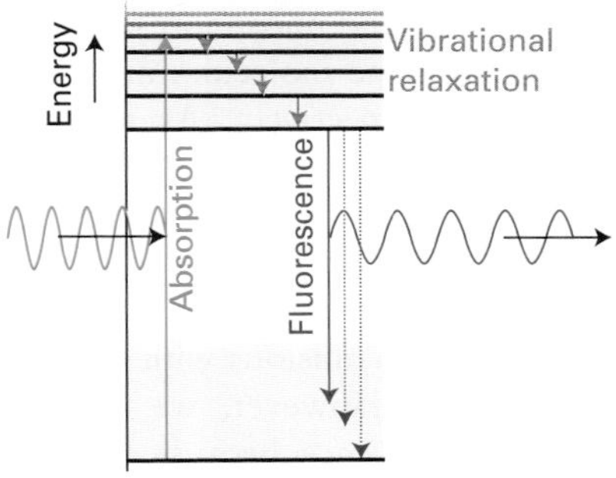

Fig. 20.16 A Jablonski diagram showing the sequence of steps leading to fluorescence. After the initial absorption the upper vibrational states undergo radiationless decay—the process of vibrational relaxation—by giving up energy to the surroundings. A radiative transition then occurs from the ground state of the upper electronic state. In practice, the separation of the ground states of the electronic states is 10 to 100 times greater than the separation of the vibrational levels.

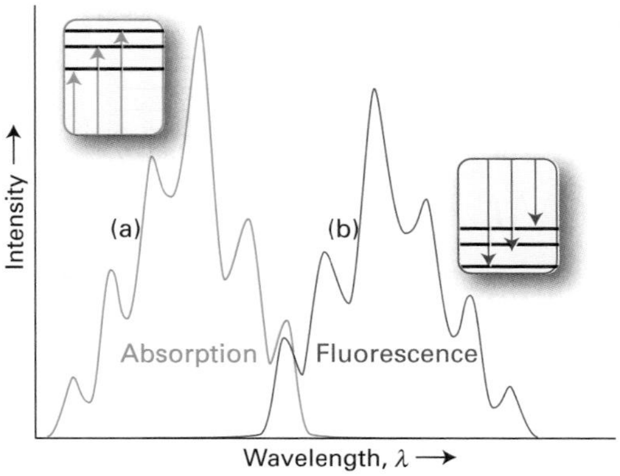

Fig. 20.17 The absorption spectrum (a) shows a vibrational structure characteristic of the upper state. The fluorescence spectrum (b) shows a structure characteristic of the lower state; it is also displaced to lower frequencies and resembles a mirror image of the absorption.

Fluorescence occurs at a lower frequency than that of the incident radiation because the fluorescence radiation is emitted after some vibrational energy has been lost to the surroundings. The vivid oranges and greens of fluorescent dyes are an everyday manifestation of this effect: they absorb in the ultraviolet and fluoresce in the visible. The mechanism also suggests that the intensity of the fluorescence ought to depend on the ability of the surrounding molecules, such as those of a solvent, to accept the electronic and vibrational quanta. It is indeed found that a solvent composed of molecules with widely spaced vibrational levels (such as water) may be able to accept the large quantum of electronic energy and so decrease the intensity of the solute's fluorescence.

20.6 Phosphorescence

Figure 20.18 is a Jablonski diagram showing the events leading to phosphorescence. The first steps are the same as in fluorescence, but the presence of a triplet state plays a decisive role. A **triplet state** is a state in which two electrons in different orbitals have parallel spins: the ground state of O_2 which was discussed in Section 14.10 is an example. The name 'triplet' reflects the (quantum mechanical) fact that the total spin of two parallel electron spins (↑↑) can adopt only three orientations with respect to an axis. An ordinary spin-paired state (↑↓) is called a **singlet state** because there is only one orientation in space for such a pair of spins. In the language introduced in Section 13.17, a triplet state has $S = 1$ and M_S has one of the three values +1, 0, and −1; a singlet state has $S = 0$ and M_S has the single value 0.

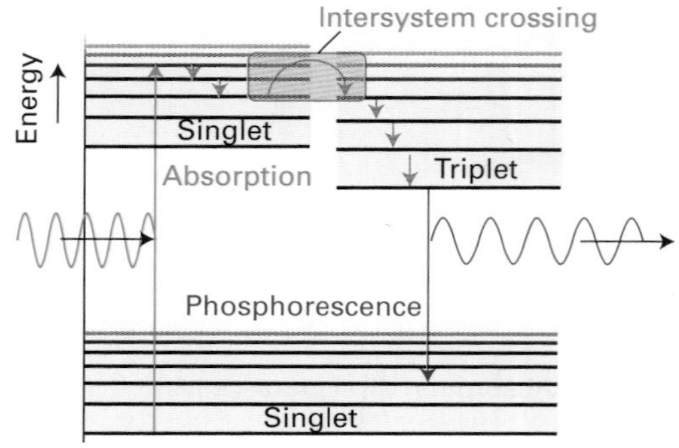

Fig. 20.18 The sequence of steps leading to phosphorescence. The important step is the intersystem crossing from an excited singlet to an excited triplet state. The triplet state acts as a slowly radiating reservoir because the return to the ground state is very slow.

The ground state of a typical phosphorescent molecule is a singlet because its electrons are all paired; the excited state to which the absorption excites the molecule is also a singlet. The peculiar feature of a phosphorescent molecule, however, is that it possesses an excited triplet state of an energy similar to that of the excited singlet state and into which the excited singlet state may convert. Hence, if there is a mechanism for unpairing two electron spins (and so converting ↑↓ into ↑↑), then the molecule may undergo **intersystem crossing** (ISC) and become a triplet state. The unpairing of electron spins is possible if the molecule contains a heavy atom, such as an atom of sulfur, with strong spin–orbit coupling (Section 13.18). Then the angular momentum needed to convert a singlet state into a triplet state may be acquired from the orbital motion of the electrons.

After an excited singlet molecule crosses into a triplet state, it continues to discard energy into the surroundings and to step down the ladder of vibrational states. However, it is now stepping down the triplet's ladder and at the lowest vibrational energy level it is trapped. The surroundings cannot extract the final, large quantum of electronic excitation energy. Moreover, the molecule cannot radiate its energy because return to the ground state is forbidden by the selection rule governing transitions: a triplet state cannot convert into a singlet state because the spin of one electron cannot reverse in direction relative to the other electron during a transition ($\Delta S = 0$ for electronic transitions). The radiative transition, however, is not totally forbidden because the spin–orbit coupling responsible for the intersystem crossing also breaks this rule. The molecules are therefore able to emit weakly and the emission may continue long after the original excited state was formed.

The mechanism of phosphorescence summarized in Fig. 20.18 accounts for the observation that the excitation energy seems to become trapped in a slowly leaking reservoir. It also suggests, as is confirmed experimentally, that phosphorescence should be most intense from solid samples: energy transfer is then less efficient and the intersystem crossing has time to occur as the singlet excited state loses vibrational energy. The mechanism also suggests that the phosphorescence efficiency should depend on the presence of a moderately heavy atom—with its ability to flip electron spins—which is in fact the case.

20.7 Quenching

A number of processes can remove, or **quench**, the excitation energy from a fluorescent or phosphorescent molecule. Examples of such processes include energy transfer, electron transfer, and photochemical reactions, reactions that are initiated by the absorption of light.

In this section we explore the rates and mechanisms of quenching, and need to gain some insight into the timescales for formation and deactivation of excited states. Electronic transitions caused by absorption of ultraviolet and visible radiation occur within 10^{-16}–10^{-15} s. We should expect, then, the upper limit for the rate constant of a quenching process, such as a first-order photochemical reaction, to be about $10^{16}\ s^{-1}$. Fluorescence is slower than absorption, with typical time constants of 10^{-12}–10^{-6} s. Therefore, the excited singlet state can initiate very fast photochemical reactions in the femtosecond (10^{-15} s) to picosecond (10^{-12} s) timescale. Typical intersystem crossing and phosphorescence time constants for large organic molecules are 10^{-12}–10^{-4} s and 10^{-6}–10^{-1} s, respectively. As a consequence, excited triplet states are photochemically important. Indeed, because phosphorescence decay is several orders of magnitude slower than most typical reactions, species in excited triplet states can undergo a very large number of collisions with other reactants before deactivation. However, we shall focus on quenching of fluorescence because important biochemical processes, such as vision (Impact 20.1) and photosynthesis (Impact 20.2), are initiated by deactivation of excited singlet states.

(a) Mechanism of decay of excited states

Consider the mechanism of deactivation of an excited singlet state in the absence of quenching. The following steps are involved:

Process	Equation	Rate
Absorption	$S + h\nu_i \rightarrow S^*$	I_{abs}
Fluorescence	$S^* \rightarrow S + h\nu_F$	$k_F[S^*]$
Intersystem crossing	$S^* \rightarrow T^*$	$k_{ISC}[S^*]$
Internal conversion	$S^* \rightarrow S$	$k_{IC}[S^*]$

in which S is an absorbing species, S* an excited singlet state, T* an excited triplet state, and $h\nu_i$ and $h\nu_F$ denote the incident and fluorescent photons, respectively. It follows that after the exciting radiation has been turned off and S* is no longer being formed,

$$\text{Rate of decay of } S^* = k_F[S^*] + k_{ISC}[S^*] + k_{IC}[S^*] = (k_F + k_{ISC} + k_{IC})[S^*] \quad (20.7a)$$

We see that the excited state decays by a first-order process, so when the light is turned off, [S*] varies with time t as

$$[S^*]_t = [S^*]_0 e^{-(k_F+k_{ISC}+k_{IC})t} = [S^*]_0 e^{-t/\tau_0} \quad (20.7b)$$

where the **observed fluorescence lifetime,** τ_0, is

$$\tau_0 = \frac{1}{k_F + k_{ISC} + k_{IC}} \quad \text{Definition} \quad \text{Observed fluorescence lifetime} \quad (20.8)$$

(Note that the observed lifetime is not simply the sum of the individual lifetimes, $\tau_F = 1/k_F$, etc.)

(b) The fluorescence quantum yield

We have seen that not every excited molecule decays by fluorescence. We therefore speak of the **quantum yield of fluorescence,** ϕ_F, which is the number of events that lead to fluorescence divided by the number of photons absorbed by the molecule in the same time interval:

$$\phi_F = \frac{\text{number of events leading to fluorescence}}{\text{number of photons absorbed}}$$

Definition Quantum yield of fluorescence (20.9a)

If each molecule that absorbs a photon fluoresces, then $\phi_F = 1$. If none does, because the excitation energy is lost before the molecule has time to fluoresce, then $\phi_F = 0$.

When both the numerator and denominator of this expression are divided by the time interval over which the events occur, we see that the fluorescence quantum yield is also the rate of fluorescence divided by the rate of photon absorption, I_{abs}:

$$\phi_F = \frac{\text{rate of fluorescence}}{\text{rate of photon absorption}} = \frac{v_F}{I_{abs}}$$

Quantum yield of fluorescence (20.9b)

We show in the following Derivation that, in the absence of a chemical reaction initiated by the excited singlet state, the quantum yield of fluorescence is

$$\phi_{F,0} = \frac{k_F}{k_F + k_{ISC} + k_{IC}} \quad \text{No quenching} \quad \text{Quantum yield of fluorescence} \quad (20.10)$$

Derivation 20.2

The quantum yield of fluorescence

Most fluorescence measurements are conducted by illuminating a relatively dilute sample with a continuous and intense beam of light. It follows that [S*] is small and constant, so we may invoke the steady-state approximation (Chapter 11) and write

$$\text{Rate of change of } [S^*] = \overbrace{I_{abs}}^{\text{rate of production of S*}} - \overbrace{k_F[S^*]}^{\text{rate of decay of S* by fluorescence}} - \overbrace{k_{ISC}[S^*]}^{\text{rate of decay of S* by intersystem crossing}} - \overbrace{k_{IC}[S^*]}^{\text{rate of decay of S* by internal conversion}}$$

$$= I_{abs} - (k_F + k_{ISC} + k_{IC})[S^*] \underbrace{=}_{\text{steady-state approximation}} 0$$

Consequently,

$$I_{abs} = (k_F + k_{ISC} + k_{IC})[S^*]$$

By using this expression and eqn 20.9b, we write the quantum yield of fluorescence as

$$\phi_{F,0} = \frac{v_F}{I_{abs}} = \frac{k_F[S^*]}{(k_F + k_{ISC} + k_{IC})[S^*]}$$

which, after cancelling the [S*], becomes eqn 20.10.

The observed fluorescence lifetime can be measured by using a pulsed laser technique. First, the sample is excited with a short light pulse from a laser using a wavelength at which S absorbs strongly. Then, the exponential decay of the fluorescence intensity after the pulse is monitored. From eqns 20.8 and 20.10, it follows that

$$\tau_0 = \frac{1}{k_F + k_{ISC} + k_{IC}} \overset{\text{multiply by } k_F/k_F}{=} \overbrace{\left(\frac{k_F}{k_F + k_{ISC} + k_{IC}}\right)}^{\phi_{F,0}} \times \frac{1}{k_F} = \frac{\phi_{F,0}}{k_F} \quad (20.11)$$

Brief illustration 20.3 The fluorescence rate constant

The fluorescence quantum yield and observed fluorescence lifetime of tryptophan in water are $\phi_{F,0} = 0.20$ and $\tau_0 = 2.6$ ns, respectively. It follows from eqn 20.11 that the fluorescence rate constant k_F is

$$k_F = \frac{\phi_{F,0}}{\tau_0} = \frac{0.20}{2.6\times10^{-9}\ \text{s}} = 7.7\times10^7\ \text{s}^{-1}$$

Self-test 20.1

A substance has a fluorescence quantum yield of $\phi_{F,0} = 0.35$. In an experiment to measure the fluorescence lifetime of this substance, it was observed that the fluorescence emission decayed with a half-life of 5.6 ns. Determine the fluorescence rate constant of this substance.

Answer: $k_F = 4.3\times10^7\ \text{s}^{-1}$

(c) The Stern–Volmer equation

A molecule in an excited state must either decay to the ground state or form a photochemical product. As shown in the following Derivation, the relation between the fluorescence quantum yields $\phi_{F,0}$ and ϕ_F in the absence and presence, respectively, of a quencher Q at a molar concentration [Q] is given by the **Stern–Volmer equation:**

$$\frac{\phi_{F,0}}{\phi_F} = 1 + \tau_0 k_Q[\text{Q}] \qquad \text{Stern–Volmer equation} \qquad (20.12)$$

This equation tells us that a plot of $\phi_{F,0}/\phi_F$ against [Q] should be a straight line with slope $\tau_0 k_Q$. Such a plot is called a **Stern–Volmer plot** (Fig. 20.19). The method is quite general and may also be applied to the quenching of phosphorescence emission.

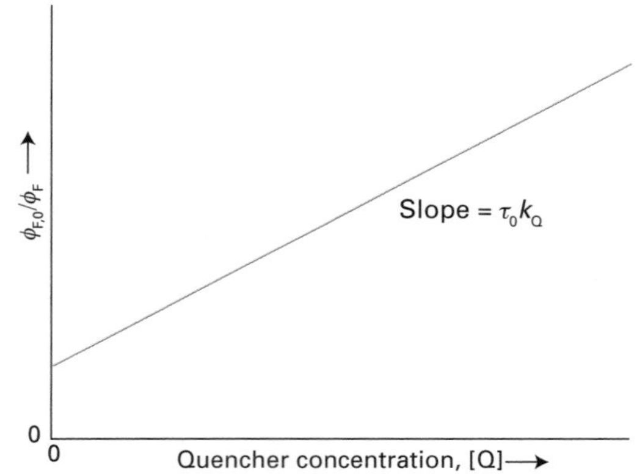

Fig. 20.19 The format of a Stern–Volmer plot and the interpretation of the slope in terms of the rate constant for quenching and the observed fluorescence lifetime in the absence of quenching.

Derivation 20.3

The Stern–Volmer equation

The addition of a quencher, Q, opens an additional channel for deactivation of S*:

Quenching: $\text{S}^* + \text{Q} \rightarrow \text{S} + \text{Q}$ Rate of quenching $= k_Q[\text{Q}][\text{S}^*]$

As in Derivation 20.2, we write the steady-state approximation for [S*] by considering all the appropriate decay processes:

$$\text{Rate of change of }[\text{S}^*] = \overbrace{I_{abs}}^{\text{rate of production of S}^*} - \overbrace{k_F[\text{S}^*]}^{\text{rate of decay of S}^*\text{ by fluorescence}} - \overbrace{k_{ISC}[\text{S}^*]}^{\text{rate of decay of S}^*\text{ by intersystem crossing}} - \overbrace{k_{IC}[\text{S}^*]}^{\text{rate of decay of S}^*\text{ by internal conversion}} - \overbrace{k_Q[\text{Q}][\text{S}^*]}^{\text{rate of decay of S}^*\text{ by quenching}}$$

$$= I_{abs} - (k_F + k_{ISC} + k_{IC} + k_Q[\text{Q}])[\text{S}^*] \underbrace{=}_{\text{steady-state approximation}} 0$$

and the fluorescence quantum yield in the presence of the quencher is

$$\phi_F = \frac{k_F}{k_F + k_{ISC} + k_{IC} + k_Q[\text{Q}]}$$

When [Q] = 0, the quantum yield is

$$\phi_{F,0} = \frac{k_F}{k_F + k_{ISC} + k_{IC}}$$

Equation 20.12 follows from

$$\frac{\phi_{F,0}}{\phi_F} = \overbrace{\left(\frac{k_F}{k_F + k_{ISC} + k_{IC}}\right)}^{\phi_{F,0}} \times \overbrace{\left(\frac{k_F + k_{ISC} + k_{IC} + k_Q[\text{Q}]}{k_F}\right)}^{1/\phi_F}$$

$$\overset{\text{Cancel } k_F}{=} \frac{k_F + k_{ISC} + k_{IC} + k_Q[\text{Q}]}{k_F + k_{ISC} + k_{IC}}$$

$$= 1 + \overbrace{\frac{1}{k_F + k_{ISC} + k_{IC}}}^{\tau_0\ (\text{eqn } 20.11)} \times k_Q[\text{Q}]$$

$$= 1 + \tau_0 k_Q[\text{Q}]$$

Because the fluorescence intensity and lifetime are both proportional to the fluorescence quantum yield (specifically, from eqn 20.11, $\tau = \phi_F/k_F$), plots of $I_{F,0}/I_F$ and τ_0/τ (where the subscript 0 indicates a measurement in the absence of quencher) against [Q] should also be linear with the same slope and intercept as those shown for eqn 20.12.

Example 20.1

Determining the quenching rate constant

The molecule 2,2′-bipyridine (**4**, bpy) forms a complex with the Ru^{2+} ion. Ruthenium(II) tris-(2,2′-bipyridyl), $Ru(bpy)_3^{2+}$ (**5**), has a strong electronic absorption at 450 nm. The quenching of the $^*Ru(bpy)_3^{2+}$ excited state by $Fe(OH_2)_6^{3+}$ in acidic solution was monitored by measuring emission lifetimes at 600 nm. Determine the quenching rate constant for this reaction from the following data:

$[Fe(OH_2)_6^{3+}]/(10^{-4}$ mol dm$^{-3})$	0	1.6	4.7	7	9.4
$\tau/(10^{-7}$ s)	6	4.05	3.37	2.96	2.17

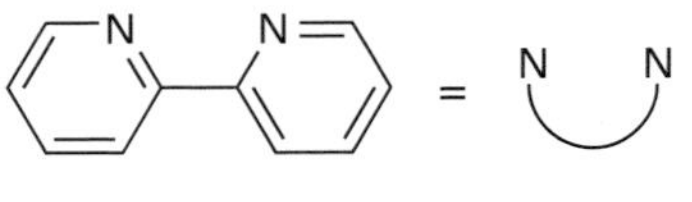

4 2,2′-Bipyridine (bpy)

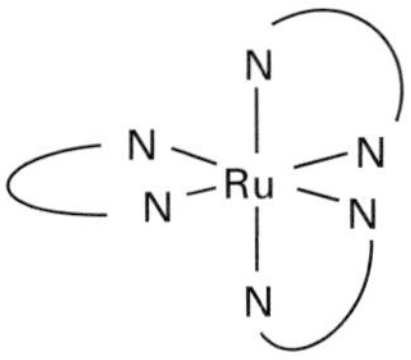

5 $[Ru(bpy)_3]^{2+}$

Strategy Rewrite the Stern–Volmer equation (eqn 20.12) for use with lifetime data; then fit the data to a straight line.

Solution Because from eqn 20.11 $\phi_{F,0}=\tau_0 k_F$ and $\phi_F=\tau k_F$, we can rewrite eqn 20.12 as

$$\frac{\phi_{F,0}}{\phi_F}=\frac{\tau_0 k_F}{\tau k_F}\overbrace{=}^{\text{Cancel }k_F}\frac{\tau_0}{\tau}\overbrace{=}^{\text{eqn 20.12}}1+\tau_0 k_Q[Q]$$

Dividing through by τ_0 gives

$$\frac{1}{\tau}=\frac{1}{\tau_0}+k_Q[Q] \qquad (20.13)$$

Figure 20.20 shows a plot of $1/\tau$ against $[Fe^{3+}]$ and the results of a fit to eqn 20.13. The slope of the line is 2.8×10^9, so $k_Q=2.8\times10^9$ dm^3 mol^{-1} s^{-1}. This example shows that measurements of emission lifetimes are preferred because they yield the value of k_Q directly. To determine the value of k_Q from intensity or quantum yield measurements, we need to make an independent measurement of τ_0.

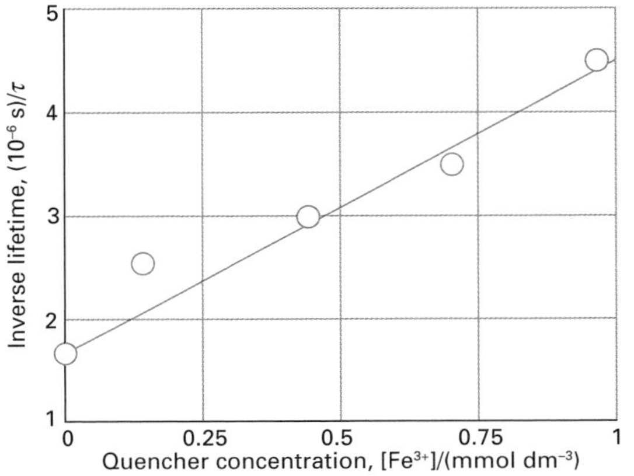

Fig. 20.20 The Stern–Volmer plot of the data for Example 20.1.

Self-test 20.2

The quenching of tryptophan fluorescence by dissolved O_2 gas was monitored by measuring emission lifetimes at 348 nm in aqueous solutions. Determine the quenching rate constant for this process from the following data:

$[O_2]/(10^{-2}$ mol dm$^{-3})$	0	2.3	5.5	8	10.8
$\tau/(10^{-9}$ s)	2.6	1.5	0.92	0.71	0.57

Answer: 1.3×10^{10} dm^3 mol^{-1} s^{-1}

Three common mechanisms for quenching of an excited singlet (or triplet) state are:

Collisional deactivation:	$S^* + Q \rightarrow S + Q$
Electron transfer:	$S^* + Q \rightarrow S^+ + Q^-$ or $S^- + Q^+$
Resonance energy transfer:	$S^* + Q \rightarrow S + Q^*$

Collisional quenching is particularly efficient when Q is an electron-rich species, such as iodide ion, which receives energy from S* and then decays non-radiatively to the ground state.

● **Brief illustration 20.4** Collisional quenching

Collisional quenching by iodide ion may be used to determine the accessibility of solvent molecules to the amino acid residues of a folded protein. For example, fluorescence from a tryptophan residue ($\lambda_{abs}\approx$ 290 nm, $\lambda_{fluor}\approx$ 350 nm) is quenched by iodide ion when the residue is on the surface of the protein and hence accessible to the solvent. Conversely, residues in the hydrophobic interior of the protein are not quenched effectively by I^-.

The quenching rate constant itself does not give much insight into the mechanism of quenching apart from suggesting that it is diffusion-controlled. However, according to the **Marcus theory** of electron transfer, which was proposed by R.A. Marcus in 1965, the rates of electron transfer (from ground or excited states) depend on:[1]

1. The distance between the donor and acceptor, with electron transfer becoming more efficient as the distance between donor and acceptor decrease.
2. The 'reorganization energy', the energy cost incurred by molecular rearrangements of donor, acceptor, and medium during electron transfer. The electron transfer rate is predicted to increase as this reorganization energy is matched more closely by the reaction Gibbs energy.

[1] For details of the Marcus theory, see our *Physical Chemistry* (2010).

3. The reaction Gibbs energy, $\Delta_r G$, with electron transfer becoming more efficient as the reaction becomes more exergonic, up to the point that $-\Delta_r G$ is equal to the reorganization energy. For example, it follows from the thermodynamic principles that lead to the electrochemical series (Chapter 9) that efficient photo-oxidation of S requires that the reduction potential of S* be lower than the reduction potential of Q.

Electron transfer can be studied by time-resolved spectroscopy. The oxidized and reduced products often have electronic absorption spectra distinct from those of their neutral parent compounds. Therefore, the rapid appearance of such known features in the absorption spectrum after excitation by a laser pulse may be taken as indication of quenching by electron transfer.

(d) Resonance energy transfer

To understand resonance energy transfer we note that in an absorption process the incident electromagnetic radiation induces a transition electric dipole moment in S. When that excited state collapses back to the ground the resulting transition dipole can induce a corresponding transition dipole moment in a neighbouring Q molecule. It does so with an efficiency, η_T, that can be expressed in terms of the fluorescence quantum yields in the absence and the presence of the quencher:

$$\eta_T = \frac{\phi_{F,0} - \phi_F}{\phi_{F,0}} \qquad \text{Definition} \quad \text{Efficiency of resonance energy transfer} \qquad (20.14)$$

According to the **Förster theory** of resonance energy transfer, which was proposed by T. Förster in 1959, for donor–acceptor (S–Q) systems that are held rigidly either by covalent bonds or by a protein 'scaffold', η_T increases with decreasing distance, R, according to

$$\eta_T = \frac{R_0^6}{R_0^6 + R^6} \qquad \text{Efficiency of energy transfer in terms of the donor–acceptor distance} \qquad (20.15)$$

where R_0 is a parameter (with units of distance) that is characteristic of each donor–acceptor pair. Equation 20.15 has been verified experimentally and values of R_0 are available for a number of donor–acceptor pairs (Table 20.2). According to the Förster theory, for a given separation, the efficiency is greatest when the emission spectrum of the donor molecule overlaps significantly with the absorption spectrum of the acceptor. In the overlap region, photons emitted by the donor can be absorbed resonantly by the acceptor for the energy gaps of the two molecules then match (Fig. 20.21).

Table 20.2
*Values of R_0 for some donor–acceptor pairs**

Donor	Acceptor	R_0/nm
Naphthalene	Dansyl	2.2
Dansyl	ODR	4.3
Pyrene	Coumarin	3.9
1.5-I-AEDANS	FITC	4.9
Tryptophan	1.5-I-AEDANS	2.2
Tryptophan	Haem	2.9

*Abbreviations: Dansyl: 5-dimethylamino-l-naphthalenesulfonic acid; FITC: fluorescein 5-isothiocyanate; ODR: octadecyl-rhodamine; 1.5-I-AEDANS: 5-(((2-iodoacetyl)amino)ethyl)amino)naphthalene-1-sulfonic acid (**6**)

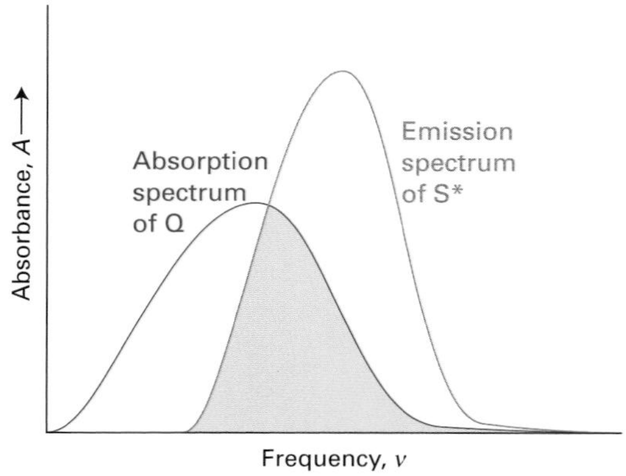

Fig. 20.21 According to the Förster theory, the rate of energy transfer from a molecule S* in an excited state to a quencher molecule Q is optimized at radiation frequencies in which the emission spectrum of S* overlaps with the absorption spectrum of Q, as shown in the shaded region.

The dependence of η_T on R forms the basis of **fluorescence resonance energy transfer** (FRET), a technique that can be used to measure distances in biological systems. In a typical FRET experiment, a site on a biopolymer or membrane is labelled covalently with an energy donor and another site is labelled covalently with an energy acceptor. The distance between the labels is then calculated from the known value of R_0 and eqn 20.15. Several tests have shown that the FRET technique is useful for measuring distances ranging from 1 to 9 nm.

Example 20.2

Interpreting the FRET technique

Consider a study of the protein rhodopsin (Impact 20.1). When an amino acid on the surface of rhodopsin was labelled covalently with the energy donor 1.5-I-AEDANS (**6**), the fluorescence quantum yield of the label decreased from 0.75 to 0.68 due to quenching by the visual pigment 11-*cis*-retinal (**2**). Using the known value of $R_0 = 5.4$ nm for the 1.5-I-AEDANS/11-*cis*-retinal pair, estimate the distance between the surface of the protein and 11-*cis*-retinal.

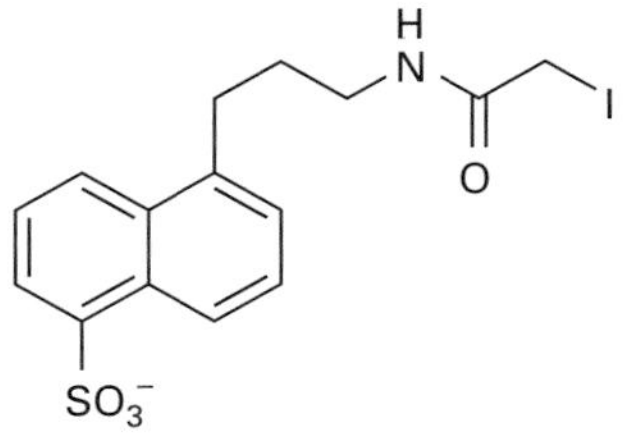

6 1.5-I-AEDANS

Strategy Use eqn 20.14 to calculate η_T from the fluorescence quantum yields. Then use eqn 20.15 and the value of R_0 to estimate R, the distance between the surface of the protein and 11-*cis*-retinal.

Solution From eqn 20.14, we calculate

$$\eta_T = \frac{0.75 - 0.68}{0.75} = 0.093$$

Before using eqn 20.15, we rearrange it:

$$\frac{1}{\eta_T} = \frac{R_0^6 + R^6}{R_0^6} = 1 + \frac{R^6}{R_0^6}$$

$$\frac{R^6}{R_0^6} = \frac{1}{\eta_T} - 1 = \frac{1-\eta_T}{\eta_T}$$

and then solve for R by taking the sixth root ($x^{1/6}$) of both sides:

$$R = R_0\left(\frac{1-\eta_T}{\eta_T}\right)^{1/6}$$

Using $R_0 = 5.4$ nm for the 1.5-I AEDANS/11-*cis*-retinal pair, we calculate

$$R = \overbrace{(5.4\text{ nm})}^{R_0} \times \left(\frac{\overbrace{1-0.093}^{\eta}}{\underbrace{0.093}_{\eta}}\right)^{1/6} = 7.9\text{ nm}$$

Therefore, we take 7.9 nm to be the distance between the surface of the protein and 11-*cis*-retinal.

Self-test 20.3

An amino acid on the surface of a protein was labelled covalently with 1.5-I AEDANS and another was labelled covalently with FITC. The fluorescence quantum yield of 1.5-I AEDANS decreased by 10 per cent due to quenching by FITC. What is the distance between the amino acids?

Answer: 7.1 nm

Impact on biochemistry 20.2

Photosynthesis

Up to 1 kW m^{-2} of radiation from the Sun reaches the Earth's surface, with the exact intensity depending on latitude, time of day, and weather. A large proportion of solar radiation with wavelengths below 400 nm and above 1000 nm is absorbed by atmospheric gases such as ozone and O_2, which absorb ultraviolet radiation, and CO_2 and H_2O, which absorb infrared radiation. As a result, plants, algae, and some species of bacteria evolved photosynthetic apparatus that captures visible and near-infrared radiation. Plants use radiation in the wavelength range 400–700 nm to drive the endergonic reduction of CO_2 to glucose, with concomitant oxidation of water to O_2 ($\Delta_r G^{\ominus} = +2880$ kJ mol^{-1}).

Plant photosynthesis takes place in the *chloroplast*, a special organelle of the plant cell. Electrons flow from reductant to oxidant via a series of electrochemical reactions that are coupled to the synthesis of ATP. In the chloroplast, chlorophylls *a* and *b* and carotenoids (of which β-carotene is an example) bind to proteins called *light-harvesting complexes*, which absorb solar energy and transfer it to protein complexes known as *reaction centres*, where light-induced electron transfer reactions occur. The combination of a light-harvesting complex and a reaction centre complex is called a *photosystem*, and plants have two: photosystem I and photosystem II.

In photosystems I and II, absorption of a photon raises a chlorophyll or carotenoid molecule to an excited singlet state. The initial energy and electron transfer events of photosynthesis are under tight kinetic control and the efficient capture of solar energy stems from rapid quenching of the excited singlet state of chlorophyll by processes that occur with relaxation times that are much shorter than the fluorescence lifetime, which is about 5 ns in diethyl ether at room temperature. Time-resolved spectroscopic data show that within 0.1–5 ps of absorption of light by a chlorophyll molecule in a light-harvesting complex, the energy hops to a nearby pigment via the Förster mechanism. About 100–200 ps later, which corresponds to thousands of hops within the complex, more than 90 per cent of the absorbed energy reaches the reaction centre. The absorption of energy from light decreases the reduction potential of special dimers of chlorophyll *a* molecules known as P700 (in photosystem I) and P680 (in photosystem II). In their excited states, P680 and P700 initiate electron transfer reactions that culminate in the oxidation of water to O_2 and the reduction of $NADP^+$ to NADPH. The initial electron transfer steps are fast and compete effectively with chlorophyll's fluorescence. For example, the transfer of an electron from the excited singlet state of P680 occurs within 3 ps. Experiments show that for each molecule of NADPH formed in the chloroplast of green plants, one molecule of ATP is synthesized. Finally, the ATP and NADPH molecules participate in the *Calvin–Benson cycle*, a sequence of enzyme-controlled reactions that leads to the reduction of CO_2 to glucose in the chloroplast.

In summary, plant photosynthesis uses solar energy to transfer electrons from a poor reductant (water) to carbon dioxide. In the process, high-energy molecules (carbohydrates, such as glucose) are synthesized in the cell. Animals feed on the carbohydrates derived from photosynthesis. The O_2 released by photosynthesis as a waste product is used to oxidize carbohydrates to CO_2. This reaction drives biological processes, such as biosynthesis, muscle contraction, cell division, and nerve conduction. Hence, the sustenance of life on Earth depends on a tightly regulated carbon–oxygen cycle that is driven by solar energy.

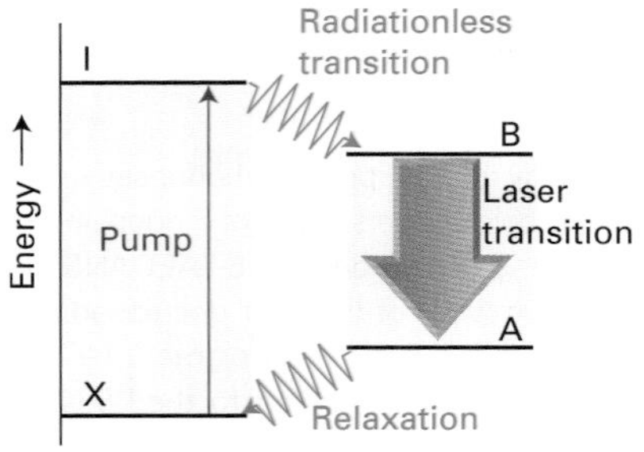

Fig. 20.22 The transitions involved in a four-level laser. Because the laser transition terminates in an excited state (A), the population inversion between A and B is much easier to achieve than when the lower state of the laser transition is the ground state.

20.8 Lasers

The word *laser* is an acronym formed from *l*ight *a*mplification by *s*timulated *e*mission of *r*adiation. As this name suggests, it is a process that depends on *stimulated* emission as distinct from the spontaneous emission processes characteristic of fluorescence and phosphorescence. In **stimulated emission**, an excited state is stimulated to emit a photon by the presence of radiation of the same frequency, and the more photons there are present, the greater the probability of the emission (for details, see Further information 20.2). To picture the process, we can think of the oscillations of the electromagnetic field as periodically distorting the excited molecule at the frequency of the transition and hence encouraging the molecule to generate a photon of the same frequency. The essential feature of laser action is the strong **gain**, or growth of intensity, that results: the more photons present of the appropriate frequency, the more photons of that frequency the excited molecules will be stimulated to form, and so the laser medium fills with photons. These photons then escape either continuously or in pulses.

(a) Requirements for laser action

One requirement for laser action is the existence of an excited state that has a long enough lifetime for it to participate in stimulated emission. Another requirement is the existence of a greater population in the upper state than in the lower state where the transition terminates. Because at thermal equilibrium the population is greater in the lower energy state, it is necessary to achieve a **population inversion** in which there are more molecules in the upper state than in the lower.

Figure 20.22 illustrates one way to achieve population inversion indirectly through an intermediate state I. Thus, the molecule is excited to I, which then gives up some of its energy non-radiatively (by passing energy on to vibrations of the surroundings) and changes into a lower state B; the laser transition is the return of B to a lower state A. Because four levels are involved overall, this arrangement leads to a **four-level laser**. One advantage of this arrangement is that the population inversion of the A and B levels is easier to achieve than when the lower state is the heavily populated ground state. The transition from X to I is caused by an intense flash of light in the process called **pumping**. In some cases the pumping flash is achieved with an electric discharge through xenon or with the radiation from another laser.

In practice, the laser medium is confined to a cavity that ensures that only certain photons of a particular frequency, direction of travel, and state of polarization are generated abundantly. The cavity is essentially a region between two mirrors, which reflect the light back and forth. This arrangement can be regarded as a version of the particle in a box, with the particle now being a photon. As in the treatment of a particle in a box (Section 12.7), the only wavelengths that can be sustained satisfy $N \times \frac{1}{2}\lambda = L$, where N is an integer and L is the length of the cavity. That is, only an integral number of half-wavelengths fit into the cavity; all other waves undergo destructive interference with themselves. In addition, not all wavelengths that can be sustained by the cavity are

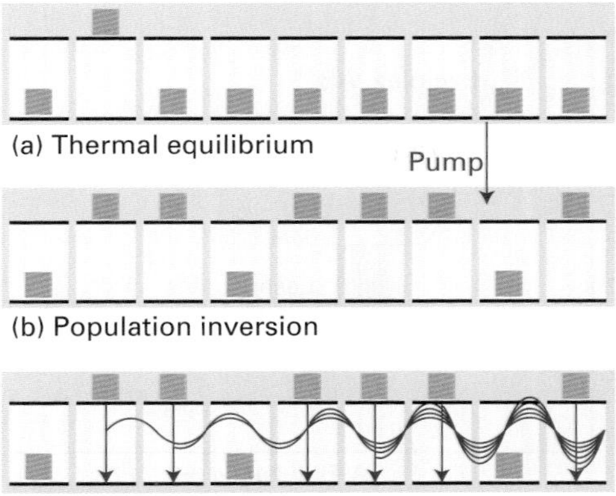

Fig. 20.23 A schematic illustration of the steps leading to laser action. (a) At thermal equilibrium, more atoms are in the ground state. (b) When the initial state absorbs, the populations are inverted (the atoms are pumped to the excited state). (c) A cascade of radiation then occurs, as one emitted photon stimulates another atom to emit, and so on. The radiation is coherent (phases in step).

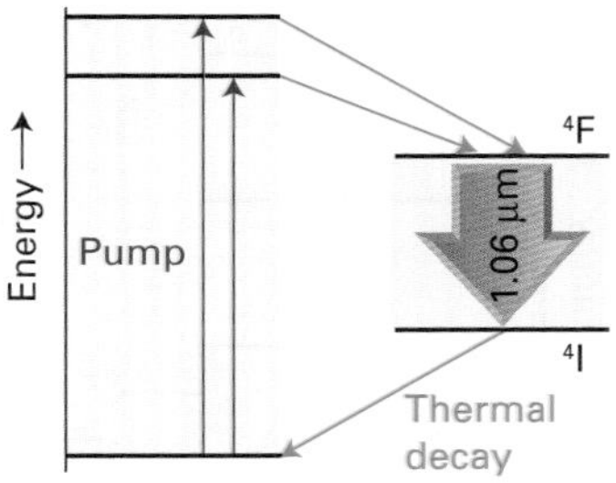

Fig. 20.24 The transitions involved in a neodymium laser. The laser action takes place between two excited states, and the population inversion is easier to achieve than in the ruby laser.

amplified by the laser medium (many fall outside the range of frequencies of the laser transitions), so only a few contribute to the laser radiation. These wavelengths are the **resonant modes** of the laser.

Photons with the correct wavelength for the resonant modes of the cavity and the correct frequency to stimulate the laser transition are highly amplified. One photon might be generated spontaneously, and travel through the medium. It stimulates the emission of another photon, which in turn stimulates more (Fig. 20.23). The cascade of energy builds up rapidly, and soon the cavity is an intense reservoir of radiation at all the resonant modes it can sustain. Some of this radiation can be withdrawn if one of the mirrors is partially transmitting.

The resonant modes of the cavity have various natural characteristics, and to some extent may be selected. Only photons that are travelling strictly parallel to the axis of the cavity undergo more than a couple of reflections, so only they are amplified, all others simply vanishing into the surroundings. Hence, laser light generally forms a beam with very low divergence. It may also be polarized, with its electric vector in a particular plane (or in some other state of polarization), by including a polarizing filter into the cavity or by making use of polarized transitions in a solid medium.

(b) Examples of lasers

A *neodymium laser* is an example of a four-level laser (Fig. 20.24). In one form it consists of Nd^{3+} ions at low concentration in yttrium aluminium garnet (YAG, specifically $Y_3Al_5O_{12}$), and is then known as a *Nd-YAG laser*. This laser operates at a number of wavelengths in the infrared, the band at 1064 nm being most common. The transition at 1064 nm is very efficient and the laser is capable of substantial power output. The power is great enough that focussing the beam on to a material may lead to the observation of **nonlinear optical phenomena**, which arise from changes in the optical properties of the substance in the presence of an intense electric field from electromagnetic radiation. A useful nonlinear optical phenomenon is **frequency doubling**, or *second harmonic generation*, in which an intense laser beam is converted to radiation with twice (and in general a multiple) of its initial frequency as it passes through a suitable material. Frequency doubling and tripling of a Nd-YAG laser produce green light at 532 nm and ultraviolet radiation at 355 nm, respectively.

In *diode lasers*, of the type used in CD players and bar-code readers, the light emission at a p–n junction (Section 17.3) is sustained by sweeping away the electrons that fall into the holes of the p-type semiconductor. This process is arranged to occur in a cavity formed by making use of the abrupt difference in refractive index between the different components of the junction, and the radiation trapped in the cavity enhances the production of more radiation. One widely used material is GaAs doped with aluminium, which produces 780 nm red laser radiation and is widely used in CD players. The new generation of DVD players that use blue rather than red laser radiation, so allowing a greater superficial density of information, use GaN as the active material.

Because *gas lasers* can be cooled by a rapid flow of the gas through the cavity, they can be used to

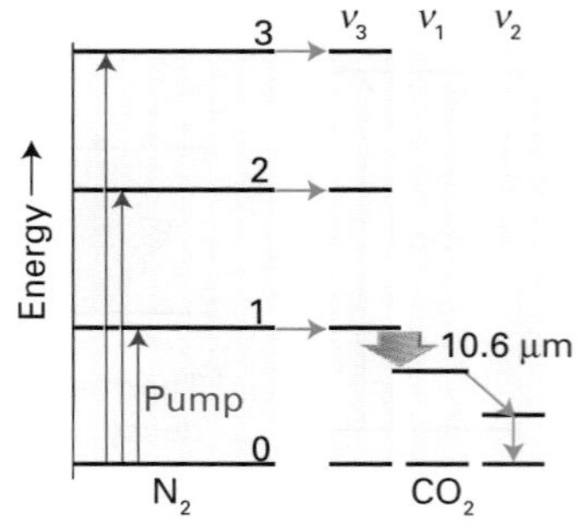

Fig. 20.25 The transitions involved in a carbon dioxide laser. The pumping also depends on the coincidental matching of energy separations; in this case the vibrationally excited N_2 molecules have excess energies that correspond to a vibrational excitation of the antisymmetric stretch of CO_2. The laser transition is from $\nu_3 = 1$ to $\nu_1 = 1$.

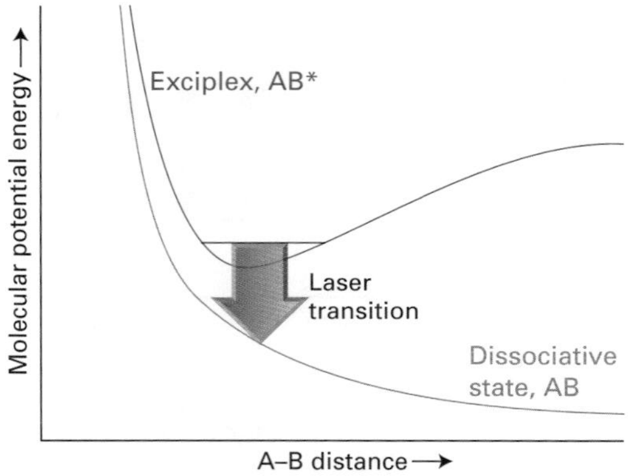

Fig. 20.26 The molecular potential energy curves for an exciplex. The species can survive only as an excited state, because on discarding its energy it enters the lower, dissociative state. Because only the upper state can exist, there is never any population in the lower state.

generate high powers. The *carbon dioxide laser*, which produces radiation between 9.2 μm and 10.8 μm, with the strongest emission at 10.6 μm, in the infrared makes use of vibrational transitions (Fig. 20.25). Most of the working gas is nitrogen, which becomes vibrationally excited by electronic and ionic collisions in an electric discharge. The vibrational levels happen to coincide with the ladder of antisymmetric stretch (ν_3, see Fig. 19.20) levels of CO_2, which pick up the energy during a collision. Laser action then occurs from the lowest excited level of ν_3 to the lowest excited level of the symmetric stretch (ν_1), which has remained unpopulated during the collisions. Such high powers of radiation can be achieved that carbon dioxide lasers can be used to cut steel for ship-building.

The population inversion needed for laser action is achieved in a more underhand way in *exciplex lasers*, for in these (as we shall see) the lower state does not effectively exist. This odd situation is achieved by forming an **exciplex**, a combination of two atoms (or molecules) that survives only in an excited state and which dissociates as soon as the excitation energy has been discarded. The term 'excimer laser' is also widely encountered and used loosely when 'exciplex laser' is more appropriate. An exciplex has the form AB* whereas an **excimer**, an excited dimer, is AA*. An example of an exciplex laser is a mixture of xenon, chlorine, and neon. An electric discharge through the mixture produces excited Cl atoms, which attach to the Xe atoms to give the exciplex XeCl*. The exciplex survives for about 10 ns, which is time for it to participate in laser action at 308 nm (in the ultraviolet). As soon as XeCl* has discarded a photon, the atoms separate because the molecular potential energy curve of the ground state is dissociative, and the ground state of the exciplex cannot become populated (Fig. 20.26).

The wavelengths of lasers can be selected in a variety of ways. *Dye lasers* have broad spectral characteristics because the solvent broadens the vibrational structure of the transitions into bands. It is possible to scan the wavelength continuously by rotating a diffraction grating in the cavity) and achieve laser action at any chosen wavelength. As the gain is very high, only a short length of the optical path need be through the dye. The excited states of the active medium, the dye, are sustained by another laser or a flash lamp, and the dye solution is flowed through the laser cavity to avoid thermal degradation (Fig. 20.27). More modern solutions are to use tuneable Ti:sapphire (for wavelengths in the range 650–1100 nm) and white light lasers.

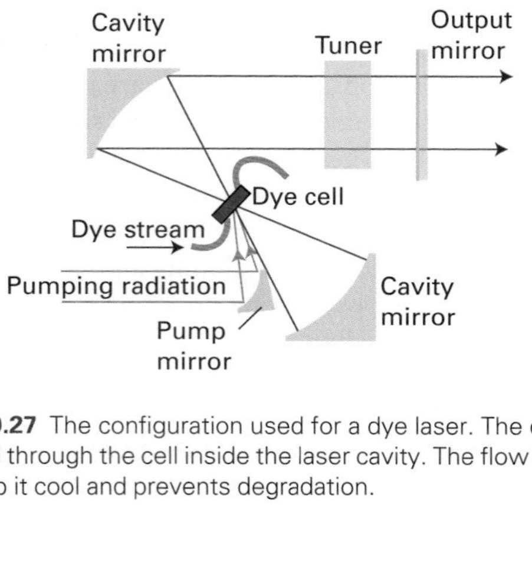

Fig. 20.27 The configuration used for a dye laser. The dye is flowed through the cell inside the laser cavity. The flow helps to keep it cool and prevents degradation.

(c) Applications of lasers in chemistry

Laser radiation has a number of advantages for applications in chemistry. One advantage is its highly monochromatic character, which enables very precise spectroscopic observations to be made. Another advantage is the ability of laser radiation to be produced in very short pulses (currently, as brief as about 1 fs): as a result, very fast chemical events, such as the individual transfers of atoms during a chemical reaction, can be followed (see Section 10.11). Intense laser radiation also reduces the time needed for spectroscopic observations. Raman spectroscopy (Chapter 19) was revitalized by the introduction of lasers because the intense beam increases the intensity of scattered radiation, so the use of laser sources increases the sensitivity of Raman spectroscopy. A well-defined beam also implies that the detector can be designed to collect only the radiation that has passed through a sample, and can be screened much more effectively against the stray scattered light that can obscure the Raman signal. Laser light can be delivered through fibre optics for use in transportable systems and focussed to such small diameters that *Raman microscopy* can be used to study sub-micrometre particles. The monochromaticity of laser radiation is also a great advantage, for it makes possible the observation of scattered light that differs by only fractions of reciprocal centimetres from the incident radiation. Such high resolution is particularly useful for observing the rotational structure of Raman lines because rotational transitions are of the order of a few reciprocal centimetres.

The large number of photons in an incident beam generated by a laser gives rise to a qualitatively different branch of spectroscopy, for the photon density is so high that more than one photon may be absorbed by a single molecule and give rise to **multiphoton processes**. Because the selection rules for multiphoton processes are different, states inaccessible by conventional one-photon spectroscopy become observable.

The monochromatic character of laser radiation allows us to excite specific states with very high precision. One consequence of state-specificity is that the illumination of a sample may be efficient in stimulating a photochemical reaction, because its frequency can be tuned exactly to an absorption. The specific excitation of a particular excited state of a molecule may greatly enhance the rate of a reaction even at low temperatures. As we saw in Chapter 10, the rate of a reaction is increased by raising the temperature because the energies of the various modes of motion of the molecule are enhanced. However, this enhancement increases the energy of all the modes, even those that do not contribute appreciably to the reaction rate. With a laser we can excite the kinetically significant mode, so rate enhancement is achieved most efficiently. An example is the reaction

$$BCl_3 + C_6H_6 \rightarrow C_6H_5\text{—}BCl_2 + HCl$$

which normally proceeds only above 600 °C in the presence of a catalyst; exposure to 10.6 μm CO_2 laser radiation results in the formation of products at room temperature without a catalyst. The commercial potential of this procedure is considerable, provided laser photons can be produced sufficiently cheaply, because heat-sensitive compounds, such as pharmaceuticals, may perhaps be made at lower temperatures than in conventional reactions.

Laser isotope separation is possible because two isotopologues (species that differ only in their isotopic composition), have slightly different energy levels and hence slightly different absorption frequencies. At least two absorption processes are required. In the first step, a photon excites an atom to a higher state; in the second step, a photon achieves photoionization from that state (Fig. 20.28). The energy separation between the two states involved in the first step depends on the nuclear mass. Therefore, if the laser radiation is tuned to the appropriate frequency, only one of the isotopologues will undergo excitation and hence be available for photoionization in the second step. An example of laser isotope separation is the photoionization of uranium vapour, in which the incident laser is tuned to excite ^{235}U but not ^{238}U. The ^{235}U atoms in the atomic beam are ionized in the two-step process; they are then attracted to a negatively charged electrode, and may be collected (Fig. 20.29).

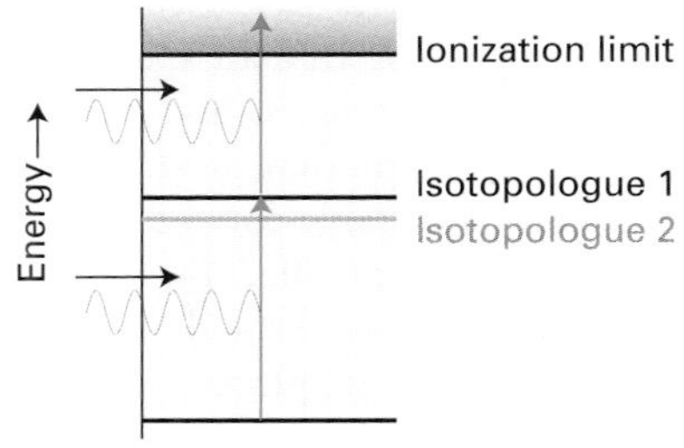

Fig. 20.28 In one method of isotope separation, one photon excites an isotopologue to an excited state, and then a second photon achieves photoionization. The success of the first step depends on the nuclear mass.

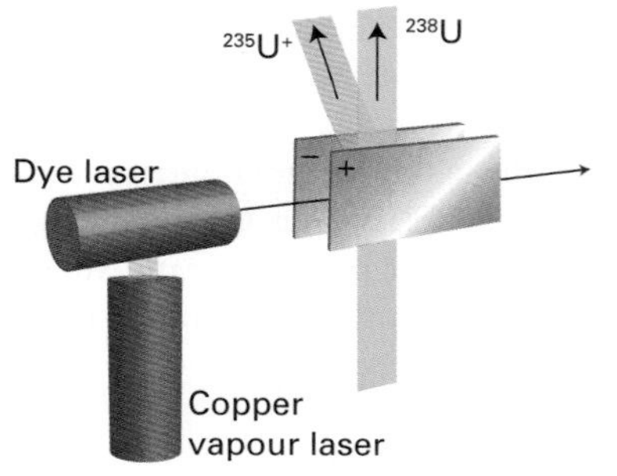

Fig. 20.29 An experimental arrangement for isotope separation. The dye laser, which is pumped by a copper-vapour laser, photoionizes the U atoms selectively according to their mass, and the ions are deflected by the electric field applied between the plates.

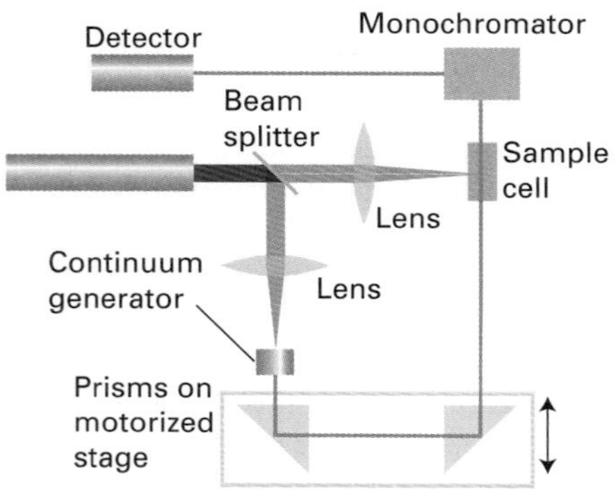

Fig. 20.30 A configuration used for time-resolved absorption spectroscopy, in which the same pulsed laser is used to generate a monochromatic pump pulse and, after continuum generation in a suitable liquid, a 'white' light probe pulse. The time delay between the pump and probe pulses may be varied.

The ability of lasers to produce pulses of very short duration is particularly useful in chemistry when we want to monitor processes in time. In **time-resolved spectroscopy**, laser pulses are used to obtain the absorption, emission, or Raman spectrum of reactants, intermediates, products, and even transition states of reactions. Lasers that produce nanosecond pulses are generally suitable for the observation of reactions with rates controlled by the speed with which reactants can move through a fluid medium. However, femtosecond to picosecond laser pulses are needed to study energy transfer, molecular rotations, vibrations, and conversion from one mode of motion to another. The arrangement shown in Fig. 20.30 is often used to study ultrafast chemical reactions that can be initiated by light. An intense but brief laser pulse, the *pump*, promotes a molecule A to an excited electronic state A* that can either emit a photon (as fluorescence or phosphorescence) or react with another species B to yield a product C. The rates of appearance and disappearance of the various species are determined by observing time-dependent changes in the absorption spectrum of the sample during the course of the reaction. This observation is made by passing a weak pulse of white light, the *probe*, through the sample at different times after the laser pulse. Pulsed 'white' light can be generated directly from the laser pulse by the nonlinear optical phenomenon of *continuum generation*, in which focussing an ultrashort laser pulse on a vessel containing a liquid such as water or carbon tetrachloride results in an outgoing beam with a wide distribution of frequencies. A time delay between the strong laser pulse and the 'white' light pulse can be introduced by allowing one of the beams to travel a longer distance before reaching the sample. For example, a difference in travel distance of $\Delta d = 3$ mm corresponds to a time delay $\Delta t = \Delta d/c \approx 10$ ps between two beams, where c is the speed of light.

Photoelectron spectroscopy

The exposure of a molecule to high-frequency radiation can result in the ejection of an electron. This **photoejection** is the basis of another type of spectroscopy in which we monitor the energies of the ejected photoelectrons. If the incident radiation has frequency ν, the photon that causes photoejection has energy $h\nu$. If the ionization energy of the molecule is I, the difference in energy, $h\nu - I$, is carried away as kinetic energy. Because the kinetic energy of an electron travelling at a speed v is $\frac{1}{2}m_e v^2$, we can write

$$h\nu = I + \tfrac{1}{2}m_e v^2 \qquad (20.16)$$

Therefore, by monitoring the speed of the photoelectron, and knowing the frequency of the incident radiation, we can determine the ionization energy of the molecule and hence the strength with which the electron was bound (Fig. 20.31). In this context the 'ionization energy' of the molecule has different values depending on the orbital that the photoelectron occupied, and the slower the ejected electron, the lower in energy the orbital from which it was ejected. The apparatus is a modification of a mass spectrometer (Fig. 20.32), in which the speed of the photoelectrons is measured by determining the

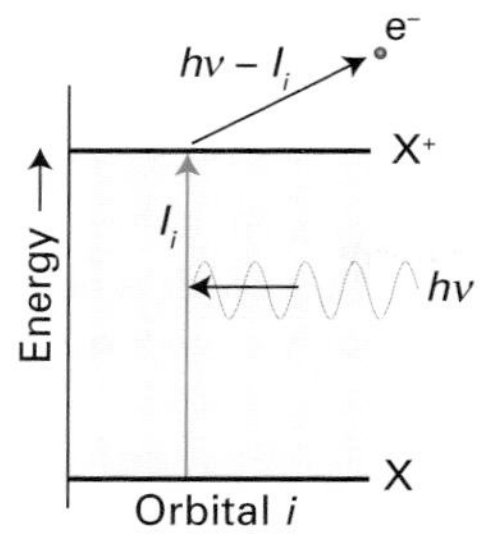

Fig. 20.31 The basic principle of photoelectron spectroscopy. An incoming photon of known energy collides with an electron in one of the orbitals and expels it with a kinetic energy that is equal to the difference between the energy supplied by the photon and the ionization energy from the occupied orbital. An electron from an orbital with a low ionization energy will emerge with a high kinetic energy (and high speed) whereas an electron from an orbital with a high ionization energy will be ejected with a low kinetic energy (and low speed).

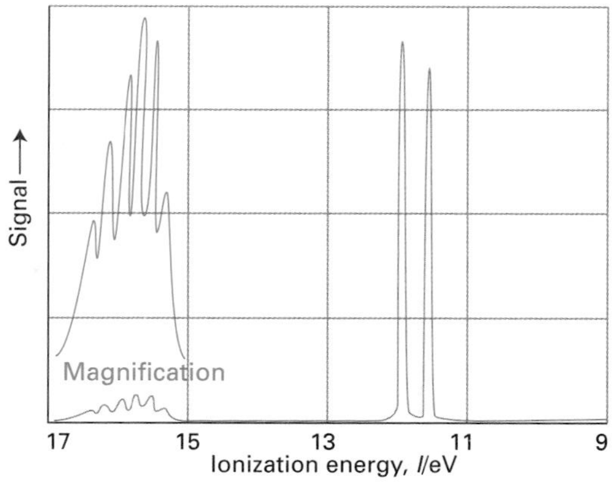

Fig. 20.33 The photoelectron spectrum of HBr. The lowest ionization energy band corresponds to the ionization of a Br lone-pair electron. The higher ionization energy band corresponds to the ionization of a bonding electron. The structure on the latter is due to the vibrational excitation of HBr^+ that results from the ionization.

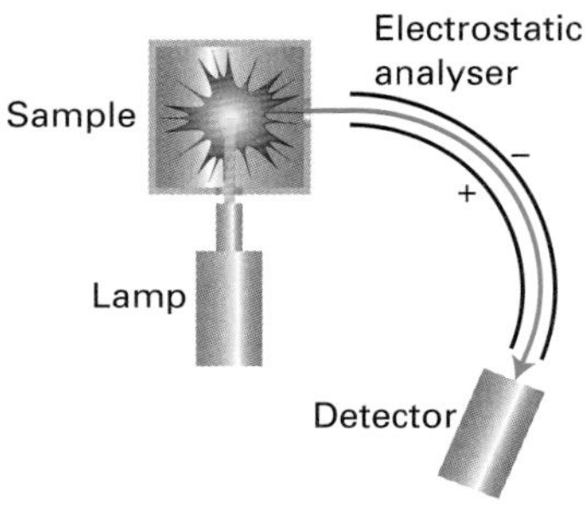

Fig. 20.32 A photoelectron spectrometer consists of a source of ionizing radiation (such as a helium discharge lamp for UPS and an X-ray source for XPS), an electrostatic analyser, and an electron detector. The deflection of the electron paths caused by the analyser depends on their speed.

strength of the electric field required to bend their paths on to the detector.

Figure 20.33 shows a typical photoelectron spectrum (of HBr). If we disregard the fine structure, we see that the HBr lines fall into two main groups. The least tightly bound electrons (with the lowest ionization energies and hence highest kinetic energies when ejected) are those in the lone pairs of the Br atom. The next ionization energy lies at 15.2 eV, and corresponds to the removal of an electron from the H–Br σ bond.

The fine structure in the HBr spectrum shows that ejection of a σ electron is accompanied by a considerable amount of vibrational excitation. The Franck–Condon principle would account for this observation if ejection were accompanied by an appreciable change of equilibrium bond length between HBr and HBr^+: if that is so, the ion is formed in a bond-compressed state, which is consistent with the important bonding effect of the σ electrons. The lack of much vibrational structure in the other band is consistent with the nonbonding role of the $Br4p_x$ and $Br4p_y$ lone-pair electrons, for the equilibrium bond length is little changed when one is removed.

Brief illustration 20.5 The photoelectron spectrum of H_2O

The highest kinetic energy electrons in the spectrum of H_2O using 21.22 eV He radiation are at about 9 eV and show a large vibrational spacing of 0.41 eV (1 eV = 8065.5 cm^{-1}). Because 0.41 eV corresponds to 3.3×10^3 cm^{-1}, which is similar to the wavenumber of the symmetric stretching mode of the neutral H_2O molecule (3652 cm^{-1}), we can suspect that the electron is ejected from an orbital that has little influence on the bonding in the molecule. That is, photoejection is from a largely nonbonding orbital.

Self-test 20.4

In the same spectrum of H_2O, the band near 7.0 eV shows a long vibrational series with spacing 0.125 eV. The bending mode of H_2O lies at 1596 cm^{-1}. What conclusions can you draw about the characteristics of the orbital occupied by the photoelectron?

Answer: The electron contributes to long-distance HH bonding across the molecule.

Further information 20.1

The Beer–Lambert law

The Beer–Lambert law is an empirical result. However, it is simple to account for its form. We think of the sample as consisting of a stack of infinitesimal slices, like sliced bread (Fig. 20.34). The thickness of each layer is dx. The change in intensity, dI, that occurs when electromagnetic radiation passes through one particular slice is proportional to the thickness of the slice, the concentration of the absorber J,

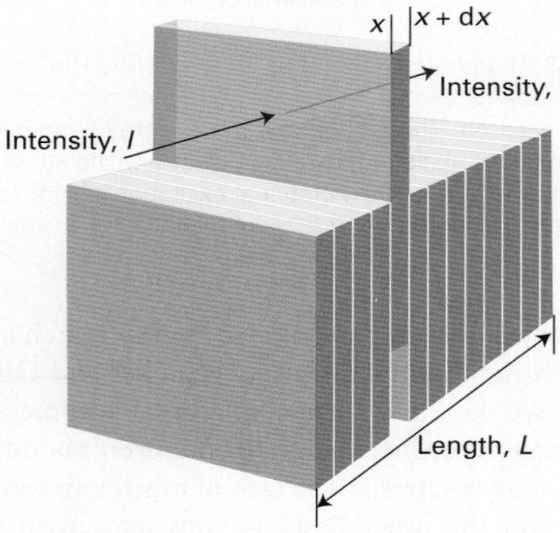

Fig. 20.34 To establish the Beer–Lambert law, the sample is supposed to be sliced into a large number of planes. The reduction in intensity caused by one plane is proportional to the intensity incident on it (after passing through the preceding planes), the thickness of the plane, and the concentration of absorbing species.

and the intensity of the incident radiation at that slice of the sample, so $dI \propto [\mathrm{J}]I dx$. Because dI is negative (the intensity is reduced by absorption), we can write

$$dI = -\kappa[\mathrm{J}]I dx$$

where κ (kappa) is the proportionality coefficient. Division of both sides by I gives

$$\frac{dI}{I} = -\kappa[\mathrm{J}]dx$$

This expression applies to each successive slice. To obtain the intensity that emerges from a sample of thickness L when the intensity incident on one face of the sample is I_0, we sum all the successive changes. Because a sum over infinitesimally small increments is an integral, we write

$$\overbrace{\int_{I_0}^{I} \frac{dI}{I}}^{\ln(I/I_0)} = -\kappa \int_0^L [\mathrm{J}]dx \overset{[\mathrm{J}]\,\text{uniform}}{=} -\kappa[\mathrm{J}]\overbrace{\int_0^L dx}^{L}$$

Therefore

$$\ln \frac{I}{I_0} = -\kappa[\mathrm{J}]L$$

Because the relation between natural and common logarithms is $\ln x = (\ln 10)\log x$, we can write $\varepsilon = \kappa/\ln 10$ and obtain

$$\log \frac{I}{I_0} = -\varepsilon[\mathrm{J}]L$$

which, on substituting $A = \log(I_0/I) = -\log(I/I_0)$, is the Beer–Lambert law, expressed as eqn 20.4.

Further information 20.2

The Einstein transition probabilities

The intensity of an absorption line is related to the rate at which energy from electromagnetic radiation at a specified frequency is absorbed by a molecule. Einstein identified three contributions to the rates of transitions between states. **Stimulated absorption** is the transition from a low-energy state to one of higher energy that is driven by the electromagnetic field oscillating at the transition frequency. He reasoned that the more intense the electromagnetic field (the more intense the incident radiation), the greater the rate at which transitions are induced and hence the stronger the absorption by the sample, so he wrote the rate of stimulated absorption as

$$\text{Rate of stimulated absorption} = NB\rho$$

where N is the number of molecules in the lower state, the constant B is the **Einstein coefficient of stimulated absorption**, and $\rho\Delta\nu$ is the energy density of radiation, the total energy in the frequency range ν to $\nu + \Delta\nu$ in a region of the electromagnetic field divided by the volume of the region, with ν the frequency of the transition. For the time being, we can treat B as an empirical parameter that characterizes the transition: if B is large, then a given intensity of incident radiation will induce transitions strongly and the sample will be strongly absorbing.

Einstein considered that the radiation was also able to induce the molecule in the upper state to undergo a transition to the lower state, and hence to generate a photon of frequency ν. Thus, he wrote the rate of this **stimulated emission** as

$$\text{Rate of stimulated emission} = N'B'\rho$$

where N' is the number of molecules in the excited state and B' is the **Einstein coefficient of stimulated emission.** Note that only radiation of the same frequency as the transition can stimulate an excited state to fall to a lower state. However, he realized that stimulated emission was not the only means by which the excited state could generate radiation and return to the lower state, and suggested that an excited state could undergo **spontaneous emission** at a rate that is independent of the intensity of the radiation (of any frequency) that is already present. Einstein therefore wrote the total rate of transition from the upper to the lower state as

$$\text{Overall rate of emission} = N'(A + B'\rho)$$

The constant A is the **Einstein coefficient of spontaneous emission.** It can be shown that the coefficients of stimulated absorption and emission are equal, and that the coefficient of spontaneous emission is related to them by[2]

$$A = \left(\frac{8\pi h\nu^3}{c^3}\right)B$$

[2] See our *Physical Chemistry* (2010) for the derivation.

The equality of the coefficients of stimulated emission and absorption implies that if two states happen to have equal populations, then the rate of stimulated emission is equal to the rate of stimulated absorption, and there is then no net absorption. The decrease in the value of A with decreasing frequency implies that spontaneous emission can be largely ignored at the relatively low frequencies of rotational and vibrational transitions, and the intensities of these transitions can be discussed in terms of stimulated emission and absorption. Then the net rate of absorption is given by

$$\text{Net rate of absorption} = NB\rho - N'B'\rho = (N - N')B\rho$$

and is proportional to the population difference of the two states involved in the transition. In Foundations 0.11 we saw that the ratio of populations of states of energies E and E' is given by:

$$\frac{N'}{N} = e^{-\Delta E/kT} \qquad \Delta E = E' - E$$

It follows that, for a constant energy difference ΔE, the population difference $(N - N')$ and the intensity of absorption increase with decreasing temperature. Also, for a specified temperature, the population difference and the intensity of absorption increase with increasing energy separation between the states.

Checklist of key concepts

☐ 1 The variation of the intensity of absorption with path length is expressed by the Beer–Lambert law.

☐ 2 An isosbestic point corresponds to a wavelength at which the total absorbance of a binary mixture is the same for all compositions.

☐ 3 The Franck–Condon principle states that because nuclei are so much more massive than electrons, an electronic transition takes place faster than the nuclei can respond.

☐ 4 A chromophore is a group with characteristic optical absorption: they include d-metal complexes, the carbonyl group, and the carbon–carbon double bond.

☐ 5 In fluorescence, the spontaneously emitted radiation ceases almost immediately (within nanoseconds) after the exciting radiation is extinguished.

☐ 6 In phosphorescence, the spontaneous emission may persist for long periods; the process involves intersystem crossing into a triplet state.

☐ 7 A Stern–Volmer plot is used to analyse the kinetics of fluorescence quenching in solution.

☐ 8 Collisional deactivation, electron transfer, and resonance energy transfer are common fluorescence quenching processes. The rate constants of electron and resonance energy transfer decrease with increasing separation between donor and acceptor molecules.

☐ 9 Laser action depends on the achievement of population inversion and the stimulated emission of radiation.

☐ 10 Applications of lasers in chemistry include Raman spectroscopy, time-resolved spectroscopy, and the study of multiphoton and state-specific processes.

☐ 11 Photoelectron spectroscopy is based on the photoejection of an electron by ultraviolet radiation or X-rays.

Road map of key equations

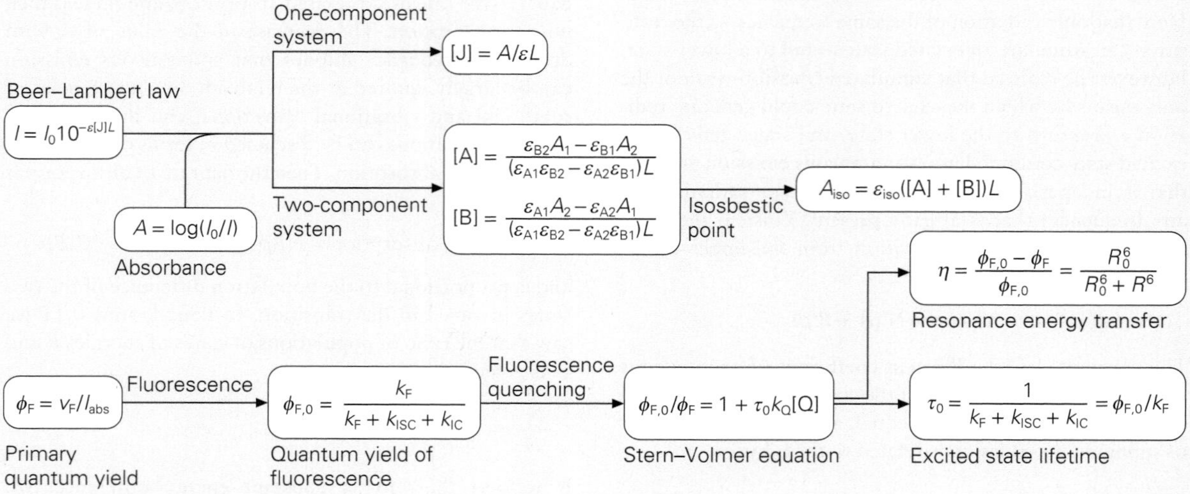

Questions and exercises

Discussion questions

20.1 Justify the form of the Beer–Lambert law. When might deviations from it be observed?

20.2 Explain the origin of the Franck–Condon principle and how it leads to the appearance of vibrational structure in an electronic transition.

20.3 Explain how colour can arise from molecules.

20.4 Describe the mechanisms of fluorescence and phosphorescence. How could you test the proposed mechanisms?

20.5 Describe the principles of laser action and the features of laser radiation that are applied to chemistry. Then, discuss two applications of lasers in chemistry.

20.6 Why is the study of fluorescence important in biology?

Exercises

20.1 An aqueous solution of a triphosphate derivative of molar mass 502 g mol^{-1} was prepared by dissolving 17.2 mg in 500 cm^3 of water and a sample was transferred to a cell of length 1.00 cm. The absorbance was measured as 1.011. (a) Calculate the molar absorption coefficient. (b) Calculate the transmittance, expressed as a percentage, for a solution of twice the concentration.

20.2 Radiation of wavelength 268 nm passed through 1.5 mm of a solution that contained benzene in a transparent solvent at a concentration of 0.080 mol dm^{-3}. The light intensity is reduced to 22 per cent of its initial value (so $T = 0.22$). Calculate the absorbance and the molar absorption coefficient of the benzene. What would be the transmittance through a cell of thickness 3.0 mm?

20.3 A *Dubosq colorimeter* consists of a cell of fixed path length and a cell of variable path length. By adjusting the length of the latter until the transmission through the two cells is the same, the concentration of the second solution can be inferred from that of the former. Suppose that a plant dye of concentration 25 μg dm^{-3} is added to the fixed cell, the length of which is 1.55 cm. Then a solution of the same dye, but of unknown concentration, is added to the second cell. It is found that the same transmittance is obtained when the length of the second cell is adjusted to 1.18 cm. What is the concentration of the second solution?

20.4 The molar absorption coefficients of two substances A and B at two wavelengths (denoted 1 and 2) are as follows: $\varepsilon_{A1} = 10.0$ dm^3 mol^{-1} cm^{-1}, $\varepsilon_{B1} = 15.0$ dm^3 mol^{-1} cm^{-1}, $\varepsilon_{A2} = 18.0$ dm^3 mol^{-1} cm^{-1}, $\varepsilon_{B2} = 12.0$ dm^3 mol^{-1} cm^{-1}. The total absorbances of a solution at these two wavelengths in a cell of length 2.0 mm were measured as 1.6 and 2.4, respectively. What are the molar concentrations of A and B in the solution?

20.5 Figure 20.35 shows the UV-visible absorption spectrum of a derivative of haemerythrin (Her) in the presence of different concentrations of CNS$^-$ ions. What may be inferred from the spectrum?

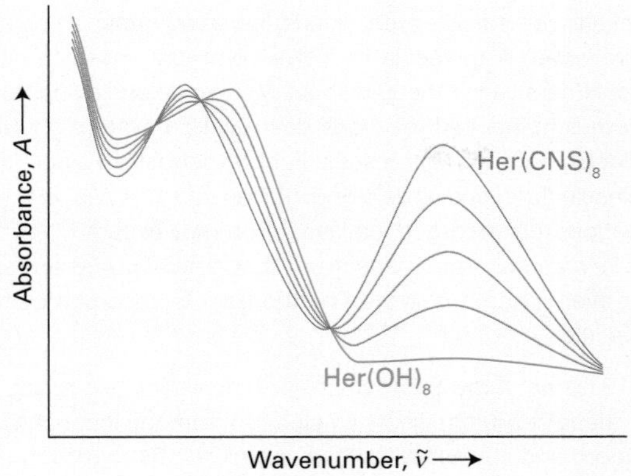

Fig. 20.35 The absorption spectra of haemerythrin in the presence of thiocyanate ions.

20.6 Suppose that you are a colour chemist and had been asked to intensify the colour of a dye without changing the type of compound, and that the dye in question was a polyene. Would you choose to lengthen or to shorten the chain? Would the modification to the length shift the apparent colour of the dye towards the red or the blue?

20.7 The compound $CH_3CH{=}CHCHO$ has a strong absorption in the ultraviolet at 46 950 cm^{-1} and a weak absorption at 30 000 cm^{-1}. Justify these features in terms of the structure of the compound.

20.8 Figure 20.36 shows the UV-visible absorption spectra of a selection of amino acids. Suggest reasons for their different appearances in terms of the structures of the molecules.

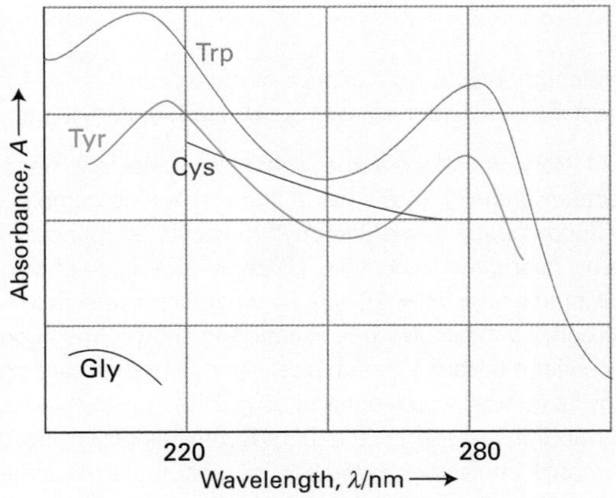

Fig. 20.36 The absorption spectra of a selection of amino acids.

20.9 The fluorescence spectrum of anthracene vapour shows a series of peaks of increasing intensity with individual maxima at 440 nm, 410 nm, 390 nm, and 370 nm followed by a sharp cut-off at shorter wavelengths. The absorption spectrum rises sharply from zero to a maximum at 360 nm with a trail of peaks of lessening intensity at 345 nm, 330 nm, and 305 nm. Account for these observations.

20.10 The line marked A in Fig. 20.37 is the fluorescence spectrum of benzophenone in solid solution in ethanol at low temperatures observed when the sample is illuminated with 360 nm light. When naphthalene is illuminated with 360 nm light it does not absorb, but the line marked B in the illustration is the phosphorescence spectrum of a solid solution of a mixture of naphthalene and benzophenone in ethanol. Now a component of fluorescence from naphthalene can be detected. Account for this observation.

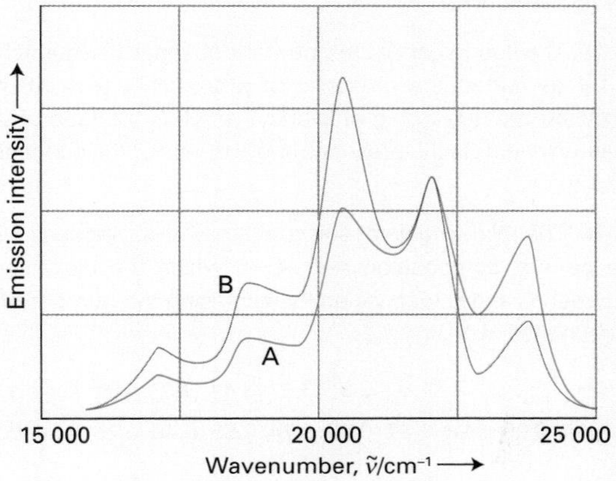

Fig. 20.37 Fluorescence spectra referred to in Exercise 20.10.

20.11 Consider a unimolecular photochemical reaction with rate constant $k_r = 1.7 \times 10^4\ s^{-1}$ that involves a reactant with an observed fluorescence lifetime of 1.0 ns and an observed phosphorescence lifetime of 1.0 ms. Is the excited singlet state or the excited triplet state the most likely precursor of the photochemical reaction?

20.12 When benzophenone is illuminated with ultraviolet radiation it is excited into a singlet state. This singlet changes rapidly into a triplet, which phosphoresces. Triethylamine acts as a quencher for the triplet. In an experiment in methanol as solvent, the phosphorescence intensity varied with amine concentration as shown below. A time-resolved laser spectroscopy experiment had also shown that the half-life of the fluorescence in the absence of quencher is 29 μs. What is the value of k_Q?

[Q]/(mol dm^{-3})	0.0010	0.0050	0.0100
I_f/(arbitrary units)	0.41	0.25	0.16

20.13 The quenching of tryptophan fluorescence by dissolved O_2 gas was monitored by measuring emission lifetimes at 348 nm in aqueous solutions. Determine the quenching rate constant for this process from the following data:

$[(O)_2]/(10^{-2}$ mol $dm^{-3})$	0	2.3	5.5	8	10.8
τ/ns	2.6	1.5	0.92	0.71	0.57

20.14 The fluorescence of a solution of a plant pigment illuminated by 330 nm radiation was studied in the presence of a quenching agent, with the following results:

[Q]/(mmol dm^{-3})	1.0	2.0	3.0	4.0	5.0
I_f/I_{abs}	0.31	0.18	0.13	0.10	0.081

In a second series of experiments, the incident radiation was extinguished and the lifetime of the decay of the fluorescence was observed:

[Q]/(mmol dm^{-3})	1.0	2.0	3.0	4.0	5.0
τ/ns	76	45	32	25	20

Determine the quenching rate constant and the half-life of the fluorescence.

20.15 The fluorescence lifetime in the absence of a quencher is 1.4 ns and in the presence of a quencher it is 0.8 ns. Calculate the quenching efficiency, which is the ratio of the quantum yields in the presence and absence of the quencher, $\phi_F/\phi_{F,0}$.

20.16 The following data refer to a family of compounds with the general composition A–B$_n$–C in which the distance R between A and C was varied by increasing the number of B units in the linker:

R/nm	1.2	1.5	1.8	2.8	3.1	3.4	3.7	4.0	4.3	4.6
η_T	0.99	0.94	0.97	0.82	0.74	0.65	0.40	0.28	0.24	0.16

Are the data described adequately by the Förster theory (eqn 20.15)? If so, what is the value of R_0 for the A–C pair?

20.17 Light-induced degradation of molecules, also called *photobleaching*, is a serious problem in *fluorescence microscopy*, in which a specimen (such as a biological cell) labelled with a fluorescent dye is observed under an optical microscope. A molecule of a dye commonly used to label biopolymers can withstand about 10^6 excitations by photons before light-induced reactions destroy its π system and the molecule no longer fluoresces. For how long will a single dye molecule fluoresce while being excited by 1.0 mW of 488 nm radiation from an argon ion laser? You may assume that the dye has an absorption spectrum that peaks at 488 nm and that every photon delivered by the laser is absorbed by the molecule.

20.18 In an X-ray photoelectron experiment, a photon of wavelength 100 pm ejects an electron from the inner shell of an atom and it emerges with a speed of 2.34×10^4 km s^{-1}. Calculate the binding energy of the electron.

20.19 The energy required for the ionization of a certain atom is 21.4 eV. The absorption of a photon of unknown wavelength ionizes the atom and ejects an electron with velocity 1.03×10^6 m s^{-1}. Calculate the wavelength of the incident radiation.

20.20 What is the kinetic energy of an electron that has been accelerated through a potential difference of 10.0 kV?

20.21 What is (a) the energy, (b) the speed of an electron that has been ejected from an orbital of ionization energy 10.0 eV by a photon of radiation of wavelength 110 nm?

20.22 In a particular photoelectron spectrum using 21.21 eV photons, electrons were ejected with kinetic energies of 11.01 eV, 8.23 eV, and 5.22 eV. Sketch the molecular orbital energy level diagram for the species, showing the ionization energies of the three identifiable orbitals.

Projects

20.23 Here we explore vision in more detail. (a) The flux of visible photons reaching Earth from the North Star is about 4×10^3 mm^{-2} s^{-1}. Of these photons, 30 per cent are absorbed or scattered by the atmosphere and 25 per cent of the surviving photons are scattered by the surface of the cornea of the eye. A further 9 per cent are absorbed inside the cornea. The area of the pupil at night is about 40 mm^2 and the response time of the eye is about 0.1 s. Of the photons passing through the pupil, about 43 per cent are absorbed in the ocular medium. How many photons from the North Star are focussed on to the retina in 0.1 s? For a continuation of this story, see R. W. Rodieck, *The First Steps in Seeing,* Sinauer (1998). (b) In the free-electron molecular orbital theory of electronic structure, the π electrons in a conjugated molecule are treated as non-interacting particles in a box of length equal to the length of the conjugated system. On the basis of this model, at what wavelength would you expect all-*trans*-retinal to absorb? Take the mean carbon–carbon bond length to be 140 pm.

20.24 Now we explore the energy and electron transfer events of photosynthesis. (a) In light-harvesting complexes, the fluorescence of a chlorophyll molecule is quenched by nearby chlorophyll molecules. Given that for a pair of chlorophyll *a* molecules $R_0 = 5.6$ nm, by what distance should two chlorophyll *a* molecules be separated to shorten the fluorescence lifetime from 1 ns (a typical value for monomeric chlorophyll *a* in organic solvents) to 10 ps? (b) The light-induced electron transfer reactions in photosynthesis occur because chlorophyll molecules (whether in monomeric or dimeric forms) are better reducing agents in their electronic excited states. Justify this observation with the help of molecular orbital theory.

21

Spectroscopy: magnetic resonance

Nuclear magnetic resonance 529

21.1 Nuclei in magnetic fields 530

21.2 The technique 532

The information in NMR spectra 533

21.3 The chemical shift 533

21.4 The fine structure 535

21.5 Spin relaxation 539

21.6 Conformational conversion and chemical exchange 541

21.7 Two-dimensional NMR 542

Electron paramagnetic resonance 543

21.8 The *g*-value 544

21.9 Hyperfine structure 545

CHECKLIST OF KEY CONCEPTS 547

ROAD MAP OF KEY EQUATIONS 548

QUESTIONS AND EXERCISES 549

One of the most widely used and helpful forms of spectroscopy, and a technique that has transformed the practice of chemistry and its dependent disciplines, makes use of an effect from classical physics. When two pendulums are joined by the same slightly flexible support and one is set in motion, the other is forced into oscillation by the motion of the common axle, and energy flows between the two. The energy transfer occurs most efficiently when the frequencies of the two oscillators are identical. The condition of strong effective coupling when the frequencies are identical is called **resonance**, and the excitation energy is said to 'resonate' between the coupled oscillators.

Resonance is the basis of a number of everyday phenomena, including the response of radios to the weak oscillations of the electromagnetic field generated by a distant transmitter. In this chapter we explore a spectroscopic application that when originally developed (and in some cases still) depends on matching a set of energy levels to a source of monochromatic radiation in the radiofrequency and microwave ranges and observing the strong absorption by nuclei and electrons, respectively, that occurs at resonance. In fact, all spectroscopy is a form of resonant coupling between the electromagnetic field and the molecules; what distinguishes magnetic resonance is that the energy levels themselves are modified by the application of a magnetic field.

Nuclear magnetic resonance

The application of resonance that we describe here depends on the fact that many nuclei possess spin angular momentum (Table 21.1). A nucleus with **nuclear spin quantum number** I (the analogue of s for electrons, and which may be an integer or a half-integer) may take $2I + 1$ different orientations relative to an arbitrary axis. These orientations are

Table 21.1

Nuclear constitution and the nuclear spin quantum number

Number of protons	Number of neutrons	I
even	even	0
odd	odd	integer (1, 2, 3, . . .)
even	odd	half-integer ($\frac{1}{2}$, $\frac{3}{2}$, $\frac{5}{2}$, . . .)
odd	even	half-integer ($\frac{1}{2}$, $\frac{3}{2}$, $\frac{5}{2}$, . . .)

distinguished by the quantum number m_I, which may take on the values $m_I = I, I-1, \ldots, -I$. A proton has $I = \frac{1}{2}$ and may adopt either of two orientations ($m_I = +\frac{1}{2}$ and $-\frac{1}{2}$). A ^{14}N nucleus has $I = 1$ and may adopt any of three orientations ($m_I = +1, 0, -1$). Spin-$\frac{1}{2}$ nuclei include protons (^{1}H), ^{13}C, ^{19}F, and ^{31}P nuclei. The state with $m_I = +\frac{1}{2}$ (↑) is denoted α, and that with $m_I = -\frac{1}{2}$ (↓) is denoted β.

The properties of magnetic fields and their interactions with matter, which lie at the core of this chapter, are summarized in The chemist's toolkit 21.1.

The chemist's toolkit 21.1 Magnetic fields

The energy of a magnetic moment $\boldsymbol{m}$ in a magnetic field $\boldsymbol{\mathcal{B}}$ is

$$E = -\boldsymbol{m}\cdot\boldsymbol{\mathcal{B}}$$

Here, $\boldsymbol{\mathcal{B}}$ is the **magnetic induction**, a measure of the strength of the magnetic field. It is reported in tesla, T, where 1 T = 1 kg s^{-2} A^{-1} (A denotes ampere). When the field lies along the z-direction,

$$E = -m_z\mathcal{B}$$

The energy of interaction of two magnetic dipoles of magnitude m_1 and m_2 lying parallel to each other in the arrangement shown in Sketch 1 is

$$E = \frac{\mu_0 m_1 m_2}{4\pi r^3}(1 - 3\cos^2\theta)$$

where μ_0 is the vacuum permeability (see inside front cover).

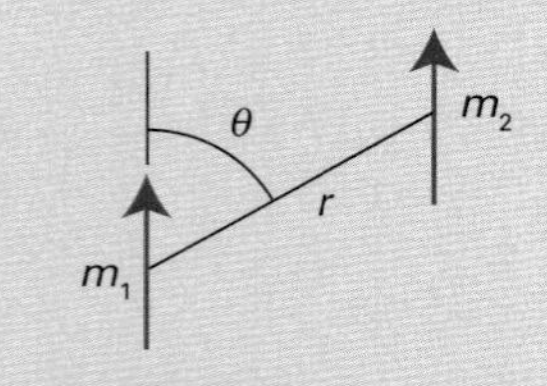

Sketch 21.1

21.1 Nuclei in magnetic fields

A nucleus with nonzero spin has a magnetic moment and behaves like a tiny magnet. The orientation of this magnet is determined by the value of m_I, and in a magnetic field $\mathcal{B}$ the $2I + 1$ orientations of the nucleus have different energies. These energies are given by

$$E_{m_I} = -\gamma_N \hbar \mathcal{B} m_I \qquad \text{Energy of nucleus} \quad (21.1)$$

where γ_N is the **nuclear magnetogyric ratio.** For spin-$\frac{1}{2}$ nuclei with positive magnetogyric ratios (such as ^{1}H), the α state lies below the β state in energy. The energy is sometimes written in terms of the **nuclear magneton,** μ_N,

$$\mu_N = \frac{e\hbar}{2m_p} = 5.051\times10^{-27}\ \text{J T}^{-1} \qquad \text{Nuclear magneton} \quad (21.2)$$

and an empirical constant called the **nuclear *g*-factor,** g_I, when it becomes

$$E_{m_I} = -g_I \mu_N \mathcal{B} m_I \qquad \text{Energy of nucleus} \quad (21.3)$$

Nuclear g-factors are experimentally determined dimensionless quantities with values typically between −6 and +6. Positive values of γ_N (and g_I) indicate that the North pole of the nuclear magnet lies in the same direction as the nuclear spin (this is the case for protons). Negative values indicate that the magnet points in the opposite direction. A nuclear magnet is about 2000 times weaker than the magnet associated with electron spin. Two very common nuclei, ^{12}C and ^{16}O, have zero spin and hence are not affected by external magnetic fields.

Table 21.2

Nuclear spin properties

Nucleus per cent	Natural abundance per cent	Spin, I	$\gamma_N/(10^7\ \text{T}^{-1}\ \text{s}^{-1})$
^{2}H (D)	0.0156	1	4.1067
^{12}C	98.99	0	–
^{1}H	99.98	$\frac{1}{2}$	26.752
^{13}C	1.11	$\frac{1}{2}$	6.7272
^{14}N	99.64	1	1.9328
^{16}O	99.96	0	–
^{17}O	0.037	$\frac{5}{2}$	−3.627
^{19}F	100	$\frac{1}{2}$	25.177
^{31}P	100	$\frac{1}{2}$	10.840
^{35}Cl	75.4	$\frac{3}{2}$	2.624
^{37}Cl	24.6	$\frac{3}{2}$	2.184

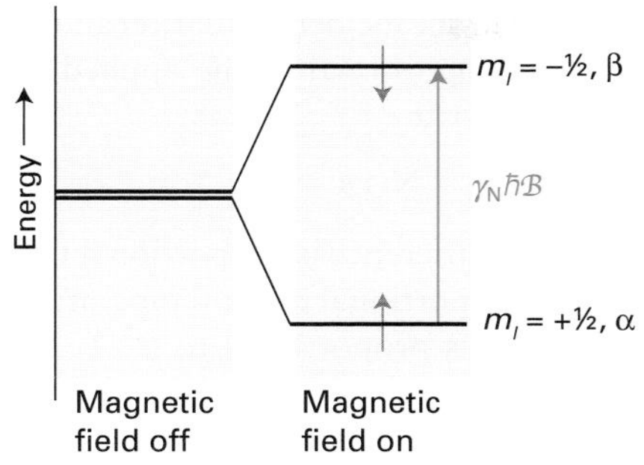

Fig. 21.1 The energy levels of a spin-1/2 nucleus (e.g. ^1H or ^{13}C) in a magnetic field. Resonance occurs when the energy separation of the levels matches the energy of the photons in the electromagnetic field.

(a) Populations

The energy separation of the two states of a spin-1/2 nucleus (Fig. 21.1) is

$$\Delta E = E_\beta - E_\alpha = -\gamma_N\hbar\mathcal{B}(-1/2) + \gamma_N\hbar\mathcal{B}(+1/2) = \gamma_N\hbar\mathcal{B} \quad (21.4)$$

As can be inferred from the discussion in Foundations 0.11, the ratio of the thermal equilibrium populations of the α and β states, N_α and N_β, is given by the Boltzmann distribution as

$$\frac{N_\beta}{N_\alpha} = e^{-\Delta E/kT} = e^{-\gamma_N\hbar\mathcal{B}/kT} \quad \text{Nuclei} \quad \text{Population ratio} \quad (21.5a)$$

We show in the following Derivation that from this relation it is possible to deduce that

$$N_\alpha - N_\beta \approx \frac{N\gamma_N\hbar\mathcal{B}}{2kT} \quad \text{Nuclei} \quad \text{Population difference} \quad (21.5b)$$

where N is the total number of spins. For nuclei with positive γ_N the α state lies below the β state, it follows from eqn 21.5 that there are slightly more α spins than β spins.

● **Brief illustration 21.1** Nuclear spin populations

For protons, $\gamma_N = 2.675 \times 10^8\ T^{-1}\ s^{-1}$. Therefore, for 1 000 000 protons in a field of 10 T at 20 °C,

$$N_\alpha - N_\beta \approx \frac{\overbrace{1\,000\,000}^{N}\times\overbrace{(2.675\times10^8\ T^{-1}\ s^{-1})}^{\gamma_N}\times\overbrace{(1.055\times10^{-34}\ J\ s)}^{\hbar}\times\overbrace{(10\ T)}^{\mathcal{B}}}{2\times\underbrace{(1.381\times10^{-23}\ J\ K^{-1})}_{k}\times\underbrace{(293\ K)}_{T}}$$

$$\approx 35$$

Even in such a strong field there is only a tiny imbalance of population of about 35 in a million.

Self-test 21.1

For ^{13}C nuclei, $\gamma_N = 6.7283\times10^7\ T^{-1}\ s^{-1}$. Determine the magnetic field necessary to induce the same imbalance in the distribution of ^{13}C spins at 20 °C.

Answer: 40 T, an unrealistically high field for an NMR spectrometer

Derivation 21.1

The population difference

To write an expression for the population difference, we begin with eqn 21.5a, written as

$$\frac{N_\beta}{N_\alpha} = e^{-\gamma_N\hbar\mathcal{B}/kT} \overbrace{\approx}^{e^{-x}=1-x+\cdots} 1 - \frac{\gamma_N\hbar\mathcal{B}}{kT}$$

where we have used the expansion $e^{-x} = 1 - x + \frac{1}{2}x^2 - \cdots$ (see The chemist's toolkit 6.1) and kept only the first two terms because x is so small. It follows that

$$\frac{N_\alpha - N_\beta}{N_\alpha + N_\beta} = \frac{N_\alpha(1 - N_\beta/N_\alpha)}{N_\alpha(1 + N_\beta/N_\alpha)} \overbrace{=}^{\text{Cancel } N_\alpha} \frac{1 - \overbrace{N_\beta/N_\alpha}^{1-\gamma_N\hbar\mathcal{B}/kT}}{1 + \underbrace{N_\beta/N_\alpha}_{1-\gamma_N\hbar\mathcal{B}/kT}}$$

$$\approx \frac{1-(1-\gamma_N\hbar\mathcal{B}/kT)}{\underbrace{1+(1-\gamma_N\hbar\mathcal{B}/kT)}_{\approx 1}} \approx \frac{\gamma_N\hbar\mathcal{B}/kT}{2}$$

Then, with $N_\alpha + N_\beta = N$, the total number of spins, we have eqn 21.5b.

(b) Resonance

If the sample is exposed to radiation of frequency ν, the energy separation of spins comes into resonance with the radiation when the frequency satisfies the **resonance condition:**

$$h\nu = \gamma_N\hbar\mathcal{B} \text{ or } \nu = \frac{\gamma_N\mathcal{B}}{2\pi} \quad \text{Nuclei} \quad \text{Resonance condition} \quad (21.6)$$

At resonance there is strong coupling between the spins and the radiation, and absorption occurs as the spins flip from the lower energy state to the upper state.

It is sometimes useful to compare the quantum mechanical and classical pictures of magnetic nuclei pictured as tiny bar magnets. A bar magnet in an externally applied magnetic field undergoes the motion called **precession** as it twists round the direction of the field (Fig. 21.2). The rate of precession is proportional to the strength of the applied field, and is in fact equal to $(\gamma_N/2\pi)\mathcal{B}$ for nuclei, which in this context is called the **Larmor precession frequency**.

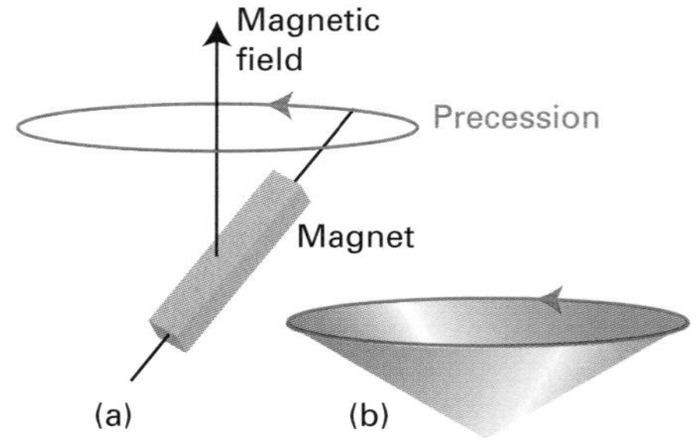

Fig. 21.2 (a) A bar magnet in a magnetic field undergoes the motion called *precession*. A nuclear spin (and an electron spin) has an associated magnetic moment, and behaves in the same way. (b) The frequency of precession is called the Larmor precession frequency, and is proportional to the applied field and the magnitude of the magnetic moment.

That is, resonance absorption occurs when the Larmor precession frequency is the same as the frequency of the applied electromagnetic field.

Brief illustration 21.2 The resonance condition

To calculate the frequency at which radiation comes into resonance with proton spins in a 12 T magnetic field we use eqn 21.6 as follows:

$$\nu = \frac{\overbrace{(2.6752\times10^{8}\ \mathrm{T^{-1}\ s^{-1}})}^{|\gamma_N|}\times\overbrace{(12\ \mathrm{T})}^{\mathcal{B}}}{2\pi} = 5.1\times10^{8}\ \mathrm{s^{-1}}$$

or 510 MHz (with 1 Hz = 1 s^{-1}).

Self-test 21.2

Determine the resonance frequency for ^{31}P nuclei, for which $\gamma_N = 1.0841\times10^{8}\ \mathrm{T^{-1}\ s^{-1}}$, under the same conditions.

Answer: 207 MHz

(c) Absorption intensities

The rate of absorption of electromagnetic radiation is proportional to the population of the lower energy state (N_α in the case of a proton NMR transition) and the rate of stimulated emission is proportional to the population of the upper state (N_β). At the low frequencies typical of magnetic resonance, spontaneous emission can be ignored because it is very slow. Therefore, the net rate of absorption is proportional to the difference in populations, $N_\alpha - N_\beta$. The intensity of absorption, the rate at which energy is absorbed, is proportional to the product of the rate of transition (the rate at which photons are absorbed) and the energy of each photon. The latter is proportional to the frequency ν of the incident radiation and therefore at resonance, proportional to the applied magnetic field, so we can write

$$\text{Intensity} \propto (N_\alpha - N_\beta)\mathcal{B} \qquad (21.7)$$

The population difference is proportional to the field and inversely proportional to the temperature, so the overall intensity is proportional to $\mathcal{B}^2/T$. It follows that decreasing the temperature increases the intensity by increasing the population difference. The intensity can also be enhanced significantly by increasing the strength of the applied magnetic field, making spectrometers operating at high fields highly desirable.

21.2 The technique

In its simplest form, **nuclear magnetic resonance** (NMR) is the observation of the frequency at which magnetic nuclei in molecules come into resonance with an electromagnetic field when the molecule is exposed to a strong magnetic field. When applied to proton spins, the technique is occasionally called **proton magnetic resonance** (^{1}H-NMR). In the early days of the technique the only nuclei that could be studied were protons (which behave like relatively strong magnets because γ_N is large), but now a wide variety of nuclei (especially ^{13}C, ^{31}P, and ^{15}N) are investigated routinely.

An NMR spectrometer consists of a magnet that can produce a uniform, intense field and the appropriate sources of radiofrequency radiation (Fig. 21.3). In simple instruments the magnetic field is provided by an electromagnet; for serious work, a superconducting magnet capable of producing fields of the order of 10 T and more is used. (A magnetic field of 10 T is very strong: a small magnet, for example, gives a magnetic field of only a few millitesla.) The

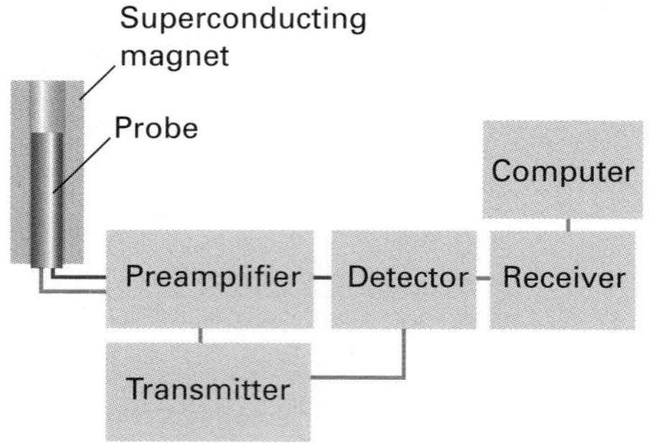

Fig. 21.3 The layout of a typical NMR spectrometer. The link from the transmitter to the detector indicates that the high frequency of the transmitter is subtracted from the high-frequency received signal to give a low-frequency signal for processing.

use of high magnetic fields has two advantages. One is that, as we have seen, the field increases the intensities of transitions. Secondly, a high field simplifies the appearance of certain spectra. Proton resonance occurs at about 400 MHz in fields of 9.4 T, so NMR is a radiofrequency technique (400 MHz corresponds to a wavelength of 75 cm).

Fourier transform NMR (FT-NMR) is the most common technique used in modern magnetic resonance. The sample is held in a strong magnetic field generated by a superconducting magnet and exposed to one or more carefully controlled brief bursts of radiofrequency radiation. This radiation changes the orientations of the nuclear spins in a controlled way, and the radiofrequency of radiation they emit as they return to equilibrium is monitored and analysed mathematically (the latter is the 'Fourier transform' part of the technique). The detected radiation contains all the information in the spectrum obtained by the earlier technique, but it is a much more efficient way of obtaining the spectrum and hence is much more sensitive. Moreover, by choosing different sequences of exciting pulses, the data can be analysed much more closely.

The information in NMR spectra

Nuclear spins interact with the *local* magnetic field, the field in their immediate vicinity. The local field may differ from the applied field either on account of the local electronic structure of the molecule or because there is another magnetic nucleus nearby.

21.3 The chemical shift

The applied magnetic field can induce a circulating motion of the electrons in the molecule, and that motion gives rise to a small additional magnetic field, $\mathcal{B}_{add}$. This additional field is proportional to the applied field, and it is conventional to express it as

$$\mathcal{B}_{add} = -\sigma\mathcal{B} \qquad \text{Shielding constant} \qquad (21.8)$$

where the dimensionless quantity σ is the **shielding constant**. We are assuming that the additional field is parallel to the applied field: more advanced treatments allow for the two not being parallel. The shielding constant may be positive or negative according to whether the induced field adds to or subtracts from the applied field. The ability of the applied field to induce the circulation of electrons through the nuclear framework of the molecule depends on the details of the electronic structure near the magnetic nucleus of interest, so nuclei in different chemical groups have different shielding constants.

Because the total local field is

$$\mathcal{B}_{loc} = \mathcal{B} + \mathcal{B}_{add} = (1-\sigma)\mathcal{B}$$

the resonance condition is

$$\nu = \frac{\gamma_N \mathcal{B}_{loc}}{2\pi} = \frac{\gamma_N}{2\pi}(1-\sigma)\mathcal{B} \qquad \text{Resonance condition} \qquad (21.9)$$

The shielding constant σ varies with the environment, so different nuclei (even of the same element in different parts of a molecule) come into resonance at different frequencies.

The **chemical shift** of a nucleus is the difference between its resonance frequency and that of a reference standard. The standard for protons is the proton resonance in tetramethylsilane, $Si(CH_3)_4$, commonly referred to as TMS, which bristles with protons and dissolves without reaction in many solutions. For ^{13}C, the reference frequency is the ^{13}C resonance in TMS, and for ^{31}P it is the ^{31}P resonance in 85 per cent $H_3PO_4(aq)$. Other references are used for other nuclei. The separation of the resonance of a particular group of nuclei from the standard increases with the strength of the applied magnetic field because the induced field is proportional to the applied field, and the stronger the latter, the greater the shift.

Chemical shifts are reported on the **δ scale**, which is defined as

$$\delta = \frac{\nu - \nu^\circ}{\nu^\circ} \times 10^6 \qquad \text{Definition} \quad \delta \text{ scale} \qquad (21.10)$$

where ν° is the resonance frequency of the standard. The advantage of the δ scale is that shifts reported on it are independent of the applied field (because both numerator and denominator are proportional to the applied field). The resonance frequencies themselves, however, do depend on the applied field through

$$\nu = \nu^\circ + (\nu^\circ/10^6)\delta \qquad (21.11)$$

A note on good practice In much of the literature, chemical shifts are reported in parts per million, ppm, in recognition of the factor of 10^6 in the definition; this is unnecessary. If you see '$\delta = 10$ ppm', interpret it, and use it in eqn 21.10, as $\delta = 10$.

Brief illustration 21.3 The δ scale 1

A nucleus with $\delta = 1.00$ in a spectrometer operating at 500 MHz, a '500 MHz NMR spectrometer', will have a shift relative to the reference equal to

$$\nu - \nu^\circ = (500 \text{ MHz}/10^6) \times 1.00 = (500 \text{ Hz}) \times 1.00 = 500 \text{ Hz}$$

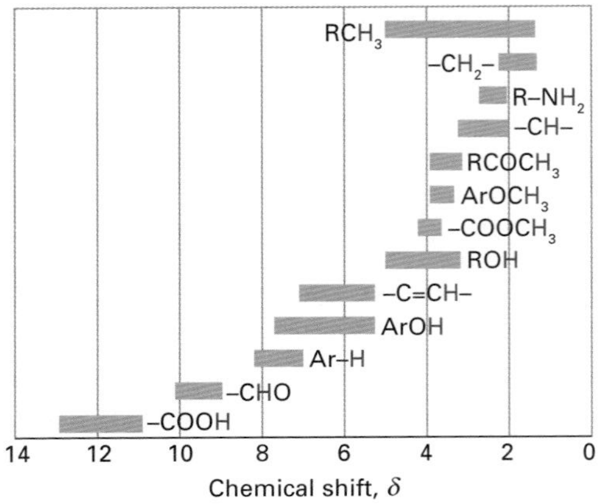

Fig. 21.4 The range of typical chemical shifts for ^{1}H resonances.

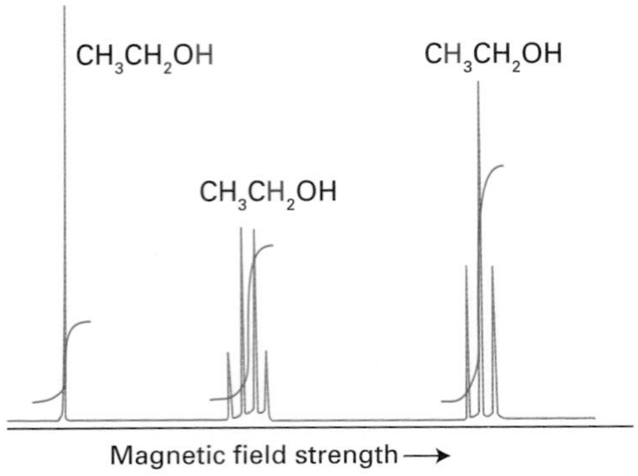

Fig. 21.5 The NMR spectrum of ethanol. The red letters denote the protons giving rise to the resonance peak, and the step-like curves are the integrated signals for each group of lines.

because 1 MHz = 10^6 Hz. In a spectrometer operating at 100 MHz, the shift relative to the reference would be only 100 Hz.

Self-test 21.3

What is the shift of the resonance from TMS of a group of nuclei with δ = 3.50 and an operating frequency of 350 MHz?

Answer: 1.23 kHz

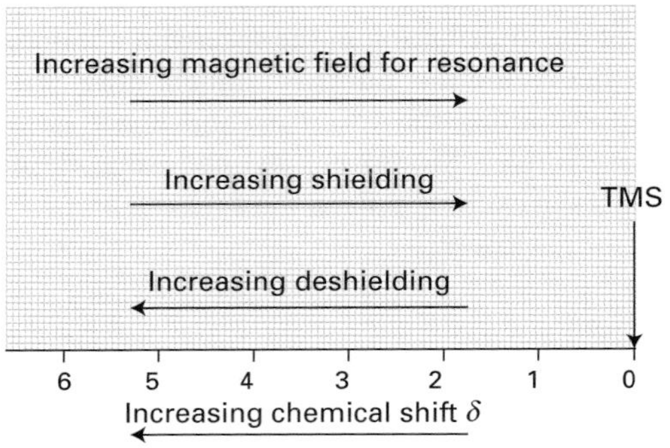

Fig. 21.6 The conventions for the display of an NMR spectrum.

If $\delta > 0$, we say that the nucleus is **deshielded;** if $\delta < 0$, then it is **shielded.** A positive δ indicates that the resonance frequency of the group of nuclei in question is higher than that of the standard. Hence $\delta > 0$ indicates that the local magnetic field is stronger than that experienced by the nuclei in the standard under the same conditions and therefore a lower applied field is needed to achieve resonance with a given radiofrequency field. Figure 21.4 shows some typical chemical shifts.

Brief illustration 21.4 The δ scale 2

The existence of a chemical shift explains the general features of the spectrum of ethanol shown in Fig. 21.5. The CH_3 protons form one group of nuclei with $\delta = 1$. The two CH_2 are in a different part of the molecule, experience a different local magnetic field, and hence resonate at $\delta = 3$. Finally, the OH proton is in another environment, and has a chemical shift of $\delta = 4$.

A note on good (or, at least, conventional) practice Traditionally, NMR spectra are plotted with δ increasing from right to left. Consequently, in a given radiofrequency the magnetic field for resonance increases from left to right (Fig. 21.6).

We can use the relative intensities of the signal (the areas under the absorption lines) to help distinguish which group of lines corresponds to which chemical group, and spectrometers can **integrate** the absorption—that is, determine the areas under the absorption signal—automatically (as is shown in Fig. 21.5). In ethanol the group intensities are in the ratio 3:2:1 because there are three CH_3 protons, two CH_2 protons, and one OH proton in each molecule. Counting the number of magnetic nuclei as well as noting their chemical shifts is valuable analytically because it helps us identify the compound present in a sample and to identify substances in different environments.

The observed shielding constant is the sum of three contributions:

$$\sigma = \sigma(\text{local}) + \sigma(\text{neighbour}) + \sigma(\text{solvent}) \quad (21.12)$$

The **local contribution,** σ(local), is essentially the contribution of the electrons of the atom that contains the nucleus in question. The **neighbouring**

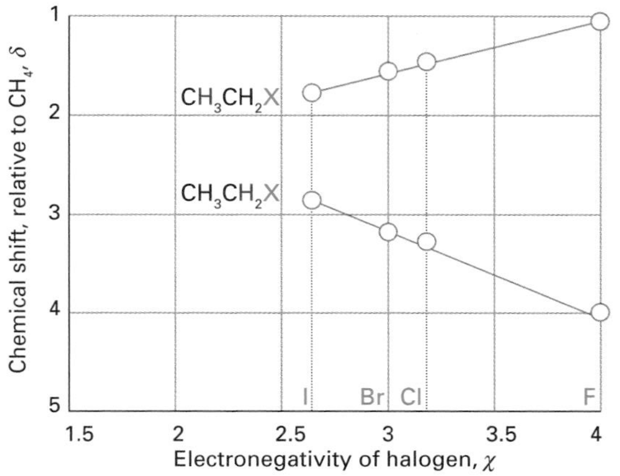

Fig. 21.7 The variation of chemical shift with the electronegativity of the halogen in the haloalkanes. Note that although the chemical shift of the immediately adjacent protons becomes more positive (the protons are deshielded) as the electronegativity increases, that of the next-nearest protons decreases.

group contribution, σ(neighbour), is the contribution from the groups of atoms that form the rest of the molecule. The **solvent contribution**, σ(solvent), is the contribution from the solvent molecules.

The local contribution is broadly proportional to the electron density of the atom containing the nucleus of interest. It follows that the shielding is decreased if the electron density on the atom is reduced by the influence of an electronegative atom nearby. That reduction in shielding translates into an increase in deshielding, and hence to an increase in the chemical shift δ as the electronegativity of a neighbouring atom increases (Fig. 21.7). That is, as the electronegativity increases, δ increases.

Another contribution to σ(local) arises from the ability of the applied field to force the electrons to circulate through the molecule by making use of orbitals that are unoccupied in the ground state. This contribution is large in molecules with low lying excited states and is dominant for atoms other than hydrogen. It is zero in free atoms and around the axes of linear molecules (such as ethyne, HC≡CH) where the electrons can circulate freely, because a field applied along the internuclear axis is unable to force them into other orbitals.

The neighbouring group contribution arises from the currents induced in nearby groups of atoms. The strength of the additional magnetic field that the proton experiences is inversely proportional to the cube of the distance r between H and X. A special case of a neighbouring group effect is found in aromatic compounds. The strong anisotropy of the magnetic

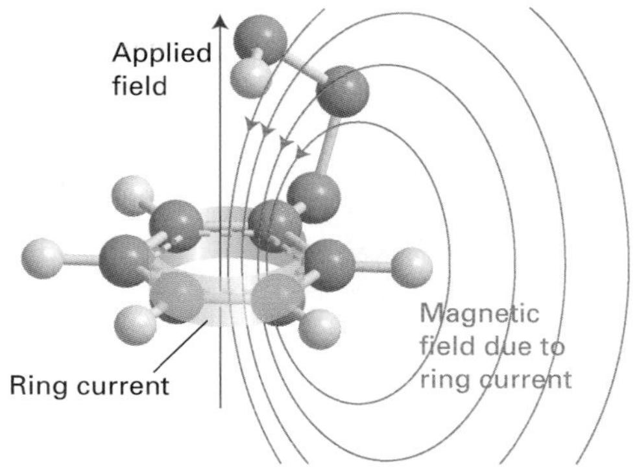

Fig. 21.8 The shielding and deshielding effects of the ring current induced in the benzene ring by the applied field. Protons attached to the ring are deshielded but a proton attached to a substituent that projects above the ring is shielded.

susceptibility of the benzene ring is ascribed to the ability of the field to induce a **ring current**, a circulation of electrons around the ring, when it is applied perpendicular to the molecular plane. Protons in the plane are deshielded (Fig. 21.8), but any that happen to lie above or below the plane (as members of substituents of the ring) are shielded.

A solvent can influence the local magnetic field experienced by a nucleus in a variety of ways. Some of these effects arise from specific interactions between the solute and the solvent (such as hydrogen-bond formation and other forms of Lewis acid–base complex formation). The magnetic susceptibility of the solvent molecules, especially if they are aromatic, can also be the source of a local magnetic field. Moreover, if there are steric interactions that result in a loose but specific interaction between a solute molecule and a solvent molecule, then protons in the solute molecule may experience shielding or deshielding effects according to their location relative to the solvent molecule (Fig. 21.9). We shall see that the NMR spectra of species that contain protons with widely different chemical shifts are easier to interpret than those in which the shifts are similar, so the appropriate choice of solvent may help to simplify the appearance and interpretation of a spectrum.

21.4 The fine structure

The splitting of the groups of resonances into individual lines in Fig. 21.5 is called the **fine structure** of the spectrum. It arises because each magnetic nucleus contributes to the local field experienced by the other nuclei and modifies their resonance frequencies. The strength of the interaction is expressed in terms of the

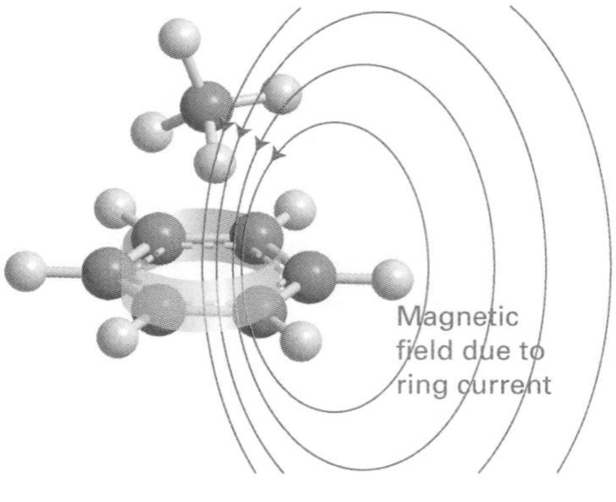

Fig. 21.9 An aromatic solvent (benzene here) can give rise to local currents that shield or deshield a proton in a solute molecule. In this relative orientation of the solvent and solute, the proton on the solute molecule is shielded.

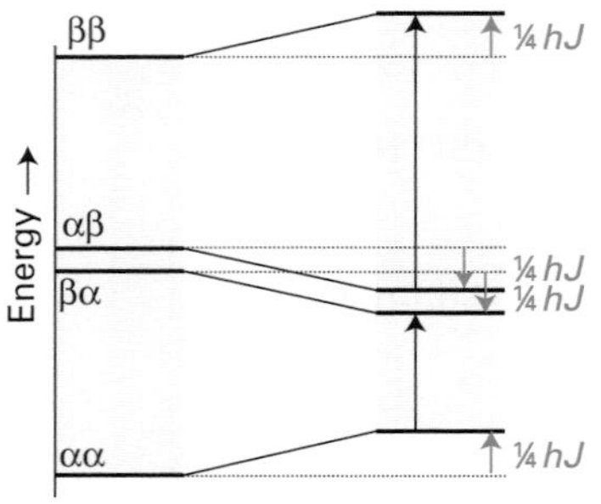

Fig. 21.10 The energy levels of a two-proton system in the presence of a magnetic field. The levels on the left apply in the absence of spin–spin coupling. Those on the right are the result of allowing for spin–spin coupling. The only allowed transitions differ in frequency by J.

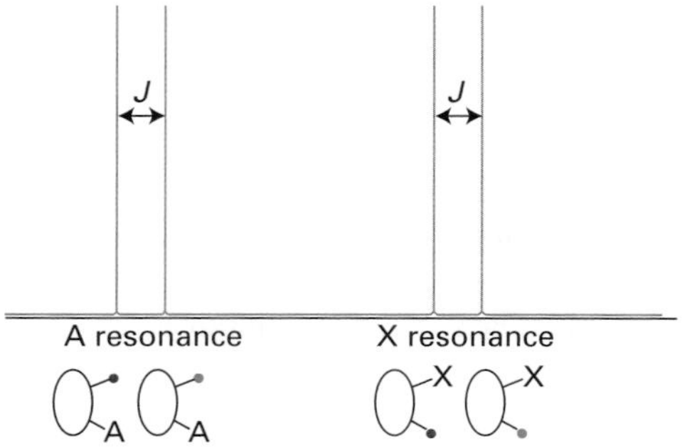

Fig. 21.11 The effect of spin–spin coupling on a NMR spectrum of two spin-½ nuclei with widely different chemical shifts. Each resonance is split into two lines separated by J. Red circles indicate α spins, green circles indicate β spins.

spin–spin coupling constant, J, and reported in hertz (Hz). Spin coupling constants are an intrinsic property of the molecule and independent of the strength of the applied field.

(a) The appearance of the spectrum

In NMR, letters far apart in the alphabet (typically A and X) are used to indicate nuclei with very different chemical shifts; letters close together (such as A and B) are used for nuclei with similar chemical shifts. Let's consider first a molecule that contains two spin-½ nuclei A and X. First, neglect spin–spin coupling. The total energy of two protons in a magnetic field $\mathcal{B}$ is the sum of two terms like eqn 21.1 but with $\mathcal{B}$ modified to $(1-\sigma)\mathcal{B}$:

$$E = -\gamma_N\hbar(1-\sigma_A)\mathcal{B}m_A - \gamma_N\hbar(1-\sigma_X)\mathcal{B}m_X$$

Here σ_A and σ_X are the shielding constants of A and X. The four energy levels predicted by this formula are shown on the left of Fig. 21.10. The spin–spin coupling energy is normally written

$$E_{\text{spin-spin}} = hJm_Am_X \qquad \text{Spin–spin coupling} \quad (21.13)$$

There are four possibilities, depending on the values of the quantum numbers m_A and m_X:

	$\alpha_A\alpha_X$	$\alpha_A\beta_X$	$\beta_A\alpha_X$	$\beta_A\beta_X$
$E_{\text{spin-spin}}$	$+\frac{1}{4}hJ$	$-\frac{1}{4}hJ$	$-\frac{1}{4}hJ$	$+\frac{1}{4}hJ$

The resulting energy levels are shown on the right in Fig. 21.10.

Now consider the transitions. When an A nucleus changes its spin from α to β, the X nucleus remains in its same spin state, which may be either α or β. The two transitions are shown in the illustration and we see that they differ in frequency by J. Alternatively, the X nucleus can undergo a transition from α to β; now the A nucleus remains in its same spin state, which may be either α or β, and again there are two transitions differing in frequency by J. As a result, the spectrum consists of a doublet of lines separated by a frequency J (Fig. 21.11).

If there is another X nucleus in the molecule with the same chemical shift as the first X (giving an AX_2 species), the resonance of A is split into a doublet by one X, and each line of the doublet is split again by the same amount by the second X (Fig. 21.12). This splitting results in three lines in the intensity ratio 1:2:1 (because the central frequency can be obtained in two ways). As in the AX case discussed above, the X resonance of the AX_2 species is split into a doublet by A.

Three equivalent X nuclei (an AX_3 species) split the resonance of A into four lines of intensity ratio 1:3:3:1 (Fig. 21.13). The X resonance remains a doublet as a result of the splitting caused by A. In general, N equivalent spin-½ nuclei split the

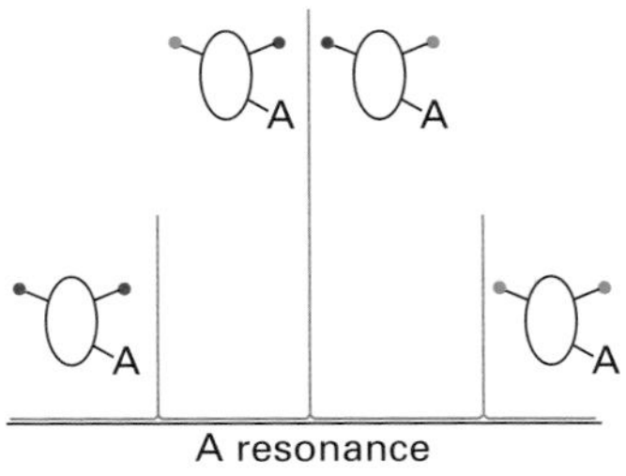

Fig. 21.12 The origin of the 1:2:1 triplet in the A resonance of an AX_2 species. The two X nuclei may have the $2^2 = 4$ spin arrangements (↑↑); (↑↓); (↓↑); (↓↓). The middle two arrangements are responsible for the coincident resonances of A.

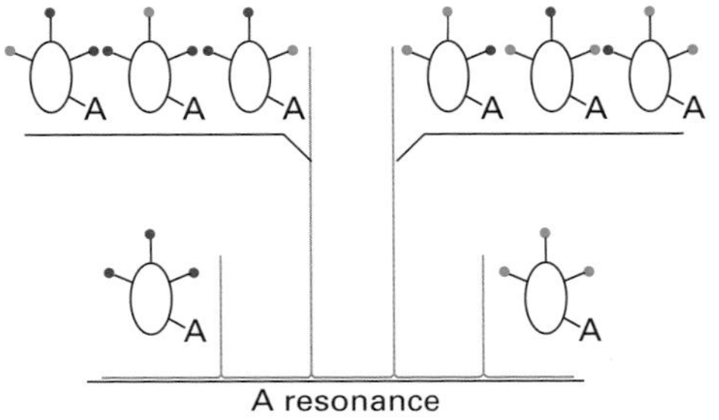

Fig. 21.13 The origin of the 1:3:3:1 quartet in the A resonance of an AX_3 species where A and X are spin-½ nuclei with widely different chemical shifts. There are $2^3 = 8$ arrangements of the spins of the three X nuclei, and their effects on the A nucleus give rise to four groups of resonances.

resonance of a nearby spin or group of equivalent spins into $N + 1$ lines with an intensity distribution given by Pascal's triangle (**1**). Successive rows of this triangle are formed by adding together the two adjacent numbers in the line above.

```
        1
      1   1
    1   2   1
  1   3   3   1
1   4   6   4   1
```

1 Pascal's triangle

Brief illustration 21.5 The fine structure in a spectrum

The three protons of the CH_3 group of CH_3CH_2OH split the single resonance of the CH_2 protons into a 1:3:3:1 quartet with a splitting *J*. Likewise, the two protons of the CH_2 group split the single resonance of the CH_3 protons into a 1:2:1 triplet. Each of these lines is split into a doublet to a small extent by the OH proton.

Self-test 21.4

What fine structure can be expected for the protons in $^{14}NH_4^+$? The nuclear spin quantum number of ^{14}N is 1.

Answer: a 1:1:1 triplet from ^{14}N

The spin–spin coupling constant of two nuclei joined by *N* bonds is normally denoted ^NJ, with subscripts for the types of nuclei involved. Thus, $^1J_{CH}$ is the coupling constant for a proton joined directly to a ^{13}C atom, and $^2J_{CH}$ is the coupling constant when the two nuclei are separated by two bonds (as in ^{13}C—C—H). A typical value of $^1J_{CH}$ is between 10^2 to10^3 Hz; the value of $^2J_{CH}$ is about 10 times less, between about 10 and 10^2 Hz. Both ^3J and ^4J give detectable effects in a spectrum, but couplings over larger numbers of bonds can generally be ignored.

Example 21.1

Interpreting an NMR spectrum

Figure 21.14 shows the ^1H-NMR spectrum of diethyl ether, $(CH_3CH_2)_2O$. What can be inferred from it?

Strategy Identify the resonances by reference to Fig. 21.4. Then decide how the fine structure should be interpreted. The integrated intensities give a clue about the numbers of protons in each chemically equivalent group. Consider why the chemical shifts are different for the two types of group and how the spectrum might change in a spectrometer operating at a higher applied field.

Solution The resonance at $\delta = 3.4$ corresponds to CH_2 in an ether; that at $\delta = 1.2$ corresponds to CH_3 in CH_3CH_2. The difference in shielding constant can be traced to the effect of the central O atom, which draws electron density towards itself. The higher value for CH_2, indicating greater deshielding, is due to the group being close to the O atom so that electron density is withdrawn from it more strongly than from the more distant CH_3 group. As we saw in Brief illustration 21.5, the fine structure of the CH_2 group (a 1:3:3:1 quartet) is characteristic of splitting caused by CH_3; the fine structure of the CH_3 resonance is characteristic of splitting caused by CH_2. The spin–spin coupling constant is $J = -60$ Hz (the same for each group). If the spectrum had been recorded with a spectrometer operating at five times the magnetic field strength, the groups of lines would have been observed to be five times further apart in frequency (but the same δ values). No change in spin–spin splitting would be observed.

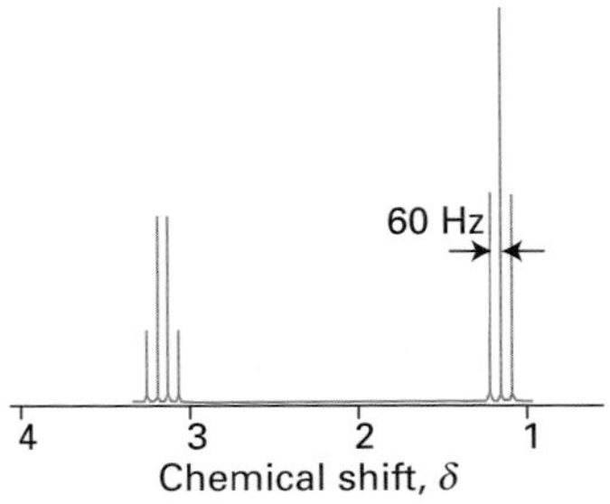

Fig. 21.14 The NMR spectrum considered in Example 21.1.

Self-test 21.5

Interpret the spectrum shown in Fig. 21.15.

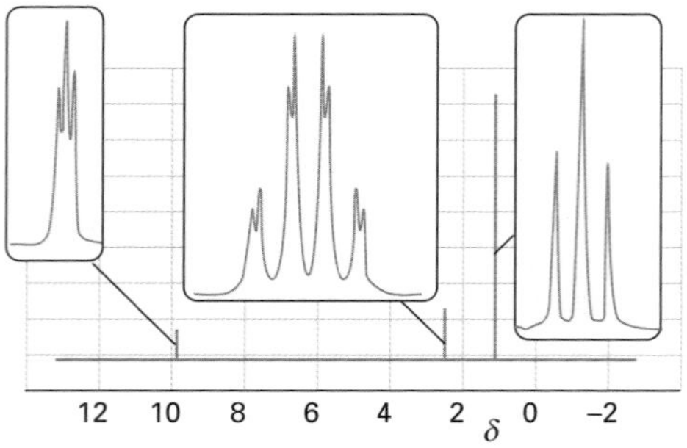

Fig. 21.15 The spectrum referred to in Self-test 21.5.

Answer: Propanal, CH_3CH_2CHO

The magnitude of $^3J_{HH}$ depends on the dihedral angle, ϕ, between the two C—H bonds (2). The variation is expressed quite well by the **Karplus equation:**

$$^3J_{HH} = A + B\cos\phi + C\cos 2\phi \quad \text{Karplus equation} \quad (21.14)$$

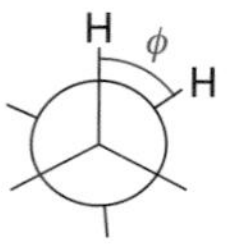

2

Typical values of A, B, and C are +7 Hz, −1 Hz, and +5 Hz, respectively.[1] Figure 21.16 shows the angular variation predicted by the equation. It follows that the measurement of $^3J_{HH}$ in a series of related compounds can be used to determine their conformations. The coupling constant $^1J_{CH}$ also depends on the hybridization of the C atom:

	sp	sp^2	sp^3
$^1J_{CH}$/Hz:	250	160	125

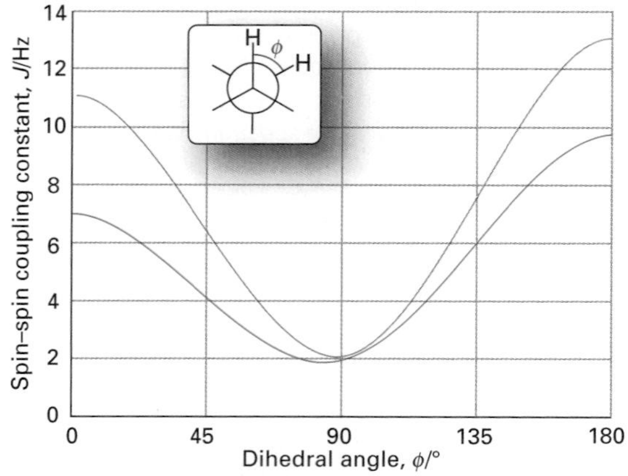

Fig. 21.16 The variation of $^3J_{HH}$ with angle, according to the Karplus equation. The orange line is for H—C—C—H and the green line for H—N—C—H.

[1] The equation is also often written in the form $^3J_{HH} = A' + B'\cos\phi + C'\cos^2\phi$.

Brief illustration 21.6 The Karplus equation

The investigation of H—N—C—H couplings in polypeptides can help reveal their conformation. For $^3J_{HH}$ coupling in such a group, A = +5.1 Hz, B = −1.4 Hz, and C = +3.2 Hz. For an α-helix, ϕ is close to 120°, which would give $^3J_{HH} \approx$ 4 Hz. For a β-sheet, ϕ is close to 180°, which would give $^3J_{HH} \approx$ 10 Hz. Consequently, small coupling constants indicate an α-helix whereas large couplings indicate a β-sheet.

Self-test 21.6

NMR experiments show that for H—C—C—H coupling in polypeptides, A = +3.5 Hz, B = −1.6 Hz, and C = +4.3 Hz. In an investigation of the polypeptide flavodoxin, the $^3J_{HH}$ coupling constant for such a grouping was determined to be 2.1 Hz. Is this value consistent with an α-helix or a β-sheet conformation?

Answer: α-helix

(b) The origin of spin–spin splitting

Spin–spin coupling in molecules in solution can be explained in terms of the **polarization mechanism,** in which the interaction is transmitted through the bonds. The simplest case to consider is that of $^1J_{XY}$, where X and Y are spin-½ nuclei joined by an electron-pair bond (Fig. 21.17). The coupling mechanism depends on the fact that in some atoms it is favourable for the nucleus and a nearby electron spin to be parallel (both α or both β), but in others it is

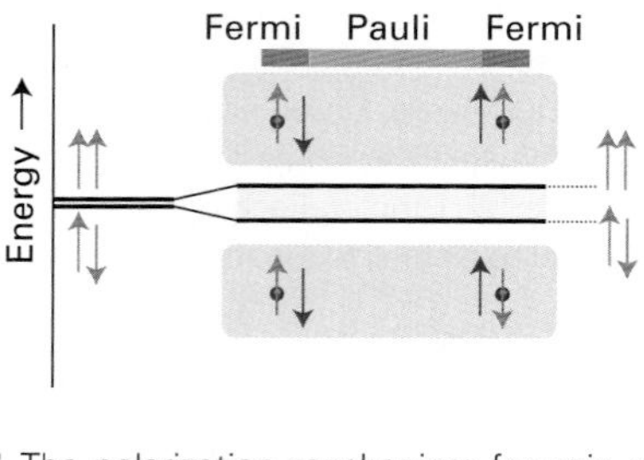

Fig. 21.17 The polarization mechanism for spin–spin coupling ($^1J_{HH}$). The two arrangements have slightly different energies. In this case, J is positive, corresponding to a lower energy when the nuclear spins are antiparallel.

favourable for them to be antiparallel (one α and the other β). The electron–nucleus coupling is magnetic in origin, and may be either a dipolar interaction (Section 15.3) between the magnetic moments of the electron and nuclear spins or a **Fermi contact interaction**, an interaction that depends on the very close approach of an electron to the nucleus and hence can occur only if the electron occupies an s orbital. We shall suppose that it is energetically favourable for an electron spin and a nuclear spin to be antiparallel (as is the case for a proton and an electron in a hydrogen atom), either $\alpha_e\beta_N$ or $\beta_e\alpha_N$, where we are using the labels e and N to distinguish the electron and nucleus spins.

If the X nucleus is α_X, a β electron of the bonding pair will tend to be found nearby (because that is energetically favourable for it). The second electron in the bond, which by the Pauli principle must have α spin if the other is β, will be found mainly at the far end of the bond (because electrons tend to stay apart to reduce their mutual repulsion). Because it is energetically favourable for the spin of Y to be antiparallel to an electron spin, a Y nucleus with β spin has a lower energy than a Y nucleus with α spin:

Low energy: $\alpha_X\beta_e \ldots \alpha_e\beta_Y$
High energy: $\alpha_X\beta_e \ldots \alpha_e\alpha_Y$

The opposite is true when X is β, for now the α spin of Y has the lower energy:

Low energy: $\beta_X\alpha_e \ldots \beta_e\alpha_Y$
High energy: $\beta_X\alpha_e \ldots \beta_e\beta_Y$

In other words, antiparallel arrangements of nuclear spins ($\alpha_X\beta_Y$ and $\beta_X\alpha_Y$) lie lower in energy than parallel arrangements ($\alpha_X\alpha_Y$ and $\beta_X\beta_Y$) as a result of their magnetic coupling with the bond electrons. That is, $^1J_{HH}$ is positive, for then hJm_Xm_Y is negative when m_X and m_Y have opposite signs.

To account for the value of $^2J_{XY}$, as in H—C—H, we need a mechanism that can transmit the spin alignments through the central C atom (which may be ^{12}C, with no nuclear spin of its own). In this case (Fig. 21.18), an X nucleus with α spin polarizes the electrons in its bond, and the α electron is likely to be found closer to the C nucleus. The more favourable arrangement of two electrons on the same atom is with their spins parallel (Hund's rule, Section 13.11), so the more favourable arrangement is for the α electron of the neighbouring bond to be close to the C nucleus. Consequently, the β electron of that bond is more likely to be found close to the Y nucleus, and therefore that nucleus will have a lower energy if it is α:

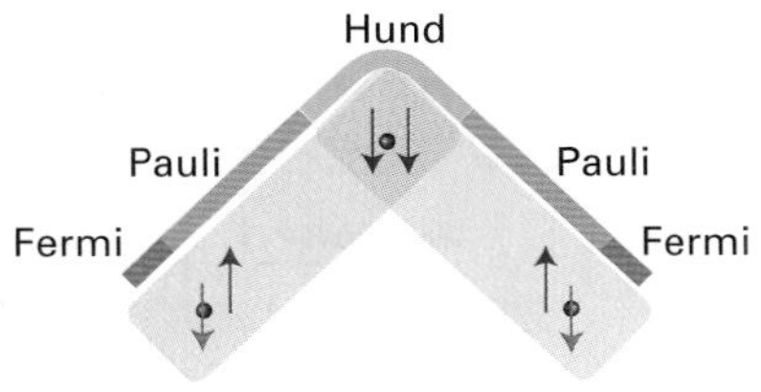

Fig. 21.18 The polarization mechanism for $^2J_{HH}$ spin–spin coupling. The spin information is transmitted from one bond to the next by a version of the mechanism that accounts for the lower energy of electrons with parallel spins in different atomic orbitals (Hund's rule of maximum multiplicity). In this case, $J < 0$, corresponding to a lower energy when the nuclear spins are parallel.

Low energy: $\alpha_X\beta_e \ldots \alpha_e[C]\alpha_e \ldots \beta_e\alpha_Y$
High energy: $\alpha_X\beta_e \ldots \alpha_e[C]\alpha_e \ldots \beta_e\beta_Y$

Low energy: $\beta_X\alpha_e \ldots \beta_e[C]\beta_e \ldots \alpha_e\beta_Y$
High energy: $\beta_X\alpha_e \ldots \beta_e[C]\beta_e \ldots \alpha_e\alpha_Y$

Hence, according to this mechanism, the energy of Y will be obtained if its spin is parallel ($\alpha_X\alpha_Y$ and $\beta_X\beta_Y$) to that of X. That is, $^2J_{HH}$ is negative for then hJm_Xm_Y is negative when m_X and m_Y have the same sign.

The coupling of nuclear spin to electron spin by the Fermi contact interaction is most important for proton spins, but it is not necessarily the most important mechanism for other nuclei. These nuclei may also interact by a dipolar mechanism with the electron magnetic moments and with their orbital motion, and there is no simple way of specifying whether J will be positive or negative.

21.5 Spin relaxation

As resonant absorption continues, the population of the upper state rises to match that of the lower state. From eqn 21.7, we can expect the intensity of the absorption signal to decrease with time as the populations of the spin states equalize. This decrease due to the progressive equalization of populations is called **saturation**.

(a) Varieties of relaxation

The fact that saturation is often not observed, especially when the radiofrequency power is kept low, must mean that there are non-radiative processes by which β nuclear spins can release energy to become α spins again, and hence help to maintain the population difference between the two sites. The non-radiative return to an equilibrium distribution of populations in a system (eqn 21.5a) is an aspect of the process called **relaxation**. If we were to imagine forming a

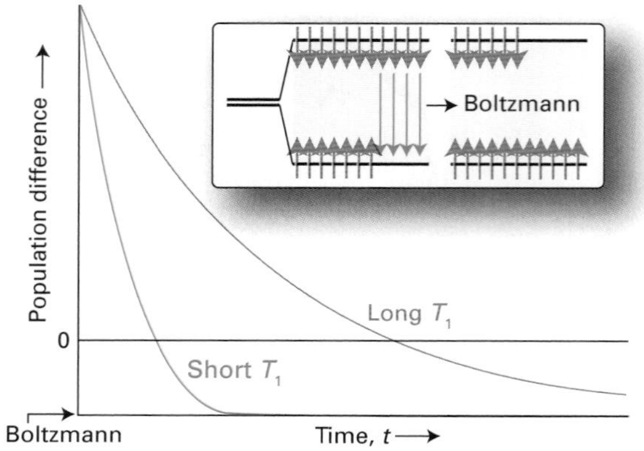

Fig. 21.19 The spin–lattice relaxation time is the time-constant for the exponential return of the population of the spin states to their equilibrium (Boltzmann) distribution.

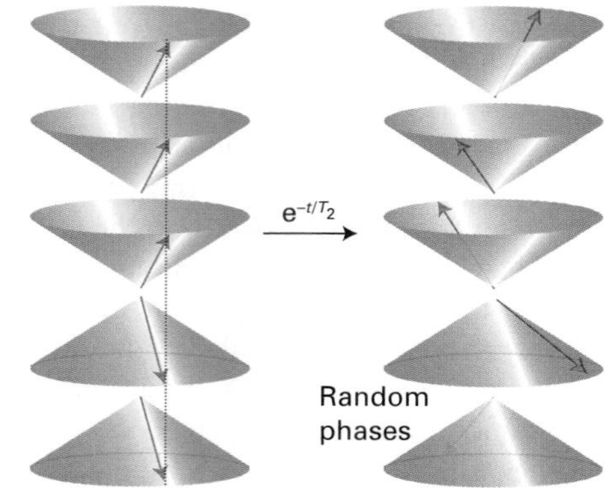

Fig. 21.20 The spin–spin relaxation time is the time constant for the exponential return of the spins to a random distribution around the direction of the magnetic field. No change in populations of the two spin states is involved in this type of relaxation, so no energy is transferred from the spins to the surroundings.

system of spins in which all the nuclei were in their β state, then the system returns exponentially to the equilibrium distribution (a small excess of α spins over β spins) with a time constant called the **spin–lattice relaxation time**, T_1 (Fig. 21.19).

However, there is another, more subtle aspect of relaxation. Consider the classical picture of magnetic nuclei (Section 21.1) and imagine that somehow we have arranged all the spins in a sample to have exactly the same angle around the field direction at an instant. If each spin has a slightly different Larmor frequency (because they experience slightly different local magnetic fields) they will gradually fan out. At thermal equilibrium, all the bar magnets lie at *random* angles round the direction of the applied field, and the time constant for the exponential return of the system into this random arrangement is called the **spin–spin relaxation time**, T_2 (Fig. 21.20). For spins to be truly at thermal equilibrium, not only is the ratio of populations of the spin states given by eqn 21.5a, but the spin orientations must be random around the field direction.

(b) Mechanisms of relaxation

What causes each type of relaxation? In each case the spins are responding to local magnetic fields that act to twist them into different orientations. However, there is a crucial difference between the two processes.

The best kind of local magnetic field for inducing a transition from β to α (as in spin–lattice relaxation) is one that fluctuates at a frequency close to the resonance frequency. Such a field can arise from the tumbling motion of the molecule in the fluid sample. If the tumbling motion of the molecule is slow compared to the resonance frequency, it will give rise to a fluctuating magnetic field that oscillates too slowly to induce transitions, so T_1 will be long. If the molecule tumbles much faster than the resonance frequency, then it will give rise to a fluctuating magnetic field that oscillates too rapidly to induce transitions, so T_1 will again be long. Only if the molecule tumbles at about the resonance frequency will the fluctuating magnetic field be able to induce transitions effectively, and only then will T_1 be short. The rate of molecular tumbling increases with temperature and with reducing viscosity of the solvent, so we can expect a dependence like that shown in Fig. 21.21.

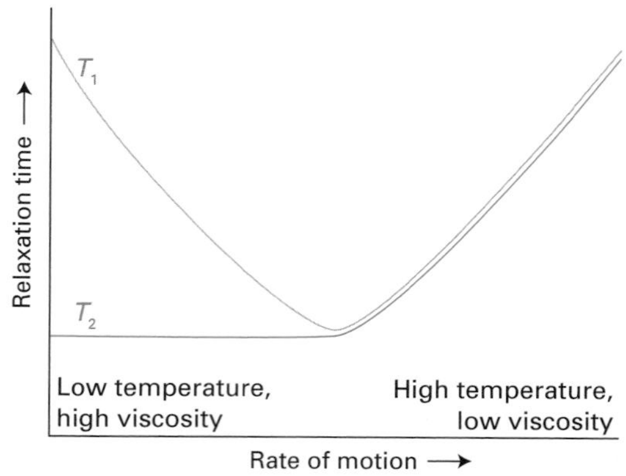

Fig. 21.21 The variation of the two relaxation times with the rate at which the molecules move (either by tumbling or migrating through the solution). The horizontal axis can be interpreted as representing temperature or viscosity. Note that at rapid rates of motion, the two relaxation times coincide.

The best kind of local magnetic field for causing spin–spin relaxation is one that does not change very rapidly. Then each molecule in the sample lingers in its particular local magnetic environment for a long time, and the orientations of the spins have time to become randomized around the applied field direction. If the molecules move rapidly from one magnetic environment to another, the effects of different magnetic fields average out, and the randomization does not take place as quickly. In other words, slow molecular motion corresponds to short T_2 and fast motion corresponds to long T_2 (as shown in Fig. 21.21). Detailed calculation shows that when the motion is fast, the two relaxation times are equal, as has been drawn in the illustration.

Spin relaxation studies, using advanced techniques that use complicated sequences of pulses of radiofrequency energy to stimulate the spins into special orientations, and then monitoring their return to equilibrium, have two main applications. First, they reveal information about the mobility of molecules or parts of molecules. For example, by studying spin relaxation times of protons in the hydrocarbon chains of micelles and bilayers it is possible to build up a detailed picture of the motion of these chains, and hence come to an understanding of the dynamics of cell membranes. Second, relaxation times depend on the separation of the nucleus from the source of the magnetic field that is causing its relaxation: that source may be another magnetic nucleus in the same molecule. By studying the relaxation times, the internuclear distances within the molecule can be determined and used to build up a model of its shape.

21.6 Conformational conversion and chemical exchange

The appearance of an NMR spectrum is changed if magnetic nuclei can jump rapidly between different environments. Consider a molecule, such as *N,N*-dimethylformamide, that can jump between conformations; in its case, the methyl shifts depend on whether they are *cis* or *trans* to the carbonyl group (Fig. 21.22). When the jumping rate is low, the spectrum shows two sets of lines, one each from molecules in each conformation. When the interconversion is fast, the spectrum shows a single line at the mean of the two chemical shifts. At intermediate inversion rates, the line is very broad. This maximum broadening occurs when the lifetime, τ (tau), of a conformation gives rise to a linewidth that is comparable to the difference of resonance frequencies, $\Delta\nu$, and both

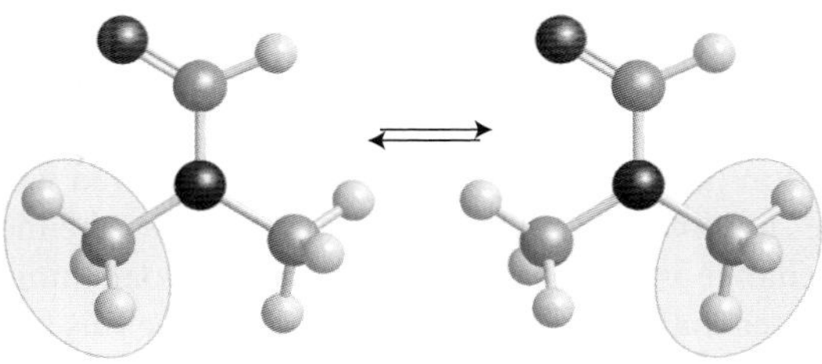

Fig. 21.22 When a molecule changes from one conformation to another, the positions of its protons are interchanged and jump between magnetically distinct environments.

broadened lines blend together into a very broad line. Coalescence of the two lines occurs when

$$\tau = \frac{2^{1/2}}{\pi\Delta\nu} \qquad \text{Coalescence criterion} \qquad (21.15)$$

Brief illustration 21.7 Line broadening

The NO group in *N,N*-dimethylnitrosamine, $(CH_3)_2N$—NO, rotates about the N—N bond and, as a result, the magnetic environments of the two CH_3 groups are interchanged. In a 600 MHz spectrometer the two CH_3 resonances are separated by 390 Hz. Coalescence will occur when the average lifetime is less than

$$\tau = \frac{2^{1/2}}{\pi \times (390\ \mathrm{s}^{-1})} = 1.2\times10^{-3}\ \mathrm{s}$$

or 1.2 ms. It follows that the signal will collapse to a single line when the interconversion rate exceeds about $1/\tau = 830\ \mathrm{s}^{-1}$.

Self-test 21.7

What would you deduce from the observation of a single line from the same molecule in a 300 MHz spectrometer?

Answer: Conformation lifetime less than 2.3 ms

A similar explanation accounts for the loss of fine structure in solvents able to exchange protons with the sample. For example, hydroxyl protons are able to exchange with water protons. When this **chemical exchange** occurs, a molecule ROH with an α-spin proton (we write this ROH_α) rapidly converts to ROH_β and then perhaps to ROH_α again because the protons provided by the solvent molecules in successive exchanges have random spin orientations. Therefore, instead of seeing a spectrum composed of contributions from both ROH_α and ROH_β

molecules (that is, a spectrum showing a doublet structure due to the OH proton) we see a spectrum that shows no splitting caused by coupling of the OH proton (as in Fig. 21.5). The effect is observed when the lifetime of a molecule due to this chemical exchange is so short that the lifetime broadening is greater than the doublet splitting. Because this splitting is often very small (a few hertz), a proton must remain attached to the same molecule for longer than about 0.1 s for the splitting to be observable. In water, the exchange rate is much faster than that, so alcohols show no splitting from the OH protons. In dry dimethyl sulfoxide (DMSO, $(CH_3)_2SO$), the exchange rate may be slow enough for the splitting to be detected.

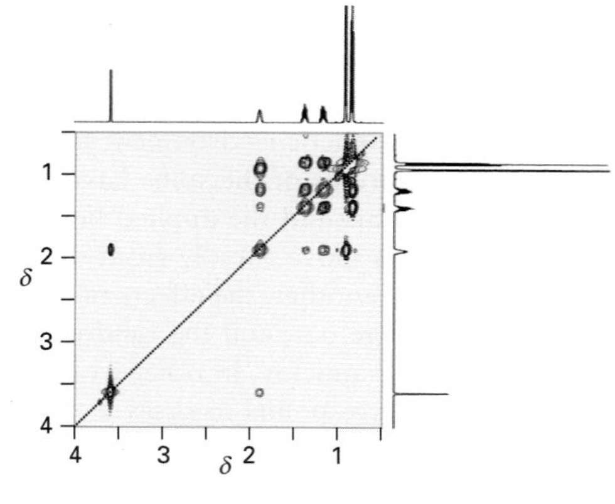

Fig. 21.23 Proton COSY spectrum of the amino acid isoleucine. The peaks on the diagonal correspond to the one-dimensional spectrum, shown along each edge.

21.7 Two-dimensional NMR

An NMR spectrum contains a great deal of information and, if many spins are present, is very complex, for the fine structure of different groups of lines can overlap. The complexity would be reduced if we could use two axes to display the data, with resonances belonging to different groups lying at different locations on the second axis. This separation is essentially what is achieved in **two-dimensional NMR**.

Much modern NMR work makes use of techniques such as **correlation spectroscopy** (COSY), in which a clever choice of pulses and Fourier transformation techniques makes it possible to determine all spin–spin couplings in a molecule. The COSY spectrum of an AX system contains four groups of signals centred on the two chemical shifts. Each group shows fine structure, consisting of a block of four signals separated by J_{AX}. The diagonal peaks are signals centred on (δ_A,δ_A) and (δ_X,δ_X) and lie along the diagonal The 'cross peaks' (or off–diagonal peaks) are signals centred on (δ_A,δ_X) and (δ_X,δ_A) and owe their existence to the coupling between A and X. Consequently, cross peaks in COSY spectra allow us to map the couplings between spins and to trace out the bonding network in complex molecules. Figure 21.23 shows a simple example of a proton COSY spectrum of isoleucine, $CH_3CH_2CH(CH_3)CH(NH_2)COOH$.

Although information from two-dimensional NMR spectroscopy is trivial in an AX system, it can be of enormous help in the interpretation of more complex spectra. For example, a complex spectrum from a synthetic polymer or a protein would be impossible to interpret in one-dimensional NMR but can be interpreted reasonably rapidly by two-dimensional NMR.

Impact on medicine 21.1

Magnetic resonance imaging

One of the most striking applications of nuclear magnetic resonance is in medicine. **Magnetic resonance imaging** (MRI) is a portrayal of the concentrations of protons in a solid object. The technique relies on the application of specific pulse sequences to an object in an inhomogeneous magnetic field (a field with strength that varies inside the sample).

If an object containing hydrogen nuclei (a tube of water or a human body) is placed in an NMR spectrometer and exposed to a *homogeneous* magnetic field (a field that has the same value throughout the sample), then a single resonance signal will be detected. Now consider a flask of water in a magnetic field that varies linearly in the *z*-direction according to $\mathcal{B}_0 + \mathcal{G}_z z$, where $\mathcal{G}_z$ is the field gradient along the *z*-direction (Fig. 21.24). Then the water protons will be resonant at the frequencies

$$\nu(z) = {}^1\!/\!_{2}\gamma_N(\mathcal{B}_0 + \mathcal{G}_z z)$$

(similar equations may be written for gradients along the *x* and *y* directions). Exposing the sample to radiation with frequency $\nu(z)$ will result in a signal with an intensity that is proportional to the numbers of protons at the position *z*. This is an example of 'slice selection', the use of radiofrequency radiation that excites nuclei in a specific region, or slice, of the sample. It follows that the intensity of the NMR signal will be a projection of the numbers of protons on a line parallel to the field gradient. The image of a three-dimensional object such as a flask of water can be obtained if the slice selection technique is applied at different orientations.

A common problem with this technique is image contrast, which must be optimized in order to show spatial variations in water content in the sample. One strategy for solving this problem takes advantage of the fact that the relaxation times of water protons are shorter for water in biological tissues than for the pure liquid. Furthermore, relaxation times from

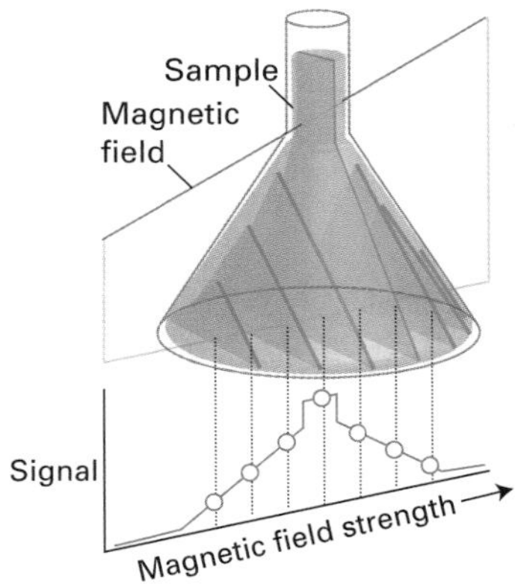

Fig. 21.24 In a magnetic field that varies linearly over a sample, all the protons within a given slice (that is, at a given field value) come into resonance and give a signal of the corresponding intensity. The resulting intensity pattern is a map of the numbers in all the slices, and portrays the shape of the sample. Changing the orientation of the field shows the shape along the corresponding direction, and computer manipulation can be used to build up the three-dimensional shape of the sample.

water protons are also different in healthy and diseased tissues. A T_1*-weighted image* is obtained by obtaining data before spin–lattice relaxation can return the spins in the sample to equilibrium. Under these conditions, differences in signal intensities are directly related to differences in T_1. A T_2*-weighted image* is obtained by collecting data after the system has relaxed extensively, though not completely. In this way, signal intensities are strongly dependent on variations in T_2. However, allowing so much of the decay to occur leads to weak signals even for those protons with long spin–spin relaxation times. Another strategy involves the use of *contrast agents*, paramagnetic compounds that shorten the relaxation times of nearby protons. The technique is particularly useful in enhancing image contrast and in diagnosing disease if the contrast agent is distributed differently in healthy and diseased tissues.

The MRI technique is used widely to detect physiological abnormalities and to observe metabolic processes. With **functional MRI**, blood flow in different regions of the brain can be studied and related to the mental activities of the subject. The special advantage of MRI is that it can image *soft* tissues (Fig. 21.25), whereas X-rays are largely used for imaging hard, bony structures and abnormally dense regions, such as tumours. In fact, the invisibility of hard structures in MRI is an advantage, as it allows the imaging of structures encased by bone, such as the brain and the spinal cord. X-rays are known to be dangerous on account of the ionization they cause; the high magnetic fields used in MRI may also be dangerous, but apart from anecdotes about the extraction of loose fillings from teeth, there is no convincing evidence of their harmfulness, and the technique is considered safe.

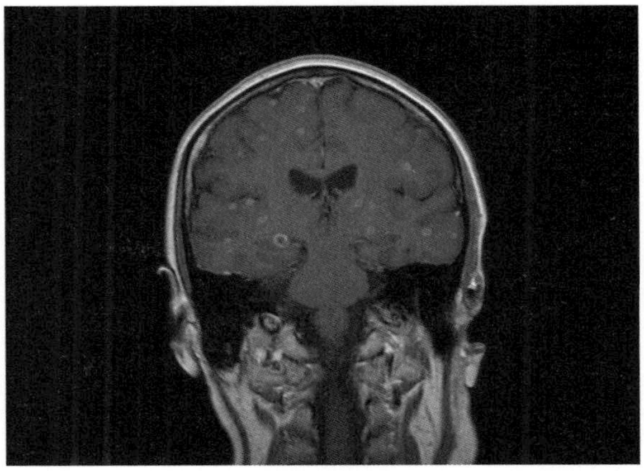

Fig. 21.25 The great advantage of MRI is that it can display soft tissue, such as in this cross-section through a patient's head. Image copyright: Dr James Holt.

Electron paramagnetic resonance

An electron (with spin quantum number $s = 1/2$) in a magnetic field can take two orientations, corresponding to $m_s = +1/2$ (denoted α or $\uparrow$) and $m_s = -1/2$ (denoted β or $\downarrow$). It possesses a magnetic moment due to its spin and this moment interacts with an external magnetic field. That is, an electron behaves like a tiny magnet with z-component

$$m_z = \gamma_e \hbar m_s \quad \text{Magnetic moment due to spin} \quad (21.16)$$

where γ_e is the **magnetogyric ratio** (or 'gyromagnetic ratio') of the electron

$$\gamma_e = -\frac{g_e e}{2m_e} \quad \text{Magnetogyric ratio} \quad (21.17)$$

and g_e is a factor, the **g-value of the electron**, which is close to 2.0023 for a free electron. The 2 comes from Dirac's relativistic theory of the electron; the 0.0023 comes from additional correction terms. It follows from the equation in The chemist's toolkit 21.1 that in a magnetic field $\mathcal{B}$ the two orientations have different energies (Fig. 21.26). These energies are given by

$$E_{m_s} = -\gamma_e \hbar \mathcal{B} m_s \quad \text{Energy of electron} \quad (21.18)$$

The energies are sometimes expressed in terms of the **Bohr magneton**

$$\mu_B = \frac{e\hbar}{2m_e} = 9.274 \times 10^{-24}\ \text{J T}^{-1} \quad \text{Bohr magneton} \quad (21.19)$$

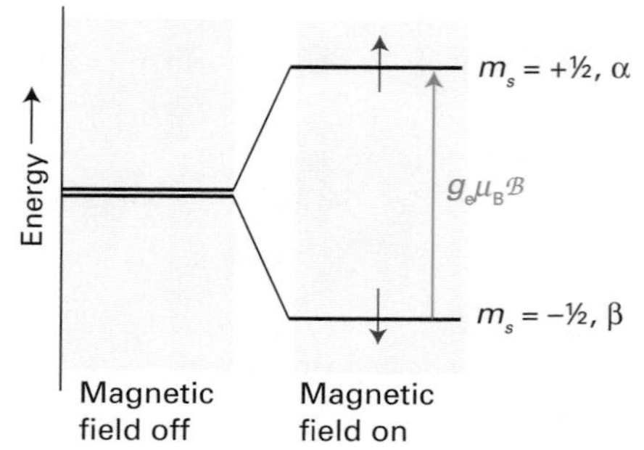

Fig. 21.26 The energy levels of an electron in a magnetic field. Resonance occurs when the energy separation of the levels matches the energy of the photons in the electromagnetic field.

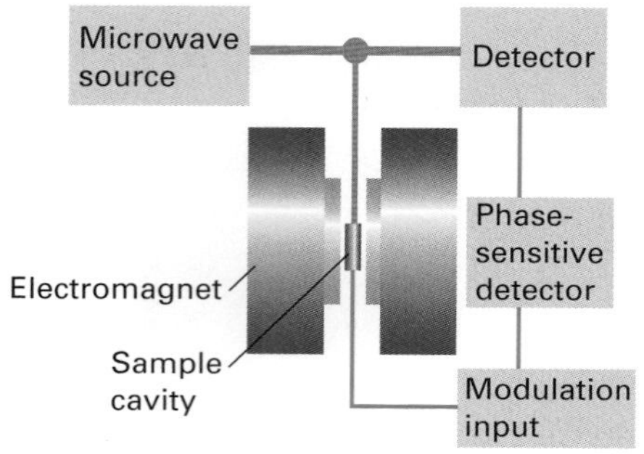

Fig. 21.27 The layout of a continuous wave EPR spectrometer. A typical magnetic field is 0.3 T, which requires microwaves of frequency 9 GHz (wavelength 3 cm) for resonance.

a fundamental unit of magnetism, for then

$$E_{m_s} = g_e\mu_B\mathcal{B}m_s \quad \text{Energy of electron} \quad (21.20)$$

It follows that the energy separation of the two spin states of an electron is

$$\Delta E = E_\alpha - E_\beta = g_e\mu_B\mathcal{B}(+^1\!/_2) - g_e\mu_B\mathcal{B}(-^1\!/_2) = g_e\mu_B\mathcal{B} \quad (21.21)$$

For an electron, the β state lies below the α state in energy and, by a similar argument to that for nuclei,

$$N_\beta - N_\alpha \approx \frac{Ng_e\mu_B\mathcal{B}}{2kT} \quad \text{Electrons} \quad \text{Population difference} \quad (21.22)$$

where N is the total number of spins.

Brief illustration 21.8 Electron spin populations

When 1000 electron spins are exposed to a 1.0 T magnetic field at 20 °C (293 K),

$$N_\beta - N_\alpha \approx \frac{\overbrace{1000}^{N}\times\overbrace{2.0023}^{g_e}\times\overbrace{(9.274\times10^{-24}\ \mathrm{J\,T^{-1}})}^{\mu_B}\times\overbrace{(1.0\ \mathrm{T})}^{\mathcal{B}}}{2\times\underbrace{(1.381\times10^{-23}\ \mathrm{J\,K^{-1}})}_{k}\times\underbrace{(293\ \mathrm{K})}_{T}}$$

$$\approx 2.3$$

There is an imbalance of populations of only about 2 electrons in a thousand.

The resonance technique for electrons in a magnetic field is called **electron paramagnetic resonance** (EPR) or **electron spin resonance** (ESR). Because electron magnetic moments are much bigger than nuclear magnetic moments, even quite modest fields can require high frequencies to achieve resonance. Much work is done using fields of about 0.3 T, when resonance occurs at about 9 GHz, corresponding to 3 cm ('X-band') microwave radiation or at about 1 T, when resonance occurs at about 35 GHz, corresponding to about 9 mm ('Q-band') microwave radiation. Electron paramagnetic resonance is much more limited than NMR because it is applicable only to species with unpaired electrons, which include radicals (perhaps prepared by radiation damage or photolysis) and d- and f-metal complexes, including such biologically active species as haemoglobin. However, it gives valuable information about electron distributions and can be used to monitor, for instance, the uptake of oxygen by haemoglobin and biological electron transfer processes.

Both Fourier transform (FT) and continuous wave (CW) EPR spectrometers are available. The FT-EPR instrument is like an FT-NMR spectrometer except that pulses of microwaves are used to excite electron spins in the sample. The layout of the more common CW-EPR spectrometer is shown in Fig. 21.27. It consists of a microwave source (a klystron or a Gunn oscillator), a cavity in which the sample is inserted in a glass or quartz container, a microwave detector, and an electromagnet with a field that can be varied in the region of 0.3 T (X-band) or 1 T (Q-band).

An EPR spectrum is obtained by monitoring the microwave absorption as the field is changed, and a typical spectrum (of the benzene radical anion, $C_6H_6^-$) is shown in Fig. 21.28. The peculiar appearance of the spectrum, which is in fact the first-derivative (the slope) of the absorption, arises from the detection technique employed (Fig. 21.29).

21.8 The *g*-value

Equation 21.21 gives the energy of a transition between the $m_s = -^1\!/_2$ and the $m_s = +^1\!/_2$ levels of a 'free' electron in terms of the *g*-value $g_e \approx 2.0023$. The magnetic moment of an unpaired electron in a

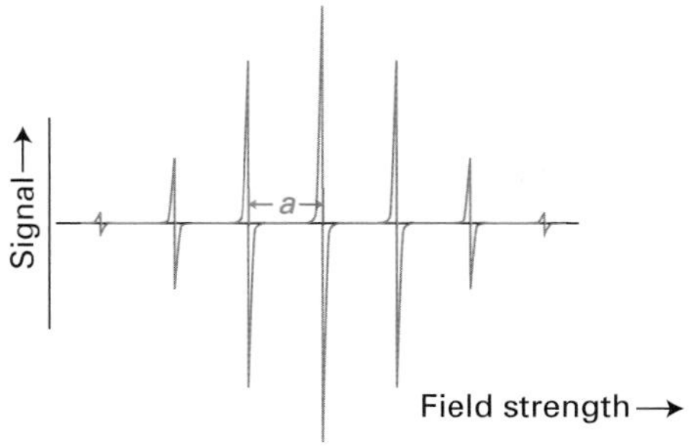

Fig. 21.28 The EPR spectrum of the benzene radical anion, $C_6H_6^-$, in fluid solution. *a* is the hyperfine splitting of the spectrum; the centre of the spectrum is determined by the *g*-value of the radical.

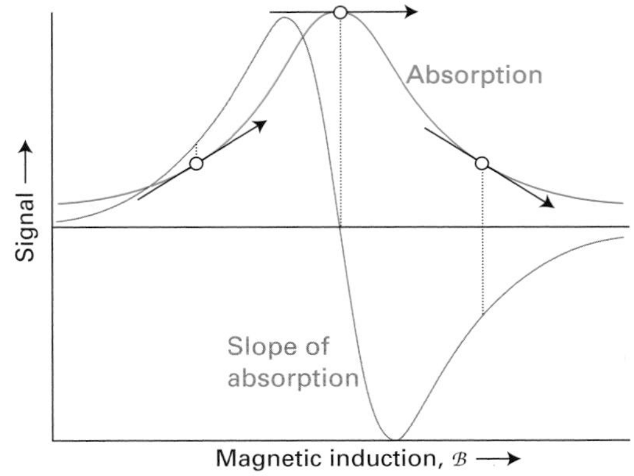

Fig. 21.29 When phase-sensitive detection is used, the signal is the first derivative of the absorption intensity. Note that the peak of the absorption corresponds to the point where the derivative passes through zero.

radical also interacts with an external field, but the *g*-value is different from that of a free electron on account of local magnetic fields induced in the molecular framework of the radical. Consequently, the resonance condition is normally written as

$$h\nu = g\mu_B\mathcal{B} \qquad \text{Resonance condition} \qquad (21.23)$$

where *g* is the empirically determined **g-value** of the radical. Many organic radicals have *g*-values close to 2.0027; inorganic radicals have *g*-values typically in the range 1.9–2.1; paramagnetic d- and f-metal complexes have *g*-values in a wider range (for example, 0 to 6).

The deviation of *g* from $g_e = 2.0023$ depends on the ability of the applied field to induce local electron currents in the radical and their transmission to the spin through spin–orbit coupling (Section 13.18). Its value gives some information about electronic structure. In that sense, the *g*-value in EPR plays a similar role to the shielding constants in NMR. However, because *g*-values differ very little from g_e in many radicals (for instance, 2.003 for H, 1.999 for NO_2, and 2.01 for ClO_2), its main use in chemical applications is to aid the identification of the species present in a sample.

Brief illustration 21.9 The *g*-value

The centre of the EPR spectrum of the methyl radical occurred at 329.40 mT in a spectrometer operating at 9.2330 GHz (in the so-called 'X-band' of the microwave spectrum). Its *g*-value is therefore

$$g = \frac{h\nu}{\mu_B\mathcal{B}} = \frac{\overbrace{(6.626\,08\times10^{-34}\ \text{J s})}^{h}\times\overbrace{(9.2330\times10^{9}\ \text{s}^{-1})}^{\nu}}{\underbrace{(9.2740\times10^{-24}\ \text{J T}^{-1})}_{\mu_B}\times\underbrace{(0.329\,40\ \text{T})}_{\mathcal{B}}}$$

$$= 2.0027$$

Self-test 21.8

At what magnetic field would the methyl radical come into resonance in a spectrometer operating at 34.000 GHz (in the so-called 'Q-band' of the microwave spectrum)?

Answer: 1.213 T

21.9 Hyperfine structure

The most important features of EPR spectra are their **hyperfine structure**, the splitting of individual resonance lines into components. In general in spectroscopy, the term 'hyperfine structure' means the structure of a spectrum that can be traced to interactions of the electrons with nuclei other than as a result of the latter's point electric charge. The source of the hyperfine structure in EPR is the magnetic interaction between the electron spin and the magnetic dipole moments of the nuclei present in the radical.

Consider the effect on the EPR spectrum of a single H nucleus located somewhere in a radical. The proton spin is a source of magnetic field, and depending on the orientation of the nuclear spin, the field it generates adds to or subtracts from the applied field. The total local field is therefore

$$\mathcal{B}_{loc} = \mathcal{B} + am_I \qquad m_I = \pm\tfrac{1}{2} \qquad (21.24)$$

where *a* is the **hyperfine coupling constant.** Half the radicals in a sample have $m_I = +\tfrac{1}{2}$, so half resonate when the applied field satisfies the condition

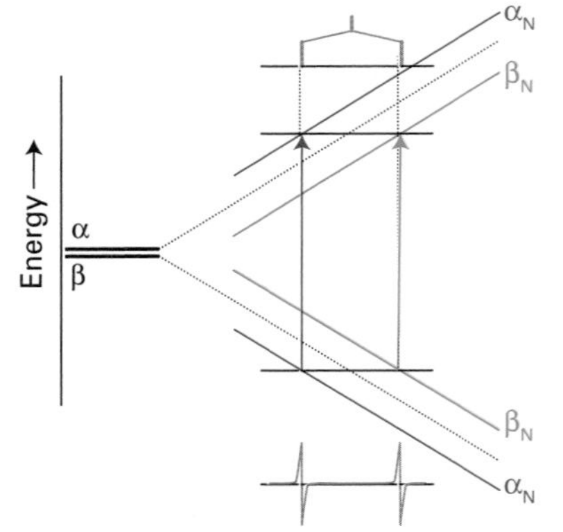

Fig. 21.30 The hyperfine interaction between an electron and a spin-½ nucleus results in four energy levels in place of the original two. As a result, the spectrum consists of two lines (of equal intensity) instead of one. The intensity distribution can be summarized by a simple stick diagram. The diagonal lines show the energies of the states as the applied field is increased, and resonance occurs when the separation of states matches the fixed energy of the microwave photon.

$$h\nu = g\mu_B(\mathcal{B} + \tfrac{1}{2}a), \quad \text{or} \quad \mathcal{B} = \frac{h\nu}{g\mu_B} - \tfrac{1}{2}a \qquad (21.25a)$$

The other half (which have $m_I = -\tfrac{1}{2}$) resonate when

$$h\nu = g\mu_B(\mathcal{B} - \tfrac{1}{2}a), \quad \text{or} \quad \mathcal{B} = \frac{h\nu}{g\mu_B} + \tfrac{1}{2}a \qquad (21.25b)$$

Therefore, instead of a single line, the spectrum shows two lines of half the original intensity separated by a and centred on the field determined by g (Fig. 21.30).

If the radical contains an ^{14}N atom ($I = 1$), its EPR spectrum consists of three lines of equal intensity, because the ^{14}N nucleus has three possible spin orientations, and each spin orientation is possessed by one third of all the radicals in the sample. In general, a spin-I nucleus splits the spectrum into $2I + 1$ hyperfine lines of equal intensity.

When there are several magnetic nuclei present in the radical, each one contributes to the hyperfine structure. In the case of equivalent protons (for example, the two CH_2 protons in the radical CH_3CH_2) some of the hyperfine lines are coincident. It is not hard to show that if the radical contains N equivalent protons, then there are $N + 1$ hyperfine lines with an intensity distribution given by Pascal's triangle (Section 21.4). The spectrum of the benzene radical anion in Fig. 21.28, which has seven lines with intensity ratio 1:6:15:20:15:6:1, is consistent with a radical containing six equivalent protons.

More generally, if the radical contains N equivalent nuclei with spin quantum number I, then there are $2NI + 1$ hyperfine lines with an intensity distribution given by modified versions of Pascal's triangle (see the exercises).

Example 21.2

Predicting the hyperfine structure of an EPR spectrum

A radical contains one ^{14}N nucleus ($I = 1$) with hyperfine constant 1.61 mT and two equivalent protons ($I = \tfrac{1}{2}$) with hyperfine constant 0.35 mT. Predict the form of the EPR spectrum.

Strategy We need to consider the hyperfine structure that arises from each type of nucleus or group of equivalent nuclei in succession. So, split a line with one nucleus, then each of those lines is split by a second nucleus (or group of nuclei), and so on. It is best to start with the nucleus with the largest hyperfine splitting; however, any choice could be made, and the order in which nuclei are considered does not affect the conclusion.

Answer The ^{14}N nucleus gives three hyperfine lines of equal intensity separated by 1.61 mT. Each line is split into doublets of spacing 0.35 mT by the first proton, and each line of these doublets is split into doublets with the same 0.35 mT splitting (Fig. 21.31). The central lines of each split doublet coincide, so the proton splitting gives 1:2:1 triplets of internal splitting 0.35 mT. Therefore, the spectrum consists of three equivalent 1:2:1 triplets.

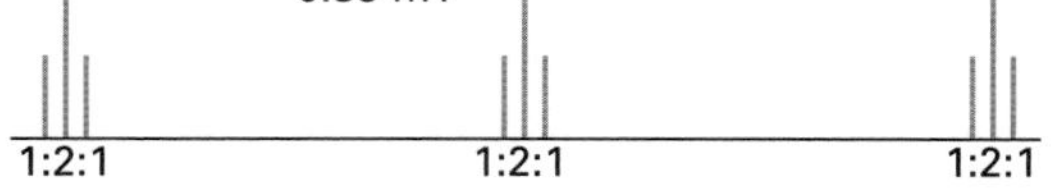

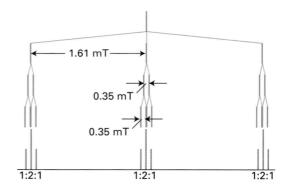

Fig. 21.31 The analysis of the hyperfine structure of radicals containing one ^{14}N nucleus ($I = 1$) and two equivalent protons.

Self-test 21.9

Predict the form of the EPR spectrum of a radical containing three equivalent ^{14}N nuclei.

Answer: Fig. 21.32

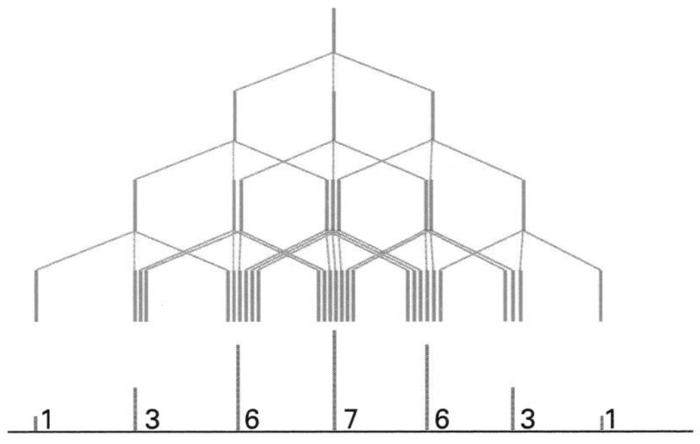

Fig. 21.32 The analysis of the hyperfine structure of radicals containing three equivalent ^{14}N nuclei.

The hyperfine structure of an EPR spectrum is a kind of fingerprint that helps to identify the radicals present in a sample. The interaction between the unpaired electron and the hydrogen nucleus responsible for hyperfine structure is either a dipolar interaction or the Fermi contact interaction described in Section 21.4. In the case of the contact interaction, the magnitude of the splitting depends on the distribution of the unpaired electron near the magnetic nuclei present, so the spectrum can be used to map the molecular orbital occupied by the unpaired electron. For example, because the hyperfine splitting in $C_6H_6^-$ is 0.375 mT, and one proton is close to a C atom with one-sixth the unpaired electron density (because the electron is spread uniformly around the ring), the hyperfine splitting caused by a proton in the electron spin entirely confined to a single adjacent C atom should be 6×0.375 mT = 2.25 mT. If in another aromatic radical we find a hyperfine splitting constant a, then the **spin density**, ρ, the probability that an unpaired electron is on the atom, can be calculated from the empirical **McConnell equation:**

$$a = Q\rho \qquad \text{McConnell equation} \qquad (21.26)$$

with Q = 2.25 mT. In this equation, ρ is the spin density on a C atom and a is the hyperfine splitting observed for the H atom to which it is attached.

Checklist of key concepts

- ☐ 1 Resonance is the condition of strong effective coupling when the frequencies of two oscillators are identical.
- ☐ 2 Nuclear magnetic resonance (NMR) is the observation of the magnetic field at which magnetic nuclei in molecules come into resonance with a radiofrequency electromagnetic field.
- ☐ 3 Electron paramagnetic resonance (EPR) or electron spin resonance (ESR) is the observation of the magnetic field at which an electron spin comes into resonance with a microwave electromagnetic field.
- ☐ 4 The intensity of an NMR or EPR transition increases with the difference in population of α and β states and the strength of the applied magnetic field (as $\mathcal{B}^2$).
- ☐ 5 The chemical shift of a nucleus is the difference between its resonance frequency and that of a reference standard.
- ☐ 6 The observed shielding constant is the sum of a local contribution, a neighbouring group contribution, and a solvent contribution.
- ☐ 7 The fine structure of an NMR spectrum is the splitting of the groups of resonances into individual lines; the strength of the interaction is expressed in terms of the spin–spin coupling constant, J.
- ☐ 8 N equivalent spin-½ nuclei split the resonance of a nearby spin or group of equivalent spins into $N + 1$ lines with an intensity distribution given by Pascal's triangle.
- ☐ 9 Spin–spin coupling in molecules in solution can be explained in terms of the polarization mechanism, in which the interaction is transmitted through the bonds.
- ☐ 10 The Fermi contact interaction is a magnetic interaction that depends on the very close approach of an electron to the nucleus and can occur only if the electron occupies an s orbital.
- ☐ 11 Relaxation is the non-radiative return to an equilibrium distribution of populations in a system with random relative spin orientations; the system returns exponentially to the equilibrium distribution with a time constant called the spin–lattice relaxation time, T_1.

☐ 12 The spin–spin relaxation time, T_2, is the time constant for the exponential return of the system into random relative orientations.

☐ 13 Coalescence of the two lines occurs in conformational interchange or chemical exchange when the lifetime, τ, of the states is related to their resonance frequency difference, $\Delta\nu$.

☐ 14 In two-dimensional NMR, spectra are displayed in two axes, with resonances belonging to different groups lying at different locations on the second axis. An example of a two-dimensional NMR technique is correlation spectroscopy (COSY), in which all spin–spin couplings in a molecule are determined.

☐ 15 The EPR resonance condition is written in terms of the g-value of the radical; the deviation of g from $g_e = 2.0023$ depends on the ability of the applied field to induce local electron currents in the radical.

☐ 16 The hyperfine structure of an EPR spectrum is the splitting of individual resonance lines into components by the magnetic interaction of the electron and nuclei with spin.

Road map of key equations

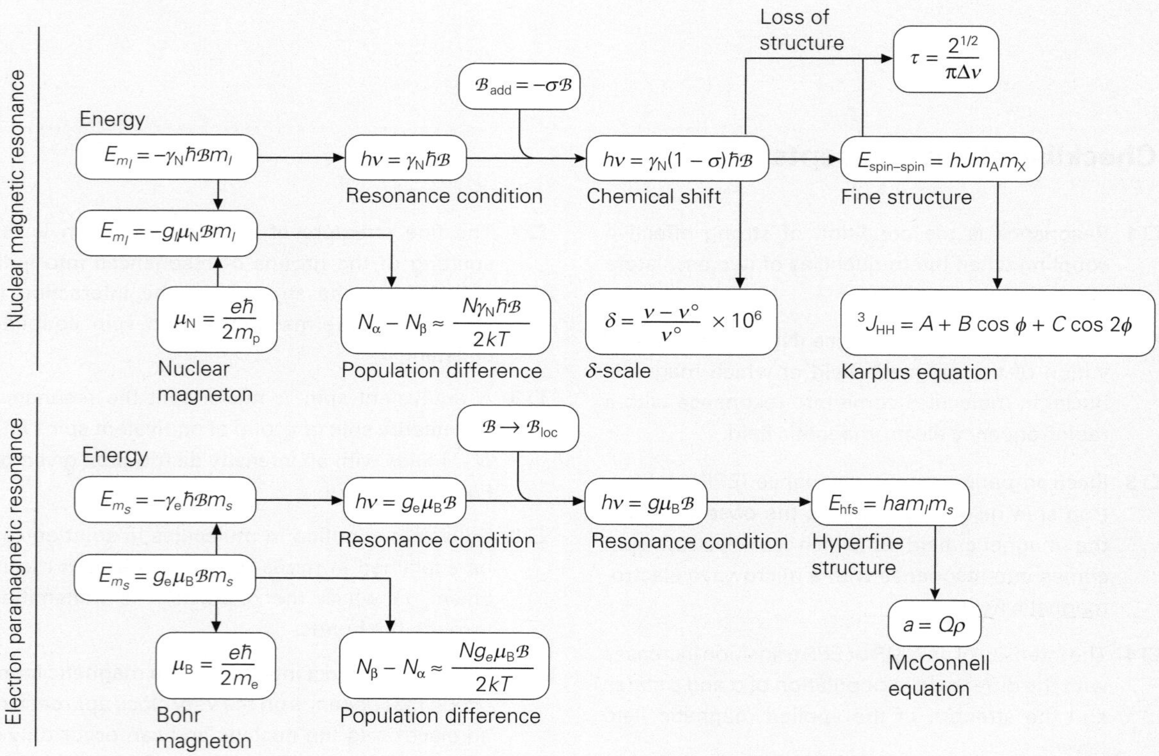

Questions and exercises

Discussion questions

21.1 Discuss the origins of the local, neighbouring group, and solvent contributions to the shielding constant.

21.2 Suggest a reason why the relaxation times of ^{13}C nuclei are typically much longer than those of ^{1}H nuclei.

21.3 Suggest a reason why the spin–lattice relaxation time of benzene (a small molecule) in a mobile, deuterated hydrocarbon solvent increases whereas that of a polymer decreases.

21.4 Discuss how the Fermi contact interaction and the polarization mechanism contribute to spin–spin couplings in NMR.

21.6 Explain how the EPR spectrum of an organic radical can be used to identify the molecular orbital occupied by the unpaired electron.

21.7 The hyperfine interaction of a π electron of an aromatic ring with a methyl group attached to the ring varies as the methyl group rotates. Suggest a mechanism for the interaction.

Exercises

21.1 Equations 21.1 and 21.3 define the *g*-factor and the magnetogyric ratio of a nucleus. Given that *g* is a dimensionless number, what are the units of γ_N expressed in (a) tesla and hertz, (b) SI base units?

21.2 The nucleus ^{33}S has $I = \frac{3}{2}$ and $\gamma_I = 2.054 \times 10^7\ T^{-1}\ s^{-1}$. Calculate the energies of the nuclear spin states in a magnetic field of 6.000 T.

21.3 The magnetogyric ratio of ^{31}P is $1.0840 \times 10^8\ T^{-1}\ s^{-1}$. What is the *g*-value of the nucleus?

21.4 Calculate the value of $(N_\alpha - N_\beta)/N$ for (a) protons, (b) carbon-13 nuclei in a field of 8.5 T.

21.5 The magnetogyric ratio of ^{19}F is $2.5177 \times 10^8\ T^{-1}\ s^{-1}$. Calculate the frequency of the nuclear transition in a field of 7.500 T.

21.6 Calculate the magnetic field needed to satisfy the resonance condition for unshielded protons in a 800.0 MHz radiofrequency field.

21.7 The resonance from a group of protons in a polypeptide is observed at $\delta = 6.33$. What is the difference in the frequency of this resonance from that of TMS in a spectrometer operating at 500.0 MHz?

21.8 The chemical shift of the CH_3 protons in acetaldehyde (ethanal) is $\delta = 2.20$ and that of the CHO proton is 9.80. What is the difference in local magnetic field between the two regions of the molecule when the applied field is (a) 1.2 T, (b) 5.0 T?

21.9 Use the information in Fig. 21.4 to state the splitting (in hertz, Hz) between the methyl and aldehydic proton resonances in a spectrometer operating at (a) 300 MHz, (b) 750 MHz.

21.10 What would be the nuclear magnetic resonance spectrum for a proton resonance line that was split by interaction with seven identical protons?

21.11 What would be the nuclear magnetic resonance spectrum for a proton resonance line that was split by interaction with (a) two, (b) three equivalent nitrogen nuclei (the spin of a nitrogen nucleus is 1)?

21.12 Repeat the analysis in Section 21.4 for an AX_2 spin-$\frac{1}{2}$ system and deduce the pattern of lines expected in the spectrum.

21.13 Sketch the appearance of the ^{1}H-NMR spectrum of acetaldehyde (ethanal) using $J = 2.90$ Hz and the data in Fig. 21.4 in a spectrometer operating at (a) 300 MHz, (b) 550 MHz.

21.14 Sketch the form of the ^{19}F-NMR spectra of a natural sample of $^{10}BF_4^-$ and $^{11}BF_4^-$.

21.15 Sketch the form of an $A_3M_2X_4$ spectrum, where A, M, and X are protons with distinctly different chemical shifts and $J_{AM} > J_{AX} > J_{MX}$.

21.16 Formulate the version of Pascal's triangle that you would expect to represent the fine structure in an NMR spectrum for a collection of *N* spin-1 nuclei, with *N* up to 5.

21.17 Formulate the version of Pascal's triangle that you would expect to represent the fine structure in an NMR spectrum for a collection of *N* spin-$\frac{3}{2}$ nuclei, with *N* up to 5.

21.18 *N*-Acetylbenzoxazepine exists as two conformational isomers. At low temperature the solution-phase ^{1}H-NMR spectrum displays two resonances separated by 119 Hz. At 325 K these two signals coalesce to form a single broad peak. What is the lifetime for interconversion of the two isomers at the higher temperature?

21.19 A proton jumps between two sites with $\delta = 2.7$ and $\delta = 4.8$. At what rate of interconversion will the two signals collapse to a single line in a spectrometer operating at 550 MHz?

21.20 Calculate the energy separation between the spin states of an electron in a magnetic field of 0.250 T.

21.21 Calculate the value of $(N_\beta - N_\alpha)/N$ for electrons in a field of (a) 0.40 T, (b) 1.2 T.

21.22 Calculate the resonance frequency and the corresponding wavelength for an electron in a magnetic field of 0.330 T, the magnetic field commonly used in EPR.

21.23 The centre of the EPR spectrum of atomic hydrogen lies at 329.12 mT in a spectrometer operating at 9.2231 GHz. What is the *g*-value of the electron in this atom?

21.24 A radical containing two equivalent protons shows a three-line spectrum with an intensity distribution 1:2:1. The lines occur at 330.2 mT, 332.5 mT, and 334.8 mT. What is the hyperfine coupling constant for each proton? What is the *g*-value of the radical given that the spectrometer is operating at 9.319 GHz?

21.25 Predict the intensity distribution in the hyperfine lines of the EPR spectra of (a) $\cdot CH_3$, (b) $\cdot CD_3$.

21.26 The benzene radical anion has $g = 2.0025$. At what field should you search for resonance in a spectrometer operating at (a) 9.302 GHz, (b) 33.67 GHz?

21.27 The EPR spectrum of a radical with two equivalent nuclei of a particular kind is split into five lines of intensity ratio 1:2:3:2:1. What is the spin of the nuclei?

21.28 Formulate the version of Pascal's triangle that you would expect to represent the hyperfine structure in an EPR spectrum for a collection of *N* spin-$\frac{3}{2}$ nuclei, with *N* up to 5.

21.29 The hyperfine coupling constants observed in the radical anions (**3**), (**4**), and (**5**) are shown (in millitesla, mT). Use the McConnell equation to map the probability of finding the unpaired electron in the π orbital on each C atom.

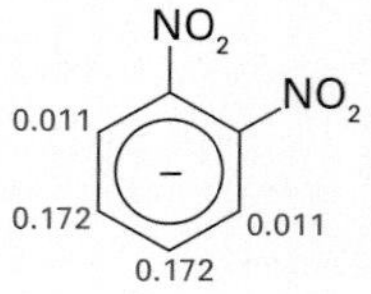

3

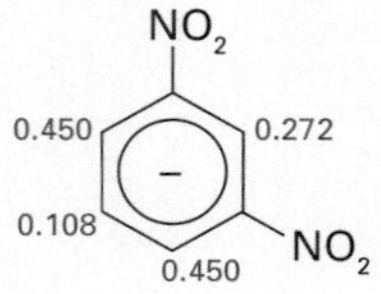

4

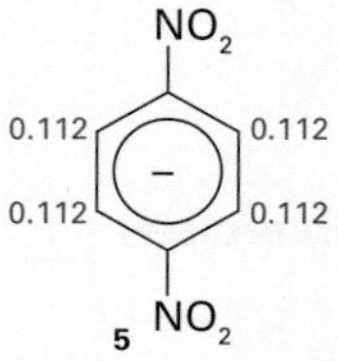

5

Projects

The symbol ‡ indicates that calculus is required.

21.30‡ Show that the coupling constant as expressed by the Karplus equation passes through a minimum when $\cos\phi = B/4C$. To do so, evaluate the first derivative with respect to ϕ and set the result equal to 0. To confirm that the extremum is a minimum, go on to evaluate the second derivative and show that it is positive.

21.31 NMR spectroscopy may be used to determine the equilibrium constant for dissociation of a complex between a small molecule, such as an enzyme inhibitor I, and a protein, such as an enzyme E:

$$\mathrm{EI} \rightleftharpoons \mathrm{E} + \mathrm{I} \qquad K_\mathrm{I} = [\mathrm{E}][\mathrm{I}]/[\mathrm{EI}]$$

In the limit of slow chemical exchange, the NMR spectrum of a proton in I would consist of two resonances: one at ν_I for free I and another at ν_EI for bound I. When chemical exchange is fast, the NMR spectrum of the same proton in I consists of a single peak with a resonance frequency ν given by $\nu = f_\mathrm{I}\nu_\mathrm{I} + f_\mathrm{EI}\nu_\mathrm{EI}$, where $f_\mathrm{I} = [\mathrm{I}]/([\mathrm{I}] + [\mathrm{EI}])$ and $f_\mathrm{EI} = [\mathrm{EI}]/([\mathrm{I}] + [\mathrm{EI}])$ are, respectively, the fractions of free I and bound I. For the purposes of analysing the data, it is also useful to define the frequency differences $\delta\nu = \nu - \nu_\mathrm{I}$ and $\Delta\nu = \nu_\mathrm{EI} - \nu_\mathrm{I}$. Show that when the initial concentration of I, $[\mathrm{I}]_0$, is much greater than the initial concentration of E, $[\mathrm{E}]_0$, a plot of $[\mathrm{I}]_0$ versus $\delta\nu^{-1}$ is a straight line with slope $[\mathrm{E}]_0\Delta\nu$ and *y*-intercept $-K_\mathrm{I}$.

21.32 Here we explore magnetic resonance imaging in more detail. (a) You are designing an MRI spectrometer. What field gradient (in microtesla per metre, $\mu\mathrm{T\,m^{-1}}$) is required to produce a separation of 100 Hz between two protons separated by the long diameter of a human kidney (taken as 8 cm) given that they are in environments with $\delta = 3.4$? The radiofrequency field of the spectrometer is at 400 MHz and the applied field is 9.4 T. (b) Suppose a uniform disk-shaped organ is in a linear field gradient, and that the MRI signal is proportional to the number of protons in a slice of width δx at each horizontal distance *x* from the centre of the disk. Sketch the shape of the absorption intensity for the MRI image of the disk before any computer manipulation has been carried out.

22

Statistical thermodynamics

There are two great rivers in physical chemistry. One is the river of thermodynamics, which deals with the relations between bulk properties of matter, particularly properties related to the transfer of energy. The other is the river of quantum theory, including spectroscopy, which deals with the structures and properties of individual atoms and molecules. These two great rivers flow together in the part of physical chemistry called **statistical thermodynamics**, which shows how thermodynamic properties emerge from the properties of atoms and molecules. The first half of this book dealt primarily with bulk properties, including thermodynamic properties. The second half has dealt with quantum theory, atomic and molecular structure. Even though through the text we have caught occasional glimpses of these great rivers flowing as one, it is in this chapter that we observe their merging.

A great problem with statistical thermodynamics is that it is highly mathematical. Many of the derivations—even the most fundamental—are beyond the scope of this text.[1] All we will see is some of the key concepts and the key results. Where possible the treatment will be qualitative.[2]

The Boltzmann distribution 551

22.1 The general form of the Boltzmann distribution 552

22.2 The origins of the Boltzmann distribution 553

The partition function 553

22.3 The interpretation of the partition function 554

22.4 Examples of partition functions 555

22.5 The molecular partition function 557

Thermodynamic properties 557

22.6 The internal energy 557

22.7 The heat capacity 559

22.8 The entropy 560

22.9 The Gibbs energy 560

22.10 The equilibrium constant 561

FURTHER INFORMATION 22.1 563

FURTHER INFORMATION 22.2 564

CHECKLIST OF KEY CONCEPTS 564

ROAD MAP OF KEY EQUATIONS 565

QUESTIONS AND EXERCISES 565

The Boltzmann distribution

In Foundations 0.11 we saw that, according to the **Boltzmann distribution**, the relative populations, N_1 and N_2, of two states of a system depends upon the absolute temperature T and the difference in their energies, ε_1 and ε_2, as

$$\frac{N_2}{N_1} = e^{-(\varepsilon_2-\varepsilon_1)/kT} \qquad \text{Boltzmann distribution} \qquad (22.1a)$$

[1] See our *Physical Chemistry* (2010) for details.

[2] In this chapter, some Examples, Self-tests, and Brief illustrations require calculus: they are marked with the symbol ‡.

where k is Boltzmann's constant, a fundamental constant with the value 1.381×10^{-23} J K^{-1}. The key conclusion is that *the relative population of the upper state decreases exponentially with its energy above the lower state.*

We also saw in Foundations 0.11 that it is common in chemical applications to use not the individual energies ε_i but energies per mole of molecules, E_i, with $E_i = N_A\varepsilon_i$, where N_A is Avogadro's constant. With $R = N_Ak$, eqn 22.1a becomes

$$\frac{N_2}{N_1} = e^{-(E_2-E_1)/RT} \qquad \text{Boltzmann distribution in terms of molar energies} \qquad (22.1b)$$

We have used this result throughout the text when making connections between thermodynamic and molecular properties, or when explaining the origins of intense spectroscopic transitions. Here we explore the Boltzmann distribution in more detail.

22.1 The general form of the Boltzmann distribution

Equation 22.1 is a special case of a more general form of the Boltzmann distribution, which tells us how to calculate the numbers of molecules in each state of a system at any temperature:

$$N_i = \frac{Ne^{-\varepsilon_i/kT}}{q} \qquad \text{Boltzmann distribution} \qquad (22.2)$$

Here N_i is the number of molecules in a state with energy ε_i, N is the total number of molecules, The term in the denominator, q, is the **partition function:**

$$q = \sum_i e^{-\varepsilon_i/kT} = e^{-\varepsilon_0/kT} + e^{-\varepsilon_1/kT} + \cdots$$

Definition Partition function (22.3)

where the sum is over all the states of the system. We shall have much more to say about q later, and see how it can be calculated and given physical meaning. At this stage it is just a kind of normalizing factor, for it ensures that the sum of all the populations is the total number of molecules in the system: that is, with N_i given by eqn 22.2, $\Sigma_i N_i = N$.

As we have emphasized when using the Boltzmann distribution earlier in the text, one very important feature of the distribution is that it applies to the populations of *states*. We have seen that in some cases (the hydrogen atom and rotating molecules are examples) several different states have the same energy. That is, some energy levels are *degenerate* (Section 12.7). The Boltzmann distribution can be used to calculate, for instance, the number of hydrogen atoms at a temperature T that have their electron in a $2p_x$ orbital. Because a $2p_y$ orbital has exactly the same energy, the number of atoms with an electron in a $2p_y$ orbital is the same as the number with an electron in a $2p_x$ orbital. The same is true of atoms with an electron in a $2p_z$ orbital. Therefore, if we want the *total* number of atoms with electrons in 2p orbitals, we have to multiply the number in *one* of them by a factor of 3. In general, if the degeneracy of an energy level (that is, the number of states of that energy) is g, we use a factor of g to get the population of the *level* (as distinct from an individual state). It is obviously very important to decide whether we wish to express the population of an individual state or the population of an entire degenerate energy level. We shall denote levels by L, so in terms of levels the Boltzmann distribution and the partition function are

$$N_L = \frac{Ng_Le^{-\varepsilon_L/kT}}{q} \qquad q = \sum_L g_Le^{-\varepsilon_L/kT}$$

General form Boltzmann distribution (22.4)

where N_L is the total number of molecules in the level L (the sum of populations of all the states of that level), g_L is its degeneracy, and ε_L is its energy.

Brief illustration 22.1 Relative populations

We saw in Section 19.1 that the rotational energy of a linear rotor is $hBJ(J+1)$ and that the degeneracy of each level is $2J+1$. Because the degeneracy of the level with $J = 2$ (and energy $6hB$) is 5 and that of the level with $J = 1$ (and energy $2hB$) is 3, the relative numbers of molecules with $J = 2$ and 1 is

$$\frac{N_2}{N_1} = \frac{\overbrace{Ng_2e^{-\varepsilon_2/kT}/q}^{\text{Boltzmann distribution}}}{\underbrace{Ng_1e^{-\varepsilon_1/kT}/q}_{\text{Boltzmann distribution}}} \overset{\text{Cancel } N \text{ and } q}{=} \frac{\overbrace{g_2}^{\text{Degeneracy, } g=2J+1}}{g_1}e^{-(\overbrace{\varepsilon_2}^{\text{Energy, } \varepsilon=hBJ(J+1)}-\varepsilon_1)/kT}$$

$$= \tfrac{5}{3}e^{-(\overbrace{6hB}^{\varepsilon_2}-\overbrace{2hB}^{\varepsilon_1})/kT} = \tfrac{5}{3}e^{-4hB/kT}$$

For HCl, $B = 318.0$ GHz, so at 25 °C (corresponding to 298 K), this ratio works out as

$$\frac{N_2}{N_1} = \tfrac{5}{3}e^{-4\times\overbrace{(6.626\times10^{-34}\text{ J s})}^{h}\times\overbrace{3.18\times10^{11}\text{ s}^{-1}}^{318\text{ GHz}}/\{\overbrace{(1.381\times10^{-23}\text{ J K}^{-1})}^{k}\times\overbrace{(298\text{ K})}^{T}\}}$$

$$= 1.36$$

There are *more* molecules in the level with $J = 2$ than in the level with $J = 1$, even though $J = 2$ corresponds to a higher energy. Each individual *state* with $J = 2$ has a lower population than each state with $J = 1$, but there are more states in the level with $J = 2$.

One important convention that we adopt (largely for convenience) is that *all energies are measured relative to the ground state*. That is, we set the ground-state energy equal to zero, even if there is a zero-point energy. For instance, the energies of the states of a harmonic oscillator are measured from zero for the ground state:

Actual energies: $\varepsilon = \frac{1}{2}h\nu, \frac{3}{2}h\nu, \frac{5}{2}h\nu, \ldots$
Our convention: $\varepsilon = 0, h\nu, 2h\nu, \ldots$

Likewise, the energies of the hydrogen atom are measured from zero for the 1s orbital:

Actual energies: $\varepsilon = -hcR_H, -\frac{1}{4}hcR_H, -\frac{1}{9}hcR_H, \ldots$
Our convention: $\varepsilon = 0, \frac{3}{4}hcR_H, \frac{8}{9}hcR_H, \ldots$

This convention greatly simplifies our interpretation of the significance of q.

22.2 The origins of the Boltzmann distribution

The conceptual basis of the derivation of eqn 22.4 is very simple. We imagine a stack of energy levels arranged like bookshelves, one above the other. Then we imagine being blindfolded and throwing balls (the molecules) at the shelves (the energy levels) and letting them land on the available shelves entirely at random, apart from one condition. That condition is that the total energy, ε, of the final arrangement must have the actual energy of the sample of matter we are seeking to describe. We should not press this analogy too far: it is intended just to provide a visualizable image and actually has little relation to the way that actual molecules become distributed! But we can proceed and conclude that, provided the temperature is above absolute zero, not all the balls are allowed to land on the bottom shelf, for that would give a total energy of zero. Some of the balls may land on the bottom shelf, but there must be others ending up on higher shelves to ensure that the total energy is ε. If we imagine throwing 100 balls at a set of shelves, then we shall end up with one particular valid distribution. If we repeated the experiment with the same number of balls, we would end up with a different but still valid distribution. If we went on repeating the experiment, we would get many different distributions, but some of them would occur more often than others (Fig. 22.1).

When this game is analysed mathematically, it turns out that the *most probable* distribution—the arrangement that turns up most often—is that given by eqn 22.4. In other words, *the Boltzmann distribution is the outcome of blind chance occupation of energy levels, subject to the requirement that the total energy has a particular value*. When we deal with about 10^{23} molecules and repeat the experiment millions of times, the Boltzmann distribution turns out to be very accurate, and we can use it with confidence for all typical samples of matter.

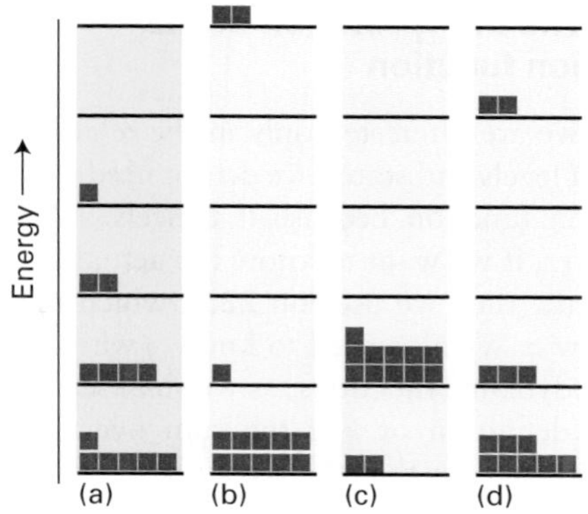

Fig. 22.1 The derivation of the Boltzmann distribution involves imagining that the molecules of a system (the squares) are distributed at random over the available energy levels subject to the requirements that the number of molecules and the total energy is constant, and then looking for the most probable arrangement. Of the four shown here, the numbers of ways of achieving each arrangement are (a) 181 180, (b) 858, (c) 78, (d) 12 870. (To calculate the number of ways, W, of arranging N molecules with N_1 in state 1, N_2 in state 2, etc., use $W = N!/N_1!N_2!\ldots$, with $n! = n(n-1)(n-2)\ldots 1$, and $0! = 1$.) The number of ways of achieving (a) is by far the greatest, so this distribution is the most probable; it corresponds to the Boltzmann distribution.

The partition function

The key concept of quantum mechanics is the existence of a wavefunction that contains in principle all the dynamical information about a system, such as its energy, the electron density, the dipole moment, and so on. Once we know the wavefunction of an atom or molecule, we can extract from it all the dynamical information possible about the system—provided we know how to manipulate it. There is a similar concept in statistical thermodynamics. The partition function, q, contains all the *thermo*dynamic information about the system, such as its internal energy, entropy, heat capacity, and so on. Our task here is to see how to calculate the partition function and how to extract the information it contains.

22.3 The interpretation of the partition function

When we are interested only in the relative populations of levels and states, we do not need to know the partition function because it cancels in eqn 22.1. However, if we want to know the actual population of a state, then we use eqn 22.2, which requires us to know q. We also need to know q when we derive thermodynamic functions, as we shall see.

The definition of q is the sum over states (not levels; remember that there may be several states of the same energy), as given in eqn 22.3. We can write out the first few terms as follows:

$$q = 1 + e^{-\varepsilon_1/kT} + e^{-\varepsilon_2/kT} + e^{-\varepsilon_3/kT} + \cdots$$

The first term is 1 because the energy of the ground state (ε_0) is 0, according to our convention, and $e^0 = 1$. In principle, we just substitute the values of the energies, evaluate each term for the temperature of interest, and add them together to get q. However, that procedure does not give much insight.

To see the physical significance of q, let's suppose first that $T = 0$. Then, because $e^{-\infty} = 0$, all terms other than the first are equal to 0, and $q = 1$. At $T = 0$ only the ground state is occupied and (provided that state is non-degenerate) $q = 1$. Now consider the other extreme: a temperature so high that all the $\varepsilon_i/kT = 0$. Then, because $e^0 = 1$, the partition function is $q \approx 1 + 1 + 1 + 1 + \cdots = N_{\text{states}}$, where N_{states} is the total number of states of the molecule. That is, at very high temperatures, all the states of the system are thermally accessible. It follows that if the molecule has an infinite number of states, then q rises to infinity as T approaches infinity. We should begin to suspect that the partition function is telling us the number of states that are occupied at a given temperature.

Now consider an intermediate temperature, at which only some of the states are occupied significantly. Suppose that the temperature is such that kT is large compared to ε_1 and ε_2 but small compared to ε_3 and all subsequent terms (Fig. 22.2). Because ε_1/kT and ε_2/kT are both small compared to 1, and $e^{-x} \approx 1$ when x is very small, the first three terms are all close to 1. However, because ε_3/kT is large compared to 1, and $e^{-x} \approx 0$ when x is large, all the remaining terms are close to 0. Therefore,

$$q = 1 + \overbrace{e^{-\varepsilon_1/kT}}^{1} + \overbrace{e^{-\varepsilon_2/kT}}^{1} + \overbrace{e^{-\varepsilon_3/kT}}^{0} + \overbrace{e^{-\varepsilon_4/kT}}^{0} + \overbrace{e^{-\varepsilon_5/kT}}^{0} + \cdots$$

and $q \approx 1 + 1 + 1 + 0 + \cdots = 3$. Once again, we see that the partition function is telling us the number of significantly occupied states at the temperature of interest. That is the principal meaning of the partition function: *q tells us the number of thermally accessible states at the temperature of interest.*

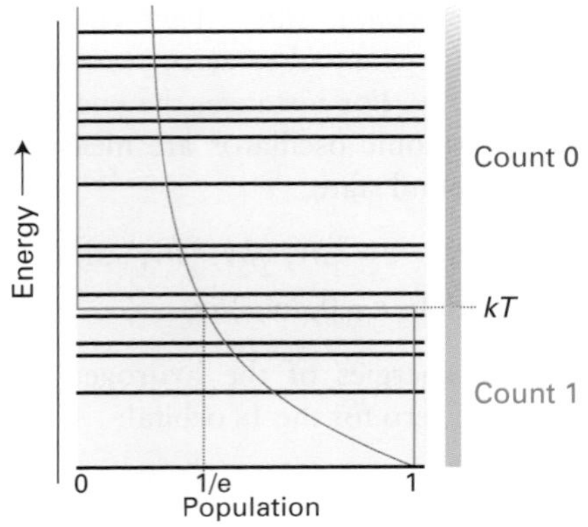

Fig. 22.2 The partition function is a measure of the number of thermally accessible states. Thus, for all states with $\varepsilon < kT$ the exponential term is reasonably close to 1 whereas for all states with $\varepsilon > kT$ the exponential term is close to 0. The states with $\varepsilon < kT$ are significantly thermally accessible.

Once we grasp the significance of q, statistical thermodynamics becomes much easier to understand. For example:

- We can anticipate, even before we do any calculations, that q increases with temperature, because more states become accessible as the temperature is raised.
- At low temperatures q is small, and falls to 1 as the temperature approaches absolute zero (when only one state, the ground state, is accessible and we are supposing that that state is non-degenerate).
- Molecules with numerous, closely spaced energy levels (like the rotational states of bulky molecules) can be expected to have very large partition functions.
- Molecules with widely spaced energy levels can be expected to have small partition functions, because only the few lowest states will be occupied at low temperatures.

Example 22.1

Calculating a partition function

The boat conformation of cyclohexane (**1**) lies 22 kJ mol^{-1} higher in energy than the chair conformation (**2**). Calculate the partition function for the cyclohexane molecule, confining attention to these two conformations only. Show how the partition function varies with temperature.

1

2

Strategy Whenever calculating a partition function, start at the definition in eqn 22.3 and write out the individual terms. Remember to set the ground state energy equal to 0. When the energies of states are given in joules (or kilojoules) per mole, replace the k in the definition of q by $R = N_A k$.

Solution There are only two states, so the partition function has only two terms. The energy of the chair form is set at 0 and that of the boat form is $E = 22$ kJ mol^{-1}. Then, with

$$\frac{E}{RT} = \frac{\overbrace{2.2\times10^4\ \text{J mol}^{-1}}^{E}}{(8.3145\ \text{J K}^{-1}\ \text{mol}^{-1})\times T} \overbrace{=}^{\text{Cancel J mol}^{-1}} \frac{2.6\ldots\times10^4}{T\ \text{K}^{-1}} = \frac{2.6\ldots\times10^4\ \text{K}}{T}$$

it follows that

$$q = 1 + e^{-(2.6\ldots\times10^4\ \text{K})/T}$$

This function is plotted in Fig. 22.3. We see that it rises from $q = 1$ (only the chair form is accessible at $T = 0$, when $(2.6 \times 10^4\ \text{K})/T = \infty$ and $e^{-\infty} = 0$) to $q = 2$ at $T = \infty$ (when $(2.6 \times 10^4\ \text{K})/T = 0$ and $e^0 = 1$; both states are thermally accessible at high temperatures). At 20 °C, $q = 1.0001$, and there is a very small proportion of molecules in the boat form.

A note on good practice Note how the units are treated in the exponent: the units of E and R cancel apart from K^{-1} in the denominator, which becomes K in the numerator (in the form 2.6×10^4 K), which will cancel the units K of T when values of the latter are introduced. You will sometimes see an expression like '$q = 1 + e^{-2.6\ldots/T}$, with T in kelvins' (or, worse, 'with T the absolute temperature'); retention of the units, as we show, is unambiguous and therefore better practice.

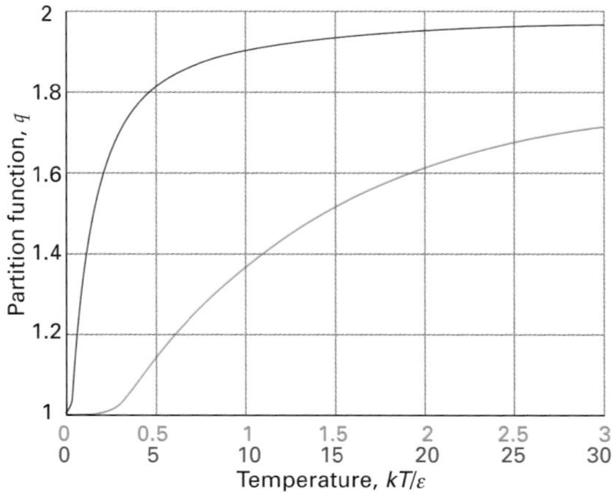

Fig. 22.3 The partition function for a two-level system with states at the energies 0 and ε. Note how the partition function rises from 1 and approaches 2 at high temperatures.

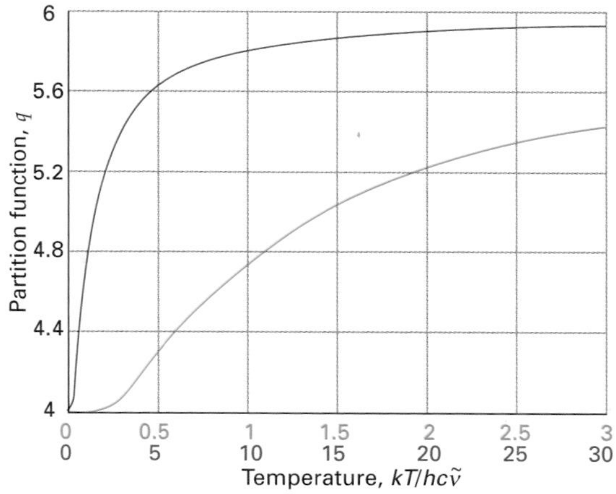

Fig. 22.4 The partition function for the system treated in Self-test 22.1. Note how q rises from 4 (when only the four states of the $^2P_{3/2}$ level are occupied) and approaches 6 (when the two states of the $^2P_{1/2}$ level are also accessible). At 20 °C, $kT/hc\tilde{\nu} = 0.504$, corresponding to $q = 5.21$.

Self-test 22.1

The ground configuration of a fluorine atom gives rise to a ^{2}P term with two levels, the $J = \frac{3}{2}$ level (of degeneracy 4) and the $J = \frac{1}{2}$ level (of degeneracy 2) at an energy corresponding to 404.0 cm^{-1} above the ground state. Write down an expression for the partition function and plot it as a function of temperature. *Hint*: The notation used here was introduced in Section 13.17. Take $E = hc\tilde{\nu}$ for the energy of the upper level. In this instance, the ground state is degenerate.

Answer: $q = 4 + 2e^{-hc\tilde{\nu}/kT}$; Fig. 22.4

22.4 Examples of partition functions

In a number of cases it is possible to derive simple closed expressions for partition functions, which are then very convenient to use for calculating various properties.

(a) The translational partition function

Suppose a molecule of mass m is confined in a flask of volume V at a temperature T, then (as shown in Further information 22.1) to a good approximation for typical containers and $T > 0$, the **translational partition function**, q^T, is

$$q^T = \frac{(2\pi mkT)^{3/2}V}{h^3} \qquad \text{Translational partition function} \qquad (22.5)$$

We see that the partition function increases with temperature, as we have come to expect. However, notice that q^T also increases with the volume of the flask. That we should expect too: the energy levels of a particle in a box become closer together as the size of the box increases (Section 12.7), so at a given temperature, more states are thermally accessible.

Brief illustration 22.2 The translational partition function

Suppose we have an O_2 molecule (of mass $32m_u$) in a flask of volume 100 cm^3 at 20 °C. Its translational partition function is

$$q^T = \left(2\pi\times\overbrace{32\times\overbrace{(1.661\times10^{-27}\ \text{kg})}^{m_u}}^{m}\times\overbrace{(1.381\times10^{-23}\ \text{J K}^{-1})}^{k}\times\overbrace{(298\ \text{K})}^{T}\right)^{3/2}$$
$$\times\frac{\overbrace{(1.00\times10^{-4}\ \text{m}^3)}^{V}}{\underbrace{(6.626\times10^{-34}\ \text{J s})^3}_{h^3}}$$
$$= 9.67\times10^{25}$$

Note that a huge number of translational states are accessible at room temperature. This result is consistent with the derivation of eqn 22.5, which assumed that the translational energy levels form a near continuum in containers of macroscopic size.

A note on good practice All the units must cancel because all partition functions are dimensionless numbers. Here, because 1 J = 1 kg m^2 s^{-2}, the units cancel as follows:

$$\frac{(\text{kg J K}^{-1}\ \text{K})^{3/2}\ \text{m}^3}{(\text{J s})^3}=\frac{(\text{kg kg m}^2\ \text{s}^{-2})^{3/2}\ \text{m}^3}{(\text{kg m}^2\ \text{s}^{-2}\ \text{s})^3}$$
$$=\frac{(\text{kg m s}^{-1})^3\ \text{m}^3}{(\text{kg m}^2\ \text{s}^{-1})^3}=\frac{\text{kg}^3\ \text{m}^6\ \text{s}^{-3}}{\text{kg}^3\ \text{m}^6\ \text{s}^{-3}}=1$$

It might seem irksome to do this cancellation explicitly, but it is a very good way of making sure that you have set up the numerical calculation correctly.

(b) The rotational partition function

The **rotational partition function,** q^R, can also be approximated when the temperature is high enough for many rotational states to be occupied. For a linear rotor it turns out (see Further information 22.1) that for heavy molecules and $T > 0$,

$$q^R = \frac{kT}{\sigma hB} \qquad \text{Large molecules} \quad \text{Rotational partition function} \qquad (22.6)$$

In this expression, B is the rotational constant (Section 19.1) and σ is the **symmetry number:** $\sigma = 1$ for an unsymmetrical linear rotor (such as HCl or HCN) and $\sigma = 2$ for a symmetrical linear rotor (such as H_2 or CO_2). The symmetry number reflects the fact that an unsymmetrical molecule is distinguishable after rotation by 180° but a symmetrical molecule is not. When evaluating q we have to count only distinguishable states, and a symmetrical molecule has fewer distinguishable states than a less symmetrical molecule. A more formal explanation is that, as shown in Section 19.2, the Pauli principle excludes certain states of symmetrical molecules, so the number of thermally accessible states is smaller by the appropriate factor than for unsymmetrical molecules, where there is no such restriction. The rotational partition function of HCl at 25 °C works out to 19.6 (see Exercise 22.11), so about 20 rotational states (not levels: remember the $(2J + 1)$-fold degeneracy of each rotational level; 20 states corresponds to about the first 4 levels) are significantly occupied at that temperature.

(c) The vibrational partition function

We show in Further information 22.1 that the **vibrational partition function,** q^V, is

$$q^V = \frac{1}{1-e^{-h\nu/kT}} \qquad \text{Vibrational partition function} \qquad (22.7)$$

Equation 22.7 is the partition function for a harmonic oscillator or any vibrating diatomic molecule. Figure 22.5 shows how q^V varies with temperature. Note that:

- $q^V = 1$ at $T = 0$, when only the lowest state is occupied.
- As T becomes high, so q^V becomes infinite because all the states of the infinite ladder are thermally accessible.
- At room temperature, and for typical molecular vibrational frequencies, q^V is very close to 1 because only the vibrational ground state is occupied (see Exercise 22.13).

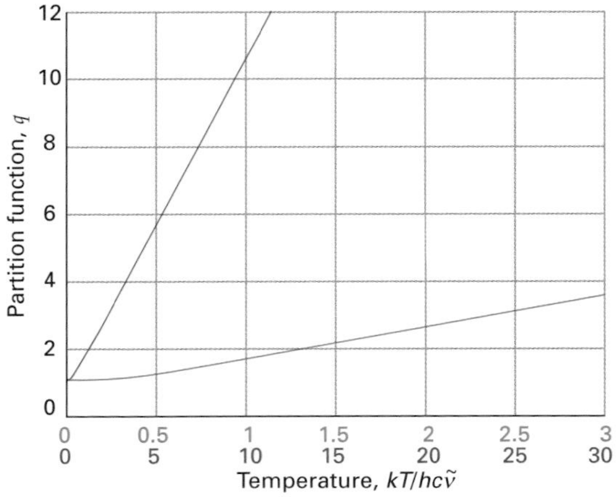

Fig. 22.5 The partition function for a harmonic oscillator. For an oscillator with $\tilde{\nu} = 1000$ cm^{-1}, at 20 °C, $kT/hc\tilde{\nu} = 0.204$, corresponding to $q = 1.01$.

We also show in Further information 22.1 that when the temperature is so high that $h\nu/kT \ll 1$, eqn 22.7 simplifies to

$$q^{\mathrm{V}} \approx \frac{kT}{h\nu} \quad \text{High-temperature limit} \quad \text{Vibrational partition function} \qquad (22.8)$$

This expression is consistent with the interpretation of q as the number of thermally accessible states: the average energy of a harmonic oscillator is kT (from the equipartition theorem, see Foundations 0.12) and the separation of energy levels is $h\nu$, so about $kT/h\nu$ states must be occupied. (Think of a ladder with rungs separated by $h\nu$: to reach kT we need a ladder with $kT/h\nu$ rungs.)

(d) The electronic partition function

No closed form can be given for the **electronic partition function**, q^{E}, the partition function for the distribution of electrons over their available states. However, for closed-shell molecules the excited states are so high in energy that only the ground state is occupied, and for them $q^{\mathrm{E}} = 1$. Special care has to be taken for atoms and molecules that do not have closed shells.

Brief illustration 22.3 The electronic partition function

The electron configuration of the ground state of Na is $[\mathrm{Ne}]3s^1$ (Section 13.11). Because there are two possible orientations of the electron spin in the 3s orbital, and both are equally likely, the ground electronic state of Na is doubly degenerate. It follows that the electronic partition function is $q^{\mathrm{E}} = g^{\mathrm{E}} = 2$.

22.5 The molecular partition function

The energy of a molecule can be approximated as the sum of contributions from its different modes of motion (translation, rotation, and vibration), the distribution of electrons, and the electronic and nuclear spin. Given that the energy is a sum of independent contributions, we show in the following Derivation that the partition function is a product of contributions:

$$q = q^{\mathrm{T}} q^{\mathrm{R}} q^{\mathrm{V}} q^{\mathrm{E}} \quad \text{Molecular partition function} \qquad (22.9)$$

where T denotes translation, R rotation, V vibration, and E the electronic contribution. The contribution from electronic spin is important in atoms or molecules containing unpaired electrons.

Derivation 22.1

Factorization of the partition function

Suppose the energy can be expressed as the sum of contributions from two modes A and B (such as vibration and rotation), and that we can write $\varepsilon_{i,j} = \varepsilon_i^{\mathrm{A}} + \varepsilon_j^{\mathrm{B}}$, where i denotes a state of mode A and j denotes a state of mode B and the sums that we will have to do are over both i and j independently. Then, the partition function is

$$q = \sum_{i,j} \mathrm{e}^{-\varepsilon_{i,j}/kT} \overset{\varepsilon_{i,j}=\varepsilon_i^{\mathrm{A}}+\varepsilon_j^{\mathrm{B}}}{=} \sum_{i,j} \mathrm{e}^{-\varepsilon_i^{\mathrm{A}}/kT-\varepsilon_j^{\mathrm{B}}/kT}$$

$$\overset{\text{Use } \mathrm{e}^{x+y}=\mathrm{e}^x\mathrm{e}^y}{=} \sum_{i,j} \mathrm{e}^{-\varepsilon_i^{\mathrm{A}}/kT}\mathrm{e}^{-\varepsilon_j^{\mathrm{B}}/kT} = \overbrace{\sum_i \mathrm{e}^{-\varepsilon_i^{\mathrm{A}}/kT}}^{q^{\mathrm{A}}}\overbrace{\sum_j \mathrm{e}^{-\varepsilon_j^{\mathrm{B}}/kT}}^{q^{\mathrm{B}}}$$

$$= q^{\mathrm{A}} q^{\mathrm{B}}$$

This argument is readily extended to three and more modes, as in eqn 22.9.

Thermodynamic properties

The principal reason for calculating the partition function is to use it to calculate thermodynamic properties of systems as small as atoms and as large as biopolymers. There are two fundamental relations we need. We can deal with First-Law quantities (such as heat capacity and enthalpy) once we know how to calculate the internal energy. We can deal with Second-Law quantities (such as the Gibbs energy and equilibrium constants) once we know how to calculate the entropy.

22.6 The internal energy

To calculate the total energy, ε, of the system, we note the energy of each state (ε_i), multiply that energy by the number of molecules in the state (N_i), and then add together all these products:

$$\varepsilon = N_0\varepsilon_0 + N_1\varepsilon_1 + N_2\varepsilon_2 + \cdots = \sum_i N_i\varepsilon_i$$

However, the Boltzmann distribution tells us the number of molecules in each state of a system, so we can replace the N_i in this expression by the expression in eqn 22.2:

$$\varepsilon = \sum_i \overbrace{\frac{N\mathrm{e}^{-\varepsilon_i/kT}}{q}}^{\text{Population of state } i} \times \overbrace{\varepsilon_i}^{\text{Energy of state } i} = \frac{N}{q}\sum_i \varepsilon_i \mathrm{e}^{-\varepsilon_i/kT} \qquad (22.10)$$

If we know the individual energies of the states (from spectroscopy, for instance), then we just substitute their values into this expression. However, there is a much simpler method—or at least a much more succinct formula—available when we have an expression for the partition function, such as those given in Section 22.4. In the following Derivation we show that the energy depends upon how the partition function q varies with temperature as

$$\varepsilon = \frac{NkT^2}{q} \times \text{slope of } q \text{ plotted against } T = \frac{NkT^2}{q} \times \frac{dq}{dT} \qquad (22.11)$$

Derivation 22.2

The internal energy from the partition function

The sum on the right of eqn 22.10 resembles the definition of the partition function, but differs from it by having the ε_i factor multiplying each term. However, we can recognize (by using the rules of differentiation set out in The chemist's toolkit 1.3) that

$$\frac{d}{dT}e^{-\varepsilon_i/kT} \overset{\text{Use } de^{f(x)}/dx=e^{f(x)}\times df/dx}{=} e^{-\varepsilon_i/kT} \times \frac{d}{dT}\left(-\frac{\varepsilon_i}{kT}\right) \overset{\text{Use } d(1/x)/dx=-1/x^2}{=} \frac{\varepsilon_i}{kT^2}e^{-\varepsilon_i/kT}$$

In other words,

$$\varepsilon_i e^{-\varepsilon_i/kT} = kT^2\frac{d}{dT}e^{-\varepsilon_i/kT}$$

With this substitution, the expression for the total energy becomes

$$\varepsilon = \frac{N}{q}\sum_i kT^2\frac{d}{dT}e^{-\varepsilon_i/kT} = \frac{N}{q}kT^2\frac{d}{dT}\overbrace{\sum_i e^{-\varepsilon_i/kT}}^{q}$$

because (in blue) kT^2 is a constant and the sum of derivatives is the derivative of the sum. Magically (or, more precisely, mathematically), the expression for the partition function has appeared, so we can write

$$\varepsilon = \frac{NkT^2}{q}\frac{dq}{dT}$$

which, because dq/dT is the slope of a graph of q plotted against T, is eqn 22.11.

The remarkable feature of eqn 22.11 is that it is an expression for the total energy in terms of the partition function alone. The partition function is starting to fulfil its promise to deliver all thermodynamic information about the system.

There is one more detail to take into account before we use eqn 22.11. Recall that we have set the zero of energy at the energy of the lowest state of the molecule. However, the internal energy of the system might be nonzero on account of zero-point energy, and the ε in eqn 22.11, and E, the corresponding molar energy, are energies *above* the zero-point energy. That is, the molar internal energy at a temperature T is

$$U_m = U_m(0) + N_A\varepsilon \qquad \text{Internal energy} \quad (22.12)$$

with ε given by eqn 22.11.

‡Example 22.2

Calculating the internal energy

Calculate the molar internal energy of a monatomic gas.

Strategy The only mode of motion of a monatomic gas is translation (we ignore electronic excitation). Therefore, substitute the translational partition function in eqn 22.5 into eqn 22.11 (using the precise mathematical form given in Derivation 22.2) and then insert the result into eqn 22.12. The partition function has the form $q = aT^{3/2}$, where a is the collection of constants

$$a = \frac{(2\pi mk)^{3/2}}{h^3}V$$

Solution First, we need the first derivative of q with respect to T:

$$\frac{dq}{dT} = \frac{d}{dT}(aT^{3/2}) \overset{dx^n/dx=nx^{n-1}}{=} \tfrac{3}{2}aT^{1/2}$$

When we substitute this result into eqn 22.11 we get

$$\varepsilon = \frac{NkT^2}{q}\times\frac{dq}{dT} = \frac{NkT^2}{aT^{3/2}}\times\tfrac{3}{2}aT^{1/2} \overset{\text{Cancel } a}{=} \tfrac{3}{2}NkT$$

The molar internal energy is obtained by replacing N by Avogadro's constant and using eqn 22.12:

$$U_m = U_m(0) + \tfrac{3}{2}N_AkT = U_m(0) + \tfrac{3}{2}RT$$

The term $U_m(0)$ contains all the contributions from the binding energy of the electrons and of the nucleons in the nucleus. The term $\tfrac{3}{2}RT$ is the contribution to the internal energy from the translational motion of the atoms in their container. We see that this treatment gives the same result obtained from use of the equipartition theorem (Foundations 0.12).

Self-test 22.2

Calculate the molar internal energy of a gas of diatomic molecules.

Answer: $U_m = U_m(0) + \tfrac{5}{2}RT$

22.7 The heat capacity

Once we have calculated the internal energy of a sample of molecules, it is a simple matter to calculate the heat capacity. It should be recalled that the heat capacity at constant volume, C_V, is defined as the slope of the plot of internal energy against temperature:

$$C_V = \frac{\Delta U}{\Delta T} \quad \text{at constant volume}$$

Therefore, all we need do is to evaluate the slope of the expression for U obtained from the partition function.

‡Brief illustration 22.4 The heat capacity

The slope of U with respect to T is actually the first derivative:

$$C_V = \frac{dU}{dT} \quad \text{at constant volume}$$

(Remember from Section 2.8 that a more sophisticated notation for this expression is $C_V = (\partial U/\partial T)_V$.) The constant-volume molar heat capacity of a monatomic gas is therefore obtained by substituting the molar internal energy, $U_m = U_m(0) + \frac{3}{2}RT$, into this expression:

$$C_V = \frac{d}{dT}(U_m(0) + \tfrac{3}{2}RT) = \tfrac{3}{2}R$$

To calculate $C_{p,m}$, we use eqn 2.17 ($C_{p,m} - C_{V,m} = R$) and obtain $C_{p,m} = \frac{5}{2}R$.

‡Self-test 22.3

Calculate the contribution to the molar constant-volume heat capacity of a two-state system, like the chair–boat interconversion of cyclohexane (Example 22.1), and show how the heat capacity varies with temperature.

Answer: $C_{V,m} = R(E/RT)^2 e^{E/RT}/(1 + e^{E/RT})^2$, Fig. 22.6

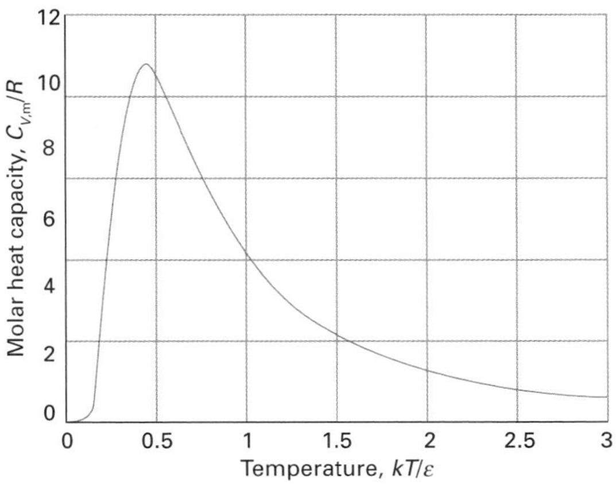

Fig. 22.6 The variation of the heat capacity of a two-level system with states at energies 0 and ε. Note how the heat capacity is zero at $T = 0$, passes through a maximum at $T = 0.417\varepsilon/k$, and approaches 0 at high temperatures.

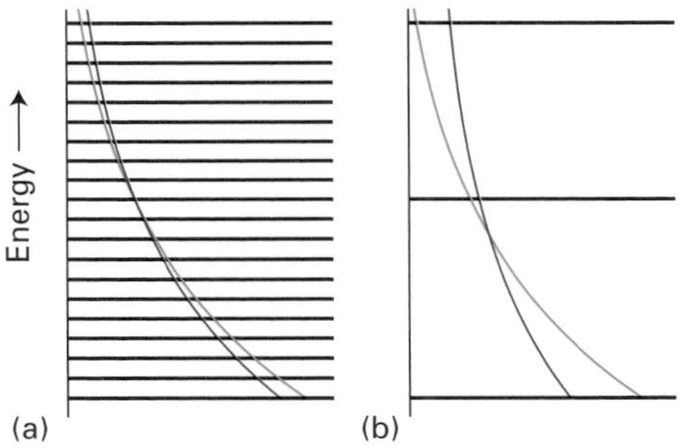

Fig. 22.7 The heat capacity depends on the availability of levels. (a) When the levels are close together, a given amount of energy arriving as heat can be accommodated with little adjustment of the populations and hence the temperature that occurs in the Boltzmann distribution. This system has a high heat capacity. (b) When the levels are widely separated, the same incoming energy has to be accommodated by making use of higher energy levels, with a consequent greater change in the 'reach' of the Boltzmann distribution, and therefore a greater change in temperature. This system therefore has a low heat capacity. In each case the green line is the distribution at low temperature and the red line that at higher temperature.

We are now ready to understand the molecular reason why different substances have different molar heat capacities. When the available energy levels are close together, a given quantity of energy arriving as heat can be accommodated with little adjustment of the populations and hence with little modification of the temperature that occurs in the Boltzmann distribution and specifies the distribution of populations. The relative insensitivity of temperature to the arrival of energy corresponds to a high heat capacity (Fig. 22.7). When the energy levels are widely separated, the arriving energy must be accommodated by making use of the high energy levels with a consequent greater 'reach' of the Boltzmann distribution and hence a greater modification of the temperature. That is, widely spaced energy levels correlate with a low heat capacity.

The translational energy levels of molecules in a gas are very close together, and all monatomic gases have similar molar heat capacities. The separation of the vibrational energies of atoms bound together in solids depends on the stiffness of the bonds between them and on the masses of the atoms. As we saw in

Chapter 12, the stronger the bond and the lighter the atoms in a bond, the larger is the separation between vibrational energy levels. As a result, solids show a wide range of molar heat capacities. Very large molecules, like polymers, have large numbers of atoms and can vibrate in many different ways. Many of these ways correspond to collective motion of many atoms, so the vibrational energies are spaced closely. Hence, heat capacities of polymers may be large.

Water, as so often, is anomalous. It is a small, rigid molecule but has a high heat capacity. The anomaly can be traced to hydrogen bonds, which link many molecules together into clusters that vibrate in numerous ways. Consequently, the vibrational energies are close together, and the heat capacity of water is larger than expected for a substance consisting of small molecules interacting weakly.

22.8 The entropy

Boltzmann showed that there is a close relation between the entropy and the partition function: both are measures of the number of arrangements available to the molecules. The precise connection for *distinguishable* molecules (those locked in place in a solid) is[3]

$$S = \frac{U - U(0)}{T} + Nk \ln q \qquad (22.13a)$$

Distinguishable particles — The entropy in terms of the partition function

The analogous term for *indistinguishable* molecules (identical molecules free to move, as in a gas) is

$$S = \frac{U - U(0)}{T} + Nk \ln q - Nk(\ln N - 1) \qquad (22.13b)$$

Indistinguishable particles — The entropy in terms of the partition function

Because we can calculate the first term on the right from q, we now have a method for calculating the entropy of any system of non-interacting molecules once we know its partition function.

Example 22.3

Calculating the entropy

Calculate the contribution that rotational motion makes to the molar entropy of a gas of HCl molecules at 25 °C.

Strategy We have already calculated the contribution to the internal energy (Self-test 22.2), and we have the rotational partition function in eqn 22.6 (with $\sigma = 1$). We need to combine the two parts. We use eqn 22.13a because we are concentrating on the internal motion (the rotation) of the molecules, not their translational motion.

Solution We substitute $U_m - U_m(0) = RT$ and $q = kT/hB$ into eqn 22.13a, and obtain (for $T > 0$)

$$S_m = \overbrace{\frac{RT}{T}}^{U_m - U_m(0)} + R \ln \overbrace{\frac{kT}{hB}}^{q} = R\left(1 + \ln \frac{kT}{hB}\right)$$

Notice that the entropy increases with temperature (Fig. 22.8). At a given temperature, the entropy is larger the smaller the value of B. That is, bulky molecules (which have large moments of inertia and therefore small rotational constants) have a higher rotational entropy than small molecules. Substitution of the numerical values gives $S_m = 3.98R$, or 33.1 J K^{-1} mol^{-1}.

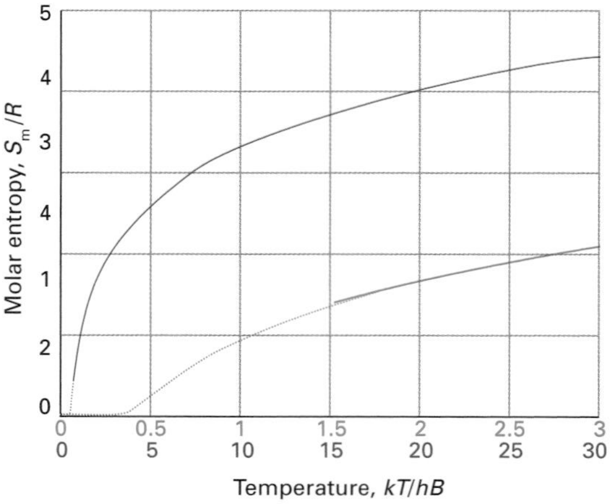

Fig. 22.8 The variation of the rotational contribution to the molar entropy with temperature. Note that eqn 22.6 is valid only for high temperatures, so the formula derived in Example 22.3 cannot be used at low temperatures (so we have terminated the curves before they become invalid). The dotted lines show the correct behaviour.

Self-test 22.4

The rotational partition function of an ethene molecule is 661 at 25 °C. What is the contribution to its molar rotational entropy?

Answer: 7.49R

22.9 The Gibbs energy

The Gibbs energy, G, was central to most of the thermodynamic discussions in the early chapters of this book, so to show that statistical thermodynamics is really useful we have to see how to calculate G from the partition function, q. We shall confine our attention to a perfect gas, because it is difficult to

[3] For a derivation, see our *Physical Chemistry* (2010).

take molecular interactions into account, and in the following Derivation we show that for a gas of N molecules

$$G - G(0) = -NkT \ln \frac{q}{N}$$

Perfect gas — The Gibbs energy in terms of the partition function (22.14)

Derivation 22.3

Calculating the Gibbs energy from the partition function

To set up the calculation, we go back to first principles. The Gibbs energy is defined as $G = H - TS$, and the enthalpy, H, is defined as $H = U + pV$. Therefore

$$G = U - TS + pV$$

For a perfect gas we can replace pV by $nRT = NkT$ (because $N = nN_A$ and $R = N_A k$), and note that at $T = 0$, $G(0) = U(0)$ (because the terms TS and NkT vanish at $T = 0$). Therefore,

$$G - G(0) = U - U(0) - TS + NkT$$

Now we substitute eqn 22.13b for S, and obtain

$$\begin{aligned} G - G(0) &= -NkT \ln q + kT(N \ln N - N) + NkT \\ &= -NkT(\ln q - \ln N) \end{aligned}$$

Then, because $\ln q - \ln N = \ln(q/N)$, we obtain eqn 22.14.

We can convert eqn 22.14 into an expression for the molar Gibbs energy. First, we write $N = nN_A$, and it becomes

$$G - G(0) = -nN_A kT \ln \frac{q}{nN_A}$$

Then we introduce the **molar partition function**, $q_m = q/n$, with units 1/mole (mol^{-1}). On dividing both sides of the preceding equation by n, we get

$$G_m - G_m(0) = -RT \ln \frac{q_m}{N_A}$$

Perfect gas — The molar Gibbs energy (22.15)

Example 22.4

Calculating the Gibbs energy

Calculate the molar Gibbs energy of a monatomic perfect gas and express it in terms of the pressure of the gas.

Strategy The calculation is based on eqn 22.15. All we need to know is the translational partition function, which is given in eqn 22.5. Convert from V to p by using the perfect gas law.

Solution When we substitute $q_m = (2\pi mkT)^{3/2}V/nh^3$ into eqn 22.15 we get

$$G_m - G_m(0) = -RT \ln \frac{(2\pi mkT)^{3/2}V}{nh^3 N_A}$$

Next, we replace V by nRT/p (notice that the ns cancel), and obtain (after a little tidying up, including writing $R = kN_A$)

$$G_m - G_m(0) = -RT \ln \left\{ \frac{(2\pi mkT)^{3/2}V}{nh^3 N_A} \times \overbrace{\frac{nN_A kT}{p}}^{V=nRT/p,\ R=N_A k} \right\} \qquad (22.16)$$

$$\overset{\text{Cancel } n, N_A}{=} -RT \ln \frac{(2\pi m)^{3/2}(kT)^{5/2}}{ph^3} = -RT \ln \frac{1}{ap}$$

$$\overset{-\ln x = \ln(1/x)}{=} RT \ln(ap) \quad \text{with} \quad a = \frac{h^3}{(2\pi m)^{3/2}(kT)^{5/2}}$$

The Gibbs energy increases logarithmically (as $\ln p$) as p increases, just as we saw in Section 5.2 (eqn 5.3).

Self-test 22.5

Ignore vibration and write the molar partition function of a diatomic molecule as $q_m^T q^R$ (see eqn 22.9). What is the molar Gibbs energy of such a gas?

Answer: As in eqn 22.16, but the natural logarithm term is $\ln(a\sigma hB/kT)p$

The only further piece of information we require is the expression for the *standard* molar Gibbs energy, for that played such an important role in the discussion of equilibrium properties. All we need to do is to use the partition function calculated at $p^\ominus$. For instance, for a monatomic gas, we use $p = 1$ bar in eqn 22.16 and obtain the standard value of the molar Gibbs energy. In general, we write

$$G_m^\ominus - G_m^\ominus(0) = -RT \ln \frac{q_m^\ominus}{N_A}$$

The standard molar Gibbs energy in terms of the standard molar partition function (22.17)

where the standard state sign on q simply reminds us to calculate its value at $p^\ominus$; to do so, we use $V_m^\ominus = RT/p^\ominus$ wherever it appears in $q_m^\ominus$. We shall see an example of that in the following section.

22.10 The equilibrium constant

We can go beyond the qualitative picture developed above by writing a statistical thermodynamic expression for the equilibrium constant. We show in Further information 22.2 that, for the equilibrium $A(g) + B(g) \rightleftharpoons C(g)$,

$$K = \frac{q_m^{\ominus}(C)N_A}{q_m^{\ominus}(A)q_m^{\ominus}(B)}e^{-\Delta E/RT}$$

The equilibrium constant in terms of the partition function (22.18)

where ΔE is the difference in molar energies between the ground state of the product and that of the reactants. This expression is easy to remember: it has the same form as the equilibrium constant written in terms of the activities (Section 7.3), but with $q_m^{\ominus}/N_A$ replacing each activity (and an additional exponential factor):

$$K = \frac{\overbrace{p_C/p^{\ominus}}^{\text{Replace with } q_m^{\ominus}(C)/N_A}}{\underbrace{(p_A/p^{\ominus})}_{\text{Replace with } q_m^{\ominus}(A)/N_A}\underbrace{(p_A/p^{\ominus})}_{\text{Replace with } q_m^{\ominus}(B)/N_A}} = \frac{q_m^{\ominus}(C)N_A}{(q_m^{\ominus}(A)/N_A)(q_m^{\ominus}(B)/N_A)}\underbrace{e^{-\Delta E/RT}}_{\text{Additional exponential factor}}$$

When we cancel the N_A, we obtain eqn 22.18.

Equation 22.18 is quite extraordinary, for it provides a key link between partition functions, which can be derived from spectroscopy, and the equilibrium constant, which is central to the analysis of chemical reactions at equilibrium. It represents the merging of the two rivers that have flowed through this text.

Example 22.5

Calculating an equilibrium constant

Calculate the equilibrium constant for the gas-phase ionization $Cs(g) \rightleftharpoons Cs^+(g) + e^-(g)$ at 500 K.

Strategy This is a reaction of the form $A(g) \rightleftharpoons B(g) + C(g)$ rather than $A(g) + B(g) \rightleftharpoons C(g)$, so we need to modify eqn 22.18 slightly, but the form to use should be clear. Analyse each species individually, and write its partition function as the product of partition functions for each mode of motion. Evaluate these partition functions at the standard pressure (1 bar), and combine them as specified in eqn 22.18. For the difference in energy ΔE, use the ionization energy of Cs(g).

Solution The equilibrium constant is

$$K = \frac{q_m^{\ominus}(Cs^+,g)q_m^{\ominus}(e^-,g)}{q_m^{\ominus}(Cs,g)N_A}e^{-\Delta E/RT}$$

Note how, in this instance, Avogadro's constant appears in the denominator: its units, mol^{-1}, ensure that K is dimensionless. The electron has translational motion, so we need its translational partition function. The spin states contribute a factor of 2 to the molecular partition function. Therefore

$$q_m^{\ominus}(e^-) = \overbrace{2}^{q^E} \times \overbrace{\frac{(2\pi m_e kT)^{3/2}V^{\ominus}}{nh^3}}^{q_m^T} \overset{V^{\ominus}=nRT/p^{\ominus}}{=} \frac{2(2\pi m_e kT)^{3/2}RT}{p^{\ominus}h^3}$$

The Cs^+ ion, a closed-shell species, has only translational freedom:

$$q_m^{\ominus}(Cs^+,g) = \frac{(2\pi m_{Cs}kT)^{3/2}RT}{p^{\ominus}h^3}$$

The partition function of the Cs atom has a translational and a spin contribution, as we saw in Section 22.5:

$$q_m^{\ominus}(Cs,g) = \overbrace{2}^{q^E} \times \overbrace{\frac{(2\pi m_{Cs}kT)^{3/2}RT}{p^{\ominus}h^3}}^{q_m^T}$$

(We are not distinguishing the masses of the Cs atom and the Cs^+ ion.) Then, with $\Delta E = I$, the ionization energy of the atom, we find

$$K = \frac{\overbrace{\{(2\pi m_{Cs}kT)^{3/2}RT/p^{\ominus}h^3\}}^{Cs^+}\times\overbrace{\{2(2\pi m_e kT)^{3/2}RT/p^{\ominus}h^3\}}^{e^-}}{\underbrace{\{2(2\pi m_{Cs}kT)^{3/2}RT/p^{\ominus}h^3\}}_{Cs}\times N_A}\times e^{-I/RT}$$

$$\overset{\text{Cancel terms}}{=} \frac{(2\pi m_e kT)^{3/2}RT}{p^{\ominus}h^3 N_A}\times e^{-I/RT}$$

$$\overset{R=N_A k}{=} \frac{(2\pi m_e kT)^{3/2}kT}{p^{\ominus}h^3}\times e^{-I/RT}$$

$$= \frac{(2\pi m_e)^{3/2}(kT)^{5/2}}{p^{\ominus}h^3}\times e^{-I/RT}$$

When we substitute the data (only the ionization energy is specific to the element), we find:

$$K = \frac{\left(2\pi\times\overbrace{9.109\times10^{-31}\text{ kg}}^{m_e}\right)^{3/2}\times\left(\overbrace{1.381\times10^{-23}\text{ J K}^{-1}\times1000\text{ K}}^{kT}\right)^{5/2}}{\left(\underbrace{10^5\text{ Pa}}_{p^{\ominus}}\right)\times(6.626\times10^{-34}\text{ J s})^3}$$

$$\times\, e^{-(3.76\times10^5\text{ J mol}^{-1})/(8.314\text{ J K}^{-1}\text{ mol}^{-1})\times(1000\text{ K})}$$

$$= 2.42\times10^{-19}$$

A note on good practice Verify that the units do in fact all cancel (use 1 J = 1 kg m^2 s^{-2} and 1 Pa = 1 kg m^{-1} s^{-2}). The K calculated by the procedure described here is the thermodynamic equilibrium constant, which for gases is expressed in terms of the partial pressures of the reactants and products (relative to the standard pressure).

Self-test 22.6

Calculate the equilibrium constant for the dissociation $Na_2(g) \rightleftharpoons 2\,Na(g)$ at 1000 K. You will need the following information about $Na_2(g)$: $B = 46.38$ MHz, $\tilde{\nu} = 159.2$ cm^{-1}, the dissociation energy is 70.4 kJ mol^{-1}, and $q_E = 2$.

Answer: 2.42

Further information 22.1

The calculation of partition functions

1. The translational partition function

We consider a particle of mass m in a rectangular box of sides X, Y, Z. Each direction can be treated independently and then the total partition function obtained by multiplying together the partition functions for each direction. The same strategy was used to write an expression for the molecular partition function by multiplying the contributions from (independent) modes of molecular motion.

The energy levels of a molecule of mass m in a container of length X are given by eqn 12.9 with $L = X$:

$$E_n = \frac{n^2h^2}{8mX^2} \qquad n = 1, 2, \ldots$$

The lowest level ($n = 1$) has energy $h^2/8mX^2$, so the energies relative to that level are

$$\varepsilon_n = (n^2 - 1)\varepsilon \qquad \varepsilon = h^2/8mX^2$$

The sum to evaluate is therefore

$$q_X = \sum_{n=1}^{\infty} \mathrm{e}^{-(n^2-1)\varepsilon/kT}$$

The translational energy levels are very close together in a container the size of a typical laboratory vessel; therefore, the sum can be approximated by an integral:

$$q_X = \int_1^{\infty} \mathrm{e}^{-(n^2-1)\varepsilon/kT}\mathrm{d}n \approx \int_0^{\infty} \mathrm{e}^{-n^2\varepsilon/kT}\mathrm{d}n$$

The extension of the lower limit to $n = 0$ and the replacement of $n^2 - 1$ by n^2 introduces negligible error but turns the integral into standard form. We make the substitution $x^2 = n^2\varepsilon/kT$, implying $\mathrm{d}n = \mathrm{d}x/(\varepsilon/kT)^{1/2}$, and therefore that

$$q_X = \left(\frac{kT}{\varepsilon}\right)^{1/2} \overbrace{\int_0^{\infty} \mathrm{e}^{-x^2}\mathrm{d}x}^{\pi^{1/2}/2}$$

$$= \left(\frac{kT}{\varepsilon}\right)^{1/2}\left(\frac{\pi^{1/2}}{2}\right) \overset{\varepsilon=h^2/8mX^2}{\widehat{=}} \left(\frac{2\pi mkT}{h^2}\right)^{1/2} X$$

The same expression applies to the other dimensions of a rectangular box of sides Y and Z, so

$$q^{\mathrm{T}} = q_X q_Y q_Z = \left(\frac{2\pi mkT}{h^2}\right)^{3/2} \overbrace{XYZ}^{V} = \left(\frac{2\pi mkT}{h^2}\right)^{3/2} V$$

where $V = XYZ$ is the volume of the box.

2. The rotational partition function

The rotational partition function of an unsymmetrical (AB) linear rigid rotor is

$$q^{\mathrm{R}} = \sum_J \overbrace{(2J+1)}^{g_J}\mathrm{e}^{-\overbrace{hBJ(J+1)}^{\varepsilon_J}/kT}$$

where the sum is over the rotational energy levels and the factor $2J + 1$ takes into account the degeneracy of the levels. When many rotational states are occupied and kT is much larger than the separation between neighbouring states, we can approximate the sum by an integral:

$$q^{\mathrm{R}} = \int_0^{\infty} (2J+1)\mathrm{e}^{-hBJ(J+1)/kT}\mathrm{d}J$$

Although this integral looks complicated, it can be evaluated without much effort by noticing that because

$$\frac{\mathrm{d}}{\mathrm{d}J}\mathrm{e}^{-hBJ(J+1)/kT} \overset{\mathrm{d}e^{af(x)}/\mathrm{d}x=ae^{af(x)}(\mathrm{d}f/\mathrm{d}x)}{\widehat{=}} -\frac{hB}{kT}(2J+1)$$

it can also be written as

$$q^{\mathrm{R}} = -\frac{kT}{hB}\int_0^{\infty}\left(\frac{\mathrm{d}}{\mathrm{d}J}\mathrm{e}^{-hBJ(J+1)/kT}\right)\mathrm{d}J$$

Then, because the integral of a derivative of a function (in blue) is the function itself,

$$q^{\mathrm{R}} = -\frac{kT}{hB}\mathrm{e}^{-hBJ(J+1)/kT}\Big|_0^{\infty} = \frac{kT}{hB}$$

For a homonuclear diatomic molecule, which looks the same after rotation by 180°, we have to divide this result by 2 to avoid double-counting of states, so in general

$$q^{\mathrm{R}} = \frac{kT}{\sigma hB}$$

where $\sigma = 1$ for heteronuclear diatomic molecules and 2 for homonuclear diatomic molecules.

3. The vibrational partition function

The energy levels of a harmonic oscillator form a simple ladder-like array (Fig. 22.9). If we set the energy of the lowest state equal to zero, the energies of the states are

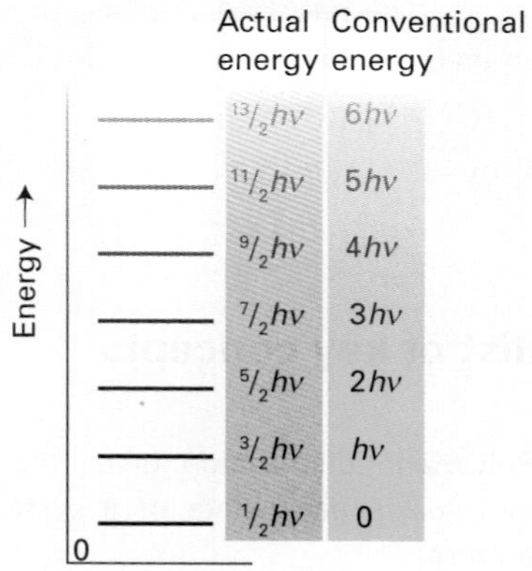

Fig. 22.9 The energy levels of a harmonic oscillator. When calculating a partition function, set the zero of energy at the lowest level, as shown on the right.

$\varepsilon_0 = 0,\ \varepsilon_1 = h\nu,\ \varepsilon_2 = 2h\nu,\ \varepsilon_3 = 3h\nu$, etc.

Therefore, the vibrational partition function is

$$q^V = 1 + e^{-h\nu/kT} + e^{-2h\nu/kT} + e^{-3h\nu/kT} + \cdots$$

$$\overset{\text{Use } e^{nx}=(e^x)^n}{=} 1 + e^{-h\nu/kT} + (e^{-h\nu/kT})^2 + (e^{-h\nu/kT})^3 + \cdots$$

According to The chemist's toolkit 6.1, the sum of the infinite series $1 + x + x^2 + \cdots$ is $1/(1-x)$, so with $x = e^{-h\nu/kT}$,

$$q^V = \frac{1}{1-e^{-h\nu/kT}}$$

which is eqn 22.7. When the temperature is so high that $h\nu/kT \ll 1$, this expression can be simplified considerably by writing $e^{-x} \approx 1 - x$ (The chemist's toolkit 6.1):

$$q^V \approx \frac{1}{1-\underbrace{(1-h\nu/kT)}_{e^{-h\nu/kT}}} = \frac{1}{h\nu/kT} = \frac{kT}{h\nu}$$

which is eqn 22.8.

Further information 22.2

The equilibrium constant from the partition function

We know from thermodynamics (Section 7.3) that the equilibrium constant for a reaction is related to the standard reaction Gibbs energy by

$$\Delta_r G^{\ominus} = -RT \ln K$$

For the reaction $A(g) + B(g) \rightleftharpoons C(g)$,

$$\Delta_r G^{\ominus} = G_m^{\ominus}(C) - \{G_m^{\ominus}(A) + G_m^{\ominus}(B)\}$$

Equation 22.17 is an expression for each of these standard molar Gibbs energies in terms of the partition function of each species, so we can write

$$\Delta_r G^{\ominus} = \left\{G_m^{\ominus}(C,0) - RT\ln\frac{q_m^{\ominus}(C)}{N_A}\right\} - \left\{G_m^{\ominus}(A,0) - RT\ln\frac{q_m^{\ominus}(A)}{N_A}\right\} - \left\{G_m^{\ominus}(B,0) - RT\ln\frac{q_m^{\ominus}(B)}{N_A}\right\}$$

The first term in each of the braces (in blue) is just the difference in ground-state energies because $G = U$ at $T = 0$, so

$$G_m^{\ominus}(C,0) - \{G_m^{\ominus}(A,0) + G_m^{\ominus}(A,0)\}$$
$$= U_m^{\ominus}(C,0) - \{U_m^{\ominus}(A,0) + U_m^{\ominus}(A,0)\} = \Delta E$$

The three logarithms can be combined to obtain

$$\ln\frac{q_m^{\ominus}(C)}{N_A} - \overbrace{\left\{\ln\frac{q_m^{\ominus}(A)}{N_A} + \ln\frac{q_m^{\ominus}(B)}{N_A}\right\}}^{\ln\{q_m^{\ominus}(A)q_m^{\ominus}(B)/N_A^2\}} \overset{\ln x - \ln yz = \ln(x/yz)}{=} \ln\frac{q_m^{\ominus}(C)N_A}{q_m^{\ominus}(A)q_m^{\ominus}(B)}$$

At this stage we have reached

$$\Delta_r G^{\ominus} = \Delta E - RT\ln\frac{q_m^{\ominus}(C)N_A}{q_m^{\ominus}(A)q_m^{\ominus}(B)}$$

The ΔE can be brought inside the logarithm by writing

$$\Delta E = -RT\ln e^{-\Delta E/RT}$$

(because $\ln e^x = x$). Therefore

$$\Delta_r G^{\ominus} = -RT\ln e^{-\Delta E/RT} - RT\ln\frac{q_m^{\ominus}(C)N_A}{q_m^{\ominus}(A)q_m^{\ominus}(B)}$$
$$\overset{\ln x + \ln y = \ln(xy)}{=} -RT\ln\left\{\frac{q_m^{\ominus}(C)N_A}{q_m^{\ominus}(A)q_m^{\ominus}(B)}e^{-\Delta E/RT}\right\}$$

All we have to do now is to compare this expression with the thermodynamic expression, $\Delta_r G^{\ominus} = -RT\ln K$, and see that the term in parentheses is the expression for K (eqn 22.18).

Checklist of key concepts

☐ 1 The Boltzmann distribution gives the numbers of molecules in each state of a system at any temperature.

☐ 2 The partition function is an indication of the number of thermally accessible states at the temperature of interest.

☐ 3 The partition function increases with temperature.

☐ 4 Molecules with numerous, closely spaced energy levels can be expected to have very large partition functions. Molecules with widely spaced energy levels can be expected to have small partition functions.

☐ **5** The molecular partition function is the product of contributions from translation, rotation, vibration, and electronic distributions.

☐ **6** The electronic partition function is $q^E = 1$ for closed-shell molecules with high-energy excited states.

☐ **7** The partition function contains all the thermodynamic information about a system, and can be used to calculate thermodynamics properties, such as the internal energy, the entropy, and the Gibbs energy.

☐ **8** The heat capacity is low when energy levels are spaced widely.

☐ **9** The equilibrium constant the distribution of molecules over the available states of a system composed of reactants and products.

Road map of key equations

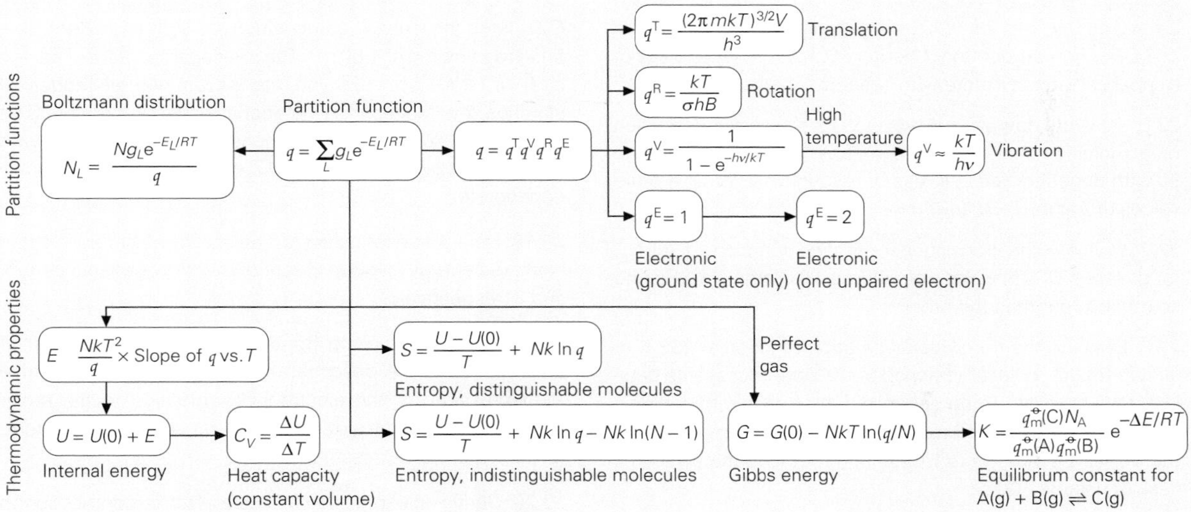

Questions and exercises

Discussion questions

22.1 Outline the principles behind the derivation of the Boltzmann distribution.

22.2 What is temperature?

22.3 Describe the physical significance of the molecular partition function.

22.4 When are particles of the same composition identical and when are they not?

22.5 Explain how the internal energy and entropy of a system composed of two levels vary with temperature.

22.6 Justify the identification of the statistical entropy with the thermodynamic entropy.

22.7 Use concepts of statistical thermodynamics to describe the molecular features that determine the magnitudes of equilibrium constants and their variation with temperature.

Exercises

22.1 Suppose polyethene molecules in solution can exist either as a single version of a random coil (that is, ignore the fact that a random coil can be achieved in many different ways) or fully stretched out, with the latter conformation 2.4 kJ mol^{-1} higher in energy. What is the ratio of the two conformations at 20 °C?

22.2 What is the ratio of populations of proton spin orientations in a magnetic field of (a) 1.5 T, (b) 15 T in a sample at 20 °C? For the energy difference, refer to Chapter 21.

22.3 What is the ratio of populations of electron spin orientations in a magnetic field of 0.33 T in a sample at 20 °C? *Hint:* For the energy difference, refer to Chapter 21.

22.4 Calculate the ratio of populations of CO_2 molecules with $J=4$ and $J=2$ at 25 °C. The rotational constant of CO_2 is 11.70 GHz. Molecular rotations are discussed in Chapter 19.

22.5 Calculate the ratio of populations of CH_4 molecules with $J=4$ and $J=2$ at 25 °C. The rotational constant of CH_4 is 157 GHz. Remember from Chapter 19 that the degeneracy of a spherical rotor in a state with quantum number J is $(2J+1)^2$.

22.6 (a) Write down the expression for the partition function of a molecule that has three energy levels at 0, 2ε, and 5ε with degeneracies 1, 6, and 3, respectively. What are the values of q at (b) $T=0$, (c) $T=\infty$?

22.7 Evaluate the translational partition function of (a) N_2, (b) gaseous CS_2 in a flask of volume 10.0 cm^3. Why is one so much larger than the other?

22.8 Evaluate the translational partition function at 298 K of (a) a methane molecule trapped in the pore of a zeolite catalyst: take the pore to be spherical with a radius that allows the molecule to move through 1 nm in any direction (that is, the *effective* diameter is 1 nm), (b) a methane molecule in a flask of volume 100 cm^3.

22.9 Evaluate the rotational partition function of HBr ($\tilde{\nu}$ = 8.465 cm^{-1}) at 298 K (a) by direct summation of the energy levels, (b) by using the high-temperature approximation, eqn 22.6.

22.10 Repeat the previous exercise at different temperatures (use mathematical software) and determine the temperature at which the approximate formula is 10 per cent in error.

22.11 Evaluate the rotational partition function at 298 K of (a) $^{1}H^{35}Cl$, for which the rotational constant is 318 GHz, (b) $^{12}C^{16}O_2$, for which the rotational constant is 11.70 GHz.

22.12 N_2O and CO_2 have similar rotational constants (12.6 and 11.7 GHz, respectively) but strikingly different rotational partition functions. Why?

22.13 Evaluate the vibrational partition function for HBr at 298 K. For data, see Table 19.2. Above what temperature is the high-temperature approximation (eqn 22.8) in error by 10 per cent or less?

22.14 A CO_2 molecule has four vibrational modes with wavenumbers 1388 cm^{-1}, 2349 cm^{-1}, and 667 cm^{-1} (the last being a doubly degenerate bending motion). Calculate the total vibrational partition function at (a) 500 K, (b) 1000 K.

22.15 The ground configuration of carbon gives rise to a triplet with the three levels $^3P_0$, $^3P_1$, and $^3P_2$ at wavenumbers 0, 16.4, and 43.5 cm^{-1}, respectively. Evaluate the partition function of carbon at (a) 10 K, (b) 298 K. Remember that a level with quantum number J has $2J+1$ states.

22.16 The ground configuration of oxygen gives rise to the three levels $^3P_2$, $^3P_1$, and $^3P_0$ at wavenumbers 0, 158.5, and 226.5 cm^{-1}, respectively. (a) Before doing any calculation, state the value of the partition function at $T=0$. (b) Evaluate the partition function at 298 K and confirm that its value at $T=0$ is what you anticipated in (a).

22.17 Calculate the molecular partition function for ethyne, C_2H_2 (with the isotopic composition ^{12}C and ^{1}H) at 298 K confined to a volume of 1.00 m^3. The rotational constant of ethyne is $\tilde{B}$ = 1.177 cm^{-1}. Ethyne has seven normal modes of vibration; the modes with vibrational wavenumbers of 3374, 1974, and 3287 are singly degenerate and those with vibrational wavenumbers of 612 and 729 cm^{-1} are doubly degenerate.

22.18 Derive an expression for the energy of a molecule that has three energy levels at 0, ε, and 3ε with degeneracies 1, 5, and 3, respectively.

22.19 The states arising from the ground configuration of a carbon atom are described in Exercise 22.15. (a) Derive an expression for the electronic contribution to the molar internal energy and plot it as a function of temperature. (b) Evaluate the expression at 25 °C.

22.20 Derive an expression for the electronic contribution to the molar heat capacity of an oxygen atom and plot it as a function of temperature. (b) Evaluate the expression at 25 °C. The structure of the atom is described in Exercise 22.16.

22.21 Why (in both thermodynamic and molecular terms) should substances with high heat capacities have high entropies?

22.22 Calculate the molar entropy of nitrogen gas at 298 K. Write the overall partition function as the product of the translational and rotational partition functions; the first excited vibrational state is sufficiently high in energy that the contribution from vibration of the molecule may be ignored at this temperature. For data, see Table 19.2.

22.23 Without carrying out an explicit calculation, explain the relative values of the standard molar entropies (at 298 K) of the following substances: (a) Ne(g) (146 J K^{-1} mol^{-1}) compared with Xe(g) (170 J K^{-1} mol^{-1}); (b) H_2O(g) (189 J K^{-1} mol^{-1}) compared with D_2O(g) (198 J K^{-1} mol^{-1}); (c) C(diamond) (2.4 J K^{-1} mol^{-1}) compared with C(graphite) (5.7 J K^{-1} mol^{-1}).

22.24 Estimate the change in molar entropy when a micelle consisting of 100 molecules disperses. Treat the transition as the expansion of a gas-like substance that initially occupies a volume V_{micelle} and spreads into a volume V_{solution}. What does this model neglect?

22.25 Calculate the standard molar Gibbs energy of carbon dioxide gas at 298 K relative to its value at $T = 0$.

22.26 Write down the expression for the equilibrium constant of the reaction $N_2(g) + 3\,H_2(g) \rightleftharpoons 2\,NH_3(g)$ in terms of the molecular partition functions of the reactants and products.

22.27 Calculate the equilibrium constant for the ionization equilibrium of sodium atoms at 1000 K.

22.28 Calculate the equilibrium constant for the dissociation of $I_2(g)$ at 500 K.

Projects

The symbol ‡ indicates that calculus is required.

22.29‡ Here we use statistical thermodynamics to calculate the internal energy and heat capacity of a system (such as the surface of an atomic solid) modelled as a collection of harmonic oscillators. (a) Derive an expression for the internal energy of a collection of harmonic oscillators. Deduce from your expression the high-temperature approximation and identify the temperature above which it is reliable. *Hint:* Substitute eqn 22.7 for the partition function into eqn 22.11 for the energy. (b) Now find an expression for the heat capacity of the oscillators and its high-temperature limit.

22.30 The first exercise in this chapter invited you to neglect the fact that a random coil can be achieved in many different ways. Repeat that exercise, allowing for this feature. Explore how the ratio of populations varies with the number of units, N, in the polymer.

Resource section

1 Quantities and units

The result of a measurement is a **physical quantity** (such as mass or density) that is reported as a numerical multiple of an agreed **unit:**

physical quantity = numerical value × unit

For example, the mass of an object may be reported as $m = 2.5$ kg and its density as $d = 1.01$ kg dm^{-3} where the units are, respectively, 1 kilogram (1 kg) and 1 kilogram per decimetre cubed (1 kg dm^{-3}). Units are treated like algebraic quantities, and may be multiplied, divided, and cancelled. Thus, the expression (physical quantity)/unit is simply the numerical value of the measurement in the specified units, and hence is a dimensionless quantity. For instance, the mass reported above could be denoted m/kg = 2.5 and the mass density as d/(kg dm^{-3}) = 1.01.

Physical quantities are denoted by italic or Greek letters (as in m for mass and Π for osmotic pressure). Units are denoted by Roman letters (as in m for metre). In the **International System** of units (SI, from the French *Système International d'Unités*), the units are formed from seven **base units** listed in Table A1.1. All other physical quantities may be expressed as combinations of these physical quantities and reported in terms of **derived units**. Thus, volume is (length)^3 and may be reported as a multiple of 1 metre cubed (1 m^3), and density, which is mass/volume, may be reported as a multiple of 1 kilogram per metre cubed (1 kg m^{-3}).

A number of derived units have special names and symbols. The names of units derived from names of people are lower case (as in torr, joule, pascal, and kelvin), but their symbols are upper case (as in Torr, J, Pa, and K). The most important units of this kind for our purposes are listed in Table A1.2. In all cases (both for base and derived quantities), the units may be modified by a prefix that denotes a factor of a power of 10. In a perfect world, Greek prefixes of units are upright (as in μm) and sloping for physical properties (as in μ for chemical potential), but available typefaces are not always so obliging. Among the most common prefixes are those listed in Table A1.3. Examples of the use of these prefixes are

$$1\ \text{nm} = 10^{-9}\ \text{m} \quad 1\ \text{ps} = 10^{-12}\ \text{s} \quad 1\ \mu\text{mol} = 10^{-6}\ \text{mol}$$

The kilogram (kg) is anomalous: although it is a base unit, it is interpreted as 10^3 g, and prefixes are attached to the gram (as in 1 mg = 10^{-3} g). Powers of units apply to the prefix as well as the unit they modify:

$$1\ \text{cm}^3 = 1\ (\text{cm})^3 = 1\ (10^{-2}\ \text{m})^3 = 10^{-6}\ \text{m}^3$$

Table A1.1
The SI base units

Physical quantity	Symbol for quantity	Base unit
Length	l	metre, m
Mass	m	kilogram, kg
Time	t	second, s
Electric current	I	ampere, A
Thermodynamic temperature	T	kelvin, K
Amount of substance	n	mole, mol
Luminous intensity	I_v	candela, cd

Table A1.2
A selection of derived units

Physical quantity	Derived unit*	Name of derived unit
Force	1 kg m s^{-2}	newton, N
Pressure	1 kg m^{-1} s^{-2}	pascal, Pa
	1 N m^{-2}	
Energy	1 kg m^2 s^{-2}	joule, J
	1 N m	
	1 Pa m^3	
Power	1 kg m^2 s^{-3}	watt, W
	1 J s^{-1}	

* Equivalent definitions in terms of derived units are given following the definition in terms of base units.

Table A1.3
Common SI prefixes

Prefix	z	a	f	p	n	μ	m	c	d
Name	zepto	atto	femto	pico	nano	micro	milli	centi	deci
Factor	10^{-21}	10^{-18}	10^{-15}	10^{-12}	10^{-9}	10^{-6}	10^{-3}	10^{-2}	10^{-1}
Prefix	k	M	G	T	P				
Name	kilo	mega	giga	tera	peta				
Factor	10^3	10^6	10^9	10^{12}	10^{15}				

Note that 1 cm^3 does not mean 1 $c(m^3)$. When carrying out numerical calculations, it is usually safest to write out the numerical value of an observable as powers of 10.

There are a number of units that are in wide use but are not a part of the International System. Some are exactly equal to multiples of SI units. These include the *litre* (L), which is exactly 10^3 cm^3 (or 1 dm^3) and the *atmosphere* (atm), which is exactly 101.325 kPa. Others rely on the values of fundamental constants, and hence are liable to change when the values of the fundamental constants are modified by more accurate or more precise measurements. Thus, the size of the energy unit *electronvolt* (eV), the energy acquired by an electron that is accelerated through a potential difference of exactly 1 V, depends on the value of the charge of the electron, and the present (2013) conversion factor is 1 eV = 1.602 177 × 10^{-19} J. Table A1.4 gives the conversion factors for a number of these convenient units.

Plans have been accepted by the overall authority responsible for maintaining units (the *Bureau of Weights and Measures*) for redefining certain quantities (particularly the kilogram and the mole), but have not yet (in 2013) been implemented. For the latest values of fundamental constants, refer to http://physics.nist.gov/constants.

Table A1.4
Some common units

Physical quantity	Name of unit	Symbol for unit	Value
Time	minute	min	60 s
	hour	h	3600 s
	day	d	86 400 s
Length	ångström	Å	10^{-10} m
Volume	litre	L, l	1 dm^3
Mass	tonne	t	10^3 kg
Pressure	bar	bar	10^5 Pa
	atmosphere	atm	101.325 kPa
Energy	electronvolt	eV	1.602 177 × 10^{-19} J
			96.485 31 kJ mol^{-1}
	calorie	cal	4.184 J

All values in the final column are exact, except for the definition of 1 eV.

2 Data section

Table 1 *Thermodynamic data for organic compounds at 298.15 K*

	M/ (g mol^{-1})	$\Delta_f H^\ominus$/ (kJ mol^{-1})	$\Delta_f G^\ominus$/ (kJ mol^{-1})	$S_m^\ominus$/ (J K^{-1} mol^{-1})†	$C_{p,m}^\ominus$/ (J K^{-1} mol^{-1})	$\Delta_c H^\ominus$/ (kJ mol^{-1})
C(s) (graphite)	12.011	0	0	5.740	8.527	−393.51
C(s) (diamond)	12.011	+1.895	+2.900	+2.377	6.113	−395.40
CO_2(g)	44.010	−393.51	−394.36	213.74	37.11	
Hydrocarbons						
CH_4(g), methane	16.04	−74.81	−50.72	186.26	35.31	−890
CH_3(g), methyl	15.04	+145.69	+147.92	194.2	38.70	
C_2H_2(g), ethyne	26.04	+226.73	+209.20	200.94	43.93	−1300
C_2H_4(g), ethene	28.05	+52.26	+68.15	219.56	43.56	−1411
C_2H_6(g), ethane	30.07	−84.68	−32.82	229.60	52.63	−1560
C_3H_6(g), propene	42.08	+20.42	+62.78	267.05	63.89	−2058
C_3H_6(g), cyclopropane	42.08	−103.85	−23.49	269.91	73.5	−2220
C_4H_8(g), 1-butene	56.11	−0.13	+71.39	305.71	85.65	−2717
C_4H_8(g), *cis*-2-butene	56.11	−6.99	+65.95	300.94	78.91	−2710
C_4H_8(g), *trans*-2-butene	56.11	−11.17	+63.06	296.59	87.82	−2707
C_4H_{10}(g), butane	58.13	−126.15	−17.03	310.23	97.45	−2878
C_5H_{12}(g), pentane	72.15	−146.44	−8.20	348.40	120.2	−3537
C_5H_{12}(l)	72.15	−173.1				
C_6H_6(l), benzene	78.12	+49.0	+124.3	173.3	136.1	−3268
C_6H_6(g)	78.12	+82.93	+129.72	269.31	81.67	−3320
C_6H_{12}(l), cyclohexane	84.16	−156	26.8		156.5	−3902
C_6H_{14}(l), hexane	86.18	−198.7		204.3		−4163
$C_6H_5CH_3$(g), methylbenzene (toluene)	92.14	+50.0	+122.0	320.7	103.6	−3953
C_7H_{16}(l), heptane	100.21	−224.4	+1.0	328.6	224.3	
C_8H_{18}(l), octane	114.23	−249.9	+6.4	361.1		−5471
C_8H_{18}(l), iso-octane	114.23	−255.1				−5461
$C_{10}H_8$(s), naphthalene	128.18	+78.53				−5157
Alcohols and phenols						
CH_3OH(l), methanol	32.04	−238.86	−166.27	126.8	81.6	−726
CH_3OH(g)	32.04	−200.66	−166.27	239.81	43.89	−764
C_2H_5OH(l), ethanol	46.07	−277.69	−174.78	160.7	111.46	−1368
C_2H_5OH(g)	46.07	−235.10	−168.49	282.70	65.44	−1409
C_6H_5OH(s), phenol	94.12	−165.0	−50.9	146.0		−3054
Carboxylic acids, hydroxy acids, and esters						
HCOOH(l), formic	46.03	−424.72	−361.35	128.95	99.04	−255
CH_3COOH(l), ethanoic	60.05	−484.3	−389.9	159.8	124.3	−875
CH_3COOH(aq)	60.05	−485.76	−396.46	178.7		
$CH_3CO_2^-$(aq)	59.05	−486.01	−369.31	86.6	−6.3	
CH_3(CO)COOH(l), pyruvic	88.06					−950
$CH_3(CH_2)_2$COOH(l), butanoic	88.10	−533.8				
$CH_3COOC_2H_5$(l), ethyl acetate	88.10	−479.0	−332.7	259.4	170.1	−2231
$(COOH)_2$(s), oxalic	90.04	−827.2			117	−254
CH_3CH(OH)COOH(s), lactic	90.08	−694.0	−522.9			−1344
$HOOCCH_2CH_2COOH$(s), succinic	118.09	−940.5	−747.4	153.1	167.3	
C_6H_5COOH(s), benzoic	122.13	−385.1	−245.3	167.6	146.8	−3227
$CH_3(CH_2)_8$COOH(s), decanoic	172.27	−713.7				
$C_6H_8O_6$(s), ascorbic	176.12	−1164.6				
$HOOCCH_2$C(OH)(COOH) CH_2COOH(s), citric	192.12	−1543.8	−1236.4			−1985
$CH_3(CH_2)_{10}$COOH(s), dodecanoic	200.32	−774.6			404.3	
$CH_3(CH_2)_{14}$COOH(s), hexadecanoic	256.41	−891.5				
$C_{18}H_{36}O_2$(s), stearic	284.48	−947.7			501.5	

Table 1 *(continued)*

	M/(g mol^{-1})	$\Delta_f H^\ominus$/(kJ mol^{-1})	$\Delta_f G^\ominus$/(kJ mol^{-1})	$S_m^\ominus$/(J K^{-1} mol^{-1})†	$C_{p,m}^\ominus$/(J K^{-1} mol^{-1})	$\Delta_c H^\ominus$/(kJ mol^{-1})
Alkanals and alkanones						
HCHO(g), methanal	30.03	−108.57	−102.53	218.77	35.40	−571
CH_3CHO(l), ethanal	44.05	−192.30	−128.12	160.2		−1166
CH_3CHO(g)	44.05	−166.19	−128.86	250.3	57.3	−1192
CH_3COCH_3(l), propanone	58.08	−248.1	−155.4	200.4	124.7	−1790
Sugars						
$C_5H_{10}O_5$(s), D-ribose	150.1	−1051.1				
$C_5H_{10}O_5$(s), D-xylose	150.1	−1057.8				
$C_6H_{12}O_6$(s), α-D-glucose	180.16	−1273.3	−917.2	212.1		−2808
$C_6H_{12}O_6$(s), β-D-glucose	180.16	−1268				
$C_6H_{12}O_6$(s), β-D-fructose	180.16	−1265.6				−2810
$C_6H_{12}O_6$(s), α-D-galactose	180.16	−1286.3	−918.8	205.4		
$C_{12}H_{22}O_{11}$(s), sucrose	342.30	−2226.1	−1543	360.2		−5645
$C_{12}H_{22}O_{11}$(s), lactose	342.30	−2236.7	−1567	386.2		
Nitrogen compounds						
$CO(NH_2)_2$(s), urea	60.06	−333.51	−197.33	104.60	93.14	−632
CH_3NH_2(g), methylamine	31.06	−22.97	+32.16	243.41	53.1	−1085
$C_6H_5NH_2$(l), aniline	93.13	+31.1				−3393
$CH_2(NH_2)COOH$(s), glycine	75.07	−532.9	−373.4	103.5	99.2	−969

Data: NBS, TDOC. †Standard entropies of ions may be either positive or negative because the values are relative to the entropy of the hydrogen ion.

Table 2 *Thermodynamic data for elements and inorganic compounds at 298.15 K*

	M/(g mol^{-1})	$\Delta_f H^\ominus$/(kJ mol^{-1})	$\Delta_f G^\ominus$/(kJ mol^{-1})	$S_m^\ominus$/(J K^{-1} mol^{-1})†	$C_{p,m}$/(J K^{-1} mol^{-1})
Aluminium (aluminum)					
Al(s)	26.98	0	0	28.33	24.35
Al(l)	26.98	+10.56	+7.20	39.55	24.21
Al(g)	26.98	+326.4	+285.7	164.54	21.38
Al^{3+}(g)	26.98	+5483.17			
Al^{3+}(aq)	26.98	−531	−485	−321.7	
Al_2O_3(s, α)	101.96	−1675.7	−1582.3	50.92	79.04
$AlCl_3$(s)	133.24	−704.2	−628.8	110.67	91.84
Argon					
Ar(g)	39.95	0	0	154.84	20.786
Antimony					
Sb(s)	121.75	0	0	45.69	25.23
SbH_3(g)	153.24	+145.11	+147.75	232.78	41.05
Arsenic					
As(s, α)	74.92	0	0	35.1	24.64
As(g)	74.92	+302.5	+261.0	174.21	20.79
As_4(g)	299.69	+143.9	+92.4	314	
AsH_3(g)	77.95	+66.44	+68.93	222.78	38.07
Barium					
Ba(s)	137.34	0	0	62.8	28.07
Ba(g)	137.34	+180	+146	170.24	20.79
Ba^{2+}(aq)	137.34	−537.64	−560.77	+9.6	
BaO(s)	153.34	−553.5	−525.1	70.43	47.78
$BaCl_2$(s)	208.25	−858.6	−810.4	123.68	75.14

Table 2 *(continued)*

	$M/(g\ mol^{-1})$	$\Delta_f H^{\ominus}/(kJ\ mol^{-1})$	$\Delta_f G^{\ominus}/(kJ\ mol^{-1})$	$S_m^{\ominus}/(J\ K^{-1}\ mol^{-1})^{\dagger}$	$C_{p,m}/(J\ K^{-1}\ mol^{-1})$
Beryllium					
Be(s)	9.01	0	0	9.50	16.44
Be(g)	9.01	+324.3	+286.6	136.27	20.79
Bismuth					
Bi(s)	208.98	0	0	56.74	25.52
Bi(g)	208.98	+207.1	+168.2	187.00	20.79
Bromine					
Br_2(l)	159.82	0	0	152.23	75.689
Br_2(g)	159.82	+30.907	+3.110	245.46	36.02
Br(g)	79.91	+111.88	+82.396	175.02	20.786
Br^-(g)	79.91	−219.07			
Br^-(aq)	79.91	−121.55	−103.96	+82.4	−141.8
HBr(g)	90.92	−36.40	−53.45	198.70	29.142
Cadmium					
Cd(s, γ)	112.40	0	0	51.76	25.98
Cd(g)	112.40	+112.01	+77.41	167.75	20.79
Cd^{2+}(aq)	112.40	−75.90	−77.612	−73.2	
CdO(s)	128.40	−258.2	−228.4	54.8	43.43
$CdCO_3$(s)	172.41	−750.6	−669.4	92.5	
Caesium (cesium)					
Cs(s)	132.91	0	0	85.23	32.17
Cs(g)	132.91	+76.06	+49.12	175.60	20.79
Cs^+(aq)	132.91	−258.28	−292.02	+133.05	−10.5
Calcium					
Ca(s)	40.08	0	0	41.42	25.31
Ca(g)	40.08	+178.2	+144.3	154.88	20.786
Ca^{2+}(aq)	40.08	−542.83	−553.58	−53.1	
CaO(s)	56.08	−635.09	−604.03	39.75	42.80
$CaCO_3$(s) (calcite)	100.09	−1206.9	−1128.8	92.9	81.88
$CaCO_3$(s) (aragonite)	100.09	−1207.1	−1127.8	88.7	81.25
CaF_2(s)	78.08	1219.6	−1167.3	68.87	67.03
$CaCl_2$(s)	110.99	−795.8	−748.1	104.6	72.59
$CaBr_2$(s)	199.90	−682.8	−663.6	130	
Carbon					
C(s) (graphite)	12.011	0	0	5.740	8.527
C(s) (diamond)	12.011	+1.895	+2.900	2.377	6.133
C(g)	12.011	+716.68	+671.26	158.10	20.838
C_2(g)	24.022	+831.90	+775.89	199.42	43.21
CO(g)	28.011	−110.53	−137.17	197.67	29.14
CO_2(g)	44.010	−393.51	−394.36	213.74	37.11
CO_2(aq)	44.010	−413.80	−385.98	117.6	
H_2CO_3(aq)	62.03	−699.65	−623.08	187.4	
HCO_3^-(aq)	61.02	−691.99	−586.77	+91.2	
CO_3^{2-}(aq)	60.01	−677.14	−527.81	−56.9	
CCl_4(l)	153.82	−135.44	−65.21	216.40	131.75
CS_2(l)	76.14	+89.70	+65.27	151.34	75.7
HCN(g)	27.03	+135.1	+124.7	201.78	35.86
HCN(l)	27.03	+108.87	+124.97	112.84	70.63
CN^-(aq)	26.02	+150.6	+172.4	+94.1	

Table 2 *(continued)*

	$M/(g\ mol^{-1})$	$\Delta_f H^\ominus/(kJ\ mol^{-1})$	$\Delta_f G^\ominus/(kJ\ mol^{-1})$	$S_m^\ominus/(J\ K^{-1}\ mol^{-1})^\dagger$	$C_{p,m}/(J\ K^{-1}\ mol^{-1})$
Chlorine					
$Cl_2(g)$	70.91	0	0	223.07	33.91
$Cl(g)$	35.45	+121.68	+105.68	165.20	21.840
$Cl^-(g)$	35.45	−233.13			
$Cl^-(aq)$	35.45	−167.16	−131.23	+56.5	−136.4
$HCl(g)$	36.46	−92.31	−95.30	186.91	29.12
$HCl(aq)$	36.46	−167.16	−131.23	56.5	−136.4
Chromium					
$Cr(s)$	52.00	0	0	23.77	23.35
$Cr(g)$	52.00	+396.6	+351.8	174.50	20.79
$CrO_4^{2-}(aq)$	115.99	−881.15	−727.75	+50.21	
$Cr_2O_7^{2-}(aq)$	215.99	−1490.3	−1301.1	+261.9	
Copper					
$Cu(s)$	63.54	0	0	33.150	24.44
$Cu(g)$	63.54	+338.32	+298.58	166.38	20.79
$Cu^+(aq)$	63.54	+71.67	+49.98	+40.6	
$Cu^{2+}(aq)$	63.54	+64.77	+65.49	−99.6	
$Cu_2O(s)$	143.08	−168.6	−146.0	93.14	63.64
$CuO(s)$	79.54	−157.3	−129.7	42.63	42.30
$CuSO_4(s)$	159.60	−771.36	−661.8	109	100.0
$CuSO_4{\cdot}H_2O(s)$	177.62	−1085.8	−918.11	146.0	134
$CuSO_4{\cdot}5H_2O(s)$	249.68	−2279.7	−1879.7	300.4	280
Deuterium					
$D_2(g)$	4.028	0	0	144.96	29.20
$HD(g)$	3.022	+0.318	−1.464	143.80	29.196
$D_2O(g)$	20.028	−249.20	−234.54	198.34	34.27
$D_2O(l)$	20.028	−294.60	−243.44	75.94	84.35
$HDO(g)$	19.022	−245.30	−233.11	199.51	33.81
$HDO(l)$	19.022	−289.89	−241.86	79.29	
Fluorine					
$F_2(g)$	38.00	0	0	202.78	31.30
$F(g)$	19.00	+78.99	+61.91	158.75	22.74
$F^-(aq)$	19.00	−332.63	−278.79	−13.8	−106.7
$HF(g)$	20.01	−271.1	−273.2	173.78	29.13
Gold					
$Au(s)$	196.97	0	0	47.40	25.42
$Au(g)$	196.97	+366.1	+326.3	180.50	20.79
Helium					
$He(g)$	4.003	0	0	126.15	20.786
Hydrogen (see also deuterium)					
$H_2(g)$	2.016	0	0	130.684	28.824
$H(g)$	1.008	+217.97	+203.25	114.71	20.784
$H^+(aq)$	1.008	0	0	0	0
$H_2O(l)$	18.015	−285.83	−237.13	69.91	75.291
$H_2O(g)$	18.015	−241.82	−228.57	188.83	33.58
$H_2O_2(l)$	34.015	−187.78	−120.35	109.6	89.1

Table 2 *(continued)*

	M/(g mol^{-1})	$\Delta_f H^\ominus$/(kJ mol^{-1})	$\Delta_f G^\ominus$/(kJ mol^{-1})	$S_m^\ominus$/(J K^{-1} mol^{-1})†	$C_{p,m}$/(J K^{-1} mol^{-1})
Iodine					
I_2(s)	253.81	0	0	116.135	54.44
I_2(g)	253.81	+62.44	+19.33	260.69	36.90
I(g)	126.90	+106.84	+70.25	180.79	20.786
I^-(aq)	126.90	−55.19	−51.57	+111.3	−142.3
HI(g)	127.91	+26.48	+1.70	206.59	29.158
Iron					
Fe(s)	55.85	0	0	27.28	25.10
Fe(g)	55.85	+416.3	+370.7	180.49	25.68
Fe^{2+}(aq)	55.85	−89.1	−78.90	−137.7	
Fe^{3+}(aq)	55.85	−48.5	−4.7	−315.9	
Fe_3O_4(s) (magnetite)	231.54	−1184.4	−1015.4	146.4	143.43
Fe_2O_3(s) (haematite)	159.69	−824.2	−742.2	87.40	103.85
FeS(s, α)	87.91	−100.0	−100.4	60.29	50.54
FeS_2(s)	119.98	−178.2	−166.9	52.93	62.17
Krypton					
Kr(g)	83.80	0	0	164.08	20.786
Lead					
Pb(s)	207.19	0	0	64.81	26.44
Pb(g)	207.19	+195.0	+161.9	175.37	20.79
Pb^{2+}(aq)	207.19	−1.7	−24.43	+10.5	
PbO(s, yellow)	223.19	−217.32	−187.89	68.70	45.77
PbO(s, red)	223.19	−218.99	−188.93	66.5	45.81
PbO_2(s)	239.19	−277.4	−217.33	68.6	64.64
Lithium					
Li(s)	6.94	0	0	29.12	24.77
Li(g)	6.94	+159.37	+126.66	138.77	20.79
Li^+(aq)	6.94	−278.49	−293.31	+13.4	+68.6
Magnesium					
Mg(s)	24.31	0	0	32.68	24.89
Mg(g)	24.31	+147.70	+113.10	148.65	20.786
Mg^{2+}(aq)	24.31	−466.85	−454.8	−138.1	
MgO(s)	40.31	−601.70	−569.43	26.94	37.15
$MgCO_3$(s)	84.32	−1095.8	−1012.1	65.7	75.52
$MgCl_2$(s)	95.22	−641.32	−591.79	89.62	71.38
$MgBr_2$(s)	184.13	−524.3	−503.8	117.2	
Mercury					
Hg(l)	200.59	0	0	76.02	27.983
Hg(g)	200.59	+61.32	+31.82	174.96	20.786
Hg^{2+}(aq)	200.59	+171.1	+164.40	−32.2	
Hg_2^{2+}(aq)	401.18	+172.4	+153.52	+84.5	
HgO(s)	216.59	−90.83	−58.54	70.29	44.06
Hg_2Cl_2(s)	472.09	−265.22	−210.75	192.5	102
$HgCl_2$(s)	271.50	−224.3	−178.6	146.0	
HgS(s, black)	232.65	−53.6	−47.7	88.3	
Neon					
Ne(g)	20.18	0	0	146.33	20.786

Table 2 *(continued)*

	M/(g mol^{-1})	$\Delta_f H^{\ominus}$/(kJ mol^{-1})	$\Delta_f G^{\ominus}$/(kJ mol^{-1})	$S_m^{\ominus}$/(J K^{-1} mol^{-1})†	$C_{p,m}$/(J K^{-1} mol^{-1})
Nitrogen					
N_2(g)	28.013	0	0	191.61	29.125
N(g)	14.007	+472.70	+455.56	153.30	20.786
NO(g)	30.01	+90.25	+86.55	210.76	29.844
N_2O(g)	44.01	+82.05	+104.20	219.85	38.45
NO_2(g)	46.01	+33.18	+51.31	240.06	37.20
N_2O_4(g)	92.01	+9.16	+97.89	304.29	77.28
N_2O_5(s)	108.01	−43.1	+113.9	178.2	143.1
N_2O_5(g)	108.01	+11.3	+115.1	355.7	84.5
HNO_3(l)	63.01	−174.10	−80.71	155.60	109.87
HNO_3(aq)	63.01	−207.36	−111.25	146.4	−86.6
NO_3^-(aq)	62.01	−205.0	−108.74	+146.4	−86.6
NH_3(g)	17.03	−46.11	−16.45	192.45	35.06
NH_3(aq)	17.03	−80.29	−26.50	113.3	
NH_4^+(aq)	18.04	−132.51	−79.31	+113.4	+79.9
NH_2OH(s)	33.03	−114.2			
HN_3(l)	43.03	+264.0	+327.3	140.6	
HN_3(g)	43.03	+294.1	+328.1	238.97	43.68
N_2H_4(l)	32.05	+50.63	+149.43	121.21	98.87
NH_4NO_3(s)	80.04	−365.56	−183.87	151.08	139.3
NH_4Cl(s)	53.49	−314.43	−202.87	94.6	84.1
Oxygen					
O_2(g)	31.999	0	0	205.138	29.355
O(g)	15.999	+249.17	+231.73	161.06	21.912
O_3(g)	47.998	+142.7	+163.2	238.93	39.20
OH^-(aq)	17.007	−229.99	−157.24	−10.75	−148.5
Phosphorus					
P(s, wh)	30.97	0	0	41.09	23.840
P(g)	30.97	+314.64	+278.25	163.19	20.786
P_2(g)	61.95	+144.3	+103.7	218.13	32.05
P_4(g)	123.90	+58.91	+24.44	279.98	67.15
PH_3(g)	34.00	+5.4	+13.4	210.23	37.11
PCl_3(g)	137.33	−287.0	−267.8	311.78	71.84
PCl_3(l)	137.33	−319.7	−272.3	217.1	
PCl_5(g)	208.24	−374.9	−305.0	364.6	112.8
PCl_5(s)	208.24	−443.5			
H_3PO_3(s)	82.00	−964.4			
H_3PO_3(aq)	82.00	−964.8			
H_3PO_4(s)	94.97	−1279.0	−1119.1	110.50	106.06
H_3PO_4(l)	94.97	−1266.9			
H_3PO_4(aq)	94.97	−1277.4	−1018.7	−222	
PO_4^{3-}(aq)	94.97	−1277.4	−1018.7	−222	
P_4O_{10}(s)	283.89	−2984.0	−2697.0	228.86	211.71
P_4O_6(s)	219.89	−1640.1			
Potassium					
K(s)	39.10	0	0	64.18	29.58
K(g)	39.10	+89.24	+60.59	160.336	20.786
K^+(g)	39.10	+514.26			
K^+(aq)	39.10	−252.38	−283.27	+102.5	+21.8
KOH(s)	56.11	−424.76	−379.08	78.9	64.9
KF(s)	58.10	−576.27	−537.75	66.57	49.04
KCl(s)	74.56	−436.75	−409.14	82.59	51.30
KBr(s)	119.01	−393.80	−380.66	95.90	52.30
KI(s)	166.01	−327.90	−324.89	106.32	52.93

Table 2 *(continued)*

	M/(g mol^{-1})	$\Delta_f H^\ominus$/(kJ mol^{-1})	$\Delta_f G^\ominus$/(kJ mol^{-1})	$S_m^\ominus$/(J K^{-1} mol^{-1})†	$C_{p,m}$/(J K^{-1} mol^{-1})
Silicon					
Si(s)	28.09	0	0	18.83	20.00
Si(g)	28.09	+455.6	+411.3	167.97	22.25
SiO_2(s, α)	60.09	−910.93	−856.64	41.84	44.43
Silver					
Ag(s)	107.87	0	0	42.55	25.351
Ag(g)	107.87	+284.55	+245.65	173.00	20.79
Ag^+(aq)	107.87	+105.58	+77.11	+72.68	+21.8
AgBr(s)	187.78	−100.37	−96.90	107.1	52.38
AgCl(s)	143.32	−127.07	−109.79	96.2	50.79
Ag_2O(s)	231.74	−31.05	−11.20	121.3	65.86
$AgNO_3$(s)	169.88	−124.39	−33.41	140.92	93.05
Sodium					
Na(s)	22.99	0	0	51.21	28.24
Na(g)	22.99	+107.32	+76.76	153.71	20.79
Na^+(aq)	22.99	−240.12	−261.91	+59.0	+46.4
NaOH(s)	40.00	−425.61	−379.49	64.46	59.54
NaCl(s)	58.44	−411.15	−384.14	72.13	50.50
NaBr(s)	102.90	−361.06	−348.98	86.82	51.38
NaI(s)	149.89	−287.78	−286.06	98.53	52.09
Sulfur					
S(s, α) (rhombic)	32.06	0	0	31.80	22.64
S(s, β) (monoclinic)	32.06	+0.33	+0.1	32.6	23.6
S(g)	32.06	+278.81	+238.25	167.82	23.673
S_2(g)	64.13	+128.37	+79.30	228.18	32.47
S^{2-}(aq)	32.06	+33.1	+85.8	−14.6	
SO_2(g)	64.06	−296.83	−300.19	248.22	39.87
SO_3(g)	80.06	−395.72	−371.06	256.76	50.67
H_2SO_4(l)	98.08	−813.99	−690.00	156.90	138.9
H_2SO_4(aq)	98.08	−909.27	−744.53	20.1	−293
SO_4^{2-}(aq)	96.06	−909.27	−744.53	+20.1	−293
HSO_4^-(aq)	97.07	−887.34	−755.91	+131.8	−84
H_2S(g)	34.08	−20.63	−33.56	205.79	34.23
H_2S(aq)	34.08	−39.7	−27.83	121	
HS^-(aq)	33.072	−17.6	+12.08	+62.08	
SF_6(g)	146.05	−1209	−1105.3	291.82	97.28
Tin					
Sn(s, β)	118.69	0	0	51.55	26.99
Sn(g)	118.69	+302.1	+267.3	168.49	20.26
Sn^{2+}(aq)	118.69	−8.8	−27.2	−17	
SnO(s)	134.69	−285.8	−256.8	56.5	44.31
SnO_2(s)	150.69	−580.7	+519.6	52.3	52.59
Xenon					
Xe(g)	131.30	0	0	169.68	20.786
Zinc					
Zn(s)	65.37	0	0	41.63	25.40
Zn(g)	65.37	+130.73	+95.14	160.98	20.79
Zn^{2+}(aq)	65.37	−153.89	−147.06	−112.1	+46
ZnO(s)	81.37	−348.28	−318.30	43.64	40.25

Data: NBS, TDOC. †Standard entropies of ions may be either positive or negative because the values are relative to the entropy of the hydrogen ion.

Table 3a *Standard potentials at 298.15 K in electrochemical order*

Reduction half-reaction	$E^{\ominus}/V$	Reduction half-reaction	$E^{\ominus}/V$
Strongly oxidizing			
$H_4XeO_6 + 2H^+ + 2e^- \rightarrow XeO_3 + 3H_2O$	+3.0	$Cu^{2+} + e^- \rightarrow Cu^+$	+0.16
$F_2 + 2e^- \rightarrow 2F^-$	+2.87	$Sn^{4+} + 2e^- \rightarrow Sn^{2+}$	+0.15
$O_3 + 2H^+ + 2e^- \rightarrow O_2 + H_2O$	+2.07	$AgBr + e^- \rightarrow Ag + Br^-$	+0.07
$S_2O_8^{2-} + 2e^- \rightarrow 2SO_4^{2-}$	+2.05	$Ti^{4+} + e^- \rightarrow Ti^{3+}$	0.00
$Ag^{2+} + e^- \rightarrow Ag^+$	+1.98	$2H^+ + 2e^- \rightarrow H$	0, by definition
$Co^{3+} + e^- \rightarrow Co^{2+}$	+1.81	$Fe^{3+} + 3e^- \rightarrow Fe$	−0.04
$HO_2 + 2H^+ + 2e^- \rightarrow 2H_2O$	+1.78	$O_2 + H_2O + 2e^- \rightarrow HO_2^- + OH^-$	−0.08
$Au^+ + e^- \rightarrow Au$	+1.69	$Pb^{2+} + 2e^- \rightarrow Pb$	−0.13
$Pb^{4+} + 2e^- \rightarrow Pb^{2+}$	+1.67	$In^+ + e^- \rightarrow In$	−0.14
$2HClO + 2H^+ + 2e^- \rightarrow Cl_2 + 2H_2O$	+1.63	$Sn^{2+} + 2e^- \rightarrow Sn$	−0.14
$Ce^{4+} + e^- \rightarrow Ce^{3+}$	+1.61	$AgI + e^- \rightarrow Ag + I$	−0.15
$2HBrO + 2H^+ + 2e^- \rightarrow Br_2 + 2H$	+1.60	$Ni^{2+} + 2e^- \rightarrow Ni$	−0.23
$MnO_4^- + 8H^+ + 5e^- \rightarrow Mn^{2+} + 4H_2O$	+1.51	$Co^{2+} + 2e^- \rightarrow Co$	−0.28
$Mn^{3+} + e^- \rightarrow Mn^{2+}$	+1.51	$In^{3+} + 3e^- \rightarrow In$	−0.34
$Au^{3+} + 3e^- \rightarrow Au$	+1.40	$Tl^+ + e^- \rightarrow Tl$	−0.34
$Cl_2 + 2e^- \rightarrow 2Cl^-$	+1.36	$PbSO_4 + 2e^- \rightarrow Pb + SO_4^{2-}$	−0.36
$Cr_2O_7^{2-} + 14H^+ + 6e^- \rightarrow 2Cr^{3+} + 7H_2O$	+1.33	$Ti^{3+} + e^- \rightarrow Ti^{2+}$	−0.37
$O_3 + H_2O + 2e^- \rightarrow O_2 + 2OH^-$	+1.24	$Cd^{2+} + 2e^- \rightarrow Cd$	−0.40
$O_2 + 4H^+ + 4e^- \rightarrow 2H_2O$	+1.23	$In^{2+} + e^- \rightarrow In^+$	−0.40
$ClO_4^- + 2H^+ + 2e^- \rightarrow ClO_3^- + H_2O$	+1.23	$Cr^{3+} + e^- \rightarrow Cr^{2+}$	−0.41
$MnO_2 + 4H^+ + 2e^- \rightarrow Mn^{2+} + 2H_2O$	+1.23	$Fe^{2+} + 2e^- \rightarrow Fe$	−0.44
$Br_2 + 2e^- \rightarrow 2Br^-$	+1.09	$In^{3+} + 2e^- \rightarrow In^+$	−0.44
$Pu^{4+} + e^- \rightarrow Pu^{3+}$	+0.97	$S + 2e^- \rightarrow S^{2-}$	−0.48
$NO_3^- + 4H^+ + 3e^- \rightarrow NO + 2H_2O$	+0.96	$In^{3+} + e^- \rightarrow In^{2+}$	−0.49
$2Hg^{2+} + 2e^- \rightarrow Hg_2^{2+}$	+0.92	$U^{4+} + e^- \rightarrow U^{3+}$	−0.61
$ClO^- + H_2O + 2e^- \rightarrow Cl^- + 2OH^-$	+0.89	$Cr^{3+} + 3e^- \rightarrow Cr$	−0.74
$Hg^{2+} + 2e^- \rightarrow Hg$	+0.86	$Zn^{2+} + 2e^- \rightarrow Zn$	−0.76
$NO_3^- + 2H^+ + e^- \rightarrow NO_2 + H_2O$	+0.80	$Cd(OH)_2 + 2e^- \rightarrow Cd + 2OH^-$	−0.81
$Ag^+ + e^- \rightarrow Ag$	+0.80	$2H_2O + 2e^- \rightarrow H_2 + 2OH^-$	−0.83
$Hg_2^{2+} + 2e^- \rightarrow 2Hg$	+0.79	$Cr^{2+} + 2e^- \rightarrow Cr$	−0.91
$Fe^{3+} + e^- \rightarrow Fe^{2+}$	+0.77	$Mn^{2+} + 2e^- \rightarrow Mn$	−1.18
$BrO^- + H_2O + 2e^- \rightarrow Br^- + 2OH^-$	+0.76	$V^{2+} + 2e^- \rightarrow V$	−1.19
$Hg_2SO_4 + 2e^- \rightarrow 2Hg + SO_4^{2-}$	+0.62	$Ti^{2+} + 2e^- \rightarrow Ti$	−1.63
$MnO_4^{2-} + 2H_2O + 2e^- \rightarrow MnO_2 + 4OH^-$	+0.60	$Al^{3+} + 3e^- \rightarrow Al$	−1.66
$MnO_4^- + e^- \rightarrow MnO_4^{2-}$	+0.56	$U^{3+} + 3e^- \rightarrow U$	−1.79
$I_2 + 2e^- \rightarrow 2I^-$	+0.54	$Mg^{2+} + 2e^- \rightarrow Mg$	−2.36
$Cu^+ + e^- \rightarrow Cu$	+0.52	$Ce^{3+} + 3e^- \rightarrow Ce$	−2.48
$I_3^- + 2e^- \rightarrow 3I^-$	+0.53	$La^{3+} + 3e^- \rightarrow La$	−2.52
$NiOOH + H_2O + e^- \rightarrow Ni(OH)_2OH^-$	+0.49	$Na^+ + e^- \rightarrow Na$	−2.71
$IAg_2CrO_4 + 2e^- \rightarrow 2Ag + CrO_4^{2-}$	+0.45	$Ca^{2+} + 2e^- \rightarrow Ca$	−2.87
$O_2 + 2H_2O + 4e^- \rightarrow 4OH^-$	+0.40	$Sr^{2+} + 2e^- \rightarrow Sr$	−2.89
$ClO_4^- + H_2O + 2e^- \rightarrow ClO_3^- + 2OH^-$	+0.36	$Ba^{2+} + 2e^- \rightarrow Ba$	−2.91
$[Fe(CN)_6]^{3-} + e^- \rightarrow [Fe(CN)_6]^{4-}$	+0.36	$Ra^{2+} + 2e^- \rightarrow Ra$	−2.92
$Cu^{2+} + 2e^- \rightarrow Cu$	+0.34	$Cs^+ + e^- \rightarrow Cs$	−2.92
$Hg_2Cl_2 + 2e^- \rightarrow 2Hg + 2Cl^-$	+0.27	$Rb^+ + e^- \rightarrow Rb$	−2.93
$AgCl + e^- \rightarrow Ag + Cl^-$	+0.22	$K^+ + e^- \rightarrow K$	−2.93
$Bi^{3+} + 3e^- \rightarrow Bi$	+0.20	$Li^+ + e^- \rightarrow Li$	−3.05

Table 3b *Standard potentials at 298.15 K in alphabetical order*

Reduction half-reaction	$E^{\ominus}$/V	Reduction half-reaction	$E^{\ominus}$/V
$Ag^+ + e^- \rightarrow Ag$	+0.80	$I_2 + 2e^- \rightarrow 2I^-$	+0.54
$Ag^{2+} + e^- \rightarrow Ag^+$	+1.98	$I_3^- + 2e^- \rightarrow 3I^-$	+0.53
$AgBr + e^- \rightarrow Ag + Br^-$	+0.0713	$In^+ + e^- \rightarrow In$	−0.14
$AgCl + e^- \rightarrow Ag + Cl^-$	+0.22	$In^{2+} + e^- \rightarrow In^+$	−0.40
$Ag_2CrO_4 + 2e^- \rightarrow 2Ag + CrO_4^{2-}$	+0.45	$In^{3+} + 2e^- \rightarrow In^+$	−0.44
$AgF + e^- \rightarrow Ag + F^-$	+0.78	$In^{3+} + 3e^- \rightarrow In$	−0.34
$AgI + e^- \rightarrow Ag + I^-$	−0.15	$In^{3+} + e^- \rightarrow In^{2+}$	−0.49
$Al^{3+} + 3e^- \rightarrow Al$	−1.66	$K^+ + e^- \rightarrow K$	−2.93
$Au^+ + e^- \rightarrow Au$	+1.69	$La^{3+} + 3e^- \rightarrow La$	−2.52
$Au^{3+} + 3e^- \rightarrow Au$	+1.40	$Li^+ + e^- \rightarrow Li$	−3.05
$Ba^{2+} + 2e^- \rightarrow Ba$	−2.91	$Mg^{2+} + 2e^- \rightarrow Mg$	−2.36
$Be^{2+} + 2e^- \rightarrow Be$	−1.85	$Mn^{2+} + 2e^- \rightarrow Mn$	−1.18
$Bi^{3+} + 3e^- \rightarrow Bi$	+0.20	$Mn^{3+} + e^- \rightarrow Mn^{2+}$	+1.51
$Br_2 + 2e^- \rightarrow 2Br^-$	+1.09	$MnO_2 + 4H^+ + 2e^- \rightarrow Mn^{2+} + 2H_2O$	+1.23
$BrO^- + H_2O + 2e^- \rightarrow Br^- + 2OH^-$	+0.76	$MnO_4^- + 8H^+ + 5e^- \rightarrow Mn^{2+} + 4H_2O$	+1.51
$Ca^{2+} + 2e^- \rightarrow Ca$	−2.87	$MnO_4^- + e^- \rightarrow MnO_4^{2-}$	+0.56
$Cd(OH)_2 + 2e^- \rightarrow Cd + 2OH^-$	−0.81	$MnO_4^{2-} + 2H_2O + 2e^- \rightarrow MnO_2 + 4OH^-$	+0.60
$Cd^{2+} + 2e^- \rightarrow Cd$	−0.40	$Na^+ + e^- \rightarrow Na$	−2.71
$Ce^{3+} + 3e^- \rightarrow Ce$	−2.48	$Ni^{2+} + 2e^- \rightarrow Ni$	−0.23
$Ce^{4+} + e^- \rightarrow Ce^{3+}$	+1.61	$NiOOH + H_2O + e^- \rightarrow Ni(OH)_2 + OH^-$	+0.49
$Cl_2 + 2e^- \rightarrow 2Cl^-$	+1.36	$NO_3^- + 2H^+ + e^- \rightarrow NO_2 + H_2O$	+0.80
$ClO^- + H_2O + 2e^- \rightarrow Cl^- + 2OH^-$	+0.89	$NO_3^- + 3H^+ + 3e^- \rightarrow NO + 2H_2O$	+0.96
$ClO_4^- + 2H^+ + 2e^- \rightarrow ClO_3^- + H_2O$	+1.23	$NO_3^- + H_2O + 2e^- \rightarrow NO_2^- + 2OH^-$	+0.10
$ClO_4^- + H_2O + 2e^- \rightarrow ClO_3^- + 2OH^-$	+0.36	$O_2 + 2H_2O + 4e^- \rightarrow 4OH^-$	+0.40
$Co^{2+} + 2e^- \rightarrow Co$	−0.28	$O_2 + 4H^+ + 4e^- \rightarrow 2H_2O$	+1.23
$Co^{3+} + e^- \rightarrow Co^{2+}$	+1.81	$O_2 + e^- \rightarrow O_2^-$	−0.56
$Cr^{2+} + 2e^- \rightarrow Cr$	−0.91	$O_2 + H_2O + 2e^- \rightarrow HO_2^- + OH^-$	−0.08
$Cr_2O_7^{2-} + 14H^+ + 6e^- \rightarrow 2Cr^{3+} + 7H_2O$	+1.33	$O_3 + 2H^+ + 2e^- \rightarrow O_2 + H_2O$	+2.07
$Cr^{3+} + 3e^- \rightarrow Cr$	−0.74	$O_3 + H_2O + 2e^- \rightarrow O_2 + 2OH^-$	+1.24
$Cr^{3+} + e^- \rightarrow Cr^{2+}$	−0.41	$Pb^{2+} + 2e^- \rightarrow Pb$	−0.13
$Cs^+ + e^- \rightarrow Cs$	−2.92	$Pb^{4+} + 2e^- \rightarrow Pb^{2+}$	+1.67
$Cu^+ + e^- \rightarrow Cu$	+0.52	$PbSO_4 + 2e^- \rightarrow Pb + SO_4^{2-}$	−0.36
$Cu^{2+} + 2e^- \rightarrow Cu$	+0.34	$Pt^{2+} + 2e^- \rightarrow Pt$	+1.20
$Cu^{2+} + e^- \rightarrow Cu^+$	+0.16	$Pu^{4+} + e^- \rightarrow Pu^{3+}$	+0.97
$F_2 + 2e^- \rightarrow 2F^-$	+2.87	$Ra^{2+} + 2e^- \rightarrow Ra$	−2.92
$Fe^{2+} + 2e^- \rightarrow Fe$	−0.44	$Rb^+ + e^- \rightarrow Rb$	−2.93
$Fe^{3+} + 3e^- \rightarrow Fe$	−0.04	$S + 2e^- \rightarrow S^{2-}$	−0.48
$Fe^{3+} + e^- \rightarrow Fe^{2+}$	+0.77	$S_2O_8^{2-} + 2e^- \rightarrow SO_4^{2-}$	+2.05
$[Fe(CN)_6]^{3-} + e^- \rightarrow [Fe(CN)_6]^{4-}$	+0.36	$Sn^{2+} + 2e^- \rightarrow Sn$	−0.14
$2H^+ + 2e^- \rightarrow H_2$	0, by definition	$Sn^{4+} + 2e^- \rightarrow Sn^{2+}$	+0.15
$2H_2O + 2e^- \rightarrow H_2 + 2OH^-$	−0.83	$Sr^{2+} + 2e^- \rightarrow Sr$	−2.89
$2HBrO + 2H^+ + 2e^- \rightarrow Br_2 + 2H_2O$	+1.60	$Ti^{2+} + 2e^- \rightarrow Ti$	−1.63
$2HClO + 2H^+ + 2e^- \rightarrow Cl_2 + 2H_2O$	+1.63	$Ti^{3+} + e^- \rightarrow Ti^{2+}$	−0.37
$H_2O_2 + 2H^+ + 2e^- \rightarrow 2H_2O$	+1.78	$Ti^{4+} + e^- \rightarrow Ti^{3+}$	0.00
$H_4XeO_6 + 2H^+ + 2e^- \rightarrow XeO_3 + 3H_2O$	+3.0	$Tl^+ + e^- \rightarrow Tl$	−0.34
$Hg_2^{2+} + 2e^- \rightarrow 2Hg$	+0.79	$U^{3+} + 3e^- \rightarrow U$	−1.79
$Hg_2Cl_2 + 2e^- \rightarrow 2Hg + 2Cl^-$	+0.27	$U^{4+} + e^- \rightarrow U^{3+}$	−0.61
$Hg^{2+} + 2e^- \rightarrow Hg$	+0.86	$V^{2+} + 2e^- \rightarrow V$	−1.19
$2Hg^{2+} + 2e^- \rightarrow Hg_2^{2+}$	+0.92	$V^{3+} + e^- \rightarrow V^{2+}$	−0.26
$Hg_2SO_4 + 2e^- \rightarrow 2Hg + SO_4^{2-}$	+0.62	$Zn^{2+} + 2e^- \rightarrow Zn$	−0.76

Index

Ab initio method, 369
Absolute entropy, 101
Absolute zero, 21
Absorbance, 236, 506
Absorption coefficient, 505, 507
Absorption intensity, 505
Acceleration, 6
Acceptable wavefunction, 295
Acceptor band, 423
Accommodation, 455
Acetylene, VB description, 350
Acid, 186
Acid buffer, 199
Acid catalysis, 276
Acid–base indicator, 200
Acidity constant, 187
  polyprotic acid, 192
Acidosis, 200
Activated complex, 254
Activated desorption, 462
Activation barrier, 251
Activation control, 272
Activation energy, 249, 253
  negative, 269
  viscosity, 275
Activation Gibbs energy, 255
Activation-controlled limit, 273
Active transport, 216
Activity, 146, 210
Activity coefficient, 146, 210
Adiabatic wall, 49
ADP, 173
Adsorbate, 449
Adsorption, 449
Adsorption isotherm, 456
Adsorption rate, 461
AEDANS, 517
Aerosol, 404
AES, 451
AFM, 453
Alkalosis, 200
Alkene, VB description, 350
Alkene isomerization, 466
Allosteric effect, 182
Allotrope, 429
Allowed transition, 326
Alloy, phase diagram, 157
α-helix, 397
Altitude, 60
Aluminium extraction, 217
Alveoli, 145
AM1, 370
Amide group, VB description, 351
Ammonia synthesis, 167
Amount of substance, 3
ampere (unit), 57, 213
Amphiphilic, 405
Amphiprotic species, 195
Amplitude, 293
Analyte, 196
ångström, 571
Angular momentum, 5, 303
  quantization, 305
  vector model, 307
Angular velocity, 5
Angular wavefunction, 318
Anharmonicity, 491
Anharmonicity constant, 492
Anion configuration, 331
Anode, 222
Antibonding orbital, 355
Antiferromagnetism, 432
Antiparallel β sheet, 398
Anti-Stokes radiation, 488
Antisymmetric stretch, 494
Approximation
  Born–Oppenheimer, 345
  orbital, 327
  steady-state, 268
Aquatic life, 144
Aragonite, 120
Arrhenius equation, 249
Arrhenius parameters, 249
Arrhenius plot, 250
Arrhenius, S., 249
Artist's colour wheel, 504
Atmosphere
  Earth's, 25
  ionic, 204, 211
  planetary, 30
atmosphere (unit), 7, 571
Atomic force microscopy, 453
Atomic orbital, 318
Atomic radius, 332
Atomic spectra, 288
Atomic weight, 4
ATP, 173
ATP synthesis, 518
Aufbau principle, 329
Auger effect, 451
Auger electron spectroscopy, 451
Autoprotolysis constant, 188
Autoprotolysis equilibrium, 187
Average molar mass, 412
Average speed, 27
Avogadro's constant, 3
Avogadro's principle, 21
Azeotrope, 154
Azimuth, 305

Balmer series, 316
Balmer, J., 316
Band gap, 422
Band structure, 495
Band theory, 421
bar (unit), 6, 571
Barometric formula, 25
Base, 186
Base buffer, 199
Base catalysis, 276
Base pair, 398
Base unit, 570
Basicity constant, 188
Beer–Lambert law, 236, 505, 506, 524
Bending mode, 494
Benzene
  elpot surface, 371
  MO description, 368
  VB description, 353
BET isotherm, 459
β-carotene, 300
β-pleated sheet, 397
β-sheet, 397
Bilayer, 407
Bilayer vesicle, 407
Bimolecular reaction, 266
Binary mixture, 24, 152
Biochemical cascade, 510
Biofuel cell, 230
Biological membrane, 407
Biological standard state, 173, 228
Bipolaron, 403
Biradical, 362
Black-body radiation, 497
Blood, buffer action, 200
Body-centred cubic, 441
Bohr effect, 208
Bohr frequency condition, 288, 316, 477
Bohr magneton, 543
Bohr radius, 319, 321
Boiling, 124
Boiling point, normal and standard, 124
Boiling-point constant, 147
Boiling temperature, 124
  effect of pressure, 121
Boltzmann distribution, 11, 531, 551
  chemical equilibrium, 176
  rotational states, 483
Boltzmann formula, 100
Boltzmann's constant, 11, 552
Bomb calorimeter, 61, 78
Bond, 345
  vibrational frequency, 309
Bond angle and hybridization, 351
Bond enthalpy, 76
Bond length determination, 486
Bond order, 362
Bond torsion, 400
Bonding orbital, 355
Born interpretation, 293
Born, M., 293
Born–Haber cycle, 426
Born–Mayer equation, 428
Born–Oppenheimer approximation, 345
Boson, 326, 338, 482
Boundary condition, 244, 295, 298
  cyclic, 304
Boundary surface, 322
Boyle's law, 19
Bragg, L., 437
Bragg, W., 437
Bragg's law, 437
Branch, 495

Branching, 280
Bravais lattice, 433
Breathing, 145
Bremsstrahlung, 436
Brønsted acids and bases, 186
Brønsted–Lowry theory, 186
Brunauer, S., 459
Buffer action, 198
Building-up principle, 329
Butadiene, MO description, 367
Butler–Volmer equation, 468

Caesium-chloride structure, 442
Cage effect, 272
Calcite, 120
calorie (unit), 571
Calorimeter, 57, 61, 78
Calorimeter constant, 57
Calorimetry, 73
Calvin–Benson cycle, 518
Capillary action, 410
Carbon dioxide
  isotherms, 34
  normal mode, 493
  phase diagram, 127
  supercritical, 125
Carbon dioxide laser, 520
Carbon monoxide
  MO description, 365
  polarity, 378
  residual entropy, 104
Carbon nanotube, 430
Carbonic acid speciation, 192
Carbonic anhydrase, 200
Carnot efficiency, 94
Casein, 405
Catalysis, 276, 463
Catalyst, 276
  effect on equilibrium, 181
Catalyst properties, 465
Catalytic activity, 465
Catalytic constant, 278
Catalytic efficiency, 279
Catalytic reforming, 466
Cathode, 222
Cation configuration, 331
CCD, 505
ccp, 440
Cell, 217
Cell membrane, 216, 408
Cell notation, 222
Cell overpotential, 472
Cell potential, 222
  variation with temperature, 229
Cell reaction, 222
Celsius scale, 10
Centrifugal distortion, 479, 486
Cesium, *see* caesium
CFC, 36
Chain carrier, 280
Chain reaction, 280
Charge, 57
Charge-coupled device, 505
Charge–dipole interaction, 380
Charge-transfer transition, 509
Charles's law, 21
Chemical amount, 3
Chemical bond, 344
  vibrational frequency, 309
Chemical equilibrium
  Boltzmann distribution, 176
Chemical exchange, 541
Chemical kinetics, 235
Chemical potential, 136
  perfect gas, 137
  pressure dependence, 137
  real solution, 146
  solute, 144
Chemical reaction, spontaneous, 105
Chemical shift, 533
  electronegativity, 535
Chemisorption, 455
Chemisorption ability, 466
Chlorofluorocarbon, 36
Chlorophyll, 517
  absorption spectrum, 504
Chloroplast, 517
Cholesteric phase, 403
Cholesterol, 408
CHP system, 230
Chromophore, 508
Chromosphere, 338
Clapeyron equation, 120
Classical mechanics, 5, 287
Classical thermodynamics, 47
Clausius–Clapeyron equation, 122
Clear sky, 26
Clebsch–Gordan series, 336
Climate change, 497
Close packing, 440
Closed system, 48
Cloudy sky, 26
CMC, 405, 411
CNDO, 370
Co-adsorption, 459
Coagulation, 409
Coalescence criterion, 541
Coefficient
  absorption, 505, 507
  activity, 146, 210
  diffusion, 273
  Einstein, 524
  extinction, 236, 505
  Hill, 185
  integrated absorption, 507
  mean activity, 210
  molar absorption, 236, 505
  osmotic virial, 151
  rate, 239
  transfer, 469
  transmission, 254
  virial, 36
Coefficient of cooling performance, 112
Coefficient of heating performance, 112
Coefficient of viscosity, 276
Coherent light, 519
Colatitude, 305
Colligative property, 147
Collision cross-section, 32, 253
Collision flux, 450
Collision frequency, 32, 251
Collision theory, 251
Collisional deactivation, 487, 515
Colloid, 403
Colorimeter, 526
Colour, 503
Colour and frequency, 504
Colour wheel, 504
Combined gas equation, 23
Combined heat and power system, 230
Combining electrode potentials, 228
Combustion, 78
Common logarithm, 55
Common-ion effect, 203
Competitive inhibition, 280
Complementary observables, 297, 307
Components, number of, 126
Compression factor, 36
Computational chemistry, 369
Concentration determination, 506
Condensation, 72
Conducting polymer, 403
Conduction, 423
Conduction band, 422
Conductivity, 213
Conductivity and mobility, 215
Conductivity cell, 213
Configuration, 100
Conformational conversion, 541
Conformational energy, 399
Conformational entropy, 397
Conjugate acid, 187
Conjugate base, 187
Conjugated polyene, 375
Consecutive reactions, 265
Consolute temperature, 156
Constant
  acidity, 187
  anharmonicity, 492
  autoprotolysis, 188
  basicity, 188
  boiling-point, 147
  Boltzmann's, 11, 552
  calorimeter, 57
  catalytic, 278
  critical, 124
  cryoscopic, 147
  dielectric, 210
  ebullioscopic, 147
  freezing-point, 147
  gas, 11, 19
  Henry's law, 143
  hyperfine coupling, 545
  Madelung, 428
  Michaelis, 277
  molar gas, 11
  normalization, 298
  Planck's, 14, 288
  rate, 239
  rotational, 478
  Rydberg, 316, 317
  solubility, 202
  solubility product, 202
  spin–spin coupling, 536
Constructive interference, 435
Contact interaction, 539
Continuous wave EPR, 544
Continuum generation, 522
Contour length, 396
Convection, 274
Conventional temperature, 70
Cooling curve, 120, 157
Cooling, spontaneous, 91
Cooper pair, 424
Cooperative binding, 182
Coordination number, 441
  ion, 442
Copolymer, 394
Corona, 338
Correlation spectroscopy, 542
Cosmic ray, 14
COSY, 542
Coulomb interaction, 210

Coulomb potential energy, 9, 210, 316
coulomb (unit), 57
Coupled reactions, 172
Covalent bond, 345
Covalent solid, 421, 429
Critical constants, 124
Critical constants, van der Waals, 39
Critical micelle concentration, 405, 411
Critical molar volume, 35
Critical point, 124
Critical pressure, 35, 124
Critical temperature, 35
  superconductivity, 424
Cross-section, 32
Cryoscopic constant, 147
Crystal diode, 478
Crystal structure, 432
Crystal system, 433
Crystallinity, 402
Crystallization, 444
Cubic close-packed, 440
Cubic system, 433
Curvature and kinetic energy, 292, 294
CW-EPR, 544
Cyclic boundary conditions, 304
Cyclic voltammetry, 471
Cytosol, 216

d block, 331
d electron, 320
d metal, 331
d orbital, 325
dalton (unit), 413
Dalton's law, 23
Daniell cell, 221
Davisson, D., 291
Davisson–Germer experiment, 291
day (unit), 571
de Broglie relation, 291, 293
de Broglie, L., 291
Deactivation, 487, 515
Debye T^3 law, 103
debye (unit), 378
Debye, P., 211, 378, 438
Debye–Hückel limiting law, 211
Debye–Hückel theory, 211, 468
Decay, 510
Defect, 450
Definite integral, 53
Degeneracy, 303, 552
  hydrogenic atom, 320
Degree of freedom, 126
Delocalization energy, 369
Delocalized electrons, 369
δ orbital, 361
δ scale, 533
Denaturation, 399
Density, 4
Density functional theory, 370
Deoxyribonucleic acid, 398
Depression of freezing point, 147
Deprotonation, 188
Derivative, 40
Derived unit, 570
Deshielded, 534
Desorption, 462
Destructive interference, 435
Detector, 505
Determinant, 367
DFT, 370
Dialysis, 405
Diamagnetic, 362, 431
Diamond, 429
Diathermic wall, 49
Dichlorobenzene, 379
Dielectric constant, 210
Differential equation, 244, 265
Differential overlap, 370
Differential scanning calorimetry, 73
Differentiation, 40
Diffraction, 289, 435
Diffraction grating, 505
Diffraction pattern, 436
Diffusion, 30, 272, 273, 282
Diffusion coefficient, 273
  temperature dependence, 275
Diffusion control, 272
Diffusion-controlled limit, 272
Diffusion equation, 274, 282
Dihelium molecule, 357
Diode junction, 424
Diode laser, 519
Dipole moment, 378
  induced, 382, 488
  vector addition, 379
Dipole–dipole interaction
  rotating dipole, 382
  stationary dipole, 381
Dipole–induced dipole interaction, 383
Diprotic acid, speciation, 193
Disperse system, 403
Dispersion interaction, 384
Dissociation, 76, 510
Dissociation limit, 510
Dissociative adsorption, 459
Distillation, 153
Distribution of molecular speeds, 28
Disulfide link, 398
DNA, 398
DNA kinetics, 265
Donor band, 423
Dopant, 423
Doppler effect, 343, 486
Doppler linewidth, 487
d-orbital contribution, 361
d-orbital occupation, 330
Double bond, 350
  chromophore, 508
Double helix, 399
Drift speed, 214
Drug design, 389
Dry cell, 217
DSC, 73
Dual-function catalyst, 466
Duality, 292
Dubosq colorimeter, 526
Dye laser, 520
Dynamic equilibrium, 120
Dynamic light scattering, 415

Eadie–Hofstee plot, 285
Ebullioscopic constant, 147
Eddy, 25
Eel, 217
Effect
  Bohr, 208
  common-ion, 204
  Doppler, 486
  hydrophobic, 386
  Joule–Thomson, 41
Effective atomic number, 332
Effective mass, 314, 490
Effective nuclear charge, 328
Effector molecule, 216
Efficiency
  catalytic, 279
  heat engine, 94
  quenching, 516
  refrigerator, 112
Effusion, 30
Einstein relation, 275
Einstein transition probability, 524
Einstein, A., 289, 524
Elastomer, 401
Electric charge, 57
Electric current, 213
Electric dipole moment, 378
Electric double layer, 408
Electric eel, 217
Electric field, 215
Electrical double layer, 467
Electrical heating, 57
Electrochemical cell, 217
Electrochemical series, 228
Electrode, 217, 220
Electrode compartment, 217
Electrode potential, 468
  combining, 228
Electrode processes, 467
Electrodialysis, 405
Electrokinetic potential, 408
Electrolysis, 217, 472
Electrolyte, 217
Electrolyte solution, 133
Electrolytic cell, 217
Electromagnetic radiation, 12
  quantized, 289
Electromagnetic spectrum, 14
Electromagnetic wave, 12
Electromotive force, 223
Electron affinity, 76, 334
Electron density, X-ray diffraction, 438
Electron diffraction, 436
Electron-gain enthalpy, 76
Electron pair and bond formation, 357
Electron paramagnetic resonance, 544
Electron spin, 325
Electron spin resonance, 544
Electron transfer, 468, 515
Electronegativity, 363
  chemical shift, 535
  dipole moment, 378
Electronic conductor, 421
electronvolt (unit), 289, 571
Electro-osmotic drag, 230
Electrophoresis, 409, 415
Electrostatic potential energy, 9
Electrostatic potential surface, 370
Elementary reaction, 266
Elevation of boiling point, 147
Eley–Rideal mechanism, 464
Elpot surface, 370
emf, *see* cell potential
Emmett, P., 459
Emulsifying agent, 405
Encounter pair, 272
End point, 202
Endergonic compound, 172
Endergonic reaction, 170
Endothermic compound, 83
Endothermic process, 50
Endothermic reaction, spontaneous, 170

Energy, 48
  defined, 8
  internal, 59
Energy density, 46
Energy level
  electron in magnetic field, 544
  harmonic oscillator, 308, 489
  hydrogenic atom, 317
  nucleus in magnetic field, 530
  particle in a box, 299
  particle on a ring, 304
  rotation, 478
  symmetric rotor, 481
Energy quantization, 295
Enthalpy, 62
  heat supplied at constant pressure, 63
  phase transition, 70
Enthalpy change, 63
Enthalpy changes, addition of, 72
Enthalpy density, 79
Enthalpy of activation, 255
Enthalpy of adsorption, 455, 458
Enthalpy of chemisorption, 456
Enthalpy of combustion, 78
Enthalpy of formation, 82
Enthalpy of fusion, 72
Enthalpy of ionization, 74
Enthalpy of mixing
  gases, 139
  ideal solution, 142
Enthalpy of physisorption, 455
Enthalpy of reaction, 80
Enthalpy of vaporization, 71
Entropy, 91
  absolute, 101
  Boltzmann's formula, 100
  conformational, 397
  defined, 92
  hydrophobic effect, 387
  low temperature, 103
  molecular interpretation, 100
  partition function, 560
  phase transition, 97
  random coil, 397
  residual, 104
  state function, 93
  variation with temperature, 96
Entropy change
  heating, 95
  irreversible process, 100
  isothermal expansion, 94
Entropy change, in surroundings, 99
Entropy of activation, 255
Entropy of fusion, 97
Entropy of mixing
  gas, 139
  ideal solution, 142
Entropy of phase transition, 97
Entropy of reaction, 105
Entropy of vaporization, 97
Entropy units, 93
Enzyme, 277
Enzyme kinetics, 235
Enzymolysis , rate of, 277
EPR, 544
Equation
  Born–Mayer, 428
  Butler–Volmer, 468
  Clapeyron, 120
  Clausius–Clapeyron, 122
  combined gas, 23
  differential, 244, 265
  diffusion, 282
  Eyring, 254
  Henderson–Hasselbalch, 197
  Karplus, 538
  McConnell, 547
  Nernst, 224
  partial differential, 310
  Schrödinger, 292
  secular, 366
  simultaneous, 367
  Stern–Volmer, 514
  van 't Hoff (equilibrium), 178
  van 't Hoff (osmosis), 150
Equation of state, 18, 37
Equilibrium
  approach to, 263
  autoprotolysis, 187
  chemical, 166
  mechanical, 7
  sedimentation, 415
  thermal, 10
  thermodynamic criterion, 167
Equilibrium composition, 174
  effect of catalyst, 181
  effect of compression, 179
  effect of temperature, 177
Equilibrium constant
  cell potential, 224
  concentration form, 176
  defined, 169
  molecular interpretation, 176
  partition function, 561, 564
  standard electrode potential, 225
  temperature dependence, 178, 262
Equipartition theorem, 12
Equivalence point, *see* stoichiometric point
Equivalent nuclei, 536
ESR, 544
Essential symmetry, 433
Ethanol, NMR spectrum, 534
Ethene
  MO description, 366
  VB description, 350
Ethyne, VB description, 350
Eutectic composition, 158
Eutectic halt, 158
Evaporation, 70
Exchange-current density, 230, 469
Excimer, 520
Excimer laser, 520
Exciplex, 520
Excluded volume, 37
Exclusion rule, 496
Exergonic compound, 172
Exergonic reaction, 169
Exothermic compound, 83
Exothermic process, 50
Expansion work, 51
Expansion, spontaneous, 91
Exponential decay, 243
Exponential function, 29
Extended Debye–Hückel law, 212
Extensive property, 4
Extinction coefficient, 236, 505
Eyring equation, 254

Fat, combustion of, 80
FEMO, 375
Femtochemistry, 255
Fermi contact interaction, 539
Fermi level, 422
Fermion, 326, 338, 482
Ferromagnetism, 431
Fibre, 402
Fick's first law, 273, 282
Fick's second law, 274, 282
Fine structure, NMR, 535
Fingerprint region, 494
First derivative, 40
First ionization energy, 333
First ionization enthalpy, 74
First Law of thermodynamics, 61
First-order reaction, 240
Flash desorption, 455
Flash photolysis, 237
Flocculation, 409
Flow method, 237
Fluorescence, 510, 511
Fluorescence lifetime, 513
Fluorescence quenching, 512
Fluorescence resonance energy transfer, 516
Fluorine molecule, MO description, 362
Flux, 273
Foam, 411
Fock, V., 331
Food and energy reserves, 80
Forbidden transition, 326
Force, 6
Force constant, 308, 489
Förster theory, 516
Förster, T., 516
Fourier synthesis, 438
Fourier transform EPR, 544
Fourier transform NMR, 533
Four-level laser, 518
Fraction deprotonated, 188
Fraction protonated, 190
Fractional composition, 193
Fractional coverage, 455
Fractional distillation, 153
Franck–Condon principle, 507
Freedom, degrees of, 126
Free-electron molecular orbital theory, 375
Freely jointed chain, 396
Freezing temperature, 125
Freezing-point constant, 147
Frequency, 13
Frequency doubling, 519
FRET, 516
Friedrich, W., 437
Fructose-6-phosphate, 166
FT-EPR, 544
FT-NMR, 533
Fuel, 79
Fuel cell, 217, 230
Functional MRI, 543
Fusion, 72

g,u classification, 356
Gallium arsenide, 424
Galvani potential, 468
Galvanic cell, 217
γ-ray, 14
Gas, 2
  kinetic model, 26
  kinetic molecular theory, 41
  liquefaction, 40
  real, 33
Gas constant, 11, 19

Gas electrode, 220
Gas laser, 519
Gas solubility, 145
Gaussian function, 29, 309
Gaussian-type orbital, 370
Gel, 404
Gerade, 356
Gerlach, W., 325
Germanium nanowire, 430
Germer, L., 291
g-factor, nuclear, 530
Gibbs energy
  activation, 255
  chemical equilibrium, 166
  defined, 106
  mixing, 138
  partition function, 560
  perfect gas, 114
  properties, 107
  variation with pressure, 114
  variation with temperature, 117
Gibbs energy of formation, 171
Gibbs energy of mixing, ideal solution, 142
Gibbs energy of reaction, 166
  electrochemical determination, 228
Gibbs phase rule, 126
Gibbs, J.W., 106
Glass electrode, 227
Glass transition temperature, 402
Global minimum, 401
Glucose-6-phosphate, 166
Gouy–Chapman model, 468
Graham's law of effusion, 31
Grain boundary, 441
Graph, 20
Graphene, 429
Graphite, 429
Graphs, plotting, 152
Gravimetry, 455
Gravitational potential energy, 9
Greenhouse effect, 497
Greenhouse gas, 494, 497
Gross selection rule, 484
Grotrian diagram, 327
Grotthus mechanism, 215
GTO, 370
Gunn diode, 478
g-value, electron, 544, 545
Gyromagnetic ratio, 543

Haemoglobin, 181, 398
Half-life, 247
Half-reaction, 218
Halley's comet, 127
Hall–Hérault process, 217
Hamilton, W., 292
Hamiltonian, 292
Hanes plot, 286
Hard-sphere potential energy, 388
Harmonic oscillator, 308, 489
Harmonic wave, 293
Harned cell, 230, 234
Harpoon mechanism, 253
Hartree, D.R., 331
Hartree–Fock procedure, 331
hcp, 440
Head group, 405
Heat, 49
  molecular interpretation, 50
  sign convention, 51
Heat capacity, 56, 64
  low temperature, 65
  partition function, 559
  relation between, 65, 68
  relation to energy levels, 559
  variation with temperature, 64
Heat engine, 93
Heat output of fuels, 79
Heat pump, 93, 94, 112
Heating, 49
Heisenberg, W., 296
Heisenberg's uncertainty principle, 296
Helium, phase diagram, 128
Helium-I and II, 128
Helix–coil transition, 286
Helmholtz layer, 468
Helmholtz plane, 468
Hemoglobin, *see* haemoglobin
Henderson–Hasselbalch equation, 197
Henry, W., 142
Henry's law, 142
Henry's law constant, 143
Hess's law, 81
Heterogeneity index, 412
Heterogeneous catalysis, 463
Heterogeneous catalyst, 276
Heterogeneous equilibrium, 202
Heteronuclear diatomic molecule, 363
Hexagonal close-packed, 440
Hexagonal system, 433
HF-SCF, 331
Highest occupied molecular orbital, 365
High-performance liquid chromatography, 126
High-temperature superconductor, 421, 424
Hill coefficient, 185
Histidine, 208
Homeostasis, 80, 200
HOMO, 365
Homogeneous catalyst, 276
Homogeneous mixture, 133
Homogenized milk, 405
HOMO–LUMO energy gap, 371
Homonuclear diatomic molecule, 356
  MO description, 361
Hooke's law, 308, 401
hour (unit), 571
HPLC, 126
HTSC, 421, 424
Hückel approximation, 422
Hückel method, 366
Hückel, E., 211, 366
Humphreys series, 341
Hund's rule, 330
Hybrid orbital, 349
Hybridization, 349
  spin–spin coupling, 538
Hybridization and bond angle, 351
Hydrodynamic radius, 215
Hydrogen atom, spectrum, 315
Hydrogen bond, 385
  fibre, 402
  polypeptide, 397
Hydrogen electrode, 220
Hydrogen fluoride molecule, MO description, 364
Hydrogen molecule
  ground state, 357
  MO description, 357
  VB theory, 346
Hydrogen molecule ion, 354
Hydrogen storage, 230
Hydrogen/oxygen cell, 230
Hydrogenic atom, 315
  degeneracy, 320
  energy levels, 317
  ground state, 321
  ionization energy, 318
  wavefunctions, 318, 319
Hydronium ion, 187
Hydrophilic, 387
Hydrophobic effect, 386
Hydrostatic pressure, 7
Hyperbola, 20
Hyperfine coupling constant, 545
Hyperfine structure, 545

Ice polymorphs, 127
Ice structure, 444
Ice-I structure, 127
Ideal gas versus perfect gas, 142
Ideal solution, 140
Ideal–dilute solution, 143
IHP, 468
Indefinite integral, 53
Independent migration of ions, 213
Indicator, 200
INDO, 370
Induced dipole moment, 382, 488
Inertia, 5
Infrared active, 494
Infrared spectrum, 495
Inhibition, 280
Inhibitor (enzyme), 280
Initial condition, 244
Initial rate, 241
Initiation, 280
Inner Helmholtz plane, 468
Instantaneous rate, 238
Insulator, 421
Integral, 53
Integral protein, 407
Integrated NMR signal, 534
Integrated absorption coefficient, 507
Integrated rate law
  first order, 243
  second order, 245
Integration, 53
Intensity
  NMR, 532
Intensity of absorption, 236
Intensive property, 4
Intercept, 20
Interference, 435
Intermediate, 265
Intermetallic compound, 424
Internal energy, 59
  partition function, 557
  state function, 60
Internal energy and enthalpy of reaction, relation between, 79
Intersystem crossing, 512
Ion channel, 216
Ion pump, 216
Ionic atmosphere, 204, 211
Ionic bond, 345
Ionic conductivity, 213
Ionic crystal, 442
Ionic mobility, 214
Ionic model, 425
Ionic strength, 212
Ionic–covalent resonance, 352

Ionization energy, 74, 318, 333, 384
ISC, 512
Isochore, 178
Isodensity surface, 370
Isolated system, 48
Isolation method, 241
Isomorphous replacement, 439
Isosbestic point, 506
Isosteric enthalpy of adsorption, 458
Isotherm, 20
Isothermal expansion, 59
Isotope separation, 31, 521
Isotopologue, 486, 521

Jablonski diagram, 511
joule (unit), 10, 570
Joule, J., 41
Joule–Thomson effect, 41

K_α radiation, 437
Kammerlingh Onnes, H., 424
Karplus equation, 538
Kekulé structure, 353
Kelvin scale, 10
Kelvin, Lord, 41
kilogram (unit), 2
Kinetic control, 270
Kinetic energy, 8
Kinetic energy and curvature, 292, 294
Kinetic energy density, 46
Kinetic model of gases, 26
Kinetic molecular theory, 26, 41
Kinetic theory of gases, 253
Kink defect, 450
Kirchhoff's law, 84
Klystron, 478
KMT, 26, 41
  criteria for validity, 33
Knipping, P., 437
Kohlrausch, F., 213
Kohlrausch's law, 213

λ-line, 128
Lamellar micelle, 407
Langmuir isotherm, 456
  co-adsorption, 459
  dissociative, 459
Langmuir–Hinshelwood mechanism, 464
Laplace equation, 410
Larmor precession frequency, 531
Laser, 518
Laser cavity, 518
Laser light scattering, 415
Lattice enthalpy, 426
Law
  Beer–Lambert, 236, 505, 506, 524
  Boyle's, 19
  Bragg's, 437
  Charles's, 21
  conservation of energy, 9, 48
  conservation of momentum, 6
  Dalton's, 23
  Debye-Hückel, 211
  Debye T^3, 103
  Extended Debye–Hückel, 212
  Fick's first, 273, 282
  Fick's second, 274, 282
  First, of thermodynamics, 61
  Graham's, 31
  Henry's, 142
  Hess's, 81
  Hooke's, 308, 401
  Kirchhoff's, 84
  Kohlrausch's, 213
  limiting, 141
  Nernst distribution, 159
  Ohm's, 213
  Raoult's, 139
  rate, 239
  Second, of thermodynamics, 92
  Stokes', 215
  Third, of thermodynamics, 102
LCAO-MO, 354
Le Chatelier's principle, 177
LEED, 452
Lennard-Jones (12,6) potential energy, 388
Level, 337
Lever rule, 155
Lewis theory, 345
Lewis, G.N., 345
Life, 108
Lifetime broadening, 487
Light, 13
  coherent, 519
Light and colour, 504
Light as particles, 290
Light-harvesting complex, 517
Light scattering, 415
Light-emitting diode, 424
Limiting law, 19, 141
Limiting molar conductivity, 213
Linde refrigerator, 41
Lindemann mechanism, 271
Lindemann, F., 271
Lindemann–Hinshelwood mechanism, 271
Linewidth, 486
Linear combination, 349
Linear combination of atomic orbitals, 354
Linear momentum, 5
Linear rotor, 478
Linear variation, 289
Linear-sweep voltammetry, 470
Lineweaver–Burk plot, 278
Liquefaction of gases, 40
Liquid, 2
Liquid crystal, 403
Liquid junction potential, 222
Liquid structure, 128
Liquid surface, 409
Liquid–liquid phase diagram, 156
Liquidus, 157
Lithium configuration, 328
litre (unit), 4, 571
Local contribution, 534
Local magnetic field, 533
Local minimum, 401
Logarithm, 55
London formula, 384
London interaction, 384
Lone pair, chromophore, 508
Long period, 331
Long-range order, 128
Low-energy electron diffraction, 452
Lowest unoccupied molecular orbital, 365
Lumiflavin, 400
LUMO, 365
Lung, 145
Lyman series, 316
Lysine, 208

Macromolecule, 394, 395
Macular pigment, 509
Madelung constant, 428
Magnetic dipole interaction, 530
Magnetic field, 529
Magnetic quantum number, 306
Magnetic resonance imaging, 542
Magnetic susceptibility, 431
Magnetization, 431
Magnetogyric ratio
  electron, 543
  nuclear, 530
Magneton, 543
Malleable, 441
Many-electron atom, 315, 331
Many-electron wavefunction, 327
Marcus theory, 515
Marcus, R., 515
Mars van Krevelen mechanism, 466
Mass, 2
Mass concentration, 134
Mass density, 4
Matrix-assisted desorption/ionization, 413
Matter, 1
Matter wave, 291
Maximum velocity (enzymolysis), 278
Maximum work, 52
Maxwell distribution of speeds, 28, 252
MBE, 430
McConnell equation, 547
Mean activity coefficient, 210
Mean bond enthalpy, 77
Mean free path, 32
Mean speed, 27
Mechanical equilibrium, 7
Mechanical work, 50
Meissner effect, 432
Melting, 72
Melting temperature, 402
Mesophase, 403
Metabolic acidosis, 200
Metabolic alkalosis, 200
Metal crystal, 440
Metal–insoluble salt electrode, 220
Metallic conductor, 421
Metallic solid, 420
Methane
  normal modes, 494
  VB description, 349
Micelle, 405
Michaelis constant, 277
Michaelis–Menten mechanism, 277
Microporous material, 467
Microwave radiation, 14
Microwave spectroscopy, 478, 484
Milk, 405
Miller indices, 434
MINDO, 370
minute (unit), 571
Mixing, Gibbs energy, 138
Mixture, 133
  binary, 24, 152
  perfect gas, 24
  phase diagram, 152
  volatile liquids, 153
MO theory, 344
Mobility, 214
Molality, 134
Molar absorption coefficient, 236, 505
Molar concentration, 2, 134
Molar conductivity, 213
Molar enthalpy, 62
Molar gas constant, 11
Molar Gibbs energy, 114

Molar heat capacity, 56
Molar internal energy, 59
Molar magnetic susceptibility, 431
Molar mass, 3
  average, 412
  osmometry, 151
Molar quantity, 5
Molar solubility, 202
Molar volume, 21
Molarity, 134
Mole fraction, 23, 135
  relation to molality, 134
Molecular beam epitaxy, 430
Molecular collisions, 32
Molecular dynamics, 390
Molecular interaction, 33
Molecular obital, 353
  criteria for formation, 360
  notation, 356
Molecular orbital theory, 344, 353
Molecular partition function, 557
Molecular potential energy curve, 346
Molecular recognition, 389
Molecular shape, 345
Molecular solid, 421, 443
Molecular spectroscopy, 477
Molecular speed, 28
Molecular vibration, 489
Molecular weight, 4
Molecularity, 266
Moment of inertia, 5, 303, 478
Momentum, 5
Monochromatic, 13
Monochromator, 505
Monoclinic, 433
Monodisperse, 412
Monolayer, 407
Monomer, 394
Monte Carlo method, 390
Morse potential energy, 374
MRI, 542
Mulliken scale, 363
Mulliken, R., 363
Multiphoton process, 521
Multi-walled nanotube, 430
MWNT, 430
Myoglobin, 181, 398

NAD^+, 184, 219
NADH, 219
NADPH, 518
Nanotechnology, 429
Nanowire, 429
Natural linewidth, 488
Natural logarithm, 55
Near infrared, 14
Negative activation energy, 269
Neighbouring group contribution, 534
Nematic phase, 403
Neodymium laser, 519
Nernst distribution law, 159
Nernst equation, 224
Network solid, 421
Neutron diffraction, 436
newton (unit), 6, 570
Newton's laws, 6
Nicotinamide, 219
Nicotine, 191
Nitric oxide, kinetics, 267
Nitrogen molecule
  MO description, 361
  VB description, 348
NMR, 532
  two-dimensional, 542
NMR spectrometer, 532
Node, 299
Non-competitive inhibition, 280
Non-electrolyte solution, 133
Non-expansion work, 107
Nonlinear optical phenomena, 519
Nonpolar molecule, 378
Nonpolarizable electrode, 470
Non-radiative decay, 510
Non-spontaneous change, 90
Normal boiling point, 124
Normal freezing point, 125
Normal melting point, 125
Normal mode, 493
Normalization constant, 298
n-to-π^* transition, 508
n-type semiconductivity, 423
Nuclear *g*-factor, 530
Nuclear magnetic resonance, 532
Nuclear magnetogyric ratio, 530
Nuclear magneton, 530
Nuclear repulsion, 347
Nuclear spin, 530
Nuclear spin quantum number, 529
Nuclear statistics, 482
Number-average molar mass, 412
Nylon, 402

Ohm's law, 213
OHP, 468
Open system, 48
Operator, 293
Orbital angular momentum, 306, 318, 335
Orbital angular momentum quantum number, 306
Orbital approximation, 327
Orbital energy variation along Period 2, 360
Orbital occupation, 330
Orbital overlap, 354
Order of reaction, 240
Ordinary differential equation, 244
Orthorhombic, 433
Oscillator, 308
Osmometry, 151
Osmosis, 149
  reverse, 152
Osmotic pressure, 149
Osmotic virial coefficient, 151
Outer Helmholtz plane, 468
Overall order, 240
Overlap and symmetry, 360
Overlap integral, 354
Overpotential, 468, 472
Oxidation, 218
Oxidation number, 218
Oxidation state, 218
Oxidizing agent, 218
Oxygen binding, 181
Oxygen molecule
  MO description, 362
  VB description, 348

p band, 422
P branch, 495
p electron, 320
p orbital, 320, 324
Packing fraction, 441
Pair distribution function, 128
Parabola, 175
Parabolic potential energy, 308, 489
Parallel band, 494
Paramagnetic, 362, 431
Parcel (of air), 25
Parity, 356
Partial charge, 363
  interacting, 377
Partial differential equation, 310
Partial molar Gibbs energy, 136
Partial molar property, 135
Partial molar volume, 135
Partial pressure, 23
Partial vapour pressure, 139
Partially miscible liquids, 155
Particle in a box, 298
Particle in a two-dimensional box, 302
Particle on a ring, 304
Particle on a spherical surface, 306
Particles as waves, 291
Partition function, 552
  electronic, 557
  entropy, 560
  equilibrium constant, 561, 564
  factorization, 557
  Gibbs energy, 560
  heat capacity, 559
  interpretation, 554
  molecular, 557
  rotational, 563, 556
  translational, 555, 563
  vibrational, 556, 563
pascal (unit), 6, 570
Pascal's triangle, 537
Paschen series, 316
Passive transport, 216
Patch clamp technique, 216
Patch electrode, 216
Path function, 55
Pauli exclusion principle, 328, 482
Pauli principle, 338, 482
Pauli, W., 328
Pauling scale, 363
Pauling, L., 363
Penetration, 328
Peptide link, 395, 397
Peptizing agent, 404
Perfect elastomer, 401
Perfect gas, 19
  chemical potential, 137
  mixing, 138
Perfect gas equation of state, 19
Perfect gas versus ideal gas, 142
Period (of oscillation), 13
Periodic trend, 332
Periodicity, 332
Peripheral protein, 407
Permittivity, 210, 377
Pfund series, 316
pH, 187
  buffer solution, 199
  determination, 227
  relation between pH and pOH, 188
  salt solution, 196
  stoichiometric point, 198
  variation of electric potential, 225
Phase, 70
Phase boundary, 118
Phase diagram, 118

alloy, 157
copper/aluminium, 163
helium, 128
hexane/nitrobenzene, 155
liquid crystal, 404
liquid–liquid, 156
mixture, 152
model protein, 164
silver/tin, 163
steel, 163
water, 127
water/nicotine, 157
water/triethylamine, 156
Phase problem, 439
Phase rule, 126
Phase stability, 114
Phase transition, 70, 113
effect of pressure, 119
Phosphatidyl choline, 407
Phosphorescence, 510, 511
Phosphoric acid, 194
Phosphorylation, 173
Photodiode, 505
Photoejection, 290, 522
Photoelectric effect, 289
Photoelectron, 290
Photoelectron spectroscopy, 522
Photoemission spectroscopy, 451
Photon, 12, 13, 289
Photon spin, 326
Photosphere, 338
Photosynthesis, 517
Photosystem I and II, 517
Physical quantity, 2, 570
Physisorption, 455
π bond, 348, 359
g,u character, 360
π-stacking, 389, 398
π-to-π^* transition, 508
pK_a, 187
pK_b, 188
relation between pK_a and pK_b, 188
pK_w, 188
Planck, M., 289
Planck's constant, 14, 288
Planetary atmospheres, 30
Plasma, 46
Plastic, 402
p–n junction, 424
Polar coordinates, 305
Polar molecule, 364, 378
Polarizability, 383, 488, 492
Polarizability volume, 383
Polarizable electrode, 470
Polarization mechanism, 538
Polarized light, 13
Polaron, 403
Polyatomic molecule
MO description, 365
VB description, 348
Polydisperse, 412
Poly-L-glycine, 398
Polymer, 394
Polymorph, 127, 429
Polypeptide, 395, 397
Polyprotic acid, 192
Population, 11, 552
nuclear spin, 531
rotational states, 483
Population inversion, 518
Position–momentum uncertainty relation, 297
Potential energy, 9
molecular interaction, 34
Powder diffraction, 438
Power series, 149
ppm, 533
Precession, 531
Precursor state, 461
Pre-exponential factor, 249, 253
Prefix (SI), 570
Pressure, 6
critical, 35, 124
curved surface, 410
Pressure jump, 264
Pressure units, 7
Primary structure, 395
Primitive cubic, 442
Principle
Aufbau, 329
Avogadro's, 21
building-up, 329
Franck–Condon, 507
Le Chatelier's, 177
Pauli, 338, 482
Pauli exclusion, 328, 482
uncertainty, 296
Probabilistic interpretation, 294
Probability density, 293
harmonic oscillator, 309
Probe, 522
Promotion, 349
Propagation, 280
Proportionality, 289
Protein, 395
combustion of, 80
Protein folding, 399
Proton magnetic resonance, 532
Proton migration, 215
Proton transfer, tunnelling, 301
Pseudo order, 241
Pseudofirst-order reaction, 241
Pseudosecond-order reaction, 241
p-type semiconductivity, 423
Pulse radiolysis, 237
Pump, 518, 522

Q branch, 495
Q-band, 544
QCM, 455
Quadratic equations, 175
Quadratic term, 12
Quantities, 2
Quantization, 10
confinement, 300
Quantized, 289
Quantum dot, 430
Quantum number
angular momentum, 305
atom, 318
introduced, 298
K, 481
magnetic, 306
nuclear spin, 529
orbital angular momentum, 306
spin magnetic, 325
spin, 325
total orbital angular momentum, 336
vibrational, 308
Quantum theory, 1, 287
Quantum yield, 513
Quark, 326
Quartz crystal microbalance, 455
Quaternary structure, 395
Quench, 512
Quenching efficiency, 516
Quenching method, 238
Quenching rate constant, 515
Quinoline, 191

R branch, 495
Radial distribution function, 322
Radial node, 322
Radial wavefunction, 318, 324
Radiation, 12
Radiative decay, 510
Radio wave, 14
Radioactive decay, 265
Radius, 305
Radius of gyration, 396
Radius ratio, 442
Radius-ratio rule, 443
RAM, 4
Ramachandran diagram, 400
Raman activity, 496
Raman effect, 488
Raman microscopy, 521
Raman shift, 488
Raman spectroscopy, 488
Random coil, 396, 397
Random walk, 273
Rankine scale, 16
Raoult, F., 139
Raoult's law, 139
Rate
absorption, 525
adsorption, 461
Rate coefficient, 239
Rate constant
definition, 239
temperature dependence, 250
units, 239
viscosity dependence, 276
Rate law, 239
catalysis, 463
chain reaction, 281
determination, 241
formulation, 267
Rate of effusion, 31
Rate-determining step, 269
Rayleigh radiation, 488
Reaction enthalpy, 80
temperature dependence, 84
Reaction Gibbs energy, 166, 516
concentration dependence, 168
electrical determination, 223
Reaction mechanism, 266
Reaction order, 240
Reaction profile, 251
Reaction quotient, 168
half-reaction, 221
Reaction rate, defined, 238
Real gas, 19, 33
Real solution, 146
Redox couple, 218
Redox electrode, 221
Redox reaction, 217
Reduced variable, 39
Reduced mass, 317, 490
Reducing agent, 218

Reduction, 218
Reference state, 82
Refrigerator, 93, 94, 112
Relation between pH and pOH, 188
Relation between pK_a and pK_b, 188
Relative atomic mass, 4
Relative molecular mass, 4
Relative permittivity, 377
Relaxation, 539
Relaxation methods, 263
Reorganization energy, 515
Repulsion, 388
Residence half-life, 462
Residual entropy, 104
Residue, 395
Resistance, 213
Resistivity, 213
Resonance, 352, 529
Resonance condition
  EPR, 545
  NMR, 531
Resonance energy transfer, 515, 516
Resonance Raman spectroscopy, 497
Resonant mode, 519
Respiratory acidosis, 200
Resting potential, 216
Restoring force, random coil, 401
Resultant dipole, 379
Retardation, 280
Retinal, 374, 509
Reverse osmosis, 152
Reverse transition, enthalpy of, 72
Reversible process, 52
Reversible work, 54
Rhodopsin, 509
Rhombohedral, 433
Ribonucleic acid, 398
Ridge, 26
Ring current, 535
RMM, 4
rms speed, 27
RNA, 398
Rock-salt structure, 442
Root-mean-square separation, 396
Root-mean-square speed, 27
Rotation, 303
Rotational constant, 478
Rotational partition function, 563
Rotational transitions, 484
Rule
  Hund's, 330
  lever, 155
  phase, 126
  radius-ratio, 443
  Trouton's, 98
Russell–Saunders coupling, 335
Rydberg constant, 316, 317
Rydberg, J., 316

s band, 422
s electron, 320
s orbital, 320
Salt bridge, 217
SAM, 452, 456
SATP, 22
Saturated solution, 202
Saturation, 539
Scanning Auger electron microscopy, 452
Scanning tunnelling microscopy, 453
SCF procedure, 369
Scherrer, P., 438
Schrödinger equation, introduced, 292
Schrödinger, E., 292
Second derivative, 40
Second harmonic generation, 519
Second ionization energy, 333
Second ionization enthalpy, 74
Second Law and life, 108
Second law of motion, 6
Second Law of thermodynamics, 92
Secondary structure, 395
Second-order reaction, 240
Secular equation, 366
Sedimentation, 414
Sedimentation equilibrium, 415
Selection rule, 326
  many-electron atom, 337
  molecular vibration, 490
  rotation, 484
  rotational Raman, 488
  vibrational Raman, 492
Self-assembled monolayer, 456
Self-assembly, 406
Self-consistent field, 331
Self-consistent field procedure, 369
Semiconductor, 421
Semi-empirical method, 369
Separation of variables, 310
SFC, 126
Shape, 345
SHE, 225
Shell, 320
Shielded, 534
Shielding, 328
Shielding constant, NMR, 533
Short-range order, 128
SI, 2
SI units, 570
siemens (unit), 213
σ bond, 347
σ electron, 354
σ orbital, 354
Sign convention, 51
Silicon junction, 424
Silicon nanowire, 430
Silver–silver chloride electrode, 220
Simultaneous equations, 367
Singlet state, 511
Single-walled nanotube, 430
Slice selection, 542
Slip plane, 441
Slope, 20
Smectic phase, 403
Soap, 405
Sodium chloride, band structure, 425
Sol, 404
Solar radiation, 497
Solid, 2
  types of, 420
Solidus, 157
Solubility constant, 202
Solubility equilibria, 202
Solubility product, 202
Solubility product constant, 202
Solubility, effect of added salt, 204
Solute, 133
Solvent contribution, 535
sp hybrid, 350
sp^2 hybrid, 350
sp^3 hybrid, 349
Sparingly soluble, 202
Speciation, 193
Specific enthalpy, 79
Specific heat, 56
Specific heat capacity, 56
Specific selection rule, 485
Spectral line, 288, 316
Spectrometer, 505
Spectrophotometry, 236
Spectroscopy, 477
  time-resolved, 522
Spectrum, 288
Speed, 5
  mean, 27
  molecular, 28
Speed of light, 13
Sphalerite structure, 443
Spherical harmonic, 306, 318
Spherical polar coordinates, 305
Spherical rotor, 482
Spin, 325
  nuclear, 530
Spin magnetic quantum number, 325
Spin quantum number, 325
Spin relaxation, 539
Spin wavefunction, 338
Spin-$^1/_2$ particle, 326
Spin–lattice relaxation, 540
Spin–orbit coupling, 337
Spin–spin coupling, hybridization, 538
Spin–spin coupling constant, 536
Spin–spin relaxation, 540
Spontaneous change, 90
Spontaneous emission, 525
Spontaneous mixing, 138
Spontaneous process, Gibbs energy, 106
Spontaneous reaction, 105
SPR, 455
Stable compound, 172
Standard ambient temperature and
    pressure, 22
Standard boiling point, 124
Standard cell potential, 224
Standard chemical potential, 137
Standard electron gain enthalpy, 76
Standard enthalpy of combustion, 78
Standard enthalpy of formation, 82, 371
Standard enthalpy of fusion, 72
Standard enthalpy of ionization, 74
Standard enthalpy of vaporization, 70
Standard entropy of reaction, 105
  electrochemical determination, 229
Standard Gibbs energy of formation, 171
Standard hydrogen electrode, 225
Standard molality, 134
Standard molar concentration, 134
Standard molar entropy, 102
Standard molar Gibbs energy, partition
    function, 561
Standard oxidation potential, 225
Standard potential, 225, 371
Standard pressure, 6, 22
Standard reaction enthalpy, 81
Standard reaction entropy, 105
Standard reaction Gibbs energy, 168, 170
  equilibrium constant, 169
Standard reduction potential, 225
Standard state, 69, 146
Standard temperature and pressure, 22
Star spectra, 338

State, 10
State function, 60
State specificity, 521
Statistical entropy, 101
Statistical thermodynamics, 48, 551
Steady-state approximation, 268
Steel, 163
Step defect, 450
Steric factor, 253
Stern, O., 325
Stern–Gerlach experiment, 325
Stern–Volmer equation, 514
Stern–Volmer plot, 514
Sticking probability, 462
Stimulated absorption, 524
Stimulated emission, 518, 524
STM, 453
Stoichiometric number, 239
Stoichiometric point, 196
Stokes radiation, 488
Stokes's law, 215
Stopped-flow method, 237
STP, 22
Strong acid, 188
Strong base, 188
Sublimation, 72
Sublimation vapour pressure, 120
Subshell, 320
Sun, structure of, 46
Superconducting magnet, 533
Superconductivity, 424
Superconductor, 421
Supercritical carbon dioxide, 125
Supercritical fluid, 36, 125
Supercritical fluid chromatography, 126
Supercritical water, 126
Superfluid helium, 128
Superposition, 296
Surface excess, 411
Surface growth, 450
Surface plasmon resonance, 455
Surface tension, 409
Surfactant, 405, 409
Sweating, 80
SWNT, 430
Symmetric rotor, 481
Symmetric stretch, 494
Symmetry and overlap, 360
Synchrotron radiation, 437
System, 48
Système International, 2
Système International d'Unités, 570

T_1-weighted image, 543
T_2-weighted image, 543
Tafel plot, 469
TDS, 462
Teller, E., 459
Temperature, 10
  conventional, 70
  critical, 35
  distinct from heat, 49
  thermodynamic, 10
Temperature jump, 263
Temperature programmed desorption, 462
Temperature–composition diagram, 152
Term symbol, 335
Termination, 280
Terrace, 450
Tertiary structure, 395
Tetragonal, 433
Tetrahedral hybrid, 349
Theorem
  equipartition, 12
  variation, 352
Theory
  Brønsted–Lowry, 186
  collision, 251
  Debye–Hückel, 211, 458
  Förster, 516
  free-electron molecular orbital, 375
  kinetic molecular, 26
  Lewis, 345
  molecular-orbital, 344, 353
  quantum, 287
  transition-state, 253
  valence-bond, 344
Thermal analysis, 120, 157
Thermal desorption spectroscopy, 462
Thermal equilibrium, 10
Thermochemistry, 47
Thermodynamic stability, 114
Thermodynamic temperature, 10
Thermodynamically stable compound, 172
Thermodynamically unstable compound, 172
Thermodynamics, 1, 47
Third body, 281
Third Law of thermodynamics, 102
Third-Law entropy, 102
Thomson, W. (Lord Kelvin), 41
Tie line, 153
Time constant, 248
Time of flight, 32
Time-resolved spectroscopy, 522
Time-resolved X-ray diffraction, 444
Titrant, 196
Titration, 196
  strong acid/strong base, 196
  weak acid/strong base, 197
  weak base/strong acid, 199
TMS, 533
tonne (unit), 571
torr (unit), 7
Total energy, 9
Total interaction, 385
Total orbital angular momentum, 335
Total orbital angular momentum quantum number, 336
TPD, 462
Trajectory, 292
Transfer coefficient, 469
Transition metal, 331
Transition probability, 524
Transition temperature, 118
Transition-state theory, 253
Translation, 298
Translational partition function, 563
Transmembrane potential, 216
Transmission coefficient, 254
Transmission probability, 301
Transmittance, 506
Transport, active and passive, 216
Triclinic, 433
Trigonal planar hybrid, 350
Triple bond, 350
Triple point, 125
Triplet state, 511
Triprotic acid, speciation, 194
Trough, 26
Trouton's rule, 98
Tunnelling, 301
Turning point, 508
Turnover frequency, 278
Two-dimensional NMR, 542
Two-dimensional square well, 302
Type I and II superconductors, 432

Ubiquitin, 74
UHV, 451
Ultracentrifugation, 414
Ultra-high vacuum, 451
Ultrapurity, 159
Ultraviolet photoemission spectroscopy, 451
Ultraviolet radiation, 14
Uncertainty broadening, 487
Uncertainty principle, 296, 307
Ungerade, 356
Unimolecular reaction, 266, 271
Unique rate, 239
Unit, 2, 570
Unit cell, 432
Unstable compound, 172
Upper consolute temperature, 156
Upper critical solution temperature, 156
UPS, 451

Vacuum permeability, 530
Vacuum permittivity, 9
Valence band, 423
Valence bond theory, 344, 346
Valence-shell electron pair repulsion model, 345
van 't Hoff equation
  equilibrium constant, 178
  osmosis, 150
van 't Hoff isochore, 178
van der Waals equation of state, 37, 38
van der Waals interactions, 376
van der Waals loops, 39
van der Waals molecule, 256
van der Waals parameters, 38
van der Waals, J., 37
Vaporization, 70
Vapour deposition, 72
Vapour diffusion, 444
Vapour pressure, 119
  variation with temperature, 122
Vapour, distinct from gas, 35
Variation principle, 364
Variation theorem, 352
VB theory, 344
Vector, 306
  addition and subtraction, 336
Vector model, 307
Velocity, 5
Vertical transition, 508
Vibration, 308, 489
Vibrational mode, 492
Vibrational partition function, 563
Vibrational quantum number, 308
Vibrational Raman spectra, 492, 496
Vibrational spectroscopy, 489
Vibrational structure, 507
Vibrational transitions, 490
Vibration–rotation spectrum, 495
Virial coefficient, 36
Virial equation of state, 37
Viscosity, temperature dependence, 275
Visible light, 14
Vision, 509
Volcano curve, 465

Voltaic cell, 217
Voltammetry, 470
Volume, 4
von Laue, M., 437
VSEPR model, 345

Water
  critical point, 126
  enthalpy of vaporization, 71
  MO description, 365
  phase diagram, 127
  photoelectron spectrum, 523
  residual entropy, 104
  vapour pressure, 119
  VB description, 348, 351
watt (unit), 570
Wave, 12
Wavefunction
  conditions on, 295
  harmonic oscillator, 309
  hydrogenic atom, 318, 319
  introduced, 292
  particle in a box, 298
  spin, 338
  VB theory, 346
Wavelength, 13, 477
Wavenumber, 13, 477
Wave–particle duality, 292
Weak acid, 188
Weak base, 188
Weather, 25
Weather map, 26
Weight of a configuration, 100
Weight-average molar mass, 412
White light, 13
Width at half-height, 486
Work, 8, 48
  isothermal, reversible, 54
  maximum, 52
  measurement, 51
  molecular interpretation, 50
  non-expansion, 107
  reversible, 54
  sign convention, 51
Work function, 290
Wrinkle, Nature's abhorrence of, 275

Xanthophyll, 509
X-band, 544
XPS, 451
X-ray, 14
X-ray crystallography, 444
X-ray diffraction, 435
  time-resolved, 444
X-ray diffractometer, 438
X-ray photoemission spectroscopy, 451

YAG laser, 519

Z-average molar mass, 413
Zeolite, 467
Zero-current cell potential, 223
Zero-point energy
  harmonic oscillator, 308
  particle in a box, 299
Zeroth-order reaction, 240
Zinc-blende structure, 443
Zone levelling, 159
Zone refining, 159